October 31–November 4, 2011
Miami, Florida, USA

Association for
Computing Machinery

Advancing Computing as a Science & Profession

MSWiM'11

Proceedings of the 14th ACM International Conference on
Modeling, Analysis, and Simulation of Wireless and Mobile Systems

Sponsored by:

ACM SIGSIM

Association for
Computing Machinery

Advancing Computing as a Science & Profession

The Association for Computing Machinery
2 Penn Plaza, Suite 701
New York, New York 10121-0701

Notice to Past Authors of ACM-Published Articles
ACM intends to create a complete electronic archive of all articles and/or other material previously published by ACM. If you have written a work that has been previously published by ACM in any journal or conference proceedings prior to 1978, or any SIG Newsletter at any time, and you do NOT want this work to appear in the ACM Digital Library, please inform permissions@acm.org, stating the title of the work, the author(s), and where and when published.

ISBN: 978-1-4503-0898-4

Additional copies may be ordered prepaid from:

ACM Order Department
PO Box 30777
New York, NY 10087-0777, USA

Phone: 1-800-342-6626 (USA and Canada)
+1-212-626-0500 (Global)
Fax: +1-212-944-1318
E-mail: acmhelp@acm.org
Hours of Operation: 8:30 am – 4:30 pm ET

ACM Order Number: 617111

MIX
Paper from responsible sources
FSC www.fsc.org
FSC® C102081

Printed in the USA

Message from the General Chair

Welcome to the 14th ACM International Conference on Modeling, Analysis and Simulation of Wireless and Mobile Systems (MSWiM). For over a decade, MSWiM has established itself as a premier conference in the areas of modeling, analysis and performance evaluation of wireless and mobile systems. The importance of these areas is becoming more evident, year-by-year, with the unprecedented ubiquity and capability of mobile systems. There is no doubt that research in these areas will remain essential to the design of efficient mobile systems and the future Internet.

This year, we have a very strong and rich program consisting of a selection of excellent regular papers. The main conference is single track, providing an excellent opportunity to meet experts and top researchers in the field. In addition, we are happy to have two outstanding distinguished keynote speakers, Professor J.J. Garcia-Luna-Aceves (University of California, Santa Cruz) and Professor Mostafa Ammar (Georgia Institute of Technology). The program also includes a distinguished invited talk by Professor Jerome Haerri (EUROCOM), short papers, poster and demo/tool sessions, and tutorials.

Co-located with the main conference, are Symposia, covering several specializations within mobile and wireless systems: Mobility management and wireless access (MobiWAC), Performance evaluation of wireless ad hoc, sensor, and ubiquitous networks (PE-WASUN) and QoS and Security of wireless and mobile networks (Q2SWinet), and two workshops covering wireless multimedia networks (WMuNeP), and measurement of wireless and heterogeneous networks (PM2HW2N). Furthermore, this year, MSWiM has added a new symposium on Design and Analysis of Intelligent Vehicular networks and Applications (DIVANet).

Many thanks go to the Chair of the Steering Committee, Professor Azzedine Boukerche, and to the members of the steering committee for creating and maintaining a clear, consistent and recognized vision for MSWiM over many years. I wish to express my gratitude -- and congratulations -- to the Program Co-Chairs Bjorn Landfeldt and Luciano Bononi, and all the TPC members, for the outstanding job they did in overseeing a thorough and rigorous review process and for selecting the technical program. They have exhibited a great deal of professionalism and hard work, making my job extremely simple.

The rest of the organizing committee also put forward a great effort, starting from the workshop chair Damla Turgut, tutorial co-chairs Abdallah Shami and Andrea M. Tonello, and, to the poster chair Xianbin Wang and demo/tools co-chairs Fei Richard Yu and Regina B. de Araujo. They all deserve our thanks as their efforts were instrumental to the success of the conference. In addition, thanks to Mirela A. M. Notare, Salil Kanhere for advertising, Samer Samarah, the registration chair and Lotzi Boloni, our local arrangements chair. Richard W. Pazzi, as always, helped out with numerous tasks during the process of organizing MSWIM 2011, and deserves a big thanks! We also extend our thanks to the external referees for providing high quality reviews within a tight schedule.

In the end, I wish to thank our main sponsor the ACM SIGSIM. I am looking forward to meeting you in beautiful Miami Beach, Florida to participate in thought-provoking, interesting and stimulating discussions, and enjoy the whole program.

Ahmed Helmy
University of Florida, USA
General Chair MSWiM 2011

Message from the Program Co-Chairs

On behalf of the Organizing Committee, it is our great pleasure to welcome you to the 14th ACM International Conference on Modeling, Analysis and Simulation of Wireless and Mobile Systems (MSWiM), held in 2011 in the exciting city of Miami Beach, Florida, USA. MSWiM has a long tradition of introducing important, high quality and timely work, and we are very pleased that this year's event maintains the high standard set in previous years.

In 2011, the call for papers attracted a record high of 190 registered papers in all areas of mobile and wireless systems of which 160 were accepted into the review process. The submitted papers came from 32 countries, and the accepted papers from 21 countries, reflecting the truly international profile of MSWiM. It is also informative to note that the five most commonly listed topics for submissions to MSWiM this year were:

- Performance evaluation and modeling;
- Wireless network algorithm and protocols;
- Sensor and actuator networks;
- Analytical models; and,
- Wireless mesh networks, mobile ad hoc networks, and VANETS.

The submissions included a large number of papers of very high quality making the selection process difficult and competitive. In the end, we selected 45 regular papers, which is an acceptance rate of 24%, and an additional 8 short papers were recommended for the program for their high quality and contribution. The MSWiM 2011 technical program also comprises two distinguished keynote addresses by Professors J.J. Garcia-Luna-Aceves, and Mostafa H. Ammar, a distinguished invited talk by Professor Jerome Haerri, a poster and demonstrator session and two tutorials. In addition, there are three one-day Symposia: MobiWAC, PE-WASUN and Q2SWinet as well as two workshops: WMuNeP and PM2HW2N co-located with the main conference.

Among the 45 regular papers, three have been selected as candidates for the best paper award at the conference and will be announced in the opening session. The papers selected are:

- "Modeling and Simulation of Service Composition in Opportunistic Networks," by Umair Sadiq, Mohan Kumar, Andrea Passarella and Marco Conti
- "Maximal Utility Rate Allocation for Energy Harvesting Wireless Sensor Networks," by Bo Zhang, Robert Simon and Hakan Aydin
- "Seamless Rate Adaptation for Wireless Networking," by Hao Cui, Chong Luo, Kun Tan, Feng Wu and Cheng Wen Chen

The winner among these three papers will be announced at the conference banquet, and will be reported in the proceedings of MSWiM 2012.

We would like to congratulate the winners of the best paper award of MSWiM 2010:

- "Dynamic Random Replication for Data Centric Storage," Angel Cuevas, Manuel Uruena and Gustavo De Veciana.

Our thanks go to Professor Azzedine Boukerche and the Advisory Board for their vision in guiding MSWiM from workshop to symposium and finally to conference status over the past 13 years. We thank Professor Ahmed Helmy, the General Chair, for sharing his advice and experience in handling the issues that arise in conference organization; his input was invaluable. A special thank you go to the Technical Program Committee as well as the external reviewers, for all the hard work in reviewing and judging the contributions of all the submitted papers.

We thank our organizing committee, as putting together a conference such as MSWiM is a team effort. Thanks to tutorial co-chairs Abdallah Shami and Andrea M. Tonello, workshop chair Damla Turgut, demo/tools co-chairs Fei Richard Yu and Regina B. de Araujo, and poster chair Xianbin Wang, for organizing the value-added sessions MSWiM attendees have come to expect. Our thanks go to Mirela A. M. Notare and Salil Kanhere for advertising the conference. Thanks are also due to Samer Samarah, our registration chair and Lotzi Boloni, our local arrangements chair. Last but not least we want to thank Richard W. Pazzi for helping out with the web site, putting together the final proceedings as publications chair and for helping out with numerous tasks during the process of organizing MSWIM 2011.

Finally, we thank our sponsors, ACM and ACM SIGSIM, for their continued support of these successful meetings.

In closing, we warmly welcome you to MSWiM 2011. We are confident that both the conference and your stay in Miami Beach will be informative, interesting, and above all enjoyable!

Bjorn Landfeldt
MSWiM'11 Program Co-Chair
University of Sydney, Australia

Luciano Bononi
MSWiM'11 Program Co-Chair
University of Bolgna, Italy

Table of Contents

Session 13: Mobile and Social Networks Modeling

Session 14: Modeling and Analysis Methodologies I

Session 15: Modeling and Analysis Methodologies II

Session 16: Positioning and Localization

Session 17: Rate Adaptation and Control

MSWIM 2011 Conference Organization

General Chair: Ahmed Helmy *(University of Florida, USA)*

Program Chairs: Bjorn Landfeldt *(University of Sydney, Australia)*
Luciano Bononi *(University of Bologna, Italy)*

Workshop Chair: Damla Turgut *(University of Central Florida, USA)*

Poster Chair: Xianbin Wang *(University of Western Ontario, Canada)*

Demo/Tools Co-chairs: Fei Richard Yu *(Carleton University, Canada)*
Regina B. de Araujo *(Universidade Federal de Sao Carlos, Brazil)*

Tutorials Co-Chairs: Abdallah Shami *(The University of Western Ontario, Canada)*
Andrea M. Tonello *(DIEGM - Universita di Udine, Italy)*

Publication Chair: Richard W. Pazzi *(University of Ottawa, Canada)*

Local Arrangements Chair: Lotzi Boloni *(University of Central Florida, USA)*

Publicity Chairs: Mirela A. M. Notare *(Barddal University, Brazil)*
Salil Kanhere *(University of New South Wales, Australia)*

Registration Chair: Samer Samarah *(University of Ottawa, Canada)*

Steering Committee Chair: Azzedine Boukerche *(University of Ottawa, Canada)*

Program Committee: Nael Abu-Ghazaleh *(State University of New York at Binghamton, USA)*
Ozgur Akan *(Koc University, Turkey)*
Jalel Ben-othman *(University of Versailles, France)*
Brahim Bensaou *(The Hong Kong University of Science and Technology, Hong Kong)*
Luciano Bononi *(University of Bologna, Italy)*
Azzedine Boukerche *(University of Ottawa, Canada)*
Claudio Casetti *(Politecnico di Torino, Italy)*
Chi Chung Cheung *(The Hong Kong Polytechnic University, Hong Kong)*
Carla-Fabiana Chiasserini *(Politecnico di Torino, Italy)*
Kaushik Chowdhury *(Northeastern University, USA)*
Marco Di Felice *(University of Bologna, Italy)*
Lorenzo Donatiello *(University di Bologna, Italy)*
Falko Dressler *(University of Erlangen, Germany)*
Stephan Eidenbenz *(Los Alamos National Laboratory, USA)*
Khalil El-Khatib *(UOIT, Canada)*
Ehab Elmallah *(University of Alberta, Canada)*

Program Committee (continued):

Ece Gelal *(University of California, Riverside, USA)*
Silvia Giordano *(University of Applied Science - SUPSI, Switzerland)*
Kartik Gopalan *(State University of New York at Binghamton, USA)*
Khaled Harras *(Carnegie Mellon University, USA)*
Xiaojun Hei *(Huazhong University of Science and Technology, PR China)*
Shengming Jiang *(South China University of Technology, PR China)*
Holger Karl *(University of Paderborn, Germany)*
Peng-Yong Kong *(Institute for Infocomm Research, Singapore)*
Jerzy Konorski *(Gdansk University of Technology, Poland)*
Bjorn Landfeldt *(University of Sydney, Australia)*
Lavy Libman *(University of Sydney, Australia)*
Renato Lo Cigno *(University of Trento, Italy)*
Francesco Lo Presti *(Universitá di Roma Tor Vergata, Italy)*
Michela Meo *(Politecnico di Torino, Italy)*
Jogesh K. Muppala *(HKUST, Hong Kong)*
Hamid Nafaa *(University College Dublin, Ireland)*
Sotiris Nikoletseas *(University of Patras and Computer Technology Institute, Greece)*
Richard W. Pazzi *(University of Ottawa, Canada)*
Ravi Prakash *(University of Texas at Dallas, USA)*
Tahiry Razafindralambo *(INRIA Lille - Nord Europe, France)*
Samer Samarah *(University of Ottawa, Canada)*
Dirk Staehle *(University of Wuerzburg, Germany)*
Violet Syrotiuk *(Arizona State University, USA)*
Mineo Takai *(University of California, Los Angeles, USA)*
Bin Tang *(Wichita State University, USA)*
Terence D. Todd *(McMaster University, Canada)*
Damla Turgut *(University of Central Florida, USA)*
Emmanouel Varvarigos *(University of Patras, Greece)*
Hongyi Wu *(University of Louisiana at Lafayette, USA)*
Yaoqing Yang *(University of Nebraska-Lincoln, USA)*
Sherali Zeadally *(University of the District of Columbia, USA)*
Jingyuan Zhang *(University of Alabama, USA)*
Jun Zhang *(The Hong Kong Polytecnic University, Hong Kong)*
Yanchao Zhang *(Arizona State University, USA)*

Additional reviewers:

Giuseppa Alfano
Onur Altintas
Constantinos Marios Angelopoulos
Baris Atakan
Saeed Bastani
A. Ozan Bicen
Thinus Booysen
Athanassios Boulis
Luca Chiaraviglio
Delia Ciullo
Carmelo Costanzo
Gabriele D'ngelo
Rui Dai
Zhonghai Deng
Jiaqi Duan
David Eckhoff
Dionysios Efstathiou
Ines El Korbi
Miguel Erazo
Milan Erdelj
Ozgur Ergul
Alessandro Finamore
Marco Fiore
Eman Fituri
Anna Faerster
Victor Govindaswamy
Quansheng Guan
Burhan Gulbahar
Bingyi Guo
Wenxuan Guo
Abdulla Hamad
Seon Yeong Han
Hanan Hassanein
Ka Lok Hung
Ganesh Iyer
Nadeem Javaid
Josep Miquel Jornet
Abdellatif Kobbane
Murat Kocaoglu
Vinay Kolar
Xia Li
Guikai Liu
Li Liu
Shichuan Ma
Leonardo Maccari
Adnan Majeed
Derya Malak
Francesco Malandrino
Choong Hock Mar
Gustavo Marfia
Moreno Marzolla
Karen Miranda
Keyvan Moghadam
Vikram Munishwar
Ghasem Naddafzadeh Shirazi
Yousof Naderi
Enrico Natalizio
Hoang Anh Nguyen
Nam Nguyen
Guoqin Ning
Mohamed Ould Cheikh
Dorin Panaitopol
Dimitra Patroumpa
Konstantinos Pelechrinis
Daniele Puccinelli
Christoforos Raptopoulos
Saquib Razak
Massimo Reineri
Marco Roccetti
Claudio Rossi
Nandakishore Santhi
Ghalib Shah
Mohamed Shazly
Mohammad Sheikh Zefreh
Paschalis Sofotasios
Christoph Sommer
Hwee-Xian Tan
Lin Tang
Chintha Tellambura
Ming-Fong Tsai
Nicholas Valler
Jason Min Wang
Li Wang
Zhijun Wang
Hirozumi Yamaguchi
Guanhua Yan
Keiichi Yasumoto

Sponsor: SIGSIM

Redefining Routing and Channel Access in Ad Hoc Networks

J.J. Garcia-Luna-Aceves*
Computer Engineering Department
University of California
Santa Cruz, CA 95064
Palo Alto Research Center
Palo Alto, CA 94304
jj@soe.ucsc.edu

ABSTRACT

There has been considerable work on the architecture and protocols of wireless networks since the inception of the ALOHA system and the DARPA packet radio network approximately 40 years ago. However, the design choices that have been made for wireless networks are derivatives of protocols designed for wireline networks and are such that: (a) channel access is either very inefficient and unable to support multimedia traffic across multiple hops, or requires complex clock synchronization to support scheduling schemes based on time division of the channel and dynamic channel reuse; (b) the routes established in wireless networks have no correlation to the transmission schedules used over the wireless links forming those routes, which reduces the efficiency with which flows can be forwarded; and (c) the mechanisms used for information dissemination in wireless ad hoc networks do not scale with the size or dynamics of the network.

In this talk, I argue that the way in which naming, addressing, routing and channel access are done in wireless networks must be redefined from the ground up in order for these networks to scale and perform much more effiiciently than they currently do. This redesign must take into account the facts that: (a) nodes do not attach to the network at static points and need not be grouped in stable sets; (b) wireless links do not operate in isolation from one another; (c) establishing transmission schedules and routes are intimately related to one another; (d) flooding must be eliminated in the signaling of routing protocols; and (e) routing and channel access protocols do not have to use static identifiers that have no relationship to their operation.

I summarize recent results on automatic incremental routing (AIR) and harmonic channel access, which follow the above design guidelines. AIR attains efficient routing in large networks without the use of traditional routing tables or flooding, and supports unicast and multicast routing using the same signaling. Harmonic channel access attains distributed transmission schedules based on circular permutations of identifiers of those entities competing for the shared channel, which need not be node identifiers.

Categories and Subject Descriptors

C.2.1 [**Network Architecture and Design**]: Wireless Networks

General Terms

Algorithms, Performance, Design

Bio

J.J. Garcia-Luna-Aceves holds the Jack Baskin Endowed Chair of Computer Engineering at the University of California, Santa Cruz (UCSC), is Chair of the Computer Engineering Department, and is a Principal Scientist at the Palo Alto Research Center (PARC). Prior to joining UCSC in 1993, he was a Center Director at SRI International (SRI) in Menlo Park, California. He has been a Visiting Professor at Sun Laboratories in Menlo Park, California, and a Principal of Protocol Design at Nokia in Mountain View, California. He received the IEEE Computer Society Technical Achievement Award in 2011 "for pioneering contributions to the theory and design of communication protocols for ad-hoc wireless networks." He is a fellow of the Institute of Electrical and Electronics Engineers (IEEE), the Association for Computing Machinery (ACM), and the American Association for the Advancement of Science (AAAS). Dr. Garcia-Luna-Aceves holds 36 U.S. patents and has published three books and more than 420 journal and conference papers. He is the co-recipient of the IEEE Fred W. Ellersick 2008 MILCOM Award for best unclassified paper entitled "On The Capacity Improvement of Multicast Traffic with Network Coding." He is also co-recipient of Best Paper Awards at the the European Wireless Conference 2010, IEEE MASS 2008, SPECTS 2007, IFIP Networking 2007, and IEEE MASS 2005 conferences, and of the Best Student Paper Award of the 1998 IEEE International Conference on Systems, Man, and Cybernetics. He received the SRI International Exceptional-Achievement Award in 1985 and 1989, and is listed in Marquis Who's Who in America and Who's Who in The World.

*This work was supported in part by the Baskin Chair of Computer Engineering

MSWiM'11, October 31–November 4, 2011, Miami, Florida, USA.
ACM 978-1-4503-0898-4/11/10.

Invited Talk

Living in the WAM Continuum: Unified Design and Operation of Wireless and Mobile Networks

Mostafa Ammar
School of Computer Science
Georgia Institute of Technology
Atlanta, GA , USA
ammar@cc.gatech.edu

Abstract

Until recently, the vast majority of research in wireless and mobile networks focused on so-called Mobile Ad Hoc Networks (MANETs), where relatively stable end-to-end paths are the norm. More recently, research has focused on a different, Intermittently Connected Network (ICN) paradigm, where stable end-to-end paths are the exception and intermediate nodes may store data while waiting for transfer opportunities towards the destination. Protocols developed for MANETs generally do not work in ICNs since the connectivity assumptions are so different. In this talk I will first give an overview of ICNs and the types of challenges involved in their design and operation. I will then discuss our work in the WAM (wireless and mobile) Continuum project which is based on the simple but powerful observation that MANETs and ICNs fit into a continuum that generalizes these two previously distinct categories. Building on this observation, our work develops a framework that goes further to scope the entire space of wireless and mobile networks. I will summarize two efforts: The first gives clues of the fundamental relationship between ICNs and MANETs by unifying the use of message ferrying in ICNs with the well-known use of connected dominating set-based routing in MANETs. The second effort aims at developing a formal WAM Continuum framework where a network can be characterized by its position in this continuum. Certain network equivalence classes can be defined over subsets of this WAM continuum and this classification can be used to inform network design and operation. I will demonstrate how the unified view enabled by our framework can be used as a systematic, formal descriptive and evaluative tool.

Categories & Subject Descriptors:

C.2.1[Network Architecture and Design]: Wireless Networks

General Terms: Algorithms, Design

Bio

Mostafa Ammar is a Regents' Professor with the School of Computer Science at the Georgia Institute of Technology. Dr. Ammar's research interests are in network architectures, protocols and services. He has contributions in the areas of multicast communication and services, multimedia streaming, content distribution networks, network simulation and most recently in disruption-tolerant networks and overlay network design. Dr. Ammar has published extensively in these areas. He served as the Editor-in-Chief of the IEEE/ACM Transactions on Networking from 1999 to 2003. He was the co-TPC Chair for the IEEE ICNP 1997, ACM Co-Next 2006 and ACM SIGMETRICS 2007 conferences. Dr. Ammar is a Fellow of the IEEE and a Fellow of the ACM.

MSWiM'11, October 31–November 4, 2011, Miami, Florida, USA.
ACM 978-1-4503-0898-4/11/10

Invited Talk

Evaluating Cooperative ITS Applications for Sustainable and Safe Mobility with iTETRIS*

Jérôme Härri
EURECOM†
Mobile Communications Department
06904 Sophia-Antipolis, France
jerome.haerri@eurecom.fr

ABSTRACT

We see today urban traffic growing faster than the capacity of road infrastructures and posing a challenge to traffic safety and to the sustainability of our mobility. In this context, cooperative mobility strategies are expected to become a key enabler for a better usage of the available road infrastructures. Considering the estimation by the American Automotive Association of the yearly cost of traffic safety to 160 billion USD, or the yearly cost of traffic congestion to 50 billion EUR by the European Commission, local road authorities are facing a strategic challenge to develop and propose cooperative ITS applications to the public.

The question is how can the actual worth of investment and effectiveness of cooperative ITS applications on city road traffic be estimated? To answer this question, new tools capable of providing realistic and large scale assessments and guidelines to road authorities are required. To this aim, we need to define an extensible architecture allowing the interactions of realistic models spanning from wireless communication to traffic and mechanical engineering, and provide a modular integration methodology of cooperative ITS applications.

In this talk, we present the outcome of the recently completed work of the European FP7 project iTETRIS during which an open, ETSI standard compliant, and flexible ITS simulation platform has been developed. We describe the modeling challenges, design choices, selected scenarios, and illustrate the benefits and functionalities of the iTETRIS platform through the integration of exemplary key cooperative ITS applications. We further discuss future evolutions, trends, and potential roles of simulations and the iTETRIS platform to reach sustainable and safe mobility resulting in economical, social and ecological wealth improvements.

*This work has been partly funded by the European Commission through FP7 ICT Project iTETRIS: An Integrated Wireless and Traffic Platform for Real-Time Road Traffic Management Solutions (No. FP7 224644). The author wishes to acknowledge the Commission for their support, as well as all project members for their technical involvements and fruitful suggestions in the content of this talk.

†EURECOM industrial members: BMW Group, Cisco, Monaco Telecom, Orange, SAP, SFR, STEricsson, Swisscom, Symantec, Thales.

MSWiM'11, October 31–November 4, 2011, Miami, Florida, USA.
ACM 978-1-4503-0898-4/11/10.

Categories and Subject Descriptors

I.6.5 [**Simulation and Modeling**]: Model Development—*Modeling methodologies*; H.4.3 [**Information Systems Applications**]: Communications Applications

General Terms

Design, Performance, Algorithms

1. BIOGRAPHY

Jérôme Härri is an Assistant Professor at the Mobile Communication Group at EURECOM, France, and conducting research in wireless vehicular networks. Previously, he led a Traffic Telematics Junior Research Group at the Institute of Telematics of the Karlsruhe Institute of Technology (KIT), Germany. His research interests are related to the investigation of cooperative ITS strategies and to the characterization of the mutual relationship between vehicular mobility and heterogeneous vehicular communication.

Jérôme has been Co-Chair of the special session on traffic telematics at IEEE PIMRC'09, has served as Demo Chair and TPC Co-chair of the IEEE Conference on Wireless Vehicular Communications (WiVeC) 2008 and 2010 respectively, and is the author of the chapter on "Vehicular Mobility Modeling for VANET" in the book "VANET: Vehicular Applications and Inter-Networking Technologies", (Eds. H. Hartenstein and K. Laberteaux) published by Wiley in 2010.

He has been involved in the EU FP7 project iTETRIS, where he contributed to the architecture and the development of ITS applications and ITS infrastructure deployment strategies. He is now involved in the French Field Operational Test (FOT) project SCORE@F and the smart grid project VELCRI with simulation-related evaluations of cooperative ITS applications based on the iTETRIS Platform. He is actively participating to standardization activities of the ETSI TC ITS and of the CAR 2 CAR Communication Consortium.

Jérôme holds a M.Sc. degree and a Dr. ès sc. degree in telecommunication from the Swiss Institute of Technology (EPFL), Lausanne, Switzerland.

Maximal Utility Rate Allocation for Energy Harvesting Wireless Sensor Networks

Bo Zhang Robert Simon Hakan Aydin
Department of Computer Science
George Mason University
Fairfax, Virginia, USA, 22030
{ bzhang3, simon, aydin }@cs.gmu.edu

ABSTRACT

There is currently tremendous interest in deploying energy-harvesting wireless sensor networks. Engineering such systems requires striking a careful balance between sensing performance and energy management. Our work addresses this problem through the design and analysis of a harvesting-aware utility-based sensing rate allocation algorithm. Based on a network utility formulation, we show that our algorithm is optimal in terms of assigning rates to individual nodes to maximize overall utility, while ensuring energy-neutral operation. To our knowledge, our work is the first optimal solution that maximizes network utility through rate assignments for tree-structured energy harvesting sensor networks. Our algorithm is fast and efficient with running time $O(N^3)$, where N is the number of nodes. We evaluate the performance, scalability, and overhead of our algorithm for various utility functions and network sizes, underlining its significant advantages.

Categories and Subject Descriptors

C.2.1 [**Network Architecture and Design**]: Distributed networks; C.2.2 [**Network Protocols**]: [Applications]

General Terms

Algorithms, Performance, Experimentation

Keywords

Sensor networks, Energy harvesting, Rate allocation

1. INTRODUCTION

The rapid introduction of new software and hardware functionalities has stimulated the development of complex wireless sensor network (WSN) applications. For this new generation of WSN applications, maximizing the value or *utility* of sensed data, as perceived by end-users, is of paramount importance. For most WSN systems, a major constraint to utility maximization is the limited energy availability of sensor nodes. One solution to this problem is to devise energy harvesting techniques to power sensor nodes at run-time [2, 4, 5]. Since the availability of environmental energy sources is often limited and highly time-varying, the goal of simple utility maximization is constrained by the need for managing unpredictable energy supplies. Our work addresses this problem by designing and analyzing a WSN-specific, energy-harvesting aware utility maximization algorithm.

Much previous work in this area assumes that application utility increases linearly with the rate at which sensor nodes sense or collect data [1, 11, 12, 13]; therefore the maximization of the network-wide data collection rate leads to utility maximization. However, for many WSN applications increasing the level of sensed data reporting only increases the utility of the application in sub-linear fashion. Consider, for instance, video or motion sampling as part of an intrusion detection system. Since humans can only move at a certain speed, sampling above a specific threshold only marginally increases the utility of the application. For this reason, our work models utility as a non-decreasing concave function, such that its rate of increase (*marginal utility*) decreases as the sensing and reporting rate increases [6, 21, 22]. The utility perceived by application end-user is the aggregate utility achieved jointly by all the nodes in the network.

With energy harvesting equipment such as solar panels or wind generators, sensor nodes harvest power from environmental sources and potentially sustain their operation perpetually. In [5], the concept of perpetual operation is formally stated as the *energy neutral condition*. This means that sensor nodes must always maintain positive energy storage levels to avoid energy depletion and hence operation interruption. A further complexity arises in multi-hop WSNs, where nodes must forward data received from other nodes in addition to their locally generated data. Given limited energy supplies, control algorithms must balance the local and external sensing and communication rates. Favoring local rates over external rates may result in high local utility, but low network-wide utility, since the rates of other nodes are throttled by the high local rate.

To solve the above issues we propose MAX-UTILITY, an *epoch*-based rate allocation algorithm. MAX-UTILITY is designed to maximize total application utility, while regulating nodes energy consumption resulted from sensing and reporting activities to guarantee energy neutral operation for every node. In our work, 'rate' is equated with packet rate, which is directly tied to sensing rate. An epoch is a time interval during which the amount of power harvested

MSWiM'11, October 31–November 4, 2011, Miami, Florida, USA.

remains relatively stable and is reasonably predictable. The length of an epoch may range from several minutes to a couple of hours, depending on the energy source type. Our approach exploits the concavity of the utility function, and a special property of tree-based networks to allocate rates to nodes as evenly as possible for achieving utility maximization, while maintaining the minimum sensing rate required by the application and the available energy and data forwarding capacity of the nodes.

We formally prove the optimality of MAX-UTILITY in the sense of utility maximization. For a system with N nodes, our algorithm has a time complexity of $O(N^3)$. We also develop a distributed version of this algorithm for use in systems without central control points. *To the best of our knowledge, this is the first optimal solution to maximize general network utility through rate assignments to individual nodes in tree-structured sensor networks, while guaranteeing energy neutrality.* A limitation of our approach is that it applies only to tree-based WSNs. However, since trees are a common logical routing structure for WSN systems [1, 3, 8], we believe our algorithm is widely applicable. While the related studies in [6, 21] target general multi-hop networks and offer innovative solutions, in general, they do not guarantee optimality. Finally, using solar harvesting traces obtained over a year-long study in [17], we have evaluated MAX-UTILITY against an alternative algorithm, under a variety of experiment settings. Our results show that MAX-UTILITY delivers superior utility improvement while ensuring energy neutral operation for all the nodes.

2. BACKGROUND AND RELATED WORK

Rate allocation for wireless sensor networks has been explored in [1, 6, 12, 13, 22]. In [1], the authors propose to maximize lexicographic rate assignments to sensor nodes. The rate assignment problem is formulated as linear optimization problem, and solved optimally by centralized and distributed algorithms. [12] proposes a rate control approach for a single energy harvesting node to achieve a series of objectives including the maximization of average sensing rate over time. These objectives are formulated as optimization problems and solved using multi-parametric control algorithms. [13, 14] propose a flow control algorithm for energy harvesting WSNs. [14] proposes an energy budgeting algorithm that defines the amount of energy a node can use for each epoch. The derived energy budget assignment is optimal in the sense that the variance of energy assigned across epochs is minimized. Using this formulation the flow control algorithm in [13] maximizes the amount of data collected over the network, given that no node consumes more energy than the assigned budget.

An implicit assumption in [1, 12, 13] is that the system utility increases linearly with the rate of nodes. However, for many applications the increase of utility slows down as rate increases (the *diminishing returns* principle). Using this observation, [6, 21, 22] model system utility as a concave and non-decreasing function of rate, and propose primal-dual based algorithms to maximize utility over the entire network. Among these works, [6] is the most related one to ours that target utility maximization for energy harvesting WSNs. However, [6, 21, 22] all assume specific utility functions that are continuously differentiable, which may limit their application for a more general class of functions. Moreover, their proposed solutions are not optimal, and can incur high control overhead and unpredictable running time, thus potentially limiting their practical implementation within resource-constrained WSN systems.

Research presented in [7, 15] formulate the Network Utility Maximization (NUM) problem for Internet congestion control. Primal-dual based algorithms are proposed to solve the problem. The issue we target shares the same structure with the NUM problem, but is for a WSN environment. The utility maximization problem has been also extensively studied for real-time embedded systems. [18] addresses utility maximization for energy-constrained systems that execute periodic real time tasks. The objective is to maximize the total utility obtained from execution of these tasks, while satisfying the deadlines of all the tasks, the tight system energy budget and the minimum system performance requirement. In [19], the authors extend the approach in [18] for solar-powered embedded systems. However, [19] only considers two different epochs in each day. [9, 10] address a similar problem as [19] but with a more complex system model. They assume a highly dynamic energy harvesting model, and assume that WSN applications have multiple discrete service and performance levels, and each has different utility values and energy demands. The problem is then to decide when to select which service level to run in order to maximize the total utility, without over-using the available energy. Finally, in [23], we propose energy management approach for maximizing the energy storage level of nodes in energy harvesting WSNs. The approach utilizes two energy saving techniques, dynamic voltage scaling and dynamic modulation scaling.

3. SYSTEM ARCHITECTURE

3.1 Environmental energy harvesting model

We assume each sensor node consists of an energy harvester head, and several energy consuming hardware units, including a CPU, a wireless transceiver, and required sensor suites. The harvester head is energy source-specific, such as a solar panel or wind generator. The energy storage unit, such as a rechargeable battery or super-capacitor, has a maximum energy capacity of Γ^{max} joules. This unit receives power from the energy harvester, and delivers power to the sensor node. We take the commonly used approach that the amount of harvested power is uncontrollable, but reasonably predictable, based on the source type and harvesting history [5, 6, 12]. To capture the time-varying nature of environmental energy, time is divided into *epochs* of length S. Harvested power is modeled as an epoch-varying function denoted by P_k^h, where k is the epoch number. P_k^h remains within each epoch k, but changes for different epochs. The time unit used for harvesting prediction is therefore one epoch. The *prediction horizon*, H is an interval containing a number of epochs during which harvesting predictions can be reasonably made. The length of H may vary depending on the type of environmental sources and prediction techniques used. Our approach does not depend on the length of H, and requires harvesting prediction for only the coming epoch.

3.2 Network and application model

We consider general WSN applications that periodically collect data from N sensor nodes populated over the target environment. The nodes are organized into a data collection tree using any tree construction protocol, such as the Collection Tree Protocol (CTP) [3]. A sensor node is denoted as V_i, and the base station is denoted as BS. Within a tree-based routing structure, at any time each node V_i is connected to BS by a single path ρ_i consisting of zero or more intermediate nodes. We use the notation $V_i \in \rho_j$ to indicate that V_i resides on a given path ρ_j.

A node V_i senses the environment and sends the resulting data in packets towards BS along the path ρ_i, at rate r_i. These packets are referred as *internal packets*. Each node V_i may also forward *external packets* at a certain rate. These are received from the set of descendant nodes $\{V_j\}$, where V_i resides on V_j's path to BS, i.e. $V_i \in \rho_j$. Therefore the total outbound traffic at V_i equals $r_i + \sum_{j:V_i \in \rho_j} r_j$. We refer to r_i as V_i's *internal packet rate*, and $\sum_{j:V_i \in \rho_j} r_j$ as its *external packet rate*.

Sensor nodes produce utility by sensing and reporting data to BS. Therefore we define the utility accrued by any node V_i as a function of its packet rate r_i, $U(r_i)$. U is a positive, non-decreasing and concave function. Then we define the aggregate utility U^{tot} accrued jointly by all N nodes in the network as:

$$U^{tot} = \sum_{i=1}^{N} U(r_i) \tag{1}$$

Our energy model assumes, without loss of generality, that sensing and processing energy costs are fixed, relatively negligible and can be ignored. We therefore consider the radio transceiver as the main sink for energy consumption. Packet routing results in two types of energy consuming activities, packet transmission and reception. We denote the energy spent on transmitting an internal or external packet as e^{tx}, where e^{tx} can be measured in advance. Packet reception also consumes energy. Our architecture assumes that reception energy is controlled by the MAC layer, using techniques such as TDMA or duty-cycled LPL approaches such as B-MAC [16]. Therefore we model per-epoch reception energy as a constant E^{rx} which is dependent on the epoch length. We can calculate, at each node V_i, the total energy consumed E_i^c by handling internal and external packets in an epoch of length S as:

$$E_i^c = e^{tx} \cdot (r_i + \sum_{j:V_i \in \rho_j} r_j) \cdot S + E^{rx} \tag{2}$$

We also assume that each node V_i has a pre-assigned per-epoch energy budget B_i. Such energy budgets can be produced by algorithms such as [9, 13, 14], based on the predictions of harvested energy for any of the epochs in the horizon. Further, related work such as [5, 6] require that a sensor node can consume no more than the amount of energy harvested in any epoch. For these models a node's allocated energy budget in any epoch is set to the amount of energy it can harvest. Any of these energy budgeting algorithms guarantees the energy-neutral operation across all epochs.

4. THE MAXIMAL UTILITY RATE ALLOCATION PROBLEM

In this section, we define our rate allocation problem for energy harvesting WSN systems. Our objective is to maximize the network utility U^{tot} (Eq. (1)), given the limited energy harvesting ability of nodes. We achieve this goal by adjusting packet rate r_i. Since the harvested power changes from epoch to epoch, the rates of nodes need to be readjusted in every epoch. We formulate this objective as an optimization problem called *Network Utility Maximization with Energy Harvesting* (*NUM-EH*) as:

$$\begin{aligned} Max \quad & U^{tot} && (3)\\ s.t. \quad & \forall V_i, && \\ & E_i^c \le B_i && (4)\\ & r_i \ge r^{min} && (5)\\ & r_i + \sum_{j:V_i \in \rho_j} r_j \le R_i^{cap} && (6)\end{aligned}$$

The optimal solution to the above problem consists of the rates for all N nodes, i.e. $\{r_1, \ldots, r_N\}$ that maximizes U^{tot}. The constraint (4) enforces that the energy consumption E_i^c of any node V_i must be smaller than the assigned energy budget B_i in an epoch. The constraint (5) enforces that the rate of any node must be higher than the minimum required value r^{min}, in order to maintain the basic service level at individual nodes. Finally, to avoid packet congestion the total packet rate at any V_i, i.e. $r_i + \sum_{j:V_i \in \rho_j} r_j$ must be smaller than V_i's packet forwarding capacity denoted by R_i^{cap}.

We observe that *NUM-EH* is a concave maximization problem with three linear constraints. This is seen from Eq. (2), where E_i^c is a linear function of r_i and the set of external rates $\{r_j\}$. We notice that this problem is a special case of the well-known network utilization maximization problem [6, 7, 15] which can be solved using primal-dual based algorithms. However, such algorithms are typically too computationally expensive for resource-constrained sensor nodes. We instead propose a polynomial-time algorithm to solve the problem optimally and cost-effectively.

5. RATE ALLOCATION ALGORITHM

In this section, we propose the algorithm MAX-UTILITY that optimally solves problem *NUM-EH*. MAX-UTILITY is applicable to arbitrary utility functions that are concave and non-decreasing. We first propose a centralized version of this algorithm that can be run on a base station. We then show how to implement MAX-UTILITY in a fully distributed fashion so resource-constrained sensor nodes can collaboratively produce optimal rate assignments.

First, we show that the constraint (4) and (6) can be combined into one single constraint as follows. Substituting the equality (2) into constraint (4) gives:

$$E_i^c = e^{tx} \cdot (r_i + \sum_{j:V_i \in \rho_j} r_j) \cdot S + E^{rx} \le B_i \tag{7}$$

yielding:

$$r_i + \sum_{j:V_i \in \rho_j} r_j \le \frac{B_i - E^{rx}}{e^{tx} \cdot S} \tag{8}$$

The right-hand side of the above inequality, $\frac{B_i - E^{rx}}{e^{tx} \cdot S}$, is a constant as B_i, E^{rx}, e^{tx} and S are all known constants. This can be combined with constraint (6) into one single constraint, $r_i + \sum_{j:V_i \in \rho_j} r_j \le CAPACITY_i$ where:

$$CAPACITY_i = Min\left\{R_i^{cap}, \frac{B_i - E^{rx}}{e^{tx} \cdot S}\right\} \tag{9}$$

We refer to $CAPACITY_i$ as the *rate capacity* of node V_i. We can re-write problem *NUM-EH* concisely as:

$$Max \quad U^{tot} = \sum_{i=1}^{N} U(r_i)$$

$$s.t. \quad \forall V_i, r_i \geq r^{min} \qquad (10)$$

$$\forall V_i, r_i + \sum_{j:V_i \in \rho_j} r_j \leq CAPACITY_i \qquad (11)$$

We refer to constraint (10) as the *min-rate constraint*, and constraint (11) as the *capacity constraint*.

5.1 The centralized version

Now we present the centralized version of algorithm MAX-UTILITY. Before the algorithm starts, BS collects two pieces of information from each node. The first is the rate capacity $CAPACITY_i$, computed locally by each node using Eq. (9). The second is the node id of parent of each node, and the BS uses the parenthood relation of nodes to derive the structure of the existing data collection tree. MAX-UTILITY allocates rates as evenly as possible to nodes, while also satisfying constraints (10-11). Given an arbitrary concave, non-decreasing functions U, this will maximize network utility U^{tot}. This property is formally given later in Proposition 1. MAX-UTILITY runs in multiple iterations, and assigns rates to a subset of nodes in each iteration. The iteration ends when rates are assigned to all N nodes.

$CAP[i]$	The remaining rate capacity of node V_i
$unassigned[i]$	The set of unassigned nodes in V_i's subtree τ_i
$\lvert unassigned[i] \rvert$	The size of $unassigned[i]$
$ASSIGNED\_SET$	The set of assigned nodes
$\tilde{R}[]$	The rate assignment derived by MAX-UTILITY
$r_c[i]$	The maximum common rate for nodes in $unassigned[i]$
V_u	The node with the least $r_c[i]$ among all the unassigned nodes

Table 1: List of notations

The algorithm has two lists of inputs and one output. The first input list contains N rate capacities $CAPACITY_i$, one for each node V_i. The second list contains N vectors, one for each V_i. Denote the subtree rooted at node V_i by τ_i. The vector of each V_i contains all the nodes in the subtree τ_i rooted at V_i. This vector is derived based on the previously discovered tree structure. For notational simplicity we re-use τ_i to represent this vector. The output is the rate assignment vector derived by MAX-UTILITY, denoted by $\tilde{R}[]$.

MAX-UTILITY uses one global variable and three per-node variables that are updated from iteration to iteration. The global variable $ASSIGNED\_SET$ is a set containing all the nodes in the network that have been assigned rates so far. As will be seen later, in each iteration MAX-UTILITY adds at least one node to $ASSIGNED\_SET$. The algorithm terminates when all N nodes are in $ASSIGNED\_SET$. Next we declare three per-node variables. First, $CAP[i]$ is the remaining capacity of node V_i, initialized to $CAPACITY_i$. Second, $unassigned[i]$ is a subset of τ_i containing any nodes in τ_i that have not yet been assigned rates. $unassigned[i]$ includes V_i itself, and is initialized to τ_i. Finally, $r_c[i]$ is the maximum common rate for nodes in $unassigned[i]$. Table 1 summarizes these notations. This construction is illustrated in Fig. 1. In the 1st iteration, when none of nodes in subtree τ_7 rooted at V_7 have been assigned, $unassigned[7]$ contains all four nodes in τ_7, $\{V_1, V_2, V_6, V_7\}$. The last variable $r_c[i]$ is the maximum common rate that can be assigned to the nodes in $unassigned[i]$. $r_c[i]$ is computed by dividing V_i's remaining rate capacity $CAP[i]$ by $|unassigned[i]|$ which is the number of nodes in $unassigned[i]$, i.e. $r_c[i] = \frac{CAP[i]}{|unassigned[i]|}$. For example, in the 1st iteration, $r_c[7] = \frac{CAPACITY_7}{|unassigned[7]|} = \frac{16}{4} = 4$.

Algorithm 1 MAX-UTILITY

```
1: - Input: {CAPACITY_i} and {τ_i}; Output: R̃[]
2: - Initialization: ASSIGNED_SET = ∅,
   ∀V_i, CAP[i] = CAPACITY_i, unassigned[i] = τ_i
3: for each node V_i in the network do
4:    If CAP[i] < r^min · |unassigned[i]|, then return ∅
5: end for
6: while |ASSIGNED_SET| < N do
7:    for any node V_i ∉ ASSIGNED_SET do
8:       r_c[i] = CAP[i]/|unassigned[i]|
9:    end for
10:   Find out V_u which has the least r_c[i] among any V_i ∉
      ASSIGNED_SET
11:   for any node V_i ∈ unassigned[u] do
12:      Set R̃[i] = r_c[u], and add V_i to ASSIGNED_SET

13:   end for
14:   for any node V_i ∉ ASSIGNED_SET do
15:      if V_u ∈ unassigned[i], i.e. V_i is V_u's ancestor then
16:         CAP[i] = CAP[i] − r_c[u] · |unassigned[u]|
17:      end if
18:      if V_u ∈ unassigned[i] then
19:         for any node V_j ∈ unassigned[u] do
20:            Remove V_j from unassigned[i].
21:         end for
22:      end if
23:   end for
24: end while
25: return R̃[].
```

MAX-UTILITY, shown in Algorithm 1, is specified as follows. Line 2 initializes $ASSIGNED\_SET$ to $\emptyset$, $CAP[i]$ to $CAPACITY_i$, and $unassigned[i]$ to τ_i. In lines 3-5, for any node V_i we check whether it has sufficient capacity to sustain the minimum required rate r^{min} for all the nodes in its subtree τ_i. If there exists a node with insufficient capacity, then we know there is no feasible rate assignment that can satisfy the min-rate constraint and capacity constraint (at V_i) at the same time. At this point MAX-UTILITY terminates immediately with empty $\tilde{R}[]$. Line 6 starts the rate assignment loop which will terminate when all N nodes are assigned. In lines 7-9, for any node $V_i \notin ASSIGNED\_SET$, we compute the maximum common rate $r_c[i]$ that can be assigned to the nodes in $unassigned[i]$ (line 8). In line 10, we find the node V_u which has the least $r_c[i]$ among any $V_i \notin ASSIGNED\_SET$. Then for any nodes in $unassigned[u]$, MAX-UTILITY assigns $r_c[u]$ to them and adds them to $ASSIGNED\_SET$ (line 11-13).

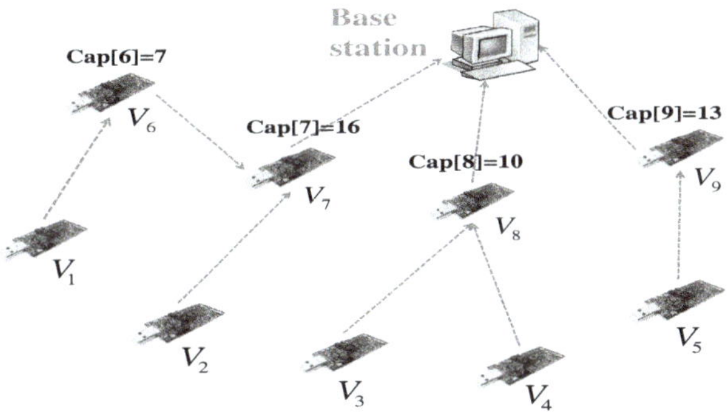

Figure 1: Example Data Collection Tree

Next, lines 14-23 update $CAP[i]$, and $unassigned[i]$ for any nodes without an assigned rate. Specifically, since any node V_i where $V_u \in unassigned[i]$ are ancestors of V_u and need to forward packets received from the nodes in $unassigned[u]$, we need to subtract $r_c[u] \cdot |unassigned[u]|$ (i.e. the total traffic that V_i received from V_u) from their $CAP[i]$ values (line 15-17). For any other nodes that are not ancestors of V_u, their $CAP[i]$ values remain the same. Moreover, any nodes $V_j \in unassign[u]$ with newly assigned rates must be removed from the $unassigned[i]$ sets of any node V_i that is an ancestors of V_u, i.e. $V_u \in unassigned[i]$ (line 18-22). The rate assignment process from line 7 to 23 continues until all N nodes are assigned.

As can be seen, in each iteration, MAX-UTILITY picks a V_u which has the least common rate $r_c[i]$ among any unassigned node V_i, and assigns $r_c[u]$ uniformly to any nodes in $unassigned[u]$, then produces a pruned tree by removing any newly assigned nodes. We refer to the node with the least common rate $r_c[i]$ among any nodes in a tree (or pruned tree) as the *critical node* of the tree. Note that in any iteration, the selected V_u is the critical node in the tree (pruned tree). The concept of critical node plays an important role in our optimality proof.

Steps of MAX-UTILITY are illustrated through an example. In Fig. 1, before the 1st iteration starts, $r_c[6] = \frac{7}{2} = 3.5$, $r_c[7] = \frac{16}{4} = 4$, $r_c[8] = \frac{10}{3}$, $r_c[9] = \frac{13}{2} = 6.5$. Line 10 picks V_8 as V_u, and line 12 assigns $\tilde{R}[3] = \tilde{R}[4] = \tilde{R}[8] = \frac{10}{3}$, and adds V_3, V_4, V_8 to $ASSIGNED\_SET$. Since this assignment does not affect any other nodes, lines 14-23 are not executed. In the 2nd iteration, V_6 is picked as V_u, so MAX-UTILITY assigns $\tilde{R}[1] = \tilde{R}[6] = 3.5$, and updates $CAP[7] = 16 - 7 = 9$, and $unassigned[7] = \{V_2, V_7\}$. V_1, V_6 are added to $ASSIGNED\_SET$. In the 3rd iteration, V_7 is picked as V_u as $r_c[7] = 9/2 = 4.5 < r_c[9] = 6.5$. MAX-UTILITY assigns $\tilde{R}[2] = \tilde{R}[7] = 4.5$, marks V_2, V_7 assigned and makes no update. In the 4th iteration, V_9 is picked as V_u, MAX-UTILITY assigns $\tilde{R}[5] = \tilde{R}[9] = 6.5$, and adds V_5, V_9 to $ASSIGNED\_SET$. Since all the nodes are now in $ASSIGNED\_SET$, the algorithm terminates and returns $\tilde{R}[] = \{3.5, 4.5, 10/3, 10/3, 6.5, 3.5, 4.5, 10/3, 6.5\}$.

5.2 Optimality of algorithm MAX-UTILITY

In Eq. (1), the network utility U^{tot} is defined as the aggregate utility accrued jointly by all the nodes, $\sum_{i=1}^{N} U(r_i)$. Since $U(r_i)$ is concave, U^{tot} can be maximized by assigning packet rates to nodes as evenly as possible, while satisfying the min-rate, and capacity constraints at any nodes. This observation is supported by the following proposition, which can be proved based on the properties of concave functions:

PROPOSITION 1. *Given n non-negative real numbers $\{r_1, \ldots, r_n\}$ where $\sum_{i=1}^{n} r_i \leq W$ (W is any non-negative constant), $\sum_{i=1}^{n} U(r_i)$ (U is a concave, non-decreasing function of r) is maximized when $r_1 = r_2 = \ldots = r_n = W/n$.*

Now we prove that the rate assignment $\tilde{R}[]$ derived by Algorithm MAX-UTILITY maximizes U^{tot}, while satisfying constraints (10-11). For the sake of contradiction, we assume that $\tilde{R}[]$ is not optimal, and that there exists an optimal assignment $R^*[]$ which differs from $\tilde{R}[]$. The network utility resulting from $R^*[]$ is denoted by $U^{tot,*}$. We will show that $R^*[] \neq \tilde{R}[]$ contradicts the optimality of $R^*[]$, hence proving that $R^*[i] = \tilde{R}[i]$ must hold for every node V_i in the network. We first propose Lemma 1, which is proved in the Appendix.

LEMMA 1. *Given an arbitrary tree-based network, the optimal rate assignment $R^*[]$ must assign equal rates to all nodes in $unassigned[u]$, where V_u is the critical node of the tree.*

Lemma 1 implies that $R^*[]$ must assign a common rate denoted as $R^*[u]$ to all the nodes in $unassigned[u]$. We now present Lemma 2, which shows that $R^*[u]$ must be equal to $r_c[u]$, i.e. the maximum common rate that can be assigned to nodes in $unassigned[u]$.

LEMMA 2. *Given an arbitrary tree-based network where V_u is the critical node in the tree, the optimal rate assignment $R^*[]$ must assign a common rate $r_c[u]$ to any nodes in $unassigned[u]$, regardless of the rates assigned to other nodes in the tree.*

The proof of Lemma 2 can be found in the Appendix. Since $\forall V_i \in unassigned[u], \tilde{R}[i] = r_c[u]$ holds (line 12), Lemma 2 implies that $\forall V_i \in unassigned[u]$, $R^*[i] = \tilde{R}[i]$ must hold. In other words $\tilde{R}[]$ derived by MAX-UTILITY is optimal for any nodes in $unassigned[u]$, where V_u is the critical node. Since assigning $R^*[u] = r_c[u]$ to any node $V_i \in unassigned[u]$ is optimal regardless of what rates are assigned to the rest of nodes, we can focus on finding the optimal rate assignment for a *partial tree* formed by disregarding any nodes in $unassigned[u]$ (i.e. removing nodes in $unassigned[u]$ from $unassigned[i]$ of any ancestors of V_u), and setting $CAP[i] = CAP[i] - r_c[u] \cdot |unassigned[u]|$ for any node V_i that is an ancestor of V_u. Note that the partial tree generated in this way contains exactly the set of nodes remaining unassigned at the beginning of the 2nd iteration of MAX-UTILITY. Also each node in the partial tree must have the same $CAP[i]$ and $unassigned[i]$ as they have in the 2nd iteration of MAX-UTILITY. Since Lemma 2 applies to an arbitrary tree, it holds for this partial tree as well. In other words, $R^*[]$ must assign a common rate $r_c[u']$ to any nodes $V_i \subset unassigned[u']$, where $V_{u'}$ is the critical node of the partial tree.

Since the structure and capacity of the partial tree is the same as the updated tree in the 2nd iteration, the critical node $V_{u'}$ must be exactly the V_u picked by MAX-UTILITY in the 2nd iteration. Thus MAX-UTILITY must also assign $r_c[u']$ to any $V_i \in unassigned[u']$. Therefore, $\forall V_i \in$

$unassigned[u']$, $\tilde{R}[i] = r_c[u']$ must hold. Since $\forall V_i \in unassigned[u']$, $R^*[i] = r_c[u']$ holds as implied by Lemma 2, we have $\forall V_i \in unassigned[u']$, $\tilde{R}[i] = R^*[i]$ must hold. That is, $\tilde{R}[]$ is optimal for any nodes in $unassigned[u']$. Again, we may disregard any nodes in $unassigned[u']$ and focus on the new partial tree formed in this way. The new partial tree has the same structure and capacities as the updated tree in the 3rd iteration of MAX-UTILITY. As we iteratively prune the tree (partial tree) and apply Lemma 2 to the generated partial trees until all the nodes are pruned except the root BS, we have proved that the $\tilde{R}[]$ derived by MAX-UTILITY is completely identical to $R^*[]$, and therefore $\tilde{R}$ is optimal for the entire original tree.

Complexity analysis:

We show that for N nodes, the centralized MAX-UTILITY runs in $O(N^3)$ time. We first focus on one iteration of the while-loop from lines 6-24. The for-loop from lines 7 to 9 has complexity $O(N)$, since there are $O(N)$ unassigned nodes in each iteration. The selection of V_u in line 10 has complexity $O(N)$ as this is equivalent to finding the minimum element among $O(N)$ $r_c[i]$ values. The rate assignment to nodes in $unassigned[u]$ in lines 11-13 has complexity $O(N)$ as well, since the size of $unassigned[u]$ is $O(N)$. The for-loop from lines 14 to 23 has complexity of $O(N^2)$. In each loop, we need to update $CAP[i]$ and $unassigned[i]$ for $O(N)$ unassigned nodes V_i. Updating $CAP[i]$ for each V_i takes $O(1)$ time (line 16), while updating each $unassigned[i]$ take $O(N)$ time as it requires removal of $O(N)$ nodes (lines 19-21). Thus, the entire for-loop completes in $O(N^2)$ time as there are $O(N)$ nodes $V_i \notin ASSIGNED\_SET$. Therefore, each iteration of MAX-UTILITY completes in $O(N^2)$ time.

Next the number of iterations of MAX-UTILITY is given by $O(N)$, as the set $ASSIGNED\_SET$ is initially empty and each iteration adds at least one node to $ASSIGNED\_SET$. Therefore, the total complexity of MAX-UTILITY is $O(N^3)$. For practical purposes we expect that the number of iterations will be small. This is because the nodes that are close to the root have more descendent nodes, hence much larger $unassigned[i]$ than the nodes in the lower levels of the tree. Therefore the high level nodes usually have small $r_c[i]$ values and are more likely to be picked as V_u. Since once these nodes are picked as V_u, all nodes in their $unassigned[i]$ sets are assigned and added to $ASSIGNED\_SET$ at once in one iteration, the set $ASSIGNED\_SET$ will grow rapidly and the expected number of iterations of MAX-UTILITY should be small. We will justify this assertion through simulation analysis.

5.3 The distributed version

We now present the distributed version of MAX-UTILITY, referred as MAX-UTILITY-D. We decompose and distribute the computations in Algorithm 1 to every node in the network. The purpose of MAX-UTILITY-D is to support systems that do not possess or make use of a single resource-rich central control point. MAX-UTILITY-D only requires a single coordinator node such as a routing tree root. The tree root can be any node in the network. The root does not perform computations but only disseminates the minimum common rate $r_c[u]$ to all other nodes.

MAX-UTILITY-D is specified as follows. Initially, each node individually computes its rate capacity $CAPACITY_i$ using Eq. (9). Then MAX-UTILITY-D starts an initialization stage during which all the nodes send an empty control packet to the root. If a node V_i receives a control packet from another node V_j, then V_i knows V_j resides in its subtree τ_i. By the end of the initialization stage, each node knows its subtree τ_i, and hence the initial $unassigned[i]$. Using $CAPACITY_i$ and $|unassigned[i]|$, V_i computes the common rate $r_c[i]$ that can be assigned to nodes in $unassigned[i]$.

Each iteration of MAX-UTILITY-D consists of two stages. The first stage determines the minimum common rate $r_c[u]$ and the critical node V_u in the tree by requiring all the nodes forward their $r_c[i]$ values to the root. In the second stage, the root disseminates $r_c[u]$ across the network, and all nodes in $unassign[u]$ receive and use $r_c[u]$ as their packet rate. In the first stage, each leaf node in the tree V_i sends its computed $r_c[i]$ towards the root. V_i's parent V_j receives $r_c[i]$, compares $r_c[i]$ to $r_c[j]$ of itself, and forwards the larger of the two upwards. V_j also temporarily stores any received $r_c[i]$. This process proceeds over the entire tree in bottom-up fashion, and finally the root discovers the minimum common rate $r_c[u]$ in the tree. Then in the second stage the root disseminates $r_c[u]$ across the tree. When a node receives $r_c[u]$ from the root, it compares $r_c[u]$ to its local r_c to identify whether it is the critical node. If it is not the critical node, then it compares $r_c[u]$ to each of the previously stored r_c values of its descendent nodes to find out who is the critical node. In this way the critical node V_u identifies itself, and commands all unassigned nodes in $unassigned[u]$ to use $r_c[u]$ as their rates for the coming epoch. This rate assignment command can be piggybacked in the packet for disseminating $r_c[u]$.

After the dissemination of $r_c[u]$ completes, all the nodes has learned the identity of the critical node V_u, either by comparing $r_c[u]$ to its own or stored r_c values, or from the rate assignment command issued by V_u. Finally, we need to update $CAP[i]$ and $unassigned[i]$ of the nodes remaining unassigned. If a node V_i is an ancestor of V_u, it has to update $CAP[i] = CAP[i] - r_c[u] \cdot |unassigned[u]|$ and then remove all nodes in $unassigned[u]$ from $unassigned[i]$. At the end of the iteration, each node discards the temporarily stored information and continues to the next iteration. In the next iteration, only the nodes remaining unassigned participate. The rate assignment iteration continues until all the nodes are assigned.

Complexity analysis:

In MAX-UTILITY-D, the tree root only disseminates $r_c[u]$. Each non-root node requires $O(N)$ comparisons in each iteration. This is because each non-root node compares its own r_c to $O(N)$ r_c values received from its descendent nodes (the 1st stage), and compares $r_c[u]$ disseminated by the root to its own r_c and $O(N)$ temporarily stored r_c. Further, updating $CAP[i]$ and $unassigned[i]$ takes $O(N)$ time for each node as described in the complexity analysis of centralized MAX-UTILITY.

Now we analyze the message complexity of the algorithm. The initialization stage requires one round of network-wide data collection. In each iteration, the determination of $r_c[u]$ requires all the nodes to report their $r_c[i]$ values, thus incurring another round of data collection. The root then announces $r_c[u]$ to all the nodes, incurring one round of network-wide data dissemination. In each data collection and dissemination round, each node sends exactly one packet. Given $O(N)$ iterations, MAX-UTILITY-D needs $O(N)$ rounds of network-wide data collection and dissemination. Finally, as mentioned in the analysis of the centralized MAX-

UTILITY, the expected number of iterations is much lower than N in practice.

6. PERFORMANCE EVALUATION

Though we have formally proved the optimality of MAX-UTILITY, we have conducted a series of simulation experiments to evaluate its performance gain and overhead. We compare our algorithm against an alternative heuristic, called *Random Rate Augmentation* (RRA). The specification of RRA is given in the Appendix. By design, the rate assignment derived by RRA is also feasible in term of satisfying constraint (10-11). Although RRA is not optimal, it achieves reasonably high utility values as it maximizes total packet rate (flow) over the tree. Other algorithms, such as the one proposed in [6] are not directly comparable to ours, since their application and network model are different.

We compare the two rate allocation algorithms under various experiment settings that consider different energy budgeting schemes, utility functions, and network sizes. Without loss of generality, we consider solar-powered sensor networks in our simulation. The solar power harvesting profile is obtained from the Hamburg University of Technology [17]. We set the length of horizon to one day (24 hours) which is further divided into 96 epochs each with length of 15 minutes. We assume two energy budgeting schemes, denoted as Strictly Energy Neutral (SEN) and Maximum Uniform Budget (MUB). The SEN scheme assigns the exact amount of harvested energy as the energy budget for any epoch. For the evening epochs with extremely low harvested energy, SEN assigns a minimum energy budget equaling $5J$ to maintain the minimum rate r^{min}. The MUB scheme computes the maximum uniform energy budget across all 96 epochs while satisfying the min-rate and capacity constraints, and assigns this budget to every epochs. The MUB scheme uses the algorithm proposed in [9]. Note that the energy budget varies across epochs if SEN scheme is used, while the budget remains constant if MUB scheme is used. We are interested in evaluating how different budgeting schemes affect the network utility. This helps system designers in choosing the right energy budgeting scheme for their applications. We consider three different concave utility functions: $log(100r+1)$ (denoted as LOG), $\sqrt{100r}$ (SQR) and $log[\sqrt{1000r}+1]$ (denoted as composite utility or CPST). We use the coefficients 100 and 1000 with r because r is commonly small. Finally, we repeat our simulation in six networks of different size, containing 25, 36, 64, 100, 169, 225 nodes respectively. We examine the accrued utilities, and the algorithm overhead as the network grows.

The sensor nodes are organized into a collection tree using the CTP protocol. The energy consumption for transmitting a packet is randomly selected from the range $(0.05-0.1)J$. This range is obtained based on the measurements in [20]. The energy storage device has capacity of 1000J. The initial energy level in the horizon is set to 500J. The minimum rate r^{min} is set to 0.01, i.e. 1 packet per 100 seconds.

6.1 Simulation results

We evaluated algorithm MAX-UTILITY and RRA in TOSSIM/TinyOS. We present our evaluation results along four dimensions: accrued network utility U^{tot}; rate (r_i) distribution among nodes; energy level variation across epochs of two selected nodes; algorithm running time and control overhead. We calculate the energy level Γ_k of a node at the end of the kth epoch using the formulas below:

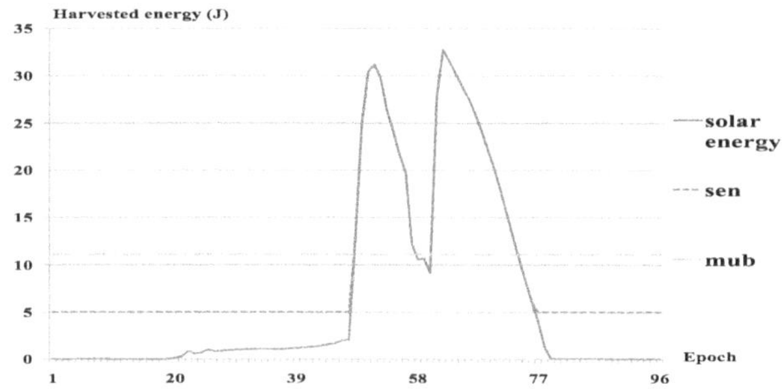

Figure 2: Solar energy harvested throughout a day.

$$\Gamma_1 = Min.\{\Gamma^{init} + P_1^h \cdot S - E_1^c, \Gamma^{max}\}$$
$$\Gamma_k = Min.\{\Gamma_{k-1} + P_k^h \cdot S - E_k^c, \Gamma^{max}\}$$

Γ^{init} is the initial energy level. Γ^{max} is the energy capacity. Recall that P_k^h is the harvested power, S is the epoch length, and E_k^c is the energy consumed in epoch k (Eq. (2)). Γ_{k-1} is the ending energy level in the previous epoch $k-1$, which is also the starting energy level in epoch k. $P_k^h \cdot S$ gives the energy harvested in epoch k.

In Fig. 2, we show the solar energy trace, along with the energy budget assignments derived by SEN and MUB schemes. As seen from the figure, the amount of solar energy reaches its peak in the afternoon due to high sunlight intensity. The line representing the energy budget assignment derived by SEN overlaps the line representing the solar energy in noon and afternoon. This is because SEN assigns the amount of harvested energy as the energy budget for any epochs, except for the evening epochs in which the harvested energy is less than 5J. Finally, MUB assigns the highest uniform budget (around 11J) for any epochs.

6.2 Network utility

In Fig. 3 and 4, we compare the utility accrued by MAX-UTILITY and RRA algorithm. We show the results for the 100-node network. In each figure we plot the utility values in all 96 epochs, for all three utility functions. In each figure, we plot six utility lines, one for each combination of the three utility functions and two rate allocation algorithms.

Fig. 3 assumes SEN energy budgeting scheme. Since SEN scheme assigns the harvested energy as the energy budget, we observe that the variations in accrued utility indicate similar pattern as the harvested energy (Fig. 2). Specifically the utility reaches its peak in the afternoon since high solar energy availability leads to high rate capacity $CAPACITY_i$ at any nodes. For any utility functions, MAX-UTILITY (OPT) achieves significantly higher utility than RRA. Specifically for SQR, the utility value increases above 500 in the afternoon if MAX-UTILITY is used. However, if RRA is used, the utility accrued never increases above 300. For the rest of the day, MAX-UTILITY achieves utility figures that are mostly above 200, while RRA achieves utility figures that are around 100.

In Fig. 4, we repeat the simulation for MUB energy budgeting. Because the MUB scheme is used, the energy budget B is uniform across different epochs, thus yielding constant rate capacities at nodes and much more stable utility per-

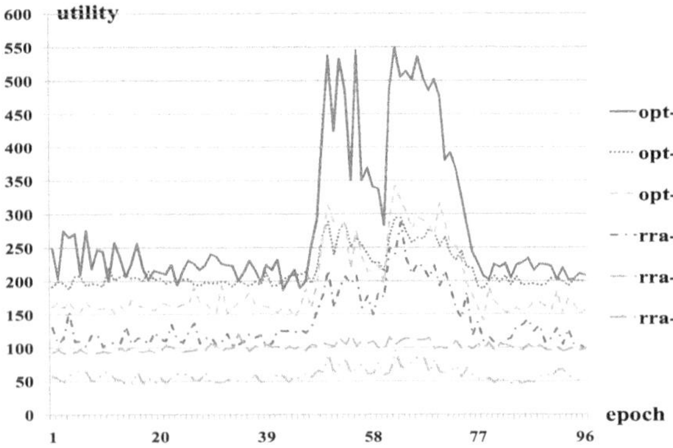

Figure 3: Utility accrued by max-utility (opt) and rra, with different utility functions, SEN budgeting.

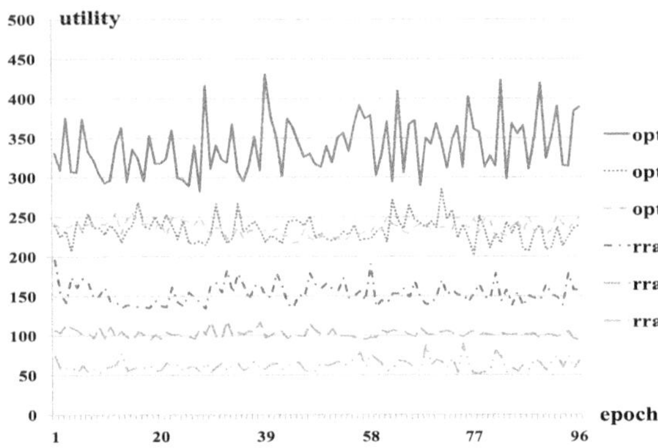

Figure 4: Utility accrued by max-utility (opt) and rra, with different utility functions, MUB budgeting.

formance in all six lines. Again, MAX-UTILITY achieves much higher utility than RRA. For SQR, the utilities accrued by MAX-UTILITY remain around 350 throughout the day, while the one for RRA remains around 150. As seen from Fig. 3 and 4, we observe that the MUB scheme maintains stable and high utility at any time and is hence suitable for the applications which desire utility stabilization. On the other hand, SEN energy allocation achieves very high utility when harvesting ability is high and is suitable for the applications where maximum utility is sought.

6.3 Rate assignment and energy storage level

In this section, we show the resulting rate assignment and energy storage levels of the two algorithms. First, we plot in Fig. 5 the distribution of rates of all the nodes in the 100 nodes network at midnight, while using the SEN scheme and the CPST utility function. From the figure, we observe that MAX-UTILITY tends to distribute rates more evenly across nodes than RRA. The rate assignment by MAX-UTILITY has standard deviation of 0.1268, while for RRA the deviation equals 0.1936. In Fig. 6(a-b) we plot the energy levels of a heavily-loaded node and a lightly-loaded node in 96 epochs. The heavily-loaded node is high in the tree, thus has higher workload and energy demand and lower energy

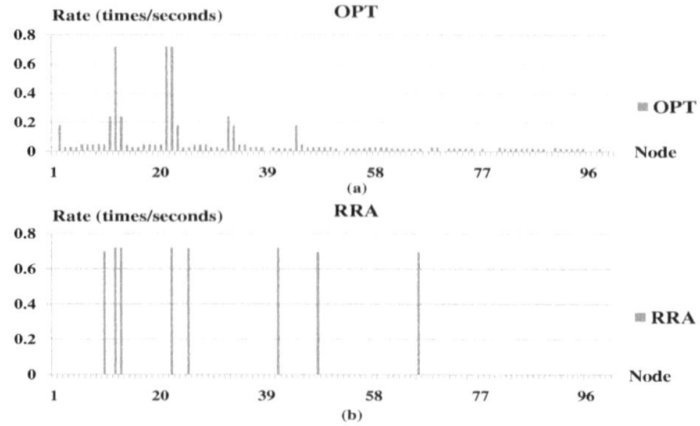

Figure 5: Distribution of rate (times/seconds) of all the nodes at midnight by (a) MAX-UTILITY; (b) RRA.

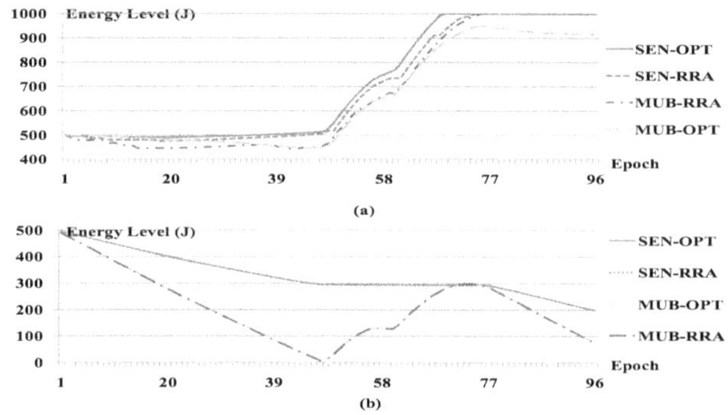

Figure 6: Variation of Energy levels over 96 epochs of (a) a lightly-loaded node; (b) a heavily-loaded node.

level. As seen from the figure, the energy neutral condition is maintained for both nodes at all times.

6.4 The impact of network size

Now we study the impact of network size on performance and overhead of MAX-UTILITY. In Fig. 7, we increase the size of the network from 25 nodes to 225 nodes, and plot the utility accrued by MAX-UTILITY. This simulation assumes SEN energy budgeting and LOG utility. We observe from the plot that the utility grows steadily as the network grows from 25 nodes to 169 nodes. As computed, the average utility is 78.10 for 25 nodes network, 109.34 for 36-nodes network, 142.84 for 64-nodes network, 203.79 for 100-nodes network, 396.54 for 169-nodes network, respectively. However, as the network size increases to 225, the utility increases to only 403.48. In other words the utility growth almost stops. This is because that, although there are potentially more utility contributors in the 225-nodes network, the achievable utility is constrained by the nodes that are close to the tree root, since the rate capacity of these nodes does not change with the network size. System designers must be aware of this phenomenon while choosing the network size and routing patterns.

We measure the actual running time of MAX-UTILITY in terms of the number of iterations shown in Table 2. Again, we assume 6 network sizes. As seen from the table, the

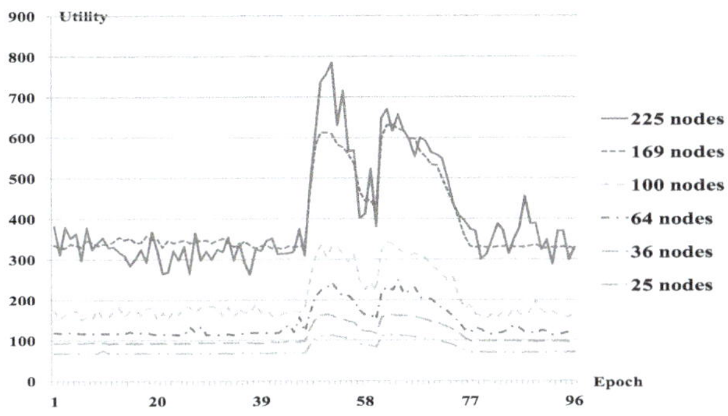

Figure 7: Total utility accrued by MAX-UTILITY in networks of different size

number of iterations does not increase with the network size. On the contrary, we observe that the 25-node network requires the largest number of iterations, 9.97, while the 64-nodes network requires the smallest number of iterations, 6.99. Based on the tree structure we observe that the number of iterations depends heavily on the number of children the tree root has. As mentioned in 5.2, the children of the root have large group of descendent nodes, hence are more likely to be picked as the critical node V_u. Once they become V_u, all their descendent nodes are added to the set $ASSIGNED\_SET$ in one iteration. This leads to the rapid growth of $ASSIGNED\_SET$.

Finally, we count the average number of control packets sent by a node in each invocation of the distributed MAX-UTILITY to be 27.32 in 100-nodes network. If the node operates at rate $r_i = 1$ packet per second then 900 data packets will be sent in an epoch. The packet overhead is then $27.32/900 \approx 3\%$. This indicates that the MAX-UTILITY-D is an entirely feasible alternative to MAX-UTILITY.

Num. of nodes	25	36	64	100	169	225
Num. of iterations	9.97	7.99	6.99	7.99	8.99	8.99

Table 2: The number of iterations of MAX-UTILITY as a function of network size

7. CONCLUSIONS

This paper addressed a utility maximization problem for energy harvesting sensor networks. Utility is defined as concave and non-decreasing function of nodes sensing rate. We proposed MAX-UTILITY, a rate allocation algorithm to maximize the total utility achieved jointly by all the nodes, while ensuring energy neutrality for any nodes. We formally proved the optimality of the algorithm, and indicated that the algorithm can derive optimal rate assignment in $O(N^3)$ time where N is the network size. Finally, extensive simulation results demonstrate its superior performance.

Acknowledgments: This work was supported, in part, by US National Science Foundation awards CNS-1016855 and CNS-0546244 (CAREER Award).

8. REFERENCES

[1] K.-W. Fan, Z. Zheng, and P. Sinha. Steady and fair rate allocation for rechargeable sensors in perpetual sensor networks. In *the 6th ACM conference on Embedded network sensor systems*, (SenSys'08), Raleigh, North Carolina, 2008.

[2] S. Farhan, S. Devyani, and P. Chou. Everlast: Long-life, supercapacitor-operated wireless sensor node. In *the 6th ACM conference on Embedded network sensor systems*, (Sensys'09), San Diego, California, 2009.

[3] O. Gnawali, R. Fonseca, K. Jamieson, D. Moss, and P. Levis. Collection tree protocol. In *the 6th ACM conference on Embedded network sensor systems*, (Sensys'05), Berkeley, California, 2009.

[4] X. Jiang, J. Polastre, and D. Culler. Perpetual environmentally powered sensor networks. In *the 4th ACM/IEEE international symposium on Information processing in sensor networks*, (IPSN'05), Los Angeles, California, 2005.

[5] A. Kansal, J. Hsu, S. Zahedi, and M. B. Srivastava. Power management in energy harvesting sensor networks. *ACM Trans. on Embedded Computing System*, 6(4):32, 2007.

[6] R. Liu, P. Sinha, and C. E. Koksal. Joint energy management and resource allocation in rechargeable sensor networks. In *the 29th IEEE Conference on Information Cmmunications*, (Infocom'10), San Diego, California, 2010.

[7] S. H. Low. A duality model of tcp and queue management algorithms. *IEEE/ACM Trans. on Networking*, 11:525–536, 2002.

[8] S. Madden, M. Franklin, J. Hellerstein, and W. Hong. Tag: A tiny aggregation service for ad-hoc sensor networks. In *the USENIX Symposium on Operating Systems Design and Implementation*, (OSDI'02), Boston, MA, 2002.

[9] C. Moser, J. Chen, and L. Thiele. Reward maximization for embedded systems with renewable energies. In *the 14th IEEE International Conference on Embedded and Real-Time Computing Systems and Applications*, (RTCSA'08), Kaohsiung, Taiwan, 2008.

[10] C. Moser, J.-J. Chen, and L. Thiele. Power management in energy harvesting embedded systems with discrete service levels. In *the 14th ACM/IEEE International Symposium on Low Power Electronics and Design*, (ISLPED'09), San Fancisco, California, 2009.

[11] C. Moser, L. Thiele, D. Brunelli, and L. Benini. Robust and low complexity rate control for solar powered sensors. In *the conference on Design, automation and test in Europe*, (Date'08), Munich, Germany, 2008.

[12] C. Moser, L. Thiele, D. Brunelli, and L. Benini. Adaptive power management for environmentally powered systems. *IEEE Transactions on Computers*, 59(4):478 – 491, 2010.

[13] D. K. Noh and K. Kang. A practical flow control scheme considering optimal energy allocation in solar-powered wsns. In *the 18th IEEE International Conference on Computer Communications and Networks*, (ICCCN'09), Washington, DC, 2009.

[14] D. K. Noh, L. Wang, Y. Yang, H. K. Le, and T. Abdelzaher. Minimum variance energy allocation for a solar-powered sensor system. In *the 5th IEEE International Conference on Distributed Computing in Sensor Systems*, (DCOSS'09), Marina Del Rey, California, 2009.

[15] D. P. Palomar and M. Chiang. A tutorial on decomposition methods for network utility maximization. *IEEE J. Sel. Areas Commun*, 24:1439–1451, 2006.

[16] J. Polastre, J. Hill, and D. Culler. Versatile low power media access for wireless sensor networks. In *the 2nd ACM International conference on Embedded networked sensor systems*, (SenSys'04), Baltimore, Maryland, 2004.

[17] C. Renner, J. Jessen, and V. Turau. Lifetime prediction for supercapacitor-powered wireless sensor nodes. In *the 8th GI/ITG KuVS Fachgespräch "Drahtlose Sensornetze"*, Hamburg, Germany, 2009.

[18] C. Rusu, R. Melhem, and D. Mossé. Maximizing the system value while satisfying time and energy constraints. In *the 23rd IEEE Real-Time Systems Symposium*, (RTSS'02), Washington, DC, 2002.

[19] C. Rusu, R. Melhem, and D. Mossé. Multi-version scheduling in rechargeable energy-aware real-time systems. *J. Embedded Comput.*, 1(2):271–283, 2005.

[20] V. Shnayder, M. Hempstead, B. rong Chen, G. W. Allen, and M. Welsh. Simulating the power consumption of large-scale sensor network applications. In *the 6th ACM conference on Embedded network sensor systems*, (Sensys'04), Baltimore, Maryland, 2004.

[21] L. Su, Y. Gao, Y. Yang, and G. Cao. Towards optimal rate allocation for data aggregation in wireless sensor networks. In *the 12th ACM International Symposium on Mobile Ad Hoc Networking and Computing*, (Mobihoc'11), Paris, France, 2011.

[22] X. Wang and K. Kar. Cross-layer rate control for end-to-end proportional fairness in wireless networks with random access. In *the 6th ACM International Symposium on Mobile Ad Hoc Networking and Computing*, (Mobihoc'05), Urbana-Champaign, Illinois, 2005.

[23] B. Zhang, R. Simon, and H. Aydin. Energy management for time-critical energy harvesting wireless sensor networks. In *the 12th International Symposium on Stabilization, Safety and Security of Distributed Systems*, (SSS'10), New York City, New York, 2010.

APPENDIX

Proof of Lemma 1:

For the purpose of contradiction, we assume that, in $R^*[]$, nodes in *unassigned*$[u]$ are assigned unequal rates $\{\mu_1, \ldots, \mu_{|unassigned[u]|}\}$. However, as shown by Proposition 1 if this assumption holds we can always further increase $U^{tot,*}$ by equalizing $\{\mu_1, \ldots, \mu_{|unassigned[u]|}\}$ to a common rate $R^*[u]=\frac{\mu_1+\ldots+\mu_{|unassigned[u]|}}{|unassigned[u]|}$. This contradicts the optimality of $R^*[]$. Therefore, *equal rate assignment to nodes in unassigned*$[u]$ *is a necessary condition for the optimality of* $R^*[]$.

We still need to show such rate equalization does not violate the min-rate and capacity constraints (Eq. (10-11)). This is justified as follow. First, the min-rate constraint is not violated, because the new common rate $R^*[u]$ must be larger than the minimum of $\{\mu_1, \ldots, \mu_{|unassigned[u]|}\}$, which is in turn larger than r^{min} (due to the feasibility of $R^*[]$). The rate capacity constraint is also satisfied. This is because V_u is the critical node with the least common rate r_c among any nodes in the tree. Thus, $r_c[u]$ cannot be larger than $r_c[i]$ of any node V_i in *unassigned*$[u]$. Further, $R^*[u] \leq r_c[u]$ must hold, since if $R^*[u] > r_c[u]$ we have $R^*[u] \cdot |unassigned[u]| > r_c[u] \cdot |unassigned[u]|$, which gives $\mu_1 + \ldots + \mu_{|unassigned[u]|} > CAP[u]$. This implies that $R^*[]$ violates the capacity constraint which contradicts the feasibility of $R^*[]$. Since $R^*[u] \leq r_c[u]$, and $r_c[u] \leq r_c[i]$, $\forall V_i \in unassigned[u]$ we know the common rate $R^*[u]$ cannot be larger than the maximally allowed common rate $r_c[i]$ of any node V_i in *unassigned*$[u]$, hence cannot violate the capacity constraint at these nodes. Finally, the capacity violation cannot happen to any nodes outside *unassigned*$[u]$ since the total packet flow at V_u does not change after the rate equalization. Therefore, we have proved that in $R^*[]$, any nodes in *unassigned*$[u]$ must use a common rate.

Proof of Lemma 2:

In Lemma 1, we have shown that in $R^*[]$ all nodes in *unassigned*$[u]$ must be assigned a common rate $R^*[u]$, where V_u is the critical node in the tree. Now we show that this common rate $R^*[u]$ must be equal to $r_c[u]$ in order for $R^*[]$ to be optimal. Recall that $r_c[u] = \frac{CAP[u]}{|unassigned[u]|}$. Notice that nodes in *unassigned*$[u]$ cannot be assigned a common rate $R^*[u]$ that is higher than $r_c[u]$ as this will exceed V_u's capacity, so $R^*[u] > r_c[u]$ cannot hold. Therefore we assume $R^*[u] < r_c[u]$. We show this assumption cannot hold either, and thereby $R^*[u] = r_c[u]$ must hold.

If $R^*[u] < r_c[u]$ holds, we assert that the rate capacity of V_u is not fully utilized. This is because $R^*[u] < r_c[u]$ means

$$R^*[u] \cdot |unassigned[u]| < r_c[u] \cdot |unassigned[u]| = CAP[u]$$

Since the left-hand side of the above inequality gives the total packet rate at V_u, if $R^*[u]$ is assigned to every nodes in *unassigned*$[u]$, this inequality implies V_u's capacity $CAP[i]$ is under-utilized.

Given that V_u's capacity is under-utilized, we make another assertion that there exists a node V_k, an ancestor of V_u, that must have its rate capacity fully utilized. This is because if all the ancestors of V_u are also under-utilized, then we can further increase $U^{tot,*}$ by increasing $R^*[u]$ by an amount of $\epsilon/|unassigned[u]|$, where $\epsilon > 0$ is the strictly positive, minimum un-utilized capacity among any ancestors of V_u. This contradicts the optimality of $R^*[]$. If there are multiple such V_k nodes, we consider the one that is in the lowest level of the tree.

Given the existence of V_k, we assert that in $R^*[]$ there must exist a node $V_j \in \tau_k$ and $V_j \notin unassigned[u]$ which has rate $R^*[j] > R^*[u]$. (Recall that τ_k is the subtree rooted at V_k.) This is because if no node in τ_k has rate higher than $R^*[u]$, then the total packet rate at V_k must be strictly smaller than $R^*[u] \cdot |\tau_k|$ ($|\tau_k|$ is the number of nodes in τ_k), and in turn no larger than $r_c[k] \cdot |\tau_k| = CAP[k]$, i.e. V_k's capacity. $R^*[u] \cdot |\tau_k| \leq r_c[k] \cdot |\tau_k|$ holds because $R^*[u] < r_c[u]$ is assumed, and $r_c[u] \leq r_c[k]$ since V_u has the least r_c among any nodes in the tree. This implies the capacity of V_k is under-utilized which contradicts our previous assertion that V_k is fully utilized. Thus we justify the existence of such V_j.

Next, given $R^*[j] > R^*[u]$, we know that the marginal utility of the concave utility function U at point $R^*[j]$ is lower than that at $R^*[u]$. Therefore we can further increase $U^{tot,*}$ by decreasing $R^*[j]$ by a strictly positive amount ϵ', and increasing $R^*[u]$ for each node in *unassigned*$[u]$ by the amount of $\frac{\epsilon'}{|unassigned[u]|}$, as long as $R^*[j] > R^*[u]$ holds. In other words, we can always increase $U^{tot,*}$ by reducing the difference between $R^*[j]$ and $R^*[u]$. This rate adjustment cannot violate the min-rate constraint as $R^*[u] > r^{min}$. The capacity constraint is also not violated, as the decrease of $R^*[j]$ does not increase the total packet rate at any nodes. The increase of $R^*[u]$ by $\frac{\epsilon'}{|unassigned[u]|}$ will not exceed the capacities of any V_u's ancestors inside τ_k, as long as ϵ' is no larger than the minimum unused capacity of any ancestors of V_u in τ_k. This is because V_k is the lowest ancestor of V_u whose capacity is fully utilized, hence any ancestors of V_u inside τ_k (i.e. below V_k) must have positive unused capacity. Further, the total packet rate at any V_u's ancestors outside τ_k does not change after the rate adjustment, so their capacity constraints cannot be violated either.

Therefore, the feasible increase of $U^{tot,*}$ contradicts the optimality of $R^*[]$, hence justifying that $R^*[u] < r_c[u]$ cannot hold, and $R^*[u] = r_c[u]$ must hold in order for $R^*[]$ to be optimal. Therefore we have proved Lemma 2.

Specification of the RRA algorithm

The RRA algorithm initially allocates r^{min} to every node, and updates the remaining capacity of nodes accordingly. If this violates the capacity constraints at some nodes, then we know there exists no feasible assignments for the given network. Otherwise, we start from that assignment and progressively improve U^{tot} by augmenting the rates of nodes with unused capacity. Specifically, based on the initial assignment, RRA randomly selects a node V_i with positive remaining capacity, and determines the maximum possible rate increment which V_i and its ancestors can afford. Then RRA adds up this increment to V_i's current rate, and zeros out its remaining capacity. RRA updates the remaining capacity of other nodes accordingly. This is the end of one round of augmentation. We then continue augmenting other nodes until no node has remaining capacity. The resulting rates of nodes is the final utility achieved by RRA.

Optimizing Push/Pull Envelopes for Energy-Efficient Cloud-Sensor Systems

Yi Xu
CISE Department
University of Florida
Gainesville, FL32611, USA
yixu@cise.ufl.edu

Sumi Helal
CISE Department
University of Florida
Gainesville, FL32611, USA
helal@cise.ufl.edu

My T. Thai
CISE Department
University of Florida
Gainesville, FL32611, USA
mythai@cise.ufl.edu

Mark Schmalz
CISE Department
University of Florida
Gainesville, FL32611, USA
mssz@cise.ufl.edu

ABSTRACT

Unlike traditional distributed systems, where the resources/needs of computation and communication dominate the performance equation, sensor-based systems (SBS) raise new metrics and requirements for sensors as well as for computing and communication. This includes sensing latency and energy consumption. In this paper, we present a performance model for SBS based on a three-tier architecture that uses edge devices to connect massive-scale networks of sensors to the cloud. In this architecture, which we call *Cloud, Edge, and Beneath* (CEB), initial processing of sensor data occurs in- and near-network, in order to achieve system sentience and energy efficiency. To optimize CEB performance, we propose the concept of *optimal push/pull envelope* (PPE). PPE dynamically and minimally adjusts the base push and pull rates for each sensor, according to the relative characteristics of sensor requests (demand side from the Cloud) and sensor data change (supply side from Beneath). We demonstrate the CEB architecture and its push/pull envelope optimization algorithm in an experimental evaluation that measures energy savings and sentience efficiency over a wide range of practical constraints. In addition, from the experiments we demonstrate that by combining PPE optimization algorithm with lazy sampling algorithm, we can achieve further energy saving.

Categories and Subject Descriptors

C.2 COMPUTER-COMMUNICATION NETWORKS. C.2.1 Network Architecture and Design. C.2.4 Distributed Systems. C.4 PERFORMANCE OF SYSTEMS.

General Terms

Algorithms, Experimentation, Measurement, Performance.

Keywords

Pervasive computing, cloud computing, energy efficiency, sentience efficiency, push pull envelope, optimization, performance

1. INTRODUCTION

Mobile and pervasive computing have recently garnered significant attention because of their potential to enable novel and attractive solutions in areas such as environmental monitoring [7], transportation enterprises [16], and health care [8]. The development of mobile and pervasive computing has benefitted from many different views of the communication and computing universe such as layered design, client-server, distributed networks, cloud-based computing, and so forth. In particular, cloud computing provides on-demand provision of computational services, allowing end users to access applications and data from a cloud on demand, anywhere in the world. Applications are hosted in terms of the *Software as a Service* (SaaS) model, whereby cloud server(s) automatically scale to meet client demand. However, cloud-based computing paradigms also exhibit theoretical and implementational limitations and disadvantages. For example, cloud-based computing is limited by a lack of depth and expressiveness. This tends to promote problems when attempting to abstract sensor and application layers, in a physically rigorous manner that must be convenient for system designers. By way of illustration, we first consider a traditional model that combines pervasive computing with the cloud computing paradigm, then contrast this with the CEB paradigm.

1.1 The Two-Layered Model: Scalability Challenges in Cloud-Based Sensing

In typical practice, cloud computing providers deliver common business applications online that can be accessed from a Web service such as a browser, while the actual programs and data are stored on servers. Clouds often appear to users as single points of access for their computing needs, and cloud-based products are usually expected to meet customers' quality of service (QoS) requirements, typically including service level agreements (SLAs). As a result, the emergence of cloud-based computing has provided ample opportunity for rejuvenation of a maturing information technology (IT) industry, in terms of a layered paradigm (*user > application > cloud > supporting devices*) with unique QoS and accounting models and practices.

In customary models of cloud-based computing, applications reside in the cloud, and devices that service the cloud's provision of applications reside in some device layer. Additionally, the emergence of pervasive sensing seeks to exploit sensors from cloud-based applications. Since the focus of this paper is on sensor computing in active pervasive spaces, we assume that the device layer includes sensors. Unfortunately, this two-layered model, shown notionally in Figure 1.1, has the following drawbacks that make it impractical for pervasive sensor computing:

- The massiveness (scope and detail) of sensor hardware in pervasive spaces makes coherent sensor management and maintenance extremely difficult. As a result, system complexity does not scale linearly with an increasing number of sensors and associated hardware devices.
- Application programmers must know many implementational details of sensor and communication hardware, in order to

successfully program the two-layer model (cloud and sensors) to realize pervasive sensing. This implies the presence of system integration knowledge, personnel, and effort, which can be expensive, time-consuming, and error-prone.

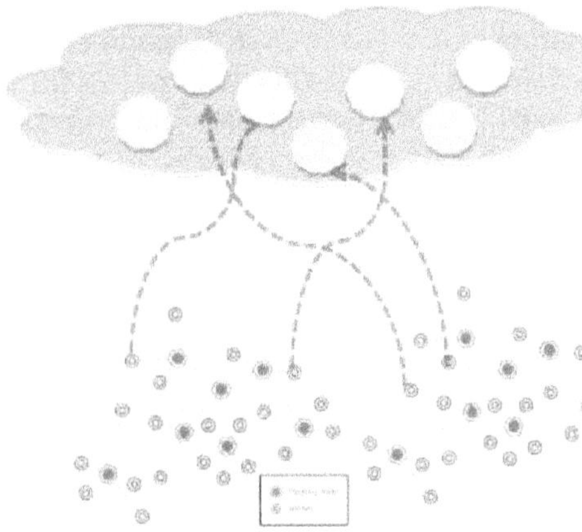

Figure 1.1. Customary two-layer model of cloud-based sensor computing

For purposes of discussion, let us suppose that this two-layer model is applied to cloud computing, where applications are in the cloud and sensors are beneath the cloud. Here, it is possible that an application directly connected to the device layer could make excessive demands on the devices, thereby overloading the devices. For example, sensors deployed in a busy highway to monitor traffic conditions can soon run out of power due to the fact that they could be overpowered when being accessed by large number of applications at the same time. This could cause excess energy consumption, with possible harm to the sensors or other objects in the device layer. Similarly, it is possible to slow down or "crash" the cloud, by having too many sensors for the cloud to process efficiently.

Further, to achieve efficiency and accuracy, we have found that the user should not be involved in dynamic interactions between the cloud and device layers. Thus, we have found an important additional flaw in the two-layer model:

- Importantly for wireless applications, the two-layer model raises significant concerns about energy consumption. In particular, applications in the cloud (by definition of cloud computing) need not know of each other's existence or functionality. Thus, when different applications request data from one sensor or group of sensors, these requests can be processed separately. This property supports redundancy of requests directed to a given sensor or subnet, which leads to redundant data acquisition that causes unnecessary energy consumption. Recall that cloud applications need not know of, or communicate among, each other. Thus, there exists no structural basis for optimizing the cloud's use of sensor hardware in terms of energy consumption, or other parameters such as field-of-view; optimization of spatial, temporal, or spectral coverage; latency hiding; and performance constraints on collaborative distributed sensing.

To opmization we propose a three- layered model (diagrammed in Figure 1.2) that has a novel layer called the *edge*, which resides between the cloud and sensor layers. We discuss the properties of this new model, as follows.

1.2 The Three-Layered Model: *Cloud*, *Edge*, and *Beneath* (CEB)

In the three-layered (CEB) model of sensor computing, sensors are partitioned into groups, each of which is managed by a single edge. In practice, an edge could be a mobile device, a computer, or a collection of sensor devices or computers. Such edge devices could be controlled by an application designed to monitor multiple conditions or activities within a particular environment. For example, the edge could be controlled by a homeowner via a security application, and be used for managing sensors deployed in his home (e.g., temperature, humidity, smoke, fire, or intrusion sensors). Similarly, sensing devices could be employed in monitoring patient health at home, in order to supply assistive services. Thus, sensors can be conveniently grouped implementationally by *type* (e.g., scalar, staring-array, spectral), *functionality* (e.g., temperature, surveillance camera), or *location* (e.g., within a building or an office complex, or outdoors).

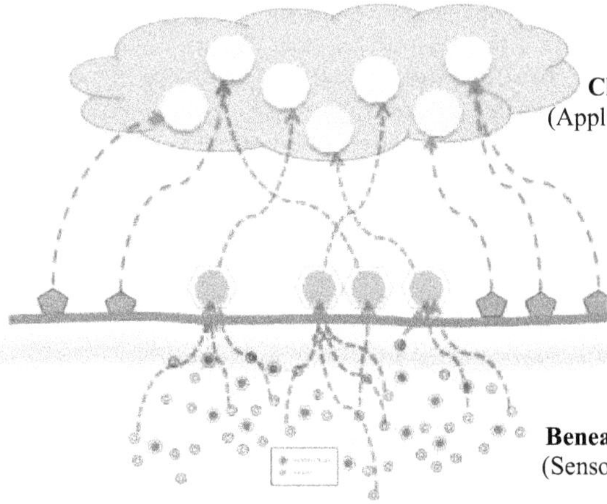

Figure 1.2. Three-layer model of sensor computing with cloud, edge, and beneath

In contrast with the traditional two-layered model illustrated in Figure 1.1, our three-layered CEB model realizes the following advantages and benefits:

- Sensor management and maintenance are performed by one or more edge(s), to which the sensors are connected. Thus, locality of control is preserved, since an edge only controls sensors located within its domain. As a result of such partitioning, scalability is significantly improved – as opposed to the two-layer model where sensors can be globally controlled by multiple competing or interacting applications (as discussed in Section 1.1).
- Edges can abstract unnecessary information about implementational or internal details of sensors, routers, and other sensing-related hardware located in the *Beneath* layer. This directly facilitates tractable models and programming of sensor configuration and control procedures, which we have previously shown to be efficient and robust in practice [17].
- Edges support staging and optimization, while allowing for many types of optimization procedures to be implemented. Edges also implement *localization* and decoupling, to achieve good software engineering practice through a loosely-coupled, layered architecture. In practice, edges can implement staging and power optimization by more

efficiently retrieving sensor data on behalf of applications, for example, by *caching*. Via data caching mechanisms, it is possible to reuse historical sensor data, thereby reducing energy consumption associated with repeated sensor data acquisition and sampling by multiple applications. In addition to energy reduction, caching could conceivably allow a reduction in the total number of sensors, as coverage could be partially managed by the caching capabilities of one or more edges.

Therefore, the introduction of the edge layer makes the three-layer architecture more tractable, better engineered, and more efficient. However, the minimization of energy consumption still remains a concern. In response to this situation, this paper describes the concept of an *optimal push-pull* (Section 2). We briefly explain a supporting event-driven programming model (Section 3), then present our optimization approaches (Section 4): In particular, our OPT-1 optimization algorithm is based on the relative characteristics of demand side (from the *Cloud* layer), while OPT-2 exploits the relative characteristics of data on the supply side (from *Beneath*). We also demonstrate experimentally determined performance gains of OPT-1, OPT-2, and their combination, in terms of practical applications.

2. Problem: Energy Efficiency of Cloud, Edge and Beneath

Recall that the two-layer model (Section 1.1) generates repeated data acquisition requests from one or more applications in the cloud to sensors in the device layer. This behavior can cause significant power consumption due to repeated sensor sampling. In the three layer model, we seek to minimize energy consumption by sentient control of sensors in the beneath (device) layer via optimization algorithms in the edge layer. Correspondingly, we have developed optimization algorithms that efficiently manage energy consumption between the cloud and edge layers.

In particular, our research shows that the updating of sensor readings can be realized via communication between cloud, edge, and beneath layers using information push and pull mechanisms. Push allows the data sink (e.g., the edge layer) to subscribe to a particular data source (e.g., a sensing device in the sensor layer), to received continuous readings at a constant rate. In contrast, information pull supports on-demand data query, such that the data sink can request, then acquire, sensor readings as one or more individual values. Advantages and disadvantages of these approaches are discussed, as follows.

2.1 The Push/Pull Envelope

When sensor data are needed at a constant rate, push requires less downlink traffic (cloud to edge, edge to sensor), since the application issues one subscription request only, then acquires the stream of values from the sensor. In contrast, pull incurs a round-trip penalty for each data query (e.g., involving steps such as request, transmit, receive, and acknowledge). However, push mode tends to be less economical when handling sporadic data requests, as an application-subscribed sensor in push mode samples and transmits a stream of data, whether or not all data values are processed by the application. In push mode, this physical transmission of the sensor value stream occurs continuously – even when the data are not needed by the application – leading to a substantial expenditure of energy. Since some sensor values are not processed, some of this energy is wasted. Thus, to achieve a design tradeoff between energy costs associated with push and pull modes, we need to know something about the application. For example, as shown in Figure 2.1 we have found that, when transitioning from MICA2 [18] to RCB [6], network communication decreases from 25 percent to 4.5 percent of total energy cost. Correspondingly sensor energy consumption due to sampling increases from 74 percent to 94 percent.

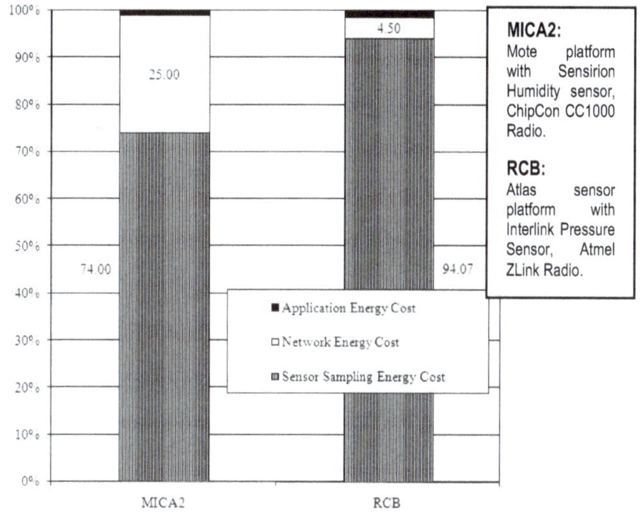

Figure 2.1. Differences in energy consumption between MICA2 and RCB (lower power, same sampling technology) for wireless sensing applications.

In response to this problem, we have developed the concept of a *push-pull envelope* (PPE), which is defined as follows.

Definition 1: An *optimal push-pull envelope* is a multidimensional sequence that determines (a) which sensors are active, and (b) which sensors should be in pull or push mode (along with push filters). The push-pull envelope minimizes work and total energy consumption in the sensor layer, as well as in the cloud, thus maximizing *sentience efficiency*.

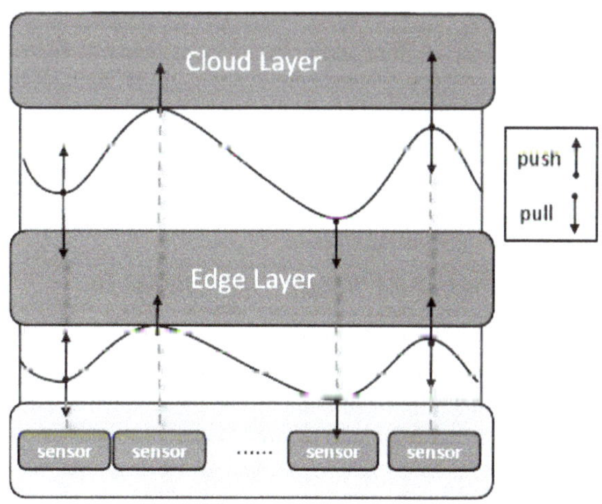

Figure 2.2. Push-pull envelope between cloud and edge, and between edge and sensors

In practice, the PPE can be thought of as a hybrid approach for achieving near-optimal energy consumption, as illustrated notionally in Figure 2.2. Architecturally, a PPE can exist between the cloud and edge layers, as well as between the edge and sensor layers. In a slightly lower-level view, a PPE can be thought of as an optimal configuration of push/pull modes for each sensor over the lifetime of service execution in the cloud. In this perspective, the PPE effectively describes an optimal execution of applications whose combined energy consumption (due to push and pull operations) is minimized. This does not imply a static view – the PPE can vary spatiotemporally. Therefore, in this paper, we seek to dynamically compute the optimal PPE between adjacent layers of the three-layer model that optimizes system energy efficiency.

Before presenting our algorithm for finding the optimal PPE, we overview an event-driven programming model that supports the programming paradigm employed in our system development of applications in the cloud.

3. Event-Driven Programming Model

An event-driven programming model is a programming paradigm whose control and data flows are determined by events. In pervasive sensor computing, an event is always associated with sensor outputs. In practice, applications developed on event-driven programming paradigms usually consist of two phases: (1) event detection, and (2) event handling.

In particular, in our model, the basic element of an application is a *rule*, which conforms to the *Event, Condition, and Action* (ECA) structure. By specifying events, conditions and actions, program-mers formulate rules that represent constraints on permissible application behavior over a range of situations. To ensure the adherence to these rules, the application constantly checks sensor data and evaluates the prespecified rules. When a rule evaluates to *true* in response to one or more events, corresponding actions (e.g. actuating a device or invoking a service) are executed, to respond to the event. Details about the design and implementation of ECA are found in our early work [15].

We have extended the event formulation structure to support composite events [15]. For example, an event associated with a single sensor only is called a *basic event*, while an event defined in terms of one or more basic events is called a *composite event*. Composite events usually require data from multiple sensors. For example, assume that a basic event e_1 is associated with a thermometer and another basic event e_2 is associated with a humidity sensor. Then a composite event could be defined based on e_1 and e_2, with the purpose of monitoring a physical event such as current weather condition.

In the next section, we present our optimization approaches to realize energy efficiency of systems that are build based on this event-driven programming model.

4. Cloud-Sensor Energy Optimization

Our research has shown that an optimal PPE can be achieved via two optimizations (called OPT-1 and OPT-2) that collaboratively determine the data delivery strategy among the three layers of our CEB sensor computing model. For each sensor, OPT-1 attempts to compute an optimal configuration of the push/pull envelope between the cloud and edge layers. Via monitoring data requests from the cloud, the edge learns the history of relative characteristics of an application's demands for each sensor. The edge then utilizes this reference knowledge to derive the optimal push/pull strategy for subsequent data deliveries that are expected to minimize energy cost. Additionally, this PPE is dynamically optimized at runtime to adapt to changing sensor sampling requests from the cloud. For each sensor, the second algorithm (OPT-2) optimizes the configuration of the push/pull envelope between the edge and beneath layers. For example, given long-term learning of the data output by a particular sensor, assume that the sensor data changes at a relatively constant rate that is less than the required data push rate calculated by OPT-1. Here, the sensor can choose a reduced data push rate, which further reduces energy consumption.

4.1 Cloud–Edge PPE Optimization

This section describes a dynamic algorithm that, for a single sensor, finds an optimal constant base push rate f^*, given the relative characteristics of sensor sampling requests from the cloud. As we have observed, requests from the cloud usually change with time. However once the arrival rate of a request surpasses a prespecified level for a given time interval, we can use base push with a constant rate to satisfy a portion of the data requests. This statement holds because push is less costly than pull if (a) the pushed data is known to be adopted by its sink (i.e., applications in the cloud), and (b) we can employ supplemental pulls to meet the unsatisfied requests. Taking advantage of this optimization opportunity, the edge adjusts the rate of base pushes to effect an overall (base + supplemental) rate change in order to find the optimal f^* that minimizes total energy cost. It is important to note that the edge continuously senses and analyzes sensor sampling requests from the cloud, to capture the relative characteristics of past requests that will support computation of f^*.

As the computation of f^* is based on a history of sensor values, our optimization approach must be able to dynamically adapt to changes in requests issued by the cloud during a specific time interval. Therefore, the edge chooses an evaluation window within which sensor sampling requests are analyzed. Our algorithm implements a simple but self-adaptive approach that selects the length W_e of this evaluation window. The following terms pertain to our discussion, in terms of a single sensor s:

$R(t_i)$ Total number of data acquisition and sampling requests for sensor s received by the edge from the cloud, over time t_i

D Sliding window at each end of which $R(t_i)$ is recorded. The length of D is d, which implies that $d = t_i - t_{i-1}$.

$R'(t_i)$ Average arrival rate of data acquisition and sampling requests from sensor s received by the edge from the cloud, within the sliding window D $[t_i - d, t_i]$, as follows:

$$R'(t_i) = \frac{R(t_i) - R(t_{i-1})}{d} \quad (1)$$

W_e Evaluation window within which multiple sensor requests from the cloud are monitored and analyzed by the edge, to provide information for computing the next f^*.

L Length of evaluation window defined as the number of sliding windows within We , i.e., $L = |We| / d$.

In summary, by analyzing the history of sampling requests from the cloud for a single senor s within the evaluation window W_e, the edge finds the optimal base push rate f^* that is expected to maximize energy savings for subsequent data queries. Therefore, the following three problems need to be solved: (1) determine the objective function and an algorithm to solve it; (2) determine when to re-evaluate f^*, and (3) specify the evaluation window W_e.

To solve Problem 1, we construct the objective function by using the energy saving rate of our mixed push-pull scheme (base push rate at f^*), to replace the pure pull scheme that is one of the customary approaches in current sensor data acquisition practice. The energy cost of a sensor node is modeled by considering both transmission and sampling [6] as major contributing factors for overall energy consumption. Accordingly, we define two energy cost coefficients: α, the energy consumption factor for one-time transmission (either sending or receiving a packet); and β, the energy cost for one sensor sampling operation. Therefore, the cost of a pull operation is $2\alpha + \beta$ (receiving query + sending data + sampling), and a push operation costs $\alpha + \beta$ (neglecting the one-time subscription). Within evaluation window W_e, the energy cost of our mixed push-pull strategy (base push rate at f^*) is

$$C_{PPE} = (\alpha+\beta)\sum_{t_i \in T}(f^* \times d) + (2\alpha+\beta)\sum_{t_i \in c_1}(R'(t) - f^*)\times d \qquad (2)$$
$$= (\alpha+\beta)\times \sum_{t_i \in (c_1 \cup c_2)}(f^* \times d) + (2\alpha+\beta)\times \sum_{t_i \in c_1}(R'(t) - f^*)\times d$$

where T represents all of the end points of sliding window D within W_e; c_1 denotes the times t_i at which $R'(t_i) > f^*$ (i.e. supplemental pull needed at sliding window $[t_{i-1}, t_i]$); and c_2 indicates the times t_i at which $R'(t_i) \le f^*$. Observe that $T = c_1 \cup c_2$.

Additionally, the energy cost by using pure pull scheme to satisfy all of the data requests is given by

$$C_{ppull} = (2\alpha+\beta)\sum_{t_i \in T}(R'(t_i)\times d) = (2\alpha+\beta)\sum_{t_i \in (c_1 \cup c_2)}(R'(t_i)\times d) \qquad (3)$$

Therefore, the objective function (energy saving rate) is defined as:

$$\max R_{save}(f^*, W_e) = \frac{C_{ppull} - C_{PPE}}{C_{ppull}} \qquad (4)$$
$$= \frac{\alpha \times f^* \times L - (2\alpha+\beta)\sum_{t_i \in c_2}(f^* - R'(t_i))}{(2\alpha+\beta)\sum_{t_i \in T} R'(t_i)}$$

where R_{save} is determined by f^* and the evaluation window W_e. The objective function is thus given by

$$\max R_{save}(f^*, W_e) = \frac{\alpha \times f^* \times L - (2\alpha+\beta)\times \frac{\sum_{t_i \in c_2}(f^* - R'(t_i))}{L} \times L}{(2\alpha+\beta)\times \frac{\sum_{t_i \in T} R'(t_i)}{L}\times L}$$
$$= \frac{\alpha \times f^* - (2\alpha+\beta)\times \frac{\sum_{t_i \in c_2}(f^* - R'(t_i))}{L}}{(2\alpha+\beta)\times \frac{\sum_{t_i \in T} R'(t_i)}{L}} \qquad (5)$$

We define the average of $R'(t)$ within evaluation window W_e as

$$P_r = \frac{\sum_{t_i \in T} R'(t_i)}{L} \qquad (6)$$

which denotes the average arrival rate of sampling requests for s received by the edge from the cloud within evaluation window W_e. We also define the average rate at which superfluous push operations happen within evaluation window W_e, as follows:

$$SPR(f^*) = \frac{\sum_{t_i \in c_2}(f^* - R'(t_i))}{L} \qquad (7)$$

It is clear that the value of $SPR(f^*)$ changes with variations in f^*.

Therefore, we can reformulate the objective function as

$$\max R_{save}(f^*, W_e) = \frac{\alpha \times f^* - (2\alpha+\beta)\,SPR(f^*)}{(2\alpha+\beta)\,P_r} \qquad (8)$$
$$= \frac{\alpha}{2\alpha+\beta}\times\frac{f^*}{P_r} - \frac{SPR(f^*)}{P_r}$$

In addition, the optimization algorithm must be able to adapt to the change of requests from the cloud (Problem 2). In our approach, after f^* is determined for subsequent data queries, the edge will continue to monitor factors that may affect PPE performance. For example, two factors in our performance evaluation algorithm are P_r and $SPR(f^*)$ (Equations 6 and 7). For either factor, once the difference between its current value and the value used to evaluate current f^* exceeds a designated threshold φ, a new round of f^* evaluation will be started. Simultaneously, the current PPE will be terminated as it might be inaccurate for processing current sampling requests.

We solve Problem 3 by initially assuming that the evaluation window W_e starts at the time when the current f^* is calculated, and ends at the time when f^* is re-evaluated. This simple approach is effective, since the length of W_e will adapt to the changing value of $R'(t_i)$. For example, if $R'(t_i)$ changes rapidly, then the performance of the current PPE will likely decrease rapidly. Therefore, the length of W_e for the next evaluation of f^* will be relatively shorter than it is when $R'(t_i)$ changes slowly. So the length of the evaluation window reflects how rapidly the evaluation of f^* adapts to changing sensor requests from the cloud. The shorter the evaluation window, the faster the evaluation of f^* adapts to the change of requests. For this reason, W_e should be shorter when $R'(t_i)$ changes rapidly. Conversely, W_e should be longer if $R'(t_i)$ changes slowly. Our experimental research results show that this approach supports the accurate tracking of changes in sensor request patterns that are typical of sensor demands by cloud applications analyzed thus far.

Given the preceding development, we summarize the proposed optimization algorithm (OPT-1), as follows.

OPT-1: Optimizing PPE between Cloud and Edge

Input: threshold φ; initial length of evaluation window L
* t_c: current system time
Algorithm:
```
Set evaluation window We = [tc - L · d, tc];
while true do
    Calculate optimal f* = FindFstar(R(tc), We);
    Record SPR'=SPR(f*); Pr'= Pr;
    Set base push rate = f* and start mixed push pull mode;
    Set t1 = tc;
    PerformCheck(f*, SPR', Pr') //quit when f* needs re-evaluation
    Set t2 = tc; We = [t1, t2];
end while
```

The supporting algorithm *FindFstar*, which optimizes f^*, is specified as follows.

Method *FindFstar*: Find the Optimal f^* Using Binary Search

```
Input: R(ti), We
Output: the base push rate f*.

Calculate R'(ti), for each ti within We.
Find min R'(ti) and max R'(ti), for ti within We .
if Rsave( min R'(ti), We ) > Rsave( min R'(ti)+Δf, We) then
      return min R'(ti) as the optimal f*;
end if
if Rsave( max R'(ti), We) > Rsave( max R'(ti)- Δf, We ) then
      return max R'(ti) as the optimal f*;
end if
fL = min R'(ti), fR = max R'(ti) ;
f' = (fL + fR)/2
while Rsave(f' - Δf, We)> Rsave(f', We) or
      Rsave(f' + Δf, We)>Rsave(f', We) do
      if  Rsave(f' - Δf, We) > Rsave(f', We) then
            fR = f'
      else fL = f'
      end if
      f' = (fL + fR)/2
end while
return f' as the optimal f*
```

Method *PerformCheck:* Monitoring Performance of f^*

```
Input: f*, SPR', Pr'
* Wc: time window within which runtime SPR(f*) and Pr are
calculated. It has fixed length, i.e., |Wc|.
* γ1: threshold used for monitoring SPR beyond which f* re-
evaluation will be triggered.
* γ2 : threshold used for monitoring Pr beyond which f* re-
evaluation will be triggered.

ReEva=false; //Initialization
while !ReEva do
   Calculate SPR(f*) and Pr within current Wc
   if  |SPR(f*) - SPR' | / SPR' > γ1 or
       | Pr - Pr' | / Pr' > γ2   then
       ReEva=true;
   end if
   Wait | Wc | time for the next time window Wc
end
```

The following proof sketch demonstrates the rationality and correctness of the function *FindFstar*. In order to apply a binary search algorithm to our process of searching for an optimal value of f^*, the objective function max $R_{save}(f^*, W_e)$ must satisfy the following conditions:

i. The optimal $f^* \in$ [min $R'(t_i)$, max $R'(t_i)$] for $t_i \in W_e$.

ii. There is only one optimal f^* for *max* R_{save} (f^*, W_e), s.t., min $R'(t_i) \leq f \leq$ max $R'(t_i)^*$.

iii. For min $R'(t_i) \leq f < f^*$, $R_{save}(f, W_e)$ increases as f increases; for $f^* < f \leq$ min R' (t_i), R_{save} (f, W_e) decreases as f increases.

Condition (i) is trivial. The proof of conditions (ii) and (iii) are as follows.

First, by calculating the derivative of (8) we get

$$\frac{d R_{save}(f^*)}{d f^*} = \frac{1}{P}\cdot\left(\frac{\alpha}{2\alpha+\beta} - \frac{d\, SPR(f^*)}{d f^*}\right) \qquad (9)$$

If we can prove (9) is a monotone decreasing function of f^*, then (ii) and (iii) will be guaranteed in that $R_{save}(f^*)$ can changes only in the following three ways as f^* increases from min $R'(t_i)$ to max $R'(t_i)$: a) always positive, b) always negative, c) first positive and then negative.

First of all, assume that functions $f = R_{save}(t)$ maps $t_i \in W_e$ to $f_i \in$ [min $R'(t_i)$, max $R'(t_i)$] for all i = 1..*n*. Sort f_i in non-decreasing order, i.e.

$$\min R'(t_i) = f_0 \leqslant f_1 \leqslant .. \leqslant f_n = \max R'(t_i) \qquad (10)$$

Partition the interval [min $R'(t_i)$, max $R'(t_i)$] into disjoint union of half-open intervals $[f_i, f_{i+1})$ where i = 0..n-1. For $f' \in [f_i, f_{i+1})$

$$SPR(f') = \frac{1}{L}\sum_{t=0}^{i}(f'-f_t) = \frac{1}{L}(i \times f' - S_i) \qquad (11)$$

Where, $S_i = \sum_{t=0}^{i} f_t$

Hence,

$$\frac{d\, SPR(f')}{d f'} = \frac{i}{L} \qquad (12)$$

For $f' \in (f_i, f_{i+1})$. Note that the derivative is not defined when $f' = f_i$, i = 1..*n*. However, the side limits (lim − and lim +) of $SPR(f')$ agrees at those special points, $f' = f_i$, i = 1..*n*-1. In other words, $SPR(f')$ is continuous within [min $R'(t_i)$, max $R'(t_i)$] and it is differentiable within [min $R'(t_i)$, max $R'(t_i)$] except at a finite set of points.

Assume that $f_a \in (f_{i_a}, f_{i_a+1})$ and $f_b \in (f_{i_b}, f_{i_b+1})$.

If $f_a < f_b$, then $i_a \leqslant i_b$. Therefore, $SRP(f_a)' \leqslant SRP(f_b)'$

So we can see that the derivative of $SPR(f^*)$ is a monotone increasing function of f^* and hence the derivative of $R_{save}(f^*)$ is a monotone decreasing function of f^* according to (9). Finally, we prove that condition (ii) and (iii) are satisfied.

4.2 Edge-Sensor Sampling Optimization

After the edge calculates the optimal base push rate f^* for a particular sensor *s* by running OPT-1, it will send f^* to *s* to set its data sampling rate. Fortunately, after receiving f^*, the sensor also has the chance to determine if it can choose an even lower sampling rate at which data is pushed to the cloud via the edge. The key concept is, if the data from *s* changes at a rate much lower than the required sampling rate f^*, then a lower push rate might sufficiently reflect the change of sensor data. We call this mode *lazy sensor sampling*. For our specific scenario where applications follow an event-driven paradigm, the base push rate of a particular sensor *s* should reflect the rate of change of the values of basic events associated with *s*. The goal of our lazy sampling algorithm (OPT-2) is to find the rate f^{**} at which *s* samples and pushes its data to the edge, using the lowest sampling rate to meet the cloud's demands for sensor data.

For a single sensor *s*, the following terms pertain:

$V(t_i)$ Number of changes (until time t_i) in the values of basic events that are associated with *s*.

D_l Sliding window at each end of which $V(t_i)$ is recorded. The length of D_l is $d_l = t_i - t_{i-1}$.

$V'(t_i)$ Average change rate of the value of basic events that are associated with s within the sliding window D_l $[t_i - d_l, t_i]$. Thus, $V'(t_i)$ is given by

$$V'(t_i) = \frac{V(t_i) - V(t_{i-1})}{d_l} \tag{13}$$

Similarly, we obtain the acceleration in the basic event change rate associated with the s, which we denote as $V''(t_i)$, as follows:

$$V''(t_i) = \frac{V'(t_i) - V'(t_{i-1})}{d_l} \tag{14}$$

Let us now describe constraints on lazy sampling. An activation condition of lazy sampling guarantees that the sensor data changes at a relatively constant rate, such that our lazy sampling algorithm will not lose track. This condition can be expressed intuitively in terms of a threshold derived from $V''(t_i)$, as follows:

$$\sum_{i=c-k}^{c} |V''(t_i)| < \tau \tag{15}$$

where t_c is the current time, τ is the threshold, and k is obtained experimentally.

If the above condition is satisfied, then our OPT-2 algorithm will compute an optimal or near-optimal rate f^{**} at which s will sample and push its data to the edge. This near-optimal sampling rate is determined by summing the predicted $V'^*(t)$ at the next time slice, toleranced by a safe margin, to compensate for any abrupt change of sensor data. The estimation of $V'^*(t)$ at the next time slice should be based on the sensor data history, with provision for temporal tolerancing, as mentioned previously. We have found that the following method for determining $V'^*(t)$ is practicable:

$$V'^*(t) = \gamma \times V'(t_c) + \gamma \times (\gamma - 1) \times V'(t_{c-1}) + \gamma \times (\gamma - 1)^2 \times V'(t_{c-2}) + \ldots + \varepsilon \sum_{i=c-k}^{c} |V''(t_i)| = \gamma \times \sum_{i=0}^{k} (\gamma - 1)^i \times V'(t_{c-i}) + \varepsilon \sum_{i=c-k}^{c} |V''(t_i)| \tag{16}$$

where γ influences how rapidly $V'^*(t)$ adapts to changes of $V'(t_i)$, and $\varepsilon \sum_{i=c-k}^{c} |V''(t_i)|$ denotes the temporal tolerance (margin of safety).

We thus summarize our lazy sampling algorithm, as follows.

OPT-2: Lazy Sampling Algorithm

Input: f^* calculated by edge through OPT-1, $V(t_i)$
Algorithm:
lazyMode=false; //Initialization
while *true* **do**
    if Condition (15) is satisfied **and** ! lazyMode **then**
        set f^{**}= *FindLazyRate* (f^*, $V(t_i)$);
        lazyMode=true;
    end if
    if Condition (15) is not satisfied **and** lazyMode **then**
        set $f^{**}=f^*$;
        lazyMode=false;
    end if
    wait for k time.
end while

Method *FindLazyRate*: **Find the sensor sampling rate f^{**}**

Input: f^*, $V(t_i)$
Output: the base push rate f^{**}.

Calculate $V'^*(t)$
if $V'^*(t) < f^*$ **then**
    return f^{**}= $V'^*(t)$ // lazy sampling mode starts
else return $f^{**}=f^*$;
end if

Importantly, to enable our optimization, caching as a supporting technology must be implemented in our three layer model.

4.3 Supporting Technology: Caching

Some events are tolerant to time fidelity in that their respective applications do not require frequent evaluation at all times. For example, when monitoring room temperature, an air conditioning system does not need to get temperature reading at a very high frequency, since the changing rate of room temperature is usually slow. In addition, an air conditioning system may need to monitor room temperature only occasionally during the spring and fall. Considering these scenarios, data of a particular sensor can be considered fresh within a certain "lifetime" after it is sampled. Taking advantage of this fact, we can employ data caching mechanisms so that historical sensor data could be reused to satisfy different requests for a common sensor data, thereby reducing energy consumption. In our model, caching are employed in all three layers (cloud, edge, and sensors) to realize energy saving.

Caching data in the cloud [19] allows different applications in the cloud (possibly located at different cloud servers) to reuse sensor data without requesting sensor data from the lower layers. Once a request is issued by an application, the cache is checked for "freshness". If it is, data will be used directly. Therefore, only those requests that cannot be satisfied by cached sensor data will be forwarded to the edges that will fetch data on behalf of applications. Similarly, caching is implemented at the edge and even sensors to further reduce energy consumption by enabling historical data reuse whenever possible.

5. Experimental Evaluation

5.1 Experimental Setup

We evaluate the performance of CEB under OPT-1, OPT-2 and their combination, by comparing their performance with that of CEB under a pure pull mechanism. We use an emulation approach where the requests from the cloud and data from the beneath are simulated by generators and in which edges and sensor nodes are simulated by software emulators. As input to our optimization algorithms, the arrival of sensor sampling requests and the generation of sensor data are simulated in the following way.

Sensor sampling request generator emulates the arrivals of requests from applications in the cloud asking for data from sensors. We simulate total number of K applications and total number of S sensors. In our experiment, we choose {K=50, S=50; K=100, S=50; K=500, S=100}. Sensors are randomly assigned to each application following a normal distribution (N (μ=2, σ^2=4)). Each application requires data from an arbitrary set of sensors out of the sensors assigned to it for execution. The time between two active phases of an application follows an exponential distribution with average arrival rate λ=1. For the generation of sensor data, we use the normal distribution. However, parameters of μ and σ^2 are varied during the course of each experiment. High σ^2 emulates the phase when the data of a sensor is changing rapidly while lower σ^2 emulates the phase when sensor data is changing slowly.

We emulated the effect of three sets of sensor platforms [Table 1] by normalizing their energy consumption coefficients. We conduct experiments on each set of coefficients to quantify the platform's receptiveness to our optimizations. Transmission energy cost includes two parts, cost of transmission between edge and sensor and cost of transmission between cloud and edge. We assume that for all three sensor platforms transmission cost between edge and cloud is unchanged.

Table 1. Normalized energy coefficients for transmission and sampling on three different sensor platforms. S1: MICA2 sensor platform with Sensirion Humidity sensor, ChipCon CC1000 Radio [18], S2: Atlas sensor platoform with Interlink Pressure Sensor, Atmel ZLink Radio [6] and S3: MicroLEAP with ECG sensor, class-2 Bluetooth 2.0 radio [20]. α_2 is derived from [21] (assume average 10 hops from edge to cloud).

Sensor platform	β: sampling	α_1 transmission between edge and sensor	α_2 transmission between cloud and edge
S1	0.21	0.36	0.43
S2	0.88	0.03	0.09
S3	0.26	0.33	0.41

5.2 Evaluation Metrics

The main performance metric, which is measured throughout all experiments, is the energy saving rate of our mixed push-pull scheme to replace the pure pull scheme i.e. R_{save}. The actual energy saving rate in all experiments is given by (19).

$$C_{ppull} = (2\alpha_1 + 2\alpha_2 + \beta) \times m_{pull} \tag{17}$$

$$C_{ppe} = 2\alpha_2 \times n_{pull1} + \alpha_2 \times n_{push1} + (2\alpha_1 + \beta) \times n_{pull2} + (\alpha_1 + \beta) \times n_{push2} \tag{18}$$

$$R_{save} = 1 - \frac{C_{ppe}}{C_{ppull}} \tag{19}$$

(17) indicates the total energy cost by using pure pull scheme in which m_{pull} means the total number of pulls (cloud pulls data from sensors) in pure pull scheme. (18) represents the total energy cost by using our mixed push pull scheme. n_{pull1} and n_{push1} means the total number of pulls and pushes respectively between edges and the cloud in our mixed push pull scheme; while n_{pull2} and n_{push2} means respectively the total number of pulls and pushes between edges and sensors in our mixed push pull scheme.

5.3 Performance Comparison Results

We compare the performance of OPT-1, OPT-2 and OPT-1 and OPT-2 combined, for sensor platforms S1 and S2. The result is recorded every T time (5 sec) as shown in Figure 5.1. (K=500, S=100).

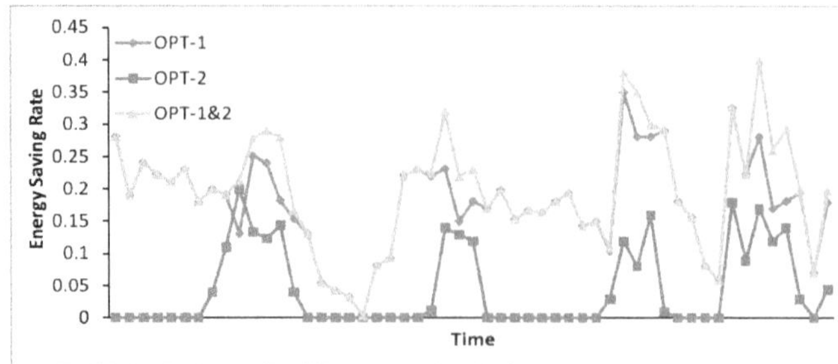

(a) Result for sensor platform S1

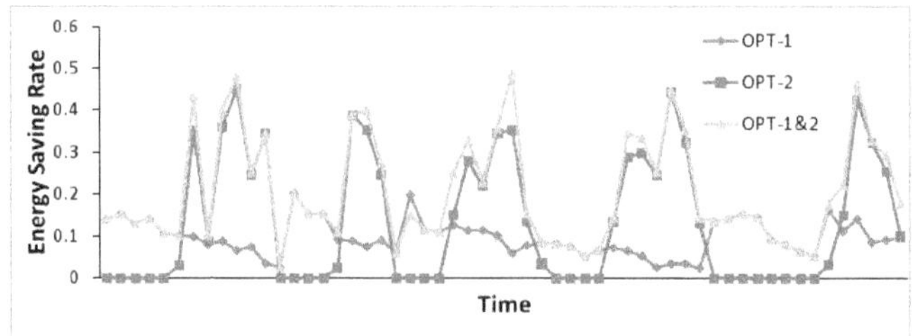

(b) Result for sensor platform S2

Figure 5.1. Energy saving rate for three optimization techniques on sensor platform S1 and S2

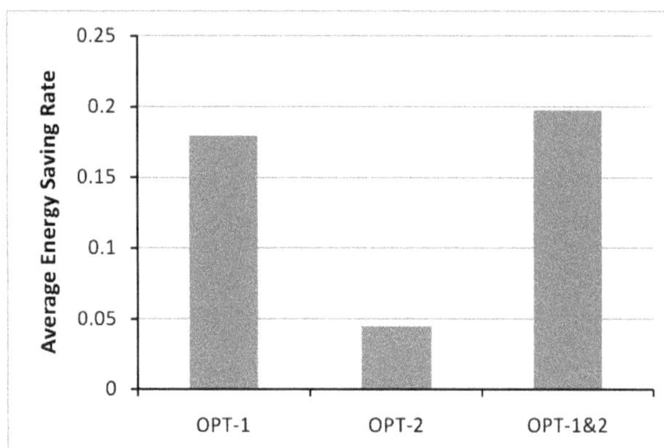

(a) Results for sensor platform S1

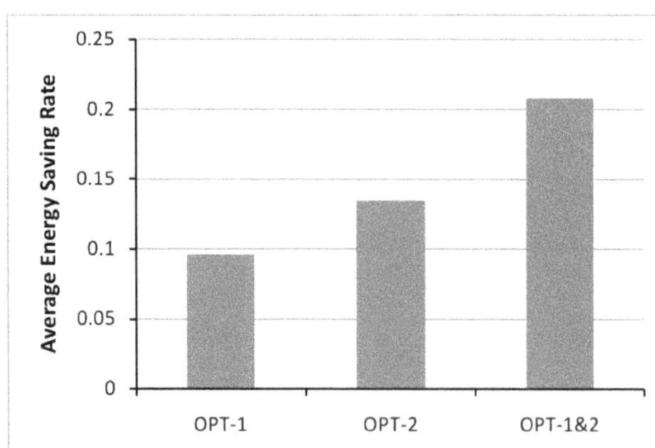

(b) Results for sensor platform S2

Figure 5.2. Average energy saving rate for three algorithms on sensor platform S1 and S2 (over 255 seconds)

From Figure 5.1. and 5.2., we can see that for sensor platform S1 whose transmission energy cost accounts for most of the energy consumption, OPT-1 shows better performance than that for sensor platform S2 where sensor sampling accounts for most of the energy consumption. This can be explained by the motivation of OPT-1, in which by replacing as many pulls with pushes the number of transmission can be reduced, leading to energy saving. On the contrary, OPT-2 shows better performance in S2 than in S1. The reason is that OPT-2 reduces the number of push as well as the number of data sampling when data of a given sensor does not change rapidly during a particular period. In Figure 5.2, we show that the combined application of OPT-1 and OPT-2 gives a superior performance than by utilizing OPT-1 or OPT-2 alone.

In addition, we measure the effects of optimization on three sets of sensor platforms and compare the results in Figure 5.3. From this result, we see that S1 adopts more pushes than S2 while it adopts less pull than S3. The reason is that in S2 supplemental pull has fewer penalties than in S1 and hence pull will be preferred as its on-demand feature.

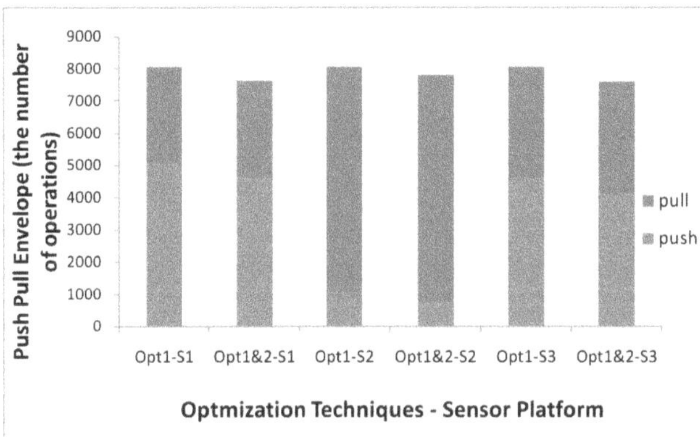

Figure 5.3. The push-pull envelope on different sensor platforms (S1-S3)

6. Related Work

Integrating sensor networks with the emerging data center cloud model of computing is becoming a popular paradigm of choice in system development for various application domains including health care [1], warehouse monitoring [4] and environmental monitoring [3]. In order to realize this integration, in [4] and [5] similar architectures are proposed to utilize virtualization as supporting technology for integrating physical sensors as services into the cloud. These architectures, however, lack an intermediate layer between the cloud and various physical devices, which raises serious scalability issues. In order to enable scalable collaborative sensor-centric applications in the cloud, [2] and [1] provided their own frameworks which respectively use Pub/Sub broker and exchange service as intermediate layer between the cloud and device layers, to preserve locality of control. However, these approaches disregard energy consumption, which could significantly affect system performance. In our work presented herein, we introduce, justify, and take advantage of a three-layer model (cloud, edge, and beneath) to enhance scalability as well as increase energy efficiency via algorithms for optimizing the push/pull envelope.

Additionally, in wireless sensor networks, special data acquisition techniques have been developed for event detection supporting real-time application execution. A typical scheme is polling [11], in which a data sink sequentially polls its underlying sensors for new data. In contrast, a bottom-up sensor-driven model [12] has also been proposed, assuming that sensors are capable of pushing data to applications when an event occurs. To improve the efficiency of data delivery and enable data sharing, messaging paradigms such as publish/subscribe [9] and push-pull [10] have been widely adopted in sensor data acquisition. Optimization techniques to balance push and pull have been extensively discussed in [10][13][14], which focus on network topology and routing algorithms. Furthermore, a new model discussed in [15] utilizes the mixed push/pull strategy and takes advantage of the optimization opportunity provided by the event structure and its time-frequency relaxations. However, none of the above approaches fit perfectly into a cloud-based sensor computing paradigm. This is especially true when one considers the massiveness of sensors and applications that tend to be invisible to each other. In order to overcome this problem, we have developed a push pull strategy based on the relative characteristics of sensor requests (demand side from the cloud) and sensor data (supply side from beneath). These two features can be easily captured in our model, and support our claims of effectiveness and significance in promoting our approach's energy efficiency.

7. Conclusion

We adopt the Cloud, Edge and Beneath (CEB) architecture to sensor data access by Cloud applications. CEB more naturally represents actual sensor system deployments in which an edge device (e.g., gateways) manages a locale of sensors. CEB allows for many optimization opportunities. In this paper, we present two optimizations. OPT-1 optimizes the proper mix of base push and supplemental pull between the cloud and the edge. OPT-2 optimizes the sampling and delivery of sensor data between the sensors and the edge, based on sensor domain value characteristics. We present both algorithms along with an experimental evaluation of their individual and combined impact. The results shows that by combining both algorithms, CEB can always achieve better performance than adopting any of the algorithms individually.

8. REFERENCES

[1] C. Rolim, F. Koch, C. Westphall, J. Werner, A. Fracalossi, G. Salvador, "A Cloud Computing Solution for Patient Data Collection in Health Care Institutions," in Proc. of ETELEMED. IEEE, 2010, pp. 95–99

[2] M. Hassan, E. Huh. "A Framework of Sensor-Cloud Integration: Opportunities and Challenges". International Conference on Ubiquitous Information Management and Communication.

[3] X.H. Le, S. Lee, T. Phan, V. La, A. Khattak, M. Han, H. Dang, M. Hassan, M. Kim, K. Koo, Y.K. Lee, E.N. Huh, "Secured WSN-integrated cloud computing for u-Life care", in: Seventh Annual IEEE Consumer Communication and Networking Conference (CCNC), January 9–12, Las Vegas, 2010, pp. 1–2.

[4] K Lee and D. Hughes. "System architecture directions for tangible cloud computing." In International Workshop on Information Security and Applications (IWISA 2010), in Qinhuangdao, China, October 22- 25, 2010.

[5] M. Yuriyama and T. Kushida, "Sensor-Cloud Infrastructure - Physical Sensor Management with Virtualized Sensors on Cloud Computing", The 13th International Conference on Network-Based Information Systems (NBiS-2010), 2010

[6] R. Bose and A. Helal, "Sensor-aware Adaptive Push-Pull Query Processing in Wireless Sensor Networks," Submitted to the 6th International Conference on Intelligent Environments - IE'10, Kuala Lumpur, Malaysia, July 19-22, 2010.

[7] Y. Kawahara, M. Minami, H. Morikawa, T. Aoyama: "Design and Implementation of a Sensor Network Node for Ubiquitous Computing Environment", In Proc. Of VTC2003-Fall 2003.

[8] A. Helal, D. Cook, M. Schmalz, "Smart Home-based Health Platform for Behavioral Monitoring and Alteration of Diabetes Patients," Journal of Diabetes Science and Technology, Volume 3, Number 1, January 2009. Pp 141-148.

[9] N. Rosa, C.Ferraz, J. Kelner, E. Souto, G. Guimarães, G. Vasconcelos, M. Vieira. "Mires: a publish/subscribe middleware for sensor networks". Personal and Ubiquitous Computing , Volume 10 Issue 1, December 2005. Springer-Verlag.

[10] X. Liu, Q. Huang, Y. Zhang. "Balancing Push and Pull for Efficient Information Discovery in Large-Scale Sensor Networks". IEEE Transactions on Mobile Computing, Volume 6, Issue 3, March 2007 Page(s):241 – 251.

[11] Z. Zhang, M. Ma, Y. Yang. "Energy Efficient Multi-Hop Polling in Clusters of Two-Layered Heterogeneous Sensor Networks". IPDPS '05: Proceedings of the 19th IEEE International Parallel and Distributed Processing Symposium (IPDPS'05), April 2005.

[12] S. Reilly and M. Haahr. "Extending the Event-based programming model to support Sensor-Driven Ubiquitous Computing Applications". Proceedings of the 2009 IEEE International Conference on Pervasive Computing and Communications (Percom'09).

[13] Z. Tao, Z. Gong, Z. OuYang, J. Xu. "Two New Push-Pull Balanced Data Dissemination Algorithms for Any-Type Queries in Large-Scale Wireless Sensor Networks". International Symposium on Parallel Architectures, Algorithms, and Networks, 2008. 7-9 May 2008 pp: 111 – 117.

[14] S. A. Hashish, A. Karmouch. "Topology-based on-board data dissemination approach for sensor network." Proceedings of the 5th ACM international workshop on Mobility management and wireless access. Chania, Crete Island, Greece, 2007. pp. 33 – 41.

[15] C. Chen, Y. Xu, K. Li, and A. Helal, "Reactive Programming Optimizations in Pervasive Computing," in proceedings of 10th Annual International Symposium on Applications and the Internet, Seoul, Korea.

[16] M. Khanafer, M. Cuennoun, H. T. Mouftah, "WSN Architectures for Intelligent Transportation Systems" Proceedings of 3rd New Technologies, Mobility and Security (NTMS), Cairo, Egypt

[17] J. King, R. Bose, H. Yang, S. Pickles and A. Helal, "Atlas – A Service-Oriented Sensor Platform, " Proceedings of the first IEEE International Workshop on Practical Issues in Building Sensor Network Applications (SenseApp 2006). Tampa, Florida, November 2006.

[18] S. Lai, J. Cao, Y. Zheng, "PSWare: a Publish/Subscribe Middleware Supporting Composite Event in Wireless Sensor Network". IEEE International Conference on Pervasive Computing and Communications, 2009. pp.1-6.

[19] Y. Cardenas, J. M. Pierson, L. Brunie: "Uniform Distributed Cache Service for Grid Computing." In: Proceedings of the International Workshop on Database and Expert Systems Applications, pp. 351–355 (2005)

[20] L.K. Au, W.H. Wu, M.A. Batalin, D.H. McIntire and W.J. Kaiser, "MicroLEAP: Energy-aware Wireless Sensor Platform for Biomedical Sensing Applications". IEEE BIOCAS2007. November 2007. pp.158-162.

[21] V. Sivaraman, A. Vishwanath, Z. Zhao, C. Russell, "Profiling Per-Packet and Per-Byte Energy Consumption in the NetFPGA Gigabit Router," IEEE INFOCOM 2011 Workshop on Green Communications and Networking

Energy Evaluations in Wireless Sensor Networks – A Reality Check

Christian Haas
Institute of Telematics
Karlsruhe Institute of Technology, Germany
christian.haas@kit.edu

Joachim Wilke
Institute of Telematics
Karlsruhe Institute of Technology, Germany
joachim.wilke@kit.edu

ABSTRACT

The development of *energy-efficient* applications and protocols is one of the most important issues in Wireless Sensor Networks (WSN). However, most publications up to now avoid time consuming realistic energy evaluations and oversimplify their evaluation with regard to energy-efficiency. This work aims at lowering the barrier for *realistic* energy evaluations. We focus on a generic application that simply transmits one packet using TinyOS Low Power Listening (LPL), which we evaluate using the WSN testbed *SANDbed*. Our results *disprove* some intuitive expectations. For example, we show that transmitting packets with a large payload can be cheaper in terms of energy consumption than a small payload. As packet transmission is part of almost any WSN application, the results shown are important to many WSN protocol evaluations. As an addition, we contribute our *lessons learned* by discussing the most important challenges and pitfalls we faced during our evaluation.

Categories and Subject Descriptors

H.4 [**Information Systems Applications**]: Miscellaneous

General Terms

Design, Experimentation, Measurement

1. INTRODUCTION

For most future applications, the nodes in a Wireless Sensor Network (WSN) have to be small and cheap. As a direct consequence, the nodes are heavily constrained regarding energy, memory, and processing resources. Top priority of a WSN application engineer is therefore to pay attention to an applications' resource efficiency, especially to energy-efficiency.

Regarding energy-efficiency, previous work has two major drawbacks. First, protocols are evaluated in isolation, i.e., neglecting the influence of other protocols and network layers. Moreover, using simple *simulator* tools is the most popular way of evaluation. Some simulators, like TOSSIM [3], omit lower network levels, ignoring influence of the radio channel or the MAC protocol. Counting data packets or data volume and then estimating energy consumption is a common way here. An improvement is emulating the sensor node combined with a cycle level energy model, as it is implemented by AVRORA [4]. However, the results still depend on the accuracy of the energy model provided. *Formal* evaluations use abstractions to model the sensor network and the environment. Even simple formal models are hard to handle, e.g., when doing formal verification. Results from more realistic evaluations are mostly acquired by *measuring* the energy consumption of a single node using an oscilloscope or by deploying a test application on a WSN *testbed.* The first approach has the disadvantage, that the energy consumption of only a single node is measured, whereas in a sensor *network* there are commonly multiple nodes and also node interactions involved. We argue, that the energy consumption of a WSN node is a complex phenomenon, that is hard to evaluate in a precise manner, because it involves more than deducing it from the communication characteristic of an application. It is therefore best evaluated using a *real* wireless sensor network.

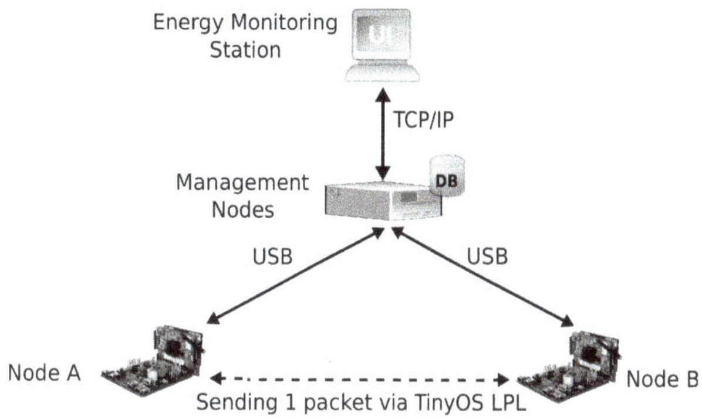

Figure 1: Experiment setup.

In this work, we use the WSN Testbed *SANDbed* as evaluation platform. It uses real hardware nodes to get trustworthy results with regard to energy consumption. Because communication is the most energy consuming task sensor nodes perform, we present evaluation results of simple communication between sensor nodes. We show how cross layer effects, especially from the MAC layer, influence energy consumption and what has to be considered to properly evaluate energy-efficiency of any WSN application. Additionally,

MSWiM'11, October 31–November 4, 2011, Miami, Florida, USA.

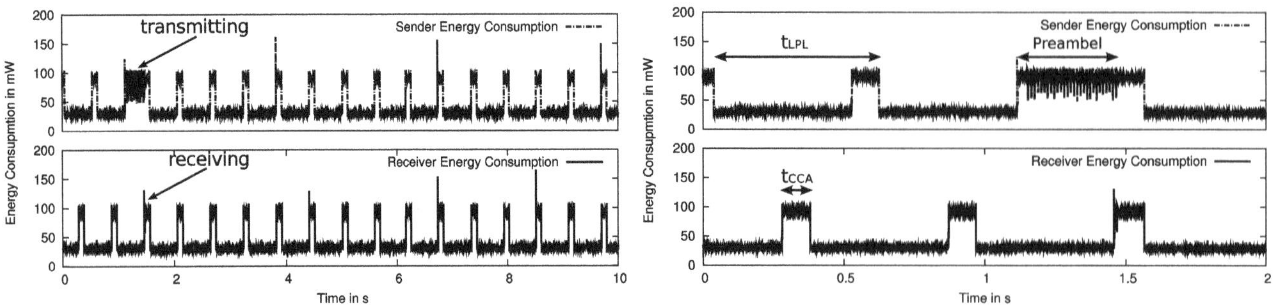

(a) Plot of a complete measurements of ten seconds.

(b) Zoomed plot of the transmission.

Figure 2: Energy consumption of two nodes (sender and receiver) while transmitting a single packet.

we contribute a set of pitfalls we faced during our evaluation and present best practices to evaluate energy-efficiency flawlessly.

2. EXPERIMENT SETUP

For a distributed measurement of energy-consumption, we use the Sensor Node Management Devices(SNMDs) at the SANDbed testbed at the KIT [2]. Each SNMD is equipped with a MICAz sensor node and its default sensor board. The SNMDs are capable of measuring the voltage and the current with a very high resolution as they provide sampling frequencies of up to 250kHz with an average measurement error below 1% [1].

We use TinyOS and enable its Low Power Listening mechanism as energy-efficient MAC protocol. The basic idea behind LPL is to periodically check the wireless medium for activity using a *Clear Channel Assessment* (CCA). The check is done every t_{LPL} milliseconds. Between these checks, the radio is powered off to save energy. If a node wants to transmit data, it continuously repeats its data packet for at least t_{LPL} milliseconds to assert the receiver had the possibility to wake up and receive the data. The length of the CCA check t_{CCA} is thus another important parameter, as the receiver must be able to detect the activity on the medium during this time slot.

Figure 1 shows the setup of our evaluation environment. We are using two measurement devices, Node A and Node B, to perform a distributed energy measurement. These two devices are SNMDs with an attached MICAz sensor node running TinyOS. The measurement devices are connected to a management node to store the evaluation results in a local database.

The application used for our reality check is very simple. One node sends exactly one data packet within a predefined time interval of 10 seconds to the other node using TinyOS LPL as MAC layer protocol. The sending time is randomly chosen by the sending node within the time interval. For evaluation purposes, we are measuring the energy consumption of both sender and receiver node. To get an idea about the influence of the LPL parameter t_{LPL}, we are varying the t_{LPL} time between 50ms and 4000ms. The length of the CCA check t_{CCA} is fixed. Further more, we are using two different payload sizes to identify differences in energy consumption. All experiments are repeated 100 times to get statistically significant results. In our evaluation later on, we choose random delays to power on nodes for every experiment run to avoid timing based effects. To cope with hardware tolerances, we are switching the sender and receiver role after 50 runs. Our experiment setup is summarized in Table 1.

Figure 2a shows the energy consumption of two nodes over time while sending and receiving a packet using TinyOS LPL. The SNMDs used to get the measurements in this plot sampled the node's current draw and voltage with 9kHz. The zoomed plot on the right side shows the transmission in detail. One can see the periodic CCA check and its duration t_{CCA} as well as the preamble of the sender. The time between two CCA checks is t_{LPL}. Having this plot in mind, it is obvious, that altering MAC parameters significantly influences energy consumption of both nodes without changing the application on top.

In each repetition we are measuring the energy consumption of sender and receiver node and whether the packet is transmitted successfully. Analyzing the above experiment setup, the sample application and related work regarding energy evaluation, one would intuitively expect the following outcome of our evaluation:

- The lower we set the LPL parameter t_{LPL}, the more energy is consumed by the two sensor nodes as a lower LPL parameter means shorter sleep intervals.
- With bigger payload sizes, one would expect a bigger overall energy consumption of the two nodes because of the higher data volume that has to be transmitted.

3. EVALUATION

In this section we analyze the influence of t_{LPL}, the influence of payload size and the influence of radio timing effects on the energy consumption of sender and receiver nodes.

MAC protocol	TinyOS LPL
Packets transmitted	1
CCA check length t_{CCA}	1600cycles ($\approx$ 11ms)
LPL parameter t_{LPL}	$50, 100, 250, 500, 750$ms $1, 1.25, 1.5, 2, 2.5, 3, 3.5, 4$s
Payload size S	$1, 90$Bytes
Experiment duration	10s
Experiment repetitions	2x 50
Sampling rate	9kHz

Table 1: Experiment setup parameters

3.1 Influence of t_{LPL}

In Figure 3, the energy consumption of sender and receiver node while transmitting 1 Byte of payload using different values for t_{LPL} is shown. The energy consumption is thereby averaged over 100 experiment repetitions. According to our evaluation, sending is always more expensive than receiving. Moreover, the energy consumption of the receiving node decreases with the size of t_{LPL}. As for the sending node, higher values of t_{LPL} do not automatically lead to lesser energy consumption. For our experiment, there is an optimal value for the t_{LPL} parameter for the sending node at approx. 2000ms. This effect can be explained with the mechanism TinyOS LPL uses for sending packets: If the t_{LPL} value is high, the sending node needs to send a preamble that covers at least t_{LPL} ms, whereas the receiver only has to perform a relatively short CCA-Check of approx. 11 ms. Therefore, the energy used for sending the preamble dominates the energy saved by powering off the radio between two CCA checks.

As our example shows, an energy-efficient MAC layer has a high influence on the overall node energy consumption. Any evaluation regarding energy consumption and communication therefore has to include a precise model for the energy consumption of the MAC layer.

3.2 Influence of radio timing effects

Figure 3 also shows the minimum and maximum values we recorded for every configuration during the 100 experiment repetitions. As stated in the previous section, the nodes started each experiment with random radio startup times, so that the CCA checks are not performed synchronously. Additionally, the sending time of the packet is randomly picked by the sending node. As can be seen, the energy consumed by the sending node has a high fluctuation, depending on the concrete sending time in conjunction with the receiving nodes CCA check. If the sending node transmits the packet shortly after the receiving node has done its last CCA check, more energy for transmitting a longer preamble is needed for the sending node.

The evaluation clearly shows that timing effects highly influence the nodes energy consumption. Also, the energy consumption of one particular experiment run can strongly deviate from the average energy consumption. In our sample application, the sending nodes energy consumption can fluctuate by ± 50 percent between the minimum and maximum value.

3.3 Influence of payload size S

As can be seen in Figure 4a, the receiving node consumes the same amount of energy for both payload sizes of 1 and 90 Byte. For the sending node, transmitting 90 Byte payload is slightly less energy consuming than transmitting 1 Byte of payload. This is caused by the *CC2420 transceiver* of the MICAz sensor node; the transceiver consumes less energy in the sending state than in listening or receiving state. With bigger payload sizes, the sending nodes transceiver is longer in the sending state than with lower payload sizes during the preamble phase of LPL.

Despite the smaller energy consumption of the bigger payload, it is clearly shown that the MAC layer parameter t_{LPL} has way more impact on the nodes energy consumption than the payload size.

Our evaluation clearly shows that an evaluation of energy consumption by simply comparing the data volume transmitted is incomplete or even leads to wrong conclusions. In our simple application, a node that is sending more data would be the slightly more energy-efficient node.

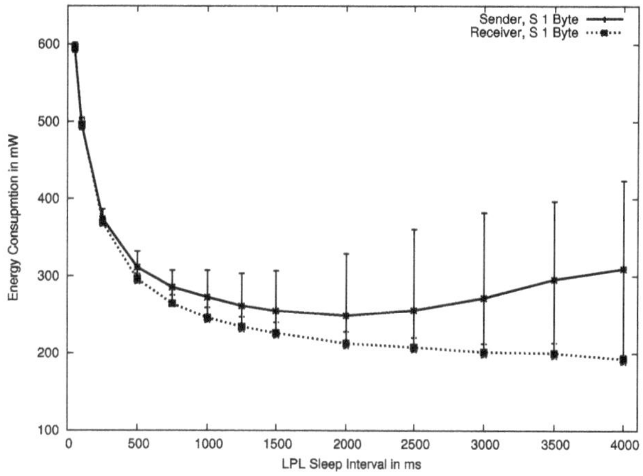

Figure 3: Influence of MAC parameter t_{LPL}.

3.4 Influence of hardware tolerances

Figure 4b shows again the influence of the t_{LPL} parameter on the nodes energy consumption. In contrast to Figure 4a the first 50 runs and the second 50 runs have been *splitted* and not aggregated as before. This way, the first part of the experiment, using node A as sender and node B as receiver can be compared with the second part where the node's roles are switched. The energy consumption differs between node A and node B by an approximately constant amount even while having the same role. In this case the energy consumption of node A is always a bit higher than that of node B.

To compare nodes with different roles with each other, roles *must* be switched to average energy consumption results. Without that, only comparisons of results measured at the *same* node are possible which constraints evaluation possibilities.

4. LESSONS LEARNED

During our evaluation, we identified several pitfalls and challenges that are important when evaluating energy-efficiency of WSN nodes. In a WSN, testbed nodes can often be powered on *simultaneously*. This is, however, no realistic behavior. In reality, nodes are powered on *asynchronously* depending on the deployment method. Moreover, clock drift and radio communication can influence the radio timing. As a result, to evaluate both the startup phase of a WSN application as well as a period of time during WSN application operation, nodes have to powered on in a *pseudo-random manner* to avoid timing based effects. This also means, that there have to be many runs of the same experiment using different seeds for the pseudo random number generator, to average results. In our evaluation, we choose random delays to power on nodes for every experiment run to avoid trapping timing based effects.

A large number of experiment runs not only consumes time, but requires the nodes to be powered off and on many times. During our experiments we noticed some weird be-

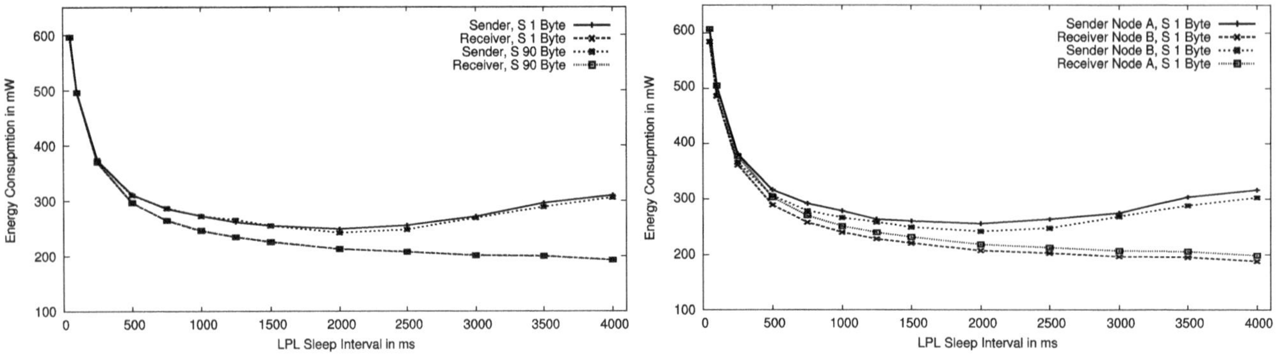

(a) Payload size S (b) Hardware tolerances

Figure 4: Influence of other parameters on energy consumption.

havior using MICAz nodes. The first one is, that the startup phase differs in energy consumption and duration from node to node. The second one is, that the current consumption of the node is often fluctuating during the very first seconds. As a result, we have put a waiting phase between powering on the nodes and starting the measurement, to avoid this fluctuating period.

Even after the startup phase, we still noticed significant offsets regarding energy consumption of different nodes while running identical operations. Further experiments showed that energy consumption of the 22 MICAz nodes in our testbed differs up to $\pm 10\%$. As a direct consequence for the evaluation, we switched the roles of the two used nodes, to average hardware tolerance effects. It is important to know about the existence of hardware tolerances and its range, as sensor nodes deployed in real environments will also have such tolerances. As a result, comparisons of absolute energy consumptions are almost impossible. Only relative comparisons seem feasible. This is, however, the important one to support design decisions as what protocol to use for a certain operation purpose.

Another issue is *when* to start or stop measurements. If an event, like a packet transmission, has to be evaluated, an event based starting and stopping of energy measurements would be the best choice. However, when evaluating *distributed events*, which even a simple packet transmission is, the challenge is how to start and stop such an event based measurement on multiple nodes simultaneously. Another possibility is to start and stop measurements on a fixed time based schedule. This is, however, only possible if the evaluated application is adapted to the evaluation's needs to make start and stop time predictable. In our evaluation we have decided to use the second approach, as our example application is simple enough to predict start time and measurement duration a priori.

Compared to a conventional simulative evaluation, a realistic energy evaluation is much more time consuming, because it runs in real time. Summarizing duration of energy measurement, node preparation (e.g., flashing nodes) and waiting times (e.g., node startup phase), the evaluation required several days of continuous operation in our testbed. It is also worth noting, that the amount of data collected was quite big. We collected several gigabytes of raw measurement data consisting of 16bit current and voltage values over time using a sampling frequency of only 9kHz.

5. CONCLUSION

Evaluating energy-efficiency of WSN protocols still remains a challenging tasks. However, the results of real measurements are worth time and effort. Our evaluation of a basic scenario revealed several interesting results. First, we disproved the common belief that energy consumption and data volume correlate. Instead we showed, that a larger data volume can even reduce energy consumption in some cases. Second, we pointed out the impact of t_{LPL} on overall energy consumption, as an example for the MAC layers influence on energy-efficiency. Last, we summarized challenges and pitfalls one should be aware of when evaluating energy-efficiency using real hardware.

Up to now, no simulator tool considers hardware tolerances and their consequences to energy consumption and network lifetime. As a result, future work will include an improved version of AVRORA that implements this behavior. We also plan to extend our scenario to include concrete transport and security protocols and other MAC layers, like S-MAC.

6. REFERENCES

[1] A. Hergenröder and J. Horneber. Facing Challenges in Evaluation of WSN Energy Efficiency with Distributed Energy Measurements. In *Proceedings of Multihop Wireless Network Testbeds and Experiments Workshop (TESTBEDS)*, Istanbul, Turkey, July 2011.

[2] A. Hergenröder, J. Wilke, and D. Meier. Distributed Energy Measurements in WSN Testbeds with a Sensor Node Management Device (SNMD). In *Workshop Proceedings of the 23th Conference on Architecture of Computing Systems (ARCS)*, pages 341–438, Hannover, Germany, Feb. 2010. VDE Verlag.

[3] V. Shnayder, M. Hempstead, B.-r. Chen, G. W. Allen, and M. Welsh. Simulating the power consumption of large-scale sensor network applications. In *Proceedings of the 2nd Conference on Embedded networked sensor systems (SenSys)*, page 188, New York, USA, Nov. 2004. ACM Press.

[4] B. L. Titzeri, K. D. Lee, and J. Palsberg. Avrora: Scalable Sensor Network Simulation with Precise Timing. In *Proceedings of the 4th International Symposium on Information Processing in Sensor Networks (IPSN)*, page 67, Piscataway, NJ, USA, 2005.

Proactive Reconfiguration of Wireless Sensor Networks

Marcel Steine[1], Cuong Viet Ngo[2], Ramon Serna Oliver[2], Marc Geilen[1], Twan Basten[1,4], Gerhard Fohler[2], Jean-Dominique Decotignie[3]
[1]Department of Electrical Engineering, Eindhoven University of Technology, The Netherlands
[2]Chair of Real-Time Systems, Technische Universtät Kaiserslautern, Germany
[3]Real-Time Software and Networking, CSEM, Neuchâtel, Switzerland
[4]Embedded Systems Institute, Eindhoven, The Netherlands
m.steine@tue.nl

ABSTRACT

Network dynamics, such as mobility and increase in network load, can influence the performance of a Wireless Sensor Network (WSN). In this paper, we introduce a method which exploits design-time knowledge of the application scenario dynamics to construct a *proactive* run-time reconfiguration approach. The approach anticipates for the impact that predefined dynamic events can have on the performance of the WSN by switching between various *modes* of operation defined at design-time. A mode defines the values for the controllable parameters of the network protocol stack. Our approach explicitly differentiates between parameters that can be adapted locally, per node, and those that should be considered globally for the whole WSN. Design-time definition of modes results in a very low run-time overhead as we only require detection of the mode to use and a low overhead synchronization to change global parameters. The approach is made robust by using a recovery approach for nodes unaware of their global mode after, for example, (re-)joining the network. Experiments with an office monitoring deployment and extensive simulations of a cow-health monitoring scenario show that our approach can easily be adopted by practical WSN deployments and results in a significant reduction in resource usage, e.g., power consumption in our examples, at a very low run-time overhead cost.

Categories and Subject Descriptors: C.2.1 [Network Architecture and Designs]: Wireless communication; C.2.4 [Computer-Communication Networks]: Distributed Systems; C.4 [Performance of Systems]: Performance attributes.
General Terms: Algorithms, Design, Performance.

1. INTRODUCTION

A Wireless Sensor Network (WSN) is a distributed network of small autonomous devices that are capable of sensing, processing and wireless communication. These devices, commonly called nodes, contain one or more sensors and are constrained by stringent resource limitations. They have to collaborate to perform a task specified by the end-user of the WSN. A typical example is a smart building scenario, where information from nodes is used to make smart decisions on, for example, energy management.

With the recent inclusion of WSNs in many different practical deployments, the impact of network dynamics, such as variation in network load and node mobility, on the behaviour of a WSN becomes more prominent. This dynamism should be taken into account when designing a WSN to ensure that the performance, i.e., Quality-of-Service (QoS), is sufficient to perform the required task, while optimizing the resource usage. The controllable parameters of the network protocols used by the nodes, for example the radio transmission power and MAC duty cycle, play an important role in how the network behaves, the resulting QoS it can provide, and the resources required. Currently, parameter values are often fixed or adapted at run-time in response to locally observed performance.

Fixing the value for the parameters at design-time, which is current practice for most WSN deployments, requires to anticipate for the worst-case dynamics of the network to ensure the required QoS at all times. This can result in a conservative selection of parameter values and QoS over-provisioning during the times the network is not experiencing its worst-case dynamics. Over-provisioning can result in a superfluous use of resources, such as the power consumed by nodes. To ensure a good performance for, for example, the monitoring of employees in an office building, we would have to configure the network for the maximum number of persons to monitor when using only single parameter values. This could mean that, for instance, the time the radio can spend in its power efficient idle state is limited to be able to deal with the worst-case amount of data. When there are only a few or no persons in the building, radio time is reserved but not needed, leading to an unnecessary use of battery power.

Adapting the parameter values at run-time provides more freedom to tune the behaviour of the WSN to the experienced dynamics and reduce this waste of resources, while QoS is still met. *Run time reconfiguration* approaches define when and how parameters should be adapted. Various specialized protocols have been developed that include reconfiguration by adapting their own parameters in response to particular changes in the locally observed performance [3, 5, 8, 15, 18, 21]. With the use of run-time techniques, such

MSWiM'11, October 31–November 4, 2011, Miami, Florida, USA.

as feedback control [3, 8, 15, 21], or machine learning [5, 18], these approaches converge to parameter values suited for the current situation. These kind of approaches are *reactive*; they only adapt parameters after a local change of performance has been observed. This may result in a (long) phase, between the occurred dynamics and required change of parameters, in which the performance of the network might be unacceptable or resources might be wasted. On the other hand, too quick responses may lead to unnecessary adaptations to temporary glitches and changes and possibly to instability. Furthermore, current approaches typically only consider adaptation of parameters that can be changed locally. Additional parameters exist with an important impact on the network performance, which should change only in a coordinated manner for all nodes in the network. Moreover, the adaptation techniques used by these specialized approaches are not generally applicable to other protocols and induce a (possibly large) run-time overhead.

Existing reconfiguration approaches are defined independent of the application scenario. The design-time knowledge of the scenario can help in the design of a run-time reconfiguration approach, by anticipating for upcoming dynamics potentially impacting the QoS of the network. The closing of an office building will indicate the presence of a lot less employees to monitor. A person starting to walk will impact the mobility of attached nodes. By detecting these kinds of events we can *proactively* adapt parameters to deal with the (upcoming) effects of the dynamics on the QoS and avoid phases with insufficient QoS or waste of resources. With design-time knowledge of the frequency of events we can furthermore more accurately assess the option of changing global parameters, which can have a strong impact on QoS, but requires synchronization overhead in exchange.

In this paper, we introduce a method to exploit design-time knowledge of application scenario dynamics to construct a proactive run-time reconfiguration approach. At design-time, the WSN designer defines modes of operation, defining the values of the controllable parameters of the network stack, to be adapted in response to expected and detectable scenario dynamics. We explicitly distinguish parameters that can be adapted locally and those that should be considered globally. We therefore define a global mode of operation, equal for every node in the network, and a local mode, which can be different between nodes. These modes are selected by keeping in mind the benefit of and overhead involved in using them at run-time.

At run-time, nodes locally observe the design-time defined events. An event either triggers the change of a local mode or is communicated to a centralized location, for example as part of a distributed detection of a global event. A centralized location, such as a sink, initiates the change of global mode by communicating to all nodes in the network. Due to the design-time definition of the modes, a simple, low overhead flooding approach suffices. Our reconfiguration approach is made robust for practical deployments by a recovery approach that allows nodes to determine the currently used global mode after, for example, (re-)joining the network or missing a request to change a global parameter.

Experiments with an office monitoring deployment and extensive simulations of a cow-health monitoring scenario show the practical applicability of the method. It is shown that, compared to using a single set of parameter values or a reactive reconfiguration approach, a significant reduction in power consumption of the nodes can be obtained for these scenarios, while the QoS is maintained.

The remainder of this paper is organized as follows. In Section 2 related work is discussed. Section 3 discusses our proactive run-time reconfiguration approach in more detail. In Section 4 we show how the method can be applied to two practical scenarios. In Section 5 we discuss the results of the integration of the proactive reconfiguration approach with the use of extensive simulations and experiments with an actual deployment. Section 6 concludes.

2. RELATED WORK

The run-time reconfiguration approach as constructed with the method suggested in this paper, differentiates itself from existing related work by the combination of its (i) proactive approach as opposed to the common reactive approach, (ii) use of both local and global parameters and (iii) independence of protocol stack and focus on practical applicability.

In our method we explicitly exploit a priori knowledge and analysis of the application scenario. This allows us to use a proactive reconfiguration approach, which anticipates on changes in application dynamics which can be deduced from observable events. In the area of real-time embedded systems the benefits of using a proactive approach is already recognized. In [6], scenarios, related to what we call modes, are defined at design-time in order to reduce resource consumption by the system by predicting in which scenario the system operates.

Existing approaches in the area of WSN are independent of the application scenario and react to the effects caused by changes in dynamic behaviour instead. These reactive run-time reconfiguration approaches can be classified as either distributed [3, 15, 21] or centralized [8]. In a centralized approach a central location, such as the sink, collects information from all nodes in the network in order to make decisions on how to reconfigure the network.

Centralized approaches do not allow for adaptation of parameters in response to frequent changes, due to the overhead involved in taking the decision and communicating it. Purely centralized approaches are therefore not suitable for most practical deployments. In a distributed approach, all nodes decide locally on how to adapt their configuration, based only on locally observed quality aspects. For example, in [3] the transmission power of a node is adapted in response to fluctuations in the quality of links or workload observed by the routing protocol. In several MAC protocols [15, 21], the duty-cycle is adapted based on the amount of traffic.

In our method we consider adapting both local and global parameters as well as the issues involved in coordinating and maintaining consistent global parameter values. Nodes (frequently) adapt their local parameters in a distributed manner, while we take a centralized approach to coordinate changes of global parameters. With our method, the overhead of the coordination can be assessed and kept under control with design-time knowledge of the frequency of events.

While many reconfiguration approaches depend on specialized adaptive protocols, of which examples are given above, our approach considers the existing protocol stack as a black box with an interface of changeable protocol parameters. This is common for design-space exploration approaches which are used to find a single set of parameter values [4, 7]. It makes our method independent of the used

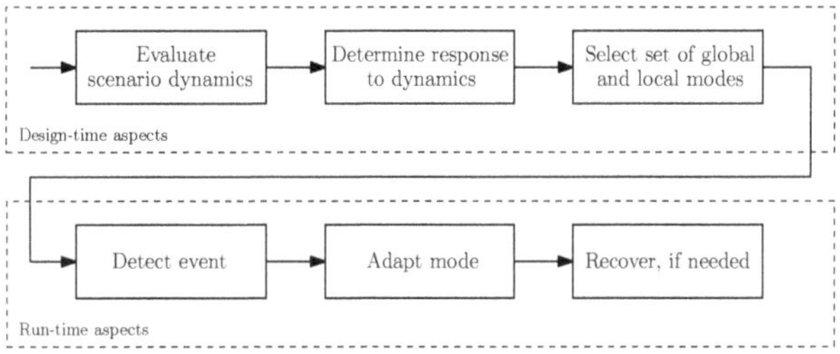

Figure 1: Method design-flow

protocol stack and allows us to change parameters for different protocol layers at the same time, increasing the applicability for new and existing WSN deployments.

The use of design-time defined modes results in a very low run-time overhead for reconfiguration, because we only need a lightweight detection of modes and switching mechanism for the global parameters. This makes it very well suitable for integration in diverse practical WSN deployments. Furthermore, the integration of a recovery approach is emphasized for the use in any practical deployment. The need of such a recovery approach is recognized [1, 14], but its current application in practice induces a large overhead. Compared to approaches as described in [1, 14] we require a smaller scale recovery of only a single (global) mode identifier instead of large code fragments, thereby limiting the overhead involved with the recovery.

3. PROACTIVE RUN-TIME RECONFIGURATION

In this section we introduce our method for designing a proactive run-time reconfiguration. We look in more detail at both the design-time and run-time aspects involved. An overview of the steps is shown in Figure 1. In general, at design-time we define different modes of operation, i.e., sets of parameter values, to be selected by the nodes at run-time, triggered by observable dynamic events, defined using knowledge of the actual environment in which the nodes are to be used. These modes are stored on the nodes locally when deployed. At run-time, we ensure that nodes are aware of the mode in which they should operate.

3.1 Design-time Mode Selection

Evaluate dynamics. The first step of our method is the classification of observable events indicating scenario dynamics that potentially lead to a change of QoS. This change may be due to the impact of the dynamics on important characteristics of the network, such as the network topology, node mobility or data being generated or handled by the nodes. From experience with actual deployments we see that these dynamics can often easily be derived using only high-level knowledge of a designer of the WSN and the target application. Detailed information on, for example, the internal functionality of the network protocols, is usually not required. For instance, when a person starts to walk after being seated, the mobility of the on-body node is expected to change. Or, at the start of a working day, the number of persons, and thereby the amount of data traffic, in an office environment is likely to increase.

Determine response. We respond to an upcoming change in the QoS by the adaptation of controllable parameters of the used network protocols. We can differentiate between two kinds of controllable parameters. A *local parameter* can be set independently for every node in the network. An example is transmission power. A *global parameter* should be set to the same value for every node in the network. An example is the radio channel or frame length for certain MAC protocols. Changing a global parameter requires synchronization to maintain proper functioning of the WSN. Since a global parameter affects the behaviour of every node in the network it often has a larger impact on the QoS compared to changing a local parameter and therefore plays an important role in run-time reconfiguration.

Many possible parameter values for the different protocols and complex cross-layer dependencies might exist. Run-time exploration of the parameter-space is hard and time-consuming and it cannot be done in a distributed fashion without knowledge about the rest of the network, which makes it not suitable for any practical (proactive or reactive) reconfiguration. We therefore determine the parameters at design-time. For this we rely on existing techniques, such as analytical approaches [4, 7], simulations [9, 11], or experience with practical WSN deployments. While our approach is independent of the design-time analysis techniques and controllable protocol parameters considered, they are important for the overall improvement obtained when using the suggested reconfiguration approach.

Determine modes. The next step is to determine to which dynamics we actually want to respond at run-time, keeping in mind the trade-off between the benefits of adaptation and the overhead incurred. The local detection of an event on a node is cheap but a distributed detection of a global event can be expensive. Similarly, adapting a local parameter is cheap but the adaptation of a global parameter is expensive. The frequency of events plays an important role in the overall configuration overhead and depends on the run-time execution of the scenario. However, at design-time we can already analyse the (expected or maximum) frequency of events in the given scenario. In a smart building scenario we know it will become night only once a day, while a person starting to walk might occur much more frequent (but maybe still infrequent enough to change a global parameter depending on the scenario).

The set of parameter values for every combination of defined dynamics, is referred to as the *mode* of operation. Based on the difference between changing local and global parameters we divide the operational mode of a node into a *local mode* and a *global mode*. The local mode defines the local parameter values, while the global mode defines the global parameter values.

After selecting the modes and defining their parameter values, the modes can be stored locally on every node, so that they can quickly be accessed by the node at run time. Note that the possible combinations of modes stored may be different for every node in the (heterogeneous) network. At any moment in time a node is required to operate according to a single combination of local and global mode.

3.2 Run-time Mode Detection

To make the nodes aware of the modes to use at a given moment in time, we have the three final steps of our method as depicted in Figure 1.

Detect event. Unless the detection of the required events is already part of the application, small procedures need to be written for the nodes to detect the events for decision making on mode changes. Events are typically detected using sensor information and/or the current time. When an event is detected that is relevant for the local mode of other nodes in the network or the selection of global modes, it should be communicated to those other nodes or a central location. For the communication of detected events to other nodes we can communicate a unique event identifier to the required nodes using the existing routing infrastructure.

Respond to event. In response to events triggering a new local mode, a node can immediately adapt its local parameters based on the locally stored mode information. For example, a node may change its transmission power without any significant overhead, assuming the protocol stack provides an interface to adapt parameters externally. For the coordinated change of a global mode we rely on a centralized location. As for the local mode change, this centralized location can decide to change the global mode based on detection of local events and/or events of one or more nodes in the network. Once decided to change the global mode, all nodes in the network should be informed.

As the parameter values to be used for every mode can be accessed locally, the centralized location only needs to communicate a predefined unique identifier of the global mode to all nodes. Because of the small amount of information to communicate, we observed that we can resort to a simple and low overhead flooding approach initiated by the centralized location to inform all nodes about the mode change. Starting from the centralized location, the mode identifier is broadcast to all of its neighbours and every node receiving the information for the first time will broadcast it thereby disseminating the information to all nodes in an iterative fashion.

Recovery approach. In a practical WSN deployment, nodes might experience heavy external interference for a longer time and may need to re-join the network. This may result in nodes being in the wrong global mode, after missing a global mode change. Nodes may also freshly join an existing network. Inconsistent global modes lead to nodes not being able to communicate, e.g., when using different radio channels, or a reduced efficiency of the communication, e.g., when using different radio sampling intervals.

Nodes should become aware of using a wrong global mode in order to start looking for the right one. When inconsistent global modes lead to nodes not being able to communicate, nodes may start to acquire the global mode after being unable to communicate for a given amount of time. In general, nodes can then cycle though their different possible global modes and send packets to their neighbours requesting for the currently used global mode. In specific cases, nodes may be able to locally observe which global mode is used and thereby avoid any communication overhead, for example by overhearing communication while changing radio frequencies in order to find the appropriate globally used radio channel.

Discovering an erroneous global mode is less obvious when nodes are still able to communicate, but less efficiently, when using different global modes. A node may not be able to make a distinction between using a wrong global mode or experiencing increased interference. In this case nodes should either regularly request the active global mode from neighbouring nodes or observe this information based on overhearing communication if possible. In this work, we want to emphasize the need of such a recovery approach for practical applications. Exploring the trade-offs involved in designing a recovery approach and the impact of the used protocol stack in more detail is considered for future work.

4. RECONFIGURATION SCENARIOS

In this section, we construct reconfiguration approaches for two practical scenarios. We start by looking at mobility monitoring of cows, based on actual end-user requirements as defined in the WASP project [20]. A second scenario is the monitoring of the location of employees as part of a smart-building scenario. We have chosen scenarios that differ in the deployment characteristics and protocol stack and we show that both allow an intuitive integration of a proactive reconfiguration approach using our method. In the next section, we discuss the experimental results of the integration for both scenarios, where we use extensive simulations for the cow-health monitoring scenario and experiments with an actual deployment for the office monitoring scenario.

4.1 Cow-health Monitoring Scenario

Health problems of cows often result in decreased mobility. Mobility monitoring leads to an early detection and treatment of health problems, improving animal well-being and lowering production losses. A WSN is proposed where a sensor node is attached to the leg of a cow [20]. For our scenario we consider the monitoring of thirty-eight cows with a single BSN [2] sensor node each in a barn of 40 by 60 meters. Twelve static BSN nodes are placed in a grid to support communication to a sink location (being a static node at one corner of the barn). When a cow is walking, acceleration is measured and sent, via the static network of nodes, to the sink for processing in order to detect mobility problems, such as limping. When a cow is stationary, the sink only needs to be informed about the posture of the cow, e.g., standing or laying, to allow activity monitoring. The goal of the deployment is to get a high delivery ratio of the information collected from the cows, while, as it is very inconvenient to regularly remove the nodes from the cows, the WSN operates as power efficient as possible.

When looking at the behaviour of the monitored dairy farm cows, we can observe that during most of the day they stay in the barn with limited mobility and twice a day, at regular times, the farmer leads them to walk to a milking parlor. This procedure can take up to two hours including the walking. With this knowledge of the application scenario we can identify several observable events of interest for reconfiguration. First of all, whether a cow starts walking or becomes stationary has an effect on both the amount of traffic as on the amount of mobility in the network. Furthermore, whether the cows are in the process of being milked or not heavily impacts the traffic load in the entire network as during this period all cows will (slowly) walk to the milking parlor. For the ease of explanation we only focus on these dynamics, but one could think of other interesting dynamics for a more fine-grained reconfiguration, such as going from day to night and vice versa or the time that cows are fed.

To find the appropriate response to the selected events we start by looking at the controllable parameters. For the protocol stack we focus on a basic TDMA MAC protocol with a fixed slot assignment. Every TDMA frame consists of an active period in which every node has a fixed slot for sending while it listens to all other nodes in the other slots. Furthermore, there is a sleep period in which all the nodes

Table 1: Mode selection for cow-health monitoring scenario

Global mode	-	*Not milking*	*Not milking*	*Milking*	*Milking*
Local mode (for mobile nodes)	-	*Stationary*	*Walking*	*Stationary*	*Walking*
Local mode (for static nodes)	-	*Static*	*Static*	*Static*	*Static*
TDMA slot-size	0.1 s	0.05 s	0.05 s	0.1 s	0.1 s
TDMA sleep-time	0 s	10 s	10 s	0 s	0s
GBR update interval	30 s	60 s	60 s	30 s	30 s
Radio TX power (mobile nodes)	-10dBm	-10dBm	-5dBm	-15dBm	-15dBm
Radio TX power (static nodes)	-15dBm	-15dBm	-15dBm	-15dBm	-15dBm

turn off their power consuming radio. To get the packets to the sink, we use Gradient Based Routing (GBR) [13], where nodes maintain their hop-count by periodic flooding initiated by the sink and only forward packets received from nodes with a higher hop-count. The controllable parameters we consider are the TDMA slot-size, used for sending and receiving packets, and sleep-time, which are both global parameters, and the radio transmission (TX) power, which is a local parameter. We furthermore consider the flooding interval of GBR, which is a parameter which can be changed by the sink node on its own, but affects the behaviour of all nodes in the network. It is therefore considered to be a global parameter.

Design-time analysis to decide on optimal values for these controllable parameters is done with the use of simulations. We implemented the scenario in OMNeT++ [11], a discrete-event simulation environment, with the use of MiXiM [9], a modeling framework created for wireless networks, to simulate with different parameter values to get an understanding of the impact of changing different parameters. Table 1 shows the values for the parameters resulting from the design-space exploration to get a delivery ratio higher than 90% at the lowest possible power consumption for the different possible combinations of modes. The table furthermore shows the values that we would use when no reconfiguration approach is applied, but just a single (worst-case) configuration is used over time, in the first column. This is the current state-of-the-art as has been implemented in the WASP project.

We can see in Table 1 that global modes are only changed for the milking period which happens two times a day. Furthermore, the change of local mode can be completely observed locally. The expected overhead of the reconfiguration is therefore negligible and we integrate all the considered modes in our reconfiguration approach.

For the (mobile) nodes to detect whether the monitored cow is changing its posture from walking to stationary and vice versa, needed to select their local mode, we use simple thresholding on the accelerometer values, obtained from the BSN node. This procedure is already part of the application and the observation of these events can therefore be reused. As milking periods start and end every day around the same time, initializing a global mode based on the wall-clock time monitored by the sink is the easiest solution. Another option would be, for example, to use a dedicated sensor informing the sink as soon as the milking parlor is used.

We observed during simulations that a single flood is enough to inform all nodes, due to the small size of the information packet, the used contention free MAC and large density of the network. Because of this and the fact that nodes remain connected to the network throughout the day, a recovery approach is not required for this scenario. For the office scenario discussed next, this plays a more prominent role.

4.2 Office Monitoring Scenario

From monitoring the location of persons in a building we could for instance automatically control the lights in every room of the building based on the presence and activities of persons, potentially reducing the power consumption compared to the manual (de-)activation of light by the employees. We can similarly imagine control of temperature, humidity or other environmental aspects, or provide feedback based on activity. This makes it an interesting scenario for a WSN deployment.

We consider a deployment with a static backbone consisting of 15 TelosB [16] nodes, which monitor light intensity and forward packets to the sink, as shown in Figure 2. To monitor the location of employees in the building, a BSN node is carried by employees, in our case 5 persons, when they start working. They send the (static) parent node to which they communicate, as an indicator of their location, to the sink every 10 seconds.

The dynamics of employees show several interesting observable events for which we could reconfigure. Change in activity of the monitored employee, e.g., walking vs. stationary, impacts the mobility of the node carried by the person. Furthermore, there is an obvious daily pattern of day and night. At night no or just a limited number of employees need to be monitored, which can drastically reduce the amount of data that is communicated within the WSN.

The protocol stack we consider consist of the default Low-Power-Listening (LPL) MAC [10], available in TinyOS [17], a basic asynchronous MAC protocol which reduces idle listening by sampling the medium at a given interval. We use a tree based routing approach where the static nodes have a fixed parent. Due to the structure of the deployment, mobile nodes can determine their closest static node (to forward packets to and as an indication for their location) by iteratively broadcasting a request for hop-count information from the static nodes in range. The considered controllable parameters are the sampling interval of the MAC protocol, a global parameter, and the frequency at which the mobile nodes check their parent for the routing protocol and the transmission power of the radio, both local parameters.

We determined the parameter values to use based on experimenting with the deployment. With the actual deployment in place we determined the impact of changing the parameters on the delivery ratio and power consumption. This resulted in the parameter value selection as shown in Table 2. The first column of the table shows the configuration we would use when using only a single configuration (and thereby configure for the worst-case). Note that for the static nodes we use only a single local mode and retain the same local parameter values at all time.

With the switch between day and night we adapt the global LPL sampling interval parameter, such that nodes are able to handle the amount of traffic experienced during

Table 2: Mode selection for office monitoring scenario

Global mode	-	*Day*	*Day*	*Night*	*Night*
Local mode (for mobile nodes)	-	*Stationary*	*Walking*	*Stationary*	*Walking*
Local mode (for static nodes)	-	*Static*	*Static*	*Static*	*Static*
LPL sampling interval (for all nodes)	500 ms	500 ms	500 ms	2000 ms	2000 ms
Routing request interval (for mobile nodes)	3 s	60 s	3 s	60 s	3 s
Radio TX power (for mobile nodes)	-15 dBm	-15 dBm	-10 dBm	-15 dBm	-10 dBm
Radio TX power (for static nodes)	-15 dBm	-15 dBm	-15 dBm	-15 dBm	-15 dBm

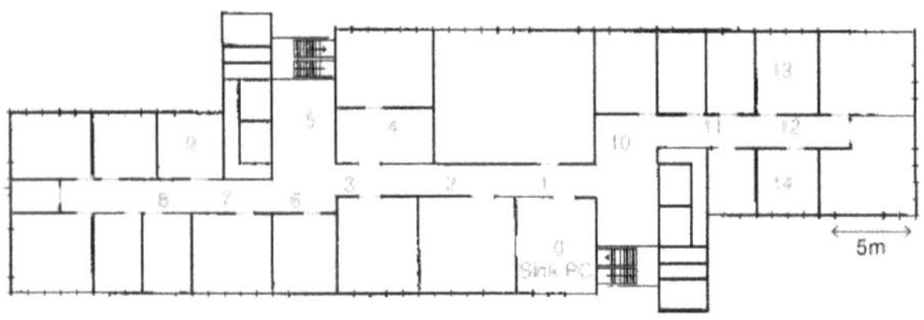

Figure 2: Deployment of 15 static TelosB nodes

that mode. While walking, the request interval is selected to be 3 seconds, which is enough to quickly respond to changing connectivity based on the distance between the nodes and walking speed. While stationary, nodes do not experience this change in connectivity and a high value of 60 seconds is selected. The transmission power is chosen such that a node can always connect to at least one static node in the monitored environment. When walking, we observed a higher interference on the sent packets, for which we compensate by a slightly higher transmission power. Note that we have chosen not to use this higher transmission power for the single configuration approach. This is because this higher transmission power is only found to be useful during the limited time a person is walking, which does not outweigh the disadvantage of having a high transmission power during the entire day. The use of modes allows for a more fine-grained exploitation of performance trade-offs.

For the (mobile) nodes to detect whether the monitored person is changing its posture, we can use the same approach as for the cow-health monitoring scenario. For the global mode to use, nodes start in a consistent global mode during the deployment and are informed about any change of the global mode afterwards. For this scenario we can initiate the change of global mode based on wall-clock time. Time is typically monitored by a computer controlling the WSN and a global mode change can be initiated by communicating to the connected sink node (via serial communication).

We observed that for our scenario, a single broadcast is often not enough to change a global mode due to the lack of redundancy in the deployment and protocols. With more retransmissions of the mode change information packet we quickly converge to successfully switching all (connected) nodes. Based on these results, we can flood a global mode request a conservative number of five times to ensure a sufficient minimal probability of connected nodes requiring to use a recovery approach. Since a global mode happens only two times a day this repeated flooding induces a negligible amount of the total traffic load during the entire day, i.e., 10 packets per node during the entire day.

As nodes (re-)joining the network are an integral part of the application, we do require a recovery approach for these nodes to determine the globally used parameter values. We implemented the recovery approach as proposed in this paper where the node cycles through the global modes while requesting their parent for the global mode if they observe they are entering the network after being disconnected. Experiments with this approach show that the time needed to find the global mode is typically less than a second when in range of the network. We did not implement a repeated check of the global mode for this scenario thereby accepting the small probability of a connected node experiencing limited connectivity if a mode change request is not received.

5. PERFORMANCE EVALUATION

In this section, we discuss results of simulations for the cow-health monitoring scenario and experiments with the office building monitoring deployment. Results show that the integration of our reconfiguration approach has a significant positive effect on the power consumption, while there is no negative effect on the delivery ratio, compared to a basic approach using a static worst-case mode.

None of the existing reactive approaches, such as [3, 5, 8, 15, 18, 21], can directly be used for the entire protocol stack and controllable parameters we consider. To see the impact on the metrics and behaviour of the network when using a proactive compared to a reactive approach we have created our own. We create the best possible reactive approach for the two scenarios, also using the optimal parameter knowledge, expressed by the modes, obtained at design-time. It adapts the parameters in response to changes in the observed network performance instead of application events.

5.1 Cow-health Monitoring Scenario

We implemented the cow-health monitoring scenario and integrated the proactive reconfiguration approach, as constructed in the previous section, for OMNeT++ [11]. The results of the integration for a small-scale WASP demonstrator are discussed in [19]. We compare our proactive approach, in terms of the (conflicting) delivery ratio and power consumption of the nodes in the network, with the approach of using a single worst-case configuration (as is the current practice for this scenario) and a reactive approach.

The considered reactive approach determines the value for the local parameters, i.e., only the transmission power, based on the locally monitored quality of sending packets to the parent node. More specific, the transmission power is increased according to the values related to the modes as soon as the observed delivery ratio to the parent is lower than 85% and lowered as soon as the ratio is higher than 95%. We respond to the periods of high data traffic, as observed when milking, by monitoring the amount of data processed at the sink. We set the global parameters, TDMA slot-size, sleep-time and GBR update interval, to the values related to the milking mode (Table 1) when it is observed at the sink that at least 80% of the cows is sending at the rate related to walking and to the not milking values otherwise.

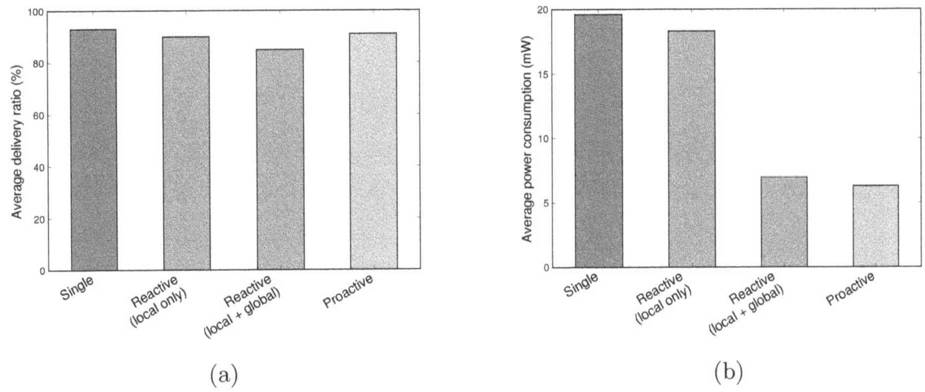

Figure 3: Impact on average (a) delivery ratio (b) power consumption for cow-health monitoring scenario

The sketched scenario is simulated for a day, in which two milking periods of two hours occur, for the three approaches, i.e., a single (worst-case) configuration, reactive (adapting only the local parameters and adapting both the local and global parameters) and proactive using modes. To have statistically reliable results, every experiment was repeated 10 times; the results showed limited variation. The average delivery ratio and the average power consumption of the nodes are shown in Figure 3. Shown results are the averages over all runs. The delivery ratio was determined by the ratio of packets sent by the mobile nodes and the number of packets received by the sink. For the power consumption we focus on the main power consuming part of the node, i.e., the radio [12, 16], which, in this case, depends on the chosen TDMA schedule.

The first thing we can observe is that for this scenario a significant reduction of power consumption, and thereby increased lifetime of the WSN, can be obtained by the introduction of the proactive approach compared to the currently used single configuration, while the delivery ratio is still maintained at a high level. When we look at the results of the reactive approach we see that if we only consider adapting local parameters, as is typically done, we see only a limited improvement, but as soon as global parameters are adapted as well, the average power consumed by the nodes reduces and is only slightly higher than the proactive approach. This shows the importance of global parameters for this scenario.

When comparing the reactive and proactive approach the most profound difference can be seen when network dynamics occur. With the reactive approach there is a delay between observing the high network load and the adaptation of the global parameters, i.e., TDMA schedule. In this period packets are lost due to the limited capacity reserved by the TDMA schedule. This only slightly impacts the delivery ratio if it is averaged over a long period, due to the limited frequency of the events. This effect is discussed in more detail for the actual deployment used for the office monitoring scenario, discussed in the next subsection.

What we observe when looking in more detail at the simulations is that in the period between observing the high network load and adaptation of the global parameters, the reactive approach first tries to compensate for the low delivery ratio by increasing the transmission power, which increases power consumption, while the solution lies in the adaptation of the TDMA schedule. Such a cross-layer view is not present in reactive approaches for individual parameters, but it is an additional advantage present in our method.

What we furthermore see is that the coordinated change of the global parameters becomes less efficient due to the high traffic and more than one retransmission is needed to successfully change. This increases the time until the required adaptation even more. Packets are lost during this transitional phase. This shows an advantage of our proactive approach; it can anticipate and adapt for the high traffic just before it actually occurs, avoiding adaptation in the period in which coordinating a mode change is more challenging and many packets can be lost.

5.2 Office Monitoring Scenario

We implemented the office monitoring scenario and integrated our reconfiguration approach, as described in the previous section, using TinyOS 2 [17]. We did a one day experiment based on the parameter values shown in Table 2. We decided to use the night mode before 9 AM, after which most employees start working, and after 7 PM, the time at which most employees have ended their working day. Because mobility and interference patterns are hard to control and differ between the different experiments, it is very hard to make a precise quantitative comparison between the different approaches. What the experiments do show is that we can successfully integrate our reconfiguration approach in a practical environment. We also observe significant positive trends for the delivery ratio and power consumption when using a proactive approach. In this subsection we furthermore look in more detail at the comparison with a reactive approach showing the benefit of proactivity compared to the delayed response of a reactive approach.

Figure 4 shows the impact of integrating our proactive approach on the average delivery ratio and average power consumption, where we differentiate between the static and mobile nodes. Unlike the simulations, practical experiments show a large difference between the performance of individual nodes, mainly due to the larger impact of external interference. The error bars show the maximum and minimum observed delivery ratio and power consumption of individual nodes. The results are based on experiments starting at 8 AM, before all monitored employees started working and 1 hour before the global mode switch, until 8 PM, after all employees stopped working and 1 hour after the global mode switch. For practical reasons we could not leave the

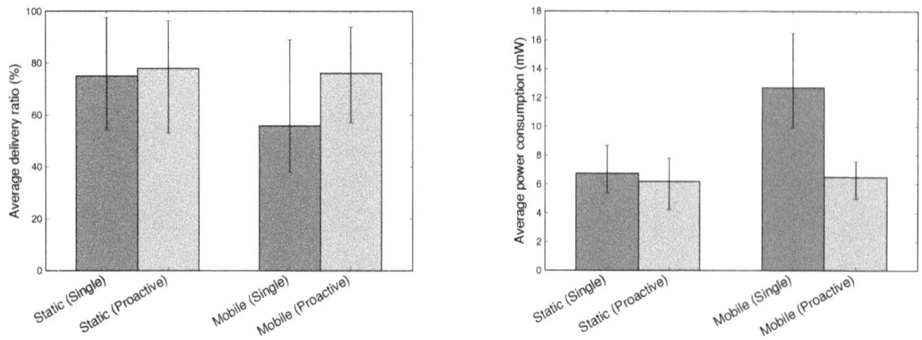

(a) (b)

Figure 4: Impact on average (a) delivery ratio (b) power consumption for office monitoring scenario

nodes over night. For the remaining 12 hours we therefore used a delivery ratio and power consumption based on experiments with only the static network in place and assumed it would be similar during the entire night. The results are based on the time they are on, which is the complete day for the static nodes and around 9 hours for the mobile nodes. The delivery ratio is based on the (application) packets sent, when the nodes were inside the building, i.e. connected to the network, in case of the mobile nodes. For the power consumption we focus on the power spent by the radio, which is derived by monitoring the time spent in every radio state, i.e., sending, receiving and idle [16].

The proactive reconfiguration resulted in a significant better power consumption for the mobile nodes, while the average delivery ratio is maintained and even improved. For the (less power constrained) static nodes no significant impact is observed due to the limited network dynamics and impact of changing parameters observed by these nodes.

Like in the cow-health monitoring scenario, when comparing with a reactive approach which is created to optimize the quality metrics averaged over a long time, it is expected to have a similar impact on the metrics due to limited network dynamism. The main difference between the reactive and proactive approach is found around periods of network dynamics, e.g., when people are moving. We did additional experiments with the deployment to zoom in on this behaviour and its effects on the performance metrics for local mode adaptation. We used a single mobile node which in 1 minute moves from one side of the corridor to the other, stays there for 2 minutes, moves to the other side in 1 minute and stays there for 2 minutes. During the experiment, the mobile node sends packets to the sink once every 5 seconds, while the static nodes only forward these packets. Results are collected after an initialization period of 1 minute.

We integrated a reactive approach in the routing protocol, which provides the parent to use for communication. The locally observable change of parent (for the mobile nodes) is a good indication of a changing topology and the routing protocol adapts local parameters in response, i.e., the transmission power (TX) is increased to -10dBm and the routing request interval reduced to 3 seconds. If no change of the parent is observed in the last 30 seconds, the activity of the protocols is relaxed by reducing TX to -15dBm and the interval to 60 seconds. With our experiment we focus on local

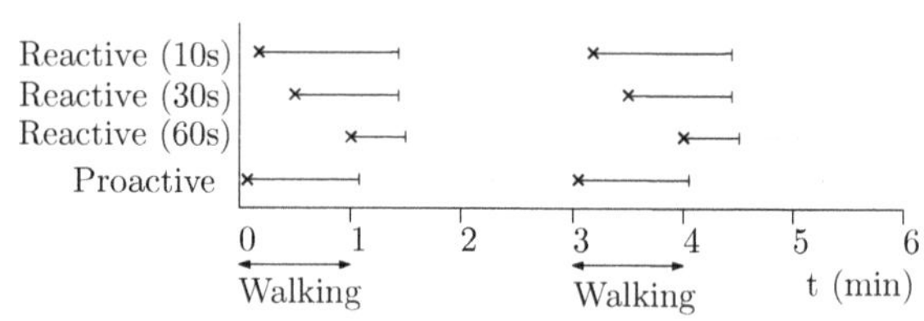

Figure 5: Adaptation responsiveness

events and retain the same global parameter values, i.e., an LPL sampling interval of 500 ms.

Figure 5 shows the reconfiguration moments in time of the proactive approach compared to the reactive approach as described above. Furthermore, it shows the results for the reactive approach when spending more time on the local analysis of the performance, i.e. reduce the request interval to 30 and 10 seconds, in order to increase the responsiveness. While the proactive approach can accurately adapt during the times it is required, using a reactive approach introduces a delay for the adaptation. After the node starts moving a delay arises until a parent change is observed. After the period of walking, there is still a small delay due to the introduced interval to be sure that the topology stopped changing. Reducing this interval can reduce the delay, but can result in a more frequent, unnecessary, adaptation for slow topology changes.

Figure 6 shows the delivery ratio and power consumption of the mobile node for the different activities, i.e., walking and stationary, when using the different adaptation approaches. The delivery ratio when walking can in this case potentially be higher than stationary due to the lower average number of hops the packets need to travel to the sink. With the reactive approach, delay in adaptation results in nodes using an unreachable parent. For the reactive approach we can therefore see a clear trade-off between the responsiveness to topology changes, which is reflected in the delivery ratio when walking, and power spent by the nodes. Putting more effort in observing the local performance, during the entire run-time, results in a better responsiveness to dynamic events, but also requires more power during the period in which no dynamics occur. With the proactive approach we actively observe the dynamics and this trade-off does not need to be made, resulting in both a better delivery ratio (when walking) and overall power consumption.

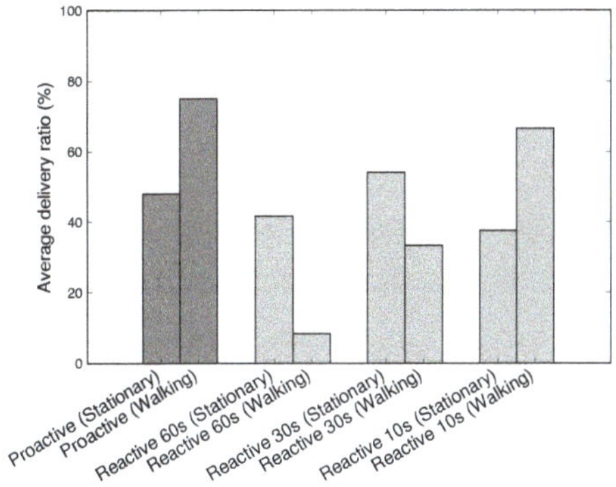

(a)

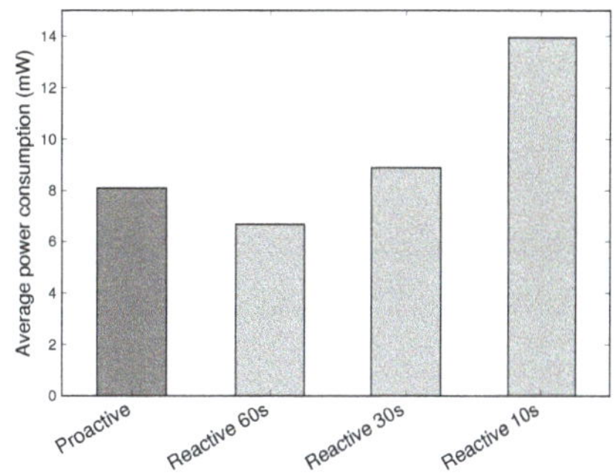

(b)

Figure 6: Impact on average (a) delivery ratio per activity (b) power consumption of the mobile node for small scale experiment

6. CONCLUSIONS

In this paper, we have introduced a proactive run-time reconfiguration approach for Wireless Sensor Networks (WSNs). This approach explicitly exploits a priori knowledge of the application scenario to guide a reconfiguration process.

At design-time, observable events signalling a change in dynamics experienced in the network and thereby potentially the Quality-of-Service (QoS) provided by the network are identified. Using existing design-time analysis techniques, the optimal response to these dynamics in terms of adapting controllable parameters is determined. Based on this analysis, different modes of operation are defined. We distinguish parameters that can be adapted locally and those that should be considered globally. We therefore define a global mode of operation, equal for every node in the network, and a local mode, which can be different between nodes. These modes are selected by keeping in mind the benefit of using them at run-time and the overhead involved.

At run-time, nodes locally observe the design-time defined events, which either trigger the change of a local mode or is communicated to a centralized location. A centralized location, such as a sink, initiates the change of global mode by communicating to all nodes in the network, for which a simple, low overhead flooding approach suffices. Our reconfiguration approach is made robust for practical deployments by a recovery approach that allows nodes to determine the currently used value for global parameters after, for example, (re-)joining the network or missing a global mode request.

We showed a proactive reconfiguration approach for practical cow-health monitoring and office monitoring scenarios. The use of a proactive approach results in a significant positive effect on the used resources, in this case power consumption, while there is no negative effect on other QoS metrics, i.e. the packet delivery ratio. Due to the controlled number of global parameter changes and design-time definition of the modes, there is only a very low run-time overhead incurred by our approach. By comparing with the most applicable reactive approach, we have seen that being proactive allows an early reconfiguration for network dynamics thereby avoiding phases in which the performance of the network is limited.

Acknowledgment

This work was in part supported by EC FP6 project IST-034963 WASP.

7. REFERENCES

[1] S. Brown and C. J. Sreenan. Software Update Recovery for Wireless Sensor Networks. *SAEL*, 29:107–125, 2010.
[2] BSN website, ICL London. http://vip.doc.ic.ac.uk/bsn/m621.html.
[3] O. Chipara et al. Real-time Power-Aware Routing in Sensor Networks. *IWQoS*, pages 83–92, 2006.
[4] K. P. Ferentinos and T. A. Tsiligiridis. Adaptive design optimization of wireless sensor networks using genetic algorithms. *Computer Networks*, 51(4):1031 – 1051, 2007.
[5] A. Forster. Machine learning techniques applied to wireless ad hoc networks: Guide and survey. *ISSNIP*, 2007.
[6] S. V. Gheorghita et al. System-scenario-based design of dynamic embedded systems. *ACM Trans. Des. Autom. Electron. Syst.*, 14(1), article 3, 45 pages, 2009.
[7] R. Hoes et al. Analysing QoS trade-offs in wireless sensor networks. *MSWiM*, pages 60–69, 2007.
[8] S. Kogekar et al. Constraint-guided dynamic reconfiguration in sensor networks. *IPSN*, pages 379–387, 2004.
[9] MiXiM website. mixim.sourceforge.net.
[10] D. Moss and P. Levis. BoX-MACs: Exploiting Physical and Link Layer Boundaries in Low-Power Networking. Technical report, Stanford Information Networks Group, 2008.
[11] OMNeT++ website. www.omnetpp.org.
[12] A. Prayati et al. A modeling approach on the TelosB WSN platform power consumption. *J. Syst. Softw.*, 83:1355–1363, 2010.
[13] C. Schurgers and M. Srivastava. Energy efficient routing in wireless sensor networks. *MILCOM*, 2001.
[14] M. Strasser and H. Vogt. Autonomous and distributed node recovery in wireless sensor networks. SASN, pages 113–122, 2006.
[15] Y. Sun et al. RI-MAC: a receiver-initiated asynchronous duty cycle MAC protocol for dynamic traffic loads in wireless sensor networks. SenSys, pages 1–14, 2008.
[16] TelosB Datasheet, Crossbow Inc. www.xbow.com.
[17] TinyOS website. www.tinyos.net.
[18] S. Tomforde et al. Adaptive Control of Sensor Networks. *ATC*, pages 77 – 91, 2010.
[19] C. Viet Ngo, R. Serna Oliver, and G. Fohler. SIMOSP: A Simple Mode Switch Protocol for Wireless Sensor Networks. Technical report, Technische Universtät Kaiserslautern, October 2010.
[20] WASP (Wirelessly Accessible Sensor Populations) project website. http://www.wasp-project.org/.
[21] W. Ye, J. Heidemann, and D. Estrin. Medium access control with coordinated adaptive sleeping for wireless sensor networks. *IEEE/ACM Trans. Netw.*, 13(3):493–506, 2004.

The Dark Side Of DEMMON: What Is Behind The Scene In Engineering Large-Scale Wireless Sensor Networks

Stefano Tennina*, Ricardo Gomes and Mario Alves
CISTER Research Center
Polytechnic Institute of Porto (ISEP/IPP)
Porto, Portugal
sota@isep.ipp.pt, rftg@isep.ipp.pt, mjf@isep.ipp.pt

Vincenzo Ciriello and Gabriella Carrozza
SESM scarl
Giugliano in Campania
Naples, Italy
vciriello@sesm.it, gcarrozza@sesm.it

ABSTRACT

Most research work on WSNs has focused on protocols or on specific applications. There is a clear lack of easy/ready-to-use WSN technologies and tools for planning, implementing, testing and commissioning WSN systems in an integrated fashion. While there exists a plethora of papers about network planning and deployment methodologies, to the best of our knowledge none of them helps the designer to match coverage requirements with network performance evaluation. In this paper we aim at filling this gap by presenting an unified toolset, i.e., a framework able to provide a global picture of the system, from the network deployment planning to system test and validation. This toolset has been designed to back up the EMMON WSN system architecture for large-scale, dense, real-time embedded monitoring. It includes network deployment planning, worst-case analysis and dimensioning, protocol simulation and automatic remote programming and hardware testing tools. This toolset has been paramount to validate the system architecture through DEMMON1, the first EMMON demonstrator, i.e., a 300+ node test-bed, which is, to the best of our knowledge, the largest single-site WSN test-bed in Europe to date.

Categories and Subject Descriptors

B.4.4 [**Performance Analysis and Design Aids**]: [Simulation, Verification, Worst-case analysis]

General Terms

Design, Experimentation

Keywords

Large Scale Wireless Sensor Networks, Toolset, Testbed

*Corresponding author.

1. INTRODUCTION

Wireless Sensor Networks (WSNs) have emerged as an infrastructure to support new classes of large-scale and dense networked embedded systems. While in the last decade there has been a plethora of scientific publications on WSNs, the vast majority focuses on protocols and algorithms (e.g. medium access control, routing, data aggregation) while only a few papers report on real(istic) applications [24].

Several relevant work on WSN architectures and solutions supported by working prototypes have been described in the literature (e.g. [5, 12, 14, 19]) and in the scope of research projects (e.g. [2, 11, 23]). However, in [28] we have shown that none of them fulfills all requirements for large-scale and dense real-time monitoring.

EMMON [1] is an ARTEMIS industry–driven project that aimed to design a WSN system architecture targeting large-scale and dense real-time monitoring applications, such as real-time pollution/noise maps of a city, or critical infrastructures monitoring (such as bridges, tunnels or the power grid). The EMMON architecture encompasses all system components: command and control graphical user interface, communication network architecture, middleware and hardware platform. It is (as far as possible) built upon standard and commercially available (COTS) technologies, maintaining as much flexibility as possible, but still guaranteeing its adequateness to fit specific applications' requirements [21]. We envisage to fulfill Quality–of–Service (QoS) requirements in an integrated fashion, considering scalability, timeliness, reliability/robustness and energy-efficiency. These QoS requirements can be found in applications [21] such as energy-efficiency in data centers [19] and infrastructures monitoring (e.g., bridges, tunnels or the power grid) [8].

Importantly, and as reported in this paper, this WSN architecture is supported by a *EM-set*, a unique and complete planning, dimensioning, simulation and analysis toolset, for deployment planning, worst case dimensioning, protocol simulation, remote programming and network sniffing. This paper outlines such a toolset and shows how it has been used to test and validate the baseline EMMON architecture by extensive simulation and experimental evaluation, including through a 300+ node test-bed [1], which is, to the best of our knowledge, the largest single-site WSN test-bed in Europe.

The remainder of the paper is organized as follows. After a brief overview of the EMMON system architecture in Section 2, the focus is put on the EM-set, encompassing network

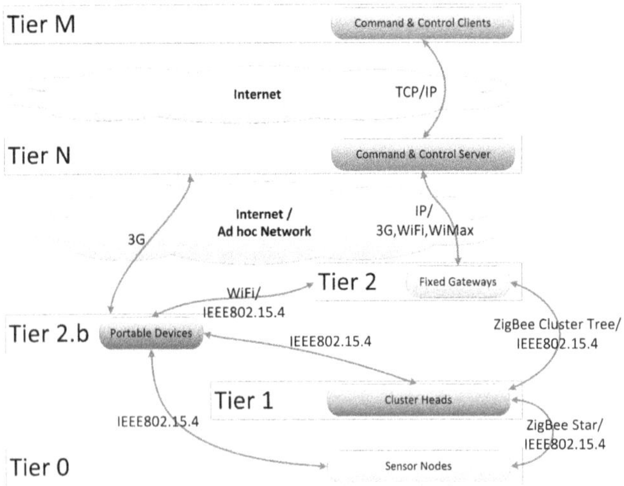

Figure 1: EMMON Hierarchical System Architecture.

deployment planning (Section 3), analytical and simulation models (Section 4), nodes programming/testing and experimental data gathering (Section 5). Finally, Section 6 illustrates some results of the first instantiation of the EMMON architecture on a physical test-bed (DEMMON1[1]) and compares them to simulation and analytical results. Concluding remarks are given in Section 7, as well as an outline of the on-going work related to the instantiation of the EMMON system architecture into several application scenarios.

2. EMMON ARCHITECTURE & TOOLSET

This section outlines the EMMON architecture and the toolset used to support it and to provide the results described in Section 6.

2.1 System Architecture

Building on the alternatives identified in [28] and to cope with scalability and QoS requirements, EMMON adopts a hierarchical, multi-tier network architecture as sketched in Fig. 1. Its main characteristics are summarized as follows.

(i) The synchronized version of the IEEE 802.15.4 MAC is used at the lowest tiers. Nodes are synchronously active or sleeping, with a dynamically adaptable duty-cycle (each cluster can operate at different duty-cycles). This enables to find the best delay/throughput vs. energy trade-off. Both best-effort (CSMA/CA, during the CAP) and real-time (GTS, during the CFP) traffic classes are supported.

(ii) WSN nodes are organized into a ZigBee-based Cluster-Tree network model [17], rooted at a gateway playing the role of the sink. A cluster-tree is a hierarchical architecture *per-se* and is adequate for convergecast traffic, as utilized in the majority of WSN applications. However, to avoid collisions between clusters while meeting all end-to-end deadlines of a predefined set of time-bounded data flows and minimizing the energy consumptions of the nodes, clusters' active portions are scheduled in a non-overlapping sequence – using the Time Division Cluster Scheduling (TDCS) [17].

(iii) We adopt tree-routing for upstream traffic, which has a negligible memory footprint and processing delay since no routing tables are needed. We also support efficient geographical-based routing of queries for the downward flow, for disseminating requests from a single root to all the nodes involved. This has a huge impact in terms of scalability, since it allows users to interact with the system through the definition of geographical objects, rather than any explicit request for raw readings from specific sensor nodes.

(iv) Beaconing inherently enables the support for time synchronization at the Data Link and Network layers. Accordingly, it enables accurate time stamping of sensor data (required by many applications), energy-efficiency through duty-cycling, cluster scheduling techniques and a contention-free MAC.

(v) Data aggregation, sensor- and data-fusion mechanisms [22] are implemented at all levels of the architecture: (v.a) at the sensor nodes (SNs), by aggregating multiple readings taken over time (temporal aggregation), before sending these data upstream; (v.b) at the cluster heads (CHs), by aggregating multiple readings coming from different sensors or children CHs (spatial aggregation), before forwarding the report upstream; (v.c) at the gateway (GW), where sensor fusion, i.e., inferring useful information through, e.g., model fitting, is done by considering multiple reports coming from the CHs, before sending them to the C&C; and (v.d) at the C&C, where a complete information can be returned to the users by allowing, e.g., a correlation of the incoming sensor reports with other available data (e.g., current traffic conditions in urban noise or air quality monitoring applications [21]).

(vi) A novel EMMON-specific middleware runs on all the elements of the system: it glues all the components together, from the C&C clients to the SNs, leading them to work properly over heterogeneous communication technologies (Fig. 1). It supports both periodic measurements reporting and event based (alarms) applications. It also greatly helps in networking and system management operations, thanks to its distribution of the intelligence as low as possible in the network's tiers.

(vii) The EMMON C&C subsystem is the interface to the end-users. It is composed by two applications, one which runs on the C&C server and uses the middleware API, and the other on the C&C clients, where the Graphical User Interface is implemented.

Finally, the baseline IEEE 802.15.4 and ZigBee protocol stack is supported by a solid critical mass, designed and implemented in synergy with the TinyOS 15.4 and ZigBee Working Groups [13].

2.2 EM-set: System Planning, Simulation, Analysis toolset

This section provides a snapshot of EM-set, the EMMON toolset, before describing each component in Sections 3 – 5.

Fig. 2 provides an overall perspective of EM-set. It is worthwhile to stress that the integration between the different components (inputs/outputs) into a single framework enables to speedup system design. In fact, the ultimate goal of this toolset is to thoroughly allow functional and non-functional performance indices evaluation under several aspects. For instance, this toolset allows to assess the scalability limits of the EMMON network through an evaluation of e.g., the end-to-end delays derived from analytical and simulation models as well as from experimental trials. The results presented in Section 6 derive from the use of *EM-set*.

[1]The acronym stands for EMMON Demonstrator, phase 1.

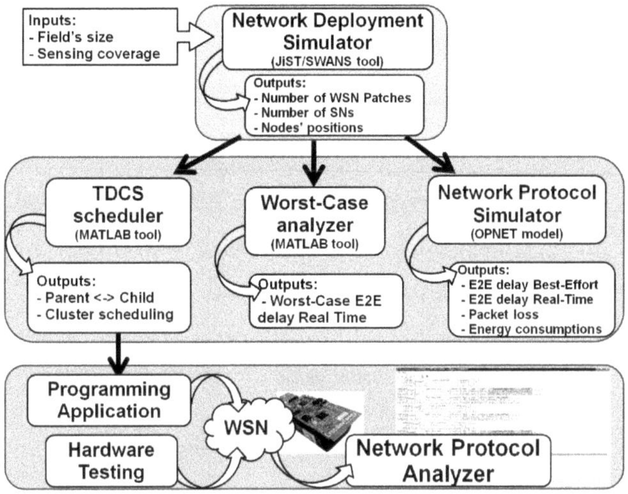

Figure 2: EM-set: EMMON toolset for network planning, dimensioning simulation analysis and test/programming.

Starting from the field area and the sensing coverage of each SN and assuming the structure of a basic WSN Patch, i.e., a GW and C CHs surrounding it, the Network Deployment Planning outputs the number of WSN Patches and SNs needed to cover that area, as well as their optimal positions. These outputs feed the TDCS scheduler, the Worst-Case analyzer and the Network Protocol simulator. These three tools enable network dimensioning and a performance evaluation. Upper bounds on the end-to-end delays of real-time traffic and end-to-end delays for both best-effort and real-time traffic classes, packet losses and network lifetime (as per node energy consumptions) are computed. The TDCS scheduler outputs the topology of the WSN Patch and the clusters' scheduling that are used to feed the Remote Programming and Testing tools (to program/test the WSN nodes via a USB tree) and to evaluate the network performance through Network Protocol Analyzers (i.e., a sniffer tool to capture the packets and a customized parser to extract the most useful statistics from the logs).

3. NETWORK DEPLOYMENT PLANNING

The Network Deployment Planning is composed by a single engine: a simulator to study the optimal deployment problem. Assuming that in EMMON there is some control over node deployment, i.e., CHs and GWs are placed in order to maximize network connectivity, the SNs are generically supposed to be scattered all over the monitored area. We built upon an Open Source tool named SiDNet-SWANS [25], built over JiST, a java based discrete event simulator. We choose JiST/SWANS because it allows to easily simulate very large scale WSNs. The results in [29] confirm that JiST/SWANS presents better time execution performance, compared to more recent simulators as NS-3 or OmNet++. Moreover, [6] shows that JiST/SWANS is really capable to simulate networks composed by a high number of nodes (up to one million) in acceptable time and using common PCs. NS-3 and OMNeT++ experiments showing similar capabilities are not available. [26] confirms the qualitative features of the JiST's simulation, comparing it with NS-2 and getting very similar performance results.

In our scenario, the deployment problem is composed by two levels: (i) inter-patch deployment, i.e., how to distribute the WSN patches on the field, and (ii) intra-patch deployment, i.e., how to place the nodes internally to the patch, in particular how to arrange the CHs and the GW. The designed tool is based on innovative deployment strategies. The inter-patch deployment organizes the field as a grid, where each cell has a specific geometric shape (e.g., square, triangular or hexagonal). This deployment strategy has already been discussed in several studies [7, 31], but we use it in a slightly different way. In those approaches, the nodes (or a set of nodes in some cases) are placed on the vertices of each cell (e.g., on the corners of the square, or on the vertices of triangles), while, in our approach, the nodes are placed inside each cell, meaning that each cell contains a WSN Patch having the GW in the middle. Currently, the designed tool provides two alternative grid deployments: square based deployment (SBD), where each WSN patch is circumscribed in a square and triangle based deployment (TBD) in which the patches are inscribed into equilateral triangles. Fig. 3 shows WSN Patches deployed in e.g., a square-shaped grid.

As far as the intra-patch deployment is concerned, the objective is to organize the deployment based on a Cluster-Tree model, as shown in Fig. 3: the diamond markers are the GWs[2], while the star markers represent the CHs[3]. The WSN Patch is logically organized according to a tree based topology, in which the GW acts as the root and each CH as node. Hence, according to the EMMON assumptions (i.e., to have some control on the GW and CHs positions) the designed deployment scheme provides the positions of CHs organized in "rings" around the GW (for all patches), while SNs (point markers in Fig. 3) are randomly scattered over the entire field. Then, nodes must organize themselves into a Cluster-Tree topology, using the IEEE 802.15.4 standard association mechanism, i.e., each node chooses its parent based on some metrics (e.g., the best link quality level). As a consequence, the deployment planning tool actually runs network simulations taking into account realistic signal propagation conditions. However, to allow for fault tolerance and recovery strategies to be implemented in the EMMON system, an additional requirement has been included, i.e., that each child should have at least two alternative candidate parents to choose from. This imposes a certain degree of redundancy on the number of CHs.

Finally, the deployment schemes are evaluated against two performance indices: (i) network connectivity and (ii) field's coverage. Connectivity refers to how much of the generated data can ultimately arrive at the C&C, from the SNs and through the CHs and the GWs. A network is considered to be fully connected if all nodes can deliver their data to the destination. According to the definition reported in [30, 32], a field is considered to be covered if all its points are within the sensing range of at least one active sensor. The sensing range R_r is defined as the maximum distance at which a sensor node can measure environmental parameters (e.g., temperature, humidity, light, etc.) and is an input to this tool. In EMMON, we are interested in checking both network connectivity and field's coverage simultaneously. Therefore we

[2]Every GW is placed in the center of each cell.
[3]Every CH surrounds a GW in each cell.

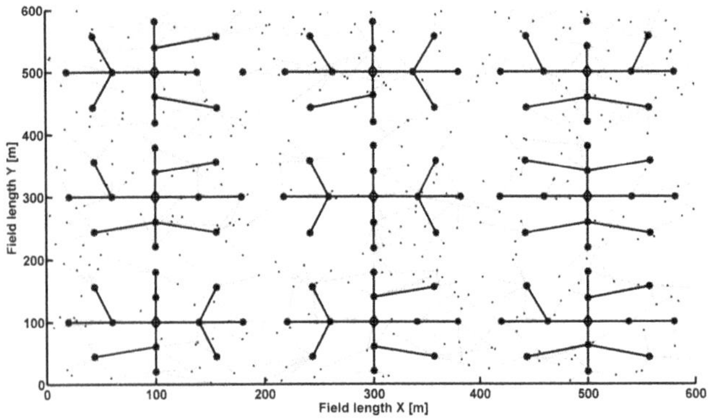

Figure 3: Example of WSN Patches with a SBD scheme to cover at least 80% of a 49 km^2 field (particular). Diamond markers are GWs; stars are CHs, distributed along 2 hops around each GW; points are SNs, randomly deployed.

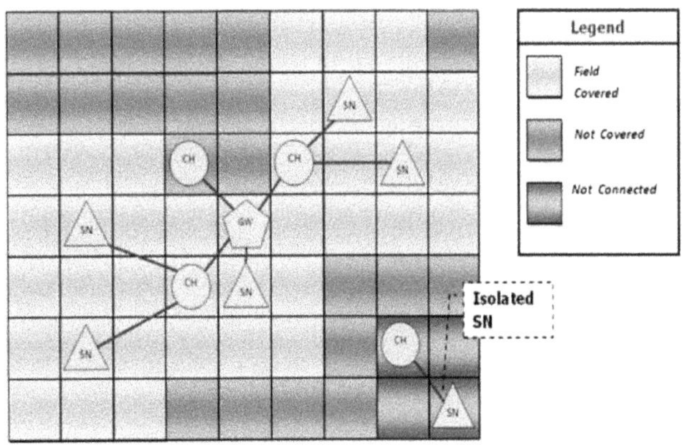

Figure 4: Field coverage analysis example.

make the latter definition more restrictive by assuming that a point is *covered* if there is at least one SN less than R_r meters away from it and that SN is connected to the network, i.e., it is able to deliver its measurements to the C&C.

Fig. 4 reports an example of this concept: the triangles, circles and pentagon represent SNs, CHs and GW, respectively, while the blue lines represent the link between the nodes. For the sake of simplicity, for this example, let us assume that the connection between the GW and the remote C&C is guaranteed at any time. So, if a SN has a path to the GW, we say that it is connected. Furthermore, let us suppose that each SN has a sensing range equal to one cell, i.e., each SN can measure environmental data only from its eight adjacent cells. Then, the green cells in Fig. 4 represent all the points covered: there is at least a SN in an adjacent cell, and the SNs covering the cell are connected. Both the red and violet cells are uncovered: the former, because they are not close to any SN; the latter, because a close SN does not have any path to the GW.

The field coverage analysis algorithm is reported in Fig. 5. It is implemented as a post-processing analysis tool and requires as inputs the results of the simulation. It is then able to output the field coverage analysis. In particular, the algorithm requires a list L_s of all the nodes positions and

- Input:
  - L_s : list of sensor positions on the field.
  - R_r : sensor reading range.
- Algorithm:
  - Field is represented as a boolean matrix F.
  - $\forall(x, y) \in L_s$ considering a square A of side $2R_r$ centred in (x,y).
  - $\forall(x_i, y_i) \in L_s$:

$$if\left(\sqrt{(x-x_i)^2+(y-y_i)^2} < R_r\right) \quad => \quad F[x_i, y_i]=1$$

  - $\text{Field coverage} = \dfrac{\#\text{elements of F equal to 1}}{\#\text{field's point}}$

Figure 5: Field coverage analysis algorithm.

considers the sensing range R_r as a parameter. The field is represented as a boolean matrix where each element denotes if the corresponding cell is covered or not. To this aim, we should consider as covered those cells that fall into a circle centered on the SN and whose radius is equal to R_r. The field coverage is then calculated as the ratio between the covered cells and the whole area.

4. PERFORMANCE ANALYSIS

Focusing on the portion of the EMMON architecture below the GW, i.e., on a *single* WSN Patch, the outputs from the network deployment simulator are translated into appropriate inputs for three tools [15]: the TDCS Scheduler, the Worst-Case Analyzer and the Network Protocol Simulator (recall Fig. 2). These tools are applied to each WSN Patch, as outlined next.

4.1 TDCS Scheduler

In the Cluster-Tree WSN model, clusters' activity is scheduled according to a time division approach, where clusters' active periods do not overlap. Beacons are messages sent by every local coordinator of any WSN Patch (i.e., the GW and the CHs) and serve to maintain the synchronization among the nodes of each cluster. This has the advantage of improving the coordination to save energy (reduce retransmissions, put the nodes to sleep and wake them up again in a synchronous fashion) and of guaranteeing a given level of QoS. However, to preserve the coordination and avoid intra-cluster collisions, the TDCS algorithm [17] is needed. This mechanism involves the definition of the Start Time values of the MAC protocol, such that the active portions of each cluster are interleaved during the inactive portion of all the others sharing the same collision domain, as shown in Fig. 6.

This design choice leads to a given upper bound on the allowed number of clusters (Γ[4]) and related duty cycles (DC). Knowing that $0 \leq SO \leq BO \leq 14$ (being SO the Superframe Order and BO the Beacon Order defined in the IEEE 802.15.4 standard), the maximum number of allowed clusters is $\Gamma = 2^{BO-SO}$ and the associated duty cycle is $DC = {}^1\!/_\Gamma = 2^{SO-BO}$. An example of these relations is shown in Fig. 7 where SO has been fixed to 4. As a consequence, under the assumption that the couple (BO,SO) is

[4]The total number of clusters in a WSN Patch must include also the GW, as a special cluster head.

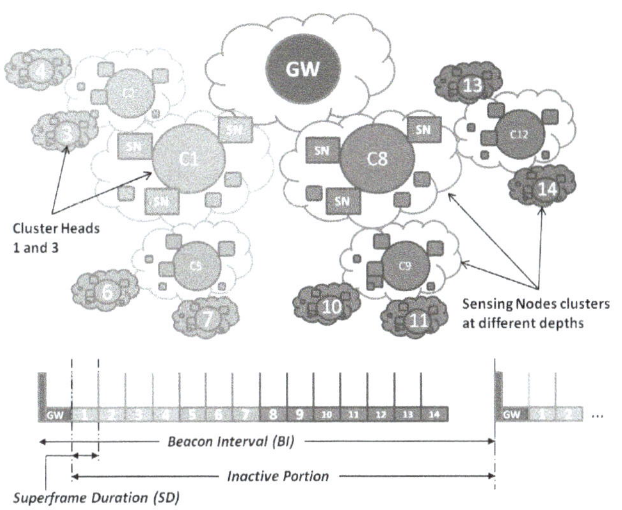

Figure 6: Time Division Cluster Scheduling (TDCS): it is assumed that every cluster has the same value of the couple (BO, SO). The inactive portion of the GW (only) is also shown, as an example.

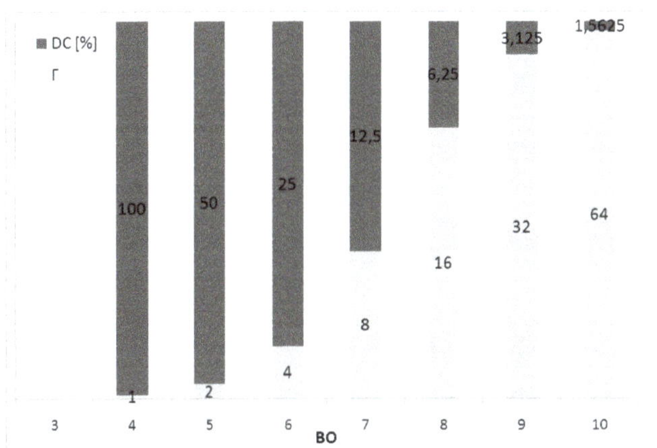

Figure 7: TDCS: upper bound on the allowed number of clusters (Γ) and related duty cycles (*DC*) for SO=4 and a subset of values of BO.

the same for every cluster in a WSN Patch, at design time, those relations help identifying the best trade-off between Γ and the related DC.

The TDCS Scheduler implements this mechanism. Given the WSN Patch characteristics (i.e., parent-child relation in the Cluster-Tree scheme), this tool (built in MATLAB) computes the minimum BO along with the start time for each cluster head to schedule its active portion (superframe), as shown in Fig. 6.

4.2 Worst-Case Analyzer

The Worst-Case Analyzer is a MATLAB tool (Fig. 8) which estimates an upper bound for the end-to-end delay of the real-time traffic in IEEE 802.15.4/ZigBee Cluster-Tree Wireless Sensor Networks, i.e., the traffic whose packets are sent during the contention free portion (GTS slots) of the superframe.

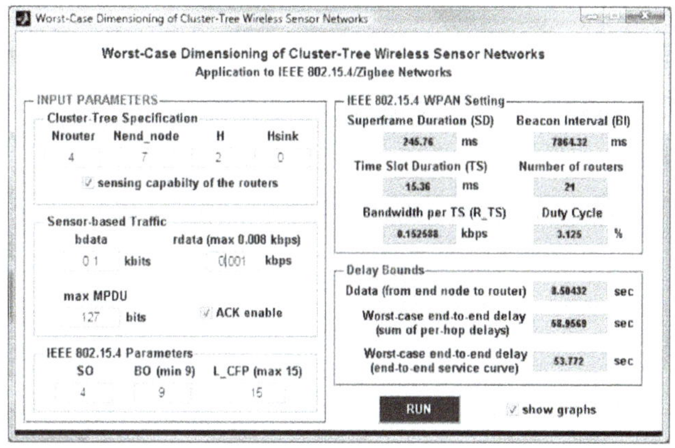

Figure 8: MATLAB tool for worst-case analysis and dimensioning.

This tool builds on the Network Calculus mathematical model, as described in [17] and enables to iteratively find the best duty-cycle vs. delay/throughput trade-off, by varying network parameters such as sensor traffic.

4.3 Network Protocol Simulator

The performance of the communication protocol is accurately evaluated using a simulator built in OPNET [3]. This is a complete simulation model for IEEE 802.15.4/ZigBee that complements a previous model [16].

Recently, several analytical and simulation models of the IEEE 802.15.4 protocol have been proposed. Nevertheless, currently available simulation models [27] for this protocol are both inaccurate and incomplete, and in particular they do not support the GTS mechanism, which is required for time-sensitive wireless sensor applications. Although a commercial software (but free for academia), OPNET Modeler was chosen due to its accuracy and to its sophisticated graphical user interface.

The idea behind this simulation model was triggered by the need to build a very reliable model of the IEEE 802.15.4 and ZigBee protocols for Wireless Sensor Networks (WSNs). The IEEE 802.15.4/ZigBee simulation model builds on the wireless module, an add-on that extends the functionality of the OPNET Modeler with accurate modeling, simulation and analysis of wireless networks. According to the overview presented in Section 2, three types of nodes are implemented, namely a GW, a CH and a SN. All types of nodes have the same internal architecture (node domain), but they differ in the available user-defined attributes.

The structure of the IEEE 802.15.4/ZigBee simulation model is presented in Fig. 9. The *physical layer* consists of a wireless radio transmitter and receiver compliant to the IEEE 802.15.4 standard running at 2.4 GHz frequency band with 250 kbps data rate. Default TelosB-related settings are used for the physical characteristics of the radio channel such as background noise and interference, propagation delay, antenna gain, and bit error rate.

The *data link layer* supports the beacon-enabled mode and implements two medium access control protocols according to the IEEE 802.15.4 standard, namely the contention-based slotted CSMA/CA and contention-free GTS. MAC payload (MSDU) incoming from the network layer is wrapped in MAC header and MAC footer and stored into two sep-

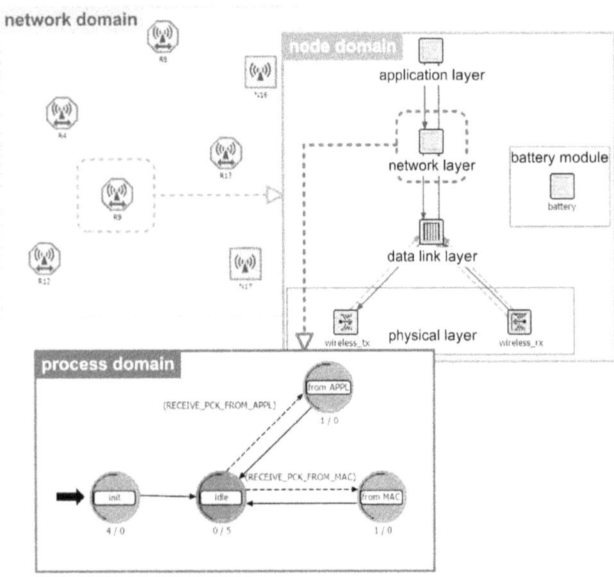

Figure 9: The structure of the IEEE 802.15.4/ZigBee Opnet simulation model.

arate FIFO buffers, namely a buffer for best-effort data frames and another buffer for real-time data frames. The frames are dispatched to the network when the corresponding CAP or CFP is active. On the other hand, the frame (MPDU) incoming from the physical layer is unwrapped and passed to the network layer for further processing. The data link layer also generates required commands (e.g., GTS allocation, deallocation and reallocation commands) and beacon frames when a node acts as GW or CH.

The *network layer* implements address-based tree routing according to the ZigBee specification. The frames are routed upward or downward along the cluster-tree topology according to the destination address by exploiting the hierarchical addressing scheme provided by ZigBee.

The *application layer* can generate unacknowledged and/or acknowledged best-effort and/or real-time data frames transmitted during CAP or CFP, respectively. There is also a battery module that computes the consumed and remaining energy levels. The default values of current draws are set to those of the widely-used MICAz [9] or TelosB [10] motes.

With the help of this simulator, we can infer (Table 1) end-to-end delays for both real-time and best-effort traffic, as well as compute network statistics such as packet loss, network throughput and lifetime, via per-node energy consumption estimation. In particular, this simulation model gets user-defined attributes (network topology and nodes parameters) through an XML file, which we automatically generate through an ad-hoc MATLAB script, which takes as inputs the outputs of the Network Deployment Simulator (i.e., nodes positions and parent-child relations, Section 3) and the windows offset defined by the TDCS scheduling tool (Section 4.1).

5. REMOTE PROGRAMMING & TESTING

Moving from the outputs of the TDCS tool, a set of appropriate MATLAB/Unix SHELL scripts are used to automatically generate header files (.h), compile the firmware of the nodes with each individual configuration and load the compiled firmware in parallel into each node, through the USB tree. Interestingly, using this tool to program in parallel the nodes over the USB tree we were able to save 66% in programming time with respect to a classical serial programming. In particular, in the DEMMON1 deployment, it took around 30 minutes to serially program a patch composed by 100 nodes, while with this tool the programming time reduced to no more than 10-12 minutes.

Finally, with the help of sniffer devices like [18] and a custom-designed log parser, built in C++ and MATLAB, EMMON specific data from the IEEE 802.15.4 frames are extracted. The outputs of the sniffer/parser (e.g., average and max delay) can then be compared with the theoretical (i.e., worst-case) and simulation (i.e., average and max) end-to-end delays.

We also designed an application to automate the testing of the hardware (USB cabling/hubs and TelosB nodes) having an element as reference (e.g., a previously tested TelosB node or USB hub). This testing tool compares the behavior of the elements under tests with the reference one. It was fundamental to early identify: (i) 19 out of 300+ TelosB with faulty humidity sensors (i.e., not usable as SNs); (ii) 2 out of 50+ USB hubs broken and (iii) a whole set of (3 m length) USB cables not compliant with the specification given to our local supplier.

6. DEMMON1 CASE STUDY

This section illustrates DEMMON1, the first demonstrator of the EMMON architecture. A comparison between simulation, analytical and experimental results validates our EMMON architecture (in terms of the interoperability among its architectural components), characterizes its performance and scalability limits and shows the usefulness of the toolset presented in this paper.

6.1 Site Layout

To validate our architecture, we performed a real deployment as shown in Fig. 10(a). 303 TelosB nodes were organized into 3 WSN Patches, with the possibility of defining different topologies by programming the nodes over a USB tree using our toolset. The GWs communicated via wired LAN to a host PC running the C&C server. The WSN Patches simultaneously operated in three distinct frequency channels. These channels were chosen to minimize the impact of the actual external interference [20]. In order to do so, a pre–deployment Electro-Magnetic Interference (EMI) analysis in the deployment site has been performed using HeatMapper [4], a free Wi-Fi coverage mapping site survey tool made by Ekahau, whose results are shown in Fig. 10(b). In particular, Fig. 10(b) clearly shows that the WiFi access points all around the deployment site were using standard IEEE 802.11 channels (namely, ch.1, ch.6 and ch.11) allowing us to assign IEEE 802.15.4 ch.15, ch.25 and ch.26 channels to our three patches, as it was expected [20].

Each WSN Patch was composed of 100 nodes organized into a 3-hops Cluster Tree model having the GW as root, 3 CHs on the first hop and 9 on the second hop for a total of 12 clusters, each of them with 8 to 10 SNs. Since all nodes were sharing the same collision domain, i.e., nodes were able to hear/interfere each other, the CHs locations were randomly chosen but the TDCS tool was used to define a scheduling among the clusters in each WSN Patch. As already men-

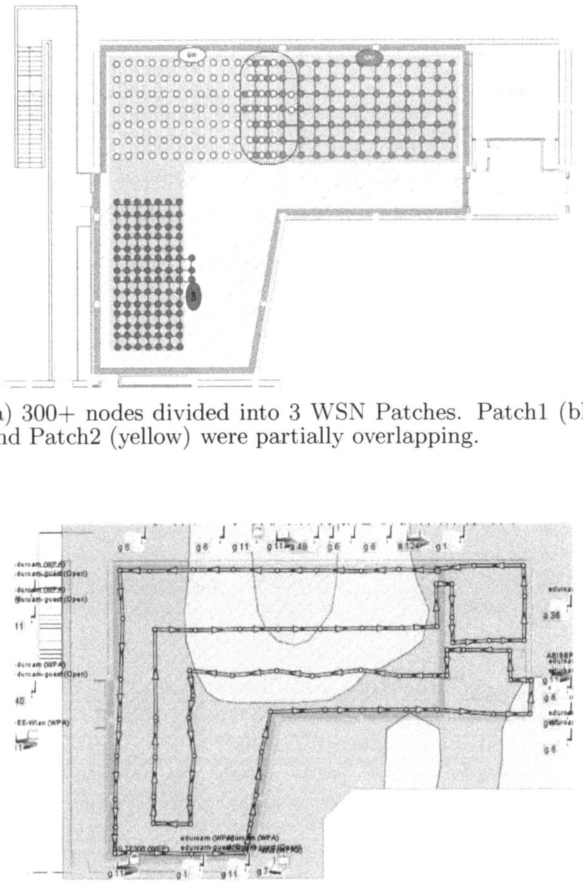

(a) 300+ nodes divided into 3 WSN Patches. Patch1 (blue) and Patch2 (yellow) were partially overlapping.

(b) ISM spectrum (namely, WLAN) interference study in the site's area [4].

Figure 10: DEMMON1 - First EMMON deployment (ISEP, Porto, Portugal).

tioned, frequency division guaranteed no interference among the nodes of different WSN Patches.

To stress our architecture, we physically overlapped some nodes belonging to two patches (Fig. 10(a)). Any query cast from the C&C involving that area runs in parallel on those nodes and data aggregation is performed first at the CHs of each patch, then at the GWs and ultimately at the C&C level. This clearly demonstrate how distributing the intelligence on the lowest tiers as possible of the system is actually implemented.

6.2 Performance Analysis

Moving from the coverage analysis results, performance limits of the communication protocol have been evaluated by focusing on the WSN Patch level, i.e., the portion of the EMMON network below the gateway. In particular, using the previously referred toolset, several scenarios were identified based on all the combinations of the input parameters in Table 1. This setup generated different network topologies with maximum depth[5] (L_m) ranging from 2 to 5 and total number of nodes in one WSN Patch ranging from 25 to 501[6] (see Table 2).

In the simulated scenarios, the nodes' application layer generates and sends upstream a packet every T seconds. Packets have a fixed size of P bytes. Both *best-effort (BE)* and *real-time (RT)* traffic classes are generated with the same packet generation ratio (nodes use the contention–based access period and the contention free period of the IEEE 802.15.4 superframe, respectively). BE traffic is intended to simulate periodic sensor reports, while RT traffic accounts for event-driven alarm notifications. As a consequence, only CHs generate RT traffic: this reflects that only CHs should reliably trigger alarm notifications towards the GW, by filtering out noisy SNs readings.

The value of SO has been fixed to 4, which means an active portion of 960 symbols, while BO has been computed for each scenario by our TDCS MATLAB tool, as well as the set of CHs start times, in order to fit with the number of clusters (Γ). We assume that every cluster in a WSN Patch has the same value of the couple (BO, SO), as in Fig. 6, and, recalling that the size of the collision domain is as large as the network size, i.e., nodes can interfere each other, the TDCS scheduler gives a conservative solution, where clusters' active portions do not overlap[7]. As a consequence, in our setup (Table 1), BO ranges between 7 and 9, corresponding to beacon interval (BI) values of ~ 2 and ~ 8 s, respectively, and duty cycles of 12.5% and 3.125%, respectively (Fig. 7).

In this paper, the end-to-end (e2e) delay for BE (e2e-BE) and RT (e2e-RT) traffic classes and the packet loss ratio are the performance indices, while other available figures (e.g., per-node energy consumptions) are not shown here, due to space constraints.

The same experiments were run live on our testbed. However, only best effort traffic was generated, using the CAP portion of the IEEE 802.15.4 MAC. The traffic was monitored through protocol analyzers with the help of some sniffer tools (like [18]) to compute the statistics for the e2e delay.

Table 2 shows an excerpt of the results related to network performance. In particular, a subset of all the scenarios is presented with an increasing level of network complexity, enabling the comparison between *simulation* and *experimental* results for the e2e-BE delay, as well as between *simulation*

Table 1: WSN Patch Validation Setup

Parameter	Range	Meaning
R_m	{2;3;4;5}	Number of children CH per parent
Γ	{5;17;21}	Number of Clusters
Σ	{5;10;15;20;24}	Number of children SNs per parent
SO	4	Superframe Order
P	137 B	NPDU size
T	2 s	Packet generation ratio

[5]Maximum number of hops between a SN and the GW.

[6]Rememer that larger system scale can be achieved through the multiple patches mechanism.

[7]The tool is ready yet to cope with scenarios where two clusters are spatially apart to not interfere each other. In this case, their active portions can be overlapped, with an evident gain in terms of system delay, due to the reduction of the beacon interval.

Table 2: Simulation, worst-case and experimental results (excerpt) from the campaign defined in Tab. 1.

Tot Nodes	Γ (clusters)	Σ (SNs per CH)	Rm	Lm	BO	BI [s]	CAP - Best Effort: Pkt Loss [%] SIMULATION	CAP - Best Effort: End-To-End Delay [s] SIMULATION avg	std	max	CAP - Best Effort: End-To-End Delay [s] EXPERIMENTAL avg	std	max	CFP - Real Time: End-To-End Delay [s] SIMULATION avg	std	max	CFP - Real Time: End-To-End Delay [s] ANALITICAL WorstCase
25	5	5	4	2	7	2.048	0.56	1.73	1.12	3.73	1.32	0.36	2.06	1.95	1.13	3.92	5.82
45	5	10	4	2	7	2.048	2.39	1.80	1.12	3.80	1.59	0.40	1.99	1.97	1.13	3.91	5.82
65	5	15	4	2	7	2.048	5.53	1.81	1.12	3.82	1.67	0.38	2.14	1.97	1.14	3.91	5.82
85	5	20	4	2	7	2.048	8.48	1.82	1.13	3.83	1.68	0.56	2.19	2.00	1.15	3.91	5.82
101	5	24	4	2	7	2.048	10.20	1.83	1.15	3.84	1.71	0.53	2.35	2.04	1.14	3.90	5.82
97	17	5	5	3	9	8.192	2.47	13.96	5.52	36.03	12.99	5.16	45.70	14.65	5.35	34.02	41.86
257	17	15	5	3	9	8.192	10.29	16.28	6.58	47.02				14.45	5.30	23.53	41.86
401	17	24	5	3	9	8.192	13.08	16.38	7.49	44.20				14.61	5.36	23.53	41.86
97	17	5	3	4	9	8.192	2.75	22.56	12.89	109.90	18.84	5.89	23.85	23.63	12.89	68.73	69.87
257	17	15	3	4	9	8.192	10.84	23.86	12.53	116.30				22.15	10.70	62.82	69.87
401	17	24	3	4	9	8.192	13.32	20.97	9.67	54.85				19.17	8.14	55.62	69.87
97	17	5	2	5	9	8.192	2.14	23.18	9.03	38.97	25.14	7.87	95.38	24.44	9.00	57.97	94.76
257	17	15	2	5	9	8.192	10.49	27.11	11.47	75.04				24.79	9.17	52.78	94.76
401	17	24	2	5	9	8.192	13.16	28.22	13.02	78.44				24.17	9.44	54.93	94.76
121	21	5	2	5	9	8.192	2.23	23.08	9.04	39.13				24.29	8.91	57.34	94.76
321	21	15	2	5	9	8.192	10.28	26.30	11.58	76.18				23.95	9.49	54.08	94.76
501	21	24	2	5	9	8.192	12.84	27.15	12.35	67.63				23.92	9.39	38.89	94.76

and *theoretical* worst case analysis for the e2e-RT. For the sake of comparison, since GTS slots are allocated only to CHs, both e2e-BE and e2e-RT delays are computed as sum of per-hop delays of messages sent from child to parent in the tree, recursively up to the GW. A column is for showing the packet loss ratio (for BE traffic only: these values account for the number of packets whose sending failed after three retransmissions). On the contrary, due to our design and setup choices (i.e., to assign GTS slots to CH children only), RT traffic experienced no packet losses.

Although experimental results are available for scenarios with up to 101 nodes, i.e., the maximum dimension of a single WSN Patch in DEMMON1 (Fig. 10(a)), overall the following conclusions can be drawn: (i) the statistics of the e2e-BE delay match the experimental ones; (ii) the analytical tool for worst case dimensioning gives an upper bound of the maximum e2e-RT delay; (iii) as expected, while the statistics of e2e-RT delay are not influenced by the clusters' size (Σ), for e2e-BE delay the impact of a more crowded network becomes not negligible; (iv) by looking at the scenarios with $\Gamma = 17$ and by averaging among the 5 values of Σ, a topology with a wider ($R_m = 5, L_m = 3$) rather than a deeper ($R_m = 2, L_m = 5$) tree shows gains in the e2e-BE and e2e-RT delays of almost 68.2% and 66.2%, respectively. Additionally, the difference in terms of packet loss for the same scenarios is negligible. Finally, these results highlight that the EMMON network architecture scales well with the number of nodes in a WSN Patch. This is one of the most important aspects of EMMON, since it shows that our hierarchical architecture can scale easily to thousands of nodes by dividing the network into patches.

6.3 Network Coverage Simulation Analysis

Due to the limited area of our deployment for DEMMON1, we did not actually apply the network planning tool as described in Section 3. However, to further assess the scalability of the proposed architecture, a set of experiments have been conducted to find the minimum number of nodes needed to cover a field of size up to 49 km^2.

Each WSN Patch fits into a square cell of length $L_{SBD} =$ 200 m (SBD scheme) or an equilateral triangle tile of length L_{TBD} = 200 m (TBD scheme), with the GWs placed in the middle, and $C = 12$ CHs evenly spread along two rings around the GW, whose radius are R_1 = 40 m and R_2 = 80 m, respectively (c.f. Fig. 3). Then, a variable number of SNs are randomly distributed in the field and every node (SN and CH) associates to a parent node following the IEEE 802.15.4 association mechanism and a simple metric based on the evaluation of the received signal strength (RSSI). The RSSI measurements are simulated by relying on the inherent radio propagation and IEEE 802.15.4 interference models offered by the simulator [25] and tuned with TelosB–like physical parameters.

Table 3: Coverage Simulation Results - SBD

Field Size [km^2]	WSN Patches	SNs to cover $\geq$ 60%	SNs to cover $\geq$ 80%
1	25	600	1250
9	225	5000	10000
49	1225	30000	50000

Given $R = 50$ m, the communication radio range common to every node in the WSN Patch and $R_r = 25$ m, the sensing range of each SN (Fig 5), the field coverage has been computed as described in Section 3. Results are reported in Table 3 and Table 4 for the SBD and TBD schemes, respectively. In particular, Table 3 reports on the numbers of network elements needed to cover at least 60% or 80% of the monitoring area, while in Table 4, assuming the same number of SNs as in Table 3, the focus is on the resulting coverage.

Comparing the two deployment schemes we can claim that: (i) SBD reaches an almost full coverage, while the TBD shows limited performance, but (ii) to allow neighbor patches to work in parallel, the SBD scheme requires that each patch runs on one of 9 distinct frequency channels (each patch has 8 neighbor patches); while the TBD requires only 4 channels (each patch has 3 neighbor patches). As a consequence, the best deployment approach is to be traded off

Table 4: Coverage Simulation Results - TBD

Field Size $[km^2]$	WSN Patches	SNs	Coverage [%]
1	77	600	39.51
1	77	1250	66.37
9	527	5000	36.88
9	527	10000	60.20
49	2627	30000	36.74
49	2627	50000	52.95

with the application requirements: e.g., in a controlled environment like in a Data Center we might allow to reserve more channels for our monitoring network, while in Urban environments only the classical IEEE 802.15.4 four channels might be available. However, both tables show that the number of WSN Patches as well as the average number of SNs per patch increase quite slowly. This confirms that the proposed architecture scales well against the size of the monitored area, with relatively stringent coverage requirements.

7. CONCLUSIONS AND FUTURE WORK

This paper outlined our toolset to support EMMON WSN system architecture for large–scale, dense and real–time embedded monitoring. It is a complete toolset for engineering these systems, encompassing deployment planning, network dimensioning, analysis, protocol simulation and nodes testing/programming: this helps the designer to have a global picture of the whole system.

This toolset has been paramount to test the EMMON baseline system architecture through extensive simulation as well as experimental evaluation, proving its feasibility and scalability. DEMMON1, i.e., the first EMMON demonstrator, is a 300+ nodes test-bed and is, to the best of our knowledge, the largest single–site WSN test-bed in Europe.

Ongoing work about the EMMON system includes the extension of its baseline architecture with security and reliability add-ons. As a consequence, parallel work will be devoted on extending the toolset to support the design and validation of such new features. For instance, sniffer and parser tools (as mentioned in Section 5) will be extended (i) to allow the synchronization among multiple distributed sniffing points, (ii) to decode ciphered packets and, finally, (iii) to inject packets in a fully controlled way, in order to test the network's fault tolerance mechanisms.

Overall, we believe that the EMMON system architecture and the related toolset presented in this paper will foster and ease the design, test and validation of WSN applications.

8. ACKNOWLEDGMENTS

This work has been partially supported by (i) National Funds, through the FCT - Portuguese Foundation for Science and Technology, under grant ref. ARTEMIS-0003-2008, (ii) the ARTEMIS Joint Undertaking, under grant agreement n. 100036, (iii) a consortium of companies consisting of Critical Software S.A., Intesys Ltd, Critical Software Technologies Ltd, SESM S.c.a.r.l., Akting Ingeniaritza S.L., Centro de Estudios e Investigaciones Técnicas de Gipuzkoa, (iv) the CONET - Cooperating Objects Network of Excellence, funded by the European Commission under FP7 with grant ref. FP7-ICT-224053.

9. REFERENCES

[1] Emmon - embedded monitoring. European Community's Seventh Framework Programme (FP7/2007-13). Grant agreement No. 217967. (http://www.artemis-emmon.eu).

[2] Wirelessly accessible sensor populations (wasp) project. FP6-IST-2005-2.5.3 Embedded Systems, Contract Number IST-034963. Duration: Sep 2006 to Oct 2010 (http://www.wasp-project.org).

[3] Open-zb: Open source toolset for ieee802.15.4 and zigbee, 2010. (http://www.open-zb.net/index.php).

[4] Ekahau heatmapper - the free wi-fi coverage mapping site survey tool, 2011. (http://www.ekahau.com).

[5] A. Arora, R. Ramnath, E. Ertin, P. Sinha, S. Bapat, V. Naik, V. Kulathumani, H. Zhang, H. Cao, M. Sridharan, S. Kumar, N. Seddon, C. Anderson, T. Herman, N. Trivedi, M. Nesterenko, R. Shah, S. Kulkami, M. Aramugam, L. Wang, M. Gouda, Y. ri Choi, D. Culler, P. Dutta, C. Sharp, G. Tolle, M. Grimmer, B. Ferriera, and K. Parker. Exscal: elements of an extreme scale wireless sensor network. In *Embedded and Real-Time Computing Systems and Applications, 2005. Proceedings. 11th IEEE International Conference on*, pages 102 – 108, 2005.

[6] R. Barr, Z. J. Haas, and R. van Renesse. Jist: an efficient approach to simulation using virtual machines: Research articles. *Softw. Pract. Exper.*, 35(6):539–576, 2005.

[7] E. Biagioni and G. Sasaki. Wireless sensor placement for reliable and efficient data collection. In *System Sciences, 2003. Proceedings of the 36th Annual Hawaii International Conference on*, page 10 pp., jan. 2003.

[8] M. Ceriotti, L. Mottola, G. P. Picco, A. L. Murphy, S. Guna, M. Corra, M. Pozzi, D. Zonta, and P. Zanon. Monitoring heritage buildings with wireless sensor networks: The torre aquila deployment. In *Proceedings of the 2009 International Conference on Information Processing in Sensor Networks*, IPSN '09, pages 277–288, Washington, DC, USA, 2009. IEEE Computer Society.

[9] I. Crossbow Technology. *MICAz Mote Datasheet.*, 2009.

[10] I. Crossbow Technology. *TelosB Mote Datasheet.*, 2009.

[11] P. Dutta, P. Aoki, N. Kumar, A. Mainwaring, C. Myers, W. Willett, and A. Woodruff. Common sense: Participatory urban sensing using a network of handheld air quality monitors (demonstration). In *Proceedings of SenSys*, pages 349–350, Berkeley, CA, November 2009.

[12] O. Gnawali, K.-Y. Jang, J. Paek, M. Vieira, R. Govindan, B. Greenstein, A. Joki, D. Estrin, and E. Kohler. The tenet architecture for tiered sensor networks. In *Proceedings of the 4th international conference on Embedded networked sensor systems*, SenSys '06, pages 153–166, New York, NY, USA, 2006. ACM.

[13] J.-H. Hauer, R. Daidone, R. Severino, J. Busch, M. Tiloca, and S. Tennina. An open-source ieee 802.15.4 mac implementation for tinyos 2.1, February 2011. Poster Session at 8th European Conference on

Wireless Sensor Networks (http://www.nes.uni-due.de/ewsn2011).

[14] T. He, S. Krishnamurthy, L. Luo, T. Yan, L. Gu, R. Stoleru, G. Zhou, Q. Cao, P. Vicaire, J. A. Stankovic, T. F. Abdelzaher, J. Hui, B. Krogh, T. U. E. S. Krishnamurthy, L. Luo, T. Yan, L. Gu, R. Stoleru, G. Zhou, and Q. Cao. Vigilnet: An integrated sensor network system for energy-efficient surveillance. *ACM Transaction on Sensor Networks*, 2:1–38, 2006.

[15] P. Jurcik. *Real-time Communication over Cluster-tree Wireless Sensor Networks.* Electrical engineering and information technology, control engineering and robotics, Department of Control Engineering Faculty of Electrical Engineering Czech Technical University in Prague Czech Republic, in collaboration with CISTER-ISEP Research Unit Polytechnic Institute of Porto Portugal, Prague, January 2010.

[16] P. Jurcik, A. Koubaa, M. Alves, E. Tovar, and Z. Hanzalek. A simulation model for the ieee 802.15.4 protocol: Delay/throughput evaluation of the gts mechanism. In *Modeling, Analysis, and Simulation of Computer and Telecommunication Systems, 2007. MASCOTS '07. 15th International Symposium on*, pages 109 –116, oct. 2007.

[17] P. Jurcik, A. Koubâa, R. Severino, M. Alves, and E. Tovar. Dimensioning and worst-case analysis of cluster-tree sensor networks. *ACM Transactions on Sensor Networks*, 7:14:1–14:47, September 2010.

[18] A. Koubaa, S. Chaudhry, O. Gaddour, R. Chaari, N. Al-Elaiwi, H. Al-Soli, and H. Boujelben. Z-monitor: Monitoring and analyzing ieee 802.15.4-based wireless sensor networks. In *6th IEEE LCN Workshop on Network Measurements, in conjunction with 36th IEEE Conference on Local Computer Networks (LCN 2011)*, Bonn, Germany, October 2011.

[19] C.-J. M. Liang, J. Liu, L. Luo, A. Terzis, and F. Zhao. Racnet: a high-fidelity data center sensing network. In *Proceedings of the 7th ACM Conference on Embedded Networked Sensor Systems*, SenSys '09, pages 15–28, New York, NY, USA, 2009. ACM.

[20] C.-J. M. Liang, N. B. Priyantha, J. Liu, and A. Terzis. Surviving wi-fi interference in low power zigbee networks. In *Proceedings of the 8th ACM Conference on Embedded Networked Sensor Systems*, SenSys '10, pages 309–322, New York, NY, USA, 2010. ACM.

[21] R. Monica and P. Braga. Operational requirements consolidated from end-users input and opinions. Deliverable D3.1, EMMON Project, ARTEMIS Joint Undertaking Call for proposals ARTEMIS-2008-1, Grant agreement no. 100036, February 2010.

[22] E. F. Nakamura, A. A. F. Loureiro, and A. C. Frery. Information fusion for wireless sensor networks: Methods, models, and classifications. *ACM Comput. Surv.*, 39, September 2007.

[23] M. Presser, A. Gluhak, D. Babb, L. Herault, and R. Tafazolli. e-sense - capturing ambient intelligence for mobile communications through wireless sensor networks. In *http://www.ist-esense.org/index.php?id=18*, 2006.

[24] B. Raman and K. Chebrolu. Censor networks: a critique of "sensor networks" from a systems perspective. *SIGCOMM Comput. Commun. Rev.*, 38:75–78, July 2008.

[25] P. Scheuermann. Sidnet-swans – a simulator and integrated development platform for sensor networks applications, 2010. (http://users.eecs.northwestern.edu/ ocg474/SIDnet.html)

[26] E. Schoch, M. Feiri, F. Kargl, and M. Weber. Simulation of ad hoc networks: ns-2 compared to jist/swans. In *Simutools '08: Proceedings of the 1st international conference on Simulation tools and techniques for communications, networks and systems & workshops*, pages 1–8, ICST, Brussels, Belgium, Belgium, 2008. ICST (Institute for Computer Sciences, Social-Informatics and Telecommunications Engineering).

[27] C. P. Singh, O. P. Vyas, and M. K. Tiwari. A Survey of Simulation in Sensor Networks. In *Proc. of the 4th International Conf. on Computational Intelligence for Modelling Control and Automation (CIMCA)*, pages 867–872, Dec. 2008.

[28] S. Tennina, M. Alves, P. G. Sousa, M. Santos, P. Braga, M. Bouroche, G. Carrozza, R. Monica, F. Pacheco, R. Gomes, A. Garg, and R. Severino. Evaluation report: evaluation of possible solutions, concepts for new communication methods. Deliverable D4.2, EMMON Project, ARTEMIS Joint Undertaking Call for proposals ARTEMIS-2008-1, Grant agreement no. 100036, 2010.

[29] E. Weingartner, H. vom Lehn, and K. Wehrle. A performance comparison of recent network simulators. In *Communications, 2009. ICC '09. IEEE International Conference on*, pages 1 –5, 14-18 2009.

[30] K. Xu, H. Hassanein, G. Takahara, and Q. Wang. Relay node deployment strategies in heterogeneous wireless sensor networks: single-hop communication case. In *Global Telecommunications Conference, 2005. GLOBECOM '05. IEEE*, volume 1, page 5 pp., nov.-2 dec. 2005.

[31] K. Xu, G. Takahara, and H. Hassanein. On the robustness of grid-based deployment in wireless sensor networks. In *Proceedings of the 2006 international conference on Wireless communications and mobile computing*, IWCMC '06, pages 1183–1188, New York, NY, USA, 2006. ACM.

[32] K. Xu, Q. Wang, H. Hassanein, and G. Takahara. Optimal wireless sensor networks (wsns) deployment: minimum cost with lifetime constraint. In *Wireless And Mobile Computing, Networking And Communications, 2005. (WiMob'2005), IEEE International Conference on*, volume 3, pages 454 – 461 Vol. 3, aug. 2005.

Ant Colony Optimization with Fuzzy Heuristic Information Designed for Cooperative Wireless Sensor Networks

Marcelo Portela Sousa
Institute for Advanced Studies in Communications, UFCG
Campina Grande, PB, Brazil
marcelo.portela@ieee.org

Waslon Terllizzie A. Lopes
Institute for Advanced Studies in Communications, UFCG
waslon@ieee.org

Marcelo Sampaio de Alencar
Institute for Advanced Studies in Communications, UFCG
malencar@ieee.org

ABSTRACT

Biologically inspired networking is becoming a very active area of research and development because methods from biology have attractive features such as scalability and resilience to changes in the environment. In this paper, the authors propose a cross-layer protocol for cooperative wireless sensor networks, inspired by the collaborative behaviour of ant groups. The fuzzy heuristic information modelling is proposed to assist the clustering and routing processes at network layer. A resultant relaying threshold is combined with an adaptive invoking of cooperative modulation diversity, at link and physical layers. Simulation results show the performance enhancement in network lifetime and packet loss rate, compared with another cross-layer cooperative system. The evaluation of energy consumption distribution and a comparison between the performance of fuzzy and crisp modelling are also provided.

Categories and Subject Descriptors

C.2.1 [**Network Architecture and Design**]: Wireless Communication

General Terms

Algorithms, Design, Performance

Keywords

Wireless Sensor Networks, Ant Colony Optimization, Fuzzy Heuristic Information, Cooperative Modulation Diversity

1. INTRODUCTION

Cluster-based protocols are successful methods for energy saving in wireless sensor networks (WSN). In each group, a cluster-head is elected as coordinator of that cluster, which collects data from other nodes, aggregates and reports it to the sink node. In clustered schemes, the cluster-head election process is a fundamental issue and impacts significantly in the network energy consumption. The clustering inspired by the organized behaviour of social ants, with the use of Ant Colony Optimization (ACO) have been developed to enhance the performance of WSNs [1–8]. However, those proposed schemes are based on single-hop networks, which waste energy and limit the transmission distance range [9].

ACO techniques utilize a graph $\mathcal{G} = (\mathcal{C}, \mathcal{L})$, in which $\mathcal{C} = \{c_1, c_2, \ldots, c_{N_c}\}$ is a finite set of components, $\mathcal{L} = \{l_{c_i c_j} | (c_i, c_j) \in \tilde{\mathcal{C}}\}, |\mathcal{L}| \leq N_c^2$ is a finite set of possible connections/transitions among the elements of $\tilde{\mathcal{C}}$, since $\tilde{\mathcal{C}}$ is a subset of the Cartesian product $\mathcal{C} \times \mathcal{C}$. ACO algorithms can be used to find minimum cost paths feasible with the constrains Ω of an optimization problem. For example, in the travelling salesman problem, $\mathcal{C}$ is the set of cities, $\mathcal{L}$ is the set of arcs connecting cities, and a solution ψ is an Hamiltonian circuit. On the other hand, in the AntNet scheme, related to a generic routing problem in communication networks, $\mathcal{C}$ and $\mathcal{L}$ correspond respectively to the sets of processing nodes and of communication links of a real network [10,11].

In ACO systems, the information collected by the ants in the search process is stored in pheromone trails, τ. The arcs also have *a priori* information, η, which is heuristic about the problem definition. If η represents a cost function related to distance, the imprecision in measurement can degrade the performance of the system [11]. Fuzzy inference systems can deal with such imprecisions and enhance the system performance [12–14].

Apart from the limited resources, the fading caused by multipath in wireless channels affect the quality of transmission and increases the Packet Loss Rate (PLR). The use of truncated Automatic Repeat Request (ARQ) decreases the PLR, but depending on the relation between the maximum number of retransmissions allowed and the channel quality, it may degrade the network lifetime due the required retransmissions [15–17].

Cooperative modulation diversity (CMD) can solve that tradeoff, by combating the fading effects without incurring waste of bandwidth or energy. Different of other recently proposed cooperative schemes, as SCA with LEACH [18], which spends too much energy in introducing redundancy by a space-time block coding, CMD rotates the angle of the signal constellation, and interleaves the transmitted component symbols [18–21].

This paper presents a cross-layer design for multi-hop cooperative wireless sensor networks, inspired by the behaviour of ants. The proposed protocol intends to increase the network lifetime and decrease the packet loss rate of WSNs. The goals are reached by an optimal election of cluster-heads, at network layer, and by the control of possible further CDM retransmissions, at link an physical layers. A relaying threshold and a truncated ARQ process guide the cross-layer operation. By simulations, the comparison with another cooperative design attests the performance superiority of the proposed protocol.

The remaining of the paper is organized as follows: Section 2 provides an overview of a classical ACO system operation related to a generic routing problem in communication networks, the AntNet system. Section 3 models the proposed clustering protocol, with the use of fuzzy heuristic information. The operation of cooperative modulation diversity is described in Section 4. Simulation results are discussed in Section 5 and the conclusions are summarized in Section 6.

2. ANTNET OVERVIEW

In AntNet, each ant searches for a minimum cost path between a pair of nodes, i and d [22]. If an ant κ is in node i, it hops to j, in accordance with a decision rule that is a function of the ant's memory, $\mathcal{M}^{\kappa}$, and of the local ant-routing table, $\mathcal{A}_i$. That table is obtained by a composition of the pheromone trails, τ_{ijd}, and of the heuristic information, η_{ijd}. Once the ant κ has completed a path, it deposits an amount of pheromone, $\Delta\tau^{\kappa}$, proportional to the goodness of the path it built. In this way, after reaching its destination node, the ant moves back to its source node, along the same path, and increases the pheromone intensity, modelled by the Assignment 1:

$$\tau_{ijd} \leftarrow \tau_{ijd} + \Delta\tau^{\kappa}. \tag{1}$$

To prevent a premature convergence to non-optimal solutions, the pheromone of outgoing connections evaporates, indicated by the Assignment 2:

$$\tau_{ijd} \leftarrow \frac{\tau_{ijd}}{(1+\Delta\tau^{\kappa})}, \forall j \in \mathcal{N}_i, \tag{2}$$

in which $\mathcal{N}_i$ is the set of node i neighbours. The relation between $\mathcal{A}_i$, and the ant's decision rule, p^{κ}_{ijd}, is given by:

$$p^{\kappa}_{ijd} = \begin{cases} \mathcal{A}_i = \dfrac{\omega\tau_{ijd} + (1-\omega)\eta_{ijd}}{\omega + (1-\omega)(|\mathcal{N}_i| - 1)}, & \text{if } j \notin \mathcal{M}^{\kappa}, \\ 0, & \text{if } j \in \mathcal{M}^{\kappa}, \end{cases} \tag{3}$$

in which $\omega \in [0, 1]$ is a weighting factor between τ_{ijd} and η_{ijd}, and the denominator is a normalization term. The ant's memory, $\mathcal{M}^{\kappa}$, indicates the set of nodes visited by the ant, and its use can avoid the occurrence of loops.

3. A NEW CLUSTERING PROTOCOL

The proposed clustering protocol is based on the behaviour of ants that need to find optimal paths from the source to a final destination. The main goal is the optimal election of cluster-heads, in each round, and control of possible further cooperative retransmissions. The protocol modifies the classical modelling of the ACO system, AntNet, since it translates the representation of vertices into edges, as well as the representation of edges into vertices. That is, each sensor node, s (vertex), in the network that uses the proposed protocol can be seen as a path (edge), ijd, in the AntNet system. Then, the election of the best sensor node as a cluster-head, by the cluster, is equivalent to a choice of the best path from a source to a final destination, by an ant.

The operation of the proposed scheme starts with the random deployment of artificial ants in the monitored region. Each cluster receives an ant, κ, that indicates the respective first elected cluster-head. In the next elections, each cluster node runs decision rules and generates special values, denoted `chance`.

For each sensor node, the respective ant, κ, travels to the final destination and moves back to the nest increasing the pheromone intensity, τ_s. In the sensor network domain, this is equivalent to each sensor node running Assignment 1. The update variable, $\Delta\tau^{\kappa}$, indicates the quality of the chosen path by the ant. This can measure how good was the operation performance of the sensor node in the previous round. In the proposed protocol, the update variable is:

$$\Delta\tau^{\kappa} = \frac{\varepsilon_s \cdot \rho_s}{\zeta_s \cdot \Gamma_s}, \tag{4}$$

in which ε_s is the residual energy of the node, ζ_s is the energy consumed by the node in the previous round, Γ_s is the total number of transmissions realized by the node in the previous round and $\rho_s = 2$, if the previous transmitted packet was successfully recovered by the next destination node. Otherwise, if even with retransmissions the previous packet was not correctly recovered, $\rho_s = 1$. Those values were determined empirically.

After the trail update, all the sensor nodes run Assignment 2, which corresponds to the pheromone trails evaporation. The next step is the processing of the heuristic information, η_s, given by the output of a fuzzy inference system, explained in the next section. By combining of τ_s and η_s, the sensor node calculates the value of the decision rule, according to Formula 3. This value indicates the probability of a path attract ants, and equivalently, indicates the probability of a sensor node elect itself as a cluster-head.

The nodes state the decision rule value as the variable `chance`. Each node advertises a message for the other candidates, with that value attached, and waits messages from other nodes. If its `chance` is higher than the `chance` from other nodes, the sensor node advertises a `cluster-head message`, which means that the sensor node elected itself a cluster-head. If a node which is not a cluster-head receives a `cluster-head message`, it selects the closest cluster-head as its coordinator and sends a message to join that cluster.

The sensor node which has the next higher value of `chance` (the second place in the election) becomes a relay candidate, which, in a possible retransmission stage, may become a vice cluster-head and cooperate in a diversity scheme. That node reaches a relaying threshold, and the only `chance` greater in that cluster is the `chance` of the elected cluster-head (the first place in the election).

3.1 Fuzzy Heuristic Information

In ACO algorithms, the heuristic information represents a local information which does not depend on the quality of previous iterations. In Ant System, an algorithm that optimizes solutions to the travelling salesman problem (TSP), $\eta_{ij} = 1/J_{ij}$, in which J_{ij} represents the distance between cities i and j [11]. In AntNet, the heuristic information is given by [22]

$$\eta_{ij} = 1 - \frac{q_{ij}}{\sum_{l \in \mathcal{N}_i} q_{il}}, \tag{5}$$

in which q_{il} is the queue length (in bits to be sent) of the link that connects the node i to its neighbour j. The set of neighbours of node i is given by $\mathcal{N}_i$. In the proposed scheme, the heuristic information η_s, represented by the variable eta, relates two other variables: local_distance and CH_dispersion.

local_distance is the sum of distances between the candidate node and other nodes within a specific radius of transmission. The greater the sum, the higher the energy to transmit the sensed data to the candidate node. The operation of this variable was proposed in the CHEF protocol [23]. However, if there are few nodes within a specific radius of transmission, the sum of distances between the candidate node and other nodes can be small. In this case one may infer, erroneously, that the node energy consumption is lower than in the case in which the nodes, in a higher number, are located closer to the candidate node. The proposed scheme overcomes this drawback, with the normalization of that sum by means of division of local_distance by the number of nodes that are within the specific radius of transmission.

CH_dispersion is the sum of distances between the candidate node and the cluster-heads within a radius of transmission. This variable is also normalized. However, the greater the sum, the higher the chance to coordinate the cluster. Besides promoting a good distribution of cluster-heads, it balances the transmission loads and the network processing. The estimation of positions and distances is done by the received signal strength intensity (RSSI), at a set-up phase.

Since eta is the combination of two variables based on imprecise measurements of distance, the fuzzy logic is well suited to represent its final processing. Fuzzy logic is a mathematical tool that relies on purely qualitative variables, in contrast to the quantitative nature of crisp values [12]. Another important advantage of fuzzy logic is that it typically requires few computational resources [24]. In WSNs, besides the uncertainty of distance measurements by RSSI, the positioning of the sensor nodes can be subjected to little changes, what can be compensated in the fuzzy inference operation.

In fuzzy logic decisions are based on fuzzy IF-THEN rules, which are used to determine the value of output variables using approximate reasoning [13]. The inference system is composed by *linguistic* variables and logical operators. The fuzzy rules used in the proposed protocol, to generate the value of the heuristic information, are presented in Table 1. Once the values of local_distance and CH_dispersion becomes small (Close) and great (Far) values, respectively, the fuzzy heuristic information presents the higher values, and thus, the greater chances of elect a cluster-head. The fuzzy variable eta is defuzzified (transformed to a crisp number) by the use of the Center of Area (CoA) method, which corresponds to calculate the centroid of the fuzzy output [23]. The membership functions adopted are formed by triangles due the requirement of processing simplicity and they are illustrated in Figures 1, 2, 3.

Table 1: Fuzzy rules used in the proposed protocol.

Rule	IF		THEN
	local_distance	CH_dispersion	eta
1	Close	Far	Very High (VH)
2	Medium	Far	High (H)
3	Far	Far	Rather High (RH)
4	Close	Medium	High Medium (HM)
5	Medium	Medium	Medium (M)
6	Far	Medium	Low Medium (LM)
7	Close	Close	Rather Low (RL)
8	Medium	Close	Low (L)
9	Far	Close	Very Low (VL)

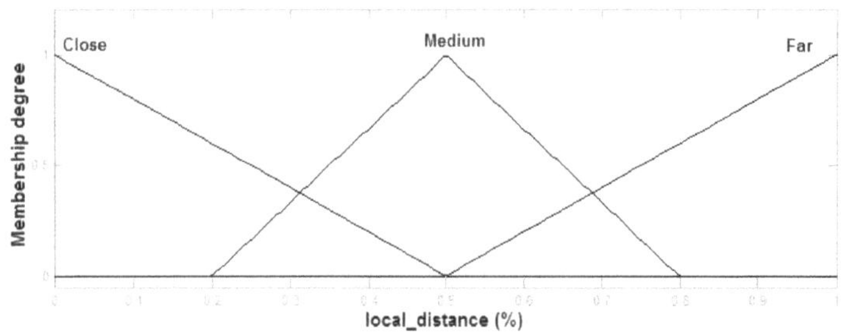

Figure 1: local_distance membership functions.

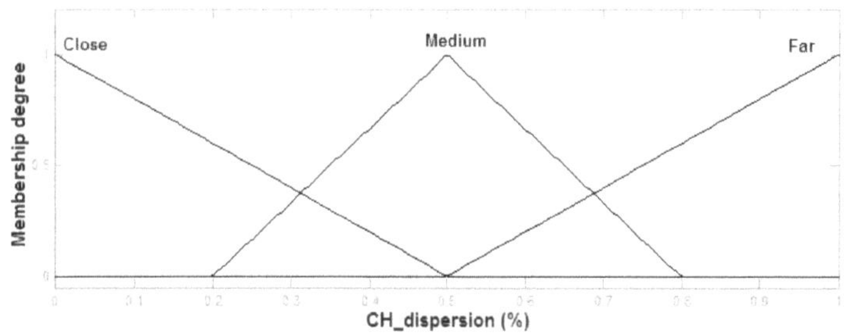

Figure 2: CH_dispersion membership functions.

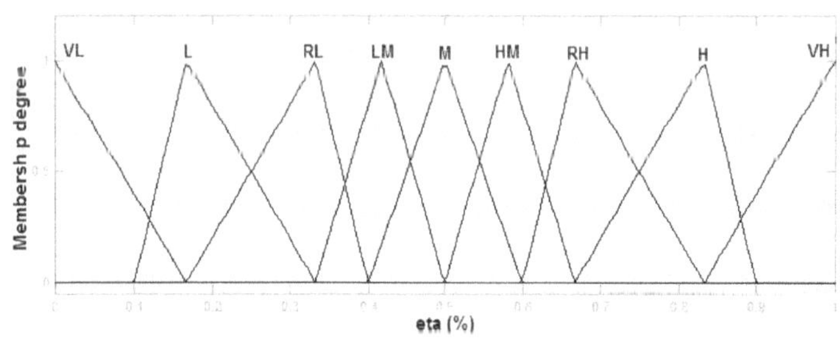

Figure 3: eta membership functions.

4. COOPERATIVE MODULATION DIVERSITY

Concerning the biological behaviour of ants, a group transmission can be seen as an efficient way to carry a large prey to the nest, since ants working as a group can carry close to ten times the carrying capacity of a solitary ant [25].

In the proposed protocol, that collaboration between ants is translated into the use of a cooperative technique between the nodes to mitigate the effects of the channel fading. Cooperative modulation diversity (CDM) exploits diversity gain in a system if each component of the transmitted signal is affected by independent channel fading. Furthermore, to achieve the maximum diversity gain, any two signal points in the system constellation must have the maximum number of distinct components. CDM is invoked in an adaptive manner, since it is needed only if transmission errors occur.

The next transmission hop from a source cluster-head can be performed in two stages: broadcast and retransmission.

4.1 Broadcast Stage

In the broadcast stage, a source cluster-head forwards its clustered message to another one, *i.e.*, the closest neighbour cluster-head and the next hop in the routing process towards the sink node. The relay candidate, indicated by the relaying threshold, receives the transmitted packet due to the broadcast nature of the wireless channel. The remaining cluster sensor nodes activate the sleep mode and save energy. The packet is transmitted by a conventional QPSK modulation scheme. The modulated signal is given by [19]

$$\begin{aligned} s(t) &= A \sum_{n=-\infty}^{+\infty} a_n p(t - nT_s)\cos(2\pi f_c t) \\ &+ A \sum_{n=-\infty}^{+\infty} b_n p(t - nT_s)\sin(2\pi f_c t), \end{aligned} \quad (6)$$

in which

$$\begin{aligned} a_n, b_n = &\ \pm 1 \quad \text{with equal probability} \\ p(t) = &\ \begin{cases} 1, & 0 \le t \le T_s \\ 0, & \text{elsewhere} \end{cases} \end{aligned} \quad (7)$$

for a carrier frequency, f_c, and a carrier amplitude, A. The transmitted packet has a CRC attached and the receiver (either a neighbour cluster-head, or the relay candidate, or the sink node) detects it. An acknowledgement is sent back to the source cluster-head. If the packet is correctly detected by the receiver, not necessarily the relay candidate, the source cluster-head remains transmitting new packets and the previous process is repeated. Otherwise, the retransmission stage begins.

4.2 Retransmission Stage

If the relay candidate receives the packet correctly, it becomes a *vice cluster-head* and, jointly with the source cluster-head, they retransmit the packet using the cooperative modulation diversity. Otherwise, a simple QPSK transmission is used again, the relay candidate activates the sleep mode and saves energy. The retransmissions continue until the packet is successfully delivered, or the number of retransmissions exceeds $N_r^{\max}$, which is a preset parameter indicating the maximum number of retransmissions allowed per packet. The value of $N_r^{\max}$ can depend on the application.

In CDM, if a QPSK constellation is rotated by a certain angle, a kind of redundancy between the two quadrature channels is introduced and the system can take advantage of the derived diversity. Then, both the source and vice cluster-heads rotate the constellation by an angle θ

$$\begin{aligned} s(t) &= A \sum_{n=-\infty}^{+\infty} x_n p(t - nT_s)\cos(2\pi f_c t), \\ &+ A \sum_{n=-\infty}^{+\infty} y_n p(t - nT_s)\sin(2\pi f_c t), \end{aligned} \quad (8)$$

in which

$$\begin{aligned} x_n &= a_n \cos\theta - b_n \sin\theta, \\ y_n &= b_n \sin\theta + b_n \cos\theta. \end{aligned}$$

The constant phase θ is selected in such a way that the squared Euclidean distance between QPSK signal constellations is maximized for both components, inphase and quadrature [19].

Quadrature components are generated and each component is independently interleaved. The signal interleavers are chosen such that after deinterleaving, the two components will be independent. An example of the interleaving process is presented in Table 2. For simplicity, the interleaving of just two symbols is presented. The first symbol transmitted by the source cluster-head has the quadrature component of the second symbol (Q_2). On the other hand, the vice cluster-head transmits a symbol with the quadrature component of the first symbol (Q_1). It can be noted that the nodes involved in the cooperative modulation transmission send just half of the total information amount, individually. The two components are then upconverted to the carrier frequency and added, using the following Expression

$$\begin{aligned} s_s(t) &= A \sum_{n=-\infty}^{+\infty} x_n p(t - nT_s)\cos(2\pi f_c t) \\ &+ A \sum_{n=-\infty}^{+\infty} y_{n-k} p(t - nT_s)\sin(2\pi f_c t), \end{aligned} \quad (9)$$

in which k is an integer representing the time delay in number of symbols introduced by the interleaving between the I and Q components.

Table 2: The cooperative interleaving process for two symbols.

	Time slot 1		Time slot 2	
Source Cluster-head	Q_2	I_1	Idle	Idle
Vice Cluster-head	Idle	Idle	Q_1	I_2

The bit error rate comparison between transmissions using simple QPSK, and QPSK with modulation diversity, for a rotation angle $\theta = 27°$, in a Rayleigh channel, is presented in Figure 4. It can be perceived that modulation diversity is more robust to the fading effects.

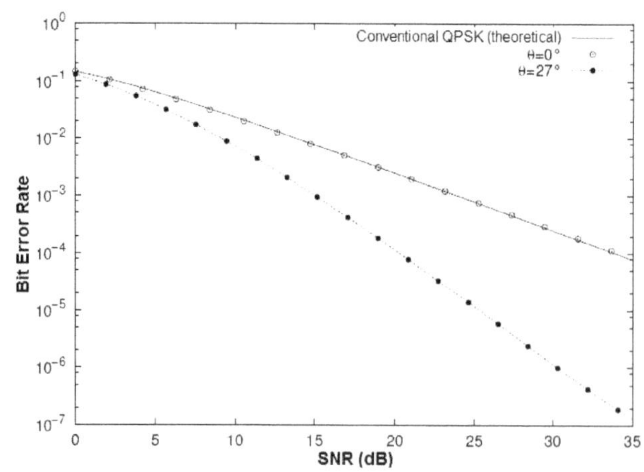

Figure 4: Bit error rate for the modulation diversity scheme.

4.3 Channel Model and Decoding System

Consider a communication channel with frequency nonselective slowly fading with a multiplicative factor representing the effect of fading and an additive term representing the Gaussian noise. The received signal is

$$r(t) = \alpha(t)s(t) + n(t), \qquad (10)$$

in which $\alpha(t)$ is modelled as zero-mean complex Gaussian process. The received signal, $r(t)$, is first downconverted to baseband. The obtained signal (equivalent lowpass) in one signalling interval is

$$r_l(t) = \alpha_n e^{-j}\phi_n s_l(t) + z(t), \quad nT_s \leq t \leq (n+1)T_s, \quad (11)$$

in which $z(t)$ represents the complex white Gaussian noise, α_n is the fading amplitude (considered constant over one symbol interval), ϕ_n is the phase shift due to the channel fading, and $s_l(t)$ corresponds to the equivalent low pass of the transmitted signal $s(t)$ [19]. With the phase shift estimation of the received signal at the sink node, and after the demodulation, the received vector is given by

$$\tilde{\mathbf{r}}_n = \alpha_n \mathbf{s}_n + \mathbf{z}_n, \qquad (12)$$

in which $\mathbf{s}_n$ is the vector representation of the transmitted signal at time nT_s, and the elements of the complex vector $\mathbf{z}_n$ are independent identically distributed Gaussian random variables with zero mean and variance $N_0/2$.

The decoded vector at the sink node, after the deinterleaving process, is

$$\mathbf{r}_n = \alpha_n x_n + \mathrm{Re}\{\mathbf{z_n}\} + j[\alpha_{n+k} y_n + \mathrm{Im}\{\mathbf{z_n}\}] \qquad (13)$$

which is then processed using symbol-by-symbol detection. The optimum demodulator computes the squared Euclidean distance between the received vector and each of the four signal vectors of the QPSK scheme and then decides in favor of the one closest to $\mathbf{r}_n$ [19].

5. SIMULATION RESULTS

In the simulations, the sensor network is composed by 100 nodes. The nodes are deployed randomly on an area of 50×50 meters. The sink node is located at the coordinates $x = 25$ and $y = 150$ meters. It is assumed that each node has an initial energy of 3 mJ. The radio dissipates $\epsilon_{\text{elec}} = 50$ nJ/bit to run the transmitter or receiver circuitry and $\epsilon_{\text{fs}} = 10$ pJ/bit/m^2, or $\epsilon_{\text{mp}} = 0.0013$ pJ/bit/m^4 for the transmitting amplifier to achieve an acceptable $\frac{E_b}{N_0}$. Consider d_0 as a specific threshold distance, given by

$$d_0 = \sqrt{\frac{\epsilon_{\text{fs}}}{\epsilon_{\text{mp}}}}. \qquad (14)$$

Thus, to transmit a l-bit message at a distance d using the radio model, the radio spends [23]

$$E_{Tx}(l,d) = \begin{cases} l \cdot (\epsilon_{\text{elec}} + \epsilon_{\text{fs}} \cdot d^2), & \text{if } d \leq d_0 \\ l \cdot (\epsilon_{\text{elec}} + \epsilon_{\text{mp}} \cdot d^4), & \text{if } d > d_0 \end{cases} \qquad (15)$$

and to receive this message, the radio spends

$$E_{Rx}(l) = \epsilon_{\text{elec}} \cdot l. \qquad (16)$$

The simulated sensor network is illustrated in Figure 5. The five cluster-heads are represented by circles and aggregate the sensed information of the other sensors, in the respective clusters. That aggregation process is illustrated in Figure 6. The figures show that by the use of the proposed scheme, the cluster-heads are well positioned, which contributes in saving overall energy.

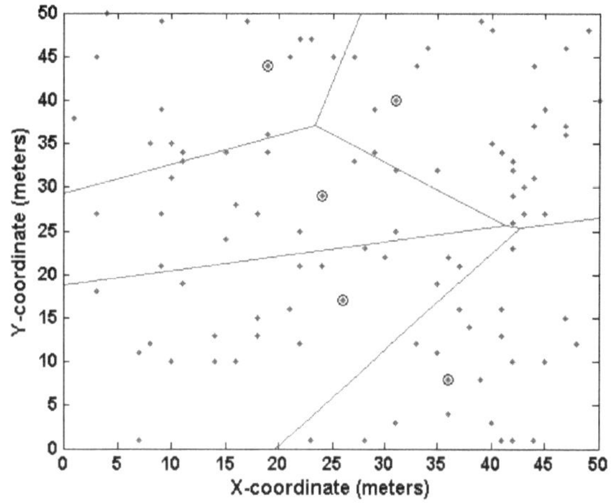

Figure 5: Five cluster-heads (represented in circles) elected by the proposed protocol. The sensor network considered in the simulations is composed by 100 nodes, randomly deployed.

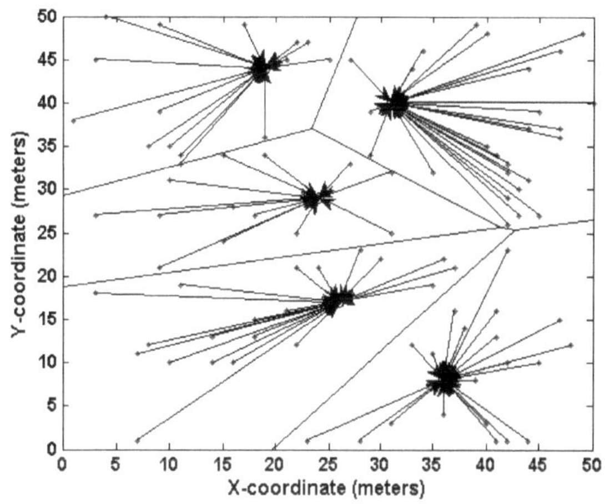

Figure 6: The aggregation process, which reduces the required amount of information to be transmitted to the sink node.

The performance evaluation was done by comparing simulation results between the proposed protocol and another cooperative cross-layer design, SCA with LEACH. The angle of rotation is selected as $\theta = 27^0$, because it is optimum for QPSK constellations [21]. Both systems use truncated ARQ and a CRC with $C = 16$ bits is assumed with a cyclic generator polynomial of $G_{CRC16}(D) = D^{16} + D^{12} + D^5 + 1$. The packet length is 120 bits and the simulations were processed in Matlab 7.

SCA with LEACH uses the LEACH protocol at the network layer and the cooperative space-time block codes at the physical layer, with the adaptive invoking of cooperative diversity, at the link layer, only if errors occur. Even with similar bit error rate behaviour [18, 26] of the modulation diversity, the amount of encoded data transmitted is twice the original message. In current simulations, two cooperative nodes are used in the SCA with LEACH oper-

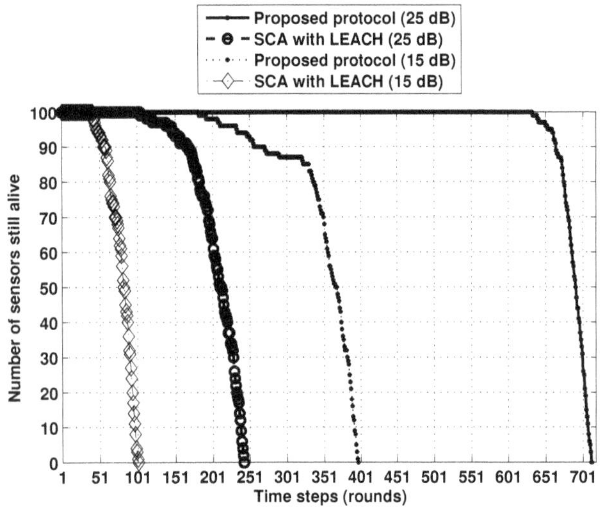

Figure 7: Amount of sensors still alive as a function of time steps. In all evaluated channel conditions, the proposed protocol extends the network lifetime, compared with SCA with LEACH.

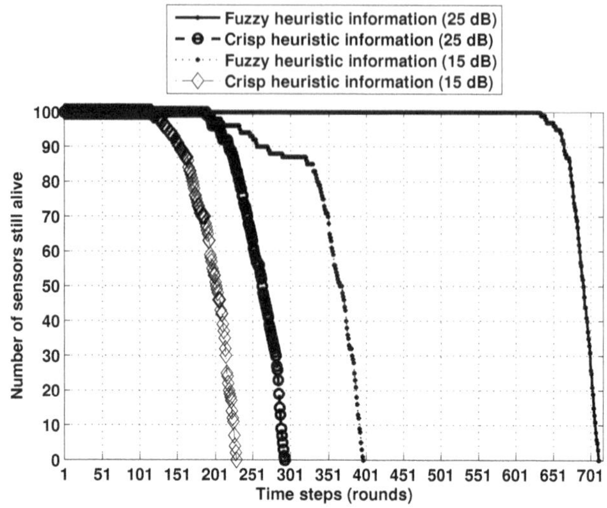

Figure 8: Performance comparison between the heuristic information modelled by a fuzzy inference and by a crisp logic. The fuzzy heuristic information improves the network performance, prolonging its operation in different channel propagation conditions.

ation. In the proposed protocol, two nodes are also used in the cooperation.

Network lifetime is the time span from the deployment to the instant when the network is considered non-functional. When a network should be considered non-functional is, however, application-specific. It can be, for example, the instant when the first sensor dies, a percentage of sensors die, the network partitions, or the loss of coverage occurs [27]. In this paper, the network lifetime measurement is based on the amount of rounds (time steps) that the first or last sensor node dies.

The first evaluation related to the network lifetime is illustrated in Figure 7, in which the amount of nodes still alive is observed as a function of the time steps (number of rounds). The performance comparison between the two simulated schemes is obtained using two different propagation conditions. As expected, for the best condition (SNR = 25 dB) the network lifetime is extended presenting higher values of first and last dead node than in the case in which the channel is degrade (SNR = 15 dB). In the poorest channel condition, the proposed protocol presents 396 rounds until the complete network inactivation, while, in the SCA with LEACH protocol, the simulated network utilizes 101 rounds to reach the same effect. In this scenario, the proposed protocol prolongs the network lifetime in approximately four times. Moreover, the proposed scheme overcomes the performance of SCA with LEACH even if different SNRs are compared.

Figure 8 presents a comparison between the performance evaluation related to the network lifetime, in which the fuzzy heuristic information and the crisp heuristic information are considered. In the crisp case, the heuristic information is obtained by the arithmetic mean between the heuristic variables `local_distance` and `CH_dispersion`. The proposed scheme, which uses the fuzzy heuristic information, presents a better performance in the considered propagation scenarios. Those results can be explained by the ability of fuzzy inference systems in dealing with the uncertainty of some measurements, specifically, the estimations of distances realized by the received signal intensity.

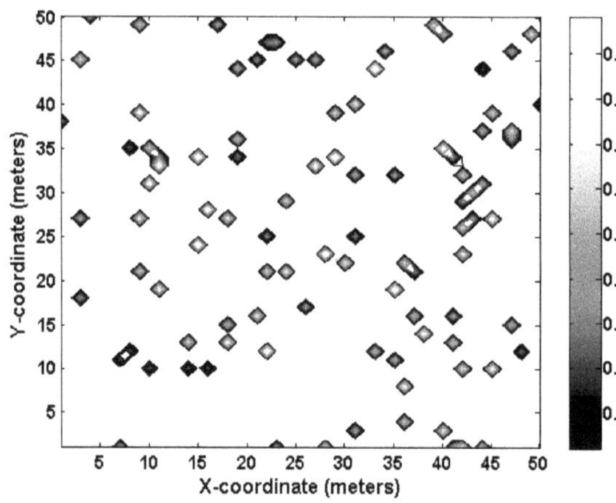

Figure 9: Residual energy distribution in the sensor network, using the SCA with LEACH protocol. An unbalanced distribution of energy can be verified. In this evaluation, the maximum number of retransmissions is four.

Figures 9 and 10 present the performance evaluation related to the energy-balancing, for the round operation of number 50. The performance of the proposed scheme (Figure 10) shows a distribution of residual energy more uniform than the SCA with LEACH protocol (Figure 9). Besides the energy-balancing evaluation, the level of node energy presented in the figures, reinforce the superiority of the proposed scheme in maximizing the network lifetime, since light colors indicate more residual energy.

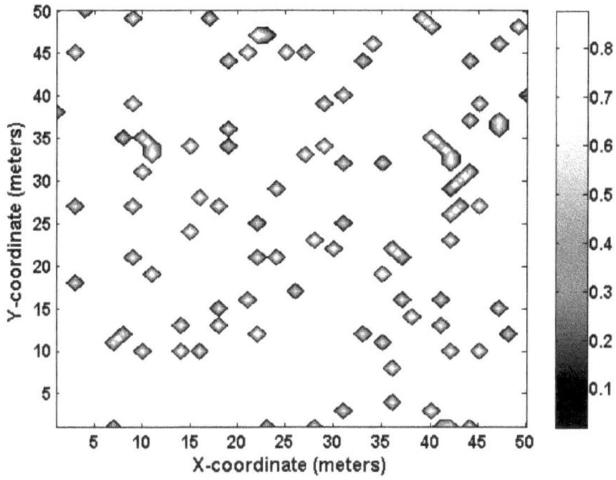

Figure 10: Residual energy distribution in the sensor network, using the proposed protocol, which presents an energy-balanced performance along the network. In this evaluation, the maximum number of retransmissions is four.

Figure 11 presents the performance comparison related to network lifetime, in which the number of rounds for the last dead node is evaluated, as a function of the channel SNR and $N_r^{\max}$. It can be noted that for all the propagation conditions, the proposed protocol presents better results than SCA with LEACH, since the network lifetime is extended for more rounds. Furthermore, related to $N_r^{\max}$, the proposed scheme presents soft and small variations, on the contrary of SCA with LEACH, which, mainly for poor propagation conditions, decreases the amount of rounds, because more energy is spent in more allowed retransmissions.

The performance evaluation related to the packet loss rate as a function the channel SNR is presented in Figure 12. For low values of SNR, the proposed protocol performance is similar to that observed for the SCA with LEACH, because the channel quality is severely degraded. However, beyond 12 dB, the PLR measured in the proposed protocol decreases more than using SCA with LEACH, which also contributes to enhance the network lifetime.

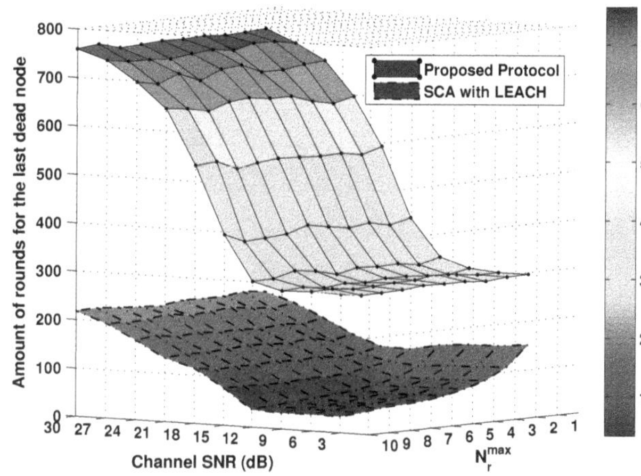

Figure 11: The lifetime performance comparison between the proposed protocol and the SCA with LEACH scheme. In all cases, the proposed protocol presents better results, with greater amount of rounds for the last dead node.

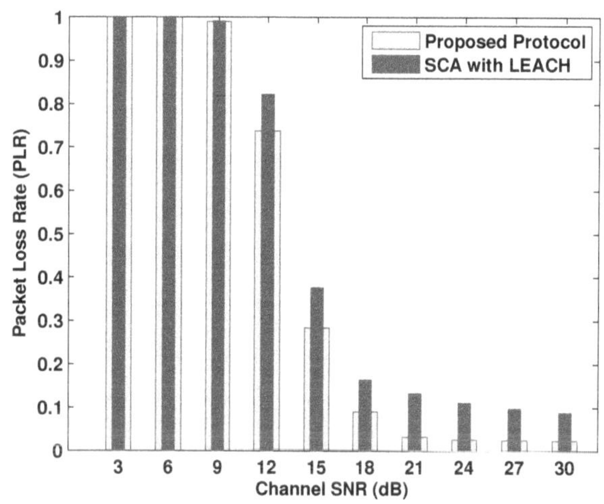

Figure 12: The packet loss rate comparison as a function of the channel quality. Under poor propagation conditions, the performance evaluation is similar, but beyond 12 dB, the proposed protocol presents lower values of PLR.

6. CONCLUSIONS

The authors presented the design and evaluation performance of a novel cooperative cross-layer protocol for multi-hop wireless sensor networks. The proposed protocol is biologically inspired in the organized behaviour of ants, in which a classical ACO system is adapted to the considered WSN constraints. The main goal is the optimal election of cluster-heads and controlling the process of possible further cooperative retransmissions.

The performance evaluation was done by comparing simulation results between the proposed scheme and another cooperative cross-layer design, SCA with LEACH. The performance metrics evaluated were the network lifetime as function of the channel SNR, maximum number of retransmissions, time steps, and of the energy balancing. The other evaluated parameter was the packet loss rate as a function of the channel SNR. In both evaluations, the proposed scheme overcomes SCA with LEACH, increasing the network lifetime and decreasing the PLR. That superiority can be explained by the energy efficiency in running the cooperative diversity by the proposed scheme, since the nodes involved in the cooperation need to transmit just half of the total symbols, individually. Furthermore, the bio-inspired cluster-head election takes into account variables related to residual energy, quality and consumption of previous transmissions, and deals with the uncertainty of distance estimations, by a fuzzy heuristic information modelling.

Future research includes the study of the interleaving depth impact on the network lifetime and transmission delay. Furthermore, the authors intend to investigate the scalability related to the number of network nodes, size of the monitored region and the relaying threshold variation.

Acknowledgement

The authors would like to thank Iecom, UFCG, CNPq for supporting this research.

7. REFERENCES

[1] S. Selvakennedy, S. Sinnappan, and Yi Shang. A biologically-inspired clustering protocol for wireless sensor networks. *Computer Communications*, 30:2786–2801, October 2007.

[2] N. Ding and P. Xiaoping. Data gathering communication in wireless sensor networks using ant colony optimization. In *IEEE International Conference on Robotics and Biomimetics (ROBIO'04)*, pages 822–827, 2004.

[3] W. Rui, L. Yan, Ye Gangqiang, Lu Chaoxia, and Pan Quan. Swarm intelligence for the self-organization of wireless sensor network. In *IEEE Congress on Evolutionary Computation (CEC'06)*, 2006.

[4] V. Vijaykumar, R. Chandrasekar, and T. Srinivasan. An ant odor analysis approach to the ant colony optimization algorithm for data-aggregation in wireless sensor networks. In *International Conference on Wireless Communications, Networking and Mobile Computing (WiCOM'06)*, sept. 2006.

[5] R. Chandrasekar and S. Misra. Introducing an aco based paradigm for detecting wildfires using wireless sensor networks. In *International Symposium on Ad Hoc and Ubiquitous Computing (ISAUHC '06)*, pages 112–117, 2006.

[6] S. Wang, X. Wang, L. Ding, D. Bi, and Z. You. Collaborative hybrid classifier learning with ant colony optimization in wireless multimedia sensor networks. In *7th World Congress on Intelligent Control and Automation (WCICA'08)*, 2008.

[7] A.-A. Salehpour, B. Mirmobin, A. Afzali-Kusha, and S. Mohammadi. An energy efficient routing protocol for cluster-based wireless sensor networks using ant colony optimization. In *International Conference on Innovations in Information Technology (IIT'08)*, 2008.

[8] C. Zhang and Q. Xu. Clustering approach for wireless sensor networks using spatial data correlation and ant-colony optimization. In *International Conference on Networks Security, Wireless Communications and Trusted Computing (NSWCTC '09)*, 2009.

[9] Y. Chen, J. Zhang, and I. Marsic. Link-layer-and-above diversity in multihop wireless networks. *IEEE Communications Magazine*, 47(2):118–124, February 2009.

[10] M. Dorigo and T. Stützle. *Ant Colony Optimization*. MIT Press, 2004.

[11] M. Dorigo and G. Caro. Ant colony optimization: a new meta-heuristic. In *Proceedings of the Congress on Evolutionary Computation (CEC'99)*, 1999.

[12] Lotfi A. Zadeh. Fuzzy sets. *Information and Control*, 8(3):338–353, 1965.

[13] Lotfi A. Zadeh. The concept of a linguistic variable and its application to approximate reasoning – I. *Information Sciences*, 8(3):199–249, 1975.

[14] M. P. Sousa, R. F. Lopes, A. Kumar, W. T. A. Lopes, and M. S. Alencar. Cooperative STBC with fuzzy election applied to surveillance wireless video sensor networks. In *IEEE Global Communications Conference (Globecom Workshops'10)*, 2010.

[15] Q. Liu, S. Zhou, and G. B. Giannakis. Cross-Layer combining of adaptive modulation and coding with truncated ARQ over wireless links. *IEEE Transactions on Wireless Communications*, 3(5):1746–1755, Sept. 2004.

[16] E. Malkamaki and H. Leib. Performance of truncated type-II hybrid schemes with noisy feedback over block fading channels. *IEEE Transactions on Communications*, 48(9):1477–1487, Sep 2000.

[17] M. P. Sousa, R. F. Lopes, W. T. A. Lopes, and M. S. Alencar. Low-energy selective cooperative diversity with ARQ for wireless image sensor networks. In *IEEE 72nd Vehicular Technology Conference Fall (VTC - Fall)*, pages 1–5, sept. 2010.

[18] M. P. Sousa, M. S. Alencar, A. Kumar, and W. T. A. Lopes. Performance evaluation of a selective cooperative scheme for wireless sensor networks. *The Sixth ACM International Symposium on Performance Evaluation of Wireless Ad Hoc, Sensor, and Ubiquitous Networks - ACM PE-WASUN*, 2009.

[19] S. B. Slimane. An improved PSK scheme for fading channels. *IEEE Transactions on Vehicular Technology*, 47(2):703 –710, May 1998.

[20] T. ur Rehman Ahsin and S. B. Slimane. Detection strategies in cooperative relaying with network coding. In *IEEE 21st International Symposium on Personal Indoor and Mobile Radio Communications (PIMRC'10)*, pages 12–17, sept. 2010.

[21] S. A. Ahmadzadeh, S. A. Motahari, and A. K. Khandani. Signal space cooperative communication. *IEEE Transactions on Wireless Communications,*, 9(4):1266–1271, april 2010.

[22] G. Caro and M. Dorigo. AntNet: distributed stigmergetic control for communications networks. *Journal of Artificial Intelligence Research*, 9:317–365, December 1998.

[23] J. Kim, S. Park, Y. Han, and T. Chung. CHEF: Cluster Head Election mechanism using Fuzzy logic in Wireless Sensor Networks. In *10th International Conference on Advanced Communication Technology (ICACT'08)*, pages 654–659, 2008.

[24] N. Baldo and M. Zorzi. Fuzzy logic for cross-layer optimization in cognitive radio networks. *IEEE Communications Magazine,*, 46(4):64 –71, April 2008.

[25] G. Montemayor and J. Wen. Decentralized collaborative load transport by multiple robots. In *IEEE International Conference on Robotics and Automation (ICRA'05)*, pages 372–377, April 2005.

[26] W. R. Heinzelman, A. Chandrakasan, and H. Balakrishnan. Energy-efficient communication protocol for wireless microsensor networks. *In Proceedings of the 33rd Annual Hawaii International Conference on System Sciences*, 2:10 pp., Jan. 2000.

[27] Y. Chen and Q. Zhao. On the lifetime of wireless sensor networks. *IEEE Communications Letters*, 9(11):976–978, November 2005.

Time-Space Correlation for Real-Time, Accurate, and Energy-Aware Data Reporting in Wireless Sensor Networks*

Leandro A. Villas[1,2], Azzedine Boukerche[2], Daniel L. Guidoni[1], Horacio A. B. F. de Oliveira[4], Regina B. Araujo[3], and Antonio A. F. Loureiro[1]

[1]UWL - Research Laboratory, Federal University of Minas Gerais, Belo Horizonte, Brazil
[2]PARADISE - Research Laboratory - SITE, University of Ottawa, Ottawa, Canada
[3]WINDIS - Research Laboratory, Federal University of São Carlos, São Carlos, Brazil
[4]LAMP - Research Laboratory, Federal University of Amazonas, Manaus, Brazil

{leandro,guidoni,loureiro}@dcc.ufmg.br, boukerch@site.uottawa.ca, regina@dc.ufscar.br and horacio@dcc.ufam.edu.br

ABSTRACT

One of the main applications in Wireless Sensor Networks (WSNs) is the accurate monitoring of extensive areas. However, due to the high density of nodes, redundant and correlated data are usually sensed by nearby nodes over time. Since saving energy is a key aspect of these networks, it is important to take advantage of these correlations to decrease communication and data exchange. However, current proposals usually results in high delays and outdated data arriving at the sink node. In this work, we go further and fully exploit both spatial and temporal correlations to perform near real-time data collection in WSNs. Also, in our proposal, called **E**fficient Data Collection **A**ware of **S**pace-**T**ime Correlation (EAST), obtained results clearly indicate that an event can be sensed with a high accuracy of more than 99.7% while still saving the residual energy of the nodes in more than 14 times when compared to the accurate data collection strategy.

General Terms

Algorithms

Categories and Subject Descriptors

C.2 [**Computer-Communication Networks**]: Network Protocols - Routing protocols

*This work is partially supported by Canada Research Chair program and NSERC. The authors also wish to thank FAPEMIG, CNPq, FAPESP, CAPES (INCT-SEC Project, processes 573963/2008-8 and 08/57870-9).

MSWiM'11, October 31–November 4, 2011, Miami, Florida, USA.

1. INTRODUCTION

Wireless Sensor Networks (WSNs) [3, 7, 15, 16] are basically guided by events. These networks are composed of a large number of small sensor devices that are able to sense these events and gather data from their surroundings, process these data, and send them toward a sink node using multihop wireless communication. The sink node, then, is responsible for sending the gathered data to a monitoring facility.

Due to their small sized and battery-operated nature, sensor nodes are usually energy-constrained devices. Thus, saving energy is one of the main concern in these networks. Energy consumption is generally associated with communication, which is often the most expensive activity in terms of energy. For that reason, algorithms and protocols designed for WSNs should consider the energy consumption in their conception [2, 11].

When sensing a physical phenomenon, nearby nodes data tend to be correlated in both time and space [10, 21]:

- *Temporal correlation:* the change pattern of current sensor readings is equal or similar to the readings observed at previous times.
- *Spatial correlation:* the change pattern of the data sensed by nearby nodes are expected to be the same or similar.

This sensing process is likely to produce a large amount of data and to take advantage of these time-space correlation to decrease communication cost is one of the key issues if energy needs to be saved. For this reason, some techniques such as spatial suppression [4, 8, 9, 19, 20] and temporal suppression [10, 13] have been explored in the literature to reduce the amount of communication required to report the gathered data, specially in applications that tolerate approximate readings. However, current solutions for time-space correlation usually introduces delays in data transmissions as well as the reception of outdated information by the sink node [5, 12, 14], which can be inappropriate for a number of real-time sensor network applications.

In this work, we go further and present a new algorithm for spatial and temporal suppression in data collection, which we refer to as the EAST (**E**fficient Data Collection **A**ware of **S**pace-**T**ime Correlation) algorithm. In our EAST algorithm, gathered data is received by the sink node in almost real-time while still taking full advantage of spatio-temporal correlations in terms of both communication cost and accuracy. In our approach, sensor nodes are grouped by using a spatial correlation approach while the leader and representative nodes execute a temporal suppression technique. The leader node generates a representative value for the whole group based on data received by the representative nodes, which is a subset of all nodes sensing the same event.

Our proposed solution dynamically adjust itself according to both network and event characteristics, which is a key contribution commonly ignored in current proposed solutions. Thus, in the EAST algorithm (i) residual energy of sensor nodes are balanced, (ii) energy consumption is reduced by eliminating redundant notifications, and both (iii) the number of representative nodes and (iv) error threshold are dynamically adjusted according to the event characteristics and accuracy requirements. We present an extensive set of experiments that show the need for new algorithms in these scenarios (specially in real-time applications) and also clearly indicate the good performance of our EAST algorithm when compared to similar proposed solutions.

The remaining of this paper is organized as follows. In the next section, we provide an overview of the existing approaches exploiting spatial and temporal correlation in WSNs. Our proposed energy-aware spatial and temporal correlation mechanism (EAST algorithm) is presented in Section 3, while detailed performance evaluation and simulation results are presented in Section 4. Finally, in Section 5 we present our conclusions.

2. RELATED WORK

In [18], the authors explore the nature of the physical phenomenon observed by a WSN to propose a new framework for developing efficient data based protocols in WSNs. The main goal of the framework is to incorporate temporal correlation among consecutive observations of the phenomenon to reduce communication cost. Spatial correlation is also explored by showing that nearby nodes tend to have the same observed data. The authors explored two ways for using their proposed framework: first, in developing efficient medium access protocols and, second, in developing reliable event data reporting in WSN.

In [5], the authors propose a framework based on temporal correlation to decrease the number of transmitted messages required to monitor a physical phenomenon. The framework uses a Self-Based Regression (SBR) algorithm [6] in order to process the observed data before sending it to the sink node. Since a high number of data is generated when sensing an event, in the proposed algorithm all of this data is stored in a buffer and, only when this buffer is full, the SBR algorithm process the whole gathered data in order to find only the representative information. The authors claim that by sending a small amount of representative data, the sink node can reconstruct the observed phenomenon without losing accuracy. The main drawback of this approach is the waiting time until the buffer fills up, in which case results in the sink node receiving only outdated information and reconstructing the past.

Another similar approach is proposed in [12, 14]. In this solution, called Spatiotemporal Clustering and Compressing Schemes (SCCS), each node stores its monitored data in a buffer and when the buffer is full, it is executed a DCA (divide and conquer) algorithm in order to consider the temporal correlation among the gathered data. The goal of the DCA is to find the minimum set of data to be transmitted to the sink node. The DCA algorithm works as follows. Considering all reading inside the buffer, the DCA creates a line between the first element and the last element. For each buffer value, the algorithm calculates the distance between this value and the created line. If the value is smaller than a predefined threshold, the solution indicates that this value was already been considered and it is not necessary to include it again in the packet that will be sent to the sink. When the value is greater than the threshold, the algorithm split the line into two lines (one line up to this value and another line up to the endpoint of the previous line). When a line is split in two lines, all the buffer values are verified again. These steps are repeated until a created line is not split in two. Also, the SCCS solution creates a cluster among the nodes that sensed the event in order to perform spatial correlation to reduce the number of transmitted messages. As mentioned before, by using a buffer to perform the temporal correlation, the waiting time to deliver the gathered data maybe inappropriate for a number of real-time sensor network applications.

3. A NEW EFFICIENT ALGORITHM FOR SPACE-TIME CORRELATION

In this section, we define our spatial and temporal correlation models and we propose the EAST algorithm. One of the key aspects of our proposed EAST algorithm is that the size of the correlation region and error threshold of readings can be changed dynamically according to the event characteristics in order to achieve the application's accuracy requirements. It results in a better use of the available energy in the nodes that are sensing the event by eliminating redundant notifications as well as by using dynamic routes and low communication overhead. These and other characteristics of our EAST algorithm are discussed in this section.

3.1 Spatial Correlation Model

As mentioned before, nearby nodes tend to detect similar values. However, this closeness (c), i.e., the Euclidean distance between the nodes that detect similar values, depends on both application requirements and event characteristics. Some critical applications are less tolerant to discrepancies in the sensed data about the observed phenomenon, requiring closer nodes to notify the sensed data (i.e., the correlation region is smaller). Other applications might be more tolerant to discrepancies, not demanding that closer nodes report the sensed data (i.e., the correlation region is wider).

DEFINITION 3.1 (CORRELATION REGION). *We define a correlation region as an area where the values sensed by the sensor nodes are considered similar (for the application). Therefore, a single reading within this region is sufficient to represent it. The size of the correlation region (c) varies according to both application and event type. For events whose characteristics change significantly at short range, the sink node should decrease the size of the correlated region, i.e., this event needs to be notified by closer nodes. For events whose characteristics do not change significantly at short range, the sink node can increase the size of the correlated region.*

3.2 Temporal Correlation Model

Sensor readings about the environment are typically periodic; consequently, the time-ordered sequence of sensed data constitutes a time series. Due to the nature of the physical phenomenon, there is a significant temporal correlation among each consecutive observation of a sensor node and gathered data are usually similar over a short-time period. Thus, in these cases, sensor nodes do not need to transmit their readings if the current reading is within an acceptable error threshold regarding the last reported reading. The sink node can just assume that any unreported data are unchanged from the previously received ones. The degree of correlation between consecutive sensor measurements might vary according to the characteristics of the phenomenon.

DEFINITION 3.2 (TEMPORAL SUPPRESSION). *Each source node keeps the last reported reading. When current reading (R_{new}) is available, R_{new} is compared against the last reported reading (R_{old}). The current reading of a source node is reported iff the given relative error threshold is greater than the temporal coherency tolerance, i.e.* $\left(\frac{|(R_{new}-R_{old})|}{R_{old}}\right) \times 100 > tct$, *where* tct *is the percentage of tolerance coherency temporal. Otherwise the value* R_{new} *is suppressed.*

3.3 Overview of the EAST Algorithm

The main idea of our proposed EAST algorithm is to manage the energy consumption of the nodes that detected an event by eliminating redundant notifications. The proposed algorithm forwards gathered data using shortest routes (in Euclidean distance) toward the sink node. Our algorithm considers the following roles to perform data routing:

- *Member Node*: A node that is currently detecting one or more events. In case its sensed data is redundant, it will not report the gathered data.
- *Representative Node*: A node detecting an event and reporting the gathered data to a coordinator representing not only itself but all nearby nodes with similar readings while still applying temporal suppression.
- *Coordinator*: A node detecting the event and responsible for gathering all of the event data sent by representative nodes. It processes the received data and send the result toward the sink node.
- *Relay*: A node that forwards data toward the sink node.
- *Sink*: The gateway between the WSN and the monitoring facility.

The proposed EAST algorithm uses shortest routes (in Euclidean distance) in two different levels for forwarding the gathered data toward the sink node. At the first level, representative nodes uses shortest routes to forward data towards the coordinator node. At the second level, the coordinator nodes use shortest routes to forward data toward the sink node. Figure 1 shows two examples of the routing structure obtained by the EAST algorithm. In this figure, a gray field indicates the event area, cells represent the regions of correlation and, lastly, the red line is the shortest route.

The proposed EAST algorithm is adaptive and the main objective is to reduce energy consumption in data gathering while still preserving the data accuracy. To achieve this goal, EAST dynamically changes the size of the correlation region and value of coherency tolerance according to the event characteristics. For this, an event is divided into cells, as depicted in Figure 1. Each cell defines a correlation region and nodes within each cell are assumed to be spatially correlated. Only one node within a cell notifies the sensed information, if and only if, the given relative error threshold is greater than the temporal coherency tolerance. This last node is the representative node of the cell. Cells are independent from each other, so the change of representative nodes in one cell does not require any reconfiguration. The alternation of a representative node in each cell is done in order to balance the energy consumption of spatially correlated nodes and the temporal suppression is performed to reduce the reporting of redundant data. Since the correlation regions are independent, the resizing of the correlation regions does not require any additional communication among the nodes within the event areas in order to compute the new cell they belong to. Furthermore, each node locally performs temporal suppression without the need for communicating with neighbors. The proposed spatiotemporal correlation approach is highly adaptive and scalable regarding events of different intensities.

The EAST algorithm is performed in three phases. In Phase 1, presented in Section 3.4, sensor nodes store the sink's position as well as their neighbors positions. Phase 2, presented in Section 3.5, consists of cluster formation, coordinator election, and the division of the event area into cells. Finally, in Phase 3, presented in Section 3.6, representative nodes are chosen and data are transmitted.

3.4 Discovery of the sink's and neighbors' positions

After the sensor nodes deployment, a data flooding is usually executed in the network in order to propagate the sink's interest. Our EAST algorithm takes advantage of this flooding to also perform the discovery and storage of the sink's and neighbors' positions. Thus, in our algorithm, the sink node starts by flooding a configuration message. The message contains four fields: `ID`, `CoordSender`, `CoordSink` and `Phenomenon_of_Interest`, where `ID` is the node identifier that retransmitted the message, `Phenomenon_of_Interest` is the application's interest (e.g. temperature higher than 25 degrees), `CoordSender` is the node's position (x_n, y_n) that relays the configuration message, and `CoordSink` is the sink's position (x_s, y_s). In this phase, sensor nodes store the received information in a table of neighbors `neighborhood`. This information will be used in Phases 2 and 3.

3.5 Cluster Formation, Leader Election, and Division of the Event Area into Cells

The second phase of our EAST algorithm only starts when in the presence of events. Thus, when an event is detected by one or more nodes, the leader election algorithm is started with the sensing nodes running for leadership (group coordinator) – this process is described in Algorithm 1. For this election, all detecting nodes are eligible (Lines 8 and 9 of Algorithm 1) and the group leader (Coordinator node) will be the node with higher residual energy. (Lines 13 and 14 of Algorithm 1). At the end of the election algorithm only one leader node exists in the group. In case of a tie, the ID parameter is used as a tie breaker. The remaining nodes detecting the same event become member nodes. At each notification, a subset of the member nodes will be representative nodes, as explained later in this section). The coordinator gathers the information collected by the representative nodes, processes the information, and sends them toward the sink node.

After the clustering, the region's event is decomposed into $\left(\frac{2r_e}{c}\right)^2$ cells, where r_e is the event maximum radius and c is the cell size (correlation region). Figure 1 shows an example in which the region's event is decomposed into 9 cells. Each cell is represented by an ordered pair (x_c, y_c). If $c = 0$, then there is no spatial correlation between nodes and all nodes in the group are representative nodes. Otherwise, each node computes the value of x_c and y_c of the cell to which it belongs to. For this computation, the node's position (x_n, y_n), the event central position (x_e, y_e), and the cell size (c) are required. Lines (17 to 26 of Algorithm 1) show this computation of x_c and y_c.

It is important to ensure that all nodes in the same cell communicate with each other. For this, we consider that the maximum size of a cell c to be the length of the triangle's leg in a right triangle when r_c is the hypotenuse, i.e., $c = r_c \cos 45°$ where r_c is the communication radius of sensor nodes. Since the cell size can vary in order to control the tradeoff between data accuracy and energy consumption, the correlation region may only vary between

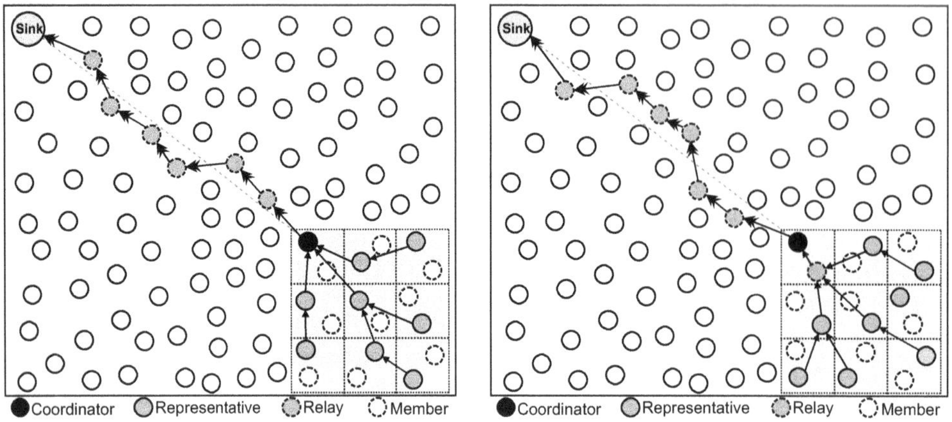

(a) Routing structure at instant i.

(b) Routing structure at instant i+1.

Figure 1: Examples of routing structure used by the EAST algorithm.

$0 \leq c \leq r_c \cos 45°$. When $c = 0$, all nodes report the sensed data (an optimal solution in terms of accuracy). For $c > 0$, only the representative node at each cell reports the sensed data.

The EAST algorithm selects a single representative node at each cell of dimensions c^2 for each notification. Figures 1(a) and 1(b) show representative nodes at different times in the event's region. These representative nodes chosen in the set of member nodes are the nodes that have higher energy residual among the nodes belonging to the same cell, which ensures the energy consumption distribution among the nodes.

3.6 Data Transmissions

After computing the cell that a node belongs, the node checks whether it is a representative node and also if the relative error threshold is greater than the temporal coherency tolerance (Line 29). If both conditions are satisfied, then the sensed value (R_{rew}) is sent toward the group coordinator. The group coordinator then processes and sends the collected information toward the sink using the shortest path. Based on its position and the sink's position, the Coordinator creates a straight line segment that connects itself to the sink. When data transmission is performed, the closest nodes to both its straight line segment and to the endpoint of this straight line will be chosen to forward the data.

If the sink needs to dynamically resize the correlation region or the coherency tolerance value in order to meet any application's requirement, it sends a new value of c and tct to the nodes of the group so they can recalculate their cells. Thus, the parameters of our proposed algorithm can be dynamically controlled by the sink node, which has a complete view of the phenomenon.

4. PERFORMANCE EVALUATION

We compared our proposed EAST algorithm to the SCCS algorithm as well as the accurate data collection strategy, which is the optimal solution in terms of accuracy. As mentioned before, in the SCCS algorithm, each node stores its monitored data in a buffer and, when the buffer is full, the node processes the data in its buffer in order to consider the temporal correlation among the monitored values and report the result to the sink node. On the other hand, in the accurate data collection strategy, every sensor is requested to report its readings to the sink node in each round of data gathering. The main purpose of this comparison is to evaluate the performance of our proposed algorithm considering the following metrics: (i) notifications, (ii) readings reported, (iii) readings per data packet, (iv) energy consumption, (v) data accuracy, and (vi) delay notification.

4.1 Methodology

The evaluation is performed through simulations by using the SinalGo version v.0.75.3 [17] simulator, a well-known network distributed algorithm simulator. In all results, curves represent average values, while error bars represent confidence intervals for 95% of confidence from 33 different instances (seeds). The simulation parameters are presented in Table 1. The event occurs in random positions and sensor nodes are randomly deployed.

Table 1: Simulation parameters

Parameters	Values
Sink node	1 (top left)
Number of nodes	1024
Number of events	1
Density (avg. neigh. number)	25
Event diameter (m)	(50, 100, 150, 200)
Correlation Region (c)	(15, 30, 45)
Temporal coherency tolerance	(0.5, 1, 2, 3, 4)
Buffer size	(25, 50, 100, 200)
Notification rate (per minute)	1
Communication radius (m)	80
Sensor field (m)	900×900
Simulation duration (days)	7

The following metrics were used for the evaluation:

- *Notifications:* Number of notifications performed by nodes that detect the event.
- *Readings reported:* Number of readings reported.
- *Readings per data packet:* Average number of readings within each packet (It is the ratio of Readings reported and Notifications).
- *Energy consumption:* The amount of energy consumed by sensors that detected the event.
- *Data accuracy:* The accuracy of the information on the observed phenomenon regarding to the original information.

Algorithm 1 EAST Algorithm.

```
▷ Variables:
tct = {Temporal Coherency Tolerance.}
R_old = ∅ {Last Reported Reading.}
Start AnnouncementInterestMessage

▷ Input:
AnnouncementInterestMessage
Action:
Stores the Neighbor's and Sink's Position
Stores the Phenomenon of Interest
[Re]Start AnnouncementInterestMessage

▷ Input:
EventDetected
Action:
node.Role := Coordinator
Send EventAnnouncementMessage
Start cellComputation

▷ Input:
msg_i = response(EventAnnouncementMessage)
Action:
if EnergyLevel(node) > EnergyLevel(msg_i) then
    node.Role := Member;
    Retransmits (EventAnnouncementMessage)
end if

▷ Input:
cellComputation timeout
Action:
x_c := 0 and y_c := 0
if (x_n − x_e) / (c/2) > 1 then
    x_c := ⌊((x_n − x_e) − (c/2)) / c⌋ + 1
    y_c := ⌊((y_n − y_e) − (c/2)) / c⌋ + 1
end if
if (x_n − x_e) / (c/2) < −1 then
    x_c := ⌊((x_n − x_e) + (c/2)) / c⌋ − 1
    y_c := ⌊((y_n − y_e) + (c/2)) / c⌋ − 1
end if

▷ Input:
DataTransmissions
Action:
R_new := sensedvalue
if node.Role = Representative and
    (|(R_new − R_old)| / R_old) × 100 > tct then
    Send R_new to Coordinator
    R_new := R_new
end if
if node.Role = Member then
    Forwards R_new to Coordinator
end if
if node.Role = Coordinator then
    Processing received R_new
    Forwards the result to Sink
    if node.Role = Relay then
        Forwards R_new to Sink {T}emporal Coherency Tolerance é foda
    end if
end if
```

- *Delay notification:* Time to deliver the gathered data.

In all evaluated cases, a number of variations of our proposed EAST algorithm were considered. First, we evaluated our EAST algorithm while exploring only the temporal correlation. We also evaluated our algorithm (EAST-15, EAST-30, and EAST-45) exploiting both temporal correlation and spatial correlation (with correlation regions of size 15, 30, and 45). For the SCCS algorithm, we considered buffer with different storage capacities of 25, 50, 100, and 200 readings.

4.2 Event Model

For our event model, we used a set of one-week environmental monitoring data collected in Amazon rainforest in Brazil [1]. These data are Celsius temperature taken at intervals of 1 minute. The samples are shown in Figure 2.

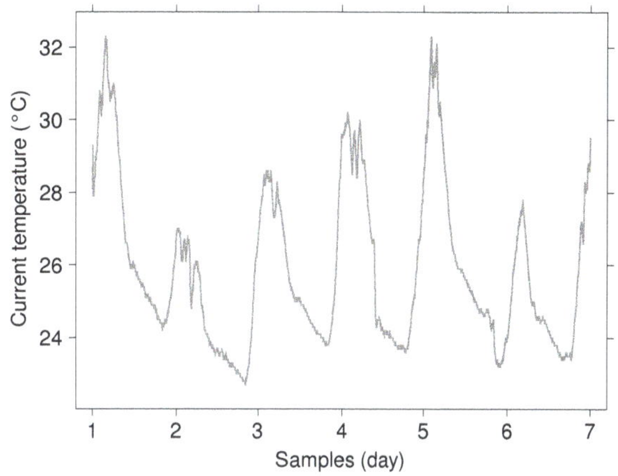

Figure 2: Data collected from Amazon rainforest.

In our application, the sink node will ask for the temperature in a region of interest. All nodes in this region will then send the data according to our proposed algorithm. To use the real-world data in our simulations, we consider the temperature in coordinate (x, y) in the event area is given by Equation 1, where T_E is temperature in event center which was obtained from real-world data set presented in Figure 2, D_E is the Euclidean distance (meters) to event center and T_D is the temperature decrease (degree Celsius per meters).

$$temperature = T_E - (D_E \times T_D) \quad (1)$$

4.3 Notifications and Readings Reported

For this analysis, the temporal coherency tolerance, buffer size, and correlation region (presented in Table 1) were all varied to evaluate their impact in the number of readings that can be eliminated by exploiting the spatial and temporal correlation.

Figure 3(a) shows that when the temporal coherency tolerance increases, the number of notifications performed by our EAST algorithm decreases while the number of notifications in the SCCS remains fixed since the readings will only be transmitted when the buffer fills up. Consequently, in SCCS, the number of notifications depends on the buffer size and not on the temporal coherency tolerance. Moreover, in our EAST algorithm, when the size of the correlation region increases (15, 30 e 45), the number of representative nodes decreases, which also decreases the number of notifications.

In Figure 3(b), we can see that the number of readings reported by our EAST algorithm is similar to the ones presented in Figure 3(a). This is because whenever the current reading is above the temporal coherency tolerance, the data is notified. Note that in most cases the EAST algorithm reports least readings than the SCCS algorithm, but for small the temporal coherency tolerance, the SCCS algorithm presents less notifications by exploring the use of a buffer in which each notification may contain more than one reading. Since the SCCS algorithm creates a line segment between the first and last reading of the buffer, this technique has good results in terms of energy consumption when the buffer size increases

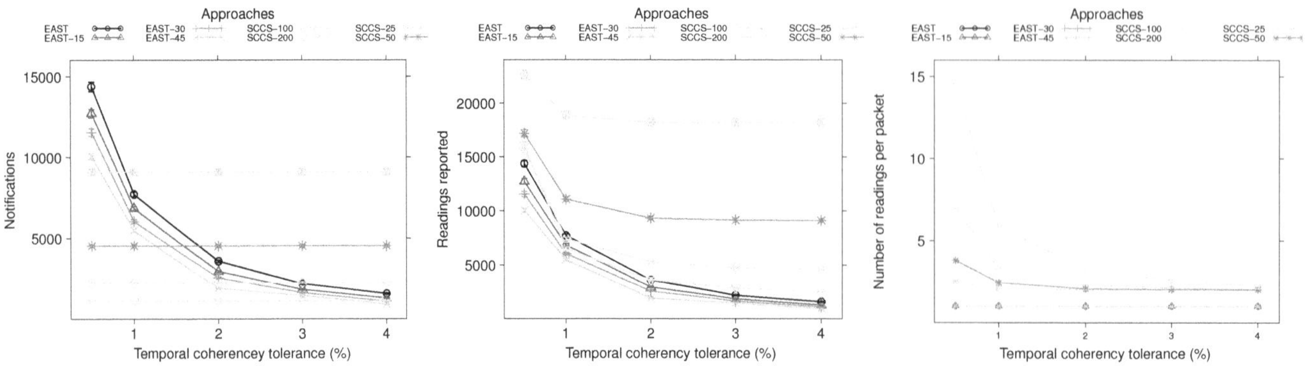

(a) Number of notifications (b) Number of readings reported (c) Readings in each data packet

Figure 3: Notifications × Readings.

(as depicted in Figure 4). In this case, it is necessary a few line segments to represent all values inside the buffer.

Figure 3(c) shows the average number of readings within each transmitted packet. As we can see, our proposed EAST algorithm, in any situation, sends only one reading within the packet. But the number of readings per packet in SCCS algorithm depends both on the capacity of the buffer and the temporal coherency tolerance. When the temporal coherency tolerance is small, the SCCS algorithm needs to split the original line segment into a high number of other line segments to represent the original values. Because of this, more readings will be necessary to represent the monitored values. It is important to point out that when we increase the buffer size, the SCCS algorithm will split the original line segment into more line segments to represent the original values, since more reading are being considered by the algorithm. On the other hand, when we increase the temporal coherency tolerance, it is not necessary to split the original line segment into a high number of line segments. For instance, when the temporal coherency tolerance is 4%, the SCCS algorithm transmits the same number of readings to represent the sensed event.

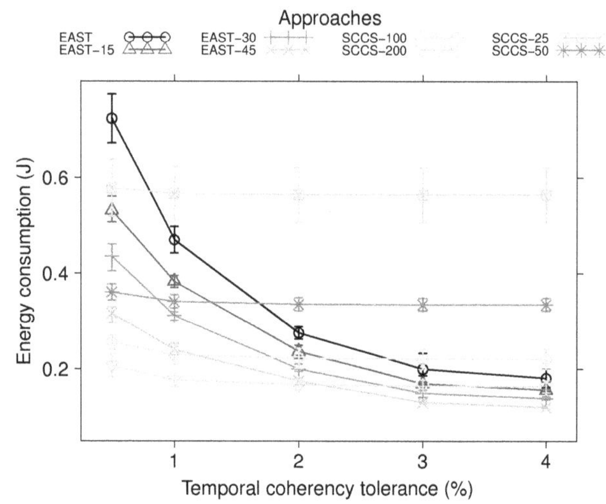

Figure 4: Average energy consumption.

4.4 Energy

Figure 4 shows the average energy consumption of the nodes within the observed phenomena area. For this analysis, the temporal coherency tolerance, buffer size, and correlation region (presented in Table 1) were all varied to evaluate their impact in the energy consumption.

As depicted in Figure 3(a), when the temporal coherency tolerance increases, the number of notifications performed by EAST decreases while the number of notifications in the SCCS remains fixed and is dependent on the buffer size, which has an impact on the data accuracy (see Figure 5). Consequently, the energy consumption also decreases in the EAST algorithm (see Figure 4). Moreover, when the size of the correlation region increases (15, 30 e 45), the number of notifications decreases. Consequently, the energy consumption also decreases, since a smaller number of sensor nodes report their readings.

The results for the accurate data collection strategy were not plotted on the graph by having a very high energy consumption. The high energy consumption in this strategy is due to the fact that all readings are notified to ensure data accuracy of 100%. On average, the accurate data collection strategy consumes $10J$, i.e., more than 14 times than to SCCS and EAST algorithms.

4.5 Data Accuracy

When both temporal and spatial correlation are exploited, the level of accuracy in information about the observed phenomenon tends to reduce. In this simulation scenario, the temporal coherency tolerance, buffer size, and correlation region were all varied to evaluate the accuracy of the readings. The phenomenon observed was the temperature, but the proposed mechanism works for any other type of phenomenon with different characteristics. We analyzed the accuracy at the sink node when computing the values for minimum, mean, and maximum temperatures.

As depicted in Figure 5, when the temporal coherency tolerance increases, the data accuracy in the EAST algorithm slowly decreases. On the other hand, in the SCCS algorithm, when the buffer capacity increases, the data accuracy decreases more quickly. It happens because when we increase the buffer size, less readings will be necessary to represent the monitored event (see Figure 3(b)). Thus, considering less values, the SCCS algorithm will not achieve good values of data accuracy.

For the evaluated scenarios, the smallest accuracy observed in our proposed algorithm was of 98.7% and the energy consumption was less than $0.2J$, while the accurate data collection strategy con-

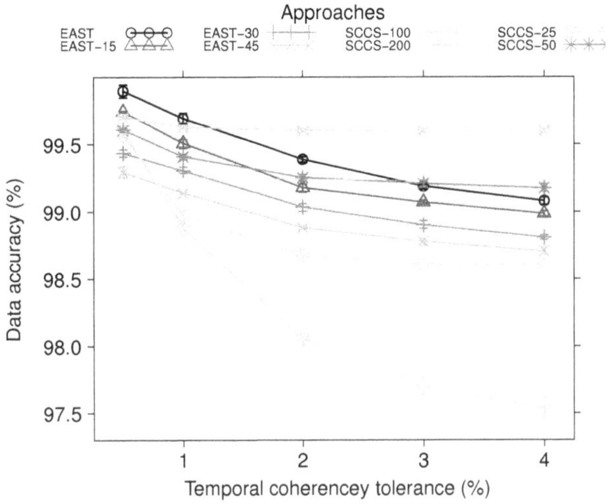

Figure 5: Data accuracy.

sumes $10J$, which indicates the advantages of using our spatial and temporal correlation techniques.

4.6 Delay notification at Each Reading

As mentioned before, the SCCS algorithm explores the use of a buffer to store the readings and, then, it is processed while exploiting temporal correlation. However, the use of a buffer has some disadvantages. The main drawback of this technique is the delay of the notification of each data. For instance, suppose the scenario of SCCS with a buffer size to store 100 readings and an interval of reading of 60 seconds. In this case, since the node will send its values only after 100 readings, it implies a delay of nearly 6000 seconds. Thus, the sink node will be notified about the event considering old readings, which is not acceptable in several WSN applications. One way to bypass this problem is to reduce the buffer size. However, as depicted in Figure 4, when the SCCS uses a small buffer size, the average energy consumption is greater than our EAST algorithm. For instance, the use of a buffer of 25 readings consumes on average 55% more energy than the in our proposed EAST algorithm. Albeit with a buffer size of 25 readings, SCCS algorithm reaches, on average, a data accuracy of 99.62% while the EAST algorithm remains on 99.14% (see Figure 5).

On the other hand, as the EAST algorithm considers the last notified reading to exploit the temporal correlation, as soon as the relative error threshold of the actual reading is greater than the tolerance of temporal coherency, the algorithm notifies the actual reading to the sink node. Because of this, every time the sensed value is beyond the error threshold the sink will be notified in real time, which is expected by most WSNs applications.

5. CONCLUSIONS AND FUTURE WORK

This work proposed the EAST algorithm, an algorithm for energy-aware data forwarding that takes full advantage of both spatial and temporal correlation mechanism to save energy while maintaining real-time, accurate data report toward the sink node.

In our proposed solution, nodes that detected the same event are grouped in correlated regions and a representative node is selected at each correlation region to gather and report data about the local state of the observed phenomenon. The temporal coherency tolerance and size of correlation region can be dynamically changed in order to achieve the required accuracy of the sensed information.

Simulation results show clearly that by using spatial and temporal correlation, the information about the event can be sensed with a high accuracy of more than 99.7% while saving the residual energy of the nodes in more than 14 times when compared to the accurate data collection strategy. The obtained results are very promising, but some advantages still need to be further exploited.

As future work we intend to consider the previous readings in the region of correlation to improve the accuracy of sensed information about the observed phenomenon. To achieve this goal, representative nodes can estimate the values of their region of correlation taking into account the time and recent readings. In addition, we intend to consider correlation regions of different sizes for an event.

6. REFERENCES

[1] Data collected of Amazon forest in Brazil, last visited in May, 2011.

[2] H. S. AbdelSalam and S. Olariu. A lightweight skeleton construction algorithm for self-organizing sensor networks. In *IEEE International Conference on Communications - ICC*, pages 1–5. IEEE.

[3] I. F. Akyildiz, W. Su, Y. Sankarasubramaniam, and E. Cyirci. Wireless sensor networks: A survey. *Computer Networks*, 38(4):393–422, March 2002.

[4] I. F. Akyildiz, M. C. Vuran, and Özgür B. Akan. On exploiting spatial and temporal correlation in wireless sensor networks. In *In Proceedings of WiOpt 2004: Modeling and Optimization in Mobile, Ad Hoc and Wireless Networks*, pages 71–80, 2004.

[5] A. Deligiannakis and Y. Kotidis. Geosensor networks. chapter Exploiting Spatio-temporal Correlations for Data Processing in Sensor Networks, pages 45–65. Springer-Verlag, Berlin, Heidelberg, 2008.

[6] A. Deligiannakis, Y. Kotidis, and N. Roussopoulos. Compressing historical information in sensor networks. In *Proceedings of the 2004 ACM SIGMOD international conference on Management of data*, SIGMOD '04, pages 527–538, New York, NY, USA, 2004. ACM.

[7] D. Estrin, L. Girod, G. Pottie, and M. Srivastava. Instrumenting the world with wireless sensor networks. In *ICASSP'01: Proceedings of the International Conference on Acoustics, Speech, and Signal Processing*, pages 2033–2036, Salt Lake City, Utah, June 2001.

[8] T. D. Le, N. D. Pham, and H. Choo. Towards a distributed clustering scheme based on spatial correlation in wsns. In *Wireless Communications and Mobile Computing Conference, 2008. IWCMC '08. International*, pages 529 –534, 2008.

[9] L. Liu, S. Member, and P. S. Yu. Asap: An adaptive sampling approach to data collection in sensor networks. *IEEE Transactions on Parallel and Distributed Systems*, 2007:1766 1783, 2007.

[10] J.-K. Min and C.-W. Chung. Edges: Efficient data gathering in sensor networks using temporal and spatial correlations. *J. Syst. Softw.*, 83:271–282, February 2010.

[11] S. Olariu, Q. Xu, and A. Zomaya. An energy-efficient self-organization protocol for wireless sensor networks. In *Intelligent Sensors, Sensor Networks and Information Processing Conference (ISSNIP)*, pages 55–60, Melbourne, Australia, December 2004. IEEE.

[12] N. D. Pham, T. D. Le, K. Park, and H. Choo. Enhance exploring temporal correlation for data collection in wsns. In *Proceedings of the IEEE International Conference on*

Research, Innovation and Vision for the Future, RIVF '08, pages 204–208. IEEE, 2008.

[13] N. D. Pham, T. D. Le, K. Park, and H. Choo. Sccs: Spatiotemporal clustering and compressing schemes for efficient data collection applications in wsns. *International Journal of Communication Systems*, 2010.

[14] N. D. Pham, T. D. Le, K. Park, and H. Choo. Sccs: Spatiotemporal clustering and compressing schemes for efficient data collection applications in wsns. *International Journal of Communication Systems*, 23:1311–1333, 2010.

[15] G. J. Pottie and W. J. Kaiser. Wireless integrated network sensors. *Communications of the ACM*, 43(5):51–58, May 2000.

[16] K. Romer and F. Mattern. The design space of wireless sensor networks. *IEEE Wireless Communications*, 11(6):54–61, December 2004.

[17] Sinalgo. Simulator for network algorithms, 2008. Distributed Computing Group - ETH-Zurich, last visited in October, 2008.

[18] M. C. Vuran, O. B. Akan, and I. F. Akyildiz. Spatio-temporal correlation: theory and applications for wireless sensor networks. *Computer Networks*, 45:245–259, June 2004.

[19] S. Yoon and C. Shahabi. Exploiting spatial correlation towards an energy efficient clustered aggregation technique (cag) [wireless sensor network applications]. In *Communications, 2005. ICC 2005. 2005 IEEE International Conference on*, volume 5, pages 3307 – 3313 Vol. 5, May 2005.

[20] J. Yuan and H. Chen. The optimized clustering technique based on spatial-correlation in wireless sensor networks. In *Information, Computing and Telecommunication, 2009. YC-ICT '09. IEEE Youth Conference on*, pages 411 –414, 2009.

[21] Y. Zhang, N. Meratnia, and P. Havinga. Outlier Detection Techniques for Wireless Sensor Networks: A Survey. *IEEE Communications Surveys & Tutorials*, 12(2):159–170, 2010.

An Adaptive and Composite Spatio-Temporal Data Compression Approach for Wireless Sensor Networks*

Azad Ali, Abdelmajid Khelil, Piotr Szczytowski and Neeraj Suri
Department of CS, TU Darmstadt, Germany
{azad, khelil, szczytowski, suri}@deeds.informatik.tu-darmstadt.de

ABSTRACT

Wireless Sensor Networks (WSN) are often deployed to sample the desired environmental attributes and deliver the acquired samples to the sink for processing, analysis or simulations as per the application needs. Many applications stipulate high granularity and data accuracy that results in high data volumes. Sensor nodes are battery powered and sending the requested large amount of data rapidly depletes their energy. Fortunately, the environmental attributes (e.g., temperature, pressure) often exhibit spatial and temporal correlations. Moreover, a large class of applications such as scientific measurement and forensics tolerate high latencies for sensor data collection. Accordingly, we develop a fully distributed adaptive technique for spatial and temporal in-network data compression with accuracy guarantees. We exploit the spatio-temporal correlation of sensor readings while benefiting from possible data delivery latency tolerance to further minimize the amount of data to be transported to the sink. Using real data, we demonstrate that our proposed scheme can provide significant communication/energy savings without sacrificing the accuracy of collected data. In our simulations, we achieved data compression of up to 95% on the raw data requiring around 5% of the original data to be transported to the sink.

Categories and Subject Descriptors

C.2.1 [**Communication Networks**]: Network Architecture and Design —*network communications, wireless communication*; I.6.5 [**Simulation and Modeling**]: Model Development—*Modeling methodologies*

General Terms

Algorithms, Design, Measurement, Performance

Keywords

spatial and temporal correlations, hierarchical clustering, energy efficiency, modeling, approximation, accuracy

*Research supported in part by HEC and DFG GRK 1362 (TUD GKMM).

1. THE PROBLEM AND THE APPROACH

In WSN deployments, sensor nodes are often distributed over the monitoring area for unattended environmental monitoring or for supervisory functions. The typical WSN functionality being (i) local event detection and reporting it to the sink, and (ii) continuous data collection by sampling the environment and sending the samples to the sink. In this paper, we deal with continuous data collection. Applications utilize continuously collected data for (a) real-time decision making, such as surveillance, or (b) delay-tolerant processing such as modeling, analysis [17] and inference [3]. In this work, we develop adaptive modeling algorithms that exploit the delay-tolerance of the data collection to maximize data compression. As examples, various scientific applications, such as, volcano monitoring [21] or eco-systems [17] [5], require detailed ambient data with high spatio-temporal sampling resolution for fine-granular understanding of the physical processes. For such applications, a WSN is essentially a spatio-temporal sampling system. High spatial resolution is reachable if we sample at a granular node level. Similarly, high temporal resolution is achieved by transmitting the samples repeatedly. Consequently, large data volumes need to be sent to the sink. Sending a message is a costly energy consuming operation and fetching more data from the network results into more message transmissions and higher energy consumption. Sensor nodes have limited computational capabilities and are powered by finite energy sources. Accordingly, highly accurate continuous data collection is a challenging problem in WSN given their energy, communication and computational constraints.

Fortunately, WSNs exhibit naturally high redundancy in spatial sampling due to the redundant sensor node deployment for connectivity and failure tolerance. Often, the sampled attributes are also temporally compressible due to temporal correlation [18]. Redundant deployment and temporal correlation allow a significant reduction of communication overhead through spatio-temporal compression. Various WSNs deployed for scientific monitoring [21], [5], [17] continuously harvest data for modeling, analysis and simulations. They generally tolerate a certain data collection latency. Hence, for such applications real-time data acquisition does not have precedence over the quality and quantity of the data. This *latency-tolerance* in reporting the data to the sink opens up fundamental design flexibility in data collection to significantly improve WSN energy efficiency. We propose a scheme to exploit the redundancies present in WSN along with the data collection delay-tolerance to maximize spatio-temporal data compression.

Our Contributions: A number of varied attempts such as aggregation, temporal compression [10] [8] [24] and spatial compression [6] [18] [23] have been proposed to reduce data volume to be transported to the sink. However, only a few composite spatio-temporal approaches, such as [18] [22] exist. These approaches exploit both spatial and temporal correlation in data, thus reaching higher data traffic reduction. However, the existing approaches are (a) partially centralized limiting local decision making for adaptability, (b) tailored to specific attribute dynamics providing no self-adaptability or (c) limited in their scope of exploiting either spatial or temporal correlation reducing compressibility.

Based on this background, we summarize our contributions as follows:

- A fully decentralized Adaptive and Composite Spatio-Temporal Compression (ASTC) technique.
- First exploitation of latency tolerance in reporting for maximal spatio-temporal compression.
- A novel model-based hierarchical clustering that adapts to attribute dynamics for spatio-temporal compression.
- Mechanisms to guarantee attribute value accuracy to be within defined bound and ensure the sample granularity to be on node level.

The paper is structured as follows. We present the system model and design requirements in Section 2. The proposed composite ASTC compression is presented in Section 3. Section 4 details the analysis and evaluation of ASTC with related work discussed in 5.

2. PRELIMINARIES

2.1 System Model

We consider a WSN composed of N static sensor nodes $S = \{s_1, s_2, \ldots, s_N\}$ and a static sink. Each sensor node periodically samples the environment attribute values every T time units. The nodes are battery powered with limited processing and storage capabilities. The WSN deployment consists of arbitrary node distributions with varying spatial node densities as per coverage requirements. We consider the clocks of nodes to be synchronized [7]. A reliable end-to-end data transport service, such as [15], to transport messages from sensor nodes to the sink and acknowledgement schemes to ensure inter-node message delivery are utilized.

Notation	Description
C_i	ith Cluster
S_i	ith Sensor node
V	Sample Value Queue
ϕ	Model Parameters
Φ_i	ith Model
Ψ	Model Cache
$m_{\#}$	Models in Ψ
ϵ	Error Threshold
$O_{\#max}$	Maximum Outliers
W_p	Prediction Window

Table 1: Notations

2.2 Design Requirements

Our objective is to maximize the spatio-temporal data compression with accuracy guarantees for continuously monitored phenomenon. Conceptually, our requirements are:

1) The sink should be able to reconstruct the signal (sample values) of each node from collected compressed data.
2) The accuracy of the reconstructed signal should be within the application-driven error bound.
3) The number of messages to report data to the sink should be minimal.
4) The algorithms should adapt to evolving attribute dynamics and be independent of network properties such as node distribution, topology and routing protocols.
5) The data compression models should be efficiently computable on resource-limited sensor nodes.

3. THE PROPOSED ADAPTIVE SPATIO-TEMPORAL COMPRESSION (ASTC)

We propose a decentralized adaptive composite compression technique while exploiting application latency tolerance. Before discussing ASTC in detail, we first describe the redundancies in sampled values and how we can conceptually exploit them for data compression.

3.1 Redundancies and Correlations in a WSN

WSN generally exhibit both spatial and temporal redundancies due to redundant node deployment and spatial and temporal correlations between the sampled values. It is also substantiated by our observation of real world data [11] where we noticed that the trends/patterns are similar for a large number of nodes in each other's physical neighborhood, as shown in Fig. 1 by gray shaded regions. This key observation suggests that we can construct a single model for the nodes (spread over a large area) that exhibit similarity in sampled values. The redundant deployment and dynamic phenomenon highlight two key aspects of correlations between nodes as (1) the nodes in closest neighborhood (1-hop distance) have the highest spatio-temporal correlation and the correlation within 1-hope cluster is fairly persistent, and (2) correlations between 1-hop clusters change over time due to changing phenomenon distribution over nodes. Hence, a certain group of 1-hop clusters may be correlated for a certain time duration and after some time, when the phenomenon has changed, another group of 1-hop clusters may be correlated, as shown in Fig. 1 at time $t + \Delta t$ the clusters belong to different regions. Hence, we propose to exploit the spatial redundancy in two stages, i.e., proactive 1-hop cluster formation that are highly correlated and then merging the 1-hop clusters to larger regions/clusters. Spatial compression in our approach differs from contemporary approaches, such as [8] [22], that form large monolithic clusters. ASTC exploits temporal redundancy by constructing simple models on 1-hop clusters.

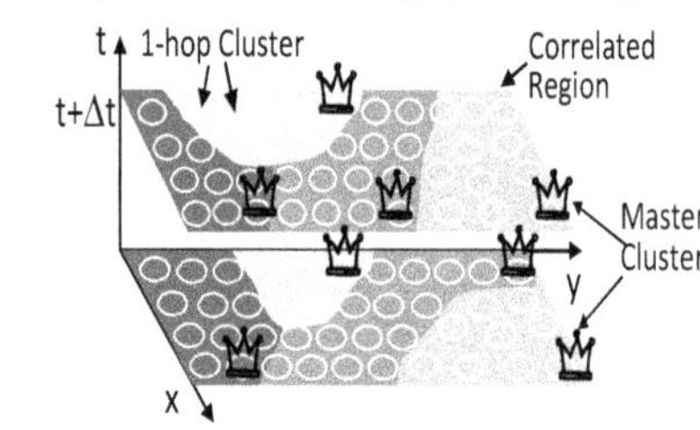

Figure 1: Phenomenon Distribution

The asymmetry in correlations between the nodes in close proximity and the nodes farther away makes the spatio-temporal compression a challenging problem. The nodes in close proximity of 1-hop distance generally exhibit persistently high correlations in sample values. However, the correlations between the 1-hop clusters are generally non-persistent and dynamic when observed over a long time window. The existing literature addresses this asymmetry by limiting the modeling to just 1-hop clusters [18], assuming the attribute dynamics to be constant [4], requiring assistance from the sink [27], [28] or by clustering based on the instantaneous values rather than models [29], [30], [8].

3.2 A Guide through the ASTC Approach

Given the nature of sensor redundancies and sensed attributes correlations in WSN, the proposed ASTC approach performs spatio-temporal compression in three stages:

- Stage 1: 1-hop cluster formation.
- Stage 2: Temporal modeling on clusters.
- Stage 3: Merging 1-hop clusters.

In Stage 1, correlated sensor nodes form small 1-hop clusters based on a short history of the attribute values to exploit strong local correlation. In Stage 2, we exploit the temporal correlations by constructing the models on a small number of clusters, referred to as *master clusters* (depicted in Fig. 1 bearing crown). Each constructed model is limited to the respective master clusters and approximates the sampled values of all the member sensor nodes of the master cluster.

In Stage 3, we propose schemes to utilize the models constructed on the master clusters to also approximate the sampled values of nodes in surrounding 1-hop clusters. The master cluster sends the model to its neighboring clusters. The cluster members fit the received model to their sampled values and accordingly either accept the model or reject the model. The clusters accepting the model merge to form a correlated region or larger clusters and further propagate the model to their neighboring clusters. Following this scheme, only a small set of the models constructed on master clusters can approximate the entire network both in space and time. The spatial compression in ASTC is a two level hierarchical clustering. The first and second hierarchy levels are formed during Stages 1 and 3 of ASTC. Stage 1 is executed only once, while Stages 2 and 3 are repeated to continuously model the sampled value and adapt to the changing dynamics. In the following sections, we detail the ASTC stages.

3.3 Stage 1: 1-hop Cluster Formation

In contrast to existing approaches, we initially construct small 1-hop clusters, instead of large monolithic clusters, to exploit local spatial correlations. The 1-hop clusters are later (Stage 3) merged to form larger clusters to model the dynamic correlations between the 1-hop clusters. Constructing and maintaining large monolithic clusters, in comparison to smaller 1-hop clusters, would incur high maintenance costs as we will elaborate in Section 3.5. As 1-hop cluster formation and inter-cluster routing is a well explored area in WSN [1], we only briefly describe them here. 1-hop clusters are formed based on the similarity of short history of the values that are transmitted by the nodes that candidate to be the cluster head. All nodes are initially eligible to be a cluster head candidate until some surrounding node announces its intent to be a cluster head as triggered by random timer. The candidate cluster head transmits a short history of the sampled values. The sensor nodes join the cluster for which the the difference between their sampled values and received sample values is within the accuracy requirement set by the application. The nodes unable to join a cluster, initiate their own cluster formation requests. Hence, a 1-hop cluster (C_i) is a set of one cluster head node (S_{CH}) and 'r' member nodes (S_i) such that they are 1-hop away from the cluster head, i.e., $C_i = \{S_{CH}, S_1, S2, \dots, S_r\}, \wedge \forall S_i \in C_i\ hopdist(S_{CH}, S_i) = 1$. Members of one cluster might be able to listen to more than one cluster heads but exclusively belong to one cluster, i.e., $C_i \cap C_j = \emptyset,\ i \neq j$. Each 1-hop cluster head discovers immediate 1-hop neighboring clusters (C_N) around it. In the discovery process they exchange the cluster Ids (same as cluster head Id) and hop distances of cluster head from the sink. They also fix the gateway nodes that are used to communicate between the two neighboring cluster heads. The 1-hop cluster formation process is performed only once. We do not explicitly refresh 1-hop clusters to maintain the correlations between the nodes within these clusters. Rather, we design Stages 2 and 3 of ASTC in an adaptive manner such that 1-hop cluster rearrangement takes place dynamically in response to the changes in the physical phenomenon, as detailed further in Section 3.5.5.

3.4 Stage 2: Temporal Modeling on Clusters

Next, we elaborate a temporal compression scheme to model the sampled values of all the member nodes within the 1-hop cluster. We emphasize that the model construction is carried out only by master clusters (selection of master clusters is explained in Section 3.5.1), while the rest of the clusters use the models constructed by the master clusters to approximate their sampled values.

DEFINITION 1. *$x(t)$ is an Autoregressive Moving Average Process $ARMA(p,q)$ of order (p,q) $p,q \in \mathbb{N}$ and $\forall$ t,*

$$x(t) = \Phi(\phi, \theta, x(t)) = \begin{array}{l}\phi_1 x(t-1) + \cdots + \phi_p x(t-p) + \\ \theta_1 z(t-1) + \cdots + \theta_q z(t-q)\end{array} \quad (1)$$

where ϕ and θ are model coefficients, $z(t)$ is white noise with mean zero and variance σ^2, denoted as $WN(0, \sigma^2)$ and $\Phi(\phi, \theta, x(t))$ (or in short Φ) denotes a model.

The use of computationally complex models to approximate generic patterns, typically observed in sampled attributes, is often not achievable on a sensor node due to (a) the limited nodes resources and (b) the unknown attribute dynamics. However, sampled values can be decomposed into short duration segments. Each segment can be dynamically modeled as $\mu + x(t)$ (where μ is the mean or linear component and $x(t)$ is the random component), allowing the nodes to adapt to unforeseen changes in the attribute dynamics without requiring prior knowledge of dynamics. Random component can be modeled by simple Autoregressive Moving Average (ARMA) models (Def. 1) or even simpler variants, i.e., Autoregressive (AR) and Moving Average (MA) models [3]. These models allow adaptability but are simple enough to be evaluated on common sensor nodes. We next describe how a master cluster constructs the model.

3.4.1 Master Cluster Model Construction

While all the sensor nodes sample the desired attribute, only the master clusters construct the models. The master cluster head maintains a queue V of n sample values to train the model. We set $q = 0$ and $p = 3$ in Eq. (1) to reduce the complexity of the model further. Eq. (1) reduces to

$$x(t) = \sum_{p=1}^{3} \phi_p x(t-p) \quad (2)$$

to construct a third order AR (AR3) model. The sampled queue is broken into mean and random components, i.e., $\mu + x(t)$. The AR3 model fitting error is minimized in mean-square error sense for the random component $(V - \mu)$ in the training queue as given by the following equation

$$\frac{\partial}{\partial \phi_p} E(t) = \frac{\partial}{\partial \phi_p} (\sum_{i=1}^{n} (x(t) - \sum_{p=1}^{3} \phi_p x(t-p)) = 0 \quad (3)$$

The predicted value based on model parameters is given by

$$\hat{V}(t) = \mu + \sum_{p=1}^{3} \phi_p \times (V(t-p) - \mu) \quad (4)$$

A model can predict only a certain number of sample values within accuracy bound referred to as *the prediction window*. Beyond the prediction window the model is rendered invalid. In order to adapt to the changing attribute dynamics a new model is constructed in the same fashion using the new sampled values in the sample queue.

Naturally, the sampled values may contain spurious and noisy values that result in modeling errors. However, accuracy of the model is important as it is used to approximate not only the master cluster member nodes sampled values but also the surrounding clusters. Hence, the cluster head involves the cluster members in model building by requesting the member nodes also to build the model. A few nodes are randomly selected to send their models to the cluster head. The cluster head broadcasts the models to the member nodes. Each member node fits each model to their sampled values and report back error for each model. The model with the minimum error is selected. The final ASTC scheme constructs a batch of models, referred to as *model cache*, instead of one model. This is detailed in Section 3.4.4.

3.4.2 Guaranteed Accuracy

We define two parameters to ensure the data reproduced from models on the sink to be within tolerable error bound, (1) an error upper bound (ϵ), and (2) the allowed maximum number of outliers $O_{\#max}$ that a node can tolerate to accept a model. ϵ is the maximum allowed error between the value predicted by the model and the actual sampled value.

$$\hat{x}(t) = \begin{cases} x(t), \text{if } |\hat{x}(t) - x(t)| > \epsilon; \\ \hat{x}(t), \text{otherwise} \end{cases} \quad (5)$$

If the model estimated value differs from the sampled value by ϵ, the sampled value is classified as an outlier (Eq. (5)). The cluster head gathers and reports the outliers to the sink to maintain accuracy within ϵ. Accordingly, $O_{\#max}$ is directly proportional to the desired accuracy level (ϵ).

3.4.3 Cluster Synchronization

We use the model constructed on the master cluster to also approximate the sampled values of surrounding cluster nodes. Though, the attribute dynamics are generally similar over a large area but change slightly from one cluster to another. Hence, when the model constructed on a master cluster is used to approximate the sample values in other clusters, the prediction window of the model can be different in different clusters. Different lengths of the prediction window results into approximating different number of samples with the same model in different clusters. This results into desynchronization of the clusters in terms of the approximated sample values. Once the clusters are desynchronized one cluster can not use the model of an other cluster (master cluster), as nodes in surrounding clusters could be approximating sample values ahead or behind the values for which the model was constructed. Hence, to avoid such desynchronization we restrict the prediction window length to a fixed value of W_p. Even if a model approximates the values within ϵ beyond W_p, the next values are considered to be in the next segment and are approximated by the next model. If the prediction window is shorter than W_p, the nodes compensate it with the outliers. If the model can not approximate the values even after allowing $O_{\#max}$ outliers, the model is rejected by the neighboring cluster node.

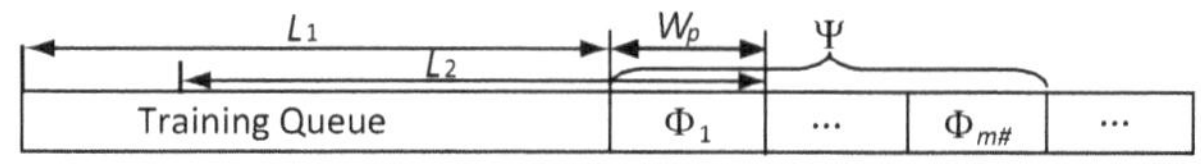

Figure 2: Model Caching

3.4.4 Model Caching

As per our design requirements (Section 2.2) we use simple models. Consequently, the prediction window is also expected to be short (in comparison to more complex models that may not be computable on sensor nodes) reducing the temporal compressibility. We assume tolerance in reporting latency. Hence, in order to increase the temporal compressibility, the master cluster heads construct a consecutive batch of models referred to as *model cache*. We denote the model cache by Ψ, which is a set of $m_{\#}$ models $\Psi = \{\Phi_1, \Phi_2, \ldots \Phi_{m\#}\}$. The model cache is constructed in the same fashion as a single model as explained in Section 3.4.1 but $m_{\#}$ models are constructed at the same time instead of just one as shown in Fig. 2. Each model in the model cache has a prediction window of W_p. Hence, to model $m_{\#}$ models the cluster head waits until $(m_{\#} - 1) \times W_p$ values are sampled in addition to the initial training queue. The first model Φ_1 is constructed from the sampled queue equal to the training length (Fig. 2, L_1). The next model Φ_2 is constructed from the next sampled attribute values (Fig. 2, L_2). It is repeated until the complete model cache Ψ has been constructed. To construct the next Ψ, additional $(m_{\#} - 1) \times W_p$ values are sampled and similarly modeled.

3.5 Stage 3: Merging 1-hop Clusters

In Stage 1, we exploited limited but strong spatial correlations to form 1-hop clusters. In Stage 2, we constructed the model caches to exploit the temporal correlation on the master clusters, achieving spatio-temporal compression. Now, we detail our scheme to extend the spatio-temporal compression beyond the master clusters. Various existing techniques use instantaneous value based monolithic clustering [1] [8] [22] to exploit the spatial correlation. They generally are very costly as they incur continuous maintenance cost. We on the other hand use models instead of instantaneous values while exploiting the spatial correlation. Large monolithic clusters based on models require tracking of all the sensor nodes that fit the entire model cache length. Moreover, the attribute dynamics may often change beyond the 1-hop cluster range. Hence, active tracking of the sensor nodes to form and maintain large clusters can require continuous message exchanges resulting in high energy consumption. Therefore, instead of forming large monolithic clusters we use a two level hierarchical clustering. The first level is 1-hop clusters (Stage 1), where generally nodes are highly correlated. The second hierarchical clustering level is constructed by the merging the 1-hop clusters that can be approximated by the same model cache to form larger clusters referred to as model cache cliques or simply cliques (Def. 2). The schemes described next (including the schemes in Stage 1 and 2) are executed on sensor nodes without assistance from the sink highlighting the fully decentralized nature of ASTC. Before we describe clique formation we discuss how the master clusters are selected and how a model cache constructed on the master cluster can be used to approximate the sampled values of the surrounding clusters nodes.

3.5.1 Master Cluster Head Selection

The master cluster merges with the surrounding clusters based on its model cache to form cliques or correlated regions. We are interested in determining the smallest number of correlated regions in the network or the smallest set of model caches that best approximates the largest network parts in order to transport maximum data in least number of messages to the sink. We formulate the problem of defining the minimum number of spatio-temporally correlated regions in terms of cliques in a network graph.

DEFINITION 2. *We define a clique as a network subgraph Q_C, a subset of the cluster heads, such that each cluster head in Q_C satisfies the model cache Ψ.*

Finding the master clusters that can construct the best model caches that form largest coverage clique is NP complete [2]. Hence, we utilize a heuristic based on the requirement of minimizing the message overhead. As the information flow direction is from the network to the sink, hence we propose the farthest cluster to be selected as the master cluster to initiate the clique formation and expand the clique in the direction of the sink by sending the model cache to the neighboring clusters. If we allow otherwise, i.e., the clusters nearer to the sink initiates the clique formation, the clique might expand in the direction away from the sink. We will have to spend additional messages to transport the accumulated information back to the sink. Hence, the heuristic biased to expand in the direction of the sink generally reduces the message cost to transport the information to the sink. Each cluster head knows its hop distance and the neighboring cluster heads hop distance from the sink as explained in Section 3.3. The cluster heads use this information to figure out the farthest cluster and hence the master cluster locally. In case of cluster heads having same number of hops, the cluster head with higher Id has the precedence. Next, we describe how a model cache constructed on master cluster (Section 3.4.4) can be used to approximate the sampled values of the nodes in the surrounding 1-hop clusters.

3.5.2 Model Cache Acceptance by 1-hop Clusters

The master cluster constructs the model cache as described in Section 3.4 and sends it to the neighboring clusters. A model cache received by a neighboring cluster must approximate the values of its member nodes within the defined error bounds (Section 3.4.2). To evaluate whether the received model cache approximates the sampled values, each member node predicts the values as:

$$\hat{V}(t) = \mu_{local} + \sum_{p=1}^{3} \phi_p \times (V(t-p) - \mu_{local}) \quad (6)$$

where V are the sample values of the node, μ_{local} is the sample mean of the local cluster head and model parameters are as receiving from the master cluster. Each cluster uses μ_{local} as we observed μ changes very quickly from one cluster to the other, while model parameters remain similar. The nodes predict the values using received model parameters using Eq. (6) and the error, $\hat{V}(t) - V(t)$, is calculated and the outliers are counted. Total count of the outliers for each model Φ in the model cache Ψ for the prediction length W_p should be less than the maximum number of allowed outliers ($O_{\#MAX}$) for model acceptance (Section 3.4.2). Hence, for each model in the model cache to be accepted, the node counts the outliers for each model as:

$$count\ |x_i(t) - \hat{x}_i(t) > \varepsilon| \le O_{\#max}, \hat{x_i} \in \Phi, j = 1...W_p \quad (7)$$

If all the models in the model cache satisfy the criteria in Eq. (7) the model cache is accepted by the node.

Algorithm 1 Model Cache Acceptance

```
1: msg.type='Ψfit';
2: msg.Ψ = Ψ;
3: timer.start();
4: broadcast(msg)
5: function msgReceive(msg)
6: add Ψ to Ψ_List;
7: function timer.expired()
8: Ψ.accepted=(Ψ_List total accepts> %r, r ∈ C)
9: broascast(Ψ.acceptance);
10: if Ψ.accepted then
11:     remove member that rejected Ψ & add the new requesting
        member
12: end if
```

In Alg. 1 and Alg. 2 we describe the model cache acceptance by the cluster head and model cache evaluation by the sensor nodes in a neighboring cluster. When a neighboring cluster head receives the model cache from the master cluster, it broadcasts the model cache (with locally calculated means) to the cluster members (Alg. 1, L. 1-4). The cluster head starts a timer to wait for the responses to be collected (Alg. 1, L. 3). Each cluster member prepares a vote, using Eq. (7), by counting outliers to either reject or accept the model cache (Alg. 2, L. 3). Each node must accept all the models in the model cache to accept the model cache. The cluster head is not notified of the acceptance of the model cache (positive vote), rather when the timer expires on the cluster head it assumes the model cache to be accepted by the particular node. The negative votes are, however, explicitly reported along with the outliers to be reported to the sink (Alg. 2, L. 5). Withholding report for positive votes saves messages because most of the nodes in the 1-hop cluster accept the model cache. The members nodes send outliers to the cluster head instead of sending them to the sink (Alg. 2, L. 6). The cluster heads reports the outliers efficiently to the sink by sending multiple outliers in one message. The cluster head counts the votes for model cache once the timer has expired (Alg. 1, L. 7-8). A model cache is accepted as a model cache for the cluster if it received acceptance from at least $k\%$ member nodes (Alg. 1, L. 8).

3.5.3 Self-Adaptive Clique Formation

So far, we have described the formation of the 1-hop clusters (Section 3.3), the model cache construction on the master clusters (Section 3.4.1) and how a model cache constructed on the master cluster can be used to approximate the values of sensors in the surrounding clusters (Section

Algorithm 2 Model Cache Evaluation

```
1: function msgReceive(msg)
2: if msg.type=='Ψfit' then
3:    vote=(solve Eq. (7) for Φ, Φ ∈ Ψ)?reject:accept
4:    if vote == reject then
5:       unicast(vote,outliers[],C_H);
6:    else if count(outliers > 0) then
7:       unicast(outliers[],C_H);
8:    end if
9: end if
```

3.5.2). Here, we describe the clique formation based on the model caches constructed on the master clusters. For clique formation the master cluster sends its model cache to its immediate neighboring 1-hop clusters. Each neighboring cluster evaluates the model cache for acceptance (Section 3.5.2). If the neighboring cluster head accepts the model cache it joins the clique and sends the model cache to its neighboring clusters and the process repeats. Hence, the clique formation is essentially controlled flooding of model cache over the cluster heads around the master cluster. The flooding stops once the model cache is rejected by the cluster heads. Next, we describe this scheme in detail.

Alg. 3 describes the expansion of the clique and the functionality of the clusters that constitute the clique and send the model caches to their neighboring clusters to further expand the clique. Initially only master cluster constitutes the clique as it initiates the clique. Alg. 4 describes the functionality of a neighboring cluster receiving the clique 'join' request to merge into the clique.

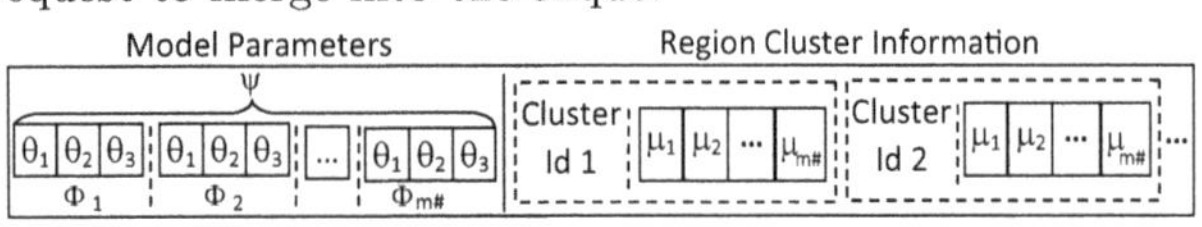

Figure 3: Join Message Payload Format

The master cluster initiates the clique formation by constructing the model cache (Section 3.4.4). Fig. 3 shows the format of the *'join' message* payload that is used by the cluster heads to propagate the list of constituting clusters and the model cache. The master cluster head adds its cluster Id, the model parameters and the sample mean values for each model in the model cache to the message payload. It sends the 'join' message to all the neighboring clusters (C_N) through the gateway sensor nodes (S_G) (Alg. 3, L. 1). Each neighboring cluster head executes the 1-hop cluster model acceptance algorithm (Alg. 4, L. 4). If the model cache is accepted, the cluster joins the clique by appending its cluster Id and the sample mean values to the message payload. The cluster joining the clique considers itself on the boundary of the clique and executes Alg. 3 to further propagate the 'join' message to its neighboring clusters, hence expand the clique (Alg. 4, L. 8-9). The neighboring clusters receiving 'join' message always notify the requesting cluster whether it is joining the clique (Alg. 4, L. 11). Each cluster head maintains a local record of its neighboring cluster with respect to their status regarding the clique. The receiving cluster heads update their local record regarding the neighboring cluster status from the clique 'join' messages during each message exchange (Alg. 4, L. 7). Once the responses from all the neighboring cluster C_N are received by the requesting clusters, it uses its local record to check if all C_N around it belong to the same clique. If all the surrounding clusters belong to the same clique, it implies that this cluster is not on the clique boundary. It notifies all cluster heads in C_N that it is not on boundary anymore through a "Not Boundary Cluster" message (NBC_{Msg}) and transfers it local list of clusters in the clique to the neighboring clusters (Alg. 3, L. 6-9). Each cluster head knows the status of each neighboring cluster whether it has joined the clique and whether it is on the clique boundary by continuously maintaining the local record and forwarding it to its neighbors.

Algorithm 3 Clique Expansion

```
1: 1HSend(Join_Msg, C_i ∧ ∀C_i∈C_N)
2: function receive(msg)
3: if msg.type='Resp_Msg' then
4:    clique.add(msg.accepted,msg.C_HR); R#++;
5:    if R#==C_N# then
6:       if ∀clique.acctance then
7:          NBC_Msg.clique=clique;
8:          1HSend(NBC_Msg, C_i ∧ ∀C_i ∈ C_N)
9:       end if
10:   end if
11: else if msg.type=='NBC_Msg' then
12:   clique.C_Hi.boundary = false ∧ C_Hi = msg.C_H;
13:   clique.remove((NBC_Msg.clique);
14: end if
```

The clique expansion continues until Alg. 2 returns positive join responses for the new clusters. A cluster head knows the clique expansion has stopped locally around it if it received the responses from all the neighboring clusters and one or more responses are negative. This cluster now stays on the clique boundary. The join message is flooded from the master cluster till the final boundary of the clique. Hence, each boundary cluster possesses the partial list of clusters forwarded during clique expansion by the clusters from the clique body. The border is traversed to accumulate the clusters list. The border traversal is initiated by a cluster possessing special token termed as the boundary traversal token (BTT). The master cluster initially assigns itself the BTT. The BTT possessing cluster forwards it to its neighboring cluster with least hops from sink if the BTT possessing cluster is no longer on the boundary.

Algorithm 4 Clique Joining by Neighboring Cluster

```
1: function receive(msg)
2: if msg.type='Join' then
3:    Resp_Msg.accept='false';
4:    Call Alg. 1 to evaluate msg.Ψ
5:    if msg.Ψ.accepted then
6:       Resp_Msg.accept='true';
7:       clique.add(msg.C_RequestCH);
8:       me.Boundary=true;
9:       Execute Alg. 3
10:   end if
11:   1HSend(Resp_Msg, C_HR)
12: end if
```

The clusters beyond the clique boundary follow the normal procedure of determining the farthest cluster to initiate and form a new clique. The cluster adjacent to the boundary of the clique exclude their neighbors that are already part of an other clique in determining the farthest cluster. Using our proposed scheme the nodes dynamically group to create a region that is spatially and temporally correlated for a given attribute for the modeled time duration. Hence, we have one model cache to be reported to the sink that represents the behavior of the spatio-temporally correlated region. During the clique formation each node in each cluster takes part in model cache acceptance, hence the data regenerated from the model caches on the sink are accurate to the level of single node. The particular discrepancies are corrected through the outliers sent by the clusters head for their respective members. We do not assume any particular distribution of the sensor nodes; therefore there can be multiple 1-hop clusters in different parts of the network that may have the maximum number of hops in the local neighborhood. We also set an upper bound on the time (T_{max}) that a cluster head can wait for the larger hop number cluster (clusters farther from sink) to initiate the clique formation. On the expiration of the wait time period the cluster head initiates clique formation. Hence, there can be multiple instances of clique formation executing in parallel. Two or more cliques formations execute mutually exclusively, i.e., a growing clique stops at the boundary of the other growing clique. We put such a restriction because the two cliques are growing based on two different model caches.

3.5.4 ASTC Iterations and Dynamic Adaptability

The second level of cluster hierarchy or the clique is a temporary entity to determine the correlated region based on the model cache and is not strictly a larger cluster in conventional sense. It does not have a cluster head and we do not maintain it. Maintaining such large cluster body will be very costly, because it is built using the model cache, which requires the clusters to agree on a sequence of values over time. Hence, cliques are reconstructed rather than being maintained. We show in Section 4.1 that the cost to construct a clique is very low. The reconstruction of the cliques allows to continuously adapt to the changing dynamics. The 1-hop cluster that were part of one clique may be part of another clique or even make their own clique in the next iteration of clique formation depending on the changes in the phenomenon. After reporting the model caches the cluster heads wait for enough data to be collected to construct the next model cache as explained in Section 3.4.4. The process of the model cache construction and the clique formation is repeated and the sink receives accurate continuous data.

3.5.5 1-hop Cluster Dynamic Rearrangement

The 1-hop cluster members usually stay correlated. However, due to changes in physical phenomenon, the correlations even at the 1-hop cluster level change and the clusters require rearrangement. We do not implement an explicit mechanism to detect such changes as it would require further message overhead. We determine such changes in the node correlations in Section 3.5.2 when $k\%$ of nodes agree with the model cache and the model cache is accepted by the cluster head due to majority vote. The cluster head evicts the remaining $r - k\%$ nodes that send the negative votes (reject the model cache). Using overlapping nature of the clusters the evicted sensor nodes wait and snoop the model caches sent by rest of the surrounding clusters. The evicted sensor node checks the model cache as if it were part of this cluster and updates the cluster head accordingly. If the model cache is accepted by the cluster head, the evicted sensor node joins the cluster. If the evicted node can not join any surrounding cluster it forms a new cluster and becomes the cluster head. Such a rare condition arises due to a new phenomenon developing in this region. The sensor nodes in this region (around the evicted sensor node) will start leaving their current cluster (as the evicted sensor node did) and will eventually form a cluster with the evicted sensor node. Consequently, the sensor nodes rearrange the 1-hop cluster to self-adapt to the environmental changes.

4. EXPERIMENTS AND DISCUSSION

We now evaluate the proposed spatio-temporal compression technique both analytically and via simulations.

4.1 Complexity and Compressibility Analysis

Our main objective is to perform maximal spatio-temporal compression with minimal number of messages without sacrificing data accuracy. The message overhead of our proposed scheme is very low. The maximum number of messages that a 1-hop cluster must exchange to process a model cache can be given by $Msgs = 2 \times \Psi_{Msg} + \sum_{S_k \in C_j, S_{k_{O\#}} > 0} Out_{Msg} + Resp_{Msg} + NBC_{Msg}$. The first message is the broadcast message (Ψ_{Msg}) to the member nodes to evaluate the model. The member nodes respond only if there is an outlier to be reported (Out_{Msg}). The cluster head reports to the requesting cluster head (through the gateway nodes) whether it accepted the model cache ($Resp_{Msg}$). If the model cache is accepted the cluster head broadcasts the model cache message again (Ψ_{Msg}) to the neighboring clusters (through gateway nodes). If the cluster is no longer on the boundary, it broadcasts the 'not-boundary-cluster message' (NBC_{Msg}) to the neighboring cluster through the gateway nodes. Hence, in the worst case the maximum number of messages a cluster head has to transmit are four. For a member node that is not the gateway node it is only one (Out_{Msg}, if there are outliers). For a member node that is also a gateway node it transmits four messages ($Out_{Msg} + \Psi_{Msg} + Resp_{Msg} + NBC_{Msg}$). The master cluster has an additional round of one message to construct a model cache.

The compression achieved within 1-hop cluster can be calculated as $Total_{bytes} = S_{\# \in C_i} \times W_p \times m_{\#}$. If there are $S_{\# \in C_i} = 7$ sensor nodes in a 1-hop cluster, the prediction window is $W_p = 15$, the number of models in Ψ are $m_{\#} = 3$, then one model cache is able to compress 315 sample values within one 1-hop cluster. For each 1-hop cluster joining the clique we achieve similar compression rate. Practically achieved compression is slightly lower due to a few outlier values (which is also bounded by the maximum number of allowed outliers $O_{\#max}$).

The computation overhead to achieve the compression is minimal. Only the master clusters construct the model cache, hence the master cluster heads have to compute matrices A and B for 3 unknowns of a linear system $AX = B$. The other nodes (either in the master cluster or the clusters joining the clique) need to solve a third order linear equation (AR3) and compare it with the sampled values.

The collection of data to construct Ψ introduces latency, which can be calculated as $Latency = m_{\#} \times W_p \times T$, where T is sampling period. Hence, Ψ and W_p are selected while keeping in consideration the maximum desired latency.

4.2 Experimental Performance Evaluation

ASTC Simulation Settings: We use a publicly available real-world data set [11], containing traces for temperature, humidity, light and voltage. We selected the temperature readings for four continuous days because, as compared to the other attributes, the temperature readings are not monotonic. The temperature continuously changes during day and night and hence provides a good opportunity to test the adaptability to the changing dynamics. The network simulations were performed in TOSSIM. The signal reconstruction at the sink is conducted using MATLAB. The main overhead of our scheme is to transport the *model caches* and the *outliers values*. Hence, we measure the efficiency by counting the total number of messages required to transport the model caches and the outliers. The network consists of 51 nodes with a data set of more than half-million data samples with the sampling period of 31 sec. We selected the AR3 model, set the training history length to 100, $O_{\#max} = 5$, $W_p = 20$, $m_{\#} = 3$ and $\varepsilon = 0.2$ (less than 1% mean error). In the following evaluation studies, one parameter is varied at a time and the others are fixed to the default values. To cover the complete spectrum for spatial and temporal compressions, we compare ASTC to its two variants. The first variant is the temporally limited case when the model cache is limited to one model only to compare to the temporally limited schemes. The second variant is the spatially limited schemes when spatial compression is limited to 1-hop cluster only. We also compare ASTC to PAQ [18] for PAQ's ability to dynamically adapt to unknown attribute dynamics. In order to compare to the schemes using no models we simulated for instantaneous value based clustering and further compare to continuous raw data collection to establish a baseline to measure the compressibility.

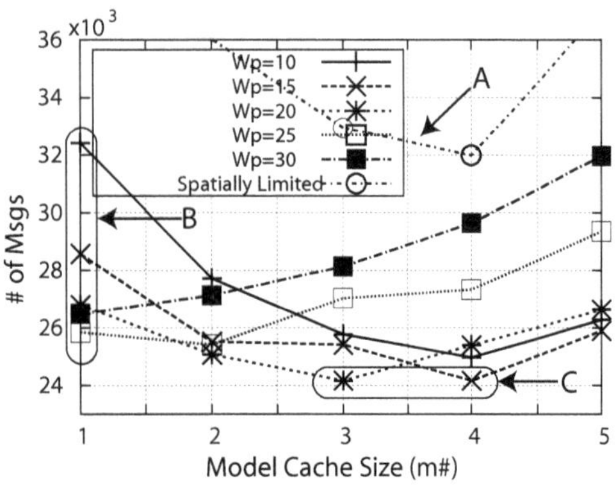

Figure 4: Message Cost for Varying W_p and $m_{\#}$

Spatio-Temporal Compression in Terms of ASTC Parameters: In order to elaborate the results we explain the spatio-temporal relations, as introduced in Section 3.1, in terms of parameters introduced in our scheme. The temporal compressibility can be controlled by changing both model cache size ($m_{\#}$) and prediction window (W_p). With each increment in $m_{\#}$ we increase one model in the model cache that compresses W_p additional samples as explained in Section 3.4.4. By increasing W_p we approximate more samples with the same model, which may not be efficient if W_p is too long and hence could result in more outliers. Hence, increasing W_p in comparison to $m_{\#}$ is not as efficient because for $m_{\#}$ we construct an additional model that approximates the new dynamics. The temporal compression for the given parameters is given by $m_{\#} \times W_p$ as shown in Section 3.4.4. The temporal compression scope influences the achievable spatial compression (hence the clique size). The short temporal compression implies the clusters have to agree for a short duration of time, which is highly likely. However, we will require more model caches to compress the complete sampled history. Long temporal compression scope can compress the whole history in fewer iterations but requires clusters to agree for a long duration of time, which results in smaller cliques. Hence, we have to find the compromise between the temporal and spatial compression to optimize the message cost. Our simulations provide a guide to choose the parameters for the optimal results.

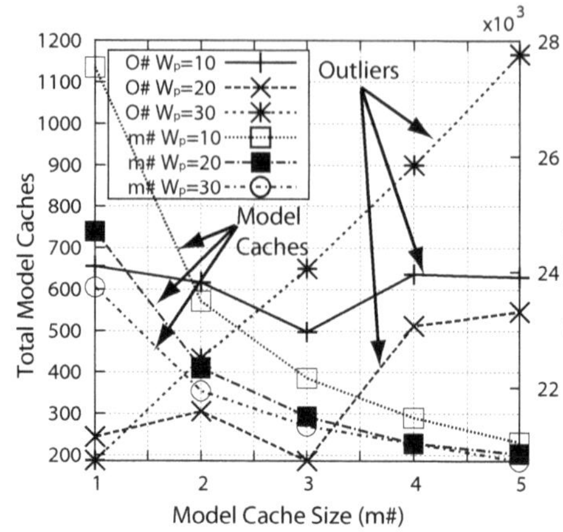

Figure 5: Total Number of Outliers and Model Caches

ASTC Efficiency and Compressibility: Fig. 4 depicts the message cost for $m_{\#}$ varied from 1 to 5 models per model cache and W_p varied from 10 to 30 approximated samples per model. Fig. 5 helps us understand the reason for changes in message cost as it depicts the generated model caches and the outliers with varying $m_{\#}$ and W_p. In Fig. 4 we initially observe a clear drop in the number of messages required to transport the data as we increase the models in $m_{\#}$, because each additional model compresses W_p more sample values. Fig. 5 confirms this by showing a clear drop in the model caches with increasing $m_{\#}$ and initial drop in the outliers for the short prediction window.

The number of messages, after the initial drop, start to increase in Fig. 4 for the shorter prediction window ($W_p = 10$ *to* 20) and model cache with the large number of models ($m_{\#} = 4$ *to* 5). This is due to (a) the limited size of 28 bytes for the message payload in TOSSIM, as we increase models we require more parameters and more messages to transmit all the parameters as shown in Section 3.5.3 (b) larger $m_{\#}$ and W_p increase the temporal compression making it difficult to approximate the cluster farther away from master cluster. Hence, it results in more outliers and smaller cliques. This is also evident for the long prediction window (30), where the message cost starts increasing very early on. Fig. 5 confirms the increase in the message cost by showing a clear increase in outliers for the large W_p and $m_{\#}$.

In Fig. 4 (Label A) we also illustrate the spatially limited case when the spatial compression is limited to only 1-hop cluster. The message cost is so high that we had to reduce the scale to show ASTC optimized results. PAQ [18] costs also lie on this line when $m_{\#} = 1$, however it is still high enough (47,215 messages) that it can not be displayed in the graph. The simulation results for instantaneous value based clustering required 109,862 messages. The continuous raw data collection resulted in 39,33,345 messages.

Fig. 4 (Label B) can also be used to study the temporally-limited case when we limit $m_{\#} = 1$ and allow clique formation. We can again see the same trend, i.e., the message cost decreases with increasing W_p as more clusters accept the model cache and increase only when the W_p gets too long (30) as it results in increased outliers and decreased spatial compression scope, resulting in smaller cliques.

ASTC has low overhead of outliers and low message cost (high compressibility). For example, in Fig. 5 for $m_{\#} = 3$ and $W_{max} = 20$ outliers on the cluster heads amount to roughly 5% of the raw data. Taking continuous raw data collection as a baseline to measure the message cost, ASTC requires around 0.6% of the messages required to transport the complete sampled history.

Parameter Selection: From the results in Fig. 4 and Fig. 5 we learn that the upper bound for $m_{\#}$ is dictated by the implementation platform as more models may require more messages due to limited payload size (in TOSSIM the maximum payload size is 28). However, W_p depends on the phenomenon. In our simulations $m_{\#}$ of size 4 and 5 and W_p 15 and 20 yielded the best results (Fig. 4, Label C).

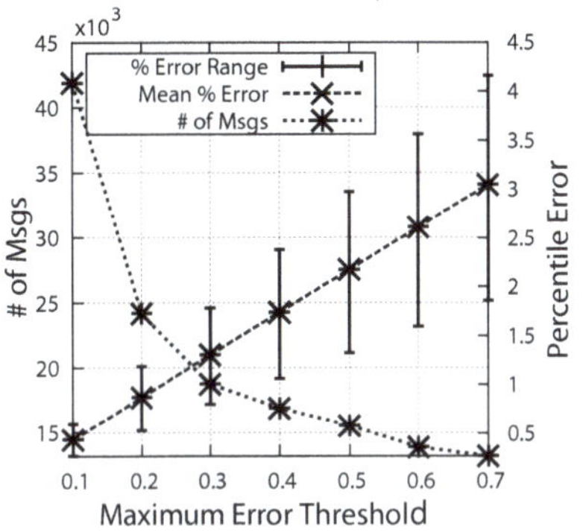

Figure 6: Total Communication Cost with Varying ε

ASTC Accuracy: Fig. 6 shows the effect of changing the error tolerance (ε) and maximum, minimum and average error induced after reconstruction. It further asserts that ASTC fulfills the accuracy requirements imposed by the user. Increasing the error threshold naturally results in less number of outliers and more spatio-temporal compressibility, hence the messages count decreases. Setting ϵ to 0.2 generates maximum 1.18%, minimum 0.53% and average error of 0.87%.

For maximum allowed outliers ($O_{\#max}$) neither large nor small values are beneficial as shown in Fig. 7. Too small $O_{\#max}$ requires strict matching of model to the sampled value, hence spatial compression is limited and smaller cliques are formed limited spatio-temporal compressibility. Allowing $O_{\#max}$ to be too large generates too many outliers requiring more messages to transport the outliers. In our simulations $O_{\#max} = 5$ yielded the best results.

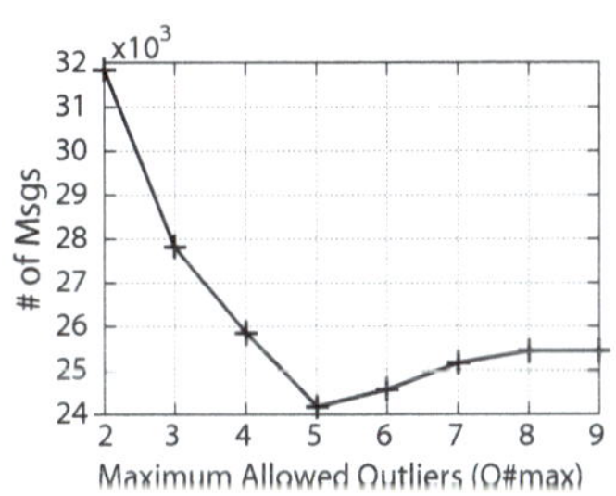

Figure 7: Total Communication Cost with Varying $O_{\#max}$

ASTC Latency: Given the sampling time of 31 sec the latency of constructing model cache in each ASTC round ranged from 310 sec for $m_{\#} = 1$ and $W_p = 10$, to the medium case of 1860 sec for $m_{\#} = 3$ and $W_p = 20$, to the extreme case of 4650 sec for $m_{\#} = 5$ and $W_p = 30$.

5. RELATED WORK

This paper focused on continuous data collection rather than event-based WSN design. Our primary objective is to spatio-temporally compress the data for continuous data collection with sensor-node granularity while exploiting delay-tolerance in data collection. There is wide range of research work in WSN for data compression, simple aggregation, suppression, filtering such as TinyDB [12] and Cougar [9], to name a few. However, there is very limited work in the area of spatio-temporal compression. We summarize some of the important research work regarding spatial and temporal compression and discuss in detail the existing spatio-temporal techniques.

Spatial Compression: The objective of spatial sampling is to collect interested attribute snapshots. The key idea behind spatial compression is to constrain neighboring sensor nodes with similar senor readings from transmitting redundant data. [13] [10] [16] [24] are a few of many spatial compression techniques relying on compressive sensing, spatially correlated models, filtering and aggregation.

Temporal Compression: The driver of temporal compression is to exploit temporal correlation present in the attribute values. The key idea in these approaches is to let every sensor node create a prediction model for its sensor readings and send the model to the sink. The sensor node should send an updated model only if the model is not valid due to changes in the signal dynamics. The approaches in [14] and [23] construct models based on Markov-chains and time series, respectively.

Spatio-Temporal Compression There is very limited research in the area of spatio-temporal compression. We have attempted to give an exhaustive account of the state of the art in the field. In [19], the authors developed a theoretical framework to model the spatial and temporal correlations in WSN. This framework enables the development of efficient medium access and reliable event transport in WSN, which exploits these advantageous intrinsic features of the WSN paradigm. This work does not focus on continuous data collection contrary to our target applications. [29], [30] and [8] form large monolithic clusters based on instantaneous values or aggregates of the values. Hence, they continuously collect data from the member nodes to check for consistency and maintain these clusters by breaking up and merging them. We on the other hand use hierarchical clustering, i.e., (1) 1-hop cluster requiring rare maintenance and (2) clique that are not maintained rather reconstructed at a minimal message cost. Additionally, cliques (clusters) in our work are formed based on models rather than instantaneous values. [18] proposes realtime data collection through continuous modeling. It is closest to our proposed scheme in terms of adaptability to changing dynamics. Hence, we have included it for comparison in simulations. However, it is limited both in spatial and temporal compressibility. The focus of our proposed scheme, in contrast to [18], are the applications that are delay-tolerant in data collection. [25] uses Kalman filter for modeling within 1-hop clusters, which incurs heavy computational cost. [26] proposes to reduce energy consumption through hierarchical aggregation. Both [25] and [26] require the node location information, which could be prohibitively expensive. Our proposed scheme works independent of the location information and uses simple models that are easily computable on a sensor node. [27] and [28] use centralized heuristics for cluster formation and nodes correlation determination. Our proposed scheme is completely distributed and does not require any information from the sink other than a few parameters initially required to setup the proposed distributed scheme as per the user requirements. [22] has two modes of operation, i.e., interactive mode is limited to spatial compression only and streaming mode performs both spatial and temporal compression. Streaming mode of [22], [20] and [4] construct probability density function (pdf) for modeling the attribute. These schemes construct pdf by collecting samples values of sensors on the sink once and are not updated afterwards. Hence, correctness of such schemes can not be guaranteed as the pdf constructed in a certain time window can not be guaranteed to be valid after attribute dynamics change limiting the use of such schemes in long term continuous monitoring. They also require expensive long training time (e.g., 15 days for [20]). We do not assume the system dynamics to be constant and construct the model dynamically based on the changes in the dynamics. Therefore, our technique can not only track the evolving attribute dynamic but also adapts in case of unexpected or unpredictable changes in the dynamics.

The existing composite approaches require the signal and its statistical properties to be constant, require location information, are partially centralized or use instantaneous values instead of models for clustering. The common factor among all the schemes is that they target real-time or immediate data collection. None of existing works exploits the latency-tolerance of many applications, thus, loose efficiency potentials. In contrast, our approach is adaptive, does not require location information, is fully decentralized, uses simple easily computable models and exploits the delay tolerance to maximize the data collection.

6. CONCLUSION AND FUTURE WORK

We have developed ASTC, a fully distributed spatio temporal data compression technique for accurate sensor data collection in WSN. ASTC dynamically adapts to approximate the monitored attribute both in space and time. The use of 'simple models batches' instead of complex monolithic models was the key idea to design a technique that adapts to the dynamic changes of the attribute. Simple model batches combined with the first exploitation of data collection delay tolerated by the application is used to maximize compression ratio and significantly reduce the message cost.

In our simulations, we were able to reconstruct the signal from spatio-temporally compressed data on the sink with granularity of a single node, mean error less than 1% and data compression to 95% of the raw data samples. In our future work, we aim to (a) tolerate node crashes and spurious data and (b) extend to multi-variate compression.

7. REFERENCES

[1] A. A. Abbasi and M. Younis. A survey on clustering algrithms for WSN. *Comp. Comm.*, 30(14-15), 2007.

[2] J. Abello et al. On maximum clique problems in very large graphs. *External Memory Algorithms*, American Mathematical Society 1999.

[3] A. Ali et al. MPM: Map based predictive monitoring for sensor networks. In *AUTONOMICS*, 2009.

[4] D. Chu et al. Approximate data collection in sensor networks using probabilistic models. In *ICDE*, 2006.

[5] I. A. Dario et al. Underwater acoustic sensor networks: Research challenges. *Ad Hoc Netw.*, 3(3), 2005.

[6] A. Deshpande et al. Model-driven data acquisition in sensor networks. In *VLDB*, 2004.

[7] Faizulkhakov, Ya. R. Time synchronization methods for wireless sensor networks: A survey. *Program. Comput. Softw.*, 33(4), 2004.

[8] B. Gedik et al. ASAP: An adaptive sampling approach to data collection in sensor networks. *IEEE TPDS*, 18(12), 2007.

[9] J. Gehrke and S. Madden. Query processing in sensor networks. *IEEE Pervasive Computing*, 3(1), 2004.

[10] A. Jindal and K. Psounis. Modeling spatially correlated data in sensor networks. *ACM TOSN*, 2(4), 2006.

[11] S. Madden. Intel lab data, 2003, http://db.cscail.mit.edu/labdata/labdata.html

[12] S. R. Madden et al. TinyDB: an acquisitional query processing system for sensor networks. *ACM TODS*, 30(1), 2005.

[13] M. Mahmudimanesh et al. Reordering for better compressibility: Efficient spatial sampling in wireless sensor networks. In *SUTC*, 2010.

[14] R. A. F. Mini et al. Prediction-based energy map for wireless sensor networks. *Ad Hoc Netw.*, 3(2), 2005.

[15] F. K. Shaikh et al. Generic information transport for wireless sensor networks. In *SUTC*, 2010.

[16] I. Solis and K. Obraczka. Isolines: efficient spatio-temporal data aggregation in senor networks. *Wire. Comm. and Mob. Comp.*, 9(3), 2009.

[17] Gilman Tolle et al. Macroscope in the Redwoods, In *SenSys*, 2005.

[18] D. Tulone and S. Madden. PAQ: time series forecasting for approximate query answering in sensor networks. In *EWSN*, 2006.

[19] M. C. Vuran et al. Spatio-temporal correlation: Theory and applications for wireless sensor networks. *Comp. Net. Journal*, 45(3), 2004.

[20] L. Wang and A. Deshpande. Predictive modeling based data collection in WSN. In *EWSN*, 2008.

[21] G. Werner-Allen et al. Monitoring volcanic eruptions with a wireless sensor networks. In *EWSN*, 2005.

[22] S. Yoon and C. Shahabi. The clustered aggregation (cag) technique leveraging spatial and temporal correlations in wireless sensor networks. *ACM TOSN*, 3(1), 2007.

[23] C. Zhang et al. Model-aided data collecting for wireless sensor networks. In *HPCC*, 2006.

[24] J. Zhao et al. Residual energy scan for monitoring sensor networks. In *WCNC*, 2002.

[25] J. Min and C. Chung EDGES: Efficient data gathering in sensor networks using temporal and spatial correlations. *JSS*, 83(2), 2010.

[26] S. J. Baek et al. Minimizing energy consumption in large-scale sensor networks through distributed data compression and hierarchical aggregation. *IEEE JSAC*, 22(6), 2004.

[27] C. Liu et al. An energy efficient data collection framework for sensor networks by exploiting spatiotemporal correlation. *IEEE TPDS*, 18(7), 2007.

[28] H. Gupta et al. Efficient gathering of correlated data in sensor networks. In *MobiHoc*, 2005.

[29] H. Jiang et al. Prediction or not? An energy-efficient framework for clustering-based data collection in wireless sensor networks. *IEEE TPDS*, 22(6), 2011.

[30] N. D. Pham et al. SCCS: Spatiotemporal clustering and compressing schemes for efficient data collection applications in WSNs. *Int. J. Comm. Sys.*, 23(11), 2010.

Monitoring Quality Optimization in Wireless Sensor Networks with a Mobile Sink

Xu Xu and Weifa Liang
Research School of Computer Science, The Australian National University
Canberra, ACT, Australia
grace.xu@anu.edu.au, wliang@cs.anu.edu.au

ABSTRACT

The exploitation of sink mobility has been proven to improve various network performance significantly, including network lifetime, data delivery latency, network connectivity, and so on. In this paper we consider a novel network model consisting of sensors, gateways, and a mobile sink, which can be applied to many realistic applications such as city traffic monitoring, patients monitoring, and forest fire surveillance. We assume that there is a roadmap in the monitoring region for the mobile sink to access, and gateways are located on roads. The mobile sink moves at a constant speed along a closed tour of roads to collect data from gateways. The travelling distance of the mobile sink per tour is bounded by a given value. Due to the limited communication time between the sink and each gateway, sometimes it is not possible for the mobile sink to collect the data generated from all sensors, consequently causing monitoring quality loss. In this paper, we study the problem by formulating it to find a closed tour for the mobile sink, such that the monitoring quality loss is minimized, subject to the tour length constraint. Since the problem is NP-hard, we propose a heuristic for it. Also, we design an energy-efficient routing protocol for data collection that balances the energy consumption among sensors. We finally conduct extensive experiments by simulation to evaluate the performance of the proposed schemes. The experiment results show the effectiveness of the proposed heuristic to optimize the network performance.

1. INTRODUCTION

Mobile sinks (MSs) have been exploited in wireless sensor networks (WSNs) to improve the network performance. The use of MSs effectively shifts the burden of data transmission from individual sensors to themselves, thus balancing the energy consumption among sensors and thereby prolonging the network lifetime [13]. Also, they enable more reliable transmission between data sources and destinations [17], improve the network throughput [18], and reduce the data delivery latency effectively [8].

MSWiM'11, October 31–November 4, 2011, Miami, Florida, USA.

In reality, however, there are several practical constraints imposed on MSs. First, mobile sinks are usually devices installed on robots, shuttles, or airplanes, which are powered by petrol or electricity and supposed to be recharged/refueled over time. Once they start moving in the network, the maximum travelling distance per tour is subject to their power sources. Thus, the length of each MS is constrained. Second, vehicles with MSs cannot travel anywhere in a monitoring area without restriction [1]. Obstacles such as ponds, bushes, buildings, and rocks should be avoided. Moreover, vehicles like city commuters can only stop at stations instead of any spot on the road. Fig. 1 is an example of a WSN deployed in a city while the mobile sink is carried on a bus starting from S. The roadmap for the bus is comprised of all roads in Fig. 1. Before the bus runs out of petrol, it is supposed to return to S. Third, the speed of a MS is usually bounded by multiple factors, such as, traffic conditions, engine limitations, and speed restrictions on highways or in suburbs. Other constraints imposed a MS include the maximum distance between its two consecutive sojourn locations, the minimum time for the MS to stay at a sojourn location [5], and the candidate sojourn locations for the MS [3].

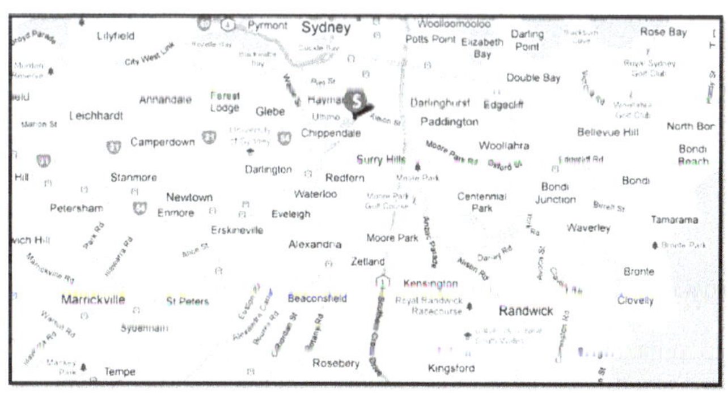

Figure 1: A roadmap for the mobile sink

In this paper, we consider a MS moving in a given roadmap M for data collection while it must return to the depot for data uploading and fuel refilling after each tour, and the tour length is bounded by L. There are some applications falling into this category. One example is a sensor network deployed in a city where the shuttle equipped with the sink has a fixed route and returns to the depot at the end of each tour. These two constraints restrict the movement of MSs.

There are some *gateways* located on each road in the roadmap, playing the roles of data relay, storing data generated from sensors temporally and transmitting data directly to the MS later. Gateways are assumed to be rechargeable and have enough buffer capacities to store data before transmitting to the MS. Data collection from a gateway only happens when the MS moves within the transmission range of the gateway. This mode of data transmission has several advantages. First, the MS does not have to visit every sensor deployed in the network, reducing the data delivery latency, especially for large scale networks. Second, having the MS visit partial nodes balances the energy consumption among sensors and the trade-off between energy conservation and delivery delay can be achieved. Third, collecting data from gateways, rather than a large number of sensors, decreases the overhead of constructing routing structures.

In this paper, we take into account the time spent on data uploading from gateways to the MS. Sometimes the MS is not able to collect data generated from all sensors. For example, if the monitoring region is large and the length constraint is relatively small, the MS can only traverse a portion of all roads in the map and visit some of the gateways. In that case, monitoring quality loss is unavoidable. Data generated from some sensors, referred to as *active nodes*, can be collected while those from others, referred to as *inactive nodes*, have to be discarded. We assume sensors densely deployed in the network, thus, the sensing data is highly spatially and temporally correlated, especially among neighboring sensors [19, 6]. The sensing data of active nodes can be used to estimate that of inactive nodes and the estimation error is used to measure the monitoring quality loss. Choosing different roads as a tour for the MS, the monitoring quality loss varies. Also, having the set of active nodes and their corresponding gateways, an appropriate routing structure is needed to enable the successful and efficient data collection. The objective of this paper is to *find a closed tour on the given roadmap for the MS starting from the depot each tour and design a routing protocol, such that the MS is able to efficiently collect data with minimized monitoring quality loss, and return to the depot at the end of each tour subject to the length constraint.*

Our main contributions are as follows. We first define the monitoring quality and formulate the quality optimization problem as a joint optimization problem. Due to its NP-hardness, we then propose a strategy to find a closed tour for the MS meeting the length constraint. We also devise an energy-efficient routing protocol for network lifetime prolongation. We finally conduct extensive experiments to demonstrate the effectiveness of the proposed algorithms.

The rest of the paper is organized as follows. Section 2 discusses the related work. Section 3 proposes the problem formulation. Section 4 and Section 5 present a novel two-stage heuristic algorithm consisting of the length-constrained tour finding and the energy-efficient data routing. Section 6 evaluates the performance of the proposed algorithm through simulations. Section 7 concludes the paper.

2. RELATED WORK

Controlled sink mobility has been extensively studied to mitigate uneven energy consumption among sensors in WSNs [1, 2, 3, 4, 13]. The basic idea is to shift the relay workload from the individual sensors to the mobile sink(s). The mobile sink(s) can either (i) visit every deployed sensor and gathers its generated data via one-hop transmission [4]; (ii) visit some nodes, referred to as sojourn locations in [13], and collect data generated from all sensors at each sojourn location; or (iii) visit gateways, referred to as *rendezvous points* in [2], and collect data generated from partial sensors at each gateway. The first mode of data collection is the most energy-efficient but it may cause long data delivery latency. In the second mode, the data delivery delay is minimized at the cost of unbalance in energy consumption among sensors. The traditional *sink neighborhood problem* [17] is not solved and still compromises the network performance. The third mode is the hybrid of the previous two. The trade-off between the data latency and the energy consumption can be achieved. Our paper falls into this category.

Previous works have discussed sink mobility based on a hybrid mode of data collection. Xing *et al.* [2] aimed to find a set of rendezvous points (RPs) with the objective to minimize the energy consumption on data collection. The optimal locations of RPs are identified subject to the constraint on data delivery latency. Rao *et al.* [7] considered the k multi-hop data collection problem with a mobile sink, where k is a configurable parameter. The desirable balance between energy efficiency and data delivery delay can be achieved by varying k. In [9, 10], data generated from remote sensors are sent to the ones more closed to the moving path of the mobile sink and the sink will pick up the data when it passes by.

In some practical applications, however, the mobility of the sink is constrained. For example, in [11], mobile sinks can only move in a given roadmap and should start from and return to the depot after finishing each tour. Assuming the mobile sinks can visit all walking paths in the roadmap, the problem in [11] is to minimize the maximum path length of mobile sinks, which is then reduced to min-max k-Chinese postman problem (MM k-CPP). Unlike their work, besides the roadmap constraint on the mobile sink, we assume the length of each tour is bounded and consider the monitoring quality optimization problem. In [1], the mobile sink is supposed to move along a fixed trajectory and visit the fixed set of gateways. In this work, the moving path for the mobile sink is to be determined from the given roadmap subject to the tour length constraint such that the monitoring quality is maximized.

3. PRELIMINARIES

Consider a wireless sensor network $G = (V \cup GW, E)$ where V is the set of low-cost sensors, GW is the set of powerful (or solar-powered), large-storage gateways, and E is the set of links connecting sensors and gateways. $n = |V|$, $n_g = |GW|$, and $m = |E|$. There is a given roadmap $M = (V_r, E_r)$ in the monitoring region for the MS, where E_r is the set of roads in the roadmap and V_r is the set of road intersections. $n_r = |V_r|$ and $m_r = |E_r|$. $V_r = \{p_1, p_2, \dots, p_{n_r}\}$. It indicates that the MS can only move along the roads in the map. Gateways are deployed on the roads and communicate with the MS when the MS moves within the transmission ranges of gateways. Each gateway in GW and each sensor in V equipped with an omni-directional antenna has a fixed, identical transmission range r. Assume the locations of GW and V are stationary and known a *priori*. There is a link between two sensors, or a sensor and a gateway, if they are within the transmission range of each other. Assume that each sensor in V has identical data generation rate r_g.

Two constraints are imposed on the MS: the MS should start from and return to the depot p_0; its total travelling distance per tour should be no longer than L. Also, the MS moves at a constant speed v.

3.1 Monitoring quality loss

Let $GW = \{g_1, g_2, \ldots, g_{n_g}\}$ be the set of gateways. The MS collects data from g_k when it moves within the intersection of the road and the transmission range of g_k. Denote by I_k the length of the intersection for gateway g_k, $1 \le k \le n_g$. Note that the longest I_k is two times of the gateway transmission radius r, i.e., $I_k \le 2 \cdot r$, for $1 \le k \le n_g$. Assuming that there is no data aggregation on gateways or sensors, the maximum amount of data that can be collected by the sink from g_k per tour is

$$data(g_k) = \frac{I_k}{v} \cdot r_k \le \frac{2 \cdot r \cdot r_k}{v}, \ 1 \le k \le n_g, \quad (1)$$

where r_k is the data transmission rate of gateway g_k. Thus, the upper bound on the number of active nodes for g_k is the number of sensors that can transmit their data to g_k, while all their data can be collected by the MS in the next tour,

$$c(g_k) = \lfloor \frac{data(g_k)}{\frac{L}{v} \cdot r_g} \rfloor \le \lfloor \frac{2 \cdot r \cdot r_k}{L \cdot r_g} \rfloor, \ 1 \le k \le n_g, \quad (2)$$

$c(g_k)$ is referred to as the *quota* of g_k.

The monitoring quality loss will occur when not all sensing data is collected by the MS in a tour. Assuming a closed tour P subject to L is found, the set of gateways deployed on the roads in P is $GW' \subset GW$. Let D_a be the set of nodes, whose generated data can be collected by the MS, referred to as *active nodes*, and $n_a = |D_a| \le \sum_{g_k \in GW'} c(g_k)$.

The monitoring quality loss is defined as the error caused by data estimation of uncollected data, using the collected ones. Assume x_{tj} is the actual reading of sensor v_j at time t, and $\widehat{x}_{tj}$ is an estimate of x_{tj}. We denote by $QL(P, D_a, V)$ the quality loss with D_a as the set of active nodes:

$$QL(P, D_a, V) = \frac{\sum_{v_j \in V - D_a} \sum_{t=1}^{L(P)/v} (x_{tj} - \widehat{x}_{tj})^2}{n \cdot L(P)/v}, \quad (3)$$

where $L(P)$ is the length of a tour P and $L(P) \le L$.

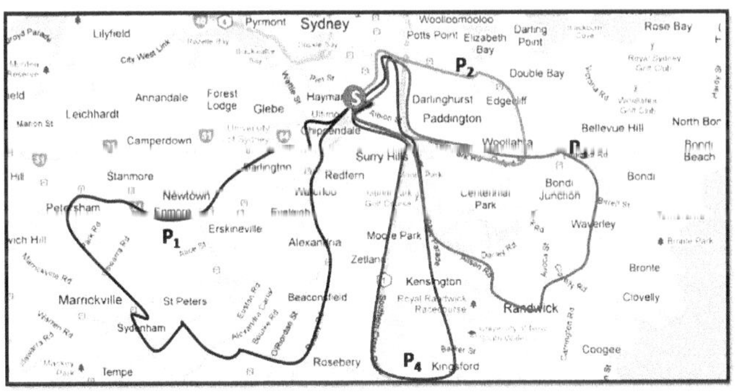

Figure 2: Tour options for the mobile sink

In Fig. 2, for instance, there are four candidate closed tours P_1, P_2, P_3, P_4. Assuming the length of P_1 exceeds L, it should be eliminated. If travelling along P_3 results in the minimum quality loss among the three tours, P_3 should be selected as the tour for the MS.

The *distance-constrained monitoring quality optimization problem* in a sensor network $G = (V \cup GW, E)$ with a mobile sink starting from depot p_0 and moving in a given roadmap $M = (V_r, E_r)$ at a constant speed v is to jointly identify active nodes, find a closed tour P subject to the length constraint L, and design an energy-efficient routing protocol for data collection, such that the mobile sink is able to efficiently collect data from active nodes with the minimum monitoring quality loss and return to p_0 at the end of each tour.

The problem in [1] is a special case of the problem concerned in our paper without the tour length constraint, where there is only one road in the roadmap and the fixed trajectory for the MS is the round trip on this road. Since the problem of data quality maximization with the MS moving along a fixed trajectory has been proven to be NP-hard [1], the problem concerned in this work is NP-hard, too.

4. FINDING A LENGTH-CONSTRAINED TOUR

Due to the NP-hardness of the problem, we propose a two-stage heuristic, which consists of finding the length-constrained tour for the MS and devising a routing protocol for data collection (see Section 5). The tour finding stage is comprised of the following three steps.

4.1 Identifying the set of candidate active nodes

As mentioned in Section 1, we need to explore the spatial correlation among data generated from all sensors before we can determine the set of active nodes.

We Assume that L is large enough so that the MS can traverse all roads in E_r. We aim to identify the set of active nodes $\mathcal{CS}$ to minimize the monitoring quality loss. To do so, we collect data generated from all sensors within a *training period* P_t. The problem is then reduced to the *quota-constrained data quality maximization problem* in [1] and the method in [1] can be adopted to determine $\mathcal{CS}$.

Nodes in $\mathcal{CS}$ are referred to as *candidate active nodes* because the constraint L has not been considered at this stage, and not all data generated from sensors in $\mathcal{CS}$ is guaranteed to be collected by the MS. With the set $\mathcal{CS}$, the monitoring quality is the best case. If $L < \sum_{e \in E_r} dist(e)$ is taken into account, where $dist(e)$ is the Euclidean distance between two endpoints of e, only nodes in a subset of $\mathcal{CS}$ are able to transmit their data to the MS, compromising the monitoring quality.

The following steps are based on this set of candidate active nodes. After the tour is found, the MS will move along it in the following *operation phase* P_o. Then the tour is to be re-identified and the training and operation phases are repeated.

4.2 Allocating candidate active nodes to gateways

Having identified the set of candidate active nodes, we first allocate them to gateways, subject to the gateway quotas. That is, partition $\mathcal{CS}$ into n_g disjoint subsets $D_1, D_2, \ldots, D_{n_g}$ for gateways subject to $c(g_k)$, $1 \le k \le n_g$, using the method in [1]. These subsets satisfy: (i) $\bigcup_{k=1}^{n_g} D_k = \mathcal{CS}, D_k \cap D_l = \emptyset$,

$1 \leq k, l \leq n_g, k \neq l$; (ii)$|D_k| \leq c(g_k)$, $1 \leq k \leq n_g$; (iii)$\sum_{k=1}^{n_g} \sum_{u \in D_k} dist(u, g_k)$ is minimized, where $dist(u, g_k)$ is the Euclidean distance between u and g_k.

We then assign each road with candidate active nodes. For $e \in E_r$, define $D(e) = \bigcup\{D_k | \text{if } g_k \text{ is on the road } e\}$. As a result, the set $\mathcal{CS}$ is partitioned into m_r disjoint sets, $D(e_1), D(e_2), \ldots, D(e_{m_r})$. And the *monitoring quality gain* of road $e \in E_r$ is obtained by using Eq.(3), denoted by $q(e)$.

4.3 Finding the distance-constrained tour

Recall that the MS starts from the depot in the roadmap, moves along roads with high quality gains, and returns to the depot, subject to the length constraint L. To find such a closed tour, we adopt the similar idea in [13]. That is, we convert this optimization problem into the distance-constrained shortest path problem, which has been extensively studied [14, 15].

We first construct a weighted, directed graph $G_M = (V_r, E_r, \omega, d)$. The weight of each edge $\langle p_i, p_j \rangle \in E_r$ is $\omega(p_i, p_j) = q(\langle p_i, p_j \rangle)$ and $d(p_i, p_j)$ is the Euclidean distance between p_i and p_j, where $p_i, p_j \in V_r$ are road intersections in roadmap M.

To find an optimal tour in $M = (V_r, E_r)$ is then reduced to find a closed path in G_M, starting from p_0 and ending at p_0, such that the total weight of the edges in the path is maximized, while the sum of the lengths of the edges in the path is bounded by L. We refer to this problem as *the distance-constrained longest path problem*, which however is NP-hard, since the well known NP-hard Hamiltonian problem [12] is a special case of the problem where no distance constraint is imposed.

We then instead propose a heuristic by converting the problem into a distance-constrained shortest path problem in another auxiliary directed graph $G'_M = (V_r, E_r, \omega', d)$. G'_M is constructed as follows. For each directed edge $\langle p_i, p_j \rangle \in E_M$,

$$\omega'(p_i, p_j) = \begin{cases} Q & \text{if } q(\langle p_i, p_j \rangle) = 0, \\ \frac{1}{q(\langle p_i, p_j \rangle)} - \rho & \text{otherwise} \end{cases} \quad (4)$$

where Q and ρ are positively large and small constants, $Q \geq q_{max}$, $0 < \rho \leq \frac{1}{q_{max}}$, and $q_{max} = \max_{e_l \in E_r}\{q(e_l)\}$. The purpose of introducing term ρ is to break the tie of two shortest paths between a pair of nodes with equal length by favoring the one with the larger quality gain. Let E_p be the set of edges in P. The distance-constrained shortest path problem in G'_M is to find a path P that consists of vertices in V_r, with p_0 as both its starting point and end point, such that $L(P) = \sum_{\langle p_i, p_j \rangle \in E_p} d(p_i, p_j) \leq L$ and $\sum_{\langle p_i, p_j \rangle \in E_p} \omega'(p_i, p_j)$ is minimized. There are several approximation algorithms for the distance-constrained shortest path problem. We modify the one by Chen *et al.* [15] for this problem, which consists of finding a feasible tour followed by local improvement to the found tour if possible.

4.3.1 Finding a feasible tour

We calculate all pairs of shortest paths for all pairs of nodes u and v in V_r, denoted by $SP(u, v)$. $P = \langle p_0 \rangle$ initially. Let $P = \langle p_0, p_1, \ldots, p_i, p_0 \rangle$ be the currently constructed path. We extend P by adding a next vertex u into P, which minimizes the total weight of the edges in the path

$$\sum_{j=0}^{i-1} \omega'(p_j, p_{j+1}) + \sum_{e_1 \in SP(p_i, u)} \omega'(e_1) + \sum_{e_2 \in SP(u, p_0)} \omega'(e_2), \quad (5)$$

subject to the constraint L

$$\sum_{j=0}^{i-1} d(p_j, p_{j+1}) + |SP(p_i, u)| + |SP(u, p_0)| \leq L. \quad (6)$$

All vertices in $SP(p_i, u)$ and $SP(u, p_0)$ should be added into P in order. The algorithm terminates when the length constraint is no longer met.

Note that $SP(p_i, u)$ and $SP(u, p_0)$ may contain vertices that have already been visited and included in P. Adding them again to P is, however, necessary. Otherwise, the MS may not be guaranteed to return to p_0 after visiting all other vertices in P. For instance, assume $P = \langle p_0, p_1, p_3, p_5, p_0 \rangle$ is the path to be extended, $SP(p_5, p_7) = \langle p_5, p_1, p_7 \rangle$, and $SP = \langle p_7, p_4, p_3, p_0 \rangle$. We now check whether adding p_7 to P still meets the L constraint. If we only consider to add un-visited vertices into P, the updated path will be $P = \langle p_0, p_1, p_3, p_5, p_7, p_4, p_0 \rangle$. Assume $L(P) \leq L$. Note that there is no edge connecting p_5 and p_7, or p_4 and p_0 in G'_M. In that case, after the MS visits p_5, it can not visit p_7 since $\langle p_5, p_7 \rangle \notin E_r$. In order to reach p_7, the MS has to visit other vertices not between p_5 and p_7 in P (p_1 in this example). Similarly, from p_4 to p_0, the MS has to visit p_3. Since $d(p_5, p_1) + d(p_1, p_7) \geq d(p_5, p_7)$ and $d(p_4, p_3) + d(p_3, p_0) \geq d(p_4, p_3)$, the actual tour length may exceed L. Therefore, we need to follow Eq.(6) to check whether the length constraint is met. And if yes, the updated path should be $P = \langle p_0, p_1, p_3, p_5, p_1, p_7, p_4, p_3, p_0 \rangle$.

4.3.2 Local improvement

Assume that a feasible tour $P = \langle p_0, p_1, \ldots, p_k, p_0 \rangle$ has been found. $E_p = \{\langle p_0, p_1 \rangle, \langle p_1, p_2 \rangle, \ldots, \langle p_k, p_0 \rangle\}$, $L(P) = \sum_{e \in E_p} dist(e) \leq L$, where $dist(e)$ is the Euclidean distance between two endpoints of e. We then perform a local improvement to the feasible solution by attempting to add more vertices to P as long as the length constraint is still met. To do so, we iteratively check whether there exists a vertex $p_j \notin P$ meeting the following conditions: (i) $\langle p_i, p_j \rangle \in E_r$ and $\langle p_j, p_{i+1} \rangle \in E_r$, where $\langle p_i, p_{i+1} \rangle \in E_p$, $i \neq 0$; (ii) $L(P) + d(p_i, p_j) + d(p_j, p_{i+1}) - d(p_i, p_{i+1}) \leq L$. If yes, add p_j into P since a better path $P = \langle p_0, p_1, \ldots, p_i, p_j, p_{i+1}, \ldots, p_k, p_0 \rangle$ is found. If there are multiple vertices meeting the length constraint, the one leading to the maximum quality gain will be added. This procedure continues until the L constraint is no longer met. The computational complexity of this local improvement is $O(n_r^2) = O(n^2)$.

$D_a = \bigcup_{e \in E_p} D(e)$ is the set of active nodes associated with path P. GW' is the set of gateways deployed on edges in E_p. The MS moves along E_p and collects the data generated from sensors in D_a by visiting gateways in GW'.

There may exist duplicate edges in the tour. In the above example, if the shortest path between p_7 and p_0 is $SP(p_7, p_0) = \langle p_7, p_1, p_3, p_0 \rangle$, the updated tour $P = \langle p_0, p_1, p_3, p_5, p_1, p_7, p_1, p_3, p_0 \rangle$, where edges $\langle p_1, p_3 \rangle$ and $\langle p_1, p_7 \rangle$ are visited twice. This will compromise the monitoring quality since the MS will not collect data when it moves along an edge which has been visited in the same tour. Consider two different tours with the same length in two roadmaps, shown in Fig. 3. P_1 is a round trip in which each edge is visited twice while in P_2, each edge is visited only once. Note that $L(P_1) = L(P_2) = 4 \cdot dist(e)$. Assuming the numbers of gateways on each edge in both P_1 and P_2 are identical, $|GW_1'| = \frac{1}{2} \cdot |GW_2'|$ and $|D_{a1}| = \frac{1}{2} \cdot |D_{a2}|$.

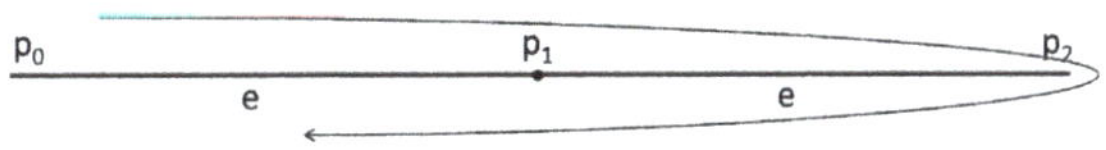

(a) P_1

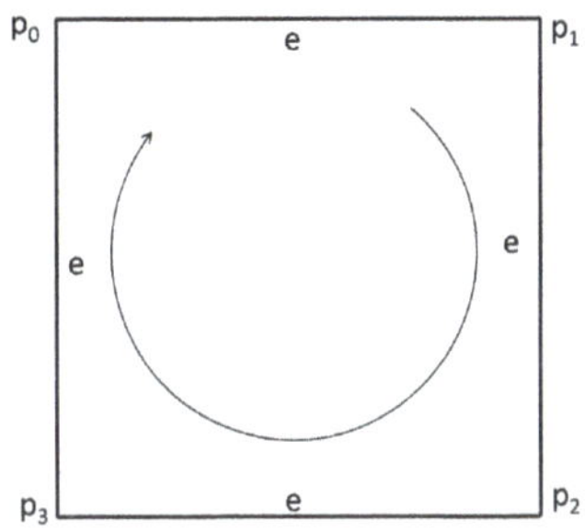

(b) P_2

Figure 3: Two different tours with the same length

5. ENERGY-EFFICIENT DATA ROUTING

Having the closed tour P, the next stage is to route data generated from active nodes to corresponding gateways energy-efficiently. In this section, we aim to devise a routing protocol for this purpose. Assume an active node $v_i \in D_a$ is allocated to gateway g_k on road $e \in E_p$, with $dist(v_i, g_k) > r$. In that case, v_i cannot send its sensing data to g_k directly, rather, it needs at least one node acting as a relay node between them. Note that such relay nodes could be active nodes allocated to the same gateway, those allocated to other gateways, or inactive nodes.

We construct a weighted graph $G_w = (GW \cup V, E, \eta)$, where $\eta(u,v) = dist(u,v)$ for each $\langle u,v \rangle \in E$. We aim to construct a routing structure consisting of $|GW'|$ trees T_k rooted at g_k, $1 \le k \le |GW'|$. The routing tree T_k should span all active nodes in D_k and the $\sum_{\langle u,v \rangle \in T_k} \eta(u,v)$ is to be minimized. The problem of finding such a routing tree is then reduced to the Steiner Tree Problem (STP), with D_k as the *terminal set*. The STP is NP-hard, and in this paper, we find a solution to the problem by modifying the approximation algorithm by Kou *et al.* [16]. The algorithm is as follows.

First, compute all pairs of shortest paths in G_w. An auxiliary complete weighted graph G_{D_k} consisting of only D_k is constructed. The weight assigned to each edge in G_{D_k} is the length of the shortest path between the two endpoints of the edge in G_w. Second, find a minimum spanning tree (MST) in G_{D_k}. Let E_{OPT} be the set of edges in the MST. A subgraph of G_w, G_k, is induced by the edges in E_{OPT}. Note that each edge in E_{OPT} corresponds to a shortest path in G_w. Third, find a MST in the subgraph G_k, and prune those branches of the tree that do not contain the node in D_k. The resultant tree T_k is an approximate Steiner tree with D_k as the terminal set. Finally, assume that the set of vertices in the Steiner tree is V_{D_k}. For each $u \in V_{D_k}$, we increase weights of its associated edges in G_w by a small constant Δ_w. As a result, the opportunity of the same node involved by more than one Steiner trees is reduced and the energy consumption among sensors can be further balanced. The found Steiner trees T_k, $1 \le k \le |GW'|$, are used as routing trees for data collection. They form the routing structure $\mathcal{F}$.

Note that T_k rooted at g_k may include nodes not in D_k, and in that case, the number of descendants of g_k, $|V_{D_k}|$, will exceed $c(g_k)$. However, g_k is still able to fulfill the data relay task since it is only supposed to store data generated from sensors in D_k and upload them to the MS when the MS moves within its transmission range. Data generated from nodes in V_{D_k} but not in D_k is discarded.

The network lifetime is defined as the time of the first sensor's failure due to the depletion of its energy [21]. We only consider the energy consumption on wireless communication including data transmission and reception [22]. Let $c(v_i)$ be the number of nodes using v_i as the relay node in $\mathcal{F}$, and let $E_c(v_i)$ be the energy consumption of v_i in a tour.

$$E_c(v_i) = d_T(v_i) \cdot E_T(v_i, v_j) + d_R(v_i) \cdot E_R(v_i) \tag{7}$$

where $E_R(v_i)$ and $E_T(v_i, v_j)$ are the amounts of energy consumed by v_i on receiving 1 bit of data and transmitting 1 bit of data to v_j; $d_T(v_i)$ and $d_R(v_i)$ are respectively the amounts of data transmitted from and received by v_i in a tour.

$$E_R(v_i) = \epsilon_{elec} \tag{8}$$

$$E_T(v_i, v_j) = \epsilon_{elec} + \epsilon_{amp} \cdot dist^2(v_i, v_j), \tag{9}$$

where ϵ_{elec} is the energy cost of processing 1 bit data and ϵ_{amp} is the transmitter amplifier. Recall that $dist(v_i, v_j)$ is the Euclidean distance between v_i and v_j.

$$d_T(v_i) = \begin{cases} r_g \cdot (c(v_i) + 1) & \text{if } v_i \in D_a, \\ r_g \cdot c(v_i) & \text{otherwise} \end{cases} \tag{10}$$

$$d_R(v_i) = r_g \cdot c(v_i) \tag{11}$$

The network lifetime is

$$\frac{IE}{max\{E_c(v_i) \mid v_i \in V\}}, \tag{12}$$

where IE is the initial energy capacity of sensors.

The heuristic including both the tour finding stage and the routing protocol designing stage is referred to as algorithm `Monitoring_Quality_Maximization`, or `MQM` for short.

6. PERFORMANCE EVALUATION

In this section, we evaluate the proposed algorithms through experiments. To evaluate the monitoring quality loss and the network lifetime, we vary network topologies, the deployment of gateways, and the length constraint L.

The roadmap in the monitoring area is shown in Fig. 4, in which there are 8 road intersections {A, B, C, D, E, F, G, H} and 13 roads. MS starts from `A` and will return to `A` at the end of each tour.

The parameter settings are listed in Table 1. We assume $I_k = 2 \cdot r$, $1 \le k \le n_g$. We adopt the energy consumption model in [20]. Also, we adopt the same way as [1] to generate sensing data in the range [0,100] and fix the ratio of training phase over the operation phase $R = P_t/P_o$ to be 10%. $P_o = 1000$ in our default simulation setting. All

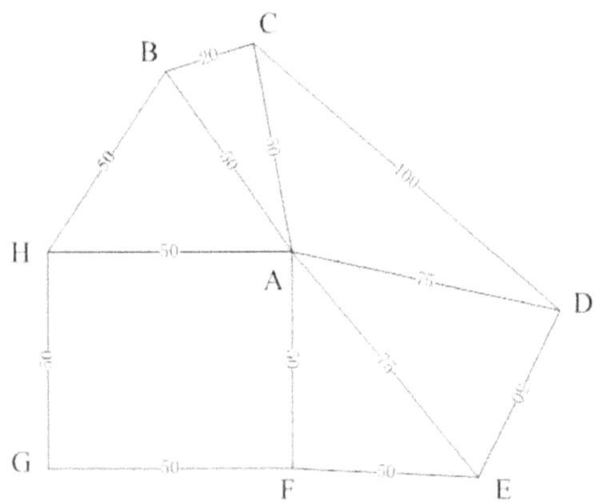

Figure 4: Monitoring region with the roadmap

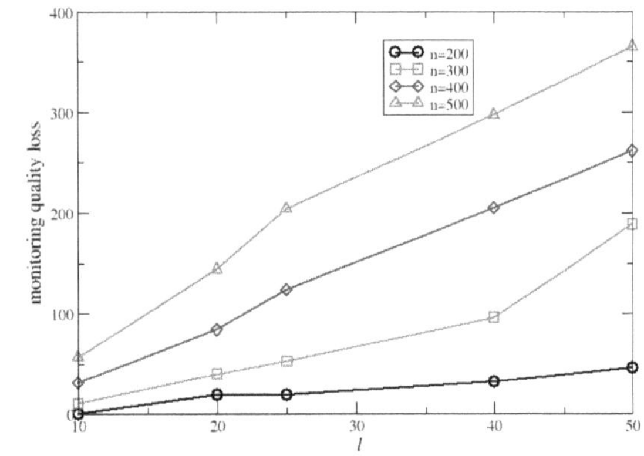

(a) Monitoring quality loss with different gaps of gateways

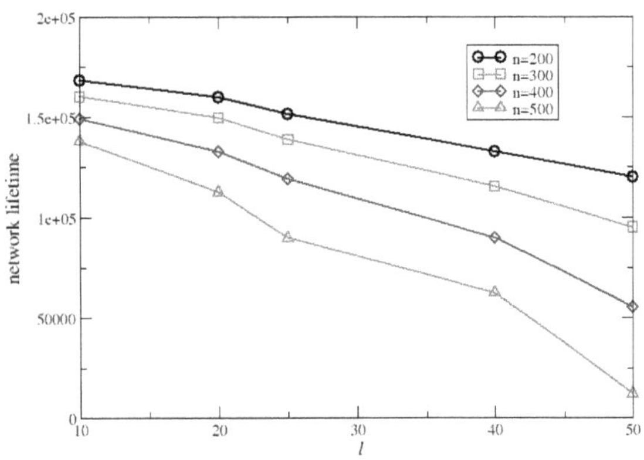

(b) Network lifetime with different gaps of gateways

Figure 5: Network performance with different gateway deployments

Table 1: Simulation Setting

v	$2\ m/s$
r_k	$100\ bit/s$
r_g	$1\ bit/s$
r	$10\ m$
IE	$100\ Jules$
ϵ_{amp}	$100\ pJ/bit/m^2$
ϵ_{elec}	$50\ nJ/bit$

experiment results are the mean of results delivered by the algorithm with 50 different network topologies and identical parameter settings.

6.1 Impact of gateway deployment on the network performance

We first evaluate the monitoring quality and the network lifetime with different gateway deployments. In practice, we may install a gateway every l meters along roads in the roadmap, or deploy gateways according to the sensor density. In our simulations, we adopt the first way of gateway deployment and refer to l as the gap of gateways.

Fixing $n = 200$ and $L = 200$, we set l to be 10, 20, 25, 40, and 50 respectively. The quota of each gateway is 10 by Eq. (2). In Fig. 5(a), the curve with $n = 200$ shows the monitoring quality loss with different values of l. It indicates that the more densely the gateways deployed (the smaller the l), the smaller the monitoring quality loss. This is because more gateways will be able to relay data from more sensors to the MS. For example, when $l = 10$, the number of gateways the MS can visit is $200m/10m = 20$ and the total number of active nodes is $20 \times 10 = 200$, which is the set of all sensors. Thus, there is no monitoring quality loss. When l is set to be 20 and 25 respectively, the number of gateways the MS can visit are $200m/20m = 10$ and $200m/25m = 8$. And the corresponding numbers of active nodes are $10 \times 10 = 100$ and $8 \times 10 = 80$. Compared with the latter case, 100 active nodes will produce less quality loss, although sensing information loss is unavoidable in both cases. The curve with $n = 200$ in Fig. 5(b) demonstrates that with a smaller l, more gateways are able to distribute the data relay load evenly among sensors, resulting in a longer network lifetime.

In reality, the value of l is determined by both the user requirement on the monitoring quality and the budget of gateways. The number of gateways is $\lfloor L/l \rfloor$ and the gateway quota is $\frac{2r \cdot r_l}{L \cdot r_g}$ referring to Eq.(2). If the data transmission rates of all gateways are identical, the total number of active nodes is $\frac{L}{l} \cdot \frac{2r \cdot r_l}{L \cdot r_g} = \frac{2r \cdot r_l}{l \cdot r_g}$. That is, the number of active nodes only depends on l. If quality loss is not acceptable, gateways should be deployed every $\frac{2r \cdot r_l}{n \cdot r_g}$ meters on roads. If the budget for gateways is limited, we should increase l at the cost of larger monitoring quality loss.

6.2 Impact of network size on the network performance

We then investigate the network performance by varying n from 200 to 500, while keeping $L = 200$. The monitoring quality loss and the network lifetime are shown in Fig. 5(a) and (b), respectively.

From Fig. 5, we note that with constant l, a larger n leads to a greater quality loss and a shorter network lifetime.

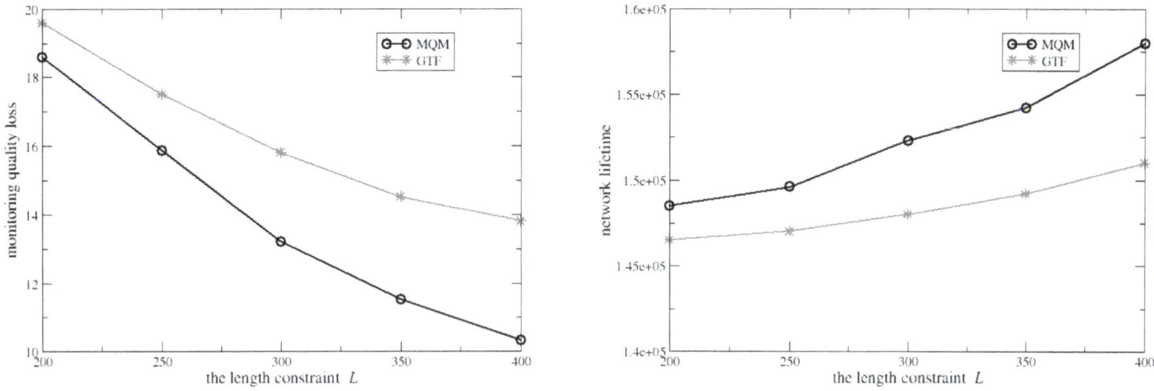

(a) Monitoring quality loss with different L

(b) Network lifetime with different L

Figure 6: Performance evaluation with different length constraints

This is because the same number of active nodes can better represent the other nodes when the network size is smaller. Also, with the increase of network size n, more energy will be consumed on the data collection, resulting in a shorter network lifetime.

6.3 Impact of tour length constraint on the network performance

We next evaluate the network performance by varying the length constraint L from 200 to 400 with the increment of 50 while fixing $l = 25$ and $n = 200$. In Fig. 6, curve `MQM` shows the monitoring quality loss and the network lifetime with various L. Generally, larger L results in better network performance.

As discussed above, the number of active nodes is irrelevant to L. However, different settings of L do impact the network performance in the following two aspects. First, in terms of data freshness. Larger L causes a longer data delivery latency. With $L = 200m$, e.g., the MS collects data generated within the last $\frac{200m}{2m/s} = 100s$ while with $L = 400m$, the collected data is generated within the last $\frac{400m}{2m/s} = 200s$. Second, in the solution domain. Larger L enlarges the domain for a better solution in terms of less monitoring quality loss, and more energy-efficient data collection. With $l = 25$, the number of active nodes is 80 regardless of the value of L. Setting L to be 200 and 400, e.g., the numbers of gateways the MS can visit are 8 and 16, with quotas 10 and 5 respectively ($8 \times 10 = 16 \times 5 = 80$). However, compared to $L = 200$, the scenario with $L = 400$ explores the roadmap with more relaxing constraint, thus is more likely to involve edges with higher monitoring quality gains and distribute the data relay load more evenly among sensors.

6.4 Performance comparison

We finally compare the performance of algorithm `MQM` with that of a greedy heuristic. The greedy heuristic consists of two stages as well. The only difference between them lies in the tour finding stage after assigning each road with a monitoring quality gain in the auxiliary graph G_M. Let $P = \langle p_0, p_1, \ldots, p_l, p_0 \rangle$ be the tour generated so far. The next vertex to be added to P is determined as follows. The algorithm first checks all un-visited neighboring vertices of p_l in G_M and adds p_k between p_l and p_0 in P if it meets: (i) $p_k \notin P$; (ii) $q(\langle l, k \rangle) = max\{q(\langle l, i \rangle) | \langle l, i \rangle \in E_r\}$; and (iii) $|SP(p_k, p_0)| + d(p_l, p_k) + L(P) - d(p_l, p_0) \le L$. If no such a vertex is found, the algorithm checks all visited neighboring vertices of p_l, finds $p_k \in P$ with the minimum $|SP(p_k, p_0)| + d(p_l, p_k) + L(P) - d(p_l, p_0)$ ($\le L$), and adds it between p_l and p_0 in P. The algorithm terminates if the L constraint is no longer met. We refer to this heuristic as algorithm `Greedy_Tour_Finding`, or `GTF` for short.

We compare the network performance delivered by algorithm `MQM` and `GTF` by fixing $n = 200$, $l = 25$, and varying L from 200 to 400 with the increment of 50. The monitoring quality loss and the network lifetime delivered by two algorithms are shown in Fig. 6(a) and Fig. 6(b) respectively. From the figures we can notice that algorithm `MQM` always outperforms algorithm `GTF` and with the increase of L, the performance gap between them becomes larger and larger.

7. CONCLUSION

In this paper, the problem of optimizing the monitoring quality using a mobile sink has been considered. The mobile sink has been assumed to move in a given roadmap for collecting the data generated from some sensors by visiting the gateways. We have formulated the problem as a joint optimization problem that aims to find a closed tour for the mobile sink to minimize the monitoring quality loss, and to design an energy-efficient routing protocol for data collection. We have proposed a novel two-stage heuristic for it. We also evaluated the performance of the proposed algorithms against another heuristic by simulation. The experimental results demonstrate that the proposed algorithms always outperform the other one.

8. REFERENCES

[1] X. Xu, W. Liang, and Tim Wark. Data quality

maximization in sensor networks with a mobile sink. *IEEE Proc. of DCOSS*, 2011.

[2] G. Xing, T. Wang, W. Jia, and M. Li. Rendezvous design algorithms for wireless sensor networks with a mobile base station. *ACM Proc. of MobiHoc*, 2008.

[3] Z. M. Wang, S. Basagni, E. Melachrinoudis, and C. Petrioli. Exploiting sink mobility for maximizing sensor networks lifetime. *IEEE Proc. of HICSS*, 2005.

[4] A. A. Somasundara, A. Ramamoorthy, and M. B. Srivastava. Mobile element scheduling with dynamic deadlines. *IEEE Trans. on Mobile Computing*, vol. 6, pp. 395-410, 2007.

[5] S. Basagni, A. Carosi, E. Melachrinoudis, C. Petrioli, and Z. Wang. Controlled sink mobility for prolonging wireless sensor networks lifetime. *Wireless Networks*, Vol. 14, pp.831-858, 2008.

[6] X. Xu, Y. Hu, J. Bi, and W. Liu. Adaptive nodes scheduling approach for clustered sensor networks. *IEEE Proc. of Symposium on Computers and Communications (ISCC)*, 2009.

[7] J. Rao and S. Biswas. Joint routing and navigation protocols for data harvesting in sensor networks. *IEEE Proc. of International Conference on Mobile Ad Hoc and Sensor Systems (MASS)*, 2008.

[8] A. S. Arun, R. Aditya, and B. S. Mani. Mobile element scheduling with dynamic deadlines. *IEEE Trans. on Mobile Computing*, vol. 6, pp. 395-410, 2007.

[9] M. Ma and Y. Yang. SenCar: an energy-efficient data gathering mechanism for large-scale multihop sensor networks. *IEEE Trans. on Parallel and Distributed Systems (TPDS)*, vol. 18, pp. 1476-1488, 2007.

[10] D. Jea, A. A. Somasundara, and M. Srivastava. Multiple controlled mobile elements (data mules) for data collection in sensor networks. *IEEE Proc. of DCOSS*, 2005.

[11] S. Tang, J. Yuan, X. Li, Y. Liu, G. Chen, M. Gu, J. Zhao, and G. Dai. DAWN: Energy efficient data aggregation in WSN with mobile sinks. *IEEE Proc. of International Workshop on Quality of Service (IWQoS)*, 2010.

[12] M. R. Garey, D. S. Johnson, and L. J. Stockmeyer. Some simplified NP-complete graph problems. *Theoretical Computer Science*, vol. 1, pp. 237-267, 1976.

[13] W. Liang, J. Luo, and X. Xu. Prolonging network lifetime via a controlled mobile sink in wireless sensor networks. *IEEE Proc. of Globecom*, 2010.

[14] R. Hassin. Approximation schemes for the restricted shortest path problem. *Mathematics of Operations Research*, vol. 17, pp. 36-42, 1992.

[15] S. Chen, M. Song, and S. Sahni. Two techniques for fast computation of constrained shortest paths. *IEEE/ACM Trans. on Networking*, vol. 16, pp. 105-115, 2008.

[16] L. Kou, G. Markowsky, and L. Berman. A fast algorithm for Steiner trees. *Acta Informatica*, vol. 15, pp. 141-145, 1981.

[17] X. Xu and W. Liang. Placing optimal number of sinks in sensor networks for network lifetime maximization. *IEEE Proc. of ICC*, 2011.

[18] L. Qiu, R. Chandra, K. Jian, and M. Mahdian. Optimizing the placement of integration points in multi-hop wireless sensor networks. *IEEE Proc. of International Conference on Network Protocols (ICNP)*, 2004.

[19] B. Gedik, L. Liu, and P. S. Yu. ASAP: An adaptive sampling approach to data collection in sensor networks. *IEEE Trans. on Parallel and Distributed Systems*, vol. 18, pp. 1766-1783, 2007.

[20] W. R. Heinzelman, A. Chandrakasan, and H. Balakrishnan. Energy-efficient communication protocol for wireless microsensor networks. *Proceedings of Annual Hawaii International Conference on System Sciences (HICSS)*, 2000.

[21] J-H Chang and L. Tassiulas. Energy conserving routing in wireless ad hoc networks. *IEEE Proc. of INFOCOM*, 2000.

[22] G. J. Pottie. Wireless sensor networks. *IEEE Proc. of Information Theory Workshop*, pp. 139-140, 1998.

Close-to-optimal Energy Balanced Data Propagation via Limited, Local Network Density Information

Azzedine Boukerche
PARADISE Research Laboratory,
University of Ottawa, Canada
boukerch@site.uottawa.ca

Dionysios Efstathiou
Research Academic Computer Technology Institute (CTI) and University of Patras, Greece
eustathi@ceid.upatras.gr

Sotiris Nikoletseas
Research Academic Computer Technology Institute (CTI) and University of Patras, Greece
nikole@cti.gr

Christoforos Raptopoulos
Research Academic Computer Technology Institute (CTI) and University of Patras, Greece
raptopox@ceid.upatras.gr

ABSTRACT

We study the problem of energy-balanced data propagation in wireless sensor networks. The energy balance property is crucial for maximizing the time the network is functional, by avoiding early energy depletion of a large portion of sensors. We propose a distributed, adaptive data propagation algorithm that exploits limited, local network density information for achieving energy-balance while at the same time minimizing energy dissipation.

We investigate both uniform and heterogeneous sensor placement distributions. By a detailed experimental evaluation and comparison with well-known energy-balanced protocols, we show that our density-based protocol improves energy efficiency significantly while also having better energy balance properties.

Furthermore, we compare the performance of our protocol with a centralized, off-line optimum solution derived by a linear program which maximizes the network lifetime and show that it achieves near-optimal performance for uniform sensor deployments.

Categories and Subject Descriptors

C.2.1 [**Network Architecture and Design**]: [Wireless Communication]

General Terms

Algorithms, Design, Performance

Keywords

Wireless Sensor Networks, Algorithms, Data Propagation, Energy Balance, Routing, Performance Evaluation

1. INTRODUCTION

We study Wireless Sensor Networks (WSNs) envisioned as very large collections of autonomous smart sensor nodes, which distributively form an ad-hoc wireless communication network. The nodes are deployed in a region of interest in order to monitor crucial events and propagate sensed data to a base station usually called the *Sink*. WSNs enable important applications ranging from smart/green building automation to environmental monitoring.

Motivation and Related Work. In WSNs it is crucial to design protocols that are energy efficient and maximize the network lifetime. Towards this goal, diverse approaches including hop-by-hop transmission techniques [5], as well as clustering algorithms [8] and power-saving modes [6] have been proposed.

All the above techniques do not explicitly take care of possible overuse of certain sensors in the network. For example, in a hop-by-hop propagation protocol where data must be forwarded to the *Sink*, the sensors in the area around the *Sink* tend to be overused creating a bottleneck area, since these nodes have to relay the data for other nodes, resulting in an unequal energy consumption. Thus, these nodes may die out very early resulting to network disconnection, although there may be still significant amounts of energy in the other sensors of the network. Also, distant transmissions in cluster head schemes, tend to overuse the cluster heads as the cost of a transmission is roughly some power of the distance between the communicating. The rotation of cluster heads may not be enough (especially in large networks with high event generation rate) to avoid their energy depletion.

Lifespan maximization can be achieved by keeping energy balance [3]. To overcome the unbalanced energy consumption, the authors in [4] proposed a mixed routing strategy where in each step the algorithm decides probabilistically and locally whether to propagate data one-hop towards the *Sink*, or to send it directly to the *Sink*. In [4], it was shown that such a randomized mixed strategy can be used to balance the energy consumption among nodes and increase the lifespan of the network. However, only homogenous topologies and uniform data generation rates are studied, while the energy efficiency itself is not explicitly addressed.

In [2] the authors propose an on-line distributed algorithm

for lifespan maximization where in each step the algorithm decides to propagate data one-hop towards the *Sink* or to send it directly to the *Sink* based however on explicit information about the energy spent by the current node and the energy spent by the nodes that are one-hop closer to the *Sink*. No topology heterogeneity is assumed and energy efficiency is only indirectly taken into account.

A nice, related but different, approach for achieving load balancing is proposed in [9], where the authors propose a non-uniform node distribution deployment to achieve energy-balanced consumption in the network. More precisely, in the (off-line) network deployment phase, they regulate the number of nodes in each sector in order to tune the ratio between the node densities in the adjacent (i+1)-th and i-th sectors. For this deployment, they also present an efficient routing scheme (q-switch) that achieves energy balance using single hop transmissions. Our protocol can be used for a broad class of network deployments and achieves energy balance in a different way, by using a combination of short, long and "bypass" transmissions.

In [1] the authors suggest exploiting mobility to lower transmission power thus indirectly leading to a more uniform energy dissipation.

The Problem. We study the problem of how to achieve energy-balanced data propagation in distributed WSNs for both uniform and heterogeneous sensor deployments. The energy balance property guarantees that the average energy spent per sensor is the same for all sensors in the network at any time during the network operation. This property is crucial for prolonging the network lifetime by avoiding early energy depletion of sensors and the non-utilization of available energy on sensors. Together with energy balance we also want (in contrast to previous work) to optimize the energy efficiency of data propagation.

Our Contribution. Our study was originally motivated by [7] where the authors define the energy balance property and propose an energy-optimal and energy balanced algorithm for the particular problem of sorting in WSNs.

We propose an on-line, distributed algorithm which takes into account limited, local density information towards balancing the energy consumption and thus increasing the lifespan of the network. Apart from the usual well-studied random uniform sensor deployments, we also study, for the first time in the context of the energy balance property, the more realistic case of a heterogeneous sensor placement by introducing a general model of non-uniform deployment. In the relevant state of the art, energy-balancing protocols have not been studied in heterogeneous node placement settings.

The experimental results show that our density-based heuristic actually works well for both uniform and non-uniform network deployments and has much better performance compared to other well-known energy balance data propagation algorithms for both sensor placement distributions and various node densities. In particular, its performance for uniform sensor deployment (as shown via comparing to an off-line optimum evaluated by a linear program) is near-optimal.

2. MODEL

We consider a plane sensor network, in which the sensors and the single *Sink* node are static. We abstract the network by a graph G(V,E), where V denotes the set of nodes (sensors), while $E \subseteq V^2$ represents the set of edges (wireless links). An edge between two nodes in the graph exists iff the distance between the corresponding sensors in the network is below the transmission range R. The distance between nodes is Euclidean, and the path length is the sum of the hops of the intermediate pairs of subsequent nodes.

Without loss of generality, we assume that network deployment area is a circle of radius D. We virtually slice the network into $M = \frac{D}{R}$ co-centric *Rings* and $N = \frac{2\pi}{\phi}$ *Slices*, where $\phi = \frac{\pi}{6}$. A sector is defined as the intersection of a specific ring and slice. For example, in Fig. 1 the network is divided into 12 Slices where each Slice contains 20 Sectors, resulting in a total of 240 sectors in the network.

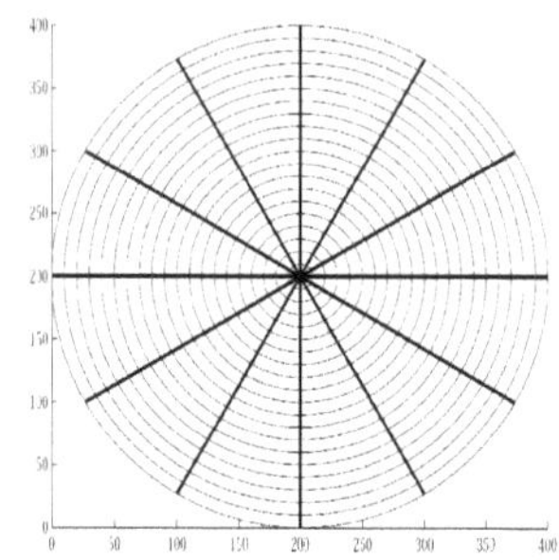

Figure 1: The sectorized network model.

Each node knowns its location, has an unique ID and belongs to exactly one Sector of a Slice. Using this information, each node can identify in which Slice and Sector it belongs to. Nodes are aware of which nodes belong to neighboring Sectors and the position of the static Sink. This information can be disseminated through a set-up phase initiated by the Sink during which the position of the Sink and the IDs of the nodes in neighboring Sectors are diffused.

Uniform deployment. In this case, we deploy the n nodes of the network uniformly at random in the network deployment area, see Fig. 2 (left). Notice that the expected density is the same for all sectors in the network.

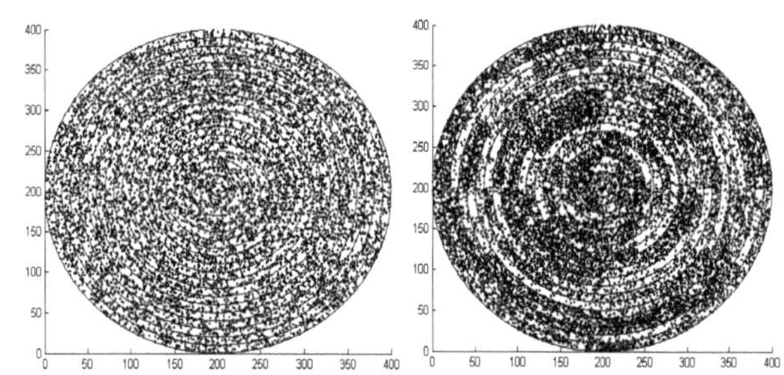

Figure 2: Uniform deployment (left) and Non-uniform deployment (right).

Non-uniform deployment. Our experiments are also performed in random instances of the following general model of non-uniform deployment: Denote by $S_{i,j}$ the sector corresponding to the intersection of slice i and ring j. Let $c > 1$ be an arbitrary constant. Each sector $S_{i,j}$ chooses independently a number $\delta_{i,j} \in [1, c]$ according to the uniform distribution $\mathcal{U}[1, c]$. We will refer to the number $\delta_{i,j}$ as the *relative density* of sector $S_{i,j}$. Clearly, values of $\delta_{i,j}$ close to 1 imply low relative density and values close to c imply high relative density. By combining the knowledge about the total number of sensors in the network n, together with the relative density $\delta_{i,j}$ and the area $A_{i,j}$ of every sector, we compute the number of nodes $n_{i,j}$ deployed in sector $S_{i,j}$ by the following formula:

$$n_{i,j} = \frac{n}{\sum_{i',j'} \frac{A_{i',j'}}{A_{i,j}} \frac{\delta_{i',j'}}{\delta_{i,j}}}.$$

where $n = \sum_{i,j} n_{i,j}$. Finally, we scatter $n_{i,j}$ nodes in the area corresponding to sector $S_{i,j}$, see Fig. 2 (right).

Notice that the fraction of the actual densities of two sectors $S_{i,j}$ and $S_{i',j'}$ is exactly $\frac{A_{i',j'}}{A_{i,j}} \frac{\delta_{i',j'}}{\delta_{i,j}}$ (which also justifies the name relative density for $\delta_{i,j}$). Furthermore, if all sectors have the same relative density (i.e. $\delta_{i,j} = \delta_{i',j'}$, for all i', j'), we get the uniform deployment.

Our protocols operate at the network layer, so we are assuming appropriate underlying data-link, MAC and physical layers. The nodes' memory is assumed limited e.g. we allow messages to piggy-back a constant number of bits of information only, encoding the position of the last node visited and the position of the *Sink*.

We consider two types of data transmission: a) single-hop transmission (cheap in terms of energy and slow) between two nodes belonging to neighboring sectors and b) direct transmission (expensive in terms of energy and fast) where the node that holds the data transmits directly to the *Sink*.

We assume that the energy spent at a sensor when transmitting data messages is proportional to the square of the transmitting distance. The size of messages is considered to be constant. We only count (for simplicity) energy spent during transmissions.

3. THE OFF-LINE OPTIMAL SOLUTION TO THE PROBLEM

Consider now a specific slice in our sensor network model. This slice contains N sectors, which we denote by $S_1, S_2, \ldots, S_N$, dropping the slice number of the sectors for the sake of simplicity. Every sector S_i contains n_i sensors and each sensor has $E_{initial}$ initial energy at the beginning of the network operation. Given a specific slice, we model the routing of messages with a linear program (LP) (see below definitions 4 - 5).

Definition 1. λ is the number of messages produced by each sensor in the network.

Definition 2. $E_{initial}$ is the initial energy of each node in the network.

Definition 3. n_i is the number of nodes in sector i, where $1 \leq i \leq N$.

Definition 4. The λ maximization problem is on input $\{E_{initial}, n_i\}$, to find the maximum number of messages that every sensor can produce with given initial energy and node n_i distribution per sector. If we multiply λ with the total number of nodes in the network, we get the maximum number of messages throughout the network that are routed to the *Sink*.

Maximize λ:

$$\sum_{j=1}^{N} f_{ij}\,(i-j)^2 \cdot R^2 \leq n_i * E_{initial}, \qquad 1 \leq i \leq N \qquad (1)$$

$$\sum_{j=0}^{N} f_{ij} = \sum_{j=1}^{N} f_{ji} + \lambda \cdot n_i, \qquad 1 \leq i \leq N \qquad (2)$$

$$f_{ij} = 0, \qquad j \notin \{0,\, i-1\}, \qquad 1 \leq i \leq N \qquad (3)$$

$$f_{ij} > 0, \qquad j \in \{0,\, i-1\}, \qquad 1 \leq i \leq N \qquad (4)$$

In Equation 1, f_{ij} is the number of messages routed from sector i to sector j while not spending more energy than available. Eq. 2 guarantees flow preservation i.e. all messages generated at or forwarded to a sector are pushed out of it. Eq. 3 and Eq. 4 allow only next-slice transmissions and direct transmissions to the sink. Equations 5 to 8 are similar. Notice that in the above linear program, λ is a variable of the LP and $E_{initial}$ is given as input, while in the following linear program $E_{initial}$ is a variable of the LP and λ is predefined.

Definition 5. The $E_{initial}$ minimization problem is on input $\{\lambda, n_i\}$, to find the minimum initial energy $E_{initial}$ which achieves the desired λ data generation rate per node in a given n_i node distribution. Thus, by multiplying the output $E_{initial}$ of the LP with the number of nodes we get the optimal energy consumption of the whole network.

Minimize $\mathbf{E_{initial}}$:

$$\sum_{j=1}^{N} f_{ij}\,(i-j)^2 \cdot R^2 \leq n_i * E_{initial}, \qquad 1 \leq i \leq N \qquad (5)$$

$$\sum_{j=0}^{N} f_{ij} = \sum_{j=1}^{N} f_{ji} + \lambda \cdot n_i, \qquad 1 \leq i \leq N \qquad (6)$$

$$f_{ij} = 0, \qquad j \notin \{0,\, i-1\}, \qquad 1 \leq i \leq N \qquad (7)$$

$$f_{ij} > 0, \qquad j \in \{0,\, i-1\}, \qquad 1 \leq i \leq N \qquad (8)$$

4. THE PROTOCOLS

4.1 The P_i Energy Balance Protocol

The authors in [4] propose an energy balance algorithm where in each step the node that holds the data decides whether to propagate data one-hop towards the *Sink* with probability p_i , or to send data directly to the *Sink* with probability 1 - p_i, where i is the current sector. The appropriate probability p_i is calculated as follows:

$$p_i = 1 - \frac{3x}{(i+1)(i-1)}$$

where x is a free (e.g. controlling) parameter and i is related to the sensor'Šs distance to the *Sink*. This randomized choice balances the (cheap but slow) one-hop transmissions with the (expensive but fast) direct transmissions to the *Sink*, which are more expensive but "bypass"the nodes lying close to the *Sink*. Note that, in most data propagation protocols, the bottleneck area around the *Sink* tends to be overused and die out early. The lifespan of the network is maximized by ensuring that the energy consumption in each slice is the same. Sensors are assumed to be randomly distributed with uniform distribution.

4.2 The E_i Energy Balance Protocol

In [2] the authors propose an on-line distributed algorithm for lifespan maximization (via energy balance) of a WSN in a data propagation scenario. Let n be a sensor. V_n is the neighborhood of node n. Node n knows the energy spent by each node in its neighborhood. Let m be the sensor of neighborhood V_n with the lowest energy spent. When n holds data it makes the following decision:

- If node n has spent more energy than m, then n sends the message to m (spending one energy unit).
- Otherwise, n sends the message directly to the *Sink* spending d^2 energy units, where d is the distance from n to the *Sink*.

4.3 Our Density-Based Protocol

The basic idea of our protocol is to take into account information about the density of the Previous and the Next Sectors in the current Slice in order to take a decision to forward data to the next Sector or directly to the *Sink*.

Definition 6. Let $d_{i,j}$ be the density of the j-th Sector in the i-th Slice.

Definition 7. If S is the current Slice and r is the current Sector of the node, let $F_{\frac{P}{N}}$ be the fraction of the density of the Previous β Sectors over the density of the Next β Sectors, which is calculated as follows:

$$F_{\frac{P}{N}} = \frac{\sum_{j=r+1}^{r+1+\beta} d_{S,j}}{\sum_{j=r-1-\beta}^{r-1} d_{S,j}}$$

We provide three options in order to examine the ratio of the densities of the Previous Sectors to the densities of the Next Sectors:

a) **Local Knowledge** ($\beta = 1$). Firstly, to take into account the density of the Previous Sector and the Next Sector only. If S is the current Slice and r is the current Sector, we compute the density ratio as follows:

$$F_{\frac{P}{N}} = \frac{d_{S,r+1}}{d_{S,r-1}}$$

b) **Partial Knowledge** ($\beta = 3$). Secondly, to take into account the density of the Previous β Sectors and the Next β Sectors; β is taken 3 as a good compromise of information and cost. If S is the current Slice and r is the current Sector, we compute the density ratio as follows:

$$F_{\frac{P}{N}} = \frac{\sum_{j=r+1}^{r+1+\beta} d_{S,j}}{\sum_{j=r-1-\beta}^{r-1} d_{S,j}}$$

c) **Global Knowledge** ($\beta = N$). Thirdly, to take into account the density of the all Previous Sectors and the all Next Sectors. If S is the current Slice, r is the current Sector and $T_{Sectors}$ is the total number of Sectors, we compute the density ratio as follows:

$$F_{\frac{P}{N}} = \frac{\sum_{j=r+1}^{T_{Sectors}} d_{S,j}}{\sum_{j=1}^{r-1} d_{S,j}}$$

We characterize the density of the Next or Previous Sectors as "high", when the average density of the Next or Previous β Sectors is M times larger than the global network density D_{Global}; in particular we chose M to be equal with 3. Similarly, we characterize the density of the Next or Previous Sectors as "low", when the average density of the Next or Previous β Sectors is M times lower than the global network density D_{Global}.

Definition 8. If S is the current Slice and r is the current Sector of the node, let $F_{Previous}$ be the fraction of the density of the Previous β Sectors over the global network density, which is calculated as follows:

$$F_{Previous} = \frac{\sum_{j=r+1}^{r+1+\beta} d_{S,j}}{D_{Global}}$$

Definition 9. If S is the current Slice and r is the current Sector of the node, let F_{Next} be the fraction of the density of the Next β Sectors over the global network density, which is calculated as follows:

$$F_{Next} = \frac{\sum_{j=r-1-\beta}^{r-1} d_{S,j}}{D_{Global}}$$

4.3.1 Our Data Propagation Scheme

The node that holds the data to be delivered to the *Sink* decides to act suitably according to the relation of densities in previous and next sectors towards the sink.

a) $\mathbf{F_{\frac{P}{N}} < (1-\alpha)}$, where α constant such that $0 < \alpha < 1$. The particular value of constant α can be chosen by the protocol designer to fine-tune performance: a large α value would lead to less energy consumption but a higher data delivery latency while a smaller α would incur a smaller data delivery latency but it would increase a the energy consumption. The experimental evaluation suggested that the value $\alpha = 0.7$ leads to an optimized trade-off. If the density of the Next β Sectors is much higher than the density of the Previous β Sectors, the node forwards data as following:

   - If both the Next and the Previous β Sectors have high densities then the node forwards data hop-by-hop, since the forward area can afford the load of the previous area.
   - If both the Next and the Previous β Sectors have low densities then the node forwards data using a probability p_1. The probability p_1 randomizes the propagation of the message one-hop towards the *Sink* or directly to the *Sink* as follows:

$$\text{Probability} = \begin{cases} \gamma_1, & \text{"Jump" to the Sink} \\ 1-\gamma_1, & \text{Hop-by-hop transmission} \end{cases}$$

In fact, a small, constant probability γ_1 where $\gamma_1 > 0$ for jumps significantly improves performance, while larger values are not appropriate due to the high energy cost of jumps. The value of $\gamma_1 = \frac{1}{30}$ has been chosen after many simulations tries suggesting that this value optimizes performance. Again, the protocol designer can freely choose γ_1 achieve particular trade-offs: a large γ_1 would lead to faster but more expensive data propagation while a small one would result to more conserving energy consumption but slower data propagation. This selection encourages a lot the hop-by-hop transmissions. The jump transmissions are expensive for the sender node but do not overload the front network area.

b) $(1-\alpha) < \mathbf{F}_{\frac{P}{N}} < (1+\alpha)$, where α a constant as in case a). This suggests a threshold behaviour for $\mathbf{F}_{\frac{P}{N}}$ around the relative density value 1. This means that if the density of the Next β Sectors and the Previous β Sectors are approximately the same, the node should forward data as follows:

- If the Previous and the Next β Sectors have high densities, comparing with the Global density then the node forwards data using a probability p_2 which randomizes the propagation of the message one-hop towards the *Sink* or by-passing the overloaded front area as follows:

  $$\text{Probability} = \begin{cases} \gamma_2, & \text{By-pass transmission} \\ 1-\gamma_2, & \text{Hop-by-hop transmission} \end{cases}$$

  As in the case a) for jumps, a similar behavior applies this case for bypasses: a very small constant bypass probability γ_2, where $\gamma_2 > 0$ a very small constant suffices to improve performance a lot, with the protocol implementor being free to select the particular value accordingly. In by-pass transmission a node chooses randomly between the left and the right diagonial sector towards the *Sink* in order to transmit its data. The value of probability $1-\gamma_2$ has been chosen to be $\frac{29}{30}$ (i.e. $\gamma_2 = \frac{1}{30}$) after fine-tuning, encouraging a lot the hop-by-hop transmissions, as the front area is dense enough to transmit the data from the backward network area. However, the by-pass transmissions are used in order to avoid the overloaded front network area.

- If the Previous and the Next β Sector have low densities, the node forwards data to the next Sector, because the front area can relay the load of the backward network area.

c) $\mathbf{F}_{\frac{P}{N}} > (1+\alpha)$, where $0 < \alpha < 1$ a constant as in cases a) and b). If the density of the Next β Sectors is smaller than the density of the Previous β Sectors, the node forwards data using a probability p_3 which randomizes the propagation of the message one-hop towards the *Sink* or by-passing the overloaded front area as follows:

$$\text{Probability} = \begin{cases} \gamma_3, & \text{By-pass transmission} \\ 1-\gamma_3, & \text{Hop-by-hop transmission} \end{cases}$$

Similarly to cases a) and b), here again a very small, arbitrarily chosen bypass probability γ_3, where $\gamma_3 > 0$ a small constant is used. The particular value of probability $\gamma_3 = \frac{19}{20}$ (i.e. $\gamma_3 = \frac{1}{20}$) has been chosen after many simulations tries suggesting that this value optimizes performance. This value of probability γ_3 encourages a lot the the hop by hop transmissions over the by-pass transmissions. The by-pass transmissions are used in order to avoid the overloaded sparse front network area.

Overall, this algorithmic design is very simple basically suggesting a threshold behaviour around the value 1 for the relation of "forward" and "backward" network densitites. Also, a probabilistic, adaptive addition of a small amount of jumps or bypass transmissions (on top of multi-hop data propagation, which is the prevailing propagation method) is able to incur significant performance gains.

5. EXPERIMENTAL EVALUATION

Experimental Setup. Our simulation environment for making the experiments is Matlab 7.9.0. We consider two types of node deployment: a) **random uniform** and b) **heterogenous**. The transmission range R is taken 10, the initial energy in each sensor is $E_{initial} = 10^6$ and we conducted experiments for various node densities ($10{\cdot}10^3$, $15{\cdot}10^3$, $20{\cdot}10^3$ and $25{\cdot}10^3$ nodes). As depicted in Fig. 1, the circle network of radius 200 is divided into 12 Slices where each Slice contains 20 Sectors.

In detail, the network area is rectangular, with length and width equal to 40 units. For statistical smoothness, we apply several times the deployment of nodes in the network and repeat each experiment 50 times. For each experiment we simulate 30.000 data propagations and the average value is taken. The statistical analysis of the findings (the median, lower and upper quartiles, outliers of the samples) demonstrates very high concentration around the mean, so in the following figures we only depict average values.

The *Sink* is placed at the center $(100, 100)$ of the circle deployment area. In simulation experiments, we focus on the following metrics: a) **data delivery latency** which is the average number of hops needed to reach the *Sink*; b) **average energy consumption** is the average energy spent per node during the network operation; c) **variation of energy consumption** which is the deviation of the energy consumed by a node from the average energy consumption in the network. This metric is useful to determine whether the energy consumption in a network is energy-balanced, e.g. as it is impossible to achieve a close to zero deviation, the lower is the deviation the more energy-balanced is the network and vice versa; and d) **success ratio** is the ratio of the total number of packets received by the sink to the number of packets generated by the whole network.

We assume an energy model, in which the energy consumed by a message transmission between two nodes is considered the square of the distance between the nodes. For the above metrics the average is taken over all sensor deployments and algorithm's repetitions.

5.1 The impact of β (network information)

In order to choose the ideal β we have conducted a detailed experimental comparison of our protocol performance for various choices of β. As described previously, β is the amount of knowledge required by our protocol. More specifically, β is the number of the Previous β and Next β Sectors of which a node needs to know their density. In Section 4.3.1, we proposed the following basic choices about β, a) local knowledge of the network ($\beta = 1$) in which a node knows the density of the neighboring Sectors only, b) partial knowledge ($\beta = 3$) where a node knows the densities of the three Previous and Next Sectors and c) global density information ($\beta = N$) where a node has global information about the densities of all the N Sectors in the Slice.

We concluded that the best choice of β is 3, as it is feasible to get distributively Partial knowledge of the neighboring Sectors and has better performance from the case of local knowledge $\beta = 1$.

5.2 Comparison of Protocols

For the uniform deployment scenario, Fig. 3 (top) shows a negative impact of our Density-Based Protocol on data delivery latency and its performance is close to P_i's results. On the other hand, the E_i Protocol is the fastest of all as it makes a lot of direct transmissions. As the density of the network increases, the E_i Protocol becomes faster while the performance of the other two protocols remains stable.

The energy-latency trade-off is more clear if you observe the Fig. 3 (middle) that illustrates the average energy consumption per node. In more detail, the Density-Based Protocol is the most energy conservative. As expected the E_i Protocol depletes the energy of the nodes quickly and its energy-efficiency gets worse as the number of the network nodes increases.

Fig. 3 (bottom) depicts the variation of energy consumption which is a metric of the energy-balance property. Our Density-based Protocol has at least 4 times lower deviation from the average energy consumption than the deviation of the other two protocols. Our protocol's behaviour improves remarkably as the network density increases. We can claim that our Density-Based Protocol prolongs the lifetime of the network more than the other two energy-balance protocols. This result indicated that knowledge of the densities of neighboring areas (or more precisely, predicting the traffic forwarded to and received from other network areas) can lead to energy balance consumption. The P_i Protocol has the worst energy-balance behavior and this can be explained by the fact that it uses only proximity information in order to decide the next-hop transmission.

For the non-uniform deployment, the protocols have the same behavior for the data latency metric as shown in Fig. 4 (top). The results for the average energy consumption metric are shown in Fig. 4 (middle). The E_i Protocol is more expensive in the case of heterogeneous deployments rather than the uniform and the most expensive of all the protocols for both deployments. Also, in the heterogeneous case, the E_i Protocol has lower variation of energy consumption, see Fig. 4 (bottom), and for smaller network densities ($10*10^3$, $15*10^3$ nodes) has even better energy-balance performance than our Density-based Protocol. However, our Density-based Protocol's energy-balance behavior improves as the network density increases and achieves a more energy-efficient balanced consumption than the E_i protocol which has similar energy-balance behavior but in a less efficient way as it spends 3 times more energy than the Density-based Protocol.

5.3 Energy-Balance Property

In this section we study the energy-balance property of the protocols. In detail, we present graphically the spatial evolution of energy dissipation in a network of 20.000 nodes after 30.000 data propagations for both random uniform (see

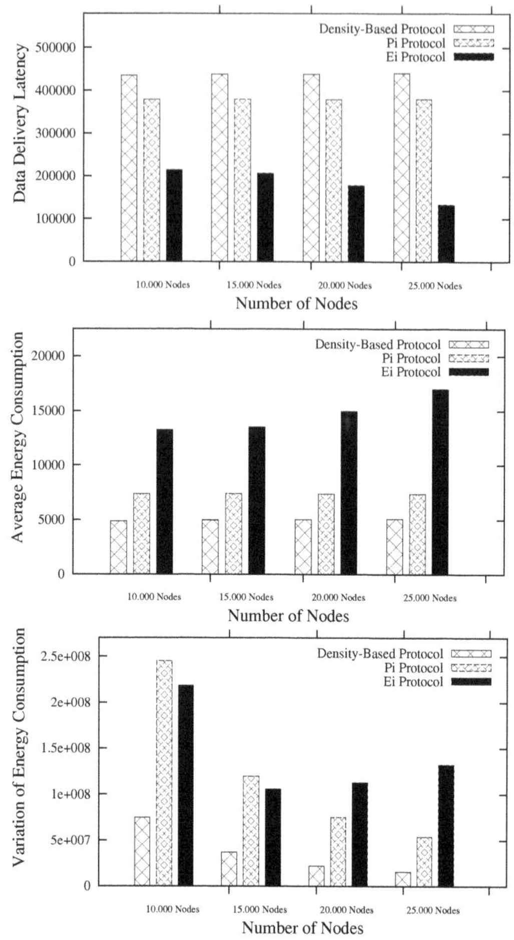

Figure 3: Experimental Comparison of the four protocols for various node densities in Uniform Network Deployment.

Fig. 5) and heterogeneous (see Fig. 6) sensor deployments. Nodes with high energy dissipation are depicted with dark colours. In contrast, nodes with high residual energy are depicted with bright colours.

We observe in Fig. 5 (left), that our Density-based Protocol achieves better energy balancing than the P_i Protocol Fig. 5 (center) and the E_i Protocol Fig. 5 (right). In detail, P_i Protocol tends to overuse the nodes that are in the middle in each sector, which can be explained by the random choice of the node of the next Sector as it is more probable to choose a node which is in the middle of the Sector area. The E_i Protocol exhausts the distant nodes as it makes many expensive jumps at the start of the network operation. On the other hand, our Density-based Protocol spends the energy of the sensors in a more balanced way.

Interestingly, in the scenario of heterogeneous node deployment depicted in Fig. 6 our Density-based protocol Fig. 6 (left) is again more energy-balanced than the P_i Protocol Fig. 6 (center) and the E_i Protocol Fig. 6 (right).

5.4 Comparing with the optimum

In this section, we compare the performance of the three energy-balance protocols (Density-based, P_i and E_i Proto-

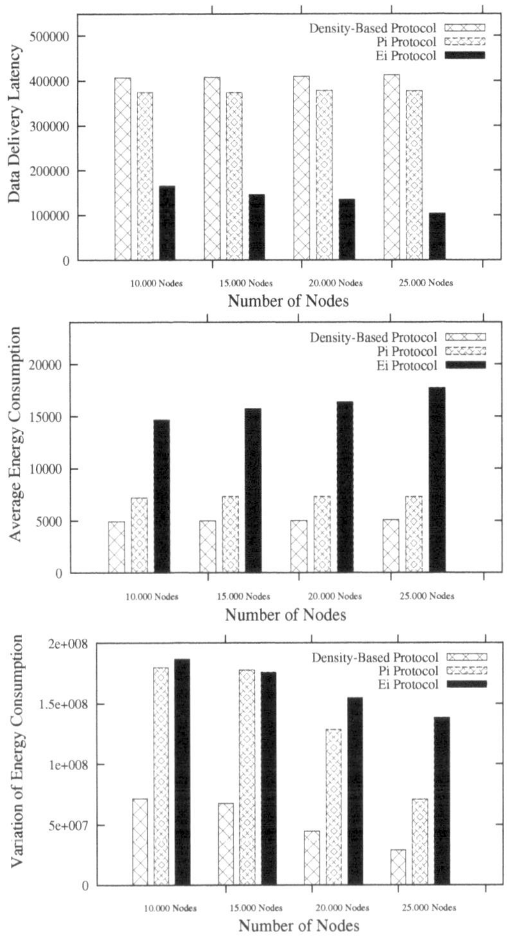

Figure 4: Experimental Comparison of the four protocols for various node densities in Non-Uniform Network Deployment.

col) with the centralized off-line optimal solution as given by the linear program described in Definition 5.

In the case of uniform deployment, Fig. 7 shows that the success ratio of Density-based Protocol is consistently higher during the network operation as the number of messages generated per node increases. The success ratio of E_i is close to our protocol's performance. On the other hand, P_i has the worst performance in terms of success ratio because it has the worse energy-balance behavior, and the nodes close the critical area around the *Sink* are depleted very early.

The energy consumption results are shown in Fig. 8. As expected, the E_i Protocol is the most expensive as it makes a lot of jumps at the start of the network operation. Our protocol is the most energy efficient of all and has close to optimal performance in terms of energy consumption. The P_i Protocol is the second expensive and it is more consuming if you take into consideration that it has very low success ratio as the network continues to operate (for $\lambda > 5$).

In the case of heterogeneous deployment, we remark that the energy consumption of the optimum solution is greatly reduced (Fig. 10), whereas the energy dissipation of other protocols is increased. As noticed in [9], you can achieve the optimum energy-balanced energy consumption in a sensor network by placing the nodes in a heterogeneous manner. If

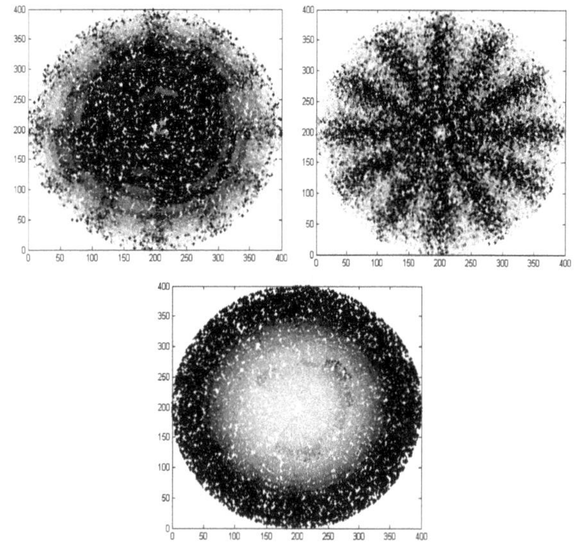

Figure 5: Energy dissipation map in uniform deployment using the Density-Based (top left), P_i (top right) and E_i Protocol (bottom).

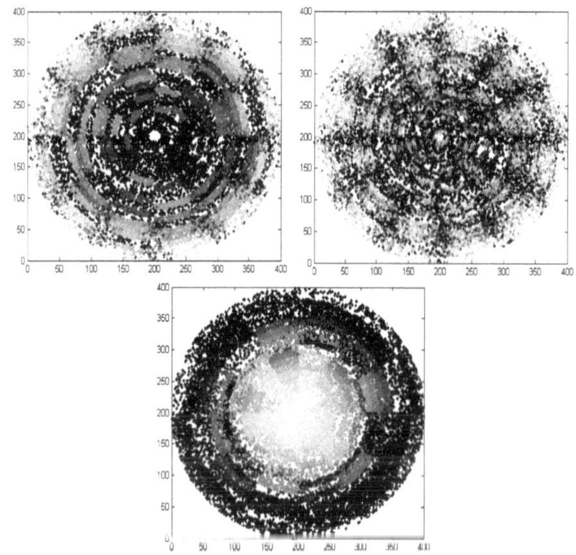

Figure 6: Energy dissipation map in non-uniform deployment using the Density-Based (top left), P_i (top right) and E_i Protocol (bottom).

you place the nodes uniformly in the area of interest, the nodes nearer to the sink carry heavier traffic loads and they will deplete their energy faster. For that reason, the optimum for the non-uniform deployment case has better performance from the optimum for the uniform deployment.

6. CONCLUSIONS/FUTURE WORK

Our approach is one of the few first that investigate the problem of maximizing the network lifetime (via energy balance) in the context of heterogeneous sensor deployment. We propose a new distributed, local and lightweight protocol that achieves improved energy balanced data propagation

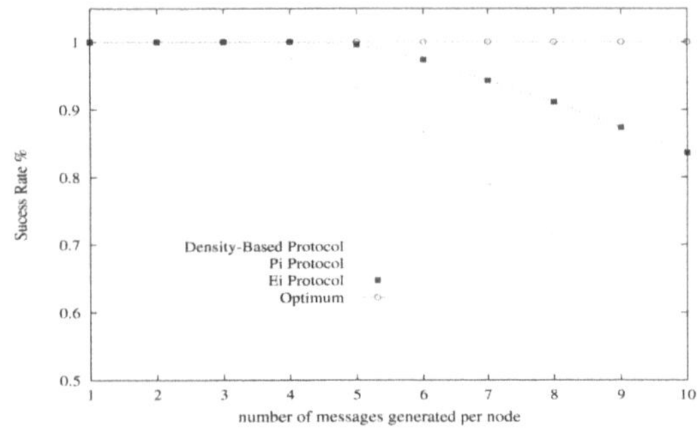

Figure 7: Success rate of the protocols and the optimum in Uniform Network Deployment.

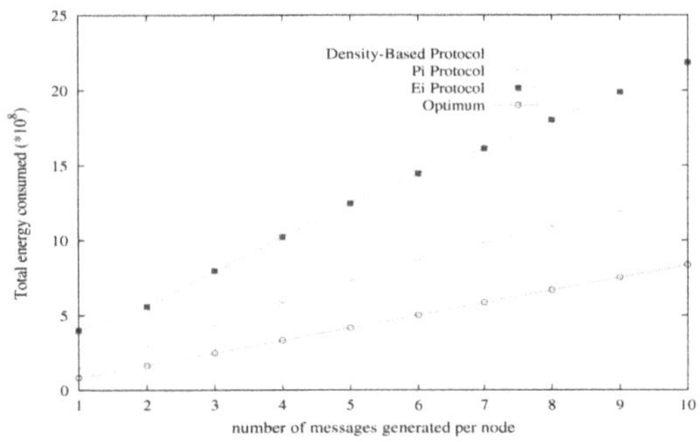

Figure 8: Energy consumption of the protocols and the optimum in Uniform Network Deployment.

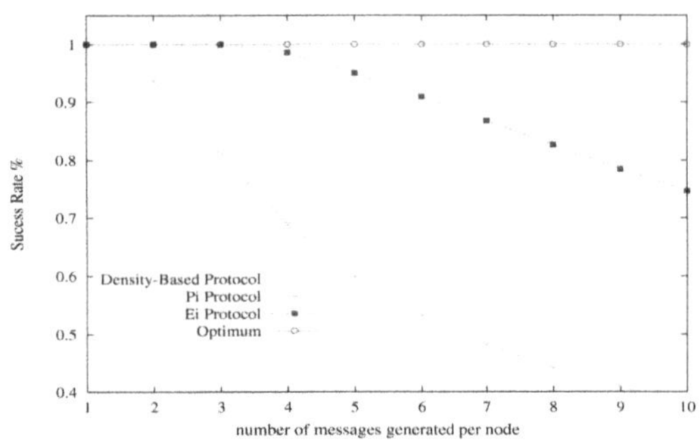

Figure 9: Success rate of the protocols and the optimum in Non-Uniform Network Deployment.

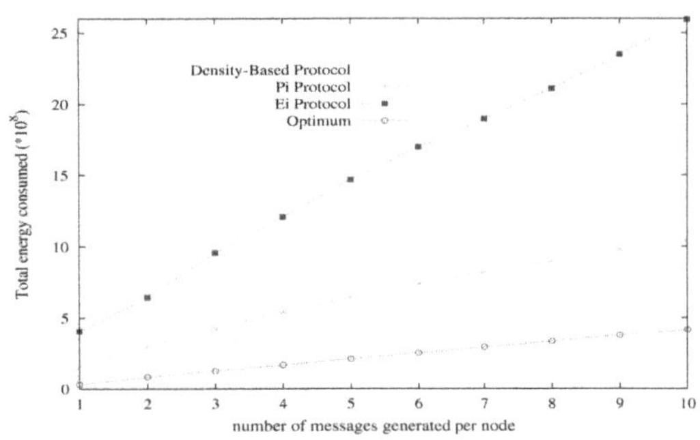

Figure 10: Energy consumption of the protocols and the optimum in Non-Uniform Network Deployment.

than other well known energy balance schemes; the main idea of our method is to exploit some limited knowledge of node density in the neighborhood of the node currently holding data under propagation. As shown by detailed simulation results, in uniform deployment our protocol has near-optimal performance (very close to the centralized off-line optimum solution by a linear program); in particular, our method is energy balanced and energy efficient.

In future work, we intend to study the problem of energy balanced data propagation in sensor networks with mobile nodes. Also, we plan to come up with a mathematically strict way of calculating the probabilities p_1, p_2 and p_3 in the Data Propagation Scheme of our Density-Based Protocol. Furthermore, we plan to implement the above mentioned energy-balanced protocols in a experimental test-bed consisting of Xbow's TelosB motes, and study their performance in real world conditions. We plan to also study the behaviour of our method in the regulated deployment of [9]. Finally, we plan to study the performance of our protocol in various propagation models where the power fall more rapidly than the square of the distance (e.g. $n = 3, n = 4$).

Acknowledgements This work has been partially supported by the EU Project HOBNET (ICT/FIRE STREP 257466).

7. REFERENCES

[1] F. Pasquale A. Clementi and R. Silvestri. Manets: High mobility can make up for low transmission power. In *ICALP*, pages 387–398, 2009.

[2] O. Powell A. Jarry, P. Leone and J. Rolim. An optimal data propagation algorithm for maximizing the lifespan of sensor networks. In *DCOSS*, volume 4026/2006, pages 405–421, 2006.

[3] S. Nikoletseas A. Jarry, P. Leone and J. Rolim. Optimal data gathering paths and energy balance mechanisms in wireless networks. In *DCOSS*, 2010.

[4] S. Nikoletseas C. Efthymiou and J. Rolim. Energy balanced data propagation in wireless sensor networks. *Wireless Networks*, 12(6):691–707, November 2006.

[5] R. Govindan C. Intanagonwiwat and D. Estrin. Directed diffusion: A scalable and robust communication paradigm for sensor networks. In *Proc. 6th ACM/IEEE International Conference on Mobile Computing MOBICOM*, 2000.

[6] S. Ganeriwal C. Schurgers, V. Tsiatsis and M. Srivastava. Topology management for sensor networks: Exploiting latency and density. In *MOBICOM*, pages 135–145, 2002.

[7] M. Singh and V. Prasanna. Energy-optimal and energy-balanced sorting in a single-hop wireless sensor network. In *Proc. First IEEE International Conference on Pervasive Computing and Comminications PERCOM*, 2003.

[8] A. Chandrakasan W. R. Heinzelman and H. Balakrishnan. Energy-efficient communication protocol for wireless microsensor networks. In *Proc. 33rd Hawaii International Conference on System Sciences HICSS*, 2000.

[9] G. Chen X. Wu and S. K. Das. Avoiding energy holes in wireless sensor networks with nonuniform node distribution. *IEEE Trans. Parallel Distrib. Syst.*, 19(5):710–720, 2008.

Minimum Latency Data Aggregation In The Physical Interference Model

Nhat X. Lam, Min Kyung An, Dung T. Huynh and Trac N. Nguyen
Department of Computer Science
University of Texas at Dallas
Richardson, Texas 75080
{nxl081000, mka081000, huynh, nguyentn}@utdallas.edu

ABSTRACT

Data aggregation has been the focus of many researchers as one of the most important applications in Wireless Sensor Networks. A main issue of data aggregation is how to construct efficient schedules by which data can be aggregated without any interference. The problem of constructing minimum latency data aggregation schedules (MLAS) has been extensively studied in the literature although most of existing works use the graph-based interference model.

In this paper, we study the MLAS problem in the more realistic physical model known as Signal-to-Interference-Noise-Ratio (SINR) where few works exist and algorithms that guarantee theoretical performances are scarce [17, 16]. We first derive an $\Omega(\log n)$ approximation lower bound for the MLAS problem in the metric SINR model. We also prove the NP-completeness of the decision version of MLAS in the geometric SINR model. This is a significant contribution as these results have not been obtained before for the SINR model. In addition, we propose a constant factor approximation algorithm whose latency is bounded by $O(\Delta + R)$ for the dual power model, where Δ is the maximum node degree of a network and R is the network radius. Finally we study the performance of the algorithms through simulation.

Categories and Subject Descriptors

C.2.1 [**Computer-Communication Networks**]: Network Architecture and Design—*Wireless communication*; F.2.2 [**Analysis of Algorithms and Problem Complexity**]: Nonnumerical Algorithms and Problems—*Geometrical problems and computations*

General Terms

Algorithms, Theory

Keywords

Data aggregation, Inapproximability, NP-Completeness, Geometric graph, Signal-to-Interference-Noise-Ratio, Approximation algorithm

1. INTRODUCTION

In Wireless Sensor Networks (WSNs), a number of small-sized sensor nodes monitor nearby environmental condition and collect data using their radio signals. The collected data is forwarded to a destination called a *sink*. This type of application is called *data aggregation*, and it is one of the most crucial applications of WSNs.

Despite the remarkable improvement over traditional sensing technology, the sensor nodes still have limited resources. Therefore, reducing energy consumption to prolong the lifetime of the network is an important issue in WSN research. In order to avoid the unnecessary retransmission using sensors' limited power, researchers have focused on assigning time slots to sensor nodes to obtain a good schedule. Due to the periodical occurrence of data collection, reducing the *latency* of the schedules, that is, constructing schedules with minimum number of time slots, has been widely investigated.

The problem of constructing minimum latency data aggregation schedules in the graph-based model has been studied by many researchers. In the collision-free model, [2] proposed a heuristic algorithm that constructs a tree with a schedule. It assigns a code and a time slot to each node to communicate with its parent node. Later, [6] claimed that there exists a schedule whose length is at most $\min\{\log_2(\frac{|C|+2}{3})+1, 3\log_3(\frac{2|C|+3}{5})+2\}$ for unit disk graphs, where C is a set of source nodes that have data to be sent to a sink node. Recently, in [13], Huang et al. introduced a nearly constant factor approximation algorithm whose latency is bounded by $23R + \Delta - 18$, where R is the network radius and Δ is the maximum node degree of the network. And in [4], Chen et al. proposed a Δ-approximation algorithm, and showed NP-hardness for the grid topologies.

While only collision was considered in those papers, in [1] the authors focused on interference as well. Given a transmission power level d for each node, the interference range is defined as αd, where $\alpha \geq 1$ is the interference factor. [1] proved an $\Omega(\log n)$ approximation lower bound in the metric model, and proposed a heuristic algorithm under the assumption that nodes can have non-uniform power levels. In [20], Xu et al. studied lower bounds for data aggregation latency for the cases $1 < \alpha < 3$ and $\alpha \geq 3$ in the uniform power model, and obtained $\max\{\frac{\Delta}{\phi}, R\}$ and $\max\{\Delta, R\}$, respectively, where $\phi = \frac{2\pi}{\lfloor \arcsin \frac{\alpha-1}{2\alpha} \rfloor}$.

Recently, several researchers have started investigating data aggregation scheduling in the more realistic physical model known as the Signal-to-Interference-Noise-Ratio (SINR) model. Unlike the graph model, in the physical model, the signal sent by a node u fades, and background noise and signals sent concurrently by other nodes can interfere with u's signal. Thus, the signal sent to node v may not be strong enough to be received and, hence, transmitted data is lost. In the SINR model, only if the SINR is beyond a certain threshold, the transmitted data can be successfully received at the receiver node.

The first investigation of the problem of constructing minimum latency data aggregation schedules in the SINR model has been done by Li et al. in [17]. Under the assumption that all nodes are initially assigned a uniform power level, [17] introduced a scheduling algorithm whose latency is bounded by $O(\Delta + R)$ which is asymptotically optimal. Later, [16] proposed a distributed algorithm that yields $O(K)$ time slots, where K is the logarithm of the ratio between the lengths of the longest and shortest links in a network. It also proposed, under the assumption that the transmission power of each node is large enough to cover the maximum node distance in the network, a centralized algorithm whose latency is $O(\log^3 n)$ which is currently best. While these studies have been concerned with data aggregation, some other researchers have focused on *scheduling* [3, 8, 9, 10, 15] and *broadcasting* [12] algorithms in the SINR model. [8] was the first study of NP-completeness of scheduling in the geometric SINR model, and [12] introduced an approximation algorithm for broadcasting in the 2-disk and SINR models.

In this paper, we continue the study of minimum latency data aggregation scheduling (MLAS) in the SINR model. We show that there is no algorithm having an approximation ratio better than $\Omega(\log n)$ in the metric SINR model. In addition, we prove the NP-hardness of the MLAS problem in the geometric SINR model. To the best of our knowledge these results have not been obtained in the literature although several researchers have designed approximation algorithms for MLAS. (Note that existing NP-hardness results hold for the graph-based interference model only.) Additionally, based on an algorithm in [17], we also introduce a constant factor approximation algorithm yielding a latency bounded by $O(\Delta + R)$ for geometric MLAS in the dual power model. Simulation results show that our algorithm produces better schedules. This algorithm can be adapted to the case where multiple power levels are present.

This paper is organized as follows. In Section 2, we describe our network models and introduce the definitions used in this paper. In Section 3, we show the $\Omega(\log n)$ approximation lower bound for MLAS problem in the metric model with uniform power, and derive the NP-hardness of the MLAS problem in the geometric model. In Section 4 we introduce a constant factor approximation algorithm for the MLAS problem in dual power model and show some simulation results. Section 5 contains some concluding remarks.

2. NETWORK MODELS

In this paper, a wireless network consists of a number of arbitrarily distributed sensor nodes. Each node is equipped with a radio frequency transceiver that sends and receives data. We model such a network as (V, D, p), where V is the set of n sensor nodes, $D : V \times V \longrightarrow R^+$ represents the distance function between nodes in V and $p : V \longrightarrow \{p_1, p_2, ...p_k\}$ is the power assignment function that assigns each node a power level $p(v) \in \{p_1, p_2, ...p_k\}$.

Considering a communication link $l(s, r)$ in this network, where s is a sender and r is a receiver, and let X be the set of other concurrently transmitting links. Adopting the physical interference model in [11], we define the $SINR$ value of the link $l(s, r)$ as follow.

$$SINR(s,r) = \frac{p(s)D(s,r)^{-\alpha}}{N + \sum_{(u,v)\in X} p(u)D(u,r)^{-\alpha}}$$

where $N > 0$ is the ambient noise and $2 \leq \alpha \leq 6$ is the path loss exponent. The receiver r can successfully receive the signal from the sender s if and only if the $SINR(s, r)$ value exceeds a given threshold $\beta \geq 1$. Thus, a node u with power $p(u)$ can send signals to only nodes that are within the distance $d \leq (\frac{p(u)}{N\beta})^{\frac{1}{\alpha}}$. We call these nodes u's neighbors. Implicitly, the network communication is modeled as a directed graph $G(V, E)$, where $E = \{(s \to r) | s, r \in V, D(s, r) \leq (\frac{p(s)}{N\beta})^{\frac{1}{\alpha}}\}$.

The minimum aggregation scheduling problem is defined as follows. A schedule is defined to be a sequence of time slots, at each of which, several nodes are scheduled to send its aggregated data to one of its neighbors, and every node can be scheduled as a sender only once. Formally, at each time slot t, we have an assignment vector $\pi_t = (l_{t_1}, l_{t_2}, ...l_{t_w})$, in which l_{t_i} is a directed link from s_{t_i} to r_{t_i} satisfying SINR threshold inequalities, i.e., all senders s_{t_i} can transmit concurrently. And a schedule, as a sequence of assignment vectors, is denoted as $\Pi = (\pi_1, \pi_2, ...\pi_L)$, where L, the length of the schedule, is also called its *latency*.

Minimum Latency Aggregation Scheduling (MLAS): Given a set $C \subset V$ of source nodes and the sink node $s \in V \setminus C$, the objective of the data aggregation scheduling problem is to find the minimum latency schedule to aggregate data from all source nodes in C to the sink node s.

Metric Model: In the metric version of the problem, we assume sensor nodes are located in a metric space (V, D, p) where D satisfies the triangle inequality.

Geometric Model: The geometric model is a special case of the metric model, where sensor nodes are deployed in the plane and D is defined as the Euclidean distance between nodes.

3. COMPLEXITY OF MINIMUM LATENCY AGGREGATION SCHEDULING

3.1 Inapproximability of Metric MLAS

In this section, we prove an $\Omega(\log n)$ approximation lower bound for the MLAS problem in the metric model. The result even holds true in the uniform power model ($k = 1$).

THEOREM 1. *There is no approximation algorithm having an approximation ratio better than $\Omega(\log n)$ for the MLAS problem in the metric model unless $NP \subseteq DTIME(n^{\log\log n})$.*

PROOF. We construct a polynomial-time approximation-preserving reduction from the *set cover* problem that is known to be hard to approximate. Let I_S be an instance of the set cover problem consisting of a collection S of subsets of a finite set E of elements. Let n and m denote the cardinalities of E and S, respectively. A solution to I_S is a subset $S' \subseteq S$ such that every element $e \in E$ is in at least one set $A \in S'$. Let OPT_S denote an optimal solution to I_S.

Given the instance I_S of the set cover problem, we construct in polynomial time an instance I_M of the MLAS problem as follows. I_M consists of $n + mn + 1$ nodes. The set of nodes V in I_M is partitioned into three subsets of nodes. The first subset contains only a single node s_n as the sink node. The second subset C_S and the third subset C_E consist of mn and n nodes, respectively. All nodes in C_E are source nodes. (See Figure 3.1.)

The set C_S is broken down into n layers of nodes. Each layer k contains m nodes, namely $s_k^1, s_k^1, \ldots\ s_k^m$. Consider the first layer of nodes in this set denoted by C_S. These m nodes in C_S correspond to the m sets in the collection S. For each node $a \in C_S$, let $S(a)$ be the set in S that a represents. Similarly, the n nodes in C_E correspond to the n elements in E of I_S. And $e(a)$ denotes the element in E that corresponds to node a in C_E.

We now define the distance function D on $V \times V$ for the metric model of the MLAS problem. Let $\tau > 0$ be some constant, and define the distance between nodes $u, v \in V$ as follows:

$$D(u,v) = \begin{cases} \tau & u = s_k^i, v = s_{k+1}^i, \\ & 1 \leq i \leq m, 0 \leq k < n \\ \tau & u = s_{n-1}^i, v = s_n, \\ & 1 \leq i \leq m \\ 2\tau & u = s_k^i, v = s_k^j, \\ & 1 \leq i < j \leq m, 0 \leq k < n \\ \tau & u = s_1^i, v \in C_E, \\ & e(v) \in S(u) \\ 2\tau & u, v \in C_E \end{cases}$$

All other distances follow from symmetry or are induced by the shortest path metric. We define the transmission power for each node as $p = \beta N \tau^\alpha$. From this definition, observe that there does not exist any direct connection between nodes in C_E, nor between nodes in different columns of C_S. Similarly, there is no direct connection between any node a in C_E and nodes in C_S whose corresponding sets do not contain $e(a)$.

It has been shown in [7, 18] that the set cover problem cannot be approximated with an approximation factor better than $\Omega(\log n)$ unless $NP \subset DTIME(n^{\log \log n})$. Let OPT_M be the optimal solution to instance I_M. The following two implications will be shown in Lemma 2 and Lemma 3 below. For some constant $\nu \geq 2$:

$$|OPT_S| \leq \nu \implies L(OPT_M) \leq n + n\nu$$
$$|OPT_S| > \ln(n)\,\nu \implies L(OPT_M) > n\ln(n)\,\nu$$

Thus, for the inapproximability factor r the following holds:

$$r \geq \frac{n\ln(n)\,\nu}{n + n\nu} \geq \frac{\ln(n)\,\nu}{\nu + 1} \geq \frac{\ln(n)}{2}$$

In addition, we have that $\ln|V| < \ln(n + nm + 1) < 2\ln(n) + c$, for some constant c. Substituting this into the above inequality concludes the proof of the theorem. □

In order to prove Lemma 2 and Lemma 3, we need the following fact.

FACT 1. *While a node in V is transmitting, no other node can send data without any interference.*

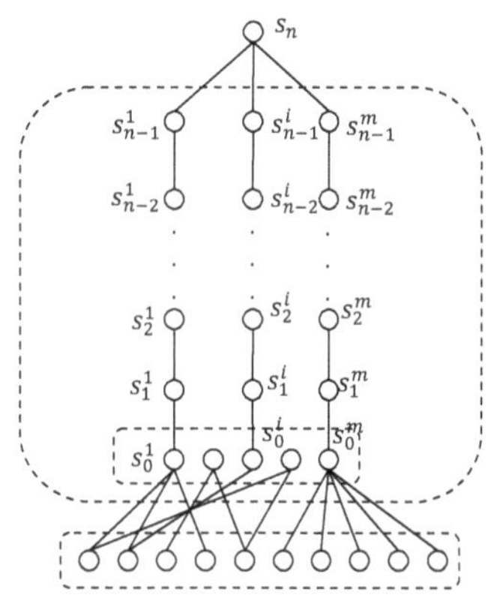

Figure 1: The corresponding metric MLAS instance

PROOF. Consider a sender node u with the transmission power $p = \beta N \tau^\alpha$. A node v can successfully receive data from u if $D(u,v)$ is $\leq (\frac{p_L}{\beta N})^{\frac{1}{\alpha}} = \tau$. Now observe that the distances between u and its neighbors are at least τ. Therefore, it cannot send data simultaneously with other nodes. Thus, the Fact 1 follows. □

LEMMA 2. *If the size of OPT_S is $\leq \nu$, then $L(OPT_M) \leq n + n\nu$.*

PROOF. Let us consider a solution of I_M constructed from OPT_S of I_S that is constructed as follows. Every node in C_E sends to its neighbor whose corresponding set is in OPT_S. Then, these selected nodes send data to the sink s_n through all intermediate nodes sequentially.

According to Fact 1, the length of this schedule is $n|OPT_S| + n$. So, if $|OPT_S| \leq \nu$, we also have $L(OPT_M) \leq n\nu + n$. □

LEMMA 3. *If the size of OPT_S is $> ln(n)\nu$, then $L(OPT_M) > n\ln(n)\nu$.*

PROOF. Observe that nodes in the first layer of C_S can only send their data to the sink node sequentially through the nodes in their column. And for each starting node in the first layer of C_S, it requires at least n steps to reach the sink node. Thus, if the size of OPT_S is $> ln(n)\nu$, then at least $\ln(n)\,\nu$ such nodes in the first layer of C_S are needed to collect data from all nodes in C_E, $L(OPT_M)$ is $> n\ln(n)\,\nu$. □

3.2 NP-Completeness of Geometric MLAS

In this section, we prove the NP-completeness of the MLAS decision problem in the geometric model for a special case, where the objective is to aggregate data from all nodes to the sink. Thus, the general version of the problem is also NP-complete.

It is straightforward that this problem is in NP. In order to prove its NP-hardness, we construct a polynomial time reduction from the *Partition* problem which was proven NP-complete in [14]. This decision problem is defined as follows. Given a finite set of distinct and positive integers, the objective is to determine if it is possible to divide this set into two subsets such that the sums of all integers in each subset are equal.

Let I_P be an instance of the Partition problem consisting of a set S of n distinct and positive integers $a_1, a_2, \ldots a_n$. Without loss of generality, assume that $a_1 < a_2 < \ldots <$

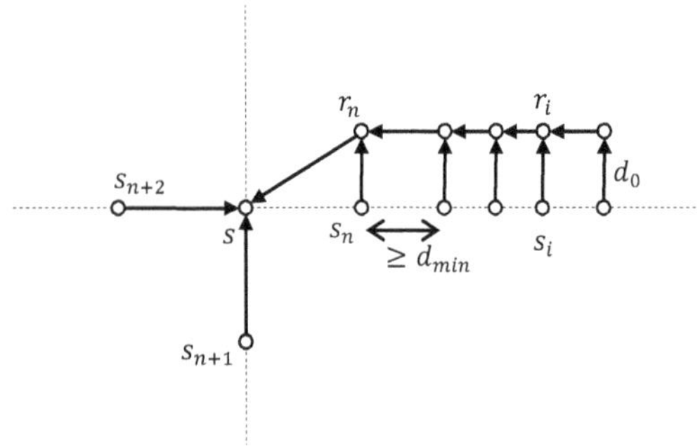

Figure 2: The corresponding geometric MLAS instance

a_n. We construct in polynomial time an instance I_M of the geometric MLAS problem as follows.

I_M consists of $2n+3$ nodes which include $2n$ nodes s_i and r_i, 2 nodes s_{n+1} and s_{n+2} and a sink node s. Letting P be some constant, we deploy these nodes on the plane at the following positions.

$$pos(s) = (0,0)$$
$$pos(s_{n+1}) = \left(0, -\left(\frac{24P}{N\beta(A^\alpha\beta+24)}\right)^{\frac{1}{\alpha}}\right)$$
$$pos(s_{n+2}) = \left(-\left(\frac{24P}{N\beta(A^\alpha\beta+24)}\right)^{\frac{1}{\alpha}}, 0\right)$$

and for all $1 \leq i \leq n$,

$$pos(s_i) = \left((\frac{P}{b_i N\beta})^{\frac{1}{\alpha}}, 0\right)$$
$$pos(r_i) = \left((\frac{P}{b_i N\beta})^{\frac{1}{\alpha}}, d_0\right)$$

where $\sigma = \sum_{i=1}^{n} a_i$, $A = \left((\frac{1}{a_n-1})^{\frac{1}{\alpha}} - (\frac{1}{a_n})^{\frac{1}{\alpha}}\right)$, $b_i = \frac{a_i A^\alpha}{12\sigma}$, and $d_0 = \left(\frac{12P\sigma}{12\sigma N\beta + nN\beta^2}\right)^{\frac{1}{\alpha}}$.

It remains to define the power of each node. Letting $d(u,v)$ denote the Euclidean distance between u and v, we define
$P(s_i) = P(s_{n+1}) = P(s_{n+2}) = P$
$P(r_i) = N\beta d(r_i, r_{i+1})^\alpha, 1 \leq i \leq n, r_{n+1} \equiv s$

FACT 2. *Let $S_i = \{s_j | 1 \leq j \leq n+1 \wedge i \neq j\}$. It holds for all $1 \leq i \leq n$ that the SINR exceeds β when node s_i is scheduled to send data to r_i concurrently with the set S_i.*

PROOF. Consider the distance between two nodes s_i and s_j, where $1 \leq i, j \leq n$.

$$d(s_i, s_j) = \left|(\frac{P}{b_i N\beta})^{\frac{1}{\alpha}} - (\frac{P}{b_j N\beta})^{\frac{1}{\alpha}}\right|$$
$$= \frac{(12P\sigma)^{\frac{1}{\alpha}}}{AN\beta}\left|(\frac{1}{a_i})^{\frac{1}{\alpha}} - (\frac{1}{a_j})^{\frac{1}{\alpha}}\right|$$
$$\geq \frac{(12P\sigma)^{\frac{1}{\alpha}}}{AN\beta}\left((\frac{1}{a_n-1})^{\frac{1}{\alpha}} - (\frac{1}{a_n})^{\frac{1}{\alpha}}\right) = (\frac{12P\sigma}{N\beta})^{\frac{1}{\alpha}}$$

Now, letting $d_{min} = (\frac{12P\sigma}{N\beta})^{\frac{1}{\alpha}}$, we have a lower bound for the SINR value of a link (s_i, r_i) as follows.

$$SINR(s_i, r_i) > \frac{Pd_0^{-\alpha}}{N + nPd_{min}^{-\alpha}} = \frac{P\frac{12\sigma N\beta + nN\beta^2}{12P\sigma}}{N + nP\frac{N\beta}{12P\sigma}} = \beta$$

□

FACT 3. *For all $1 \leq i \leq n$, s_i can send data only to r_i.*

PROOF. From Fact 2, we can see that s_i can send data to r_i. We now show that s_i cannot send data to other nodes.

Observe that with $\alpha \geq 2$ and $a_n \geq 2$ we have $A^\alpha < 0.086$. Hence, for b_i it holds that $0 < b_i < 1$ for all $1 \leq i \leq n$. Therefore, s_i cannot send data to s since $Pd(s_i, s)^{-\alpha} = b_i N\beta < N\beta$.

Finally, we see that all other nodes have a distance at least d_{min} away from s_i. Thus, s_i cannot send data to any such node u since $Pd(s_i, u)^{-\alpha} \leq Pd_{min}^{-\alpha} = \frac{N\beta}{12\sigma} < N\beta$. □

FACT 4. *s_{n+1} and s_{n+2} can send data only to s.*

PROOF. Observe that s_{n+1} (s_{n+2}) can send data to s since $Pd(s_{n+1}, s)^{-\alpha} = \frac{N\beta(A^\alpha\beta+24)}{24} > N\beta$. But s_{n+1} (s_{n+2}) cannot send data to any node s_i since

$$Pd(s_{n+1}, s_i)^{-\alpha} \leq Pd(s, s_i)^{-\alpha} < N\beta$$

Next we claim that s_{n+1} and s_{n+2} cannot communicate with each other. This holds true because $Pd(s_{n+1}, s_{n+2})^{-\alpha} = \frac{N\beta(\frac{A^\alpha\beta}{24}+1)}{2^{\frac{\alpha}{2}}} < \frac{N\beta(0.004\beta+1)}{2} < N\beta$. □

FACT 5. *When a node r_i is sending to r_{i+1} ($r_{n+1} \equiv s$), no other node can send without interference.*

PROOF. Assume that there exists a node u sending data at the same time with the link (r_i, r_{i+1}) and this node causes an interference $I_u > 0$ at r_{i+1}. We have $SINR(r_i, r_{i+1}) \leq \frac{P(r_i)d(r_i, r_{i+1})^{-\alpha}}{N+I_u} < \frac{P(r_i)d(r_i, r_{i+1})^{-\alpha}}{N} = \beta$. This means that r_i cannot successfully send data to r_{i+1}. □

FACT 6. *It holds for all $1 \leq i < n$ that r_i can aggregate data to s through r_{i+1} only.*

PROOF. Due to the power of r_i, it is obvious that r_i cannot send data to any nodes s_j and r_j for all $j > i+1$, nor to node s_{i+1}. In addition, Fact 3 shows that s_i can send data to r_i only. Although, r_i may be able to communicate with nodes on its right- hand side, in order to reach s on its left-hand side it must send data to r_{i+1}. □

LEMMA 4. *I_P has a solution if and only if I_M has a $n+2$ latency aggregation schedule.*

PROOF. First, we will show that every aggregation schedule of I_M requires at least $n+2$ time slots. To this end, observe that due to Fact 5 and Fact 6, every solution to I_M must contain n time slots, in each of which, r_i is scheduled to send its data to r_{i+1}. Moreover, due to Fact 4 and Fact 5, every solution must contain 2 additional time slots in which s_{n+1} and s_{n+2} send their data to the sink s. Thus, the lower bound $n+2$ follows. We now return to the proof of Lemma 4. For the only-if direction, assume that (S_1, S_2) is a partition of S where $\sum_{a \in S_1} a = \sum_{a \in S_2} a$. We show how to construct an $(n+2)$-latency schedule to aggregate data from all nodes to s. Together with the $n+2$ lower bound, it follows that this schedule is an optimal solution to I_M.

In the first time slot, we schedule all nodes s_i, that correspond to integers a_i in S_1, to send data to r_i. Also in this time slot, we schedule s_{n+1} to send data to s.

All remaining nodes s_i, that correspond to integers a_i in S_2, are assigned to send data to r_i in the second time slot.

And also in this time slot, s_{n+2} is scheduled to send its data to s.

The ith time slot of next n time slots is used for r_i to send its data to r_{i+1}, where $r_{n+1} \equiv s$. These n time slots do not have any interference since each contains only one link. Thus, we only need to consider the first two time slots. Due to Fact 2, it is guaranteed that the $SINR$ value at all receivers except s exceeds β. Therefore, we only need to check the $SINR$ value at s in these two time slots.

Since the analyses of these two time slots are similar, we just consider the first one. We have

$$SINR(s) = \frac{\frac{N\beta(A^\alpha\beta+24)}{24}}{N+\sum_{a_i\in S_1} Pd(s,s_i)^{-\alpha}} = \frac{\frac{N\beta(A^\alpha\beta+24)}{24}}{N+\frac{A^\alpha N\beta}{12\sigma}\sum_{a_i\in S_1} a_i} = \beta$$

This means that all links in the first time slot can successfully transmit their data.

For the if direction, we show that if I_M has an $n+2$ latency aggregation schedule, then I_P has a solution. To this end, observe that due to the $n+2$ lower bound and Fact 5, in order to obtain a $(n+2)$-latency schedule, all nodes s_i must be scheduled with either s_{n+1} or s_{n+2}. This means that we have to split the set of all nodes s_i into two different time slots such that in each time slot the interference experienced at s is exactly $\frac{A^\alpha N\beta}{24}$. It is straightforward to see that the partition corresponding to this schedule yields a solution to I_P. □

THEOREM 5. *The MLAS decision problem in the geometric model is NP-complete.*

PROOF. In Lemma 4, we have proved that Partition is polynomial-time reducible to MLAS. As MLAS is easily seen to be in NP, it is therefore NP-complete. □

4. CONSTANT FACTOR APPROXIMATION ALGORITHMS

In this section, we introduce a constant factor approximation algorithm for the geometric MLAS problem in the dual power model, and then provide a small improvement to get better schedules.

In practice, the links (s,r), where $d(s,r) = (\frac{p(s)}{N\beta})^{\frac{1}{\alpha}}$, are not good candidates to be scheduled concurrently. We are more interested in links (s,r) where $d(s,r) \leq (\frac{\min_{u\in\{s,r\}} p(u)}{\gamma N\beta})^{\frac{1}{\alpha}}$, for some constant $\gamma > 1$. In designing the constant-factor approximation algorithm for MLAS problem, we assume that the undirected graph $G'(V,E')$ where $E' = \{(s,r)|d(s,r) = (\frac{\min_{u\in\{s,r\}} p(u)}{\gamma N\beta})^{\frac{1}{\alpha}}\}$, is connected. We further make the traditional assumption that $\alpha > 2$ [11].

4.1 Data Aggregation Tree Construction

In the dual power model, each node is assigned to one of the low power level p_l or the high power level $p_h = p_l \cdot \delta$, where $\delta > 1$. We use v_h to denoted a node that is assigned power level p_h, also called *high power node*, and v_l to denote a node that is assigned power level p_l, also called *low power node*. We use $d_h = (\frac{p_h}{\gamma N\beta})^{\frac{1}{\alpha}}$ and $d_l = (\frac{p_l}{\gamma N\beta})^{\frac{1}{\alpha}}$ to denote the maximum link lengths in the graph $G'(V,E')$ generated by v_h's and v_l's , respectively.

The algorithm starts by partitioning a network into cells each of which has diagonal lengths d_h or d_l. This induces two grids where the upper-left corners have coordinates $(1,1)$. Let us denote the cells whose diagonal length is d_h by **HC** (**HighCell**), and the cells whose diagonal length is d_l **LC** (**LowCell**). Each **HC** (**LC**) is denoted by a Cell-ID **HC**(i,j) (**LC**(l,m)), if its upper-left corner has coordinates (i,j) ((l,m)).

We now construct a data aggregation tree on G' as done by Li et al. in [17]. [17] chooses a *center* node c, and constructs a breadth first search (BFS) tree (cf. [5]) on G' rooted at the node c in order to get a latency bound in terms of the network radius R rather than its diameter. A node v can be chosen as the center node if the distance from v to the most distant node from v is minimum. Once the data is aggregated to the center node c, it sends the aggregated data to the sink node via a shortest path. This takes only $O(R)$ time slots. Based on the BFS tree, [17] finds a Maximal Independent Set (MIS) on G' using an algorithm in [19]. (A subset $V' \subseteq V$ of a graph $G = (V,E)$ is said to be *independent* if for any vertices $u,v \in V'$, $(u,v) \notin E$. An independent set is said to be *maximal* if it is not a proper subset of another independent set.) We call the nodes in an MIS *dominators*, and the others *dominatees*. The MIS constructed by [19] guarantees that the distance between any pair of its complementary subsets is *exactly two hops*. Therefore, in each **LC** we can have at most 1 dominator, and in each **HC** we can have at most $\lceil\delta^{\frac{1}{\alpha}}\rceil^2$ dominators (See Lemma 8). To obtain a Connected Dominating Set (CDS) of G', [17] connects the dominators using some *connectors* that were originally dominatees. (A *dominating set* (DS) is a subset $V' \subseteq V$ such that every vertex v is either in V' or adjacent to a vertex in V'. A DS is said to be *connected* (CDS) if it induces a connected subgraph.) Since there may exist dominatees that are not connected to the CDS, [17] connects each of the remaining dominatees to its neighboring dominator that has the smallest hop-distance to the root of the BFS tree. We denote the newly formed tree by T_{Agg}, and use it as the data aggregation tree in our algorithm.

4.2 Data Aggregation Scheduling

Once the data aggregation tree is obtained, a number of iterations are performed to find a schedule. The scheduling is based on two constant values $\lceil H_h\rceil$ and $\lceil H_l\rceil$ which guarantee that two high power senders (low power senders) can transmit data without any interference if they are at least $\lceil H_h\rceil$ ($\lceil H_l\rceil$) cells apart from each other. The details are contained in Algorithms 1 and 2.

Our algorithm is based on [17], and extended to fit the dual power model. In the first phase, it schedules the nodes in T_{Agg} starting with the dominatees so that they can send data successfully to their upper level dominators (Steps 4–6 in Algorithm 1). After all dominatees are scheduled, only dominators and connectors are left. In the second phase (Steps 7–14 in Algorithm 1), all dominators and connectors are scheduled by first assigning time slots to all dominators at level i so that they can send data successfully to the connectors at level $i-1$ (Steps 8–10 in Algorithm 1). Algorithm 1 then assigns time slots to the connectors (Steps 11–13 in Algorithm 1). Steps 7–14 are repeated for scheduling the remaining dominators and connectors level by level until all dominators and connectors are scheduled.

Since our algorithm is developed for the dual power model, we again divide the phases as follows:

1. **Phase 1**: dominatees to dominators
   (a) High power dominatees to dominators
   (b) Low power dominatees to dominators

Algorithm 1 Data Aggregation Scheduling

Input: A set V of nodes with two power levels p_h and p_l
Output: Length of schedule

Partition the network into **HC**s and **LC**s each of which has diagonal length $(\frac{p_h}{\gamma N\beta})^{\frac{1}{\alpha}}$ and $(\frac{p_l}{\gamma N\beta})^{\frac{1}{\alpha}}$, respectively.
Construct T_{Agg} using an algorithm in [17].
Set the starting time slot $t \leftarrow 1$
Let $S_h \subseteq V$ be the set of high power dominatees, and $S_l \subseteq V$ be the set of low power dominatees.
$t \leftarrow$ **HighPowerScheduling**(S_h, t)
$t \leftarrow$ **LowPowerScheduling**(S_l, t)
for $i = R$ to 2 **do**
  Let $S_h^i \subseteq V$ be the set of high power dominators, and $S_l^i \subseteq V$ be the set of low power dominators at level i in T_{Agg}.
  $t \leftarrow$ **HighPowerScheduling**(S_h^i, t)
  $t \leftarrow$ **LowPowerScheduling**(S_l^i, t)
  Let $S_h^{i-1} \subseteq V$ be the set of high power connectors, and $S_l^{i-1} \subseteq V$ be the set of low power connectors at level $i-1$ in T_{Agg}.
  $t \leftarrow$ **HighPowerScheduling**(S_h^{i-1}, t)
  $t \leftarrow$ **LowPowerScheduling**(S_l^{i-1}, t)
end for
return $t - 1$

Algorithm 2 High(Low)PowerScheduling Algorithm

Input: A set S of high (low) power nodes and a starting time slot t
Output: Time slot t

while $S \neq \emptyset$ **do**
  Pick one high (low) power node $v \in S$ as a sender in each **HC** (**LC**). Let $S' \subseteq S$ be the set of such senders.
  for $t_1 = 0, ..., \lceil H_h \rceil - 1$ and $t_2 = 0, ..., \lceil H_h \rceil - 1$ ($t_1 = 0, ..., \lceil H_l \rceil - 1$ and $t_2 = 0, ..., \lceil H_l \rceil - 1$) **do**
    Let $Z \subseteq S'$ be the set of nodes with Cell-ID **HC**(x, y) (**LC**(x, y)) such that $t_1 = x \bmod \lceil H_h \rceil$, and $t_2 = y \bmod \lceil H_h \rceil$ ($t_1 = x \bmod \lceil H_l \rceil$, and $t_2 = y \bmod \lceil H_l \rceil$)
    for each node $w \in Z$ **do**
      Assign time slot t to w
    end for
    $t \leftarrow t + 1$, $S \leftarrow S \setminus S'$, $S' \leftarrow S' \setminus Z$
  end for
end while
return t

2. **Phase 2**:
   (a) **Sub-phase 2.1**: dominators to connectors
      i. High power dominators to connectors
      ii. Low power dominators to connectors
   (b) **Sub-phase 2.2**: connectors to dominators
      i. High power connectors to dominators
      ii. Low power dominators to connectors

While scheduling, Algorithm 1 uses Algorithm 2 as a subroutine to assign the same time slots to high (low) power nodes if they are $\lceil H_h \rceil$ ($\lceil H_l \rceil$) cells away from others.

4.3 Analysis of Algorithm 1

In this subsection, we analyze Algorithm 1 and bound the latency of the schedule produced by the algorithm. We first prove that any two high power nodes (low power nodes) can send data without any interference if they are $\lceil H_h \rceil$ ($\lceil H_l \rceil$) cells apart. Note that while scheduling, high power nodes and low power nodes are not scheduled to send data at the same time.

LEMMA 6. *For SINR threshold $\beta \geq 1$, path loss exponent $\alpha > 2$, background noise $N > 0$, and some constant $\gamma > 1$, let $H_h = (\frac{p_h \cdot 2\pi}{N(\alpha-2)(\gamma-1)})^{\frac{1}{\alpha-2}} \cdot (\frac{\gamma N\beta}{p_h})^{\frac{1}{\alpha}} + 1$. Then any two high power nodes that are at least $\lceil H_h \rceil$ cells away from each other can send data simultaneously without interference.*

PROOF. We bound H_h under the assumption that all nodes transmit with power level p_h since high power nodes are not scheduled at the same time with low power nodes in Algorithm 1.

In order to bound the shortest distance between any two high power nodes that can send data simultaneously, we first consider the following worst case for a sender node i. As node i is trying to send its data to a receiver node j, the possible longest distance between i and j is $(\frac{p_h}{\gamma N\beta})^{\frac{1}{\alpha}}$. In order that the receiver j receives data from the sender i successfully, for all other sender nodes that are sending data simultaneously, the following must be satisfied:

$$\frac{p_h \cdot ((\frac{p_h}{\gamma N\beta})^{\frac{1}{\alpha}})^{-\alpha}}{N + \sum_{u \notin \{i,j\}} p_h \cdot d(u,j)^{-\alpha}} \geq \beta$$

which implies

$$\frac{\sum_{u \notin \{i,j\}} p_h \cdot d(u,j)^{-\alpha}}{N(\gamma - 1)} \leq 1 \qquad (1)$$

Let d be the shortest distance between receiver j and one of the other senders transmitting data at the same time as i. Then,

$$\frac{p_h}{N(\gamma-1)} \cdot \sum_{u \notin \{i,j\}} d(u,j)^{-\alpha} \leq \frac{p_h}{N(\gamma-1)} \cdot \int_d^{\infty} 2\pi r \cdot r^{-\alpha} dr = \frac{p_h \cdot 2\pi}{N(\gamma-1)} \cdot \frac{d^{2-\alpha}}{\alpha-2}$$

From inequality (1), we get

$$\frac{\sum_{u \notin \{i,j\}} p_h \cdot d(u,j)^{-\alpha}}{N(\gamma-1)} \leq \frac{p_h \cdot 2\pi}{N(\gamma-1)} \cdot \frac{d^{2-\alpha}}{\alpha-2} \leq 1$$

Now, we bound the shortest distance d satisfying the following inequality:

$$\frac{p_h \cdot 2\pi}{N(\gamma-1)} \cdot \frac{d^{2-\alpha}}{\alpha-2} \leq 1$$

We have

$$d^{2-\alpha} \leq \frac{N(\gamma-1)(\alpha-2)}{p_h \cdot 2\pi} \Rightarrow d \geq \left(\frac{p_h \cdot 2\pi}{N(\gamma-1)(\alpha-2)}\right)^{\frac{1}{\alpha-2}}$$

Let $D := (\frac{p_h \cdot 2\pi}{N(\gamma-1)(\alpha-2)})^{\frac{1}{\alpha-2}}$. Then D is a lower bound for d.

Next, we bound the number of cells between j and another sender that is closest to j, say i', as follows. Assume a straight line between i' and j, and relay nodes with the power level p_h on the line. Let h be the number of cells between i' and j. As $D \le h \cdot (\frac{p_h}{\gamma N \beta})^{\frac{1}{\alpha}}$, we have $D \cdot (\frac{p_h}{\gamma N\beta})^{-\frac{1}{\alpha}} \le h$. Therefore, the receiver and another sender should be at least $D \cdot (\frac{\gamma N\beta}{p_h})^{\frac{1}{\alpha}}$ cells away, and any two senders should be at least

$$H_h = D \cdot \left(\frac{\gamma N\beta}{p_h}\right)^{\frac{1}{\alpha}} + 1 = \left(\frac{p_h \cdot 2\pi}{N(\gamma-1)(\alpha-2)}\right)^{\frac{1}{\alpha-2}} \cdot \left(\frac{\gamma N\beta}{p_h}\right)^{\frac{1}{\alpha}} + 1$$

cells apart. Thus, if any two high power nodes that are at least $\lceil H_h \rceil$ cells away from each other, they can send data simultaneously without any interference. □

Using a similar argument as in the proof of Lemma 6, we can set H_l as follows.

LEMMA 7. *Let $H_l = (\frac{p_l \cdot 2\pi}{N(\alpha-2)(\gamma-1)})^{\frac{1}{\alpha-2}} \cdot (\frac{\gamma N\beta}{p_l})^{\frac{1}{\alpha}} + 1$ for a SINR threshold $\beta \ge 1$, path loss exponent $\alpha > 2$, background noise $N > 0$, and some constant $\gamma > 1$. Then any two low power nodes that are at least $\lceil H_l \rceil$ cells apart can send data simultaneously without interference.*

Next, we prove that the latency of a schedule found by Algorithm 1 is bounded by $O(\Delta + R)$. We need the following lemmas.

LEMMA 8. *The number of dominators in an* **HC** *is at most $\lceil \delta^{\frac{1}{\alpha}} \rceil^2$.*

PROOF. Without loss of generality, consider **HC**(1, 1), and let us find the number of **LC**s that are covered by the **HC**. Notice that the number of **LC**s whose Cell-ID is **LC**$(i, 1)$ within the **HC** is $\lceil ((\frac{p_h}{\gamma N\beta})^{\frac{1}{\alpha}}/\sqrt{2})/((\frac{p_l}{\gamma N\beta})^{\frac{1}{\alpha}}/\sqrt{2}) \rceil = \lceil \delta^{\frac{1}{\alpha}} \rceil$. Therefore, an **HC** can cover at most $\lceil \delta^{\frac{1}{\alpha}} \rceil^2$ **LC**s. Since there is at most one dominator in an **LC**, the number of dominators in an **HC** is at most $\lceil \delta^{\frac{1}{\alpha}} \rceil^2$. □

LEMMA 9. *The number of low power connectors in an* **LC** *is at most 20.*

PROOF. Consider a dominator in an **LC**. In Figure 4.3, the cells are **LC**s, the diameter of the inner circle is $(\frac{p_l}{\gamma N\beta})^{\frac{1}{\alpha}}$, and the diameter of outer dashed circle is $3(\frac{p_l}{\gamma N\beta})^{\frac{1}{\alpha}}$. Since the low power connectors connect dominators in CDS, the number of low power connectors in the inner white circle cannot exceed the number of dominators in the shaded part. And, as seen in the Figure 4.3, the number of **LC**s covered by the shaded part is at most 20. Since the number of dominators in an **LC** is at most 1, the number of dominators in the shaded part is at most 20. Therefore, the number of low power connectors in an **LC** is also at most 20. □

LEMMA 10. *The number of high power connectors in an* **HC** *is at most $21\lceil \delta^{\frac{1}{\alpha}} \rceil^2$.*

PROOF. As in the proof of Lemma 9, consider a dominator in an **HC**. In Figure 4.3, suppose the cells are **HC**s. The diameter of the inner circle is $(\frac{p_h}{\gamma N\beta})^{\frac{1}{\alpha}}$, and the diameter of the outer dashed circle is $3(\frac{p_h}{\gamma N\beta})^{\frac{1}{\alpha}}$. As the number of **HC**s covered in the outer dashed circle is at most 21, and each **HC** can have at most $\lceil \delta^{\frac{1}{\alpha}} \rceil^2$ dominators (See Lemma 8), it follows that the number of high power connectors in one **HC** is at most $21\lceil \delta^{\frac{1}{\alpha}} \rceil^2$. □

LEMMA 11. *For any node at most $\omega = \frac{2^\alpha \delta^2 (d_l)^\alpha}{\beta} - 1$ neighboring nodes can send data simultaneously in dual power model.*

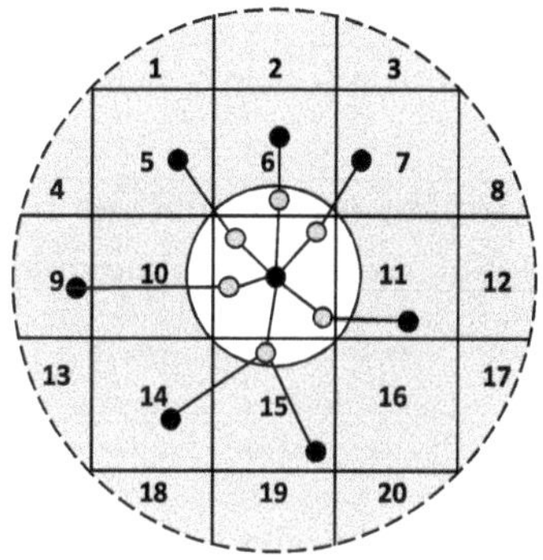

Figure 3: Black nodes represent dominators, and gray nodes represent connectors.

PROOF. Similar to the proof in [17], we prove this by contradiction. Let l be the length of the shortest link between any two of the neighbors. Assume that one of the end points of the link is a high power node v_h, and it is sending its data to another end point v_r. Let S_j be the set of all the other concurrently-transmitting senders and assume that all nodes in S_j are low power nodes so that the $SINR$ of v_r is largest. Now suppose that $|S_j| = \omega' \ge \frac{2^\alpha \delta^2 (d_l)^\alpha}{\beta}$. Then, the $SINR$ of v_r is $SINR(v_r) = \frac{p_h \cdot l^{-\alpha}}{N + \sum_{v_l \in S_j} \frac{p_l}{d(v_l, v_r)^\alpha}} \le \frac{p_l \cdot \delta}{N + \sum_{v_l \in S_j} \frac{p_l}{d(v_l, v_r)^\alpha}} < \frac{p_l \cdot \delta}{\sum_{v_l \in S_j} \frac{p_l}{d(v_l, v_r)^\alpha}} \le \frac{\delta}{\sum_{v_l \in S_j} \frac{1}{d(v_l, v_r)^\alpha}} \le \frac{\delta}{\omega'(2d_h)^{-\alpha}} = \frac{2^\alpha \delta^2 (d_l)^\alpha}{\omega'} \le \beta$. Thus, v_h cannot send data without interference, a contradiction! □

COROLLARY 12. *If Δ be the maximum node degree in a network, then every data aggregation schedule has at least $\max\{\Delta/\omega, R\}$ time slots.*

THEOREM 13. *Algorithm 1 produces data aggregation schedules whose latency is bounded by $O(\Delta + R)$, and it is a constant-factor approximation algorithm.*

PROOF. First consider Phase 1 of Algorithm 1. There exist at most Δ dominatees in each **HC** and **LC**. Therefore, data from all high power dominatees can be aggregated to their dominators in $\Delta \cdot \lceil H_h \rceil^2$ time slots, and data from all low power dominatees in $\Delta \cdot \lceil H_l \rceil^2$ time slots. Thus, Phase 1 takes $\Delta \cdot (\lceil H_h \rceil^2 + \lceil H_l \rceil^2)$ time slots.

Next consider Sub-phase 2.1. Since there exist at most $\lceil \delta^{\frac{1}{\alpha}} \rceil^2$ dominators in each **HC** and at most 1 dominator in each **LC**, data from all dominators can be aggregated to their connectors in at most $\lceil \delta^{\frac{1}{\alpha}} \rceil^2 \cdot \lceil H_h \rceil^2 + \lceil H_l \rceil^2$ time slots. For Sub-phase 2.2, since there exist at most $21 \cdot \lceil \delta^{\frac{1}{\alpha}} \rceil^2$ connectors in each **HC** and at most 20 connectors in each **LC**, data from all connectors can be aggregated to their dominators in at most $21 \cdot \lceil \delta^{\frac{1}{\alpha}} \rceil^2 \cdot \lceil H_h \rceil^2 + 20 \cdot \lceil H_l \rceil^2$ time slots. As these two Sub-phases are iterated R times, Phase 2 takes at most $R(22 \cdot \lceil \delta^{\frac{1}{\alpha}} \rceil^2 \cdot \lceil H_h \rceil^2 + 21 \cdot \lceil H_l \rceil^2)$ time slots.

Finally, observe that $H_h = (\frac{p_l \cdot \delta \cdot 2\pi}{N(\alpha-2)(\gamma-1)})^{\frac{1}{\alpha-2}} (\frac{\gamma \beta N}{p_l \cdot \delta})^{\frac{1}{\alpha}} + 1 = (\frac{p_l \cdot 2\pi}{N(\alpha-2)(\gamma-1)})^{\frac{1}{\alpha-2}} (\frac{\gamma\beta N}{p_l})^{\frac{1}{\alpha}} \cdot \delta^{\frac{1}{\alpha-2}} \cdot \delta^{-\frac{1}{\alpha}} + 1 = (H_l - 1) \cdot \delta^{\frac{2}{\alpha(\alpha-2)}} + 1$.

Thus the latency of a schedule found by Algorithm 1 is $\Delta(\lceil H_h \rceil^2 + \lceil H_l \rceil^2) + R(22\lceil \delta^{\frac{1}{\alpha}} \rceil^2 \lceil H_h \rceil^2 + 21\lceil H_l \rceil^2) = \Delta(\lceil (H_l - 1)\delta^{\frac{2}{\alpha(\alpha-2)}} + 1 \rceil^2 + \lceil H_l \rceil^2) + R(22\lceil \delta^{\frac{1}{\alpha}} \rceil^2 \lceil (H_l - 1)\delta^{\frac{2}{\alpha-2}} + 1 \rceil^2 +$

$21\lceil H_l \rceil^2) = O(\Delta + R)$. This is a constant factor approximation ratio, because any schedule requires $\geq \max\{\Delta/\omega, R\}$ time slots. □

4.4 Improvement on Algorithm 1

In this subsection, we provide an improvement on Algorithm 1. Since Algorithm 1 schedules the nodes level by level based on T_{Agg}, some nodes in different levels cannot be scheduled at the same time even though they are far apart. Therefore, we modify Algorithm 1 to schedule such nodes more efficiently. The details are contained in Algorithm 3. Instead of scheduling the nodes level by level, Algorithm 3 schedules all leaves of T_{Agg} at each iteration. In Algorithm 3, T_r denote a subtree of T_{Agg} rooted at the sink node s. The leaves of T_r contain all data aggregated up to round $r-1$.

Algorithm 3 Improved Data Aggregation Scheduling

Input: A set V of nodes with two power levels p_h and p_l
Output: Length of schedule

```
Run Steps from 1 to 3 in Algorithm 1.
Let S_h ⊆ T_0 be the set of high power dominatees, and
  S_l ⊆ T_0 be the set of low power dominatees.
Run Steps from 5 to 6 in Algorithm 1.
r ← 1
while T_r ≠ {s} do
    Z_r ←leaves of T_r − {s}
    Let S_h be the set of high power nodes in Z_r, and S_l
    be the set of low power nodes in Z_r.
    t ← HighPowerScheduling(S_h, t)
    t ← LowPowerScheduling(S_l, t)
    r ← r + 1
end while
return t − 1
```

Next, we show that Algorithm 3 also has the same latency bound as Algorithm 1.

THEOREM 14. *Algorithm 3 produces data aggregation schedules whose latency is $O(\Delta + R)$, and it is a constant factor approximation algorithm.*

PROOF. The proof is similar to the proof of Theorem 13, and is omitted. □

4.5 Experimental Results

In this subsection, we compare the performance of Algorithms 1 and 3 on the same networks. In our simulation, networks were generated randomly in the Euclidean plane where the number of nodes are n = 300, 400, and 500, respectively. The sink node s is randomly picked among the nodes that are randomly deployed on an area of size 100×100. In order to build the communication graph G', we first found the minimally required high power level p_h to make the graph G' connected. If there exists any node v that can still maintain connectivity using p_l, then v is assigned power level p_l. For each n, we generate 50 different networks, and average the lengths of the schedules produced by the algorithms over the networks generated.

Tables 1 and 2 show the performance of Algorithms 1 and 3 with $\alpha = 5$ and $\beta = 1$, where A1 denotes Algorithm 1, and A3 Algorithm 3. As seen in Table 1, when δ becomes larger (i.e., the number of high power nodes becomes larger in our simulation), the latency also increases. Table 2 shows that as

Table 1: Latency of Algorithms 1 and 3 with $\gamma = 1.1$

n	300		400		500	
δ	A1	A3	A1	A3	A1	A3
10	271.84	257.20	343.66	315.40	400.72	375.50
10^2	279.56	253.90	361.92	322.70	429.98	369.10
10^3	292.66	273.20	380.88	347.40	452.90	398.70
10^4	295.98	284.20	389.68	366.20	464.22	422.80

Table 2: Latency of Algorithms 1 and 3 with $\delta = 10$

n	300		400		500	
γ	A1	A3	A1	A3	A1	A3
1.1	271.84	257.20	343.66	315.40	400.72	357.50
2	235.20	211.50	290.40	249.70	330.96	273.64
3	231.30	207.20	285.40	244.70	322.30	262.80
4	230.10	205.30	276.30	234.00	314.90	254.60
5	229.20	204.30	276.20	233.80	314.90	254.50

γ becomes larger (i.e., as we consider many shorter links), the latency becomes smaller. The results show that Algorithm 3 performs better than Algorithm 1 in most cases. It also can be observed that as the networks become denser with larger γ, Algorithm 3 performs much better than Algorithm 1.

5. CONCLUSIONS

In this paper, we have studied the problem of constructing minimum latency data aggregation schedules (MLAS) in the SINR model. We have shown that there is no approximation algorithm for MLAS having an approximation ratio better than $\Omega(\log n)$ in the metric model unless $NP \subseteq DTIME\left(n^{\log \log n}\right)$. We have also shown that NP-completeness holds for the MLAS decision problem in the geometric model. In addition, we have proposed constant factor approximation algorithms for the dual power model whose latency is bounded by $O(\Delta + R)$. We have studied the performance of the algorithms through simulation. The question as to whether geometric MLAS is APX-hard remains an open problem.

6. REFERENCES

[1] M. K. An, N. X. Lam, D. T. Huynh, and T. N. Nguyen. Minimum data aggregation schedule in wireless sensor networks. In *Proc. of 23rd Intern. Conf. on Computer Applications in Industry and Engineering. ISCA*, pp. 257-363, Nov. 2010.

[2] V. Annamalai, S. Gupta, and L. Schwiebert. On tree-based convergecasting in wireless sensor networks. In *Wireless Communications and Networking*, Vol 3, pp. 1942 -1947, Mar. 2003.

[3] C. Avin, Z. Lotker, F. Pasquale, and Y.-A. Pignolet. A note on uniform power connectivity in the SINR model. In *Algorithmic Aspects of Wireless Sensor Networks*, pp. 116-127, 2009.

[4] X. Chen, X. Hu, and J. Zhu. Data gathering schedule for minimal aggregation time in wireless sensor networks. *Int. J. Distrib. Sen. Netw.*, 5:321–337, July 2009.

[5] T. H. Cormen, C. E. Leiserson, R. L. Rivest, and C. Stein. *Introduction to Algorithms*. MIT Press and McGraw-Hill, third edition, 2009.

[6] H. Du, X. Hu, and X. Jia. Energy efficient routing and scheduling for real-time data aggregation in WSNs. *Comput. Commun.*, 29:3527–3535, Nov 2006.

[7] U. Feige. A threshold of ln(n) for approximating set cover. *Journal of the ACM*, 41(5):314–318, 1998.

[8] O. Goussevskaia, Y. A. Oswald, and R. Wattenhofer. Complexity in geometric SINR. In *MobiHoc 2007*, pp. 100-109, Sep 2007.

[9] O. Goussevskaia, R. Wattenhofer, M. Halldorsson, and E. Welzl. Capacity of arbitrary wireless networks. In *INFOCOM 2009*, pp. 1872 -1880, April 2009.

[10] J. Grönkvist and A. Hansson. Comparison between graph-based and interference-based STDMA scheduling. In *MobiHoc 2001*, pp. 255-258, 2001.

[11] P. Gupta, S. Member, and P. R. Kumar. The capacity of wireless networks. *IEEE Trans. on Information Theory*, 46:388–404, 2000.

[12] S.-H. Huang, P.-J. Wan, J. Deng, and Y. Han. Broadcast scheduling in interference environment. *IEEE Trans. on Mobile Computing*, 7(11):1338–1348, 2008.

[13] S.-H. Huang, P.-J. Wan, C. Vu, Y. Li, and F. Yao. Nearly constant approximation for data aggregation scheduling in wireless sensor networks. In *INFOCOM 2007*, pp. 366-372, May 2007.

[14] R. Karp. Reducibility among combinatorial problems. In R. Miller and J. Thatcher, editors, *Complexity of Computer Computations*. Plenum Press, pp. 85–103, 1972.

[15] E. Lebhar and Z. Lotker. Unit disk graph and physical interference model: Putting pieces together. In *Proc. of 2009 IEEE Intern. Symposium on Parallel & Distributed Processing*, pp. 1-8, 2009.

[16] H. Li, Q. S. Hua, C. Wu, and F. C. M. Lau. Minimum-latency aggregation scheduling in wireless sensor networks under physical interference model. In *Proc. of 13th ACM Intern. conf. on Modeling, analysis, and simulation of wireless and mobile systems*, pp. 360–367, 2010.

[17] X.-Y. Li, X. Xu, S. Wang, S. Tang, G. Dai, J. Zhao, and Y. Qi. Efficient data aggregation in multi-hop wireless sensor networks under physical interference model. In *Proc. of 6th IEEE Conf. on Mobile Adhoc and Sensor Systems*, pp. 353-362, Oct. 2009.

[18] C. Lund and M. Yannakakis. On the hardness of approximating minimization problems. *J. ACM*, 41(5):286–293, 1994.

[19] P.-J. Wan, K. Alzoubi, and O. Frieder. Distributed construction of connected dominating set in wireless ad hoc networks. In *INFOCOM 2002*, Vol 3, pp 1597-1604, 2002.

[20] X. Xu, X.-Y. Li, X. Mao, S. Tang, and S. Wang. A delay-efficient algorithm for data aggregation in multihop wireless sensor networks. *IEEE Trans. on Parallel and Distributed Systems*, 22(1):163 –175, 2011.

On Realization of Reliable Link Layer Protocols with Guaranteed Sustainable Flows for Wireless Communication

Sohraab Soltani
Department of Computer Science
University of Texas
Tyler, TX, 75799 USA
ssoltani@uttyler.edu

Hayder Radha
Department of Computer Science and Engineering
Michigan State University
East Lansing, MI, 48824 USA
radha@egr.msu.edu

ABSTRACT

Despite major developments in link-layer design to address reliability issues associated with the wireless communication (in a presence of heavy noise), these efforts fall short on many fronts. This includes a clear demonstration regarding the viability of a truly reliable link-layer capable of providing a minimal level of guaranteed sustainable flows for the higher layers. In this paper, we present an analytical and experimental study to design and implement a reliable wireless link-layer that provides sustainable flow control. We develop an experimental platform using Software Radio Defined (SDR) technology with the Universal Software Radio Peripheral (USRP) frontend to capture and measure the behavior of an error process imposed on a wireless channel. Next, we design a Reliable And StablE (RASE) link-layer protocol to provide reliability (by achieving optimal throughput) and stability (by ensuring a sustainable traffic flow) for realtime and non-realtime wireless communications. We then incorporate the RASE protocol into the SDR-USRP platform to investigate the level of throughput and realtime stability achieved in comparison with the IEEE802.11 ARQ and the FEC-based HARQ protocols. We demonstrate experimentally that RASE provides 20%-50% improved reliability. In addition, realtime video communication experiments show a 2-8dB PSNR gain in playback quality.

Categories and Subject Descriptors

C.2.2 [**Computer Systems Organization**]: Computer-Communication Networks

General Terms

Algorithm,Design, Reliability, Performance

Keywords

Wireless Network, Link-layer Protocol, Stability, Reliability

1. INTRODUCTION

Despite the unprecedented success and proliferation of wireless LANs over the past decade, there are a few major shortcomings in the underlying link-layer protocols of well-established wireless systems. These shortcomings are exacerbated as the level of heterogeneity and high-bandwidth requirements of emerging applications increase dramatically. In particular, popular wireless link-layer protocols, such as the retransmission ARQ-based approach (employed by the IEEE 802.11 standard suite) are designed to achieve some level of reliability by dropping corrupted packets at the receiver and by performing one or more retransmission attempts.

The key issues associated with the current design paradigm of the 802.11 link-layer protocol are 1) *Inefficient reliability*: The 802.11 ARQ approach discards a distorted packet and request for a retransmission even when there is only a *single bit error* in the packet. This operation is performed until a packet is received error-free or a maximum number of retransmission attempts is reached. Hence, the potential throughput of 802.11 systems could improve significantly; 2) *Largely absent stability*: The ARQ-based link-layer and other recent protocols do not address the stability aspect of data communications in terms of maintaining a sustainable flow, which is critical for a dynamic and heterogeneous wireless environment.

Many leading research efforts [3]–[18] have highlighted the inefficiencies of the current 802.11 link-layer protocol. They propose a variety of remedy solutions to present new link-layer and cross-layer protocols that utilize corrupted packets (or *partial packets*) instead of discarding them [6]–[17]. In addition to Hybrid ARQ (HARQ) based methods [11]–[16], examples of recent efforts include Partial Packet Recovery (PPR) [17], packet combining [3]–[10], Cross-Layer Design with Side-information (CLDS) [22–24], and Automatic Code Embedding (ACE) [18]. Some of these approaches, such as PPR and packet combining, exploit physical layer information regarding the quality of individual bits to improve the probability or recovering corrupted packets. Others, such as CLDS, and ACE, utilize information available in current 802.11 link-layer protocols in conjunction with error correcting codes to recover corrupted packets. The ACE protocol [18], the only proposed solution that addresses reliability and stability, fails to demonstrate its viability via a true and practical design and implementation. Other methods, especially the ZIPTX [8] protocol, provide some evidence of a working system, but virtually ignore any aspects of stability.

In this paper, we analytically and experimentally investigate the impact of reliability and stability on wireless link-layer communication where we design, implement, and test a "reliable and stable" protocol in a real-system settings. First we develop an experimental wireless platform using Software Radio Defined (SDR) technology (which is implemented in GNU Radio toolkit [26]) with Universal Software Radio Peripheral (USRP) frontend [27] to a) measure the behavior of an error process of a real wireless channel conditions; and b) implement error control protocols at the link-layer that employ and manipulate received corrupted packets. We propose a Reliable And StablE (RASE) link-layer protocol to provide reliability (which achieves optimal throughput) and sustained flow control for realtime and non-realtime wireless link-layer communications. This protocol is built on the ACE framework [18] to demonstrate its viability in ensuring reliability and stability via a true and practical design and implementation. A major contribution of this work is to demonstrate the efficiency of the link-layer in providing a reliable and sustainable wireless communication in real-system settings. To that end, we utilize the GNU Radio SDR-USRP platform to investigate the achievable throughput and realtime stability of RASE in comparison to the IEEE802.11 ARQ and the FEC-based HARQ protocols. In this work, all analyses regarding the wireless channel noise condition and protocol performance measurements are based on a complete realization and are not a trace-based analysis. Our contributions can be summarized as follows:

- We develop a SDR-USRP platform to measure and investigate real error conditions at the link-layer.
- We develop RASE for wireless link-layer communication to demonstrate the viability of a truly reliable link-layer design capable of providing a minimal level of guaranteed sustainable flows.
- We use GNU-USRP based platform to implement, test, and conduct comparative performance analysis of RASE and other contemporary link-layer protocols.

For reliability, the experimental results show that RASE outperforms other protocols regardless of the channel condition. For instance, the goodput and throughput achieved by RASE are respectively 20%-50% and 10%-30% higher than the IEEE802.11 ARQ and HARQ protocols. We design and implement realtime wireless video communication using H.264/AVC JM17.0 codec [30] in conjunction with the GNU-USRP based platform to measure the flow sustainability of video traffic. RASE provides a 2-8dB PSNR improvement in video playback quality in comparison with other protocols over varying wireless channel noise conditions.

The remainder of the paper is organized as follows: We present the SDR-USRP experimental platform in Section 2. The proposed RASE protocol is presented in Section 3. In Section 4, we evaluate the performance of RASE. Section 5 concludes the paper.

2. SDR-USRP PLATFORM

Current wireless network devices such as Peripheral Component Interconnect (PCI) wireless adaptor cards for computer desktops (e.g., Linksys WMP54G) and notebooks (e.g., Linksys WPC54G), USB network adaptors (e.g., Linksys WUSB54G), and Wireless CompactFlash Card Adapters for PDAs (e.g., Linksys WCF54G) are utilized to establish communication channels among users in different wireless networks. By design, these devices use IEEE802.11 standard to access wireless channels, perform contention and error control at the Physical (PHY) and MAC layers. The current design of the IEEE802.11 prevents PHY layer to pass distorted packets (packets with unverified checksum) to the higher layers. This introduces a major challenge for this work since our objective is to utilize link-layer corrupted packets to i) capture and measure the

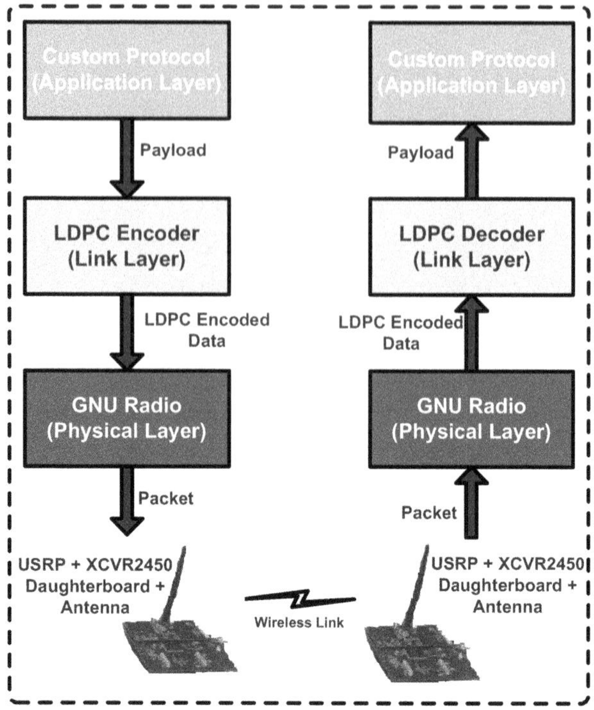

Figure 1: The layered architecture of the GNU-USRP based experimental setup.

behavior of wireless error processes; and ii) implement error control protocols at the link-layer which deploy partial packets. To tackle this problem, we need to modify the configuration of these devices to pass the distorted packets to the MAC layer. However, since the off-the-shelf wireless devices are not open source products, the modification of the source code is not possible.

In this work, we use Software-Defined Radio (SDR) technology, a radio communication system where hardware components (e.g., mixers, filters, amplifiers, modulators/demodulators, detectors. etc.) are instead implemented using software. In particular, GNU Radio package [26] is used as the software-defined radio system frontend on sender and receiver nodes in our platform. Since GNU Radio is an open-source software development toolkit, we can easily reconfigure and modify its building blocks. Fig. 1 shows the layered architecture of our experimental platform. At the highest level, *Application Layer* resides to generate data packets and pass them down to the link-layer[1]. At the link-layer, the data packets are encoded and encapsulated into the link-layer packets. These packers are then sent to the PHY layer (i.e., GNU Radio) to be transmitted over the wireless channel. We use the Universal Software Radio Peripheral (USRP) [27] as a frontend of GNU Radio. This is because USRP has an open design, with freely available schematics and drivers and can be integrated with GNU Radio.

USRP can simultaneously receive and transmit on two antennas. Daggerboards mounted on URSP provide flexible, fully integrated RF front-ends. In this paper, we employ RF *XCVR2450* transceiver daughterboards which covers both the ISM band at 2.4 GHz and the entire 4.9 to 5.9 GHz band, including the public safety, UNII, ISM, and Japanese wireless bands.

Our experimental platform consists of two USRP devices: one

9 [1]It is important to note we are interested in point-to-point link-layer wireless communication thus the modification of the TCP, Network and medium access control is irrelevant to this work

Table 1: List of the parameters and corresponding values in our platform.

Frequency	2.4-2.5 GHz, 4.9-5.9 GHz
PN Sequence Chip Rate	4 - 32 M chips / sec
PN Sequence Length	3 - 4095 chips
Receiver Gain	0 - 92 dB
Resolution	31.25 - 250 nsec
Transmitted Power	-2.75 dBm
ADC Sample Rate	64 MSa / sec
Modulation Scheme	GMSK
Bursty Mode	Disabled

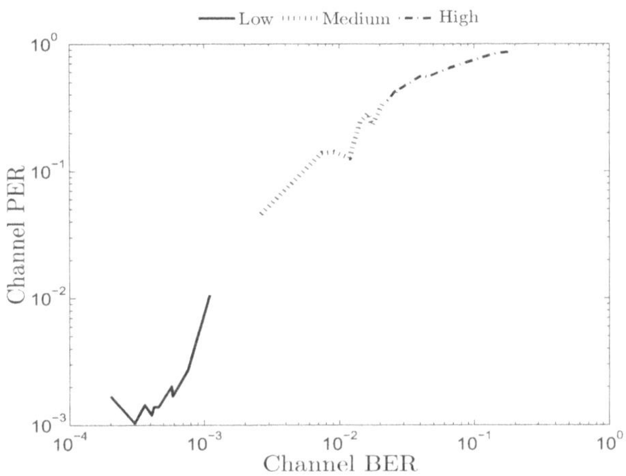

Figure 2: The log-log plot of PER with respect to BER

connected to a desktop PC which we refer to as the Data-Host (DH) PC and the other connected to the Data-Ghost (DG) laptop computer. In our experiments, the application layer in DH and DG respectively generate data and acknowledgment packets. We set up the wireless communication between DH and DG using the parameters listed in Table 1.

Using the GNU-USRP based wireless platform, we conduct several communication scenarios to capture the behavior of wireless error in the presence of fading, interference and mobility. To measure the amount of error introduced in every packet during a transmission over a specific channel, DH transmits a predefined 1500 byte payload in every transmission. Accordingly, DG captures the received packet and performs XOR operations with the predefined payload. Through this, we can measure the number of byte (and bit) errors introduced in each packet. We characterize wireless channel condition with the average Bit Error Rate (BER) $\bar{\epsilon}$ and the average Packet Error Rate (PER) $\bar{\delta}$ where

$$\bar{\epsilon} = \frac{\text{\# of distorted bits during a session}}{\text{Total \# of transmitted bits}}$$
$$\bar{\delta} = \frac{\text{\# of packets with at least one distorted bit}}{\text{Total \# of transmitted packets}}.$$

Our analysis consists of measuring 100 wireless channel conditions which are categorized in three groups: i) *Fading factor*: In this set of scenarios, the DG laptop is located within certain distances of DH. The distance between DG and DH governs the impact of fading on the wireless communication. In particular, the error conditions for channels where DH and DG are relatively close to each other (e.g., 1-2m) are remarkably better than those with DG and DH are farther; ii) *Interference factor*: To capture the impact of an

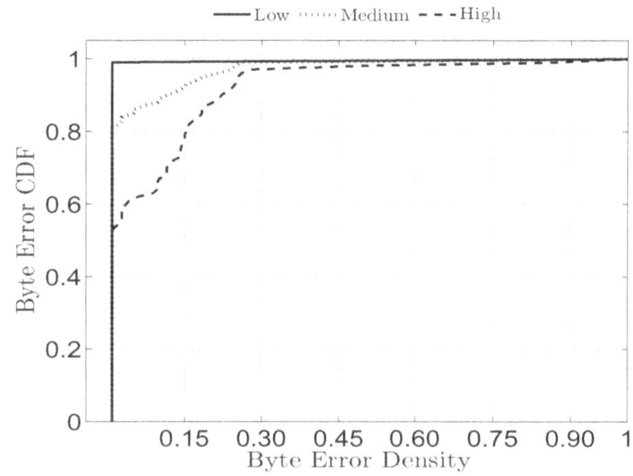

Figure 3: Byte error CDF for channel clusters

interference on a wireless channel condition, we establish wireless communications in the presence of other wireless networks. As more active wireless nodes are presented in the environment, the likelihood of packet collisions increases leading to frequent packet errors due to the interference; iii) *Mobility factor*: The DH transmits data packets to the DG laptop which is a mobile node and changes its location frequently. In these experiments, the mobility is conducted within a room and the hallway (next to the room where DH resides); however the wireless connection never fails; all the transmitted packets are received by DG.

Fig. 2 shows the log-log plot of the average PER with respect to the average BER of the channels collected in our experiments. In general channel error conditions are grouped into three clusters: i) *Low*: The wireless channels are mostly error-free during the communication with BER $\bar{\epsilon} \leq 0.001$ and PER $\bar{\delta} \leq 0.010$; ii) *Medium*: The wireless error is partially impacted by fading, interference, and mobility. This communication still contains error-free transmissions however the likelihood of packet exposures to errors is with BER $0.001 \leq \bar{\epsilon} \leq 0.02$ and PER $0.01 \leq \bar{\delta} \leq 0.35$; iii) *High*: These channels are constantly affected by the wireless noise thus the probability of packet error is high resulting in wireless communications with BER $0.02 \leq \bar{\epsilon} \leq 0.18$ and PER $0.35 \leq \bar{\delta} \leq 0.86$. To investigate the level of packet distortion in each cluster, Fig. 3 illustrates the cumulative density function of packet byte errors with respect to the likelihood of byte distortions for each cluster. For *Low* cluster, the amount of byte distortion is very low (e.g., less than 1%). For *Medium* and *High* clusters 30% and 60% of transmitted bytes contain errors.

The empirical error density captured by the SDR-USRP Platform indicates a possible impact of the wireless distortion on the reliability and stability of a wireless communication.

3. THE RASE PROTOCOL

In this section, we present a Reliable And StablE (RASE) protocol built on the Automatic Code Embedding (ACE) framework [18] to demonstrate its viability in providing reliability (which achieves optimal throughput) and sustained flow control for realtime and non-realtime wireless link-layer communication via a true and practical design and implementation. The design architectures of the RASE protocol is illustrated in Fig. 4. The RASE protocol has two components: i) *LDPC Encoder/Decoder* and ii) *Channel Stare Prediction*.

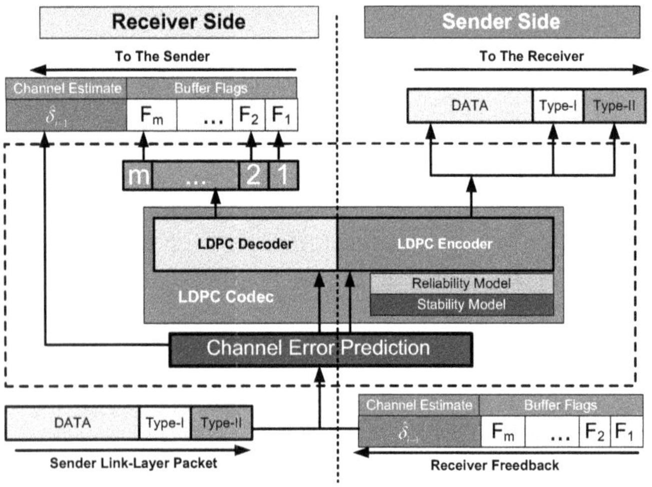

Figure 4: The design architecture of the RASE protocol.

3.1 LDPC Encoder/Decoder

The RASE sender encapsulates *new* codewords into link-layer packets by encoding the payload of every packet with LDPC codes. More specifically, every network-layer datagram received by the link-layer is encoded by the LDPC encoder component and the RASE link-layer headers are concatenated to this encoded datagram to generate a link-layer frame. The encoding process is as follows: consider the ith network-layer datagram passed down to the link-layer. The LDPC encodes k_i data bits of the payload with x_i parity bits constructing a datagram codeword $C_i(k_i, x_i)$. The RASE sender uses puncturing to identify and remove a subset of parity bits from the original codeword, and the unpunctured bits are sent to the receiver. We refer to these (x_i) parity symbols as *type-I parity*. For the systematic case this could be envisioned, as a truncated codeword containing the original data and some embedded parity (i.e., type-I parity) bits. If the decoding (at the receiver) with the original type-I parities is unsuccessful, additional parity bits (*type-II parity*) are then sent, piggybacking on the next packet to be transmitted. This additional parity is concatenated to the already received codeword until successful decoding is achieved.

In [18], we proved that under the ACE framework, there exist only one optimal code embedding rate under which the reliability and stability of link-layer traffic is ensured. Recall that the operational code embedding rate measures the fraction of data bits that are embedded in a codeword. For instance a codeword $C_i(k_i, x_i)$ is generated based on the code rate $R_i = \frac{k_i}{k_i + x_i}$. This finding is repeated in the following Lemma:

LEMMA 1. *An optimal solution for code embedding rate that ensures reliability and stability in wireless transmission over a channel with BER ϵ_i is a unique solution and is given by:*

$$R_i = 1 - \frac{\epsilon_i}{\alpha(\epsilon_i)}, \quad \epsilon_i \ll \alpha$$

where $\alpha(\epsilon_i)$ is error-correcting capability of a decoder.

Lemma 1 determines the optimal code embedding rate for a general decoder with error-correcting capability $\alpha(\epsilon_i)$ (when the channel BER is ϵ_i). In the implementation of RASE, the LDPC encoder uses $\alpha(\bar{\epsilon}_i)$ to measure the expected error-correcting capability of the LDPC soft decision. Because the RASE sender uses a prediction of the channel BER $\bar{\epsilon}_i$ as an estimate of the expected error in the ith transmission. Therefore, an *operational* optimal code embedding rate for parity allocation is

$$R_i = 1 - \frac{\bar{\epsilon}_i}{\alpha(\bar{\epsilon}_i)}. \tag{1}$$

An encoded frame received by the PHY layer is passed up to the RASE link-layer receiver. The LDPC decoder component of the RASE receiver performs decoding operations to retrieve the original network-layer datagram. If the decoding is successful, the packet is decapsulated (the link-layer headers will be removed from the packet) and the corresponding datagram is sent up to the network-layer. Otherwise, the link-layer frame is stored in a RASE buffer until type-II parity bits arrive for that frame. The receiver has a finite buffer which can accommodate up to m corrupted messages waiting for recovery. If a newly corrupted packet finds all rooms in the buffer occupied, it does not enter the buffer and is dropped. The status of the receiver is reported to the transmitter via certain flags in an acknowledgment message called *buffer flags*.

An accurate prediction/estimation of the channel condition has a critical impact on the overall performance of the RASE protocol. The reason is two folded: i) The RASE LDPC encoder uses an estimate of the channel BER to compute an optimal code rate to ensure reliable and sustainable flow (see Equation (1)); ii) The RASE LDPC decoder performs LDPC soft decision decoding [28] by computing the log-likelihood ratios (LLRs) for the received bits based on an estimate of the prevailing channel conditions. More specifically, the RASE protocol uses an iterative belief propagation LDPC decoder [20] which attempts to correct errors in the received packet iteratively using an estimate of the channel BER. This estimate is provided by the *Channel State Prediction* component of the RASE protocol described in the following section.

3.2 Channel State Prediction

The channel condition is predicted based on the receiver feedback (provided to the sender by the acknowledgment packet (ACK)). Let $N_1, \cdots, N_m$ represent the error status of the last m packets. Recall that the distribution of the number of errors in a packet (denoted by random variable N) carries the following probability densities,

$$\begin{aligned} P[N=0] &= 1-\bar{\delta} \\ P[N=j] &= \frac{\bar{\delta}}{1-(1-\bar{\epsilon})^n}\binom{n}{j}\bar{\epsilon}^j(1-\bar{\epsilon})^{n-j}. \end{aligned}$$

Using these probability values and the m observations, RASE formulates the following likelihood equation to predict the channel PER and BER,

$$\begin{aligned} L(\bar{\delta},\bar{\epsilon}|N_1,\cdots,N_m) &= \left(1-\bar{\delta}\right)^{\sum_{k=1}^{m} I[N_k=0]} \\ &\times \left[\frac{\bar{\delta}}{1-(1-\bar{\epsilon})}\right]^{\sum_{k=1}^{m} I[N_k>0]} \\ &\times \prod_{k=1}^{m}\binom{n}{N_k}\bar{\epsilon}^{N_k}(1-\bar{\epsilon})^{n-N_k} I[N_k>0]. \end{aligned}$$

By solving this equation (using log-likelihood method), RASE uses the following formulas to predict channel BER and PER,

$$\bar{\epsilon} = \frac{\sum_{k=1}^{m} N_k I[N_k>0]}{n\sum_{k=1}^{m} I[N_k>0]} \tag{2}$$

$$\bar{\delta} = \frac{\sum_{k=1}^{m} I[N_k>0]}{n}. \tag{3}$$

Fig.5 illustrates a short communication example consisting four

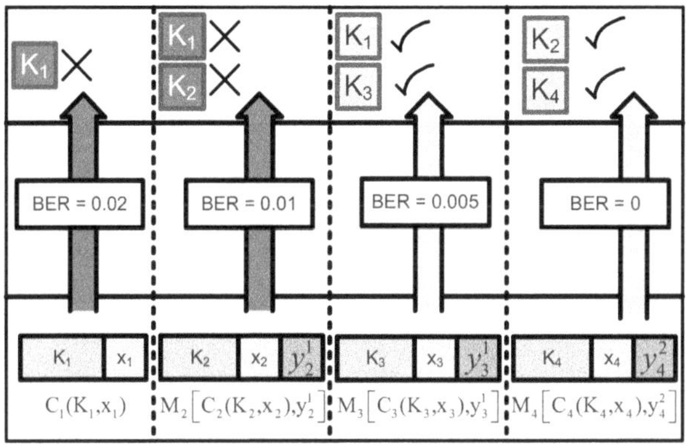

Figure 5: A short wireless link-layer communication example for RASE.

transmission intervals $\tau_1, \cdots, \tau_4$. In τ_1, a codeword $C_1(K_1, x_1)$ is sent. A receiver that fails to decode C_1, stores C_1 in its buffer and requests for additional parity bits (hereafter type-II parity bits) for C_1. In τ_2, the transmitter sends $M_2 = [C_2(K_2, x_2), y_2^1]$. The receiver uses x_2 to decode C_2 and employs type-II parity bits y_2^1 (y_i^j denote additional parity for $C_j, j < i$ transmitted in τ_i) in addition to x_1 to decode C_1. The receiver fails to decode C_1 and C_2. In τ_3, the sender transmits $M_3 = [C_3(k_3, x_3), y_3^1]$. In τ_3, the receiver successfully decodes C_3 using x_3 and $C_1(k_1, x_1 + y_2^1 + y_3^1)$. Consequently, the receiver requests for type-II parity bits for C_1. Accordingly, in τ_4, the sender transmits $M_4 = [C_4(K_4, x_4), y_4^2]$. The receiver decodes C_4 using x_4 and $C_2 = (K_2, x_2 + y_4^2)$ successfully. Therefore, the data packets $K_1, \cdots, K_4$ are delivered to the higher layer. This example demonstrates the efficiency of RASE in providing reliability and stability in wireless link-layer communication. In the next section, we use our SDR-USRP based platform to implement and test the performance of RASE in comparison with other link-layer protocols in real-system settings.

4. EXPERIMENT

Using the SDR-USRP platform, we implemented four link-layer protocols namely: IEEE802.11 ARQ, Hybrid ARQ type-I (HARQ-I), Hybrid ARQ type-II (HARQ-II), and the RASE protocol. These protocols reside in *link-layer* section of the layered architecture shown in Fig 1 and are implemented using C++ modules and glued in SDR-USRP platform with Python.

4.1 Reliability Analysis

The main objective in ensuring reliability for wireless communication is to maximize the bandwidth utilization of wireless medium while delivering error-free information packets to the destination. In this section, we determine the reliability achieved by the IEEE802.11 ARQ, Hybrid ARQ type-I (HARQ-I), Hybrid ARQ type-II (HARQ-II), and the RASE protocol in terms of the average *goodput* and *throughput*. *Goodput* is the ratio of the total number of new data bits delivered correctly (to the higher layers) to the total number of transmitted bits during a session. The *throughput* measures the fraction of error-free packets received at the link-layer. These measures represent the performance of a protocol "per channel use". So a protocol with average goodput and throughput close to "one" ensures a better reliable wireless transmission.

Fig. 6 illustrates the average goodput and throughput over different wireless conditions represented by the channels average BERs and PERs. Over good channel conditions all protocols achieve

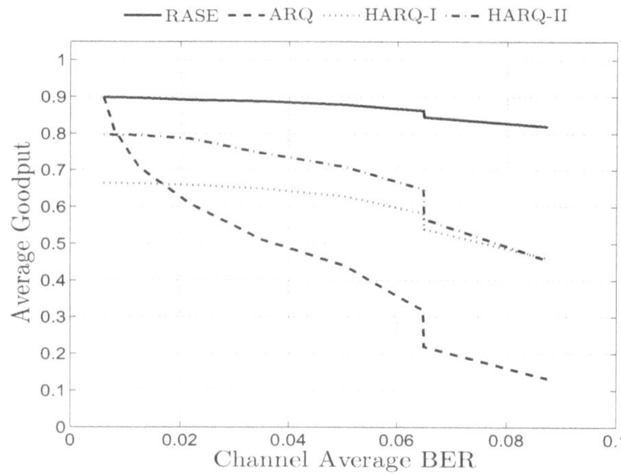

9(a) Goodput vs. Channel BER

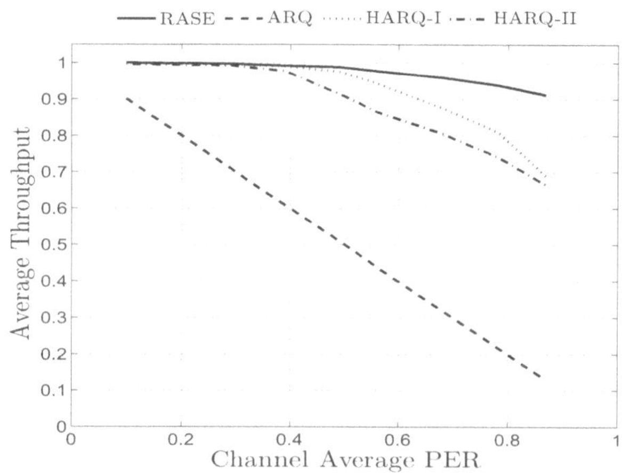

9(b) Throughput vs. Channel PER

Figure 6: Performance gain under different link-layer protocols over varying channel conditions.

more than 90% reliability. However, the goodput and throughput under the IEEE802.11 ARQ protocol drops significantly over more realistic channel conditions. For instance, IEEE802.11 ARQ achieves only 70% goodput over a channel with average BER 0.01. It is important to note that the throughput achieved under the ARQ mechanism has the inverse relationship with channel BER. HARQ-I and HARQ-II demonstrate relatively a better performance than IEEE802.11 ARQ, however, the goodput gain over these protocols is remarkably lower than their corresponding throughput gain. This stems from the fact that HARQ-I and HARQ-II protocols utilize substantial portion of channel bandwidth to transmit parity bits for data error correction. Therefore although the throughput increases (since more packets are delivered correctly), the goodput decreases since many of the corrected packets contain parity bits which are unusable for the higher layers. The RASE protocol achieves similar reliability as other protocols over good channel conditions and it successfully retains noticeably high goodput and throughput over channels with higher error conditions. RASE gains 85% goodput (93% throughput) over a channel with average BER 0.06 and PER 0.4 while IEEE802.11 ARQ achieves around 35% (60%), HARQ-I 55% (80%) and HARQ-II 65% (90%) reliability per channel use.

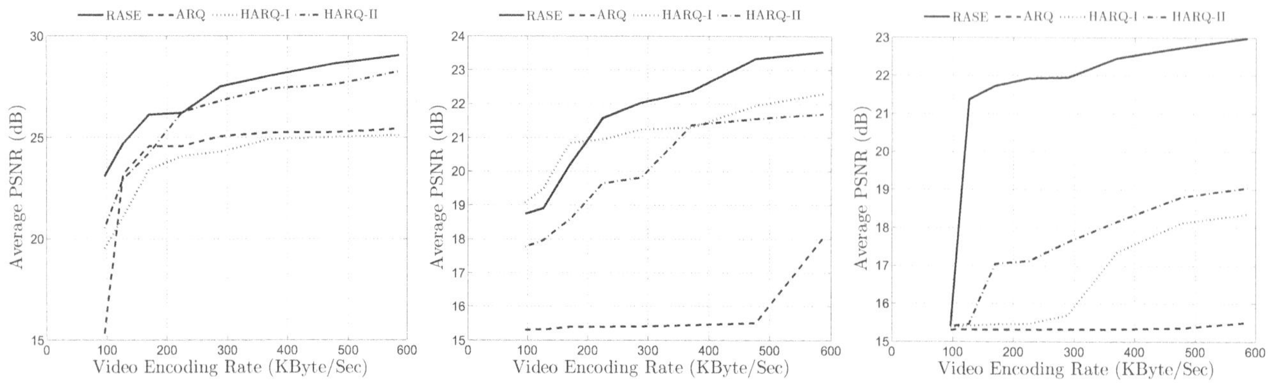

9(a) BER: 0.002 & PER: 0.02 9(b) BER: 0.007 & PER: 0.12 9(c) BER: 0.014 & PER: 0.22

Figure 7: Stefan PSNR gain over different channel conditions with respect to varying video encoding rates.

4.2 Realtime Video Stability Analysis

In this section we conduct several realtime video experiments to determine the level of guaranteed sustainable flow of the wireless communication provided by the link-layer protocol. The realtime video quality is an adequate representation of system stability since the lack of timely arrival of video information at the video decoder will result in frame freezing, skipping and/or playback jitter leading to an unsatisfactory user experience.

Often in realtime multimedia application (such as video streaming and video conferencing), a user can choose the video playback quality based on the available bandwidth and the latency of the Internet connection. In our experiment, we use the H.264/AVC JM17.0 codec [30] for video encoding/decoding operations. We configure the H.264/AVC JM17.0 codec to achieve uniform playback quality by setting the quantization parameter (QP) to a specific value for each frame. Under this setting, QP dictates the number of bits allocated to each video frame leading to variable bit rate (VBR) video encoding. That is QP (and consequently video encoding rate) determines the video playback quality. Accordingly, the link-layer sender should deliver video data at or above particular rate specified by the video encoder. The value of video encoding rate determines the video decoder tolerable delay. Those packets missing their deadlines are unusable by the decoder which leads to the degradation of video quality.

Our realtime experiment setup is as follows: at the application layer, the H.264/AVC JM17.0 codec [30] encodes video packets. Specifically, the Network Abstraction Layer Units (NALU) of H.264 standard is utilized to encode the video stream. Traditionally each NALU is embedded in MPEG-2/RTP-IP/H.32x/UDP packets. We choose each NALU packet (at the application layer) to be embedded within a single UDP packet (at the transport layer) along with an incremental index for packet sorting at the decoder. Under this setup, the video stream is partitioned into fixed size video slices equal to the size of Maximum Transmission Unit (MTU) including the UDP packet header size and the index used for sorting. The video packets are then transferred to wireless link-layer where they enter the link-layer buffer. Next, the link-layer protocol delivers each video packet to the GNU radio for transmission over a wireless channel using USRP boards. Each transmitted video packet distorted based on the channel condition arrives at the receiver. At this point, the link-layer protocol attempts to deliver each video packet before the video decoder deadline expires. To measure the video playback quality, we use the average PSNR value which is a function of the mean square error (MSE) between the values of the original and decoded Y frame pixels,

$$PSNR = 10log_{10}\left[\frac{255^2}{||\mathbf{y_{org}} - \mathbf{y_{dec}}||_2^2}\right].$$

In case of a frame loss, we use the variance of the pixels to calculate the frame PSNR as follows:

$$PSNR_l = 10log_{10}\left[\frac{255^2}{||\mathbf{y_{org}} - \overline{\mathbf{y_{org}}}||_2^2}\right].$$

where $\overline{\mathbf{y_{org}}}$ represents the mean of the values of the original frame pixels and $||.||_2$ is the $\mathcal{L}^2$-norm.

4.2.1 Quality Selection Stability

In this section, we measure the impact of a realtime video transmission in different qualities on the wireless communication stability. The sender encodes Stefan CIF sequence using different encoding rates where the corresponding video packets are delivered to the video decoder using IEEE802.11 ARQ, HARQ-I, HARQ-II, and RASE protocols. In particular, we choose eight quality videos produced by setting QP to 20-32 with step size two where each QP value corresponds to a specific video encoding rate. The video sequence encoded under each QP setting is transmitted over 10 channels with different wireless conditions. Therefore, for each protocol 80 and overall 320 experiments were conducted.

Fig. 7 illustrates the PSNR gain with respect to different video encoding rates (i.e., different video quality) over various channel conditions. For the higher video encoding rate, a user experiences a better video playback quality. This is intuitive since the video decoder is using more information packets to retrieve the original video. Over channels with good conditions (e.g., BER 0.002), all four protocols produce a relatively acceptable playback quality where RASE outperforms with more than 2dB PSNR gain. However, for channels with higher presence of noise, the video quality under IEEE802.11 ARQ and HARQ-I is lower than HARQ-II and RASE. This stems from the fact that the IEEE802.11 ARQ and HARQ-I protocols use retransmissions which introduce delay in delivering video packets to the destination. Therefore, over channels with high likelihood of packet distortion, frequent retransmissions will result in delivering of video information after the realtime video deadline expires. RASE achieves a remarkably improved performance than HARQ-II because RASE always sends *new* video packets to the destination causing the decoder not to starve (wait) for the video information. For instance, at video encoding rate of 500 Kbyte/Sec, RASE gains 7dB (7dB), 2dB (5dB) and 1dB (4dB)

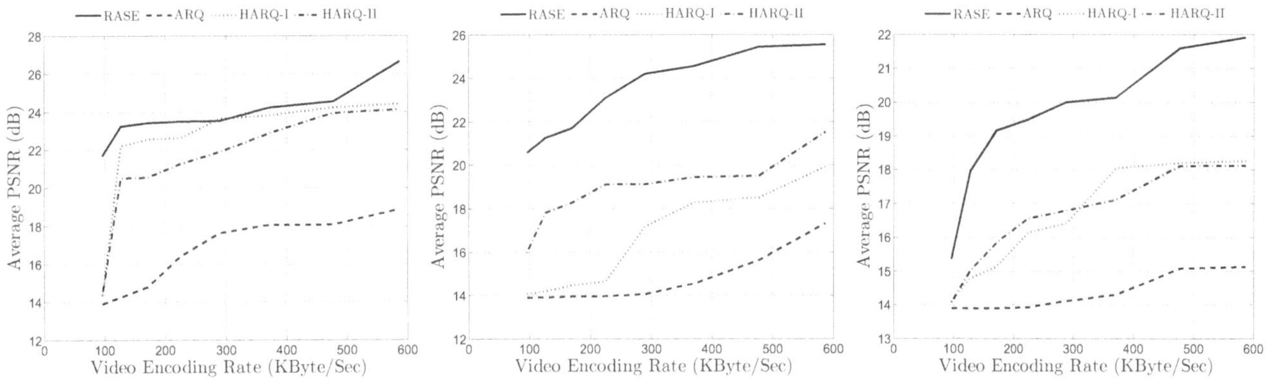

9(a) BER: 0.007 & PER: 0.12 9(b) BER: 0.014 & PER: 0.22 9(c) BER: 0.024 & PER: 0.35

Figure 8: Foreman PSNR gain over different channel conditions with respect to varying video encoding rates.

more than IEEE802.11 ARQ, HARQ-I and HARQ-II respectively over channels with BER 0.007 (0.014).

4.2.2 Scene Change Stability

In practice, not only the playback quality but also the nature of the video has an impact on the overall user experience. For a video sequence with a frequent scene change (where each frame represents different scene) it is vital that *all* video frames are delivered and decoded on time. This is important since a single frame skipping, loss or jitter in video is sufficient to deteriorate the system performance from the user viewpoint. To determine the stability performance of the link-layer protocols for this type of video, we use *Foreman CIF* sequence which contains noticeable number of scene changes. Similar to the *Stefan CIF* experiment, we measure the video playback quality with respect to different video encoding rates and varying wireless channel conditions through 320 experimental setups.

Fig. 8 illustrates the PSNR gain of playback quality where the RASE protocol achieves higher video quality. This is due to the fact that RASE delivers video data bits rapidly such that all video frames are decoded with high quality regardless of the variations in channel conditions. For instance, at video encoding rate of 300 Kbyte/Sec, RASE PSNR gain is respectively 6dB, 7dB, and 10dB higher than HARQ-II, HARQ-I, and IEEE802.11 ARQ over a channel with BER 0.014.

To determine the impact of wireless channel noise variations on frame quality, in Fig. 9 we measure the impact of channel BER and PER variations on video playback quality for a fixed video encoding rate. Specifically, Fig 9(a) illustrates the PSNR gain of the video quality encoded with rate 450Kbyte/Sec with respect to varying channel BERs. For low values of BER all protocols performs relatively well close to ideal frame quality. However, as BER increases, the performance of IEEE802.11 ARQ and HARQ-I dramatically decays and falls below HARQ-II. At the same time RASE maintains the high visual quality. For transmissions where BER is more than 0.02, RASE achieves respectively 8dB, 5dB, and 4dB PSNR gain over IEEE802.11 ARQ, HARQ-I and HARQ-II. Fig. 9(b) also demonstrates similar performance gain with respect to variations in channel PER where RASE outperforms other protocols regardless of channel conditions. In all cases, we observe that over channel conditions with very high presence of noise (with BER and PER more than 0.03 and 0.4), the video quality deteriorates significantly.

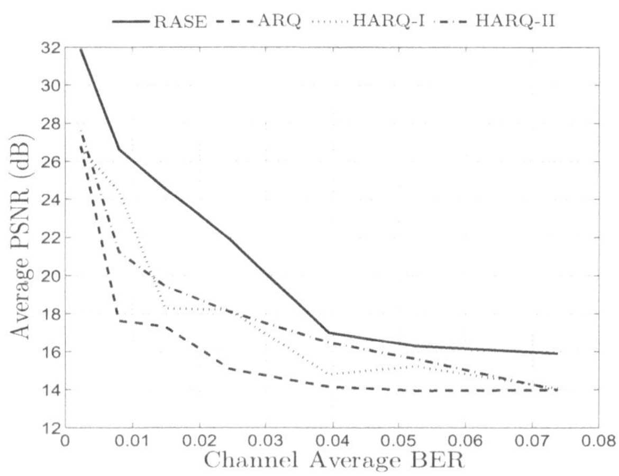

9(a) Video rate: 450 Kbyte/Sec

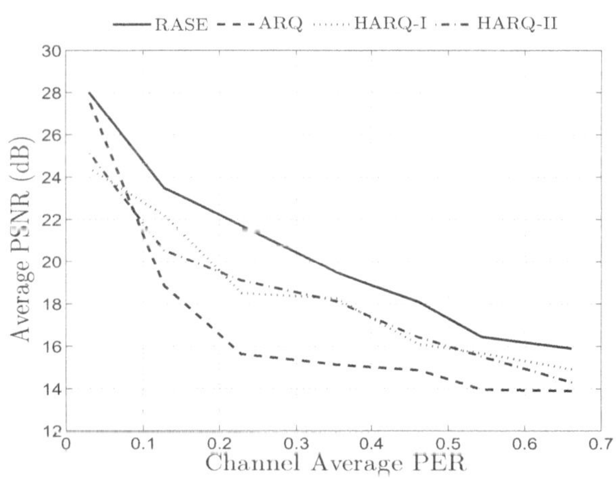

9(b) Video rate: 600 Kbyte/Sec

Figure 9: Foreman PSNR gain over varying channel conditions.

5. CONCLUSIONS

In this paper, we presented an analytical and experimental study on the impact of wireless errors in providing reliable and sustainable traffic flow at the link-layer. We presented analytical models to formulate reliability and stability achieved under the ARQ- and

FEC-based link-layer protocols. We then presented a Reliable And StablE (RASE) protocol to provide reliability and sustained flow control for realtime and non-realtime wireless link-layer communication. This protocol is built on the ACE framework [18] to demonstrate its viability via a true and practical design and implementation. By developing a SDR-USRP based experimental platform, we measured and captured the error behavior of realistic wireless noise conditions. More importantly, we implemented, tested, and conducted comparative performance analyses of RASE, IEEE802.11 ARQ, HARQ-I and HARQ-II based link-layer protocols. These evaluations clearly demonstrate the efficacy of RASE to achieve significant reliability while sustaining high quality video information flow throughout a session over various real-system settings.

6. REFERENCES

[1] S. S. Karande and H. Radha, "Hybrid Erasure-Error Protocols for Wireless Video," IEEE Transactions on Multimedia, vol. 9, no. 2, Feb 2007.

[2] IEEE Computer Society LAN MAN Standard Committee, "Wireless LAN Medium Access Control (MAC) and Physical Layer (PHY) Specifcations," IEEE Std. 802.11-1999, New York, 1999.

[3] J. G. Kim and M. M. Krunz, "Delay analysis of selective repeat ARQ for a Markovian source over a wireless channel," *IEEE Trans. Veh. Technol.*, vol. 49, no. 5, pp. 1968Ű1981, Sep. 2000

[4] T. W. A. Avudainayagam, J.M. Shea and L. Xin. Reliability Exchange Schemes for Iterative Packet Combining in Distributed Arrays. *Proc. of the IEEE WCNC*, volume 2, pages 832-837, 2003.

[5] S. S. Chakraborty, E. Yli-Juuti, and M. Liinaharja. An ARQ Scheme with Packet Combining. *IEEE Comm. Letters*, 1998.

[6] H. Yomo, S. S. Chakraborty, and R. Prasad, "IEEE 802.11 WLAN with Packet Combining", International Conference on Computer and Device 2004 (CODEC-04), January, 2004, Kolkata, India

[7] Grace Woo, Pouya Kheradpour, Dawei Shen, and Dina Katabi, "Beyond the Bits: Cooperative Packet Recovery Using Physical Layer Information," ACM MOBICOM, 2007.

[8] K. C. Lin, N. Kushman, and D. Katabi. Ziptx: Harnessing partial packets in 802.11 networks. In MobicomŠ08, September 2008.

[9] E. C. Strinati, S. Simoens, and J. Boutros, "Performance evaluation of some Hybrid ARQ schemes in IEEE 802.11a Networks", *IEEE VTC*, 4(4):2735- 2739, 2003

[10] G. Caire and D. "Tuninetti. The throughput of Hybird-ARQ protocols for the Gaussion collision channel". *IEEE Trans. Inform. Theory*, July 2001.

[11] S. Lin and P. S. Yu, "A hybrid ARQ scheme with parity retransmission for error control of satellite channels," IEEE Trans. Commun., vol. 30, pp. 1701Ű1719, July 1982.

[12] Y. Wang and S Lin, "A modified selective-repeat type-II hybrid ARQ system and its performance analyses," *IEEE Transactions on Communications* 31(5), pp. 124-133, 1983.

[13] G. Caire and D. Tuninetti, "The throughput of Hybird-ARQ protocols for the Gaussion collision channel", *IEEE Trans. Inform. Theory*, 47:1971Ű1988, July 2001.

[14] Y. Jin,G. Bai,P. Zhang,H. Shen, and J. Wang, "Performance Evaluation of a Hybrid FEC/ARQ for Wireless Media Streaming, IEEE International Conference on Circuits and Systems for Communications, 2008.

[15] , Z. Li,and T. Herfet, "MAC layer multicast error control for IPTV in wireless LANs",IEEE Transactions on Broadcasting, vol.55, num.2, pages 353-362, 2009.

[16] J. C. Bolot, S. Fosse-Parisis, and D. Towsley, "Adaptive FEC-based error control for internet telephony," in Proc. IEEE INFOCOM Š99, 1999, vol. 3, pp. 1453Ű1460.

[17] K. Jamieson and H. Balakrishnan. "PPR: Partial Packet Recovery for Wireless Networks". In ACM SIGCOMM, Kyoto, Japan, August 2007.

[18] S. Soltani, K. Misra, and H. Radha, "On Link-Layer Reliability and Stability for Wireless Communication," ACM MOBICOM, 2008.

[19] S.B. Wicker and V.K. Bhargava. Reed-Solomon Codes and Their Applications. *IEEE Press*, 1994.

[20] W. E. Ryan, "An introduction to LDPC codes," August 2003.

[21] S. M. Ross, "Introduction to Probability Models", Acad. Press. 2003.

[22] L. Rizzo, "Effective erasure codes for reliable computer communication protocols," Comput. Commun. Rev., vol. 24, no. 2, pp. 24Ű36, Apr. 1997.

[23] A., Argyriou, "Cross-layer error control for multimedia streaming in wireless/wireline packet networks", IEEE Trans. Multimedia, vol.10,num.6, pages 1121-1127, 2008,

[24] S. Shakkottai, T. Rappaport, and P. Karlsson, "Cross- Layer Design for Wireless Networks," IEEE Commun. Mag., vol. 41, no. 10, Oct. 2003, pp. 74Ű80.

[25] Y. Tian, K. Xu, and N. Ansari. TCP in Wireless Environments: Problems and Solutions. *IEEE Radio Communication*, 43(3): S27-S32, March 2005.

[26] E. Blossom, "GNU radio: tools for exploring the radio frequency spectrum", *Linux Journal* vol. 2004, no. 10, pp. 4, 2004.

[27] D. Shen, "Tutorial 4: The USRP Board", University of Notre Dame, 2005.

[28] J. Hagenauer and P. Hoecher. "A Viterbi Algorithm with Soft-Decision Outputs and its Applications". In IEEE GLOBECOM, 1989.

[29] "LDPC Software", `http://www.cs.utoronto.ca/radford/ldpc.software.html`

[30] H.264/AVC Software Coordination `http://iphome.hhi.de/suehring/tml/`

Interference Cancellation-based RFID Tags Identification

Raju Kumar, Thomas F. La Porta
Computer Science and Engineering
Pennsylvania State University
University Park, PA 16802
{rajukuma,tlp}@cse.psu.edu

Gaia Maselli, Chiara Petrioli
Computer Science Department
University of Rome "La Sapienza"
Via Salaria 113, 00198, Rome, Italy
{maselli,petrioli}@di.uniroma1.it

ABSTRACT

In this paper we investigate interference cancellation to faster identify tags in RFID networks. We explore how interference cancellation can be applied to ALOHA and tree-based identification schemes, its limitations, the extent of achievable improvements, and the overhead incurred to obtain effective gains. Analytical and simulation results show that for an ALOHA-based scheme interference cancellation allows us to identify nearly 23% of tags without directly interrogating them. This speeds up tag identification (over 20% faster) while producing little overhead. For a tree-based scheme nearly 50% of the tags are identified by exploiting interference cancellation, resulting in an improvement of the identification rate of over 20%. Finally, we propose an enhancement of the tree-based scheme with interference cancellation that achieves a further identification speed up of 50%.

Categories and Subject Descriptors

C.2 [**Computer-Communication Networks**]: Network Protocols

General Terms

Performance

1. INTRODUCTION

Radio Frequency IDentification (RFID) tags [20] are used to identify objects onto which they are attached. These tags are read by an RFID reader, thereby enabling a range of tasks such as identification, tracking, monitoring, etc. Passive RFID tags use the energy derived from the RFID reader transmissions to transmit data using FSK backscatter modulation [16].

The purpose of an RFID network is often labeling and automation of inventory management in diverse environments. Each RFID tag has a globally unique electronic product code (EPC) or ID. Specifications for RFID, like EPCglobal [1] and ISO/IEC 18000 [7] allow IDs to be up to 256 bits long (including some checksum bits). In a dense deployment of RFID tags, rapid identification of all tags within communication range of the reader is an imperative. This requires all tags to transmit their IDs to the reader, which may lead to tag-tag collisions when replying to the reader queries.

In this paper, we consider the case of a single reader querying a set of passive tags. Several schemes, referred to as anti-collision mechanisms, have been proposed to make tag identification efficient. Two of the prominent categories of these schemes [18] are *ALOHA-based* and *tree-based*. ALOHA-based schemes allocate frames comprised of slots. Each unrecognized tag transmits in a randomly selected slot. Only those tags that transmit alone in a slot are recognized. As long as tags remain to be recognized, more frames are issued. In contrast, tree-based schemes query nodes based on a specific criteria such as prefix of ID or generated random numbers. Tags are recursively split into subgroups until single tag groups are obtained and thereafter identified.

With existing solutions, when collisions occur at a reader due to the concurrent transmission by multiple tags, no tags are recognized and the signal received is of little use. Previous work [11] showed that almost 50% of the identification time is wasted in collisions. In this paper, we propose to use *interference cancellation (IC)* to utilize the signal of collisions to improve the tag identification rate. For example, let tags T_1 and T_2 transmit concurrently and the combined signal received at the reader be Y_{12}. Existing anti-collision schemes require all colliding tags to retransmit later. Let tag T_1 transmit next so that the signal received at the reader is Y_1. If instead of requiring tag T_2 to retransmit as well, the *residual signal* $Y_{12} - Y_1$ is demodulated, T_2 may be recognized depending on the signal-to-noise ratio (SNR) of this residual signal. We refer to tags recognized due to interference cancellation as "inferred" tags in the rest of this work.

In this paper we make the following contributions:

- We apply interference cancellation to two anti-collision mechanisms representative of ALOHA-based and tree-based schemes: Tree Slotted ALOHA (TSA) [3] and Query Tree (QT) [13]. The interference cancellation-based protocols are termed ICTSA and ICQT, respectively. We investigate the performance of ICTSA and ICQT analytically and through simulation showing the gain in identification rate they obtain. Our results show that interference cancellation enhances identification rate by up to 20% for both TSA and QT. We also determine the efficacy of interference cancellation by measuring the fraction of tags inferred and the number of interference cancellations attempted per tag (called

MSWiM'11, October 31–November 4, 2011, Miami, Florida, USA.

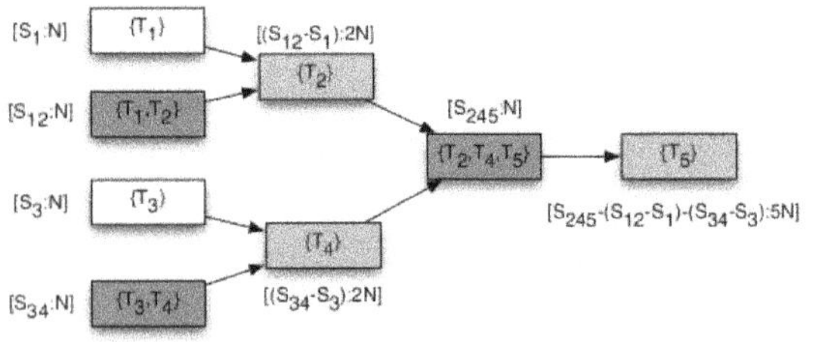

Figure 1: Example of interference cancellation.

overhead). ICTSA infers up to 23% of the tags by just imposing a 0.34 overhead per inferred tag. When applied to QT, interference cancellation infers 50% of the tags while requiring an average overhead of 1.22 per tag.

- We propose an enhancement of ICQT. This enhanced version of the protocol, termed E-ICQT, exploits failures in applying interference cancellation to skip reader queries that would lead to collisions or no replies. E-ICQT increases the identification rate up to 50% with respect to ICQT, and an overall 80% with respect to QT.

The rest of this paper is organized as follows. Section 2 provides background information on interference cancellation for RFID. We then present the details of ICTSA and ICQT and analyze their identification rate performance in Section 3. Simulation results are presented in Section 4. Section 5 introduces and investigates E-ICQT. Section 6 presents related works and Section 7 concludes the paper.

2. BACKGROUND

In this section we provide details of interference cancellation for RFID tags. We use and extend the notation of Zhang et al. [25]. Let T_i denote a tag, its ID, as well as the application level packet that it transmits to an RFID reader. The packet T_i is turned into X_i by the channel coding and modulation layer. After the transmission is successfully received as signal Y_i by the reader it is demodulated back into packet T_i. Let x, y, and t denote a symbol from packets X, Y and T, respectively. The following relation holds:

$$y_i = n + h_i x_i$$

where h_i is the complex path loss coefficient for the channel from tag T_i to the reader and n is the time-dependent noise at the reader (assumed to be complex Gaussian with unit variance [25]). When a set of tags T transmit concurrently, the signal received at the reader is $n + \sum_i h_i x_i$ where $h_i x_i$ is the signal received by reader from tags T_i in the set T. Since tag identification is generally accomplished in a few seconds [11], the tags and the reader can be assumed to be nearly stationary during the identification process. As a result, the noise and the transmission power of a tag at the reader can be assumed to be relatively stable.

According to Fuschini et al. [5], the bit error rate (BER) of the transmission of a tag when received at the reader is given by

$$BER = \frac{1}{2}\text{erfc}\left(\frac{|V_0| \cdot m}{2 \cdot \sqrt{2} \cdot \sigma}\right) \quad (1)$$

where V_0 is the voltage at the reader, σ is the standard deviation of the Additive White Gaussian Noise (AWGN), and m is the modulation index. As factors like multipath or electromagnetic coupling due to nearby objects become more prominent, m decreases from an ideal value of 0.8 to a low value of 0.2. Let the average power of the noise at the reader (based on clear channel assessment) be NdB. Let the average power of tag T_i transmission at the reader be S_i dB. (Unless stated explicitly, a signal does not contain noise and all power levels are in dB.) Based on Equation (1) the packet error rate (PER) for 56 and 96 bit IDs [10, 11] goes from 1 to 0 over a small range of 4dB. With larger packet sizes, this range of SNR in which PER transitions from 1 to 0 decreases. Due to this sharp transition, we assume that when the SNR $S_i - N$ is greater than or equal to a threshold τ, upon demodulation the checksum of the tag will match and the ID is correctly received. At lower SNRs, whether due to signal degradation or collision, the packet received is erroneous and the checksums of the tag IDs do not match.

Figure 1 shows an example of the interference cancellation approach used here. Each node represents either a successful transmission, a collision, or an inferred signal, and is annotated with the relevant set of tags, the corresponding signal, and noise powers. When the difference of the signals of $\{T_1, T_2\}$ and $\{T_1\}$, referred to as *residual signal*, is demodulated, it may lead to tag T_2 being recognized. The power of this residual signal is $S_{12} - S_1$. However, its noise component is $2N$ since subtracting signals does not cancel noise; instead, noise components are added [24]. As a result, the SNR of the residual signal is $(S_{12} - S_1) - 2N$. An important consequence of the increasing of noise is that the SNR of successively inferred packets using interference cancellation decreases. Since the SNR of the residual signal must be greater than or equal to the threshold τ, this limits the extent of the possible interference cancellation. For example, in Figure 1 the residual signal for tag T_5 has the power $S_{245} - (S_{12} - S_1) - (S_{34} - S_3)$ while its noise component is $5N$. The SNR of this residual signal may not be high enough to be demodulated correctly.

Note that this kind of interference cancellation is based on difference of signals from *different* transmissions. In contrast, capture effect [9] and successive interference cancellation [6] use the difference of signals *derived* from the *same* transmission. To demonstrate the effectiveness of interference cancellation as a stand-alone technique for improving tag identification in this paper we do not consider capture effect or successive interference cancellation.

3. APPLYING INTERFERENCE CANCELLATION

In this section we introduce interference cancellation-based Tree Slotted ALOHA and Query Tree protocols, termed ICTSA and ICQT, respectively. We also analyze their performance in terms of tag identification rate. In order to provide an upper bound on the gain obtainable through interference cancellation, we analyze the considered identification schemes under the assumption that a native or residual signal can be decoded even if the SNR is below the required threshold.

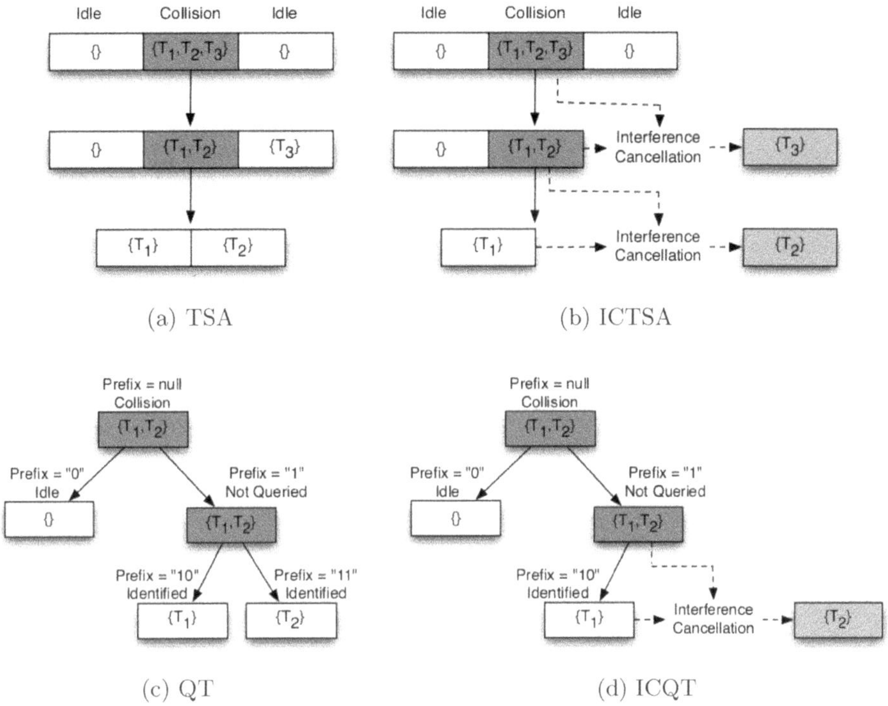

Figure 2: Application of interference cancellation to TSA and QT protocols.

3.1 ICTSA

Tree Slotted ALOHA (TSA) [3] initially allocates a frame comprised of slots. Each tag randomly picks one slot in the frame. For each slot that results in a collision of tags, TSA assigns another frame, the *child frame*. Only the tags that collide in a slot transmit in this new frame. A frame can have multiple children and these children frames can have subsequent children frames. A child only has a single parent frame (more accurately, a parent slot in the parent frame). As a result, TSA creates a tree of frames to recognize tags. Consider the example in Figure 2(a) where tags T_1, T_2, and T_3 collide in the second slot of a frame. Hence a child frame is allocated to resolve this collision. T_1 and T_2 again collide in this child frame and need a third frame to recognize them. If interference cancellation is applied to this example, the difference between the signal of collision of tags T_1, T_2, and T_3 from the first frame and that of T_1 and T_2 from the second frame can be used to recognize tag T_3 (Figure 2(b)). Similarly, interference cancellation can be used to infer tag T_2 in the third frame. Note that if a new tag is recognized due to interference cancellation using signals of a slot in a parent frame and that of the initial slots of the child frame, the child frame is terminated even if slots remain in the frame. This is because if any other tag in addition to the inferred tag was yet to transmit in the remainder of the child frame, interference cancellation would have failed. Thus, for this example, interference cancellation reduces the number of slots required from 8 to 6 i.e., a gain of 33% in terms of tag identification rate.

More generally, a collision of tags in a slot provides a signal Y. In the child frame allocated to resolve this collision, after every non-idle slot, a cumulative signal Y_c is obtained by combining the signals of all non-idle slots up to this time. After each non-idle slot (except the last slot in a frame), interference cancellation is attempted on the residual signal $Y - Y_c$. If demodulating this residual signal leads to a tag being recognized, the rest of the child frame is truncated since interference cancellation will not be successful if any other tags remain to be discovered. If interference cancellation does not lead to a tag being recognized, it could be due to either multiple remaining tags, zero remaining tags, or low SNR; there is no way to differentiate among these outcomes. Hence in this case the remainder of the frame is not terminated.

TSA Analysis: A critical issue with TSA (and ICTSA) is to optimally set the size of each frame. A constant frame size may lead to either too many idle slots or too many collisions. We first determine how to set the frame size to maximize the rate at which tags are recognized by TSA.

Let the number of tags to transmit in the frame to be allocated be n and the number of slots in the frame be $s(n)$. Since tags independently and uniform randomly select the slot in which to transmit, the probability of k out of n tags transmitting in one of the $s(n)$ slots in a frame is

$$p(s(n), n, k) = \binom{n}{k} \left(\frac{1}{s(n)}\right)^k \left(1 - \frac{1}{s(n)}\right)^{n-k} \quad (2)$$

Interestingly, an idle slot may not be the same size as a non-idle slot. Khandelwal et al. [8] note that when no response is received for a delay corresponding to 10 bits, the reader terminates a slot and indicates the beginning of the next slot. Let β be the ratio of the size of an idle slot to that of a non-idle slot. β depends on several factors - number of bits in the ID, modulation used, delay before terminating an idle slot, etc. Different values of β have been used in prior work [8, 11, 17]. We adopt the values of β used in La Porta

et al. [11] - $\beta = 0.03$ for ALOHA-based schemes and 0.13 for tree-based schemes. Since contemporary research has often considered the case of all slots being the same duration i.e., $\beta = 1$, we present results for $\beta = 1$ as well.
Given β, the rate at which tags are recognized by TSA in a frame is

$$R(s(n)) = \frac{p(s(n), n, 1) \cdot s(n)}{s(n) - (1-\beta) \cdot p(s(n), n, 0) \cdot s(n)} \quad (3)$$

where $p(s(n), n, 1) \cdot s(n)$ is the number of tags recognized in the frame and $p(s(n), n, 0) \cdot s(n)$ is the number of idle slots. When $\beta = 1$, $s(n) = n$ leads to maximum tag recognition rate i.e., the number of slots in a frame should be the same as the number of tags that transmit in the frame. When $\beta = 0.03$, $s(n) = 4.4n$ maximizes this rate (as also observed in La Porta et al. [11]). *For brevity, hereon we adopt the notation that* $\{A, B\}$ *will indicate that the value corresponding to* $\beta = 1$ *is* A *and that for* $\beta < 1$ *is* B. Hence for TSA, $s(n) = \{n, 4.4n\}$ maximizes the tag recognition rate. Note that the frame size is larger for smaller value of β because large frames can be used to decrease collisions, while incurring an insignificant penalty of more idle slots.
Let the number of slots in the tree of frames to recognize n tags be $X(n)$. Let the root frame consist of $s(n)$ slots. For collision of $k \geq 2$ tags, TSA is recursively invoked. Hence the number of slots in a TSA tree of frames is given by the number of slots in the root frame plus the number of slots in children frames of collision slots i.e.,

$$X(n) = \begin{cases} 1 & \text{if } n = 1, \\ s(n) + s(n) \cdot \sum_{k=2}^{n} p(s(n), n, k) \cdot X(k) & \text{if } n > 1. \end{cases} \quad (4)$$

where $s(n)$ is the number of slots in the root frame. $s(n) \cdot p(s(n), n, k)$ is the number of slots that lead to collision of k tags, each of which requires $X(k)$ slots to identify these colliding tags. Note that $X(n)$ is recursively defined.
Similarly, the number of silent slots in this tree of frames is

$$Y(n) = \begin{cases} s(n) - 1 & \text{if } n = 1, \\ p(s(n), n, 0) \cdot s(n) + \\ s(n) \cdot \sum_{k=2}^{n} p(s(n), n, k) \cdot Y(k) & \text{if } n > 1. \end{cases} \quad (5)$$

where $p(s(n), n, 0) \cdot s(n)$ is the number of idle slots in the root frame. Note that when $n = 1$, the number of idle slots in a frame is $s(n) - 1$.
Hence the rate of recognizing tags in TSA i.e., the ratio of the number of tags to the total identification delay in terms of non-idle slots is

$$R = \frac{n}{X(n) - (1-\beta) \cdot Y(n)} \quad (6)$$

As a result, for TSA, $s(n) = \{n, 4.4n\}$ maximizes the tag recognition rate and results in $R = \{0.43, 0.81\}$.
ICTSA Analysis: Now, lets consider the impact of augmenting TSA with interference cancellation. Let $r(s(n), n, k, j)$ (Equation (7)) be the probability that interference cancellation truncates the current frame with k slots remaining such that j tags transmit in the $(k+1)$-th last slot. The total number of ways in which n tags can transmit in $s(n)$ slots is $s(n)^n$. Out of these, the number of ways in which 1 tag transmits in the last k slots while j tags transmit in the $k+1$-th last slot and the remaining tags transmit in the remaining slots is $\binom{n}{1} \cdot \binom{n-1}{j}$. But 1 tag can transmit in k slots in $\binom{k}{1}$ ways. Moreover, these tag groups can be ordered in 3! ways, and only one is the case we consider. Finally, the $n-1-j$ tags in the first $s(n)-(k+1)$ slots can transmit in $(s(n)-(k+1))^{(n-1-j)}$ ways. This leads us to the expression in Equation (7).
The total number of slots required with interference cancellation is $X'(n)$ (Equation (8)) where $j \in [1, n-1]$ is the number of tags that collide in the $(k+1)-th$ last slot. The expected number of slots in the frame saved due to the frame being truncated is $\sum_{k=1}^{s(n)-1} \sum_{j=1}^{n-1} (r(s(n), n, k, j) \cdot k)$. Note that to enable interference cancellation in the root frame, the signal of all transmitting tags together needs to be available. This can be made available at the cost of an initial frame with only 1 slot. Hence the number of total slots required is $X'(n) + 1$.
The number of silent slots that occur in the tree of frames is $Y'(n)$ (Equation (9)). First, the number of idle slots when a frame is not truncated is $p(s(n), n, 0) \cdot s(n)$. But when a frame is truncated, the number of these idle slots that are saved is $\sum_{k=1}^{s(n)-1} \sum_{j=1}^{n-1} r(s(n), n, k, j) \cdot (k-1)$.
Note that an optimal $s(n)$ for TSA might not be optimal for ICTSA. A small value of $s(n)$ will lead to more collisions while a large value will offer less opportunity for interference cancellation. We arrive at the optimal $s(n)$ for ICTSA by parameter exploration of the expression for its tag identification rate. With interference cancellation, $s(n) = \{0.8n, 3n\}$ leads to maximum rate of recognizing tags $R' = \{0.52, 0.87\}$. Thus interference cancellation improves TSA rate of recognizing tags by $\{21\%, 6\%\}$.

3.2 ICQT

Upon request from a reader, tree-based protocols [4, 14] recursively split tags into two sub-groups until single-tag groups are obtained. Query Tree (QT [13]) protocol queries tags based on a prefix. When a tag ID matches the queried prefix, it replies with the remainder of its ID. Essentially, QT corresponds to exploring the binary tree of the possible prefixes. Another tree-based scheme, Binary Splitting [15], splits tags based on binary random numbers generated by tags. Given a uniform distribution of IDs though, QT behaves like a deterministic version of the Binary Splitting protocol.
A prefix query p may lead to a collision in QT. If so, subsequently, prefix "$p0$" and "$p1$" are scheduled for querying. If "$p0$" is queried and no reply is received, it is apparent that query prefix "$p1$" will lead to a collision. As a result, instead of "$p1$", "$p10$" and "$p11$" are scheduled for querying next. When augmented with this optimization, QT protocols are sometimes referred to as Query Tree Improved (QTI) [13].
Consider the example in Figure 2(c). Let tag T_1 ID be "10..." while that of T_2 be "11...". When the first query for *null* prefix is transmitted, both tags reply. Next the query prefix is "0" and neither of the tags reply to this query. This indicates that querying prefix "1" will lead to a collision. Hence prefix "10" is queried next and only tag T_1 replies and is identified. Next, prefix "11" is queried and only tag T_2 replies and is identified.
Now, consider the case when interference cancellation is applied to this example (Figure 2(d)). Once tag T_1 is recognized, an interference cancellation is attempted between the

$$r(s(n),n,k,j) = \frac{1}{3!}\frac{1}{s(n)^n}\binom{n}{1}\binom{n-1}{j}\binom{k}{1}(s(n)-(k+1))^{(n-1-j)} \tag{7}$$

$$X'(n) = \begin{cases} 1 & \text{if } n = 1, \\ s(n) - \sum_{k=1}^{s(n)-1}\sum_{j=1}^{n-1}(r(s(n),n,k,j)\cdot k) + s(n)\cdot\sum_{k=2}^{n} p(s(n),n,k)\cdot X'(k) & \text{if } n > 1. \end{cases} \tag{8}$$

$$Y'(n) = \begin{cases} p(s(n),n,0)\cdot s(n) & \text{if } n = 1, \\ p(s(n),n,0)\cdot s(n) - \sum_{k=1}^{s(n)-1}\sum_{j=1}^{n-1}(r(s(n),n,k,j)\cdot(k-1)) + s(n)\cdot\sum_{k=2}^{n} p(s(n),n,k)\cdot Y'(k) & \text{if } n > 1. \end{cases} \tag{9}$$

Figure 3: ICTSA analysis

signal received and the signal for the collision corresponding to *closest ancestor node*. This is because the parent of a node may not correspond to an actual query due to the optimization component of QT explained earlier. If the interference cancellation is successful, tag T_2 is recognized. Thus, in this example interference cancellation reduces the number of queries from 4 to 3 i.e., an identification rate gain of 33%.

In general, when a collision is observed for a query prefix "*p*", prefix "*p*0" is queried next. If a reply is received for this query, an interference cancellation attempt is made using the signals for reply to queries for prefixes "*p*" and "*p*0". If a tag can be recognized from this residual signal, prefix "*p*1" need not be queried. If a tag is not recognized from the residual signal, because of low SNR of the signal or either zero or multiple tags being involved in the residual signal, prefix "*p*1" is queried.

QT Analysis: Assuming a uniform distribution of IDs among the tags to be identified, the probability of k out of n tags replying to a query when all n tags replied to the prefix query corresponding to the parent node in the query tree is

$$p(n,k) = \binom{n}{k}/2^n \tag{10}$$

Let $X(n)$ denote the number of nodes i.e., queries, in the query tree for QT to identify n tags. Hence

$$X(n) = \begin{cases} 1 & \text{if } n \leq 1, \\ 1 + p(n,0)(X(0) + (X(n)-1)) + \\ \sum_{k=1}^{n} p(n,k)(X(k) + X(n-k)) & \text{if } n > 1. \end{cases} \tag{11}$$

where $X(k)$ and $X(n-k)$ are the number of nodes in the left and right subtree, respectively. Note that for each silent node on the left subtree, the corresponding node on the right subtree is bypassed i.e., not queried, due to the optimization aspect of QT. This leads to the $(X(n)-1)$ component.

Now, the number of silent nodes $Y(n)$ in the query tree is

$$Y(n) = \begin{cases} 1 & \text{if } n = 0, \\ 0 & \text{if } n = 1, \\ \sum_{k=0}^{n} p(n,k)(Y(k) + Y(n-k)) & \text{if } n > 1. \end{cases} \tag{12}$$

Hence, the rate of recognizing tags in QT is $R \simeq \{0.38, 0.44\}$.

ICQT Analysis: When QT is enhanced with interference cancellation, the number of nodes in the query tree does not include the right child of any node that splits into $(n-1)$ tags in the left subtree and only one tag in the right subtree. Hence the number of nodes in the query tree is

$$X'(n) = \begin{cases} 1 & \text{if } n \leq 1, \\ 1 + p(n,0)\cdot(X'(0) + (X'(n)-1)) + \\ p(n,n-1)\cdot X'(n-1) + \\ \sum_{k=1,k\neq n-1}^{n} p(n,k)\cdot(X'(k) + X'(n-k)) & \text{if } n > 1. \end{cases} \tag{13}$$

where the element corresponding to $k = n-1$ results in successful interference cancellation and occurs with probability $p(n, n-1)$.

Note that interference cancellation does not reduce silent nodes in the query tree. As a result, the number of silent nodes in the query tree for ICQT is $Y'(n) = Y(n)$. Subsequently, the rate of recognized tags in ICQT is $R' \simeq \{0.46, 0.56\}$. This leads to a gain $\{21\%, 27\%\}$ for tag identification rate.

4. RESULTS

We characterize three elements of a tag identification protocol - average number of tags identified per unit time (referred to as *rate*), average fraction of tags identified due to interference cancellation (referred to as *inferred fraction* (IF)), and average number of interference cancellations per tag (referred to as *overhead* (Ovh)). Since the slot duration depends on the modulation used, number of bits in the ID, etc., unit time is considered the same as the duration of a non-idle slot. The IDs of the tags are assumed to be uniformly distributed.

Wu et al. [22] contend that the signal strength of a tag transmission depends mainly on four important factors: The reader transmission power, the distance between tag and reader, relative orientation of reader and tag radios, and the material around a tag. Since tags may be deployed over a large area, the range of signal strengths of tag transmissions may be large. In the results presented here the signal strengths received at the reader are selected uniformly and randomly between 25dB and 50dB. The noise level is set to 5dB. For a signal to be correctly decoded the SNR is required to be greater than $\tau = 18$dB. Since the range of signal strengths and the SNR threshold have an impact on the results of ICTSA and ICQT, we also present results for rate and inferred fraction of variants of ICTSA and ICQT in case of an ideal channel where the SNR does not impact reader reception. There variants are referred to as OICTSA and OICQT in the pictures. They serve the purpose of bounding the gains achievable through interference cancellation.

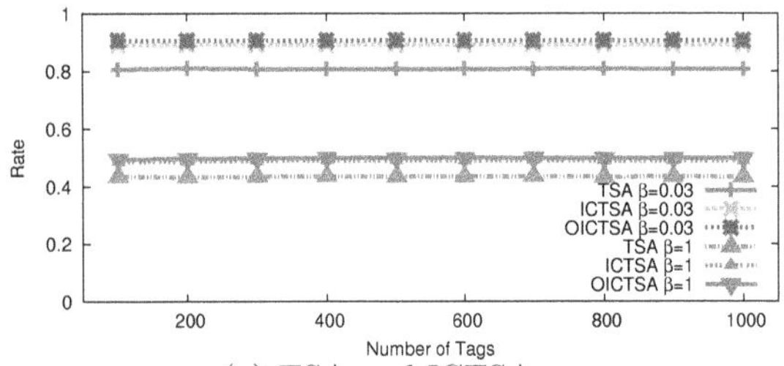

(a) TSA and ICTSA

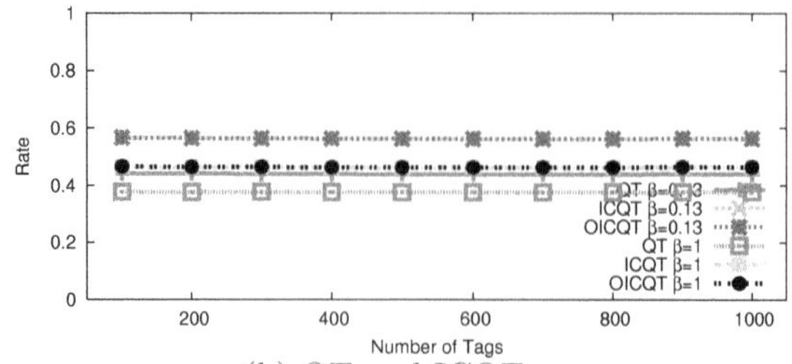

(b) QT and ICQT

Figure 4: Rate.

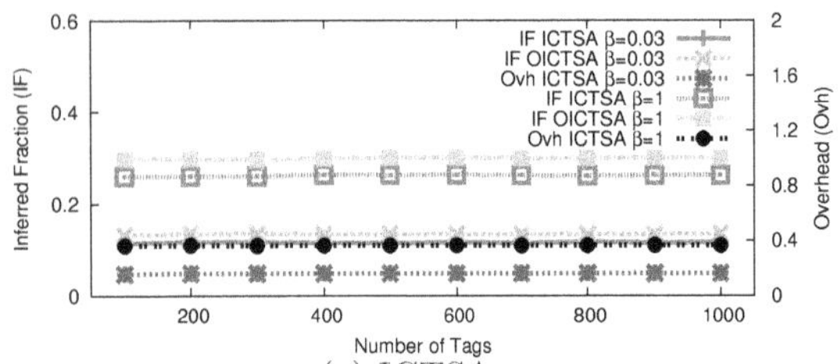

(a) ICTSA

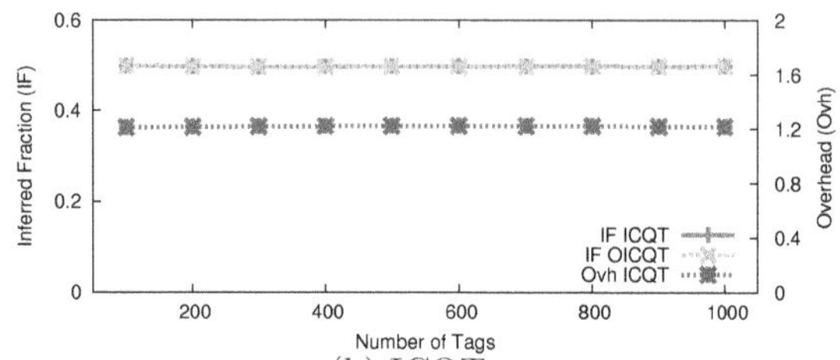

(b) ICQT

Figure 5: Inferred fraction and overhead.

Clearly, the overhead as defined here metric does not affect OICTSA and OICQT.

All results presented in this section have been obtained through simulations by averaging over 100 different runs.

For the TSA protocol and its interference cancellation based counterparts we suppose to know the number of tags to be identified. Tag cardinality estimation methods [10, 11] allow an accurate measure of the number of tags to recognize in a fraction of the time required to identify them. This allows us to set an initial optimal frame size.

Figure 4(a) shows that the tag identification rate increases from $\{0.43, 0.81\}$ to $\{0.49, 0.89\}$ to $\{0.50, 0.91\}$ for TSA to ICTSA and OICTSA. Hence interference cancellation can bring rate gains of up to $\{16\%, 12\%\}$. Note that the identification rates for $\beta = 0.03$ case are higher than that for the $\beta = 1$ case because the optimal frame size is much larger for $\beta = 0.03$ (Section 3) and the cost of idle slots is negligible.

Figure 4(b) shows the identification rate for QT and its variants. Interference cancellation improves the identification rate from $\{0.38, 0.44\}$ to $\{0.46, 0.56\}$ to $\{0.46, 0.56\}$ for QT to ICQT to OICQT. Though the number of idle and non-idle slots in a query tree is independent of β, a smaller β leads to a reduced identification delay and as a consequence, a higher identification rate. Again, the maximum gain achieved by interference cancellation in terms of identification rate is bounded by $\{23\%, 28\%\}$.

Figure 5(a) shows the inferred fraction and overhead for when interference cancellation is applied to TSA. $\{23\%, 9\%\}$ of the tags are inferred by ICTSA while OICTSA infers $\{27\%, 10\%\}$ of the tags. Again, $\beta = 1$ leads to more interference cancellation primarily due to the optimal frame size being smaller. This is because when $\beta = 0.03$, it is fairly inexpensive to increase the size of a frame and pay the minuscule penalty of excessive idle slots. As a consequence, this leads to an attrition of scope for interference cancellation. The overhead of ICTSA is only $\{0.34, 0.12\}$ interference cancellations per tag.

Figure 5(b) shows the inferred fraction and overhead for QT based schemes. Since QT schemes do not employ the concept of a frame, the inferred fraction and overhead are independent of β. ICQT and OICQT lead to nearly 50% of the tags to be identified due to interference cancellation and the overhead for ICQT is limited to 1.22 interference cancellations per tag.

First, note that ICTSA inferred fraction is less than that for ICQT. This is because only one interference cancellation can be successful per frame for ICTSA while one interference cancellation per internal node in the query tree can be successful for ICQT. Second, the overhead of ICTSA is less than that for ICQT because the number of interference cancellation attempts per frame is nearly the same as the number of non-idle slots in the frame. But the number of non-idle slots is less than the number of tags that transmit in the frame. On the other hand, each internal node in query tree for ICQT results in a collision and presents a candidate for interference cancellation. Hence the overhead of interference cancellation is higher than that for ICTSA.

5. ENHANCED ICQT

In this section we show how for some protocols, such as ICQT, we can use the outcome of interference cancellation to skip some colliding and idle queries, thus further improving the tag identification rate. We term the resulting protocol E-ICQT.

Let us consider a query "x" that causes a collision (see Figure 6). According to ICQT "x" is followed by a query "$x0$" performed by a subset of the tags that generated the collision in the query "x". After this second query, the reader attempts to apply interference cancellation. If the application is successful, it leads to the identification of a tag without querying it. If interference cancellation is not successful, it is because of one of the following two cases. The first case happens when the set of tags responding to "$x0$" is the same that responded to "x". In this case the new query "$x1$" would be idle, and therefore can be skipped. If the set of tags responding to "$x0$" is smaller than the set of tags that

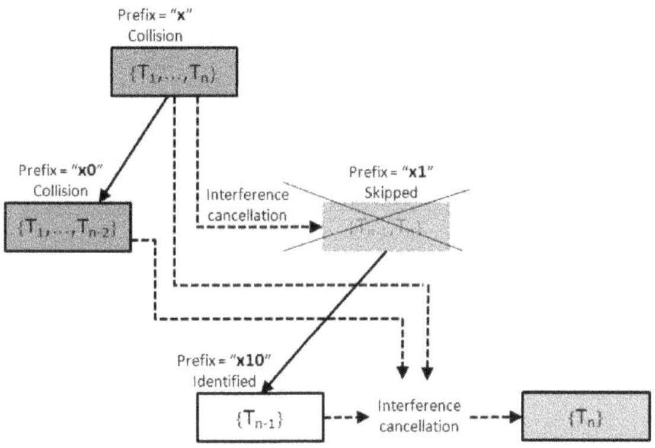

Figure 6: E-ICQT example.

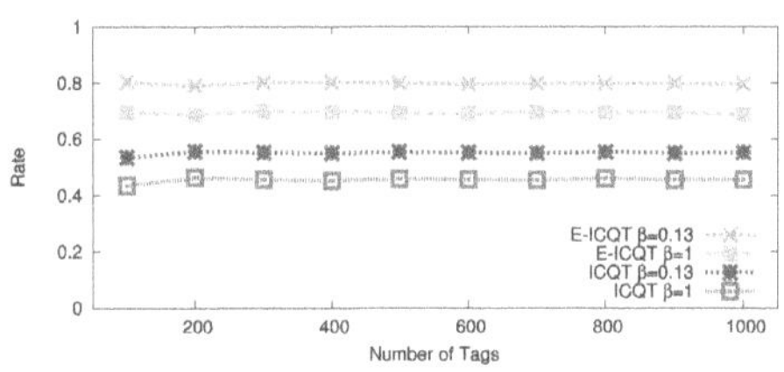

Figure 7: Identification rate for E-ICQT.

responded to "x" (with the difference between the two sets being greater than one tag) the new query "$x1$" would generate a collision. ICQT says that this collision can be used (along with its left son in the tree) to apply interference cancellation at the next tree level. However, we observe that the information provided by such collision can be retrieved by taking the difference between the collisions in "$x1$" father and left brother. As a result, query "$x1$" can be skipped and query "$x10$" can be executed instead. Through this simple reasoning E-ICQT allows us to halve the number of colliding queries, and to reduce the number of idle queries.

Figure 7 shows that the tag identification rate increases from $\{0.46, 0.56\}$ for ICQT to $\{0.69, 0.80\}$ for E-ICQT. Therefore, the improvement over ICQT is up to $\{50\%, 42\%\}$ and, even more remarkably, the improvement over QT is $\{81\%, 81\%\}$. We note that the inferred fraction for E-ICQT is the same as that of ICQT (i.e., 50%). In other words, E-ICQT does not impact on the interference cancellation process. It only smartly exploits known information for reducing time consuming colliding and idle queries without imposing extra overhead, which remain the same of ICQT (i.e., 1.22).

As a matter of fact, interference cancellation can be unsuccessful also because of low SNR (Section 2). However, in order to evaluate the effectiveness of the E-ICQT protocol, here we have considered scenarios with an ideal channel. Also the tag that is eventually not recognized through interference cancellation is queried at the next level. Future work will study the impact of physical layer issues on the E-ICQT protocol performance.

6. RELATED WORK

The application of interference cancellation to identify RFID tags has been proposed previously. A preliminary attempt in presented in [24], that provides an early glimpse into how to apply interference cancellation to a tree-based anti-collision protocol.

A deeper study is presented in [23], where physical layer network coding is applied to a basic Framed Slotted Aloha (FSA) protocol. FSA issues consecutive frames to which all unrecognized tags participate, picking a random slot. The idea in [23] is to apply interference cancellation between a collision slot and one or more single slots. Specifically, they remove from the mixed signal (e.g., $Y_1Y_2 \dots Y_k$) of a collision of k tags, the signals of $k-1$ tag IDs (e.g., $Y_1Y_2 \dots Y_k - (Y_1 + Y_2 + \dots + Y_{k-1})$), received in different single slots. In our work, we apply interference cancellation also when the signal to be subtracted is generated by a collision of multiple tags. The application is successful every time the second group of tags is the same as the first group except for one tag (e.g, $Y_1Y_2 \dots Y_{k-1}$). Another difference with [23] is that our interference cancellation approach is much more lightweight. Applying interference cancellation to FSA requires to *guess* all but one of the tags that collide in a slot to identify a new tag. In other words, whenever a new tag is identified, it may be useful to solve a previous collision. However, the reader does not know in which of the previous collision slots this tag has transmitted. It also does not know with which and with how many other tags the new tag has collided. Therefore, the reader has to attempt an interference cancellation among every collision that happened in earlier frames and every subset of recognized tags, which incur a significant computational overhead.

Our investigation is more general, targeting both ALOHA-based and tree-based identification protocols. The schemes we consider, TSA and QT, provide higher identification rate than FSA. Furthermore, instead of guessing à la FSA, in both our ICTSA and ICQT we use a *parental* relationship between tag collisions and identifications, making the application of interference cancellation immediate, and curtailing the computational overhead significantly. The need of guessing is bypassed by subtracting the signal of a collision or identification directly from the signal of the slot or of the query at the parent level (Section 3).

We finally mention that interference cancellation is not the only technique that can be used for improved identification rate. Other methods, such as capture effect [9] and successive interference cancellation (SIC) [6] can be exploited. Lai et al. [12] observe a pitfall in the application of capture effect to a tree-based protocol. In these protocols, once a tag is recognized based on its reply to a query, the corresponding node in the query tree is marked as a leaf and its descendants are not explored. By using capture effect, tags can be recognized in spite of collisions with other tags. As a consequence, such colliding tags are never detected. Lai et al. [12] propose an efficient method to address this issue.

Tseng et al. [19] partition tags into groups based on the signal strength of tag transmissions. For instance, tags surrounding a reader are broken into four groups. Tags in the first and third group form a partition and the remaining groups form the other partition. To recognize tags belonging to any partition, DFSA is used. By making tags with disparate signal strengths part of the same group, SIC [6] and multi-user detection [21] can be used to recognize tags. The authors report a 5% increment in the tag identification rate. Bhanage et al. [2] use SIC to reduce error rate during tag recognition.

7. CONCLUSIONS

RFID anti-collision schemes do not take advantage of the signal received when tag-tag collisions occur. In this paper we propose to use interference cancellation to use these collision signals to expedite the tag identification process. Our results show that interference cancellation can infer 23% and 50% of the tags when applied to the previously proposed schemed TSA and QT, respectively. In addition, the tag identification rate induced by interference cancellation applied to TSA and QT improves by 20%. Our investigation and results lead us to propose a simple enhancement to ICQT that further improves the identification rate for this tree-based protocol.

8. ACKNOWLEDGEMENTS

This work was supported by the National Science Foundation CNS-0916171 and CHIRON European project. CHIRON is a project of the European Joint Undertaking on Embedded Systems ARTEMIS, co-funded by the EU and by National Authorities including MIUR. CHIRON: Contract n. 100228.

9. REFERENCES

[1] EPC R© Radio-Frequency Protocols Class-1 Generation-2 UHF RFID Protocol for Communications at 860 MHz–960 MHz Version 1.1. 0, 2006.

[2] G. Bhanage, Y. Zhang, Y. Zhang, W. Trappe, and R. Howard. RollCall: the design for a low-cost and power efficient active RFID asset tracking system. In *EUROCON*, 2007.

[3] M. Bonuccelli, F. Lonetti, and F. Martelli. Instant collision resolution for tag identification in RFID networks. *Elsevier, Ad Hoc Networks*, (5), 2007.

[4] J. Capetanakis. Tree algorithms for packet broadcast channels. *IEEE Transactions on Information Theory*, 25(5), 1979.

[5] F. Fuschini, C. Piersanti, F. Paolazzi, and G. Falciasecca. On the efficiency of load modulation in RFID systems operating in real environment. *IEEE Antennas and Wireless Propagation Letters*, 7, 2008.

[6] D. Halperin, T. Anderson, and D. Wetherall. Taking the sting out of carrier sense: interference cancellation for wireless lans. In *ACM MobiCom*, 2008.

[7] S. Hong, H. Zhang, I. Song, K. Chang, D.-B. Shin, and H.-S. Lee. ISO/IEC 18000-7 based on rfid multi-hop relay system. In *IEEE ISCIT*, 2009.

[8] G. Khandelwal, K. Lee, A. Yener, and S. Serbetli. ASAP: a MAC protocol for dense and time constrained RFID systems. *EURASIP Journal on Wireless Communications and Networking*, 2, 2007.

[9] J. Kim and J. Lee. Capture effects of wireless CSMA/CA protocols in Rayleigh and shadow fading channels. *IEEE Transactions on vehicular technology*, 48(4), 1999.

[10] M. Kodialam and T. Nandagopal. Fast and reliable estimation schemes in RFID systems. In *ACM MobiCom*, 2006.

[11] T. F. La Porta, G. Maselli, and C. Petrioli. Anticollision protocols for single-reader rfid systems: Temporal analysis and optimization. *IEEE Transactions on Mobile Computing*, 10:267–279, February 2011.

[12] Y.-C. Lai and L.-Y. Hsiao. General binary tree protocol for coping with the capture effect in rfid tag identification. *IEEE Communications Letters*, 14(3), March 2010.

[13] C. Law, K. Lee, and K.-Y. Siu. Efficient memoryless protocol for tag identification. In *ACM DIALM*, 2000.

[14] J. Massey. Collision resolution algorithms and random-access communication. *Multi-Users Communication Networks, CISM Courses and Lectures*, (256), 1981.

[15] J. Myung, W. Lee, J. Srivastava, and T. K. Shih. Tag-splitting: Adaptive collision arbitration protocols for rfid tag identification. *IEEE Transactions on Parallel and Distributed Systems*, 18(6), june 2007.

[16] V. Sarangan, M. Devarapalli, and S. Radhakrishnan. A framework for fast RFID tag reading in static and mobile environments. *Computer Networks*, 52(5), 2008.

[17] B. Sheng, Q. Li, and W. Mao. Efficient continuous scanning in rfid systems. In *IEEE INFOCOM*, 2010.

[18] D. Shih, P. Sun, D. Yen, and S. Huang. Taxonomy and survey of RFID anti-collision protocols. *Computer communications*, 29(11), 2006.

[19] D. Tseng and Z. Lin. Anti-collision algorithm with the aid of interference cancellation and tag set partitioning in radio-frequency identification systems. *Communications, IET*, 3(1), 2009.

[20] N. Vaidya and S. R. Das. Rfid-based networks: exploiting diversity and redundancy. *SIGMOBILE Mob. Comput. Commun. Rev.*, 12(1), 2008.

[21] S. Verdu. *Multiuser detection*. Cambridge Univ Pr, 1998.

[22] N. Wu, M. Nystrom, T. Lin, and H. Yu. Challenges to global RFID adoption. *Technovation*, 26(12), 2006.

[23] M. Zhang, T. Li, S. Chen, and B. Li. Using analog network coding to improve the rfid reading throughput. In *Proceedings of the 2010 IEEE 30th International Conference on Distributed Computing Systems*, ICDCS '10, pages 547–556, Washington, DC, USA, 2010. IEEE Computer Society.

[24] N. Zhang and B. Vojcic. Binary search algorithms with interference cancellation RFID systems. In *IEEE MILCOM*, 2005.

[25] S. Zhang, S. Liew, and L. Lu. Physical layer network coding schemes over finite and infinite fields. In *IEEE Globecom*, 2008.

Modeling and Performance Analysis of DMAC for Wireless Sensor Networks

Tao Zheng
School of Computer Science
University of Oklahoma
Norman, Oklahoma
tao@ou.edu

Sridhar Radhakrishnan*
School of Computer Science
University of Oklahoma
Norman, Oklahoma
sridhar@ou.edu

Venkatesh Sarangan
Chennai Innovation Labs
Tata Consultancy Services
Chennai, India
venkatesh.sarangan@tcs.com

ABSTRACT

DMAC is a popular data gathering MAC protocol for wireless sensor networks. It employs staggered sleep-awake schedules to enable continuous data forwarding along a data gathering tree, resulting in reduced end-to-end delays and energy consumption. In this paper we present a generalized model for the DMAC protocol and for this generalized model we have analyzed end-to-end delay and energy consumption with respect to the source node for both constant bit rate traffic and stochastic traffic following a Poisson process. The stochastic traffic scenario is modeled as a discrete time Markov chain and expressions for state transition probabilities, average delay and average energy consumption are developed and it is shown that these can be evaluated numerically. Simulations are carried out with various parameters and the results are in line with the analytical results. Both analysis and simulations indicate that DMAC is more suitable for constant bit rate traffic than stochastic traffic.

Categories and Subject Descriptors

C.4 [**Performance of Systems**]: Modeling techniques

General Terms

Performance, Theory

Keywords

Wireless sensor networks, DMAC, Modeling, Markov chain

1. INTRODUCTION

Wireless sensor networks are multi-hop ad-hoc networks consisting of a large number of sensor nodes communicating through wireless medium. There have been growing research interest in wireless sensor networks due to their potentials for a broad range of applications, such as environmental monitoring, health monitoring, military surveillance, inventory tracking, homeland security.

Energy saving is one of the key challenges in wireless sensor networks due to infeasible battery replacement in most situations. Many MAC protocols have been proposed to save the unnecessary energy consumption due to idle listening by putting sensor nodes into sleep. SMAC [15], TMAC [4], BMAC [11] and XMAC [2] are some examples of the energy saving MAC protocols. DMAC [8] is another energy saving MAC protocol specifically designed for low-rate data gathering in wireless sensor networks. It employs *staggered* sleep-awake schedules to enable continuous data forwarding along a data gathering tree rooted at the sink node. Contention is reduced because the active periods are now separated. Additional active periods can be added to a node with little overhead through the *more data* flag when a node has multiple packets to send. A *data prediction* mechanism is used to add additional receiving and sending slots in order for other child nodes to send data in a timely manner. *More-To-Send* packets are used to avoid collision caused by interference between nodes on different branches of the data gathering tree.

Most of the work on energy saving MAC protocols for wireless sensor networks had taken the pragmatic and experimental approaches, with no analytical model presented to provide insights on how their protocols perform. Only a few recent work attempted to model and analyze some of the MAC protocols. Yang and Heinzelman had proposed Markov models for both SMAC [13] and XMAC [14]. In [13], two Markov models were proposed to evaluate the throughput of SMAC with and without retransmissions. The models were validated through simulations with varying number of sensor nodes, queue capacity, packet arrival rate and contention window. Similar approach is used to evaluate the throughput of XMAC in [14]. Both analytical models match simulation results within a 5% range. Even though the lack of analytical work on energy saving MAC protocols for wireless sensor network, quite a lot of Markov models were proposed to analyze the IEEE 802.11 [1, 3, 6] and the IEEE 802.15.4 [7, 5, 12, 9, 10]. Bianchi [1] proposed a Markov model to analyze the saturation throughput of the IEEE 802.11 distributed coordination function, which inspired most of the analytical work coming after it.

To the best of our knowledge, there is no analytical model proposed for evaluating the performance of DMAC. In this paper, we present a generalized model for DMAC and analyze its performance under both CBR traffic and stochastic

*Author to whom all correspondence should be sent.

MSWiM'11, October 31–November 4, 2011, Miami, Florida, USA.

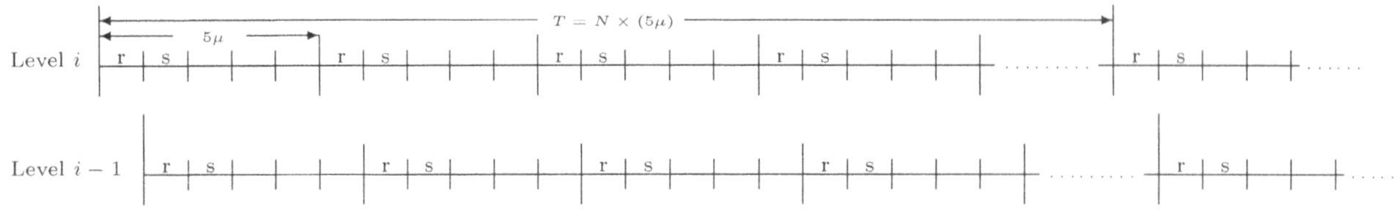

Figure 1: Time slots in DMAC

traffic following a Poisson process. The stochastic traffic scenario is modeled as a discrete time Markov chain and the analytical results are obtained using numerical methods. The consistency between the analytical results and the simulation results shows that the analytical approach can be used as a complementary tool for performance analysis on DMAC.

This paper is organized as follows. In Section 2, we present a generalized model for DMAC. We analyze the average delay and average energy consumption for both CBR traffic and stochastic traffic in section 3. In section 4, we present and compare the analytical results and the simulation results. We conclude and state our future work in section 5.

2. A GENERALIZED MODEL

As shown in Figure 1, DMAC is a time-slotted MAC protocol in which time is divided into small time slots. All the time slots have equal length and is long enough to send out one data packet along with control packets. Let μ be the length of such a time slot. In DMAC, a node will only send one data packet every 5 time slots in order to avoid collisions, so every 5 consecutive time slots can be virtually grouped into a time frame. The first time slot in a 5μ time frame is a receiving slot and labeled by r in Figure 1, and the second slot in a 5μ time frame is a sending slot and labeled by s in Figure 1. The other 3 time slots are all sleeping slots since the node knows the downstream nodes along a multi-hop path will forward the data packet in the next 3 time slots. Additional receiving/sending slots will be held in the subsequent 5μ time frames if (1) a *more data* flag is set in the current data packet because the sender has more data to send or (2) a sender has lost channel contention but overheard an *ACK* packet from its parent or (3) a node on a different branch is sending an explicit *MTS* packet.

A cycle, denoted by T, contains N number of 5μ time frames. That is, $T = N \times (5\mu)$. In the DMAC paper, $N = 4$. The receiving/sending time slots of a node at level $i-1$ is staggered with respect to level i by shifting one time slot to the right. Assume E_r is the energy consumed in the receiving time slot and E_s is the energy consumed in the sending time slot. Also assume that the energy consumed in the sending time slot is 0 if there is no data to be sent.

3. ANALYSIS ON DMAC

In this section, we analyze the average delay and average energy consumption at a single source node under both CBR and stochastic traffic following a Poisson process. For the sake of simplicity, we ignore the random delay introduced by contention and assume all the sending slots consume the same amount of energy in presence of data and same for all the receiving slots, however sending slots and receiving slots may consume different amount of energy.

3.1 CBR Traffic

CBR traffic can be regarded as a deterministic process. Let T_i be the time moment at which the ith packet arrives at a node, where $i = 0, 1, 2, ...$, and Δt denote the time interval between two consecutive data packets. Then we have

$$T_i = T_0 + i \times \Delta t \tag{1}$$

For the sake of simplicity, let's assume $T_0 = 0$ and $\Delta t = M\mu$ where $M = 1, 2, ...$ That is, $T_i = iM\mu$.

3.1.1 Case 1: $M \geq 5N$

This case can be regarded as $M = 5Nc + \delta$ with $c = 1, 2, 3...$ and $0 \leq \delta < 5N$. In this case, the traffic load is so light that only the first receiving and sending slots in a T cycle are likely to be waken up to receive or send packets.

$$\begin{aligned} T_i &= i\mu(5Nc + \delta) & (2)\\ &= (ic + \left\lfloor \frac{i\delta}{5N} \right\rfloor)5N\mu + [i\delta - \left\lfloor \frac{i\delta}{5N} \right\rfloor 5N]\mu \\ &= (ic + \left\lfloor \frac{i\delta}{5N} \right\rfloor)5N\mu + \Delta_i & (3) \end{aligned}$$

where $\lfloor X \rfloor$ is the floor integer of a real number X and

$$\Delta_i = (i\delta - \left\lfloor \frac{i\delta}{5N} \right\rfloor 5N)\mu \tag{4}$$

Let D_i denote the delay of the ith packet at the node, then

$$D_i = \begin{cases} \mu - \Delta_i, & \text{if } 0 \leq \Delta_i \leq \mu \\ (5N+1)\mu - \Delta_i, & \text{if } \mu < \Delta_i < 5N\mu \end{cases} \tag{5}$$

Let $\tilde{D}$ denote the average delay over a large number of data packets. If $\delta = 0$, it is obvious that $\tilde{D} = \mu$. For $\delta \neq 0$, let $\frac{\delta}{5N} = \frac{m}{n}$ where m,n are positive integers and coprime. It is trivial to show that Δ_i is periodic with a period n since n is the smallest integer such that $\Delta_{i+n} = \Delta_i$. To calculate $\tilde{D}$, it is sufficient to calculate the average delay of all the n packets in a single period instead. And $0 \leq \Delta_i \leq \mu$ is now equivalent to $\Delta_i = 0$ or $\Delta_i = \mu$. That is,

$$\frac{i\delta}{5N} - \left\lfloor \frac{i\delta}{5N} \right\rfloor = 0 \tag{6}$$

or

$$\frac{i\delta}{5N} - \left\lfloor \frac{i\delta}{5N} \right\rfloor = \frac{1}{5N} \tag{7}$$

From Equation (6), we know that $\frac{i\delta}{5N} = \frac{im}{n}$ must be an integer and hence $i = Bn$ where $B = 0, 1, 2, ...$ That also means that within a period n, there is exactly one such i

satisfying Equation (6). From Equation (7), we have

$$\frac{i\delta}{5N} = C + \frac{1}{5N}$$
$$i = \frac{5CN+1}{\delta}, \quad (8)$$

where C is such an integer that $\frac{5CN+1}{\delta} = \lfloor \frac{5CN+1}{\delta} \rfloor$. Let A be the number of such C's also satisfying $i < n$, which can be computed numerically. Now the overall average delay $\tilde{D}$ at a source node can be expressed as

$$\begin{aligned}\tilde{D} &= \frac{1}{n}\left[\sum_{i=0}^{n-1}((5N+1)\mu - \Delta_i) - 5N(A+1)\mu\right] \\ &= \frac{\mu}{n}\left[n(5N+1) - 5N(A+1) - \sum_{i=0}^{n-1}\left(i\delta - 5N\left\lfloor\frac{im}{n}\right\rfloor\right)\right] \\ &= \frac{\mu}{n}\left[n(5N+1) - 5N(A+1) - \frac{\delta n(n-1)}{2} \right. \\ &\quad \left. + \frac{5N(m-1)(n-1)}{2}\right] \quad (9)\end{aligned}$$

where we have utilized the formula for positive integers m and n which are coprime,

$$\sum_{i=1}^{n-1}\left\lfloor\frac{im}{n}\right\rfloor = \frac{1}{2}(m-1)(n-1) \quad (10)$$

The average energy consumed within a T cycle will be:

$$\begin{aligned}\tilde{E} &= \frac{1}{nc+m}\left[(nc + \left\lfloor\frac{n\delta}{5N}\right\rfloor)E_r + nE_s\right] \\ &= \frac{1}{nc+m}[(nc+m)E_r + nE_s] \quad (11)\end{aligned}$$

where $(nc+m)$ is the number of receiving time slots up to the reception of the nth packet. Since only one packet can be sent in a cycle $T = 5N\mu$, only n number of sending time slots really wake up to send the n number of packets.

3.1.2 *Case 2:* $5 <= M < 5N$

Let i_k and $i_{k'}$ denote the $i_k th$ and $i_{k'} th$ packets starting from 0. They are also the first packets in the kth and $k'th$ T cycles to be buffered and delayed to their next T cycles, respectively. That is, i_k and $i_{k'}$ will be the smallest integer satisfying the following condition:

$$i_k M\mu - [5Nk'\mu + \mu + 5(i_k - 1 - i_{k'})\mu] > 0, \quad (12)$$

wherein $i_k M\mu$ is the arrival time of the $i_k th$ packet, $i_k - i_{k'}$ is the number of packets buffered and sent out before the $i_k th$ packet. This condition basically states that the arrival time of the $i_k th$ packet must come after the beginning of the last sending time slot.

Equation (12) can be reduced to

$$i_k > \frac{5Nk' - 5i_{k'} - 4}{M-5} \quad (13)$$

Since i_k is the smallest integer satisfying the above inequality, it can be expressed as

$$i_k = \left\lceil\frac{5Nk' - 5i_{k'} - 4}{M-5}\right\rceil \quad (14)$$

It is obvious that the initial condition $i_1 = 1$. Let K be the first T cycle such that $i_K = 5NK/M$. If such a K exists, then the process follows a periodic pattern. All the T cycles thereafter will repeat the first K number of T cycles. Such a K can be numerically computed through the following iterations:

$$\begin{aligned}&\textit{initial condition: } k' = 1 \textit{ and } i_1 = 1 \\ &\textit{iteration: } i_k = \left\lceil\frac{5Nk' - 5i_{k'} - 4}{M-5}\right\rceil \\ &\qquad k = k' + \left\lceil\frac{i_k M - 5Nk'}{5N}\right\rceil \quad (15) \\ &\qquad k' = k \\ &\textit{stop condition: } i_k = 5Nk/M \textit{ or } k \textit{ is very large.}\end{aligned}$$

Let S be the set of k's obtained through the iterations, then the average delay at a node can be expressed as

$$\begin{aligned}\tilde{D} = \frac{M\mu}{5NK}\sum_{k\in S}&\left[\sum_{j=1}^{i_k - i_{k'}}(5(j-1)+1) + \sum_{j=i_k}^{\lfloor\frac{5Nk}{M}\rfloor}(5Nk - jM)\right. \\ &\left. - \sum_{j=\lceil\frac{5Nk'}{M}\rceil}^{i_k - 1}(jM - 5Nk')\right]. \quad (16)\end{aligned}$$

$\tilde{D}$ can be calculated as i_k can be computed through the iterations.

The average energy consumed within a T cycle can be expressed as

$$\begin{aligned}\tilde{E} &= \frac{1}{K}\left[\frac{5NK}{M}(E_r + E_s) + KE_r\right] \\ &= \frac{5N}{M}(E_r + E_s) + E_r. \quad (17)\end{aligned}$$

$\frac{5NK}{M}$ number of packets are sent out in K number of T cycles, each packet consumes $E_r + E_s$ amount of energy. An extra receiving time slot in every T cycle is used before going to sleep.

3.1.3 *Case 3:* $M < 5$

This case represents a saturated and unstable state. The arrival time of the ith packet is $T_i = iM\mu$ and the sending time of the ith packet is $(i \times 5\mu + \mu)$. Therefore the average delay upon the sending of the kth packet is

$$\tilde{D} = \frac{1}{k}\sum_{i=0}^{k}[i(5-M)+1]\mu. \quad (18)$$

The average energy consumption within a T cycle is $\tilde{E} = N(E_r + E_s)$.

3.2 Stochastic Traffic

Now let us assume the packet arrivals follow a Poisson process $\{N(t) : t \geq 0\}$ with a rate parameter λ, where $N(t)$ is the number of packets that have arrived up to time t. Then the probability of k number of packets arriving in time interval $(t, t+\Delta t]$ is given by a Poisson distribution:

$$P[N(t+\Delta t) - N(t) = k] = \frac{e^{-\lambda\Delta t}(\lambda\Delta t)^k}{k!} \quad (19)$$

where $k = 0, 1, 2....$ Poisson process is memoryless, which means that the number of packets arriving after time t is independent of the number of packets arriving before time t.

X_{ij} denotes the state of a sensor node at the end of every T cycle, where $i = 0, 1, 2, ..., N$ and $j = 0, 1, 2, ...$ are two discrete valued random variables. Random variable i denotes that the first i number of 5μ time frames in the T cycle are used for receiving and sending. Random variable j denotes that there are j number of packets buffered at the end of the T cycle. Then the random variables $\{X_{ij}\}$ forms a Discrete Time Markov Chain since the next state only depends on the current state and not on the past. Let $P_{ij,mn}$ denote the state transition probability from state X_{ij} to X_{mn}. Let $D_{ij,mn}$, $E_{ij,mn}$ be the average delay and the energy consumption incurred at the state X_{ij}, respectively. Note that $D_{ij,mn}$, $E_{ij,mn}$ only take into account the effects of the packets arrived or buffered at the current T cycle and exclude the effects of the packets buffered at the previous T cycle. A few cases need be considered here.

3.2.1 *Case 1:* $0 \leq m < N$ *and* $m < j$

It is obvious that in this case the state transition probability $P_{ij,mn} = 0$. If $j < N$, then all the j packets buffered at the end of the previous T cycle are supposed to be sent out at the first j number of 5μ time frames of the current T cycle. If $j \geq N$, then the first N number of packets will be sent out in the current T cycle. Therefore, $m \geq j$ or $m = N$ is the necessary condition for having a non-zero state transition probability $P_{ij,mn}$. All these states in this case should be excluded in the calculation of the average delay or the average energy consumption.

3.2.2 *Case 2:* $0 \leq j < m < N$

The state transition probability can be expressed as:

$$\begin{aligned} P_{ij,mn} &= e^{-\lambda\mu(5(m-1)+1)} \frac{[\lambda\mu(5(m-1)+1)]^{(m-j)}}{(m-j)!} \times I_{j,m} \\ &\times e^{-\lambda\mu(5(N-m+1)-1)} \frac{[\lambda\mu(5(N-m+1)-1)]^n}{n!} \\ &= e^{-\lambda\mu(5N)} \frac{[\lambda\mu(5(m-1)+1)]^{(m-j)}}{(m-j)!} \times I_{j,m} \\ &\times \frac{[\lambda\mu(5(N-m+1)-1)]^n}{n!} \end{aligned} \tag{20}$$

where $I_{j,m}$ is an integral depending on j and m:

$$\begin{aligned} I_{j,m} = & \frac{(m-j)!}{(5(m-1)+1)^{(m-j)}} \int_0^{5(j-1)+1} \int_{\tau_1}^{5j+1} \int_{\tau_2}^{5(j+1)+1} \\ & \cdots \int_{\tau_{m-j-2}}^{5(m-3)+1} \int_{\tau_{m-j-1}}^{5(m-2)+1} d\tau_{m-j} d\tau_{m-j-1} ... d\tau_3 d\tau_2 d\tau_1 \end{aligned} \tag{21}$$

for $0 < \tau_1 < \tau_2 < \tau_3 < ... < \tau_{m-j-1} < \tau_{m-j} < 5(m-2)+1$.

In this case, in addition to the j number of packets from the previous T cycle, there must be another $(m-j)$ number of packets arriving during the first $5(m-1)+1$ time slots. The probability for that to occur is given by the terms before the $I_{j,m}$ term in Equation (20). However, this does not guarantee that all those $(m-j)$ packets will be sent out as supposed to be. There are further restrictions on the arrival time of the $(m-j)$ packets in order for them to be sent. The probability of having proper arrival timings among those $(m-j)$ packets is given by the term $I_{j,m}$, which we will explain in details shortly. Furthermore, there should be another n number of packets coming and being buffered during the rest of the T cycle, whose probability is given by the terms after the term $I_{j,m}$.

In order to understand the term $I_{j,m}$, first let's recall the order statistics property of a Poisson process. For a Poisson process $\{N(t) : t \geq 0\}$, suppose we are given that for a fixed t, $N(t) = n$. Let T_i be the arrival time of the ith event, where $i = 1, 2, ..., n$. Let $u_1, u_2, ..., u_n$ be $i.i.d$ random variables each having a uniform distribution over $(0, t)$ and $U_{(1)}, U_{(2)}, ..., U_{(n)}$ be their order statistics. Then the conditional joint probability density function of $(T_1, T_2, ..., T_n)$ given that $N(t) = n$ is the same as the joint probability density function of $(U_{(1)}, U_{(2)}, ..., U_{(n)})$. That is,

$$\begin{aligned} & f_{T_1,T_2,...,T_n}((\tau_1, \tau_2, ..., \tau_n)|N(t) = n) \\ = & \begin{cases} \frac{n!}{t^n}, & \text{for } 0 < \tau_1 < \tau_2 < ... < \tau_n < t \\ 0, & \text{otherwise} \end{cases} \end{aligned} \tag{22}$$

Without loss of generality, we assume $\mu = 1$. According to the order statistics property in Equation (22), the conditional joint probability density function of the $(m-j)$ packets is $\frac{(m-j)!}{(5(m-1)+1)^{(m-j)}}$, for $0 < \tau_1 < \tau_2 < \tau_3 < ... < \tau_{m-j-1} < \tau_{m-j} < 5(m-1)+1$. Furthermore, in order for the first packet among the $(m-j)$ packets to be sent out, it must arrive no later than the $(5(j-1)+1)th$ time slot. That is, $0 < \tau_1 \leq 5(j-1)+1$ must be satisfied so that a *more data* flag will piggyback the jth data packet being sent out. As a result, the next receiving/sending time slots will stay awake to send more data. The second packet must come after τ_1 but no later than the $(5j+1)th$ time slot, i.e., $\tau_1 < \tau_2 \leq 5j+1$, and so forth. Integrating the conditional joint probability density function over all those ranges will give us the conditional joint probability for all the $(m-j)$ packets to be sent out.

The average delay $D_{ij,mn}$ can be expressed as

$$\begin{aligned} D_{ij,mn} = & e^{-\lambda\mu(5N)} \frac{[\lambda\mu(5(m-1)+1)]^{(m-j)}}{(m-j)!} \frac{\mu}{(m-j+n)} \\ & \times \left\{ A_{j,m} + I_{j,m} B_{m,N} + I_{j,m} \sum_{k=1}^{n} (5(k-1)+1) \right\} \\ & \times \frac{[\lambda\mu(5(N-m+1)-1)]^n}{n!} \end{aligned} \tag{23}$$

where $A_{j,m}$ is an integral depending on j and m:

$$\begin{aligned} A_{j,m} = & \frac{(m-j)!}{(5(m-1)+1)^{(m-j)}} \int_0^{5(j-1)+1} \int_{\tau_1}^{5j+1} \int_{\tau_2}^{5(j+1)+1} \\ & \cdots \int_{\tau_{m-j-2}}^{5(m-3)+1} \int_{\tau_{m-j-1}}^{5(m-2)+1} \sum_{k=1}^{m-j} (5(j+k-2)+1-\tau_k) \\ & \cdot d\tau_{m-j} d\tau_{m-j-1} ... d\tau_2 d\tau_1 \end{aligned} \tag{24}$$

for $0 < \tau_1 < \tau_2 < ... < \tau_{m-j-1} < \tau_{m-j} < (5(m-2)+1)$, and $B_{m,N}$ is an integral depending on m and N:

$$\begin{aligned} B_{m,N} = & \frac{n!}{(5(N-m+1)-1)^n} \int_{5(m-1)+1}^{5N} \int_{\tau_1}^{5N} \cdots \int_{\tau_{n-2}}^{5N} \\ & \cdot \int_{\tau_{n-1}}^{5N} \sum_{k=1}^{n} (5N - \tau_k) d\tau_n d\tau_{n-1} ... d\tau_2 d\tau_1 \end{aligned} \tag{25}$$

for $5(m-1)+1 < \tau_1 < \tau_2 < ... < \tau_{n-1} < \tau_n < 5N$.

Inside the curly bracket of Equation (23), the $A_{j,m}$ term is associated with the delays of the $(m-j)$ number of packets received and to be sent in the current T cycle. The $B_{m,N}$ term is related to the sleep delays of the n number of buffered packets up to the end of the current T cycle. The last term is associated with the delays of the n number of buffered packets with respect to the beginning of the next T cycle. The $B_{m,N}$ term and the last term together give the delays of the n number of buffered packets.

3.2.3 *Case 3:* $0 \leq j = m < N$

The state transition probability can be expressed as:

$$\begin{aligned} P_{ij,mn} &= e^{-\lambda\mu(5(j-1)+1)} e^{-\lambda\mu(5(N-j+1)-1)} \\ &\times \frac{[\lambda\mu(5(N-j+1)-1)]^n}{n!} \\ &= e^{-\lambda\mu(5N)} \frac{[\lambda\mu(5(N-j+1)-1)]^n}{n!} \end{aligned} \tag{26}$$

A node must stay awake at the receiving and sending time slots of the first $5(j-1)+1$ number of time slots in the current T cycle in order to send all the j number of packets buffered during the previous T cycle. To end up at a state X_{mn} with $m = j$, there must be no packet arriving by the end of the first $(5(j-1)+1)$ time slots. The probability for that to happen is $e^{-\lambda\mu(5(j-1)+1)}$ according to the Poisson distribution in the Equation (19), wherein $k = 0$ and $\Delta t = (5(j-1)+1)\mu$. Also to end up at a state X_{mn}, there must be another n number of packets arriving during the rest of the T cycle after the first $(5(j-1)+1)$ time slots. The probability for that to happen is again given by the Poisson distribution in the Equation (19), wherein $k = n$ and $\Delta t = [5(N-j+1)-1]\mu$.

The delay $D_{ij,mn}$ in this case can be expressed as:

$$\begin{aligned} D_{ij,mn} &= e^{-\lambda\mu(5N)} \frac{[\lambda\mu(5(N-j+1)-1)]^n}{n!} \frac{\mu}{n} \\ &\times \left\{ C_{j,N} + \sum_{k=1}^{n} (5(k-1)+1) \right\} \end{aligned} \tag{27}$$

where $C_{j,N}$ is an integral depending on j and N:

$$\begin{aligned} C_{j,N} &= \frac{n!}{(5(N-j+1)-1)^n} \int_{5(j-1)+1}^{5N} \int_{\tau_1}^{5N} \cdots \int_{\tau_{n-2}}^{5N} \\ &\cdot \int_{\tau_{n-1}}^{5N} \sum_{k=1}^{n} (5N-\tau_k) d\tau_n d\tau_{n-1} ... d\tau_2 d\tau_1 \end{aligned} \tag{28}$$

for $5(j-1)+1 < \tau_1 < \tau_2 < ... < \tau_{n-1} < \tau_n < 5N$.

Inside the curly bracket of Equation (27), the $C_{j,N}$ term is associated with the sleep delays of the n number of buffered packets up to the end of the current T cycle. The last term is related to the delays of the n number of buffered packets with respect to the beginning of the next T cycle.

3.2.4 *Case 4:* $0 \leq j < m = N$

The state transition probability can be expressed as:

$$P_{ij,mn} = e^{-\lambda\mu(5N)} \frac{[\lambda\mu(5N)]^{(N-j+n)}}{(N-j+n)!} \times I_{j,N} \tag{29}$$

where $I_{j,N}$ is an integral depending on j and N:

$$\begin{aligned} I_{j,N} &= \frac{(N-j+n)!}{(5N)^{(N-j+n)}} \int_0^{5(j-1)+1} \int_{\tau_1}^{5j+1} \int_{\tau_2}^{5(j+1)+1} \\ &\cdots \int_{\tau_{N-j-2}}^{5(N-3)+1} \int_{\tau_{N-j-1}}^{5(N-2)+1} \int_{\tau_{N-j}}^{5N} \int_{\tau_{N-j+1}}^{5N} \\ &\cdots \int_{\tau_{N-j+n-1}}^{5N} d\tau_{N-j+n} ... d\tau_{N-j+1} d\tau_{N-j} ... d\tau_3 d\tau_2 d\tau_1 \end{aligned} \tag{30}$$

for $0 < \tau_1 < \tau_2 < ... < \tau_{N-j-1} < \tau_{N-j} < ... < \tau_{N-j+n-1} < 5N$.

In this case, in addition to the j number of packets from the previous T cycle, there must be another $(N-j+n)$ number of packets arriving during the current T cycle. The first $(N-j)$ ones among the $(N-j+n)$ number of packets will be sent out resulting in a state X_{Nn}. The probability to have $(N-j+n)$ number of packets coming during the current T cycle is given by the terms before the $I_{j,N}$ term in Equation (29). Again, this does not guarantee that its first $(N-j)$ packets will be sent out as supposed to be. There is further restriction on the arrival time of the $(N-j+n)$ packets in order for the first $(N-j)$ packets to be sent. The probability of having proper arrival timings among those $(N-j+n)$ packets is given by the term $I_{j,N}$.

Without loss of generality, we again assume $\mu = 1$. According to the order statistics property in Equation (22), the conditional joint probability density function of the $(N-j+n)$ packets is $\frac{(N-j+n)!}{(5N)^{(N-j+n)}}$, for $0 < \tau_1 < \tau_2 < ... < \tau_{N-j-1} < \tau_{N-j} < ... < \tau_{N-j+n-1} < 5N$. Furthermore, in order for the first packet among the $(N-j+n)$ packets to be sent out, it must arrive no later than the $(5(j-1)+1)th$ time slot. That is, $0 < \tau_1 \leq 5(j-1)+1$ must be satisfied. The second packet must come after τ_1 but no later than the $(5j+1)th$ time slot, i.e., $\tau_1 < \tau_2 \leq 5j+1$. It repeats until $\tau_{N-j-1} < \tau_{N-j} \leq 5(N-2)+1$. After that, $\tau_{k-1} < \tau_k < 5N$, for $k = N-j+1, N-j+2, ..., N-j+n$. Integrating the conditional joint probability density function over all those ranges will give us the conditional joint probability for all the $(N-j+n)$ packets to be sent out.

The delay $D_{ij,mn}$ can be expressed as

$$\begin{aligned} D_{ij,mn} &= e^{-\lambda\mu(5N)} \frac{[\lambda\mu(5N)]^{(N-j+n)}}{(N-j+n)!} \frac{\mu}{(N-j+n)} \\ &\times \left\{ F_{j,m} + I_{j,N} \sum_{k=1}^{n} (5(k-1)+1) \right\} \end{aligned} \tag{31}$$

where $F_{j,m}$ is an integral depending on j and m:

$$\begin{aligned} F_{j,m} &= \frac{(N-j+n)!}{(5N)^{(N-j+n)}} \int_0^{5(j-1)+1} \int_{\tau_1}^{5j+1} \int_{\tau_2}^{5(j+1)+1} \\ &\cdots \int_{\tau_{N-j-2}}^{5(N-3)+1} \int_{\tau_{N-j-1}}^{5(N-2)+1} \int_{\tau_{N-j}}^{5N} \int_{\tau_{N-j+1}}^{5N} \cdots \int_{\tau_{N-j+n-1}}^{5N} \\ &\cdot \left(\sum_{k=1}^{N-j} (5(j+k-2)+1-\tau_k) + \sum_{k=N-j+1}^{N-j+n} (5N-\tau_k) \right) \\ &\cdot d\tau_{N-j+n} ... d\tau_{N-j+1} d\tau_{N-j} ... d\tau_3 d\tau_2 d\tau_1 \end{aligned} \tag{32}$$

for $0 < \tau_1 < \tau_2 < ... < \tau_{N-j} < ... < \tau_{N-j+n} < 5N$.

The first summation term of $F_{j,m}$ in Equation (32) is associated with the delays of the $(N-j)$ number of packets received and to be sent in the current T period. The second summation term of $F_{j,m}$ is related to the sleep delays of the n number of buffered packets up to the end of the current T cycle. Inside the curly bracket of Equation (31), the last term is associated with the delays of the n number of buffered packets with respect to the beginning of the next T cycle.

3.2.5 *Case 5:* $N = m \leq j \leq N + n$

The state transition probability can be expressed as:

$$P_{ij,mn} = e^{-\lambda\mu(5N)} \frac{[\lambda\mu(5N)]^{(N-j+n)}}{(N-j+n)!} \tag{33}$$

In this case, more than N number of packets are buffered by the end of the previous T cycle, but only N out of them can be sent out. The other $(j - N)$ number of packets stay in the buffer. To end up with a state at X_{Nn}, there must be another $(n - j + N)$ number of packets coming during the current T cycle. The probability for that to occur is given by the Equation (19), wherein $k = n-j+N$ and $\Delta t = 5N\mu$.

The delay $D_{ij,mn}$ can be expressed as

$$D_{ij,mn} = e^{-\lambda\mu(5N)} \frac{[\lambda\mu(5N)]^{(N-j+n)}}{(N-j+n)!} \frac{\mu}{(N-j+n)} \times \left\{ G_{j,m} + \sum_{k=j-N+1}^{n} (5(k-1)+1) \right\} \tag{34}$$

where $G_{j,m}$ is an integral depending on j and m:

$$G_{j,m} = \frac{(N-j+n)!}{(5N)^{(N-j+n)}} \int_0^{5N} \int_{\tau_1}^{5N} \int_{\tau_2}^{5N} \cdots \int_{\tau_{N-j+n-1}}^{5N} \cdot \sum_{k=1}^{N-j+n} (5N - \tau_k) d\tau_{N-j+n} ... d\tau_3 d\tau_2 d\tau_1 \tag{35}$$

for $0 < \tau_1 < \tau_2 < ... < \tau_{N-j+n} < 5N$.

Inside the curly bracket of Equation (34), The $G_{j,m}$ term is associated with the delays of the $(N - j + n)$ number of packets arrived at the current T cycle up to the end of the current T cycle. The second term is associated with the delays of the $(N - j + n)$ buffered with respect to the beginning of the next T cycle.

3.2.6 *Case 6:* $m = N, j >= 2N$

This case represents a saturated and unstable state. It occurs only when the traffic load exceeds the channel capacity. Since no equilibrium state in this case, Markov model is no longer applicable. Therefore, we assume the traffic load is less than the channel capacity and hence ignore this case.

The state transition probabilities and the delay can be summarized as Equation (36) and Equation (37), respectively.

Let π_{ij} be the equilibrium probability that the system is in the state X_{ij}, then the global balance equations for the state X_{ij} can be expressed as

$$\pi_{ij} \sum_{m=0}^{N} \sum_{n=0}^{M} P_{ij,mn} = \sum_{m=0}^{N} \sum_{n=0}^{M} \pi_{mn} P_{mn,ij} \tag{38}$$

The sum of all the probabilities of leaving a state should always be 1, *i.e.*,

$$\sum_{m=0}^{N} \sum_{n=0}^{M} P_{ij,mn} = 1 \tag{39}$$

Therefore, Equation (38) can be reduced as:

$$\pi_{ij} = \sum_{m=0}^{N} \sum_{n=0}^{M} \pi_{mn} P_{mn,ij} = \sum_{n=0}^{M} \sum_{m=0}^{N} \pi_{mn} P_{mn,ij} \tag{40}$$

Since the state transition probability $P_{mn,ij}$ is independent of m, Equation (40) can be written as

$$\pi_{ij} = \sum_{n=0}^{M} \left(P_{0n,ij} \sum_{m=0}^{N} \pi_{mn} \right) \tag{41}$$

Let's define

$$\hat{\pi}_n = \sum_{m=0}^{N} \pi_{mn} \tag{42}$$

for $n = 0, 1, 2...$, then Equation (41) becomes

$$\pi_{ij} = \sum_{n=0}^{M} P_{0n,ij} \hat{\pi}_n \tag{43}$$

We further have

$$\hat{\pi}_j = \sum_{i=0}^{N} \pi_{ij} = \sum_{i=0}^{N} \sum_{n=0}^{M} P_{0n,ij} \hat{\pi}_n = \sum_{n=0}^{M} \left(\sum_{i=0}^{N} P_{0n,ij} \right) \hat{\pi}_n \tag{44}$$

In matrix notation, this becomes, with $\hat{\boldsymbol{\pi}}$ the column vector with elements $\hat{\pi}_n$,

$$\hat{\boldsymbol{\pi}} = \boldsymbol{Q} \hat{\boldsymbol{\pi}} \tag{45}$$

or

$$(\boldsymbol{I} - \boldsymbol{Q}) \hat{\boldsymbol{\pi}} = \boldsymbol{0} \tag{46}$$

The determinant of $\boldsymbol{I} - \boldsymbol{Q}$ will be 0. To have a unique non-zero solution of $\hat{\boldsymbol{\pi}}$, replace one of the equation in Equation (46) with the normalization equation:

$$\sum_{n=0}^{M} \hat{\pi}_n = 1 \tag{47}$$

Now we have reduced a set of $(M + 1) \times (M + 1)$ equations into a set of $(M + 1)$ equations instead. Solving Equation (46) together with Equation (47) is much easier than solving Equation (40). After we have $\hat{\boldsymbol{\pi}}$, we can obtain the probability π_{ij} that the system is in the state X_{ij} from Equation (43).

The average sleep delay incurred in the state X_{mn} can be expressed as

$$\tilde{D}_{mn} = \frac{\sum_{i=0}^{N} \sum_{j=0}^{M} \pi_{ij} D_{ij,mn}}{\pi_{mn}} \tag{48}$$

The overall average delay is

$$\tilde{D} = \frac{\sum_{m=0}^{N} \sum_{n=0}^{M} \pi_{mn} \tilde{D}_{mn}}{1 - \hat{\pi}_0} \tag{49}$$

Assume the sink node is at level 0, then the average end-to-end delay for a source node at level L will be

$$D_{end-end} = \tilde{D} + (L - 1)\mu \tag{50}$$

The energy consumption $E_{ij,mn}$ can be expressed as

$$E_{ij,mn} = \begin{cases} 0, & \text{for } 0 \leq m < N, m < j \\ E_r, & \text{for } m = j = n = 0 \\ (m - j + n)(E_r + E_s), & \text{otherwise} \end{cases} \tag{51}$$

$$P_{ij,mn} = \begin{cases} 0 & \text{if } 0 \le m < N \text{ and } m < j \\ e^{-\lambda\mu(5N)} \frac{[\lambda\mu(5(m-1)+1))]^{(m-j)}}{(m-j)!} \times I_{j,m} \times \frac{[\lambda\mu(5(N-m+1)-1))]^{n}}{n!} & \text{if } 0 \le j < m < N \\ e^{-\lambda\mu(5N)} \frac{[\lambda\mu(5(N-j+1)-1))]^{n}}{n!} & \text{if } 0 \le j = m < N \\ e^{-\lambda\mu(5N)} \frac{[\lambda\mu(5N)]^{(N-j+n)}}{(N-j+n)!} \times I_{j,N} & \text{if } 0 \le j < m = N \\ e^{-\lambda\mu(5N)} \frac{[\lambda\mu(5N)]^{(N-j+n)}}{(N-j+n)!} & \text{if } N = m \le j \le N+n \end{cases} \tag{36}$$

$$D_{ij,mn} = \begin{cases} 0 & \text{if } 0 \le m < N \text{ and } m < j \\ e^{-\lambda\mu(5N)} \frac{[\lambda\mu(5(m-1)+1))]^{(m-j)}}{(m-j)!} \frac{[\lambda\mu(5(N-m+1)-1))]^{n}}{n!} \frac{\mu}{(m-j+n)} & \\ \quad \times \left\{A_{j,m} + I_{j,m}B_{m,N} + I_{j,m}\sum_{k=1}^{n}(5(k-1)+1)\right\} & \text{if } 0 \le j < m < N \\ e^{-\lambda\mu(5N)} \frac{[\lambda\mu(5(N-j+1)-1))]^{n}}{n!} \frac{\mu}{n} \times \left\{C_{j,N} + \sum_{k=1}^{n}(5(k-1)+1)\right\} & \text{if } 0 \le j = m < N \\ e^{-\lambda\mu(5N)} \frac{[\lambda\mu(5N)]^{(N-j+n)}}{(N-j+n)!} \frac{\mu}{(N-j+n)} \times \left\{F_{j,m} + I_{j,N}\sum_{k=1}^{n}(5(k-1)+1)\right\} & \text{if } 0 \le j < m = N \\ e^{-\lambda\mu(5N)} \frac{[\lambda\mu(5N)]^{(N-j+n)}}{(N-j+n)!} \frac{\mu}{(N-j+n)} \times \left\{G_{j,m} + \sum_{k=j-N+1}^{n}(5(k-1)+1)\right\} & \text{if } N = m \le j \le N+n \end{cases} \tag{37}$$

Similarly, the average energy consumed in the state X_{mn} can be expressed as

$$E_{mn} = \frac{\sum_{i=0}^{N}\sum_{j=0}^{M}\pi_{ij}P_{ij,mn}E_{ij,mn}}{\pi_{mn}} \tag{52}$$

The overall average energy consumption in a T cycle is

$$\tilde{E} = \sum_{m=0}^{N}\sum_{n=0}^{M}\pi_{mn}E_{mn} \tag{53}$$

4. PERFORMANCE EVALUATION

This section presents and compares the numerical and simulation results for both CBR and stochastic traffic.

4.1 Numerical Evaluation

Even though the integrals in Equation (36) and Equation (37) are not closed-form expressions, they all can be evaluated numerically with little efforts. For example, the $I_{j,m}$ can be computed through a small MATLAB function as follows:

```
function ans = Ijm(j,m)
    syms f t;
    f=1;
    for i=m-j-1:-1:1
        f=int(f,t,t,5*(j+i-1)+1);
    end

    if(j>0)
        f=int(f,t,0,5*(j-1)+1);
    else
        f=int(f,t,0,1);
    end
    ans=factorial(m-j)*double(f);
    ans=ans/((5*(m-1)+1)^(m-j));
```

All the other integrals in Equation (36) and Equation (37) can be evaluated numerically in similar ways. Table 1 lists the state transition probabilities for $N = 4$ as stated in the DMAC paper [8]. The packet arrival rate $\lambda = 2$ packets/second and the time slot $\mu = 0.00967$ second. Column j lists the state transition probabilities from the state X_{0j} to every other state, for $j = 0, 1, 2, 3, 4$. The summation of each column is equal or very close to 1 as expected. The state probabilities are listed in Table 2 and it is easy to verify that all the state probabilities add up to 1.

Table 1: State transition probabilities (N=4)

	(0,0)	(0,1)	(0,2)	(0,3)	(0,4)
(0,0)	0.6792	0	0	0	0
(0,1)	0.2496	0	0	0	0
(0,2)	0.0459	0	0	0	0
(0,3)	0.0056	0	0	0	0
(0,4)	0.0005	0	0	0	0
(1,0)	0.0131	0.6792	0	0	0
(1,1)	0.0048	0.2496	0	0	0
(1,2)	0.0009	0.0459	0	0	0
(1,3)	0.0001	0.0056	0	0	0
(1,4)	0.0000	0.0005	0	0	0
(2,0)	0.0001	0.0131	0.6792	0	0
(2,1)	0.0000	0.0036	0.1839	0	0
(2,2)	0.0000	0.0005	0.0249	0	0
(2,3)	0.0000	0.0000	0.0022	0	0
(2,4)	0.0000	0.0000	0.0002	0	0
(3,0)	0.0000	0.0014	0.0788	0.6792	0
(3,1)	0.0000	0.0002	0.0137	0.1182	0
(3,2)	0.0000	0.0000	0.0012	0.0103	0
(3,3)	0.0000	0.0000	0.0001	0.0006	0
(3,4)	0.0000	0.0000	0.0000	0.0000	0
(4,0)	0.0000	0.0002	0.0122	0.1445	0.6792
(4,1)	0.0000	0.0001	0.0031	0.0405	0.2627
(4,2)	0.0000	0.0000	0.0004	0.0060	0.0508
(4,3)	0.0000	0.0000	0.0000	0.0006	0.0066
(4,4)	0.0000	0.0000	0.0000	0.0000	0.0006
Total	1.0000	1.0000	1.0000	1.0000	0.9999

4.2 Simulation Results

We have carried out the simulations on a line topology with 11 nodes using the ns-2. The data packet size is 70 bytes and the simulation time is set to 60 minutes. Table 3 shows some parameters used in both the DMAC paper [8] and in our simulations as well.

Figure 2 compares the numerical and simulation results on average end-to-end delays and energy consumption with varying arrival intervals for both CBR and stochastic traffic.

Figure 2(a) shows the numerical results accurately match the simulation results for CBR traffic with intervals equal or larger than 10μ. An interesting periodic and symmet-

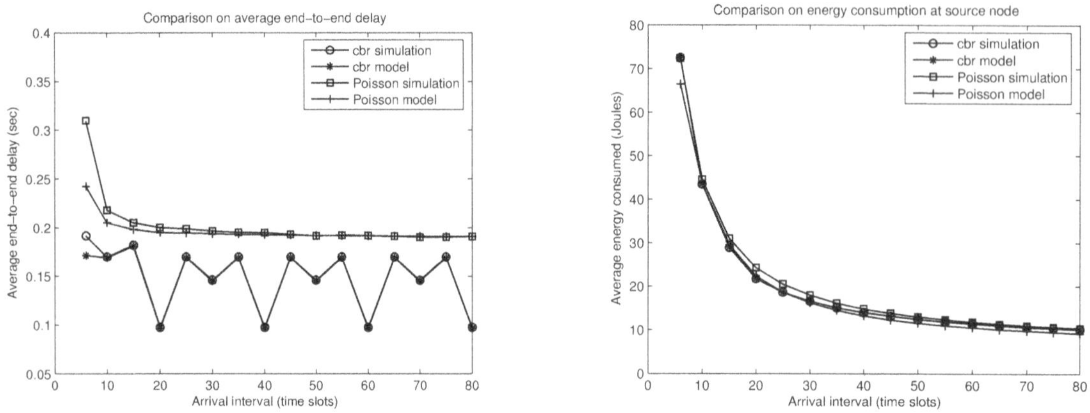

(a) The comparison of delay (b) The comparison of energy consumption

Figure 2: Comparison with different arrival intervals

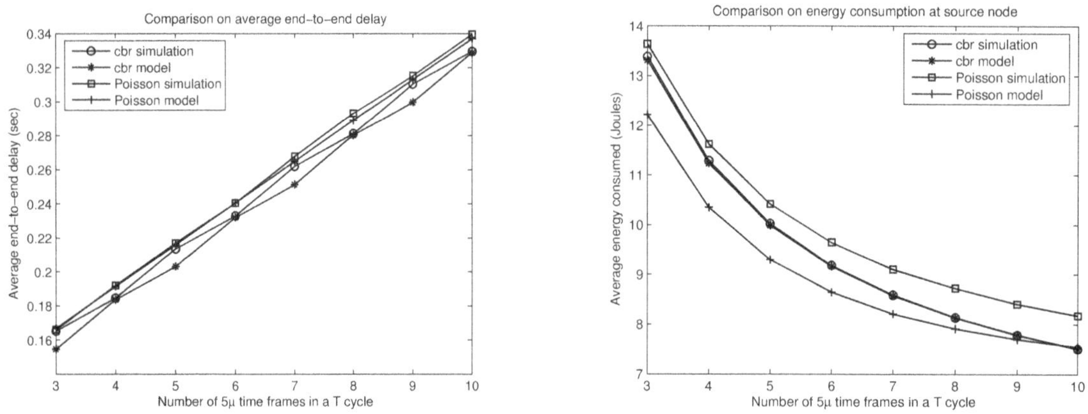

(a) The comparison of delay (b) The comparison of energy consumption

Figure 3: Comparison with different number of 5μ time frames in a T cycle

Table 2: Probabilities at state X_{ij} (N=4)

	0	1	2	3	4
0	0.4735	0.1740	0.0320	0.0039	0.0004
1	0.1798	0.0661	0.0121	0.0015	0.0001
2	0.0343	0.0093	0.0013	0.0001	0.0000
3	0.0077	0.0013	0.0001	0.0000	0.0000
4	0.0018	0.0005	0.0001	0.0000	0.0000
$\hat{\pi}_j$	0.6971	0.2512	0.0456	0.0055	0.0005

Table 3: Parameters used in the simulations

Parameter	IEEE 802.11
transmission power	660 mW
receiving power	395 mW
idle power	35 mW
data rate	2 Mbps
range	250 m

ric pattern is observed for CBR traffic for the selected time intervals within the same range. The numerical results for stochastic traffic are also close to the simulation results for large arrival intervals. As the interval is getting smaller, the stochastic model requires more states in order to have an accurate prediction, which explains the growing gap between the numerical and simulation results for stochastic traffic. Larger delays are observed for stochastic traffic at the selected arrival rates compared to CBR traffic, which indicates DMAC adapts better to CBR traffic than stochastic traffic at those arrival rates. The absence of the periodic pattern in stochastic traffic reveals that stochastic traffic is less sensitive to the changes of intervals than CBR traffic is.

In Figure 2(b), the numerical results are also close to the simulation results when the arrival interval is large. The absence of oscillation in energy consumption for CBR traffic indicates that the energy consumption is less sensitive to the changes of intervals than the delay is.

Figure 3 shows the effect of varying number of 5μ time frames in a T cycle on average end-to-end delays and energy consumption. The traffic interval is fixed to 0.6 seconds. Figure 3(a) shows the numerical results of the delay match the simulation results pretty well for both CBR and stochastic traffic. The delay increases as the number of 5μ time frames increases for both traffic conditions.

Figure 3(b) shows the energy consumption decreases as the number of 5μ time frames in a T cycle increases for both traffic conditions. Even though stochastic traffic does not match its numerical results with its simulation results as well as CBR traffic does, the same tendency is still observed. From Figure 3(a) and 3(b), we can also see the tradeoff between the delay and the energy consumption with the selection of different number of 5μ time frames in a T cycle.

5. CONCLUSIONS AND FUTURE WORK

In this paper, we proposed generalized models of the DMAC protocol for both CBR traffic and stochastic traffic following a Poisson process. A discrete-time Markov chain is used to model the stochastic traffic scenario. The average delay and energy consumption at a source node are expressed and can be evaluated numerically. The close match between the numerical results and the simulation results validates the correctness of the models. The models provide insight on the adaptivity of DMAC under CBR and stochastic traffic conditions and indicate that DMAC only adapts varying traffic conditions to a limited extent due to the low channel utilization and DMAC adapts CBR traffic better than the stochastic traffic. The stochastic model can also be applied to scenarios with multiple source nodes. As future work, we plan to extend the numerical evaluations and simulations to include scenarios with multiple source nodes and draw comparisons.

6. REFERENCES

[1] G. Bianchi. Performance analysis of the ieee 802.11 distributed coordination function. *IEEE JSAC*, 18(3):535–546, 2000.

[2] M. Buettner, G. V. Yee, E. Anderson, and R. Han. X-mac: a short preamble mac protocol for duty-cycled wireless sensor networks. In *SenSys'06*, pages 307–320, 2006.

[3] P. Chatzimisios, A. C. Boucouvalas, and V. Vitsas. Ieee 802.11 packet delay - a finite retry limit analysis. In *GLOBECOM'03*, pages 950–954, 2003.

[4] T. V. Dam and K. Langendoen. An adaptive energy-efficient MAC protocol for wireless sensor networks. In *SenSys'03*, pages 171–180, Los Angeles, Nov. 2003.

[5] B. Gao, C. He, and L. Jiang. Modeling and analysis of ieee 802.15.4 csma/ca with sleep mode enabled. In *ICCS'08*, pages 6–11, Nov. 2008.

[6] Z. Hadzi-Velkov and B. Spasenovski. Saturation throughput - delay analysis of ieee 802.11 dcf in fading channel. In *ICC'03*, 2003.

[7] J. He, Z. Tang, H. H. Chen, and S. Wang. An accurate markov model for slotted csma/ca algorithm in ieee 802.15.4 networks. *Communications Letters, IEEE*, 12(6):420 –422, June 2008.

[8] G. Lu, B. Krishnamachari, and C. S. Raghavendra. An adaptive energy-efficient and low-latency mac for data gathering in sensor networks. In *IPDS'04*, page 224a, 2004.

[9] J. Misic, S. Shafi, and V. B. Misic. Performance of a beacon enabled ieee 802.15.4 cluster with downlink and uplink traffic. *IEEE Trans. Parallel Distrib. Syst.*, 17:361–376, April 2006.

[10] P. Park, P. Soldati, C. Fischione, and K. H. Johansson. A generalized markov chain model for effective analysis of slotted ieee 802.15.4. In *MASS'09*, pages 130–139, 2009.

[11] J. Polastre, J. Hill, and D. Culler. Versatile low power media access for wireless sensor networks. In *SenSys'04*, pages 95–107, 2004.

[12] S. Pollin, M. Ergen, S. C. Ergen, B. Bougard, L. V. D. Perre, F. Catthoor, I. Moerman, and P. Varaiya. Performance analysis of slotted carrier sense ieee 802.15.4 medium access layer. In *GLOBECOM'06*, 2006.

[13] O. Yang and W. Heinzelman. Modeling and throughput analysis for smac with a finite queue capacity. In *ISSNIP'09*, pages 409–414, Dec. 2009.

[14] O. Yang and W. Heinzelman. Modeling and throughput analysis for xmac with a finite queue capacity. In *GLOBECOM'10*, pages 1–5, Dec. 2010.

[15] W. Ye, J. Heidemann, and D. Estrin. An energy-efficient MAC protocol for wireless sensor networks. In *INFOCOM'02*, pages 1567–1576, New York, June 2002.

Effects of Carrier Sense Modeling on Wireless Network Simulation Results

Athanassios Boulis
NICTA, Networks research group
13 Garden St,
Eveleigh, NSW, Australia
+61293762070
athanassios.boulis@nicta.com.au

Yuriy Tselishchev
NICTA, Networks research group
University of Sydney, School of IT
13 Garden St,
Eveleigh, NSW, Australia
yuri.tselishchev@nicta.com.au

ABSTRACT

We investigate the effect of radio modeling on the simulation results of low-power wireless networks. In particular we focus on the modeling of carrier sensing and state transition delays and describe how these aspects of low-power radios are captured in the Castalia simulator. These aspects are usually neglected by network-level simulators yet they can have a significant effect on the higher level simulation results. We show these effects by exploring a simple simulation scenario: a clique of N nodes -all in range with each other- with N-1 transmitter nodes sending packets to one receiver node, employing the simple CSMA/CA MAC protocol. Varying the sending rate and number of nodes, we show that the throughput at the receiver node can greatly vary depending on the assumptions we make about the radio. We also show that simple decisions at the MAC level, such as the back-off duration and the method of transmitting buffered packets, have a much more pronounced effect on performance when the radio is modeled more accurately.

Categories and Subject Descriptors

C.2.1 [Network Architecture and Design]: Wireless Communication, I.6.0 [Simulation and Modeling]: General

General Terms

Performance, Experimentation

Keywords

Radio model, wireless network simulation, carrier sense modeling

1. INTRODUCTION

Simulation is extremely important for the evaluation of wireless networks, especially low-power large-scale wireless networks such as wireless sensor networks (WSN), where real hardware experimentation in realistic scenarios is often prohibitively costly. Indeed the WSN community has relied heavily on simulation to propose and evaluate solutions at every layer of a WSN system. Researchers have used various platforms for their simulation needs from custom simulation environments to more mature simulators such as ns-2[1], TOSSIM[2], J-Sim[3], and MiXiM[4]. Accurate modeling of all aspects of a complex system, such as a WSN, is very challenging. Even for the aforementioned mature simulators it is affected by issues such as developer resources, simulator scope, and performance.

However, one part that has been consistently neglected is the radio subsystem. For example, researchers had to wait until 2007 to have a basic 802.11 radio model for ns-2 [5], and until 2008 to have a basic 802.15.4 ns-2 radio model. Before these models, only a rudimentary SNR threshold was applied without consideration to packet lengths, modulation, scheme, and preambles. MiXiM uses similar models to current ns-2, while J-Sim and TOSSIM are still using simpler radio models, applying some variations of the SNR threshold and probabilistic packet delivery. Despite the improvements in some of the simulators, another aspect of radios has not received the appropriate attention: delay modeling. Changing radio states does incur some delay, and more importantly, sensing a busy channel incurs a *variable* delay, depending on radio state, signal sensed, and radio design parameters.

In this paper we show the importance of such modeling using the Castalia simulator. Castalia is another mature simulator for WSN (and body area networks), developed at NICTA since 2006 and released publicly as open-source software since 2007. Castalia grew out of the need to have a WSN-specific simulator offering the most accurate modeling available in the research community, starting with the communication models (i.e., channel and radio models). In 2010, with the release of version 3.0 we have incorporated a more detailed radio model, which addresses delays more realistically, among other features. The importance of such modeling is shown in a carefully constructed, yet simple, simulation scenario where the influence of several other simulation aspects (most notably the wireless channel) is eliminated: N-1 nodes send packets at a constant rate to a receiver node. All the nodes are in range with each other, achieving perfect communication when no interference is present. Employing CSMA/CA MAC, one would expect that the received packets would increase as the rate and number of senders increases, reach a maximum point dependent on the radio's data rate, and then plateau. Instead we see great variability on the received packets curves, (e.g. different maxima points according to number of nodes, decreasing performance with increasing rate) depending on characteristics of the radio and some MAC operational parameters. The variability is due to different numbers of collisions happening, despite the use of CSMA/CA. The collisions occur because CSMA is imperfect in the present of delays. These delays are not due to the propagation of signals as modeling of traditional long-distance wireless networks dictates, but due to the operation of digital radios.

The rest of the paper is organized as follows: section 2 briefly describes Castalia's radio model focusing on the carrier sense modeling, section 3 describes the simulation scenario used throughout the paper and explains how collisions can happen, while section 4 discusses and explains a wide variety of simulation results, showcasing the importance of radio modeling. Finally, section 5 concludes the paper.

2. CASTALIA'S RADIO MODEL

The radio model in Castalia operates with event messages received from the wireless channel. The messages signify either the start or the end of a transmission. The start transmission message carries information about the signal's ID, received signal strength, modulation, carrier frequency, and encoding. The ending signal carries only the ID. Therefore the transmission of a packet has duration (i.e. the transmission time) and it is not a single event in the simulator. As a result, we can have signals from multiple sources overlapping and interfering with each other. At every signal change (i.e., transmission start or end message received) Castalia calculates the Signal Interference and Noise Ratio (SINR) of every signal and based on modulation type and some other radio parameters the bit errors on every signal are computed. Finally based on encoding and bit errors during the total time of packet transmission it is decided if a packet is received or not.

This model is similar to the advanced PHY models found ns-2 and MiXiM. A major difference however is the way we handle the total received power, calculate the Received Signal Strength Indicator (RSSI), and consequently perform Clear Channel Assessment (CCA). Instead of instantly return the current total received power, or return it after a constant delay, we model what takes place in real low-power transceivers [7]: when the radio is in RX/listening mode it integrates the total received power over a number of radio symbols. The result of this integration is the RSSI value. RSSI is then compared to a programmable threshold to determine if the channel is clear or not. Figure 1 depicts this.

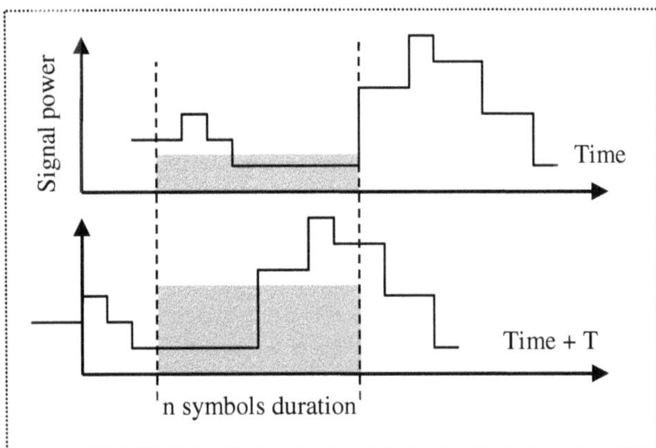

Figure 1: RSSI is integration over x symbols

Furthermore, there is a RSSI valid flag. When transitioning from the TX or the sleep state to the RX state, the radio has to wait the number of symbols it takes to calculate RSSI before asserting the RSSI value to be valid. The implication of these facts is that the time it takes to declare the channel busy when a signal starts being transmitted, depends on the number of RSSI symbols, the strength of the signal, the CCA threshold, and finally how long was the radio in RX mode before the start of the signal.

Apart from the CCA variable delay, Castalia also accounts for constant transition delays between radio states. A full description of Castalia's models and parameters can be found in its User's Manual.

3. SIMULATION SCENARIO

As already mentioned, the simulation scenario we examine has N-1 nodes sending to one receiver node. The path loss from any node to any other node is set to 82dB (using a path loss map file), the transmission power is set to 0dBm, while the sensitivity of the radio is -95dBm. This simple setup has the advantages of "black or white" events, i.e., either there is a guaranteed packet reception when just one signal is present, or there is a guaranteed collision when 2 or more signals are overlapping. We cannot have packet errors due to low SNR, or packet receptions despite (weak) interference, which are usually possibilities with Castalia's complex modeling. Thus we can focus on the radio effects of interest. The simulation time is 10 secs.

The sending nodes employ non-persistent CSMA/CA MAC to avoid collisions. That is, when a node has a packet in the buffer to transmit it asks the radio if the channel is clear (CCA). If the reply is "RSSI not ready", it asks again after RSSI integration time. If the reply is "channel clear" it transmits all packets waiting in the buffer. If the reply is "channel busy" it backs off a random amount of time (upper limit dependent on MAC parameters) and performs again CCA when that time has passed.

We are using 130-byte packets (100 bytes application payload + 30 bytes of all layers overhead). Whenever we change a radio or MAC operating parameter, we measure the performance (packets successfully received) at varying conditions: sending rate [10..100]pks/sec, number of nodes [2..10]. Each performance point is the average result from 20 seed sets. Thus a graph like the one in figure 2 needs 9×10×20 = 1800 simulation runs.

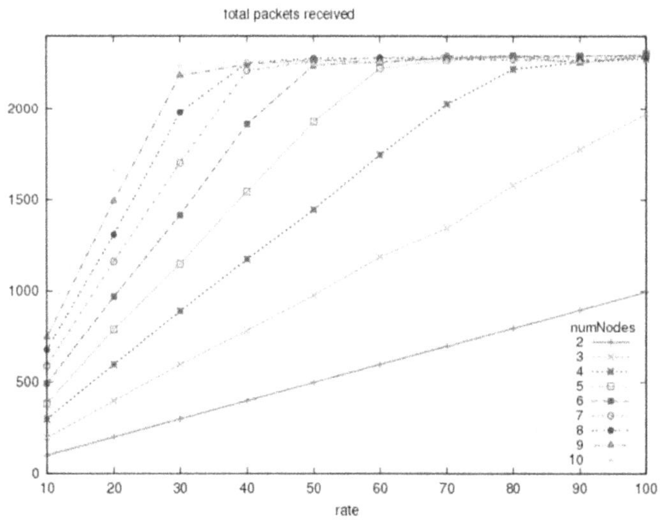

Figure 2: Performance when negligible collisions occur.

If all delays are negligibly small (short transitions, detect signal in just 1 symbol) then a small back-off time is sufficient to avoid most collisions. A small back-off time is an efficient back-off time since it decreases the channel idle time (i.e., all nodes backing off). Figure 2 depicts the performance (packets received at the receiver node during full simulation time) in such a case. As expected, the performance increases with increasing rate and increasing number of nodes till it plateaus close to the theoretical limit (the 250Kbps radio can receive up to 2403 packets in 10 secs).

It is interesting to examine how delays affect collisions. Imagine a node that has been transmitting (TX), it stops to sense the channel and then starts transmitting again. What is the earliest possible time and the latest possible time that another node can start a colliding transmission? Figure 3 presents the answer graphically.

In the figure, **TR** in the state transition time (Tx→Rx or Rx→Tx), **Rs** is the Time for the RSSI register to become valid (=RSSI symbols), **Rc** is the time for the RSSI register to fall below the CCA threshold (channel clear). Rc depends on how strong was the signal received. If the signal was X times stronger than the CCA threshold then it would take (1 - 1/X) of the total RSSI symbols to get cleared. **Rn** is the time for the RSSI register to go beyond the CCA threshold (channel busy). Rn depends on how strong is the signal being received. If the signal is X times stronger than the CCA threshold then it will take 1/X of the total RSSI symbols to assert that the channel is busy. Note that Rs = Rn + Rc.

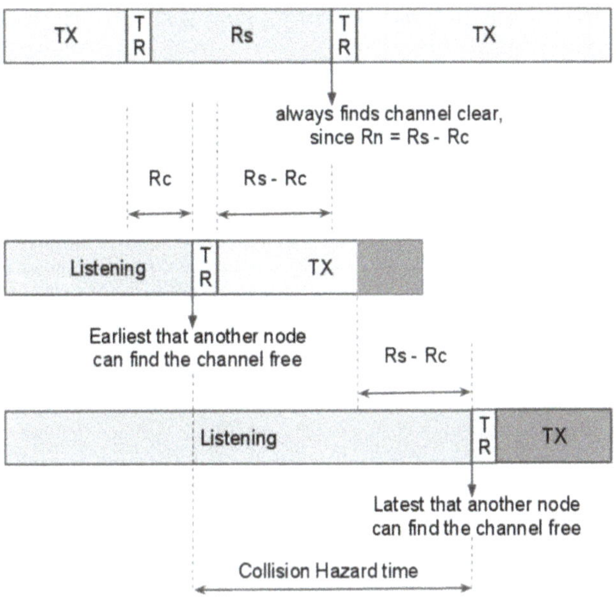

Figure 3: Colliding transmissions in CSMA/CA

The collision hazard time (the interval that a node deciding to TX will cause a collision) is TR + 2(Rx-Rc) = TR + 2Rn. Consider also the simpler case where a listening node decides to transmit in a free channel. In this case, a second node can only decide to TX after the first node making the collision hazard time TR + Rn. In both cases there is a simple linear relationship between delays and collision hazard time, which in turn, together with the back-off parameters and number of nodes wanting to transmit, affects the probability of collision.

4. SIMULATION RESULTS

In this section we will see concretely how radio parameters relating to the carrier sense delay, affect the performance. Moreover we will see how certain MAC parameters affect the performance when the radio delays are not negligible. It is important to mention that the default back-off scheme is to always choose a random time in the interval [0..16]msec (note also that a packet needs 4.13msec to be transmitted).

We start with the effect of RSSI symbols which usually vary between 8 and 64 in digital transceivers. Figure 4 shows the performance when RSSI integration is 16 symbols and the CCA threshold is -83dBm (i.e., 1dB below the received signal strength, which means we need approximately ¾ of the RSSI symbols to pass the threshold). We can immediately notice the changes from figure 3. Even though some of the general trends remain, the curves are clearly separated at the maximum performance level they can reach. More nodes means a smaller number of maximum packets received because there is a greater probability for collisions.

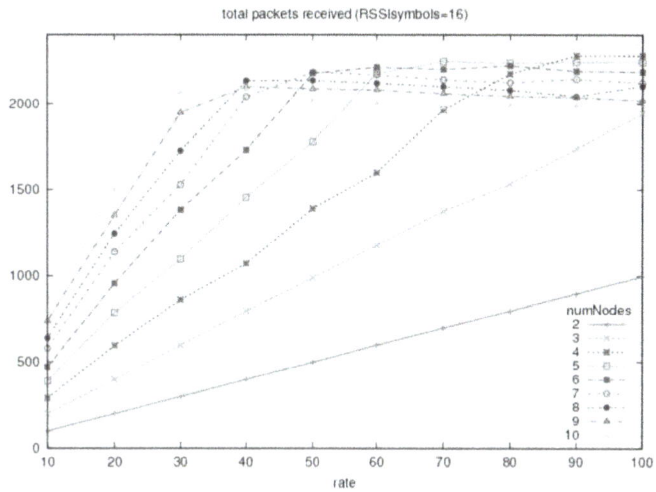

Figure 4: Performance for 16 RSSI symbols, CCA -83dBm

The same effect is reinforced when we increase the RSSI symbols to 32, and thus increase the delay to sense a busy channel.

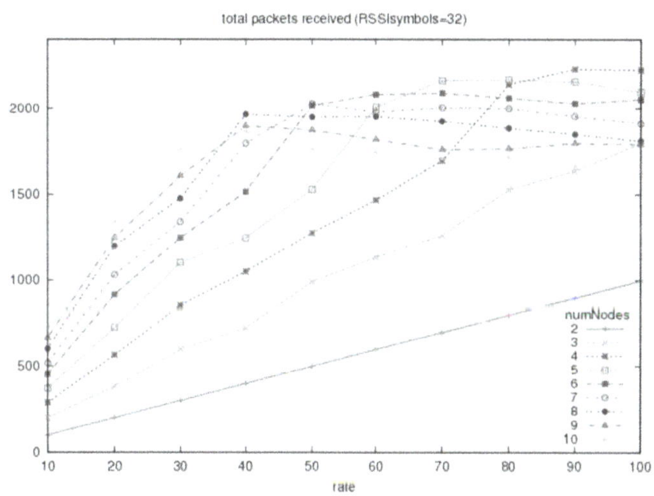

Figure 5: Performance for 32 RSSI symbols, CCA -83dBm

We can also vary the CCA threshold to affect the time/delay we need to sense a carrier. For example if the CCA threshold is 12dB below the receiving signal strength, we only need 1/16 of the RSSI symbols to go beyond the threshold and find the channel busy. Figure 6 shows the performance curves for various CCA thresholds for the case of 10 nodes. We clearly see that the lower the threshold, the smaller the delay to sense the carrier and the fewer the collisions.

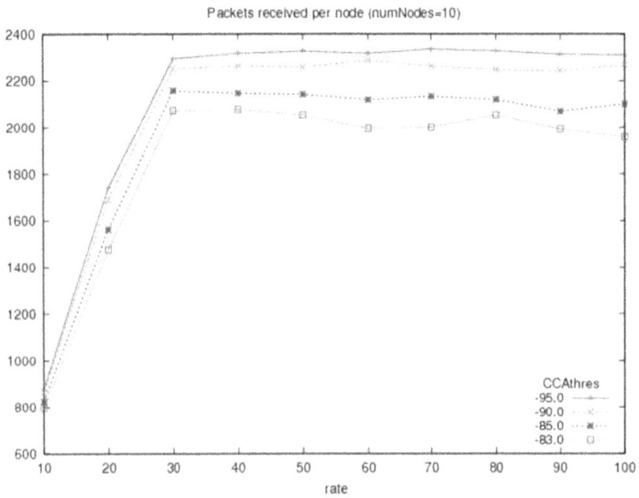

Figure 6: Performance for various CCA thresholds, 10 nodes, RSSI symbols=16

Apart from varying radio parameters to reveal the effect of delays in the higher level simulation results, we can also vary some MAC parameters to achieve the same goal. For example, consider the back-off time. If the delays were negligible then we would want to use a small back-off time so we can be efficient (minimize the amount of time that nodes are backing off without any node transmitting). For example, with our constant back-off scheme, make the upper limit of the back-off window, 2ms, instead of 16ms. However, making the back-off time short means that more nodes will be contending for the channel in a unit of time, which in turn increases the probability of collisions.

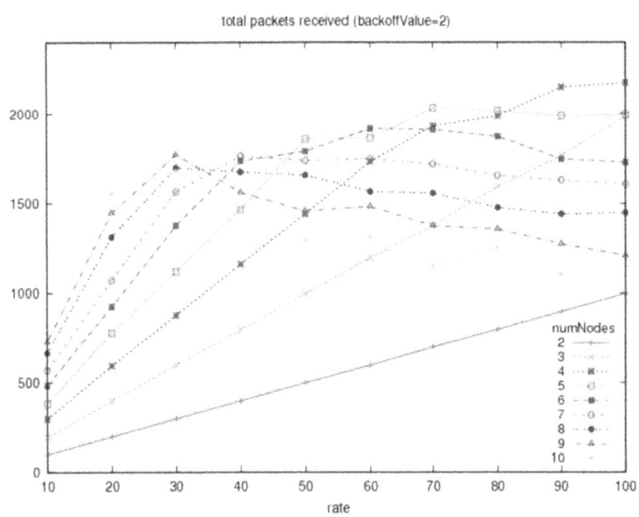

Figure 7: Performance with short average back-off time

Figure 7 shows this effect. We notice the clear separation of curves, as more nodes increase the probability for collisions. We also note that after the maximum performance is reached for each curve, there is a small performance decrease as the rate increases. This is again due to the increased collisions occurring as the nodes have more packets to transmit and more nodes are likely to contend each time.

A similar effect to the short back-off can be observed when we employ p-persistent CSMA/CA. With p-persistent CSMA a node continuously senses the channel and when found free it decides to transmit with probability p. Since most low cost radios so not have an interrupt mechanism to report a clear channel, we implement persistence with fast CCA polling (with a period of RSSI symbols). Figure 8 shows results for p=0.1

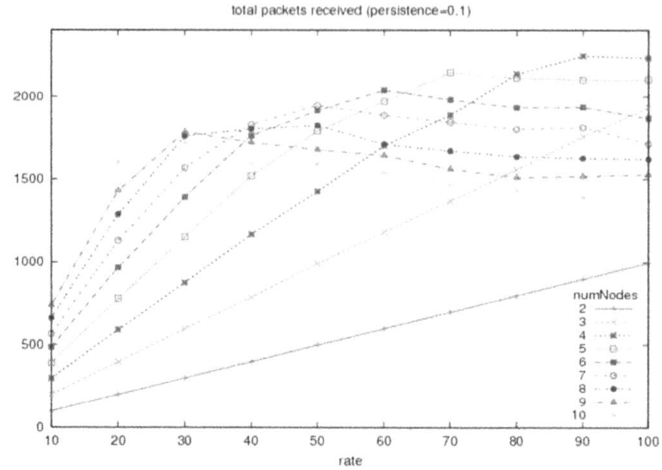

Figure 8: Performance for 0.1-persistent CSMA/CA

We can also examine back-off schemes different to the constant one describes so far. For example, in high contention situations, it is common to increase the back-off window if a node repeatedly senses the channel busy. In the simple multiplicative back-off scheme, a node draws its back-off randomly from the window [0..16x]ms where x is the number of consecutive times the channel has been found busy before this attempt. As contention increases the back-off windows increase to regulate it, and avoid excessive collisions.

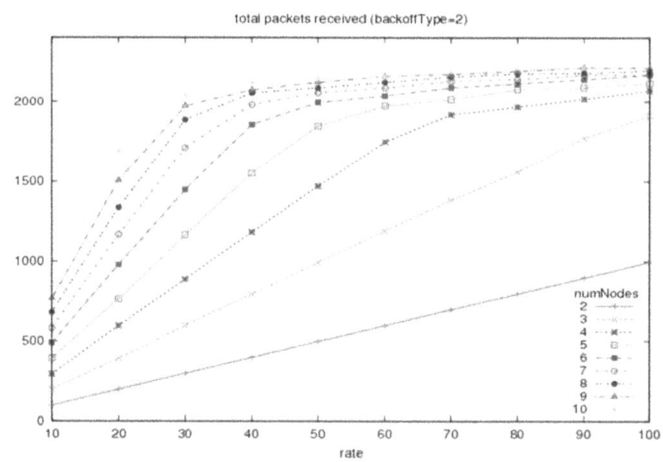

Figure 9: Performance for an increasing back-off window

Figure 9 shows the results for such a back-off scheme. It is interesting to note that the trend of previous cases is inverted with higher number of nodes achieving higher performance. This implies that the scheme is effective in avoiding considerable collisions but also suffers from often leaving the channel unused.

Finally, it is interesting to examine the performance in another MAC operational option, which is usually neglected. As mentioned in earlier sections, the MAC has a buffer to store packets received from the application. Although this in not explicitly mentioned in CSMA/CA it is a practical reality of all

MAC protocols. It was hinted earlier that when the channel is found free all packets in the buffer are transmitted. This is the more efficient option but it is not the only option. A more fair option that avoids hogging the channel by any particular node is to transmit only one packet when the channel is free. If a node has more packers in its buffer it will have to re-acquire the channel. When no radio delays are modeled, the one-by-one approach is as efficient to the alternative and guaranteed to be fairer. However, if we incorporate delays (RSSIsymbols=16, CCA=-83dm, backoff=[0..16]) a dramatically different picture is drawn. The major effect is the sharp drop in performance as the rate increases.

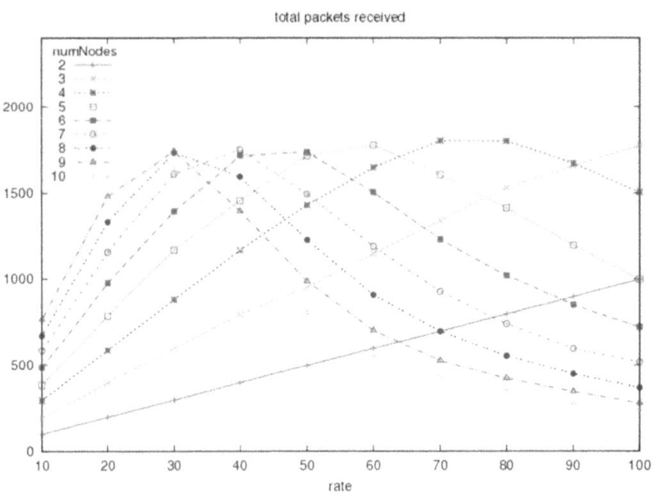

Figure 10: Performance when packets transmitted one-by-one

This effect was initially puzzling but simulation traces examination revealed the cause: A node will not relinquish the channel until all packets in its buffer are transmitted, even in the "one-by-one" case (i.e., node performs carrier sense after each packet transmission). The explanation for this is shown in figure 3. There is no possibility for a previously transmitting node to find the channel busy. It can only collide with another node. Once a collision between nodes occurs, then the two nodes *lock into a time pattern* and collide all packets in their buffer. The greater the rate, the more packets collide. This pattern locking can be broken by simply introducing a small random offset ([0..1ms]) before every decision to sense the channel and transmit. Figure 11 shows the results of this improvement. Now the curves reach a maximum performance and sustain it much like figure 4. The overall performance is worse since there is more transition overhead, but the results are fairer across node, a fact that can be also seen by the smoother curves of figures 10 and 11 compared to figures 4-8. Due to the modeling of radio delays, an interesting performance behaviour has been revealed, and a simple protocol enhancement can improve it.

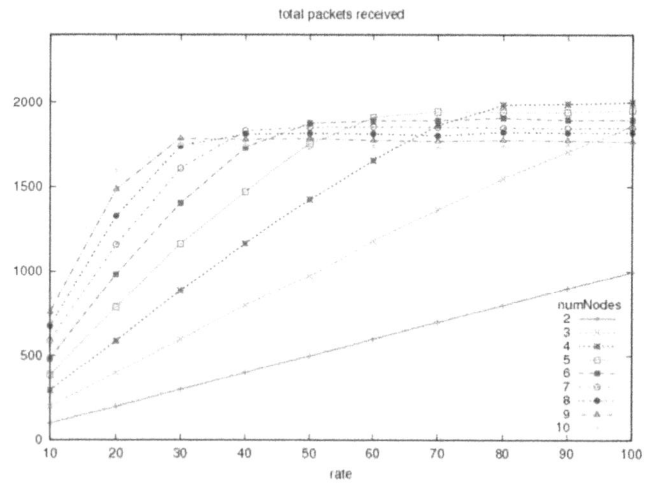

Figure 11: Performance one-by-one + random TX offset

5. CONCLUSION

In this paper we showed that the usually-neglected variable delay to perform carrier sense and the state transition delays, can have a significant impact on the throughput of a clique of nodes sending packets using CSMA/CA. Moreover, simple MAC decisions that are usually left undefined (e.g., transmit all packets in the buffer when channel found free, or just one packet) have a much more pronounced impact when radio delays are taken into account.

6. REFERENCES

[1] Ns-2 simulator website: http://www.isi.edu/nsnam/ns/

[2] Levis, P., Lee, N., Welsh, M., and Culler, D. "TOSSIM: accurate and scalable simulation of entire TinyOS applications," in proceedings of the 1st international Conference on Embedded Networked Sensor Systems (SenSys '03), Los Angeles, CA, USA, November 5-7, 2003

[3] Sobeih, A., Chen, W. P., Hou, J. C., Kung, Lu. C., Li, N., Lim, H., Tyan, H. Y. and Zhang, H., "J-sim: A simulation environment for wireless sensor networks," in Annual Simulation Symposium, pages 175–187, 2005

[4] MiXiM simulator website: http://mixim.sourceforge.net/

[5] Chen, Q. et al, "Overhaul of IEEE 802.11 Modeling and Simulation in NS-2", in proceedings of the 10th ACM International Conference on Modeling Analysis and Simulation of Wireless and Mobile Systems (MSWiM '07), Chania, Greece, October 22-26, 2007

[6] Castalia simulator website: http://castalia.npc.nicta.com.au

[7] CC2420 transceiver datasheet: http://focus.ti.com/docs/prod/folders/print/cc2420.html

A Novel Neighbor Discovery Protocol for Ultraviolet Wireless Networks

Leijie Wang and Yiyang Li
Dept. of Electrical Engr.
University of California
Riverside, CA 92521, USA
{lewang,yiyli}@ee.ucr.edu

Zhengyuan Xu
Dept. of Electronic Engr.
Tsinghua University
Beijing 100084, P.R. China
xuzy@tsinghua.edu.cn

Srikanth V. Krishnamurthy
Dept. of Computer Science
and Engr.
University of California
Riverside, CA 92521, USA
krish@cs.ucr.edu

ABSTRACT

Ultraviolet (UV) communication is an attractive option for tactical networks or environmental monitoring. The underlying UV PHY layer has unique characteristics that render previously proposed higher layer protocols for RF communications inappropriate or inefficient. Neighbor discovery is an important functional component of a UV ad hoc wireless network. While there has been some work in UV PHY layers, there is very limited work in network study. In this paper, we propose a new neighbor discovery protocol for this setting; unlike prior protocols, our approach alleviates the negative effects of random access based collisions by choosing a leader that arbitrates the discovery process. Without prior knowledge of the number of nodes in the network, the approach facilitates neighbor discovery in a fast, fair and efficient manner. We perform extensive simulations with a realistic UV PHY layer and demonstrate that the approach reduces the required neighbor discovery time by as much as 90 %. We also examine the impact of various system parameters that can be especially useful to UV network and system designers.

Categories and Subject Descriptors

C.2.5 [**Computer-Communication Networks**]: Local and Sensor Networks

General Terms

Design, Performance

Keywords

Neighbor discovery, optical wireless communications, ultraviolet

1. INTRODUCTION

Ultraviolet or UV communications are especially attractive in applications such as environment monitoring and tactical battlefield communications [1]. Recently, there has been significant progress in terms of building advanced low cost, low power and small size light emitting diodes (LEDs) that operate in the deep UV part of the spectrum. Further, high gain photomultiplier tubes (PMTs) that can be used as receivers have contributed to the rapidly increased interest towards the growth of UV communications. Atmospheric scattering allows the establishment of non-line of sight (NLOS) links with UV; UV photons are scattered multiple times between transmitters and receivers. Thus, even if there is no direct line-of-sight (LOS) path, communications can be established. This feature provides significant advantages for the establishment of communications in various complex scenarios where LOS communications are infeasible. Other properties that make UV communications attractive are that a UV link is resistant to RF jamming, and the probability of the signals being intercepted is very low at long distances due to the unique power decay profile of UV signals.

In the applications envisioned, nodes are likely to be deployed in an ad hoc manner. A node is unlikely to have a priori information about its neighbors or the network topology. As discussed later, the problem of neighbor discovery is non-trivial since UV communications are likely to use directional transmissions; the node should realize not only the identities of its neighbors but also the best direction in which to transmit in order to effectively communicate with each such neighbor. Line-of-sight UV signals are prone to blockage by obstacles; fortunately, nodes can establish possibly multiple NLOS links using which they can communicate.

Our objective in this paper, is to design and evaluate an efficient neighbor discovery protocol that (a) allows the fast and efficient discovery of neighbors, (b) can be applied without any prior knowledge of the network topology or configuration, and (c) allows each node to determine the various directions that can be used to communicate with each of its neighbors and *rank* these directions in terms of their effectiveness.

While there have been prior efforts on neighbor discovery for RF wireless networks, they cannot be directly applied in the UV context. First, the unique UV characteristics are different from the propagation effects experienced on an RF channel. Many of the prior approaches (discussed later) rely on GPS or other enhancements which are unlikely to be available in the application scenarios of interest. Secondly, almost all of the previous approaches only discover neighbors using LOS links; neither do they consider NLOS links nor do they provide any assessment on the relative goodness of these links. There has been very limited work on neighbor discovery in the UV context [7]. However, this approach takes significant time to operate and is inefficient since it uses random access as the basis for neighbor discovery. We propose a much more efficient protocol that can speed up the neighbor discovery process dramatically.

The rest of the paper is organized as follows. We present related

MSWiM'11, October 31–November 4, 2011, Miami, Florida, USA.

work on neighbor discovery in Section 2. In Section 3, we introduce the propagation characteristics of UV communications. We review the credit-collection based neighbor discovery protocol proposed in [7] in Section 4 to motivate design of our new protocol; we then describe our basic protocol design based on the use of a leader to arbitrate neighbor discovery. In Section 5, we consider unfairness aspects of the neighbor discovery procedure suffered and propose modifications to improve the fairness. In Section 6, we discuss how each node determines when to terminate neighbor discovery, without a priori knowledge about the number of its neighbors. We conclude in Section 7.

2. RELATED WORK

Neighbor discovery has been previously studied in the context of RF wireless networks; in particular, the focus has been on cases where directional communications are the norm. Most of the proposed approaches use random access as a basis for neighbor discovery. In [3], the authors examine the frequency with which a node should transmit neighbor discovery probes in order to maximize the probability of discovering a neighbor within a preset time period. They find that this probability is influenced by the density of nodes in the network (the number of neighbors of a node) and the beamwidth used for transmissions. A TDMA based neighbor discovery algorithm is proposed for 3-D RF wireless networks in [4]; the neighbor discovery is based on establishing a handshake between the neighbors that are trying to discover each other. However, the approach assumes that each node is equipped with a Global Positioning System (GPS) device. The authors suggest that nodes begin with using low powers so as to discover *close by* neighbors in interference reduced settings; the power is gradually increased to discover neighbors that are further away. Luo et al. analyze neighbor discovery in a CDMA-like system; they assume that each node is aware of its neighbors' identities and their spreading codes a priori. An energy efficient neighbor discovery protocol is proposed in [5]. Jakllari et al. propose an integrated neighbor discovery and MAC protocol for an ad hoc network with directional antennas [9]. In their work, they explicitly consider mobility unlike in the other efforts. Aloha-based neighbor discovery protocols for RF wireless networks are comprehensively analyzed in [2]. The authors propose a collision detection mechanism to improve performance. The algorithms proposed also allow nodes to determine when to terminate neighbor discovery, without any prior knowledge of their neighborhoods.

None of the above approaches can be directly applied in a ultraviolet wireless network due to its inherent physical (PHY) layer properties. Specifically, almost all of these protocols assume the existence of LOS links; they do not consider the possibility of transmitting in different directions to establish NLOS links (as is possible with UV) and thus, do not provide a relative ranking of these links in terms of the quality of communications. Second, they do not account for the inherent properties of the UV scattering channel (discussed later). Finally, many of them implicitly assume either prior knowledge of a node's neighborhood or specialized equipment such as GPS.

The first work (the only work to the best of our knowledge) on neighbor discovery in UV wireless networks is found in [7]. A credit-collection protocol is designed to rank the transmission directions of a node for each neighbor, based on packet losses on each directional link. This protocol requires a large number of packet transmissions to achieve stable results and it takes a relatively long time to terminate. Our goal here is to design a much faster protocol for neighbor discovery to obtain the same information about neighbors as in [7].

3. UV LINK AND TRANSCEIVER MODELS

In this section, we describe the propagation characteristics of UV communications. UV signal delivery is realized via the propagation of photons between the transmitter (made up of LEDs) and the receiver (referred to as the PMT as discussed earlier). The UV photons are scattered one or more times after being emitted by the LEDs. The atmospheric scattering is governed by a stochastic process that one has no control on, or exact knowledge of how many times the photons are scattered or in which direction they are scattered in the atmosphere. Note here that before the photons are captured by the PMT, their propagation direction might change multiple times due to what is defined as multiple scattering. Stated otherwise, a LOS link is not mandatory for UV communication although a LOS link can provide relatively lower path loss as has been verified in [10]. Although the path loss model we adopt in this paper considers multiple-scattering effects, the general propagation trends that we describe are consistent with single-scattering so that we can use geometric demonstrations to facilitate understanding. Figure 1 depicts a typical NLOS UV communication link. As long as there is a *common volume* between the transmission beam cone and the receiving field of view (FOV), a communication link can be potentially established.

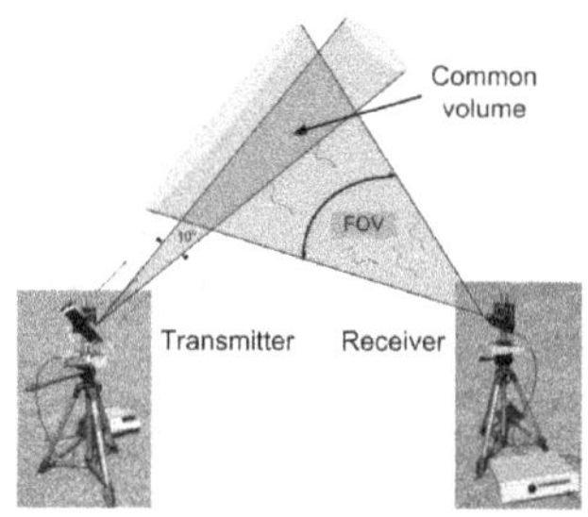

Figure 1: Common (overlap) volume between the transmission beam and the receiving field of view (FOV) enables a NLOS link.

Some path loss models have been proposed under the assumption that photons are scattered only once, which is *approximately* correct when the *optical length* is relatively short. The optical length is defined as the product of actual distance between the transmitter (Tx) and the receiver (Rx) and what is referred to as the scatter coefficient [10]. In our work, we adopt the path loss model which captures both single-scattering and multiple-scattering effects. More importantly, the model can capture arbitrary transmission directions from the LEDs. Thus the scenarios we study are not limited to the case where the transmission beam axis and the FOV axis are on the same plane (coplanar case). We introduce an additional angle to describe the noncoplanar case as shown in Fig. 2. The path loss model is developed in [8] by fitting a curve to experimentally obtained data on a real UV test-bed. A typical transceiver configuration is presented in Fig. 4. The receiver is designed to point vertically and the transmitter can orient itself any direction by turning on the corresponding LEDs on specific *facets*. The corresponding link loss model is adopted throughout this paper.

The empirical model is presented in mathematical form below:

$$L = \xi r^{\alpha} e^{b\varphi}. \tag{1}$$

The values of the path loss factor ξ, the path loss exponent α and the off-axis angle exponent factor b are related to the transmission beam divergence, the FOV and the Tx/Rx pointing angles. Extensive experiments on a real UV Physical layer test-bed have been conducted to obtain these parameters. In our simulations below, we

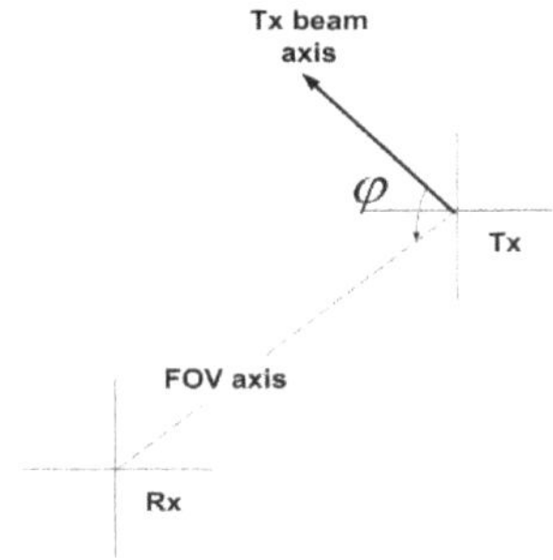

Figure 2: Transmission beam axis and FOV axis are located on the different planes. The off-axis angle φ is introduced to describe the offset between these two planes.

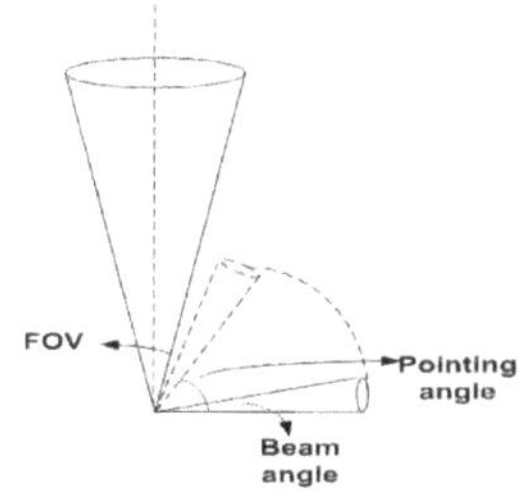

Figure 3: Beam angle, receiver FOV and pointing angle.

set these parameters to be those associated with the case where, the transmission pointing angle is equal to 10°. They are, $\xi = 5 \times 10^9$, $\alpha = 0.4$, $b = 8.7$, and $\varphi \in [0, \pi]$.

As is typical with UV transceivers, we assume that the Tx and the Rx are co-located on a communication node. Multiple transmitters are simultaneously activated to enable omni-directional transmissions. Each Tx consists of one or more LEDs, which are mounted on the side facets of a node as shown in Fig. 4. The LEDs installed on each facet transmit the same information-bearing signal when they are turned on. Transmitters are switched on or off digitally; no mechanical motion is needed to steer beams. The Rx is deployed on the top of the node to achieve omni-directional reception, i.e., the receiving pointing angle is 90° and Rx vertically faces upwards. A practical transceiver is not limited to the configuration in Fig. 4. The number of directions could be varied adaptively based on different application requirements and optical device settings.

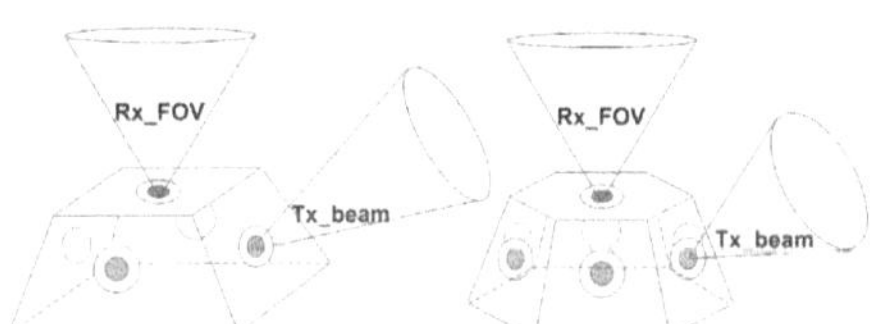

Figure 4: A UV transceiver with 4 (left) and 6 (right) transmission directions. Dots on the side facets are the directional LED transmitters; dots on the top facet are the omni-directional photon detectors.

With this design, the transceiver consists of multiple **directional transmitters** and an **omni-directional receiver**. The multiple transmitters can function individually to provide directional communication links. In other words, each target receiver can be associated with *one* best direction in terms of the lowest path loss. The transmitters can also work simultaneously to achieve omni-directional transmissions, which is useful when one needs to realize a broadcast function.

As previously discussed, atmospheric scattering cannot be controlled artificially. Some photons emitted by the Tx might be scattered back to its co-located Rx. This is defined as backscattering, which results in self-interference. Li et al. have conducted experiments to verify that as long as the Tx pointing angle is relatively small to make the beam depart from the FOV, the self-interference is negligible [8]. The basis for this conclusion is that in such scenarios, there is no (or very little) overlap between the Rx-FOV and the co-located Tx-beam as shown in Figs. 3 and 4. Since the transceiver is free of self-interference, spatial full-duplex communications are feasible with above the transceiver configuration. This feature greatly improves the throughput of UV communications and is a differentiating factor compared to RF communications (typically half-duplex).

4. OUR LEADERSHIP-BASED NEIGHBOR DISCOVERY PROTOCOL

In this section, we first elaborate the previously proposed neighbor discovery protocol for a UV network in [7]. We motivate and then, detail the design of our approach.

The first and the only neighbor discovery protocol (to our best knowledge) for a UV network is proposed in [7]. It makes no assumptions about a priori knowledge, such as the number of neighbors or the relative positions of neighbors. The basic procedure is described below:

- The Rx of every node is in a standby state all the time; it can receive anytime thanks to the full-duplex capability.
- A node randomly chooses a direction to send a packet after waiting for an exponentially distributed period.
- The transmitted packet could be either a request packet or a feedback packet. The request packet contains information about the request sender's ID and the ID of the direction that is being currently used. The feedback packet is composed of feedback sender's ID, request sender's ID, the IDs of the directions, and the total counts of received request packets in corresponding directions.
- When a node receives a request packet from a sender, it schedules a feedback packet to be returned to that sender. If no request packet from other nodes is received within a fixed pre-determined time period, the node will send out a request packet in a randomly chosen direction.
- Each node counts the number of received feedback packets and categorizes them into different groups based on the information carried in such packets. To elucidate, once request sender receives a feedback packet in return, it is aware of the recipient of the request, as well as the corresponding direction in which the request was received. The sender then adds a *credit* to that direction.
- A node ranks the directions for each of its neighbors, based on the number of credits accumulated for each direction. The direction with the highest number of credits is the best direction for that neighbor.

As may be evident from the above description, a node needs to collect a sufficiently large number of credits in order to accurately rank the directions towards a given neighbor. Since the primary mode of communication is random access, collisions occur. This can skew the results since small sample sets may not yield accurate estimates of the best direction. In other words, this results in short term inaccuracies. Convergence to correct results takes a relatively long period of time.

Given this limitation, and the importance of expeditious neighbor discovery, we seek to design an interference-free environment for the process. Our goal here is to enable the discovery of neighbors and the best direction for communications with each neighbor with a *one-time packet exchange*. We later show that our method (proposed below) drastically improves the performance of neighbor discovery in terms of the time taken, compared to the method proposed in [7].

Our proposed approach: In a nutshell, our proposed approach relies on sequential neighbor discovery. A single node (the leader) initially performs neighbor discovery; after that node is done, the leadership is passed on to a second node and so no. We call our approach the "leadership-based neighbor discovery" protocol. In more detail, our approach consists of the following steps:

- Initially, all the nodes are in a standby state. They only receive (listen) packets but do not transmit anything.
- A pre-chosen leader node (could be randomly chosen) will then initiate the neighbor discovery process by sending a request packet in a randomly chosen direction. This request packet includes the ID of the sender and the direction in which it is transmitted.
- The leader waits for a fixed time period T, and then switches to the next direction to send the next request packet. It repeats the process until it performs packet transmissions in all possible directions.
- When a neighbor node receives a request packet from the leader node, it calculates the path loss experienced. Note that this is possible because there is no other concurrent transmission. (This is a key difference from the method in [7]; the path loss cannot be obtained since transmissions are subject to interference.)
- Each such neighbor waits for a period t; this period is randomly chosen in $[0, T]$. After this period, the neighbor randomly chooses a direction using which it transmits a feedback packet back to the leader. The feedback packet contains the ID of responding node and the path loss estimated using the request packet. The reception of the feedback packets is not guaranteed because contention exists among different neighbors that attempt sending feedback. However as long as the length of the feedback packets is much shorter than maximal waiting time T, it is likely to receive many of these feedback packets with a high probability.
- After the process is complete, the leader node can rank the directions for each discovered neighbor found based on the path loss information contained in feedback packets. It then randomly selects a successor from the nodes in its neighbor list. It sends a notifier packet to the selected successor node, using the best direction recorded for this node. The successor then performs neighbor discovery. All other nodes are made aware, that the neighbor discovery *token* has now been passed on to a different node.

It is possible that the notifier packet gets lost; to account for this, if no request packets are heard from the successor for a period $2T$, the former leader node will re-select a new successor. The process may have to be repeated in extreme cases until a successful leadership transfer is achieved. The steps described above are formalized with the pseudo code below:

```
K = total no. of directions;
Leader:
while rotation not finished yet do
  if time T is up then
    ND_request.source_ID=self_id;
    ND_request.direc_ID=uniformly_choose(K);
    send ND_request;
    set T;
  else
    if receive ND_feedback then
      neighbor_ID=ND_feedback.source_ID;
      prev_direc=ND_feedback.prev_direc;
      PathLoss=ND_feedback.path_loss;
      PathLoss.node[neighbor_ID].direc[prev_direc]=PathLoss;
    end if
  end if
end while
# of discovered neighbors is summarized here;
while i < # of discovered neighbors do
  order PathLoss.node[i].direc[*];
end while
select leader=uniformly_choose(# of discovered neighbors);
send Notifier packet to next leader;
Other nodes:
if receive ND_request then
  dest_ID=ND_request.source_ID;
  prev_direc=ND_request.direc_ID;
  PathLoss=Tx_power/Rx_power;
  set t=uniformaly_choose(T);
  if time t is up then
    ND_feedback.source_ID=self_id;
    ND_feedback.prev_direc=prev_direc;
    ND_feedback.path_loss=PathLoss;
    send ND_feedback;
  end if
end if
```

Evaluations of our approach: We evaluate the performance of our neighbor discovery protocol via extensive simulations performed using OPNET version 16.0 [12]. The simulation settings are listed in Table 1. The chosen transmission power corresponds to the UV transmission range (approximately 50m in Fig. 5). In all of the following simulations, we assume that the transceiver has 6 directions for transmission operations. The more the directions, the larger the neighbor table maintained and the longer the neighbor discovery process takes. The relation between the number of directions and the time consumed by the neighbor discovery process is discussed later. We adopt the channel model from Section 3 to characterize signal propagation. By default, the chosen parameters correspond to the scenario when the transmission pointing angle is equal to $10°$. To recall, these parameters are: $\xi = 5\times10^9$, $\alpha = 0.4$, $b = 8.7$. The SIR threshold is set to be 1dB; the signal is detectable if the power is larger than the total interference power by this value. The sizes chosen for the three kinds of packets are consistent with that by IEEE 802.11 [11]; we choose this since protocols for UV networks are yet to be standardized. Each node is placed randomly in a network area of $1 \times 10^4 \mathrm{m}^2$. The total number of nodes in the network is 10. All the simulation curves are computed as averages of over 10 runs.

We compare the newly proposed protocol (labeled leadership-based) with the previously proposed protocol from [7] (labeled credit-collection) in Fig. 5. We would like to point out here that this comparison is with the fastest version of the algorithm proposed in [7]. The expected value of the waiting interval λ for credit-collection is set to be 10^{-3}s. The uniformly distributed waiting interval T with the proposed protocol is set to be 2×10^{-3}s; this ensures that the mean is 10^{-3}s and thus, the settings are consistent in both cases. In Fig 5, The leadership-based algorithm leads to the sharp speed up in the neighbor discovery process. It saves about 92% of the time

Table 1: Parameters settings in simulations

Transmission power	4 mW
Number of directions	6
Transmission pointing angle	10°
SIR threshold	1 dB
Collision model	Physical (accumulative) model
Traffic pattern	10000 pkt/sec
Request packet size	32 bits/pkt
Feedback packet size	80 bits/pkt
Notifier packet size	32 bits/pkt
Network size	100 m by 100 m
Number of nodes	10

consumed in neighbor discovery, if one were to have a requirement that every node find 70% of its neighbors in the network. It saves about 97% of the time consumed in neighbor discovery if this requirement is stricter and every node has to find about 80% of its neighbors. The credit collection protocol consumes a lot of time to stabilize. These results suggest that interference has a dominant influence on neighbor discovery in UV networks; using a relatively interference-free approach drastically speeds up the neighbor discovery process.

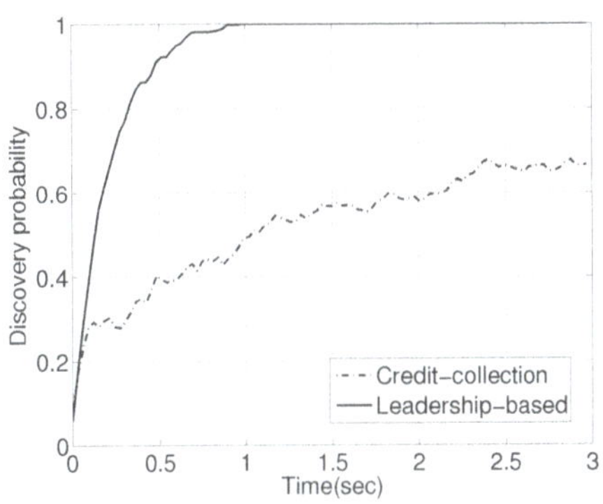

Figure 5: Comparison between leadership-based algorithm and credit-collection algorithm proposed in [7].

5. IMPROVING FAIRNESS

Although not explicitly stated earlier, we wish to point out here that a node can assume the role of the leader multiple times. We expect that the neighbor discovery process is a background process that is continuously executed; this is essential since the neighborhoods of nodes could change either due to failures, environmental changes (obstacles causing links to fail) or mobility. In this section, we consider the fairness of our approach in terms of the number of times a node gets an opportunity to become the leader.

To examine the fairness of our approach, we perform a simulation experiment, wherein we run the discovery process until all nodes discover almost all of their neighbors (> 99 %). The total simulation time was 3 seconds and node '0' was arbitrarily chosen to be the initial leader. Table 2 shows the number of times that each node is chosen to be the leader (referred to as *leader records*). We immediately see that certain nodes have a better chance in neighbor discovery; in other words, an imbalance is observed. Node 1 is chosen to be the leader 29 times; however, node 8 is only chosen 12 times. Table 3 shows the leader records up to the first 0.54 seconds of the simulation. In this period, the percentage of neighbors discovered is on average 95% and 7 of the nodes had found all of their neighbors. We observe that node 0 and node 4 assumed the role of the leader 6 times each; many other nodes had the opportunity only twice. We also indicate in this table, the percentage of neighbors discovered within this time. We observe that node 8 has discovered all of its neighbors even though it has been a leader just twice. Most nodes find all of their neighbors as long as they become a leader more than twice. This observation is encouraging in that, a node only has to serve as a leader a few times, in order for it to find its neighbors. If a node is chosen as a leader too often, it is likely to waste time and resources. Thus, it appears meaningful to provide equal opportunities of being a leader to all nodes in the network.

Table 2: Records of being the leader (time length = 3 secs)

Find 100% neighbors					
Node ID	0	1	2	3	4
Leader counts	22	29	14	13	18
Node ID	5	6	7	8	9
Leader counts	21	22	18	12	25

Table 3: Records of being the leader (time length = 0.54 secs)

Node ID	0	1	2	3	4
Leader counts	6	4	2	3	6
Neighbor percentage	1	1	0.75	0.89	1
Node ID	5	6	7	8	9
Leader counts	5	3	2	2	5
Neighbor percentage	1	1	0.81	1	1

A modified approach to improve fairness: In order for all nodes to equally share the leadership duties, intuitively it would seem that nodes need to have a *count* of how many times each has served as a leader. Then, the leadership role can be passed onto the neighbor who has served the least number of times. This will ensure that the number of times that each node serves as the leader will be roughly equivalent.

To achieve this goal, we incorporate what we call *a leader table* in the notifier packet. This enlarges the size of notifier packets compared to what was used earlier (as in Table 1). Specifically, the size of the packet will increase by $(\#of\ nodes \times 8\ bits)/packet$. The first leader node (also called trigger node), say node 0, will create this table and mark itself as being the leader once. Once the notifier packet is received by the next leader node (say node 1), a new entry is added into the table to record that node 1 served as the leader once. This leader table is updated at each leadership transfer instance. In this way, we ensure that the current leader node possesses the most up-to-date leader table. Each leader node will choose the next leader with the least *experience* of being the leader. If multiple nodes are qualified to be the next leader, one of them is randomly chosen. The pseudo code below describes the process in detail.

K = total no. of directions;
Leader:
extract Leader_table from Notifier packet;
while rotation not finished yet **do**
  if time T is up **then**
    ND_request.source_ID=self_id;
    ND_request.direc_ID=uniformly_choose(K);
    send ND_request;
    set T;

else
if receive ND_feedback **then**
neighbor_ID=ND_feedback.source_ID;
prev_direc=ND_feedback.prev_direc;
PathLoss=ND_feedback.path_loss;
PathLoss.node[neighbor_ID].direc[prev_direc]=PathLoss;
end if
end if
end while
# of discovered neighbors is summarized here;
while i < # of discovered neighbors **do**
order PathLoss.node[i].direc[*];
end while
select next_leader=Min(Leader_table[neighbor_ID]);
send Notifier packet to next_leader;
Other nodes:
if receive ND_request **then**
neighbor_ID=ND_request.source_ID;
prev_direc=ND_request.direc_ID;
ND_feedback.path_loss=Tx_power/Rx_power;
t=uniformly_choose(T);
if t is up **then**
ND_feedback.source_ID=self_id;
ND_feedback.prev_direc=prev_direc;
send ND_feedback;
end if
end if

Evaluating the modified approach: Next, we present evaluations of our modified approach. Tables 4 and 5 show the number of times that each node in the network serves as a leader, when the overall percentage of neighbors discovered is approximately 100% and 95%, respectively. The results are from one randomly chosen realization. We observe that the variance in the counts (in Table 4) is relatively small; specifically, it is 0.44 compared to 26.84 in Table 2 with the original scheme. We see from Table 5 that all the nodes have been the leader, the same number of times; this means that the approach is fair. Both Table 3 and Table 5 reflect scenarios where each node found approximately 95% of its neighbors, successfully. However, the time incurred with the latter algorithm is only 0.42s; this represents a 22% reduction in time compared to the original approach. This reduction is a direct artifact of the increased efficiency with the modified scheme; nodes that have previously served as leaders and have found their neighbors do not waste time in serving as leaders again.

Table 4: Records of being the leader (time length = 3 secs)

Find 100% neighbors					
Node ID	0	1	2	3	4
Leader counts	20	19	20	20	18
Node ID	5	6	7	8	9
Leader counts	19	19	19	20	20

Table 5: Records of being the leader (time length = 0.42 secs)

Node ID	0	1	2	3	4
Leader counts	3	3	3	3	3
Neighbor percentage	0.89	1	1	1	0.94
Node ID	5	6	7	8	9
Leader counts	3	3	3	3	3
Neighbor percentage	1	1	1	0.71	1

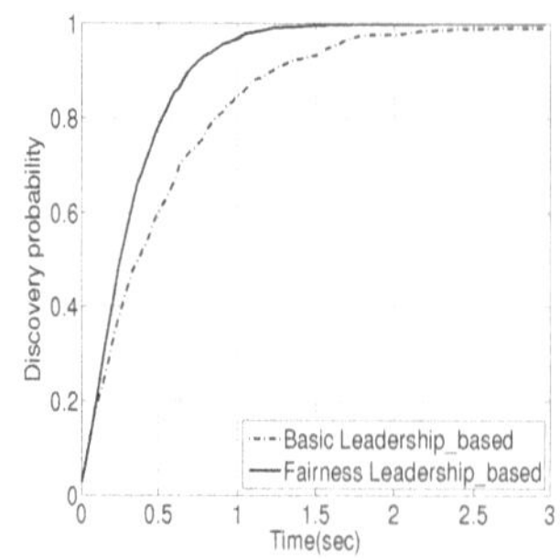

Figure 6: Comparison between schemes with or without fairness consideration.

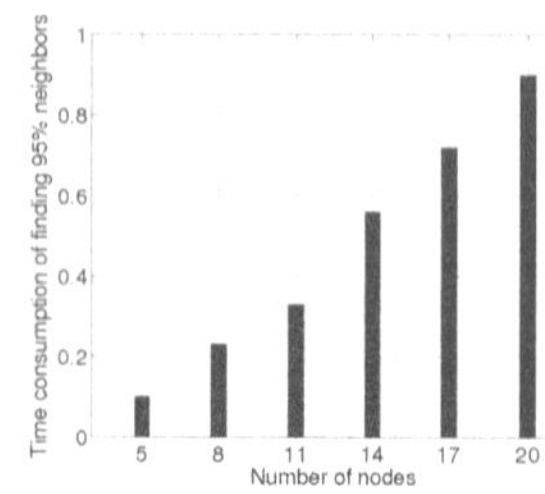

Figure 7: The time taken for neighbor discovery varies with number of nodes. The time shown is when 95% of neighbors are discovered.

Figure 6 depicts how the modified scheme improves the performance of neighbor discovery. We choose a 20-node scenario and the area of deployment is still 100 m$\times$100 m. The maximum transmission range and transmission power are 50 m and 0.4 mW, respectively; these are the same as in Table 1. From Fig. 6. We see that the time needed by every node for finding 95% its neighbors on average, with the basic leadership-based algorithm is about 1.62 seconds; this time is reduced to about 0.9 seconds with modified fair scheme. The modified scheme brings about a 45% performance improvement *after* taking into account the increase in packet transmission time due to the enlarging of the notifier packet.

Finally, we examine how the time for neighbor discovery varies as we vary the number of nodes in the network. Figure 7 shows how the time required to discover 95% of the neighbors changes (on average), as we vary the number of nodes in the network. We retain the coverage area to be 100 m$\times$100 m for each simulation point represented in the figure. The notifier packet length is also set to be identical to that in the scenario with 20 nodes; we do this to isolate the impact of node density from the incurred overhead changes from the notifier packets. Figure 7 shows that the more the nodes in the network, the longer it takes for neighbor discovery as one might expect. More importantly, one can conclude that the time consumption is *proportional* to the number of nodes. The relationship is almost linear as seen in the figure.

As described earlier, our leadership-based algorithm attempts to construct an interference-free environment for the leader node to find its neighbors. However, with an increase in node numbers or network size, the time taken for neighbor discovery will increase. We further increase the node density (see Fig. 8) and we find that the time taken grows almost linearly versus the number of nodes. For very dense networks, as compared with the credit-collection algorithm, we found that the savings in terms of time get larger with increased node density. We use a curve fitting method to obtain extrapolated results for the leadership-based (solid line) and the credit-collection (dotted line) protocols, respectively. The curve fitting expression for the solid line is $0.4264 \times \exp(0.02799n)$ and the expression for the dotted line is $0.4937 \times \exp(0.03880n)$, where n stands for the number of nodes in the network. The slopes for both curves after differentiation of these expressions are obtained as $0.4264 \times 0.02799 \times \exp(0.02799n)$ and $0.4937 \times 0.03880 \times \exp(0.03880n)$, respectively. For a fixed value of n, the slope of the curve for the credit-collection protocol is always larger than the one for the newly proposed leadership-based protocols. Thus, one might expect the difference between two curves to grow when the number of nodes increases; this can be observed in Fig. 8.

In order to cover all the directions seamlessly, the number of transmitters installed should be inversely proportional to the transmission beam divergence of each Tx. For example, if the beam divergence is 60°, the number of transmitters should at least be 6 to cover all the directions. With more number of transmitters installed, the transmission beam divergence of each Tx can be reduced. Assuming that the transmission power is kept the same, small beam divergence results in longer transmission range and lower interference to other nodes. However, a large number of transmitters also implies that the time necessary for the neighbor discovery increases because the nodes have to search and rank more directions. Figure 9 shows the relationship between the time required and the number of transmitters. Each direction corresponds to one transmitter individually; for e.g. the plot corresponding to 15 directions represents the scenario where the transceiver is equipped with 15 facets and 15 transmitters. From this figure, we conclude that the more directions, the longer the neighbor discovery process. For the four cases compared in Fig. 9, in order to find 95% of neighbors on average, it takes 0.9 secs, 1.4 secs, 1.92 secs and 2.39 secs, for the cases with 6 directions, 9 directions, 12 directions, and 15 directions, respectively.

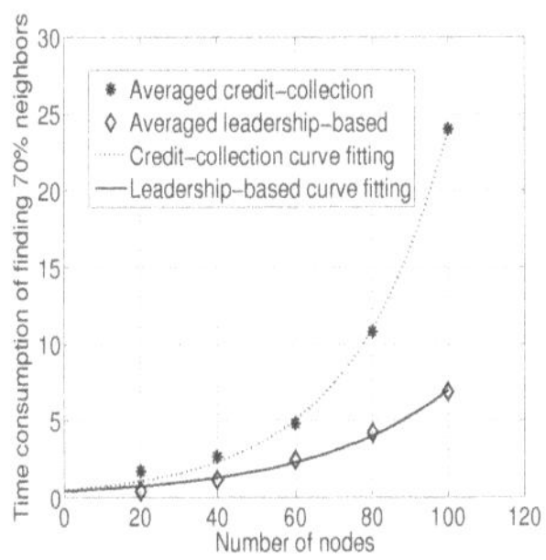

Figure 8: The time taken for neighbor discovery varies with number of nodes – a comparison of the leadership-based algorithm and the credit-collection algorithm. The time shown is when 70% of the neighbors are discovered, on average.

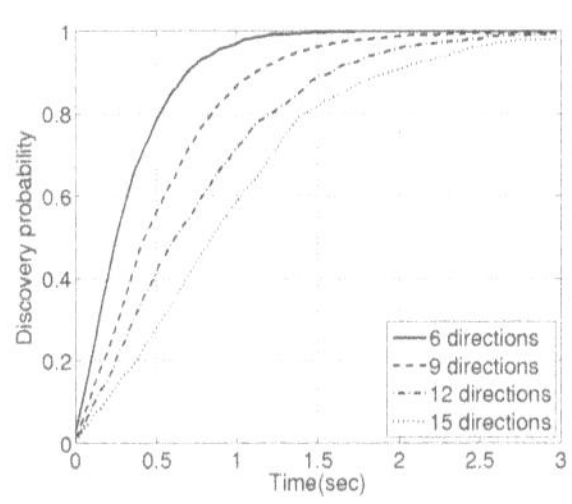

Figure 9: Performance with different numbers of directions. 20 nodes in total are distributed in the network.

6. TERMINATING NEIGHBOR DISCOVERY

In most application scenarios, it is important to bootstrap the network within a relatively short period. It is unlikely that a deployer will be able to pre-configure nodes with the specific topology; moreover, link qualities are likely to vary temporally. Our goal here is to determine when a neighbor discovery process can be considered to have terminated.

The scheme for terminating neighbor discovery: Achieving the above goal is far from trivial. Due to multi-user interference, feedback packets may collide, leading to information loss. As a consequence, a leader node is unlikely to realize all its neighborhood information by simply serving as the leader once. Thus, as discussed earlier, a node assumes the role of the leader multiple times. At each instance, it updates its neighborhood information with any data that is newly collected. If a node is unable to obtain new information upon becoming the leader, it is likely that it has already obtained its complete neighborhood information. In fact, the more times this happens, the more confidence one has in this conclusion. Thus, we make a modified rule that if a node does not acquire new information for a certain consecutive number of instances when it serves as a leader, it assumes that it does not need to be a leader anymore. It *marks* itself as being done with neighbor discovery. If all nodes have marked themselves as being done, then the procedure terminates. The information on which of the nodes are marked as above is recorded in a termination table. This table is exchanged in the notifier packets; thus, it is globally disseminated. We discuss this table later.

Each node maintains a counter which is initialized to '0'. Whenever it has acted as a leader but has not obtained any new neighborhood information, it increments this counter. If it does obtain new information, the counter is reset. When the counter value reaches a pre-defined threshold N, the node marks itself. When it transfers leadership, it updates the termination table to indicate that it is done with neighbor discovery. A simple bit corresponding to the nodes' IDs can be set to indicate that it is done. In other words, the termination table can simply be a bitmap which is equal in size to the number of nodes in the network.

With the information in the termination table, the current leader has the most up-to-date information and can make a timely and efficient decision on which node to choose as the leader next. Note here that this can be combined with the fairness criterion described earlier; however, we use this as the explicit criteria in the remainder of the section.

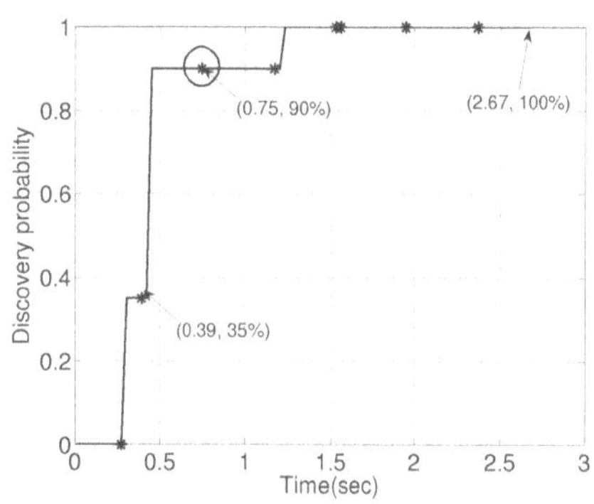

Figure 10: Illustrating the termination process. Solid line shows how the percentage of discovered neighbors increases with time; '*'s represent the instances when the node starts serving as the leader.

Evaluations: Figure 10 shows the percentage of neighbors discovered with time for a single realization from the perspective of an arbitrarily chosen node in the network. We simulate the 20-node scenario and other simulation settings are the same as described in Section 5. In this figure, we see that the node under consideration found 35% of its neighbors after the first instance of being the leader; this percentage increased to 75% after its second instance of being the leader. However, as shown by the circle in the plot, after the node served as the leader for the third time, there was no new information obtained; note however that, the neighbor discovery was not complete for the node. Although the counter increased by 1, it was reset after the node served as a leader for the fourth time, since new information was acquired. The neighbor discovery procedure finally ended at 2.67 seconds because the counter threshold was reached. At that point, the node had indeed discovered all of its neighbors.

Table 6 shows the total leader counts *after* every node found 100% of its neighbors. We set the termination counter threshold mentioned above to be 2. Interestingly, from the Table 6 we ob-

Table 6: Leader records after finding 100% neighbors

Node ID	0	1	2	3	4
Leader counts	6	4	5	7	4
Node ID	5	6	7	8	9
Leader counts	5	4	5	6	6
Node ID	10	11	12	13	14
Leader counts	5	6	6	4	2
Node ID	15	16	17	18	19
Leader counts	7	7	7	6	7

serve that all the nodes except the one with node ID of 14 served as the leader without any information update, more than 2 times. This is because, node 14 was the final node in the neighbor discovery process; unless its counter was set to 2, the process could not be terminated. However, node 14 could be assigned to being a leader only by one of its neighbors. In this case, nodes attempted to search for 14 and interim also served as leaders again (the leadership token was passed around). Finally, the token was passed onto node 14 and the process terminated.

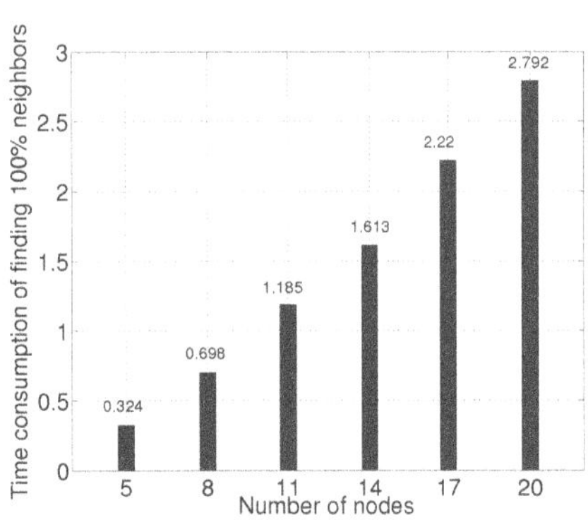

Figure 11: Illustration of the performance of the termination process with different node densities. All the nodes in the network find 100% of their neighbors.

Figure 11 shows the performance of the protocol with the termination procedure with six scenarios, with different node densities. We see the termination time is proportional to the node density as one may expect.

7. CONCLUSIONS

Neighbor discovery in UV networks is an important function that allows nodes to bootstrap the topology and begin operations. While there have been a plethora of proposals for neighbor discovery in the RF context, they are inapplicable in the UV context given its unique PHY characteristics.

In this paper, we propose a novel neighbor discovery protocol for UV networks. The key idea is to eliminate interference by arbitrating the discovery process by using leaders; a leader is the only node that is allowed to perform neighbor discovery at any given time. The leadership role is passed on from node to node. We demonstrate that our protocol is extremely effective in performing neighbor discovery via extensive simulations that incorporate the UV PHY layer. In particular, the time needed for neighbor discovery process is decreased by as much as 90% as compared to the only previously proposed neighbor discovery scheme for UV networks. We also provide extensive evaluations to capture the impact of various network and system parameters on the performance of our protocol.

8. REFERENCES

[1] Z. Xu and B. M. Sadler. Ultraviolet communications: potential and state-of-the-art. *IEEE Commun. Magazine*, 46(5), May 2008.

[2] S. Vasudevan, D. Towsley, R. Khalili, and D. Goeckel. Neighbor discovery in wireless networks and the coupon collector's problem. In *ACM Mobicom*, 2009.

[3] S. Vasudevan, J. Kurose, and D. Towsley. On neighbor discovery in wireless networks with directional antennas. In *IEEE INFOCOM*, March 2005.

[4] G. Pei, M. Albuquerque, J. Kim, D. Nast, and P. Norris. A neighbor discovery protocol for directional antenna networks. In *IEEE MILCOM*, October 2005.

[5] M. J. McGlynn and S. A. Borbash. Birthday protocols for low energy deployment and flexible neighbor discovery in ad hoc wireless networks. In *Mobihoc*, September 2001.

[6] J. Luo and D. Guo. Neighbor discovery in wireless ad hoc networks based on group testing. In *46th Annual Alerton Conf.*, September 2008.

[7] Y. Li, L. Wang, Z. Xu, and S. V. Krishnamurthy. Neighbor discovery for ultraviolet ad hoc networks. In *IEEE JSAC*, 2011 (accepted with revision).

[8] Y. Li, J. Ning, Z. Xu, S. V. Krishnamurthy, and G. Chen. Uvoc-mac: A MAC protocol for outdoor ultraviolet networks. In *IEEE ICNP*, October 2010.

[9] G. Jakllari, W. Luo, and S. Krishnamurthy. An integrated neighbor discovery and MAC protocol for ad hoc networks using directional antennas. *IEEE Trans. on Wireless Communications*, 6(3), 2007.

[10] G. Chen, Z. Xu, H. Ding, and B. M. Sadler. Path loss modeling and performance trade-off study for short-range none-line-of-sight ultraviolet communication. *Optics Express*, 17(5), March 2009.

[11] Wireless LAN medium access Control (MAC) and physical layer (PHY) specifications. http://standards.ieee.org/getieee802/download/802.11-2007.pdf.

[12] Opnet User's Documentation. http://www.opnet.com.

Analysis of Link Break Detection using HELLO Messages

Zainab R. Zaidi
National ICT Australia (NICTA)
zainab.zaidi@nicta.com.au

Marius Portmann
National ICT Australia (NICTA)
The University of Queensland, Australia
marius.portmann@nicta.com.au

Wee Lum Tan
National ICT Australia (NICTA)
weelum.tan@nicta.com.au

ABSTRACT

HELLO messages are widely used for neighbor discovery in routing protocols for wireless multihop networks. In this paper, we provide mathematical and experimental proofs that the current strategy of declaring a link down based on observations that a specific number of HELLO messages are lost, is not a correct strategy. In our mathematical analysis, we characterized the error bursts over wireless links by extending the Gilbert bit error model, and validate our model using HELLO loss data from our wireless mesh testbed. Our analysis shows that error burst lengths follow a geometric distribution, i.e. the probability of additional losses in the burst does not depend on the observed losses. We propose an alternative link break detection strategy where a link break is declared when the observed mean error burst length for the link is longer than the mean route recovery time. Our testbed results show that running the AODV routing protocol with our proposed strategy yields better throughput and lower control message overhead, compared to the case where AODV relies on its default settings of declaring link breaks when two consecutive HELLO messages are lost.

Categories and Subject Descriptors

C.2.2 [**Network Protocols**]: Routing protocols; G.3 [**Probability and Statistics**]: Distribution functions

General Terms

Experimentation; Performance

1. INTRODUCTION

Although a lot of work has been done to develop routing protocols for wireless multihop networks that could yield reasonable performance over uncertain wireless media, they are generally observed to be under-performing in practical scenarios. This is mostly because various protocol parameters are not tuned to suit the operational settings. A classic example is the HELLO messaging used for neighborhood discovery in various routing protocols, e.g. AODV (Ad hoc On-demand Distance Vector) [13], OLSR (Optimized Link State Routing) [9], etc. Two parameters are associated with the HELLO protocol: U: the number of HELLO messages received before a link is declared *UP*, and D: the number of HELLO messages lost before a link is declared *DOWN* [1].

While the parameter U is usually selected as 1, it is not immediately clear what should be the right value for D. In practical multihop networks, wireless link quality may fluctuate randomly (due to interference, fading, etc.). As such, setting D to a low value may lead to poor performance since it will lead to unnecessary declarations of link breaks and subsequent new route discoveries. On the other hand, setting D to a large value is also not a good solution as valuable time would be lost if the link is under a long error burst. These practical issues usually do not arise in simulation experiments where simplistic error models are used for wireless channels.

Not much work has been reported in the literature to optimize the parameters of the HELLO protocol. A default value of $D = 2$ is selected in the original AODV routing protocol based on simulation results [13]. D is termed as the *allowed-hello-loss* in [13]. In a follow-up work [3], field experiments were carried out with three nodes equipped with 802.11b radios for $D = 2$ and $D = 3$. The experiment results show that $D = 2$ gives better throughput in a mobile setting while $D = 3$ is better in a static setting. A detailed literature survey is presented in Section 5.1.

In this paper, our goal is to come up with a method to select an optimum value of D which would maximize link utilization and throughput. We first develop an analytical model for describing error bursts over a wireless link and validate it using the HELLO loss data from our wireless mesh testbed. The error burst distribution is found by extending the well-known Gilbert bit error model for bursty channels [7]. Our results show that error burst lengths are geometrically distributed. As a consequence, the probability of observing additional losses does not depend on the observed losses. This indicates that the D parameter (or *allowed-hello-loss*) should not be used as an indicator for broken links. In this paper, we propose an alternative strategy of characterizing each link in terms of its observed mean error burst length, and declaring a link as broken only when it is greater than the average route recovery time of the network. Using this approach, results from AODV routing experiments on our wireless mesh testbed show throughput improvements and lower control message overhead compared to the case with AODV using the default value of $D = 2$.

MSWiM '11, October 31–November 4, 2011, Miami, Florida, USA.

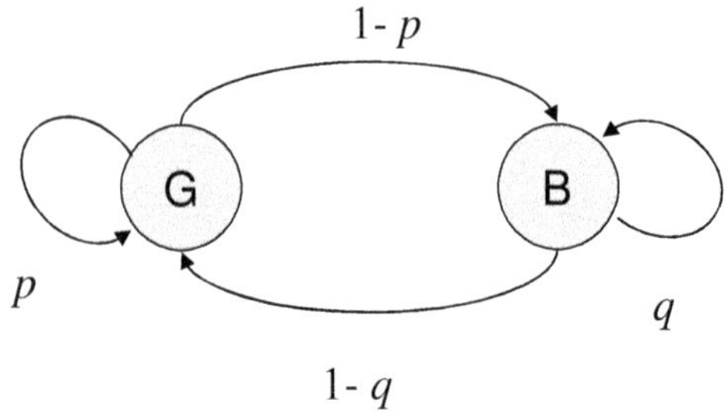

Figure 1: The basic Gilbert model for bit errors over a bursty channel.

The paper is organized as follows: Section 2 presents the packet error model which leads to a geometric distribution for error burst lengths. Section 3 presents the proposed strategy for declaring link breaks, Section 4 presents validation results for the burst length distribution and the performance evaluation of our new strategy of link break detection. Section 5 recaps previous work in this area and finally Section 6 concludes the paper.

2. PACKET BURST ERROR MODEL

Most of the work on error modeling over wireless channels focus on bit level statistics. There are some instances where packet/frame level modeling are discussed but the papers, in general, talk about the appropriateness of particular mathematical models in certain contexts. They do not provide explicit expressions for the distribution of error burst lengths which could be used for the selection of optimal values for the parameter D in our case. A detailed literature survey is presented in Section 5.2.

In this section we extend the most widely used bit error model for bursty channels, i.e. the Gilbert model [7]. While prior works have extended the Gilbert model before (cf. Section 5.2), in this paper, our novel contributions are in terms of deriving an exact closed-form expression for the distribution of error burst length and its validation with real HELLO loss data from our wireless mesh testbed.

2.1 Extending the Gilbert bit error model

The basic Gilbert model [7] is shown in Fig. 1. The system (channel) could be in either of two states: G (*Good*) or B (*Bad*). States G and B have different error probabilities, with the error probability in B much larger than that in G. We have simplified the model by assuming that there is no error in G and when in B, there is 100% error. This assumption greatly simplifies our analysis and our validation study shows that the simplified model is sufficient to model the burstiness of errors in HELLO packets. The transition probabilities $P(G|G) = p$ and $P(B|B) = q$ are the only remaining unknowns in the simplified Gilbert model. The transition probability matrix for the Gilbert model is then defined as:

$$A = \begin{pmatrix} p & 1-q \\ 1-p & q \end{pmatrix}$$

Let P_n be the steady-state probabilities of being in states G and B, i.e., $P_n = [P(G) \;\; P(B)]'^{1}$.

$$P_n = AP_n = \begin{pmatrix} \frac{1-q}{2-p-q} \\ \frac{1-p}{2-p-q} \end{pmatrix} \quad (1)$$

Suppose the event that the ith packet of a data sequence is correctly received is denoted by C_i and the event of the reception of ith packet in error is represented by E_i. We assume that each packet contains N bits. According to the Gilbert model of Fig. 1, the dependence between two consecutive packets in a sequence is actually the dependence between the last bit of the preceding packet and the first bit of the following packet. If the preceding packet is in error, the last bit could be either in state G or B. However, if the preceding packet is correctly received, its last bit can only be in G state with probability 1, i.e.,

$$P(C_{i+1}|E_i) = HP_n, \quad (2)$$

$$P(C_{i+1}|C_i) = p^N, \quad (3)$$

where $H = [p^N \;\; (1-q)p^{N-1}]$. Here the effects of forward error correction (FEC) are ignored. Also, in the context of HELLO packets, FEC is only used for packet headers and not for the body of the message. Nevertheless, our validation study with real HELLO messages from wireless mesh testbed indicates that this assumption does not have great impact on the applicability of our model in practical scenarios.

Similarly, the probability that the $i + 1^{th}$ packet is in error is as follows:

$$P(E_{i+1}|E_i) = 1 - P(C_{i+1}|E_i) = 1 - HP_n, \quad (4)$$

$$P(E_{i+1}|C_i) = 1 - P(C_{i+1}|C_i) = 1 - p^N. \quad (5)$$

According to [10, 16], a first-order Markov chain is sufficient to model packet errors over a bursty channel. Using the conditional probabilities in Eqns. (2) to (5), we can actually build a two-state model similar to the Gilbert model to represent packet errors as shown in Fig. 2. The two states consists of state E, representing the event E_i defined above, and state C, representing a correctly delivered packet defined as the event C_i previously. It is important to note that the transition probabilities for the Gilbert model and the packet error model are not necessarily stationary for practical channels.

2.2 Error burst length distribution

Let BL represent the error burst length in terms of number of packets, so we have,

$$P(BL = k) = P(E_i \cdots E_{i+k-1}C_{i+k}) \quad (6)$$

$$= P(E_i)P(E_{i+1}|E_i) \cdots P(E_{i+k-1}|E_{i+k-2}) \cdot P(C_{i+k}|E_{i+k-1}),$$

$$= HP_n[1 - HP_n]^k \quad (7)$$

Thus, Eqn. (7) shows that the error burst length follows a geometric distribution. The mean error burst length can then be found analytically as follows:

$$E[BL] = \frac{1 - HP_n}{HP_n} \quad (8)$$

Throughout this paper, we use $E[BL]$ to denote the mean error burst length in terms of consecutive losses of HELLO messages. We interchangeably use the units of $E[BL]$ as

$^{1\prime}$ represents matrix transpose.

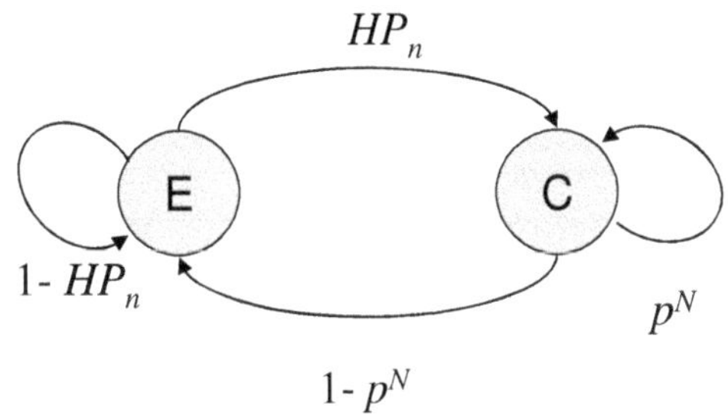

Figure 2: A two-state packet error model for wireless links.

either number of HELLO packets or seconds, as one HELLO message is typically transmitted every second. The mean error burst length is very useful in estimating the geometric distribution parameters for real data, i.e.,

$$HP_n = \frac{1}{1 + E[BL]}. \tag{9}$$

Section 4.2 presents model validation results using the HELLO loss data from our wireless testbed links where we have very strong evidence that the above geometric distribution model of packet burst errors is indeed a practical and realistic model. In our experimental study, we have calculated the burst lengths in terms of consecutive lost HELLOs for various wireless links in our testbed in both directions. Using the error burst data, we estimated the mean error burst length and the geometric distribution parameters using (9). We have used statistical tools to match the empirical distribution of the error burst length with a geometric distribution with the same mean.

3. LINK BREAK DETECTION STRATEGY

3.1 Link characterization

efore discussing the implications of our packet error model on the selection of optimal values for the parameter D, an interesting issue is to look at various link quality parameters and their ability to sufficiently characterize the packet loss behavior of a link. The BER (Bit Error Rate) and PER (Packet Error Rate) parameters are widely used to characterize the quality or strength of a wireless link. However, we will show in this section that they are not sufficient measures, especially when used for routing purposes, as there may be very different mean error burst lengths associated with the same BER or PER, depending on the parameters in the error model. Furthermore, wireless links with longer error bursts, on the average, should be treated differently from the ones with shorter error bursts, when selecting a route.

BER is the steady-state probability of being in state B in the Gilbert model, i.e.

$$\text{BER} = \frac{1-p}{2-p-q}. \tag{10}$$

For a specific BER, there could be multiple solutions for the pair of parameters p and q. Besides conforming to the above equation, p and q should also be legitimate probability values, i.e., $0 \leq p, q \leq 1$. A range of feasible solutions is shown by straight lines in Fig. 3 for a specific BER, shown as labels on the lines. For a link with BER $= 0.01$, p has a higher value, while it does not matter much what the value of q is. Similarly, for a very bad link, q has a higher value irrespective of the value for p.

As shown in Fig. 3, the highest variability in values for p and q is observed for middle-range values of BER. This variability has huge effects on the error burst length distribution. Two links with the same BER can have drastically different distributions of error burst length, from close to single packet errors to very long mean error burst length as shown in Fig. 4. Fig. 4 shows the mean error burst lengths for combinations of p and q in Fig. 3. Each curve is labeled with the specific BER value. For all values of BER shown in Fig. 4, the mean error burst length seems to be zero or very small for values of $p \geq 0.97$.

Re-writing the expected value of error burst length in

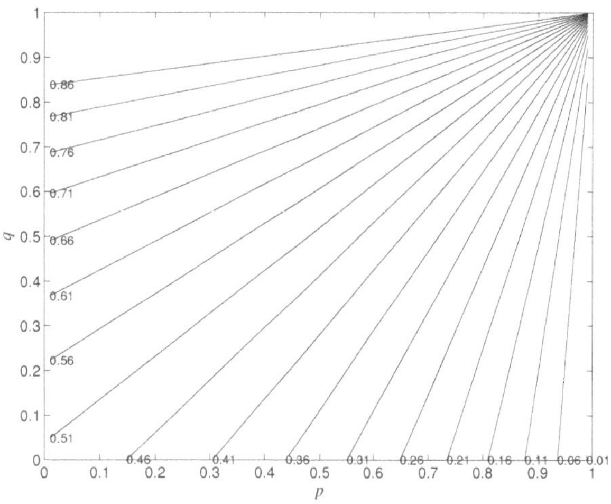

Figure 3: Feasible solutions for specific BER in terms of the transition probabilities p and q.

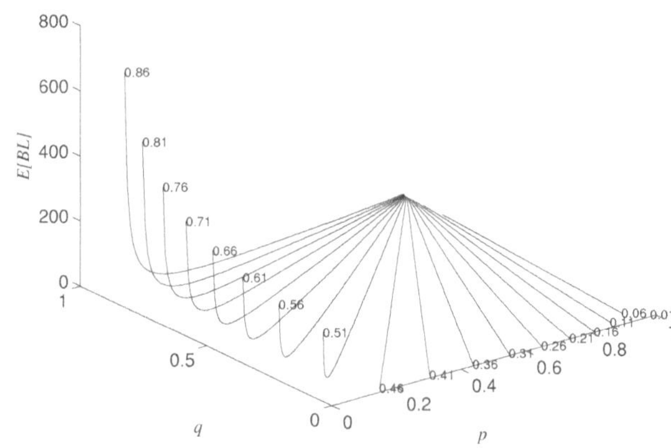

Figure 4: Mean error bursts length w. r. t. p and q for different BER.

terms of BER, i.e. replacing

$$HP_n = [p^N \quad (1-q)p^{N-1}] * \begin{bmatrix} \frac{1-q}{2-p-q} \\ \frac{1-p}{2-p-q} \end{bmatrix}$$
$$= (1 - \text{BER})p^{N-1}$$

in Eqn. (8), we get:

$$E[BL] = \frac{1}{(1-\text{BER})p^{N-1}} - 1 \qquad (11)$$

From the expression above, we can see that for the same BER, if p is larger, the mean error burst length will be smaller as shown in Fig. 4. The mean error burst length increases with BER and also when p becomes smaller. Thus, we can have links with the same BER but different mean error burst length, depending on the probability p of remaining in the G (*Good*) state in Fig. 1.

PER is the steady-state probability of being in state E of the packet error model given in Fig. 2, i.e.

$$\text{PER} = \frac{1-p^N}{1+HP_n-p^N} = \frac{(1-p^N)(1+E[BL])}{1+(1-p^N)(1+E[BL])}.$$

Similarly as above, we can re-write this expression to show the mean error burst length in terms of the parameters PER and p. As before, we can show that two links with the same PER can have different mean error burst length, depending on the parameter p. As such, links specified in terms of BER or PER alone do not give complete information about the packet burst errors.

3.2 Link break detection

Ideally, an optimum value of D in the HELLO protocol would be one that maximizes the probability of observing longer error bursts given that D losses have been observed. However, using Eqn. (7), the conditional probability of observing k additional losses when D losses are already observed, is found to be independent of D, i.e.

$$P(BL \geq D + k | BL \geq D) = (1 - HP_n)^k = P(BL \geq k) \quad (12)$$

This result is a consequence of the memoryless property of a geometric distribution and is in complete contradiction to the way we handle link break declarations in HELLO protocol. Observing some HELLO losses does not change the uncertainty of future losses over a wireless link.

It makes sense to leave a wireless link under fade to look for a better alternative in order to minimize the time of inactivity or the lost time. If the link enters the error state or state E of Fig. 2 and is not declared broken, the average lost time $E[T_L]$ will be equal to $E[BL]$, i.e. the mean error burst length. On the other hand, if a link is declared broken as soon as it enters the error state, the average lost time will t_r, where t_r is the recovery time in seconds. Recovery time includes the time to notify the source of the traffic, as well as initiating and completing a new route discovery process.

In this paper, we propose to declare a link break if its mean error burst length is longer than the average time required to recover from a link failure, i.e. if $E[BL] > t_r$, the link is declared down. $E[BL]$ is considered here in units of seconds. The periodic HELLO messages will be used to estimate $E[BL]$ within a time window for every link. A very short time window will not allow us to observe all probable error burst lengths, while a very long time window will not allow $E[BL]$ to adapt quickly to the changes in the wireless medium. We selected the averaging time window to be 100 seconds in our testbed experiments; as the typical mean t_r is usually under 2 seconds for our network, a window of 100 seconds is more than enough to have error burst lengths greater than t_r if there is any. Our experiment results show that this strategy works better in terms of yielding higher throughput than the default strategy of using $D = 2$ in the AODV routing protocol. Details are given in Section 4.3. We also note that estimating $E[BL]$ requires an initial setup time in order to compute the first value of $E[BL]$ over the entire time window. The rest of the values can be updated using a sliding window approach and hence our strategy does not induce any delays in declaring link breaks when compared to the default strategy of using $D = 2$.

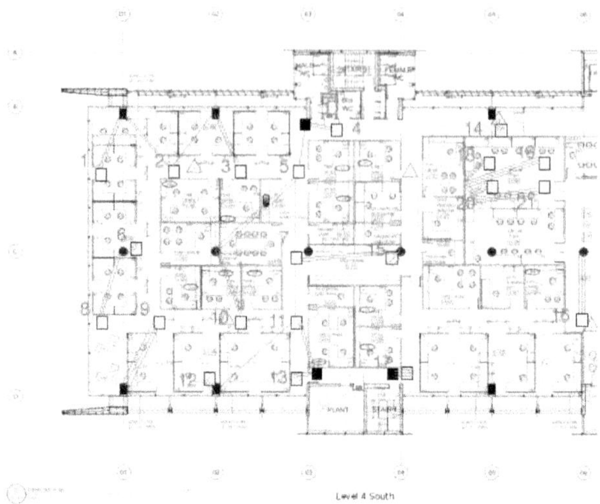

Figure 5: Layout of a section of NICTA indoor wireless mesh testbed.

4. EVALUATION

This section discusses our experimental results in validating the packet loss model, and the performance comparison of the AODV routing protocol in delivering packets with our new strategy of link break detection and with the default scheme of using $D = 2$.

4.1 Wireless mesh testbed

We have used NICTA's wireless mesh testbed for our experiments (http://mytestbed.net/). The NICTA wireless mesh testbed is composed of 40 static nodes installed across three floors at the NICTA office building in Sydney. A layout of a section of the testbed is shown in Fig. 5. In Fig. 5, the yellow (or light grey in greyscale) squares denote the location of the mesh nodes and the numbers show their respective IDs. The black squares and straight lines are irrelevant for our purpose. They show the power outlets in the building with which the mesh boxes are attached to.

Each node consists of a Ubiquity SR2 wireless card supporting the IEEE 802.11a/b/g standard. We have installed

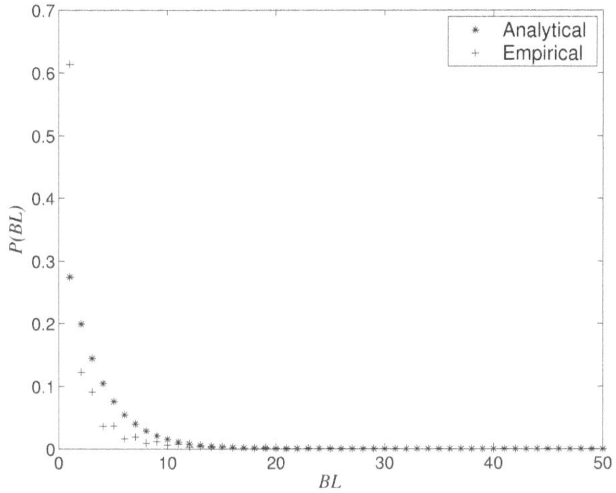

Figure 6: Empirical and analytical distributions for error bursts over link $16-7$.

the AODV-UU implementation[2] of the AODV routing protocol on our testbed nodes. In all our experiments, we have used the default configuration of MadWiFi driver for the wireless interfaces and configured the nodes to use 802.11a channel 56.

4.2 Model validation

Model validation using real HELLO loss data is the critical component of this study. We have performed experiments over several links in the NICTA wireless mesh testbed spanning multiple days. The links included in our study ranged from very good, where only a small fraction of HELLO messages are lost, to very bad, where only a few HELLO messages are received over a long period of time.

In these experiments, the AODV-UU routing daemons run on various combinations of two or more nodes, and statistics about reception of HELLO messages are collected. An example is shown in Fig. 6 where empirical and analytical distributions are drawn for error bursts over link 16-7, i.e. the link between node 16 and node 7 in Fig. 5. The analytical distribution is given in Eqn. (7) where the parameter HP_n is estimated from the data, i.e. $\hat{HP_n} = (1 + \hat{\bar{BL}})^{-1}$, where $\hat{\bar{BL}} = \sum_i BL_i/N_o$, and N_o is the number of observed error bursts. The empirical distribution can be found as

$$P_{emp}(BL = k) = \frac{1}{N_o}\sum_{i=1}^{N_o} \mathbf{1}_{\{BL_i = k\}}$$

There are several statistical goodness-of-fit tests defined to estimate the quality of fit of the samples to a particular distribution. We start with Pearson's χ^2 test. This test is very sensitive to the number of observations, and for links with longer error bursts, we need many more observations to have a reasonable test result. Two other goodness-of-fit tests which can work with smaller sample sizes are also used in our experiments. The test statistics are given as follows:

[2] http://aodvuu.sourceforge.net

Table 1: Model validation results

Link	PER (%)	$E[BL]$	Trial length (days)	Result H_0 accepted by
2-3	0.8	0.0081	3	All tests at $\alpha = 0.2$
3-2	0.9	0.0091	3	
4-2	21.8	0.2791	1	
2-4	29.24	0.4137	5	
17-9	0.48	0.0048	5	
17-5	65.55	1.9025	1	
15-17	0.42	0.0042	1	
17-15	0.31	0.0031	1	
16-17	0.15	0.0015	1	
17-16	0.34	0.0034	1	
7-4	51.61	1.1429	5	CVM and AD tests at $\alpha = 0.2$
2-17	91.23	10.4444	1.5	
9-4	53.94	1.1709	5	
4-9	58.18	1.3915	5	
9-17	41.14	0.699	5	
5-17	86.72	6.5295	1	
7-16	68.44	2.1686	1	
16-7	42.71	0.7454	1	
7-17	70.86	2.4323	1	
17-7	57.26	1.3396	1	
17-2	99.71	461.667	1.5	AD at $\alpha = 0.2$
4-7	50.46	1.0205	1	CVM at $\alpha = 0.05$ and AD at $\alpha = 0.005$
5-4	50.13	1.0067	5	
2-7	50.17	1.0067	1	CVM and AD tests at $\alpha = 0.2$ with truncation
7-2	50.23	1.0091	1	
25-27	50.07	1.0027	2	
27-25	50.16	1.0065	2	

1. **χ^2 test:**

$$\chi^2 = \sum_k \frac{(\text{expected}_k - \text{observed}_k)^2}{\text{expected}_k},$$

where $\text{expected}_k = N_o \times P(BL = k)$ and $\text{observed}_k = \sum_{i=1}^{N_o} \mathbf{1}_{\{BL_i = k\}}$.

2. **Cramér-von Mises (CVM) test** [2]:

$$W^2_{N_o} = N_o \sum_k \left[\{F_{emp}(BL = k) - F(BL = k)\}^2 \times P(BL = k)\right],$$

where $F(BL = k) = P(BL \leq k)$.

3. **Anderson-Darling (AD) test** [2]:

$$W^2_{N_o} = N_o \sum_k \left[\frac{\{F_{emp}(BL = k) - F(BL = k)\}^2}{F(BL = k)(1 - F(BL = k))} \times P(BL = k)\right].$$

We also use truncation with the CVM and AD tests where the statistics is summed for the values of k in the interval $[1, \min\{k \geq \text{max observed BL}; (1 - F(BL = k))^3 \leq 10^{-4}/N_o\}]$. The bootstrap method is used to calculate the

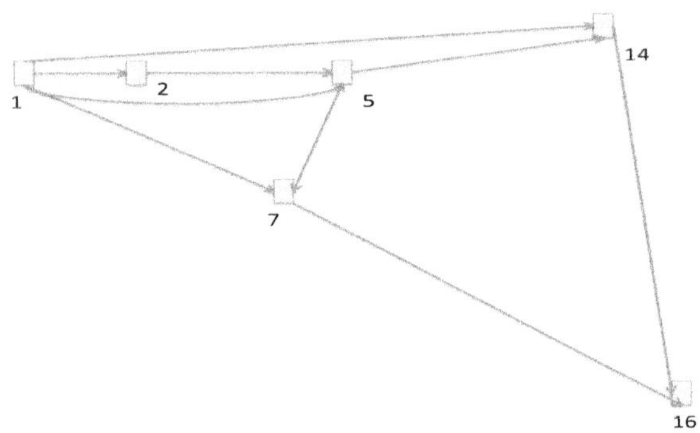

Figure 7: Available paths between node 1 and node 16 in one of our experiments.

critical values for the statistics. The bootstrap method uses 1000 trials and generates N_o samples in each trial [2]. The hypothesis test is defined as

$$H_0 : P_{emp}(BL = k) = P(BL = k) \quad (13)$$
$$H_1 : P_{emp}(BL = k) \neq P(BL = k)$$

The summary of the validation results is shown in Table 1. For 25 links, out of a total of 27 links in our validation study, H_0 was accepted at $\alpha = 0.2$ by the CVM and AD tests with proper truncation to avoid tail behavior. For links from node 4 to node 7, and from node 5 to node 4, H_0 was accepted by the CVM test at $\alpha = 0.05$ and the AD test at $\alpha = 0.005$.

For 10 out of the 27 links, we are able to accept H_0 at $\alpha = 0.2$ by the χ^2 test as well. In general, these links have smaller mean error burst length and our experiment duration was long enough to get sufficient samples suitable for a χ^2 test. Moreover for 23 links, we are also able to accept H_0 by the CVM and AD tests without the use of truncation defined above. The links in our experiment varied in quality. The minimum average error burst length observed was 0.001, in terms of the number of consecutive HELLO losses, with PER = 0.2% for the link from node 16 to node 17. The maximum average error burst length observed was 461.67 with PER of 99.7% for the link from node 17 to node 2. The HELLO loss data over these 27 links were collected for different durations ranging from 1 to 5 days, as shown by the *Trial length* column in Table 1.

4.3 Performance evaluation

To compare the default setting of $D = 2$ in the AODV routing protocol with our new strategy of declaring a link break when $E[BL] > t_r$, where t_r is the mean route recovery time, we selected a few groups of nodes in our testbed which have at least two paths between the source node and the destination node. For example, Fig. 7 shows the available paths between the source node 1 and the destination node 16, via nodes 2, 5, 7, and 14. In our experiments, these paths consist of a different number of hops, ranging from a single hop to five hops across the different groups of nodes. A 54 Mbit/s UDP flow is transmitted between the source node and the destination node within each group of nodes for 10 minutes using the *Iperf* tool. We selected UDP traffic to avoid the artifacts of the congestion control mechanism of TCP flows.

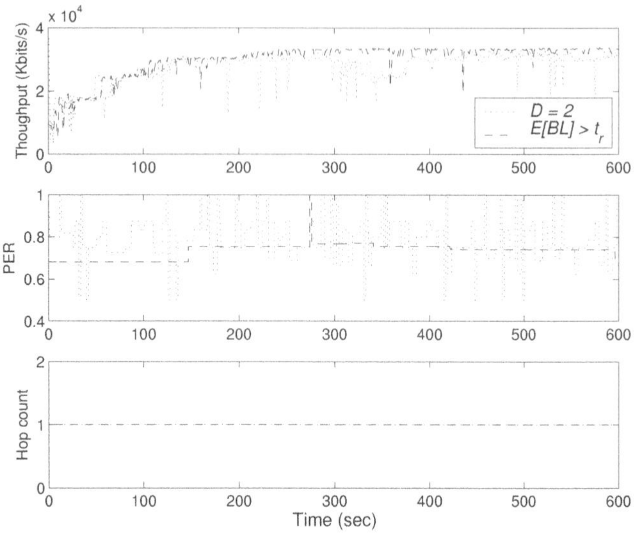

Figure 8: Throughput comparison using the default setting of $D = 2$ and the new strategy of declaring a link broken when $E[BL] > t_r$.

We ran AODV-UU with its default parameters on all the nodes for our baseline experiment. For the alternate strategy of link break detection, we modified the AODV-UU code to save the history of HELLO messages reception within the specified averaging window, which is set to 100 seconds in these experiments. Whenever a new HELLO message is received from a neighbor node, the receiver recalculates $E[BL]$ using the received HELLO messages history. If $E[BL] > t_r$, the value of D is set to 1 so that the next time the link is in error or enters the E state of Fig. 2, the link will be declared down. On the other hand, if $E[BL] \leq t_r$, we keep D at its initial value, which is 100 in our experiments. This implementation maintains the 'link break' mechanism of AODV-UU which is triggered by the parameter D, although it is not a very accurate implementation of our strategy. This is because by updating $E[BL]$ only when a HELLO message is received, we may miss an error burst with length less than the initial value of D until it finishes. Furthermore, the link is assumed to be active throughout such a burst. Nevertheless, we remark that this inaccuracy degrades the performance of our strategy and an accurate implementation can only enhance the throughput performance reported in our results. For all of our experiments, we calculated t_r from the log files of our baseline experiments. The alternate strategy of link break detection requires a startup time to estimate $E[BL]$, so we started UDP flows in all the experiments after 2 minutes of starting the AODV deamon on the nodes.

A sample result is shown in Fig. 8, where the throughput in each second is plotted for the default setting of $D = 2$, and for the case where D is set to 1 or 100 based on the computed $E[BL]$. The middle and bottom graphs show the average packet error rate (PER) using only HELLO losses and the hop count of the particular path for each second. The set of nodes in this experiment consists of nodes 2, 4, and 5. The source is selected to be node 2 and the destination is node 4. In this experiment, paths selected by AODV

Table 2: Throughput comparison.

Source-Destination	Nodes in the experiment	Hops	Average throughput (Mbits/s)		Average RREQ/RREP overhead (No. of packets)	
			$D = 2$	$E[BL] > t_r$	$D = 2$	$E[BL] > t_r$
1-16	1, 2, 5, 7, 14, 16	2-4	3.4	4.68	1341	568.67
16-5	5, 7, 12, 16, 17	2-3	2.92	3.04	1094	496
28-32	$28, \cdots, 32$	1-3	11.02	12.9	925	262.33
2-4	2, 4, 5	1	26.32	28.48	424.67	12

always consist of one hop as shown in the bottom graph, although an alternate path of two hops via node 5 could also be selected. The middle graph shows the average packet error rate (PER) calculated from the loss of HELLO messages for the active path. PER is averaged over the interval when a path is active. Interestingly, the more stable plot of PER for $E[BL] > t_r$ shows that paths remain active for a longer duration with our strategy of link break detection compared to that of $D = 2$. The top graph shows the throughput performance. Averaged over 10 minutes, the mean throughput using our strategy is 29.96 Mbits/s, while with the default setting of $D = 2$, the mean throughput is 27.65 Mbits/s.

Table 2 shows the average throughput for all of our experiments, with the averaging performed over the length of the experiment (which is set at 10 minutes) and also over the number of experiments. We conducted three trials for each experiment. Table 2 also shows the average control message overhead of Route-Request (RREQ) and Route-Reply (RREP) packets transmitted in the experiments, with the averaging performed over the number of trials. We have counted all RREQ and RREP packets transmitted from all nodes, from the log files. Since our strategy, in general, yields more stable paths, the control message overhead of route discovery is also considerably low. The more stable paths, resulting from the use of our strategy, are also responsible for the more stable PER plot in the middle graph of Fig. 8. Interestingly, for the last experiment listed in Table 2 where a one hop path was always selected, $D = 2$ was responsible for reporting a link break so many times that a huge number of RREQ broadcasts were needed to find the destination which is only a hop away.

5. RELATED WORK

5.1 Selection of D or allowed-HELLO-loss

Not much work has been reported in the literature to optimize the parameters of the HELLO protocol. The original AODV paper [13] presented a simple simulation-based results to show that $D = 2$ is the best choice. The paper presented route acquisition latency and goodput results for $D = \{1, \cdots, 4\}$ for networks with 50 and 100 nodes. $D = 2$ is selected as the best choice despite it having the worst latency for the 50-nodes network and not achieving the best goodput for both 50-nodes and 100-nodes networks. In a follow-up work [3], field experiments were carried out with three 802.11b radios for $D = 2$ and $D = 3$. The experiments were performed in two modes: static (all nodes are stationary) and mobile (one node moves further away). The results show that $D = 2$ gives better throughput in a mobile setting while $D = 3$ is better in a static setting.

Some other recent works include the performance analysis of neighborhood discovery via HELLO messaging [1, 12]. Their approach utilizes Markov chain modeling to derive the steady-state probabilities of link detection. However, the studies do not consider the drawbacks of using a larger D, i.e., lower throughput when the link remains inactive for long period.

5.2 Packet and bit error models

While a lot of work has been done in characterizing errors in bursty channels in order to evaluate the performance of error correction and detection coding schemes, not much has been done to validate these models for wireless multihop networks using 802.11 radios. In the context of wireless mesh networks, a lot of research is focussed on modeling the effects of hidden terminals on packet delivery while ignoring all other causes of errors [4].

The classic approach of modeling bit errors using a 2-state Markov chain is given by Gilbert [7], where the *Good* state has zero error rate and the *Bad* state has non-zero error probability. Gilbert's model is generalized by Elliott to introduce a smaller error rate in the *Good* state compared to that of the *Bad* state [5]. This model is further generalized to include N states partitioned into k *Good* states and $N - k$ *Bad* states in [6]. In [6], expressions are also given for error free *run* lengths and error *burst* lengths as products of state transition probabilities in the Gilbert-Elliott model.

Several enhancements have also been suggested for the basic Gilbert model which are more realistic for specific scenarios, e.g. in [11], the Gilbert-Elliott model is enhanced and compared for 802.11 wireless links. The enhancement involved making the transition probability between *Good* and *Bad* states to be dependent on the current bit [11]. An alternate model is presented in [14] where independent errors are considered separately from correlated errors for general packet transmission systems.

While the enhanced versions of the Gilbert model promise more accuracy, they are mathematically more complicated and it is difficult to derive closed-form and simple expressions for required statistics such as probability of error burst length, etc. We started with the simplified form of the Gilbert model in this paper and our strategy was to use more complicated versions of the model only when validation with real data fails. The simplified Gilbert model, used in our paper, has also been successfully used to describe both the packet loss characteristics in Internet video transmission [8], and the cell discard model for MPEG video transmission in ATM networks [15]. Other efforts include the characterization of packet or block errors by extending the Gilbert model [10, 16] in a similar fashion as in [6]. In [16], mutual information arguments are used to show that a first-order time-homogeneous Markov chain is sufficient to represent packet errors in a bursty channel. However, in [10], it is shown that this can be achieved only by using non-stationary, time-inhomogeneous Markov chain. Simulations based on the Gilbert model are used to verify the accuracy of packet characterization in [16] and [10].

The packet loss model developed in this paper is in accordance with the previous results [6, 10, 16]. The use of the simplified Gilbert model helped us to come up with the closed-form expressions for the distribution of error burst length and the transition probabilities for the packet level model, as functions of the parameters in the Gilbert model. We also stepped forward and validated the model using real HELLO loss data from our wireless mesh testbed, thus making a genuine case for its application in practical scenarios.

6. CONCLUSION

In this paper, we have shown that declaring a link break when a specific number of HELLO message losses are observed is not a correct strategy. Our mathematical analysis yields a geometric distribution for error burst length for HELLO message losses. This result is also validated using real HELLO loss data from our wireless mesh testbed. For a geometric distribution, the probability of observing additional losses does not depend on the observed losses, and hence D or *allowed-hello-loss* is an irrelevant parameter for the HELLO protocol. Instead, we propose to characterize a link based on its observed mean error burst length and when it is found to be more than the mean route recovery time for the network, it is better to declare a link break in order to improve link utilization and minimize loss of time. Our testbed experiments show that improved throughput and lower control message overhead performance are achievable using our new strategy, compared to the default setting of HELLO messaging in the AODV routing protocol.

As a future extension to the current study, we intend to look into the discrepancy in HELLO message reception rate and data delivery rate over a wireless link and how it affects the performance of our new strategy for link break detection.

Acknowledgement

The authors like to acknowledge the assistance of Ms. Shachi Shastry, Dr. Thierry Rakotoarivelo, and Mr. Christoph Dwertmann in setting up the experiments over the NICTA wireless mesh testbed.

7. REFERENCES

[1] BARAS, J. S., TABATABAEE, V., PURKAYASTHA, P., AND SOMASUNDARAM, K. Component based performance modeling of the wireless routing protocols. In *Proceedings IEEE ICC Ad Hoc and Sensor Networking Symposium 2009* (June 2009), pp. 1–6.

[2] C. BRACQUEMOND, E. C., AND GAUDOIN, O. A comparative study of goodness-of-fit tests for the geometric distribution and application to discrete time reliability. `http://www-ljk.imag.fr/SMS/ftp/BraCreGau02.pdf`. [Online; accessed 11-May-2011].

[3] CHAKERES, I., AND BELDING-ROYER, E. The utility of hello messages for determining link connectivity. In *Wireless Personal Multimedia Communications, 2002. The 5th International Symposium on* (2002), vol. 2, pp. 504 – 508 vol.2.

[4] EGELAND, G., AND ENGELSTAD, P. A model for the loss of hello-messages in a wireless mesh network. In *Communications (ICC), 2010 IEEE International Conference on* (May 2010), pp. 1 –6.

[5] ELLIOTT, E. Estimates of error rates for codes on burst-noise channels. *Bell Systems Technical Journal 42* (September 1963), 1977–1997.

[6] FRITCHMAN, B. A binary channel characterization using partitioned markov chains. *Information Theory, IEEE Transactions on 13*, 2 (April 1967), 221–227.

[7] GILBERT, E. Capacity of a burst-noise channel. *Bell Systems Technical Journal 39* (September 1960), 1253–1265.

[8] GIROD, B., STUHLMÜLLER, K., LINK, M., AND HORN, U. Packet loss resilient internet video streaming. In *in Proceedings of SPIE Visual Communications and Image Processing '99* (January 1999), pp. 833–844.

[9] JACQUET, P., MUHLETHALER, P., CLAUSEN, T., LAOUITI, A., , QAYYUM, A., AND VIENNOT, L. Optimized link state routing protocol for ad hoc networks. In *Proceedings of the IEEE INMIC 2001* (August 2001), pp. 62–68.

[10] JIAO, C., SCHWIEBERT, L., AND XU, B. On modeling the packet error statistics in bursty channels. In *Local Computer Networks, 2002. Proceedings. LCN 2002. 27th Annual IEEE Conference on* (November 2002), pp. 534 – 541.

[11] KÖPKE, A., WILLIG, A., AND KARL, H. Chaotic maps as parsimonious bit error models of wireless channels. In *INFOCOM 2003. Twenty-Second Annual Joint Conference of the IEEE Computer and Communications. IEEE Societies* (March-3 April 2003), vol. 1, pp. 513 – 523 vol.1.

[12] MEDINA, A., AND BOHACEK, S. A performance model of neighbor discovery in proactive routing protocols. In *Proceedings of the 7th ACM workshop on Performance evaluation of wireless ad hoc, sensor, and ubiquitous networks* (New York, NY, USA, October 2010), PE-WASUN '10, ACM, pp. 66–70.

[13] PERKINS, C. E., AND ROYER, E. M. Ad hoc on-demand distance vector routing. In *Proceedings of the 2nd IEEE Workshop on Mobile Computing Systems and Applications* (February 1999), pp. 90–100.

[14] SARKER, J., HASSAN, M., AND HOSSAIN, A. Statistical modelling of correlated and independent errors in packet data transmission systems. In *Personal Wireless Communications, 1997 IEEE International Conference on* (December 1997), pp. 499 –504.

[15] ZHANG, L., CHOW, D., AND NG, C. H. Cell loss effect on qos for mpeg video transmission in atm networks. In *Communications, 1999. ICC '99. 1999 IEEE International Conference on* (June 1999), vol. 1, pp. 147 –151 vol.1.

[16] ZORZI, M., AND RAO, R. On the statistics of block errors in bursty channels. *Communications, IEEE Transactions on 45*, 6 (June 1997), 660 –667.

GDE: A Distributed Gradient-Based Algorithm for Distance Estimation in Large-Scale Networks

Qingzhi Liu
ES, EEMCS, TU Delft
the Netherlands
q.liu-1@student.tudelft.nl

Andrei Pruteanu
ES, EEMCS, TU Delft
the Netherlands
a.s.pruteanu@tudelft.nl

Stefan Dulman
ES, EEMCS, TU Delft
the Netherlands
s.o.dulman@tudelft.nl

ABSTRACT

Today, wireless networks are connecting most of the devices around us. The scale of these systems demands for novel techniques to maintain availability for various services such as routing, localization, context detection etc. Distance estimation is one of their most important building blocks. The majority of current algorithms, presumes knowledge about node position via systems such as GPS. While for some application scenarios this approach is feasible, for a lot of cases it suffers from frequent unavailability and high costs in terms of energy consumption. The main contribution of this paper is the introduction of a novel distributed algorithm called *GDE*, for the estimation of distances in large-scale wireless networks. *GDE* is a mechanism which estimates distances between nodes based solely on local interactions. The evaluation by means of simulations shows that *GDE* succeeds in estimating the distance between nodes in both static and mobile scenarios with considerably high accuracy, even under the influence of different kinds of environment parameters, such as node density, node speed, spatial node distribution, multicast percentage, etc.

Categories and Subject Descriptors

C.2.4 [**Distributed Systems**]: Distributed applications

General Terms

Algorithms, Design

Keywords

hop count, distance estimation, distributed algorithm

1. INTRODUCTION

Nowadays, large-scale wireless networks are connecting most of the devices around us. Along with ubiquity comes also the problem of tremendous scale. The current approaches that enable services such as routing, data dissemination, context detection, etc. require new, radical changes, in order to assure scalability. When location information is needed, most of the current algorithms presume that node position information via GPS is available. For some application scenarios, such an approach is feasible (e.g., outdoors). Unfortunately, for most of the cases, GPS is either not available (e.g., indoors) or it is inaccurate and most likely consumes a lot of power. For all these reasons, there is a need for ways to estimate distances between nodes only based on local interaction with no additional information such as the one provided by GPS. In this paper, we introduce *a gradient-based distance estimation algorithm*, called *GDE*, to estimate distances using very simple, local interactions between nodes. The algorithm is based on a hop-count gradient that is smoothed to decreases the estimation error [6]. Additionally, we introduce two error compensation methods in order to improve its accuracy. Our approach is suited also for mobile networks and deployments having non-uniform spatial distributions.

The paper is a theoretical study on the feasibility of such an approach for large-scale, dynamic systems. We validate our models via simulations with various environment parameter setups, such as node density, node speed, propagation model, etc. The testing results show that our algorithm has good performance in distance estimation accuracy. The rest of the paper is organized as follows. Section 4 presents the *GDE* algorithm for a static, uniform deployment. Section 5 presents strategies making the algorithm work in real deployments. An in-depth simulation-based analysis of *GDE* for various parameter setups is presented in Section 6. We conclude the paper and give insights about the future work in Section 7.

2. RELATED WORK

Distance estimation for wireless systems is one of the most important building blocks for various services such as routing [7], clustering [9], localization [3, 8], etc. For the case of geographic routing, nodes use locations as their addresses, and forward packets (when possible) in a greedy manner towards the destination. However, this approach demands for the nodes to be aware of their location information. While this is a feasible assumption in some settings (e.g., nodes have GPS modules attached), there are many cases (e.g., indoors) where node location information is either not available or it requires additional expensive equipments. For clustering algorithms [2] that require node position information, again, all nodes are assumed to be aware of their location information. To optimize for power, the GPS module has to be turned on for a very short time during bootstrapping. This approach does make sense when the nodes are either static or there are small changes of the network topology. When nodes are always moving for example, the localization system has to be turned on continuously, thus leading to a high power consumption.

Besides GPS, other methods suitable in particular for indoor deployments, make use of the wireless signal strength [1]. The main problem this class of algorithms is facing is the lack of accurate models for indoor radio propagation. Issues arise from various sources such as multi-path effects, reflection, and fading effects. Most of the time, there is limited correlation between the received signal strength indication and the real distance between devices.

The algorithm [6] serves as a starting point for our work and builds a self-organized, global coordinate system on top of a ad-hoc wireless sensor network. It relies only on distributed, simple computations and local communication. The distance estimation results can be used to calculate an estimated position, and the algorithm adapts to failures and additions of nodes. A remarkable aspect uncovered in the described scenarios (i.e., the presence of obstacles) is that greedy routing performs better when using virtual coordinates than when using the actual geographic coordinates.

3. EXISTING GRADIENT ALGORITHMS

In this section we briefly introduce the details of the related work, mainly the algorithm presented in [6]. Suppose all nodes are static and uniformly distributed in a deployment area, and the transmission range of each node is r. The node density is defined as the average number of nodes in the local transmission range of each node. Some nodes, called *seeds*, are placed at fixed positions, and take a constant counter value. Each node, except the seeds, listens to its neighbors and selects the largest counter value, and then rewrites the local counter value decreased by one. So the counter values of all nodes in the deployment area vary uniformly: the nodes closer to the seed nodes have larger counter values, and the nodes that are further from the seed nodes have smaller counter values. We refer to this variation trend the *gradient*. Figure 1(a) shows the resulting gradient when a straight line of seed nodes are vertically placed at the left side of the deployment area, and each node in the figure is colored according to its own counter value.

One can notice in Figure 1(a) that the obtained gradient is not smoothed but looks rather like stairs. This is because each node calculates the local counter value using only *the largest counter value* in its own transmission range, so based on this strategy, some adjacent nodes will finally take the same local counter values. We call the nodes with the same counter values a *wave* in the gradient. Because the width of each wave depends on the maximum number of hops from the seeds, we call this gradient the *hop-count gradient*.

3.1 Smoothed Gradient

The distance estimation produced by the hop-count gradient is a multiple of the width of the gradient wave ω. It produces an average error of approximately 0.5ω [6]. To improve the accuracy, the local neighborhood gradient values can be used to smooth the hop-count gradient. To simplify the equations, the transmission area of every node is approximated to a square with side length $2r$. Figure 1(b) shows the transmission area of node i. By a we denote the integer

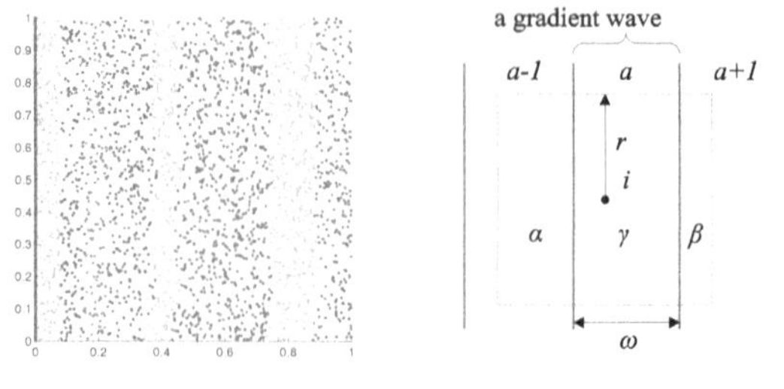

(a) (b)

Figure 1: (a). Hop-count gradient triggered by a vertical line of seeds (left side of the area). (b). Adjacent hop-count gradient waves around node i.

hop-count gradient value of node i. α, β and γ are the ratio of the three areas in various hop-count gradient waves to the size of the transmission area as seen by node i. Because the nodes are uniformly distributed in the area, α, β and γ are supposed to be equal to the rates between the number of nodes in the adjacent three areas to the total number of nodes in the transmission range respectively. Suppose the width of each wave ω equals the transmission range r of the node. The smoothed gradient value of node i is:

$$\begin{aligned} G_\alpha &= (a-1)r + (r - 2r\alpha) \overset{r=1}{=} a - 2\alpha \\ G_\beta &= ar - (r - 2r\beta) \overset{r=1}{=} a + 2\beta - 1 \\ G_{\alpha\beta} = G_{SMG} &= \frac{G_\alpha + G_\beta}{2} \overset{r=1}{=} a - \alpha + \beta - \frac{1}{2} \end{aligned} \quad (1)$$

Another approach to calculate the smoothed gradient is to average the gradient values of the neighbor nodes [6]. The average gradient value is G_a (in which n_l is the number of nodes in the transmission range of node i) and the smoothed gradient value is G_{SMG}:

$$G_{avg} = \frac{n_l\alpha(a-1) + n_l\gamma a + n_l\beta(a+1)}{n_l} = a - \alpha + \beta \quad (2)$$

$$G_{SMG} = G_{avg} - \frac{1}{2} = a - \alpha + \beta - \frac{1}{2} \quad (3)$$

4. GDE ALGORITHM

In this section we introduce the main contribution of this paper, the *GDE* algorithm. We introduce successively a series of improvements that make *GDE* a suitable algorithm to run in a distributed dynamic environment.

4.1 Gradient Width Reduction

Suppose node i has transmission range r and node k has the largest distance to node i among all the nodes in the transmission range of node i. The number of nodes in the area is not infinite, thus there is high chance that the node k is not exactly on the border of the transmission range of node i, thus the distance between node k and node i is smaller than the transmission range r. This implies that the width of the hop-count gradient wave becomes smaller than r, next referred to as *width reduction of gradient waves*.

Figure 2 shows the phenomenon of width reduction of gradient waves. Suppose node n is at the border p_{bu} of hop-count gradient wave $a-1$, and node m at the border p_b of hop-count gradient a can just discover node n. Because the node density is not infinite, there is a high chance that

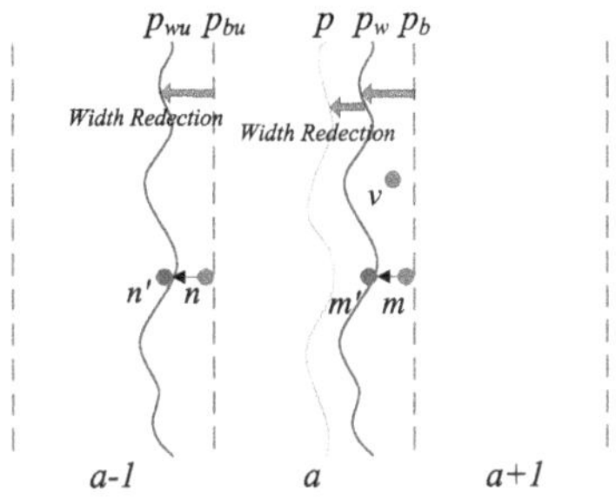

Figure 2: Width reduction of gradient waves.

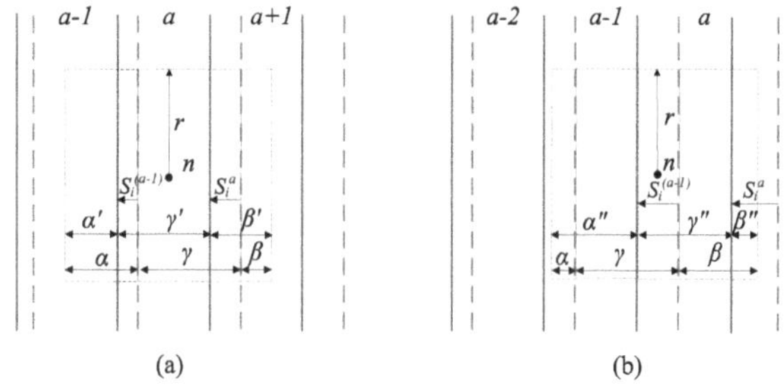

Figure 3: Average gradient for reduced width waves.

there is no node on the line of border p_{bu}, and the nodes that are closer to the border p_{bu} are on the line p_{wu}. So the real border of the hop-count gradient wave $a-1$ moves from p_{bu} to p_{wu}. Suppose node n is replaced at position n'. The last node that can discover the node n' has to move from position m to m'. The border of the hop-count gradient wave a changes from p_b to p_w, because of the movement of the previous border from p_{bu} to p_{wu}. The nodes positioned between p_b and p_w, such as node v, can no longer discover the nodes with hop-count gradient value $a-1$, so updates the hop-count gradient value to $a+1$.

At the same time, because of the nodes missing around border p_w, the nodes that are closer to the border p_w are on the line p. So the real border of the hop-count gradient wave a moves from p_b to p. As a result, the width of the hop-count gradient wave becomes smaller. At the same time, the movement of the border always influences the border positions of all the next hop-count gradient waves, so the movement distance of borders, such as p_b to p in hop-count gradient wave a, becomes larger and larger.

Suppose the estimated physical distance corresponding to a communication hop is d_{hop}. Equation 4 (taken from [4]) shows the relation between the real estimated distance and the node density n_l.

$$d_{hop} \overset{r=1}{\equiv} (1+e^{-n_l} - \int_{-1}^{1} e^{-\frac{n_l}{\pi}(\arccos t - t\sqrt{1-t^2})} dt) \quad (4)$$

Let $\Delta = 1 - d_{hop}$ be the width reduction of each hop-count gradient wave. As explained, the movement distance of borders has a cumulative characteristic and is related to the hop-count gradient value. The cumulative reduced width of the node with hop-count gradient value a is $S_i = (a-1)\Delta$ and the gradient width reduction value is $G_{si} = G_{SMG} - S_i$.

4.2 Averaging for Reduced Gradient Width

The width reduction of the gradient wave influences the result of the smoothed gradient. Each node in the one hop-count gradient wave could be placed at two positions. The first one is shown in Figure 3(a), where node n is not in the reduced width area of the gradient wave. The second place is shown in Figure 3(b), where node n is in the reduced width area of the gradient wave.

In the first case, as shown by Figure 3(a), the percentages of discovered nodes in α, β, γ become α', β', γ', as shown in Equation 5. In the second case, as shown by Figure 3(b), the percentages of discovered nodes in α, β, γ become α'', β'', γ'', as shown by Equation 6.

$$\begin{aligned}
\alpha' &= \frac{2r\alpha - S_i^{a-1}}{2r} \overset{r=1}{\equiv} \alpha - \frac{1}{2}(a-1)\Delta \\
\beta' &= \frac{2r\beta + S_i^{a}}{2r} \overset{r=1}{\equiv} \beta + \frac{1}{2}a\Delta \\
\gamma' &= \frac{2r\gamma + S_i^{a-1} - S_i^{a}}{2r} \overset{r=1}{\equiv} \gamma - \frac{1}{2}\Delta
\end{aligned} \quad (5)$$

$$\begin{aligned}
\alpha'' &= \frac{2r\alpha + (r - S_i^{a-1})}{2r} \overset{r=1}{\equiv} \alpha + \frac{1}{2} - \frac{1}{2}(a-1)\Delta \\
\beta'' &= \frac{2r\beta - (r - S_i^{a})}{2r} \overset{r=1}{\equiv} \beta - \frac{1}{2} + \frac{1}{2}a\Delta \\
\gamma'' &= \frac{2r\gamma - (r - S_i^{a-1}) + (r - S_i^{a})}{2r} \overset{r=1}{\equiv} \gamma - \frac{1}{2}\Delta
\end{aligned} \quad (6)$$

If the node is outside the reduced width area, the average gradient value is $G_{avg}^{out} = a - 2\alpha + \frac{1}{2} + (a\Delta - \frac{1}{2}\Delta)$ and the smoothed gradient value is $G_{si}^{out} = G_{avg}^{out} - \frac{1}{2} - (a-1)\Delta - \frac{1}{2}\Delta$. If the node is inside the reduced width area, the average gradient value is $G_{avg}^{in} = a - 2\alpha - \frac{1}{2} + a\Delta - \frac{1}{2}\Delta$ and the smoothed gradient value is
$G_{si}^{in} = G_{avg}^{in} - \frac{1}{2} - (a-1)\Delta - \frac{1}{2}\Delta + 1$.

Due to the fact that nodes do not have position information and cannot tell whether they are inside or outside the reduced width area, we compute the expected average value of the gradients G_{si}^{out} and G_{si}^{in}. $R_{in} = \frac{(r - rd_{hop})}{r} \overset{r=1}{\equiv} 1 - d_{hop}$ represents the percentage of the reduced width of one hop-count gradient wave normalized to r. $R_{out} = 1 - R_{in} \overset{r=1}{\equiv} d_{hop}$ is the percentage of the not-reduced width of one hop-count gradient wave, normalized to r. The gradient value is given by $G_{esi} = G_{si}^{out} R_{out} + G_{si}^{in} R_{in}$.

4.3 Overestimated Width Reduction

The gradient calculation in Section 4.1 uses the expected average value d_{hop}. But this expected average value will overestimate the width reduction of gradient waves. Suppose the gradient wave border moves as shown in the Figure 4(a). If the border moves from p_b to p, and node n is located in the area $(p + d_{hop})^+$, which means node n cannot discover the nodes on the border p, then the calculated hop distance d_{hop} is the expected width of hop-count gradient wave. But the real border that moves to p can not be a straight line, but a curved line like p_r. Node m' may move to the position of node m, to be inside the transmission range of the node n. As a result, node n can discover some of the nodes on the border p. This implies that the reduced width Δ is overestimated.

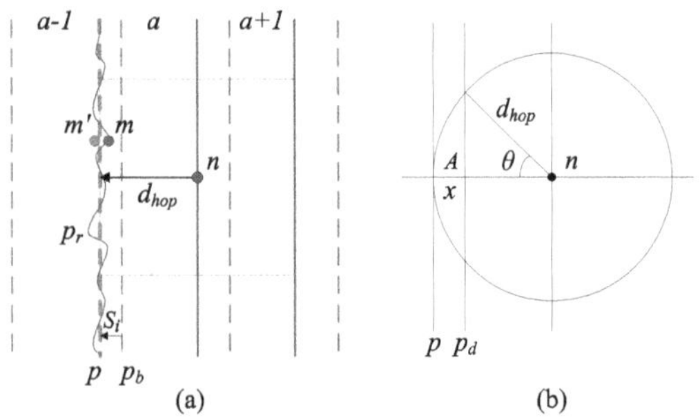

Figure 4: (a). Overestimation of gradient width reduction. (b). Expected reduced wave width.

Assume the shape of the gradient waves is parallel lines, the node density is ρ and the transmission range of each node is r. The probability that k nodes are located in the area of size h is given by $\Pr(k \in h) = \frac{(\rho h)^k}{k!} e^{-\rho h}$ [6].

For example, in Figure 4(b), node n has d_{hop} as the real communication hop distance, which is given in Equation 4. Let p be the gradient wave border using d_{hop}, p_d the real border of the gradient wave, X is the overestimated reduced width, and A is the overlap area of the overestimated reduced width area and the transmission range. The probability that there are no nodes in the area A equals the probability that the gradient border p will have a shift larger than x: $\Pr(0 \in A) = \Pr(X > x) = e^{-\rho A}$.

Therefore, the probability distribution function of the overestimated reduced width X is given by:
$F(x) = \Pr(X \leq x) = 1 - e^{-\rho A}$.

Area A is computed using Equation 7 and the probability density function of X is shown in Equation 8. The expected overestimated reduced width Δr of the one hop-count gradient wave can be written as Equation 9. a being the hop-count gradient value of the node, the gradient value for each node, the basic GDE value, can be calculated as:

$$A = 2[\pi r^2 \frac{\arccos\frac{r-x}{r}}{2\pi} - \frac{(r-x)\sqrt{r^2-(r-x)^2}}{2}] \quad (7)$$

$$f(x) = \mathrm{F}'(x) \overset{r\equiv 1}{=} \frac{\rho - 2x^2 + 4x - 1}{e^{(x-1)\sqrt{2x-x^2}+\rho\arccos(1-x)}\sqrt{2x-x^2}} \quad (8)$$

$$\Delta r = E(x) = \int_0^1 x f(x) dx = \int_0^1 x \frac{\rho - 2x^2 + 4x - 1}{e^{(x-1)\sqrt{2x-x^2}+\rho\arccos(1-x)}\sqrt{2x-x^2}} dx \quad (9)$$

$$G_{GDE} = G_{si}^{out} R_{out} + G_{si}^{in} R_{in} + (a-1) d_{hop} \Delta r \quad (10)$$

4.4 Error Reference Value

This section introduces an error reference value, which can be used as a benchmark for the testing results of the distance estimation error.

If node density ρ tends to $+\infty$, then Δ and Δr converge to 0, as shown by Equations 11 and 12, which means gradient width reduction problem disappears for large node densities.

$$\lim_{\rho\to+\infty} \Delta \overset{r\equiv 1}{=} 1 - \lim_{\rho\to+\infty} d_{hop} = 0 \quad (11)$$

$$\lim_{\rho\to+\infty} f(x) = 0 \Rightarrow \lim_{\rho\to+\infty} \Delta r = 0 \quad (12)$$

The number of nodes n in the area of size h follows a *Poisson* distribution with parameter λ (λ=ρh). If node density ρ goes to $+\infty$, then the probability distribution of n nodes in the area of size h tends to follow a *Normal* distribution with a mean λ and a deviation of λ (Equation 13). Therefore, if node density ρ goes to $+\infty$, the number of nodes in various hop-count gradient waves of a transmission range obeys a *Normal* distribution (Equation 14), in which n_α and n_β are the number of nodes in section α and β, and S_α and S_β are the area size of section α and β.

$$n \sim P(\lambda) = \frac{\lambda^n}{n!} e^{-\lambda} \overset{\lambda\to+\infty}{\Rightarrow} n \sim N(\lambda, \lambda) \quad (13)$$

$$n_\alpha \sim N(\rho S_\alpha, \rho S_\alpha),\ n_\beta \sim N(\rho S_\beta, \rho S_\beta) \quad (14)$$

The linear combination of *Normal* distributions also follows a *Normal* distribution. The computed gradient value using Equation 10 follows a *Normal* distribution (Equation 15), in which n_l is the average number of nodes in the transmission range of a node.

$$\lim_{\rho\to+\infty} G_{GDE} = \lim_{\rho\to+\infty} (a - \frac{n_\alpha}{n_l} + \frac{n_\beta}{n_l} - \frac{1}{2}) \sim N(a - \frac{\rho S_\alpha}{n_l} + \frac{\rho S_\beta}{n_l} - \frac{1}{2}, \frac{\rho(S_\alpha+S_\beta)}{n_l^2}) \quad (15)$$

If nodes are distributed on a grid and node density goes to infinity, then by using Equation 16 the gradient is computed as G_{GDE}^g. If nodes in the area are uniformly distributed and node density goes to infinity, then using Equation 17 the gradient is computed as G_{GDE}^u.

$$G_{GDE}^g \overset{\rho\to+\infty}{=} a - \frac{\rho S_\alpha}{n_l} + \frac{\rho S_\beta}{n_l} - \frac{1}{2} \quad (16)$$

$$G_{GDE}^u \overset{\rho\to+\infty}{=} a - \frac{n_\alpha}{n_l} + \frac{n_\beta}{n_l} - \frac{1}{2} \quad (17)$$

If all nodes are uniformly distributed, the error becomes ΔG, which is the absolute difference value between the gradient that is computed when nodes are distributed on a grid and the gradient computed when nodes are uniformly distributed (Equation 18), in which $S_l = \pi r^2$.

$$\Delta \mathrm{G} = G_{GDE}^u - G_{GDE}^g = (\frac{S_\alpha}{S_l} - \frac{n_\alpha}{n_l}) + (-\frac{S_\beta}{S_l} + \frac{n_\beta}{n_l}) \quad (18)$$

Equation 14 shows the probability distribution of the number of nodes n_α and n_β. While ΔG is the linear combination of the *Normal* distributions of n_α and n_β, the ΔG also follows a *Normal* distribution. The expected absolute error of distance estimation $E(|\Delta G|)$ is:

$$\Delta \mathrm{G} \sim N(\frac{S_\alpha}{S_l} - \frac{\rho S_\alpha}{n_l}, \frac{\rho S_\alpha}{n_l^2}) + N(-\frac{S_\beta}{S_l} + \frac{\rho S_\beta}{n_l}, \frac{\rho S_\beta}{n_l^2}) \overset{n_l = S_l \rho}{\sim} N(0, \frac{\rho S_\alpha + \rho S_\beta}{n_l^2}) \overset{S_\alpha + S_\beta = \frac{1}{2} S_l}{\sim} N(0, \frac{1}{2n_l}) \quad (19)$$

$$E(|\Delta G|) = \frac{1}{\sqrt{\pi n_l}} \quad (20)$$

It can be seen that as the node density tends to infinity, the error tends to zero (case not met in a reality). This result is the error of distance estimation in the ideal case, which can be used as a benchmark for the experimental results.

5. ALGORITHM PARAMETERS

So far we discussed about algorithms in more or less idealized conditions. However, for real deployments, wireless systems have to cope with many problems such as mobility, communication failures, non-uniform node distribution, influence on data dissemination speed by communication paradigms (unicast vs. multicast), etc. In this section we analyze all these aspects and see how the algorithm adapts to such conditions.

5.1 Node Mobility

In the previous sections we described an algorithm for the case when the nodes are static. In this section, we introduce a compensation strategy for the case in which the nodes are mobile (i.e., we consider the Random Walk mobility model assuming a discrete time model).

In Figure 5(a), p_b is the border of a hop-count gradient wave. Node m can travel from position p to p' and p'' in one time step. The last node that can discover the border p_b changes from node n to node n', which enlarges the width of the hop-count gradient wave in the gradient. A strategy to compensate the enlarged width in the distance estimation value is needed.

We use the expected moving distance of one node in a given time step E_{dc} to compensate the enlarged width value, which equals the movement distance from p_b to p''. The translation of the hop-count gradient into x and y coordinates is shown by Figure 5(b), in which y coordinate is parallel to the line of the seed nodes.

When nodes move with various speeds, the average expected moving distance for a node in one time step is D which makes an angle θ with the y axis (V being the projection on it). Due to the fact that all the possible movement directions are symmetrical with respect to the x coordinate, only the movement for $y > 0$ is considered (θ ranging from $-\pi/2$ to $\pi/2$). The probability density function of V is given by Equation 21. Since we need the absolute expected moving distance of one node, the expected value is integrated from 0 to D as shown by Equation 22.

$$\left.\begin{array}{c}\theta \sim U(-\frac{\pi}{2}, \frac{\pi}{2}) \\ V = D\sin\theta\end{array}\right\} \Rightarrow f(v) \overset{-D\leq v<D}{=} \frac{1}{\pi\sqrt{D^2 - v^2}} \tag{21}$$

$$E_{dc} = E(v) = \int_0^D v f(v) dv = \int_0^D \frac{v}{\pi\sqrt{D^2 - v^2}} dv \tag{22}$$

The enlarged width of each gradient wave is accumulating from one hop-count gradient to the next. For a node with hop-count gradient value a and transmission range r, the mobility compensation value C_d and the gradient value G_i is given by Equations 23 and 24.

$$C_d = [(a-1)d_{hop}E_{dc}]/r \tag{23}$$

$$G_{GDE}^{comp\,mob} = G_{GDE} + C_d \tag{24}$$

5.2 Multicast Communication

GDE algorithm does not require each node to communicate with all its neighbors. Multicast can be used for each node to decrease the number of messages in the network. Suppose the nodes are uniformly distributed in the area. If each node only selects a fraction of its neighbors as the communication targets, then the selected nodes can also construct a network. The only difference to the original network is that the new network has lower node density. According to the formulas in Section 4, *GDE* can make distance estimation in the network with different node densities. Therefore, each node can use multicast to communicate with neighbors. The equations of *GDE* algorithm introduced in Section 4 can still be used, the only difference being that the node density n_l needs to be adjusted with the percentage of neighborhood size considered.

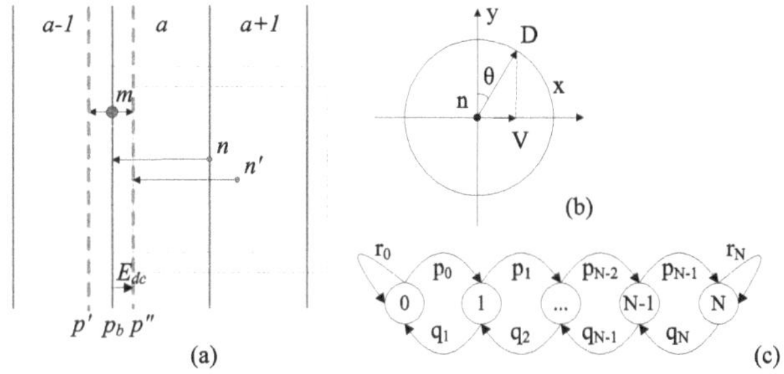

Figure 5: (a). Influence of node mobility to the border of the hop-count gradient wave. (b). Expected moving distance. (c). Moving distances of each time step of a round in the Markov Chain.

5.3 Round Length

GDE assumes a discrete time model, with un-synchronized communication rounds of size P. Each node acts once in each round, in a different time slot. The rounds are not synchronized and are being used to mimic the asynchronous nature of communication. Suppose all nodes follow a Random Walk mobility model, the transmission range is r and the expected moving distance at one time step is E_{dc}. We model the motion segments with a Markov Chain in order to calculate the expected moving distance (see Figure 5(c)). In the model we only consider nodes moving in one dimension (i.e., towards the next hop-count gradient). Suppose that state j, $j = 0, ..., N$, is the state that describes a node moving forward for a distance $j \cdot E_{dc}$. p_j represents the probability that the node moves further from the original position, while q_j the probability of the node to move closer to the original position. Let N be the maximum number of moving steps in one round of length P. Let r_0 be the probability that a node does not move out of the initial state and is equal to $\frac{1}{2}$, and let p_0 be the probability that a node moves out of the initial state and is equal to $\frac{1}{2}$. In the inner processes, the backward and forward motion probabilities of a node in one time step are the same, so the node moves to the next state or to the previous state with $p_j = q_j = \frac{1}{2}$. For the final state N, due to the fact that there are nodes that can move to the previous state or start a new calculation round, let r_N be the probability that a node starts a new round (equal to $\frac{1}{2}$), and q_N be the probability that a node moves backward (equal to $\frac{1}{2}$). Due to the fact that there is a large number of nodes, the probability of each steady state can be used as the probability that a node stays in that state of P steps. Considering the mathematical properties of the Random Walk [5] , the model that describes the steady states π_j ($0 \leq j \leq N$) is given by Equation 25. One can notice that all states converge to the same probabilities.

Algorithm 1: Calculate Gradient by Propagation

select $\{SN\}$
calculate D_{hop}, E_{sc}, E^{p}_{dc}
$E_c = (1 - D_{hop}) + E_{sc} + E^{p}_{dc}$
for each element $\{SN\}$
  if $\text{HCG}_i = (\text{HCG} - 1)$
    sum $(PE_{c\_i} + E_{c\_i})$
  end if
end for
PE_c =**average**$(PE_{c\_i} + E_{c\_i})$
$G^{adp}_{GDE} = G_{SMG} - PE_c$

$$\begin{aligned}
&\pi_0 = r_0\pi_0 + q_1\pi_1 \overset{r_0=q_1=\frac{1}{2}}{\Rightarrow} \pi_0 = \pi_1 \\
&\pi_N = p_{N-1}\pi_{N-1} + r_N\pi_N \overset{p_{N-1}=r_N=\frac{1}{2}}{\Rightarrow} \pi_N = \pi_{N-1} \\
&\pi_j = \frac{\prod_{m=0}^{j-1}\frac{p_m}{q_{m+1}}}{1+\sum_{j=1}^{N}\prod_{m=0}^{j-1}\frac{p_m}{q_{m+1}}} \overset{p=q=\frac{1}{2}}{=} \frac{1}{1+N} \quad (1 \leq j < N)
\end{aligned} \tag{25}$$

The node in state 0 does not affect the gradient of nodes in the next hop-count gradient. For this reason, we ignore state 0. For each state in the Markov Chain, it represents the expected moving distance from the original place. For state j, the moving distance is $j \cdot E_{dc}$. Because every steady state has the same probability as proved by Equation 25, the expected value for the moving distance of all states can be calculated by Equation 26. It can be found that if the length of the round becomes 1, Equation 26 is the same as Equation 22. The compensation value for the round length is given in Equation 27, and the computation for the gradient value is given in Equation 28.

$$E^{p}_{dc} = \frac{\sum_{j=1}^{P} jE_{dc}}{P} = \frac{(1+P)E_{dc}}{2} \tag{26}$$

$$C^{p}_{d} = [(a-1)d_{hop}E^{p}_{dc}]/r \tag{27}$$

$$G^{comp\,round}_{GDE} = G_{GDE} + C^{p}_{d} \tag{28}$$

5.4 Integrated Parameters Influence

For real deployments, the node distribution is not uniform. Furthermore, node speed, the neighborhood interaction and the communication periods may vary. Therefore, we introduce an algorithm that can calculate the gradient value for the nodes with various deployment conditions as shown in Algorithm 1. Each node communicates with parts of the neighbors to calculates its own gradient values as in Section 4, and propagates average values of the gradient reduced width value Δ, the static compensation value E_{sc} and the dynamic compensation value E_{dc}, from itself and the nodes in the previous hop-count gradient wave, to the nodes belonging to the next hop-count gradient wave. Each node uses the received width reduction values and compensation values to calculate its own gradient value.

For Algorithm 1, *HCG* is the hop-count gradient value, $\{SN\}$ is the subset of the neighbors to communicate, D_{hop} is the real one hop distance, E_{sc} is the compensation for the static case which equals Δr as shown in Equation 9, E^{p}_{dc} is the compensation for the mobile case given by Equation 26. E_c is the sum of reduced width $(1 - D_{hop})$, E_{sc}, and E_{dc}. PE_c is the sum of E_c for the nodes in all the previous hop-count gradients. PE_{c_i}, E_{c_i} and HCG_i are the PE_c, E_c and HCG values of node i in the neighbor subset $\{SN\}$. G_{SMG} is the gradient value computed using the simple smooth algorithm and G^{adp}_{GDE} is the final computed gradient value.

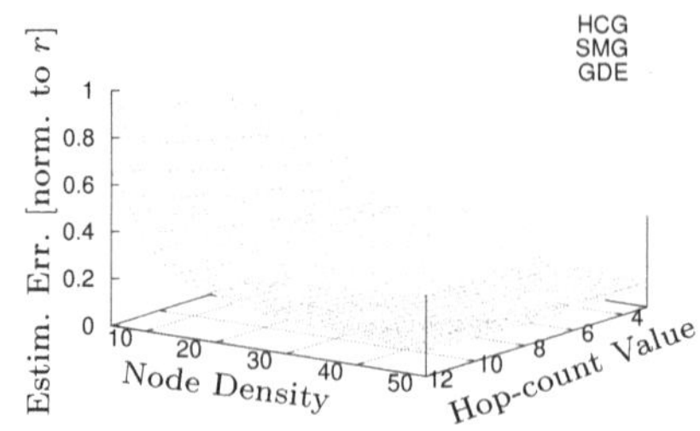

Figure 6: Absolute error for various node densities and hop-count gradient values.

6. GDE ALGORITHM ANALYSIS

To validate our algorithms, we run simulations in Matlab. The nodes are placed randomly in a square area of 1000×1000 meters. The transmission range r of each node is 80 meters. The seed nodes are placed on the left border of the square. The resulting hop-count gradient looks like parallel waves. For the dynamic case, each node moves a step in every second. We run experiment 20 times for each testing point. Each node makes a distance estimation from itself to the line of seed nodes. The error for each node is computed as $\varphi = \frac{|\xi - \theta|}{r}$, in which ξ is the estimated distance to the line of seed nodes using different algorithms, θ is the real distance towards the line of seed nodes, r is the transmission range of each node and φ is the error normalized to the transmission range r. The Hop-Count Gradient (*HCG*) and Smoothed Gradient (*SMG*) algorithms calculate the gradient values G_{HCG} and G_{SMG}, and the estimated distance is calculated by $\xi_{HCG} = G_{HCG} \cdot d_{hop}$ and $\xi_{SMG} = G_{SMG} \cdot d_{hop}$ respectively. The Gradient-based Distance Estimation (*GDE*) algorithm uses algorithms provided in the previous sections to compute the gradient value G_{GDE}, and the estimated distance is calculated by $\xi_{GDE} = G_{GDE} \cdot r$.

6.1 Node Density

With the first set of experiments we want to find what are the effects of node density and the network diameter on the accuracy of the algorithm. All nodes are static and uniformly distributed. Each node broadcasts locally to their neighbors.

The qualitative analysis about the average absolute error of the distance estimation of *HCG*, *SMG* and *GDE* algorithms is given out in Figure 6 for various node densities and network diameters. On the x axis we see the node density and, on the y axis the network diameter. The z coordinate is the average absolute error. The results show that the *HCG* has almost no decrease of the error after node density 15. The distance estimation error using *SMG* and *GDE* algorithms decreases when node density n_l increases. The error of *GDE* is always smaller than the one of *SMG*. When node density n_l is constant, and the network diameter increases, the errors of both *SMG* and *GDE* show a small increase. This is mainly caused by the cumulative effect of the distance estimation error of the nodes in the previous

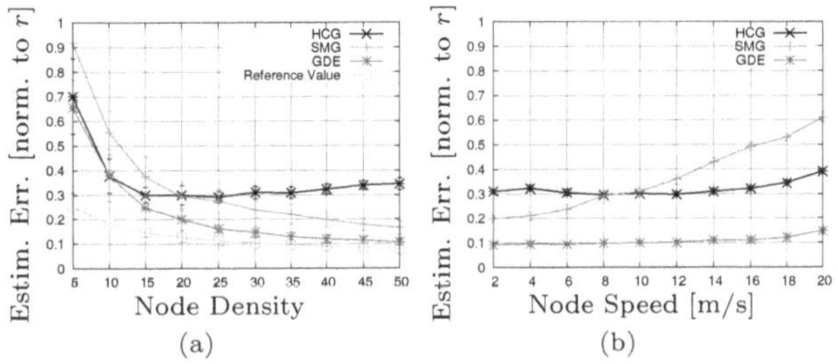

Figure 7: (a). Absolute error for various node densities. (b). Absolute error for various node speeds.

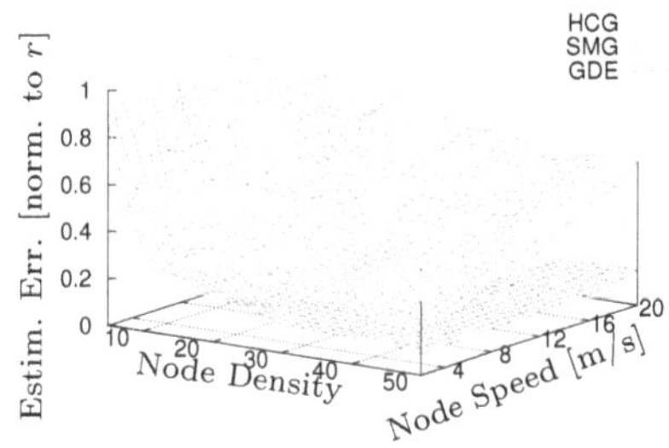

Figure 8: Absolute error for various node densities and node speeds.

hop-count waves. Still, the error of *GDE* is much smaller than the error of *HCG*. Finally, the increasing error rate of *GDE* is smaller than the one of *SMG*.

The quantitative analysis about the average absolute error of the three algorithms is given out in Figure 7(a) for various node densities. The network diameter is 12. The *GDE* method reduces the error of *HCG* and *SMG* significantly for densities higher than 15. The standard deviations of the errors for both *SMG* and *GDE* decrease when the average node density increases. The standard deviation of the estimation error for *GDE* is always smaller than the one of *SMG* for node density higher than 15. For an average node density 30 the average estimation error for *HCG*, *SMG* and *GDE* are 0.31, 0.24 and 0.15. While the standard deviation is 0.025, 0.036 and 0.020 respectively. Figure 7(a) also shows out the error reference value according to Equation 20. As the node density increases, the error of *GDE* algorithm tends closer to the error reference value.

6.2 Node Speed

The second set of experiments is intended to quantify the effects various node speed levels and node densities has on the accuracy of *GDE*. Nodes are uniformly distributed, and use broadcast to communicate with their neighbors. They use the Random Walk model with network diameter 12.

The qualitative analysis about the average absolute error of the distance estimation of *HCG*, *SMG* and *GDE* algorithms is given out in Figure 8 for various node speeds and densities. The x axis shows the node density, the y axis the node speed, and the z axis the average absolute error. On the y axis, the speed is increased in step of 2 m/s. The estimation error of *GDE* decreases as node density increases. On the other hand, it increases slightly when the average node speed increases. Also, when the node speed increases, the error of *HCG* and *SMG* increases. For node speeds higher than 8 m/s, the error of *SMG* increases greatly and is even larger than the one of *HCG*. This is mainly caused by the fact that *SMG* is designed for the static environments and it does not consider the influence of node mobility.

The quantitative analysis about the average absolute error of the three algorithms is given out in Figure 7(b) for various node speeds. Node density is set to 50. The network diameter is 12. As the speed increases, the error of *GDE* increases slightly, but *SMG* increases significantly. When the node speed is 20 m/s, the average error of distance estimation by *GDE* is 0.148, while the error levels for *HCG* and *SMG* are 0.391, 0.610 respectively. The standard deviation for *SMG* is 0.019. The standard deviation for *HCG* and *GDE* are 0.011 and 0.010 respectively. The test results show that the *GDE* algorithm has a good performance in a dynamic environment. *HCG* and *SMG* are not suitable in a dynamic environment especially at high node speed.

6.3 Spatial Node Distribution

We check next the influence of different spatial node distribution to the accuracy of the algorithms. Nodes follow again a Random Walk mobility pattern with speed of 2 m/s. The seed nodes are placed on the left border of the square. This time the placement of the nodes is not uniformly distributed. Instead, the y coordinates of nodes are uniformly distributed from 0 to 1000 meters, while the x coordinates of nodes are distributed following a normal distribution $|N(0, 0.5) \cdot 1000|$, with maximum value 1000 and minimum value 0. So the field that is closer to the seed nodes has higher node density, and the field that is further to the seed nodes has lower node density. The *GDE* gradient is calculated using the algorithm shown in Section 5.4.

In Figure 9(a), the x axis represents the node density and the y axis the average distance estimation error. As shown by the graph, the errors of *HCG* and *SMG* increase significantly. For *GDE*, the average error of distance estimation is quite low, which is around 0.150 with node density 20 and 0.089 with node density 80. Thus, *GDE* is also suited for deployments with non-uniform node distributions.

6.4 Multicast Percentage

For wireless systems where nodes follow duty-cycled sleep schedules, nodes that transmit can only be heard by a subset of their neighbors. The percentage of nodes that take part determines the speed of information diffusion. For this experiment, nodes are uniformly distributed with an average density of 50. The network diameter is 12. The nodes use a Random Walk model with speed of 2 m/s.

In Figure 9(b), the x coordinate shows the multicast percentage, while the y coordinate shows the error level. Each node randomly selects a subset of its neighbors at each time step. The multicast percentage varies from 0.1 to 1.0. The error of both *SMG* and *GDE* algorithms decreases significantly. *GDE* always computes a better estimate than *SMG*. This is given in both the average error and the standard deviation. The estimation error of *HCG* decreases from the multicast percentage 0.1 to 0.5 and shows almost no improvement afterward. For an average multicast percentage of 0.6, the error and standard deviation of *GDE* are 0.122 and 0.009 respectively, and the values are almost the same as the results of Section 6.1 with node density 30. This experiment shows that even if each node cannot communicate with all its neighbors, *GDE* still shows good performance.

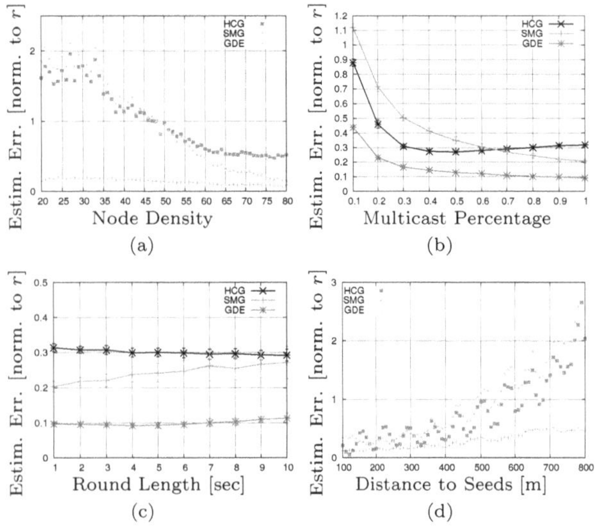

Figure 9: (a). Abs. error for normal distribution of nodes. (b). Abs. error for various multicast percentages. (c). Abs. error for various round lengths. (d). Abs. error for integrated parameters.

6.5 Round Length

As previously described, *GDE* uses a discrete time model, with un-synchronized communication rounds of size P. In this section, we test the influence of the length of the communication round to the accuracy of distance estimation. For this experiment, nodes are uniformly distributed, with an average node density of 50. They use broadcasting to communicate with their neighbors, and the network diameter is 12. In Figure 9(c), on the x axis is shown the round length in seconds, while on the y axis the estimation error. As the round length increases, the error and standard deviation of *SMG* increases significantly. On the other hand, the error and standard deviation of *GDE* shows a much smaller increase from 0.097 and 0.006 to 0.113 and 0.014. The test results show that *GDE* algorithm offers good distance estimation without requiring very frequent communication with neighbor nodes.

6.6 Adaptive Behavior

Previous experiments considered some simplified assumptions. In real deployments though, nodes are not uniformly distributed, node speed varies, local interaction is multicast communication, etc. In this section, we test all these aspects together. We consider nodes normally distributed, as in Section 6.3. The mobility model is Random Walk with node speed following a normal distribution with mean value of 2 m/s. The multicast percentage depends on the local node density. Nodes interact with a minimum 15 nodes to a maximum 30 nodes. The random variable of the number of neighbors for each node follows a Binomial distribution. For each neighbor, the probability of making successful communication is $p_s = (0.2 \cdot n_l + 15)/n_l$ (n_l is the node density). The communication round length P changes according to the node density as $P = |(p_s \cdot n_l)/15 - 1| \cdot 9 + 1$.

Figure 9(d) shows the experiment results. The x axis shows the distance to the seed nodes while y axis shows the error normalized to the transmission range r. It can be seen that, as the distance to the seed nodes increases, the distance estimation error also increases. When the distance is larger than 300 meters, the estimation error of *SMG* is even larger than that of *HCG*. The estimation error of the two algorithms increases significantly from 300 to 800 meters. As expected, *GDE* generally shows a small estimation error, with only a small increase from 0.120 to 0.456. So *GDE* can make acceptable distance estimation in the environment with integrated parameters.

7. CONCLUSIONS

Distance estimation in wireless networks is of tremendous importance for a lot of applications. The majority of the algorithms presume knowledge about node position via systems such as GPS. While this approach is feasible for some application scenarios, for a lot of cases it suffers from frequent unavailability and high costs in terms of energy consumption. For this reason, in this paper we introduce a novel distributed algorithm called *GDE* for the estimation of distances in large-scale wireless networks. It is a mechanism which estimates distances between nodes using solely local interactions. The models and the theoretical evaluations by means of simulation show that *GDE* succeeds in evaluating the distances between nodes in both static and mobile scenarios with considerably high accuracy, even under the influence of some environment parameters. As future work, we plan to extend the analysis to more realistic deployment conditions such as real human mobility models and realistic propagation models.

8. REFERENCES

[1] A. Awad, T. Frunzke, and F. Dressler. Adaptive distance estimation and localization in wsn using rssi measures. In *Proceedings DSD 2007*, pages 471 –478.

[2] G. Gupta and M. Younis. Load-balanced clustering of wireless sensor networks. In *Proceedings IEEE ICC 2003*, volume 3, pages 1848–1852. IEEE.

[3] T. He and all. Energy-efficient surveillance system using wireless sensor networks. In *Proc. of the 2nd International Conf. on Mobile Systems*. ACM, 2004.

[4] L. Kleinrock and J. Silvester. Optimum transmission radii for packet radio networks or why six is a magic number. In *Proc. of the IEEE National Telecom. Conference*, volume 4, pages 1–4, 1978.

[5] P. V. Mieghem. *Performance Analysis of Communications Networks and Systems*. Cambridge University Press, New York, NY, USA, 2005.

[6] R. Nagpal, H. Shrobe, and J. Bachrach. Organizing a global coordinate system from local information on an ad hoc sensor network. In *Information Processing in Sensor Networks*, pages 553–553. Springer, 2003.

[7] A. Rao and all. Geographic routing without location information. In *Proceedings Mobicom 2003*, pages 96–108. ACM.

[8] M. Rudafshani and S. Datta. Localization in wireless sensor networks. In *Proceedings IPSN 2007*, pages 51–60. ACM.

[9] O. Younis and S. Fahmy. Heed: a hybrid, energy-efficient, distributed clustering approach for ad hoc sensor networks. *IEEE Transactions on Mobile Computing*, pages 366–379, 2004.

Modeling and Simulation of Service Composition in Opportunistic Networks

Umair Sadiq, Mohan Kumar
University of Texas at Arlington
Arlington, TX, USA
umair.sadiq@mavs.uta.edu,
mkumar@uta.edu

Andrea Passarella, Marco Conti
IIT-CNR, Pisa, Italy
{a.passarella,m.conti}@iit.cnr.it

ABSTRACT

Pervasive networks formed by users' mobile devices have the potential to exploit a rich set of distributed service components that can be composed to provide each user with a multitude of application level services. However, mobile and pervasive networks suffer from intermittent connectivity, disconnections and partitions, such that opportunistic networking techniques are required to enable communication. This poses novel challenges to service composition techniques. While several works have discussed middleware and architecture for service composition in well-connected wired networks and in stable MANET environments, the underlying mechanism for selecting and forwarding service requests in the significantly challenging networking environment of opportunistic networks has not been addressed. The problem comprises three stages: i) selecting an appropriate service sequence set out of available services; ii) forwarding service inputs to the device hosting the next service in the composition; and iii) routing final service outcomes back to the requester. The proposed algorithm derives efficiency and effectiveness by taking into account the service load and location of devices providing the services, as well as intermittent connectivity, to select a particular service set. Through extensive simulations on real and synthetic traces, we show that by using only local knowledge collected in a distributed manner, performance close to a real-time centralized system can be achieved.

Categories and Subject Descriptors

C.2.1 [**Computer Communication Networks**]: Network Architecture and Design—*store and forward networks, wireless communication*

General Terms

Algorithms, Design, Performance

Keywords

Distance/Location Modeling, Opportunistic Networks, Service Provisioning

1. INTRODUCTION

In recent years, the number of multi-functional, personal smart devices has been increasing at a high rate. The possibility of such devices coming within communication range of each other is enhanced by the presence of multiple embedded radios. While, such opportunistic contacts between pairs of such devices have been exploited by the opportunistic networking paradigm [16, 19, 25], exploiting resources to execute remote services is an area of research that has received attention in recent times [17, 18]. The creation of a services-rich environment by effectively making services available on each device, accessible to applications on other devices in opportunistic environments is a major challenge. This paper investigates efficient service composition in opportunistic networks where services are available on devices in physical proximity of a user. Such environments include parks, malls, streets in a city or other social gathering places where citizens interact. In particular, an efficient mechanism for selecting services to complete a request and forward service parameters to corresponding devices in an opportunistic network is proposed.

Heterogeneous resources available on devices can be abstracted as services to simplify the interface and have platform independence. Services from multiple devices can be composed to provide enhanced functionality [13]. For example, to complete a task like '*take a picture* and share it on a *map* in my *current location*', services offered by a camera, GPS receiver and map provider need to be composed [2]. Service composition will be applicable in several areas - pervasive healthcare, intelligent transportation systems, crisis management, etc., [5].

When two devices are within communication range, there exists an opportunity to exploit each other's services. If the contact period is sufficiently long, nodes can utilize each other's resources to execute remote services. On the other hand, if the contact time is insufficient, service parameters can be exchanged during the contact, service can be executed offline and service outputs transmitted to the requester using opportunistic contacts. In Passarella et al. [18], multiple copies of a request are initiated to minimize delay in service execution and to increase robustness but only one-hop communication (i.e. a requester waits until a direct contact with service provider) is used. Also, a few other works propose efficient schemes and use multi-hop paths for service discovery and invocation in opportunistic networks by using a set of proxies [15] or location of the user [24]. However, composition of multiple services is not considered in these works. For composition of services, a number of middleware frameworks and architectures are proposed but these do not discuss the actual forwarding mechanism [2,7,13]. Routing is considered in [8,28] by modeling the expected colocation time between nodes based on past interactions. However, service composition fails if participants do not remain directly connected. In contrast in this paper, service composition is performed using multi-hop paths that can relay information to the service providers and back to the service requestor even when an end-to-end connected path does not exist. Furthermore, alternative paths are considered to increase the success of composition.

The problem of routing requests for service composition is different from traditional routing in two key aspects: i) service load at destination needs to be taken into account; and ii) future forwarding path needs to be considered for the remaining services that need to be composed in a sequence. Our algorithm selects a service sequence set by taking into account the service load and temporal distances between nodes. These values are measured in a distributed manner by using only opportunistic contacts. Temporal distance provides a measure of relative location of other nodes. The elegance of our solution lies in the fact that while a particular service sequence is selected by considering the temporal distance between devices, the actual routing scheme used to forward service request and parameters can be different. For example, to meet a service request of *encrypting and compressing* a file, our algorithm chooses a sequence of services (e.g., an encryption service and then a compression service) that can be composed such that these services are provided at devices in proximity of the service requester. Then, an underlying routing scheme is used to forward service request to the device providing encryption, encrypted result to the device providing compression and then final result back to the service requester.

Extensive simulation results are provided to demonstrate the performance of the proposed method in terms of service composition success, composition length, number of hops, and delays. The performance is analyzed for a range of service density, number of nodes, service request rates, request timeout durations, and routing mechanisms. It is shown that: i) composition is better than searching for the exact single service match; ii) use of multi-hop paths improves performance (service completion rate and delay) significantly as compared to one-hop direct forwarding; and iii) using only local knowledge (collected from opportunistic contacts) about load at and temporal distance from other nodes, performance close to a centralized system (exact load and temporal distances between nodes are known) can be achieved.

In rest of the paper, the terms devices and nodes are used interchangeably. The paper is organized as follows: Section 2 provides the system description. Temporal distances and service loads are modeled in Section 3. Service composition mechanism and algorithm is described in Section 4 followed by simulation results and analysis in Section 5. Section 6 discusses related work and other opportunistic forwarding schemes in literature and Section 7 concludes with directions for future work.

2. SYSTEM DESCRIPTION

2.1 Service model

We employ a service graph that represents which services can be combined in a sequence based on their input requirements. This can be done by using service ontologies and semantics in a hierarchical manner [13]. For simplicity, we represent a service based on its input and output type to generate a service graph i.e. two services can be composed if output type of one is same as required input to the other. For n_d input/output types, possible services are of type s_{xy} such that input type $x \in [1, n_d - 1]$ and output type $y \in [x+1, n_d]$. This representation illustrates services with different levels of functionality. A service s_{xy} is of functionality k if $y = x + k$. Suppose, s_{12} and s_{23} represent services *decompress* and *decrypt* respectively, then service s_{13} represents a service of higher functionality that can perform both decryption and decompression. A service with high functionality like s_{14} provides same transformation that is otherwise possible by composition of two or more services with lower functionalities (e.g., $\{s_{12}, s_{23}, s_{34}\}, \{s_{12}, s_{24}\}$ or $\{s_{13}, s_{34}\}$). Services with varying functionalities are used to demonstrate the tradeoff between selecting an exact service (that can be further away in the network) versus composition of smaller services (that might be available in proximity of service requester). To account for task and device heterogeneity, execution time of each service is distributed exponentially on the devices. The performance of service composition is explored by varying the repetition of each service i.e. the number of nodes providing that service.

2.2 Mobility model

It has been shown that real mobility traces show a power law distribution of Inter-Contact Times (ICTs) [3] in contrast to previously assumed exponential distribution. Since random waypoint and random direction models lead to exponential Inter-Contact Times [23], they are not suitable to represent nodes' mobility. In addition, it has been shown that real mobility traces have power-law flight-lengths and pause-time distributions [20]. Therefore, in our analysis and simulation, we generate synthetic traces using levy walk mobility model [20]. Levy walk represents a more realistic scenario in comparison with [18] where each node is equally likely to meet any other node.

Mobility plays a central role in the selection and forwarding of service requests in an opportunistic network. Network environment is made heterogeneous by varying the mobility characteristics of nodes. Based on levy walk, though each node always follows the same step length, pause time distribution and speed for itself, these parameters are varied across the network for other nodes. This is reflected in Table 2 that shows the values for slow and fast moving nodes. In addition to synthetic traces, results on real mobility trace from a State Fair [21] are also analyzed. The corresponding results support applicability of our proposed algorithm in homogenous (nodes with similar mobility characteristics) as well as heterogeneous (nodes with varied mobility characteristics) environments. The focus of this paper is to provide services that are in physical proximity of a user. Therefore, results obtained in schemes [6,12] for social networks, where delays are to the order of hours and days, are not valid for our application scenario.

2.3 Forwarding Schemes

There are several forwarding schemes proposed in the literature on opportunistic networks. These are discussed in Section 6. The following forwarding schemes have been found to work best in dynamic environments: i) use of timer transitivity (TT) [25] or temporal distance between nodes in homogenous (nodes with similar mobility characteristics) environments; and ii) use of encounter rate (EBR) [16, 26] in heterogeneous environments (nodes with varied mobility characteristics). In addition, we have proposed a modified version of timers (MT) [22] that works efficiently over small forwarding paths. In our simulation studies, we compare performance of the proposed selection and forwarding mechanism for service composition under 4 forwarding schemes - direct or one-hop, TT, EBR, and MT.

3. RESOURCE MODELING

3.1 Disconnected paths

In opportunistic networks, nodes providing a service may not be directly connected to the service requested. This leads our investigation into modeling the physical distance between nodes when these are devices carried by people. In [14], Kim et al. show that the magnitude of typical displacement of a mobile node is given by $r = t^{\gamma/2}$ where r^2 (denoted $M(t)$ in [14]) is the mean square displacement of a node at time t from its position at time 0 and $\gamma \in [0.5, 2]$. Then, the distance d between two nodes when they start from same position at time 0 is given by $d = 2r\sin(\theta/2)$ where $\theta \in [0, 2\pi]$ is the angle between their displacements and is uniformly distributed in the range. Therefore, expected displacement is given by

$$E[d] = 2r\frac{1}{2\pi}\int_0^{2\pi} \sin(\theta/2)d\theta = \frac{4r}{\pi} = 4t^{\gamma/2}/\pi. \quad (1)$$

Interesting result is that the physical distance between nodes is related to the separation in time i.e. time since they moved out of each other's transmission range. For this reason, we use similar timers as proposed in [25] and study properties of the shortest temporal distances between nodes. Several works have used the intuition behind this idea to approximate the physical distance between nodes and characterize the temporal distance [1, 9, 10, 22, 25, 27]. A shortest temporal distance between two disconnected nodes is the minimum time some sequence of contacts can relay information from one node to the other [27].

3.2 Shortest temporal distance

In essence, shortest temporal distance (t_{ij}) provides the minimum time (using epidemic forwarding) it would take for some information to travel from node i to node j. Therefore, to compare efficiency of an underlying forwarding scheme to send requests to different nodes in network, we simply compare this lower bound on the propagation time. For example, node i can select a service provider node j or k based on which of these has smaller distance (t_{ij} or t_{ik}). Since, no centralized infrastructure exists in an opportunistic network to provide t_{ij} and t_{ik}, the approximations t_{ji} and t_{ki} are used as they can be measured at node i in a distributed manner by using simple timers [Section 4.1]. We analyze the shortest temporal distances maintained at nodes in a real mobility trace from State Fair. In this paper, the term temporal distance always refers to the shortest temporal distance. It is found that even though actual nodes in the shortest path may be different, the temporal distance t_{ij} is approximately same as t_{ji}. This is demonstrated in Figure 1. We measure the absolute difference $|t_{ij} - t_{ji}|$ for all node pairs i,j after

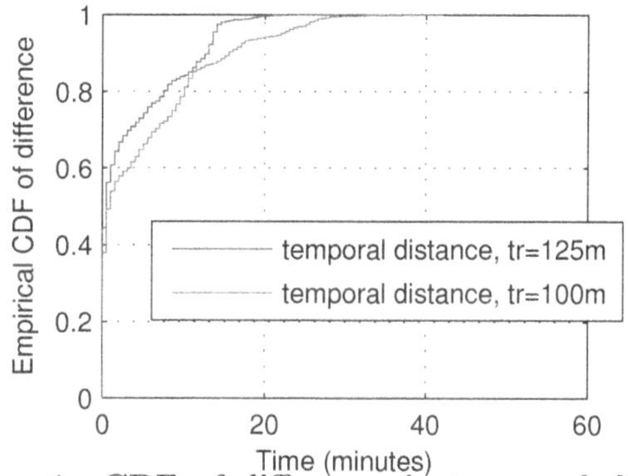

Figure 1: CDF of difference in temporal distance $|t_{ij} - t_{ji}|$ for all node pairs i, j at different transmission ranges (tr)

every 30 sec in a 90 minute duration of State Fair mobility trace [21] with 18 users. Figure 1 shows the CDF of these values. Since real mobility trace is used, different levels of connectivity are investigated by changing the transmission range tr from 100m to 125m. At higher level of connectivity (tr=125m), differences are smaller than at lower level (tr=100m). Also as shown in Figure 1, differences between temporal distances are less than 5 minutes for 50% and less than 10 minutes for around 80% of pairs at all times which justifies the use of approximation t_{ij} for t_{ji}.

3.3 Service loads

Each node provides one or more services, and maintains a FIFO queue for all service requests. Thus, even though a requester may be connected to a service provider, the request may take a long time to complete because service provider is busy. Therefore, one needs to take into account the current service load at the destination in order to compose services in minimal time. This can be done by sharing service load of all nodes in a distributed manner [Section 4.1]. Our service selection algorithm takes into account the service load at destination together with the temporal distance estimate. Though, statistical mechanisms can be used to predict the future service loads and provide better estimates, our current analysis uses a moving time-window average for the load estimate l at each node with a decaying factor α of 0.5. Load in current window l_{cw} is calculated by the total number of pending requests multiplied by the expected service execution time. Then estimate of load is updated by following rule: $l = \alpha l_{cw} + (1 - \alpha)l_{old}$ where l_{old} is the load value l in previous time window. We use a time window of 30s in our analysis.

4. SERVICE COMPOSITION

In this section, we describe how the information about temporal distance, service loads, and available services can be shared among nodes in the network. The link costs on a service graph are updated based on temporal distance between nodes and their current service loads. A particular service composition is selected based on the shortest path in this graph. Overview of the key algorithm steps is also provided in Figure 2.

4.1 Estimation of temporal distance and load

We are interested in temporal distance between nodes. An estimate of this distance at some time t_0 can be used to

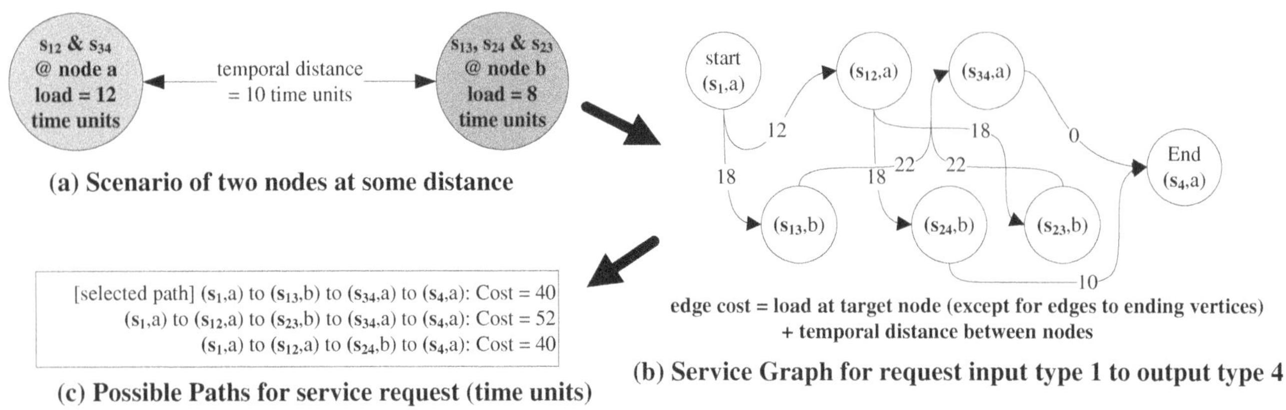

Figure 2: Overview of Composition Selection Algorithm: (a) an example scenario, (b) construction of service graph, and (c) possible paths for request completion

approximate it for a later time that is not too long after t_0. In order to estimate temporal distance from other nodes in a distributed manner, each node keeps a timer for other nodes in networks with an update rule as defined hereafter. Nodes also share the load values with rest of the network. This is in contrast with centralized estimates of load that are used in [18]. Each node maintains a load value for other nodes in network. Let $t_a(i)$ denote the time elapsed since node a last made contact with node i, where $t_a(a)$ is always set at zero. Also, let $l_a(i)$ denote the estimate for load of node i at node a. Local timer values for each node are incremented after every time unit (e.g., 30s, 60s etc. depending on devices). When node a comes into contact with some other node b, it updates its timers and loads estimates for all nodes from whom node b has a smaller temporal distance. This rule is defined as follows: $\forall i \neq a$ if $t_b(i) < t_a(i) - t_{av}$, then set

$$t_a(i) := t_b(i) + t_{av}$$
$$l_a(i) := l_b(i)$$

where t_{av} is the measure of distance between two nodes when they are within each other's transmission range. Every node performs this update when it comes into contact with another node. Note that $t_a(i)$ is same as temporal distance t_{ia} and is used to approximate t_{ai} at node a as described in Section 3.2. The value of t_{av} is a small constant greater than zero but less than or equal to one time unit (increment by which local timers are updated) i.e. $t_{av} \in (0, 1]$. Note that it's a different definition of t_{av} from that in [25] where the value of average time required to travel between the two nodes is included. A small value of t_{av} is used as we are interested in the time a request will take in forwarding and not the physical distance between nodes. Value of $t_{av} > 0$ creates a gradient in a connected subset of nodes such that forwarding paths with smaller number of hops are selected. Based on this smaller value of t_{av} we have proposed a forwarding scheme modified timer (MT) [22] in which messages are forwarded to nodes with smaller timer from the destination. MT is used as the default forwarding scheme in all simulations and is found to perform efficiently when temporal distances are small (under 15min).

4.1.1 *Service advertisement and controlling distribution radius*

By keeping track of other nodes through use of timers, the service composition mechanism retains information only about nodes that are in close vicinity. This makes our approach scalable for large networks, as nodes have a direct mechanism to filter unnecessary information about devices that are further away. Thus, each time two nodes come into contact, they only update the timers and service load estimates for those remaining nodes in network whose temporal distances (implied proximity of location) are smaller than a certain threshold. Similar timers can directly be used to control epidemic forwarding of description of services that other nodes provide to a limited space. As network size grows, nodes only maintain a fixed set of values to efficiently select and forward service requests.

4.2 Service Graph

Each device maintains a local service graph $G = (V, E)$ based on its view of rest of the network. A simple case is described in Figure 2, where services provided by two devices a and b are used to construct a service graph. The number of devices, services (including repetitions) and input/output types are represented by N, N_s and n_d respectively. The service graph G has two types of vertices $V = \{V_1, V_2\}$, where a vertex $v \in V_1$ is a device and service pair such that $|V_1| = N_s$, and vertex $v \in V_2$ is a device and input/output type pair such that $|V_2| = n_d$. The two types of the vertices are shown in Figure 2b: i) (s_{12}, a) is a vertex of type V_1, represents that service s_{12} is provided at node a; and ii) (s_1, a) is a vertex of type V_2, represents that there is an input/output of type 1 (denoted s_1) entering/exiting an application at node a. Vertices of type V_2 are only maintained for the device constructing the service graph.

A directed edge (u, v) between two vertices u and v exists if (i) input/output type of first vertex $u \in V_2$ is same as service input of second vertex $v \in V_1$ e.g., $(s_1, a) \to (s_{12}, a)$; (ii) service output of first vertex $u \in V_1$ is same as service input of second vertex $v \in V_1$ e.g., $(s_{12}, a) \to (s_{24}, b)$; or (iii) service output of first vertex $u \in V_1$ is same as input/output type of second vertex $v \in V_2$ e.g., $(s_{24}, b) \to (s_4, a)$. The cost of an edge in cases (i) and (ii) is the sum of temporal distance between devices of corresponding vertices and the load at the device of second vertex. For example, cost of $(s_{12}, a) \to (s_{24}, b)$ is 18 which is sum of temporal distance from a to b (10) and load on b (8). However, the cost of an edge in case (iii) is simply the temporal distance between devices of corresponding vertices as it is the cost of routing final results back to the service requester. Nodes of type V_2 are needed to take into account the temporal distances: (a) from the service requestor to the node providing the first service; and (b) from the node providing the last service to the requestor.

4.3 Algorithm

Our final algorithm is to compute a shortest path on a service graph based on a service request (input type and desired final output type). From the collected information about service loads, temporal distances, and services provided at the neighboring nodes, a device creates the service graph. In Figure 2, a simple case is described for a service request from input type 1 to output type 4 at device a. Using the service load and temporal distance estimate, the edge cost is computed from each service output to compatible service inputs. From the final graph, shortest path is computed from the starting vertex (s_1, a) to ending vertex (s_4, a) by Dijkstra's Algorithm. In this case the shortest path is $(s_1, a) \rightarrow (s_{13}, b) \rightarrow (s_{34}, a) \rightarrow (s_4, a)$ i.e. the next service s_{13} is to be run at device B. This information is given to the forwarding algorithm to route service request from device A to device B. After execution of this service in the sequence, the device hosting the service (device B) finds the next service in the path by re-computing the shortest path for pending request (input type 3 to output type 4) and gives the destination address to the underlying forwarding scheme. This is done until the final results are routed back to the service requestor. The path is re-computed after execution of each service in the sequence as the network topology can change in that duration to provide alternate paths that are more feasible. However, one may choose to follow the same path computed at the service requestor if the network topology does not change very fast.

The service graph has a total of $N_s + n_d$ vertices. Since, the cost of edges from services on one device to services on another is same, service graph requires only $O(N^2)$ updates instead of $O(N_s^2)$. The actual cost of update is even lower, as information about vertices experiencing changes are updated. In a large network, the size of graph can be controlled by retaining information of devices that are in close vicinity. One run of Dijkstra's Algorithm ($O((N_s+n_d)^2)$) finds shortest path from one input type to all possible output types i.e. it finds composition paths of all service requests that have the same input type. Thus, at most, a node only requires to run Dijkstra's Algorithm n_d times to compute composition path for any number of pending service requests.

4.4 Levels of awareness

Note that even though any forwarding scheme can be used to forward service requests efficiently, the selection of service set (and consequently the destination nodes for forwarding) still needs to take into account the likelihood of delivery. Even though our algorithm is independent of any forwarding scheme, selection of a particular service set has a significant impact on performance (service completion rate and delay) of any forwarding algorithm. The selection of a particular service set can be made with different levels of awareness about rest of the network. More specifically, the edge cost between vertices of service graph is based upon the knowledge of temporal distance and service loads. These levels are summarized in Table 1 for node a where $l_a(i)$ represents estimate for load of *node i* at node a and $\widehat{t_i(j)}$ represents estimate for temporal distance $t_i(j)$ at node a. Note that, timers maintained by node a only provide $t_a(i)$, therefore, different levels of awareness are used to estimate $t_i(j)$ for $i, j \neq a$.

Table 1: Settings for different levels of awareness

Level	$\widehat{t_a(i)}, \widehat{t_i(a)}$	$\widehat{t_i(j)} : i, j \neq a$	$l_a(i)$
minimal	1	1	0
local	$t_a(i)$	$t_a(i) + t_a(j)$	$l_a(i)$(delayed)
global	$t_a(i)$	$t_i(j)$(delayed)	$l_a(i)$(delayed)
perfect	$t_a(i)$	$t_i(j)$	$l_a(i)$

At the very basic level it is assumed that a node knows about services being provided in the environment but does not have an estimate of its temporal distances from particular devices. This is reflected as *minimal* in Table 1 as it requires no housekeeping.

The next level of *local* knowledge is by use of timers as described in Section 4.1. A node has knowledge of its temporal distance from other nodes in the network (e.g., $t_a(i)$ for node a). A node also knows about the latest service load at other nodes, which is forwarded to it in a distributed manner [Section 4.1] and therefore incurs some delay. This level requires exchanging $O(N)$ values ($t_a(i), l_a(i)$ $\forall i$ at node a) per contact. In order to update link cost between services provided at other nodes, a node needs to be aware of temporal distances between other nodes (i.e. $t_i(j) : i, j \neq a$). However, this information is not available in *local knowledge*. Therefore, a node uses its own timer values for approximation as shown in Table 1. For example, node a uses $(t_a(i) + t_a(j))$ as an approximation for temporal distance $t_i(j)$ because $|t_a(i) - t_a(j)| < t_i(j) < t_a(i) + t_a(j)$. Even though this approximation (an upper bound on $t_i(j)$) is not accurate, it still gives an idea of proximity of a device. Hence, the performance results using local knowledge are fairly close to ones achieved by perfect knowledge.

The next higher level of *distributed global* knowledge is when a node also receives timers of other nodes in a distributed manner i.e. the estimates of temporal distances between all nodes which requires exchanging $O(N^2)$ values ($t_i(j), l_a(i)$ $\forall i, j$ at node a) per contact. However, this information about temporal distance between other nodes is still not up-to-date because of propagation delay. Therefore, we further make comparison with a *perfect* system which assumes centralized knowledge of temporal distances and service loads i.e. all information is available to nodes without any delay.

5. SIMULATION AND ANALYSIS

Performance of service selection algorithm is evaluated for types for levels of awareness, forwarding schemes, repetition of services, request rates, and node densities. Extensive simulations are run on real as well as synthetic mobility traces. The real mobility trace has been collected from participants that carry GPS receivers which log position at 30 second intervals. These traces are collected in five different environments, but we show representative results from a State Fair [21]. In order to make a comparison with suitable number of nodes, track logs from each day are considered to be a separate user. These logs have durations from around one to ten hours each day. We truncate all logs to 90 minutes during which 18 users record their location. Synthetic mobility traces are generated using Levy Walk mobility model [20] with 20 (10 slow, 10 fast moving) and 40 (30 slow, 10 fast

Table 2: Levy mobility trace used for simulation

Description	slow	fast
Exponent of step length distribution	1.6	
Exponent of pause time distribution	0.6	
Velocity of node	1m/s	5m/s
Maximum step length	5m	25m
Minimum pause duration	30s	7.5s
Maximum pause duration	600s	60s
Simulation area	500×500m	

Table 3: Simulation Parameters

Description	Range [**default value**]
Number of services	[**20**]
Number of service providers	[**20**], 40
Repetition of each service	1, [**2**], 3, 4
Request timeout	10, [**15**], 20, 30
Request rate per node	0.2, [**0.4**], 0.67, 1/min
Trace duration	[**10 hours**]
Transmission range of device	[**100m**]

moving) nodes. Details about trace settings are provided in Table 2.

Simulation setup: All plots show the average of five simulation runs and the error bars are plotted on the minimum and maximum values in those runs. Table 3 summarizes the generic simulation parameters. In all following plots, default values of these parameters are used unless it is mentioned otherwise under the plot. Parameters used for forwarding schemes are EBR-Encounter rate window=10 minutes, t_{av}=10min for TT and 0.5min for MT. A total of seven input/output types are used to create $C_2^7 = 21$ unique services. Service s_{17} is dropped to leave 20 unique services that are provided at 20 nodes in the network. Note that for a service repetition level of 3, all services are provided by 3 unique nodes (randomly selected) such that every node provides exactly 3 services. Service requests are only generated for services with $k \geq 4$ and the average execution time of service is set to 30s. Each service request times out after 15 minutes so that previous service requests do not overload the system. In all simulation results, service requests are not generated in the last 15 minutes (duration of request timeout) of the trace duration. We consider a simulation analysis including both a transient and a steady regime phase. The transient regime lasts for approximately the initial 30 minutes, after which all nodes acquire information about network resources and system enters the steady regime.

5.1 Direct and multi-hop forwarding

Figure 3 shows the service completion rate in the Levy Walk mobility trace. In all these cases *local* level of awareness is used to construct service graph. The search for a single service that matches the request exactly only fulfills 40% of the requests. However, when the condition is relaxed to compose multiple services to meet the requirements, about 50% of service requests are completed. As a further improvement, when multi-hop forwarding paths are used to forward requests and service outputs, 70% requests are completed. This shows the significant improvement that service

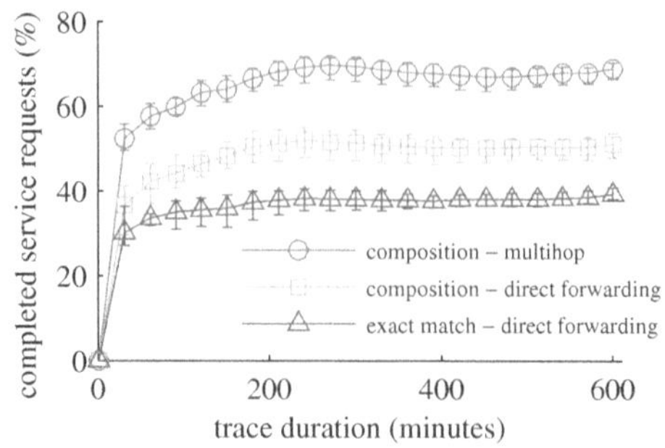

Figure 3: Higher service completion in multi-hop composition

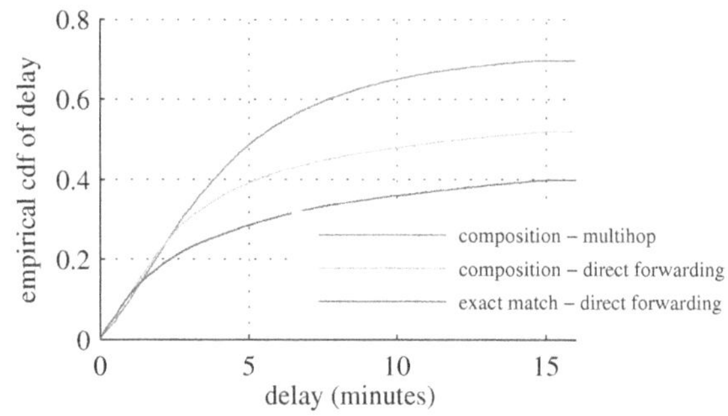

Figure 4: Lower delay in multi-hop composition

composition with multi-hop forwarding paths can have over other mechanisms. While such mechanisms are discussed in [8,28], our algorithm tolerates disconnections and utilizes paths that are formed over time. As a result, in addition to utilizing services at devices with directly connected paths to a node, ones that are present in the nearby vicinity can be invoked at the expense of slightly higher delays. However, in opportunistic networks, multi-hop forwarding still yields lower delays as compared to direct one-hop forwarding. This is shown in Figure 4 where 50% of compositions are made within 5 minutes using multi-hop forwarding, whereas it takes 12 minutes in direct forwarding. In case when only exact match is searched in the network, only 40% requests are completed after 15 minutes. The CDF plot (proportion of samples less than a value) is based on the aggregate delay per service from five simulation runs. As seen from Figure 3, the percentage of completed services increases sharply in the first 30 minutes (transient regime) and then becomes steady. In the transient regime the system starts with no initial service loads. Therefore, delays of only those compositions are considered that are generated once the system is well inside the steady state regime. Figure 4 and later plots show delays of only those service requests that are generated after first two hours of simulation.

5.2 Levels of awareness

While there is significant improvement in service completion rate and delays when some knowledge about temporal distances from other nodes is used, there is not much difference in the performances of perfect, globally aware and local schemes. This is clear from Figure 5a where in a range of service densities, completion rate of perfect, distributed global and local levels of awareness is within 3% of each other. Figure 5b shows the delays when each service is repeated twice i.e. two different nodes provide that service. The local scheme has delays between the globally aware and

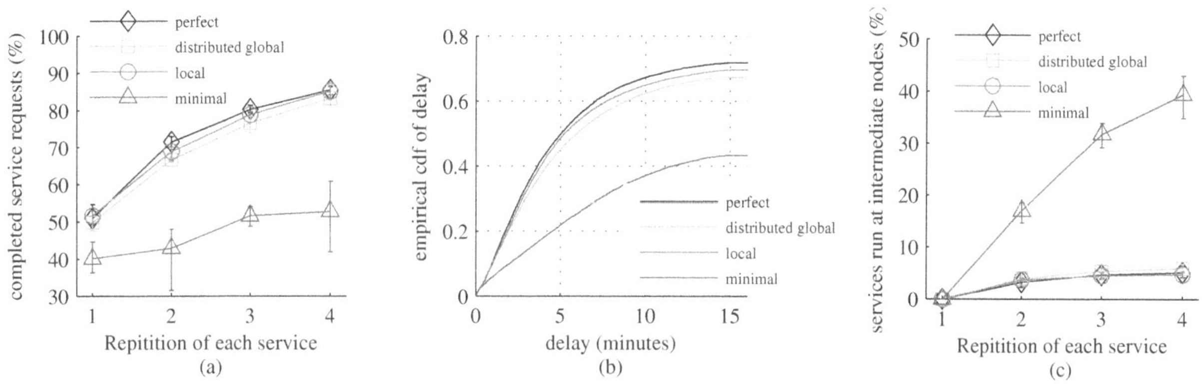

Figure 5: Levy Walk: With different levels of awareness (a) service completion, (b) delay, and (c) services run at intermediate nodes

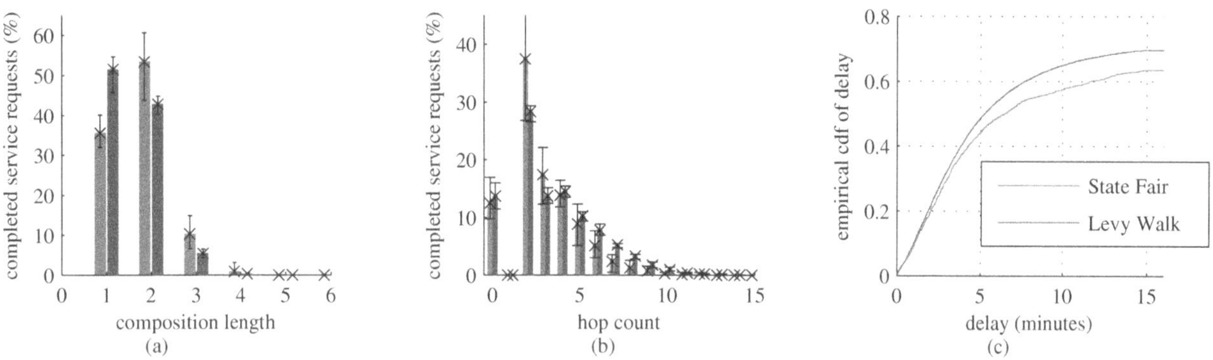

Figure 6: State Fair [red] and Levy Walk [blue] mobility trace: (a) composition length, (b) hop count, and (c) delay profiles for multi-hop service composition

perfect scheme as shown in Figure 5b. One explanation for this behavior is that approximation for temporal distance between other nodes is better than delayed information of actual temporal distance between those nodes. Also, by the time a service is forwarded and executed on a device, the network topology changes so much that it does not make much difference if local knowledge is used or perfect. Since, all three schemes have similar performance; we select the scheme with local knowledge for further analysis as it is more light weight (requires exchanging $O(N)$ temporal distance and load values per contact instead of $O(N^2)$).

In Figure 5c, we validate the selection of correct devices to compose all services. At different service densities, percentage of services that are run at intermediate nodes while they were in route to some other device is considered (i.e. when required service is opportunistically found at a device different from one selected by the algorithm). While it is quite high around 40% when minimal knowledge about network is available, it is reduced to under 5% when some knowledge about network topology is used. This validates that our service selection algorithm selects the correct devices to execute services in a composition.

5.3 Composition length, delay, and hop profiles

Figure 6 shows the detailed profile of multi-hop service composition for State Fair and Levy Walk mobility trace. Most compositions are comprised of 1 or 2 services and then there are some of three, and four services as shown in Figure 6a. In our analysis, services are distributed such that every service is provided by at least one node in the network. However, as described earlier, restricting the completion to the exact match degrades performance. Therefore, all compositions (around 60%) of length more than 1 that are shown in Figure 6a are optional, and are made to achieve higher service completion rates and minimal delays. However, these compositions come at the expense of a few extra hops as shown in Figure 6b. Most of the times (about 80%) complete services are composed by using less than 5 hops. Around 15% of requested services are found at the device itself. This causes a hop count of zero as shown in Figure 6b. If the service is not found at the requesting devices, service completion incurs at least 2 hops (one to send request and one to receive results). Figure 6c shows the empirical CDF of the time taken from service request to receiving composed results, routed back from the last node in the composition path. Around 60% of completed service requests have a delay of less than 10 minutes whereas close to 50% of these are completed in less than 5 minutes. Each service is repeated twice in this run. For higher number of repetitions, delays are lower and percentage of completed services reaches close to 100% (85% for four repetitions, see Figure 5a). This means that for, reasonable service densities, applications that can tolerate delays to the order of 5 to 10 minutes, opportunistic networks can provide a reasonable service completion rate in an environment similar to State Fair. Results in State Fair and Levy Walk are fairly similar which means that evaluations based on Levy Walk traces are close to real mobility traces.

5.4 Effect of forwarding scheme, request load, request timeout, node and service density

Figures 8 shows the service completion rate of 3 different underlying forwarding schemes used in the service selection

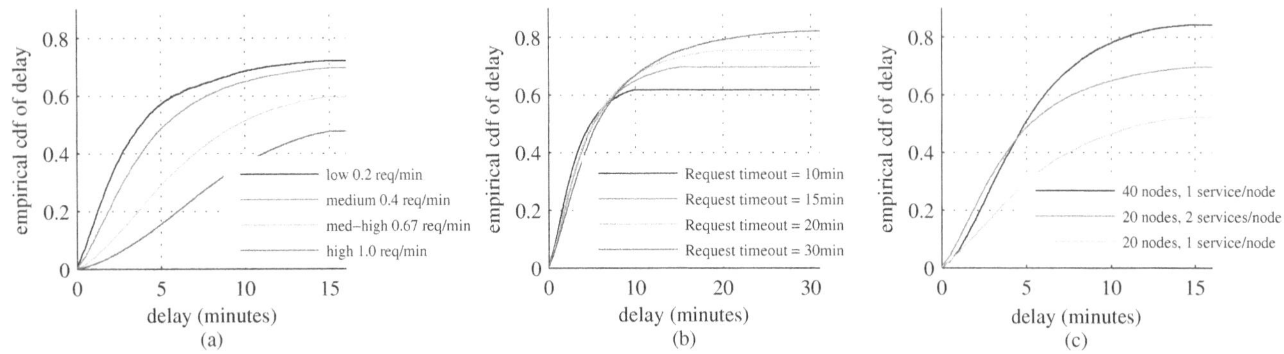

Figure 7: Levy Walk: Higher service completion and lower delay at (a) low loads, (b) long request timeout, and (c) high node, service density for multi-hop service composition

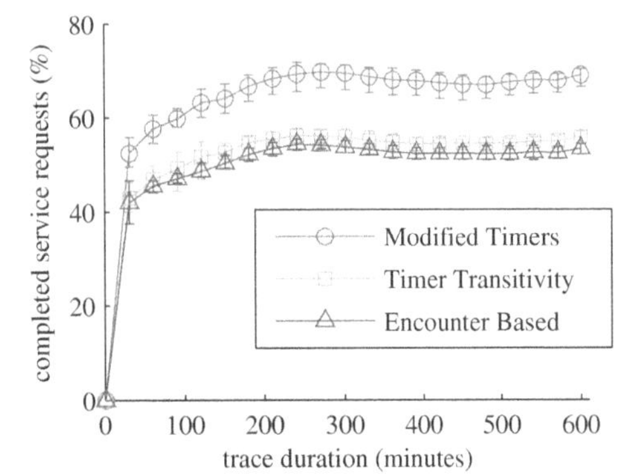

Figure 8: MT with better service completion

algorithm. Only single copy of each request is forwarded. MT shows better delivery rate (about 30-40% higher than other schemes). Network performance is also analyzed under different service load conditions and levels of connectivity. Requests rates are varied from 0.2 to 1 request per minute per node. Figure 7a shows that service completion rate decreases under high loads. At high node/service density Figure 7c shows that almost 85% of all requests can be composed. Also, when a service request does not timeout for a longer time, more requests can be completed [Figure 7b]. Thus, while service compositions can be made in a moderately connected network, our algorithm adapts well in sparse scenarios with higher service request loads and short request timeout duration. Note that variation in load greatly changes the delay profile under 5 min as shown in Figure 7a. This is because higher loads result in longer service queues and services take longer to execute even when a connected path to service provider exists. Whereas variation in request timeout changes the delay profile at over 10 min as shows in Figure 7b. This is because longer request timeouts keep the service requests in the queue which are otherwise dropped without completion. As a side effect, average service load at nodes increases slightly. For this reason, 40% of compositions are made in 4.5 min with request timeout of 30min but only in 3.5min with timeout of 10 min due to change in average load at nodes. Figure 7c shows that average delays are lower when 40 services are distributed in 40 nodes instead of 20 nodes even though the number of generated requests is double in a network with 40 nodes. This is because the network with 40 nodes is more dense and connected.

6. RELATED WORK

Brief overview of related works in single service provisioning [15,17,18,24] and service composition [2,7,8,13,28] was presented in the introduction. In this section, we discuss the highlights of these works in greater detail. There has been a significant research on semantics and ontologies to describe services and compositions (see [4] for survey and issues of service composition in mobile environments). Services to be composed can either be strictly defined in the request (static composition) [11], or can be computed at run-time (dynamic composition) based on the request [13]. Dynamic composition looks for alternate sets of services in the current environment that can be composed to provide the required functionality. Services are modeled as directed attributed graphs [13], to find possible compositions. We leverage this work for service representation to find possible compositions and explore mechanism to execute such compositions in an opportunistic network. Passarella et al., [17,18] investigated optimal policy for service execution in opportunistic networks. An optimal policy is derived for the level of replication to receive service results in minimal time but only a single service is requested and the service requester waits until a direct contact with service provider. In contrast, the algorithm presented in this paper uses a single instance of service request to compose multiple services, collects all information in a distributed manner, and uses multi-hop forwarding paths to upload requests to and download results from service provider. Since, multihop forwarding paths are used, next we describe the existing schemes to forward data in opportunistic networks that are applicable to our analysis.

Several schemes for forwarding messages in opportunistic networks have been proposed in the literature. Two key metrics used in open environments where participating nodes may not have social interaction history include, *time since last encounter* with destination [9] and *rate of encounter* [16,26] . Time since last encounter provides good location approximation in dense mobile networks [10] but has a slow warm up time (i.e. it takes a long period of time before every node has made at least one contact with every other to make an estimate of its location). To reduce warm up time, Spyropoulos et al. [25] presented a timer transitivity (nodes exchange timer values for all destinations). Whereas, use of encounter rate provides good delivery rates in heterogeneous environments by identifying nodes that are highly mobile.

7. CONCLUSIONS

In this paper, we proposed a novel algorithm for service composition in opportunistic networks. With the proposed algorithm, mobile users can benefit from a larger set of services available locally in an environment. Proposed service composition makes efficient service selections from devices that are in close proximity. Key insight used in the solution is that temporal distance between devices provides a direct measure for reachability of nodes when an end-to-end connected path does not exist. This way, based on a service graph, a composition sequence can be selected to meet the requirements of service request while any routing scheme can be used to forward service parameters.

Through extensive simulations on real and synthetic mobility traces, we conclude that using only local knowledge about temporal distances and service loads, a large percentage of services can be composed in an opportunistic network using multi-hop paths between devices. In order to complete a request, searching for possible compositions is better than looking for a single service that matches the request exactly. Also, use of multi-hop forwarding paths is more efficient than one-hop direct forwarding. Performance of service composition improves (more requests are completed with lower delays) under high service density, high node density, long request timeout duration and low request rate. Future work will explore how additional information about a node's mobility and context can be combined with temporal distances to improve successful compositions in minimal time. Also, analysis of replicated service requests at different stages of composition to increase robustness will be investigated. Replication strategies can be devised based on proximity and number of devices providing the required service.

8. ACKNOWLEDGMENTS

The research work presented in this paper was carried out with support from the US National Foundations Grants ECCS-0824120, CSR 0834493 and European Commission under the FIRE SCAMPI (FP7-IST-258414), FET-AWARE RECOGNITION (FP7-IST-257756) Projects. The authors would like to thank Sidra Bashir for help in running simulations.

9. REFERENCES

[1] H. Cai and D. Y. Eun. Aging rules: what does the past tell about the future in mobile ad-hoc networks? In *Proc. ACM MobiHoc*, 2009.

[2] I. Carreras, L. Bassbouss, D. Linner, H. Pfeffer, V. Simon, E. Varga, D. Schreckling, J. Huusko, and H. Rivas. Bionets: Self evolving services in opportunistic networking environments. In *Bioinspired Models of Network, Information, and Computing Systems*, pages 88–94. 2010.

[3] A. Chaintreau, P. Hui, J. Crowcroft, C. Diot, R. Gass, and J. Scott. Impact of human mobility on opportunistic forwarding algorithms. *IEEE Trans. Mobile Comput.*, 6:606–620, 2007.

[4] D. Chakraborty, A. Joshi, T. Finin, and Y. Yesha. Service composition for mobile environments. *Mobile Networks and Applications*, 10:435–451, 2005.

[5] M. Conti and M. Kumar. Opportunities in opportunistic computing. *Computer*, 43(1):42–50, 2010.

[6] E. M. Daly and M. Haahr. Social network analysis for routing in disconnected delay-tolerant manets. In *Proc. ACM MobiHoc*, 2007.

[7] O. Davidyuk, I. Sanchez, and J. Riekki. CADEAU: Supporting Autonomic and User-Controlled Application Composition in Ubiquitous Environments. In *Pervasive Computing and Communications Design and Deployment: Technologies, Trends, and Applications*. 2010.

[8] L. Del Prete and L. Capra. Reliable discovery and selection of composite services in mobile environments. In *Proc. IEEE EDOC*, 2008.

[9] H. Dubois-Ferriere, M. Grossglauser, and M. Vetterli. Age matters: efficient route discovery in mobile ad hoc networks using encounter ages. In *Proc. ACM MobiHoc*, 2003.

[10] M. Grossglauser and M. Vetterli. Locating nodes with ease: last encounter routing in ad hoc networks through mobility diffusion. In *Proc. IEEE INFOCOM*, 2003.

[11] X. Gu, K. Nahrstedt, and B. Yu. Spidernet: an integrated peer-to-peer service composition framework. In *Proc. IEEE HPDC*, pages 110–119, 2004.

[12] P. Hui, J. Crowcroft, and E. Yoneki. Bubble rap: social-based forwarding in delay tolerant networks. In *Proc. ACM MobiHoc*, 2008.

[13] S. Kalasapur, M. Kumar, and B. A. Shirazi. Dynamic service composition in pervasive computing. *IEEE Trans. Parallel Distrib. Syst.*, 18:907–918, 2007.

[14] S. Kim, C.-H. Lee, and D. Y. Eun. Superdiffusive behavior of mobile nodes and its impact on routing protocol performance. *IEEE Trans. Mobile Comput.*, 9:288–304, 2010.

[15] N. Le Sommer, R. Said, and Y. Mahéo. A proxy-based model for service provision in opportunistic networks. In *Proc. MPAC*. ACM, 2008.

[16] S. Nelson, M. Bakht, and R. Kravets. Encounter-based routing in dtns. In *Proc. IEEE INFOCOM*, 2009.

[17] A. Passarella, M. Conti, E. Borgia, and M. Kumar. Performance evaluation of service execution in opportunistic computing. In *Proc. 13th ACM MSWIM*, 2010.

[18] A. Passarella, M. Kumar, M. Conti, and E. Borgia. Minimum-delay service provisioning in opportunistic networks. *IEEE Trans. Parallel Distrib. Syst.*, PP(99), 2011.

[19] L. Pelusi, A. Passarella, and M. Conti. Opportunistic networking: data forwarding in disconnected mobile ad hoc networks. *IEEE Communications Magazine*, 44(11):134–141, 2006.

[20] I. Rhee, M. Shin, S. Hong, K. Lee, and S. Chong. On the levy-walk nature of human mobility. In *Proc. IEEE INFOCOM*, 2008.

[21] I. Rhee, M. Shin, S. Hong, K. Lee, S. Kim, and S. Chong. CRAWDAD trace. `http://crawdad.cs.dartmouth.edu/ncsu/mobilitymodels/GPS/NC_State_Fair`, July 2009.

[22] U. Sadiq and M. Kumar. ProxiMol: Proximity and mobility estimation for efficient forwarding in opportunistic networks. In *Proc. IEEE MASS*, 2011.

[23] G. Sharma and R. Mazumdar. Scaling laws for capacity and delay in wireless ad hoc networks with random mobility. In *Proc. IEEE ICC*, 2004.

[24] N. L. Sommer and S. B. Sassi. Location-based service discovery and delivery in opportunistic networks. In *Proc. ICN*, 2010.

[25] T. Spyropoulos, K. Psounis, and C. S. Raghavendra. Efficient routing in intermittently connected mobile networks: the single-copy case. *IEEE/ACM Trans. Netw.*, 16(1):63–76, 2008.

[26] T. Spyropoulos, T. Turletti, and K. Obraczka. Routing in delay-tolerant networks comprising heterogeneous node populations. *IEEE Trans. Mobile Comput.*, 8(8):1132–1147, 2009.

[27] J. Tang, M. Musolesi, C. Mascolo, and V. Latora. Characterising temporal distance and reachability in mobile and online social networks. *SIGCOMM Comput. Commun. Rev.*, 40(1):118–124, 2010.

[28] J. Wang. Exploiting mobility prediction for dependable service composition in wireless mobile ad hoc networks. *IEEE Trans. on Services Comput.*, 4:44–55, 2011.

Motion-based Routing for Opportunistic Ad-hoc Networks

Weihan Wang
Department of Electrical and Computer Engineering
University of Toronto
weihan@eecg.toronto.edu

Cristiana Amza
Department of Electrical and Computer Engineering
University of Toronto
amza@eecg.toronto.edu

ABSTRACT

In this paper, we introduce novel motion-based routing protocols for opportunistic packet relay in ad-hoc opportunistic networks. Unlike existing ad-hoc routing protocols, which require global knowledge about the mobility patterns of other nodes, thus raising scalability and privacy concerns, our routing protocols store and use only local motion information, such as the current direction of the local node. Such information is easily obtainable by any modern GPS-equipped mobile device.

We design and evaluate a series of protocols using different motion information, ranging from simple, easy to capture metrics i.e., speed and direction, to more complex metrics, i.e., the past or expected trajectory of the local node. These metrics allow for increasing degrees of routing accuracy at a correspondingly higher cost in terms of memory and computation. However, all of our routing protocols have constant complexity in the size of the network, and none of them rely on nodes sending or storing information about other forwarding nodes. This makes our schemes ideal for large-scale networks and non-community networks, where membership is very dynamic.

Our simulation study using real-world traces shows that our motion-based schemes perform comparably to or considerably better than existing complex, global-knowledge-based protocols with significant less utilization on memory, CPU, and bandwidth.

Categories and Subject Descriptors

C.5.3 [**Computer System Implementation**]: Portable devices; C.2.2 [**Computer - Communication Networks**]: Network protocols—*routing protocols*

General Terms

Simulation, Modeling, Performance

Keywords

Mobile Computing, Opportunistic Ad-hoc Networks, Routing Protocol

1. INTRODUCTION

Opportunistic ad-hoc networks [7, 29], support wireless communication within some type of spontaneously forming, self-organizing networks of mobile users. In such opportunistic, localized networks, usually some type of specialized wireless communication is either desirable or absolutely necessary. For example, communication may be performed opportunistically on Bluetooth because of its low-power consumption and availability at short range. Alternatively, Bluetooth, or some other specialized wireless technologies may be used together with a certain type of vehicle, e.g., cars or buses, or in regional networks unreachable by standard telecommunication technologies.

In these opportunistic ad-hoc networks, the mobile users themselves are commonly contributors to the network infrastructure as packet forwarders on behalf of other nodes. Mobile devices relay packets to neighboring devices as they come in contact with them, by taking advantage of communication opportunities that arise in the course of user mobility, until the packet eventually reaches its destination. Since mobile devices become increasingly pervasive, almost anyone with a Bluetooth device in her or his pocket becomes a potential participant in the forwarding process. We call these type of routing protocols based on mobility opportunities delay-tolerant routing (DTR).

In order for DTR to be widely accepted and deployed, two pre-requisites must be met for mobile devices involved in packet transfers: i) users should be required to reveal little private information and ii) resource consumption, for message transfers, computation and memory, required by the routing protocol should be kept as low as possible.

Many existing DTR protocols rely on revealing and aggregating detailed private statistics to other nodes, including meeting records, connection histories, and even details of buffered packets [6, 19, 24, 2, 22, 4, 15], which might raise privacy concerns. Furthermore, in order to calculate routes, these protocols commonly require each node to store the learned topology of other nodes in the whole system, such as the link state and meeting history. Thus, the cost of such DTR algorithms is $O(\sqrt{n})$ at best, which makes these solutions unsuitable for memory constrained devices or large-scale networks. In addition, computation and dissemination of such global information may consume considerable power and bandwidth, often crucial resources in mobile environments. Finally, these protocols require continuous observation, and some even rely on a priori knowledge, of other nodes to learn system topologies. Hence, they work well only for *community networks*, where membership is static and nodes are coordinated. In contrast, in *non-community networks*, such as, vehicles on a highway, or people in a shopping center, where the ad-hoc network is organized on-the-fly, and membership is very dynamic, or cannot be accurately established, these existing protocols may not work efficiently, or at all.

In this paper, we observe that, in addition to Bluetooth, GPS-capabilities are becoming mainstream on mobile devices; various alternative techniques for outdoor and indoor positioning, including WiFi and RFID localization and GSM fingerprinting, are also maturing [12, 17, 32, 25]. Based on this observation, our key idea is to use information about node movement in order to compute the *utility*, which reflects the node's probability of delivering any given packet to its known destination. Intuitively, nodes moving towards the intended packet destination are likely to successfully deliver the particular packet. Moreover, frequent visitors to the destination's neighborhood have a high probability to deliver the respective packet as well. We therefore propose a series of motion-based routing protocols using different utility functions, corresponding to the above intuitions: i) velocity-based protocols and ii) trajectory-based protocols that infer utilities from either velocity or past/expected trajectories, respectively. Each protocol in the above categories exploits increasingly more metrics, allowing for varying levels of trade-off between routing accuracy and costs in terms of memory and computation.

All the proposed protocols compute utilities locally and require no motion data about other forwarding nodes. This provides three advantages over existing DTR protocols. First, the complexity of our algorithms is *constant* with respect to the size of the network, making our protocols scalable and viable for very large networks, such as an ad-hoc network spanning the mobile users in a whole city. Second, not requiring storing and using information about other nodes makes our protocols ideal for non-community networks. Third, privacy is better preserved than previous approaches as a side-effect of localized computation.

We run simulations using two real-world traces, one consisting of more than 30 buses and the other with 300 campus wireless users, and compare our protocols with two existing global-knowledge-based DTR protocols: *MV* and *RAPID*. They are chosen as they are supposed to outperform protocols that exploit less knowledge. *MV* [6] is a classical protocol using global meeting histories to compute probabilities of delivery. *RAPID* [2] exchanges meeting histories and also metadata about each packet to estimate per-packet delay. According to previous work [2], *RAPID* significantly outperforms three existing protocols including *MaxProp* [4], *Spray and Wait* [28], and *PRoPHET* [22].

We adopt a novel metric, called *Relative CDF*, which we believe is more accurate than the traditional combination of delivery ratio and delay, as well as the more traditional metrics, to compare the performance of our protocols with the above state-of-the-art protocols. We also compare resource utilization of all protocols.

Our study show that our algorithms are resilient to inaccurate location information. With orders of magnitude lower costs in various computing resources, our algorithms perform comparable to *RAPID* and significantly better than *MV*, even in community networks, which is the environment providing the most fair comparison since they have bounded network sizes and allow global statistics to be accumulated, which are necessary for global-knowledge-based protocols such as *RAPID* and *MV* to work efficiently.

The rest of this paper is organized as follows. § 2 describes the system model and protocol design. § 3 presents the utility functions we use, the key component of our protocols. § 4 discusses privacy and other issues. § 5 presents evaluation methodology and results. Related work is summarized in § 6. § 7 concludes the paper and presents directions for future work.

2. PROTOCOL OVERVIEW

2.1 System model and assumptions

We model an opportunistic ad-hoc network as a dynamic set of mobile nodes. Nodes may join and leave the network at any time. Communications are based on pair-wise contacts: two nodes in the network are called *neighbors* and can transfer data packets bidirectionally if within communication range. Each packet has a single destination node and can be delivered from the source to the destination directly or via intermediate nodes. Naming and addressing for mobile nodes is an orthogonal issue beyond the scope of this paper; therefore, as others before us e.g., Burns et al. [6], we assume that the address for the destination of a packet can be obtained by orthogonal means.

For example, destinations for packets could be stationary nodes, part of an infrastructure which can be assumed to be always on and whose addresses, and locations can be easily obtained. Specifically, in many applications, destinations are knwon data collectors in sensor networks [27], gateways to the Internet [11], or packet relays to out-of-band channels. Finally, much like geocasting [14], applications leveraging our motion-based routing may transmit packets towards a specific region and then broadcast them within that region.

We also assume that nodes always know their own locations through localization systems. The buffer size on each node dedicated to packet forwarding is limited and when the buffer is full, old packets have to be dropped in order to receive new packets. Network bandwidth is also limited; the amount of bytes that can be transmitted during a contact is finite and proportional to the duration of the contact. Therefore, protocols must wisely transmit important data first.

2.2 Protocol operations

Our protocols operate as follows. When a node A meets another node B, A first delivers the packets that are destined to B, and then exchanges metadata to compute the *utility function* for the packets remained in its buffer. The utility function $U(n, d)$ of node n w.r.t. node d is defined to reflect the protocol's belief on n's probability to visit d's neighborhood region in the near future. Node A sorts the packets in decreasing order of *marginal utilities* gained by forwarding packets to B, i.e. $U(B, d) - U(A, d)$, with d the destination of a packet in question. All packets with *positive* marginal utilities are transferred from A to B in that order and removed from A's buffer. Transmission ends when all these packets have been forwarded or the transfer opportunity ends.

When node B receives a forwarded packet destined to d, it delivers the packet if d is among its immediate neighbors. Otherwise, it calculates $U(n,d)$ for every neighbor n and forwards the packet to a neighbor with the highest utility and which is higher than $U(B,d)$. This neighbor in turn performs the same operation. Such iterative process is necessary to pass packets on to the best node in the same cloud as, but not directly neighboring, the initiating node. The process continues until the receiving node finds no better candidate than itself. In this case, the node stores the packet in its own buffer, and when the buffer is full, drops the oldest packet, i.e. the packet with the longest time since creation.

Only one copy for each packet is kept in the network. Alternatively, nodes can always retain the packet after forwarding it to other nodes, or retain it only when observing from various facts that the probability of delivery is too low. For example, the packet's maximum utility among neighbors is below a threshold, or the node has not met others for a while indicating a sparse neighborhood network. In § 6, we summarize existing replication as well as buffer management techniques directly applicable to our protocols.

3. THE UTILITY FUNCTION

Information about a node's movement, such as, its current direction or the locations it visited in the past, can be used to predict the regions the node is going to visit. Once it is in vicinity with another node, it is likely to deliver packets to the latter using direct contacts or through intermediate nodes between them. Hence, the utility function $U(n,d)$ uses n's motion information to computes a *utility* which reflects n's probability of visiting d's neighborhood region in the near future, and therefore the likelihood of n delivering packets to d. We propose several utility functions using different motion information, each realizing a distinct motion-based protocol. *Velocity-based functions* compute utilities from n's current velocity, and *trajectory-based functions* use either n's past or expected trajectory. These functions exploit increasingly more information about n's movement for better routing performance at higher costs in computation and storage. Note that the functions support both 2D coordinates and 3D coordinates as in underwater and interplanetary networks.

None of the utility functions rely on information about *any* other forwarding nodes. The only information they require besides n's own movement is the destination's *current* geographical location, if the destination is mobile. Ideally, a packet should move towards the location where it meets the destination at the time of delivery but not where the destination dwells at present. However, this requires estimating packet delay and predicting locations based on the delay which involves non-trivial computation and potential inaccuracy.

3.1 Velocity-based utility functions

A node's current velocity is a hint to its future movement: the higher speed and less yaw of a node approaching the other node, the more likely it is to reach the latter in the imminent future. We propose three velocity-based utility functions. By convention, we refer to *velocity* as a vector whose magnitude is *speed*. Illustrated in Figure 1, we denote node n's velocity as $\vec{v}_n$ or $\vec{v}$ when used unambiguously, the vector from n to node d's current location as $\vec{d}_{n,d}$ or $\vec{d}$,

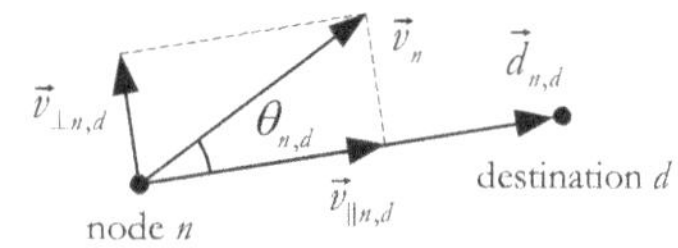

Figure 1: Notations. We omit subscripts when used unambiguously.

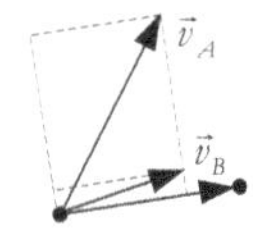

Figure 2: Node A and B have the same $\vec{v}_{\parallel}$ but B has a lower $\vec{v}_{\perp}$.

and the angle between $\vec{v}$ and $\vec{d}$ as $\theta_{n,d}$ or θ. In practice, $\vec{v}$ can be always estimated from most recent readings from n's localization system.

3.1.1 Velocity projection (U_{vp})

It is defined as the projection of $\vec{v}$ on $\vec{d}$. Hence, a node moving faster along $\vec{d}$ gains a higher utility. Note that the utility becomes negative when moving against the destination.

$$U_{\text{vp}}(n,d) = |\vec{v}| \cdot \cos\theta$$

3.1.2 Velocity cotangent (U_{vcot})

U_{vp} ignores $\vec{v}_{\perp}$, $\vec{v}$'s perpendicular components to $\vec{d}$. However, as illustrated in Figure 2, when two nodes have identical $\vec{v}_{\parallel}$ but different $\vec{v}_{\perp}$, the one with a smaller $|\vec{v}_{\perp}|$ is more likely to reach d. A better utility function shall increase as $|\vec{v}_{\perp}|$ decreases given the same velocity projection. As one of many approaches, we divide U_{vp} by $\sin\theta$, which is equivalent to multiplying $|\vec{v}|$ and $\cot\theta$. In the final equation, we use the absolute value of $\sin\theta$ to guarantee a positive utility when moving towards the destination and negative otherwise. We avoid an infinite utility by keeping the denominator above a small constant ε.

$$U_{\text{vcot}}(n,d) = |\vec{v}| \cdot \frac{\cos\theta}{\max(|\sin\theta|, \varepsilon)}$$

3.1.3 Estimated time to arrive (U_{eta})

The distance to the destination also contribute to the probability as well as the delay of packet delivery. We compute U_{eta} as the reciprocal of the estimated travel time to the destination's radio coverage based on n's speed along $\vec{d}$, and similar to U_{vcot}, divide the result by $\max(|\sin\theta|, \varepsilon)$. Hence, the final equation is equivalent to dividing U_{vcot} by the distance. As the time to arrive would make no sense if the node is departing from the destination (i.e. $\cos\theta \leq 0$), we simply use U_{vcot} in this case. Assuming the radius of radio range is R, the utility function is defined as follows:

$$U_{\text{eta}}(n,d) = \begin{cases} \dfrac{U_{\text{vcot}}(n,d)}{|\vec{d}| - R}, & \cos\theta > 0; \\ U_{\text{vcot}}(n,d), & \text{otherwise.} \end{cases} \quad (1)$$

3.2 Trajectory-based utility functions (U_{tp} and U_{tf})

Either past or future trajectories can be used to estimate probabilities of delivery. We denote the utility function based on history trajectories as U_{tp}. Leguay et al. point out that users' mobility pattern is not uniformly random but can be often characterized by power-law distributions [19]. That is, a user visits few places frequently with long stay while visiting other places rarely with short stay. If a node's past trajectory is known, we can therefore expect that its future movement follows similar patterns, and use this data to predict the probability of reaching specified regions in the future. However, this approach would be disadvantageous if mobility patterns are uniform or changing quickly over time.

Future trajectories can be accurately obtained if node schedules are known in advance. This is the case for most public transit systems. Additional information like traveling history and road condition forecasts may help refining the information. We refer to the utility function using future trajectories as $U_{\mathtt{tf}}$.

We assume that nodes store trajectories as a time series of sample points sampled at a fixed time interval t. Depending on the pace the system evolves, the interval can be as short as one second or as long as several days. We define the *individual utility* of each sample point to be negatively correlated to its distance to the destination. Sample points within the destination's radio range are more advantageous than other points because direct delivery is guaranteed when the node is at these points. We therefore define the individual utility of node n's ith sample point as

$$\hat{U}(n,d,i) = \begin{cases} \beta R^{-\lambda}, & |\vec{d}_{n_i,d}| < R; \\ |\vec{d}_{n_i,d}|^{-\lambda}, & \text{otherwise.} \end{cases} \tag{2}$$

where R is the radius of radio range, $|\vec{d}_{n_i,d}|$ is the sample point's distance to d, the exponent λ is a positive number controlling how fast the utility declines as distance increases, and the *boosting factor* β is a constant greater than one to encourage sample points within the radio range. The overall utility is then the weighted sum over all sample points:

$$U_{\mathtt{tp,tf}}(n,d) = \sum_{i=1}^{S} \omega(n,i)\hat{U}(n,d,i) \tag{3}$$

with S the total number of n's sample points and $w(n,i)$ the weight of the ith sample point reflecting the importance of the point. There are numerous methods to assign weights, which could be dependent on application semantics. For instance, heavier weights can be given to sample points lying in venues such as central stations and meeting rooms where users are likely to meet destination nodes or better candidates as packet carriers. In this paper, we employ *exponential moving average* to give higher importance to more recent sample points, as described below.

Supposing the sample point at index one is the earliest, we set $\omega(n,i) = \omega_0^{S-i}$ when using past trajectories. The constant ω_0 which is referred to as the *smoothing factor* satisfies $0 < \omega_0 < 1$, so that the smaller the value, the quicker old data discounts. The rationale behind this scheme is that more recent data predict future mobility patterns better than older data. When applied to future trajectories, the weight becomes $\omega(n,i) = \omega_0^i$ so that sample points in the immediate future are given more importance. It also makes sense in that timely delivery of packets counts more on node movement in the near future than in the far future. As a special case, setting $\omega_0 = 1$ will treat all sample points equally. Finally, The summation in Equation 3 ignores the sample points whose weight is below a threshold, in order to save computation as well as storage for insignificant sample points. We term this threshold as the *cutoff weight*.

In practice, trajectory-based functions incur higher costs than velocity-based ones in computation and storage, rendering them infeasible for extremely resource constrained devices. $U_{\mathtt{tf}}$ is unavailable if no future schedule is known; $U_{\mathtt{tp}}$ is also excluded if users are not willing to be tracked by their own devices for any reasons. We will study the performance and costs of these utility functions through experiments.

4. DISCUSSION

Because utilities are determined solely from local information (in addition to the destination's location), our protocols are highly scalable, with *constant* complexities in time, space, and communication with respect to the size of the network. In contrast, the complexities of existing DTR protocols grow at least $O(\sqrt{n})$ with the system size, making them infeasible for large-scale networks. The considerable amount of power, memory, and bandwidth consumed by these protocols also impose heavy burden to resource constrained mobile devices. However, DTR must keep resource utilization as low as possible in order to be widely accepted in today's mobile environments. Moreover, algorithms relying on global state require a priori knowledge about and continuous statistics over other nodes, which works well in community networks but not in non-community networks where the number of participants is unbounded and what a node meets most are strangers, making meeting histories and other global statistics unable to accumulate.

As a byproduct of localized computation, our protocols preserve privacy for forwarding nodes better than existing protocols whose computation is based on synthesis of private information about other nodes such as meeting histories and buffer state [6, 19, 24, 2, 22, 4, 15]. In our protocols, nodes compute their own utilities without information about other forwarding nodes, and the only peers that require a node A's utility values besides A itself are its immediate neighbors. When a neighbor, say B, requests A's utilities, rather than sending A's motion details to B, A only replies with a ready-made utility corresponding to a destination given by B. Nodes therefore exchange only high-level utility values rather than low-level private information. However, by repeatedly querying a node with well-crafted destination addresses, one can still infer private information through interpolation. This is an area for future exploration.

The above method of "delegating" utility computation also reduces bandwidth usage for trajectory-based protocols: only a single value rather than the whole set of sample points are transmitted. As another optimization, we can save repetitive computation and transmission by caching utilities for a short while on both sending and receiving nodes. If the underlying network supports broadcast, the third optimization is to broadcast when replying with utilities. Nodes receiving the broadcast may cache these values for later use.

5. EVALUATION

In this section, we first present a novel technique for network performance analysis (§ 5.1). After describing evaluation methodology (§ 5.2), we present evaluation results in the last five subsections, each answering one of the following questions:

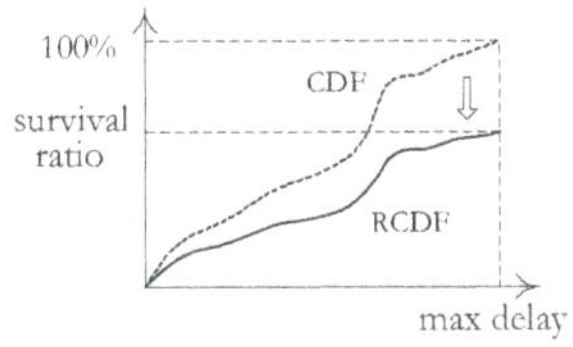

Figure 3: Converting CDF to RCDF

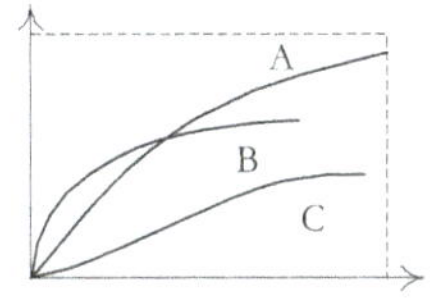

Figure 4: Comparing RCDF

- How to tune parameters for trajectory-based protocols?
- How do different protocols perform against each other?
- How sensitive are our protocols to localization errors?
- How do the protocols perform against existing ones?
- Do our protocols consume less resources than others?

5.1 Relative CDF (RCDF)

Two commonly used performance metrics for opportunistic ad-hoc networks are delivery ratio, i.e. the number of delivered packets versus the number of generated packets, and packet delay distribution, or one of its summaries like arithmetic mean or median. However, if two systems have different delivery ratios, their delay distributions are distorted and thus comparing delay distribution is biased. A system might attain worse delay distribution only because it delivers more packets. We also argue that delivery ratio is affected by delay and thus is not an independent metric. Given other factors being the same, a system with longer delay would suffer lower delivery ratio because of the time lag of the initial packets being delivered. In the following text, we describe a metric that we call Relative Cumulative Distribution Functions or RCDF to address these issues by defining *survival ratio* and combining it with delay distribution. RCDF is generally applicable to networks with packet loss.

Definition Let **A**, **D** and **I** respectively be the set of generated packets, delivered packets, and in-transit packets. Given the CDF $f(x)$ over packet delay x, the RCDF $f'(x) \equiv Sf(x)$, where the *survival ratio* S is a constant satisfying

$$\frac{|\mathbf{D}|}{|\mathbf{A}| - |\mathbf{I}|} \leq S \leq \frac{|\mathbf{D}| + |\mathbf{I}|}{|\mathbf{A}|}. \quad (4)$$

The idea of RCDF is to normalize delay distribution against all rather than received packets, and to use S to represent packets that are not lost. It is important to understand that in addition to delivered packets, *S also includes a portion of in-transit packets that will be successfully delivered in the future.* Figure 3 illustrates the transformation from CDF to RCDF.

In general cases, the exact value of S cannot be expressed in a closed form. It can be approximated as delivery ratio in some situations. Readers are referred to the appendix for the deduction of the bounds in Equation 4 and approximation. Figure 4 shows three hypothetic RCDF curves. The x-coordinate at the end of each curve corresponds to the system's maximum delay. System A and B is said to be absolutely advantageous over system C. Although the overall survival ratio of A is higher than B, B achieves more packets at the low-delay end. Whether A is better than B depends on end-user preferences. In this paper, we plot RCDF using mean of the two bounds as survival ratio. When comparing with previous work, we also include the traditional delivery ratio and delay metrics.

	diesel	*campus*
# mobile nodes	34	300
coverage	49 × 20 km	1.6 × 1.1 km
duration	80 days	45 days
radio range	10 m (Bluetooth Class 2)	
bandwidth	2.1 Mbps payload, duplex (Bluetooth 2.0)	
packet arrival	Exponential(1) min./packet/node	
packet size	1 KB	
buffer size	10 MB	
cutoff weight	1×10^{-6}	

Table 1: Default parameters for simulation

5.2 Methodology

We develop a custom discrete event simulator which, along with protocol implementations, is publicly available [1]. Since our interest is in comparing application-layer protocols, the simulator assumes ideal, lossless radio channels. Network parameters are set according to typical values in Bluetooth. Packets are continuously generated from each mobile node in a poisson process, with destinations being randomly selected fixed nodes. Table 1 summarizes the parameters used in our simulation.

We find two real-world, wireless mobility traces annotated with geographical information, namely the DieselNet trace and the Dartmouth trace. The DieselNet trace (*diesel*) [5] records bus movement around the UMass campus and the surrounding county. Each bus's log records the time and GPS coordinates when meeting other buses. We restore the bus's movement by linear interpolating these data points.

The Dartmouth campus trace (*campus*) [16] is an extensive dataset collected in the Wi-Fi network of Dartmouth College. It records over three years of mobility of around 13,000 wireless devices. Since running the whole trace is impossible on today's hardware, Leguay et al. [20] selected a subset of 15 days between January 26, 2004 to March 11, 2004 as users in this period "make an intensive and regular use of the network." The authors randomly chose 300 devices from this period for their study. We adopt the same subset in our paper. In *campus*, the log for each device records the names of the access points (APs) the device has been associated with, annotated with time. APs' geographic data is stored in a separate database. We restore the device's movement by assuming it is collocated with the AP at the beginning of each association, and linear interpolating these positions (APs with unknown locations are ignored). APs are dispersed among 185 buildings. Because our work focuses on outdoor communications where DTR is most useful, we avoid indoor networking which raises fundamentally different problems by moving all the APs in the same building to their geometric center.

Packets are continuously generated from each node in a poisson process. Packet destinations are randomly chosen points in the map which is passed through by at least one node during the simulation. Finally, we run each experi-

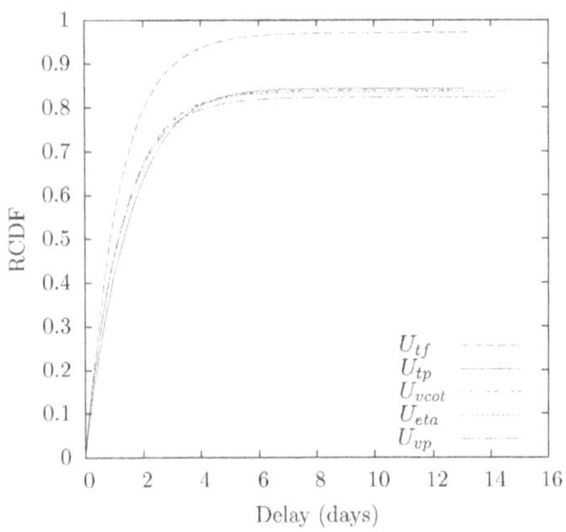

Figure 5: Comparing utility functions in *diesel*

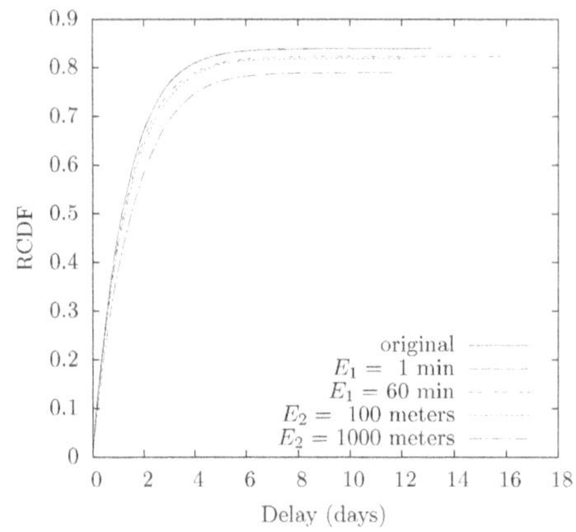

Figure 7: Sensitivity to localization errors in *diesel*

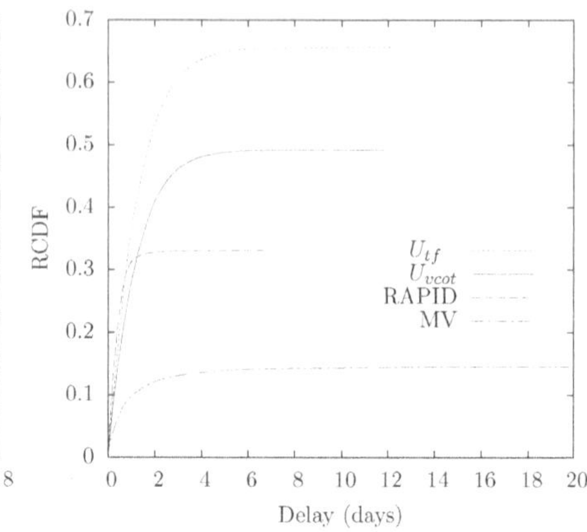

Figure 9: Performance against existing work in *diesel*

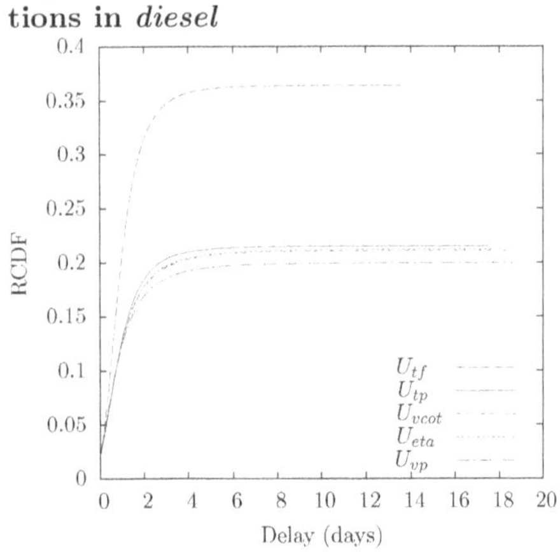

Figure 6: Comparing utility functions in *campus*

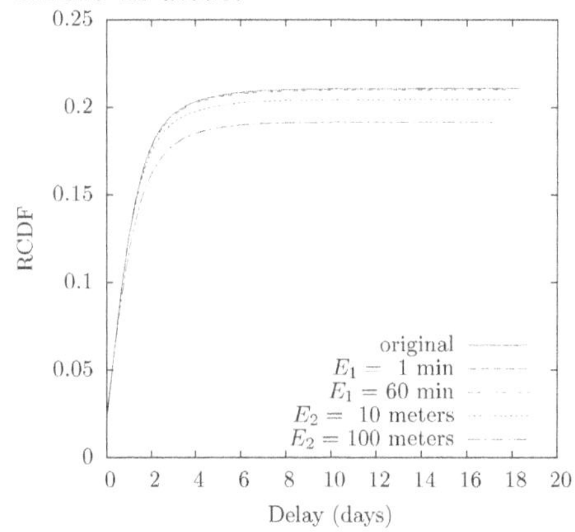

Figure 8: Sensitivity to localization errors in *campus*

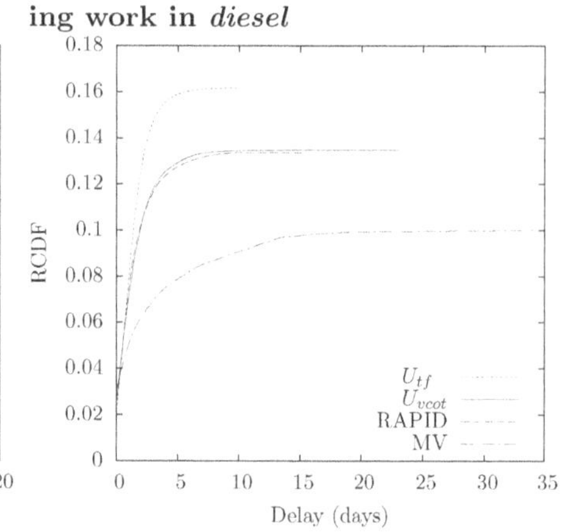

Figure 10: Performance against existing work in *campus* with 100 nodes

ment three times with different random seeds and report the average.

5.3 Parameter tuning for trajectory-based utility functions

There are four parameters for $U_{\texttt{tp}}$ and $U_{\texttt{tf}}$, the sampling interval t, the smoothing factor ω_0, the exponent λ, and the boosting factor β. We study the impact of each parameter in that order, by varying its value while setting others to a default.

The sampling interval controls the algorithms' granularity. Finer granularities result in better performance but higher demand on storage and computation. Experiments in our particular scenarios show that performance improves as t decreases, but the improvement becomes marginal when t is below several minutes.

A smaller ω_0 discounts faster the sampling points farther away the present. Proper assignment of ω_0 depends on the speed in which the system evolves. A small value shall be chosen for a fast-pacing system. For our particular scenarios, we find a value between 0.95 and 0.995 gains maximum performance. Treating all points equally, i.e. $\omega_0 = 1$, leads to low performance, which justifies the need of moving average.

According to Equation 2, λ controls how fast the utility recedes as the distance from sample points to the destination increases. A larger λ speeds up the recession and thus stresses more the preference of forwarding packets to visitors of the destination's close neighborhood. However, as desirable as it seems, if all packets are flooded to these visitors, their buffers might be overloaded and start dropping many packets. In this sense, a smaller λ spreads load more evenly leading to fewer lost packets. Our experiments show that performance peaks at $\lambda = 8$ or so.

The boosting factor β encourages nodes with direct connections to the destination. Same as λ, one may prefer a small β to distribute load evenly from direct contacts to indirect ones. The optimality of β also depends on λ. A large λ already emphasizes closeness and a small β may suffice. Our experiments suggest that when $\lambda = 1$, the best value of β is above 5,000; it reduces to 10 or so when $\lambda = 8$.

5.4 Comparing utility functions

We compare the performance of all the proposed utility functions. To calculate velocity for velocity-based functions, nodes read their location once per second and calculate velocity from two most recent readings. The parameters for $U_{\texttt{tp}}$ and $U_{\texttt{tf}}$ are set to $t = 10$ min, $\omega_0 = 0.95$, $\lambda = 8$, and $\beta = 10$. Results are shown in Figure 5 and 6. Readers may notice that the *campus* trace generally leads to less survival ratio than *diesel* (Figure 5 to 10). We believe this is caused by the less mobility of *campus* users whose movements tend to be infrequent and localized if compared to those in the *diesel* trace. Meanwhile, we strive to preserve accuracy of the simulation instead of manipulating configurations for better but less realistic results.

Jain et al. [13] found that a DTR protocol using more knowledge is expected to outperform the one with less information. Our results confirm their findings: $U_{\mathtt{tf}}$ significantly outperforms other protocols because it possesses the most complete information–nodes' future movement; others only conjecture such information using history or current information. The performance difference between $U_{\mathtt{tf}}$ and $U_{\mathtt{tp}}$ depends on how well past trajectories predict future trends. The better such *predictability*, the smaller the difference is expected to be. Similarly, the predictability using the current velocity determines the difference between $U_{\mathtt{tf}}$ and velocity-based protocols. Although the two predictabilities are generally different, they happen to be comparable in our particular experiment.

Velocity-based protocols perform closely to each other. $U_{\mathtt{vcot}}$ exploits more information and thus is slightly more effective than $U_{\mathtt{vp}}$. $U_{\mathtt{eta}}$ shows no more advantageous than $U_{\mathtt{vcot}}$, because the radio range in our experiment is so small compared to the arenas' dimensions (Table 1), that the difference between two neighboring nodes in their distances to a destination is usually negligible. In this case, $U_{\mathtt{eta}}$ degenerates to $U_{\mathtt{vcot}}$. We expect a higher performance of $U_{\mathtt{eta}}$ in long-range radio networks.

5.5 Sensitivity to localization errors

Motion-based protocols rely on localization systems to obtain the nodes' locations. In this subsection, we examine the protocols' resilience against localization errors. We model error distance from each node's estimated location to its actual position as a random variable following a Gaussian distribution with a mean of 0 and a standard deviation of E meters.

Figure 7 and 8 show the results of a representative utility function, $U_{\mathtt{vcot}}$. Under highly pessimistic localization errors, the degradation on survival ratio is less than ten percent in each scenario (we use larger errors in *diesel* considering that vehicles move faster than people and the area it covers is much bigger than *campus*). Such robustness against location inaccuracy can be explained by the *distance effect* [3], which states that nodes farther away from the target are less sensitive to location errors than closer nodes. Meanwhile, a packet is very likely to be delivered when it has already been forwarded to a node close to the destination.

5.6 Comparing performance with existing protocols

We compare our work with *MV* and *RAPID*, two global-knowledge-based DTR protocols. We choose them for fair comparison because intuitively, they are expected to outperform other protocols using less knowledge. As a tradeoff however, the former's runtime cost is also expected to be higher than the latter, magnifying the advantage of our protocols when comparing resource utilization.

MV [6] is a classical protocol exploiting meeting histories for routing and buffer management. Because *MV* does not specify how global knowledge is disseminated, we assume *perfect* global knowledge on each node which gives *MV* ideal performance. Because *MV* does not support mobile destinations, we plant 30 fixed packet receivers at random locations in each trace and when generating packets, select destinations only among these receivers. *RAPID* [2] is an intentional routing algorithm aiming to optimize a specific metric. It "significantly outperforms existing protocols" [2] including *MaxProp* [4], *Spray and Wait* [28], and *PRoPHET* [22]. In *RAPID*, nodes exchange meeting histories as well as per-packet statistics and use them to optimize average delay. For fairness, we disable acknowledgements as other protocols do not have this feature. Adding it should improve all protocols equally. Two champions of motion-based protocols are $U_{\mathtt{tf}}$, the best one given future schedules, and $U_{\mathtt{vcot}}$, a representative one when no such information exists.

	Time	Space	Comm.
Motion-based	$O(b)$	$O(1)$	$O(b)$
MV	$O(bn)$	$O(n^2)$	$O(n^2)$
RAPID	$O(bn^3)$	$O(n^2+bn)$	$O(bn)$

Table 2: Comparing complexities

Table 2 summarizes the complexities of the protocols under comparison. It shows the time complexity of computation on each node, the space required on each node to store metadata, and the communication overhead during each contact. In the table, n denotes the number of nodes and b the number of packets buffered in a node. All the complexities of motion-based protocols are constant with the size of the network, whereas the complexities of *MV* and *RAPID* grow either exponentially or linearly. Empirical study on algorithm overhead is presented in the next subsection.

Both *diesel* and *campus* represent community networks, which are the environment providing the most fair comparison. In non-community networks, the advantages of our motion-based protocols would be greatly magnified, since global statistics are difficult to accumulate in such networks but such statistic is required by global-knowledge-based protocols to work efficiently. It is also because the number of nodes in a non-community network is unbounded, leading to a sharp increase in the complexities of these protocols. In fact, *RAPID* exhausts the memory of our 3.5 GB-memory computer system with the *campus* trace (the JVM on which our simulator runs needs all of the heap to reside in physical memory). We have to reduce the number of nodes 100 for the rest of the paper. Comparison with *MV* using the original *campus* trace is reported in a separate document [1].

Results are shown in Table 3 in terms of the traditional delivery ratio and average delay. With much lower complexities, motion-based protocols outperform *MV* in both scenarios and outperform *RAPID* in *campus*. In *diesel* however, our protocols achieve higher delivery ratio but lower delay, which generates a tie. Better presentation of the results using RCDF is shown in Figure 9 and 10. Analysis on RCDF shows that in terms of survival ratio, *RAPID* in *diesel* (Figure 9) gains 3.7% and 7.0% more low-delay packets than $U_{\mathtt{tf}}$ and $U_{\mathtt{vcot}}$, respectively. However, the gain is largely offset by the overall ratio: $U_{\mathtt{tf}}$ and $U_{\mathtt{vcot}}$ achieves 98.4% and 48.7% higher survival ratio than *RAPID*. In the 100-node *campus* trace (Figure 10), *RAPID* performs almost identically to $U_{\mathtt{vcot}}$.

5.7 Comparing resource utilization with existing protocols

We measure and compare four types of resource utiliza-

	diesel		*campus (100 nodes)*	
	delivery	delay	delivery	delay
$U_{\mathtt{tf}}$	64.4%	105.0	11.8%	73.5
$U_{\mathtt{vcot}}$	47.1%	97.7	6.6%	77.6
RAPID	32.6%	34.3	7.4%	117.5
MV	13.9%	106.0	4.1%	256.6

Table 3: Comparing delivery radio and average delay (in 10^3 sec)

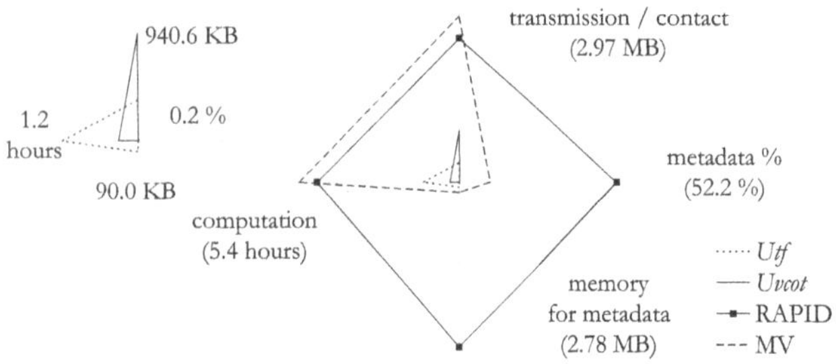

Figure 11: Resource utilization in *diesel* using the radar plot. Each axis depicts one type of resource utilization. The inset on the left is the magnified view of the plot's central area. In both views, numbers show the maximum value seen on each axis.

tion: i) the maximum amount of memory each node consumes to store metadata. We assume that basic data types including float points and packet identifiers are four-byte long. ii) communication overhead, i.e. the percentage of bandwidth used for metadata exchange. iii) the total amount of data (payload plus metadata) transmitted during each contact. iv) CPU time each node spends in computing utilities. Similar to our protocols, a key component of *MV* and *RAPID* is also the computation of a utility function. Peripheral components such as housekeeping and statistics are inexpensive and less meaningful to compare. To obtain the CPU time, we run the single-thread simulator on an Intel Xeon 2.8 GHz machine and record the time spent in executing utility functions. Each of the above metrics is reported as an average over nodes or contacts. For fairness, caching and broadcasting optimizations (§ 4) are disabled from our protocols. Finally, as mentioned earlier, readers shall notice that better performance of global-knowledge-based protocols at the cost of higher overhead, which (unavoidably) overestimates our protocols' advantage in resource utilization of many DTR protocols other than *MV* and *RAPID*.

Results are given in Figure 11 and 12 using *radar plots*, in which the total resource utilization of a scheme roughly corresponds to the area of that scheme. The results confirm the complexity analysis in Table 2. (Due to a larger constant factor in the complexity, *MV* spends more CPU time than *RAPID* despite of its lower complexity. The same situation is for *RAPID*'s communication overhead.) Our protocols consume far less resources than other protocols in all cases. If comparing $U_{\mathtt{vcot}}$ with *RAPID*, the two protocols with the closest performance, $U_{\mathtt{vcot}}$'s CPU time, memory footprint, and communication overhead are several, up to five, orders of magnitude less than *RAPID*. $U_{\mathtt{vcot}}$'s total data transmission is also significantly lower than *RAPID*.

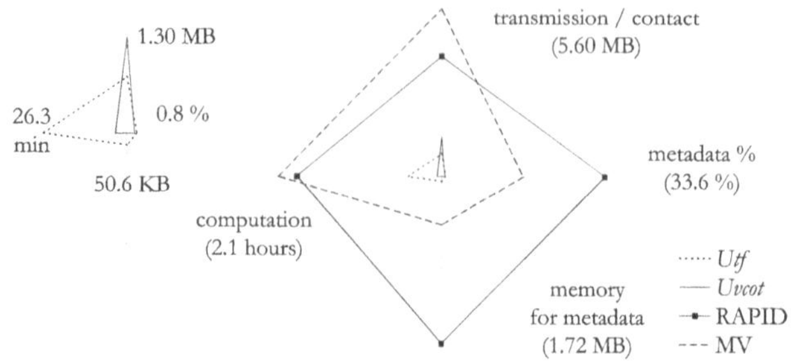

Figure 12: Resource utilization in *campus* with 100 nodes

6. RELATED WORK

6.1 Routing protocols

Numerous geographical routing protocols have been proposed in the context of mobile ad-hoc networks (MANETs) [23, 14]. These schemes forward packets to the next hop that is "closer" to the destination. Our work fundamentally differs from them for two reasons. First, MANET protocols assume end-to-end paths and may fail when no such path exists, whereas DTR protocols must cope with opportunistic connections. Second, because of this difference in requirement, geographical routing infers static network topology from node locations, and in turn find a path out of the topology, whereas our motion-based protocols infer mobility pattern from node movement, and exploit the mobility to realize packet forwarding.

DTR protocols either flood packets blindly or use certain knowledge about the network to limit flooding and to direct packet flows. Commonly used knowledge includes past or scheduled contacts, connectivities, and energy level. Such information is often disseminated across the network so that nodes can learn others' probabilities of delivery or calculate their own probabilities or both. Zhang et al. [31] provides a detailed survey on existing DTR protocols. The knowledge exploited by our protocols is the physical motion of the local node. However, the information is not propagated in order to preserve privacy and resources.

As a rudimentary work, LeBrun et al. [18] propose to use geographical information for DTR. Their protocol computes the *expected closest distance* to the destination based on the heading of a node, and decides forwarding using this distance. We go further by designing a complete protocol framework, proposing utility functions that exploit varying levels of motion information, and taking more factors into account including privacy, design tradeoffs, and mobile destinations. Furthermore, we provide empirical study and compare performance and overhead against state-of-the-art protocols with real-world traces.

Leontiadis and Mascolo [21] propose to take advantage of suggested routes from the vehicles' navigation system to select vehicles that are more likely to deliver packets to fixed destinations. Their protocol is similar to ours in the use of estimated time of arrival. However, their approach highly relies on integration with navigation systems as well as the availability of road maps. Moreover, intensive computation required by the approach also limits its applicability on power-constraint devices.

6.2 Replication and buffer management

Packets may be replicated in the network to increase their chances to reach destinations or to tolerate faults. Replication has been widely studied in the past. In particular, network coding in DTR [30] has become a promising approach to attain maximum network flow and fault tolerance using information theory. These efforts are orthogonal to and can be integrated with our work.

Our protocols drop the oldest packets when buffers full, while other dropping policies are also possible [8, 10]. More sophisticated buffer management schemes incorporate acknowledgements [4], application semantics [26], or client interests [9]. These schemes would improve certain performance metrics while calling for supports from the upper layers of the network stack.

7. CONCLUSIONS

We introduce routing protocols for opportunistic ad-hoc networks that estimate probabilities for packet delivery based on current velocity or past/expected trajectories. As an important property, these protocols employ only local information for routing decisions, which enables three advantages over existing DTR protocols: privacy protection, scalability with network sizes, and support for non-community networks. Our algorithms use increasingly more information to gain better routing accuracy at the cost of higher resource utilization in terms of memory and computation. Simulation with real-world traces show that our protocols achieve comparable or better performance while consuming far less resources than state-of-the-art protocols using global meeting histories and per-packet statistics. We also propose RCDF as a more accurate metric than the traditional combination of delivery ratio and delay for network performance analysis. It encodes both delay distribution and survival ratio in a single graphical representation.

We would like to investigate these issues in the future:

- Routing across building floors raises interesting problems. A node moving towards a destination on a different floor is unlikely to reach it if the elevator is behind the node. The protocol might also need to consider the effect of non-line-of-sight transmission through walls and floors.
- Coexisting with other protocols and switching to them on appropriate opportunities. For example, CAR [24] uses a traditional MANET protocol if an end-to-end link exists.

8. REFERENCES

[1] Additional resources. `http://www.eecg.toronto.edu/~weihan/motion`.

[2] A. Balasubramanian, B. Levine, and A. Venkataramani. DTN routing as a resource allocation problem, 2007.

[3] S. Basagni, I. Chlamtac, and V. Syrotiuk. Geographic messaging in wireless ad hoc networks, Jul 1999.

[4] J. Burgess, B. Gallagher, D. Jensen, and B. N. Levine. MaxProp: Routing for vehicle-based disruption-tolerant networks, April 2006.

[5] J. Burgess and B. N. Levine. CRAWDAD trace umass/diesel/transfer/spring2006 (v. 2007-12-02). Downloaded from `http://crawdad.cs.dartmouth.edu/umass/diesel/transfer/spring2006`.

[6] B. Burns, O. Brock, and B. N. Levine. MV routing and capacity building in disruption tolerant networks, 2005.

[7] A. Chaintreau, P. Hui, J. Crowcroft, C. Diot, R. Gass, and J. Scott. Pocket Switched Networks: Real-World mobility and its Consequences for Opportunistic Forwarding, 2005.

[8] J. Davis, A. Fagg, and B. Levine. Wearable computers as packet transport mechanisms in highly-partitioned ad-hoc networks, 2001.

[9] P. Eugster and R. Guerraoui. Probabilistic multicast, 2002.

[10] P. Eugster, R. Guerraoui, A.-M. Kermarrec, and L. Massoulie. Epidemic information dissemination in distributed systems, 2004.

[11] S. Guo, M. H. Falaki, E. A. Oliver, S. U. Rahman, A. Seth, M. A. Zaharia, and S. Keshav. Very low-cost internet access using KioskNet, 2007.

[12] J. Hightower and G. Borriello. Location systems for ubiquitous computing, 2001.

[13] S. Jain, K. Fall, and R. Patra. Routing in a delay tolerant network, 2004.

[14] X. Jiang and T. Camp. A review of geocasting protocols for a mobile ad hoc network, 2002.

[15] E. Jones, L. Li, J. Schmidtke, and P. Ward. Practical routing in delay-tolerant networks, Aug. 2007.

[16] D. Kotz, T. Henderson, and I. Abyzov. CRAWDAD trace dartmouth/campus/movement/01_04 (v. 2005-03-08). Downloaded from `http://crawdad.cs.dartmouth.edu/dartmouth/campus/movement/01_04`.

[17] A. LaMarca et al. Place Lab: Device positioning using radio beacons in the wild, 2005.

[18] J. LeBrun, C.-N. Chuah, D. Ghosal, and M. Zhang. Knowledge-based opportunistic forwarding in vehicular wireless ad hoc networks, 2005.

[19] J. Leguay, T. Friedman, and V. Conan. DTN routing in a mobility pattern space, 2005.

[20] J. Leguay, T. Friedman, and V. Conan. Evaluating mobility pattern space routing for DTNs, 2006.

[21] I. Leontiadis and C. Mascolo. GeOpps: Opportunistic Geographical Routing for Vehicular Networks. In *Proceedings of the IEEE Workshop on Autonomic and Opportunistic Communications. (Colocated with WOWMOM07)*, Helsinki, Finland, June 2007. IEEE Press.

[22] A. Lindgren, A. Doria, and O. Schelén. Probabilistic routing in intermittently connected networks, 2004.

[23] M. Mauve, A. Widmer, and H. Hartenstein. A survey on position-based routing in mobile ad hoc networks, 2001.

[24] M. Musolesi, S. Hailes, and C. Mascolo. Adaptive routing for intermittently connected mobile ad hoc networks, 2005.

[25] L. M. Ni, Y. Liu, Y. C. Lau, and A. P. Patil. LANDMARC: Indoor location sensing using active RFID, 2004.

[26] J. Pereira, L. Rodrigues, and R. Oliveira. Semantically reliable multicast: Definition, implementation, and performance evaluation, 2003.

[27] T. Small and Z. J. Haas. Resource and performance tradeoffs in delay-tolerant wireless networks, 2005.
[28] T. Spyropoulos, K. Psounis, and C. S. Raghavendra. Spray and wait: an efficient routing scheme for intermittently connected mobile networks, 2005.
[29] J. Su, A. Chin, A. Popivanova, A. Goel, and E. de Lara. User mobility for opportunistic ad-hoc networking, 2004.
[30] Y. Wang, S. Jain, M. Martonosi, and K. Fall. Erasure-coding based routing for opportunistic networks, 2005.
[31] Z. Zhang. Routing in intermittently connected mobile ad hoc networks and delay tolerant networks: overview and challenges, 2006.
[32] Z. Zhou, J.-H. Cui, and A. Bagtzoglou. Scalable localization with mobility prediction for underwater sensor networks, April 2008.

APPENDIX

In order to derive Formula 4, we divide $\mathbf{I}$ into two subsets, $\mathbf{I_s}$ for packets that will be successfully delivered in the future, and $\mathbf{I_l}$ for those will be lost. Survival ratio S is then

$$S \equiv \frac{|\mathbf{D}| + |\mathbf{I_s}|}{|\mathbf{A}|} \quad (5)$$

As the best case scenario, packets are dropped at the earliest stage of the system and packets already in the system will never be lost. Therefore,

$$|\mathbf{I_s}| \leq |\mathbf{I}|. \quad (6)$$

In the worst case, the system will lost most packets from $\mathbf{I}$. Statistically, the loss rate *within* $\mathbf{I}$ cannot be greater than the overall loss rate:

$$1 - \frac{|\mathbf{I_s}|}{|\mathbf{I}|} \leq 1 - S = 1 - \frac{|\mathbf{D}| + |\mathbf{I_s}|}{|\mathbf{A}|} \quad (7)$$

Combining Equation 5, 6 and 7 yields to Formula 4. Note that RCDF is the delay distribution over $\mathbf{D}$ and $\mathbf{I_s}$ combined. Although RCDF is actually derived from $\mathbf{D}$, because $|\mathbf{D}|$ is assumed to be sufficiently large, we can regard RCDF as an unbiased statistic over $\mathbf{I_s}$ as well.

If $|\mathbf{I}| \ll |\mathbf{A}|$ and $|\mathbf{D}|$, survival ratio can be approximated as

$$S \approx \frac{|\mathbf{D}|}{|\mathbf{A}|} = \textit{delivery ratio}.$$

It is useful when expensive or impossible to obtain $|\mathbf{I}|$. The last condition holds for most experiments that run sufficiently long. However, we observe that in opportunistic adhoc networks, the condition is not always true due to huge buffer sizes deployed in the network leading to large system capacities.

Multipath Opportunistic RPL Routing over IEEE 802.15.4

Bogdan Pavković
Grenoble Institute of Technology
CNRS Grenoble Informatics Laboratory UMR 5217
Grenoble, France
pavkovic@imag.fr

Fabrice Theoleyre
CNRS, LSIIT
University of Strasbourg
Strasbourg, France
theoleyre@unistra.fr

Andrzej Duda
Grenoble Institute of Technology
CNRS Grenoble Informatics Laboratory UMR 5217
Grenoble, France
andrzej.duda@imag.fr

ABSTRACT

We consider the problem of running RPL on top of the IEEE 802.15.4 MAC layer—the two layers operate over two different structures, a directed acyclic graph in the case of RPL and a cluster-tree for IEEE 802.15.4. We propose to adapt the cluster-tree of IEEE 802.15.4 so that it can efficiently work coupled with RPL. Nodes in our modified cluster-tree can associate with several parent nodes by taking advantage of an adequate organization of superframes at the MAC layer. Building on this modified MAC layer, we define an opportunistic forwarding scheme that extends RPL with the possibility of forwarding packets over multiple paths. Instead of always using a preferred parent, a node opportunistically forwards packets through other parents as long as their routes towards the sink are better. We take advantage of the opportunistic forwarding to support higher-priority delay-sensitive alarms that need to arrive in sink before a given deadline along with low-intensity monitoring data considered as best-effort. We compare our opportunistic version of RPL to its basic version through detailed simulations in terms of packet delivery ratio, incurred delay, and overhead.

Categories and Subject Descriptors

C.2.1 [**Computer-Communication Networks**]: [Network Architecture and Design - Wireless Communication]

General Terms

Experimentation, Algorithms, Performance

Keywords

Wireless Sensor Networks, RPL, QoS, IEEE 802.15.4, opportunistic routing, multi-path

1. INTRODUCTION

We consider Wireless Sensor Networks (WSN) supporting IP connectivity and running over low-duty IEEE 802.15.4 wireless links. Such networks will lead to the development of the future Internet of Things and enable large deployments of sensors in various domains (smart homes, smart cities, smart grids, environmental sensing, critical infrastructure surveillance, etc.).

IP connectivity in sensor networks mainly relies on two IETF standards: 6LoWPAN [1] and RPL (Routing Protocol for Low power and Lossy Networks) [2]. 6LoWPAN enables IPv6 networking over low-power wireless networks thus bridging the gap between 802.15.4 and IP in a simple way—it defines a header compression and a fragmentation mechanism for IPv6 to run over IEEE 802.15.4 radio links.

RPL is a Distance Vector routing protocol that builds a DODAG (Destination Oriented Directed Acyclic Graph) anchored at a border router of a sensor network. Each node selects at most three parent nodes. It considers the best parent node (in the sense of some metric) as a preferred one and uses it for packet forwarding to the border router. Alternate parents provide backup routes to the border router making the network robust to unexpected changes in radio connectivity. However, the network structure built by RPL is different from the topology required in multi-hop networks with the 802.15.4 MAC layer—a *cluster-tree*. This latter topology allows a hierarchical organization in which a node can only select and associate with just one parent node. So, the main problem of running RPL on top of the 802.15.4 MAC layer is to make the two different structures work together.

In this paper, we propose to adapt the cluster-tree of IEEE 802.15.4 so that it can efficiently work coupled with RPL. Nodes in our modified cluster-tree can associate with several parent nodes by taking advantage of an adequate organization of superframes at the MAC layer. Building on this modified MAC layer, we define an opportunistic forwarding scheme that extends RPL with the possibility of forwarding packets over multiple paths. Instead of always using a preferred parent, a node opportunistically forwards packets through other parents as long as their routes towards the sink are better. We take advantage of the opportunistic forwarding to support higher-priority delay-sensitive alarms that need to arrive in sink before a given deadline along with low-intensity monitoring data considered as best-effort. We compare our opportunistic version of RPL to its basic version through detailed simulations in terms of packet delivery ratio, incurred delay, and overhead. Our scheme results in improved packet delivery, shorter delays, while keeping almost the same overhead.

MSWiM'11, October 31–November 4, 2011, Miami, Florida, USA.

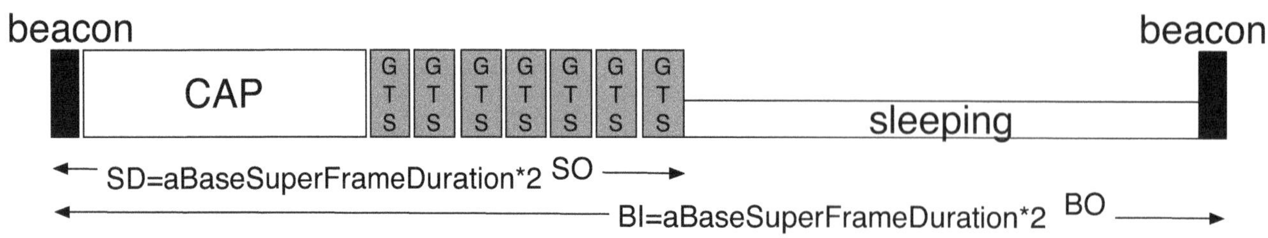

Figure 1: The superframe structure in IEEE 802.15.4

2. RELATED WORK

We briefly review here the work related to IEEE 802.15.4 and RPL.

2.1 IEEE 802.15.4

IEEE 802.15.4 defines PHY and MAC layers for wireless sensor networks [3]. In the non-beacon mode, it implements a classical CSMA-CA approach in which a receiver periodically initiates communication (the period is implementation dependent) and the PAN (wireless Personal Area Network) coordinator has to remain awake to receive frames from its children and cannot save energy. This feature increases the overhead and has a significant impact on the end-to-end delay.

IEEE 802.15.4 also offers the beacon-enabled mode in which a coordinator periodically sends a beacon to delimit its superframe (cf. Figure 1). The Contention Access Period (CAP) follows the beacon: all associated nodes may send a packet according to a slotted CSMA-CA. The superframe finishes after the Contention Free Period (CFP): nodes reserve dedicated Guaranteed Time Slots (GTS). In the beacon mode, nodes wake-up just before a beacon transmitted by the coordinator to save energy—a node may sleep during the inactive period if $BO > SO$. We consider the beacon mode in this paper, because it leads to better energy savings.

IEEE 802.15.4 requires a cluster-tree topology anchored at the PAN coordinator for multihop operation [4]. Once a node associates with a coordinator, it begins to periodically send beacons to maintain its own superframe and control the access to the channel of the associated nodes. Thus, different nodes may have overlapping superframes, which may create collisions between beacons and data frames. To alleviate this problem, two main approaches exist in the literature:

- *Beacon-Only Period (BOP)*: at the beginning of each superframe, nodes reserve a period of several slots with each slot able to contain a beacon [5] (cf. Figure 2). Interfering coordinators should choose different BOP slots. This method only solves collisions between beacons: it is only suitable for low-intensity traffic.
- *Superframe Scheduling*: nodes implement a distributed scheduling algorithm so that interfering coordinators maintain non-overlapping superframes [6][7]. Although such a collision-free scheduling is more complex, it results in better capacity.

An experimental comparison of both techniques [8] showed that Superframe Scheduling outperforms BOP in terms of the number of delivered packets: if two interfering coordinators do not have overlapping superframes, the number

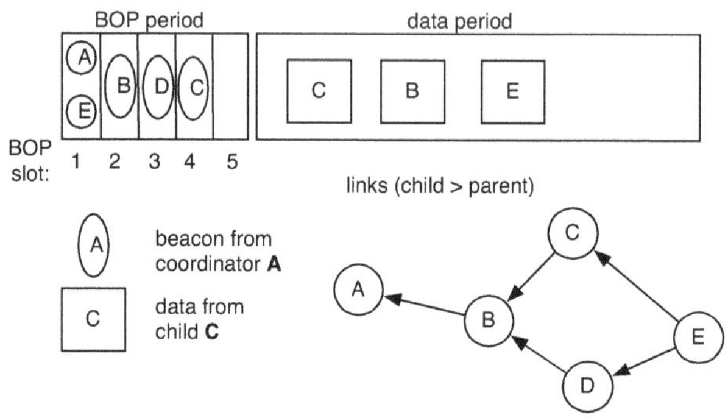

Figure 2: Beacon-Only-Period superframe organization

of collisions between `data` packets and between `beacons` is reduced. Nevertheless, its parameters (BO) should be carefully set to avoid excessive battery consumption.

2.2 Routing in multihop wireless networks

RPL has recently emerged as the standard for routing in wireless sensor networks. It defines a framework for constructing a routing topology and efficiently forwarding packets while saving energy.

The topology is based on a Destination-Oriented Directed Acyclic Graph (DODAG) anchored in the sink (the PAN coordinator in the IEEE 802.15.4 jargon). This kind of structure is particularly suitable for *convergecast* traffic: the sink is the destination for all data packets. Each node maintains its *rank* towards the sink and includes its value in periodical DIO (DODAG Information Object) messages. This rank represents more or less its *depth* in the DODAG.

To avoid loops, an objective function determining the rank is defined so that it monotonically decreases toward the sink. A simple way proposed by ROLL for computing the rank is to closely track ETX (Expected Transmission Count). This metric estimates the average number of transmissions required to send a data packet to a neighbor. By summing up the ETX along the path toward the sink, a node may obtain the cumulative ETX, strictly decreasing and consequently forbidding loops. RPL uses trickle timers to adapt the overhead incurred by the maintenance of a DODAG to the topology stability—when it is stable, nodes increase the period for sending routing advertisements.

RPL is currently under extensive implementation and experimentation work so little results are published so far.

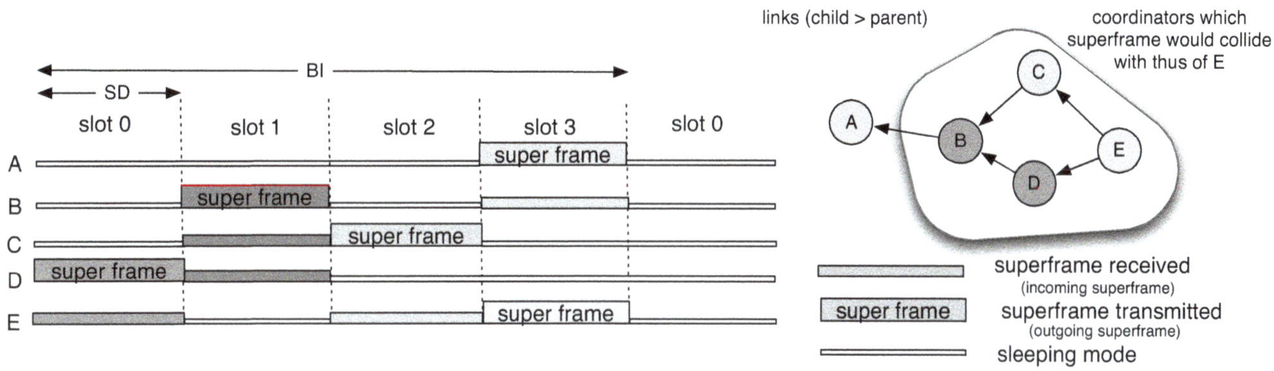

Figure 3: Scheduling of 802.15.4 superframes

Vasseur et al. experimentally evaluated RPL and showed that the use of trickle timers decreases the number of control traffic necessary to construct and maintain a stable DODAG [9]. They also observed the amount of control traffic in the case of the general repair. Clausen et al. pointed out some under specifications of the protocol, especially concerning downward traffic and point-to-point routing [10].

Since nodes in a wireless sensor network are prone to failures, RPL builds a DODAG with several paths towards the sink so that each node maintains several parents: if the preferred one fails, the node can switch to another one. The existence of several paths may improve resilience and performance [11], help to guarantee QoS [12], or allow to deal with real-time traffic [13].

Instead of providing backup routes, opportunistic routing dynamically chooses the next hop on a per packet basis [14]. Since the transmission conditions are highly time-varying, a node may estimate the link quality and the best next hop just before the transmission.

To the best of our knowledge, no work so far considered the joint RPL and IEEE 802.15.4 operation.

3. MULTIPATH OPPORTUNISTIC RPL OVER IEEE 802.15.4

Our objective is to integrate RPL and IEEE 802.15.4 to enable QoS multipath routing and improve packet delivery before a deadline, while minimizing overhead and energy consumption. We assume that there are two types of traffic in the network:

1. **low intensity monitoring data** that can be considered as best-effort;
2. **higher-priority delay-sensitive alarms** that need to arrive in sink before a given deadline.

We want to provide support for such service differentiation over RPL and multi-hop wireless network by taking advantage of multiple paths.

The first problem to consider is the tree structure of IEEE 802.15.4: in the current standard, a node can only associate with one parent to avoid loops. Indeed, all the nodes at a given depth in a tree maintain their superframe simultaneously. To take advantage of several parents in the DODAG structure, we need to allow a node to associate with several parents in the IEEE 802.15.4 structure (we consider this problem below in Section 3.1). Then, we propose an extension to RPL for using several parents in an opportunistic way—as soon as a node can send a frame to a parent in a synchronized way, it will do it, which increases the probability of packet delivery and reduces delay (cf. Section 3.2).

3.1 Support for Multiple Paths in IEEE 802.15.4

A node joining an IEEE 802.15.4 cluster-tree needs to associate with a neighbor coordinator. It may implement an active scan method: it sends a solicitation to retrieve the information of the cluster-tree from listening neighbors. This method only works when nodes never sleep, which is not the case of the applications we are targeting. In a passive scan, a node waits for `beacons` from a neighbor coordinator that define a superframe even if it is not yet associated with the cluster-tree. Then, an `association-request` / `ack` / `data-request` / `association-reply` / `ack` exchange is required to finalize the association. Thus, a node can associate with one parent during a given superframe. In the original IEEE 802.15.4, an incoming superframe directly follows an outgoing superframe (i.e. the superframes of a node and its children are immediately consecutive). To avoid loops, a node should only associate with a parent closer to the PAN coordinator. Thus, there is no way of maintaining several parents in the original IEEE 802.15.4.

To implement a DAG structure maintaining several parents, we have to appropriately schedule superframes: a node should be able to receive `beacons` from different neighbors during non overlapping superframes. We assume that all the nodes have the same BO and SO values (duration and interspacing between superframes). Consequently, a schedule of superframes corresponds to a TDMA assignment with timeslots of a fixed length. We propose to extend the algorithm proposed by Muthukumaran *et al.* [7] with a random assignment of slots to interfering nodes in a given neighborhood.

3.1.1 Detailed Schedule Construction

We assume that only 1-hop and 2-hop neighbors may interfere. To avoid collisions, we have to schedule superframes so that no pair of interfering coordinators use the same slot. The number of these slots depends on the BO and SO parameters: BO is the repetition period of superframes during which we can place at most N_{slots} superframes:

$$N_{slots} = 2^{BO-SO} \quad (1)$$

We propose a simple algorithm that each node executes to find a valid slot to transmit its superframes:

1. Each node retrieves the slots used by its 2-neighbors (i.e. interfering coordinators). A node considers only nodes with at least one child since other coordinators maintain an empty superframe and do not create collisions for `data` frames;

2. A node randomly selects one free slot.

While Muthukumaran *et al.* schedule the slots to be immediately consecutive, we rather chose a random assignment approach for the following reasons:

- it presents a trade-off to reduce the delay for both upload and download directions (Muthukumaran *et al.* focus only on the upload case);

- it accelerates convergence: two interfering nodes have little chance to choose the same slot when they have the same parent.

The resulting structure has an interesting property: a node can access the channel during several superframes coordinated by different parent neighbors, which provides the basis for multipath forwarding supported by RPL.

3.1.2 Distributed version

To obtain the list of interfering nodes, each node maintains a neighborhood table. It piggybacks in its `beacons` the list of neighbors and the slots they have already chosen. Thus, a node has just to select one free slot to avoid collisions. This procedure can be safely integrated in the association procedure of IEEE 802.15.4: a new coordinator chooses its slot just after having received its `association-reply` scheduling immediately its future superframes.

As long as a node does not have children, it may safely change its slot without impacting the MAC performance: there is no domino effect. Besides, colliding `beacons` often avoid children to associate with interfering coordinators. In other words, a coordinator without any child means perhaps that its slot collides with another one. Thus, a coordinator reapplies the assignment rule to choose its slot at the end of each of its superframes, except for an `association-request` during the current superframe (i.e. at least one child is associated).

3.1.3 Example

Figure 3 illustrates the principle of slot assignment. We assume here for the sake of simplicity that the interfering range is twice the radio range so that B and E cannot choose the same superframe. If we consider that each superframe slot is a different color, we face a classical graph coloring problem. The reader can notice that the number of slots in this example should be at least 4 since B, C, D, and E interfere with each other.

Let us focus on node E assuming that all other coordinators have already chosen a slot for their superframes. E will collect `hellos` from C and D, and will be able to construct the list of its 2-neighbors (interfering nodes). Finally, E has just to randomly select one slot not yet present in this list. In our case, it will select slot 0 since it is the only one remaining free.

3.2 Deadline Oriented Opportunistic RPL over IEEE 802.15.4

As we aim at service differentiation of best-effort and time-sensitive traffic, we propose to adapt RPL so that it can take into account delay and packet delivery before a deadline. We only consider convergecast (multipoint-to-point) traffic.

We consider three classes of service:

min-delay: critical packets for which we need to minimize the end-to-end delay without concerns for energy consumption

deadline: alarm packets to deliver before deadline D while minimizing energy consumption

best effort: packets that do not require any guarantee, but their forwarding needs to take into account energy consumption.

3.2.1 Notation

We adopt the following notation:

- N: a node that forwards a packet
- $deadline(p)$: deadline associated with packet p
- t: current time
- $d(N)$: hop distance between node N and the sink
- $slot(t)$: current superframe slot (node N has just received a beacon in this slot so it will be able to forward it)
- $slot(NH)$: slot used by the superframe of node NH
- $PDR_{bcn}(NH)$: beacon packet delivery ratio for neighbor NH
- $PDR_{data}(NH)$: data packet delivery ratio over the link to neighbor NH;
- $t_{tx_{data+ACK}}$: time needed for data and acknowledgement frame transmissions
- $queue(NH)$: queue of packets scheduled for transmission during the superframe of NH
- $STEP$: a constant to extend the local time budget if no parent can satisfy it at the first attempt

3.2.2 Taking into account deadlines with RPL

When a node generates a packet, it assigns a deadline according to the class of service it belongs to. Nodes maintain a queue of packets ordered by their deadlines. When a node has a packet to forward, it waits for a successful `beacon` reception from one of its parents. Then, it needs to decide to transmit its packets during the current superframe or later if another parent offers better performance (e.g. smaller energy consumption, better reliability).

The node extracts the first packet from its queue: if the deadline is elapsed, the packet is simply dropped and the next packet is extracted. Then, it must find the next hop that guarantees the deadline. It assumes the time before the deadline can be uniformly shared among the nodes in the route. Thus, the transmission has to meet the local *time budget* constraint:

$$budget = \frac{deadline(p) - t}{d(N)} \quad (2)$$

When a packet is at node N, the delay before the packet is correctly transmitted to the next hop NH depends on:

1. the delay until the superframe of NH starts while taking into account the average number of superframes to wait in case of beacon losses:

$$D_{sframe} = SD * \mid slot(NH) - slot(t) \mid + BI * max\left(0, \frac{1}{PDR_{bcn}(NH)} - 1\right) \quad (3)$$

   where SD denotes the superframe duration while BI represents the time separating two beacons. For the currently received beacon, this delay is zero, since the node can immediately try to send the packet.

2. the average delay until NH correctly receives the packet, it is estimated via the packet probability delivery ratio:

$$D_{tx} = \frac{t_{tx_{data+ACK}}}{PDR_{data}(NH)} \quad (4)$$

Finally, nodes need to satisfy the following deadline constraint:

$$budget \geq D_{sframe} + D_{tx} \quad (5)$$

As several candidate next hop nodes may satisfy the deadline constraint, the protocol will choose the best one based on the cumulative Expected Transmission Count (ETX) routing metric: it represents the cumulative number of packet transmissions required to reach the sink, also used as the node *rank* in the DODAG. It should be noted that per link ETX is simply calculated as inverse value of measured $PDR_{data}(NH)$ (link to neighbor NH). A node chooses among all the possible next hops the node with the lowest cumulative ETX so packets will experience the smallest delay.

If a node cannot satisfy the budget constraint, it reconsiders all the parents for an increased time budget hoping that the packet will benefit from shorter delay further in the network. We gradually extend the budget up to two times the initial value given by Eq. 2 ($STEP$ and $relax$ parameters in Algorithm 1)

Algorithm 1: Does a packet has to be transmitted in the current superframe?

```
src ← waitBeacon();
nexthopcandidate ← ∅
if empty(queue) or end(superframe, src) then
  return false
else
  repeat
    p ← getFirstPacket(queue);
    if (p.deadline ≤ t) then
      DropPacket(p);
    end if
  until (p.deadline > t)
  budget ← computeHopBudget(p, src.hops + 1);
  relax ← 0;
  while (nexthopcandidate = ∅) and (relax < 2 * budget) do
    for neigh NH do
      D_sframe ← computeDelaySuperframe(NH.sframe, NH.pdr);
      D_tx ← computeExpectedTransmissionTime(NH.pdr);
      if (budget + relax > D_sframe + D_tx) then
        nexthopcandidate ← nexthopcandidate + {NH}
      end if
    end for
    relax += budget * STEP;
  end while
  if (src = getBestETX(nexthopcandidate)) then
    return true
  else
    return false
  end if
end if
```

4. PERFORMANCE EVALUATION

We have compared our opportunistic version of RPL to its basic version in terms of packet delivery ratio, incurred delay, and overhead through detailed simulations. Both protocols take advantage of the 802.15.4 superframe scheduling so that even the basic version of RPL can dynamically adapt the choice of the preferred parent node in function of performance.

For the sake of simplicity, we implemented a centralized coloring solution to assign slots for each superframe as described in Section 3.1.1. However, we may use any scheduling algorithm creating a Directed Acyclic Graph such that the distributed version described in Section 3.1.2.

4.1 Simulation setup

We have used the WSNet/Worldsens event-driven simulator for large scale wireless sensor networks [15]. We have ported the Contiki RPL implementation [16] to WSNet. We used the IEEE 802.15.4 implementation in beacon-enabled mode [17]. We have additionally implemented the superframe scheduling mechanism.

The simulations have considered 10 different topologies and randomly deployed up to 256 nodes in a square area 400 x 400 m. To make the simulations as close as possible to the reality, we have not adopted the Unit Disk Graph assumptions commonly used in the literature, but rather the Rayleigh propagation model and the parameters of the IEEE 802.15.4 radio.

We have only considered low intensity traffic with the av-

Simulated area	400m x 400m
Number of nodes	up to 256
Traffic type, rate	periodic, 1/7.5 minutes
Simulation duration	50000 s
SO	3, 4, 5
Deadline	360s, 180s

Table 1: Simulation parameters

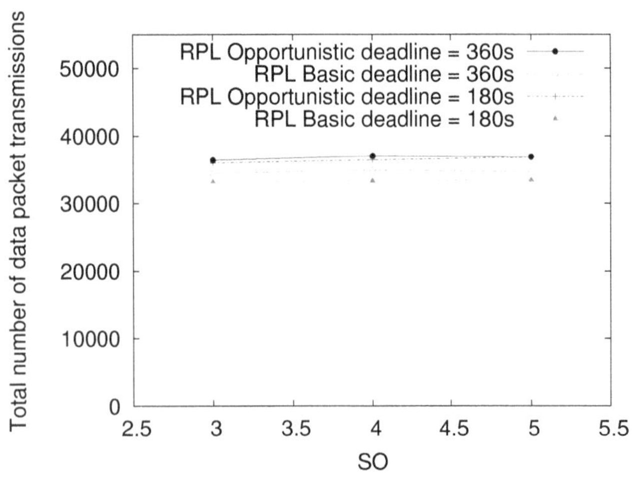

(a) total number of transmitted packets

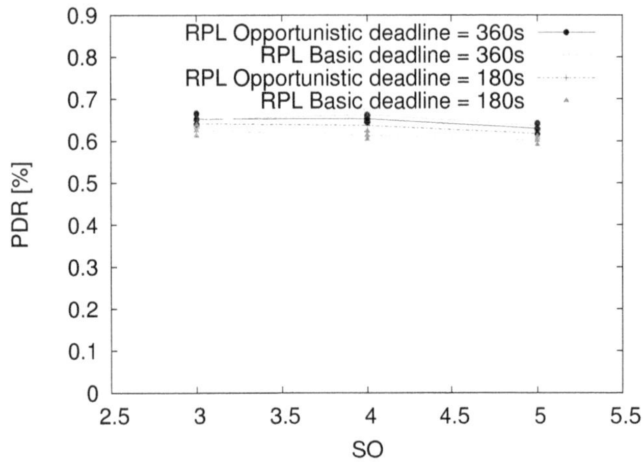

(b) PDR for *all* packet types

Figure 4: General comparison of basic and opportunistic RPL

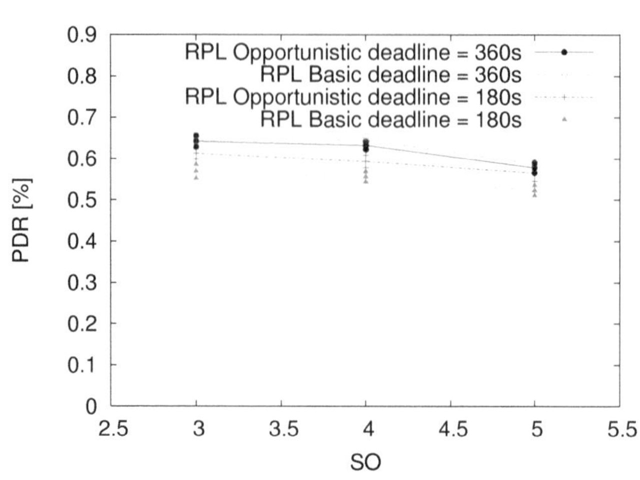

(a) PDR

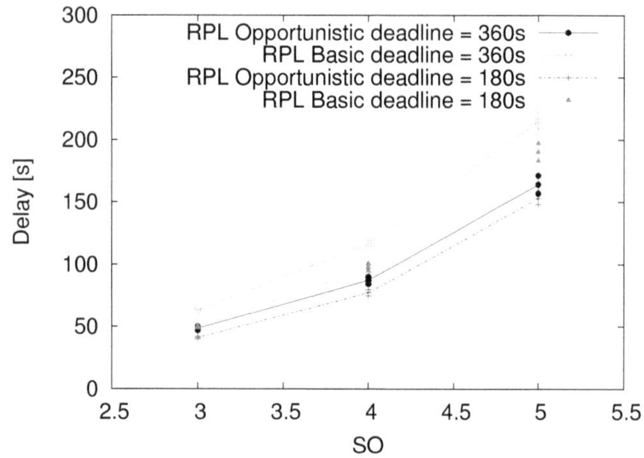

(b) delay

Figure 5: Comparison of basic and opportunistic RPL, *deadline* traffic type

erage interval between data of 7.5 minutes. We have empirically established this value to avoid performance degradation of IEEE 802.15.4 under heavy traffic. We divide the traffic into three classes with the respective proportions: 70%-20%-10% (best-effort, min-delay, deadline). We vary the SO parameter from 3 to 5 and choose the BO parameter so that the number of superframe slots is sufficient to avoid colliding superframes (Eq. 1). We run a simulation for each of the topologies during 50.000 s and average the results over multiple runs to obtain 95% confidence intervals. Table 1 summarizes the important simulation parameters.

4.2 Result analysis

Figure 4(a) presents the total number of transmitted packets for the basic and opportunistic RPL. We measure the raw number of packets transmitted by the MAC layer, i.e. a data frame transmitted for the first time or retransmitted after a failure. The IEEE 802.15.4 MAC layer drops a frame when the number of retransmissions exceeds 3 or the number of Clear Chanel Assessments exceeds 4. Packet are also dropped if the deadline is missed. At the end of the simulation, these values are summed up for all generated packets. Both protocols generate the same fixed amount of application data packets and none of them is destroyed before the end of the simulation.

We can notice that our opportunistic solution results in a slightly greater number of transmitted packets (9%). This increase may come from better performance: since less packets are dropped, this mechanically results in more transmissions at the MAC layer. A larger overhead also comes from the forwarding rule: if the deadline is short, the node will privilege the forwarding delay compared to minimizing the number of transmissions (i.e. ETX). This aggressive decision would privilege short deadlines, but also negatively impacts the number of transmissions.

Figure 4(b) presents the packet delivery ratio for all packet types. As soon as the deadline becomes more critical, the fact that we use alternative parents results in a higher PDR. Less packets get dropped because of the short packet deadline, since the fact that we have missed a beacon from our preferred parent could still be caught up with the reception of beacons from some alternative parents. In the same situation, the basic RPL version needs to wait for the whole

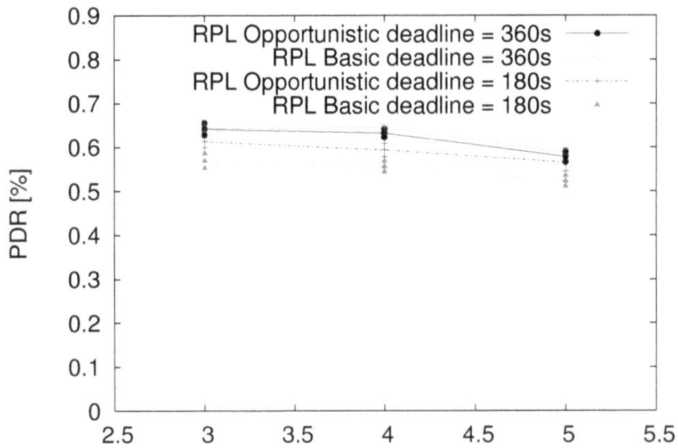

(a) PDR

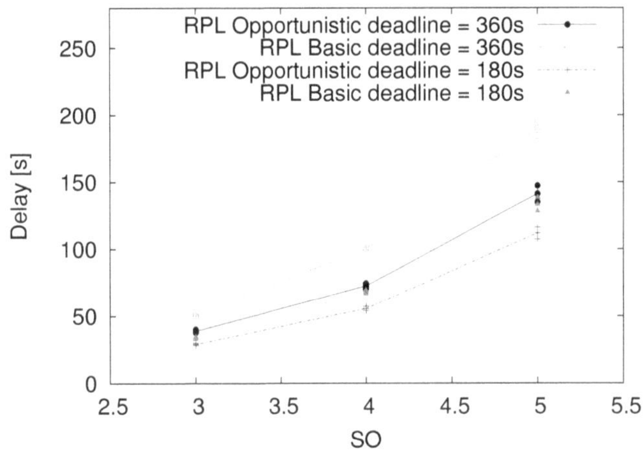

(b) delay

Figure 6: Comparison of basic and opportunistic RPL, *min-delay* traffic type

inter-beacon period (BI) and thus there is a risk that the short packet deadline incurs more packet drops.

We have considered above the performance of the tested protocols from the global point of view. Let us analyze performance with respect to the QoS delay requirements of *min-delay* and *deadline* data packet types. We can notice a similar behavior as previously for both types of traffic when it comes to PDR (Figures 5(a) and 6(a)) and for the experienced delay (Figures 5(b) and 6(b)). If we consider delay, it is clear that our opportunistic scheme exhibits much shorter delay than the basic RPL version due to the possibility of interchangeably using alternative parents. With respect to the PDR performance, our opportunistic scheme presents a real gain when we deal with packets with harsh deadline constraints.

Finally, we can notice an interesting property of our opportunistic approach that directly comes from its forwarding policy even if we do not show it in the numerical results. Our opportunistic scheme spreads traffic over parents by not only selecting the preferred parent as the next hop, but the alternative ones as well. In realistic scenarios with a limited battery capacity and limited queue lengths, this may appear as the primary concern since it would increase the overall network lifetime and prevent packet drops due to full queues. We plan to include this kind of realistic constraints in the future work.

5. CONCLUSION AND PERSPECTIVES

We have presented a scheme allowing the coexistence of two structures in emerging IP enabled wireless sensor networks: RPL routing and IEEE 802.15.4 MAC. We have modified the cluster-tree operation of IEEE 802.15.4 to support RPL DODAG so that nodes are able to follow multiple parents and use them for traffic forwarding when needed. Our simple opportunistic routing scheme benefits from an interesting feature: traffic is spread more uniformly over all possible parents instead of going through the preferred one. It results in improving the network lifetime and optimizing fairness to avoid quick energy depletion. Our solution achieves slightly better results with respect to end-to-end packet reliability (PDR) and delay while keeping almost the same amount of generated traffic.

In the future, we plan to validate the proposed scheme on an experimental testbed and take into account additional constraints of limited packet buffers. We also want to explore how reliability can improve with the use of more advanced multi-path routing schemes or by adapting the parameters of IEEE 802.15.4 to accommodate higher traffic loads.

6. ACKNOWLEDGEMENT

Authors would like to thank Nazim Abdeddaim for providing his source code implementation of the IEEE 802.15.4 protocol. His comments were of great help as well while adapting the code to work with RPL.

This work was partially supported by the French National Research Agency (ANR) project ARESA2 under contract ARESA2 ANR-09-VERS-017 and the Competitive Clusters Minalogic and System@tic under contracts FUI SensCity and SAMEE.

7. REFERENCES

[1] G. Montenegro, N. Kushalnagar, J. Hui, and D. Culler. Transmission of IPv6 Packets over IEEE 802.15.4 Networks. RFC 4944, IETF, sept. 2007.

[2] T. Winter (Ed.), P. Thubert (Ed.), and RPL Author Team. RPL: IPv6 Routing Protocol for Low Power and Lossy Networks. Internet Draft draft-ietf-roll-rpl-19, work in progress.

[3] IEEE 802.15.4 2006 standard, `http://www.ieee802.org/15/pub/TG4.html`.

[4] F. Cuomo, E. Della Luna, E. Cipollone, P. Todorova, and T. Suihko. Topology Formation in IEEE 802.15.4: Cluster-Tree Characterization. In *PerCom*. IEEE, 2008.

[5] IEEE 802.15 WPAN Task Group 4b (TG4b). `http://www.ieee802.org/15/pub/TG4b.html`, September 2004.

[6] Anis Koubâa, André Cunha, Mário Alves, and Eduardo Tovar. TDBS: a Time Division Beacon Scheduling Mechanism for ZigBee Cluster-Tree Wireless Sensor Networks. *Real-Time Systems*, 40(3):321–354, 2008.

[7] Panneer Selvan Muthukumaran, Rodolfo de Paz, Rostislav Špinar, and Dirk Pesch. MeshMAC: Enabling Mesh Networking over IEEE802.15.4 Through Distributed Beacon Scheduling. In *International Conference on Ad Hoc Networks (AdHocNets)*, Niagara Falls, Ontario, Canada, September 2009.

[8] B.C. Villaverde, R. De Paz Alberola, S. Rea, and D. Pesch. Experimental Evaluation of Beacon Scheduling Mechanisms for Multihop IEEE 802.15.4 Wireless Sensor Networks. In *International Conference on Sensor Technologies and Applications (SENSORCOMM)*, pages 226–231. IARIA, July 2010.

[9] J. P. Vasseur J. Tripathi, J. C. de Oliveira. A Performance Evaluation Study of RPL: Routing Protocol for Low Power and Lossy Networks. *Conference on Information Sciences and Systems (CISS)*, 2010.

[10] Thomas Clausen and Ulrich Herberg. Some Considerations on Routing in Particular and Lossy Environments. *IETF Workshop on Interconnecting Smart Objects with the Internet*, 2011.

[11] Deepak Ganesan, Ramesh Govindan, Scott Shenker, and Deborah Estrin. Highly-Resilient, Energy-Efficient Multipath Routing in Wireless Sensor Networks. *ACM SIGMOBILE Mobile Computing and Communications Review*, 5(4):10–24, October 2001.

[12] Xiaoxia Huang and Yuguang Fang. Multiconstrained QoS Multipath Routing in Wireless Sensor Networks. *Wireless Networks*, 14(4):465–478, 2008.

[13] B. Yahya and J. Ben-Othman. An Energy Efficient and QoS Aware Multipath Routing Protocol for Wireless Sensor Networks. In *Conference on Local Computer Networks (LCN)*, pages 93–100. IEEE, oct. 2009.

[14] Haitao Liu, Baoxian Zhang, Hussein T. Mouftah, Xiaojun Shen, and Jian Ma. Opportunistic Routing for Wireless Ad Hoc and Sensor Networks: Present and Future Directions. *IEEE Communications Magazine*, 2009.

[15] A. Fraboulet, G. Chelius, and E. Fleury. Worldsens: Development and Prototyping Tools for Application Specific Wireless Sensors Networks. In *IPSN'07 Track on Sensor Platforms, Tools and Design Methods (SPOTS)*, Cambridge, Massachusetts, USA., April 2007. ACM.

[16] Jeonggil Ko, Joakim Eriksson, Nicolas Tsiftes, Stephen Dawson-Haggerty, Andreas Terzis, Adam Dunkels, and David Culler. ContikiRPL and TinyRPL: Happy Together. In *Proceedings of the workshop on Extending the Internet to Low power and Lossy Networks (IPSN 2011)*, Chicago, IL, USA, April 2011.

[17] Nazim Abddedaim and Fabrice Theoleyre. Implementation of a WSNet Module to Simulate the IEEE 802.15.4 Beacon-Enabled Mode in Multihop Topologies. Research Report 00590853 `http://hal.archives-ouvertes.fr/hal-00590853`, HAL, May 2011.

Controlled Potential-Based Routing for Large-Scale Wireless Sensor Networks

Daichi Kominami
Graduate School of Information Science and Technology
Osaka University
1-5 Yamadaoka, Suita
Osaka 565-0871, Japan
d-kominami@ist.osaka-u.ac.jp

Masashi Sugano
School of Comprehensive Rehabilitation
Osaka Prefecture University
3-7-30 Habikino, Habikino
Osaka 583-8555, Japan
sugano@rehab.osakafu-u.ac.jp

Masayuki Murata
Graduate School of Information Science and Technology
Osaka University
1-5 Yamadaoka, Suita
Osaka 565-0871, Japan
murata@ist.osaka-u.ac.jp

Takaaki Hatauchi
Fuji Electric Co., Ltd.
1 Fuji, Hino
Tokyo 191-8502, Japan
hatauchi-takaaki@fujielectric.co.jp

ABSTRACT

Improving the scalability of wireless sensor networks is an important task, and toward this end, much research on self-organization has been conducted. However, the problem remains that much larger networks based on pure self-organization cannot be guaranteed to behave as desired. In this paper, we propose a controlled potential-based routing protocol. This protocol is based on a novel concept: a "controlled self-organization scheme", which is a self-organization scheme accompanied by control from outside the system. This scheme ensures desired network behavior by controlling a portion of nodes operated in self-organization. Through simulation experiments with a multi-sink network, we show that traffic loads can be equalized among heterogeneously distributed sink nodes, and moreover, that load balancing among the relay nodes can bring about a 138% extension of network lifetime.

Categories and Subject Descriptors

C.2.2 [COMPUTER COMMUNICATION NETWORKS]: Network Protocols—*Routing protocols*

General Terms

Algorithm, Design, Performance, Reliability

Keywords

Controlled Self-Organization, Potential-Based Routing, Diffusion Equation, Load Balancing

1. INTRODUCTION

Recent advances in wireless and micro-electromechanical system (MEMS) technologies have made it possible to develop extremely small sensor nodes that have wireless communication facilities, and as a result, considerable attention has been directed at wireless sensor networks (WSNs). Wireless sensor networks consist of a large number of sensor nodes and sink nodes, and these networks can be used for a wide range of applications free of any need for a fixed network infrastructure, such as a monitoring of phenomena over a wide area.

Many challenges must be met to realize a WSN, and one of these is to improve scalability of routing. When dealing with thousands of sensor nodes, the wireless channel is occupied by the exchange of routing information, which consumes more energy and more bandwidth. Within such networks, it is impractical to give numerous nodes unique identifiers (IDs) and exchange all routing information among them. This stands in contrast to IP networks, where an ability to access an arbitrary node is required. Another scalability problem is that heavy loads are placed on nodes neighboring the sink node, as those nodes, as a consequence of many-to-one (or many-to-some) communication, must forward data from other sensor nodes in addition to their own data.

In order to reduce the amount of routing information that needs to be exchanged throughout the network, the emphasis is on self-organization, that is, entirely local decision-making by each node based only on local information. Self-organization has good scalability, adaptability, and robustness [4], which are important properties for sensor networks. These properties are brought about by numerous interactions among the local-level components of a system without external or centralized control process. Conversely, pure self-organization has the following problems because of its bottom-up design [12] in which only local information is used.

- Guaranteeing an optimal operation is difficult.
- Confirming an operation on the entire network is difficult.
- Convergence speed after an environmental change is slow.

Since each entity in self-organization makes a decision on the basis of local information, local optimization is easy to achieve, while a global optimal operation of the network is hard to guarantee. This local decision-making based on the local interaction in large systems derives emergence, however as is generally known, controlling emergence is not realistic. Then, once a self-organized system converges to act undesirably, it remains unchanged. In [12], the author suggests use of an observer/controller architecture, where an observer and a controller are responsible for correcting the system behavior. In the case of optimal operation in self-organized routings, since each node selects its next hop based on only local information, it is not possible to deal with network problems from a macro viewpoint; one would not be able to alleviate excess concentrations of communication load induced by an irregular node distribution density.

To resolve these problems, we propose a controlled self-organized routing, which is based on a novel scheme called "controlled self-organization" [2]. This scheme rests on a controller which is responsible for ensuring the desired behavior of the system, for guaranteeing a high degree of system perform, and for encouraging a convergence of the state of the system. That is, the controller makes the self-organized system *manageable* by controlling some of the components within the self-organization scheme to accomplish the above objectives.

In this paper we adopt a potential-based routing [1, 5, 6, 8–11, 13, 15, 16] as a self-organized routing and apply controlled self-organization scheme to it. In other words, we propose a *controlled potential-based routing (CPBR)*. In potential-based routing schemes, nodes have a scalar value called "potential", and a next hop is based on only a node's potential and its neighbors' potentials. Here, a node calculates its own potential from the potentials, the number of hops from the sink node, or the remaining energy of itself or its neighbors. Basically, the smaller a hop count to the sink node is, the lower is the potential value assigned to it. Therefore, if a node simply transmits data toward one of its neighbor nodes having a smaller potential than its own, the data will eventually reach the sink node.

On the reduction of the load on nodes neighboring the sink node, multiple sink nodes are deployed over the network, with data obtained by the sink nodes transmitted to a data server (users or applications can then access the data from the server as necessary). Here, each sensor node does not necessarily select only one sink node as its destination. When potential-based routing is applied to multi-sink networks, each node only has to forward data according to potentials: the data will eventually reach a sink node. Thus, potential-based routing can be straightforward to use on multi-sink sensor networks.

Because the potential-based routing is based on self-organization, it has above-described some problems, such as concentrations of communication load provoked by an inhomogeneous distribution density of sensor or sink nodes. In CPBR, sink nodes act as controllers, according to network manager's requests, that adjust their own potentials to construct a desired potential field. We assume that multiple sink nodes are connected to one another on a high-bandwidth wired network so that they can exchange various information periodically and instantaneously; for example, the number of received data packets and the remaining energy of their neighbor nodes. Sink nodes control their own potential on the basis of this information so that even as sensor nodes decide their own potential from local information, a preferable potential field is constructed over the entire network.

In order to realize CPBR, we begin by introducing a potential-field construction from the diffusion equation in Section 2. In the section, we first show how sensor nodes construct a potential field and select a next hop according to it. We then extend this method of construction to include local optimization. In Section 3, we describe controlled potential-based routing intended for overall optimization; then in the following section, we present the results of a simulation. Finally, we conclude our paper in Section 5.

2. POTENTIAL-BASED ROUTING

2.1 Related Work

Various works related to potential-based routing have been conducted [1, 5, 6, 8–11, 13, 15, 16]. Such efforts are classified into following two types.

1. Physics-knowledge based schemes
2. Hop-count based schemes

As for physics-knowledge based schemes, the focus has been on analogies between electrostatic fields and sensor networks [5, 6] and on analogies between electrical circuits and sensor networks [10]. In these studies, potentials are assigned at sensor nodes by solving a Poisson equation. With the solution obtained from the equation, a load-balancing potential field is constructed. However, to solve the equation, certain location information is needed for the nodes. However, obtaining and exploiting location information assume the availability of GPS receivers or some other means, and thus the cost of producing such nodes becomes much higher. By extension, the cost for deploying sensor nodes also rises, which is a potentially critical problem for constructing a large scale network. In a sense, this is also a scalability problem. A related difficulty is that it might not be feasible to provide GPS receivers indoors, in rooms below ground level, within heavily forested areas, or at other such locations with limited or obstructed satellite coverage.

As for hop-count based schemes, nodes calculate their own potential essentially from their hop count from the sink nodes [8, 11, 15, 16]. In references [8, 11, 15], nodes also use their own remaining energy and that of their neighbor nodes for load balancing. An effective data aggregation mechanism supported by potential-based routing was proposed in [16], where local queue-length information is used to calculate potentials. Although the proposed routing schemes exhibit good performance, location information is required in the schemes of references [9, 11]. Also, the parameters used to calculate potentials were insufficiently examined and evaluated in references [8, 15], and thus the difference between their proposals and mere minimum hop routing along with remaining energy information is not clear. Moreover, the most important problem is that all the above-mentioned studies offer no mechanism for guaranteeing intended network operation. To resolve this critical problem, we propose controlled potential-based routing (CPBR). Our CPBR utilizes a physics-knowledge based scheme inspired by thermal diffusion phenomenon as well as [13], however, our CPBR does not require location information. We focus on an analogy between the heat conduction from a heat source and the potential conveyance from a sink node. In CPBR, sensor nodes have to change their own potential according to the potential of the sink nodes. Making use of the diffusion equation that describes heat conduction, CPBR allows diffusion of the potentials of sink nodes throughout the entire sensor network.

2.2 Potential Field Construction from the Diffusion Equation

The diffusion equation is shown by the partial differential equation (1), which provides magnitude ϕ of the diffusing quantity at time t at position $\mathbf{x}$.

$$\frac{\partial \phi(\mathbf{x}, t)}{\partial t} = D \triangle \phi(\mathbf{x}, t), \quad (1)$$

where D is the diffusion rate and takes a positive value. By discretizing this equation and regarding ϕ as a potential, it becomes possible to construct a potential field based on self-organization where the behavior is governed by only local information.

2.2.1 Discrete Diffusion Equation

Node n calculates its own potential at time step $t+1$ (denoted by $\phi(n, t+1)$) based on the discrete diffusion equation (2). In equation (2), $N_a(n)$ denotes a set of nodes neighboring node n. It can be noted from the equation, that location $\mathbf{x}$ is cleared and the potential of node n is obtained from the latest potentials of $nb(n)$ and its own last potential. In order to calculate potentials, nodes must periodically inform their neighbor nodes of their own potentials.

$$\phi(n, t+1) = \phi(n, t) + D(n) \sum_{k \in N_a(n)} \{\phi(k, t) - \phi(n, t)\}. \quad (2)$$

In discrete equation (2) (derived from continuous equation (1)), it can be thought that $D(n)$ is a parameter that changes the magnitude of influences by potentials of the neighbor nodes. An important point should be noted: potentials may oscillate when $D(n)$ is large. For the solution of the problem, we consider the case where node n has an only one neighbor node m. Equation (2) can thus be replaced by $\phi(n, t+1) = D(n)\phi(m, t) + (1 - D(n))\phi(n, t)$, which represents an internal/external division of the points on the number line. In the following, we consider the case of $\phi(n, t) < \phi(m, t)$.

In the case $0 < D(n) < 1$:

After node n receives the potential of node m, the following inequality is satisfied: $\phi(n, t) < \phi(n, t+1) < \phi(m, t)$. Repeating this procedure, potentials of node n and node m approach and converge between $\phi(n, t)$ and $\phi(m, t)$. In this case, node n's potential remains smaller than node m's potential.

In the case $1 \leq D(n) < 2$:

After node n receives the potential of node m, the following inequality is satisfied: $\phi(m, t) < \phi(n, t+1) < 2\phi(m, t) - \phi(n, t)$. Repeating this procedure, the potentials of node n and node $m$approach and converge, but the relationship between the magnitude of node n's potential and node m's potential is indefinite.

In the case $2 \leq D(n)$:

After node n receives the potential of node m, the following inequality is satisfied. $2\phi(m, t) - \phi(n, t) \leq \phi(n, t+1)$. Repeating this procedure, potentials of node n and node m remain unchanged or diverge. Moreover, the magnitude relationship between node n's potential and node m's potential is indefinite.

For the diffusion of potentials, it is preferable that $D(n)$ satisfies the following expression: $0 < D(n) < 1$. In the general case (i.e., when there exist multiple neighbor nodes), we set $D(n)$ to $\frac{\alpha}{|N_a(n)|}$, where $|N_a(n)|$ is the number of neighbor nodes of node n. As a result, it can be considered that each node has been influenced by the potential of essentially only one node. We then set α to the value between 0 and 1 in order to keep the potential from oscillating.

2.2.2 Boundary Condition

As an initial condition, the potentials of sensor nodes are set to zero. To construct a potential field from equation (2), we utilize a Dirichlet boundary condition to specify the sink nodes' potentials:

$$\forall d \in N_s, \ \phi(d, t) = \Phi(d), \quad (3)$$

where N_s is a set of sink nodes and $\phi(d, t)$ is the potential of sink node d at time step t. $\Phi(d)$ (≤ 0) is the potential control function, which we discuss later in Section 3. By the nature of the diffusion equation, this boundary condition is insufficient because the potentials of all nodes will arrive at much the same value as the potential of the sink node eventually. We thus define another boundary condition that has to be satisfied by nodes at the edge of the network:

$$\forall n \in N_{edge}, \ \phi(n, t) = 0, \quad (4)$$

N_{edge} is a set of nodes at the edge of the network, and node n is an element of the set that satisfies any of the following conditions (5) or (6):

$$\forall k \in N_a(n), \ H(n) > H(k), \quad (5)$$

$$\forall k \in N_a(n), \ \{H(n) \geq H(k)\} \cap \{D_{id}(n) = D_{id}(k)\}. \quad (6)$$

Here, $H(n)$ is the minimum hop count of node n from a nearby sink nodes and $D_{id}(n)$ is the ID of the sink node. Nodes append H and D_{id} into their ID, which is transmitted periodically to let their neighbor know their H and D_{id}. Sink nodes set their H to zero and D_{id} to their own ID. When node n receives an ID from node m, node n updates $H(n)$ and D_{id}. If $H(m) + 1$ is larger than $H(n)$, node n sets $H(n)$ to $H(m) + 1$ and also sets $D_{id}(n)$ to $D_{id}(m)$. When $H(m) + 1$ equals $H(n)$, node n changes $D_{id}(n)$ to $D_{id}(m)$ with a probability of 0.5.

The expression (5) cannot define the potentials of nodes at the network edge when two or more nodes with the same hop count exist at the network edge. Instead, we use expression (6) for that case. By using D_{id}, nodes in the middle portion of two sink nodes are prevented from mistakenly deciding that they are at the edge of the network. Because D_{id} represents the closest sink node, D_{id} does not coincide among neighboring nodes in the middle portion of two sink nodes.

2.2.3 Local Optimization

In this section, we present the construction of a potential field where nodes can select the best next hop locally. To do this, we add a term ρ on the right-hand side of the discrete diffusion equation (2).

$$\phi(n, t+1) = \phi(n, t) + D(n) \sum_{k \in N_a(n)} \{\phi(k, t) - \phi(n, t)\} + \rho(n, t), \quad (7)$$

where $\rho(n, t)$ is a variable indicating the influence of node n on the potential field at time step t (a larger ρ is associated with lower probability that node n is selected as a next hop and vice versa). Here, we show load balancing based on the residual energy with $\rho(n, t)$.

Node n increases $\rho(n, t)$ when the residual energy of node n is smaller than the average of that of the neighbor nodes whose hop count equals $H(n)$. Residual energy is contained in an ID and transmitted periodically.

The algorithm for deciding $\rho(n, t)$ is as follows, and is executed each time a potential is received.

1. Node n extracts the average residual energy of the neighbor nodes that have the same hop count as node n at time step t (denoted by $E_{avg}(n, t)$) and compares $E_{avg}(n, t)$ with own residual energy at time step t (denoted by $E_{rem}(n, t)$).
    - If $E_{rem}(n, t) > E_{avg}(n, t)$, $\rho(n, t)$ is set to zero.
    - If $E_{rem}(n, t) \leq E_{avg}(n, t)$, it follows step 3..
2. $E_{dif}(n, t)$ is the difference of energy between node n and its neighbors at time step t and assigned to $E_{avg}(n, t) - E_{rem}(n, t)$.

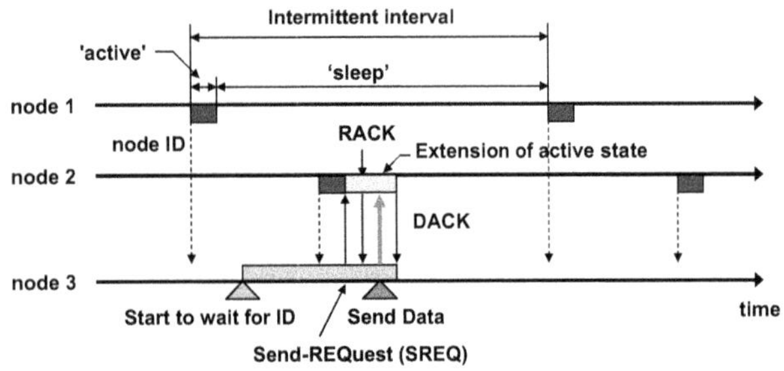

Figure 1: Intermittent receiver-driven data transmission (IRDT) protocol, which is one of the receiver-driven duty cycling MAC protocols such as RI-MAC [14]; we apply this protocol to the potential-based routing

- If $E_{dif}(n,t) < E_{dif}(n,t-1)$, $\rho(n,t)$ is unchanged.
- If $E_{dif}(n,t) \geq E_{dif}(n,t-1)$, $\delta(n,t)$ is added to $\rho(n,t)$, where $\delta(n,t)$ is the difference between $\phi(n,t)$ and average potential of the neighbor nodes whose hop count is same as node n at time step t.

3. Finally, $\rho(n,t)$ is set to $\frac{\rho(n,t)}{|N_a(n)|}$.

The procedure 3. is to suppress the dependence for the number of neighbor nodes on potential. This reduces dependency from our routing on the density of the network.

2.2.4 MAC Layer Protocol and Potential Dissemination

Another major challenge in wireless sensor network research is energy efficiency. For energy efficiency in wireless sensor networks, consideration of duty cycling MAC in which wireless nodes sleep and wake up periodically is required. Then, we use an *intermittent receiver-driven data transmission (IRDT)* protocol for MAC layer, which aims to save energy and get high reliability [7]. Note that our routing protocol is not limited to IRDT and it is applicable also in other underlying protocols. In IRDT, each receiver sends its own ID to inform other nodes that it is ready to receive a data packet (Figure 1). A sender node waits for a receiver ID, and when it acquires an ID from an appropriate receiver, it establishes a link with it by returning an SREQ message. After obtaining an acknowledgement for the SREQ (RACK), the sender transmits a data packet and terminates the communication upon receipt of an acknowledgement for the data (DACK). In this way, a sender node can communicate with one or more receivers flexibly, which can improve the communication reliability and save considerable energy. Because IRDT is an asynchronous MAC protocol that does not require synchronization, it is considered to have scalability.

By having the potential and node ID be periodically transmitted in IRDT, a node can inform its neighbor nodes of its potential. This simple modification produces little overhead. Note that because IRDT uses a duty cycling mechanism, where each node periodically cycles between an awake state and a sleep state, it is not necessarily that a transmitted potential be received by a node within the range of the communication. Each node wakes up and waits to receive potentials for a period of T_p at intervals of T_i. We refer to this period as the "sampling period", and to this interval as the "sampling interval". A sampling period should be longer than the interval of the periodic ID transmission in IRDT to ensure neighbor potentials are known by a node.

Potentials of neighbor nodes are managed in a soft-state manner. In other words, if a node receives a potential from a neighbor node during a sampling period, the node stores the potential; otherwise, the node deletes the information about the neighbor node. The procedure for calculating a potential is shown below.

During a sampling period T_p:

1. If a node receives a potential, it returns its own potential. After returning its potential, it calculates its own potential according to equation (2).
2. A node that receives, and that was intended to receive, the potential returned in step 1. also calculates its own potential.

While waiting for an ID for data transmission:

1. If a node receives an ID, it returns an SREQ containing its own potential. After returning the SREQ, it calculates its own potential.
2. A node that receives, and that was intended to receive, the potential returned in step 1., also calculates its own potential.

Immediately after a sampling period T_p:

1. Potentials of nodes whose potentials have not been updated for a period of T_i are deleted. After this process, the node calculates its own potential.

2.3 Routing in Potential Field

Here, consideration of both routing protocol and MAC protocol (in particularly, duty cycling MAC protocol) is important for energy efficiency. In IRDT, a node that has data to send waits for an ID from an appropriate node [7]. When the node receives an appropriate ID, it forwards the data to the sender of the ID. In our potential-based routing, a potential is transmitted along with an ID. A sender waits for a potential and when it receives a potential, it decides whether to forward data. Most of the proposed potential-based routing schemes use only one receiver that has the minimum potential; however, in duty-cycling MAC protocols, most the energy is consumed while the sender nodes wait for a receiver to awaken. Because energy efficiency is crucial in a sensor network, our potential-based routing uses multiple receivers. This can reduce energy consumption and improve reliability. When node n receives a potential from node r and it is not greater than its own potential, node n always returns an SREQ to node r.

3. CONTROLLED POTENTIAL-BASED ROUTING

In CPBR, multiple sink nodes share information and decide their own potentials for constructing a desired potential field. Sink nodes exchange information at regular time intervals T_m for the purpose of control. We call this information "metric value" (denoted by m). We aim here at balancing the load on sink nodes and sensor nodes.

- **Load balancing of sink nodes**
  For this purpose, sink nodes control their potential to maintain a uniform number of received data packets. The metric value here is the number of data packets received by sink node d (denoted by $N_{dt}(d)$).
- **Load balancing of sensor nodes**
  In the network, the nodes particularly neighboring the sink nodes frequently relay data and thus consume more energy. Thus, the metric value is needed to be based on the sum of the remaining energy of sink node d's neighbors (denoted by $P_{nb}(d)$). To make the energy consumption of the neighbor nodes of sink node d equal, sink nodes control their potential to equalize the decrease rate of $P_{nb}(d)$ (denoted by

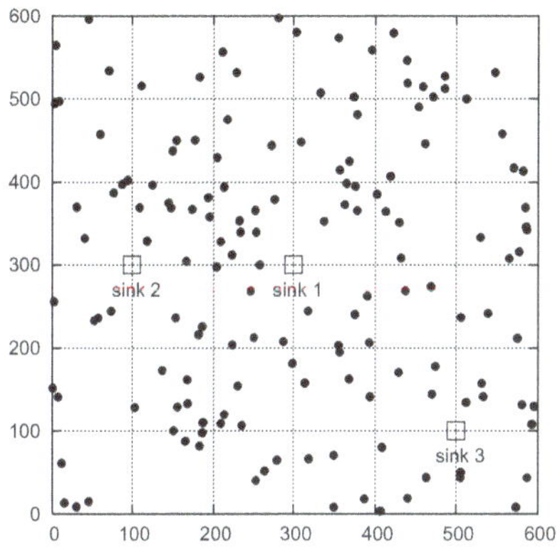

Figure 2: Network model (3 sink nodes)

Table 1: Parameter settings

Parameter	Value
λ	0.003 packet/s/node
Current consumption (TX)	20 mA
Current consumption (RX)	25 mA
Current consumption (SLEEP)	0 mA
T_i	100 s
T_m	50 s or 500 s
Φ_{init}	-30
Φ_{min}	-90
Φ_{max}	0
θ	0.5

$P_{rt}(d)$). As the metric value, the energy consumption rate, $\frac{P'_{nb}(d)-P_{nb}(d)}{P_{nb}(d)}$, is used, where $P'_{nb}(d)$ is the last $P_{nb}(d)$.

The potential of sink node d, $\phi(d)$, is given by the potential control function $\Phi(d)$. $\Phi(d)$ is decided with the following algorithm.

1. Sink nodes set their potential to the initial value Φ_{init}:

$$\Phi(d) = \Phi_{init} \ (\Phi_{init} < 0). \tag{8}$$

2. Sink nodes calculate $\bar{m}$, which is the average of the metric value. For example, $\bar{m}$ for $N_{dt}(d)$ (denoted by $\overline{m_N}$ for convenience) is defined by equation (9) and $\bar{m}$ for $P_{rt}(d)$ (denoted by $\overline{m_P}$) is defined by equation (10):

$$\overline{m_N} = \frac{\sum_{d \in N_s} N_{dt}(d)}{\sum_{d \in N_s} 1}, \tag{9}$$

$$\overline{m_P} = \frac{\sum_{d \in N_s} \{P'_{nb}(d) - P_{nb}(d)\}}{\sum_{d \in N_s} P_{nb}(d)}. \tag{10}$$

3. The potential of sink node d is given according to expression (11):

$$\Phi(d) = \begin{cases} \Phi(d) * (1 - \theta \frac{m(d)-\bar{m}}{\bar{m}}) & (m(d) \geq \bar{m}) \\ \Phi(d) * (1 + \theta \frac{\bar{m}-m(d)}{\bar{m}}) & (m(d) < \bar{m}), \end{cases} \tag{11}$$

where θ is a constant $(-1 < \theta < 1)$. The change in potential can be larger when it is away from the mean value; conversely, the change can be smaller when it is closer to the mean value. At this time, so as to avoid an aberrant value of potential, the potential is taken to be within a range decided beforehand, $[\Phi_{min}, \Phi_{max}]$.

4. SIMULATION RESULTS

We evaluate the impact of CPBR through computer simulation by using an event-driven simulation program we made written in Visual C++. The network model is a square (length of each side: 600 m) in which 150 sensor nodes are randomly deployed and 3 sink nodes (sink 1, sink 2, and sink 3) are set at points (300, 300), (100, 300), and (500,100), respectively as described in Figure 2. The communication range of each node is 100 m. We assume that data packets are generated by each sensor node according to a Poisson process with intensity λ and are sent to the sink node by multihop relay. The simulation commences after an initializing phase, in which each node sufficiently exchanges its potential with neighbor nodes. The simulation ends at 20,000 seconds. The interval of ID transmission is 1.0 s, and T_p is also set to 1.0 s. Other parameters are set as shown in Table 1. Note that θ decides the rate of increase and decrease of potential. Here, a positive value is used for θ to increase Φ when m is larger than $\bar{m}$ and vice versa. For T_m, the interval of the potential control, two values are used. T_m is set to 500 s when the metric value is the number of received data packets because the frequency of data generation ($= \lambda$) is not too high, and T_m of 50 s is used for the energy consumption rate (in this case, T_m depends on T_i, the standby interval for receiving potentials). When a node receives a potential during T_p, it immediately returns its own potential together with its remaining energy. Since this sampling period occurs at intervals of T_i, the remaining energy at neighbor nodes is updated within $\frac{T_i}{2}$ on average. Therefore, the potentials of the sink nodes can be comparatively controlled over a short cycle.

4.1 Load Balancing of Sink Nodes

Potential control based on the numbers of received data, $N_{dt}(d)$, can balance the load of the sink nodes. Simulation results are shown in Figure 3. In this figure, the y-axis indicates the number of data packets received by each sink node during T_m ($500\ s$) with 95% confidential interval. When sink nodes do not control potentials (we call it "autonomous" for simplicity), the number of received data packets remains mostly unchanged over time. With potential control of sink nodes (we call it "controlled" for simplicity), the number of received data packets (over 3 sink nodes) converges to a nearly identical value. This identical value equals $\lambda T_m \frac{N_n}{N_s}$ $(= 75)$, where N_n is the number of sensor nodes. Convergence time is about 10000 s, which indicates that 20 times control causes the number of received data packets to converge. When considering the operating time of an actual sensor network system, which can be in units of years, we note that convergence within a realistic time is possible.

Figure 4 shows the number of received data packets in a network where 5,000 sensor nodes and 100 sink nodes are randomly deployed. The network field shapes a square (length of each side: 3500 m). Also in this case, potential control works properly without a significant change in convergence time; however, there are some issues. The deviation is larger than that of Figure 3(b). This is an inevitable result of a self-organization mechanism when a network grows in scale.

Next, we examine the adaptivity of CPBR to the heterogeneous density of sensor nodes. In Figure 5, we show the impact of potential control when there is a difference in density of sensor nodes: We use a square field, with a 600 meter long side, where the right half of the field has double node density of the left half. CPBR can

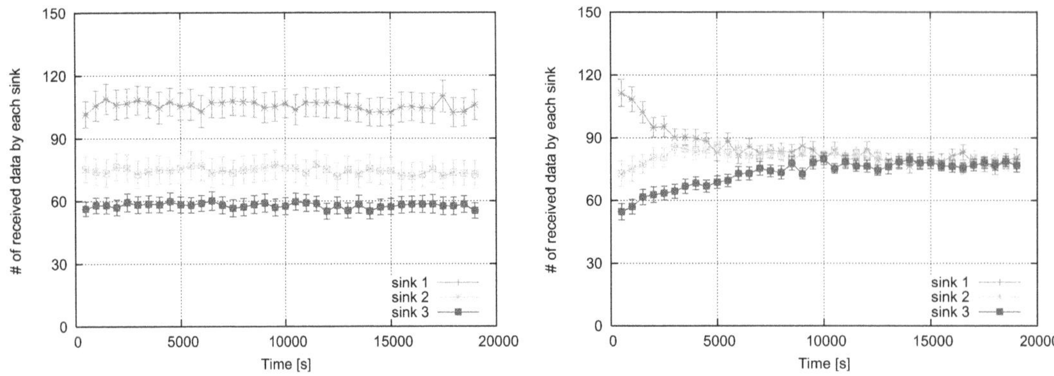

(a) Autonomous (b) Controlled

Figure 3: Potential control based on the number of received data packets (3 sinks)

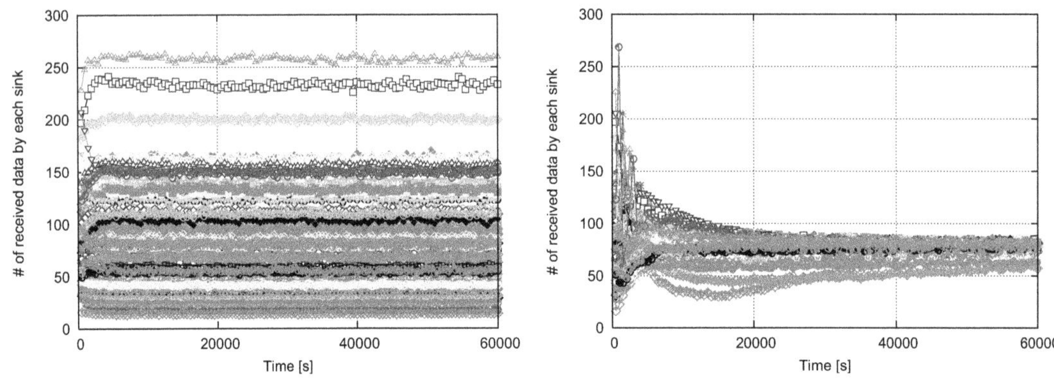

(a) Autonomous (b) Controlled

Figure 4: Potential control based on the number of received data packets (100 sinks)

attain good convergence of the number of received data packets also in this case. This indicates our proposed potential control can adaptively accommodate heterogeneous density of sensor nodes.

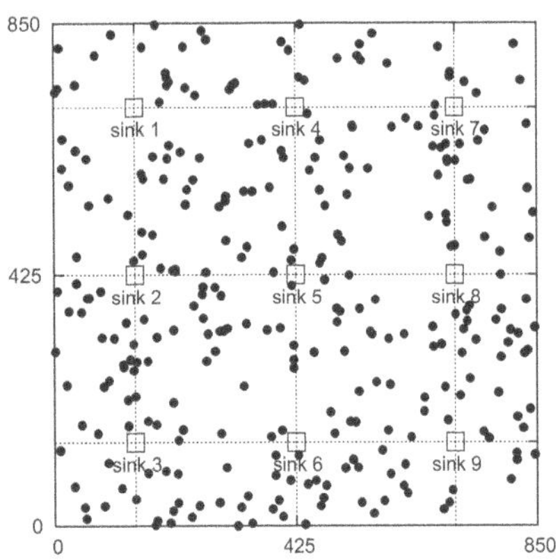

Figure 6: Network model (9 sink nodes)

We also demonstrate the robustness of CPBR against sink node failures and additions. Here, we randomly deployed 300 sensor nodes over a square-shaped network (length of each side: 850 m), and placed 9 sink nodes at locations (142, 708), (142, 425), (142, 142), (425, 708), (425, 425), (425, 142), (708, 708), (708, 425), and (708, 142) (denoted by sink 1 to sink 9 respectively) as shown in Figure 6. Here, the number of sink nodes is varied as follows.

- Six sink nodes (2, 4, 6, 7, 8, and 9) are active at 0 second.
- After four hours, two sink nodes (6 and 9) break down.
- Eight hours after that failure, two sink nodes (1 and 3) are added.
- At twelve hours into the simulation, a sink node (5) is added.

Results of the simulation are shown in Figure 7. Comparing Figure 7(a) and Figure 7(b), we find that CPBR can control potentials adequately after failures and additions of sink nodes; thus, CPBR is robust against sink node failures and additions.

4.2 Load Balancing of Sensor Nodes

Balancing the load of sensor nodes based on the residual energy is expected to be accomplished by potential control based on energy consumption rate. Figure 8 is a simulation result where potential control based on $P_{rt}(d)$ is used. The network model is same as that of Figure 2. The y-axis of the figure indicates the energy consumption of each node with 95% confidential interval in 6-hour simulation, while the x-axis represents each node sorted by

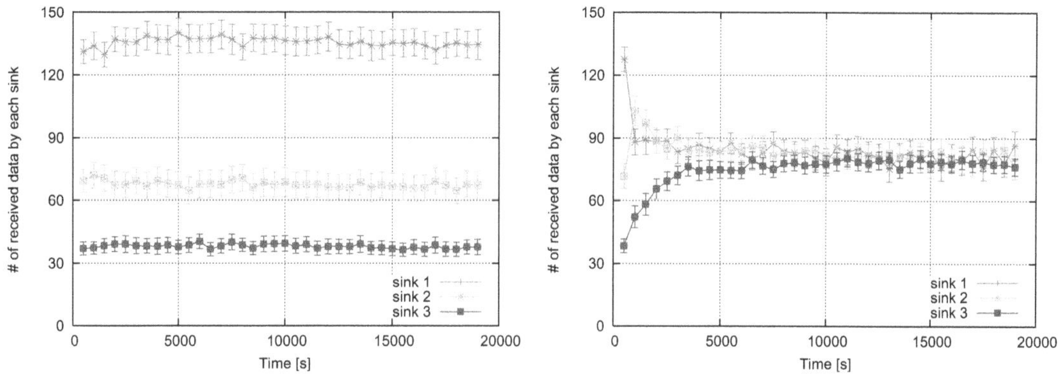

(a) Autonomous (b) Controlled

Figure 5: Potential control based on the number of received data packets (heterogeneous node density)

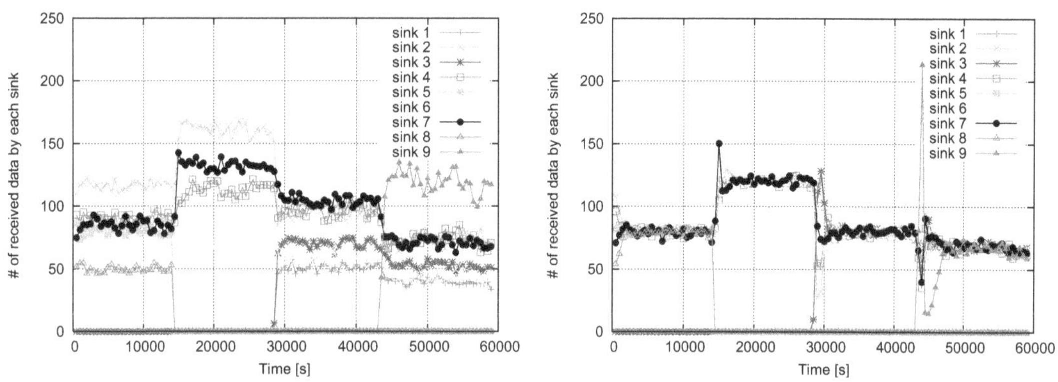

(a) Autonomous (b) Controlled

Figure 7: Potential control based on the number of received data packets (failures and additions of sink nodes)

descending order. In our potential-based routing, relay load is concentrated on the node that has the minimum potential among the neighbors of the sink node. Hence, once a potential field has been constructed, the relay load remains concentrated on a specific node (as apparent in Figure 8(a)). In this figure, because the number of received data packets at sink 1 is largest, the energy consumption of the heaviest loaded node is larger. Figure 8(c) indicates that potential control can reduce the energy consumption of that node. This is because the number of neighbor nodes at each sink nodes is nearly equal; therefore, the number of data packets received by the sink nodes are also nearly equal. If the number of neighbor nodes differs considerably, the bias of energy consumption may grow even more than the result shown in Figure 8(a). In any situation where there is a major difference among the number of neighbor nodes, the local load-balancing mechanism described in Section 2.2.3 can substantially reduce the energy consumption of the heaviest loaded node as shown in Figures 8(b) and 8(d). The results of CPBR with a local load-balancing mechanism are shown in Figure 8(d). While the total energy consumption rises due to an increase of detours, a 58.0 % reduction in the energy consumption of the heaviest loaded node was attained comparing with the result shown in Figure 8(a). Given that energy is consumed only by communication, we find that CPBR with load balancing acted to extend the time until the first node runs out of energy by 138% ($= \frac{0.58}{1-0.58} \times 100$).

Figure 9 shows the network lifetime based on the number of alive nodes and network lifetime based on reachability to sink nodes. There are various definitions for network lifetime in sensor networks depending on applications [3] and in this paper we use following two simple definitions.

1. The time until the first node depletes its energy (alive node).
2. The time until 20% of nodes lose reachability to sink nodes (80% reachability).

For evaluating network lifetime, we set the battery of sensor nodes to comparatively small value (1.0 mAh) and simulation time is set to longer time than battery lifetime. Comparing "controlled with local load balancing" with "autonomous" in Figure 9(a), the time until the first node depletes its energy is prolonged to more than twice as remarked above. In terms of the time guaranteeing 80% reachability, that of CPBR with load balancing ("controlled with local load balancing") is 11.5% shorter than that of default potential-based routing (autonomous) because CPBR increases total energy consumption. However, the time of CPBR ensuring 100% reachability is 2.14 times longer than that of default.

5. CONCLUSION

In a controlled self-organization scheme intended to ensure desired network behavior, one or more controllers control a portion

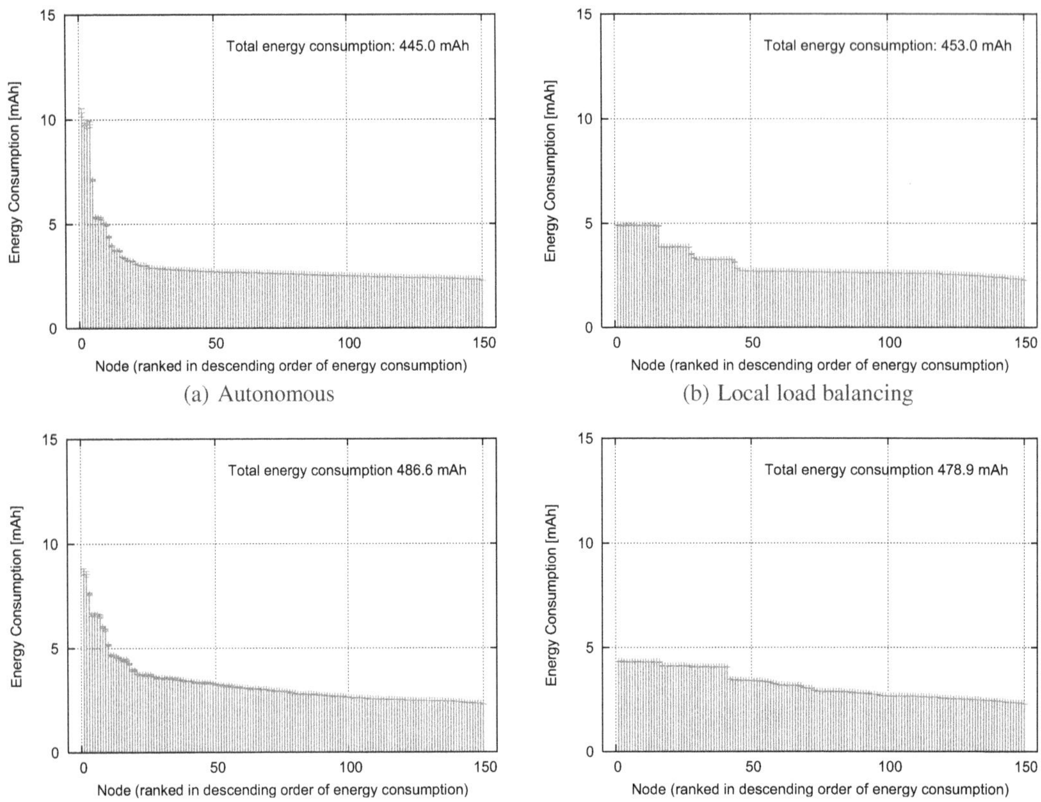

(a) Autonomous (b) Local load balancing

(c) Controlled (d) Controlled with local load balancing

Figure 8: Potential control based on neighbors' energy consumption rate (energy consumption distribution)

of nodes operated in self-organization with centralized control, distributed control, or some other control scheme. In this paper, we proposed controlled potential-based routing (CPBR), which is based on a controlled self-organization scheme. In this scheme, sensor nodes calculate their own potential in a self-organized manner while sink nodes control their own potentials according to network manager's demands by distributed control so as to construct a desired potential field. Our CPBR is over IRDT protocol, however, it is not limited to IRDT and it is applicable also in the sensor network where other MAC protocols were adopted. Through computer simulation, we show that load balancing of the sink nodes can be attained in diverse situations with potential control based on the number of data received at each sink nodes. We also show that CPBR, with potential control based on the energy consumption rate, can extend the time until the first node deplete its energy by 138 %. Presently in our laboratory, investigations into the variations in system behavior and in convergence time as network size becomes increasingly large are underway. Additionally, on the subject of our routing scheme, further evaluation on robustness and detailed study on the dependence on node density are our future work.

Acknowledgment

This research was supported in part by "Global COE (Centers of Excellence) Program" of the Ministry of Education, Culture, Sports, Science and Technology, Japan, and "Grant-in-Aid for Scientific Research (A) 21240004 and (C) 23500097" of the Japan Society for the Promotion of Science (JSPS) in Japan

6. REFERENCES

[1] A. Basu, A. Lin, and S. Ramanathan. Routing Using Potentials: A Dynamic Traffic-Aware Routing Algorithm. In *Proceedings of the 2003 Conference on Applications, Technologies, Architectures, and Protocols for Computer Communications*, pages 37–48, Aug. 2003.

[2] J. Branke, M. Mnif, C. Müller-Schloer, H. Prothmann, U. Richter, F. Rochner, and H. Schmeck. Organic Computing–Addressing Complexity by Controlled Self-Organization. In *Second International Symposium on Leveraging Applications of Formal Methods, Verification and Validation (ISoLA)*, pages 185–191, 2006.

[3] I. Dietrich and F. Dressler. On the Lifetime of Wireless Sensor Networks. *ACM Transactions on Sensor Networks*, 5(1):1–39, Feb. 2009.

[4] F. Dressler. *Self-Organization in Sensor and Actor Networks*. Wiley, 2007.

[5] S. Jung, M. Kserawi, D. Lee, and J.-K. K. Rhee. Distributed Potential Field Based Routing and Autonomous Load Balancing for Wireless Mesh Networks. *IEEE Communications Letters*, 13(6):429–431, 2009.

[6] M. Kalantari and M. Shayman. Design Optimization of Multi-Sink Sensor Networks by Analogy to Electrostatic Theory. In *Proceedings of the IEEE Wireless Communications and Networking Conference (WCNC)*, pages 431–438, Apr. 2006.

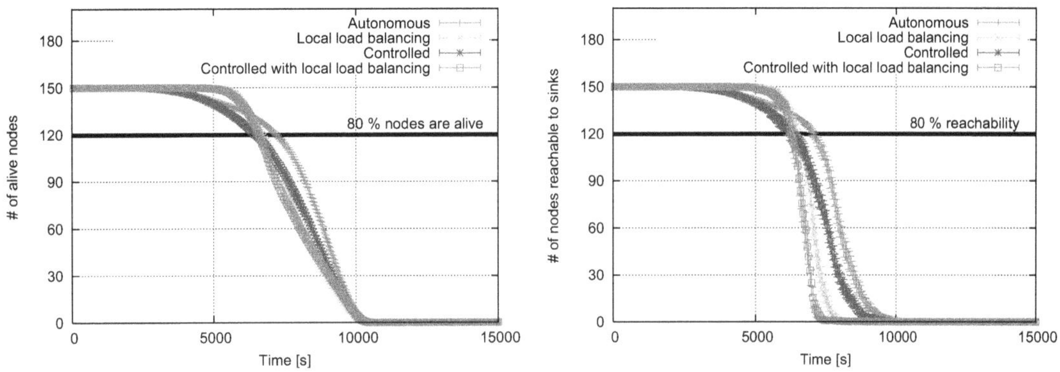

(a) Alive node

(b) 80% reachability to sink nodes

Figure 9: Potential control based on neighbors' energy consumption rate (network lifetime)

[7] D. Kominami, M. Sugano, M. Murata, and T. Hatauchi. Energy-Efficient Receiver-Driven Wireless Mesh Sensor Networks. *Sensors*, 11(1):111–137, Jan. 2011.

[8] P. Kumar, J. Kuri, P. Nuggehalli, M. Strasser, M. May, and B. Plattner. Connectivity-Aware Routing in Sensor Networks. In *Proceedings of the International Conference on Sensor Technologies and Applications (SENSORCOMM)*, pages 387–392, Oct. 2007.

[9] J. Li, S. Ji, H. Jin, and Q. Ren. Routing in Multi-Sink Sensor Networks Based on Gravitational Field. In *International Conference of Embedded Software and Systems (ICESS)*, pages 368–375, July 2008.

[10] H. Liu, Z.-L. Zhang, J. Srivastava, and V. Firoiu. PWave: A Multi-Source Multi-Sink Anycast Routing Framework for Wireless Sensor Networks. *NETWORKING 2007. Ad Hoc and Sensor Networks, Wireless Networks, Next Generation Internet*, pages 179–190, 2010.

[11] Y. Liu and W. K. G. Seah. A Scalable Priority-Based Multi-Path Routing Protocol for Wireless Sensor Networks. *International Journal of Wireless Information Networks*, 12(1):23–33, Jan. 2005.

[12] C. Müller-Schloer, H. Schmeck, and T. Ungerer. *Organic Computing–A Paradigm Shift for Complex Systems*. Birkhäuser, 2011.

[13] H. Ochiai and H. Esaki. Message Routing on Potential-fields in Forwarding-based DTNs. In *Proceedings of the 3rd International Conference on Ubiquitous Information Management and Communication (ICUIMC)*, pages 185–193, Jan. 2009.

[14] Y. Sun, O. Gurewitz, and D. B. Johnson. RI-MAC: A Receiver-Initiated Asynchronous Duty Cycle MAC Protocol for Dynamic Traffic Loads in Wireless Sensor Networks. In *Proceedings of the 6th ACM conference on Embedded network sensor systems (Sensys)*, pages 1–14, Nov. 2008.

[15] C. Wu, R. Yuan, and H. Zhou. A Novel Load Balanced and Lifetime Maximization Routing Protocol in Wireless Sensor Networks. In *Proceedings of the IEEE Vehicular Technology Conference (VTC)*, pages 113–117, May 2008.

[16] J. Zhang, Q. Wu, F. Ren, T. He, and C. Lin. Effective Data Aggregation Supported by Dynamic Routing in Wireless Sensor Networks. In *Proceedings of the IEEE International Conference on Communications (ICC)*, pages 1–6, May 2010.

A Self-Adaptive Routing Paradigm for Wireless Mesh Networks Based on Reinforcement Learning

Maddalena Nurchis
IMT Lucca, Italy
m.nurchis@imtlucca.it

Raffaele Bruno,
Marco Conti
IIT - CNR, Pisa, Italy
{r.bruno,m.conti}@iit.cnr.it

Luciano Lenzini
Dept. of Information Eng.,
University of Pisa, Italy
l.lenzini@iet.unipi.it

ABSTRACT

Classical routing protocols for WMNs are typically designed to achieve specific target objectives (e.g., maximum throughput), and they offer very limited flexibility. As a consequence, more intelligent and adaptive mesh networking solutions are needed to obtain high performance in diverse network conditions. To this end, we propose a reinforcement learning-based routing framework that allows each mesh device to dynamically select at run time a routing protocol from a pre-defined set of routing options, which provides the best performance. The most salient advantages of our solution are: *i*) it can maximize routing performance considering different optimization goals, *ii*) it relies on a compact representation of the network state and it does not need any model of its evolution, and *iii*) it efficiently applies Q-learning methods to guarantee convergence of the routing decision process. Through extensive ns-2 simulations we show the superior performance of the proposed routing approach in comparison with two alternative routing schemes.

Categories and Subject Descriptors

C.2.2 [**Computer-Communication Networks**]: Network Protocols—*Routing protocols*; C.4 [**Performance of Systems**]: Modeling techniques.

General Terms

Performance, Algorithms.

Keywords

Wireless mesh networks, unicast routing, opportunistic routing, reinforcement learning, performance evaluation.

1. INTRODUCTION

In recent years wireless mesh networks (WMNs) have attracted increasing attention due to their advantages over traditional wireless networking technologies. In particular, WMNs allow a fast, easy and cost-effective network deployment in a large variety of environments. Furthermore, WMNs are able to provide connectivity services even when fixed-network deployments are limited or not available at all. For instance, WMNs have been used as a communication infrastructure for wireless ISPs in rural or under-developed areas, municipal access networks, and neighbourhood communities, as well as to provide wireless communications among first responders in emergency situations, or backhaul services for large-scale wireless sensor networks [4]. Such diversity in applications and usage scenarios for WMNs generates a pressing demand for the development of mesh networking solutions characterized by a *high degree of flexibility and adaptability*.

Among the several issues that may affect the overall performance of a WMN, traffic routing is one of the most critical. Thus, there is a large body of studies that have proposed different routing strategies to cope with the challenges due to multi-hop operations, unreliable wireless links, unpredictable channel interference, location-dependent contention, uncertain traffic demands, QoS requirements, etc.. At one end of the spectrum there are *unicast routing* solutions. Many routing protocols in this class select minimum-cost path according to a given routing metric (e.g., [10] and references herein). Alternatively, path computation can be formulated as an optimization problem (e.g, [11, 15]). On the other end of the spectrum there are *opportunistic routing* solutions [6], which broadcast each packet and implement forwarding decisions in a hop-by-hop fashion. In this way, the routing protocol can leverage transmissions that unexpectedly reach far nodes, thus minimizing the number of hops needed to deliver a packet to the destination.

In this paper we are not concerned with the analysis of which are the scenarios where one specific routing solution for WMNs shows better performance than another specific routing solution. On the contrary, the central question addressed in this paper is: *How to design self adaptive mesh networking solutions that autonomously and on-the-fly decide upon the best routing protocol for a traffic flow from a set of supported algorithms, given an estimate of the current network state?* In principle an exact answer to this question could be obtained through a comprehensive routing model characterizing the impact of routing primitives, traffic patterns, network topologies and link characteristics on the network performance. As a matter of fact, theoretical analysis of the capacity of wireless mesh networks has received much attention in recent years (e.g., [16] and references therein). However, these studies are either limited

MSWiM'11, October 31–November 4, 2011, Miami, Florida, USA.

to asymptotic bounds or based on centralized and computationally complex models. Furthermore, existing results cannot be applied to heterogeneous environments, where each node may use a different routing strategy.

To answer the above questions, in this paper we propose a novel routing architecture that exploits *reinforcement learning* to allow each node to autonomously choose the best forwarding strategy from a pre-defined set of routing protocols as network conditions (e.g., traffic loads or link qualities) change. The idea of using machine learning techniques to solve specific routing problems in ad hoc networks is not new (see [9] for a survey). However, to the best of our knowledge there is very little research in applying reinforcement learning for the optimal composition of different routing strategies. The rationale behind the use of reinforcement learning is that it enables the design of decision agents that learn through *trial-and-error* interactions with a dynamic environment how to behave in order to maximize the reward associated to their actions [13]. For the purpose of evaluation, in this paper we design an intelligent agent able to combine a traditional unicast routing protocol with an opportunistic routing scheme. A recent paper [3] has investigated how to combine on-demand routing and delay-tolerant routing[1] in intermittently-connected ad hoc networks. However, the solution proposed in [3] is basically an extension of the AODV routing protocol [12] to cope with network partitions. On the contrary, the approach proposed here is more general because it allows the combination of multiple and arbitrary routing algorithms without making any assumption on the characteristics of the underlying network.

In the following, we first formulate the network state space for our decision problem with the objective of identifying the minimum number of state variables required to estimate the routing efficiency. Second, we define a set of rules to instruct the learning agent how to behave in order to maximize its reward. It is important to note that the reward function is used to formalize the goal of the learning problem. For instance, the reward function can give higher value to policies that improve throughput performance, reduce delays or increase reliability. This results in a highly configurable system able to customize the routing behaviour to different application objectives and QoS requirements. Finally, the tool we utilize to solve the learning problem is the *Q-learning algorithm*, which is a dynamic programming method that works by continuously improving its estimates of the value of particular actions at particular states [14]. The key properties of Q-learning are: *i*) it is a *model-free method*, i.e., it does not require any model of the system, *ii*) it converges with probability one to the optimal policy under general assumptions, and *iii*) it is very easy to implement using lookup tables. On the negative side, it may converge slowly to the optimal policy depending on the state size. Thus, to improve efficiency and speed up convergence we rely on compact network-state representations and smart initialization of action-value functions.

To confirm the effectiveness of the proposed routing approach, we present the results of an extensive simulation study using ns-2. In this study we compare the performance of three routing protocols: *i*) OLSR [7], which is one of the most representative and popular proactive routing protocols for WMNs; *ii*) PacketOPP [5], which is a recently proposed probability-based opportunistic routing protocol; and *iii*) Hybrid, a novel routing protocol obtained by applying the proposed reinforcement learning approach to dynamically combine OLSR and PacketOPP. Our results show that routing hybridization can lead to significant throughput improvements over single routing schemes in a broad range of network and traffic conditions.

The rest of this paper is organized as follows. Section 2 overviews the principles of reinforcement learning, and introduces the Q-learning algorithm. Section 3 describes the proposed routing architecture and the design of the routing learning agent. In Section 4 we report the results of our simulations. Section 5 concludes the paper with final remarks.

[1]Delay-tolerant routing is a different type of opportunistic routing, where transmission opportunities are generated by node mobility rather than wireless diversity.

2. BACKGROUND ON REINFORCEMENT LEARNING

Reinforcement Learning (RL) is a popular machine learning technique, which allows an agent to automatically determine the optimal behaviour to achieve a specific goal based on the positive or negative feedbacks it receives from the environment after taking an action [13]. More formally, let us assume that the interactions between the agent and the environment occur at a sequence of discrete time instants t. Following the same notation as in [13], the learning problem can be formulated by defining:

- The *state* $s_t \in S$ of the environment as observed by the agent.
- The *action* $a_t \in A(t)$ chosen by the agent.
- The probabilistic *reward* $r_{t+1}ß = R(s_t, a_t)$, which measures the goodness of taking action a_t in state s_t.
- The *state transition* function $T(s_t, a_t, s_{t+1})$, which provides the probability of making a transition from state s_t to state s_{t+1} after performing action a_t. .

Typically, in machine learning it is assumed that the state transition probabilities satisfy the Markov property, i.e., they are independent of any state or action previous to time t. In this case, the environment where the agent operates is described through a Markov Decision Process (*MDP*), and several optimized learning algorithms have been studied for this class of environments[2]. More specifically, let us denote with π a *policy*, which defines the probability $\pi(s, a)$ that the learning agent takes action a when in state s. In some cases, a policy can be a simple deterministic function, but arbitrarily sophisticated policies are possible. In general, the main goal of the learning agent is to find the optimal policy providing the maximum long-term reward. To express this in a mathematical setting we define the *expected return* W_t as a function of the sequence of rewards received after time step t. In the case of learning tasks that are continuos a typical formulation for the return functions is as follows

$$W_t = \sum_{k=0}^{\infty} \gamma^k r_{t+k+1} , \quad (1)$$

where γ is a parameter ($0 \leq \gamma < 1$) called *discount rate*, used to determine the present value of future rewards. If

[2]It is important to note that RL can also deal with non-Markovian environments [13]. However, Markov property is a good approximation for many network characteristics observed over long time intervals.

$\gamma = 0$, the agent will behave so as to maximize its immediate reward, even if this would imply a lower long-term return.

The solution of any RL problem can be formulated mathematically in a MDP perspective and under the discounted infinite horizon optimality model described in (1) by introducing the concept of *optimal value* $V^*(s)$ for each state $s \in S$. More precisely, $V^*(s)$ is defined as the expected return if the agent starts at state s and then executes the optimal stationary policy π^*. From the previous definition it follows that

$$V^*(s) = \max_{\pi} E\left(\sum_{t=0}^{\infty} \gamma^t r_t\right) . \quad (2)$$

In the MDP case, it is a consolidated theoretical result that the optimal value function (2) is unique and it can be computed with closed-form equations unique and it can be computed with closed-form equations if $R(s, a)$ and $T(s, a, s')$ are known [13]. However, the main idea behind RL is that it is possible to obtain the optimal policy for an MDP environment even when the model of the environment is not known or difficult to learn. Thus, several *model-free* RL methods have been developed, which are based on the concept of *action-value functions* [13]. More formally, the optimal action-value function $Q^*(s, a)$ is the expected return if the agent starts at state s, executes action a and follows the optimal policy π^* hereafter. A fundamental development in the context of model-free RL methods is the *Q-learning algorithm* [14], which updates the action-value function using the following rule

$$Q(s, a) = Q(s, a) + \alpha \left[r + \gamma \max_{a'} Q(s', a') - Q(s, a)\right] , \quad (3)$$

where s' and r are the state and reward, respectively, obtained after performing action a in state s, and α is a positive step-size parameter determining the *learning rate*. It has been shown that under general assumptions[3] the Q-learning algorithm will asymptotically converge with probability one to the optimal action-value function Q^* independently of the agent's behaviour, i.e., of the policy π being followed [14].

The great advantage of Q-learning algorithm is that it is typically easier to implement than other RL techniques. On the negative side, it may converge slowly to the optimal policy due to the so-called *exploration* and *exploitation* issue. Basically, when in state s the learning agent should exploit its accumulated knowledge of the best policy to obtain high rewards, but it must also explore actions that it has not selected before to find out a better strategy. Moreover, each action must be tried many times to gain a reliable estimate of its expected reward. In Section 3.2 we will discuss suitable action selection strategies for the application scenario addressed in this work.

3. ROUTING PROTOCOL DESIGN

In this section we first provide a high-level overview of our routing framework, and then we describe in detail the proposed learning-based routing system.

3.1 Routing architecture

Figure 1 shows the main components in the node architecture. A key assumption in our routing framework is that

[3]The most important condition is that all state-action pairs are visited infinitely often.

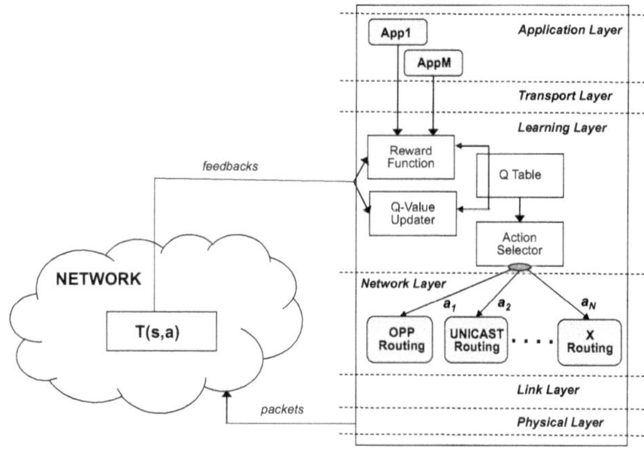

Figure 1: Node architecture

each mesh node implements a number of pre-defined routing algorithms, and it can decide on-the-fly which protocol to activate. The decision on which routing strategy to use can be made at different levels of granularity. For instance, the source node may decide the best routing algorithm on a flow-level basis. A finer granularity in the optimization process would be obtained by using a combination of routing algorithms for each connection, for instance dividing the packet flow into batches and choosing the best routing algorithm on a batch-level basis.

The core component in the architecture depicted in Figure 1 is the reinforcement learning agent implementing the Q-learning algorithm. Such agent operates between the transport layer and the network layer. Specifically, the agent receives *feedback signals* from the network, which are used both to estimate the network state, and to compute the (positive or negative) reward generated by the last action taken by the agent. Such feedback signals can be in the form of observed network quantities, or explicit messages sent by the destination and intermediate nodes. It is important to note that the reward function formalizes the goal of the agent and it is application dependent. For instance, in case of applications generating bulk data transfers, throughput is the most important performance metric and the reward should be a function of the measured connection throughput. In case of delay-sensitive traffic alternative formulations for the reward would be needed. From the knowledge of network states and transition rewards, the agent can update the Q-value for the last transition using formula (3). Then, the Q-values are represented by a two-dimensional lookup table indexed by state action pairs. Note that lookup tables are very easy to implement, although the memory requirements may be prohibitive in case of huge state spaces.

Finally, there is the action selector, which decides the fraction of packets in a flow that should be forwarded using one of the available routing protocols. The simplest solution would be to always select the routing protocol that has provided the highest throughput up to the time of the agent decision. However, there is a trade-off between short-term gains and long-term rewards. Furthermore, the agent has to execute each state-action pair infinitely often in order to guarantee the convergence to the optimal Q^* matrix. In following section suitable exploitation strategies will be defined.

3.2 Network state, reward and actions

In the rest of this section, to simplify the protocol presentation, we assume that each mesh node implements two different routing strategies, namely, an unicast routing protocol and an opportunistic routing protocol.

In our model, an *action taken by the source is one possible choice for the* η_i *parameter*, defined as the fraction of packets in batch i to be sent using the unicast routing protocol (obviously, the remaining $(1-\eta_i)$ packets are sent using the opportunistic routing protocol). In other words, the learning agent at the source node dynamically adapts the portion of traffic sent using unicast routing or opportunistic routing depending on its view of the network state and the received network feedbacks. More specifically, let ρ_i^U be the throughput of packets from batch i delivered using the unicast routing protocol, while let ρ_i^O be the throughput of packets from batch i delivered using the opportunistic routing protocol[4]. Thus, $\rho_i = \rho_i^U + \rho_i^O$ is the total throughput for packets of batch i measured at the destination. In addition, let η_i' be the fraction of the overall throughput at the destination for batch i due to the unicast transmissions. More formally, this can be expressed as $\eta_i' = \rho_i^U / \rho_i$. Now, we can introduce the concept of *routing efficiency* ϵ_i, a 3-valued variable defined as follows

$$\epsilon_i = \begin{cases} 1 & \eta_i' > \eta \\ 0 & \eta_i' = \eta \\ -1 & \text{otherwise} \end{cases} . \tag{4}$$

In other words, the ϵ_i quantity measures if the packets delivered using the unicast routing protocol suffer from less, the same number, or more losses than packets of the same batch delivered using the opportunistic routing protocol. Then, the *network state* s_i *is the two-dimensional tuple* $\{\eta_i, \epsilon_i\}$. Note that our objective is not to obtain global network information, but to define an approximate compact representation of the network state to infer the impact of agent's actions on the performance of each traffic flow.

The last element of our reinforcement learning system is the *reward* $R(s_i, a_i)$ *gained when in state* s_i *the agent performs action* a_i. Intuitively, for bulk data transfers the immediate reward should depend on the total throughput obtained at the end of the transmitted batch of packets. More precisely, let us assume that the source receives at the end of the transmission of batch i a feedback message with the ρ_i^U and ρ_i^O measures[5]. Then, the agent chooses a new value for the proportion η_{i+1} between unicast and opportunistic transmissions in the next batch $(i+1)$ (see Section 3.3 for a detailed discussion on the action selection policies). At the end of the $(i+1)$th batch, the agent will receive a new feedback message and it will calculate the reward as follows:

$$R(s_i, a_i) = \rho_{i+1} + \omega(\rho_{i+1} - \rho_i) , \tag{5}$$

where $0 < \omega < 1$, is a constant parameter. As expressed in formula (5), the reward consists of two contributions. The first term is intuitive: the higher the measured throughput, the higher is the reward gained by the source node. However, the agent should also assign higher value to actions that provide an increase in the flow throughput, and penalize actions that decrease the throughput. Thus, the reward should also depend on throughput gain $(\rho_{i+1} - \rho_i)$[6].

Whenever the source node receives a feedback message from the destination or a neighbour node with the information needed to compute the reward associated to its last action, it updates its estimates of the action values using formula (3). It is important to note that for learning problems in stationary environments it is usual to set the learning rate α as a decreasing function of time. The reason is that at the beginning of the agent's life the rate α should be large enough to quickly overcome initial conditions. As the estimates of the optimal action values converge, decreasing the rate α mitigate unnecessary fluctuations. However, the routing problem considered in this work is intrinsically *non-stationary* because the network conditions can change over time. In such cases it is useful to give more weight to recent rewards than past ones to guarantee a fast adaptation to dynamic network bahaviours. This is achieved by using a *constant* learning rate (in our implementation $\alpha = 0.5$, which provides a good tradeoff between reactivity and stability). For the same reason, we also set the discount rate γ equal to zero, to give more importance to recent rewards than to future ones, which may be associated to a network with changed conditions.

Before concluding this section, it is important to explain how our learning agent reduces the number of admissible state-action pairs to explore. First of all, we restrict ourselves the discrete η values. More formally, let $(K+1)$ denote the maximum number of positive and equally distributed values that η can take in the range $[0, 1]$. By definition it follows that:

$$\eta = k\delta \quad \text{with} \quad k = 0, 1, 2, \ldots, K , \tag{6}$$

where $\delta = 1/K$. Furthermore, for each η, we can have three different conditions for the efficiency parameter (i.e., $-1, 0, 1$). This implies that the overall size of the state space S_t is $3\times(K+1)$. In our implementation we have set $K=10$, which means that the fraction of packets in a batch that is transmitted using unicast routing changes with a step size equal to 10 percent. Our results indicate that larger K values have negligible impact on system performance.

As described in Section 3.1, in our model an action is the selection of a new η value for the successive batch of packets. In principle, after each packet batch, the learning agent could select any of the admissible values of η in the range $[0, 1]$. To restrict the number of possible state-action pairs we assume that in state $\{\eta_i, \epsilon_i\}$ the learning agent can only increase/decrease the η_i parameter by δ, or maintain it constant. In other words, the agent's actions can only cause transitions to adjacent states. This means that the total number of state-action pairs per each flow destination that we should rate with a Q-value is bounded by $27 \times (K+1)$. Note that this number is small enough to make acceptable the use of lookup tables for implementing the Q-learning algorithm.

3.3 Action selection policy

A very important learning procedure is the action selection policy. In following simulations three intuitive and quite popular policies will be tested.

[4]We assume that a tiny header is added to the traditional routing header containing information such as the batch id and the type of routing algorithm used to forward that packet.

[5]For the sake of simplicity we assume that the feedback signal is not delayed. In Section 3.4 we discuss on the impact of delayed feedbacks, which is the normal case in real-world networks.

[6]In our simulations $\omega = 0.5$

- *greedy*: The greedy method is the simplest action selection rule because, at time t it always chooses the *greedy* action a^*, namely the action with the maximum estimated action-value function. More formally this can be expressed as follows:

$$Q(s_t, a^*) = \max_a Q(s_t, a) \; , \tag{7}$$

where s_t is the network state at time t. This method is guaranteed to maximize the immediate reward but it ignores the exploration of actions with non-maximum Q value, although they could lead to action selections with better long-term return.
- $\varepsilon-$*greedy*: A simple alternative to the pure greedy strategy is a *near*-greedy selection methods. More precisely, the learning agent will behave most of the time greedily, selecting an action with maximum expected value with probability $(1-\varepsilon)$, but with a small probability ε instead selects an action uniformly over the set A_t, and independently of the action-value estimates. The advantage of an ε-greedy method is that it guarantees that every action will be sampled an infinite number of times as the time goes to infinity, which is the necessary condition for the convergence of the Q-learning algorithm. However, the advantage of ε-greedy over greedy methods highly depends on the network behaviour. For instance, if the variance of reward samples is very low, exploration may not be necessary to find the optimal action.
- *softmax*: One drawback of the ε-greedy method is that it has the same probability of choosing worst-case actions as well as best-case actions when in exploration phase. An obvious solution for this issue is to use softmax action selection rules, which rank and weight actions according to their value. The most common softmax function used in reinforcement learning to convert values in action probabilities is the following [13]:

$$\pi(s_t, a_t) = \frac{exp[Q(s_t, a)]}{\sum_{a' \in A_t} exp[Q(s_t, a')]} \; . \tag{8}$$

Based on (8), there can be a great difference between the selection probabilities of different actions depending on their values.

3.4 Practical issues

In previous section we have described the general design of our routing framework. But for the protocol to be practical, there are additional challenges, which we discuss in detail below.

- *Delayed feedbacks*: The destination is responsible for collecting the throughput measurements and sending them to the source. The simplest approach is to piggyback such information in dedicated end-to-end messages sent by the destination at the end of the batch transmission. In any case, the source's agent cannot receive the feedback signal (due to end-to-end transmission delays) in time for selecting the η value for the successive batch. In addition, sending a feedback message after each received bath would generate an excessive overhead. To address this issue, in our implementation the destination generates a feedback message only after receiving m consecutive batches of packets. Furthermore, this allows the agent to apply averaging to the per-batch throughput measurements, thus mitigating fluctuations due to transient conditions. Then, the source uses the same η value for at least $m+1$ consecutive batches. Our results indicate that $m=10$ provides a good tradeoff between responsiveness, stability of throughput estimates and efficiency.
- *Packet losses*: Packets can get lost in the network for several reasons. Generally, unicast routing protocols are totally unaware of these losses that are recovered through mechanisms implemented at different layers of the protocol stack (e.g., using layer-2 or layer-4 retransmissions). On the contrary, most of the opportunistic routing protocols directly retransmit lost packets. Since the η parameter is a measure of the routing efficiency for individual packet transmissions, retransmitted packet should are not included in such computation.
- *Implementation issues*: There are several details to take into account during practical implementation. First of all, we must decide a suitable size for the batch. Note that, differently from other routing schemes that operate on batches of packets, such as ExOR [2], it is not necessary that the source collects a full batch of packets before starting the forwarding process. In our implementation, the batch is a virtual concept used to facilitate the learning process, and to easily allocate each packet to one of the two routing options so as to respect the η fraction. Thus, each packet is forwarded as soon as it reaches the head of the transmission buffer. In our implementation, the batch size is set equal to $(K+1)$ packets.
A second concern regards the measurement of the routing efficiency. In real-world networks the instantaneous throughput may be fluctuating due to a variety of causes (e.g., transient changes of link quality due to fading variations, burstiness of packet arrivals, randomness of channel accesses, etc.). Consequently, it may be difficult to obtain stable estimates of $\epsilon=0$, as defined in (4). For these reasons, in our implementation we adopt an extended definition of routing efficiency as follows:

$$\epsilon_i = \begin{cases} 1 & \eta'_i/\eta > 1+\beta \\ 0 & |\eta'_i - \eta| \leq \beta\,\eta \\ -1 & \text{o.w.} \end{cases} \; . \tag{9}$$

In other words, relationship (9) expresses that unicast and opportunistic routing have the same efficiency if the relative difference between η' and η values is less or equal than β. In our implementation, $\beta=0.1$, which is sufficient to absorb small fluctuations of throughput measurements.
Finally, since Q-learning is an averaging method that improves its estimates of the average value of action-state pairs as new actions are taken, it is dependent on the initial values of the action-value estimates. In general, it is advantageous to use optimistic initial values to encourage exploration even if greedy actions are selected most of the time. Consequently, in our implementation we start a 2-second probing phase at the beginning of each new connection to get an initial estimate of the efficiency of unicast routing. More precisely, let us assume that in the initial probing phase the traffic is sent using only unicast routing (i.e., $\eta=0$). In addition, let $\widetilde{\phi}^U$ be the offered load per unit time measured at the source and $\widetilde{\rho}^U$ the unicast throughput measured at the destination at the end of the probing phase. Then, $\widetilde{\eta} = \widetilde{\rho}^U/\widetilde{\phi}^U$ is a rough estimate of the efficiency of unicast transmissions. Thus, a good way

for calculating an initial guess for the values of the action-state function is to assume that states with $\eta > \widetilde{\eta}$ would not significantly contribute to increase the system reward. More formally, this is equivalent to say that $Q(s,a) = \widetilde{\rho}^U$ for all $s = \{\eta, \epsilon\}$ such that $\eta \leq \widetilde{\eta}$, and $Q(s,a) = 0$ otherwise.

4. PERFORMANCE EVALUATION

A comprehensive simulation study has been conducted to compare the performance of the proposed self-adaptive routing protocol, hereafter simply called *Hybrid*, against two alternative routing schemes. We first show the gains in terms of throughput improvements in single flow scenarios. Then, we analyze scenarios with multiple flows and unreliable channels.

4.1 Simulation environment

To carry out the following simulation we use the *ns-2* network simulator, which implements the full protocol stack for multi-hop wireless networks. We consider wireless mesh networks of 25 nodes, each of which is equipped with one omnidirectional radio antenna. These static nodes are placed randomly in a 500m×500m area. Concerning the physical layer characteristics, to make more realistic simulations we use the *Shadowing* propagation model instead of the classical TwoRay propagation model, because it permits to describe the received power at a certain distance as a random variable. The following results have been obtained by setting the path loss exponent equal to 2 and the shadowing deviation to 4, which are typical values for outdoor environments [1]. If not otherwise stated the receiving threshold for the network interface is set in such a way to ensure a 95% correct reception rate at the distance of 100m. Concerning the MAC layer, we use the 802.11 DCF scheme with a fixed transmission rate equal to 11 Mbps. Moreover, we have disabled the RTS/CTS access method, since this is the default setting in most wireless networks.

As motivated in Section 1, for the purpose of evaluation we use three routing protocols. First, we use OLSR [7] as a representative of proactive link-state routing protocols, which discovers and then disseminates link state information throughout the mesh network. Individual nodes use this topology information to compute next hops for all destinations in the network using shortest hop forwarding paths and ETX routing metric [8]. Regarding the opportunistic routing option, we use the PacketOPP protocol [5], a lightweight opportunistic routing algorithm able to select at each hop, and at run-time, the candidate forwarders that can maximize the opportunistic throughput gain. Finally, the third protocol is Hybrid, which is a dynamic combination of OLSR and PacketOPP obtained by applying the framework described in Section 3. In the following simulations, to generate the data traffic we use constant bit-rate UDP flows. We do not use TCP-controlled data transfers because traditional TCP would experience unacceptable performance degradations with opportunistic routing due to the frequent out-of-order packet deliveries [2].

4.2 Numerical results

4.2.1 *Single flow*

First, we evaluate the performance of our solution under a single flow scenario. To this end, Figure 2 shows a box-and-whiskers plot depicting the quartiles of the distribution of flow throughput for OLSR, PacketOPP and Hybrid with different action-selection strategies. The results shown in Figure 2 have been obtained as follows. First, we randomly select a node pair in the network as source and destination of an UDP flow, which sends packets with a payload of 1000 bytes at an offered rate of 1 Mbps. Then, ten statistically independent runs are repeated with this node-pair selection to obtain average values and confidence intervals for the flow throughput. Since throughput performance of individual flows are highly variable, we repeat the same set of simulations with 80 different node pairs to collect a number of samples sufficiently large to obtain good estimates of quartile ranges.

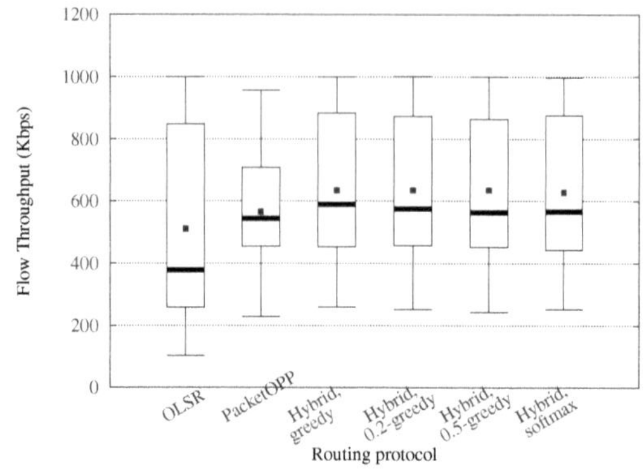

Figure 2: Boxplots of throughput between 80 node pairs in a 25-node random network for different routing schemes. Squares represents the mean throughputs.

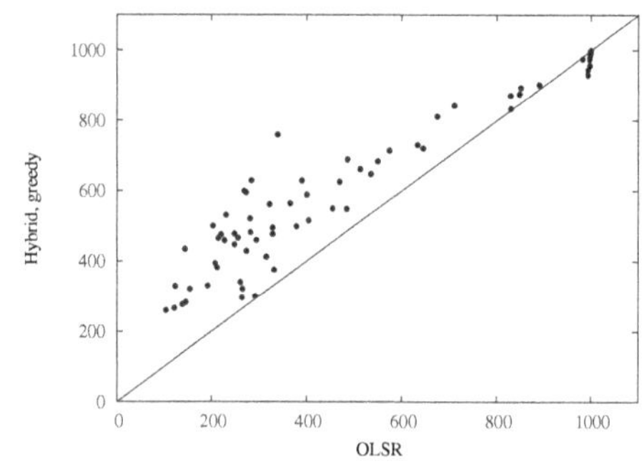

Figure 3: Scatter plot showing the relationship between the throughput of individual flows obtained with OLSR and Hybrid (greedy).

Several important considerations can be derived from the above diagram. First, if we consider the min-max interval traditional unicast routing shows the largest range, and in particular the lowest minimum throughput. By inspecting the traces we discovered that unicast routing achieves very low throughput for source-destination pairs that are separated by many hops. In these cases PacketOPP can take advantage of multiple forwarding nodes and more opportunities to use long links providing throughput gains of a factor of two or more. On the other hand, due to its self-adaptive characteristics, Hybrid protocol is able to select the routing option most suitable for both particularly disadvantaged flows and flows with the highest throughputs. The second observation regards the statistical dispersion of throughput values, i.e., the range between the first quartile and the third quartile, also known as *interquartile range*. It is important to note that a high variability of the throughput

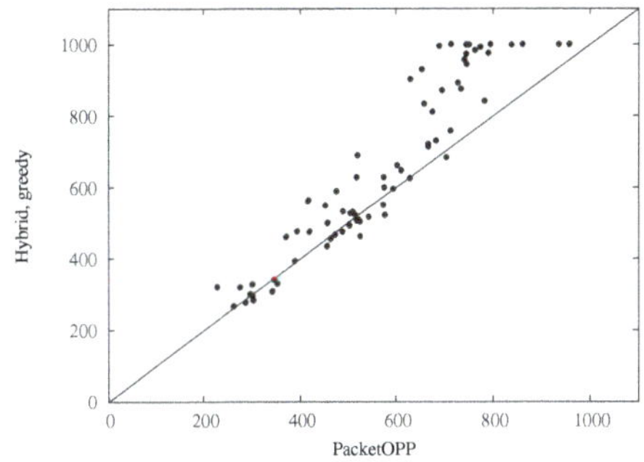

Figure 4: Scatter plot showing the relationship between the throughput of individual flows obtained with PacketOPP and Hybrid (greedy).

obtained by individual node pairs is unavoidable due to the significant differences in the length of shortest routes connecting such pairs, which span from single hop traditional routes to routes that have six hops. However, Figure 2 indicates that OLSR has the largest variability, with many flows having low throughputs. This results in a quite low median throughput. On the other hand, PacketOPP obtains a median throughput 50% higher than OLSR. Despite that, the throughput distribution for PacketOPP is quite concentrated around the median value, which means that only a few flows can achieve high throughputs. On the contrary, Hybrid protocol shows the best performance in terms of interquartile range, which is more shifted towards high throughputs. In addition, Figure 2 shows that Hybrid protocol behaves almost equally well with all the considered policies. However, the pure greedy strategy behaves slightly better the the other considered strateies (i.e., ε-greedy methods and softmax method), as it provides the highest median throughput. In the following, all the reported results (except for convergence times) are obtained using the greedy version of our Hybrid scheme.

To better understand the primary reason of throughput gains provided by Hybrid protocol over the other routing strategies, Figure 3 shows the throughput relationship between OLSR and Hybrid with a greedy action-selection policy for the 80 node pairs considered in Figure 2, while Figure 4 shows the relationship between PacketOPP and Hybrid. Figure 3 indicates that there is a large number of node pairs with low unicast throughputs that obtain a significant gain when using Hybrid protocol. Specifically, for half of the node pairs Hybrid outperforms OLSR by more than 50%. Moreover, the node pairs with the highest throughput (mainly pairs with single hop routes) are not affected by the use of Hybrid routing. This means that the overhead introduced by Hybrid (i.e., feedback messages and additional information in packet headers) have a negligible impact. On the contrary, Figure 4 indicates that the throughput gain of Hybrid over PacketOPP occurs mainly for node pairs with mid to high throughput values, while it provides similar performance for node pairs with lower throughputs. To clarify the routing selection behaviour, Figure 5 shows the probability mass function (PMF) of the average η value for the 80 considered node pairs and a greedy action-selection policy. To plot such curve we first compute the average fraction of traffic that was delivered using unicast routing for each individual node pair, and the we derive the distribution of such measurement over all the flows. Interestingly we can

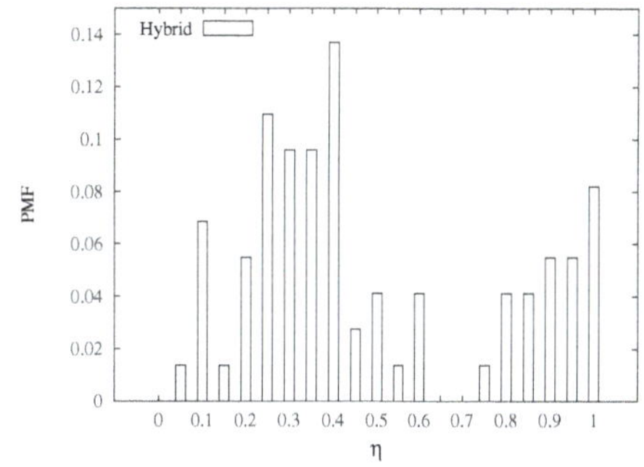

Figure 5: Probability mass function of the average η value per node pair.

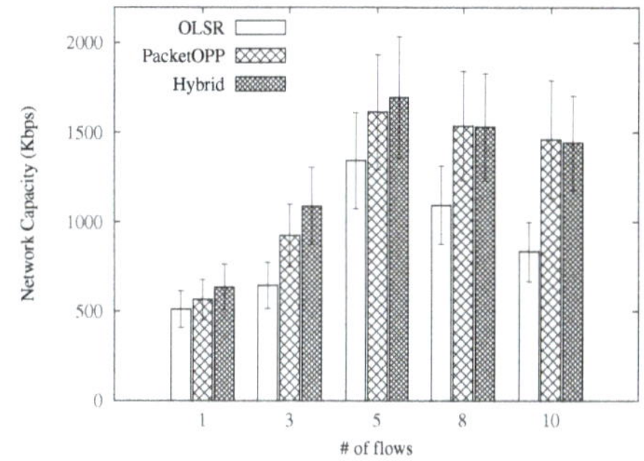

Figure 6: Comparison of network capacity for different touring protocols with varying number of simultaneous flows.

observe that the PMF of the η parameter shows two principal modes, i.e., two distinct peaks (local maxima). More precisely, most of the flows converge towards η values in the range $[0.25, 0.4]$, which means that these flows use a combination of unicast and opportunistic transmissions, but give preferences to opportunistic routing. Then, there is a second smaller group of flows that converge towards η values close to one, which are mostly node pairs that are separated by a few hops. Interestingly, there is also a non-null probability that a flow uses only opportunistic routing, i.e., $\eta \leq 0.1$.

4.2.2 *Impact of the number of flows*

We now turn to consider more scaled scenarios. Basically, we repeat the same simulations discussed in the previous section, but increasing the number of parallel data transfers active in the network. Figure 6 reports average values and 95% confidence intervals for the network capacity (i.e., the sum of the throughputs of individual flows) versus the number of flows. These results are obtained repeating each test with twenty different combinations of selected flows. Note that flow patterns are chosen in such a way that each node can be at most the source/destination of one flow. The plot indicates that Hybrid significantly outperforms OLSR in all the considered scenarios, with throughput gains ranging from 25% to 55%. On the other hand, Hybrid provides smaller improvements with respect to PacketOPP for low numbers of flows (up to 18%), and attains comparable performance for higher numbers of flows. It is important to note that the primary objective of our solution is not to define a new routing protocol, but to show how existing routing paradigms can be adaptively combined to optimize the network performance in the widest range of scenarios. In fact, a specific routing protocol can achieve better performance than another routing protocol in one scenario, but worse

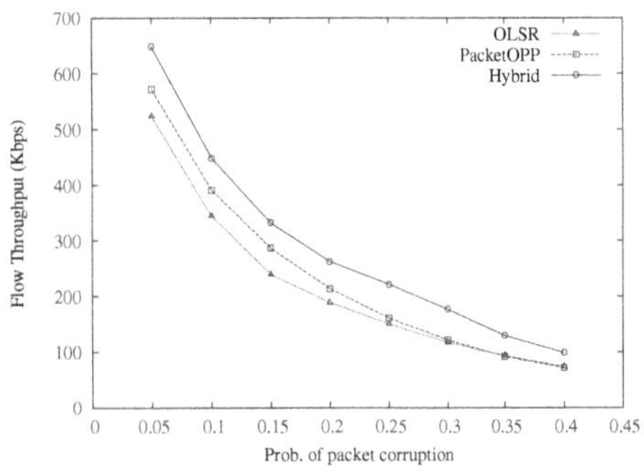

(a) One flow

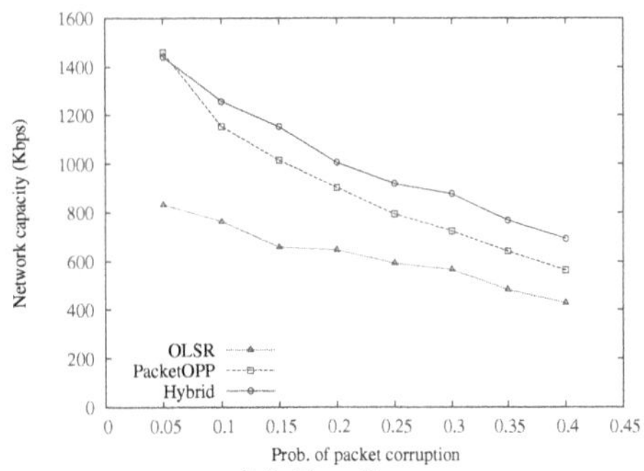

(b) Ten flows

Figure 7: Comparison of throughput performance for different routing protocols with varying shadowing intensity.

in another one. Our routing framework is capable of discovering the best routing option for every scenario using a minimal knowledge of the network behaviour and limited overheads.

4.2.3 Impact of shadowing intensity

We investigate the impact of the shadowing intensity on the routing performance. More precisely, we keep constant the path loss exponent (equal to 2) and the shadowing deviation (equal to 4), but we vary the receiver sensitivity threshold to change the probability of packet distortion at the distance of 100m. Figure 7(a) shows the average throughput obtained in the case of a single flow, while Figure 7(b) shows the network capacity with ten randomly-selected simultaneous flows. We do not report results for packet-corruption probabilities higher than 0.4 because most of the links are too unreliable to allow OLSR to discover shortest paths between all node pairs. The results confirm that Hybrid efficiency is not affected by the channel unreliability. It is also interesting to observe that Hybrid has similar performance than PacketOPP when the probability of packet corruption is low, while it provides up to 20% throughput gains in more challenged environments. Moreover, unicast routing provides the worst network capacity in all the considered scenarios.

5. CONCLUSIONS

There are many possible application scenarios related to wireless mesh networks. Thus, it is very difficult to design a general routing solution fitting the QoS needs of different applications. Motivated by this, in this paper we have proposed and evaluated a self-adaptive routing framework for wireless mesh networks, which enables the dynamic and on-the-fly combination of multiple routing strategies to maximize routing performance under arbitrary target objectives. The proposed framework relies on efficient machine learning techniques, which allows us to tackle the complexity of the problem optimization without requiring to know how system performance depend on routing primitives and network conditions. Finally, using ns-2 simulations we have shown the superior performance of a hybrid routing protocol obtained through the combination of two legacy schemes.

6. REFERENCES

[1] D. Aguayo, J. Bicket, S. Biswas, G. Judd, and R. Morris. Link-level Measurements from an 802.11b Mesh Network. *SIGCOMM Comput. Commun. Rev.*, 34(4):121–132, 2004.

[2] S. Biswas and R. Morris. ExOR: opportunistic multi-hop routing for wireless networks. *SIGCOMM Comput. Commun. Rev.*, 35(4):133–144, October 2005.

[3] J. Boice, J. Garcia-Luna-Aceves, and K. Obraczka. Combining on-demand and opportunistic routing for intermittently connected networks. *Ad Hoc Networks*, 7:201–218, 2009.

[4] R. Bruno, M. Conti, and E. Gregori. Mesh Networks: Commodity Multihop Ad Hoc Networks. *IEEE Commun. Mag.*, 43(3):123–131, March 2005.

[5] R. Bruno, M. Conti, and M. Nurchis. Opportunistic packet scheduling and routing in wireless mesh networks. In *IFIP Wireless Days'10*, 2010.

[6] R. Bruno and M. Nurchis. Survey on diversity-based routing in wireless mesh networks: Challenges and solutions. *Computer Communications*, 33(3):269–282, February 2010.

[7] T. Clausen and P. Jaquet. Optimized Link State Routing Protocol (OLSR). RFC 3626, October 2003.

[8] R. Draves, J. Padhye, and B. Zill. Routing in Multi-Radio, Multi-Hop Wireless Mesh Networks. In *Proc. of ACM MobiCom*, pages 114–128, Philadelphia, PE, USA, Sept. 26–Oct. 1 2004.

[9] A. Forster. Machine Learning Techniques Applied to Wireless Ad-Hoc Networks: Guide and Survey. In *IEEE ISSNIP'07*, December 3–6 2007.

[10] M. Genetzakis and V. Siris. A Contention-Aware Routing Metric for Multi-Rate Multi-Radio Mesh Networks. In *Proc. IEEE SECON 2008*, pages 242–250, San Francisco, CA, USA, June 16–20, 2008.

[11] X.-Y. Li, A. Nusairat, Y. Wu, Y. Qi, J. Zhao, X. Chu, and Y. Liu. Joint throughput optimization for wireless mesh networks. *IEEE Transactions on Mobile Computing*, 8(7):895–909, July 2009.

[12] C. Perkins, E. Belding-Royer, and S. Das. Ad hoc On-Demand Distance Vector (AODV) Routing. RFC 3561, July 2003.

[13] R. Sutton and A. Barto. *Reinforcement Learning: An Introduction.* The MIT Press, March 1998.

[14] C. Watkins and P. Dayan. Q-learning. *Machine Learning*, 8:279–292, 1992.

[15] J. Wellons, L. Dai, Y. Xue, and Y. Cui. Augmenting predictive with oblivious routing for wireless mesh networks under traffic uncertainty. *Computer Networks*, 54:178–195, 2010.

[16] P. Zou, X. Wang, and R. Rao. Asymptotic Capacity of Infrastructure Wireless Mesh Networks. *IEEE Trans. Mobile Comput.*, 7(8):1011–1024, August 2008.

Adaptive Overhead Reduction via MEWMA Control Charts

Kahkashan Shaukat, Douglas C. Montgomery, and Violet R. Syrotiuk
CIDSE, Arizona State University, Tempe, AZ 85287-8809
{kshaukat,doug.montgomery,syrotiuk}@asu.edu

ABSTRACT

In an effort to reduce overhead in proactive protocols, rather than use a global period for transmission of control information, we propose to transmit based on changes in a node's local conditions. For OLSR, a proactive link-state routing protocol, we use a significant change in the number of edges in a node's two-hop neighbourhood and topological graph as indicators of topology change warranting an update. When the associated *exponentially weighted moving average* (EWMA) charts and multivariate EWMA chart signal out-of-control, a node transmits a HELLO, a topology control (TC), or both HELLO and TC messages. We use `ns-2` simulations to compare OLSR to our TS-OLSR, and other variants of OLSR. We find that TS-OLSR obtains a statistically significant reduction in overhead compared to the other protocols while maintaining the packet delivery ratio. The approach is general and may find use in other applications requiring response to changes in local conditions.

Categories and Subject Descriptors

C.2.2 [**Network Protocols**]: Routing Protocols.

General Terms

Performance, Measurement, Experimentation.

Keywords

Control overhead, statistical process control, EWMA charts.

1. INTRODUCTION

Most protocols for computer networks incur control overhead in their operation. This includes negotiation, such as a handshake in IEEE 802.11 to arbitrate access to the medium, for channel or rate selection, among others. It also includes the distribution of information, such as the broadcast of link-state in the Internet's OSPF routing protocol.

MSWiM'11, October 31–November 4, 2011, Miami, FL, USA.

Link-state routing protocols have also been developed for *mobile ad hoc networks* (MANETs). A MANET is a collection of mobile wireless nodes that self-organize with no centralized control or fixed infrastructure. The OLSR, or *optimized* link-state routing protocol [1] uses *multi-point relay* (MPR) sets to implement the broadcast efficiently. An MPR set for a node is a subset of its one-hop neighbours that reaches all its two-hop neighbours.

MPRs are selected through the exchange of HELLO messages, which provide each node a view of its two-hop neighbourhood. MPRs also forward *topology control* (TC) messages; these provide each node with sufficient topology information to compute shortest path routes to all destinations.

OLSR is *proactive*, transmitting control information periodically. Because it is unlikely that the network conditions are identical at every node, it is unnecessary for each node to use the same period of transmission. A large period results in a poor packet delivery ratio; a large number of packets are dropped due to out-of-date routes. A small period transmits too frequently consuming bandwidth that can otherwise be used for data. Instead, the transmission of a control message should depend on the network conditions experienced at a node. When conditions change rapidly more frequent transmission may be necessary; when they change slowly less frequent transmissions may suffice.

In this paper, we explore such an adaptive approach for the transmission of control information with the goal of reducing overhead of the protocol. *Statistical process control* (SPC) is used to monitor a characteristic so that corrective measures can be taken when performance degrades. Most often process performance depends on multiple characteristics. A MANET may be viewed as a process and we measure its performance by the packet delivery ratio.

To monitor small shifts in one or more characteristics, *exponentially weighted moving average* (EWMA) or *multivariate* EWMA (MEWMA) charts are used [2]. These charts work well when the monitored characteristics are independently distributed; otherwise they may generate too many false alarms. To effectively use the control charts, autocorrelation is removed using *time series analysis* (TSA) [3].

We propose *Time Series OLSR* (TS-OLSR) to jointly monitor the number of edges in the two-hop neighbourhood and topological graphs at each node. An MEWMA control chart is used to decide when to transmit the corresponding HELLO and TC messages. An inherent difficulty with MEWMA charts is how to interpret an out-of-control signal because an MEWMA may generate a signal even when the individual

EWMA charts do not. Therefore individual EWMA charts are also used.

ns-2 simulations are conducted to compare the performance of TS-OLSR to OLSR [1], and other variants that seek to reduce control overhead including A-OLSR [4] and Adaptive OLSR [5]. TS-OLSR obtains a statistically significant reduction in control overhead compared to the other protocols. It transmits 29.55–54.80% less control overhead compared to OLSR.

The sequel is organized as follows. §2 summarizes previous work on reducing control overhead in OLSR, and defines EWMA and MEWMA charts for statistical process control. §3 provides an overview of ARIMA models using the Box-Jenkins method while §4 describes the construction of ARIMA models in TS-OLSR. §5 presents the simulation results. Conclusions and future directions are in §6.

2. RELATED WORK

2.1 Reducing Control Overhead in OLSR

Several ideas to reduce control overhead in OLSR have been proposed. In Clustered OLSR (C-OLSR) [6], overhead is reduced by partitioning the network into clusters and restricting the propagation of TC messages to a cluster. MPRs at the cluster level generate and forward inter-cluster topology information. In Hierarchical OLSR (HOLSR) [7], nodes use a "fisheye" technique where TC messages from nodes further away are received at larger intervals.

Some approaches to reduce overhead in OLSR incorporate the use of local conditions. In the Fast-OLSR protocol [8], a node switches from *default* to *fast-OLSR* mode when changes in its neighbourhood exceed a threshold. Fast-OLSR nodes send Fast-HELLO messages at a higher frequency to help maintain network connectivity. Adaptive OLSR [5] extends Fast-OLSR by the addition of a *fast-response* mode. A node switches from default to fast-response mode on the receipt of a Fast-HELLO message; Fast-response nodes also send Fast-HELLO messages. Generally, a fast-OSLR or default node is not selected as an MPR. A fast-OLSR node restricts the size of its MPR set to two. Adaptive OLSR (A-OLSR) [4] takes a different approach; it monitors betweenness (a centrality measure) of the two-hop neighbourhood of a node using a Shewhart control chart. An out-of-control signal reflects a change in the topology causing a TC message to be sent.

2.2 Statistical Process Control

Statistical process control (SPC) is a collection of statistical techniques to monitor the performance of a process to determine if it has shifted away from some nominal operating state (the *in-control* state).

Control charts are the basic tools used in monitoring. To detect small deviations from the in-control state, an *exponentially weighted moving average* (EWMA) control chart is used. An EWMA chart is defined as:

$$\begin{aligned} z_t &= \lambda x_t + (1-\lambda) z_{t-1}, \text{ where } z_0 = 0 \quad (1)\\ CL &= \mu_0 \pm K\sqrt{\frac{\lambda}{2-\lambda}}\frac{\sigma_0}{\sqrt{n}} \end{aligned}$$

where x_t is the characteristic being monitored, $0 < \lambda \leq 1$ is a smoothing constant, and n is the number of samples. The computation of the upper and lower *control limit* (CL) use μ_0 and σ_0, the mean and standard deviation of x_t. The parameters λ and K can be tuned to achieve a specific in-control *average run length* (ARL) [9,10]. When the monitored characteristic falls between the control region bounded by the CLs, the process is in-control. Otherwise it is out-of-control and corrective measures are required.

A multivariate EWMA (MEWMA) chart [11] is an extension of the EWMA chart and defines vectors of EWMAs. The design of MEWMA charts is a generalization of the approach to design EWMA charts [9, 10].

3. MODEL BUILDING METHODOLOGY

One assumption underlying control chart use is that the observations are independent. The charts are not effective if the monitored characteristic exhibits even low levels of correlation. They may signal too many false alarms if the data are positively correlated; they may not signal at all if the data are negatively correlated. We therefore first describe the methodology for removing correlation in modelling.

3.1 Definitions

3.1.1 Autocorrelation Function (ACF)

Given a series $x_1, x_2, \ldots, x_n$, the autocorrelation between observations k intervals apart is measured by the *lag k sample autocorrelation coefficient* ($k < n$) [3]:

$$r_k = \frac{\sum_{i=1}^{n-k}(x_i - \bar{x})(x_{i+k} - \bar{x})}{\sum_{i=1}^{n}(x_i - \bar{x})^2}; r_0 = 1. \quad (2)$$

Here, x_i and x_{i+k} are the observations at time i and $i+k$, and $\bar{x}$ is the mean of the n values. The autocorrelation coefficients are plotted against the lags. If more than 5% of them fall outside the 95% confidence interval, the data is serially correlated, and the correlation needs to be removed.

3.1.2 Partial Autocorrelation Function (PACF)

The partial autocorrelation function, ϕ_{kk} for a stationary process is defined as a function of the autocorrelations r_k of the process [3] and has the distinctive feature that $\phi_{kk} = 0$ for $k > p$ in an autoregressive process of order p. ϕ_{kj} denotes the jth coefficient in an autoregressive process of order k, so that ϕ_{kk} is the last coefficient in

$$\phi_{k1} r_{j-1} + \phi_{k2} r_{j-2} + \ldots + \phi_{kk} r_{j-k} = r_j. \quad (3)$$

3.2 ARIMA Models

A general autoregressive integrated moving average model of order (p, d, q), or $ARIMA(p, d, q)$, has three parts: the *autoregressive* (AR), *integrated* (I), and *moving average* (MA) models. The order of the model specifies how far back in time to look. The integrated model is used when the series is non-stationary, i.e, it has no constant mean or standard deviation. The ACF and PACF are used in conjunction with the time series to find the order of the autoregressive and moving average models. The objective is to find a model that adequately describes the time series yet contains relatively few parameters [3].

An $ARIMA(p, d, q)$ is given by $\Phi_p(B)\nabla^d x_t = \Theta_q(B)\epsilon_t$. Here, $\Phi_p(B)$, ∇^d, and $\Theta_q(B)$ is the AR, I, and MA model, respectively; B is the backward shift operator defined as $Bx_t = x_{t-1}$.

$\Phi_p(B)$, an autoregressive model of order p, is given by:

$$x_t = \xi + \phi_1 x_{t-1} + \phi_2 x_{t-2} + \ldots + \phi_p x_{t-p} + \epsilon_t. \quad (4)$$

The current observation x_t is regressed on p previous observations of the time series with coefficients $\phi_1, \phi_2 \ldots, \phi_p$. ξ is a constant coefficient, and ϵ_t is a random error at time t. Using the backward shift operator, Eqn. (4) can be written as $\Phi_p(B)x_t = \xi + \epsilon_t$.

$\Theta_q(B)$, a moving average model of order q, is given by:

$$x_t = \mu + \epsilon_t - \theta_1\epsilon_{t-1} - \theta_2\epsilon_{t-2} - \ldots - \theta_q\epsilon_{t-q}. \quad (5)$$

Similarly, using the backward shift operator, Eqn. (5) can be written as $x_t = \mu + \Theta_q(B)\epsilon_t$.

∇^d is the *dth* difference, i.e., taking d successive differences of the characteristic being modelled. The first difference (∇^1) is $\nabla x_t = x_t - x_{t-1}$, the second difference is $\nabla^2 x_t = x_t - 2x_{t-1} + x_{t-2}$, and so on.

Once the order (p, d, q) of the $ARIMA$ model is identified, and model checking finds the model adequate, the residuals obtained after fitting the model are independent and random. The residuals may then be used with SPC techniques to monitor the process.

3.3 The Box-Jenkins Method

The Box-Jenkins method [3] is a modelling technique for time series; the output of the obtained model may be used in monitoring. Finding an adequate ARIMA model using the Box-Jenkins method is an iterative four step process: 1) detect stationarity, 2) determine the order of the model, 3) estimate parameters, and 4) check the model.

Stationarity can be assessed from a run sequence plot that displays data in a time sequence. Long runs of increases and decreases in the series indicate non-stationarity. To remove non-stationarity, a new series comprised of differences from the first series is computed. Successive differencing may be required. ACF and PACF plots are also useful in eliminating non-stationarity.

Once a stationary time series is obtained, the order of the $ARIMA(p, d, q)$ is determined. The ACF and PACF are used to estimate the values of p and q. The ACF and PACF are plotted to identify a tentative model by comparing the observed patterns with known patterns [3] in Table 1.

Table 1: Behaviour for stationary models [12].

Model	ACF	PACF
AR(p)	Tails-off	Cuts-off after lag p
MA(q)	Cuts-off after lag q	Tails-off
ARMA(p,q)	Tails-off	Tails-off

Once the order of the ARIMA process is determined, estimation methods such as maximum likelihood or ordinary least squares are used to estimate the model parameters. We use `Minitab` [13] to estimate the models using ordinary least squares estimation and also for model checking.

In case the model is not adequate, the method iterates. If multiple models are built with similar adequacy, the model with lower order is selected.

4. ARIMA MODELS FOR OLSR

OLSR transmits overhead in the form of HELLO and TC messages consisting of two-hop neighbourhood and topological information, respectively. Since our goal is to reduce the control overhead in OLSR, we are interested to determine when each of these message types *need* to be transmitted.

We start by studying two time series at each node. The first series, computed every HELLO_INTERVAL ($2\,s$), is the number of edges in a node's two-hop neighbourhood graph. The second series is the number of edges in the topological graph derived from a node's routing table, computed every TC_INTERVAL ($4\,s$ instead of the default $5\,s$ so that the characteristics coincide more often). Each routing table entry at node i is a tuple (d, j) specifying the next-hop neighbour j to forward the packet to reach destination d. The topological graph contains the edges (i, j), and edge (j, d) indicating that d is reachable through j for the entry.

We divide the model building process into an off-line part to obtain an initial ARIMA model, and an online part to tune the model. Once tuned, the models may be used to control the transmission of HELLO and TC messages.

4.1 Off-line Model Building

Off-line model building is used to determine an adequate model and to remove non-stationarity and serial correlation. We proceed using the Box-Jenkins method. Simulations of OLSR are carried out in `ns-2` [14] using the parameters in Table 3. Figure 1 shows a representative time series for the number of links in the two-hop neighbourhood and topological graphs of a node moving $2\,m/s$.

Both time series are non-stationary; they show a large number of successive increases and decreases. Figure 2 shows the ACF for the two-hop neighbourhood and topological links time series; values exceeding the 95 % confidence interval are enclosed in a dotted ellipse. Figure 2(a) decays very slowly. Thus, the series is serially correlated and the correlative structure needs to be removed. The differenced series has a high standard deviation of 12.29 (see Figure 3(a)). The corresponding ACF in Figure 3(b) tails-off and then becomes larger at higher lags. Thus, the correlative structure needs to be removed further.

Figure 4(a) shows the second differenced time series for the two-hop neighbourhood links and Figure 4(b) shows the corresponding ACF. The standard deviation reduces to 8.32. The third differencing increases the standard deviation to 15.97 so further differencing is not necessary. The ACF in Figure 4(b) tails-off after lag two, indicating an autoregressive model of order 2. To decide the order of the autoregressive moving average model, the PACF of the second difference of two-hop neighbourhood links is studied.

A similar study of the successive differences of the topological links shows that a single differencing provides a stationary series; the ACF cuts-off at lag one (see Figure 5).

The partial autocorrelation functions for the second difference of two-hop neighbourhood links and first difference of topological links is given in Figures 6(a) and (b), respectively. Considering the value at lag 4 as an outlier, the PACF for two-hop neighbourhood links cuts-off after lag 1 while the PACF for topological links tails-off. Thus, the ARIMA model for two-hop neighbourhood links has no moving average part whereas the ARIMA model for the topological links has a moving average part of order 1.

To summarize, for the two-hop neighbourhood links, second differencing achieves stationarity; the I part of the ARIMA has order 2. The ACF tails-off after lag 2 indicating an AR part of order 2. But, the PACF cuts-off after lag 1 which means that the ARIMA has no MA part; the order of the AR part is 1 (see Table 1). The ARIMA model for the two-hop neighbourhood links is of order (1,2,0). Single differencing

Figure 1: Time series of 2-hop neighbourhood and topological links of a representative node moving $2\,m/s$.

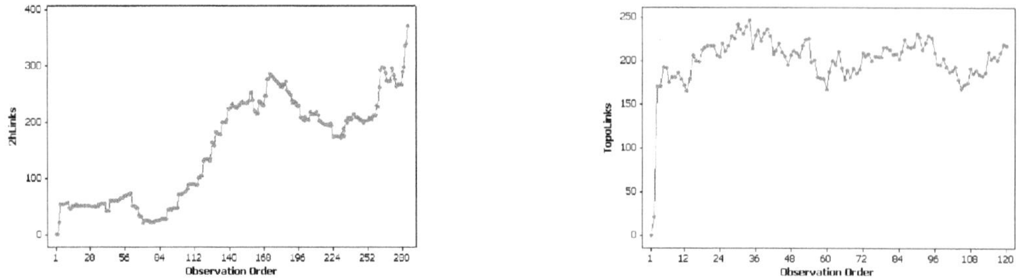

(a) Two-hop neighbourhood links

(b) Topological links

Figure 2: ACF of 2-hop neighbourhood and topological links of a representative node moving $2\,m/s$; the dotted ellipses show values exceeding the $95\,\%$ confidence interval.

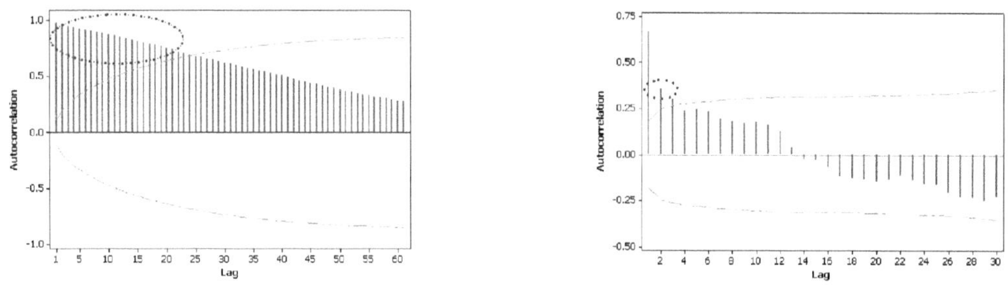

(a) ACF of 2-hop neighbourhood links

(b) ACF of topological links

Figure 3: ∇^1 and ACF of ∇^1 of the 2-hop neighbourhood links of a representative node moving $2\,m/s$.

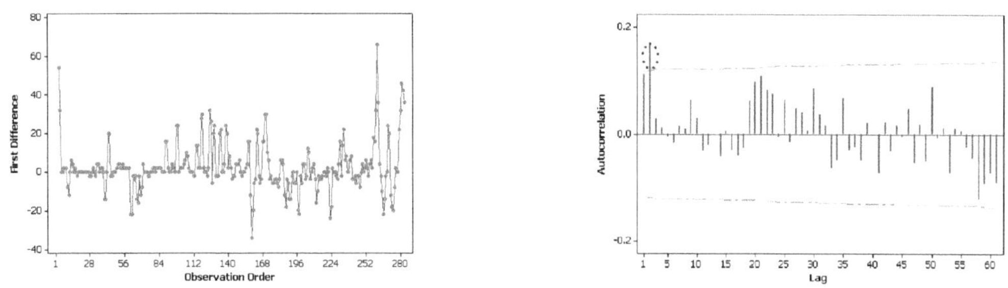

(a) ∇^1 of 2-hop neighbourhood links

(b) ACF of ∇^1 of 2-hop neighbourhood links

Figure 4: ∇^2 and ACF of $= \nabla^2$ of the 2-hop neighbourhood links of a representative node moving $2\,m/s$.

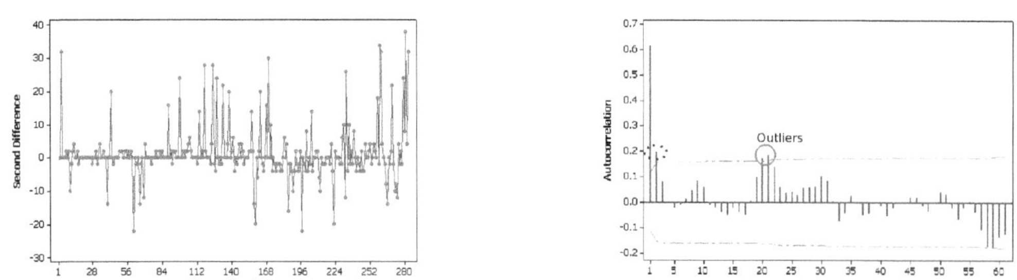

(a) ∇^2 of 2-hop neighbourhood links

(b) ACF of ∇^2 of 2-hop neighbourhood links

Figure 5: ∇^1 and ACF of ∇^1 of the topological links of a representative node moving $2\,m/s$.

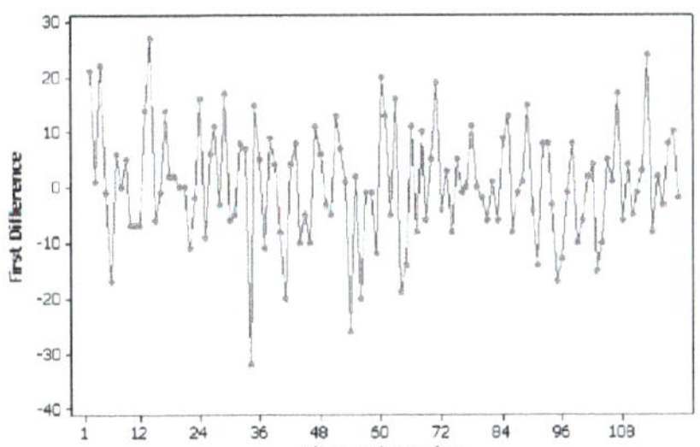

(a) ∇^1 of topological links

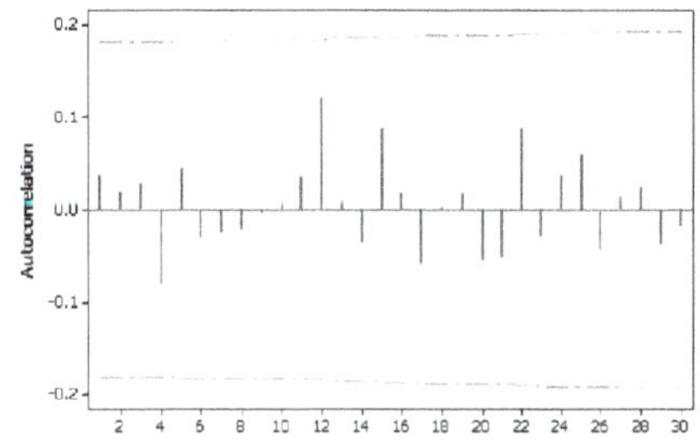

(b) ACF of topological links

Figure 6: PACF of 2-hop neighbourhood and topological links of a representative node moving $2\,m/s$.

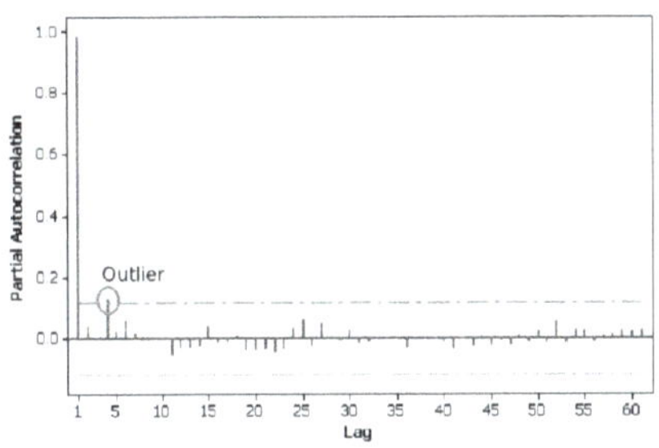

(a) PACF of 2-hop neighbourhood links

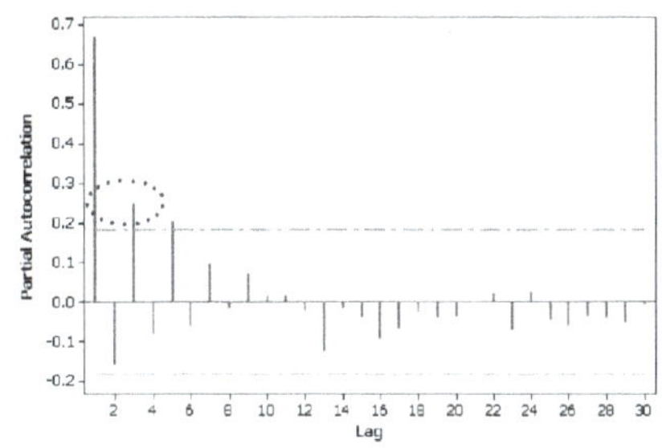

(b) PACF of topological links

obtains a stationary time series for the topological links; the ACF cuts-off at lag 1, and the PACF tails-off. Thus, the ARIMA model for the topological links is of order (0,1,1).

`Minitab` [13] is used to estimate the ξ and ϕ parameters for the two-hop neighbourhood links, and the θ parameter for the topological links. The ARIMA(1,2,0) model for the two-hop neighbourhood links is $x_t = 0.079 + 1.457x_{t-1} + 0.086x_{t-2} - 0.543x_{t-3} + \epsilon_t$, and the ARIMA(0,1,1) model for the topological links is $x_t - x_{t-1} = 1.827 + \epsilon_t - 0.037\epsilon_{t-1}$.

The residuals, ACF and PACF of the residuals, normal plot of residuals, and the fitted values versus residuals of the ARIMA(1,2,0) are shown in Figure 7. The residuals in Figure 7(a) look random, i.e., the serial correlation has been removed; since the ACF and PACF in Figures 7(b) and (c) do not exceed the 95 % confidence interval, the residuals do not show any correlation. The normal plot in Figure 7(c) follows the normal curve closely, another indication that the residuals are random. Even though the residuals versus the fitted values in Figure 7(e) show high variance, the model is acceptable. Similarly, the ARIMA(0,1,1) is an adequate model for the topological links (figures not included).

The Ljung-Box test [15] is also performed on the residuals of the ARIMA(1,2,0) and ARIMA(0,1,1) models. The χ^2 statistics together with the degrees of freedom and p-values are presented in Table 2. For the ARIMA(1,2,0) model at lag 12, the χ^2 value is 2.8, and the corresponding p-value of 0.904 indicates a 90.4% probability that the data is random; the model is adequate. The χ^2 and p-values for the ARIMA(0,1,1) model also indicate that it is adequate.

Simulation studies with node speeds 5, 10, 15 and 20 m/s are also performed. Off-line modelling for each speed shows

Table 2: Ljung-Box statistics for selected models.

ARIMA(1,2,0)				
Lag	12	24	36	48 8
χ^2	2.8	10.9	22.1	29.0
Degrees of Freedom	9	21	33	45 5
p-Value	0.904	0.927	0.879	0.950
ARIMA(0,1,1)				
Lag	12	24	36	48
χ^2	3.7	8.1	16.2	21.3
Degrees of Freedom	9	21	33	45
p-Value	0.960	0.997	0.996	0.999

that second differencing and single differencing is required to remove non-stationarity in the two-hop neighbourhood and topological series, respectively. Off-line modelling for higher speeds also shows that the order of the AR and MA models may change though they never exceed two. In some cases, both the ARIMA models have non-zero p and q values.

4.2 Online Model Building

`gretl` [16], a gnu open-source library, is used to perform the time series analysis during online modelling. There are six possible ARIMA models up to order 2 and each one is built for each characteristic; the Ljung-Box statistics for 24 lags of each model is checked. The model that provides the highest probability of the residuals being random is then used in the simulation.

With serial dependence in the data removed, the residuals are ready to be used with MEWMA and EWMA charts to

Figure 7: Time series plot of residuals, ACF and PACF of residuals, normal plot of residuals, and residuals versus fitted values of 2-hop neighbourhood links of a representative node moving $2\,m/s$.

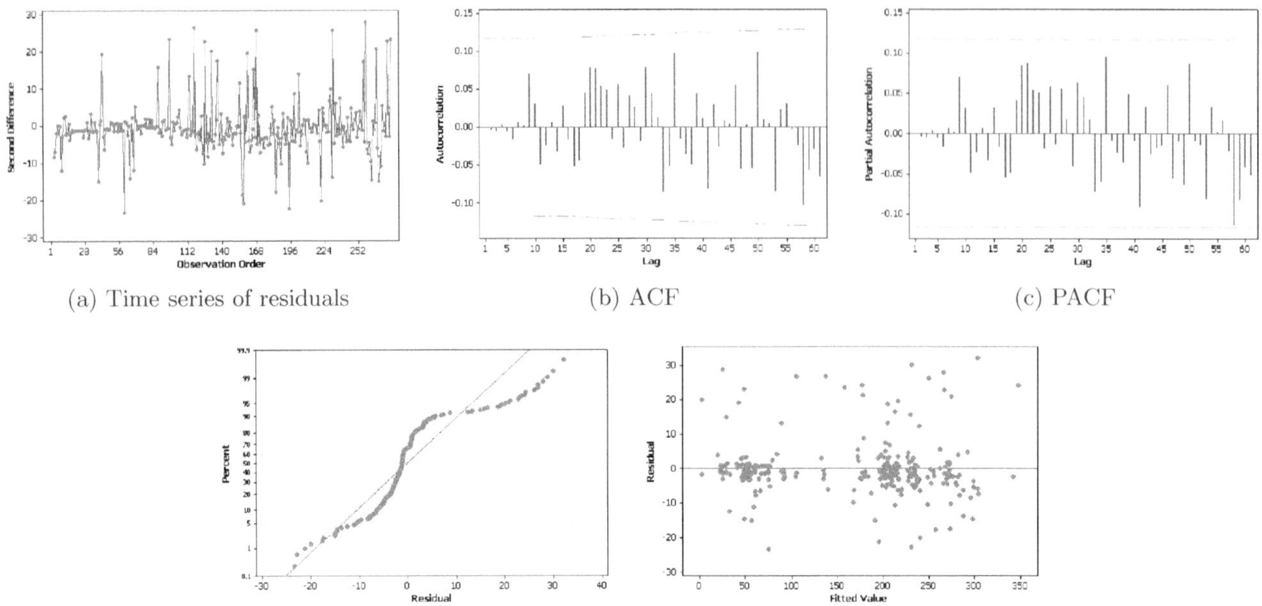

(a) Time series of residuals (b) ACF (c) PACF

(d) Normal plot of residuals (e) Residuals versus fitted values

determine when to transmit control packets. The MEWMA uses the average value of the two-hop neighbourhood links collected in the last two time intervals as one of its control characteristics; the other is the topological links. When the MEWMA generates an out-of-control signal, both HELLO and TC messages are transmitted. The simultaneous use of EWMA charts together with the MEWMA chart is two-fold: they help determine which characteristic generated the signal, and generate network activity. In case the MEWMA does not generate an out-of-control signal, the EWMA charts help decide when to send appropriate control information.

4.2.1 Tuning EWMA and MEWMA parameters

In industrial processes, the in-control ARLs are large because false alarms are a major quality control issue; it requires stopping the process to take corrective action. In our application, false alarms only generate the transmission of control information. Indeed, to some extent false alarms are beneficial because even when the system is in-control some overhead is required to keep the routes up-to-date.

The values of K and h_4 for the EWMA and MEWMA chart, are tuned using the methodology in [9] and [17], respectively. Simulations are performed to find the in-control ARL values for the EWMA and MEWMA charts when λ and r equal 0.05. $h_4 = 0.93$ corresponds to a MEWMA chart with in-control ARL of 8, while $K = 0.63$ and $K = 0.49$ provide an in-control ARL of 8 and 5 for the two-hop neighbourhood and the topological links EWMA charts, respectively.

5. SIMULATION RESULTS

5.1 Simulation Set-up

Simulations using the `ns-2` network simulator [14] with extensions for wireless mobility are performed to observe the performance of TS-OLSR compared to OLSR, Adaptive OLSR, and A-OLSR. Since `gretl` requires the sample size to be at least 100 to perform the time series analysis, the first $400\,s$ of the simulation is used to collect the sample. Data transmission starts at $400\,s$ and continues to the end of the simulation. The packet delivery ratio (PDR) and total control overhead (CO), made up of TC and HELLO messages, are measured during this time.

Table 3: Simulation parameters.

Parameter	Values
Simulator	`ns-2 version 2.29`
OLSR implementation	`um-olsr-0.8.8`
MAC protocol	IEEE 802.11b
Confidence interval	95%
Simulation area	$1000 \times 1000\,m^2$
Nodes	50
Transmission range	$250\,m$
Channel bandwidth	$2\,Mbps$
Simulation duration	$800\,s$
Traffic type	Constant bit rate (CBR)
Packet arrival rate	$10\,pkts/s$
Packet size	$512\,bytes$
Mobility model	Stationary random waypoint
Node speed	$2, 5, 10, 15,$ and $20\,m/s$
Number of flows	$1, 2, 5$ 10
λ and r	0.05

5.2 Results

An MEWMA chart using $h_4 = 0.93$ and $r = 0.05$ to jointly monitor the two-hop neighbourhood links and the topological links is shown in Figure 8. When the T_i^2 values are greater than 0.93 the MEWMA chart generates an out-of-control signal. The signal implies that both the number of two-hop neighbourhood and topological links has changed such that corrective action is required. The corrective action is to transmit a HELLO and a TC message.

Figure 8: MEWMA: node speed $2\,m/s$, ARL = 8.

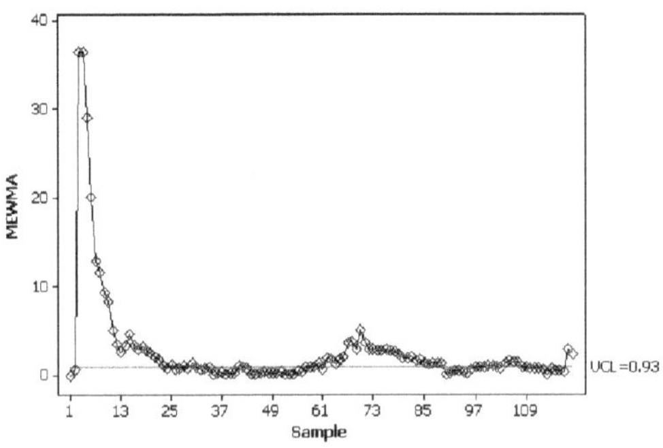

Because it is unclear which of the characteristics caused the MEWMA chart to signal, EWMA charts are used for this purpose. An EWMA chart for the residuals of the two-hop neighbourhood and topological links is given in Figure 9. When both the MEWMA and the EWMA charts signal at the same time, only a message corresponding to the EWMA chart is transmitted.

OLSR, Adaptive OLSR, A-OLSR, and TS-OLSR are simulated using the parameters in Table 3. The results are presented in Figure 10 and Table 4. Figure 10(a) shows the PDR; all three protocols perform better than OLSR at speed $2\,m/s$. At higher speeds A-OLSR and TS-OLSR show reduced performance by approximately 2%; the reduction is not statistically significant. Adaptive OLSR shows an increase in PDR by 3.37–14.20% for higher speeds; its highest achieved PDR is observed at $15\,m/s$ speed.

Figure 10(b) shows the CO. TS-OLSR outperforms OLSR by sending 29.55–54.80% less control overhead where the savings increase as speed increases. A-OLSR shows a 12.20–47.67% decrease in CO; its lowest savings occur at $5\,m/s$ which then increases as speed increases. Adaptive OLSR shows a similar trend; its decrease ranges from 15.39–58.45%. TS-OLSR outperforms all the other protocols up to speed $15\,m/s$, Adaptive OLSR has the highest CO reduction at $20\,m/s$ but it achieves this reduction by restricting the size of the MPR sets to two in *Fast-OLSR* mode.

Breaking out the total CO into TC and HELLO messages we find that TS-OLSR sends 27.25–56.83% fewer TC messages than OLSR; while A-OLSR saves 14.07–48.29%. Adaptive OLSR sends 24.74–82.74% fewer TC messages. The large savings for Adaptive OLSR again comes from its *Fast-OLSR* mode where it restricts its MPR set to two. The trend is similar for these protocols to their corresponding savings in CO.

Since A-OLSR uses betweenness to control the transmission of TC messages only, it has the same number of HELLO messages as OLSR. Adaptive OLSR sends 4.69–155.3% more HELLO messages compared to OLSR; the number of messages increase with speed. The number of HELLO messages increase due to Adaptive OLSR's sending Fast-HELLO messages at a higher frequency than the HELLO messages. TS-OLSR sends 36.95–43.37% less HELLO messages compared to OLSR; the savings decrease with increase in speed.

Table 4: Percentage difference between OLSR and its variants for a single flow with $10\,pkts/s$ arrival rate for node speeds 2, 5, 10, 15, and $20\,m/s$. Negative/positive values represent percentage decrease/increase.

Speed (m/s)	OLSR		A-OLSR	
	PDR	CO	PDR	CO
2	89.02	82090.90	+1.38	-27.75
5	68.85	88596.40	-1.15	-12.20
10	62.95	101679.40	+0.53	-22.48
15	48.83	112004.60	-1.80	-43.24
20	46.37	114815.80	-1.53	-47.67
Speed (m/s)	**Adaptive OLSR**		**TS-OLSR**	
	PDR	**CO**	**PDR**	**CO**
2	+0.75	-20.54	+1.51	-29.55
5	+3.37	-15.39	-0.29	-42.43
10	+7.17	-24.39	-1.92	-47.13
15	+14.20	-44.91	-1.30	-51.59
20	+13.04	-58.45	-2.06	-54.80

Simulations with 2, 5, and 10 flows gave similar results and are not provided here. An interesting aspect of the CO, TC, and HELLO message plots (TC and HELLO message plots not included here) for TS-OLSR is the flatness of the curve. This is due to the fact that the serial correlation of the data has been removed adequately and the usage of the same in-control ARLs for all speeds. The A-OLSR EWMA chart is tuned on node speed and Adaptive OLSR changes its message transmission scheme based on mobility.

6. CONCLUSIONS AND FUTURE WORK

Rather than transmitting HELLO and TC messages periodically as in OLSR, each node in TS-OLSR uses out-of-control signals of the MEWMA and EWMA control charts derived from the time series of the number of links in its two-hop neighbourhood and topological graph to decide when to transmit the corresponding control messages. By using control charts in this manner, TS-OLSR reduces control overhead by 29.55–54.80% compared to OLSR with an insignificant (0.29–2.06%) decrease in packet delivery ratio.

Adaptive OLSR saves slightly more control overhead than TS-OLSR at $20\,m/s$; this may be attributed to Adaptive OLSR restricting its MPR set to a size of two for higher mobility scenarios. Adaptive OLSR shows a significant increase in packet delivery ratio but, unlike TS-OLSR, it changes the protocol. This suggests that incorporating Fast-OLSR messages into TS-OLSR in high mobility scenarios may yield even more gains.

Overall, our results indicate that the use of time series analysis, ARIMA modelling, and control charts are effective tools for the adaptive transmission of control overhead.

We have used an MEWMA chart to monitor the two-hop neighbourhood and topological links. It may be interesting to see how TS-OLSR behaves if two equivalent one-sided MCUSUM charts are used to monitor positive and negative deviations from the mean. Deviation on one side may have a different meaning than deviation on the other side and appropriately tuned MCUSUMs may provide insight on what actions are suitable when they generate signals.

Figure 9: EWMA chart ($\lambda = 0.05$) for residuals of 2-hop neighbourhood and topological links, respectively, of a representative node moving $2\,m/s$.

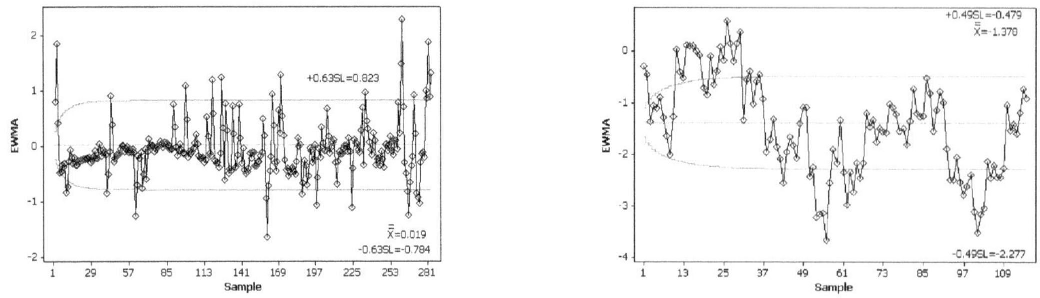

(a) 2-hop neighbourhood residuals, $K = 0.63$

(b) Topological residuals, $K = 0.49$

Figure 10: Comparison of OLSR and its variants: one data flow with arrival rate of $10\,pkts/s$.

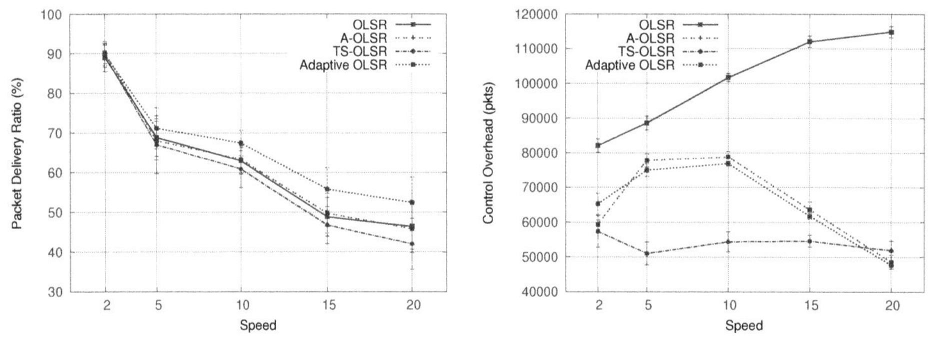

(a) Packet delivery ratio

(b) Total control overhead

7. REFERENCES

[1] T. Clausen and P. Jacquet, "RFC 3626: Optimized link state routing protocol," October 2003.

[2] D. C. Montgomery, *Introduction to Statistical Quality Control.* John Wiley & Sons, Inc., 6th ed., 2008.

[3] G. E. P. Box, G. M. Jenkins, and G. Reinsel, *Time Series Analysis: Forecasting & Control.* Prentice Hall, 3rd ed., February 1994.

[4] K. Shaukat and V. R. Syrotiuk, "Using monitoring to control a proactive routing protocol," *Ad Hoc & Sensor Wireless Networks*, vol. 6, no. 3–4, pp. 299–319, 2008.

[5] L. Qin and T. Kunz, "Adaptive MANET Routing: A Case Study," in *Springer LNCS*, vol. 5198, pp. 43–57, September 2008.

[6] F. J. Ros and P. M. Ruiz, "Cluster-based OLSR Extensions to Reduce Control Overhead in Mobile Ad Hoc Networks," in *Proc. of the Int'l Conf. on Wireless Communications and Mobile Computing*, pp. 202–207, 2007.

[7] L. Ming, G. Zhao, G. Xie, and X. Kuang, "HOLSR: A novel routing scheme of ad hoc wireless networks for pervasive computing," in *Proc. of the 2nd Int'l Conf. on Pervasive Computing and Applications (ICPCA'07)*, pp. 661–666, July 2007.

[8] H. Badis and K. A. Agha, "Scalable Model for the Simulation of OLSR and Fast-OLSR Protocols," in *Proc. of the IFIP-TC6 Mediterranean Workshop on Ad-Hoc Networks (MedHocNet'03)*, June 2003.

[9] S. V. Crowder, "Design of Exponentially Weighted Moving Average Schemes," *Journal of Quality Technology*, vol. 1, July 1989.

[10] J. M. Lucas and M. S. Saccucci, "Exponentially Weighted Moving Average Control Schemes: Properties and Enhancements," *Technometrics*, vol. 32, no. 1, pp. 1–12, 1990.

[11] C. A. Lowry, W. H. Woodall, C. W. Champ, and S. E. Rigdon, "A Multivariate Exponentially Weighted Moving Average Control Chart," *Technometrics*, vol. 34, no. 1, pp. 46–53, 1992.

[12] D. C. Montgomery, L. A. Johnson, and J. S. Gardiner, *Forecasting and Time Series Analysis.* McGraw-Hill Inc., 2nd ed., 1990.

[13] Minitab, "Minitab Software for Quality Improvement." http://www.minitab.com, 2011.

[14] The University of California, Berkeley, "The Network Simulator – `ns-2`."

[15] G. M. Ljung and G. E. P. Box, "On A Measure of Lack of Fit in Time Series Models," *Biometrika*, vol. 65, no. 2, pp. 297–303, 1978.

[16] Gnu Regression, Econometrics and Time-series Library. `http://gretl.sourceforge.net/`.

[17] S. E. Rigdon, "A Double-Integral Equation for the Average Run Length of a Multivariate Exponentially Weighted Moving Average Control Chart," *Statistics & Probability Letters*, vol. 24, no. 4, pp. 365–373, 1995.

Towards Optimal Broadcast in Wireless Networks

Zygmunt J. Haas and Milen Nikolov
Wireless Networks Laboratory (WNL),
Cornell University
Ithaca, NY 14853
{zjh1, mvn22}@cornell.edu http://wnl.ece.cornell.edu

ABSTRACT

Broadcast is a fundamental operation in networks, especially in wireless Mobile Ad Hoc NETworks (MANET). For example, some form of broadcasting is used by all on-demand MANET routing protocols, when there is uncertainty as to the location of the destination node, or for service discovery. Being such a basic operation of the networking protocols, the importance of efficient broadcasting has long been recognized by the networking community. Numerous papers proposed increasingly more efficient implementation of broadcasting, while other studies presented bounds on broadcast performance. In this work, we present a new approach to efficient broadcast in networks with dynamic topologies, such as MANET, and we introduce a new broadcasting algorithm for such networking environments. We evaluate our algorithm, showing that its performance comes remarkably close to the corresponding theoretical performance bounds. Furthermore, we compare the performance of the proposed algorithm with other recently proposed schemes, especially in the context of various mobility settings.

Categories and Subject Descriptors

C.2.1 [**Computer Communication Networks**]: Network Architecture and Design - *Network topology, Distributed networks, Wireless communication*; C.2.2 [**Network Protocols**]: Routing Protocols

General Terms

Algorithms, Design, Performance

Keywords

broadcast; stochastic routing; probabilistic broadcasting; efficient flooding, Minimum Connected Dominating Set, heuristics,

1. INTRODUCTION

Broadcast is a fundamental network operation allowing a source node to send a message to all other nodes in the network. In the context of *Mobile Ad Hoc Networks* (*MANET*), where topology can change rapidly, where all communications are carried over a wireless medium, and where the network nodes are limited in energy and computational power, an efficient broadcast mechanism is exceptionally important for the overall improvement of the network performance.

MSWiM'11, October 31–November 4, 2011, Miami, Florida, USA.

For instance, among the various proposed MANET routing protocols (e.g., AODV, DSR, OLSR, TRBF, ZRP), a prominent sub-group, referred to as *on-demand* or *reactive* routing protocols, is designed based on the philosophy that the discovery of a route in the network should be done only when there is an actual need to route traffic. The route discovery mechanism in on-demand routing protocols relies on some variant of broadcasting to locate a path between a source and a destination nodes. As another example, in highly reconfigurable topologies, where the lifetime of the routes may be shorter than the duration of a communication (especially connection-oriented communication) broadcast, by itself, could be used as a routing mechanism. Yet in other scenarios, when data dissemination to all the nodes in a MANET is needed, broadcast is an obvious solution.

The required features of a broadcast algorithm in the context of wireless reconfigurable networks are that:

(1) it reaches *all* the network nodes;

(2) it transmits the message as few times as possible (or, equivalently, reduces the number of times that the broadcast message is received by a network node, optimally to only once);

(3) each node requires only locally restricted information (e.g., only the knowledge of the k-hop topology for small value of k);

(4) it proceeds so that changes in topology during the broadcast propagation do not significantly affect (1) and (2) above.

Being such an essential network operation, it is not surprising that the importance of an efficient broadcast implementation has been widely accepted by the networking community. Unfortunately, few of the so-far-proposed approaches to the broadcast problem satisfy all the above requirements. Flooding-based protocols violate (2) above, as they lead to the notorious "broadcast storm" problem [1]. Probabilistic schemes, often based on percolation [3, 4, 5, 6, 7, 8, 9] do not satisfy (1) above.[1] Efficient backbone-based algorithms have also been proposed. They rely on finding minimum connected dominating sets (MCDS), which are constructed by identifying dominating sets, maximal independent sets, or Steiner trees prior to broadcast. Such algorithms do not tolerate dynamic topologies well [19, 20, 21, 22], thus are incompliant with (4). Furthermore, some of these algorithms are centralized [29, 30], thus violating (3). One recent example of a broadcast algorithm that satisfies (1) – (4) has been described in [25], though the algorithm requires the knowledge of the nodes' geographical position. (Section 7 provides more details and references to related work in the literature.)

[1] To ensure full coverage, a stochastic scheme would need to transmit with probability close to 1, in which case the scheme converges to flooding.

In this paper we describe, model, and simulate a novel approach to network broadcasting. The resulting algorithm satisfies the features (1)−(4) without relying on nodes' positional information. Furthermore, the algorithm is comparatively simple to implement, relying on the intuitive greedy heuristic to tackle the, otherwise, NP-hard problem [32] of finding the Minimum Connected Dominating Set (MCDS) in a Unit Disk Graph. The algorithm experimentally outperforms other schemes recently published in the literature, achieving close to the optimal performance of the broadcast operation for key metrics such as the number of transmissions. Significantly, the algorithm performs well even in highly dynamic scenarios, in the context of realistic mobility models such as the Gaussian Markov Mobility Model (GMMM) and the Reference Point Group Mobility (RPGM) model.

2. SYSTEM MODEL

The network model consists of N equal-capability nodes with unique IDs, randomly distributed in 2D plane.[2] The transmission range of all nodes is R [meters]. Two nodes are referred to as *1-hop neighbors* (or simply as *neighbors*) and can communicate directly if the distance between them is less than R. Thus, the network can be modeled as a Unit Disk Graph (UDG).

On the MAC layer, links are bidirectional and nodes share a single wireless channel. We assume a *perfect* MAC layer to eliminate other effects (e.g., congestion), so that the broadcast performance metrics reflect only the algorithmic efficiency. This also allows a meaningful comparison of various broadcast schemes, as different schemes could be differently affected by a particular MAC layer protocol and the choice of its parameters.

The system operation is time-slotted, and the network nodes are assumed to be coarse-grain synchronized. The latter is a standard assumption of many distributed algorithms and can be implemented in variety of ways. For instance, distributed, control-message-based methods for coarse-grain synchronization in multi-hop wireless networks would suffice; such schemes have been studied extensively in the literature (e.g., [28]). Recent advances in radio technologies could also be utilized [27]. Finally, all nodes are cooperative and trust each other.

3. SOLUTION DEMYSTIFIED

As hinted above, the problem of finding the most efficient broadcast scheme is to a large extent equivalent to finding an approximation of the MCDS in a Unit Disk Graph which satisfies (1) − (4). The *Time-Sequence* (TS) scheme and its variants, as proposed in this paper, aim at this goal through the use of a *greedy* heuristic. We start by providing a number of definitions:

Definition 1: A *broadcast session* is the operation (including all related events) of delivering a message, created at one node – the *source* – to all the other network nodes.

Definition 2: A *covered node* is a node that has received the broadcast message.

The source node of a broadcast session is always covered. A node that has not received the broadcast message at a particular time is referred to as an *uncovered* node at that time.

[2] However, the results in this paper apply to 1D and to 3D networks, as well.

Definition 3: The *residual coverage* (RC) of a covered node s at a particular time, referred to as $RC(s)$, equals the number of its 1-hop uncovered neighbors at that time.

Definition 4: We define S as the set of all the network nodes, C as the set of all covered nodes at a particular time, and Q as the set of nodes that have already transmitted the message at a particular time. We further define $NE(s)$ as the set of all the neighbors of the node s, where $s \in S$. We note that at any time, $Q \subseteq C \subseteq S$ and that $|S|=N$.

We represent the network as a connected graph $G=(S, E)$, where E is the set of all the links that connect any two nodes in S. Also, we label the broadcast source node as s_0.

Consider first the following centralized greedy scheme, starting with the initial set of covered node $C=\{s_0\}$ and $Q=\emptyset$. The source node transmits first, covering its neighbors: $C = \{s_0 \cup NE(s_0)\}$, $Q=\{s_0\}$. An "oracle" greedily chooses a node, s_1, from the set C-Q with the largest RC value to broadcast next; i.e., $\forall s \in (C\text{-}Q)$, $RC(s) \leq RC(s_1)$. After s_1 broadcasts, $C \leftarrow C \cup NE(s_1)$ and $Q \leftarrow Q \cup \{s_1\}$. This operation of selecting the next node to transmit as the node with the largest RC from the set C-Q is repeated until all the network nodes are covered; i.e., until $C=S$, at which time the algorithm terminates. (The total number of transmissions in a broadcast session equals $|Q|$ at the algorithm termination time.) Furthermore, the choice of node s_i to transmit in the ith iteration allows maximizing the number of covered nodes during the ith transmission. This, in turn, would intuitively <u>tend</u> to minimize the total number of transmissions during the operation of the algorithm. Of course, the algorithm does not guarantee such a minimum, as in some cases choosing a node with smaller RC value first could, in fact, result in finding nodes with much larger RC value later, so that the overall number of transmissions is smaller. In [31], the authors presented in more details the failure of a similar greedy scheme in finding MCDS in general graphs.

Notwithstanding the above, restricting the network topology to a Unit Disk Graph relaxes the problem and the above centralized greedy heuristic finds on the average a rather close approximation of an MCDS,[3] as is demonstrated in Section 5. Of course, the challenge, similarly to other efficient MCDS approximation schemes, is to implement the "oracle" in a distributed manner; i.e., ordering nodes' transmissions based on their RC values, while utilizing only local topological information.

The following TS scheme distributively approximates the transmissions' order by allowing nodes with larger RC values to transmit before transmissions of nodes with smaller RC values. The timing of transmissions is enforced by a specifically structured sequence of time-slots, which are locally available to each node, and by a scheduling procedure, which is locally executed at each node. The scheduling procedure is based on associating a particular value of RC threshold with each time-slot and on restricting transmissions in a time-slot only to nodes with RC values not smaller than the time-slot's RC threshold.

Note that the goal of the TS scheme is not to control access to the wireless channel, akin the TDMA scheme, for instance. Namely, a time-slot in the TS scheme may accommodate concurrent transmissions of multiple nodes in the network. Section 4.3 discusses how, via the structure of the time-slot sequences in the TS scheme and the operation of the proposed algorithms, potential

[3] This is often the case with the greedy algorithm in various other applications as well.

collisions and interference on the MAC and the physical layers can be avoided during the transmissions of the broadcast message.

4. THE TIME-SEQUENCE SCHEME

We start by defining a *Time Sequence,* and by presenting its structure and properties.

4.1 The Time-Sequence Structure

Let T be an ordered collection of vectors $\{T_x, T_{x-1}, \ldots, T_1\}$, where $T_i = (u_i, m_i, l_i)$ and where $u_i, m_i,$ and $l_i \in \mathbb{N}^+$ for all i. The parameters u, m, and l are referred to as *upper*, *middle,* and *lower* values. Let T be the output of **Algorithm 1** given input u.

Proposition 1: $|T| = \dfrac{u(u+1)}{2}$.

Proof: The output, T, of **Algorithm 1** can be arranged in the following isosceles triangle with sides u: the triangle consists of u levels, where the last level (level 1) comprises u vectors:

(u, u, u) -- {level u: $l = u$}

(u, u, u-1) **(u, u-1, u-1)** ----------------------- {level u-1: $l = u$-1}

⋮ ⋮

⋮ ⋮

(u, u, 1) (u, u-1, 1) … **(u, 1, 1)** -------------- {level 1: l=1}

The number of vectors at the ith level equals $1 + u - i$. The total number of vectors is, then: $\sum_{i=1}^{u} u(u+1)/2$. □

The vectors in bold are called *edge slots* and form the "base" of the isosceles triangle.

Next, let $A = (S, I)$ be a set system with $S = \{s_1^{p_1}, s_2^{p_2}, \ldots, s_n^{p_n}\}$, $p_i \in \mathbb{N}^+$ for all $i \in \{1, \ldots, x\}$, and let I be the collection of subsets of S, such that $I = \{S_1, S_2, \ldots, S_x\}$ and $S_i = \{s_j^{p_j} : p_j \geq m_j \geq l_j \geq 1\}$ given vector $T_j = (u_j, m_j, l_j)$ in T (note that $|I| = |T|$). Define a binary order relation on S, so that $s_j^{p_j} \leq s_i^{p_i} \Leftrightarrow p_j \leq p_i$.

Definition 5: An element $s_j^{p_j}$ is called *admissible in T_i*, iff $s_j^{p_j} \in S_i$.

A minimal admissible element in T_k, denoted by $min\ (T_k)$, is $s_j^{p_j}$ such that $s_j^{p_j} \leq s_i^{p_i}$ for all $s_i^{p_i}$ admissible in T_k. Consider the ordered collection of vectors T^{λ} at the level λ of T ($\lambda \in \{1, \ldots, u\}$), and the smallest element of all minimal admissible elements at this level, denoted $inf(T^{\lambda})$. Now, let L be the sequence $(inf(T^{l}), inf(T^{l-1}), \ldots, inf(T^{1}))$.

Proposition 2: The sequence L is decreasing.

Proof outline. The result follows trivially from the definition of I, the ordering of T, the order relation defined on S, and the definition of $inf(T^{\lambda})$.□

Let M be the sequence $(min\ (T_k), min\ (T_{k-1}), \ldots, min\ (T_1))$, where T_i's, $i \in \{1, \ldots, k\}$, are in T^{λ} for some λ.

Proposition 3: The sequence M is decreasing.

Proof outline: Similarly to **Proposition 2**, the proof follows trivially from the definition of I, the ordering of T, the order relation defined on S, and the definition of $min\ (T_k)$. □

```
Algorithm 1: Constructing the vector set T
Input: u
/* initialization */
upper := u
middle: = u
lower := u
T := ∅
T1 := ( upper, middle, lower )
T := T1
/* Computing the T vectors */
while middle > 1:  // Generating the next vector
   if middle == lower :
      lower := lower – 1
      mid := upper
      next_vector: = ( upper, middle, lower )
   else
   if middle > lower :  // lower remains the same
      middle := middle – 1
      next_vector := ( upper, middle, lower )
   T := T+next_vector
```

We can now define the structure of the Time Sequence (*TS*). Suppose time is slotted. Every x consecutive time-slots (naturally ordered in time) are mapped one-to-one to the vectors in the ordered collection T above. That is, each time-slot, t_j is uniquely associated with a vector in T as follows: $t_1 \leftrightarrow T_x = (u, u, u)$, $t_2 \leftrightarrow T_{x-1} = (u, u, u\text{-}1)$, …, $t_x \leftrightarrow T_1 = (u,1,1)$, where $x = 0.5u(u+1)$ from **Proposition 1.** A network node is admissible in t_j if it is admissible in T_{x+1-j} This association "wraps around"; i.e., in general for $j \geq 1$, $t_j \leftrightarrow T_{kx+1-j}$ where $k = \lceil j/x \rceil$. The time sequence (*TS*) is the ordered collection of the time-slots, together with their corresponding vectors.

4.2 Distributing the Time-Sequence Scheme

As defined above, let S be the set of all the network nodes. Each node j is assigned its value $p_j = RC(j)$. More specifically, $S = \{n_1^{RC_1}, n_2^{RC_2}, \ldots, n_N^{RC_N}\}$, where $|S| = N$. Note that as time goes by from a time-slot to the next, the values $RC(j)$ may change.

(Revisited) Definition 1: A *broadcast session* consists of all the events, starting from the transmission of the message by the source node and ending when the broadcast algorithm terminates.

Also, for now, we will assume that the broadcast session finishes in less than or equal to |T| time-slots. Then, we define the following rule:

Broadcast Rule: At every time-slot, t_i, a node considers broadcasting only if it has not broadcasted earlier in this session and it is admissible in T_i at the beginning of the time-slot t_i

Thus, **Proposition 2** implies that nodes with smaller *RC* values would not be able to broadcast (i.e., be admissible) at the beginning of the time sequence *TS*, but will potentially be admissible in later timeslots (i.e., at lower levels). And reversely, only nodes with larger *RC* values would be admissible and be able to broadcast in earlier timeslots (i.e., at higher levels) of the *TS*. Furthermore, **Proposition 3** implies that within given level of the *TS*, nodes with smaller *RC* values would again be admissible only in later timeslots at this level, and nodes with larger *RC* values would be admissible earlier in the given level.

For illustration purpose, consider the case of $u = 4$:

(4,4,4) ---{TS level 4: l=4}
(4,4,3) **(4,3,3)** -------------------------------------{TS level 3: l=3}
(4,4,2) (4,3,2) **(4,2,2)**-----------------------------{TS level 2: l=2}
(4,4,1) (4,3,1) (4,2,1) **(4,1,1)** --------------------{TS level 1: l=1}

Example 1. The output of **Algorithm 1** for $u = 4$

In this example, in the top *TS* level (level 4) only nodes with RC ≥ 4 are allowed to transmit. In the next *TS* level (level 3), first nodes with $RC \geq 4$ and then nodes with $RC \geq 3$ will be allowed to transmit. In level 2, first nodes with $RC \geq 4$, then nodes with $RC \geq 3$, and finally nodes with $RC \geq 2$ will transmit. In the last level, first nodes with $RC \geq 4$, then nodes with $RC \geq 3$, then nodes with $RC \geq 2$, and finally nodes with $RC \geq 1$ (all nodes with at least one uncovered neighbor) will be allowed to transmit.

In summary, the structure of the *TS* in conjunction with the **Broadcast Rule** has admissibility property allowing for repetitive assignment of larger transmission priority to nodes with larger *RC* values as compared to nodes with smaller *RC* values. Hence, by the virtue of the admissibility property of the **Broadcast Rule**, the greedy "oracle" scheme is emulated.

4.2.1 Distributing the TS-based Broadcast

The above time-sequence structure is computed locally by every node in the network, so that covered nodes can determine, based on their current value of *RC*, the earliest time-slot (i.e., the phase of the time sequence) at which they should first attempt to transmit the message, thus satisfying the **Broadcast Rule.** Next, we describe the operation of the scheme in more details.

4.2.1.1 Network Deployment

Upon network deployment, **Algorithm 1** is run locally by each node.[4] The input to the algorithm, the parameter u, is fixed and set up administratively and network-wide at the time of deployment. The value of u should be judiciously chosen; a too small value of u does not allow to separate in time the transmissions of nodes with different values of *RC*, thus losing the ability to assign larger priority to nodes with larger *RC* values, while a too large value of u results in many empty time-slots, thus leading to an unnecessarily long broadcast session. Section 5.1.2 addresses the choice of the u parameter in more details, discussing its effect on the scheme's performance.

Preamble	*Broadcast Field*

Figure 1. The format of a time-slot.

Time-slots have the format shown on Fig. 1. The *Broadcast Field* is fixed to be the maximum duration needed to transmit the broadcast message. The *Preamble* is used to transmit control information between adjacent nodes, and its duration is negligible compared to the total length of the time-slot. The timestamp of the initial broadcast by the source is embedded within the *Preamble,* is propagated by the broadcasting nodes, and serves by the nodes to determine the current phase of the *TS*.

4.2.1.2 Node Scheduling

The source node broadcasts the message at the beginning of the first *TS*. As the broadcast propagates throughout the network, any node upon receiving the broadcast message for the first time determines the current *TS* level and the current time-slot within the *TS*. This is achieved by subtracting the initial broadcast timestamp (in the message's *Preamble*) from the node's current local time, and by dividing the difference by the duration of a time-slot. Knowing the generic *TS* structure and the number of time-slots that passed since the beginning of the *TS,* a node is able to determine the m and l values of the current time-slot.

Said differently, as the broadcast message propagates through the network, it prompts the receiving nodes to calculate the *TS* structure and to identify the current phase within the *TS* structure.

After a receiving node calculated the current time-slot within the *TS,* the node runs **Algorithm 2** to schedule its transmission for a future time-slot. It is easy to verify that **Algorithm 2** complies with the **Broadcast Rule**. For instance, if m of the current *TS* time-slot is lower than the node's *RC* value, **Algorithm 2** schedules the node's transmission for the next immediate time-slot. Otherwise, a further future time-slot is assigned to the node. A node is never scheduled to transmit if its $RC = 0$ (i.e., it is not admissible in any time-slot).

Algorithm 2: Node i self-scheduling
Input: RC_i; (u_{ct}, m_{ct}, l_{ct}) // *vector associated with the current time-slot*

```
/* initialization */
rc := RC_i
upper := u_ct
middle := m_ct
lower := l_ct
/*if RC_i is larger than the current value of middle, the node
broadcasts in the next time-slot */
if rc > middle :
   broadcast_timeslot := next_timeslot
else
/* if RC_i is not large enough for the current priority, the node's
future broadcast time-slot is determined by the value of RC_i */
if rc ≤ middle AND rc ≥ lower :
   if (u_ct, m_ct, l_ct) is edge slot :
      if lower > 1 :
         broadcast_timeslot := ( upper, rc, lower-1 )
      else
         broadcast_timeslot := ( upper, rc , 1 )
   else
      broadcast_timeslot := ( upper, rc, lower )
else
/* if RC_i is even less than the value of lower, the node is
pushed to broadcast in a later level of the TS triangle; the
level depends on RC_i */
if rc < lower AND rc ≥ 1 :
   broadcast_timeslot := ( upper, rc, rc )
```

The value of a node's *RC* can change[5] between the time at which it had scheduled itself for transmission and the beginning of its scheduled-for-transmission time-slot, thus possibly rendering the node inadmissible in its scheduled time-slot. To avoid transmission in an incorrect time-slot, a node re-computes its *RC* value during the *Preamble* time of its scheduled-for-transmission timeslot and checks if it is still admissible in this time-slot. If so, it transmits the message during the *Broadcast Field* time. Else, it reschedules itself by employing the **Algorithm 2** again.

4.2.1.3 Residual Coverage Computation

The *RC* value of a node is needed prior to the node scheduling or rescheduling itself for transmission. This is done by a locally

[4] Practically, **Algorithm 1** should be run periodically.

[5] Due to transmissions of other nodes or due to mobility.

executed protocol – a version of a "Neighbor Discovery" protocol. Specifically, to compute its *RC* value, node *i* broadcasts a *Coverage Request* packet, *CReq*, to all its 1-hop neighbors. The *CReq* packet contains *i*'s ID. Upon receiving the *CReq* packet, each one of *i*'s uncovered neighbors replies with a *Coverage Reply* packet, *CRep*, which contains the neighbor's ID and *i*'s ID. Node *i* then counts the number of such replies (identified by its own ID in the *CRep* packets) during a time approximately equal to the duration of the *Preamble*. This simple protocol can be further improved in a variety of ways, but that is beyond the scope of this paper.

4.3 Variants of TS-based Broadcast Schemes

For clarity, let's summarize the basic *TS*-based broadcasting scheme. All nodes run **Algorithm 1** to construct the *TS*. When a node (re)broadcasts the message, all of its previously uncovered neighbors mark themselves as covered, compute their *RC*, and run **Algorithm 2** to schedule their broadcast time-slots. Just before its scheduled time to transmit (during the *Preamble* of the scheduled-for-transmission time-slot), a node re-computes and updates its *RC*. If the node's current value of *RC* has decreased (but $RC > 0$), the node determines its new time-slot assignment by re-running **Algorithm 2.** If, at any time, the *RC* value of a node equals 0, the node will never be scheduled for broadcast in this session.

We refer to this basic scheme as the *Naïve Time Sequence Scheme* (*NTSS*)[6]. In what follows, we present two variants of the NTSS, each utilizing varying degree of the neighborhood topology knowledge.

4.3.1 1-Hop Time Sequence Scheme (TSS-1Hop)

Suppose node *i* is about to transmit[7] in the current time-slot t_j. Then node *i* checks during the *Preamble* of the time-slot t_j whether any of its 1-hop neighbors are scheduled to broadcast within t_j as well. This check does not necessitate any additional transmissions, as the node can determine whether a particular neighbor is scheduled to transmit in t_j, if it receives the *CReq* message from that neighbor during the *Preamble* time. If more than one neighboring node is scheduled for t_j, the node with the largest *RC* is selected to transmit in t_j. The rest of the neighboring nodes reschedule themselves to transmit in the next time-slot.

To accommodate the TSS-1Hop operation, the *CReq* and the *CRep* control packets should also include the respective sender's *RC* value, in addition to the nodes' IDs. We omit here further details of this simple protocol.

4.3.2 2-Hop Time Sequence Scheme (TSS-2Hop)

In the TSS-2Hop scheme, a node *i* which is scheduled to transmit in the current time-slot t_j checks if there are either 1-hop or 2-hop neighbors scheduled to transmit within t_j as well. Among the set of all the 1-hop and 2-hop neighbors, the one with the largest *RC* is selected to transmit; the rest are rescheduled to the next time-slot.

Here, the control packets passed in the *Preamble* need further augmentation. The *CReq* packet contains now two IDs. Node *i* broadcasts a *CReq* packet to all its 1-hop neighbors, with the first ID set to *i* and the second ID to *null*. Each (1-hop) neighbor *n*, upon receipt of such a *CReq* packet, retransmits it to its own neighbors, with the first ID unchanged and the second ID set to *n*. Each (1-hop) neighbor of *n* (and, therefore, 2-hop neighbor of the node *i*), which is scheduled to transmit in this time-slot, replies to *n* with its *CRep* packet with its *RC* value and with the IDs unchanged. The node *n* then collects all such replies, determines the largest *RC* value from among the received *RC* values and its own *RC* value, and send its *CRep* with this largest value of *RC* to the node *i*. We omit here further details of this simple protocol.

[6] As the name indicates, the NTSS possess some deficiency, which will be cured by the other variants of the scheme.

[7] I.e., that the current value of RC(*i*) matches the time-slot t_j.

4.4 Sample Execution of NTSS

Scheduled Nodes	*RC*/Scheduled Timeslot
a	2 / (4,2,2)
b	3 / (4,3,3)
c	1 / (4,1,1)
d	1 / (4,1,1)

Step 1

Scheduled Nodes	rc/Scheduled Timeslot
a	2 / (4,2,2)
c	0 / (4,1,1)
d	1 / (4,1,1)
f	1 / (4,1,1)
g	0 / -
h	0 / -

Step 2

Scheduled Nodes	rc/Scheduled Timeslot
c	0 / (4,1,1)
d	1 / (4,1,1)
f	0 / (4,1,1)
g	0 / -
h	0 / -
e	0 / -
k	2 / (4,2,1)

Step 3

Step 4: All nodes are covered after timeslot (4,2,1). The time-sequence is exhausted at (4,1,1), where as well no nodes have RC > 0, and the algorithm terminates.

Figure 2. In the first timeslot, the source node *S* broadcast the message. Nodes *a*, *b*, *c*, and *d* receive it, mark themselves as covered, compute their RC, and schedule themselves to broadcast At **Step 1**, according to **Algorithm 2,** since node *b* has 3 uncovered neighbors it is scheduled for timeslot (4,3,3). Node *a* is scheduled similarly for (4,2,2), *c*, and *d* for (4,1,1). At **Step 2**, during the third timeslot (4,3,3) node *b* rebroadcasts the message. Nodes *g*, *h*, and *f* become newly covered and are added to the scheduled nodes list: *f* is scheduled for (4,1,1); *g* and *h* have $RC = 0$ and are not scheduled. Node *b* is removed from the list. At **Step 3**, in the sixth timeslot (4,2,2), node *a* check its *RC*. Since its *RC* remains the same, node *a* rebroadcasts the message. Nodes *e* and *k* become newly covered. Node k has two uncovered neighbors: *i* and *j*. It is scheduled for timeslot (4,2,1). Finally, at **Step 4**, during the preamble of the ninth timeslot node *k* has not changed its *RC* and rebroadcasts. All nodes are covered at this point. Hence, *c*, *d*, and *f* do not broadcast

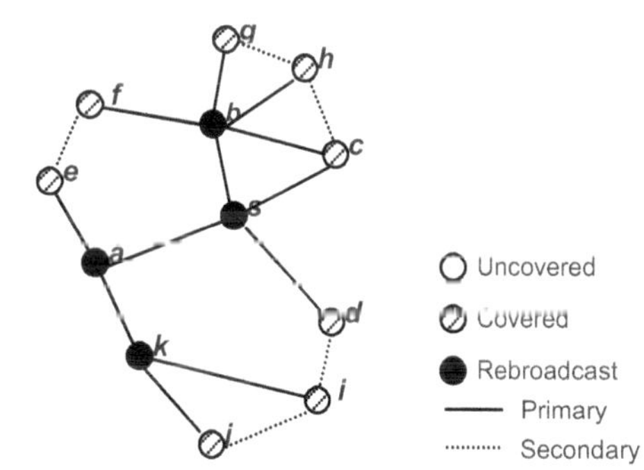

Fig 3: A network topology example, where {*s,b,a,k*} form an MCDS of the particular network graph. All nodes are covered after the broadcast session completes.

Consider the network of nodes shown on Fig. 3. For simplicity of presentation, only the NTSS variant of the *TS* scheme is discussed, which suffices to demonstrate the main mechanisms of the *TS* scheme. Suppose $u = 4$. Then, **Algorithm 1** constructs the *TS* as shown in **Example 1.** The sample run of NTSS is shown in Fig. 2, with the resulting network coverage depicted in Fig. 3.

Note that the *TS* structure allows **Algorithm 2** to give priority to nodes with higher *RC* that are scheduled later during the broadcast session. For instance, because of its larger *RC* value, node *k* will broadcast before nodes *c*, *d*, or *f*, albeit they were scheduled earlier. Ultimately, this eliminates the transmissions of nodes *c*, *d*, and *f*. Fig. 3 shows the network state after the broadcast session is completed. Edges between rebroadcast nodes, and between rebroadcast and covered nodes are labeled as primary, while edges between covered nodes as secondary.

4.5 Correctness of the TS-based Schemes

In the following proof of correctness, the network topology is assumed to be static during the broadcast session and the network graph is assumed to be connected.

Proposition 4: All *TS*-based schemes terminate in finite amount of time and guarantee full coverage of the network.

Proof (outline): Per **Algorithm 2**, for all *TS* schemes, a node is not scheduled to broadcast unless its *RC* is strictly greater than zero. Whenever a node *n* broadcasts, all of it neighbors receive the broadcast message and are marked as covered. Hence, the *RC* value of node *n* decreases to zero, and node *n* is not admissible in any future time-slot. Therefore, a node does not broadcast more than once during the execution of the algorithm. Since there is a finite number of nodes in the network, the algorithm terminates in a finite number of time steps.

Now, suppose that after a *TS*-based algorithm's termination there is at least one node, *D*, that is not covered. Since the network graph is connected, there exist at least one path from the source node, *S*, to the destination node *D*. Then, because *D* has not received the message, there are at least two neighboring nodes *X* and *Y* along this path, such that *X* has received the message and *Y* has not (note: *X* might be *S* and *Y* might be *D*). Therefore, *X*'s *RC* was greater than zero at algorithm's termination. This contradicts all *TS* schemes' termination condition. (Whenever a node is covered it computes its residual coverage. As long as node's residual coverage is greater than zero it is always scheduled to broadcast according to **Algorithm 2**.) Thus, by contradiction, all *TS*-based schemes cover all the network nodes. □

5. PERFORMANCE EVALUATION

We investigate the performance (as defined by a number of metrics) of various broadcast algorithms in three distinct network topology models: (1) a static network topology; (2) a mobile topology with *independent* mobility patterns of the nodes; and (3) a mobile topology with *correlated* (group) mobility pattern.

We compared the performance of the TS-based schemes with *position* aware *Responsibility Based Scheme* (*RBS*), suggested by Khabbazian and Bhargava [25]. In [25] authors show that RBS outperforms a few well known broadcast algorithms, such as the *Edge Forwarding* [13] algorithm, for example. Another broadcast protocol simulated here, the *Bordercast,* is the route discovery mechanism in the *Zone Routing Protocol* (*ZRP*) ([26]).

Bordercast relies only on local topological information to select the nodes to relay the broadcast message. To include an algorithm that constructs a *backbone* structure prior to the broadcast session we selected *Funke's* algorithm [21], which was shown to obtain one of the best approximation ratios to the size of the MCDS (i.e., |MCDS|), outperforming leading algorithms such as Wan's algorithm [20]. Finally, for comparison, we simulated Liu's algorithm [11] – a *node forwarding* algorithm that relies on 1-Hop positional information. Both RBS and Funke's algorithms are among the most efficient broadcast schemes in the literature.

In all the experiments, the simulation area is a 200[m] by 200[m] square; the inner square area is of dimensions (200 - R)[m] by (200 - R)[m], where R[m] is the transmission radius of all nodes and was set to R = 25[m]. The number of nodes in the network was varied to investigate the schemes' performance at different node densities. To eliminate the effects of the physical and MAC layers on the broadcast algorithms' performance, we assume no signal interference, no fading, and a "perfect" MAC Layer. All the broadcast algorithms were implemented in a JAVA discrete event simulator.

5.1 Static Network Topology

5.1.1 Transmission Complexity

The number of transmission during a broadcast session (i.e., "transmission complexity"), a crucial metric for an efficient broadcast algorithm, is investigated on Fig. 4. We were interested in understanding how our distributed algorithms compare with the centralized greedy approximation of the MCDS.

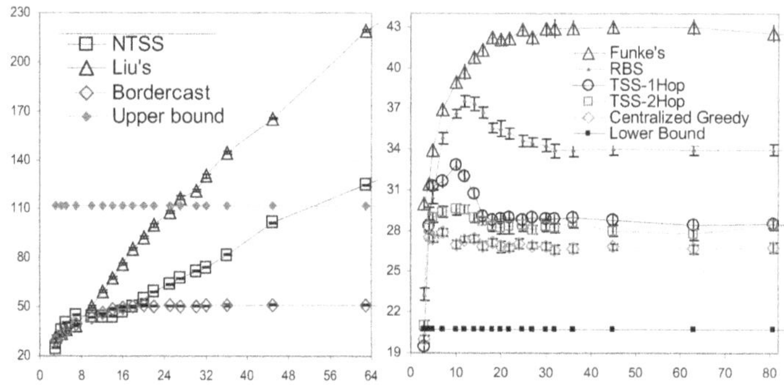

Figure 4. Number of transmissions (y-axis) vs. average node degree (x-axis) for different broadcast algorithms.

As an additional comparison of the transmission complexity, we calculated the number of transmissions using the *Linear Hexagon Coverage* technique, which provides a close approximation to the minimum number of transmissions needed to cover the entire network area, assuming sufficient node density. The upper bound of transmission complexity should be interpreted as the maximal number of transmissions that would be required by any broadcast algorithm that avoids duplicate coverage of nodes and, hence, is density independent. Due to space limitations, we omit here the derivation and an elaborate discussion of the bounds in Fig. 4.

The TSS-1Hop algorithm performs only slightly worse than the TSS-2Hop algorithm and mildly worse than the centralized greedy scheme.

Both, the NTSS and the Liu's algorithms are not density independent (surpass the upper bound). The NTSS algorithm does not perform well, because often a node is covered multiple times by transmissions of its 1-hop neighbors which occur in the same time-slot, thus resulting in redundant transmissions. This is especially common in time-slots towards the end of a broadcast session. The number of such transmissions during a simulation run is

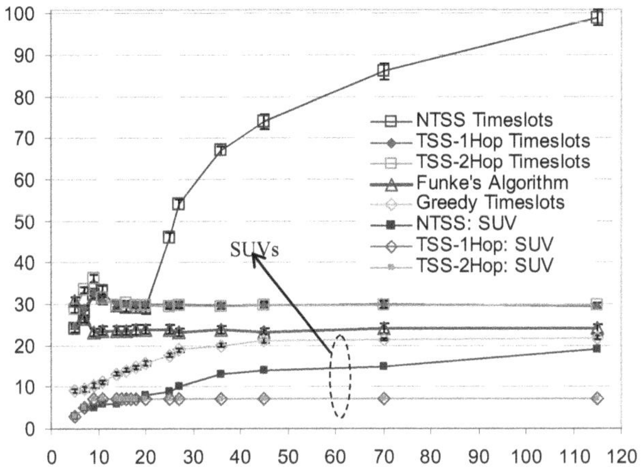

Figure 5. Delay (number of timeslots) (y-axis) vs. average node degree (x-axis). (SUV denotes Sequence Upper Values (the parameter *u*) for which the delay of *TS* schemes was measured.)

proportional to the number of nodes in the network and, hence, the transmission complexity continues to increase with the node density.

5.1.2 Delay

The time delay – the number of time-slots needed to complete a broadcast session – is presented in Fig. 5. The delay performance of the three *TS*-based schemes was obtained with *u* set to the smallest value, such that the number of broadcast transmissions is still close to optimal. These smallest values of *u* were determined via simulations; it is worth noticing that beyond average node degree of 15, these *u* values for TSS-1Hop and TSS-2Hop remain essentially constant with respect to the network density. Thus the upper value could be fixed prior to network deployment, so that the resulting time delay and the number of broadcast messages remains close to the algorithm's optimum behavior. The results for TSS-1Hop and TSS-2Hop hint that a good practice would be to set *u* to the average node degree for networks of node degrees up to 20. For networks of larger node degrees, *u* is largely density- independent and can be simply fixed at 20.

In general, there is a tradeoff between the two performance metrics: the number of transmissions during a broadcast session and the duration of a broadcast session. Intuitively, it would be beneficial to set *u* to a large value for finer-grain resolution of priorities among the nodes with different *RC* values and, thus, for a better approximation to the greedy "oracle" scheme. However, large values of *u* have the disadvantage of increasing the delay, since the *TS* grows longer (quadratic in *u*, from **Proposition 1**) and large number of initial time-slots go by empty, as no nodes are then admissible

5.2 Changing Network Topology

For all of the mobility experiments presented here, each data point was obtained after a simulation warm-up time of 1000 [s] in which the mobility model could converge to a stationary node distribution.

5.2.1 Gauss-Markov Node Mobility

In the Gauss-Markov mobility model (GMMM) [33, 34], the time is split into time intervals. At the beginning of the kth time interval, nodes' velocity is updated based on their velocity in the $(k-1)^{th}$ time interval and according to the following rule: $v_k = \alpha v_{k-1} + (1-\alpha)\bar{v} + \sqrt{(1-\alpha^2)}\, v_{x_{n-1}}$. Here, v_{k-1} is the velocity (speed and direction) of a node in the previous time interval, $v_{x_{n-1}}$ is a Gaussian random variable, $\bar{v}$ is the mean value of the velocity, and α is a parameter that determines the degree of which the current velocity depends on the previous velocity (the amount of "memory"). As α approaches 1, nodes' motion is constant; as α approaches 0 nodes' motion becomes more random. At the end of each interval, the position of a node in the network area is updated according to its velocity during this interval.

Velocity and position update interval	0.2 [s]
Velocity standard deviation	0.75 [m/s]
Velocity mean	Varied
Alpha	0.75

Table 1: The parameters of the Gauss-Markov Mobility Model.

Fig. 6 depicts the number of transmissions generated by the four investigated broadcast algorithms. As the average speed of the nodes increases, the number of broadcast transmissions saturates for all algorithms. This is an artifact of the finite network area – beyond a certain speed (about 20-30 [m/s] in this case), the locations of the nodes in the next time interval become random with respect to the locations in the previous time interval. This is equivalent to formation of a completely new random placement of nodes in each time interval, and increasing the speed further does not alter the performance of the algorithms anymore.

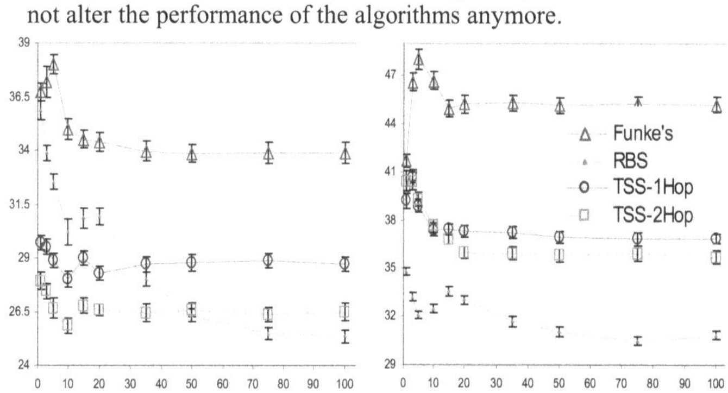

Figure 6. GMMM – number of transmissions (y-axis) vs. average speed [m/s] (x-axis); 200 and 500 nodes (*left*) and (*right*) respectively.

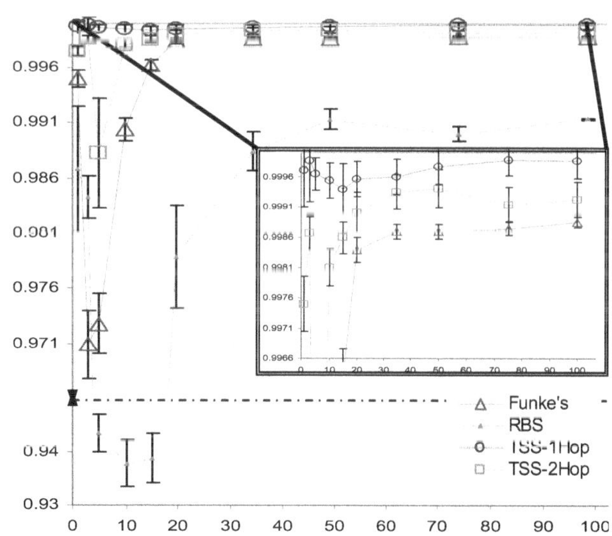

Figure 7. GMMM – the fraction of network nodes covered (y-axis) vs. average speed [m/s] (x-axis); 200 nodes.

However, the number of transmissions prior to this saturation point depends on the specific mechanisms of the broadcast algorithm in question. For instance, the decrease in the number of transmissions prior to saturation as mobility increases for the *TS*-based schemes results from the scheduling policy of these algorithms: the larger their residual coverage is, the sooner the nodes transmit. As mobility increases, there is a large chance that a node would move into an area containing larger concentration of uncovered nodes, thus increasing its *RC* and, overall reducing the total number of transmissions. Furthermore, the fraction of nodes covered increases as well, as shown in Fig. 7.

As opposed to the static case, where a broadcast session is bound to cover all the network nodes as long as the network is connected, with mobility there is a chance that some nodes will not be covered. Fig. 7 demonstrates the coverage (fraction of the network nodes reached during a broadcast session). The performance of all algorithms saturates beyond some mobility threshold. In the mobility region of 10-20 [m/s] and for smaller densities (200 nodes in the network), all algorithms are affected by disconnections in the graph during the broadcast session. Comparatively, for larger densities, all algorithms perform substantially better (the figure is omitted due to space limitations).

The notable drop in performance of the RBS scheme in Fig.10 for a range of low speeds is due to the specific RBS mechanisms: node *A* broadcasts if among its neighbors there is an uncovered node (call it node *B*), and node *B* is such that the distance between *A* and *B* is smaller than the distance between *B* and any other of the covered neighbors of *A*. For very low velocity, RBS performs well and covers almost the entire network; however, as the velocity increases, node *A*'s estimation of the distance between *B* and *A*'s neighbors becomes less accurate. Though, as the velocity of nodes increases further, RBS benefits from the fact that *A* could move into areas where only a small number of nodes is covered.

5.2.2 Group Mobility

Since independent individual node movement may be unrealistic for certain scenarios (e.g., hikers, tourist groups, military platoon, etc.), we have also simulated a group mobility model, which is

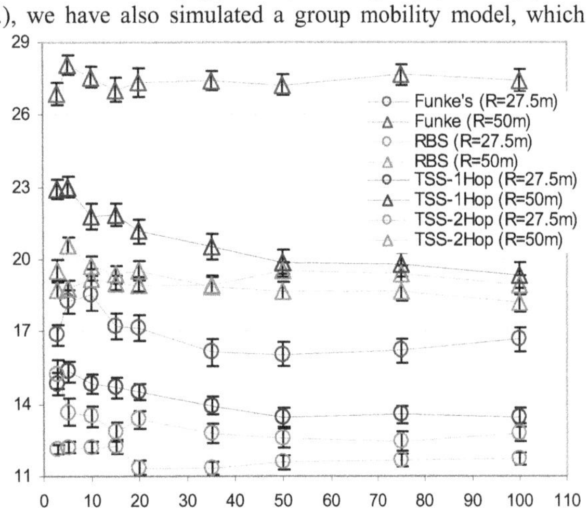

Figure 8. RPGM – number of transmissions (y-axis) vs. average speed [m/s] (x-axis); reference point (RP) radius equals 27.5[m] (*circles*) and 50[m] (*triangles*).

based on the *Reference Point Group Mobility* (*RPGM*) model described in [35]. The implemented version of RPGM follows the *Nomadic Community Mobility* model described in [34]. In this model the network nodes form groups that move together from one reference point (*RP*) in the simulation area to another. The RPs have a radius of influence and the group of nodes is free to move randomly within this radius. The simulation parameters (such as the group sizes and the number of groups in the network) used here were borrowed from [34] and [36] for the Nomadic and the RPGM models, respectively. The number of nodes was set to 200.

Pause time for mobility around the RP	0.3 [s]
Nodes velocity update frequency	0.5 [s]
Maximum reference point pause time	4 [s]
RP radius	Varied
Average group size	20% *of network size*
Nodes mean velocities	Varied

Table 2: The parameters of the RPGM mobility model.

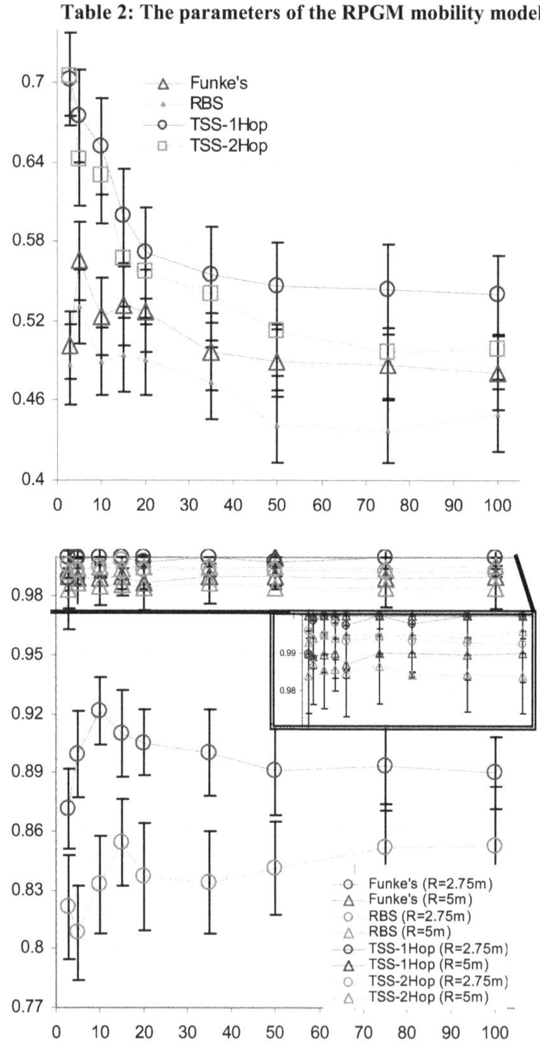

Figure 9. RPGM – fraction of nodes covered (y-axis) vs. average speed [m/s] (x-axis): (*top*) reference point (RP) radius equals 12.5[m]; (*bottom*) RP radius equals 27.5[m] and 50[m].

Figures 8 and 9 depict the number of transmissions and the network coverage. The RP radius (labeled as R in the Fig. 9), is varied to investigate the effects of node clustering about the reference point. As can be observed from these figures, the smaller the radius is, the more frequent are the disconnections and, consequently, the performance of the algorithms degrades.

As in the GMMM case, in the RPGM case too, the number of transmissions and the network coverage saturate as the network mobility increases. However, in the RPGM case there is a minimal decrease in the number of transmissions with respect to the increase in mobility; e.g., for 12.5[m] the decrease in the number of transmission is largely due to the highly clustered and disconnected network topology.

Overall, in dynamic scenarios, such as those when mobility is introduced, algorithms which rely on local topology information, such as TSS-1Hop and TSS-2Hop, outperform algorithms such as RBS and Funke's.

6. DISCUSSION

In this paper, we introduced a novel scheme for broadcasting in wireless networks that mimics in performance of the centralized greedy algorithm. This is accomplished through distributive prioritization of transmissions based on nodes' residual coverage (*RC*) and the particularly designed Time Sequence (*TS*) to schedule the nodes' transmissions. The basic NTSS scheme was improved to eliminate a major source of inefficiency – multiple coverage of nodes by more than one transmission – which resulted in the TSS-1Hop and the TSS-2Hop schemes. We proved the schemes' correctness (i.e., guaranteeing full coverage in finite time).

Through simulations and based on two metrics – the transmission complexity and the delay – we compared the performance of our schemes with other leading broadcasting schemes, initially in static networks. The TSS-1Hop and the TSS-2Hop schemes outperform all other schemes with respect to the number of message transmissions. Also, the TS-based schemes do not require additional equipment, such as GPS. Furthermore, this performance is achieved with bounded latency, and is independent of network density.

Next, we considered two types of mobility scenarios: individual node and group mobility. We showed experimentally that the TS-based schemes outperform the other schemes with respect to the network coverage by the broadcast transmissions.

Our study allows examination of the basic tradeoffs in the design of the broadcasting protocols, such as the tradeoff between transmission complexity and delay. It is worth noticing that the TS-based schemes facilitate control of this tradeoff through a single parameter only – the *u* parameter.

In addition, the current work invites interesting questions regarding the TS schemes performance that remain to be further studied. For instance, schemes' behavior in the presence of multiple simultaneous broadcast sources could be considered; the algorithms' performance given errors in the process of RC computation should also be investigated.

7. RELATED WORK

The problem of efficient broadcasting has been extensively studied in the technical literature. The initial simple concept of flooding evolved to more sophisticated schemes through building optimal network subgraphs. All through, the main algorithmic challenge has been to reduce the number of transmitted messages needed to reach all the network nodes.

Among the major shortcomings of pure flooding are the resulting large transmission complexity and the notorious broadcast storm [1]. The *Scalable Broadcast Algorithm* [2] alleviates somewhat this problem utilizing 1-Hop neighbor information.

A different approach is taken by probabilistic broadcast protocols that associate some (re)transmission probability to each node receiving the broadcast message. Schemes exploring such mechanisms were suggested in [3, 4, 5, 6, 7, 8, 9].

The interest in probabilistic broadcasting schemes is due to their inherent low transmission overhead, low processing complexity, and high tolerance of frequent and rapid topological changes. Balancing these benefits, though, is the disadvantage of inability to guarantee full network coverage.

In contrast, deterministic broadcast algorithms innately guarantee full network coverage (assuming ideal MAC layer). In the deterministic scheme of Multipoint Relaying proposed in [10], the set of retransmitting neighbor nodes is reduced from the set of all neighbors to the minimum subset of neighbors that cover the same area as that covered by the original set. This approach is an example of the "minimum forward-node set" strategy, and works such as [11, 12, 13, 14] provide approximate solutions to this NP-hard problem. To avoid the transmission of the list of forwarding nodes along with the broadcast message, a technique of self-pruning [15, 16] has been proposed.

The forward-node set and, consequently, the self-pruning problems can essentially be viewed as the task of solving the NP-hard "Minimum Connected Dominating Set" (MCDS) problem [13]. Several studies [18, 19, 20, 21, 22] have attempted to tackle the problem by constructing a communication backbone prior to the broadcast initiation. These schemes can sometimes dramatically reduce the number of transmissions. Nevertheless, as shown in [23], they do not tolerate well frequent network topological changes. For volatile communication environments, an approach to dynamically construct CDS is a better alternative. Works such as [24] and [25] offer potential solutions. However, further research is required to study the scalability of these approaches.

A number of studies have attended to the theoretical bounds of broadcast and related information dissemination mechanisms. For instance, [37, 38] provide bounds on gossiping, broadcast, and rumor spreading in general networks. Influential works by Peleg et al. in the spirit of [39] demonstrate important lower bounds on broadcast in radio networks.

8. ACKNOWLEDGMENTS

This work was supported by the NSF grant numbers ANI-0329905 and CNS-0626751 and by the AFOSR contract number FA9550-09-1-0121/Z806001. We would also like to thank the anonymous reviewers for their constructive critique and suggestions regarding this manuscript.

9. REFERENCES

[1] S.-Y. Ni, Y.-C. Tseng, Yuh-Shyan Chen, and J.-P. Sheu, "The Broadcast Storm Problem in a Mobile Ad Hoc Networks", In Proc. ACM/IEEE MobiCom, 1999, pp. 151 - 162

[2] W. Peng, and X.C. Lu, "On the reduction of broadcast redundancy in mobile ad hoc networks," In Proc. ACM/IEEE MobiHoc, 2000, pp. 129–130

[3] Z. J. Haas, J. Y. Halpern, and L. Li, "Gossip-Based Ad Hoc Routing", In Proc. IEEE INFOCOM, 2002, pp. 1707-1716

[4] Y. Sasson, D. Cavin, and A. Schiper, "Probabilistic broadcast for flooding in wireless mobile ad hoc networks," in Proc. IEEE WCNC, 2003, pp. 1124-1130
[5] D. Scott and A. Yasinsac, "Dynamic probabilistic retransmission in ad hoc networks," ICWN, 2004, pp. 158--164, CSREA Press
[6] J. Cartigny, D. Simplot, and J. Carle, "Stochastic flooding broadcast protocols in mobile wireless networks," Tech. Report LIFL, Univ. Lille1, 2002-03, 2002
[7] J. Kim, D. J. Scott, and A. Yasinsac. "Probabilistic broadcasting based on coverage area and neighbor confirmation in mobile ad hoc networks," In Proc. IEEE Globecom, 2004
[8] A. Keshavarz-Haddad, V. Ribeiro, and R. Riedi, "Color-based broadcasting for ad hoc networks," WiOpt, 2006
[9] S. Pleisch, M. Balakrishnan, K. Birman, and R. van Renesse, "MISTRAL: Efficient Flooding in Mobile Adhoc Networks," ACM MobiHoc 2006.
[10] A. Qayyum, L. Viennot, and A.Laouiti, "Multipoint relaying for flooding broadcast messages in mobile wireless networks," HICSS, 2002, Hawaii.
[11] H. Liu, P. Wan, X. Jia, X. Liu, and F. Yao. "Efficient Flooding Scheme Based on 1-hop Information in Mobile Ad Hoc Networks," IEEE INFOCOM, 2006
[12] W. Lou and J. Wu. "Double-covered broadcast (DCB): A simple reliable broadcast algorithm in MANETs," IEEE INFOCOMM, 2004
[13] Y. Cai, K. Hua, and A. Phillips, "Leveraging 1-Hop Neighborhood Knowledge for Efficient Flooding in Wireless Ad Hoc Networks," IPCCC, 2005, pp. 347-354
[14] G. Călinescu , I. I. Mandoiu , P. Wan , A. Z. Zelikovsky, "Selecting forwarding neighbors in wireless ad hoc networks," MONET, 2004, v.9 n.2, p.101-111
[15] M. Sun, W. Feng, and T. Lai, "Broadcasting in Ad Hoc Networks Based on Self-Pruning," IEEE GLOBECOM, 2001, pp. 2842-2846,
[16] Jie Wu , Fei Dai, "A Generic Broadcast Protocol in Ad Hoc Networks Based on Self-Pruning," IPDPS, 2003, p.291.
[17] J. Wu and H. Li, "On Calculating Connected Dominating Set for Efficient Routing in Ad Hoc Wireless Networks," Third Int'l Workshop DIAL-M, 1999, pp. 7-14,
[18] A. Keshavarz-Haddad, V. Ribeiro, and R. Riedi, "DRB and DCCB: Efficient and Robust Dynamic Broadcast for Ad Hoc and Sensor Networks," In Proc. IEEE SECON, 2007
[19] R. Misra, C. Mandal, "Minimum Connected Dominating Set Using a Collaborative Cover Heuristic for Ad Hoc Sensor Networks," IEEE Trans. Parallel and Distributed Sys, vol.21, no.3, pp.292-302, Mar. 2010
[20] P.-J. Wan, K. M. Alzoubi, O. Frieder, "Distributed Construction of Connected Dominating Set in Wireless Ad Hoc Networks", IEEE INFOCOM 2002, pp 1597-1604.
[21] S. Funke, A. Kesselman, U. Meyer, and M. Segal, "A Simple Improved Distributed Algorithm for Minimum CDS in Unit Disk Graphs", ACM Trans. Sensor Networks, vol. 2, no. 3, pp. 444-453, 2006
[22] B. Gao, Y. Yang, H. Ma, "An Effective Distributed Approximation Algorithm for Constructing Minimum Connected Dominating Set in Wireless Ad Hoc Networks," ICCIT, 2004, p.658-663
[23] N. Meghanathan, A. Farago, "On the stability of paths, Steiner trees and connected dominating sets in mobile ad hoc networks, Ad Hoc Networks," vol. 6, no. 5, pp. 744-769, Jul 2008
[24] I. Stoimenovic, "Comments and Corrections to 'Dominating Sets and Neighbor Elimination-Based Broadcast Algorithms in Wireless Networks", IEEE Trans. Parallel and Distributed Systems, vol. 15, no. 11 Nov. 2004
[25] Khabbazian, M., Bhargava, V. "Efficient Broadcasting in Mobile Ad Hoc Networks," IEEE Trans. on Mobile Comp., Vol.8, No.2, Feb. 2009
[26] Z. J. Haas and R. Barr, "Density-Independent, Scalable Search in Ad hoc Networks," IEEE PIMRC, 2005
[27] X. Wang, R. Dokania, A. Apsel, "PCO Based Synchronization for Ad-Hoc Duty-Cycled Impulse Radio Sensor Networks", Special issue on cognitive sensor networks of IEEE Sensors Journal, 2010.
[28] R. Solis, V. Borkar, and P. R. Kumar, "A new distributed time synchronization protocol for multihop wireless networks," In Proc. IEEE CDC, 2006, pp. 2734-2739.
[29] X. Cheng, X. Huang, D. Li, W. Wu, and D.-Z. Du, "A polynomial-time approximation scheme for the minimum-connected dominating set in ad hoc wireless networks," Networks, 42, 2003, 202--208.
[30] H. Hunt, III , M. Marathe , V. Radhakrishnan , S. Ravi , D. Rosenkrantz , R. Stearns, "NC-approximation schemes for NP- and PSPACE-hard problems for geometric graphs," Journal of Algorithms, v.26 n.2, p.238-274, Feb. 1998
[31] S. Guha and S. Khuller, "Approximation Algorithms for Connected Dominating Sets," Algorithmica, vol. 20, pp. 347-387, 1998.
[32] B.N. Clark, C.J. Colbourn, and D.S. Johnson, "Unit Disk Graphs, " Discrete Mathematics, Vol. 86, 1990, pp. 165-177.
[33] B. Liang and Z. Haas, "Predictive Distance-Based Mobility Management for PCS Networks," In Proc. IEEE INFOCOM, 1999.
[34] T. Camp, J. Boleng and V. Davies, "A survey of mobility models for ad hoc network research." Wireless Communications & Mobile Computing 2 5 (2002), pp. 483–502
[35] Hong X, Gerla M, Pei G, Chiang C. "A group mobility model for ad hoc wireless networks." In Proc. ACM MSWiM, 1999.
[36] F. Bai, N. Sadagopan, and A. Helmy, "IMPORTANT: A framework to systematically analyze the impact of mobility on performance of routing protocols for adhoc networks," In Proc. IEEE INFOCOM, 2003, pp. 825-835
[37] D. Liu , M. Prabhakaran," On Randomized Broadcasting and Gossiping in Radio Networks," In Proc.COCOON, 2002, pp.340-349
[38] D. Alistarh , S. Gilbert , R. Guerraoui , M. Zadimoghaddam, "How efficient can gossip be? (on the cost of resilient information exchange)," In Proc. ICALP, 2010
[39] N. Alon , A. Bar-Noy , N. Linial , D. Peleg, "A lower bound for radio broadcast," Journal of Computer and System Sciences, v.43 n.2, p.290-298, Oct. 1991

A Greedy Reclaiming Scheduler for IEEE 802.11e HCCA Real-Time Networks

Anna Lina Ruscelli
a.ruscelli@sssup.it

Gabriele Cecchetti
g.cecchetti@sssup.it

Giuseppe Lipari
g.lipari@sssup.it

Antonia Mastropaolo
antonella.mastropaolo@gmail.com

ReTiS Lab. - TeCIP
Scuola Superiore S. Anna
Pisa, Italy

ABSTRACT

The IEEE 802.11e standard introduces Quality of Service (QoS) support for wireless local area networks and suggests how to design a tailored HCF Controlled Channel Access (HCCA) scheduler. However the reference scheduling algorithm is suitable to assure service guarantees only for Constant Bit Rate traffic streams, whereas shows its limits for Variable Bit Rate traffic. Despite the numerous alternative schedulers proposed to improve the QoS support for multimedia applications, in the case of VBR traffic satisfactory real-time performance has not been yet achieved.

This paper presents a new scheduling algorithm, *Unused Time Shifting Scheduler* (UTSS). It integrates a mechanism for bandwidth reclaiming into a HCCA real-time scheduler. UTSS assigns the unused portion of each transmission opportunity to the next scheduled traffic stream. Thanks to such feature, traffic variability is absorbed, reducing the waste of resources. The analytical evaluation, corroborated by the simulation results, shows that UTSS is suitable to reduce the delay experienced by VBR traffic streams and to increase the maximum burstiness sustainable by the network.

Categories and Subject Descriptors

C.2.1 [**Computer-Communication Networks**]: Network Architecture and Design—*Wireless Communication.*

General Terms

Algorithms, Performance.

Keywords

Quality of Service, real-time scheduling algorithms, wireless LAN.

MSWiM'11, October 31–November 4, 2011, Miami, Florida, USA.

1. INTRODUCTION

The support for multimedia applications (like VoIP, multimedia streaming, online gaming, High Definition TV etc.) is an important challenge in the context of *Wireless Local Area Networks* (WLAN). The reference IEEE 802.11 standard [17] lacks the support of the *Quality of Service* (QoS) as requested by the multimedia applications, for instance, in terms of guaranteed bandwidth and bounded delay, jitter and packet loss. Indeed the earlier release IEEE 802.11b [18] standard was designed for best effort data transmissions. Hence the IEEE 802.11e amendment [19], has been produced to offer QoS support. In particular it introduces service differentiation, as required by multimedia applications. Service differentiation is possible thanks to two new *Medium Access Control* (MAC) functions that improve the functions existing in IEEE 802.11b. *Hybrid coordination function Controlled Channel Access* (HCCA) adds a parameterized QoS at the polling mechanism, whereas *Enhanced Distributed Channel Access* (EDCA) supplies prioritized QoS at the contention-based medium access. However numerous theoretical and simulative studies [1,4,5,8,14,35] highlighted that the reference scheduler has good performance only with *Constant Bit Rate* (CBR) traffic whereas it is not suitable to provide temporal guarantees to *Variable Bit Rate* (VBR) traffic. Thus numerous scheduling algorithms have been proposed as alternative to the reference one in order to overcome its poor performance due to the fixed values the HCCA function assigns to the transmission parameters.

In this paper we present a novel scheduler named *Unused Time Shifting Scheduler* (UTSS) which integrates a mechanism for bandwidth reclaiming into a HCCA real-time scheduler. The mechanism assigns the portion of *Transmission Opportunity* (TXOP) unused by polled stations to the next scheduled traffic stream. Therefore a dynamic computation of the current transmission time is introduced. Such dynamic computation lets the scheduler to reassign the unused bandwidth, for dealing with rate variations of VBR traffic. We will show through analytical study and simulation that UTSS improves the provided QoS, being able to assure lower access delay and to absorb the variability of VBR traffic streams, without wasting of resources.

The rest of the paper is organized as follows: in Section 2 we summarize some relevant works that improve the HCCA reference scheduler. In Section 3, the proposed scheduler is described. In Section 4 some properties of the scheduler are analytically evaluated, whereas Section 5 show its performance through simulations. Finally, in Section 6 we draw our conclusions.

2. HCCA SCHEDULING ALGORITHMS

The IEEE 802.11e standard defines a non-mandatory *reference* HCCA scheduler and the guidelines for the computation of protocol parameters by taking into account *QoS Station* (QSTAs) requirements. In particular the reference scheduler computes *Service Interval* (SI), i.e. the polling period, and *Transmission Opportunity* (TXOP), i.e. the transmission duration, as fixed values. *SI* is computed as a unique value for all admitted QSTAs with the aim to globally meet their temporal service expectations: its value has to be less than the beacon interval, thus each QSTA is polled at least once during the beacon interval, and less than the minimum *Maximum Service Interval* (MSI), thus polling period constraints of all traffic streams are respected. $TXOP$ is computed as the maximum time to transmit at the minimum physical rate Γ_i the total amount of bits that can arrive during *SI*:

$$N_i = \left\lceil \frac{SI \cdot R_i}{L_i} \right\rceil \qquad TXOP_i = max\left(\frac{N_i \cdot L_i}{\Gamma_i}, \frac{M_i}{\Gamma_i}\right) + O,$$

where M_i is the maximum *MAC Service Data Unit* (MSDU) size, i.e. 2304 bytes, L_i is the nominal MSDU size, R_i is the mean data rate and O is the transmission overhead due to interframe spaces, `ACK`, `CF-Poll`.

The admission control test for deciding whether to admit a new stream in the HCCA reference scheduler is:

$$\frac{TXOP_{k+1}}{SI} + \sum_{i=0}^{k} \frac{TXOP_i}{SI} \leq \frac{T - T_{CP}}{T} \leq 1.$$

where k is the number of admitted streams, $k+1$ indexes the newly admitted stream, T is the beacon interval and T_{CP} is the EDCA duration.

SI and $TXOP$ are recomputed only if a new *Traffic Stream* (TS) arrives. Their values are based on *worst case* conditions; this produces a too stringent admission control test and a not-optimal resource management. Moreover, since *SI* is the same of all admitted QSTASs and $TXOP$ is globally assigned to a QSTA, all different TSs are polled with the same period and are served with the same computation time. This makes the reference scheduler suitable to serve CBR traffic but unable to efficiently adapt the resource management to VBR TSs, as highlighted by numerous studies and evaluations about the reference scheduler [8, 14, 29, 34].

Many scheduling algorithms alternative to the reference one have been proposed to improve the HCCA QoS provisioning [10, 12, 28, 34] and few of them are focused on real-time support [5], i.e. on temporal guarantees. In the following some real-time schedulers that, at the best of our knowledge, have introduced significant contributions are summarized. They range over different approaches including queue length models, feedback-based schemes, bandwidth reclaiming methods, exploitation of EDCA resources using the IEEE 802.11e *HCCA-EDCA Mixed Mode* (HEMM) mode that permits a QSTA to jointly use both the HCCA and the EDCA MAC mechanisms. In particular we will deeply describe some algorithms based on the concept of deadlines that is used by the proposed UTSS scheduler.

Fair HCF (FHCF) [1] aims to improves the fairness of both CBR and VBR traffic and the delay performances assigning variable $TXOPs$ by means of a mathematical model of the uplink TSs queues length, used to estimate the global packet delay. It distinguishes between the packet queuing delay, influenced by the variations in packet size and data rate, and the waiting time delay, defined as the interval between the packet arrival time and the QSTA polling time.

The *Feedback Based Dynamic Scheduler* (FBDS) [3] deals on a closed loop feedback control to restore the balance about the right packets queue delivery by bandwidth recovering, limiting the maximum delay. $TXOPs$ are dynamically assigned according to queue length estimation at the beginning of the *Controlled Access Phase* (CAP) through a discrete time model, corrected by the use of the actual queue length information sent by each QSTA; *SI* remains fixed.

The model of the channel when both HCCA and EDCA modes are used, presented in [22], shows that incrementing the portion of HCCA increases the medium utilization of large WLAN in saturation conditions and the determinism in the channel control. Instead large EDCA networks are affected by growing collisions that degrade their performance.

In [32] the economic model used to efficiently manage elastic traffic over EDCA and HCCA functions shows how the CW_{min} parameter, the CSMA/CA scheme, and the RTS/CTS procedure affect channel congestion and throughput. Moreover it finds the optimal value of the HCCA-EDCA ratio by means of optimization techniques.

The *Adaptively Tuned HCF* (AT-HCF) algorithm [23] dynamically tunes the HCCA and the EDCA durations to the different type of traffic, until they converge to the optimal values to improve the throughput of the overall system.

The *Overboost* local node scheduler [31] exploits EDCA bandwidth to integrate the HCCA activity to limit the delay experienced by the traffic streams waiting for the next HCCA polling time and improves the network performance. Before the contention period begins, Overboost moves the TSs traffic exceeding the assigned HCCA $TXOP$ transmission time from the HCCA queue to the higher priority Access Category EDCA queue. This scheduler can collaborate with any type of centralized HCCA scheduling algorithm that continues to manage admission control, scheduling parameters, and polling list.

The use of deadlines introduces timing constraints and their management is tailored to meet temporal requirements. The *Scheduling Estimated Transmission Time - Earliest Due Date* (SETT-EDD) [13] algorithm uses a token bucket scheme of time units or $TXOP$ timer to vary their $TXOP$ over time according to the node requirements. *SI* of each node is computed taking into account its traffic profile, while *Earliest Deadline First* (EDF) [26] determines the polling order. Variable $TXOP$ and *SI* enhance the scheduler flexibility and lead to significant reduction in average transmission delay and packet loss ratio.

In [9] a timer-based scheduler computes the transmissions deadlines as the smallest between the downlink and uplink ones, and then schedules traffic streams according to EDF algorithm.

Real-Time HCCA (RTH) scheduler [7] manages $TXOP$ as critical section, which cannot be interrupted by higher priority flows, by using *Stack Resource Policy* (SRP) algorithm [2]. Then an EDF-based algorithm schedules TS transmissions. It is composed by two activities: the offline one executes admission control, computes transmission parameters and scheduling timetable, and the online one schedules traffic streams transmissions.

Adaptive Resource Reservation Over WLANs (ARROW) [33] dynamically computes each $TXOP$ by taking into account the actual different buffered TSs data at the beginning of the polling. Moreover, MSI is bounded in order to ensure no deadline miss and delay requirements. Finally *Earliest Due Date* (EDD) [21] deadlines scheduling manages the QSTAs polling list.

The *Application-Aware Adaptive HCCA Scheduler* [20], derived from ARROW, distinguishes uplink and downlink schedulers, while the *Earliest Deadline First* (EDF) algorithm [26] defines the QSTAs polling order taking into account the stations deadlines. The uplink scheduler assigns each QSTA a minimum and a maximum

SI, adapted to the application and network conditions and to the buffered traffic.

Wireless Capacity Based Scheduler (WCBS) [6] dynamically updates QSTA polling list using EDF algorithm and uses static and dynamic parameters to adapt the transmissions scheduling to the TSs characteristics. During the admission control a pair of static parameters, that do not change during normal conditions, are assigned to each TS_i taking into account its *Traffic Specification* ($TSPEC_i$): the budget Q_i, i.e. the maximum transmission time during a period ($TXOP_i$), and P_i, the service interval (SI_i). The ratio $U_i = Q_i/P_i$ is the utilization factor of the stream, i.e. its bandwidth. Therefore the admission control test is defined as:

$$\frac{TXOP_{k+1}}{SI_{k+1}} + \sum_{i=0}^{k} \frac{TXOP_i}{SI_i} \leq \frac{T - T_{CP}}{T} \leq 1. \quad (1)$$

Instead, the dynamic parameters characterize each TS_i during the scheduling phase: the remaining time c_i assigned to TS_i during the next transmission, the absolute deadline d_i before the budget Q_i has to be exhausted, the next time p_i when TS_i will be polled if it has no more data to transfer or it has exhausted its $TXOP$, and the stream *state* (*transmitting*, *active*, *polling*, *idle*).

To the best of our knowledge the first mechanism using bandwidth reclaiming in IEEE 802.11e is [24] where open-loop strategies, based on the use of TSPEC, and closed-loop strategies, that use the information about the queue sent by stations to the QAP, are compared. In particular, two different *max-min fairness* algorithms are proposed with the aim reduce the delay produced by HCCA in the case of VBR traffic by recovering spare resources through a proportional controller. A weighted proportional function assigns an addition amount of resources proportionally to the traffic class and to the buffer length. *MAXMin Fair-Adaptive* (MMF-A) assumes fixed SI, whereas *MAXMin Fair-Adaptive with Rescheduling* (MMF-AR) considers dynamic SI values. The non linearities of the system that can affect the mentioned works are overcame by [30] with an optimal controller, based on Model Predictive Control [11], that dynamically assigns resources in order to empty buffers of the stations and reduce packet loss. Moreover in [27] both cited proportional controller as optimal controller algorithms are applied to multi-class traffic with different priorities and to heterogenous traffic in order to find the resource assignment tailored to maximize the throughput, reducing the packet loss experienced by the different types of traffic.

In [25] the authors propose a bandwidth reclaiming scheme for the IEEE 802.11 PCF function with *Weighted Round Robin* (WRR) scheduling policy. It determines when the unused transmission time can be either used to advance the next polling opportunity or assigned to the Contention Period. Moreover the proposed algorithm modifies the WRR polling list in order to put the stations with higher probability of generating unused time at the end of the list. This rearrangement aims to reduce the number of reclaimed stations but it makes the solution not extensible to HCCA function, where the polling order is strictly related to real-time guarantees.

3. UNUSED TIME SHIFTING SCHEDULER

Despite of many QoS scheduling algorithms enhancing the reference one, satisfactory performance for VBR traffic with real-time guarantees has not yet been achieved. For instance, EDF-based algorithms are well performing in the case of CBR traffic streams, whereas they are not able to follow the variations of VBR traffic. Due to the traffic variability when the instant data rate drops down, a QSTA transmits data for an amount of time shorter than the assigned $TXOP$. In this case the *QoS Access Point* (QAP), listening the idle channel for a time longer than a *Short Interframe Space* (*SIFS*), assumes the control of the medium and polls the next station. Therefore the unused time is lost.

Unused Time Shifting Scheduler (UTSS) aims to face off this problem by reclaiming the unused transmission time and provides a shortcut to have instantaneous dynamic $TXOP$, without modify the admission control (see Section 4).

UTSS integrates the scheduling rules of WCBS scheduler illustrated in Section 2 by keeping the same admission control and scheduling algorithm of WCBS. In Fig 1, the Admission Control of the WCBS scheduler tests the feasibility condition, then it computes both the static and dynamic parameters and finally the newly admitted TS is enqueued in the EDF-ordered list.

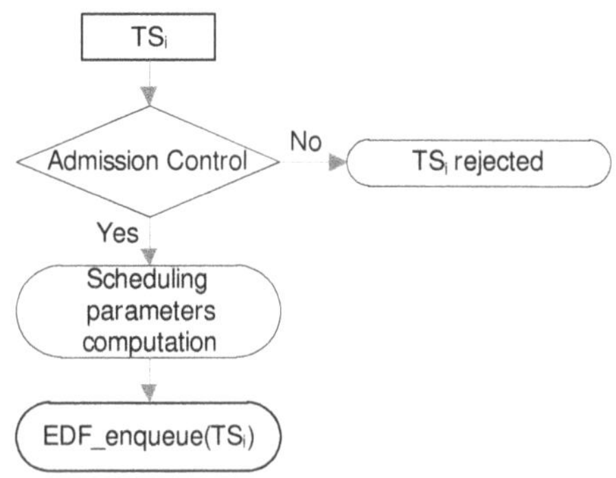

Figure 1: Admission control test.

UTSS adds a further scheduling rule during the computation of the dynamic parameters by modifying the $TXOP$ if some unused time is available. Fig 2 describes the relationship between WCBS and UTSS. The scheduler, after the extraction of the next

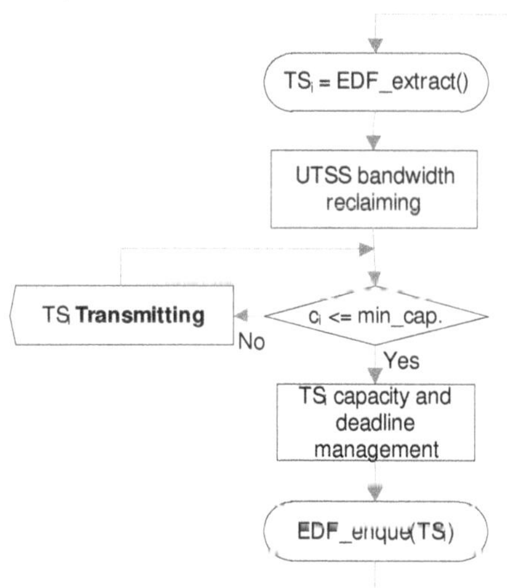

Figure 2: EDF + UTSS scheduling cycle.

TS_i from the EDT queue, performs the UTSS bandwidth reclaiming mechanism that affects the assigned $TXOP_i$. Then, if there is enough capacity to transmit, TS_i is scheduled for transmission. Otherwise WCBS operates replenishing the capacity and, if necessary, postponing the deadline. Therefore, while WCBS uses a constant $TXOP$ assigned during the admission control phase, UTSS can make this parameter variable each time TS is scheduled for transmission. In particular, each time a QSTA does not use the full allocated $TXOP$, UTSS assigns this unused time to the next scheduled TS extracted from the EDF queue during the current CAP. We adopt the following notation:

$t_{end} = t_p + TXOP$: ending time of $TXOP$, when it is completely exhausted (t_p is the polling time);

T_{spare} : spare time of $TXOP$, computed as difference between t_{end} and the time when QSTA has actually finished its transmission.

T_{spare} is computed every time QAP polls a QSTA and $T_{spare} \neq 0$ when a station ends its transmission before t_{end}. Current T_{spare} is then added to the assigned $TXOP_i$ of the next polled station QSTA$_i$ that will receive a new $TXOP_i'$ computed as follow:

$$TXOP_i' = \begin{cases} TXOP_i & \text{if } T_{spare} = 0 \\ TXOP_i + T_{spare} & \text{if } T_{spare} > 0 \end{cases}$$

Fig. 3 better details UTSS. After the extraction from the EDF

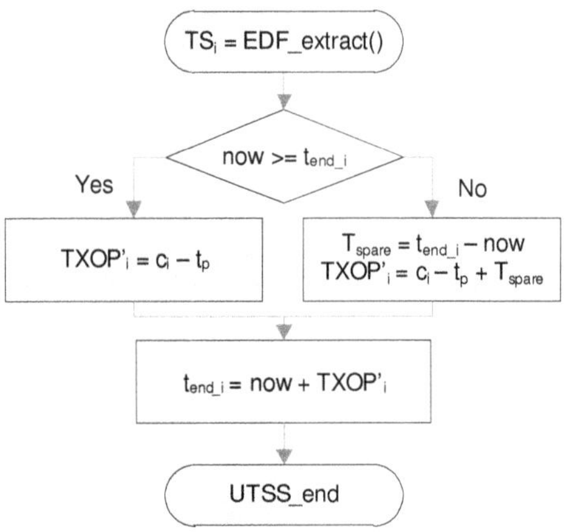

Figure 3: UTSS mechanism.

queue of the next TS to transmit, UTSS verifies if some spare time from previous transmissions is available. In such case, it adds T_{spare} to the TXOP computed by WCBS ($TXOP_i = c_i - t_p$). Then the scheduler updates the estimated transmission ending time.

Hence next scheduled traffic streams can transmit their data not only exploiting the correspondent $TXOP$ (assigned during the admission control phase) or the remaining capacity, if they have been already served, but also by using the surplus T_{spare}. Fig. 4 shows a simplified scheduling example where WCBS and UTSS behaviors are compared. In WCBS the unused bandwidth of TS$_2$ is wasted, whereas with the help of UTSS TS$_3$ transmits its data during also the unused time of previously polled TSs.

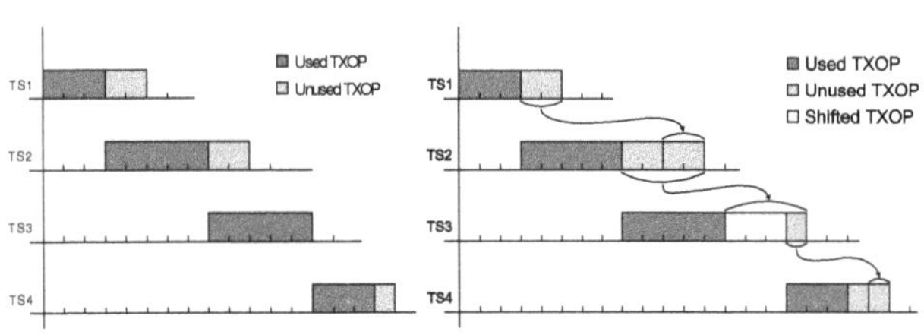

(a) WCBS (b) UTSS

Figure 4: Scheduling example.

In this way it is possible to reclaim all the unused $TXOP$ portions. The total amount of T_{spare} can satisfy the temporal requirements of a highly variable traffic with a temporary load greater than the mean value used to calculate $TXOP$. Hence T_{spare} could be useful to absorb traffic peaks which characterize highly variable bit rate applications.

4. SCHEDULING ANALYSIS

This section analyzes the UTSS algorithm in order to highlight its effects on the centralized HCCA scheduler. The objective is to assess whether UTSS changes the admission control feasibility condition. We investigate from the analytical point of view how the added UTSS mechanism impacts on the admission control performed by the overall scheduler. Moreover, we check if the accumulation of this unused time can rise or not deadline miss. Then an upper bound of the maximum acceptable T_spare is computed. Finally to study the behavior of the overall scheduler in presence of bursts of traffic, an expression of the maximum tolerable burstiness is provided.

THEOREM 1. *The UTSS mechanism does not not affect the validity of the admission control feasibility test.*

PROOF. We write Eq. 1 as follow:

$$T_{CAP} + T_{CP} = \sum_{i=0}^{k} T_i + T_{CP} \leq H$$

where T_{CAP} is the time assigned during the hyperperiod to HCCA, T_{CP}, is the time assigned to EDCA, and T_i is the $TXOP$ of each QSTA$_i$.

Without loss of generality we can assume that, $QSTA_1$ does not use its whole $TXOP_1$, thus its unused time $T_{spare_1} \neq 0$. We can highlight that, when UTSS reclaims unused time and assigns T_{spare}, the following relationship holds:

$$\sum_{i=0}^{k} T_i' = T_{eff_1} + T_2 + T_{spare_1} + \ldots + T_k$$
$$= T_{eff_1} + T_2 + T_1 - T_{eff_1} + .. + T_k$$
$$= \sum_{i=0}^{k} T_i \leq H - T_{CP}.$$

where T_i' is the new T_i computed taking into account T_{spare}, and T_{eff_1} is the used portion of $TXOP_1$. □

THEOREM 2. *The UTSS mechanism of T_{spare} assignment does not raise deadline miss.*

PROOF. We distinguish two cases: 1) assignment of T_{spare} during a CAP, and 2) assignment of T_{spare} derived from the last polled QSTA of a CAP.

Case 1 During a CAP, the assignment of T_{spare} does not miss the deadline of the polled QSTAs. In fact the MAC scheduling algorithms manage the temporal sequence of QSTAs access to the medium by the use of fixed interframe space intervals (SIFS and *PCF Interframe Space* PIFS), that set the waiting time between consecutive polling of QSTAs listed in the polling queue. These strict rules avoid the presence of idle blocking time between polled stations transmissions, as in general it could happen in real-time systems, where the addition of a further slot of transmission time could jeopardize the real-time behavior of the next process to execute, raising a deadline miss. Instead, due to the IEEE 802.11e MAC scheduling rules, the use of T_{spare} only anticipates the transmission of a polled QSTA, without impacting on the next polled QSTA behavior, as shown in Fig. 4.

Case 2 When T_{spare} is derived from the last polled QSTA during a CAP, see Fig. 5a, two additional cases exist. If a T_{spare} recovered from the transmission of the last QSTA in the current CAP is shifted to the subsequent *Contention Period* (CP), the beginning of the following CAP is not affected since in this case there is only a redistribution of time portions between *Contention Free Period* (CFP) and CP of the same hyperperiod, see Fig. 5b.

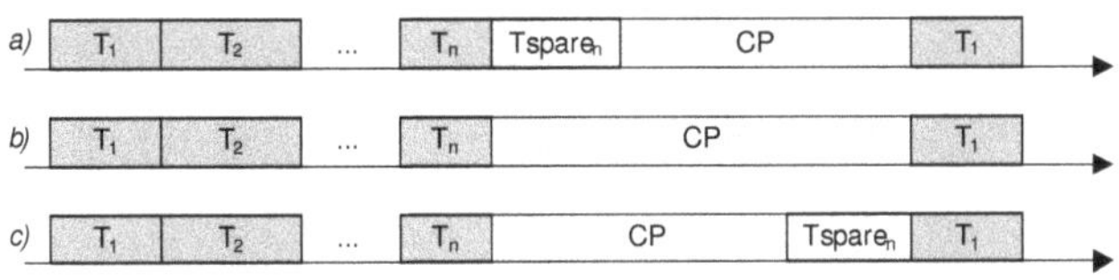

Figure 5: Example of T_{spare} propagation.

Instead, if we choose to propagate the T_{spare} allocation to the next CAP, by assigning the transmission time of the last polled QSTA to the first QSTA at the beginning of the next CAP, there is only an early polling of this QSTA of the same time interval, see Fig. 5c, whereas all other QSTAs are polled by respecting their Delay Bounds, thus without deadline miss. Moreover this choice increases the algorithm fairness by handling all QSTAs, included the first one of the new CAP, in the same way. □

In order to avoid unpredictable growing of T_{spare}, especially due to accumulation of a large number of unused portion of $TXOP$s when the T_{spare} propagation across consecutive CAP is enabled, we can try to set un upper bound Θ suitable to meet the deadlines.

PROPOSITION 1. *The upper bound Θ of T_{spare} suitable to avoid deadline miss is equal to $\Theta = d_i - t_{end_i} + \delta$.*

PROOF. In the case of one-hope propagation, i.e. in the case of propagation of T_{spare} to the consecutive QSTA:

$$T_i \le T_i' \le T_i + T_{i-1}$$

where

$$T_i' = T_i + T_{spare_i},$$

whereas if all previous QSTAs do not use their $TXOP$,

$$0 < T_{spare_i} < \sum_{j=1}^{i-1} T_j$$

In particular T_{spare} is made available after that QSTA has informed QAP that has no data to send by responding to `CF-Poll` frame with a `CF-Null` frame. This handshake requires a time interval equal to $\tau = t_{SIFS} + t_{NULL} + t_{SIFS}$, thus the general expression is:

$$\begin{aligned} 0 \le & T_{spare_i} \\ \le & \sum_{j=1}^{i-1} (T_j - \tau) \\ \le & \sum_{j=1}^{i-1} (T_j) - (i-1)\tau. \end{aligned}$$

Finally, in order to avoid deadline miss, we can assume to accept t_{spare_i} if and only if it respects its upper bound Θ:

$$0 \le T_{spare_i} \le d_i - t_{end_i} + \delta = \Theta$$

where d_i is the absolute deadline of the polled QSTA$_i$ and δ is an offset. □

Since the proposed mechanism aims to introduce flexibility in the $TXOP$ assignation at each polling by reclaiming the unused allocated resources, it could positively impact on the service provided to QSTA with VBR traffic. In fact VBR traffic makes $TXOP$ not suitable since it is computed during the admission control considering mean value parameters. This leads to poor network performance. A metric useful to characterize the *burstiness* traffic with high rate variability is the *burstiness factor* (B), defined as *the ratio between mean data rate r during a long time interval and peak data rate r' during the activity interval.* Its value ranges from 1, in the case of CBR traffic, to 0, as the peak rate increases.

THEOREM 3. *The maximum traffic burstiness of a QSTA that receives T_{spare} and tolerable by the system is increased and the new burstiness factor B_{max} is equal to:*

$$B_{max} = \frac{t_{end_i} - t_{p_i}}{T_{spare} + t_{end_i} - t_{p_i}} \le B.$$

where B is the burstiness factor without using UTSS.

PROOF. The use of $T_{spare} \neq 0$ by the next polled QSTA allows more enqueued traffic to be dispatched and the capacity to be increased and absorb peaks in the TS. Thus, with regard to the assigned $TXOP$, using T_{spare} has the same effect of instantaneously increasing the delivery rate. The portion of traffic dispatched using T_{spare} is $\varphi = r \cdot T_{spare}$. Considering that, in general, $QSTA_i$ cannot use the whole T_i, then

$$\tau \le t_{end_i} - t_{p_i} = T_{eff_i} \le T_i$$

and r' is equal to

$$r_i' = \frac{\varphi_i + r(t_{end_i} - t_{p_i})}{t_{end_i} - t_{p_i}} = \frac{r(t_{spare} + t_{end_i} - t_{p_i})}{t_{end_i} - t_{p_i}}.$$

Finally the maximum burstiness B_{max} tolerable by the system, taking advantage by the use of the UTSS mechanism, is:

$$B_{max} = \frac{t_{end_i} - t_{p_i}}{t_{spare} + t_{end_i} - t_{p_i}} \le B. \quad (2)$$

□

This result is confirmed by the performance evaluation about end-to-end delay, throughput and discarded packets, see Section 5, where UTSS is evaluated in presence of VBR traffic.

5. PERFORMANCE ANALYSIS

This section evaluates the performance of the proposed UTSS scheduling algorithm versus WCBS and reference schedulers in order to analyze the effect produced by the introduction of the new mechanism. In particular the analysis takes into account the utilization efficiency of the network, the mean access delay, and the discarding rate of queued packets with expired delay bound. The result is preceded by a description of the simulation tools, their settings, the traffic models and the considered scenario.

5.1 Simulation settings and traffic model

Performance is evaluated through simulation using *ns-2* network simulator [16]. It is assumed that QSTAs communicate directly without hidden node problem. Thus RTS/CTS mechanism, MAC level fragmentation and multirate support are disabled. The Physical layer is specified in the IEEE 802.11g standard which uses OFDM (*Orthogonal Frequency-Division Multiplexing*) as mandatory modulation scheme; its parameters are listed in Table 1.

Parameters	*Value*	*Parameters*	*Value*
SIFS (μs)	10	PLCP header (b)	24
DIFS (μs)	28	Preamble (b)	72
PIFS (μs)	19	Data Rate ($Mbit/s$)	54
Slot Time (μs)	9	Basic Rate ($Mbit/s$)	1

Table 1: MAC/PHY simulation parameters.

The presented results have been obtained simulating independent replication of 700 s with a warm-up time of 100 s until the 95% confidence interval is reached for each measure.

The network scenario chosen for simulations is composed by seven QSTA and one QAP. Each QSTA transmits *one* uplink TS which is received by the QAP. Each TS has a *Traffic Specification* (TSPEC) different from the others TS. In particular one station sends VoIP traffic encoded with G.729A codec, one station transmits a video conference and five stations transmit video streaming applications. One station send data traffic with SDU of 1500 bytes through legacy *Distributed Coordination Function* (DCF). The data station operates in asymptotic condition, i.e. it is always backlogged in order to saturate the channel. The parameters of the VoIP traffic are shown in Table 2.

Parameter	*Value*	*Parameter*	*Value*
Frame size *(B)*	10	Payload size *(B)*	20
Frame per packet	2	IP/UDP/RTP	
Period *(s)*	0.02	Header size *(B)*	40
Data rate *(kb/s)*	24	SDU size *(B)*	60

Table 2: G.729A VoIP traffic stream parameters.

Video streaming traffic has been generated using pre-encoded high quality MPEG4 trace files of 60 minutes each from the Internet archive of traces [15]. Such traces are: *Jurassic Park* (VS1), *Silence of the lambs* (VS2), *Mr. Bean* (VS3), *Die hard III* (VS4), *Robin Hood* (VS5). The video conference (VC) session has been represented by the pre-encoded *LectureHQ-Reisslein* trace file.

Parameter	VC	VS1	VS2
Mean frame size (B)	3800	3800	2900
Maximum frame size (B)	11386	11386	22239
Period (s)	0.040	0.040	0.040
Mean data rate (kb/s)	770	770	580
Maximum data rate(kb/s)	3300	3300	4400
Parameter	VS3	VS4	VS5
Mean frame size (B)	2900	3500	4600
Maximum frame size (B)	15251	16960	16550
Period (s)	0.040	0.040	0.040
Mean data rate (kb/s)	580	700	910
Maximum data rate(kb/s)	310	3400	3300

Table 3: Video streaming and video conference parameters.

5.2 Efficiency analysis

Here we evaluate the efficiency of UTSS, intended as a measure of how well it utilizes the network resources, and we compare the result with the one of the other considered schedulers.

The first efficiency parameter is the *null rate* defined as the number of `CF-Null` packets sent by a QSTA in response to a `CF-Poll`, when it has no traffic to transmit. In Fig. 6 UTSS and WCBS have similar values of null rate for each traffic stream: this is because

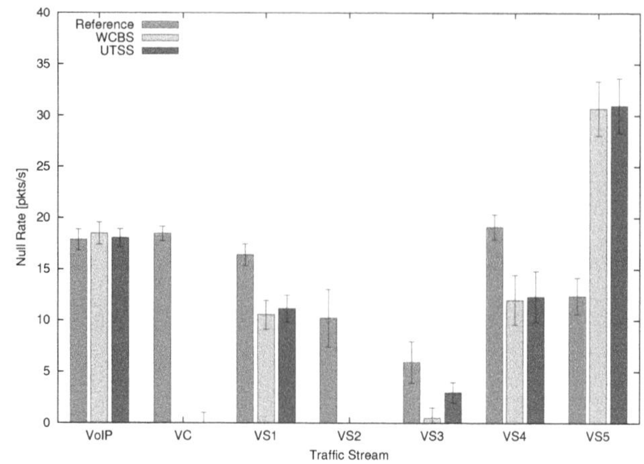

Figure 6: Null Rate.

they use the same polling interval. Often the null rate of these schedulers is better than the reference one since they poll the stations taking into account the different SI of each TS: such feature lets these schedulers to perform a zero null rate with some TS (e.g. VC and VS2 in Fig. 6). In the case of UTSS, even if the polling instant can be advanced, the polling interval variation is negligible and it does not affect the average null rate.

Fig.7 shows the polling interval used by the schedulers with each TS. The reference scheduler uses a unique value of the polling interval for all the TSs, as reported in Sec. 2, which is less than the minimum MSI of the all admitted TS. Instead, WCBS and UTSS poll each TS_i using its SI_i.

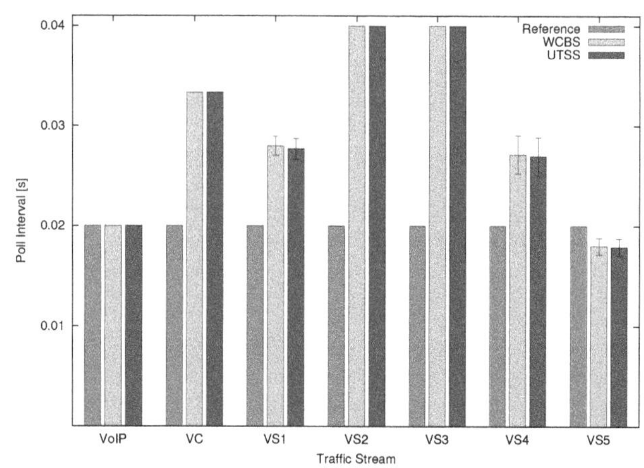

Figure 7: Poll Interval.

Finally Fig. 8 shows that UTSS does not significantly affect the throughput, thus network utilization is roughly the same.

5.3 Delay analysis

The access delay is defined as the time elapsed from when the frame reached the MAC layer until when the frame is successfully acknowledged. Fig. 9 shows that mean value of the access delay of UTSS is improved with respect to WCBS, in particular for TS with higher VBR, as VS2 and VS3. This confirms the obtained analytical results about deadline miss (see Theorem 2). Compared to the reference scheduler UTSS performs worse when serving TSs

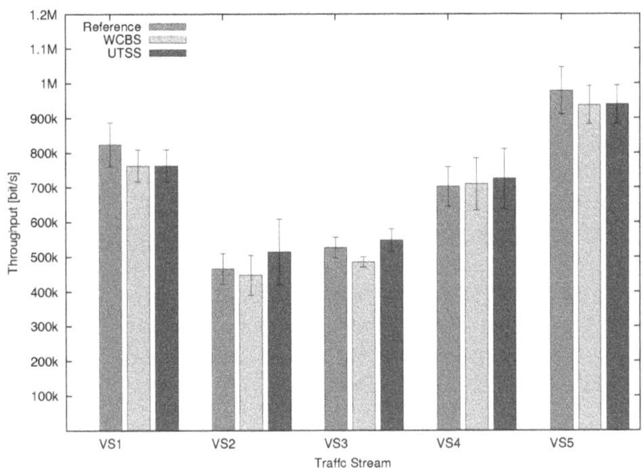

Figure 8: Throughput of VBR stations.

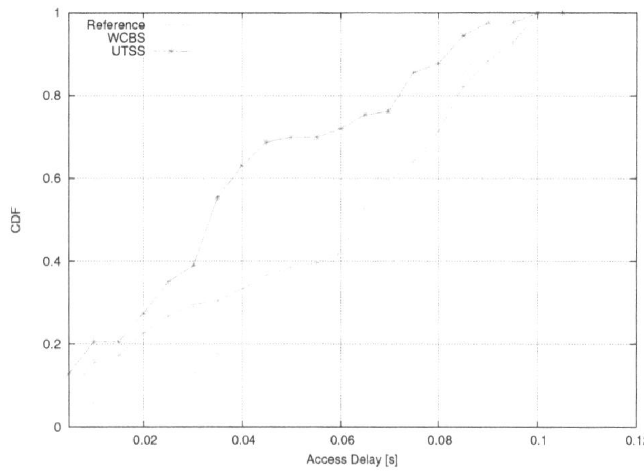

Figure 10: Access Delay CDF of VS3 Traffic Stream.

with less variable bit rate. This result is confirmed by the behavior of other EDF-based schedulers [5].

Looking at the *Cumulative Distribution Function* (CDF) of the access delay, Fig. 10 shows that after a time interval of 0.05 s UTSS allows 70% of transmitted packets whereas WCBS is limited to only 30%.

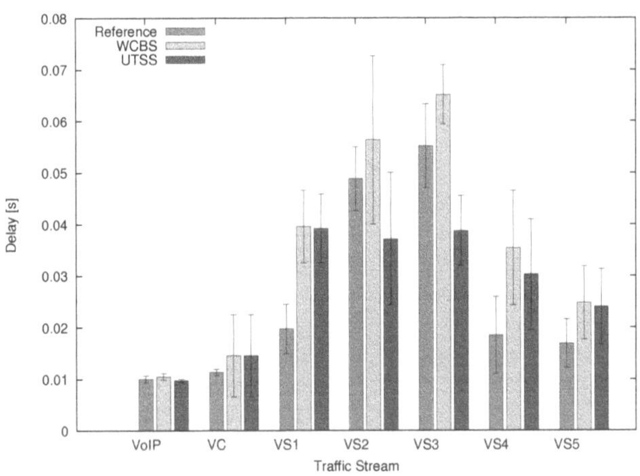

Figure 9: Mean Access Delay

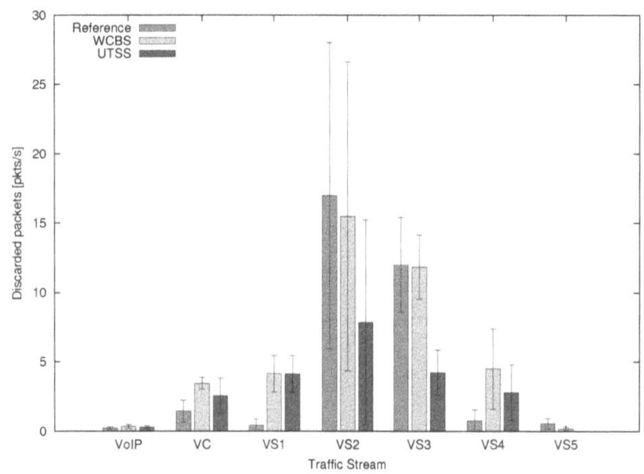

Figure 11: Discarded packets.

This is the most important result that highlights how UTSS is suitable for improving the real-time behavior of WCBS in the case of VBR traffic.

5.4 Packets drop analysis

Here we evaluate the amount of dropped packets from the queues due to expiration of *Delay bound* traffic parameter. Such parameter is chosen taking into account the length of the play-out buffer of a typical consumer device. For this reason this analysis aims to high light the differences between the schedulers and not the absolute value performed by each single scheduler.

In Fig. 11 the number of dropped packets is improved by UTSS when the considered TS is highly variable. This is due to the fact that UTSS, while it recovers the unused time from the previous transmission, it reduces the waiting time of the scheduled TS. In particular, the VS3 TS experiences an improvement up to 60% with respect to WCBS and reference scheduler. This result confirms the analytical consideration about the burstiness (see Theorem 3), which demonstrates that UTSS is suitable to absorb data rate peaks.

6. CONCLUSIONS

In this paper we have presented *Unused Time Shifting Scheduler*, a novel HCCA scheduler for IEEE 802.11e networks that is suitable for improving performance of a HCCA real-time scheduler, like WCBS, by reclaiming the unused portions of transmission opportunity intervals assigned for traffic streams transmission.

This mechanism is effective especially when high variable bit rate traffic stream are involved, such as in the case of video streaming transmissions.

The analytical evaluation shows that it efficiently uses the wireless medium in presence of different types of traffic, and increases the burstiness tolerable by the network.

These results has been validated through simulations showing that UTSS improves the performance of the existing schedulers, i.e. WCBS and reference scheduler, in terms of null rate, throughput and access delay.

7. REFERENCES

[1] P. Ansel, Q. Ni, and T. Turletti. FHCF: A simple and efficient scheduling scheme for IEEE 802.11e wireless lan. *Mobile Networks and Applications*, 11(3):391–403, June 2006.

[2] T. P. Baker. Stack-based scheduling for real-time processes. *Real-Time Syst.*, 3(1):67–99, 1991.

[3] G. Boggia, P. Camarda, L. G. Grieco, and S. Mascolo. Feedback-based control for providing Real-Time services with the IEEE 802.11e MAC. *IEEE/ACM Trans. Net.*, 2(15):323–333, Apr. 2007.

[4] G. Cecchetti and A. L. Ruscelli. Performance evaluation of Real-Time schedulers for HCCA function in IEEE 802.11e wireless networks. In *Proc. ACM Q2SWinet*, Vancouver, Canada, Oct. 2008.

[5] G. Cecchetti and A. L. Ruscelli. Real-Time support for HCCA function in IEEE 802.11e networks: a performance evaluation. *Security and Communication Networks*, 2010. Special Issue on Security for QoS Assured Wireless Networks.

[6] G. Cecchetti, A. L. Ruscelli, and F. Checconi. W-CBS: a scheduling algorithm for supporting QoS in IEEE 802.11e. In *Proc. QSHINE*, pages 1–7, Vancouver, British Columbia, Canada, 2007. ACM.

[7] C. Cicconetti, L. Lenzini, E. Mingozzi, and G. Stea. Design and performance analysis of the Real-Time HCCA scheduler for IEEE 802.11e WLANs. *Computer Networks*, 51(9):2311–2325, 2007.

[8] J. Cowling and S. Selvakennedy. A detailed investigation of the IEEE 802.11e HCF reference scheduler for VBR traffic. In *Proc.13th ICC*, pages 453–459, Chicago, US, Oct. 2004.

[9] Y. Fan and C. Huang. Real-Time traffic sheduling algorithm in WLAN. In *Proc. 4GMF*, San Diego, CA, 2005.

[10] H. Fattah and C. Leung. An overview of scheduling algorithms in wireless multimedia networks. *IEEE Wireless*, 9(5):76–83, Oct. 2002.

[11] C. Garcia, D. Prett, and M. Morari. Model predictive control: theory and practice: a survey. *Automatica*, 25(3):335–348, 1989.

[12] A. Grilo, M. Macedo, and M. Nunes. A service discipline for support of IP QoS in IEEE 802.11 networks. In *Proc. PWC*, Laapenranta, Finland, Aug. 2001.

[13] A. Grilo, M. Macedo, and M. Nunes. A scheduling algorithm for QoS support in IEEE 802.11e networks. *IEEE Wireless Comm.*, 10(3):36–43, June 2003.

[14] A. Grilo and M. Nunes. Performance evaluation of IEEE 802.11e. In *13th IEEE PIMRC*, volume 1, pages 511–517, Lisboa, Portugal, Sept. 2002.

[15] http://traces.eas.asu.edu/, 2005.

[16] http://www.isi.edu/nsnam/ns/. *Network Simulator 2*, 1995.

[17] IEEE802.11. Wireless LAN medium access control (MAC) and physical layer (PHY) specification, 2007.

[18] IEEE802.11b. Wireless LAN medium access control (MAC) and physical layer (PHY) specification: Higher-speed physical layer extension in the 2.4 GHz band, 1999.

[19] IEEE802.11e. 802.11e-2005 IEEE standard for information technology telecommunications and information exchange between systems local and metropolitan area networks specific requirements part 11: Wireless LAN medium access control (MAC) and physical layer (PHY) specifications: Amendment 8: Medium access control (MAC) quality of service enhancements, 2005.

[20] I. Inanc, F. Keceli, and E. Ayanoglu. An adaptive multimedia QoS scheduler for IEEE 802.11e wireless LANs. In *Proc. IEEE ICC*, Istanbul, Turkey, June 2006.

[21] J. R. Jackson. Scheduling a production line to minimize maximum tardiness. Research Report 43, University of California, Los Angeles, CA, 1955. Management Science Research Project.

[22] C. Kuan and K. Dimyati. Utilization model for HCCA/EDCA mixed mode in IEEE 802.11e. *ETRI Journal*, 29(6):829–831, Dec. 2007.

[23] W. K. Lai, C. Shien, and C.Jiang. Adaptation of HCCA/EDCA ratio in IEEE 802.11 for improved system performance. *Int. J. on Innovative Computing, Inf. and Contr.*, 5(11(B)):4177Ű–4188, Nov. 2009.

[24] P. Larcheri and R. Lo Cigno. Scheduling in 802.11e: Open-loop or closed-loop? In *Proc. IFIP WONS*, Les Ménuires, France, Jan. 2006.

[25] J. Lee, I. Shin, G. Park, W. Song, J. Kim, and J. Hong. An efficient bandwidth reclaim scheme for the integrated transmission of real-time and non real-time messages on the WLAN. In *Proc. ICCS*, volume 4490, pages 925–932, Beijing, China, May 2007.

[26] C. L. Liu and J. W. Layland. Scheduling algorithms for multiprogramming in a hard-real-time environment. *J. ACM*, 20:46–61, Jan. 1973.

[27] R. Lo Cigno, L. Palopoli, and A. Colombo. Analysis of different scheduling strategies in 802.11e networks with multi-class traffic. In *Proc. IEEE LCN*, Dublin, Ireland, Oct. 2007.

[28] S. Lu, V. Bharghavan, and R. Srikant. Fair scheduling in wireless packet networks. *IEEE/ACM Trans. Net.*, 7(4):473–489, Aug. 1999.

[29] S. Mangold, S. Choi, P. May, O. Klein, G. Hiertz, and L. Stibor. IEEE 802.11e wireless LAN for quality of service. In *Proc. European Wireless*, volume 1, pages 32–39, Florence, Italy, Feb. 2002. invited paper.

[30] L. Palopoli, R. Lo Cigno, and A. Colombo. Control and optimization of HCCA 802.11e access scheduling. In *Proc.IEEE CDC*, New Orleans, LA, Dec. 2007.

[31] A. L. Ruscelli, G. Cecchetti, A. Alifano, and G. Lipari. Enhancement of QoS support of HCCA schedulers using EDCA function in IEEE 802.11e networks. *Ad Hoc Networks*, In Press, 2010. Special Issue on Recent advances in analysis and deployment of IEEE 802.11e and IEEE 802.11p protocol families.

[32] V. A. Siris and C. Courcoubetis. Resource control for the EDCA and HCCA mechanisms in IEEE 802.11e networks. In *WiOpt*, pages 22–27, Boston, Massachusetts, 2006.

[33] D. Skyrianoglou, N. Passas, and A. K. Salkintzis. ARROW: An efficient traffic scheduling algorithm for IEEE 802.11e HCCA. *IEEE Transaction on Wireless Communications*, 5(12):3558–3567, Dec. 2006.

[34] S.-L. Tsao. Extending earliest-due-date scheduling algorithms for wireless networks with location-dependent errors. In *52nd IEEE VTC-Fall*, volume 1, pages 223–228, Boston, MA, Sept. 2000.

[35] S. Wang and A. Helmy. Performance limits and analysis of contention-based IEEE 802.11 MAC. In *Proc. IEEE LCN*, Tampa, FL, Nov. 2006.

Modeling Aggregation Convergecast Scheduling Using Constraints

Evandro de Souza
Computing Science Department
University of Alberta, Canada
evandro@ualberta.ca

Ioanis Nikolaidis
Computing Science Department
University of Alberta, Canada
nikolaidis@ualberta.ca

ABSTRACT

We consider the transmission scheduling problem for aggregation convergecast in wireless sensor networks. All transmissions required for a completed data collection to the sink node are to be scheduled in a single TDMA frame. Concurrent transmissions are possible within each TDMA slot as long as no collisions occur at the intended recipients of each transmission. We observe that previously proposed heuristics to solving the aggregation convergecast scheduling problem rely on a decomposition into two phases. The first phase is the construction of an aggregation tree, typically a Shortest Path Tree (SPT), and the second phase is sheduling transmissions. We express the constraints of the aggregation convergecast problem in a manner compatible with a constraint programming solver, and extract solutions for small sized instances. The, admittedly sparse, evidence suggests that the resultant aggregation trees is almost never SPTs and point to alternative considerations that would be useful when constructing aggregation convergecast scheduling heuristics.

Categories and Subject Descriptors

C.2.1 [**Network Architecture and Design**]: Network topology, Wireless communication; C.2.2 [**Network Protocols**]: TDMA, Routing protocols

General Terms

Algorithms

Keywords

Convergecast, Scheduling, Constraint Programming

1. INTRODUCTION

A commonplace application for wireless sensor networks (WSNs) is the collection of values (e.g., one measurement from each sensor) and their multi-hop transmission to a sink node. The underlying protocol is termed *convergecast.* To execute convergecast, it is useful to route packets along logical paths from sensors to the sink, and to enforce a transmission schedule, such that concurrent transmissions do not result in collisions. A particular variety of *convergecast* is *aggregation convergecast* whereby data collected from sensors are summarized/aggregated along the way of being forewarded to the sink. Aggregation has the potential to reduce the overall packet traffic, and hence provides better energy efficiency for executing aggregation queries. There are several combination of paths to aggregate the data. The union of those paths can be represented by a spanning tree, which we call the *aggregation tree.* Determining the spanning tree and its corresponding optimal makespan schedule has been shown to be an NP-hard problem [1].

The existing approximation heuristics decompose the problem into two steps: a spanning tree construction, followed by scheduling, using some heuristic [1, 4, 5]. Shortest Path Trees (SPT) has been the favorite aggregation tree. However, as of today, there has been no clear indication as to whether SPTs could result in an optimal schedule. The difficulty of course for making any statement about the potential of SPTs is that the quality of the outcome of all proposed approximations is a function of *both* the aggregation tree *as well as* of the second "stage," i.e., of the schedule construction heuristic (*given* an aggregation tree). In other words, it has been unclear whether the SPT alone or the subsequent scheduling phase is to be blamed when inefficient schedules are produced.

In this paper, we attempt to elucidate the tradeoffs behind the selection of the optimal tree. To this end, we do not restrict the family of aggregation trees in any way, and we do not perform any approximations, thus allowing us to observe if indeed the SPTs appear in better, or even optimal, schedules than what the current literature produces. This paper was conceived as an exercise in presenting the aggregation convergecast problem in a standardized Constraint Satisfaction Problem (CSP) form that ensures no misunderstandings about the nature of the underlying optimization objective and constraints. In Section 2 we provide an overview of aggregation convegecast, and in Section 3 we introduce the corresponding constraint satisfaction model. Section 4 describes our findings and Section 5 provides our conclusions.

2. AGGREGATION CONVERGECAST

One typical example of aggregation convergecast is presented on Figure 1. The sink is the top node, to where all

MSWiM'11, October 31–November 4, 2011, Miami, Florida, USA.

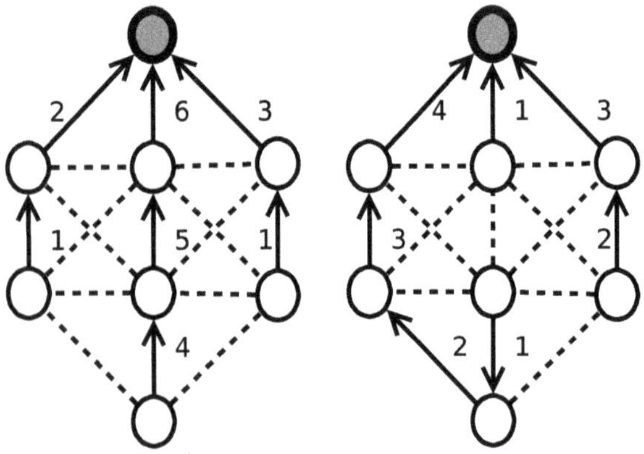

(a) SPT Schedule (b) Optimal Schedule

Figure 1: Schedule solution for example 8-node graph (numbers indicate time slot).

other nodes send their data. Figure 1(a), shows a solution using SPT. The number next to the arrows represents the time slot when a link is allowed to transmit. This solution uses a time frame length of 6 time slots. Even for this small example, it is possible to obtain a better schedule, as presented on Figure 1(b), where the schedule length is only 4 time slots, but using a non-SPT topology as aggregation routing path.

Previous works use SPT as the aggregation tree. In [1], the algorithm SDA follows the shortest data aggregation philosophy. It aggregates data along shortest paths towards the sink, incrementally constructing smaller and smaller shortest path trees. The algorithm DAS [5] is based on a aggregation tree using connected dominating sets to form a virtual backbone of a sensor network. The WIRES algorithm [4] uses a heuristic to obtain a variation of the SPT that produces better schedules, however it is still a SPT-based solution.

The number of the transmissions for an alternative to the SPT will be the same as those of the SPT because aggregation convergecast already minimizes the number of transmissions (only one transmission will be performed by each node). However an SPT does not inherently account for the interference caused by node transmissions which, as we will see, impacts, as expected, the schedule length. So, it is preferable to select an aggregation tree that minimizes the effects of the interference, and this tree is very likely not to be an SPT.

3. MODELING PROCESS

Constraint Programming (CP) is a formalism for modeling problems as a Constraint Satisfaction Problems [3]. A model consists of a number of variables, each of which has an associated domain of values. Constraints are specified on these variables restricting the set of values they can take. The objective of the problem is to find out if each of these variables can be assigned a value from its domain in such a way that all the constraints are satisfied. These three components, the variables, their values and the constraints, form the model of the problem to be solved.

The WSN is modeled as an undirected graph $G = (V, E)$, where V is the set of nodes and E is the set of edges representing all communications links available in the transmission range of the nodes. The number of nodes in the graph is represented by n. A fixed node $s \in V$ is designated as $sink$. The graph G and the $sink$ node are the inputs for the CP solver. LSP represents the number of hops of the longest shortest path to the $sink$. If each graph edge has weight 1, then LSP will represent the heaviest (across all nodes) shortest path from a node u to the sink. In this model, node u represents any node $\in V$ or the transmitter node on a communication link, while v represents the receiver node. $N(u)$ represents the (one-hop) neighbor set of node u.

3.1 The Constraints

The model to represent the aggregation convergecast is composed of several variables. The variable PP is defined for each node and represents the parent of the node in the definition of the tree. Each node must have only one parent. Each node transmits only once and only to its parent. The variable RR is defined for each node and represents the number of hops to the $sink$ using the selected tree. This variable is used to ensure that loops are avoided. The variable SH is defined for each node and represents the time slot in which the node transmits. The variable CC is defined for each time slot and represents the number of concurrent transmissions in the slot. CC is used to optimize the search process. The minimum time frame length is the final objective, so a minimization process is necessary, i.e., not just the generation of a feasible solution. The optimization criteria is represented by the variable $makespan$, which stands for the length of the TDMA schedule. It is trivial to observe that the value of SH variables need not be larger than n (the number of nodes) since this is the maximum number of time slots necessary when no concurrent transmission are possible. By the same token, the values of RR, PP and $makespan$ are upper bounded by number of nodes n.

$$PP[u] = v \;\Rightarrow\; RR[u] = RR[v] + 1 \tag{1}$$

$$PP[u] \neq v \; \begin{cases} u \rightarrow v \notin E \\ u \neq sink \end{cases} \tag{2}$$

The above constraints are the topology constraints and have the objective of limiting the communication graph, such that the final results is a spanning tree of the original graph, rooted at the $sink$ node. Constraint 1 specifies that the rank of the node u must be the rank of its parent node v plus one. Constraint 2 restricts the nodes that can be parent of node u to only those which have a direct link to it.

$$PP[u] = v \;\Rightarrow\; SH[u] \neq SH[j] \; \begin{cases} j \in N(v)\backslash\{u\} \\ u \rightarrow v \in E \\ v \rightarrow j \in E \end{cases} \tag{3}$$

Constraint 3 captures the requirement that there can be no interference at the receiver node, based on the link activation model [2]. As before, the restriction is only valid when a specific link $u \rightarrow v$ is valid (i.e., if the edge exists).

$$\left.\begin{matrix} RR[u] = 0 \\ PP[u] = u \end{matrix}\right\} for \;(u = sink) \tag{4}$$

$$SH[sink] > SH[u] \quad \{u \in V\backslash\{sink\}\} \tag{5}$$

$$SH[sink] \geqslant LSP \tag{6}$$

$$PP[u] = v \Rightarrow SH[u] < SH[v] \quad (7)$$

This set of constraints is related to the aggregation convergecast application. Constraints 4 and 5 define the unique role of the *sink* node. It is the beginning of the rank count, parent of itself, and it must be scheduled (this is a pseudo-tranmsission to itself acting as the end delimeter of the schedule) after any other node on the tree. Constraint 6 restricts the domain of the schedule of the sink node, by ensuring that it cannot be smaller than the longest shortest path. Constraint 7 captures the requirement that a parent node can only transmit after it has received transmissions from all its children.

$$CC[t] \geq CC[t+1] \quad (8)$$

The above constraint deals with the fact that there are multiple possible feasible solutions for the same schedule length. The difference between then is not relevant to the optimization process, but searching all solutions wastes a great amount of computation. One way to avoid this exhaustive search is to introduce this constraint that enforces the number of transmission per slot to be monotonically decreasing in all valid solutions.

4. RESULTS

Using the model presented in the previous section, it is possible to obtain solutions for the aggregation convergecast problem. Sample graphs were generated by randomly placing nodes on a square region with constant and uniform transmission range, forming Unit Disk Graphs (UDG). Groups of 10, 15, 20 and 25-node graphs were produced. On each group, the graph density ranges from 0.250 to 0.425 approximately. The graph density is defined as the ratio of the number of links on the graph and the maximum possible number of links (for a completely connected graph the with same number of nodes). Using graphs with different densities allows us to analyse scenarios with different levels of interference. Also, the aggregation tree size is defined as $\sum_{i=1}^{n} h_{(i,sink)}$, where h is the number of hops from node i to the *sink* node using the selected tree. That is, the aggregation tree size is the sum of each node's hop-count to the sink along the aggregation tree. It is trivial to show that the smallest aggregation tree size for a graph is achieved by its SPT.

Table 1: Minimal Tree Size for 15-node graphs

Density	WIRES	SPT Size	CP	Min TS
0.2571	8	42	7	43
0.2762	9	48	8	49
0.2952	7	32	6	34
0.3238	8	36	7	36
			6	41
0.3524	9	34	8	35
			7	36
0.3714	8	34	7	36
0.4000	9	27	8	28
			7	29

Some of the existing (two-phase) aggregation convergecast scheduling heuristics are used for comparison purposes. The Balanced Shortest Path Tree (BSPT) [4] is used as the aggregation tree for all examples. For the second phase, three heuristic-based scheduling algorithms were selected: SDA [1], DAS [5] and WIRES [4].

CP solvers have the downside of spending a large amount of time to obtain an optimal solution, mainly in the search phase. As the number of constraints and candidate trees is related to the number of nodes and links, the time required for obtaining results can be prohibitive. Therefore, the CP results presented may not be optimal, but nonetheless they required fewer time slots than the heuristics produced.

The results obtained are depicted in Figure 2. Note that the results obtained using the CP model are closer to the minimal schedule possible (expressed by the LSP, used here to illustrate the lower bound for the convergecast scheduling problem) for the given graphs.

Figure 3 presents the tree size difference between BSPT, and the tree obtained from the best result using the CP model. It is noticeable that tree size obtained with the CP model is larger than the SPT, suggesting that it is not a good strategy to use SPT for this application. Table 1 depicts the minimum schedule length for 15-node graphs possible to obtain when we relax the tree size to be larger than the minimum possible size (SPT). Even though this conclusion seems intuitive, it is remarkable that the algorithms in the literature continue using SPT as base for their aggregation convergecast tree.

The difference between schedule length obtained by the previous aggregation convergecast algorithms compared to better solutions from the CP model is related to the characteristics of the tree selected for the aggregation path. Shortest path trees use the minimal number of hops from a node to the *sink* node, independent of whether doing so might prove to be detrimental to the scheduling phase. This is typically the case when some bottleneck node is present on the graph, or specially when cliques are present. What we noticed is that SPT tends to use the clique internal edges to reach the nodes inside a clique because it produces, in most cases, the minimal path length to the *sink*. However, when the clique–internal edges are used, the nodes belonging to the same clique can not transmit in parallel. Even worse, if all clique–internal links are directed toward a single node (because it is the shortest path to the *sink*), the number of time slots cannot be reduced to less than the number of nodes in that clique. The final effect is that the overall number of time slots on the schedule cannot be smaller than the maximal clique on that graph.

Smaller schedules, found by the CP model, are possible by selecting links directed away from cliques (not using most of the clique internal edges) or bottleneck nodes (lots of incoming edges), increasing the size of the tree in order to avoid the serialization of the transmissions of nodes belonging to the same clique. This is the reason for the surprising relation between the results in Figure 2 and Figure 3. Whereas the tree can be significantly costlier (for example, the 25-node tree for 0.3 density presents the most pronounced difference) a shorter schedule can be achieved, even if the schedule is shorter by a couple of slots. One has to take into account that a couple slots difference might not appear as a small difference in the absolute sense, but relatively speaking it can represents a 10% (or even more) improvement over the schedule length produced by the best heuristic. Of course, due to the computation needs of the constraint solver, we cannot confirm if a similar proportional improvement is possible for larger networks.

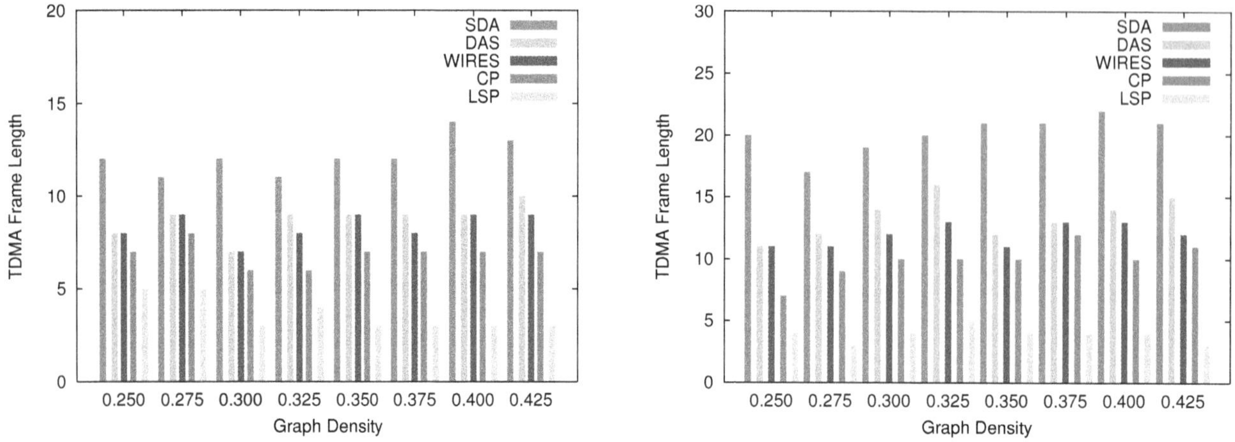

(a) 15-Node Graphs (b) 25-Node Graphs

Figure 2: Frame Length Comparison

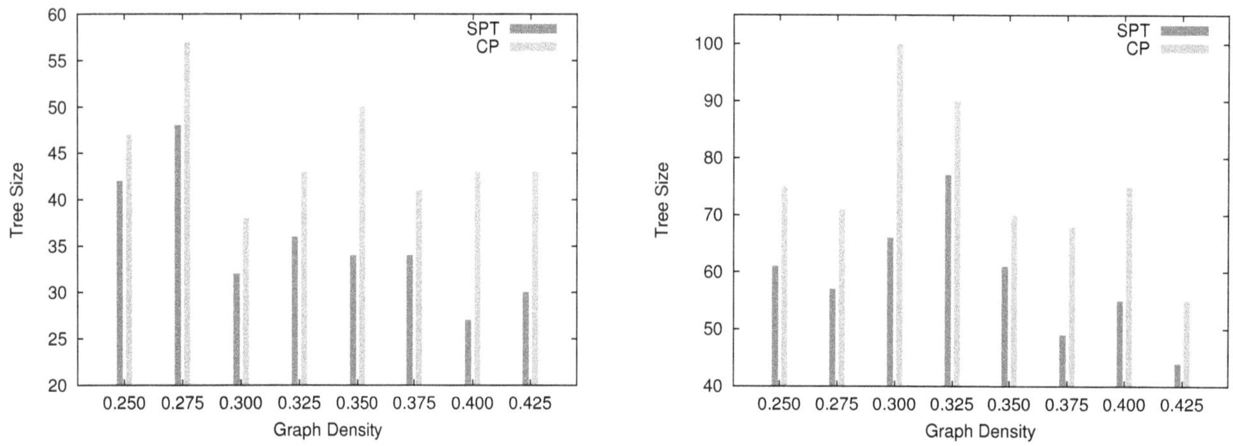

(a) 15-Node Graphs (b) 25-Node Graphs

Figure 3: Tree Size Comparison

5. CONCLUSIONS

We model aggregation convergecast scheduling as a Constraint Satisfaction Problem with the objective of obtaining the shortest possible schedule for a TDMA frame that allows the complete collection of data to the sink. Constraints related to the topology, the interference, and the application logic, as well as their relationships are modeled. We observe that SPT is very likely *not* the best choice for aggregation tree. While computational limitations do not allow us to derive results for large networks, the produced results provide valuable insights as to the "shape" that the optimal schedules and corresponding aggregation trees possess. Using non-SPT topologies that avoid the use of internal links to large cliques, has the potential to produce shorter schedules, improve the number of concurrent transmissions and hence the throughput of the solution.

6. ACKNOWLEDGMENTS

This work has been partly supported by a grant from the Natural Sciences and Engineering Research Council (NSERC).

7. REFERENCES

[1] X. Chen, X. Hu, and J. Zhu. Minimum data aggregation time problem in wireless sensor networks. In *Mobile Ad-hoc and Sensor Networks*, volume 3794 of *Lecture Notes in Computer Science*, pages 133–142. Springer Berlin / Heidelberg, 2005.

[2] I. Chlamtac, A. Farago, and H. Y. Ahn. A topology transparent link activation protocol for mobile cdma radio networks. *Selected Areas in Communications, IEEE Journal on*, 12(8):1426 –1433, Oct. 1994.

[3] R. Dechter. *Constraint Processing*. Morgan Kaufmann, 340 Pine Street, San Francisco, 2003.

[4] B. Malhotra, I. Nikolaidis, and M. Nascimento. Aggregation convergecast scheduling in wireless sensor networks. *Wireless Networks*, 17:319–335, 2011.

[5] B. Yu, J. Li, and Y. Li. Distributed data aggregation scheduling in wireless sensor networks. In *INFOCOM 2009, IEEE*, pages 2159 –2167, 2009.

Adaptive Real-Time Query Scheduling for Wireless Sensor Networks

Moutaz Saleh Mustafa Saleh
Department of Computer Science & Engineering
College of Engineering, Qatar University
2713 Doha, Qatar
moutaz.saleh@qu.edu.qa

ABSTRACT

With the recent evolution of high data rates in real-time sensor network applications, there is always an increasing demand for high performance query services in such networks. To meet this demand, we propose an adaptive approach for scheduling real-time query transmissions in wireless sensor networks (WSNs). Our purpose is to schedule multiple real-time queries efficiently by maximizing the overall network throughput while meeting the queries' deadlines. The proposed scheduler allows multiple conflict-free queries to execute concurrently in order to achieve maximum throughput, besides the scheduler works preemptively to satisfy the real-time queries' deadline constrains. The simulation results showed that our proposed scheduler can be adopted to effectively increase network throughput and eliminate query prioritization inversion.

Categories and Subject Descriptors

C.2.1 [**Computer-Communication Networks**]: Network Architecture and Design – *wireless communication, network communications.*

General Terms

Algorithm, Management, Performance.

Keywords

Real-time Query; Scheduling; Preemptive; Priority; Throughput; Conflict-free; WSN.

1. INTRODUCTION

Recent years have witnessed a necessary need towards wireless sensor networks (WSNs) that must support real-time data collection at high data rates. Examples of such systems include, patient monitoring [1], emergency response [2], and health monitoring [3]. These systems, however, pose significant challenges. First, the system must handle various types of traffic with different deadlines. Thus, a real-time communication protocol should provide effective prioritization between different traffic classes while meeting their respective deadlines. Second, the system must support high throughput since it may generate high volumes of traffic. Therefore, it is important for the system to achieve predictable and bounded end-to-end latencies.

Many WSN's real-time applications use query services to periodically collect data from sensors to a base station. Real-time applications refer to those performance critical applications that require bounded service latency. There exists a class of WSN applications that is real-time, and requires bounded latency on data delivery. Since the nodes communicate over a shared medium, it is possible that if multiple transmissions overlap in time, some of them will collide. This problem is alleviated by either sensing the medium for possible ongoing transmissions to avoid collisions, or by carefully timing and scheduling the transmissions so that no collision occurs. The collision avoidance algorithms that attempt to schedule transmissions are referred to as transmission scheduling algorithms. While contention based collision avoiding algorithms offer a simpler and more dynamic solution to the medium access problem, schedule based algorithms can provide deterministic service delay bounds that is crucial for real-time applications where the transmission delays should be known and bounded.

In this paper, we present a novel Real-Time Query Scheduling (RTQS) approach designed to meet the communication needs of high data rate real-time WSN applications. Our scheduling approach takes advantage of the common properties of WSN query services to construct a concurrent scheduler for executing conflict-free query transmissions. This can maximize the overall system throughput. Also, the scheduler works adaptively with priority-driven preemption to satisfy the real-time queries' deadline constrains. Specifically; the proposed scheduling scheme has the following attractive features: first, it achieves high query throughput by allowing concurrent execution of conflict-free query transmissions. Second, it eliminates query prioritization inversion introduced by traditional non-preemptive query scheduling. Third, it enhances the query worst-case response time of the ordinary preemptive scheduling. Finally, the associated schedulability analysis bridges the gap between the WSN queries and real-time scheduling theories.

The rest of this paper is organized as follows: Section 2 introduces related works on WSNs query scheduling and then Section 3 describes query, network, and scheduler models. Section 4 details the design and analysis of the proposed scheduling scheme. After exhibiting simulation results in Section 5, Section 6 concludes this paper.

MSWiM'11, October 31–November 4, 2011, Miami, Florida, USA.

2. RELATED WORK

In-network Query processing techniques have been widely accepted for on-line sensory data management in WSNs [4] [5]. Instead of pulling all the sensory data from nodes into a central server to process, these techniques make the sensor nodes to cooperatively process the queries from users within the network, hence can increase the flexibility and reduce the network traffic of WSNs. Scheduling schemes have also been proposed to guarantee the efficiency and reliability of in-network sensor query processing [6] [7] [8] [9] [10]. These scheduling schemes are able to coordinate the timings of nodes to avoid communication collisions and make receivers active when a transmitter transmits packets towards them. Figure 1 shows the operation of query processing in WSN.

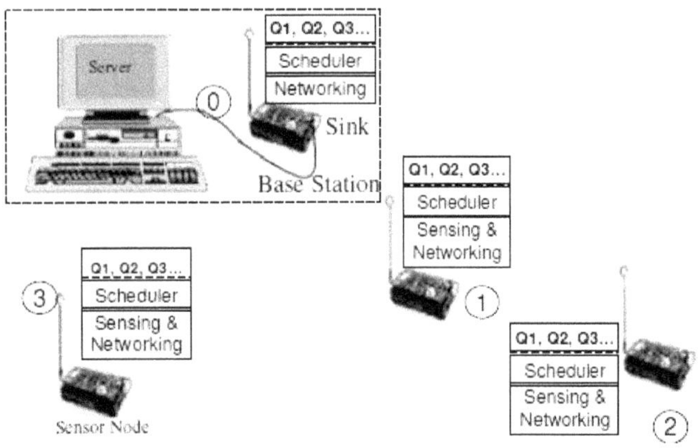

Figure 1. Query Processing in WSN

For instance, TDMA scheduling is attractive for high data rate sensor networks because it is energy efficient and can provide higher throughput than CSMA/CA protocols under heavy load. Two types of TDMA scheduling problems have been investigated in the literature: node scheduling and link scheduling. In node scheduling, the scheduler assigns slots to nodes whereas, in link scheduling, the scheduler assigns slots to links through which pairs of nodes communicate. Early TDMA scheduling protocols were designed for static or uniform workloads [11][12][13][14]. Such approaches are not suitable for dynamic applications with variable and non-uniform workloads. Several recent TDMA protocols can adapt to changes in workload. A common method to handle variable workloads is to have nodes periodically exchange traffic statistics and then adjust the TDMA schedule based on the observed workload [11][15][16]. However, exchanging traffic statistics frequently may introduce communication overhead.

In [17], a novel query scheduling technique called DCQS is presented. Instead of assigning slots to each node or link, slots are assigned to transmissions based on specific communication patterns and temporal properties of queries in WSNs. This approach allows achieving high throughput and low latency. Also, DCQS adapts to changes in workloads by exploiting explicit query information provided by the query service. Furthermore, it features a local scheduling algorithm that can accommodate changes in query rates and completions without explicitly reconstructing the schedule. In spite of all these advantages for DCQS, it ignores the quality requirements from different applications and users queries with different quality preferences are equally treated and are processed one by one according to their arrival order. As a result, those queries with high quality requirements, real-time queries, are likely to be unsatisfied, whereas the lower request queries that arrive earlier than the high request queries may be overly satisfied. Hence, for real-time query scheduling, the research work in [18] proposes a Slack-stealing Query Scheduling (SQS) algorithm which adapts preemption decisions to improve throughput while still meeting real-time queries' deadlines.

3. SYSTEM MODELS

3.1 Query Model

RTQS assumes a common query model in which source nodes produce data reports periodically. This model fits many applications that gather data from the environment at user specified rates. A query l is characterized by the following parameters: a set of sources, a function for network aggregation [19], the start time S_l, a query period P_l, a query deadline D_l, and a static priority. A new query instance is released in the beginning of each period to gather data from the WSN. We use $I_{l,u}$ to refer to the uth instance of query l whose release time is $R_{l,u} = S_l + u \cdot P_l$. The priority of an instance is given by the priority of its query. If two instances have the same query priority, the instance with the earliest release time has higher priority. For each query instance a node i need $W_l[i]$ slots to transmit its aggregated data report to its parent.

A query service works as follows: a user issues a query to a sensor network through a base station, which disseminates the query parameters to all nodes. To facilitate data aggregation, the query service constructs a routing tree rooted at the base station as the query is disseminated. The execution of a query instance entails network data aggregation. Accordingly, each non-leaf node waits to receive the data reports from its children, produces a new data report by aggregating its data with the children's data reports, and then sends it to its parent. We assume that there is a single routing tree that spans all nodes and it is used to execute all queries. This assumption is consistent with the approach adopted by existing query services [19]. During the lifetime of the application the user may issue new queries, remove queries from execution, or change the parameters of existing queries. RTQS is designed to support dynamic queries efficiently.

3.2 Network Model

RTQS models a WSN as an Interference-Communication (IC) graph as shown in Figure 2. The IC graph has all nodes as vertices and has two types of directed edges: communication and interference edges. A communication edge ab indicates that a packet transmitted by a may be received by b. A subset of the communication edges forms the routing tree used for data aggregation. An interference edge ab indicates that a's transmission interferes with any transmission intended for b even though a's transmission may not be correctly received by b. The IC graph is used to determine if two transmissions can be scheduled concurrently. We say that two transmissions, ab and cd are conflict-free and can be scheduled concurrently if a, b, c, and d are distinct and ad and cb are not communication/interference edges. Each node needs to know only its incoming/outgoing communication and interference edges. To do so, Zhou et al in [20], present a Radio Interference Detection (RID) mechanism that can be used by WSN nodes in IC graph to determine its adjacent communication and interference edges. Also, clocks are assumed to be synchronized as clock synchronization is a fundamental service in WSN since many applications must time-stamp their sensor readings to infer meaningful information about the observed events [2].

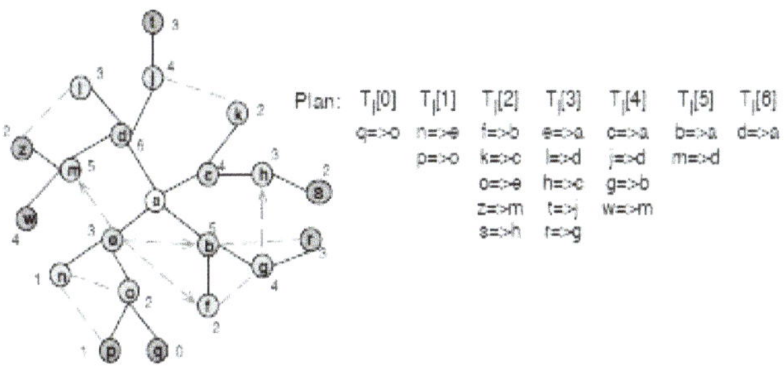

Figure 2. Example of IC Graph and its Plan

4. QUERY SCHEDULING

RTQS is designed to achieve high data throughput and differentiated query latencies through prioritized concurrent conflict-free transmission scheduling. The adopted approach relies on using two components: a planner and a scheduler. The planner constructs a plan for executing all the instances of a query. A plan is an ordered sequence of steps, each comprised of a set of conflict-free transmissions. The scheduler that runs on every node determines the time slot in which each step in a plan is executed. To improve the throughput, the scheduler may execute steps from multiple query instances in the same slot as long as they do not conflict with each other. To summarize, RTQS works as follows: (1) When a query is submitted, RTQS identifies a plan for its execution. Multiple queries be executed using the same plan, and therefore RTQS may reuse a previously constructed plan for the new query. When no plan is reused, the planner constructs a new one. (2) RTQS determines if a query meets its deadline using a schedulability analysis which is performed on the base station. If the query is schedulable, the parameters of the query are disseminated; otherwise, the query is rejected. (3) At run-time the scheduler executes all admitted queries.

A plan has two properties: First, it respects the precedence constraints introduced by data aggregation: a node is assigned to transmit in a later step than any of its children. Second, each node is assigned in sufficient steps to transmit its entire data report. We use $T_l[i]$ to denote the set of transmissions assigned to step i $(0 \leq i < L_l)$ in the plan of query l, where L_l is the length of the plan. To facilitate in-network aggregation, a node waits to receive the data reports from all its children before transmitting the aggregated data report to its parent. Therefore, to reduce the query latency, the planner assigns the transmissions of a node with a larger depth in the routing tree to an earlier step in the plan. This strategy reduces the query latency because it reduces the time a node waits for the data reports from all its children. We also assume that two queries belong to the same query class if they can be executed according to the same plan. This can be defined as follows: Let $W_l[i]$ be the number of slots node i needs to transmit its data report to its parent for an instance of query l. If the planner constructs a plan for a query l, the same plan can be reused to execute a query h if $W_l[i] = W_h[i]$ for all nodes i. Since queries with different temporal properties and aggregation functions may share a same plan, a WSN may only need to support a small number of query classes. This allows RTQS to amortize the cost of constructing a query plan over many queries and effectively reduces the overhead.

The scheduler performs a query instance according to its query plan and improves the query throughput by overlapping the transmissions of multiple instances such that: (1) all steps executed in a slot are conflict-free. (2) the steps of each plan are executed in order: if step $I_{l,u}.i$ is executed in slot s_i, step $I_{l,u}.j$ is executed in slot $s_j < s_i$ then $I_{l,u}.j < I_{l,u}.i$. This ensures that the precedence constraints required by aggregation are preserved. Additionally, the scheduler maintains a record of the start time, period, and priority of all admitted queries. This can help in determining the step numbers in which the host node is assigned to transmit or receive in each plan together with the plan's length. The scheduler considers the released instances in the order of their priority and executes all steps that don't conflict in given slot s. For high performance, this operation should be done with low time complexity. Accordingly, the concept of Minimum Step Distance (MSD) is used [6]. The MSD $\Delta(l, h)$ is the smallest step distance between $I_{l,u}$ and $I_{h,v}$ such that the two steps $I_{l,u}.i$ and $I_{h,v}.j$ may be executed concurrently without any conflict. The MSD $\Delta(l, h)$ depends on the IC graph and the plans of l and h.

Here, we consider two common RTQS algorithms: Non-preemptive Query Scheduling (NQS) and Preemptive Query Scheduling (PQS). In NQS, once a query instance is started then it can't be preempted. For that, the earliest time at which an instance $I_{l,u}$ may start is after the previous instance $I_{h,v}$ completes step Δ-1. Since the execution of $I_{l,u}$ and $I_{h,v}$ cannot be preempted, if we enforce the MSD between the start of the two instances then their concurrent execution is conflict-free for their remaining steps since steps $I_{l,u}.i=x$ and $I_{h,v}.j=x+\Delta$ are executed in the same slot. Therefore, to guarantee that a non-preemptive scheduler executes conflict-free transmissions in each slot, it suffices to enforce a MSD of Δ between the start times of any two instances. Indeed, NQS enforces a MSD of at least Δ between the start times of any two instances by starting an instance in two cases: (1) when there is no instance being executed and (2) when the step distance between the two instances is larger Δ.

Obviously, NQS has the drawback of introducing priority inversion. To eliminate such drawback, PQS is introduced with a key feature which preempts the instances that conflict with the execution of a higher priority instance. PQS maintains a run queue and a release queue which are keyed by the query instance priority. When a new instance is released, it is added to the release queue. On the other hand, when an instance is started for execution, it is added to the run queue. PQS starts an instance $I_{l,u}$ ($I_{l,u}$ € release queue) in two cases: (1) If the next step $I_{l,u}.i$ can be executed concurrently with all instances in the run queue without conflict i.e. MSD of Δ between the start times of any two instances. (2) If $I_{l,u}$ has higher priority than all the instances in the run queue. In the last case, all running instances that conflict the execution $I_{l,u}$ will be preempted. We mention here that when an instance is started it is moved from the release queue to the run queue. Oppositely, when an instance is preempted, it is moved from the run queue to the release queue. Now, by eliminating priority inversion, PQS achieves lower latencies for high priority instances when compared with NQS. However, the query throughput is lower because it allows less overlap in the execution of instances. This exemplifies the tradeoff between prioritization and throughput in query scheduling. In the next section, we will introduce our proposed scheduling algorithm to effectively combine this tradeoff.

5. QUERY SCHEDULING

Our proposed scheduler combines the benefits of NQS and PQS in that it improves query throughput while meeting all deadlines. The design of this scheduler is based on the observation that preemption lowers throughput, and hence it should be used only when necessary for meeting deadlines. We define $C_{l,u}$ as the number of slots a lower priority instance u of query l has completed during its execution. Besides, a higher priority instance v of query h may start execution in any slot in the interval $[r_{h,v}, r_{l,u}+\Delta]$, where $r_{h,v}$ is the release time of $I_{h,v}$ (the vth instance of query h), $r_{l,u}$ is the release time of $I_{l,u}$ (the uth instance of query l), *and* Δ is the MSD between $I_{l,u}$ and $I_{h,v}$. The scheduler postpone the start of the higher priority instance $I_{h,v}$ if the lower priority instance $I_{l,u}$ satisfies the condition of having $(\Delta - C_{l,u} + P_{h,v}) \leq D_{h,v}$, where $P_{h,v}$ is the execution period of the higher priority instance $I_{h,v}$, and $D_{h,v}$ is the maximum allowable deadline for the same higher priority instance $I_{h,v}$. The advantage for this approach is that it opportunistically avoids preemption and the related throughput reduction when allowed by query deadlines. This highlights that our proposed adaptive query scheduler (AQS) can adapt preemption decisions to improve throughput while meeting all queries' deadlines. To clearly show the efficiency of the proposed scheduler, we present here the worst-case response time analysis for NQS, PQS, and proposed AQS. The worst-case response time of a query is the maximum query latency of any of its instances.

5.1 Analysis of NQS

Since NQS is a non-preemptive scheduling algorithm, to compute the response time of a query l we must compute the worst-case interference of higher priority instances and the maximum blocking time of l due to the non-preemptive execution of lower priority instances. From the NQS operation, an instance is blocked for at most $\Delta-1$ slots, and a higher priority instance interferes with a lower priority instance for at most Δ slots. Accordingly, the response time R_l of query l is the sum of its plan's length L and the maximum delay W_l that any query instance experiences before it is started.

5.2 Analysis of PQS

In PQS, higher priority instance cannot be blocked by a lower priority instance. Also, after an instance completes Δ steps, no newly released instance will interfere with its execution because their step distance would be at least Δ which allowing them to execute concurrently. To compute the response time of a query l we split its instance I_l into two parts: a preemptable part of length Δ and non-preemptable part of length $L - \Delta$. Higher priority instances may interfere with I_l only during its preemptable part. Thus, the response time of a query l is the sum of response time of the preemptable part and the length of the non-preemptable part.

5.3 Analysis of AQS

To compute the worst-case response time of a query l we split again a query instance I_l into two parts: a preemptable part and a non-preemptable part. Under PQS, the preemptable part is Δ slots. In contrast, with AQS, an instance I_l may utilize from a higher priority instance I_h at least n slots. Thus, the length of the preemptable part is at most $\Delta - n$ slots; and the length of the non-preemptable part is therefore $L - (\Delta - n)$ slots which is lower compared to both NQS and PQS. This is for sure a noticeable improvement for the AQS operation.

6. SIMULATION AND RESULTS

Our simulation settings are similar to 802.11b radios. This is because we are interested in high data rate applications, and sensor nodes used for such applications often adopt high-bandwidth radios. Indeed, this is also reasonable since several real-world sensor systems use 802.11b interfaces to meet their bandwidth requirements. Accordingly, the network bandwidth is 2Mbps. The communication range is 125m. The power consumed for transmitting and receiving a packet is 1.6W and 1.4W, respectively. The size of a packet is 2KB, of which 20 bytes are used for packet headers. Based on packet size and bandwidth we computed the slot size to be 8.16ms. The queue size is 10 packets. Each experimental run takes 200s.

In the beginning of the simulation we construct the IC graph and the routing tree. The IC graph is constructed similarly to the method described in. The routing tree is constructed as follows. The node closest to the center of the topology is selected as the base-station and is the root of the routing tree. The base station initiates the construction of the routing tree by flooding setup requests. A node may receive setup requests from multiple nodes and selects the node with the latest depth as its parent. Each node in the routing tree performs in-network aggregation when executing a query. We assume that each aggregated data report fits in a single packet. The queries issued involve all nodes in the network. In all experiments the queries belong to the same query class.

The results presented in this section are the average of multiple runs for accuracy. All experiments are performed in a 750m × 750m area divided into 75m × 75m grids in which a node is placed at random. We simulate three queries with high, medium and low priorities. The query priorities are determined based on their deadlines such that the tighter the deadline, the higher the priority. Eventually, we compare the performance of our proposed AQS algorithm against both NQS and PQS with respect to the response time. The response time of a query instance is the time between its release time and completion time, i.e., when the base station receives the last data report for that instance. The average response times of NQS, PQS, and AQS when scheduling queries with high Q_H, medium Q_M and low Q_L priorities are shown in Figures 3, 4, and 5 respectively.

By observing the plotted figures, NQS meets the deadline requirements only when Q_H deadline is large enough to be scheduled. NQS misses Q_H deadline when it is tight due to the priority inversion under non-preemptive scheduling. This indicates that NQS is unsuitable for high priority queries with tight deadlines. Interestingly, under AQS, the response time of Q_H changes depending on its deadline. As the deadline becomes tighter, the response time of Q_H also decreases and remains below the deadline. We also see an increase in the response times of the lower priority queries as Q_H deadline is decreased. This is because as Q_H deadline decreases the lower priority queries may interferes high priority queries. This shows that AQS adapts effectively based on query deadlines. Moreover, note that AQS provides smaller latencies for the lower priority instances than PQS. This is because AQS has a higher throughput than PQS since it uses preemption only when it is necessary for meeting packet deadlines.

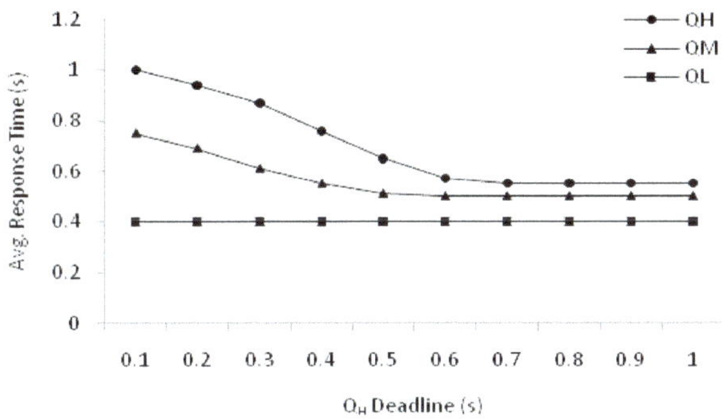

Figure 3. NQS Response Time

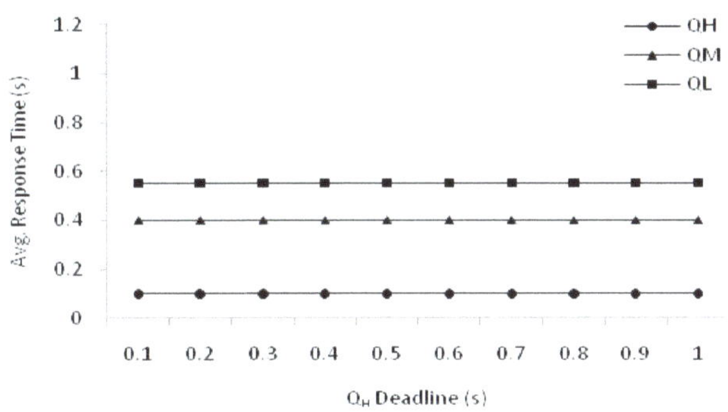

Figure 4. PQS Response Time

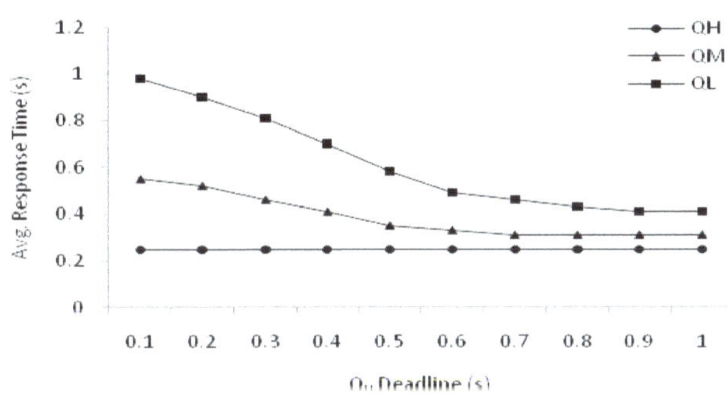

Figure 5. AQS Response Time

7. CONCLUSIONS

This paper presents an adaptive query scheduler (AQS) approach designed to meet the deadline needs of real-time queries in WSN. The proposed AQS approach allows multiple conflict-free queries to execute concurrently in order to achieve maximum throughput, besides it works preemptively to satisfy the real-time queries' deadline constrains. This scheduler combines the benefits of NQS and PQS in that it improves query throughput while meeting all deadlines. The design of this scheduler is purely based on the observation that preemption lowers throughput, and hence it should be used only when necessary for meeting deadlines. When comparing AQS with both NQS and PQS, the simulation results showed that AQS can be adopted to effectively increase network throughput and eliminate query prioritization inversion.

8. ACKNOWLEDGMENTS

My great thanks to Qatar University in supporting an ideal research environment.

9. REFERENCES

[1] Lorincz, K., Malan, D. J., Fulford-Jones, T. R. F., Nawoj, A., Clavel, A., Shnayder, V., Mainland, G., Welsh, M., and Moulton, S. 2004.. Sensor networks for emergency response: Challenges and opportunities. *IEEE Pervasive Computing*, vol. 3(4): pp.16–23.

[2] Mangharam, R., Rowe, A., Suzuki, R., and Rajkumar, R. 2006. Voice over sensor networks. *In RTSS 06.*

[3] Lynch, J. P. and Loh, K. J. 2006. *A Summary Review of Wireless Sensors and Sensor Networks for Structural Health Monitoring*. The Shock and Vibration Digest, 38(2):91–128.

[4] Madden, S., Franklin, M. J., Hellerstein, J. M., and Hong, W. 2005. Tinydb: an acquisitional query processing system for sensor networks. *ACM Trans. Database Syst.*, vol. 30(1): pp. 122–173.

[5] Yao, Y. and Gehrke,, J. 2003. Query processing in sensor networks. I*n proceedings of CIDR.*

[6] Cao, Q., Abdelzaher, T. F., He, T., and Stankovic, J. A. 2005. Towards optimal sleep scheduling in sensor networks for rare-event detection. *In proceedings IPSN.*

[7] Hohlt, B., Doherty, L., and Brewer, E. A., 2004. Flexible power scheduling for sensor networks. *In proceedings IPSN.*

[8] Sadagopan, G. Lu, N., Krishnamachari, B., and Goel, A. 2005. Delay efficient sleep scheduling in wireless sensor networks. *In proceedings of IEEE INFOCOM.*

[9] Sichitiu, M. L. 2004. Cross-layer scheduling for power efficiency in wireless sensor networks. *In proceedings of IEEE INFOCOM.*

[10] Wu, H., Luo, Q., and Xue, W. 2006. Distributed cross-layer scheduling for in-network sensor query processing. *In PerCom. IEEE Computer Society*, pp. 180–189.

[11] Cidon, I. and Sidi, M. 1989. Distributed assignment algorithms for multihop packet radio networks. *IEEE Transactions on Computer.*, vol. 38, no. 10.

[12] Ephremides, A. and Truong, T. 1990. Scheduling broadcasts in multihop radio networks. *IEEE Transactions on Communications*, vol. 38, no. 4.

[13] Ramaswami, R. and Parhi, K. K. 1989. Distributed scheduling of broadcasts in a radio network,. *In proceedings of IEEE INFOCOM.*

[14] Arikan, E. 1983. *Some complexity results about packet radio networks*. NASA STI/Recon Technical Report, vol. 83.

[15] Rajendran, V., Obraczka, K., and Garcia-Luna-Aceves, J. J. 2003. Energyefficient collision-free medium access control for wireless sensor networks. *In proceedings of SenSys.*

[16] Bao, L. and Garcia-Luna-Aceves, J. J. 2001. A new approach to channel access scheduling for ad hoc networks. I*n proceedings of MobiCom.*

[17] Chipara, O., Lu C., and Stankovich, J. A. 2006. Dynamic conflict free query scheduling for wireless sensor networks. *In proceedings of ICNP.*

[18] Chipara, O., Lu, C., and Roman G-C, 2007. Real-time query scheduling for wireless sensor networks. *In proceedings of 28th IEEE International Real-Time Systems Symposium.*

[19] Madden, S., Franklin, M. J., Hellerstein, J. M., and Hong TAG, W. 2002. A tiny aggregation service for ad-hoc sensor networks. In proceedings of OSDI.

[20] Zhou, G., He, T., Stankovic, J. A., and Abdelzaher, T. F. 2005. RID: radio interference detection in wireless sensor networks. In proceedings of IEEE INFOCOM.

Channel Sensing Strategy for Channel Load Estimation

Brian Sung Chul Choi
University of California, Los Angeles
Computer Science Department
schoi@cs.ucla.edu

Mario Gerla
University of California, Los Angeles
Computer Science Department
gerla@cs.ucla.edu

ABSTRACT

The emergence of multi-channel wireless networks and cognitive radio networks has rendered *dynamic channel selection* an important task, and many existing channel selection schemes factor in the amount of wireless activities that take place in each channel, or *channel load*, to achieve load balance and maximize the utilization of wireless resources. In such environments, a monitoring node must sense the channel to estimate the channel load, yet when the node is equipped with a single radio interface, in which case the lone interface must be used for both channel sensing and data communication, there can be only a fixed amount of time allotted to channel sensing. In this paper, we show that a careful scheduling of channel sensing is needed to improve the accuracy of channel load estimation, and based on our findings, we devise a sensing strategy that minimizes the estimation error. Evaluation shows that our scheme can reduce the relative estimation error by as much as 40% in a heavily loaded channel environment.

Categories and Subject Descriptors

C.4 [**Performance of Systems**]: Measurement techniques; C.2.3 [**Network Operations**]: Network monitoring

General Terms

Algorithms, Measurement

Keywords

Channel Load, Wireless Networks

1. INTRODUCTION

In most modern wireless networks, there is a notion of a *channel*, which is often defined to be a frequency range in a wireless spectrum. For example, IEEE 802.11b and g standards define 11 channels in the 2.4-GHz ISM band, each with the bandwidth of 20MHz. Having the spectrum divided into multiple channels lets several networks coexist in the same physical area without interfering with one another.

All nodes that are willing to communicate among themselves must be operating in a common channel. In many traditional infrastructure networks, the channel to use is configured upon network creation, and such a channel selection remains fixed semi-permanently. However, following the success of wireless networks, such a fixed-channel scheme is shown to lead to a bad utilization of wireless resources [3]. Also, the development of more advanced networks such as multi-channel networks [7, 17, 19] and cognitive radio networks [4, 10] has rendered *dynamic channel selection* an important task.

Many existing channel selection schemes rely on the measurement of how much wireless activity exists in each channel, which we define as the *channel load*. Channel load is considered, for example, in achieving channel load balance [6, 11], route management in multi-hop networks [20], and channel selection in open spectrum access networks [4, 10].

Our interest is in the method of sensing a channel and estimating the channel load. In particular, we focus on measuring *airtime utilization*, which is a common channel load metric chosen in many of the previous work [5, 10, 16, 18] and in the IEEE 802.11k standard [1, 15]. Airtime utilization[1] is the fractional duration of time during which the wireless medium is sensed to be busy[2], and it combines transmissions at different rates in a meaningful way and captures wireless traffic that may be external to the network, thus serves as a suitable channel load metric. Airtime utilization is not only used in selecting channels, but in many other contexts such as link rate adaptation [2, 8].

Most existing work that utilizes airtime utilization either uses a secondary scanning radio, with which one can perfectly determine the airtime utilization [4], or use a channel sensing session of an arbitrarily chosen duration, without considering the sensing accuracy [10]. In this paper, we show that channel load estimation can be error-prone and can be made more accurate by choosing the right sensing duration and number of sensing sessions. We then devise a sensing strategy, such that the accuracy of channel load estimation is maximized. We make the following contributions in this paper:

- We characterize how channel sensing parameters, namely sensing session duration and the number of sensing

[1] We use the terms *airtime utilization* and *channel load* interchangeably in this paper.

[2] For example, a node may conclude the medium is busy if the received signal strength is above a certain threshold.

MSWiM'11, October 31–November 4, 2011, Miami, Florida, USA.

sessions, affect the accuracy of channel load measurements both analytically and via simulation, using a bandwidth-limited model in which a wireless node is equipped with only one radio interface and can tune to one channel at a time.

- We formulate an optimization problem with which channel sensing parameters can be chosen such that the accuracy of channel load estimation is maximized, given the total sensing duration and channel switching delay as contraints, and evaluate the properties and performance of this strategy.

In Section 2, we present the system model that is used in this paper, and we discuss and characterize how different sensing parameters affect the accuracy of sensing results in Section 3. Section 4 overviews the sensing strategy we adopt based on the analysis from Section 3, and we evaluate the properties and performance of this strategy in Section 5. We review the related work in Section 6 and conclude in Section 7.

2. SYSTEM MODEL

2.1 Packet Arrival Process

Let $C = \{1, 2, \cdots\}$ be the set of all channels in the system. We assume every packet transmission in the system incurs the same transmission delay, T_p, for simplicity. We define packet arrivals to each channel to be packet arrivals to all user nodes using the channel in a collision domain. As commonly done in other work, we use a Poisson process of the rate λ_i to model the packet arrivals at each node i. Therefore, the packet arrivals to the channel with k users can also be viewed as a Poisson process of the rate λ, where $\lambda = \sum_{i=1}^{k} \lambda_i$.

Let $N(t)$ be the number of packets that have arrived in the time range $[0, t]$ in a certain channel, which is a Poisson process of rate λ. Then for any $t, s > 0$,

$$P_n(t) = Pr(N(t+s) - N(s) = n) = \frac{e^{-\lambda t}(\lambda t)^n}{n!} \quad (1)$$

where $P_n(t)$ is the probability of n packets arriving in the duration of length t. Given this definition, the probability of no packet arriving in $[0, t]$ is:

$$P_0(t) = e^{-\lambda t} \quad (2)$$

We assume perfect media access control, where there is no protocol-incurred delay; that is, when there are two packets in the *channel queue* (a virtual queue containing all packets queued at k nodes using the same channel) waiting to be transmitted, the second packet's transmission will immediately follow the first one without any delay or exchange of control messages.

2.2 Sensing Models

2.2.1 Ideal sensing model

In the *ideal sensing model*, each node has a dedicated radio that continuously monitors all $|C|$ channels in the system simultaneously. In this case, there is no extra sensing overhead that affects network performance adversely. If we let T_s denote the duration of a *sensing session*, in the ideal sensing model, we can say $T_s >> T_p$. The airtime utilization

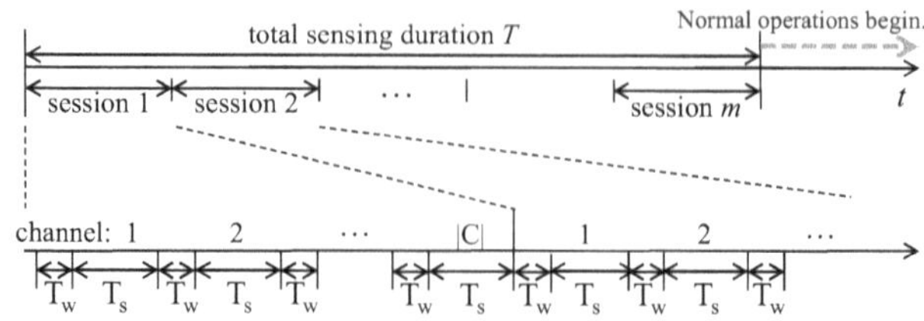

Figure 1: Channel scanning of total sensing duration T, split into m sessions.

thus converges to $\min(1, \frac{\mathbb{E}[N(T_s)]\cdot T_p}{T_s})$; i.e., it is the fractional amount of busy time, which is equal to the number of packets multiplied by the transmission delay of each packet, in a lengthy sensing session. Using the packet arrival model presented in the previous subsection, $\mathbb{E}[N(T_s)] = \lambda T_s$. Therefore, we define the *true channel load* in channel c, denoted A_c, as:

$$A_c \triangleq \min(1, \frac{\lambda T_s T_p}{T_s}) = \min(1, \lambda T_p) \quad (3)$$

A_c is capped at 1 for an obvious reason. When $A_c = 1$, we say the channel is *saturated*.

2.2.2 Bandwidth-limited model

In reality, however, the ideal sensing model is difficult to realize. Due to hardware limitations, a radio can tune to a limited number (often 1) of channels at a time. Further, a wireless node usually does not have the luxury of having a dedicated interface for monitoring channels. Therefore, in a more realistic, *bandwidth-limited sensing model*, we constrain the environment by limiting a radio interface to tune to one channel at a time, and assume that this radio is also used for data communication. To compute the channel load, therefore, a node must halt its communication, monitor the channel for a finite duration T_s and compute a *channel load estimate* (or simply an *estimate*, denoted $\bar{A}_c$), and then continue its communication after this channel sensing procedure is complete. $\bar{A}_c$ is the fraction of busy periods during the sensing session of T_s.

As a result, T_s can no longer be indefinitely large, because T_s directly translates to sensing overhead (i.e., longer the sesing session, less time there is for actual data communication). Yet reducing T_s may hurt the accuracy of channel load estimation. In the extreme case when $T_s << T_p$, it may declare the channel to be completely occupied if it overlaps with a packet transmission, even if this is a rare packet transmission in the channel, or empty if no packet is being transmitted for the short duration of T_s, even for a heavily-loaded channel. This naturally leads to sensing duration vs. sensing accuracy trade-off, and one of the goals in this paper is to characterize this trade-off and learn to strike the balance between the two.

In this model, therefore, channels in C are *scanned* individually, one after another, as illustrated in Figure 1. Each sensing *session* lasts T_s and incurs a *channel switching delay* of T_w to switch from one channel to another. This sequential monitoring of channels is repeated m times, and the $\bar{A}_c$ values obtained from individual sensing sessions are averaged to enhance estimation accuracy.[3] The *total channel sens-*

[3]By the definition of a Poisson process, the number of events

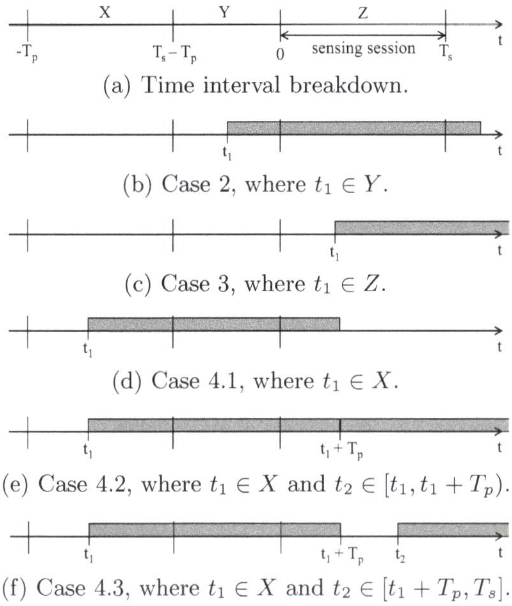

(a) Time interval breakdown.

(b) Case 2, where $t_1 \in Y$.

(c) Case 3, where $t_1 \in Z$.

(d) Case 4.1, where $t_1 \in X$.

(e) Case 4.2, where $t_1 \in X$ and $t_2 \in [t_1, t_1 + T_p)$.

(f) Case 4.3, where $t_1 \in X$ and $t_2 \in [t_1 + T_p, T_s]$.

Figure 2: Different scenarios that can occur during a sensing session of length $T_s \leq T_p$.

ing duration T, which is spent to sense all channels in C, is therefore $m|C|(T_w + T_s)$.

2.3 Computing Expected $\bar{A}_c$

In this subsection, we analytically compute the expected estimated channel load, $\mathbb{E}[\bar{A}_c]$. One can easily see $\mathbb{E}[\bar{A}_c] = \lambda T_p$ if $T_s >> T_p$, converging to the true channel load.

We focus on the expected value of an estimate for small T_s values, particularly for $0 \leq T_s \leq T_p$. With a small T_s, a monitoring node may not observe any packet transmission, observe a packet transmission partially, or observe a packet transmission for the entire sensing session. Suppose a sensing session takes place in the time interval $[0, T_s]$, and no packet is queued at time $-T_p$. We can list possible cases as follows:

Case 1: In the first case, there may be no packet arrival in the interval $[-T_p, T_s]$ (i.e., no prior packet transmission at 0 and no packet arrives in $[0, T_s]$), such that $A_c = 0$. This occurs with the probability $P_0(T_p + T_s)$.

For the remaining cases, we divide the time interval $[-T_p, T_s]$ into three subintervals X, Y, and Z, where $X = [-T_p, T_s - T_p)$, $Y = [T_s - T_p, 0)$, and $Z = [0, T_s]$, as shown in Figure 2(a). Also, let t_1 denote the time at which the first packet arrives in $[-T_p, T_s]$.

Case 2: If $t_1 \in Y$, then $\bar{A}_c = 1$ (i.e., the channel appears saturated). This occurs with the probability $P_0(T_s) \cdot (1 - P_0(T_p - T_s))$ (i.e., no packet arrival in X, and at least one packet arrival in Y).

Case 3: If $t_1 \in Z$, then $\bar{A}_c = \frac{T_s - t_1}{T_s}$. Note that $1 - P_0(t)$ is the probability of at least one packet arriving in $[0, t]$, which is equivalent to $Pr(t_1 \leq t)$, thus serves as the cumulative distribution function for the distribution $p_1(t)$, which we define as the probability of the first packet arriving at t since the time 0. Therefore, we differentiate $1 - P_0(t)$ to obtain:

$$p_1(t) = \lambda e^{-\lambda t} \tag{4}$$

which happens to be exponentially distributed. With $p_1(t)$ defined, we can compute the probability of Case 3 occurring to be $P_0(T_p) \cdot p_1(t_1)$.

Case 4: The final case occurs if $t_1 \in X$, which we can divide into three subcases:

Case 4.1: No more packets arrive afterwards, in which case $\bar{A}_c = \frac{t_1 + T_p}{T_s}$. This occurs with the probability $p_1(t_1 - (T_p)) \cdot P_0(T_s - t_1)$.

Case 4.2: The next packet arrives at $t_2 \in [t_1, t_1 + T_p]$ (i.e., during the first packet transmission), in which case $\bar{A}_c = 1$. This occurs with the probability $p_1(t_1 - (-T_p)) \cdot (1 - P_0(T_p))$.

Case 4.3: The next packet arrives at $t_2 \in (t_1 + T_p, T_s]$ (i.e., after the first packet transmission), in which case $\bar{A}_c = \frac{(t_1 + T_p) + (T_s - t_2)}{T_s}$. This occurs with the probability $p_1(t_1 - (-T_p)) \cdot P_0(T_p) \cdot p_1(t_2 - (t_1 + T_p))$.

One can easily show the exhaustiveness of these cases by summing the probabilities. See Figures 2(b) to 2(f) for the illustration of each case listed above.

We now combine the above cases to calculate $\mathbb{E}[\bar{A}_c]$ as a function of T_s and T_p, given that $0 \leq T_s \leq T_p$, as follows:

$$\begin{aligned}
&\mathbb{E}[\bar{A}_c] = \\
&e^{-\lambda T_s} \cdot (1 - e^{-\lambda(T_p - T_s)}) \cdot (1) \\
&+ \int_Z e^{-\lambda T_p} \cdot \lambda e^{-\lambda t_1} \cdot (\frac{T_s - t_1}{T_s})\, dt_1 \\
&+ \int_X \lambda e^{-\lambda(t_1 + T_p)} \cdot e^{-\lambda(T_s - t_1)} \cdot (\frac{t_1 + T_p}{T_s})\, dt_1 \\
&+ \int_X \lambda e^{-\lambda(t_1 + T_p)} \cdot (1 - e^{-\lambda T_p}) \cdot (1)\, dt_1 \\
&+ \int_X \int_{t_1 + T_p}^{T_s} \lambda e^{-\lambda(t_1 + T_p)} \cdot e^{-\lambda T_p} \cdot \lambda e^{-\lambda(t_2 - (t_1 + T_p))} \cdot \\
&\qquad \frac{(t_1 + T_p) + (T_s - t_2)}{T_s}\, dt_2 dt_1
\end{aligned} \tag{5}$$

We can also obtain the variance (and the standard deviation) by computing $\mathbb{E}[\bar{A}_c^2]$ (we can simply rewrite the above expression with each $\bar{A}_c$ value squared), the presentation of which we omit. It turns out, as presented in the following section, the analysis using (5) is sufficient to reveal the properties we are interested in. We use Monte Carlo simulation for T_s values greater than T_p, where the number of scenarios grows to the point where writing a closed-form expression like (5) is no longer practical, to confirm the analysis results.

3. ANALYSIS

3.1 Basic Results

For all analyses presented in this section, we fix T_p at 1. (Note that when $T_p = 1$, the true channel load A_c becomes $\min(1, \lambda)$). We only consider λ values between 0 and 1, because the channel is saturated with traffic for λ's greater than 1, which is all the information that the monitoring node is interested in; that is, figuring out the actual $\lambda \geq 1$ is not

in two different time intervals $[t_1, t_2]$ and $[t_3, t_4]$, $N(t_2) - N(t_1)$ and $N(t_4) - N(t_3)$, are independently distributed. We thus assume that $\bar{A}_c$ values from different sensing sessions are independent, even though there are cases in which they are dependent when more than one packet is queued in the channel queue during the sensing sessions.

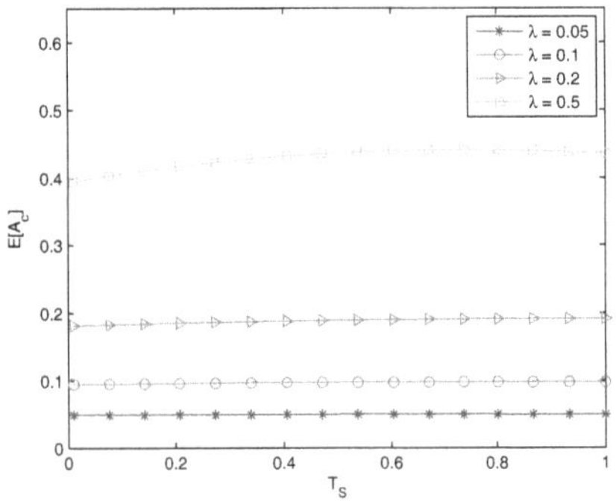

(a) $\mathbb{E}[\bar{A}_c]$ for various λ values computed using (5).

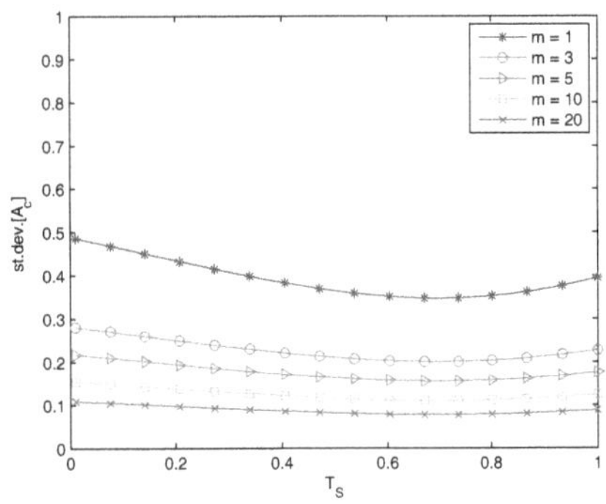

(b) $St.dev.(\bar{A}_c)$ for various m values, $\lambda = 0.5$.

Figure 3: $\mathbb{E}[\bar{A}_c]$ and $St.dev.(\bar{A}_c)$.

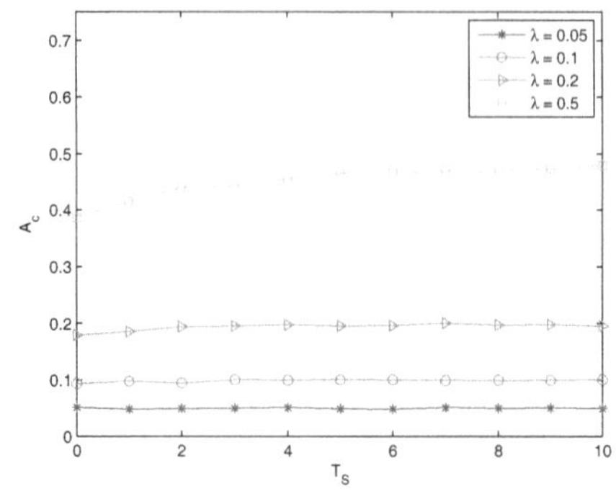

(a) Single sensing session ($m = 1$) with varying T_s.

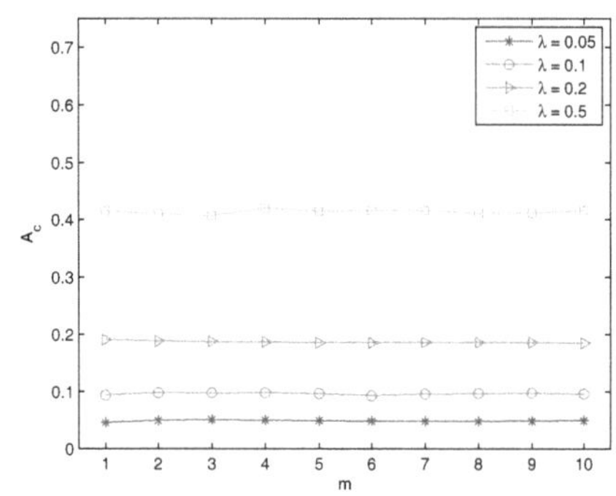

(b) m sessions of $T_s = 1$ and data fusion.

Figure 4: Single vs. multiple sessions.

of the monitoring node's interest. It should also be noted that such complete saturation does not occur often [16].

Figure 3(a) plots $\mathbb{E}[\bar{A}_c]$ computed using (5) for four different λ values, 0.05, 0.1, 0.2, and 0.5, for $0 \leq T_s \leq T_p$ and $m = 1$. For low λ values, $\bar{A}_c$ is expected to be very close to A_c even for a small T_s. Yet as λ increases, we see that it starts to underestimate. When $\lambda = 0.5$, for instance, it is underestimated by 14% when $T_s = T_p$. This is because for a higher λ, there are more packet transmissions, thus there is a better chance that the channel appears saturated during the sensing session, possibly with some packets getting queued (i.e., the session session coincides with a temporary "burst" of packet arrivals). As $\bar{A}_c$ does not capture the queueing effect and is capped at 1 for all these saturation cases regardless of how many packets are queued during the sensing session, it results in the reduction of the value of $\mathbb{E}[\bar{A}_c]$. One can see, however, that $\mathbb{E}[\bar{A}_c]$ approaches the λ value as T_s increases. Apparently, the only remedy for such underestimation is to lengthen the duration of the sensing session, T_s, such that the effect of such bursty packet arrivals is amortized.

Although $\mathbb{E}[\bar{A}_c]$ is close to A_c, there is a high variation in the estimates. We plot the standard deviation of $\bar{A}_c$ values with λ fixed at 0.5 in Figure 3(b). The estimate obtained in a single sensing session ($m = 1$) can be highly variant, with the standard deviation being higher than 0.4 for most T_s values, making $\bar{A}_c$ almost as variant as a random value. Such variation can be reduced with a longer T_s, or by taking multiple estimates ($m > 1$) and averaging them. Let $\bar{A}_c(i)$ be the ith estimate of the channel load. Mathematically,

$$St.dev.(\frac{1}{m}\sum_{i=1}^{m}\bar{A}_c(i)) = \frac{St.dev.(\bar{A}_c(1))}{\sqrt{m}} \quad (6)$$

where we assume that $\bar{A}_c(i)$'s are independently and identically distributed. This shows that an increase in m helps reduce variance and makes $\bar{A}_c$ converge to A_c, but with diminishing returns. Figure 3(b) shows the standard deviation for various m values.

3.2 Making Estimates Less Variant

In the previous subsection, we have shown that even for a small sensing duration T_s, we can expect to correctly estimate the channel load, yet the estimates can be highly variant. Two parameters we can adjust to tune this variation are T_s and m, and our goal in this subsection is to examine the effect of each parameter.

In Figure 4, we show plots of average $\bar{A}_c$ that we have obtained from Monte Carlo simulations – each data point is obtained from 10000 simulation runs. In each simulation run, we use a Poisson spike generator for packet generation, with a FIFO channel queue, and compute the estimated channel load. In Figure 4(a), we plot the average $\bar{A}_c$ obtained by varying T_s from 0 to 10, with m fixed at 1. It is confirmed that as T_s is increased, the underestimation problem becomes less visible, even for a high λ value like 0.5. In another case (Figure 4(b)), we fix each session to last T_p (i.e., $T_s = T_p = 1$), and vary m from 1 to 10. When T_s is relatively short as in this case, the underestimation problem

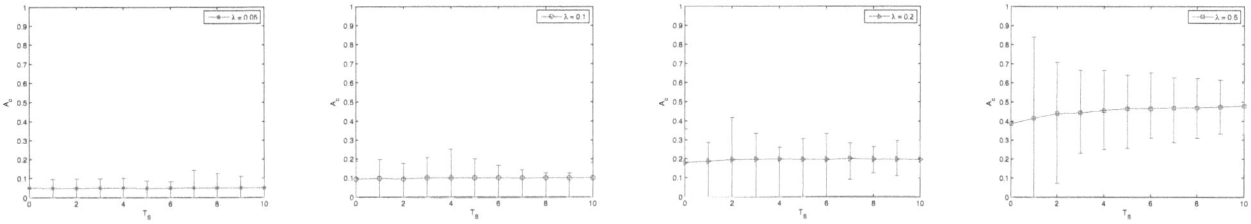

(a) $\lambda = 0.05$ (b) $\lambda = 0.1$ (c) $\lambda = 0.2$ (d) $\lambda = 0.5$

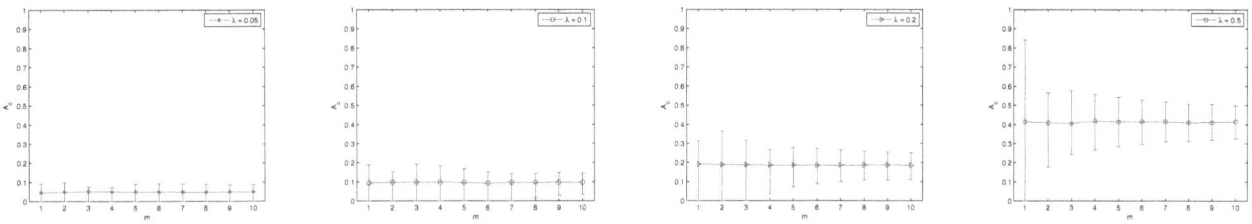

(e) $\lambda = 0.05$ (f) $\lambda = 0.1$ (g) $\lambda = 0.2$ (h) $\lambda = 0.5$

Figure 5: (a) to (d): average $\bar{A}_c$ values obtained in a single sensing session with varying T_s, (e) to (h): average $\bar{A}_c$ values obtained in multiple sensing sessions, each of length $T_s = 1$.

is not mitigated at all even with a high m, for the reason we discussed in the previous subsection.

While increasing m does not help correct the underestimation problem, it helps reduce the variation in estimates significantly. In Figure 5, we re-plot the graphs of $\bar{A}_c$ averages separately for different λ's with error bars. In Figures 5(a) to (d), we fix m at 1 and vary T_s, and in Figures 5(e) to (h), we fix T_s at $T_p = 1$ and vary m. For each data point, we show the 25th-percentile and 75th-percentile $\bar{A}_c$ values to show how much the estimates vary. It is shown that for low λ values, the variation in estimates is relatively low, as the channel is idle most of the time, not making much room for under or overestimation. As λ is increased, however, we begin to see more variation in the estimates, such that when $\lambda = 0.5$ and $T_s < 3$, we see that estimates are spread widely, stretching to more than half of the entire airtime utilization (0 to 1) range.

The two strategies of lengthening T_s and using $m > 1$ both reduce the amount of spread, but evidently the latter strategy is more effective. For example, in Figure 5(h), we see that the 25th and 75th-percentile values fall within ± 0.13 of the average for $m \geq 4$ (after which the benefits come with diminishing returns as expected). For the equivalent T_s in the single session strategy, the offsets stretch as much as ± 0.25, as shown in Figure 5(d).

In summary, we make the following observations in our analysis:

- The total sensing time spent for each channel must be larger than T_p for the estimates to be reasonably close to the true channel load.
- For a fixed total sensing duration T, increasing T_s (or reducing m) results in a more accurate average $\bar{A}_c$, as increasing m (or reducing T_s) results in the underestimation of A_c.
- The variation in estimates more quickly diminishes with an increased m than with an increased T_s.

4. CHOOSING T_s AND m

Based on the observations that we have made in the previous section on the expected $\bar{A}_c$ and its variation, we devise a sensing strategy, where the sensing parameters, namely T_s and m, are chosen to minimize the channel load estimation error.

Here, we assume that the total sensing duration T and channel switching delay T_w are given as constraints. T depends on how much time the underlying communication protocol allows for channel sensing, and is directly related to sensing overhead. How T is chosen and how often channel sensing is performed to adapt to the changing environment depend on how much accuracy the network requires and how dynamic the channel environment is, respectively, but these discussions are out of this paper's scope.

Let an error function $\epsilon(m, T_s, \lambda)$ be defined as follows:

$$\epsilon(m, T_s, \lambda) = \sqrt{\frac{\sum_{i=1}^{n}(A_c - \bar{A}_c(i))^2}{n}} \quad (7)$$

where $A_c = \min(1, \lambda T_p)$ as given before, n is the number of estimates, and $\bar{A}_c(i)$ is the ith estimate. In words, $\epsilon(m, T_s, \lambda)$ is the root mean square deviation of estimates given m, T_s, and λ, and captures both the underestimation effect and the variation in estimates. Given $\epsilon(\cdot)$, we can find the optimal m^* and its corresponding T_s^* as follows.

$$(m^*, T_s^*) = \underset{(m, T_s)}{\arg\min} \int_0^\infty p(\lambda) \cdot \epsilon(m, T_s, \lambda)\, d\lambda \quad (8)$$

where $p(\lambda)$ is a predetermined distribution of λ for channels in C. In words, we want to find a pair of m and T_s values that minimize the expected error in the estimates. $p(\lambda)$ can easily be obtained by keeping the history of channel load estimates and computing statistics out of it.

Given the constraint $T = m|C|(T_w + T_s)$, which we can solve for T_s as follows,

$$T_s = \frac{T - m|C|T_w}{m|C|} \quad (9)$$

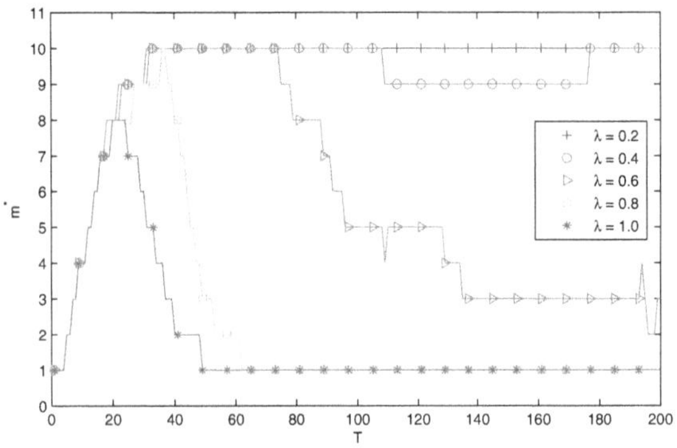

Figure 6: m^* values for various T.

we can simplify (8) to involve only one parameter, m:

$$m^* = \arg\min_m \int_0^\infty p(\lambda) \cdot \epsilon(m, \frac{T - m|C|T_w}{m|C|}, \lambda)\, d\lambda \quad (10)$$

As it is implausible to derive ϵ analytically, we empirically derive it with discretized λ values ($\lambda_1 < \lambda_2 < ...$) using Monte Carlo simulations. The discrete approximation of (10) is then:

$$\bar{m^*} = \arg\min_m \sum_{\lambda_i} p(\bar{\lambda}_j) \cdot \epsilon(m, \frac{T - m|C|T_w}{m|C|}, \lambda_j) \quad (11)$$

where $p(\bar{\lambda}_j)$ is a probability mass function defined as $p(\bar{\lambda}_j) = \int_{\lambda_{j-1}}^{\lambda_j} p_\lambda \, d\lambda$.

5. EVALUATION

In this section, we evaluate the parameter selection scheme from the previous section. For performance evaluation, we have empirically computed $\epsilon(\cdot)$ for various m, T_s, and λ values using (7) by running 10000 samples from simulation (i.e., $n = 10000$) and created an *error table*, such that $\epsilon(\cdot)$ computation is a simple table lookup, with T_s and λ approximated to the nearest discretized values. We have used T_p = 1ms and T_w = 0.2ms in all evaluation presented in this section.

We first examine how $\bar{m^*}$, the optimal number of sensing sessions that T is split into to minimize the estimation error, obtained by our strategy (using (11)), changes based on the actual channel load. We focus on $\bar{m^*}$ values only, because T_s^* is simply a function of $\bar{m^*}$ with a fixed T (as in (9)). We generate samples of λ values from a Gaussian distribution with the standard deviation of 0.1 and a varying mean, such that a channel load is assumed to be relatively static over time with a small amount of jitter, while varying T from 0 to 200ms. We have set 10 as the maximum m value.

The results are shown in Figure 6. We see that for a small T, our optimization decides that it is advantageous to perform as many sensing sessions as possible to make $\bar{A}_c$ converge to A_c quickly, as there is not enough room to increase T_s to begin with. ($\bar{m^*}$ increases proportionally to T with very small T values, since increasing m also leads to more channel switching overhead, leaving no time for sensing.) For lightly loaded channels ($\lambda \leq 0.4$), the underestimation problem is not serious, thus $\bar{m^*}$ is set high for most T values.

When the channels are more heavily loaded ($\lambda \geq 0.6$), the underestimation problem becomes more pronounced, such that it is better to increase the duration of each sensing session than to increase the number of sensing sessions (plots for $\lambda = 0.8$ and 1.0 are stacked onto one another at $m = 1$ for most T values). This is demonstrated in the graph – as T is increased beyond a certain level, our strategy selects a low $\bar{m^*}$ for high λ's, such that it can balance the amount of underestimation in heavy-load cases and the variation in the estimates.

We next quantify the amount of benefits we get from applying our sensing strategy. As done in the previous analysis, we assume the channel load is static with a random jitter, and have obtained each data point after performing 10000 simulation runs and avearging. In Figure 7, we give the plots of average errors in the estimates obtained by using various numbers of sensing sessions (m = 1 to 10), given a varying channel load (λ = 0.4, 0.6, 0.8, and 1.0). T is set to 200ms.

When $\lambda = 0.4$, it is apparent that the error is reduced as m is increased, and compared to the average error one gets without applying multiple sensing sessions (i.e., $m = 1$), the error is reduced by 13.4%. When $\lambda = 0.6$, however, there is no huge advantage either way – this is the point where the underestimation error and the variation in estimates strike a balance. As the channel load is further increased, we see the reverse effect, where performing multiple sensing sessions is no longer helpful. For λ of 0.8 and 1.0, using the optimal $\bar{m^*}$ reduces the error by 29.6% and 39.7%, respectively, when compared to the worst case. However, regardless of the optimization we perform, the estimation error tends to increase as the channel load increases, suggesting that in an environment where channels are consistently heavily loaded, the total sensing duration T must be increased accordingly to reduce the estimation error.

6. RELATED WORK

With the success of wireless communications, it has become important to reuse the wireless resources, as there is more than one network trying to access the wireless spectrum in most environments [3]. Channel division serves this purpose, and wireless networks that can utilize more than one channel, or *multi-channel networks*, have drawn much research interests [12, 17, 21].

At the same time, a new wireless access paradigm called *white space networking* has emerged recently. In this scheme, when a portion of a band that is dedicated to certain services or devices (*primary* users) is found to be idle, such a portion (called a *white space*) can be opportunistically accessed by the users that are not licensed to use the band (*secondary* users) as long as they do not interfere with the primary users. This paradigm naturally creates a dynamic multi-channel environment, consisting of channels that may not be contiguous or of the same size, the availability of which depends on the changing behavior of primary users. In white space networking, therefore, secondary users use cognitive radios [14] to constantly monitor the spectrum, in order to detect primary users and other secondary users, to identify white spaces and share them with other secondary users.

In the existing work in multi-channel and white space networking, channels are often dynamically configured based on the observed traffic load on each channel, for which accurate channel load estimation is critical. For example, Bahl et al. [4] have used channel load to compute the *expected share* of the channel, which is the metric that secondary

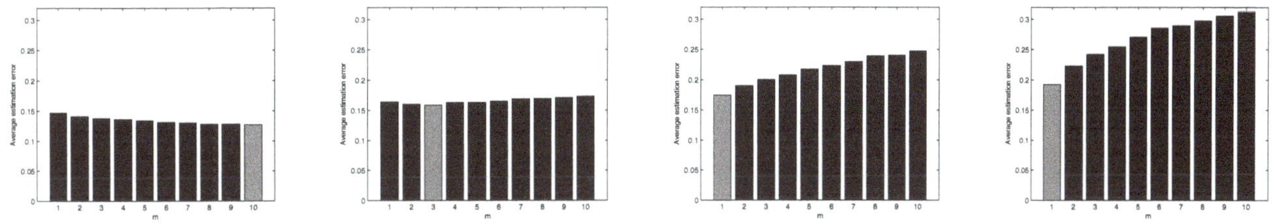

(a) $\lambda = 0.4$ (b) $\lambda = 0.6$ (c) $\lambda = 0.8$ (d) $\lambda = 1.0$

Figure 7: Average errors in the estimates for varying channel load. The red bar in each plot is the average error we get when the optimal number of sensing sessions, $\bar{m}^*$, is used.

users use to choose the white space where there exists the least amount of activity. Channel load has been used for purposes other than channel assignment, such as identifying the cause of packet losses, which is an important for rate adaptation. For instance, Jardosh et al. [8] and Acharya et al. [2] have found airtime utilization to be a useful metric in deciding whether packet losses are due to congestion or random wireless losses.

Most work that requires channel load measurements, however, assumes either that a wireless node is equipped with more than one radio interface [4,6,20], or that the channel load can be accurately measured by monitoring the wireless spectrum for a short duration [5,10], the assumptions that we have relaxed in this work.

Spectrum monitoring for channel load estimation is similar to spectrum sensing performed by cognitive radios, where the goal is to *detect* the presence of primary users. In the cognitive radio domain, there have been a lot of proposals in the topic of scheduling spectrum sensing to meet detectability requirements [9] and to maximize the throughput while still meeting such requirements [13,22]. The channel load estimation problem explored in this paper differs from the cognitive radio problem in that it is about quantifying network activity, as opposed to answering a yes-or-no question of device presence.

7. CONCLUSIONS AND FUTURE WORK

In this paper, we have shown how careful scheduling of channel sensing is important in accurately assessing the channel load, which is the amount of wireless activity that takes place in a channel, presented a method of finding optimal sensing parameters, and showed its effectiveness. We have shown via simulation that our scheme can reduce the estimation error as much as 30.7% in a heavy load channel environment.

In this paper, we have observed that when channels are more heavily loaded, the average error in the load estimates increases regardless of how one schedules the channel sensing. This observation naturally leads to our next plan of devising a method of scheduling channel sensing given a specific bound on the average error. Further, we admit that our work serves rather as a cornerstone for future work, which must deal with a more realistic wireless environment where the channel load varies over time. We are interested in characterizing the actual channel behavior via measurements, verify the findings in this paper in various environments where the simplifying assumptions we made in this work (e.g., Poisson packet arrivals) may be invalid, and figure out how often and how long channel sensing must be performed.

8. REFERENCES

[1] IEEE 802.11k-2008 – Amendment 1: Radio resource measurement of wireless LANs.

[2] P. A. Acharya, A. Sharma, E. M. Belding, K. C. Almeroth, and K. Papagiannaki. Rate adaptation in congested wireless networks through real-time measurements. *IEEE Transactions of Mobile Computing*, June 2010.

[3] A. Akella, G. Judd, S. Seshan, and P. Steenkiste. Self-management in chaotic wireless deployments. *Wireless Networks*, 13(6), December 2007.

[4] P. Bahl, R. Chandra, T. Moscibroda, R. Murty, and M. Welsh. White space networking with Wi-Fi like connectivity. In *ACM SIGCOMM'09*, Barcelona, Spain, August 2009.

[5] B. S. C. Choi, H. Im, K. C. Lee, and M. Gerla. CCH: Cognitive channel hopping in vehicular ad hoc networks. In *IEEE VTC'11-Fall*, San Francisco, CA, September 2011.

[6] W.-C. Hung, K. Law, and A. Leon-Garcia. A dynamic multi-channel MAC for ad hoc LAN. In *21st Biennial Symposium on Communications*, Kingston, Canada, June 2002.

[7] N. Jain, S. R. Das, and A. Nasipuri. A multichannel CSMA MAC protocol with receiver-based channel selection for multihop wireless networks. In *ICCCN'01*, Scottsdale, AZ, October 2001.

[8] A. P. Jardosh, K. N. Ramachandran, K. C. Almeroth, and E. M. Belding-Royer. Understanding congestion in IEEE 802.11b wireless networks. In *IMC'05*, Berkeley, CA, October 2005.

[9] H. Kim and K. G. Shin. In-band spectrum sensing in IEEE 802.22 WRANs for incumben protection. *IEEE Transactions on Mobile Computing*, 9(12):1766–1779, December 2010.

[10] W. Kim, S. Y. Oh, M. Gerla, and J.-S. Park. CoCast: Multicast mobile ad hoc networks using cognitive radio. In *Milcom'09*, Boston, MA, October 2009.

[11] P. Kyasanur and N. H. Vaidya. Routing and interface assignment in multi-channel multi-interface wireless networks. In *IEEE WCNC'05*, New Orleans, LA, March 2005.

[12] P. Kyasanur and N. H. Vaidya. Capacity of multichannel wireless networks: Impact of channels,

interfaces, and interface switching delay. Technical report, UIUC Department of Electrical and Computer Engineering, August 2006.

[13] Y.-C. Liang, Y. Zeng, E. C. Peh, and A. T. Hoang. Sensing-throughput tradeoff for cognitive radio networks. *IEEE Transactions on Wireless Communications*, 7(4):1326–1336, April 2008.

[14] J. Mitola and G. Q. Maguire. Cognitive radio: making software radios more personal. *IEEE Personal Commnuications*, 6(4):13–18, August 1999.

[15] E. A. Panaousis, P. A. Frangoudis, C. N. Ververidis, and G. C. Polyzos. Optimizing the channel load reporting process in IEEE 802.11k-enabled WLANs. In *IEEE LANMAN'08*, Chij-Napoca, Transylvania, September 2008.

[16] R. Raghavendra, J. Padhye, and R. Mahajan. Wi-Fi networks are underutilized. Technical report, Microsoft Research, August 2009.

[17] A. Raniwala and T. Chiueh. Architecture and algorithms for an IEEE 802.11-based multi-channel wireless mesh network. In *IEEE Infocom'05*, Miami, Florida, March 2005.

[18] M. Rodrig, C. Reis, R. Mahajan, D. Wetherall, and J. Zahorjan. Measurement-based characterization of 802.11 in a hotspot setting. In *ACM SIGCOMM'05 Workshops (E-WIND)*, Philadelphia, PA, August 2005.

[19] J. So and N. H. Vaidya. Multi-channel MAC for ad hoc networks: Handling multi-channel hidden terminals using a single transceiver. In *ACM MobiHoc'04*, Roppongi, Japan, September 2004.

[20] J. So and N. H. Vaidya. Load-balancing routing in multichannel hybrid wireless networks with single network interface. *IEEE Transactions on Vehicular Technology*, 56(1):342–348, January 2007.

[21] A. P. Subramanian, H. Gupta, and S. R. Das. Minimum-interference channel assignment in multi-radio wireless mesh networks. In *IEEE SECON'07*, San Diego, CA, June 2007.

[22] X. Zhou, Y. G. Li, Y. H. Kwon, and A. C. K. Soong. Detection timing and channel selection for periodic spectrum sensing in cognitive radio. In *IEEE GlobeCom 2008*, New Orleans, LA, November 2008.

Simulation Models for the Performance Evaluation of Spectrum Sharing Techniques in OFDMA Networks

Luca Anchora*, Marco Mezzavilla§, Leonardo Badia*†§, Michele Zorzi†§
* IMT Institute for Advanced Studies, Piazza S. Ponziano 6, 55100 Lucca, Italy
† Consorzio Ferrara Ricerche, via Saragat 1, 44122 Ferrara, Italy
§ Dept. of Information Engineering (DEI), via Gradenigo 6B, 35131 Padova, Italy
luca.anchora@imtlucca.it, {mezzavil,badia,zorzi}@dei.unipd.it *

ABSTRACT

Cooperation in wireless networks is an important means to improve the resource utilization efficiency. It finds an interesting application in the context of spectrum sharing, where multiple wireless users put their licensed frequency bands in common in order to achieve a better resource usage. Due to the complexity of the problem, mathematical analysis is typically focused on simple scenarios. However, we believe that, in order to obtain a concrete proof of concept of the sharing paradigm, it is mandatory to assess its performance in realistic situations, i.e., with a larger number of nodes and a wider range of applications. Therefore, the support of a proper simulation environment is fundamental for high-quality applied research. In this paper we present and evaluate an original extension of the well known ns-3 network simulator which focuses on multiple operators of the most up-to-date cellular scenarios, i.e., the Long Term Evolution of UMTS employing OFDMA multiplexing. We describe the software architecture that enables the spectrum sharing and, in particular, allows operators to interact in order to agree on a spectrum division. A sample sharing policy is given as well, and a detailed simulation campaign is run to validate the proposed architecture, assess its efficiency, and evaluate the simulation time related to scenarios with an increasing number of nodes.

Categories and Subject Descriptors

C.2.1 [**Network Architecture and Design**]: Network communications, wireless communication; C.2.2 [**Network Protocols**]: Protocol verification; I.6.5 [**Model Development**]: modeling methodologies; I.6.8 [**Types of Simulation**]: Discrete event

General Terms

Design; Performance; Verification

Keywords

Spectrum sharing; ns-3; LTE; multi-operator cellular network.

*This work was supported by the European Commission under the FP7 EU project SAPHYRE, grant agreement no. 248001.

MSWiM'11, October 31–November 4, 2011, Miami, Florida, USA.

1. INTRODUCTION

The application of game theoretic concepts to wireless networking is becoming widespread within the relevant scientific literature. Several topics related to radio resource management have been approached by means of game theoretic tools which, until few years ago, were considered typical of economics, but have by now become common knowledge among wireless engineers [17]. One of the main reasons for the success of these techniques in the study of wireless networks is that game theory is well suited to approach problems involving both multiple players with different objectives and a general scarcity of resources, characteristics which are often found in radio networks. Actually, spectrum availability problems are due more to the inefficient usage by the licensed users rather than to a real lack of available frequencies [13]. Thus, it is the selfishness of the actors that makes radio resource access problems even more acute, which motivates a proper game theoretic analysis where *egoistic* players are given incentives to *cooperate*.

In particular, early attempts at using game theory within wireless scenarios were mainly considering *cognitive networks*, a model that, since its early characterization [19], involves a distinction between primary and secondary users; these terms are also frequently encountered in game theoretic duopoly analysis à la Stackelberg [21], i.e., involving a player moving first, and another reacting subsequently. A more recent trend of analysis involves a general paradigm of collaborative usage of the wireless spectrum, where different agents are no longer framed as owner or opportunistic unlicensed (i.e., primary and secondary, respectively) users, but rather an egalitarian approach is used. In this context, the focus is more on similarly-minded players, which can be thought of as network operators, which desire to share a portion, or possibly all, of their licensed frequencies for common wireless access, if they envision a gain in doing so, which can be a larger number of users served, a wider network coverage, or any similar benefit. When realistic models for the physical layer are considered, such a collaborative approach is found to be advantageous over competition among operators due to multi-user and frequency diversity [14].

Several studies hinted that spectrum sharing may be beneficial for all the involved players if a collaborative access to the wireless resource is achieved [8, 15, 16]. However, the practical scenario considered in the analysis is often limited to small networks with few transmitter-receiver pairs, most of the times just two, i.e., a total of four nodes. Instead, we believe that a precise performance evaluation of a detailed network is key to get a clear understanding of the usefulness of the sharing concept in wireless scenarios. It is also evident that the performance assessment of a complex system such as a wireless network comprising dozens of nodes cannot be performed through an exact analysis. However, it is fairly common in the scientific community to resort to network simulation instruments; among these, the network simulator-3 (ns-3) tool [4] is well known and is currently considered as one of the most advanced and modular. It is entirely open source and its features span the entire protocol stack, from the physical layer up to the application. Such a modularity has been improved with respect to its previous version ns-2, which, while fairly accurate in the medium access and networking layers, was not sufficiently accurate in the characterization of the wireless channel, at least with the high level of detail required when dealing with spectrum sharing issues. The properties of modularity and entire bottom-up representation of the protocol stack provided by ns-3 are particularly appealing for our purposes, since the analysis of spectrum sharing, while involving physical and datalink layers, implies important consequences in protocol design at higher layers as well, thus being an inherently cross-layer problem. These reasons motivate our choice to employ an existing implementation [20] within ns-3 of the Long Term Evolution (LTE) of the Universal Mobile Telecommunications System (UMTS) [1].

The main contribution of the present paper is to introduce a novel software extension of this ns-3 version to characterize spectrum sharing scenarios where cooperation is established among multiple operators, each with a considerable number of nodes. To realize this enhancement, original software structures are introduced; in particular, as will be discussed in the following, a class describing a virtual *frequency market* has been inserted in the simulator structure. This class implements the functionalities of a virtual arbitrator, and does not represent a physical entity of the network, but rather it determines the sharing policy of the frequencies belonging to the common pool. In other words, its role is to abstract the set of rules agreed by the operators when determining the shared portion of the spectrum. In particular, two main sharing meta-policies are available, namely *orthogonal* and *non-orthogonal* sharing. In the former case, the frequencies of the shared pool still remain into exclusive usage of exactly one operator, although not necessarily the one that detains the legal property of the access on that frequency. In the latter, also simultaneous access on the same frequency is possible. In both cases, the arbitrator structure is required to give an abstract representation of every other sharing policy detail, such as priority rules among the operators in case of conflicting assignments. It is worth noting that the definition of efficient sharing policies is out of the scope of the present paper. For the sake of simplicity, we focus on orthogonal sharing, which is immediate to describe and does not require to detail any power control policy for shared frequencies. However, as the code developed is entirely modular, an extension to non-orthogonal sharing would be straightforward. Moreover, to simplify the game theoretic analysis, only competitive sharing will be modeled, leaving the issue of identifying efficient and collaborative sharing mechanisms for future work.

Besides introducing the details of the software extensions implemented within ns-3, this paper also provides the results of an extensive simulation campaign meant to assess the effectiveness of the simulator as a benchmark for testing spectrum sharing algorithms. A simple sharing algorithm is used, and the evaluation of the modified version of the simulator in terms of computational requirements is given as well. The results confirm the ability of such a software instrument to give realistic assessments of the usefulness of spectrum sharing, and at the same time motivate further efforts with game theoretic approaches to implement efficient sharing algorithms where collaborative sharing is sought.

The rest of this paper is organized as follows. In Section 2 we review related works on simulation platforms for spectrum sharing analysis. In Section 3 we describe the system model, detailing the theoretical rationale behind the spectrum sharing characterization, while in Section 4 we discuss the modifications applied to the software architecture. In Section 5 we outline the simulation scenario and in Section 6 we present numerical results to validate our proposed contribution; we finally conclude in Section 7.

2. STATE OF THE ART

The availability of a suitable simulation platform for testing protocols and algorithms is quite important, in particular for all those scenarios where the mathematical analysis becomes complex or cannot produce a solution in closed form. One immediate solution, which is often used in the literature, is to develop a basic single-purpose simulator, written from scratch and specific to the scenario under investigation. However, such a solution often violates important requirements in scientific work, i.e., generality and reproducibility of the results. Conversely, several standard code libraries and simulation tools have been developed to support researchers in their work. Some of these instruments are general purpose (e.g., SIMLIB [6], MATLAB [3]) and need customization to the particular context that is to be evaluated; others are more application specific and are meant to simulate the behavior of particular systems, e.g., OMNET++ [5].

For the case of computer networks, one of the most used tools in the research community is the Network Simulator ns [4], whose latest version is ns-3. It is an open source, free software managed by an active community of developers. The whole Internet suite protocol stack is implemented together with the most important protocols at the transport, network, and datalink layers. Therefore, many different network scenarios can be created and simulated. One of the last implemented modules realizes LTE cellular networks [20]. The introduced framework enables the creation of Base Stations (called eNodeBs, or eNBs) and user terminals (called UEs) which can communicate with the eNBs. Most of the functionalities of the physical channel and medium access have been implemented, while some of them are still empty or a sample code has been provided, giving the programmer the opportunity to introduce and test new algorithms. This paper aims at extending this basic framework by introducing the multi-cell scenario and allowing eNBs to share part of their frequencies in the downlink direction. This situation is particularly interesting when the eNBs are managed by different cellular network operators.

Although the problems of interference channels and spectrum sharing have been addressed in several papers, e.g., [11, 12], the scenario of inter-cell spectrum sharing has been considered in a small number of them so far, and even fewer papers have focused on multi-operator networks. However, since in current network deployments the coexistence of multiple operators in adjacent areas is quite common, it is sensible to investigate the efficiency of the spectrum division policies adopted in common practice. The interest in this area has increased during the last years and has been involving not only researchers, but also telecommunication companies and regulatory bodies.

A first simple concept of spectrum sharing has been introduced and analyzed in [7]. Base Stations try to face their incoming requests first by using their initial spectrum, and then by exploiting frequencies not used by the others. Two algorithms for resource allocation are presented and evaluated, but the presence of a centralized network is assumed, together with a coordinating unit that manages the whole network. In [10, 18] the authors introduce the concept of resource sharing in broadband cellular networks and show its impact on achievable capacity and packet delay. In this case, the resources shared among the different cells are the time slots (time division multiple access is employed), and operators use their allocated slots to transfer data to their mobiles. While in [18] sharing is seen only as a "last resort" solution, in [10] a new way of implementing radio networks is explored where mobiles are always connected to the best base station, regardless of whether it belongs to their home operator or not. This point is quite far from the implementation that we present in this work, where the resource shared is the band and mobile terminals are always connected to the their home operator.

Another paper where the inter-operator spectrum sharing context is taken into consideration is [9], where a game theoretic analysis is given for a cognitive context where operators are classified into primary and secondary. This is slightly different from the system modeled in our simulator, as described in the next section, where eNBs are not supposed to have sensing capabilities and such a hierarchy is not present.

3. SYSTEM MODEL

We focus on the problem of spectrum sharing in OFDMA networks, with particular reference to the LTE standard [1]. As seen in the introduction, when discussing spectrum sharing policies it is important to clarify the orthogonality of the access scheme in the pool of common frequencies, where "orthogonality" means "impossibility of simultaneous usage by more than one operator." Given that the non-orthogonal approach would require a lengthy discussion about the convergence of the contention for shared frequencies, and the description of a power control mechanism for the users (i.e., the eNBs), we will limit the following discussion to the orthogonal sharing case. Note that, in any event, this choice is made only for the sake of simplicity and is not restrictive as the software modules developed are entirely transparent to the orthogonality property, and they can be promptly extended to work under non-orthogonal sharing almost without any modification. Therefore, from this point on, we will assume that eNBs share orthogonally the pool of common frequencies so that each frequency resource can be assigned to at most a single operator (which, in turn, will use it for one of its UEs) within an allocation time slot, in our case corresponding to the LTE subframe duration, i.e., 1 ms.

Therefore, this work focuses on the definition of a modular framework developed to test different solutions and efficiently evaluate the performance in terms of throughput and execution time, observing scenarios with an increasing number of UEs and sharing percentages. The resulting software can be used as a validation platform for several sharing policies, possibly derived within a game theoretic analysis. In the following, we will show sample results for orthogonal competitive sharing. However, given the modular nature of the simulator, more complex game theoretic approaches can be framed, even resorting to dynamic games, Stackelberg games and so on [17,21].

The proposed framework can be divided into three parts. First of all, the spectrum usage parameters must be provided, i.e., physical details such as the center frequency, the channel bandwidth, the sharing percentage, and so forth (see Section 3.1). Then, local scheduling and resource allocation algorithms must be executed in each eNB in order to generate an allocation map that represents the proposed serving scheme, as detailed in Section 3.2. Finally, a virtual market is in charge of collecting the local allocation maps and derive the serving schemes that must be adopted by each eNB, according to the chosen contention solving policy, as will be illustrated in Section 3.3.

3.1 Spectrum management

Once the physical parameters have been determined, the eNBs select the set of frequencies on which they plan to interoperate. The policy behind such a cooperation agreement is out of the scope of the present paper, as it is more related to the economic agreement between the operators and their business models. However, along with different allocation and coordination techniques, it represents an interesting research topic and, thanks to this contribution, various approaches can be quantitatively evaluated. Figure 1 shows the scheme adopted to define the system sharing capabilities. According to the selected bandwidth percentage to be shared, the eNBs will allow partial access to UEs belonging to other domains.

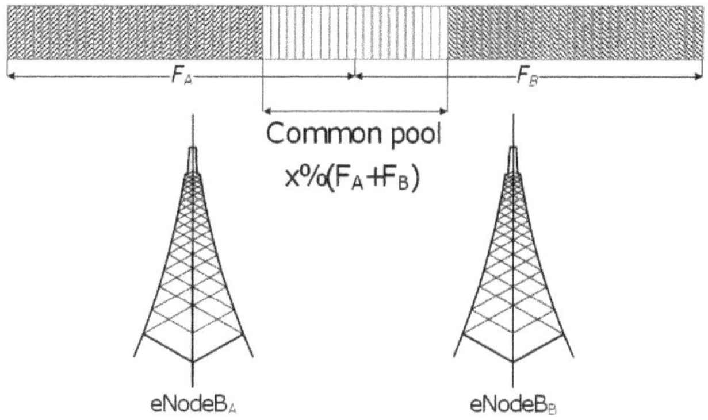

Figure 1: Spectrum sharing

3.2 Intra-cell allocation

The cell capabilities are fully characterized when the physical components have been defined. Then, a joint scheduling and resource allocation algorithm is needed to design a proper downlink transmission scheme. However, the focus of this paper is just on the integration of the proposed spectrum sharing framework for LTE systems into a simulation tool, ns-3. Conversely, the definition and the analysis of efficient game theoretic schemes which can be fed to this simulator are not directly investigated here, but are left for future work, possibly within a game theoretic context.

For what concerns the scope of this paper, two basic algorithms have been implemented and compared: on one hand, *max throughput* represents an allocation scheme for which the resources are allocated to the best UEs, without taking into account fairness among users. On the other hand a fair approach, denominated *fairness*, is proposed: the available system resources are equally distributed among the users, thus lowering the overall throughput, but increasing the average level of satisfaction of each UE. Figure 2 depicts a sample scenario, where 10 UEs and 10 resources, hereinafter referred as resource blocks (RBs), are considered. By selecting the first approach, *max throughput*, all the available resources are allocated to the UEs with the best channel quality indicator (CQI), discussed later in Section 6. Thus, by exploiting multiuser diversity, the system throughput can be very high. However, UEs with lower CQIs will never be served. Therefore, an additional technique has been introduced, i.e., the *fairness* mechanism which, as visible in the figure, will provide service to all the registered UEs. As per the previous case, each RB is allocated to the best UE, but each user cannot get more than a fixed amount of resources. This threshold is given by

$$TH = \left\lceil \frac{N_{RB}}{N_{UE}} \right\rceil, \quad (1)$$

where N_{RB} represents the total number of RBs, and N_{UE} is the num-

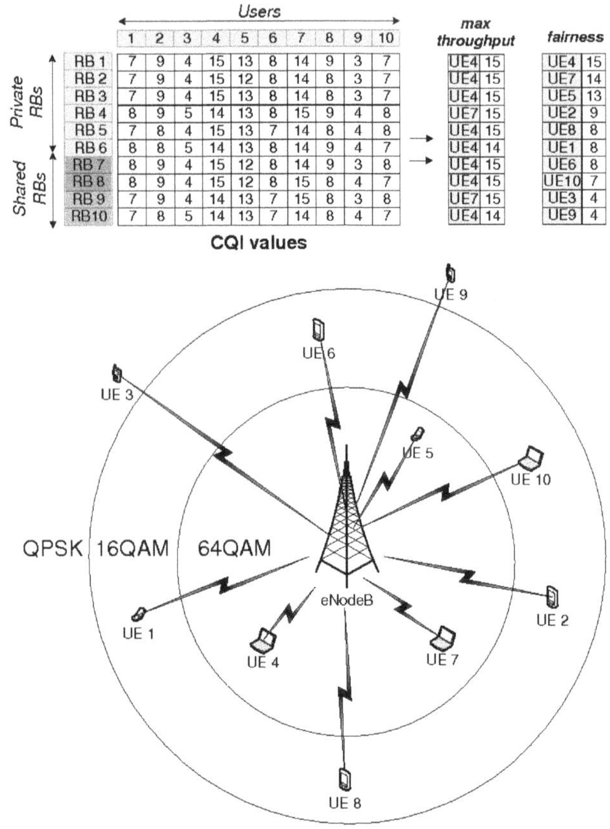

Figure 2: Intra-cell allocation

ber of registered UEs requesting admittance in the system. In the proposed example, from equation (1) the threshold is equal to 1, so all the UEs will be allocated a single RB.

3.3 Inter-cell coordination

The sharing contention policy is implemented in a separate module, here called *virtual market*. The relevant class (we refer to an Object-Oriented Programming, or OOP, paradigm) implements an arbitration rule which defines how the operators bargain the access to the common portion of the spectrum. Any complex strategy can be implemented within this class, possibly involving further extensions. In particular, this may be the place where to implement, in an entirely modular manner, some procedures inspired by game theoretic principles. Each eNB, after generating its own allocation map, sends it to the *virtual market* that gathers all the cells' allocation information and rearranges the allocation map according to the sharing policy. For the sake of simplicity, in this paper we propose immediate implementations of scheduling and resource allocation algorithms, as well as a simple procedure to handle the contentions among operators. Each eNB is assigned a *priority* value per frequency subchannel, defined as

$$PR_{eNB_j,RB_{pool,i}} = \begin{cases} p, & RB_{pool,i} \in F_{eNB_j} \\ 1-p, & \text{otherwise} \end{cases}, \qquad (2)$$

where $j \in \{1,...,m\}$ represents the eNB identifier, m is the total number of eNBs involved in the sharing process, $p \in [0,1]$ is the priority level given to the eNB, $F_{eNB_j} = \{RB_{j,1},...,RB_{j,n_j}\}$, n_j is the total number of RBs available at eNB$_j$, and $RB_{pool,i} \in F_{eNB_j} \cup \ldots \cup F_{eNB_m}$. In other words, shared resources are assigned based on these priority levels; obviously, the UEs associated to eNB$_j$ will always have higher priority than all other competing users. In our paper, the proposed approach is even simpler: we assume $p = 1$ and $m = 2$, so an eNB will assign to its UEs the shared resources belonging to the *competitor* eNB, referred to as eNB$_c$, only if these are not allocated to UEs belonging to eNB$_c$. Thus, when multiple players request the same resource, only the one with the highest priority will get it. The others end up with no assignment, which is in general inefficient.

We stress that this general strategy is not given as an optimal allocation, which ought to be derived from a (game) theoretic perspective. Rather, such an intentionally non-optimized (and actually inefficient) policy serves to show the effectiveness of our software implementation. Moreover, it can be thought of as a characterization of the inefficient Nash equilibria in the games with *competitive* sharing, while the goal of spectrum sharing should rather be a *collaborative* assignment of frequencies. Thus, our reference allocation policy correctly reflects that, if the whole common pool is shared competitively, in the long run only inefficient and unfair allocations will be achieved. However, we also remark that more efficient solutions derived through game theory, either available in the literature or originally developed, can be tested and validated within the modular framework proposed in this paper, so as to determine the choice that better suits the operator needs.

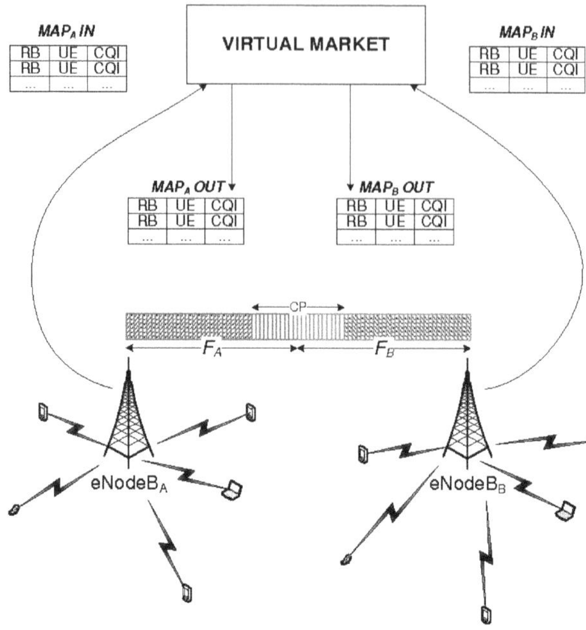

Figure 3: Inter-cell coordination

4. NS-3 LTE EXTENSION

The reference implementation of LTE to which we have applied our modifications is the one presented in [20] and included in the current release of the ns-3 simulator. Our extension introduced two main features, i.e., the implementation of multi-cell multi-operator scenarios and the definition of inter-operator downlink spectrum sharing policies. In this way, we have prepared a framework that can be used as is or extended again to simulate a broader category of scenarios. This is made possible by the extreme modularity of ns-3. It is also worth mentioning that our extension is entirely backward compatible with previous versions of ns-3.

4.1 Multi-cell multi-operator scenario

The definition of a multi-cell scenario requires first of all the definition of a separate object of the class *LteHelper* for each cell. Such an object contains a reference to the eNB and all its UEs and therefore manages the creation and configuration of all the members of a cell (e.g., registration of a UE). Different cells are managed by different *LteHelpers*.

A further modification that was required with respect to [20] regards the management of the time by each eNB. The class *LtePhy* is the base class for modeling eNB and UE physical layer. Then *EnbLtePhy* and *UeLtePhy* are derived classes that implement particular features of the physical layer for the two types of nodes, such as transmission and reception of signals on the wireless channel. The *LtePhy* class has in its private fields two *static* counters, one for the frame index and another for the subframe index within the current frame. They are incremented every time a new frame/subframe is started, a functionality that is implemented by the *EnbLtePhy* class, methods *StartFrame* and *StartSubFrame*, since it is up to the eNB to decide when to start the new frame/subframe. In a multi-cell scenario there are many eNBs, each with its own *EnbLtePhy*, and all these counters need to be incremented. Therefore, two possible solutions are available: either only an eNB increments those counters or each one of them has its own counter and increments it independently. In our implementation we have chosen the latter, thus each eNB has its private view of the time index. In our implementation, they are all synchronized, hence they start each (sub)frame at the same time, but this implementation choice does not prevent further more realistic extensions where the eNBs are non-synchronized.

4.2 Downlink spectrum sharing

Regarding the implementation of the inter-cell downlink spectrum sharing, several modifications to the base model have been written. First of all, we made eNBs aware of the additional subchannels they can use for downlink resource allocation. The original implementation assigns to each *EnbLtePhy* and *UeLtePhy* a vector of subchannels which represents the available resources they can use. In our implementation we have associated to each node an extended vector containing not only the subchannels originally assigned to it, but also those that the other eNBs are willing to share (calculated as a percentage of the original spectrum size) together with the subchannel priority access information. This vector is the set of frequencies that is actually used by the resource allocator of the eNB. The way it is used depends on the scheduling and allocation policy implemented. In particular, to customize these functionalities, it is sufficient to write a new class which extends the *PacketScheduler* class, thereby inheriting its methods, and to override the method *DoRunPacketScheduler*, i.e., the routine called at the beginning of each subframe when a new set of packets must be selected for transmission.

As a further point, we have implemented the communication and trading mechanisms among the eNBs for the sharing of the common pool. Each eNB calculates its allocation map independently, according to an internal scheduling and resource allocation policy. Then, a virtual entity has been introduced to implement the exchange of the maps and the resolution of the conflicts. In a real system, this phase requires that the eNBs communicate (e.g., through a backhaul) and agree on a final allocation map to which all of them must adhere. This virtual entity is an object defined as an instance of the class *VirtualMarket*; at the beginning of each subframe, it receives the resource allocation maps proposed by all the eNBs (competitors) and decides the final map according to some policy. Developers can implement whatever policy they need, by just modifying that class or extending it and overriding the method that deals with contention resolution, i.e., *GetAllocationMap*. The *VirtualMarket* has a collection of eNB entities, which can communicate with it through its public interface. In Figure 4 an example of such a communication is shown by means of a sequence diagram, which is also able to catch the temporal dimension of the activity. The particular communication protocol shown is the one described in the previous section for conflict resolution. An iteration is shown since every time a competitor cannot use a subchannel for some UEs (i.e., it loses the contention), it is invited to reschedule those UEs on other free resources (if any).

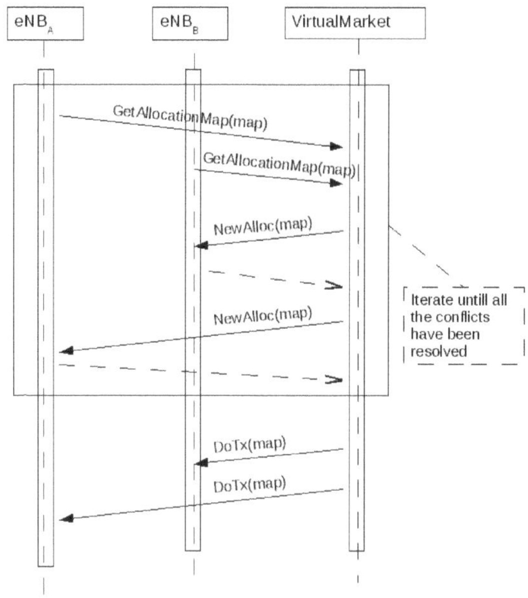

Figure 4: Sequence diagram for allocation conflict resolution

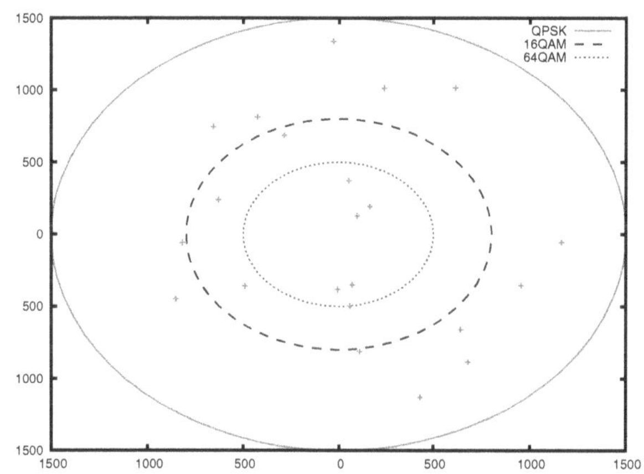

Figure 5: Deployment of the UEs around the eNB

5. SIMULATION SCENARIO

In order to test the software architecture that we have implemented, and to validate the sample sharing algorithm proposed, we have run some simulations. The aim of this phase is not to test the performance of the algorithm itself since, as already said, the focus of this paper is on the architectural extension of the simulator. The algorithm that we have implemented is meant to be just an example to show how things work, so it is not expected to be the optimal so-

CQI	Modulation	ECR	Spectral Efficiency	TB
1	QPSK	0.0762	0.15	24
2	QPSK	0.1172	0.23	40
3	QPSK	0.1885	0.38	60
4	QPSK	0.3008	0.6	100
5	QPSK	0.4385	0.88	144
6	QPSK	0.5879	1.18	196
7	16QAM	0.3691	1.48	248
8	16QAM	0.4785	1.91	322
9	16QAM	0.6016	2.41	402
10	64QAM	0.4551	2.73	452
11	64QAM	0.5537	3.32	554
12	64QAM	0.6504	3.9	654
13	64QAM	0.7539	4.52	756
14	64QAM	0.8525	5.12	856
15	64QAM	0.9258	5.55	936

Table 1: LTE MCS

Parameter	Value
Center Frequency	2.0 GHz
Channel Bandwidth	10 MHz
Subcarrier Bandwidth	15 kHz
Doppler frequency	200 Hz
$RB_{bandwidth}$	180 kHz
$RB_{subcarriers}$	12
$RB_{OFDMsymbols}$	14
eNodeB TX power	43 dBm
Noise figure (F)	2.5
Noise spectral density (N_0)	-174 dBm/Hz
Macroscopic pathloss	$128.1+(37.6\cdot\log_{10}(R))$ dB
Shadow fading	log-normal ($\mu=0,\sigma=8$ dB)
Multipath	Jakes model
Wall penetration loss	10 dB
Simulated interval	2000 subframes
Frame duration	10 ms
TTI	1 ms

Table 2: Main system parameters

lution. We are more interested in the performance and the usability of the simulator itself. In the following we present the results of a simulation campaign conducted with the extended framework for spectrum sharing in ns-3.

The scenario consists of two eNBs positioned in the same field, both with a coverage of 1500 m. An increasing number of UEs, characterized by low mobility, are registered to each station, and are uniformly distributed within the associated eNB coverage area. Figure 5 depicts an example of user distribution resulting after the execution of a simulation run with 22 UEs.

As mentioned in the previous sections, each user perceives a different quality of the channel according to its position and other minor factors. Hence, an ideal channel is established between the UE and the eNB, used for the transmission of the CQIs associated to each RB. In fact, thanks to this information, the eNB can select an adequate Modulation and Coding Scheme (MCS). As reported in Table 1, LTE technology provides 15 different option schemes [2], where ECR stands for Effective Code Rate, and represents the robustness of the selected coding scheme. Hence, each MCS determines the transport block (TB) size that results from

$$TB_{CQI}=\frac{RB_{subcarriers}\cdot RB_{OFDMsymbols}\cdot ECR_{CQI}}{TTI}, \quad (3)$$

where $RB_{subcarriers}$, $RB_{OFDMsymbols}$, and TTI, that represent the number of subcarriers per RB, the number of OFDM symbols per RB, and the scheduling time respectively, are provided in Table 2, together with the main system parameters.

The simulation campaign is executed to investigate the reliability of the proposed framework, in terms of theoretical capacity, aggregate throughput, and simulation time. In fact, as will be extensively detailed in Section 6, the system performance behavior follows the trend that we expected: on the one hand, increasing the number of UEs in the system corresponds to a throughput increase, while on the other hand increasing the sharing percentage induces a smooth decrease of the system throughput, according to the simple conflict resolution approach implemented. More specifically, the performance metrics taken into consideration are:

—**Shannon Capacity**, which represents the maximum theoretical throughput in communication systems. It is defined as the maximum of the mutual information between the input and the output of the channel, and is given by

$$C=B\cdot\log_2(1+SNR). \quad (4)$$

—**System Throughput**, which represents the aggregation of the data rates delivered to all UEs, and is computed as

$$T=\frac{\sum_{i=1}^{N}TB_{RB_i}}{TTI}, \quad (5)$$

where TB_{RB_i} represents the transport block size referred to the ith RB, and N is the total number of RBs available in the system.

—**Simulation time**, which represents the execution time of each set of simulation runs. As expected, it grows with the number of UEs and with the sharing percentage because of the higher computational complexity needed to process a larger number of operations. The reference machine is a desktop computer with a Pentium 4 CPU, 1 GB RAM and running GNU/Linux Ubuntu 10.04 as the operating system.

6. NUMERICAL RESULTS

Figures 6–7 show the performance in terms of capacity and throughput achieved by each cell when the *max throughput* allocation is used. As expected, the actual throughput value is significantly below the channel capacity, which represents the theoretical limit achievable with such a channel condition. The actual amount of data transmitted depends on the ECR and is upper bounded by the Shannon capacity. However, the behavior of both capacity and throughput as functions of the sharing percentage for different numbers of users is qualitatively similar, meaning that they differ only by a scaling factor due to the use of real coding and modulation schemes. A first conclusion that might be drawn from these figures is that, if spectrum sharing is performed in a *competitive* manner, there is no gain for the operators in sharing their frequencies. In fact, the higher the sharing percentage, the more likely the resource conflicts. Due to the lack of collaboration among the operators, they simply try to get the best frequencies, whereas other less appealing resources are wasted. This is a typical phenomenon of *non cooperative* game theory, i.e., an inefficient Nash equilibrium as a result of the selfishness of the players [17]. The situation is made worse by the constraint that the private subchannels of one eNB cannot be accessed by the other eNBs, and so some resources might be unused. These are the main reasons for the convexity of the curves: the presence of a minimum for a 50% sharing indicates

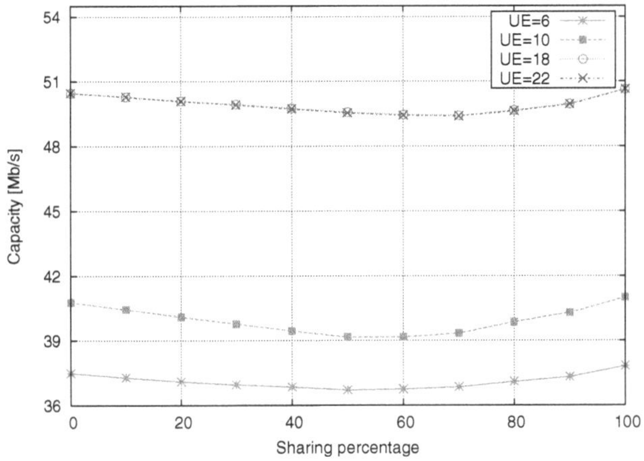

Figure 6: Cell capacity of the *max throughput* allocation

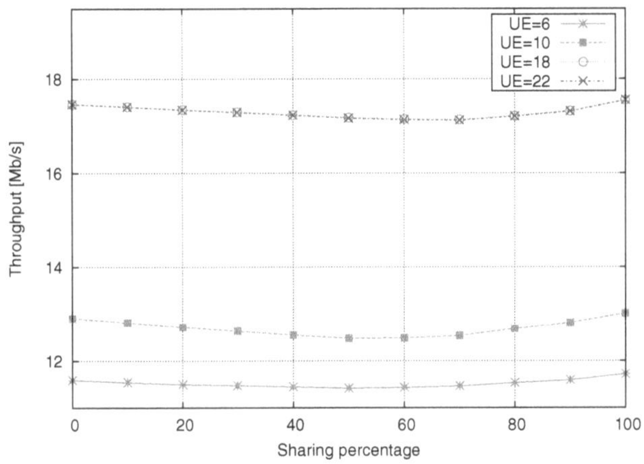

Figure 7: Cell throughput of the *max throughput* allocation

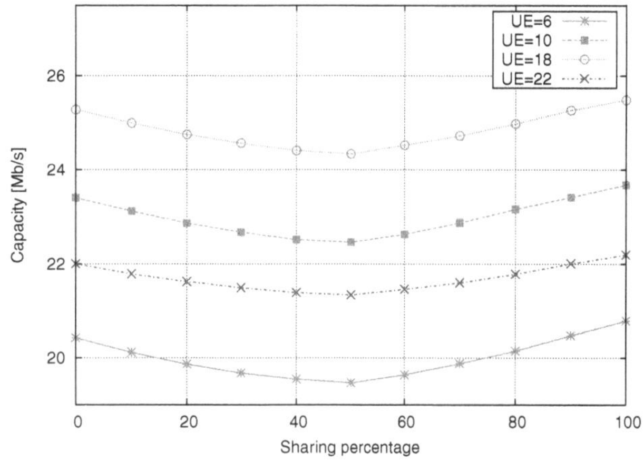

Figure 8: Cell capacity of the *fairness* allocation

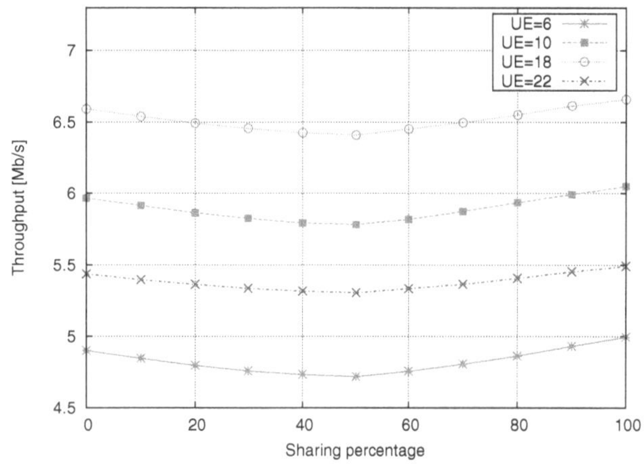

Figure 9: Cell throughput of the *fairness* allocation

the situation of maximum waste. Although the focus of this paper is just on the software framework, we see as a promising extension the identification of efficient game theoretic strategies for inducing the operators to achieve a *collaborative* sharing, thus improving the allocation efficiency. The modularity of our software allows for a prompt insertion of such strategies.

Another important intuition that can be gained from those figures is that both performance indices increase with the number of users. This is an effect of the increased multiuser diversity: the greater the number of UEs, the higher the probability that for each subchannel there is at least one of them with a good CQI. However, this increase is significant only when the number of users is low. When more users are in the system, the marginal improvement due to multiuser diversity becomes lower since for almost all the subchannels there is a user with good CQI. Thus, the curves seem to saturate around 18 users.

To sum up, the results validate the reliability of our model in spite of an inefficient sharing policy, that was not the scope of this paper. Thanks to the modularity introduced, the contention technique can be adapted to different needs, and in particular to pursue a cooperative sharing, where system capacity and throughput increase when the spectrum sharing percentage becomes higher.

Our framework also enables a comparative analysis among different allocation approaches. Therefore, we can compare the *max throughput* allocation with the other approach, i.e., the *fairness* allocation, that aims at scheduling all the users, not just the best ones. As expected, applying a fair scheduling scheme results in a decrease of the system throughput. This is shown in Figures 8 and 9, where it can be noted that the aggregate data rates obtained are roughly halved with respect to the *max throughput* approach. Moreover, it is highlighted that in the *fairness* allocation both performance indices do not always improve when the number of users increases. In fact, when the number of users is increased, the fairness constraint is harder to satisfy and may actually lead to an overall decrease of the system capacity and throughput.

Finally, the execution time resulting from a wide range of simulations is analyzed. As shown in Figure 10, the obvious complexity increase with respect to the increase of the number of UEs and spectrum sharing percentage is reflected in the graphs. A higher number of UEs requires more memory and computational resources to store and manage all those objects and thus a higher execution time. On the other hand, a greater number of shared resources implies more contention and thus more iterations of the conflict resolution algorithm. Execution times also increase for higher sharing, since the simulator has a higher number of degrees of freedom. Moreover, we remark that the tracing option was enabled in order to log

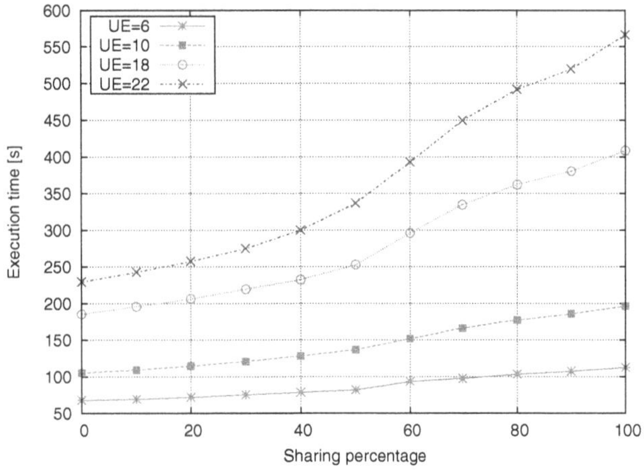

Figure 10: Simulation time

the performance indices and calculate statistics. Disk accesses are quite time consuming and can slow down the execution by more than 10 times the normal duration. However, in spite of all these points, the computational complexity scales almost linearly with the number of users, and can thus be considered acceptable for realistic and detailed simulation campaigns.

7. CONCLUSIONS

In this paper we outlined and evaluated a framework for spectrum sharing mechanisms within an LTE implementation of ns-3. The resulting software has been thoroughly tested to evaluate its correctness and reliability in achieving spectrum sharing functionalities. The results have been satisfactory under all aspects, showing that our proposed extension can serve as a concrete tool to evaluate resource sharing mechanisms in next generation wireless networks.

Future work involves the implementation and exploration of *non-orthogonal* sharing mechanisms, where multiple players are allowed to operate on the shared frequencies. Such an extension implies also the introduction of power control mechanisms to harmonize the transmission among different sources. Finally, submission of the developed software for official release in ns-3 is also planned for the near future.

8. REFERENCES

[1] 3GPP Evolved Universal Terrestrial Radio Access (E-UTRA); Overall description; Stage 2, 3GPP TS 36.300.

[2] 3GPP Tech. Specif. Group Radio Access Network; Conveying MCS and TB size via PDCCH, 3GPP TSG-RAN WG1 R1-081483.

[3] Matlab, available at: *http://www.mathworks.com/products/matlab/*.

[4] ns-3 simulator, available at: *http://www.nsnam.org/*.

[5] Omnet++ simulator, available at: *http://www.omnetpp.org/*.

[6] Simlib simulator, available at: *http://www.fit.vutbr.cz/~peringer/SIMLIB/*.

[7] B. Aazhang, J. Lilleberg, and G. Middleton. Spectrum sharing in a cellular system. In *Proc. IEEE ISSSTA*, pages 355–359, 2004.

[8] L. Badia, M. Levorato, F. Librino, and M. Zorzi. Cooperation techniques for wireless systems from networking perspective. *IEEE Wireless Commun. Mag.*, 17(2):89 – 96, Apr. 2010.

[9] M. Bennis, S. Lasaulce, and M. Debbah. Inter-operator spectrum sharing from a game theoretical perspective. *EURASIP J. on Advanc. in Sign. Proc.*, 2009:1–12, 2009.

[10] M. Bennis and J. Lilleberg. Inter base station resource sharing and improving the overall efficiency of B3G systems. In *Proc. IEEE VTC Fall*, pages 1494–1498, 2007.

[11] A. B. Carleial. Interference channels. *IEEE Trans. Inf. Theory*, IT-24(1):60–70, Jan. 1978.

[12] R. Etkin, A. Parekh, and D. Tse. Spectrum sharing for unlicensed bands. *IEEE J. Sel. Areas Commun.*, 25(3):517–528, April 2007.

[13] S. Geirhofer, L. Tong, and B. Sadler. Dynamic spectrum access in the time domain: Modeling and exploiting white space. *IEEE Commun. Mag.*, 45(5):66–72, May 2007.

[14] E. Larsson and E. Jorswieck. Competition versus cooperation on the MISO interference channel. *IEEE J. Sel. Areas Commun.*, 26(7):1059–1069, Sept. 2008.

[15] A. Leshem and E. Zehavi. Cooperative Game Theory and the Gaussian Interference Channel. *IEEE J. Sel. Areas Commun.*, 26(7):1078–1088, Sept. 2008.

[16] P. Liu, Z. Tao, S. Narayanan, T. Korakis, and S. S. Panwar. CoopMAC: A cooperative MAC for wireless LANs. *IEEE J. Sel. Areas Commun.*, 25(2):340 – 354, Feb. 2007.

[17] A. MacKenzie and L. DaSilva. *Game Theory for Wireless Engineers*. Morgan and Claypool, San Rafael, CA, 2006.

[18] G. Middleton, K. Hooli, A. Tölli, and J. Lilleberg. Inter-operator spectrum sharing in a broadband cellular network. In *Proc. IEEE ISSSTA*, pages 376–380, 2006.

[19] J. Mitola III and G. Maguire Jr. Cognitive radio: making software radios more personal. *IEEE Personal Commun. Mag.*, 6(4):13–18, 1999.

[20] G. Piro, N. Baldo, and M. Miozzo. An LTE module for the ns-3 network simulator. In *Proc. SIMUTools*, March 2011.

[21] J. Zhang and Q. Zhang. Stackelberg game for utility-based cooperative cognitive radio networks. In *Proc. ACM MobiHoc*, 2009.

Closed-Form Approximation of the Energy Detection Performance over Generalized Fading Channels

Oluwatobi Olabiyi, Annamalai Annamalai

Center of Excellence for Communication Systems Technology Research
Department of Electrical and Computer Engineering

Prairie View A&M University, TX 77446

engr3os@gmail.com, aaannamalai@pvamu.edu

ABSTRACT

Efficient spectrum sensing is one of the most crucial components in the entire operation of cognitive radio rests. Blind sensing of spectral-holes using energy detectors is one of the solutions that have been proposed for enabling opportunistic spectrum access. This article revisits the problem of energy detection of an unknown deterministic signal over a myriad of fading environments. We proposed yet another novel approach (based on the approximate finite series representations of the generalized Marcum Q-function of real order in conjunction with the derivatives of the moment generating function of signal-to-noise ratio) to analyze the performance of maximal-ratio combining (MRC) and square-law combining (SLC) energy detectors with independent but non-identically distributed (i.n.d) fading statistics, including Rice and mixed-fading channels. Our analytical framework is also capable of treating the Nakagami-m channels with non-integer fading severity indices as well as half-odd integer values for the time-bandwidth product u. Many of these cases were either intractable with the classical probability density function/contour integral approaches, or that heretofore had resisted simple/computationally efficient solutions. Comparison is also made with other series solutions that have been presented in literature and selected numerical results are provided for the receiver operating characteristic (ROC) of MRC and SLC diversity energy detectors over Rice fading channel.

Categories and Subject Descriptors

C.2.1 Network Architecture & Design: *wireless communications*; C.2.3 Network Operations: *network monitoring*; C.4 Performance of Systems: *modeling techniques*, *performance attributes*.

General Terms

Theory, Performance, Reliability, Design, Algorithms.

Keywords

Energy Detection; Receiver Operating Characteristics (ROC); Opportunistic Wireless Spectrum Access; Diversity Receiver.

1. INTRODUCTION

Among the various known spectrum sensing techniques, blind sensing using energy detectors is perhaps the simplest (i.e., low complexity) and more importantly, the "secondary users" do not require any unauthorized details or a priori knowledge of the "primary user" transmissions (i.e., the energy detector of a secondary user treats the received primary user transmission as an unknown deterministic signal). While energy detection of the received signal waveform over an observation time window may be more versatile than the cyclo-stationary feature detection approach, its performance (reliability) is severely limited by multipath fading. Therefore, performance characterization of energy detectors with diversity reception over different wireless fading environments is of practical interest.

Urkowitz [1] showed that detection problem of an unknown deterministic signal over a flat band-limited Gaussian noise channel using an energy detector boils down to a binary hypothesis-testing problem, and the statistics of the decision variable is chi-square distributed, irrespective of whether the process model is low-pass or band-pass. The analysis of performance of energy detection using probability density function (PDF) approach for deriving detection probability with single-channel and diversity receptions over Rayleigh, Rice and Nakagami-m environments can be found in [2]-[6] and have been well reviewed in [13] and [14]. In [7], the authors by using "exponential-type" contour integral representation of the generalized Marcum Q-function was able to derive closed-form solutions for maximum ratio combining (MRC) diversity over i.i.d Nakagami-m fading (with positive integer mL, where L denotes the diversity order) and the i.i.d Rice fading environment (although the final solution is in the form of an infinite series, whose summation terms require the evaluation of two separate higher order derivatives of a product term [7, eq. (27)]). Also, the final formula obtained on diversity detector over Rice fading channels appears very complicated and none of the above mentioned articles (with exception of [13] and [14]) have treated the i.n.d fading statistics case (in environments other than Rayleigh fading), or have derived for diversity energy detectors in Nakagami-m channels that is applicable to any arbitrary real value of fading severity index.

These problems were addressed in [13] and [14] where the authors presented unified solutions for evaluating P_d across different fading channels. However, the final expression though much simpler than [7] contains infinite series which brings up the issues of bounds and truncation error. In contrast, in this article we rely on a finite series approximate representations of the generalized Marcum Q-function along with the derivatives of the moment generating function (MGF) of combiner output SNR (in closed-form) to develop two unified expressions for over generalized fading channels. These compact yet general formulas are then used to analyze the performance of MRC and square law combining (SLC) energy detectors under general settings (including i.n.d fading statistics and mixed-fading scenarios). In addition, our framework place no restriction on system parameters and even for the special case of i.i.d Rice fading with MRC diversity, our solution is very concise and considerably simpler to program compared to a more complicated expression derived in [7, eq. (27)].

The remainder of this article is organized as follows. The system model and the derivation of unified expression for single channel detection are discussed in Section II. Section III considers both MRC and SLC diversity energy detectors. Selected numerical results are provided in Section V, while Section VI summarizes the key results.

2. SYSTEM MODEL AND SINGLE CHANNEL DETECTION

For the purpose of brevity and consistency, the notations similar to [6], [7], [13], and [14] are also used in this article.

The detection of the existence of the unknown deterministic signal $s(t)$ by the receiver is a binary hypothesis test as shown in (1) [1],

$$y(t)=\begin{cases} n(t) & :H_0 \\ hs(t)+n(t) & :H_1 \end{cases} \tag{1}$$

where $s(t)$ is the unknown deterministic signal waveform, $n(t)$ is the noise waveform (white Gaussian random process), H_0 is the hypothesis 0 (i.e., no $s(t)$ present) and H_1 is the hypothesis 1 (i.e., $s(t)$ present).

Therefore, a sample from noise process n_i is a Gaussian random variable with zero mean and $N_{01}W$ variance; $n_i \sim N(0, N_{01}W)$ [1], where N_{01} is one sided noise power spectral density, W is one-sided bandwidth.

Thus the probability of detection (P_d) and the probability of false alarm (P_f) of an unknown deterministic signal in additive white Gaussian noise have been shown to be given by [7].

$$P_d = Q_u(\sqrt{2\gamma},\sqrt{\lambda}) \tag{2}$$

$$P_f = \frac{\Gamma(u,\frac{\lambda}{2})}{\Gamma(u)} \tag{3}$$

where $u = TW$ is the time-bandwidth product, λ is the energy detection threshold, $Q_u(.,.)$ is the uth order generalised Marcum-Q function and $\Gamma(.,.)$ is the upper incomplete gamma function which is defined by $\Gamma(a,x)=\int_x^{\infty} t^{a-1}e^{-t}dt$ and $\Gamma(a,0)=\Gamma(a)$. P_f is the same over any fading channel since there is no γ in (3). In the other sense P_d has to be averaged over different fading channels and diversity combining.

The generalized Marcum Q-function is given by [12, Eq. (1)],

$$\begin{aligned} Q_u(\alpha,\beta) &= \int_\beta^\infty x\left(\frac{x}{\alpha}\right)^{u-1} e^{-\frac{1}{2}(x^2+\alpha^2)} I_{u-1}(\alpha x)dx \\ &= 1-\int_0^\beta x\left(\frac{x}{\alpha}\right)^{u-1} e^{-\frac{1}{2}(x^2+\alpha^2)} I_{u-1}(\alpha x)dx \end{aligned} \tag{4}$$

where $I_v(.)$ is the modified Bessel's function of the first-kind with order v.

In [11], the authors derived an approximate finite polynomial representation of Bessel function of first kind whose corresponding modified Bessel function of first kind of order v is asymptotically the same as that of the infinite series in [9, eq. (8.445)] and is given by

$$I_v(z) \cong \sum_{k=0}^{w} \frac{\Gamma(w+k)w^{1-2k}}{k!\Gamma(w-k+1)\Gamma(v+k+1)}\left(\frac{z}{2}\right)^{v+2k} \tag{5}$$

where $0 < z < 2w$. Substituting (8) into (7) and after few algebraic manipulations, we obtain two different representation of Marcum Q-function in (5) and are given by

$$Q_u(\sqrt{2\gamma},\sqrt{\lambda}) \cong 1-\sum_{k=0}^{w} \frac{\Gamma(w+k)w^{1-2k}G\left(u+k,\frac{\lambda}{2}\right)}{k!\Gamma(w-k+1)\Gamma(u+k)} e^{-\gamma}\gamma^k \tag{6}$$

and

$$Q_u(\sqrt{2\gamma},\sqrt{\lambda}) \cong \sum_{k=0}^{w} \frac{\Gamma(w+k)w^{1-2k}\Gamma\left(u+k,\frac{\lambda}{2}\right)}{k!\Gamma(w-k+1)\Gamma(u+k)} e^{-\gamma}\gamma^k \tag{7}$$

where in this case, $0<\sqrt{2\gamma\lambda}<2w$, $G(.,.)$ and $\Gamma(.,.)$ denote the lower and upper incomplete gamma function, which are defined by $G(a,x)=\int_0^x t^{a-1}e^{-t}dt$ and $\Gamma(a,x)=\int_x^\infty t^{a-1}e^{-t}dt$ respectively.

It is important to note that (9) and (10) are approximate closed form version of [13 Eq. (10) and (11)] and are slightly more general than the infinite series in [10, eq. (4)] and [8, eq. (4.74)] because u is not restricted to positive integers. Interestingly, both (9) and (10) are in a "desirable" exponential form that would facilitate the fading averaging problem in a "unified" manner using the MGF-derivative approach.

The average detection probability over a generalized fading environment can be evaluated by averaging (9) over the PDF of channel SNR $f(\gamma)$, and it is given by

$$\overline{P}_{d\ Gen} = 1-\sum_{k=0}^{w} \frac{1}{k!}\frac{\Gamma(w+k)w^{1-2k}G(u+k,\frac{\lambda}{2})}{\Gamma(w-k+1)\Gamma(u+k)}\int_0^\infty \gamma^k e^{-\gamma} f(\gamma)d\gamma \tag{8}$$

Using the Laplace transform of a derivative property, it is straight forward to show that (9) can be simplified as

$$\overline{P}_{d\ Gen} = 1-\sum_{k=0}^{w} \frac{(-1)^k}{k!}\frac{\Gamma(w+k)w^{1-2k}G(u+k,\frac{\lambda}{2})}{\Gamma(w-k+1)\Gamma(u+k)}\phi_\gamma^{(k)}(s)\Big|_{s=1} \tag{9}$$

where $\phi_\gamma^{(k)}(s)=\frac{\partial^k \phi_\gamma(s)}{\partial s^k}$ and $\phi_\gamma(s)$ denotes the MGF of channel SNR in a specified stochastic channel model. Similarly, another generic expression for the average detection probability can be derived from (10), viz.,

$$\overline{P}_{d\ Gen} = \sum_{k=0}^{w} \frac{(-1)^k}{k!}\frac{\Gamma(w+k)w^{1-2k}\Gamma(u+k,\frac{\lambda}{2})}{\Gamma(w-k+1)\Gamma(u+k)}\phi_\gamma^{(k)}(s)\Big|_{s=1} \tag{10}$$

Interestingly, the kth derivative of the MGF of SNR for all common fading channel models for an L branch i.i.d MRC receiver are summarized in [13 Table 1]. Note also that the special case of single-channel reception can be obtained by setting $L = 1$. For instance, the average detection probability in Nakagami-m and Rice channels can be computed efficiently using (12) and (13), respectively:

$$\overline{P}_{d\ Nak} = 1-\sum_{k=0}^{w} \frac{1}{k!}\frac{\Gamma(w+k)w^{1-2k}G(u+k,\frac{\lambda}{2})}{\Gamma(w-k+1)\Gamma(u+k)}\frac{\Omega^k m^m \Gamma(m+k)}{(m+\Omega)^{m+k}\Gamma(m)} \tag{11}$$

$$\begin{aligned} \overline{P}_{d\ Ric} &= 1-\sum_{k=0}^{w} \frac{1}{k!}\frac{\Gamma(w+k)w^{1-2k}G(u+k,\frac{\lambda}{2})}{\Gamma(w-k+1)\Gamma(u+k)}\frac{\Omega^k (k!)^2(1+K)}{(1+K+\Omega)^{1+k}} \\ &\times \exp\left(\frac{-K\Omega}{1+K+\Omega}\right)\sum_{i=0}^{k}\frac{1}{(i!)^2(k-1)!}\left(\frac{K(1+K)}{1+K+\Omega}\right)^i \end{aligned} \tag{12}$$

Before proceeding further, we want to point out that other than the intuitiveness of our approach, to the best of our knowledge only the proposed solution leads to a unified expression for computing

$\overline{P_d}$ over a myriad of wireless fading environments, while preserving the simplicity of a single summation series (as opposed to more complicated contour integrations and/or restrictive parameters [7]); without imposing any restrictions on the channel fading or system parameters and not plagued by the convergence and truncation error of the infinite series representation employed in [4], [13] and [14].

3. DETECTION WITH DIVERSITY SCHEMES

3.1 Pre-detection MRC Detector

The MRC output SNR is simply the sum of the instantaneous SNRs from all diversity branches, i.e., $\gamma_{MRC} = \sum_{l=1}^{L}\gamma_l$ where L denotes the diversity order. Since the signals from the L independent diversity branches are being combined before sampling, [7] has shown that the decision statistics Y_{MRC} under H_1 follows $\chi_{2u}^2(\epsilon)$ non-centrality parameter $\sum_{l=1}^{L}2\gamma_l$ while the decision statistics Y_{MRC} follows χ_{2u}^2 under H_0. Hence, the conditional detection probability can be evaluated as

$$\begin{aligned} P_{d,MRC} &= Q_u(\sqrt{2\gamma_{MRC}},\sqrt{\lambda}) \\ &= 1-\sum_{k=0}^{\infty}\frac{\gamma_{MRC}^k e^{-\gamma_{MRC}}}{k!}\frac{\Gamma(w+k)w^{1-2k}G(u+k,\frac{\lambda}{2})}{\Gamma(w-k+1)\Gamma(u+k)}, \end{aligned} \quad (13)$$

while the false alarm probability is identical to (6). The average detection probability with MRC receiver can be evaluated in a similar fashion to our development leading to (10) and (11), but with an exception that $\phi_{\gamma_{MRC}}^{(k)}(s)$ is now required. For the special case of i.i.d fading statistics, the corresponding average detection probabilities in Nakagami-m and Rice fading are shown in (15) and (16) respectively:

$$\overline{P_d}_{MRC-Nak} = 1-\sum_{k=0}^{w}\frac{1}{k!}\frac{\Gamma(w+k)w^{1-2k}G(u+k,\frac{\lambda}{2})}{\Gamma(w-k+1)\Gamma(u+k)}\frac{\Omega^k m^{Lm}\Gamma(Lm+k)}{(m+\Omega)^{Lm+k}\Gamma(Lm)} \quad (14)$$

$$\begin{aligned} \overline{P_d}_{MRC-Ric} &= 1-\sum_{k=0}^{w}\frac{1}{k!}\frac{\Gamma(w+k)w^{1-2k}G(u+k,\frac{\lambda}{2})}{\Gamma(w-k+1)\Gamma(u+k)}\frac{\Omega^k k!(L+k-1)!(1+K)^L}{(1+K+\Omega)^{L+k}} \\ &\times \exp\left(\frac{-KL\Omega}{1+K+\Omega}\right)\sum_{i=0}^{k}\frac{1}{i!(L+i-1)!(k-i)!}\left(\frac{KL(1+K)}{1+K+\Omega}\right)^i \end{aligned} \quad (15)$$

For the more general case of i.n.d branch statistics, we obtain the MGF of MRC output SNR as $\phi_{\gamma_{MRC}}(s) = \prod_{l=i}^{L}\phi_i(s)$. In this case, we can derive a closed-form expression for $\phi_{\gamma_{MRC}}^{(k)}(s)$ using the Leibniz differentiation rule, viz.,

$$\begin{aligned} \phi_{\gamma_{MRC}}^{(k)}(s) &= \sum_{n_1=0}^{k}\sum_{n_2=0}^{n_1}\cdots\sum_{n_{L-1}=0}^{n_{L-2}}\binom{k}{n_1}\binom{n_1}{n_2}\cdots\binom{n_{L-2}}{n_{L-1}}\phi_1^{(k-n_1)}(s) \\ &\times \phi_2^{(n_1-n_2)}(s)\cdots\phi_L^{(n_{L-1})}(s) \end{aligned} \quad (16)$$

The corresponding average detection probability can be obtained by substituting (19) into (12) or (13). It is important to highlight that (19) circumvents the need for partial fraction expansion of the MGF, and can be easily implemented on most mathematical software packages such as MATHEMATICA and MAPLE.

3.2 Post-detection SLC Detector

In the square-law combining receiver, the output decision is combined after sampling. The decision statistics Y_{SLC} is thus the sum of L i.i.d χ_{2u}^2 under H_0 and the sum of L i.i.d $\chi_{2u}^2(\epsilon)$ under H_1, as defined by [3][6]

$$Y_{SLC} = \sum_{l=1}^{L}Y_l \sim \begin{cases} \chi_{2Lu}^2 & : H_0 \\ \chi_{2Lu}^2(\epsilon_{SLC}) & : H_1 \end{cases} \quad (17)$$

where $\epsilon_{SLC} = \sum_{l=1}^{L}\epsilon_l = \sum_{l=1}^{L}2\gamma_l = 2\gamma_{SLC}$. Hence the false alarm and the detection probabilities in AWGN are given by

$$P_{f,SLC} = \frac{\Gamma(Lu,\frac{\lambda}{2})}{\Gamma(Lu)} \quad (18)$$

$$P_{d,SLC} = Q_{Lu}(\sqrt{2\gamma_{SLC}},\sqrt{\lambda}) = 1-\sum_{k=0}^{w}\frac{\gamma_{SLC}^k e^{-\gamma_{SLC}}}{k!}\frac{\Gamma(w+k)w^{1-2k}G(u+k,\frac{\lambda}{2})}{\Gamma(w-k+1)\Gamma(u+k)} \quad (19)$$

Then, the average detection probability with SLC diversity over a generalized fading channel can be computed as

$$\overline{P_d}_{Gen} = 1-\sum_{k=0}^{\infty}\frac{(-1)^k}{k!}\frac{\Gamma(w+k)w^{1-2k}G(u+k,\frac{\lambda}{2})}{\Gamma(w-k+1)\Gamma(u+k)}\phi_{\gamma_{SLC}}^{(k)}(s)\bigg|_{s=1} \quad (20)$$

where $\phi_{\gamma_{SLC}}(s) = \phi_{\gamma_{MRC}}(s) = \prod_{l=i}^{L}\phi_i(s)$ since $\gamma_{SLC} = \gamma_{MRC}$. For instance, the k-th derivative of the MGF of MRC output SNR listed in Table 1 can be directly used in (21) to compute the average detection probabilities with i.i.d L-branch SLC diversity in a myriad of fading environments.

4. NUMERICAL RESULTS

Since the value of w in (12) and (13) is only lower bounded by $\sqrt{\gamma\lambda/2}$, we fist investigate the values of w that provide four figure accuracy for different detection parameters. The allow us to compare the performance with other series representation in [13] and [14]. The results for different system parameter changes are obtained from MATLAB are tabulated in Table 1. The 'NN' results indicate when MATLAB is not able to give a result, in fact it has been observed that MATLAB can only give result with maximum number of terms, N_{max}=85 due to division of very large or very small numbers . It could be observed that both (12) and (13) suffer from lower accuracy even at large value of N. This is because the form of (12) and (13) makes the accuracy of the series less dependent on number of terms, N but with compromise on accuracy. However, accuracy of order 10^{-2} or 10^{-3} can be easily obtained and sustained over wide range of parameter changes. When compared with other series form, they perform worse than [13, Eq. (10) and (11)] and [14, Eq. (13)] but better than [14, Eq (11)]. Interesting, they are more stable than all other series in [13] and [14]. Also, similar to the observation in [13] the series form (12), with lower incomplete gamma function is more accurate than (13) with upper incomplete gamma function. To illustrate the performance of energy detectors using (12), we generate complementary ROC curves (the average probability of miss detection $P_m = 1-\overline{P_d}$ versus the false alarm probability P_f) for different receiver structures and different fading environments. Fig. 1 illustrates the complementary ROC curves for different detection configuration over Rice channels. It shows that the performance of energy detector improves with increasing values of the mean channel SNR Ω, Rice factor K and reduced time-bandwidth product u, as expected. Also, it indicates that energy detection can be greatly improved with diversity reception and employing MRC diversity detection gives better performance than the corresponding SLC diversity detection. Although not shown here, similar performance is observed for Nakagami-m channel. The trend and the accuracy of the obtained results are in agreement with [13] and [14].

Table 1: **The required number of terms to achieve four digits accuracy**

Channel	SNR = 5dB Pf = 0.0001 u = 1	SNR = 5dB Pf = 0.0001 u = 5	SNR = 5dB Pf = 0.01 u = 1	SNR = 5dB Pf=0.01 u = 5	SNR = 10dB Pf = 0.0001 u = 1	SNR = 10dB Pf = 0.0001 u=5	SNR = 10dB Pf = 0.01 u = 1	SNR = 10dB Pf = 0.01 u = 5
$\lvert E_{Nak} \rvert$	m=1, N_{12}=67, N_{13}=NN m=3, N_{12} =67, N_{13}=65	m=1, N_{12}=NN, N_{13}=83 m=3, N_{12}=77, N_{13}=45	m=1, N_{12}=37, N_{13}=NN m=3, N_{12}=41, N_{13}=80	m=1, N_{12}=59, N_{13}=NN m=3, N_{12}=60, N_{13}=71	m=1, N_{12}=80, N_{13}=NN m=3, N_{12}=NN, N_{13}=NN	m=1, N_{12}=NN, N_{13}=NN m=3, N_{12}=NN, N_{13}=NN	m=1, N_{12}=37, N_{13}=NN m=3, N_{12}=41, N_{13}=NN	m=1, N_{12}=69, N_{13}=NN m=3, N_{12}=80, N_{13}=NN
$\lvert E_{Ric} \rvert$	K=4, N_{12}=67, N_{13}=65	K=4, N_{12}=78, N_{13}=46	K=4, $_{12}$=41, N_{13}=81	K=4, N_{12}=61, N_{13}=72	K=4, N_{12}=NN, N_{13}=NN	K=4, N_{12}=NN, N_{13}=NN	K=4, N_{12}= 40, N_{13}=NN	K=4, N_{12}=, N_{13}=NN
$\lvert E_{Nak,3} \rvert$	m=1, N_{12}=NN, N_{13}=NN m=3, N_{12}=NN, N_{13}=NN	m=1, N_{12}=NN, N_{13}=NN m=3, N_{12}=NN, N_{13}=NN	m=1, N_{12}=42, N_{13}=NN m=3, N_{12}=43, N_{13}=NN	m=1, N_{12}=80, N_{13}=NN m=3, N_{12}=NN, N_{13}=NN	m=1, N_{12}=63, N_{13}=NN m=3, N_{12}=43, N_{13}=NN	m=1, N_{12}=NN, N_{13}=NN m=3, N_{12}=NN, N_{13}=NN	m=1, N_{12}=23, N_{13}=NN m=3, N_{12}=24, N_{13}=NN	m=1, N_{12}=53, N_{13}=NN m=3, N_{12}=34, N_{13}=NN
$\lvert E_{Ric,3} \rvert$	K=4, N_{12}=NN, N_{13}=NN	K=4, N_{12}=NN, N_{13}=NN	K=4, N_{12}=43, N_{13}=NN	K=4, N_{12}=NN, N_{13}=NN	K=4, N_{12}=46, N_{13}=NN	K=4, N_{12}=NN, N_{13}=NN	K=4, N_{12}=24, N_{13}=NN	K=4, N_{12}=37, N_{13}=NN

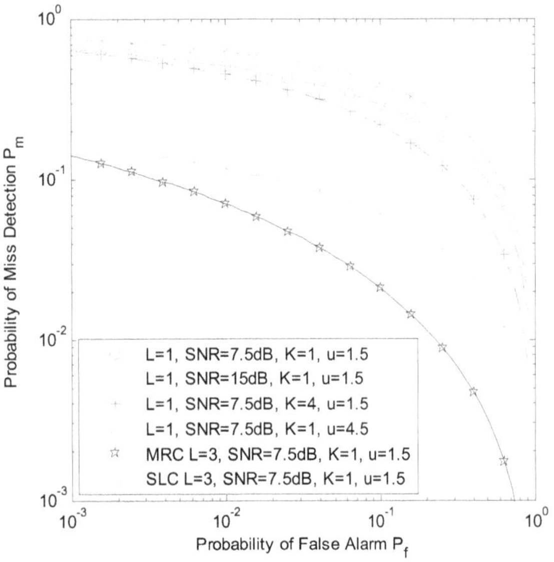

Fig. 1 Complementary ROC curves different detection configuration over Rice fading environments.

5. CONCLUSIONS

In this article, the problem of detecting an unknown deterministic signal by using a diversity energy detector has been revisited. In particular, we illustrate a new approach (based on the finite series representations of the generalized Marcum Q-function for real order and the derivatives of the MGF of combiner output SNR) to compute the desired average probability of detection over generalized fading channels. Our proposed MGF-derivative methods give similar performance as [13] and [14] and permits new results to be obtained for several cases that heretofore had resisted solution in a simple form. Moreover, our method has low computational complexity since our finite series expression does not have convergence problem. We have also demonstrated the versatility of our approach by applying it to several fading distributions and two different diversity receiver structures (MRC and SLC receivers) with i.i.d and i.n.d fading statistics. These results could be readily used in deciding the diversity order and/or the energy threshold value required to achieve a specified false alarm rate for different operating scenarios of a cognitive radio network (dynamic spectrum access).

ACKNOWLEDGMENT

This work is supported in part by funding from the US Army Research Office, National Science Foundation and Clarkson Aerospace/US Air Force Research Laboratory.

REFERENCES

[1] H. Urkowitz, "Energy Detection of Unknown Deterministic Signals," *Proceedings of the IEEE,* vol. 55, pp. 523-531, Apr 1967.

[2] V.I. Kostylev, "Energy Detection of a Signal with Random Amplitude," *in Proc. IEEE ICC' 02*, vol. 3, pp. 1606 – 1610, Apr-May 2002.

[3] Fadel F. Digham, Mohamed-Slim Alouni and Marvin K. Simon, "On the Energy Detection of Unknown Signals Over Fading Channels," *IEEE Trans. Commun.*, vol 55, no.1,pp.21-24, Jan. 2007.

[4] S.P. Herath, N. Rajatheva, C. Tellambura,"On the Energy Detection of Unknown Deterministic Signal over Nakagami Channels with Selection Combining," *in Proc. IEEE CCECE'09*, pp. 745–749, May 2009, Canada.

[5] S. Herath, N. Rajatheva, "Analysis of Equal Gain Combining in Energy Detection for Cognitive Radio over Nakagami Channels," *in Proc. IEEE GLOBECOM'08*, Dec. 2008, U.S.A.

[6] S. Herath, N. Rajatheva, "Analysis of Diversity Combining in Energy Detection for Cognitive Radio over Nakagami Channels," *in Proc. IEEE ICC'09*, June 2009, Germany.

[7] S. Herath, N. Rajatheva, C. Tellambura, "Unified Approach for Energy Detection of Unknown Deterministic Signal in Cognitive Radio over Fading Channels" *in Proc. IEEE ICC'09*, June 2009, Germany.

[8] M. K. Simon and M-S Alouini, *Digital Communication over Fading Channels*, New York: Wiley, 2 edition, 2005.

[9] I. S. Gradshteyn and I. M. Ryzhik, *Table of Integrals, Series and Products*, 5th ed., San Diego, CA: Academic, 1994.

[10] G. M. Dillard, "Recursive Computation of the Generalized Q-function," *IEEE Trans. Aero. Elec. Sys.*, vol. AES-9, July 1973, pp. 614–615.

[11] L- L. Li, F. Li and F. B. Gross, A New Polynomial Approximation for J_m Bessel Functions, *Elsevier Journal of Applied Mathematics and Computation,* Vol. 183, pp. 1220 1225, 2006.

[12] A. H. Nuttall, "Some Integrals Involving the Q_M-Function,," *IEEE Trans. Information Theory,* Vol. 21,p 95-96, January 1974.

[13] A. Annamalai, O. Olabiyi, S. Alam, O. Odejide, and D. Vaman, "Unified Analysis of Energy Detection of Unknown Signals over Generalized Fading Channels," *Proc. IEEE IWCMC'2011* , Istanbul, Sept, 2011.

[14] O. Olabiyi, and A. Annamalai, "Further Results On the Performance of Energy Detector over Generalized Fading Channels," *to appear in Proc. IEEE PIMRC'2011*, Toronto, Canada.

Efficient Evaluation of Area under the ROC Curve of Energy Detectors over Fading Channels

Oluwatobi Olabiyi, Shumon Alam, Olusegun Odejide, Annamalai Annamalai

Center of Excellence for Communication Systems Technology Research
Department of Electrical and Computer Engineering
Prairie View A&M University, TX 77446, USA

engr3os@gmail.com, femiodejide@yahoo.com, shalam@pvamu.edu, aaannamalai@pvamu.edu

ABSTRACT

In this article, we derive closed-form expressions for the area under the receiver operating characteristics (ROC) curve (hereafter, referred to as AUC metric) of energy detectors with/without diversity receiver over a myriad of fading environments. Our "unified" expression for the AUC performance metric only requires the knowledge of the derivatives of the moment generating function (MGF) of fading signal-to-noise ratio (SNR) at the receiver output. Aside from developing a new analytical framework for evaluating the AUC of energy detectors, we show that our approach also facilitates the treatment of a more generalized/realistic propagation environment (i.e., not restricted to Rayleigh/Nakagami-m channels). Moreover, the AUC analysis enables comparisons between various overlapping ROC curves to be performed easily. Different factors affecting the spectrum sensing performance were also analyzed and our key results and findings are useful in the design of cognitive radio networks.

Categories and Subject Descriptors

C.2.1 Network Architecture & Design: wireless communications; C.2.3 Network Operations: network monitoring; C.4 Performance of Systems: modeling techniques, performance attributes.

General Terms

Theory, Performance, Reliability, Design, Algorithms.

Keywords

Energy Detection; Diversity Receiver; Area under the Receiver Operating Characteristics Curve; Opportunistic Spectrum Access.

1. INTRODUCTION

The emerging technology of cognitive radio has created a paradigm shift in the design of wireless system where radio can now adapt their operating behavior to take advantage of unused spectrum. One of the main requirements of this system is to ensure that the incumbent (i.e. primary or licensed user) is not interrupted by the activity of these cognitive radios. Therefore the new radio must be capable of determining the presence or absence of an incumbent before spectrum usage. The challenge of identifying unused spectrum has become fascinating topic within research community and is an important part of IEEE802.22 standard for cognitive radio system. Energy detector has been identified as one of the prospective solutions to this problem due to the simplicity of its implementation [1].

The characterization of energy detectors has been well researched in literature with aid of ROC curves. ROC curves help to visualize, organize and select classifiers based on their performance. ROC has long been used for characterizing the system performance especially in signal detection to show the tradeoff between detection and false alarm rates [2], [8], [10], [11], [12], [13]. Also, ROC analysis is extensively used in making decisions in medicine and industry [3], [9]. For energy detection, ROC is a plot representing probability of detection (P_d) versus false alarm probability (P_f). Alternatively, it is a plot of missed detection probability (P_m=1-P_d) versus P_f (referred to as complementary ROC) [2].

Although the ROC curves fully characterize the performance of an energy detector, it is sometimes difficult to compare detectors' performance using their ROC curves especially when the curves overlap. Therefore, it is desirable to have a single figure of merit that can give distinction between two detectors. Such a measure is the area under the ROC curve (AUC) which varies between 0.5 (poor performance) and 1 (good performance) [5]. AUC is a good measure of detection capability and it represents the probability that choosing the correct decision by the detector is more likely than choosing the incorrect decision [5]. This statistical analysis is closely related to Wilcoxon ranking statistics [5] and it has been shown that AUC is the continuous equivalent of Wilcoxon ranking statistics [5]. Bounds, asymptotic expression and limiting value of AUC can also be found in [6], and [7]. Although, AUC is a useful parameter for illustrating detector's performance, limited research work regarding AUC is found in the communication areas. In [4], closed-form expressions for the AUC of an energy detector under the Nakagami-m fading channel can be found. However the derived expressions are complex and limited to Nakagami-m channel and cannot be generalized to other fading channels. To the best of our knowledge only [4] has applied this method to the analysis of performance of energy detector.

In this paper, we develop a simple, compact and closed form solution to the problem of performance analysis of energy detector using AUC. First, our final expression is much simpler than the final expression in [4] (with hypergeometric functions). Second, the nature of our final expression provides intuitive behaviour of energy detector in fading and non-fading channels. More so, since the instantaneous AUC is in the desirable exponential form, the averaging over different fade distributions is straight forward. Ultimately, our derivations culminated into k^{th} order derivative of the moment generating function (MGF) of the fading channel. This allows us to easily generalise average AUC over fading channels. The analysis has been extended to the diversity scenarios and simple closed form solutions have been

MSWiM'11, October 31 - November 4, 2011, Miami, Florida, USA

derived for both MRC and SLC diversity detectors. In addition, our work generalizes Nuttall's previous result in [14, (8)].

2. SYSTEM MODEL AND THE AREA UNDER ROC CURVE (AUC)

The detection of the received signal y(t) at the detector at time *t*, assuming flat fading can be modelled as a binary hypothesis (H_0 and H_1) [2]. Hypothesis H_0 means that the signal is absent whereas H_1 implies that the signal is present. y(t) can be modelled as

$$y(t) = \begin{cases} n(t) & : H_0 \\ hs(t) + n(t) & : H_1 \end{cases} \quad (1)$$

where h is the channel gain, $s(t)$ is the transmitted signal and $n(t)$ is the AWGN. The probability of detection (P_d) and the false alarm (P_f) of the energy detector can be written by [8]

$$P_d = Q_u(\sqrt{2\gamma}, \sqrt{\lambda}) \quad (2)$$

where $u = TW$ is the time-bandwidth product, λ is the energy detection threshold, and $Q_u(.,.)$ is the generalised (u^{th} order) Marcum-Q function.

$$P_f = \frac{\Gamma(u, \frac{\lambda}{2})}{\Gamma(u)} \quad (3)$$

where $\Gamma(a,x) = \int_x^{\infty} t^{a-1} e^{-t} dt$ is the upper incomplete gamma.

The AUC provides better insights of a detector's performance with respect to various parameters due to the fact that AUC varies from 0.5 to 1 as the energy-threshold λ varies from ∞ to 0 for energy detector [9]. The AUC can be expressed in term of instantaneous SNR γ value as

$$A(\gamma) = \int_0^1 P_d(\gamma, \lambda) dP_f(\lambda) \quad (4)$$

Since $P_f(\lambda)$ and $P_d(\gamma,\lambda)$ are both functions of the threshold λ, the threshold averaging method [9] can be used in evaluating AUC. Since $P_f(\lambda)$ varies from 0→1 as λ varies from ∞→0, Eq. (4) can be re-written by [4]

$$A(\gamma) = -\int_0^{\infty} P_d(\gamma, \lambda) \frac{\partial P_f(\lambda)}{\partial \lambda} d\lambda \quad .(5)$$

From (3), we can write

$$\frac{\partial P_f(\lambda)}{\partial \lambda} = -\frac{\lambda^{u-1} e^{-\frac{\lambda}{2}}}{2^u \Gamma(u)} \quad (6)$$

Substituting (6) and into (5) we obtain

$$A(\gamma) = \frac{1}{2^u \Gamma(u)} \int_0^{\infty} \lambda^{u-1} e^{-\frac{\lambda}{2}} Q_u(\sqrt{2\gamma}, \sqrt{\lambda}) d\lambda \quad .(7)$$

This, after change of variable substitution of $x = \sqrt{\lambda}$ could be written as

$$A(\gamma) = \frac{1}{2^{u-1} \Gamma(u)} \int_0^{\infty} x^{2u-1} e^{-\frac{1}{2}x^2} Q_u(\sqrt{2\gamma}, x) dx \quad (8)$$

Eq. (8) is evaluated as shown in Appendix A and using the result in (A.7), we obtain,

$$A(\gamma) = 1 - \sum_{v=0}^{u-1} (1/2)^{u+v} e^{-\gamma/2} L_v^{u-1}(-\gamma/2) \quad (9)$$

where $L_v^u(r) = \sum_{k=0}^{v} \binom{v+u}{v-k} \frac{(-r)^k}{k!}$ denotes the generalized Laguerre polynomial. For the special case of u=1, (A.7) reduces to Nuttall's previous result [14, Eq. (8)]. Moreover, (9) can be further compactly written as

$$A(\gamma) = 1 - \sum_{v=0}^{u-1} \sum_{k=0}^{v} \binom{v+u-1}{v-k} \left(\frac{1}{2}\right)^{k+v+u} \frac{1}{k!} \gamma^k e^{-\frac{1}{2}\gamma} \quad (10)$$

The simplicity of (10) is not yet found in any literature and it circumvents the need for sophisticated software packages to evaluate special hypergeometric function in [4]. It will be shown in the later section how (10) simplifies the averaging of $A(\gamma)$ over the probability distribution function (PDF) of fading channels.

3. AVERAGE AUC OVER GENERALIZED FADING CHANNELS

3.1 No Diversity case

The average AUC, $\overline{A}$, over generalized fading channels can be evaluated by averaging (8) over the PDF of the channel SNR, $f_\gamma(\gamma)$ yielding

$$\overline{A} = \int_0^{\infty} A(\gamma) f_\gamma(\gamma) d\gamma \quad (11)$$

Substituting (10) into (11) yields

$$\overline{A}_{gen} = 1 - \sum_{v=0}^{u-1} \sum_{k=0}^{v} \binom{v+u-1}{v-k} \left(\frac{1}{2}\right)^{k+v+u} \int_0^{\infty} \frac{\gamma^k e^{-\frac{1}{2}\gamma}}{k!} d\gamma \quad (12)$$

Similar to our dervation in [16], it is straight forward to show that (12) can be expressed as,

$$\overline{A}_{gen} = 1 - \sum_{v=0}^{u-1} \sum_{k=0}^{v} \binom{v+u-1}{v-k} \left(\frac{1}{2}\right)^{k+v+u} \frac{(-1)^k}{k!} \phi_\gamma^{(k)}(s) \Big|_{s=\frac{1}{2}} \quad (13)$$

where $\phi_\gamma^{(k)}(s) = \frac{\partial^k \phi_\gamma}{\partial s^k}$ and $\phi_\gamma(s)$ is the MGF of different stochastic fading channels. The kth derivatives of the MGF of common fading channels are listed in [13, Table 1]. Eq. (13) is the simplest form average AUC over generalized fading channel that has been derived in literature.

For example, for Nakagami-m and Rician fading channel, the average AUC is given respectively by

$$\overline{A}_{Nak} = 1 - \sum_{v=0}^{u-1} \sum_{k=0}^{v} \binom{v+u-1}{v-k} \left(\frac{1}{2}\right)^{k+v+u} \frac{1}{k!} \frac{\Omega^k m^m \Gamma(m+k)}{(m+0.5\Omega)^{m+k} \Gamma(m)} \quad (14)$$

$$\overline{A}_{Ric} = 1 - \sum_{v=0}^{u-1} \sum_{k=0}^{v} \binom{v+u-1}{v-k} \left(\frac{1}{2}\right)^{k+v+u} \frac{\Omega^k k! (1+K)}{(1+K+0.5\Omega)^{1+k}} \times \exp\left(\frac{-0.5K\Omega}{1+K+0.5\Omega}\right) \sum_{i=0}^{k} \frac{1}{(i!)^2 (k-i)!} \left[\frac{K(1+K)}{1+K+0.5\Omega}\right]^i \quad (15)$$

These expressions are simple and in closed form and they do not involve infinite series.

3.2 Diversity Based Detection

3.2.1 Maximum Ratio Combining (MRC)

MRC is a linear combining method in which the signals from all diversity branches are first co-phased and weighted in proportional to their SNR before summing. Although, this requirement made MRC an undesirable combining technique due to the need for complete channel information of each branch,

ROC analysis for MRC is found in various literatures. The output SNR, γ_{MRC} of the MRC combiner is the sum of the SNRs on all branches, i.e., $\gamma_{MRC} = \sum_{l=1}^{L}\gamma_l$ (where L is the number of diversity branches). The statistics of the decision variable of a MRC follows χ^2_{2u} distribution under H$_0$ and $\chi^2_{2u}(\gamma_{MRC})$ under H$_1$ Hypothesis [13]. Therefore, the P_d and the P_f at the MRC output for AWGN channels can be evaluated by (2) and (3) respectively. Thus without any change, using the MGF in [13, Table 2] for L-branch diversity, closed form Average AUC can be readily derived from (13).

For examples, closed form and concise expression for average AUC of MRC diversity detectors over the Nakagami-m and Rician fading channels are given respectively by.

$$\bar{A}^{MRC}_{Nak} = 1-\sum_{v=0}^{u-1}\sum_{k=0}^{v}\binom{v+u-1}{v-k}\left(\frac{1}{2}\right)^{k+v+u}\frac{1}{k!}\frac{\Omega^k m^{mL}\Gamma(mL+k)}{(m+0.5\Omega)^{mL+k}\Gamma(mL)} \quad (16)$$

$$\bar{A}^{MRC}_{Ric} = 1-\sum_{v=0}^{u-1}\sum_{k=0}^{v}\binom{v+u-1}{v-k}\left(\frac{1}{2}\right)^{k+v+u}\frac{\Omega^k (L+k-1)!(1+K)^L}{(1+K+0.5\Omega)^{L+k}} \quad (17)$$

$$\times\exp\left(\frac{-0.5KL\Omega}{1+K+0.5\Omega}\right)\sum_{i=0}^{k}\frac{1}{i!(L+i-1)!(k-i)!}\left[\frac{KL(1+K)}{1+K+0.5\Omega}\right]^i$$

3.2.2 Square-Law combing (SLC):

The SLC diversity scheme provides good alternative to MRC but with reduced accuracy due to increase in degree of freedom. Since the output decision is combined in this scheme, the decision variable is the sum of L independent and identically distributed (i.i.d.) variables. The statistics of the decision variable of a SLC follows χ^2_{2u} distribution under H$_0$ and $\chi^2_{2Lu}(\varepsilon_{SLC})$ under H$_1$ Hypothesis [13]. Where the non-centrality $\epsilon_{SLC} = \sum_{l=1}^{L}\epsilon_l = \sum_{l=1}^{L}2\gamma_l = 2\gamma_{SLC}$; hence the P_d and the P_f can be expressed by (2) and (3) respectively by replacing u with Lu.

Similar to the analysis of MRC diversity based detection, the closed form average AUC for the SLC diversity for Nakagami-m and Rician fading channels are given by (18) and (19) respectively.

$$\bar{A}^{SLC}_{Nak} = 1-\sum_{v=0}^{uL-1}\sum_{k=0}^{v}\binom{v+uL-1}{v-k}\left(\frac{1}{2}\right)^{k+v+uL}\frac{1}{k!}\frac{\Omega^k m^{mL}\Gamma(mL+k)}{(m+0.5\Omega)^{mL+k}\Gamma(mL)} \quad (18)$$

$$\bar{A}^{MRC}_{Ric} = 1-\sum_{v=0}^{uL-1}\sum_{k=0}^{v}\binom{v+uL-1}{v-k}\left(\frac{1}{2}\right)^{k+v+uL}\frac{\Omega^k (L+k-1)!(1+K)^L}{(1+K+0.5\Omega)^{L+k}} \quad (19)$$

$$\times\exp\left(\frac{-0.5KL\Omega}{1+K+0.5\Omega}\right)\sum_{i=0}^{k}\frac{1}{i!(L+i-1)!(k-i)!}\left[\frac{KL(1+K)}{1+K+0.5\Omega}\right]^i$$

4. NUMERICAL RESULTS

First, we show how the AUC works as the single figure of merit and useful it could be especially when the ROC curves overlaps. We later show the behavior of AUC with variation in different detection parameter and diversity detection schemes. Fig. 1 shows six different complementary ROC curves with three overlapping pairs. Looking at the ROC curve, it is difficult to determine with of the detection scenario within the pairs gives better performance across the ranges of false alarm probability. Using the average AUC expressions proposed in this article, each curve directly maps to a single average AUC value which makes the comparison much easier. The computed average AUC value for the curves in Fig. 1 are 0.9272, 0.9167, 0.9878, 0.9803, 0.9943, 0.9936, corresponding to curves C1 to C6.

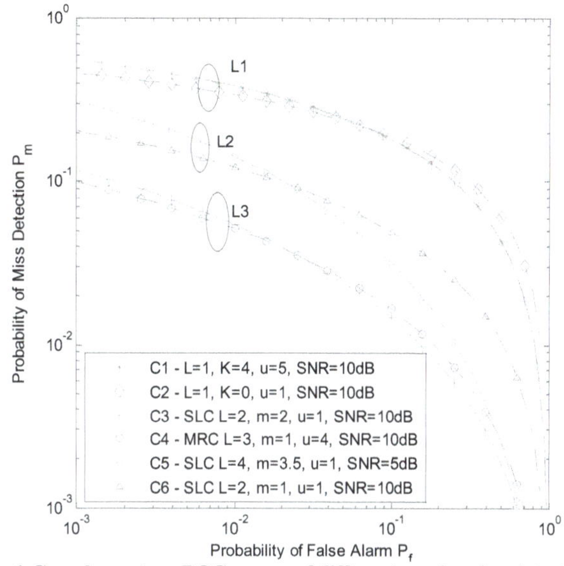

Fig. 1 Complementary ROC curves of different overlapping detection configuration.

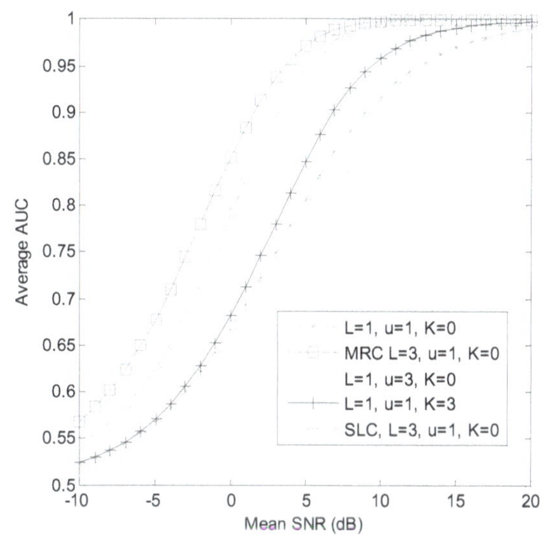

Fig. 2 Average AUC versus Mean SNR with detection parameters and different diversity schemes over Rice channel.

This implies that the average AUC can be a single figure of merit that characterizes the performance of an energy detector over the range of received signal SNR. Also, these values are in agreement with the statistical interpretation of ROC curves showing that pairs L3 performs better than L2 and L2 better than L1. In Fig. 2, the AUC characterization is investigated over the Rice fading channel with different parameter changes and diversity schemes. The trend shown by average AUC is similar to that of ROC curve in that the detector's performance improves with increasing received SNR, channel fading indices, and diversity and it also improves with reduced number of samples, u. Similarly, the figure shows that MRC diversity based detection performs better than the SLC in agreement with ROC curves in [13]. Although not shown here, similar behavior is observed for Nakagami-m channel. It is worth to mention here that, each point on any curve of Fig. 2 corresponds to an ROC curve for each SNR value. Therefore the AUC is able to provide more insight than ROC curve which is limited to specific average SNR.

5. CONCLUSIONS

In this paper we have derived a novel, simple, compact and closed form expression for the analysis of AUC that requires only $u(u+1)/2$ and $uL(uL+1)/2$ computational terms for MRC and SLC respectively. As the value of u and L are generally small, our expression provides very efficient way of evaluating the AUC. Also, our results are all in closed form and can be easily generalize over different fading channel distributions even with diversity schemes. The final expression result into kth order derivative of MGF of SNR of the channel which is readily available in literature. This is far more efficient than the approach adopted in [4] which requires complicated derivations for different channels and diversity systems. These results could be readily used to choose right parameters for energy detector receivers in cognitive radio system.

ACKNOWLEDGMENT

This research was supported in part by funding from the US Army Research Office, National Science Foundation and Clarkson Aerospace/US Air Force Research Laboratory.

APPENDIX A

In order to evaluate (8), we need to solve the following integral which does not exist in [12] and [14],

$$I = \int_0^\infty x^{2u-1} e^{-\frac{1}{2}p^2x^2} Q_u(b, ax)\,dx \qquad \text{(A.1)}$$

Using the generalized Marcum-Q function given by [14, (1)], $Q_u(b,ax) = \int_{ax}^{\infty} t\left(\frac{t}{b}\right)^{u-1} e^{-\frac{1}{2}(t^2+b^2)} I_{u-1}(bt)\,dt$, (A.1) becomes

$$I = \int_0^\infty x^{2u-1} e^{-\frac{1}{2}p^2x^2}\left[\int_{ax}^{\infty} t\left(\frac{t}{b}\right)^{u-1} e^{-\frac{1}{2}(t^2+b^2)} I_{u-1}(bt)\,dt\right]dx \qquad \text{(A.2)}$$

where $I_u(.)$ is the modified Bessel function of first kind of order u. Eq. (A.2) could be re-written (after change of the order of integration) as,

$$I = \int_0^\infty \frac{t^u}{b^{u-1}} e^{-\frac{1}{2}(t^2+b^2)} I_{u-1}(bt)\left[\int_0^{t/a} x^{2u-1} e^{-\frac{1}{2}p^2x^2}\,dx\right]dt \qquad \text{(A.3)}$$

Using [12, (8.350-1)] we obtain

$$I = \frac{2^{u-1}e^{-\frac{1}{2}b^2}}{b^{u-1}p^{2u}} \int_0^\infty t^u e^{-\frac{1}{2}t^2} I_{u-1}(bt)\, G\left(u, \frac{p^2t^2}{2a^2}\right)dt \qquad \text{(A.4)}$$

where $G(.,.)$ is the lower incomplete Gamma function.

For integer u, using [12, (8.352.1)], we obtain

$$I = \frac{2^{u-1}e^{-\frac{1}{2}b^2}(u-1)!}{b^{u-1}p^{2u}}\left[\int_0^\infty t^u e^{-\frac{1}{2}t^2} I_{u-1}(bt)\,dt - \sum_{m=0}^{u-1}\frac{1}{m!}\left(\frac{p^2}{2a^2}\right)^m \times \int_0^\infty t^{u+2m} e^{-\left(\frac{1}{2}+\frac{p^2}{2a^2}\right)t^2} I_{u-1}(bt)\,dt\right] \qquad \text{(A.5)}$$

Applying [12, (6.631.1)] and [12, (8.406.3)], we obtain

$$I = \frac{2^{u-1}e^{-\frac{1}{2}b^2}(u-1)!}{p^{2u}}\left[{}_1F_1\left(u;u;\frac{b^2}{2}\right) - \left(\frac{a^2}{a^2+p^2}\right)^u \sum_{m=0}^{u-1}\left(\frac{p^2}{a^2+p^2}\right)^m \times \frac{1}{m!}\frac{\Gamma(u+m)}{\Gamma(u)}\,{}_1F_1\left(u+m;u;\frac{a^2b^2}{2(a^2+p^2)}\right)\right] \qquad \text{(A.6)}$$

which after applying [12, (9.212.1)] and the identity [12, (8.972.1)] with few algebraic manipulation we obtain

$$I = \frac{2^{u-1}(u-1)!}{p^{2u}}\left[1 - \left(\frac{a^2}{a^2+p^2}\right)^u e^{-\frac{p^2b^2}{2(a^2+p^2)}} \times \sum_{m=0}^{u-1}\left(\frac{p^2}{a^2+p^2}\right)^m L_m^{u-1}\left(\frac{-a^2b^2}{2(a^2+p^2)}\right)\right] \qquad \text{(A.7)}$$

where $L_m^u(r) = \sum_{i=0}^{m}\binom{m+u}{m-i}\frac{(-r)^i}{i!}$ [12, (8.970.1)], is the generalized Laguerre polynomial. Also note that for the special case of $u = 1$, (A.7) reduces to Nuttall's previous result [14, (8)].

REFERENCES

[1] K. B. Letaief, W. Zhang, "Cooperative Communications for Cognitive Radio Networks.'' *Proc. IEEE*, vol. 97, no. 5, pp 878-893, May.

[2] H. Urkowitz, "Energy Detection of Unknown Deterministic Signals," *Proc. IEEE,* vol. 55, pp. 523-531, Apr 1967.

[3] J. A. Swets, R.M. Dawes, and J. Monahan, "Better Decisions Through Science," *Scientific American*, vol. 283, no. 4 pp. 82-87, Oct. 2000.

[4] S. Atapattu, C. Tellambura, and H. Jiang "Analysis of Area under the ROC Curve of Energy Detection," *IEEE Transactions on Wireless Communication*, vol. 9, no. 3, pp. 1216-1225, March 2010.

[5] J. A. Hanley and B. J. Mcneil, "The Meaning and Use of the Area under a Receiver Operating Characteristics (ROC) Curve," *Radiology*, vol. 143, no. 1, pp. 29-36, Apr. 1982..

[6] J. H. Shapiro, "Bounds on the Area under the ROC Curve," *J. Opt. Soc. Am. A,* vol. 16, no. 1, pp. 1963-1968, Jan 1999.

[7] H. H. Barrett, C. K. Abbey, and E. Clarkson, "Objective Assessment of Image Quality—III: ROC Metrics, Ideal Observers, and Likelihood-Generating Functions," *J. Opt. Soc. Am. A*, vol. 15, no. 6, pp. 1520–1535, June 1998.

[8] F. Digham, M. Alouini, and M. Simon, "On the Energy Detection of Unknown Signals over Fading Channels," *IEEE Trans. on Comm.*, vol. 55, no. 7, pp.1407-1416, July 2007.

[9] T. Fawcett, "An Introduction to ROC Analysis," *Pattern Recognition Letter*, vol. 27, no. 8, pp. 861–874, June 2006.

[10] S. Herath, N. Rajatheva, C. Tellambura,"On the Energy Detection of Unknown Deterministic Signal over Nakagami Channels with Selection Combining," *Proc. IEEE CCECE'09*, pp. 745–749, May 2009.

[11] M. K. Simon and M-S Alouini, *Digital Communication over Fading Channels*, New York: Wiley, 2nd Edition, 2005.

[12] I. S. Gradshteyn and I. M. Ryzhik, *Table of Integrals, Series and Products*, 5th Edition, San Diego, CA: Academic, 1994.

[13] A. Annamalai, O. Olabiyi, S. Alam, O. Odejide and D. Vaman, "Unified Analysis of Energy Detection of Unknown Signals over Generalized Fading Channels," to appear in *Proc. IEEE IWCMC 2011 Conference*, Istanbul, July 2011.

[14] A. H. Nuttall, "Some Integrals Involving the Q_M-Function,," *IEEE Trans. information Theory*, vol. 21, pp. 95-96, 1974.

Designing Optical Metro and Access Networks for Future Cooperative Cellular Systems

Thorsten Biermann
DOCOMO Euro-Labs
Landsberger Str. 312
80687 Munich, Germany
biermann@docomolab-euro.com

Luca Scalia
DOCOMO Euro-Labs
Landsberger Str. 312
80687 Munich, Germany
scalia@docomolab-euro.com

Holger Karl
University of Paderborn
Pohlweg 47-49
33098 Paderborn, Germany
holger.karl@uni-paderborn.de

ABSTRACT

Using Coordinated Multi-Point (CoMP) transmission and reception techniques poses challenging latency and capacity requirements on the backhaul network infrastructure of cellular access systems. On a small scale, these requirements can be fulfilled by using upcoming optical technologies like Wavelength Division Multiplexing (WDM) Passive Optical Networks (PONs). On a metro scale, however, it is unclear to which extent CoMP is feasible.

For a metro-wide cellular network deployment, we propose different backhaul network architectures and analyze their capability of fulfilling CoMP requirements. The analysis shows a trade-off between the ability of reusing existing metro network infrastructure and the area that can be covered using CoMP techniques. Based on this, we provide not only an understanding of which architecture approach fits best for a certain scenario, but also point out necessary hardware upgrades required for supporting CoMP in a desired target area.

Categories and Subject Descriptors

C.2.1 [**Network Architecture and Design**]: Network communications, Network topology, Wireless communication

General Terms

Theory, Design

Keywords

Wireless, Cooperation, CoMP, Joint processing, Backhaul, Optical, Cellular networks, Model

1. INTRODUCTION

The spread of smartphones has changed the habits of mobile users, who are nowadays more and more interested in multimedia content and social network services. To address the high bandwidth requirements of such services, requirements for future cellular networks, like 3GPP's Long Term Evolution (LTE)-Advanced, have been defined [1]. Wireless speeds of 1 Gbit/s per cell are expected.

One of the technologies envisioned to achieve this goal is Coordinated Multi-Point (CoMP) transmission and reception. In a CoMP system, multiple cells (or, equivalently, sectors) cooperate to improve the wireless service quality. In the general case, cooperating cells belong to different Base Stations (BSs), thus requiring interaction between the BSs involved in the cooperation.

Different types of cooperation are possible. Referring to current 3GPP activities, joint scheduling [2], joint beamforming [3] and joint processing [4] are techniques which are under discussion for LTE-Advanced. These techniques achieve different performance gains in terms of throughput [5–7] and have a different impact on the wireline network infrastructure of the mobile network. As the part of the wireline network infrastructure that connects the BSs to the core network plays an important role in the feasibility of BS cooperation, we use an own name for this part. It will be called *backhaul network* in the remainder of this paper.

We focus on Joint Processing (JP), which is promising among the proposed CoMP techniques due to the expected high gains. In a downlink JP system, user data is shared among different cells from where it is simultaneously transmitted to the User Equipment (UE). Constructive interference at the UE is ensured by synchronization and precoding of user data. An illustration of JP is shown in Figure 1.

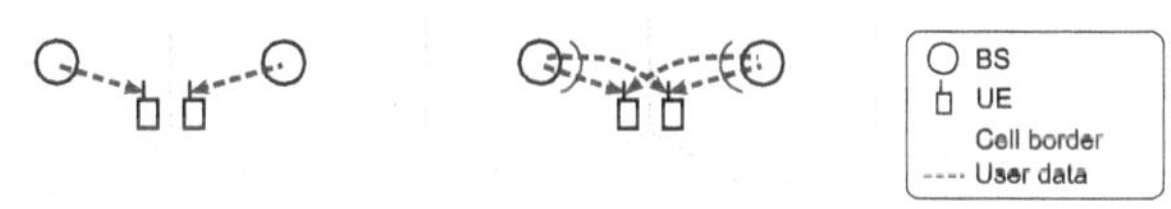

(a) Traditional transmission (no JP). (b) BSs use JP to jointly serve the UEs.

Figure 1: Downlink JP for two adjacent cells serving two UEs that are at the cell borders.

One BS usually operates multiple cells/sectors. In case JP (or any other CoMP technique) is applied only among cells that belong to the same BS, the backhaul network is not involved in the data transfer arising from cooperation. However, when CoMP spans over cells belonging to different BSs, the backhaul network is involved. User data as well

MSWiM'11, October 31–November 4, 2011, Miami, Florida, USA.
Copyright 2011 ACM 978-1-4503-0898-4/11/10 ...$10.00.

as signaling and Channel State Information (CSI) need to be shared between different BSs, thus introducing not only additional load in the network but also stringent delay requirements. These delay requirements arise from the short validity of the measured CSI, which is on the order of 1 ms. Details on the requirements will follow in Section 2.

In small-scale hot-spot scenarios, like airports or shopping malls, the requirements can be fulfilled by deploying Remote Radio Head (RRH) solutions where there is just a single, big BS unit with many remote antennas connected via optical fiber. Alternatively, when multiple BSs are distributed over a small area, like with pico cells, they can be interconnected using optical technologies, like Wavelength Division Multiplexing (WDM) Passive Optical Networks (PONs) [8]. This technology provides high link capacity and introduces only low latency for traffic that stays within a PON.

When using cooperation techniques on a larger scale, e.g., to support JP among adjacent macro cells, BSs will be involved that are connected to different PONs. In this case, inter-BSs cooperation is challenging as user, CSI, and signaling data need to go one layer up in the wireline network hierarchy, i.e., via a metro or core network, before reaching the destination BS. As a consequence, a significantly larger part of the mobile network is involved in exchanging additional traffic with strict latency constraints (Figure 2). Queuing and processing delay at intermediate nodes like Internet Protocol (IP) routers limit the cooperation feasibility.

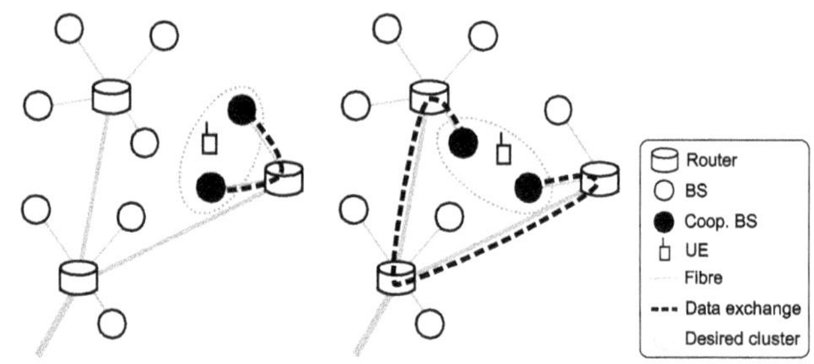

(a) Only local data exchange necessary, e.g., via a PON.

(b) Data exchange via metro network required.

Figure 2: JP cooperation scenarios and impact on the mobile backhaul infrastructure.

In our previous work, we addressed these limitations by predicting the set of BSs which are eligible for cooperation based on the current backhaul network conditions [9, 10]. Based on these predictions, only those BSs eventually cooperate that are able to fulfill the backhaul network requirements. In this paper, rather than dealing with these limitations, we design a cost-effective backhaul network architecture which inherently enables JP on a metro scale.

We have developed models of different backhaul network architectures to check whether they are able to support JP. We evaluate backhaul architecture implementations which either reuse existing metro ring infrastructures (Section 4) or are designed from scratch to include the requirements of JP (Section 5). For each of these architectures, we evaluate the area that can be covered such that JP is feasible. During the evaluation, we vary various system parameters and identify trade-offs between the different architectures.

The results of our analysis show that there exists an inevitable trade-off between reusing current infrastructures, enabling cooperation between all adjacent cells, and the limits of upcoming WDM PON technologies. Based on these insights, we consider possible enhancements to improve JP feasibility. Furthermore, we propose and analyze an architecture in Section 6 that reuses existing infrastructures to a large extent but also exploits advantages of future optical technologies to provide a high coverage area in which JP can be applied among arbitrary BSs.

2. JOINT PROCESSING

This section describes how JP works and points out the resulting strict requirements on the backhaul network.

2.1 Mechanism

In a JP system, multiple BSs simultaneously transmit downlink user data such that the BSs' signals create constructive interference at the UE. This increases the experienced Signal To Interference and Noise Ratio (SINR) and hence improves service quality. A similar idea can be applied in the uplink where multiple BSs simultaneously receive data from a single UE.

Figure 3 illustrates a downlink JP scenario. We assume a distributed JP implementation where each BS locally encodes the raw user data based on the precoding weights that have been calculated jointly. This approach is complementary to the centralized one in which data is encoded at a central location and sent, e.g., in form of IQ samples, to the other BSs. This approach is less convenient than the distributed one since transferring the encoded data via the backhaul requires 10-20 times higher data rates than the transmission of the raw data, which happens in the distributed approach.

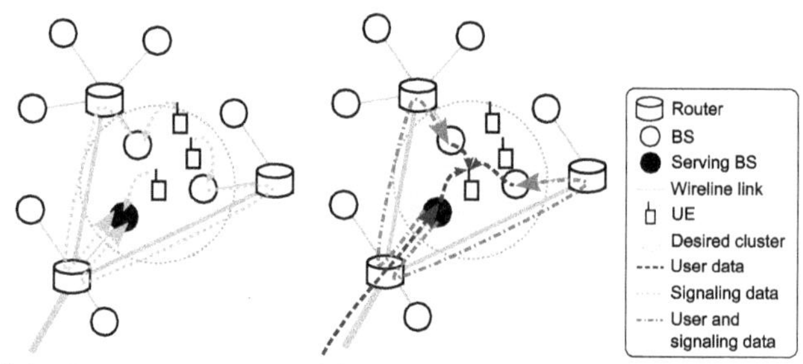

(a) Phase 1: UEs collect CSI and send it to the BSs. They forward it to the serving BS for processing.

(b) Phase 2: Serving BS forwards user data to cluster members (incl. scheduling, precoding info).

Figure 3: JP and corresponding data exchange.

In the first phase (Figure 3a), each cooperating BS collects the CSI measured by their UEs and sends it to a central point that coordinates the JP procedure, e.g., the serving BS of the UE that needs cooperation. This CSI exchange is required to permit Multi-User MIMO (MU-MIMO) processing when calculating the precoding information. In a second phase (Figure 3b), the precoding information is sent together with the raw user data to each cooperating BS where the simultaneous wireless transmission takes place eventually.

2.2 Requirements on backhaul network

CSI is only valid for a limited time as the wireless channel properties can change quickly. Typically, CSI is valid for

1 ms, which corresponds to the duration of an LTE subframe and which is the scheduling interval in LTE. This means that for JP in the downlink, the duration from the point in time when the CSI is measured at UEs to the point in time when the cooperating BSs simultaneously transmit must be below 1 ms. Simply speaking, the whole procedure described in Section 2.1 needs to be completed within 1 ms. Hence, link latencies and processing delay, e.g., at routers between cooperating BSs, play a very important role.

Note that the delay requirements of other CoMP techniques, like joint beamforming or joint scheduling, are similar. Hence, our considerations are not only valid for JP but also for CoMP schemes that do not exchange user data.

For JP in the uplink, latency constraints are less strict as there is no deadline for CSI validity within which a sending operation has to happen. Received data copies and corresponding CSI can be processed at any time.

The second limiting factor is the increased capacity requirement on the BSs' backhaul links. In LTE-Advanced, the downlink wireless peak data rate per BS sector is 1 GBit/s. This implies that for a foreseen 6 sector BS deployment, up to 6 GBit/s need to be sent wirelessly. From the backhauling perspective, the amount of user data which needs to be transported to a 6-sector BS is around 50 % of the wireless data rate, i.e., 3 GBit/s, removing wireless coding and other overhead [11].

Considering upcoming optical backhauling technologies, 2.5 GBit/s capacity provided by one WDM channel [8] already will not be enough to serve a single *non-cooperating* BS. Furthermore, as shown in Figure 3b, BSs that use JP have to exchange additional signaling and user data. Hence, additional backhaul capacity is required, which can be provided by assigning multiple WDM channels per JP-capable BS. Alternatively, increasing the WDM channel capacity to 10 GBit/s, which will be available in the near future, helps to satisfy the high requirements.

3. MODELING ASSUMPTIONS

Our study concentrates on the latency requirements that need to be fulfilled for JP. For different metro-scale backhaul architectures, we analyze the maximum area that can be covered by these architectures while still fulfilling the latency requirements of JP between BSs connected via these architectures. This is done by creating mathematical models for each of the architecture candidates, which will be described in detail in the following sections. The assumptions that all models have in common are listed in the following.

Passive Optical Networks (PONs): We use PONs [8] to connect the BSs to the wireline network infrastructure. This technology is the cheapest way to provide enough capacity for future JP-enabled systems.

PONs have a tree topology. A node called Optical Line Terminal (OLT) defines the root of the tree and is connected via a fiber, which is split up to multiple Optical Network Units (ONUs) using an Array Waveguide Grating (AWG) in case of WDM PONs or a passive splitter in case of Time Division Multiplexing (TDM) PONs. A router or L3 switch that is colocated with the OLT controls incoming and outgoing traffic to/from BSs that are connected to the ONUs.

In our scenario, deployed PONs have a splitting factor of $sf_{PON} = 32$, i.e., each OLT is connected to 32 ONUs via the remote node. In the evaluations, the splitting factor only affects the calculated number of PONs that are required to cover a certain area. Other properties are independent and are valid for any type of PON.

Optical links: All backhaul network links use optical technology. Their link latency is given by $s \cdot \frac{1.45}{c}$, where s is the link length, c is the speed of light, and 1.45 the refraction index of the fiber. This corresponds to a typical Single-Mode Fiber (SMF) setup.

IP processing: IP routers in the network need to make IP forwarding decisions. This introduces a delay of $d_{\mathrm{IP}} = 0.1$ ms [12,13], which also contains the queuing delay in the router. Notice that using technologies like Multiprotocol Label Switching (MPLS) does not lower the queuing delay [14]. In practice, both IP and MPLS show similar performances in terms of overall packet processing delay. Hence, all following results based on d_{IP} are also valid for MPLS.

Wireless parameters: We assume an MU-MIMO setup where the overall delay that is required for the wireless processing (Multiple Input Multiple Output (MIMO) processing, traversing the transceiver chains, etc.) is estimated at $d_{\mathrm{wls}} = 0.1$ ms. Furthermore, the maximum tolerable delay from measuring CSI at the UE until the point in time when the corresponding cooperative transmission takes place is $d_{\max} = 1$ ms (Section 2).

BS deployment: We assume an urban BS deployment scenario with an inter-BS distance of 1000 m such that each BS wirelessly covers an area of $A_{\mathrm{BS}} = 0.785\,\mathrm{km}^2$.

Metrics: In the following three sections, we analyze for three different metro-wide backhaul architectures the area A that can be covered using JP. The size of A changes for the architectures due to their different topologies and technologies used to interconnect the BSs, which in turn influences to which degree latency requirements can be fulfilled or not. For a network operator it is appealing to maximize A as JP can be applied dynamically for all UEs within this area, which improves their service quality.

We also focus on the shape of the covered area. It is relevant for network operators as JP will be deployed mainly in areas with high traffic demand due to economical reasons. Therefore, different architectures might be better suited for certain deployment scenarios due to the shape of their coverage area, e.g., because they match the shape of a city center.

Architecture parameters: In addition to the aspects introduced in this section, each of the architectures has some specific parameters that only pertain to one architecture. We will introduce these aspects later where they are needed.

4. RING INFRASTRUCTURES

In the past, optical metro rings have been deployed to bridge the gap between the core and access network as the reach of access network technologies were limited to a few kilometers. For mobile operators, metro rings have been beneficial as they provide redundant high speed connectivity for transporting data to and from the core network.

Today, the access part is often a bottleneck and will not be able to provide enough capacity for future wireless technologies like LTE-Advanced. On the other hand, operators would like to reuse existing infrastructures as much as possible due the high deployment costs. A possible solution to deal with these new requirements is to use PON technology while keeping the optical ring infrastructure as it is nowadays. The resulting architecture is illustrated in Figure 4; we refer to it as the *ring architecture* in the following.

The ring architecture consists of the metro ring and multi-

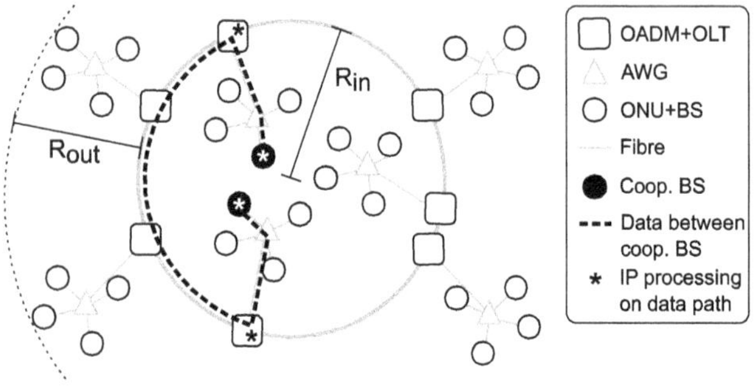

Figure 4: Ring architecture and delays that occur

ple PONs whose OLTs are connected to the ring using Optical Add-Drop Multiplexers (OADMs). The BSs are directly connected to the ONUs of the PONs.

For two BSs that apply JP, data exchange either happens directly within the same PON or via two PON and the ring in case they are connected to the same metro ring (example in Figure 4).

4.1 Ring architecture model

We assume that physical layer tunnels are available on the ring infrastructure. In this case, IP/MPLS processing is only necessary at the ring ingress and egress OADM and at the ONUs/BSs that are cooperating. The data will pass transparently through the intermediate hops. Furthermore, data from one OADM to another one on the optical ring is always sent via the shortest path, exploiting the bidirectional nature of the ring. This has the effect that the maximum distance that has to be covered is half of the ring's circumference.

In practice, not all BSs located outside the ring are required to cooperate. Because of their large distance, the benefit of JP is limited, thus making the cooperation, at least from the JP point of view, unappealing. Nonetheless, other forms of cooperation, like coordinated scheduling, could be deployed. To address this, we define an outer cooperation factor f_{out} which determines the fraction of the ring that needs to be able to cooperate. Later, we evaluate values of 0.25 and 0.5. This means that outside of the ring, BSs connected to different PONs can cooperate if the OLTs/OADMs of their PONs are located within the same half (0.25) of the ring or all need to cooperate (0.5).

Inside the ring, usually all BSs need to be able to cooperate as BSs/ONUs of different PONs are very close to each other. Analogously to f_{out}, this corresponds to an inner cooperation factor f_{in} of 0.5.

The radius of the optical ring is called R_{in}, which also equals the reach of PONs within the ring. Outside the ring, the PON reach R_{out} will be larger because $f_{\text{out}} < f_{\text{in}}$. Figure 4 illustrates R_{in} and R_{out}.

To calculate the coverage area of the ring architecture, which is the area of the circle with radius $R_{\text{in}} + R_{\text{out}}$ in Figure 4, we first calculate the accumulated IP and wireless processing delay d_{proc} that occurs between two BSs applying JP. It consists of $8 \cdot d_{\text{IP}}$ (4 times in each direction; Figure 4) and of d_{wls}.

$$d_{\text{proc}} = 8 \cdot d_{\text{IP}} + d_{\text{wls}} \tag{1}$$

The difference between $d_{\max}$ and d_{proc} remains for the propagation time in the optical fiber. Hence, the maximum possible fiber length s_{fiber} can be calculated according to Equation 2.

$$s_{\text{fiber}} = (d_{\max} - d_{\text{proc}}) \cdot \frac{c}{1.45} \tag{2}$$

As s_{fiber} is the fiber length for the full round trip, the fiber distance between two cooperating BSs must be below $s_{\text{fiber,BS}} = 0.5 \cdot s_{\text{fiber}}$. This distance must cover two times the PON reach R_{in} and the fraction f_{in} of the optical ring, such that $s_{\text{fiber,BS}} = 2 \cdot R_{\text{in}} + f_{\text{in}} \cdot 2 \cdot \pi \cdot R_{\text{in}}$. Based on this, we calculate R_{in} according to Equation 3 and R_{out} according to Equation 4.

$$\begin{aligned} R_{\text{in}} &= \frac{s_{\text{fiber,BS}}}{2 \cdot (1 + f_{\text{in}}\pi)} = \frac{s_{\text{fiber}}}{4 \cdot (1 + f_{\text{in}}\pi)} = \\ &= \frac{5}{29} c \frac{d_{\max} - 8d_{\text{IP}} - d_{\text{wls}}}{1 + f_{\text{in}}\pi} \end{aligned} \tag{3}$$

$$\begin{aligned} R_{\text{out}} &= \frac{s_{\text{fiber,BS}} - f_{\text{out}} 2\pi R_{\text{in}}}{2} = \frac{s_{\text{fiber}} - f_{\text{out}} 4\pi R_{\text{in}}}{4} = \\ &= \frac{5}{29} c (d_{\max} - 8d_{\text{IP}} - d_{\text{wls}}) \frac{1 + (f_{\text{in}} - f_{\text{out}})\pi}{1 + f_{\text{in}}\pi} \end{aligned} \tag{4}$$

Based on the two reaches R_{in} and R_{out}, the overall coverage area of the ring architecture is calculated according to Equation 5. The required number of PONs to cover this area is derived in Equation 6.

$$A = \pi \cdot (R_{\text{in}} + R_{\text{out}})^2 \tag{5}$$

$$n_{\text{PON}} = \frac{A}{A_{\text{BS}} \cdot \text{sf}_{\text{PON}}} \tag{6}$$

4.2 Results

Based on the presented model and assumptions, the resulting architecture and its coverage area is illustrated in Figure 5 for two typical parameter configurations. The plot is intended to give a first impression on the shape and size of the coverage area and to compare them to other architecture candidates later on.

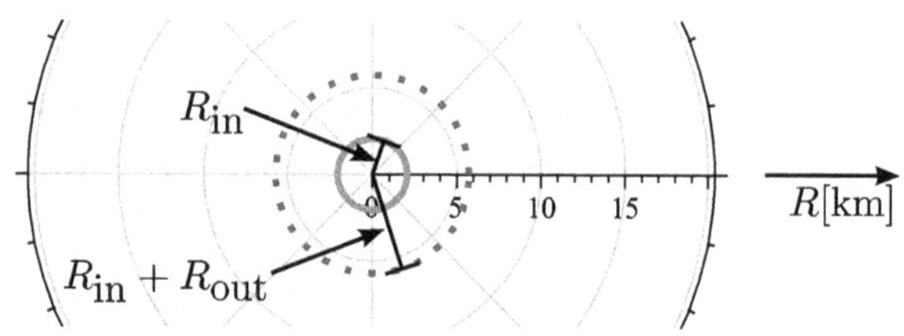

Figure 5: Size and shape of ring architecture; $d_{\text{IP}} = 0.1$ ms, $f_{\text{out}} = 0.25$. The solid circle marks the optical fiber ring, the dashed one shows the maximum range of the outer PONs.

We also evaluated the ring architecture while varying the parameters d_{IP} and f_{out}. The resulting plot is shown in Figure 6. The left y axis (black) is scaled in km for R_{in} and R_{out} and the right y axis (gray) is scaled in km^2 for A. Additionally, another y axis shows the number of PONs that are required to cover the corresponding area, implementing the linear relation of Equation 6.

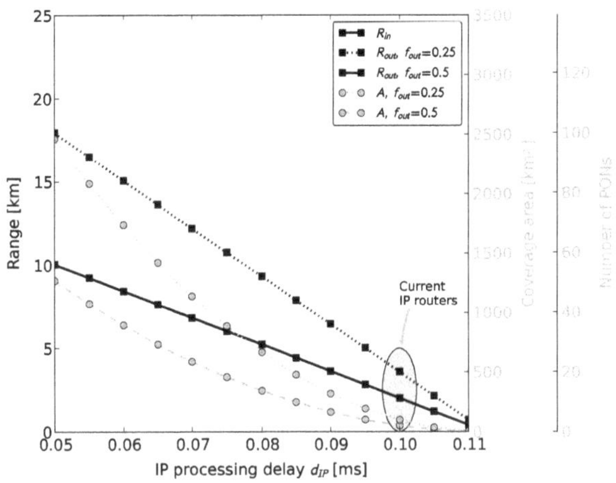

Figure 6: Feasible ring architecture size depending on d_{IP} and f_{out}

The plot shows the expected two basic behaviors: decreasing d_{IP} and f_{out} leads to a larger coverage area A and hence a higher number of PONs.

Looking at the absolute values reveals that with today's IP routing equipment ($d_{\text{IP}} = 0.1\,\text{ms}$) JP is only feasible up to a ring radius of $R_{\text{in}} \approx 2\,\text{km}$. Depending on f_{out}, the overall area that can be covered by the ring is roughly between $100\,\text{km}^2$ and $130\,\text{km}^2$, which corresponds to 4 to 5 PONs in an urban BS deployment.

The coverage area achieved by this standard ring architecture is not satisfying for a mobile network operator. For example, to cover the area of Berlin (about $900\,\text{km}^2$) with $f_{\text{out}} = 0.25$, 10 ring structures would be required, 12 for New York (about $1200\,\text{km}^2$), and 22 for Tokyo (about $2200\,\text{km}^2$).

4.3 Improvements

As the propagation delay in the fiber cannot be reduced easily, the way to have more time for signal propagation is to reduce the processing delay in the network nodes. As the major delay comes from IP/MPLS processing and queuing on the path between cooperating BSs, we propose several approaches to reduce this delay.

The first approach is to reduce the IP/MPLS processing delay in general at all locations (OADM/OLT, ONU/BS), e.g., by improving the hardware components. This leads to significant improvements (Figure 6) but is more difficult to realize in reality as hardware components need to be exchanged not only at central locations in the network but also in the field.

Another approach is to remove the IP/MPLS processing at the OADMs/OLTs. This can be achieved, e.g., by bypassing the traffic between cooperating BSs on the optical layer such that there is no delay caused by IP/MPLS processing. Then, Equation 1 changes to Equation 7, i.e., the total IP/MPLS delay is halved.

$$d_{\text{proc}} = 4 \cdot d_{\text{IP}} + d_{\text{wls}} \tag{7}$$

The same effect occurs if IP/MPLS processing is removed at the ONU/BS locations. This, however, has the same disadvantage as the described reduction of the IP/MPLS processing delay in general: equipment needs to be exchanged in the field. Furthermore, no IP/MPLS processing at the ONU complicates reusing the same PON infrastructure also for, e.g., Fiber To The Home (FTTH), and attaching multiple BSs to a single ONU. This is a clear disadvantage of removing IP/MPLS processing at the ONU locations.

Furthermore, the queuing delay can be reduced by increasing the link speeds. This is difficult in the access part of the network, i.e., in the PONs, but feasible on the links that form the metro ring.

The effects of the proposed approaches are summarized in Figure 7. The plot contains the case with IP processing at all locations (IP_{OLT}, IP_{ONU}), only at the ONUs (IP_{ONU}), and without any IP processing (no IP).

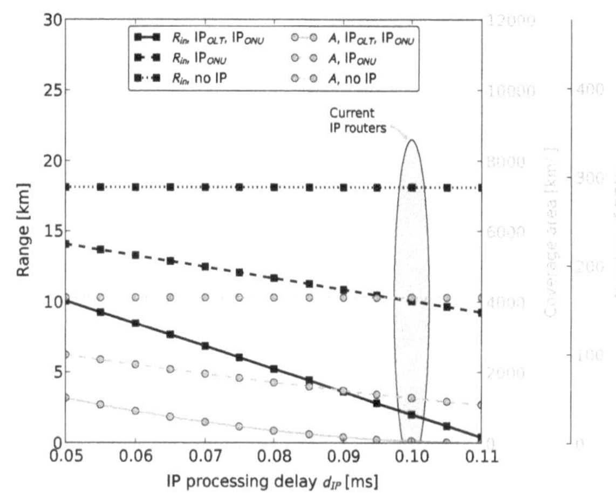

Figure 7: Feasible ring architecture size depending on d_{IP}, assuming proposed IP processing improvements at OLT and ONU; $f_{\text{out}} = 0.5$

Removing IP processing at the OADM/OLT locations leads to an increase of the ring coverage area by a factor of about 25. This means that with today's IP router equipment at the ONU/BS locations, an area of roughly $1270\,\text{km}^2$ can be covered with 50 PONs.

From the coverage perspective, this improvement is a big step. Unfortunately, equipment for implementing this step at the OADMs/OLTs is not available yet. Current OADMs support only a limited number of optical channels (on the order of 100) that are transported via the ring. This will not be enough to support separate channels for connecting each pair of cooperating BSs.

5. STAR INFRASTRUCTURES

In the past, metro networks have been deployed to cover the distance between the core network and the customer as the reach of access network technologies were limited to a few kilometers. With new access network technologies, like WDM PONs, much higher distances up to multiple tens of kilometers can be covered. This allows to directly connect the long-reach access network to the core network without the need for an intermediate metro infrastructure.

Using WDM PONs, which are able to provide a long reach and capacities that are high enough to support JP, implementing a single network covering the old metro and access part is straightforward. The resulting architecture looks like a star and is illustrated in Figure 8.

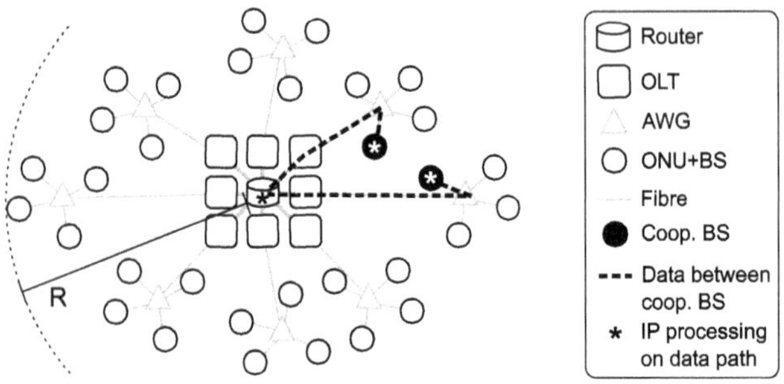

Figure 8: Star topology and delays that occur

As the OLTs are now all colocated in a single location, there is no need for a separate router for each of them as it was the case in the metro ring architecture in Section 4. All OLTs can share a single, big router which reduces IP/MPLS processing and queuing.

5.1 Star architecture model

All OLTs and the IP/MPLS router are colocated. This means that there is no optical propagation delay between the OLTs and the router.

The PON reach, i.e., the radius of the area that is covered by the star architecture is called R (Figure 8).

Calculating the coverage area of the star architecture, which is the area of the circle with radius R in Figure 8, is similar to the calculation procedure in the ring architecture. Here, the accumulated processing delay just consists of $6 \cdot d_{\mathrm{IP}}$ and of d_{wls}. The reduced IP/MPLS processing is a result of the single, centralized IP/MPLS router (cf. Figure 8) at the OLTs.

$$d_{\mathrm{proc}} = 6 \cdot d_{\mathrm{IP}} + d_{\mathrm{wls}} \tag{8}$$

The resulting maximum possible fiber length s_{fiber} is calculated exactly as already described in Equation 2.

As for the ring architecture, s_{fiber} is the fiber length for the full round trip. Similarly, the fiber distance between two cooperating BSs must be below $s_{\mathrm{fiber,BS}} = 0.5 \cdot s_{\mathrm{fiber}}$ again. This distance must cover two times the PON reach R in the star architecture. Consequently, R can be calculated according to Equation 9.

$$\begin{aligned} R &= \frac{s_{\mathrm{fiber,BS}}}{2} = \frac{s_{\mathrm{fiber}}}{4} = \\ &= \frac{5}{29} c(d_{\max} - 6 d_{\mathrm{IP}} - d_{\mathrm{wls}}) \end{aligned} \tag{9}$$

The overall coverage area of the star architecture is calculated according to Equation 10. The number of required PONs are derived according to Equation 6.

$$A = \pi R^2 \tag{10}$$

5.2 Results

We did the same evaluation for the star architecture that we have done for the ring architecture. Now, however, the results only depend on the parameter d_{IP}.

Based on the model and assumptions, the star architecture's coverage area is illustrated in Figure 9 for the typical IP/MPLS and queuing delay. Compared to the legacy metro ring architecture in Figure 5, a clear increase in the coverage area is visible.

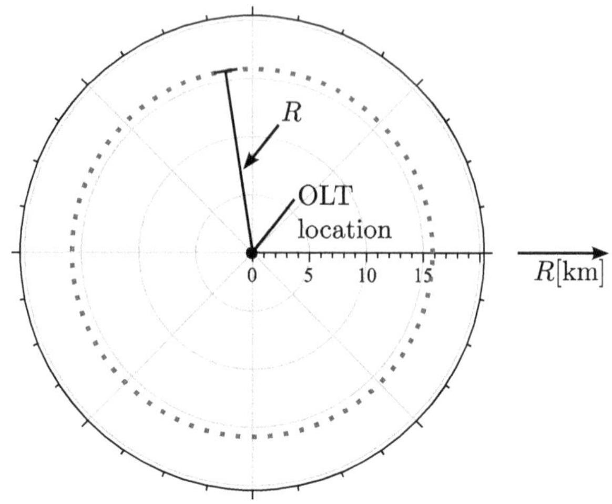

Figure 9: Size and shape of star topology; $d_{\mathrm{IP}} = 0.1\,\mathrm{ms}$. All ONUs are colocated at the origin. The dashed circle shows the maximum range of the PONs.

The coverage area depending on the IP/MPLS delay d_{IP} is shown in Figure 10 ($\mathrm{IP_{OLT}}$, $\mathrm{IP_{ONU}}$).

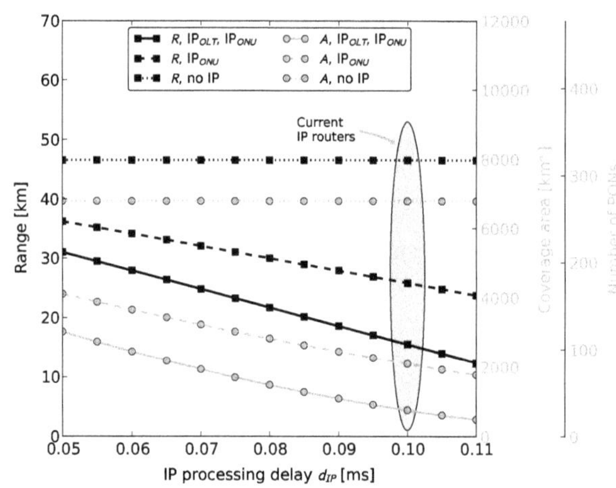

Figure 10: Feasible star architecture size depending on d_{IP}.

With today's IP/MPLS router equipment, the star architecture can cover an area of approximately $750\,\mathrm{km}^2$ ($R \approx 15.5\,\mathrm{km}$) with 30 PONs. This advantage compared to the ring architecture ($130\,\mathrm{km}^2$) with current hardware components is a result of the colocation of all OLTs. This enables to have one single, big IP/MPLS router, also colocated with the OLTs, which allows to remove one IP/MPLS processing step compared to the distributed ring architecture.

5.3 Improvements

For the star architecture, the same improvements can be considered that have been proposed for the ring architecture. Besides reducing the IP/MPLS processing delay in general or removing it at the ONUs/BSs, which both have the same serious disadvantages as in the ring architecture (cf. Section 4.3), it is easier to remove the IP/MPLS processing at a small amount of central locations. As all OLTs and the IP/MPLS router are colocated in a single data center, modifying, e.g., all OLTs can be done with low effort. The resulting coverage area using such a bypassing technique at the OLTs is shown in Figure 10 (IP_{ONU}). The plot also contains the case without any IP/MPLS processing in the network (no IP) for comparison.

The plot shows that bypassing the IP/MPLS router in the OLT data center for traffic between cooperating BSs leads to significant improvements of the architecture coverage area. The area that can be covered increases from $750\,\text{km}^2$ to $2100\,\text{km}^2$; an improvement by the factor of 2.8.

6. MERGING RING AND STAR CONCEPT

The distributed ring architecture with OADMs on the metro ring to which PONs are connected is not a solution for the future. Without improving hardware components and without extensive link capacity over-provisioning, requirements of JP cannot be fulfilled on a metro scale.

On the other hand, the star architecture approach with centralized OLTs and a single IP/MPLS router colocated with the OLTs provides a very good performance already with today's hardware components. However, fiber deployment costs are between 15,000 and 17,000 USD per kilometer on average [15]. These costs make it very unlikely to completely replace existing infrastructures, like available optical metro rings. They need to be reused as far as possible.

In line with that, we propose an architecture that combines the idea of colocating all OLTs and using a single, big router (star architecture) with existing ring fiber (ring architecture). This reduces the required time for IP/MPLS processing and queuing at intermediate nodes, which can be spent for propagation within the fiber to extend the coverage area. Existing pipes and fiber of metro rings are used to connect the centralized OLTs to the AWGs. We call this the *star-via-ring architecture*. An example is given in Figure 11.

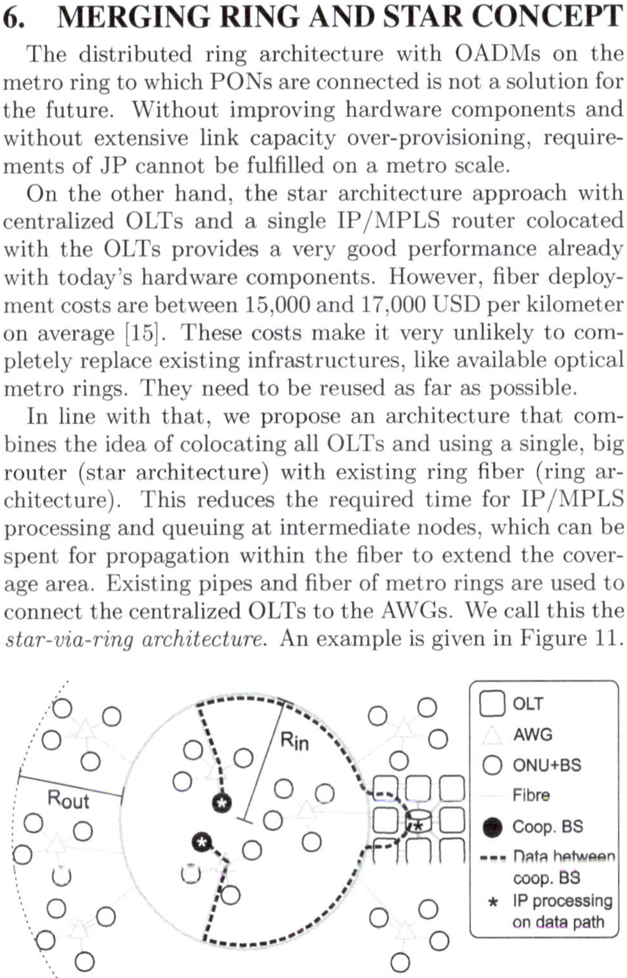

Figure 11: Star-via-ring topology and delays that occur. Multiple parallel PON fibers on the ring are drawn as one thick line.

Unlike the previous two architectures, this one is not symmetric. The location of the colocated OLTs on the ring will play an important role for the coverage area, as it will be shown in the next section.

6.1 Star-via-ring architecture model

The accumulated IP/MPLS and wireless processing delay d_{proc} that occurs between two BSs applying JP is identical to the star architecture. It consists of $6 \cdot d_{\text{IP}}$ (3 times in each direction, cf. Figure 11) and of d_{wls}.

The difference between $d_{\max}$ and d_{proc} again remains for propagation in the fiber and leads to the maximum possible fiber length s_{fiber}. Its calculation is identical to the previous architectures (Equation 2). The fiber length between two BSs must be below $s_{\text{fiber,BS}} = 0.5 \cdot s_{\text{fiber}}$, as s_{fiber} is the full round-trip fiber length.

As for the star architecture, we assume that all BSs should be able to cooperate in the star-via-ring architecture. Starting from this assumption, we can calculate the maximum possible radius of the fiber ring R_{in}. The distance $s_{\text{fiber,BS}}$ needs to cover two times the inner PON reach R_{in} and two times half of the optical ring, such that $s_{\text{fiber,BS}} = 2 \cdot R_{\text{in}} + 2 \cdot \pi \cdot R_{\text{in}}$. This is the worst case when two BSs cooperate that are attached to PONs whose fiber leaves the ring at the opposite of the OLTs' location (left side in Figure 11). Hence, we calculate R_{in} according to Equation 11.

$$\begin{aligned} R_{\text{in}} &= \frac{s_{\text{fiber,BS}}}{2 \cdot (1+\pi)} = \frac{s_{\text{fiber}}}{4 \cdot (1+\pi)} \\ &= \frac{5}{29} c \frac{d_{\max} - 6d_{\text{IP}} - d_{\text{wls}}}{1+\pi} \end{aligned} \tag{11}$$

The location where a PON fiber leaves the ring plays an important role. The worst case is the opposite location of where the OLTs are colocated. Here, half of the ring has to be traversed and the remaining part, which is equal to R_{in}, can be used to depart to the outside of the ring. When moving the point where the fiber leaves ring closer to the colocated OLTs, a larger part of the available fiber distance can be spent to depart from the ring. In Figure 11, this would result in a larger R_{out} on the right side compared to the left side, as the OLTs are colocated on the right side. Hence, the outer radius depends on α, which is the angle between the location of the colocated OLTs, the center of the ring, and the location where the PON fiber leaves the ring. We measure this angle in both directions from the colocated OLTs, i.e., $0 \leq \alpha \leq \pi$. It is illustrated in Figure 12.

To calculate the outer radius $R_{\text{out}}(\alpha)$, we look at the worst case where the cooperating BS A's PON leaves the ring at the opposite of the colocated OLTs and a second BS B's PON leaves at angle α. The total available fiber length $s_{\text{fiber,BS}}$ needs to cover half the ring and R_{in} (from the OLT to BS A) and the fraction of the ring depending on α and $R_{\text{out}}(\alpha)$ (from the OLT to BS B). This is contained in Equation 12 and leads to $R_{\text{out}}(\alpha)$ in Equation 13.

$$s_{\text{fiber,BS}} = R_{\text{in}} + \frac{1}{2} 2\pi R_{\text{in}} + \frac{\alpha}{2\pi} 2\pi R_{\text{in}} + R_{\text{out}}(\alpha) \tag{12}$$

$$\begin{aligned} R_{\text{out}}(\alpha) &= s_{\text{fiber,BS}} - R_{\text{in}} \cdot (1 + \alpha + \pi) = \\ &= \frac{s_{\text{fiber}}}{2} - R_{\text{in}} \cdot (1 + \alpha + \pi) = \\ &= \frac{5}{29} c (d_{\max} - 6d_{\text{IP}} - d_{\text{wls}}) \frac{1 + \pi - \alpha}{1+\pi} \end{aligned} \tag{13}$$

Based on the two reaches R_{in} and $R_{\text{out}}(\alpha)$, the overall coverage area of the star-via-ring architecture is calculated according to Equation 14 using polar integration.

$$\begin{aligned} A &= \int_0^{\pi} (R_{\text{in}} + R_{\text{out}}(\alpha))^2 \, \mathrm{d}\alpha = \\ &= \frac{25}{2523} c^2 (6d_{\text{IP}} + d_{\text{wls}} - d_{\max})^2 \cdot \frac{\pi(12 + 6\pi + \pi^2)}{(1+\pi)^2} \end{aligned} \tag{14}$$

6.2 Results

The evaluation of the star-via-ring architecture is similar to one of the ring and star architecture.

Based on the model and assumptions, the star-via-ring architecture's coverage area is shown in Figure 12 for the typical IP/MPLS and queuing delay of $d_{\mathrm{IP}} = 0.1\,\mathrm{ms}$.

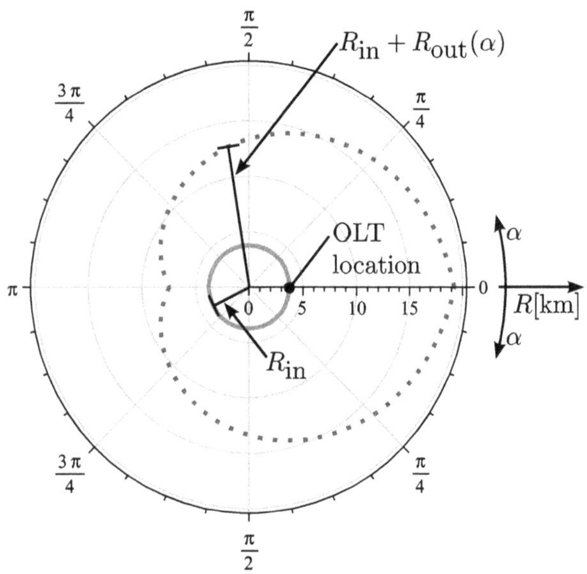

Figure 12: Size and shape of star-via-ring topology; OLTs are all colocated on the right side of the ring, $d_{\mathrm{IP}} = 0.1\,\mathrm{ms}$. The solid circle marks the optical fiber ring, the dashed one shows the maximum range of the outer PONs.

The most visible change in the coverage area is the asymmetry compared to the fully symmetric, circle-shaped coverage areas of the ring and ring architecture. The coverage area is heart-shaped now, which is caused by the colocation of all OLTs in one location on the fiber ring (at $\alpha = 0$ in Figure 12).

The coverage area depending on the IP/MPLS delay d_{IP} is shown in Figure 13 ($\mathrm{IP_{OLT}}$, $\mathrm{IP_{ONU}}$).

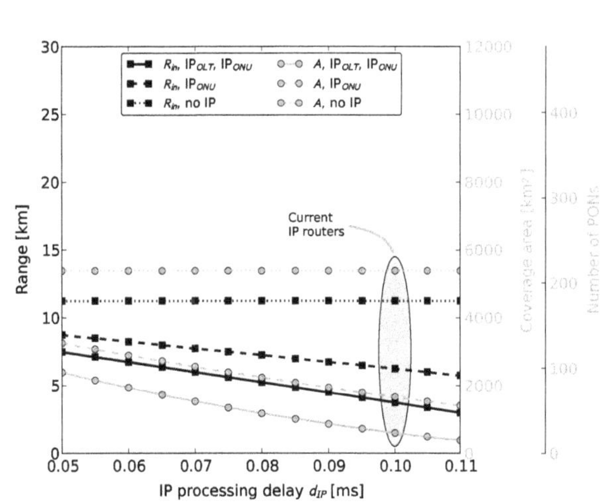

Figure 13: Feasible star-via-ring architecture size depending on d_{IP}.

Assuming today's IP/MPLS router equipment, the star-via-ring architecture is able to cover an area of approximately $600\,\mathrm{km}^2$ ($R_{\mathrm{in}} \approx 3.7\,\mathrm{km}$, $3.7\,\mathrm{km} \leq R_{\mathrm{out}} \leq 15.5\,\mathrm{km}$) with 24 PONs. This advantage compared to the ring architecture ($130\,\mathrm{km}^2$) is, as for the star architecture case, a result of colocating all OLTs and using a single IP/MPLS router. Compared to the star architecture ($755\,\mathrm{km}^2$), which is optimal in terms of coverage area as the fiber can be deployed without indirections, the star-via-ring architecture looses due to the fiber's "forced detour" via the ring. Nevertheless, it achieves 79 % of the star's coverage area and 607 % of the distributed ring architecture while being able to reuse existing fiber. The coverage areas of the three architectures are summarized in Figure 14. For better comparability, the shape of the star-via-ring architecture has been aligned according to both the architectures' optical fiber rings (Figure 14a) and to the OLT locations (Figure 14b).

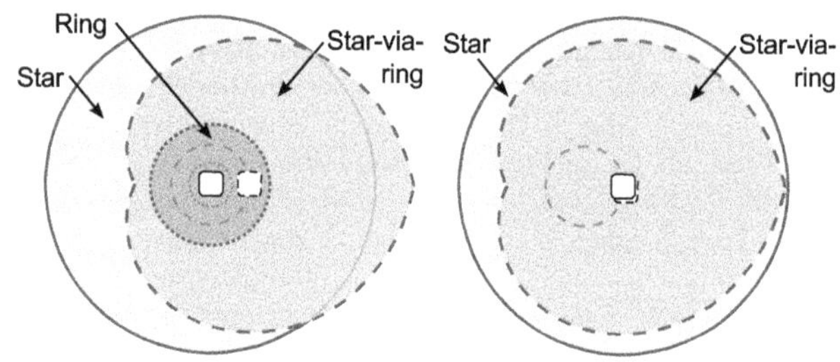

(a) Centered around optical fiber rings' midpoint. (b) Centered around colocated ONUs' location.

Figure 14: Coverage of ring (dotted), star (solid), and star-via-ring architecture (dashed); $d_{\mathrm{IP}} = 0.1\,\mathrm{ms}$, $f_{\mathrm{out}} = 0.25$. Fiber rings (ring and star-via-ring architecture) are drawn in the same line styles in red color. Colocated ONUs (star and star-via-ring architecture) are marked by the ONU symbol in the same line style.

6.3 Improvements

Applying the improvements already discussed for the star architecture (reducing IP/MPLS processing and queuing delay in general, at the ONUs/BSs, or at the OLTs) is possible for the star-via-ring architecture, too. The effects have been summarized in Figure 13.

7. CONCLUDING REMARKS

Addressing the backhaul requirements of future mobile access systems is not simply a matter of upgrading the networking hardware. Based on the results provided in this paper, the support of demanding CoMP mechanisms like JP goes beyond the adoption of high-capacity access transport technologies like WDM PONs.

We have shown that latency is the major concern for inter-BSs cooperation. On metro scale, 1 ms delay for BS-to-BS data exchange is easily exceeded.

Considering this requirement, a first conclusion we have drawn is that JP becomes feasible as soon as queuing and IP processing delay between cooperating BSs is limited, e.g., by restricting the scope in which JP is used. In small deployment scenarios, like public hot-spots and buildings, the cooperation is limited to BSs connected to the same PON infrastructure, thus avoiding to traverse the routers present in the higher layers of the wireline backhaul network hierarchy.

Deploying JP for macro cells on metro scale requires taking into account a major part of the wireline backhaul network infrastructure. Our latency analysis has shown that reusing today's ring infrastructures is feasible only with improved system components and optical bypassing techniques, which are not available yet. The star architecture, on the other hand, allows to use JP in large areas (up to 2100 km^2, e.g., to cover Tokyo) even with today's technology.

In between, the proposed star-via-ring architecture offers a reasonable trade-off between coverage area and reuse of existing ring fiber infrastructure.

Notice that for network planning, not only the area that a certain architecture can cover is important. What plays a crucial role is also the amount of population that a certain architecture can reach, to not waste money for upgrading infrastructure in areas where no revenues can be expected from that. Hence, it not only matters *where* people live but also *how* people are distributed in a certain territory. For example, if the distribution of the population is non-uniform, then the asymmetry of the star-via-ring architecture's coverage shape can be exploited to cover the most densely populated areas at reasonable CAPEX costs. On the other hand, the ring architecture is appealing in uniformly-populated small to medium-sized cities with an area of less than 100 km^2. As an example, Figure 15 overlays the population density distribution of Berlin's metro area[1] to the coverage areas of the three architectures. The star-via-ring architecture is the most appealing solution as it covers most population while reusing an existing fiber ring.

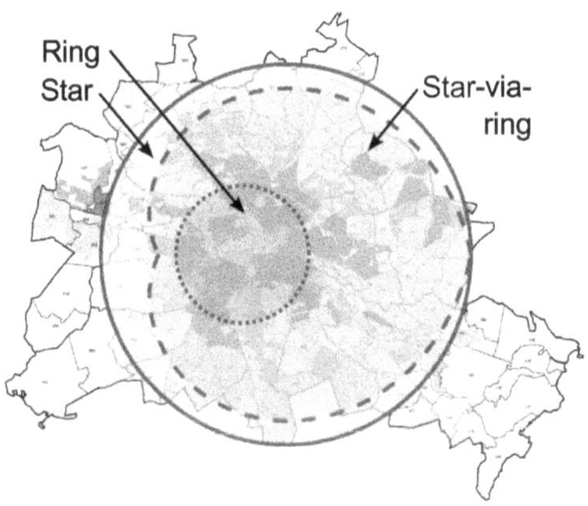

Figure 15: Population in Berlin's metro area (darker color indicates higher density) with overlaid architecture coverage areas (all drawn in the same scale).

The models presented in this paper can also be used as a tool for mobile network operators to check to which extent current wireline network deployments support CoMP techniques. This is done by feeding arbitrary network parameters to the model that represent the used hardware, topology, packet processing and queuing delay at the routers, etc.

8. REFERENCES

[1] ITU-R Rep. M.2134. Requirements related to technical performance for IMT-advanced radio interface(s), 2008.

[1]The population density map was taken from `http://www.stadtentwicklung.berlin.de/planen/basisdaten_stadtentwicklung/atlas/de/bevoelkerung.shtml`.

[2] Wei Yu, Taesoo Kwon, and Changyong Shin. Multicell coordination via joint scheduling, beamforming and power spectrum adaptation. In *Proc. IEEE International Conference on Computer Communications (INFOCOM)*, 2011.

[3] J. He and M. Salehi. Low-complexity coordinated interference-aware beamforming for MIMO broadcast channels. In *Proc. IEEE Vehic. Techn. Conf. (VTC-Fall)*, pages 685–689, 2007.

[4] Young-Han Nam, Lingjia Liu, Yan Wang, C. Zhang, Joonyoung Cho, and Jin-Kyu Han. Cooperative communication technologies for LTE-advanced. In *Proc. IEEE International Conference on Acoustics Speech and Signal Processing (ICASSP)*, pages 5610–5613, 2010.

[5] V. Jungnickel, L. Thiele, T. Wirth, T. Haustein, S. Schiffermüller, A. Forck, S. Wahls, S. Jaeckel, S. Schubert, H. Gäbler, and Others. Coordinated multipoint trials in the downlink. In *Proc. IEEE Broadb. Wirel. Access Workshop (BWAWS)*, 2009.

[6] V. Jungnickel, M. Schellmann, L. Thiele, T. Wirth, T. Haustein, O. Koch, W. Zirwas, and E. Schulz. Interference-aware scheduling in the multiuser MIMO-OFDM downlink. *IEEE Comm. Mag.*, 47(6):56–66, June 2009.

[7] S. Venkatesan, H. Huang, A. Lozano, and R. Valenzuela. A WiMAX-based implementation of network MIMO for indoor wireless systems. *EURASIP Journal on Advances in Signal Processing*, 2009.

[8] K. Grobe and J.-P. Elbers. PON in adolescence: from TDMA to WDM-PON. *IEEE Comm. Mag.*, 46(1):26–34, 2008.

[9] T. Biermann, L. Scalia, J. Widmer, and H. Karl. Backhaul design and controller placement for cooperative mobile access networks. In *Proc. IEEE Vehic. Techn. Conf. (VTC-Spring)*, 2011.

[10] T. Biermann, L. Scalia, C. Choi, H. Karl, and W. Kellerer. Backhaul network pre-clustering in cooperative cellular mobile access networks. In *Proc. IEEE World of Wireless Mobile and Multimedia Networks (WoWMoM)*, 2011.

[11] Peter Croy. LTE backhaul requirements: A reality check, Feb. 2011. Whitepaper, Aviat.

[12] Adrian Popescu and Doru Constantinescu. Measurement of one-way transit time in IP routers. Technical report, Blekinge Institute of Technology (BTH), Blekinge, Sweden, 2005.

[13] N. Hohn, K. Papagiannaki, and D. Veitch. Capturing router congestion and delay. *IEEE/ACM Transactions on Networking*, 17(3):789–802, 2009.

[14] S. Nakazawa, H. Tamura, K. Kawahara, and Y. Oie. Performance analysis of IP datagram transmission delay in MPLS: Impact of both the number and the bandwidth of LSPs of layer 2. *Proc. IEEE International Conference on Communications (ICC)*, pages 1006–1010.

[15] T. Cohen and R. Southwood. Extending open access to national fibre backbones in developing countries. In *International Telecommunication Union (ITU) 8th Global Symposium for Regulators*, 2008.

Modeling the Effect of Transmission Errors on TCP Controlled Transfers over Infrastructure 802.11 Wireless LANs

Subhashini Krishnasamy
Dept. of Electrical Communication Engg.
Indian Institute of Science, Bangalore
subhashini.kb@ece.iisc.ernet.in

Anurag Kumar
Dept. of Electrical Communication Engg.
Indian Institute of Science, Bangalore
anurag@ece.iisc.ernet.in

ABSTRACT

There have been several studies on the performance of TCP controlled transfers over an infrastructure IEEE 802.11 WLAN, assuming perfect channel conditions. In this paper, we develop an analytical model for the throughput of TCP controlled file transfers over the IEEE 802.11 DCF with different packet error probabilities for the stations, accounting for the effect of packet drops on the TCP window. Our analysis proceeds by combining two models: one is an extension of the usual TCP-over-DCF model for an infrastructure WLAN, where the throughput of a station depends on the probability that the head-of-the-line packet at the Access Point belongs to that station; the second is a model for the TCP window process for connections with different drop probabilities. Iterative calculations between these models yields the head-of-the-line probabilities, and then, performance measures such as the throughputs and packet failure probabilities can be derived. We find that, due to MAC layer retransmissions, packet losses are rare even with high channel error probabilities and the stations obtain fair throughputs even when some of them have packet error probabilities as high as 0.1 or 0.2. For some restricted settings we are also able to model tail-drop loss at the AP. Although involving many approximations, the model captures the system behavior quite accurately, as compared with simulations.

Categories and Subject Descriptors

C.2.5 [**Computer Communication Networks**]: Local and Wide Area Networks - *Access schemes*

General Terms

Performance

Keywords

WLANs, performance analysis, modeling TCP over 802.11 CSMA/CA, channel errors

1. INTRODUCTION

IEEE 802.11 supports multiple data transmission rates with different modulation schemes at the PHY layer. The aim of rate adaptation algorithms is to exploit this multi-rate capability to obtain the best throughput in different channel conditions. Some recent rate adaptation algorithms ([2, 17]) estimate the packet error rate periodically and adjust the transmission rates by comparing the error rate with some thresholds. Others ([6, 1]) use a combination of factors such as estimates of the throughput and, the probability of success, to adapt the rates. A model that predicts the throughputs obtained at different data transmission rates and different packet error rates would be very useful in providing insights in setting the threshold parameters in [2, 17] or in deciding the transmission rates and the retry counts in the retry chain in [6, 1]. We address the problem of estimating the throughputs obtained by the stations when each of them download TCP files at a constant PHY rate and experience some packet error rate on the wireless channel.

We consider an IEEE 802.11 infrastructure WLAN with a single Access Point (AP). The stations (STAs) can experience different channel conditions and bit-error-rates depending on their locations, the RF propagation environment, the external interference, and the PHY with which they are associated. We consider the problem of modeling TCP controlled file downloads from a server on the wired LAN. We focus on downloads, as studies have shown that a large fraction of wireless access traffic tends to be downlink in nature.

Related Literature.

There is a vast literature on performance modeling of the MAC protocol in IEEE 802.11 WLANs under various traffic scenarios. There have been many models ([25, 15]) which use Bianchi's fixed point analysis [5] of saturated nodes in a single-cell network. Bruno *et al.* [8] studied TCP file downloads by modeling the interaction of TCP with a p-persistent IEEE 802.11 MAC protocol by using Markov chain analysis for the number of stations that are contending for the wireless channel. In [16], TCP file downloads over the standard IEEE 802.11 DCF was modeled by using the saturation analysis ([5, 15]) assuming that the AP buffer is infinite. However, in all these papers, the authors assume ideal channel conditions and that frames can be received in error only due to collision. Some researchers have recently extended the 802.11 saturation analysis to include channel errors [9, 12, 20]. Chatzimisios *et al.* [9] perform saturation analysis assuming that all stations are associated at the same rate and experience the same BER. Multi-rate and different BER for each station with saturated queues is considered in [12, 20].

On the other hand, there have been many models which analyze the performance of TCP over wireless links by considering random packet losses [18, 14]. However, these papers consider point-to-point wireless links. Mare *et al.* [21] analyze the effect of channel errors in an IEEE 802.11 WLAN for the simple case of single TCP long file transfer. To the best of our knowledge, the interaction between packet losses, CSMA/CA random access and TCP has not been analyzed in a multi-station infrastructure WLAN.

Contributions of this paper.

We study the impact of channel errors on TCP performance in a more general framework where there are multiple STAs with different channel conditions and bit-error rates, downloading long TCP files from the wireline LAN. It is known that, without channel errors, all TCP controlled file transfers obtain equal throughputs [8, 16]. With different channel error rates, however, the TCP window processes of the connections are affected differently, and it is expected that there would be throughput unfairness and also a reduction in the total network throughput. We provide an analytical model that yields insights into these performance issues.

Our analysis proceeds by iterating between two sub-models. One is an extension of the Markov regenerative model in [8, 16] that provides the STAs' throughputs given the probability that the head-of-the-line (HOL) packet at the AP belongs to a particular STA. These HOL probabilities are then obtained from the analysis of the TCP window process by focusing on the sharing of the AP buffer by the various TCP connections. In modeling the congestion window processes, we initially assume that the AP buffer is infinite and, therefore, packet drops are only due to channel errors and not due to buffer overflow. We then extend this model to the finite buffer case. Even though many approximations are made during the course of our analysis, we find that we are able to capture the essential factors that govern the performance, and the results from our analysis compare very favorably with the simulations. It is observed that MAC layer retransmissions help in avoiding critical throughput unfairness up to fairly large error probabilities (almost 10%) but with a slight decrease in the total network throughput. But at higher error rates, STAs with higher packet loss probabilities get lower throughput due to two reasons: a) Inability of TCP to recognize losses due to channel errors, thus, causing a reduction in the windows of the STAs with higher loss probabilities, and b) FIFO queuing at the AP buffer.

2. THE SETTING AND MODELING ASSUMPTIONS

We consider an IEEE 802.11 WLAN in which the stations (STAs) are associated with a single Access Point (AP). Each STA is downloading a large file from a "server" which is on the high-speed LAN to which the AP is connected. The STAs may experience different channel conditions and therefore, are subject to different bit error rates (BER), yielding different packet error rates (PER).

We briefly describe how DCF of the 802.11 MAC protocol works: For every packet, a node begins with a uniformly distributed random backoff of mean b_0 slots (b_0 = CWmin/2) and defers the backoff when it senses a busy medium. At the end of the backoff, if the channel is idle for DIFS time, it transmits the packet. If the transmission was unsuccessful, it retries the frame with a random backoff of increased mean. We denote the mean backoff duration (in slots) after the kth attempt for a frame by b_k ($b_k = \min(2^k \times b_0, \text{CWmax}/2)$). The node tries to retransmit the frame until it is successful or the maximum retry limit is reached. The maximum possible number of attempts for an RTS or for a short frame whose size is less than or equal to the RTS threshold, is called the dot11ShortRetryLimit, denoted by K_s. For a long frame of length greater than the RTS threshold, the maximum possible number of attempts is called the dot11LongRetryLimit, denoted by K_l. After the number of attempts for a frame reaches its maximum (K_s or K_l according to the size of the frame), the packet is discarded [3]. Table 1 shows the various parameters used in the analysis and simulation.

Parameter	Symbol	Value
PHY data rate	C_d	11 Mbps
Control rate	C_c	2 Mbps
PLCP preamble time	T_P	144 μs
PHY Header time	T_{PHY}	48 μs
MAC Header size	L_{MAC}	34 bytes
RTS Frame size	L_{RTS}	20 bytes
CTS Frame size	L_{CTS}	14 bytes
MAC ACK Header size	L_{ACK}	14 bytes
IP Header	L_{IPH}	20 bytes
UDP Header	L_{UDPH}	8 bytes
TCP Header	L_{TCPH}	20 bytes
TCP ACK Packet size	$L_{TCP-ACK}$	40 bytes
Payload Data size	L	1460 bytes
Slot time	δ	20 μs
DIFS time	$DIFS$	50 μs
SIFS time	$SIFS$	10 μs
EIFS time	$EIFS$	308 μs
Min. Contention Window	CW_{min}	31
Max. Contention Window	CW_{max}	1023
Short retry limit	K_s	7
Long retry limit	K_l	4

Table 1: Various parameters used in simulation and analysis

2.1 System Model

Let M be the total number of STAs indexed by $1 \leq j \leq M$. Each STA can be associated with the AP at a different packet error rate. For simplicity, we assume that the error rates belong to a finite set (this is not restrictive as this set can have as many error rates as the number of STAs). Let n_i be the number of STAs that experience packet error rate, ε_i, $1 \leq i \leq C$, where C is the number of error rates. We say that the set of these n_i STAs belong to "Class" i, denoted by $\mathscr{C}_i$. Since we are concerned with TCP controlled downloads, the AP transmits TCP data packets. A TCP data packet transmitted by the AP to an STA of Class i fails due to channel errors (given that the packet did not suffer a collision) with probability ε_i independent of anything else.

2.2 Approximations

We make the following approximations in our model:

A1 Since, for TCP file downloads, the STAs have to send only TCP ACK packets, which are small (40 bytes), the probability that they are received in error at the AP is also small. Therefore, we assume that *errors when the STAs transmit can be ignored.*

A2 Since TCP (for the undelayed ACK case) requires equal number of packets to be transmitted by the AP (data packets) and the STAs (ACK packets), and DCF is packet fair, it should be true that a very small number of STAs must be contending for the channel most of the time [8, 16]. Therefore, the probability that the HOL packet at the AP belongs to one of these contending STAs is also small and, we assume that every successful transmission from the AP generates a packet at a previously empty STA. Furthermore, when an STA receives data from the AP, it is assumed that it immediately generates an ACK to be queued for transmission. From the above arguments we make the approximation that *an STA can have at most one TCP ACK in its buffer* [8, 16, 4].

A3 Since the server is connected to the AP on a high speed Ethernet, it is assumed (as in [16, 8, 4]) that the AP is the bottleneck for the TCP transfers and it always has packets in its buffer to contend for the channel. Moreover, we can assume that *all the packets in the TCP window of any connection are in the AP buffer.* This can be explained as follows: Since the server is on the same LAN as the AP, the number of packets "in flight" between the server and the AP can be ignored. Also, since the STAs can have at most one TCP ACK in their buffers (as seen in A2), the number of packets in the STA buffers is very small relative to the TCP windows.

A4 *The RTS and CTS packets that have not collided are not received in error.* This assumption is justified since the RTS and CTS packets are small and are sent at a low physical bit-rate.

2.3 Assumptions

Apart from the above approximations, we also make the following assumptions:

M1 The "delayed ACK" mechanism is disabled.

M2 When RTS-CTS is enabled, RTS/CTS mechanism is used to transmit data packets whereas ACK packets are transmitted by the Basic Access mode.

3. SATURATION ANALYSIS

In this section, we deviate from the setting described in Section 2 to characterize the channel contention process when all the nodes in the WLAN are saturated (infinitely backlogged queues). Results from this analysis are used in Section 4 to obtain the stationary distribution of the number of contending STAs and the error rate of the HOL packet in the AP buffer, when the STAs are doing TCP file downloads.

We extend the single-cell model with saturated queues ([5, 15]) to include channel errors. In the single-cell model, all nodes are in the carrier-sensing range of each other. We consider the scenario in which there are n STAs and an AP, all of which are infinitely backlogged and use the same backoff parameters. Due to Approximation A1, we assume that only one node (the AP) experiences poor channel condition when it transmits, while the STAs all transmit in perfect channel conditions. Each packet sent by the AP is received in error with probability ε. DCF works such that the channel is idle when all the nodes are in the backoff state and such intervals alternate with activity on the channel (a transmission or a collision). During backoff periods, time is slotted; countdown of the backoff timers and attempts by the nodes are made at the slot boundaries. Let the backoff slots be indexed by t ($t \geq 1$). We denote the long run average transmit attempt rate of a node by β, i.e., if $A(t)$ is the number of attempts by a node until slot t, then β for that node is defined as.

$$\beta := \lim_{t\to\infty} \frac{A(t)}{t} \tag{1}$$

As in [5, 15], we assume that the node attempts at every slot with probability β independent of anything else. Let β_a and β_s be the average attempt rate of the AP and the STAs respectively. We note that, in this saturated setting, the attempt rates of all the STAs are equal due to symmetry (all of them transmit with zero error probability). Similarly, define $D(t)$ for a node, as the number of attempts by the node until slot t that have resulted in collision, and $Q(t)$ as the number of attempts until slot t in which there was no collision but the transmitted frame was received in error. Then, we define γ as the long run fraction of attempts by the node that have resulted in collision, and α as the long run fraction of attempts that have failed either due to collision or erroneous reception, i.e., for a given node,

$$\gamma := \lim_{t\to\infty} \frac{D(t)}{A(t)} \tag{2}$$

and

$$\alpha := \lim_{t\to\infty} \frac{D(t)+Q(t)}{A(t)} \tag{3}$$

We denote the collision and failure probabilities for the AP and the STAs by γ_a, γ_s, α_a and α_s. Since an attempt by an STA fails only due to collisions, the failure probability for an STA, α_s, is the same as the collision probability, γ_s.

We extend the fixed point analysis in [15] to include channel errors. In [20], the authors calculate the saturation throughput obtained when different STAs experience different channel errors. In calculating the attempt and failure probabilities, they do not differentiate between the MAC retry limits for long and short frames. We show that, unlike the case where there are no channel errors, the analysis proceeds differently for the following two cases: (1) RTS-CTS is disabled and all the packets are transmitted by the Basic Access mode (which is the same as in [20]), (2) The AP uses RTS/CTS mechanism to transmit its packets while the STAs transmit in the Basic Access mode. It can be shown that, in both the cases the fixed point equations have a unique solution for β_a and β_s for any ε and n [24].

3.1 RTS-CTS Disabled

In this case, the data frames are retransmitted until the number of attempts reaches dot11ShortRetryLimit, K_s. With α_a and α_s as defined above, the attempt probabilities of the AP (β_a) and the STAs (β_s) can be written as in [15].

$$\beta_a = G(\alpha_a) \tag{4}$$

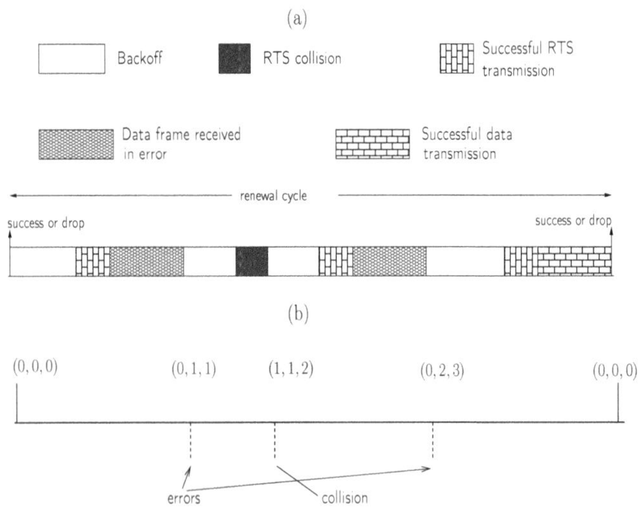

Figure 1: (a) Backoff periods and transmission attempts at the AP. Failures could be due to RTS collisions as well as channel errors. (b) Changes in the state vector at the end of activity periods. Each attempted packet starts a new backoff cycle with state $(0, 0, 0)$

$$\beta_s = G(\gamma_s) \tag{5}$$

where

$$G(\gamma) = \frac{1 + \gamma + \gamma^2 + \cdots + \gamma^{K_s - 1}}{b_0 + b_1\gamma + b_2\gamma^2 + \cdots + b_{K_s-1}\gamma^{K_s-1}} \tag{6}$$

The collision probabilities for the AP and the STAs are given by

$$\gamma_a = 1 - (1 - \beta_s)^n \tag{7}$$

$$\gamma_s = 1 - (1 - \beta_a)(1 - \beta_s)^{n-1} \tag{8}$$

The failure probability for the AP can be written as

$$\alpha_a = \gamma_a + (1 - \gamma_a)\varepsilon = 1 - (1 - \beta_s)^n(1 - \varepsilon) \tag{9}$$

Equations 4, 5, 8 and 9 can be solved to obtain β_a and β_s for a given ε and n.

3.2 RTS-CTS Enabled

In this case, the data frames are retransmitted until the number of attempts reaches dot11LongRetryLimit, K_l and the RTS frames can be attempted a maximum of K_s times. Evolution of transmission attempts at the AP is shown in Fig. 1(a). As in the RTS-CTS disabled case, the instants at which the packet is transmitted successfully or discarded are taken as renewal instants (see [15]). Applying the Renewal Reward theorem to this renewal process, the attempt probability, β_a is given by

$$\beta_a = \mathbb{E}[A_a]/\mathbb{E}[B_a]$$

where $E[A_a]$ is the expected number of attempts by the AP in a renewal cycle and $E[B_a]$ is the expected total backoff duration (in slots) at the AP before the frame is either transmitted successfully or discarded. To obtain these quantities, it suffices to know the backoff periods and the outcome of the transmission (collision, erroneous transmission or successful transmission) and thus, the time periods when the channel is active can be discarded. To calculate $\mathbb{E}[A_a]$ and $\mathbb{E}[B_a]$, we maintain in every cycle, a three-dimensional state vector (i, j, k) which changes at the end of every channel activity in the cycle (see Fig. 1(b)). Here, i represents the number of RTS collisions after the last successful RTS transmission, j is the number of data frame transmission failures since the beginning of the cycle and k is the total number of failures (collisions or errors) since the beginning of the cycle. Every cycle begins with the state $(0, 0, 0)$. Let $\mathbb{E}[A^{(i,j)}]$ be the expected number of attempts in the cycle after the current RTS frame has collided i times and the data frame transmission has failed j times, i.e., $\mathbb{E}[A^{(i,j)}]$ is the expected number of attempts in the cycle after the busy period that ends with the state (i, j, k) for any k. Then, $\mathbb{E}[A_a] = \mathbb{E}[A^{(0,0)}]$ and

$$\mathbb{E}[A^{(i,j)}] = 1 + \gamma_a\mathbb{E}[A^{(i+1,j)}] + \varepsilon(1 - \gamma_a)\mathbb{E}[A^{(0,j+1)}]$$
$$\text{for all } 0 \leq i \leq K_s - 1, \quad 0 \leq j \leq K_l - 1$$

$$\mathbb{E}[A^{(K_s,j)}] = \mathbb{E}[A^{(i,K_l)}] = 0 \qquad \text{for all } i, j$$

Similarly, let $\mathbb{E}[B^{(i,j,k)}]$ be the expected backoff duration in the cycle after the busy period that ends with state (i, j, k). Then, $\mathbb{E}[B_a] = \mathbb{E}[B^{(0,0,0)}]$ and

$$\mathbb{E}[B^{(i,j,k)}] = b_k + \gamma_a\mathbb{E}[B^{(i+1,j,k+1)}] + \varepsilon(1 - \gamma_a)\mathbb{E}[B^{(0,j+1,k+1)}]$$
$$\text{for all } 0 \leq i \leq K_s - 1, \quad 0 \leq j \leq K_l - 1, \quad k$$

$$\mathbb{E}[B^{(K_s,j,k)}] = \mathbb{E}[B^{(i,K_l,k)}] = 0 \qquad \text{for all } i, j, k$$

Thus, $\beta_a = \mathbb{E}[A_a]/\mathbb{E}[B_a] = \tilde{G}(\gamma_a, \varepsilon)$ is a function of γ_a and ε, and is obtained from the equations above. Since the STAs transmit the TCP ACKs in the Basic Access mode, $\beta_s = G(\gamma_s)$. The expressions for γ_a, γ_s, and α_a are the same as in the RTS/CTS disabled case. Equations 7, 8, $\beta_a = \tilde{G}(\gamma_a, \varepsilon)$ and $\beta_s = G(\gamma_s)$ can be solved to obtain β_a and β_s.

Comparisons between the estimates from the fixed point analysis and simulation in QualNet validate the model accuracy [24] for both the cases, i.e., when RTS/CTS is disabled and enabled.

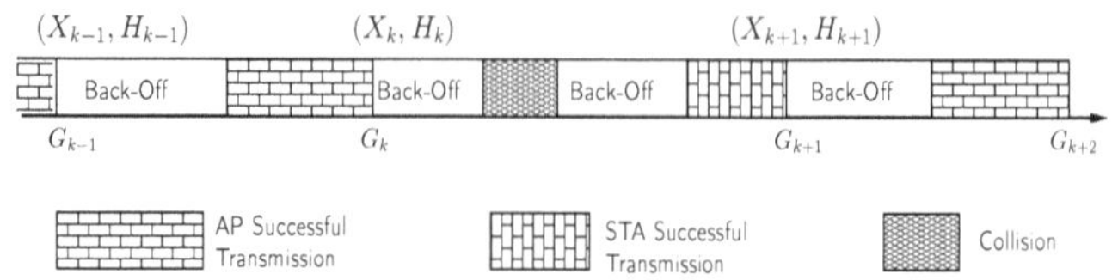

Figure 2: Evolution of channel activity: successful transmissions end at instants G_k

4. TCP CONTROLLED FILE DOWNLOADS

4.1 Model for DCF Based Medium Sharing

We now consider the case when there are M STAs that are downloading large files from the local server. Taking the assumptions in Section 2 into consideration, we can develop a stochastic model akin to the ones in [16, 4]. Consider the aggregate process of channel contention and activity (see Fig. 2). Since we have long file transfers and no external arrivals, the states of the AP and the STA queues can change only at the instants when there are successful transmissions or packet discards. Due to MAC retransmissions, we note

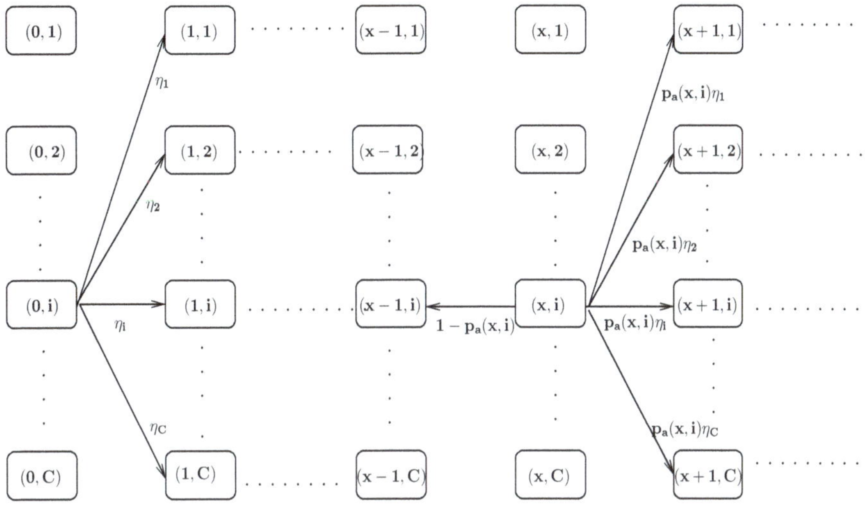

Figure 3: Transition probability diagram for the process $\{(X_k, H_k), k \geq 0\}$ embedded at epochs $G_k, k \geq 0$

that the fraction of attempts that result in a packet discard is negligibly small as compared to the fraction of attempts that result in a successful transmission. This can be verified analytically and numerically. Therefore, we make the following approximation:

A5 The states of the AP queue and the STAs' queues change only at the end of successful transmissions.

Let G_k, $k \geq 0$, denote the instants when a successful transmission ends. The successful transmission could be a data packet from the AP or a TCP ACK from one of the STAs. Let X_k denote the number of contending or *active* STAs, and H_k denote the class to which the HOL (Head Of Line) packet in the AP buffer belongs during the kth *cycle* $[G_k, G_{k+1})$. By A5, these quantities change only at the end of the cycle, i.e., at G_{k+1}. We say that a packet belongs to Class i if the destination STA of the packet belongs to Class i.

A6 We assume that in a cycle if there are x active STAs and the HOL packet in the AP buffer is from Class i, then the attempt and failure probabilities are the same as those in the saturated case with $n = x$ saturated STAs and the AP frame error probability, $\varepsilon = \varepsilon_i$ ([16, 4]).

Thus, the attempt probability and failure probability for all the nodes in the cycle $[G_k, G_{k+1})$ depend only on (X_k, H_k). Therefore, the time interval between the successes, $G_{k+1} - G_k$ depends only on (X_k, H_k). At any success instant, by Approximation A2, the number of active STAs decreases by one if the successful transmission was from an STA and increases by one if the successful transmission was from the AP. Define η_i as the fraction of packets served by the AP that belong to Class i. In other words, η_i is the fraction of the aggregate TCP throughput that is obtained by STAs of Class i.

$$\eta_i := \lim_{n \to \infty} \frac{1}{n} \sum_{m=1}^{n} I_{\{C(m)=i\}} \tag{10}$$

where $C(m)$ is the class to which the mth packet served by the AP belongs. Based on 10 (see also [4]), we assume that

A7 When a successful transmission is due to the AP, the next HOL packet is of Class i with probability η_i independent of anything else.

Therefore, it can be seen that the process, $\{(X_k, H_k), k \geq 0\}$ is a DTMC with state space $\mathcal{S} = \{(x, i) : x \geq 0, 1 \leq i \leq C\}$ where, C is the number of error classes. Note that we allow the number of active STAs to exceed M. This does not affect the accuracy of the results as, for M sufficiently large, the stationary probability of the number of active STAs exceeding M is very small for this DTMC (see Approximation A2). Also, since $G_{k+1} - G_k$ depends only on (X_k, H_k), $\{((X_k, H_k), G_k), k \geq 0\}$ forms a Markov renewal sequence [13].

The DTMC, $\{(X_k, H_k), k \geq 0\}$ has the following transition probabilities,

$$\Pr\{(X_{k+1}, H_{k+1}) = (x+1, j) | (X_k, H_k) = (x, i)\} = p_a(x, i)\eta_j \tag{11}$$

$$\Pr\{(X_{k+1}, H_{k+1}) = (x-1, i) | (X_k, H_k) = (x, i)\} = 1 - p_a(x, i) \tag{12}$$

where $p_a(x, i)$ is the probability that when the state is (x, i), the next successful transmission is from the AP. This is the same as the probability that the AP transmits successfully given that there is a successful transmission in a slot (see Equation 13). The transition probability diagram is shown in Fig. 3 and the matrix is of the form shown in Equation 14

$$\mathbf{P} = \begin{pmatrix} \mathbf{0} & \mathbf{A_0} & \mathbf{0} & \mathbf{0} & \dots \\ \mathbf{S_1} & \mathbf{0} & \mathbf{A_1} & \mathbf{0} & \dots \\ \mathbf{0} & \mathbf{S_2} & \mathbf{0} & \mathbf{A_2} & \dots \\ \vdots & \vdots & \vdots & \vdots & \ddots \end{pmatrix} \tag{14}$$

where $\mathbf{A}_x$ and $\mathbf{S}_x$, $x \geq 0$ are non-negative matrices of order C. It can be observed that the DTMC is a level dependent Quasi-Birth-and-Death (LDQBD) process [22], where the number of active STAs represents the *level* and the class of the HOL packet represents the *phase*. Since $p_a(x, i) \in (0, 1)$, the right-hand side of 11 and 12 are positive if $\eta_j > 0$, for all j and therefore, P is irreducible if $\eta_j > 0$, for all j. It can be proved using the Foster-Lyapunov stability criterion [11] that the process is positive recurrent.

THEOREM 1. *The LDQBD $\{(X_k, H_k), k \geq 0\}$ is positive recurrent.*

PROOF. *Let $V(x, i) = x$ and $\mathcal{B} = \{(x, i) : 0 \leq x \leq 1, 1 \leq$*

$$p_a(x,i) = \frac{\beta_a(x,i)(1-\beta_s(x,i))^x(1-\varepsilon_i)}{\beta_a(x,i)(1-\beta_s(x,i))^x(1-\varepsilon_i) + x\beta_s(x,i)(1-\beta_s(x,i))^{x-1}(1-\beta_a(x,i))} \quad (13)$$

$i \le C\}$. Then, $\mathbb{E}[V(X_{k+1},H_{k+1}) - V(X_k,H_k)|(X_k,H_k) = (x,i)] = 2p_a(x,i) - 1 < 0$, for all $x \ge 2$ and for all i. Also, since $p_a(x,i)$ is a decreasing function of x and $2p_a(x,i)-1 < \infty$ for $x = 0, 1$, we can find constants ϵ and b such that, $\mathbb{E}[V(X_{k+1},H_{k+1}) - V(X_k,H_k)|(X_k,H_k) = (x,i)] \le -\epsilon + bI_{\{i\in\mathcal{B}\}}$ for all $i \in \mathcal{S}$. Therefore, by the Foster-Lyapunov stability criterion, the process $\{(X_k,H_k), k \ge 0\}$ is positive recurrent and a stationary distribution exists. □

Thus, there exists a $\boldsymbol{\pi}$ that solves the system of equations $\boldsymbol{\pi} = \boldsymbol{\pi}\mathbf{P}$ and $\boldsymbol{\pi}\mathbf{e} = 1$, where $\mathbf{e}$ is a column vector of ones. The vector $\boldsymbol{\pi}$ is partitioned by the *levels* (number of active STAs) into sub-vectors of length C such that $\boldsymbol{\pi} = (\boldsymbol{\pi}_0, \boldsymbol{\pi}_1, \boldsymbol{\pi}_2, \dots)$ and the elements of $\boldsymbol{\pi}_x$ are the stationary probabilities for the states in level x, i.e., $\boldsymbol{\pi}_x(i)$ which is the ith element of $\boldsymbol{\pi}_x$ is the stationary probability that the process, $\{(X_k,H_k), k \ge 0\}$ takes the value (x,i). Theorem 12.1.1 in [19] proves that when the process is irreducible and positive recurrent, the stationary distribution has the matrix-product form, $\boldsymbol{\pi}_{x+1} = \boldsymbol{\pi}_x\mathbf{R}_x, x \ge 0$ where $\mathbf{R}_x, (x \ge 0)$ are square matrices of order C. $\mathbf{R}_x$ can be computed recursively using the equation

$$\mathbf{R}_x = \mathbf{A}_x + \mathbf{R}_x\mathbf{R}_{x+1}\mathbf{S}_{x+2}, \qquad x \ge 0 \quad (15)$$

The stationary probabilities can, therefore, be obtained by solving the following system of equations for $\boldsymbol{\pi}_0$: $\boldsymbol{\pi}_0 = \boldsymbol{\pi}_0\mathbf{R}_0\mathbf{S}_1$ and $\boldsymbol{\pi}_0(\sum_{x=0}^{\infty}\prod_{n=0}^{x-1}\mathbf{R}_i)\mathbf{e} = 1$. To compute the matrices, $\mathbf{R}_x$, we use the algorithm by Bright and Taylor [7] in which the transition probability matrix is truncated at some large level, N. Having obtained the stationary distribution, we can find, 1) the TCP throughput, $\Theta_i^{(TCP)}$ obtained by a STA belonging to Class i (Section 4.2), and 2) the probability, p_i that a packet belonging to Class i is discarded/dropped by the AP after its Maximum Retry Limit is reached (Section 4.3), as functions of the packet error rates, $\boldsymbol{\varepsilon} := (\varepsilon_1, \varepsilon_2, \dots, \varepsilon_C)$ and $\boldsymbol{\eta} := (\eta_1, \eta_2, \dots \eta_C)$.

4.2 TCP Throughput

Every successfully transmitted packet from the AP contributes to the aggregate throughput from the server to the STAs. Let $Z_k \in \{0,1\}$ denote the number of AP successes in the kth cycle and $U_k = G_{k+1} - G_k$ denote the length of the kth cycle. Let $Z(t)$ denote the number of packets served by the AP in the interval $(0,t)$. Since $\{((X_k,H_k),G_k), k \ge 0\}$ forms a Markov renewal sequence, by Markov regenerative analysis [13], the aggregate network TCP throughput, $\Theta^{(TCP)}$ is given by,

$$\Theta^{(TCP)} := L \lim_{t\to\infty} \frac{Z(t)}{t} \overset{a.s.}{=} L\frac{\sum_{x,i}\pi_x(i)p_a(x,i)}{\sum_x \pi_x(i)\mathbb{E}_{(x,i)}U} \quad (16)$$

where $\mathbb{E}_{(x,i)}U$ is the expected cycle time when the state is (x,i) (see [24]) and L is the payload size (assumed to be the same for all the TCP connections). Let the TCP throughput of each STA belonging to Class i be Θ_i^{TCP}. Since η_i is the ratio of the total throughput obtained by STAs of Class i to the total throughput obtained by all STAs,

$$\Theta_i^{(TCP)} = \frac{\eta_i\Theta^{(TCP)}}{n_i} \quad (17)$$

Thus, given $\boldsymbol{\eta}$ and the error rates, $\boldsymbol{\varepsilon}$, the stationary distribution, $\boldsymbol{\pi}$, can be obtained as described in Section 4.1, and the throughput of any STA can be obtained from Equation 17. The method to obtain $\boldsymbol{\eta}$ by fixed point iteration is shown in Section 4.5.

4.3 Obtaining the Drop Probability

4.3.1 RTS-CTS Disabled

In the Basic Access mode, a packet is dropped if it could not be transmitted successfully even after attempting to transmit K_s times. Define α_i as the fraction of AP attempts that result in failure when the HOL packet is of Class i. Let $\Phi_i(t)$ and $A_i(t)$ denote the number of failures and the number of attempts by the AP with HOL packet of Class i until slot t. Again, using Markov regenerative analysis, we can write

$$\alpha_i := \lim_{t\to\infty}\frac{\Phi_i(t)}{A_i(t)} \overset{a.s.}{=} \frac{\sum_x \pi_x(i)\mathbb{E}_{(x,i)}\Phi_i}{\sum_x \pi_x(i)\mathbb{E}_{(x,i)}A_i} \quad (18)$$

where $\mathbb{E}_{(x,i)}\Phi_i$ denotes the expected number of failed attempts (due to collisions and errors) by the AP in a cycle given the HOL packet at AP is of Class i and the number of active STAs is x, and $\mathbb{E}_{(x,i)}A_i$ is the expected number of attempts by the AP in a cycle given the HOL packet at AP is of Class i and the number of active STAs is x. Expressions for these can be obtained from renewal arguments [24]. With the following approximation, it can be seen that the drop probability, p_i, is given by $\alpha_i^{K_s}$.

A8 Every attempt to transmit a packet of Class i fails with probability α_i independent of anything else.

4.3.2 RTS-CTS Enabled

When RTS-CTS is enabled, a packet is dropped either if the number collisions of an RTS frame reaches K_s or if the packet is received in error K_l times. Define γ_i as the fraction of AP attempts that result in collision when the HOL packet belongs to Class i. Let $\Psi_i(t)$ and $A_i(t)$ denote the number of RTS collisions and the number of attempts at the AP with HOL packet of Class i in the time interval $(0,t)$. Then,

$$\gamma_i := \lim_{t\to\infty}\frac{\Psi_i(t)}{A_i(t)} \overset{a.s.}{=} \frac{\sum_x \pi_x(i)\mathbb{E}_{(x,i)}\Psi_i}{\sum_x \pi_x(i)\mathbb{E}_{(x,i)}A_i} \quad (19)$$

where $\mathbb{E}_{(x,i)}\Psi_i$ denote the expected number of attempts by the AP that result in RTS collisions in a cycle given the HOL packet at AP is of Class i and the number of active STAs is x. We assume that,

A9 Every attempt to transmit an RTS frame corresponding to a packet of Class i results in collision with probability γ_i independent of anything else.

As described, a packet can be dropped if is received in error K_l times, or if the RTS transmitted each time before the data frame is sent collides K_s times. The probability that the packet is dropped due to collisions in the kth transmission attempt of the data frame is $\left((1-\gamma_i^{K_s})\varepsilon_i\right)^{k-1}\gamma_i^{K_s}$ $(1 \le k \le K_l)$, while the probability that the packet is dropped due to the packet being received in error K_l times is equal to $\left((1-\gamma_i^{K_s})\varepsilon_i\right)^{K_l}$. Thus, the total probability that a packet of Class i is dropped is given by,

$$p_i = \gamma_i^{K_s} \sum_{k=1}^{K_l} \left((1-\gamma_i^{K_s})\varepsilon_i\right)^{k-1} + \left((1-\gamma_i^{K_s})\varepsilon_i\right)^{K_l} \quad (20)$$

4.4 Determining η by TCP Window Analysis

We have seen in Section 4.3 that, given ε, the drop probabilities, $\mathbf{p} = (p_1, p_2, \ldots, p_C)$ is a function, $\mathbf{F}(.)$ of η, both when RTS/CTS is disabled and when it is enabled. In this section, we show how to obtain the AP HOL probabilities, η, given $\mathbf{p}$, thus providing the ingredients of fixed point iteration which will be described in Section 4.5.

Recall that η_i (as defined in Equation 10) is the fraction of packets served by the AP that belong to Class i. To calculate $\eta = (\eta_1, \eta_2, \ldots \eta_C)$, we consider the congestion window evolution of the TCP connections. As mentioned in Approximation A3 in Section 2, we assume that all the packets of the window are in the AP buffer for every TCP connection. We also initially assume that the AP buffer size is large enough to avoid buffer overflows. Later, in Section 5, we extend the analysis to finite buffer size. To analyze the window evolution, the concept of *rounds* similar to that in [23] is used. A round ends when all the packets present in the AP buffer at the beginning of the round have been either transmitted successfully or discarded. The end of a round is the beginning of the next round. Consider the process $\{\mathbf{W}_k, k \geq 0\} := \{(W_k^{(1)}, W_k^{(2)}, \ldots, W_k^{(M)}), k \geq 0\}$, where $W_k^{(l)}$ denotes the number of packets of connection l in the AP buffer at the beginning of round k.

A10 It is assumed that an HOL packet of Class i is dropped by the AP with probability p_i independent of all other events.

A11 The loss recovery due to packet loss in any round is completed by the end of that round and, therefore, at the beginning of a round, the number of packets of any connection in the AP buffer is equal to the congestion window of that connection in the server.

Since the TCP window evolves according to packet losses, by A10 and A11, the process $\{\mathbf{W}_k, k \geq 0\}$ is a DTMC with a finite state space, $\mathcal{W} = \{0, 1, 2, \ldots, W_{max}\}^M$ (W_{max} is the maximum receive window). If the DTMC is irreducible, let ν be its stationary distribution (for the window evolution models that we consider in this paper, the DTMC turns out to be irreducible).

To calculate η_i, note that $\sum_{j\in\mathscr{C}_i} W_k^{(j)}$ packets of Class i and a total of $\sum_{l=1}^{M} W_k^{(l)}$ packets are served in the round k. Define Y_k as the total number of packets served in all the rounds before the beginning of round k. Then, $Y_{k+1} - Y_k = \sum_{l=1}^{M} W_k^{(l)}$ and, $\{(\mathbf{W}_k, Y_k), k \geq 0\}$ forms a Markov renewal sequence. Therefore, by Markov regenerative analysis [13],

$$\eta_i = \lim_{n\to\infty} \frac{1}{n} \sum_{m=1}^{n} I_{\{C(m)=i\}} \quad (21)$$

$$\stackrel{a.s.}{=} \frac{\sum_{j\in\mathscr{C}_i} \sum_{\mathbf{w}\in\mathcal{W}} \nu(\mathbf{w}) w_j}{\sum_{j=1}^{M} \sum_{\mathbf{w}\in\mathcal{W}} \nu(\mathbf{w}) w_j} \quad (22)$$

$$= \frac{n_i \mathbb{E}_\nu W_i}{\sum_{l=1}^{C} n_l \mathbb{E}_\nu W_l} \quad (23)$$

where $\mathbf{w} = (w_1, w_2, \ldots, w_M)$ in Equation 22, and $\mathbb{E}_\nu W_i$ is the stationary expected value of the window of a STA of Class i. Since $\mathbb{E}_\nu W_i$ is a function of p_i, η can be calculated as a function, $\mathbf{T}(.)$ of $\mathbf{p}$.

When the AP buffer is infinite, losses are only due to channel errors and thus, by A10, the window evolution of any connection is independent of that of other connections, i.e., $\{W_k^{(j)}, k \geq 0\}$ is independent of $\{W_k^{(l)}, k \geq 0\}$ for all $j, l \in \{1, 2, \ldots, M\}$. Denote the stationary distribution of the DTMC, $\{W_k^{(j)}, k \geq 0\}$, $j \in \mathscr{C}_i$ by ν_i. We calculate $\mathbb{E}_\nu W_i$ by considering a stochastic model for the evolution of the window of a TCP Reno connection when the probability that a packet is dropped is p_i. We make the following assumptions:

A12 The probability of more than one loss in a round is negligible.

A13 It is assumed that if there is a loss, recovery always happens by fast retransmit and fast recovery so that the connection is always in congestion avoidance phase.

In the congestion avoidance phase, the congestion window increases by one if all the packets in the window are acknowledged. Therefore, at the end of a round, the window of a connection increases by one if all the packets belonging to that connection have been transmitted successfully by the AP, else the window reduces to half of its previous value. For drop probability p, the window evolution process, $\{V_k, k \geq 0\}$ has the state space $\{1, 2, \ldots, W_{max}\}$ and transition probabilities given by

$$\Pr\{V_{k+1} = w+1 | V_k = w\} = (1-p)^w$$

$$\Pr\{V_{k+1} = \lceil (w/2) \rceil | V_k = w\} = 1 - (1-p)^w$$

The stationary distribution for this DTMC can be computed numerically which can be used to calculate ν_i and $\mathbb{E}_\nu W_i$ in Equation 23 for a given drop probability, p_i.

4.5 The Fixed Point Iteration

As seen in Section 4.3, $\mathbf{p} = \mathbf{F}(\eta)$. Further, in Section 4.4, we obtained $\eta = \mathbf{T}(\mathbf{p})$. Hence, we can obtain η by solving the fixed point equation, $\eta = \mathbf{T}(\mathbf{F}(\eta))$. We obtained η from the fixed point iteration, $\eta^{(k+1)} = \mathbf{T}(\mathbf{F}(\eta^{(k)}))$. It has been observed that $\eta^{(k)}$ converges to the same solution irrespective of the initial value but we have not been able to prove convergence to a unique solution. The value of η thus obtained can be used to calculate the stationary probabilities of the LDQBD and throughputs of the TCP connections using Equations 16, 17 .

5. EXTENSIONS TO FINITE AP BUFFER

Until now, we assumed that the size of the buffer at the AP is so large that for the given W_{max}, there are no packet drops due to buffer overflow (tail drop losses). But with TCP window scale option, this assumption is not valid as W_{max} is very large. We now consider the effects of finite buffer size at the AP.

Let the size of the buffer at the AP be B packets. It can be seen that the Markov chain model (Section 4) for channel contention is still valid when the AP buffer is finite. But the fraction of packets of each connection served by the AP, η, varies with B. Thus, we need to develop a model for

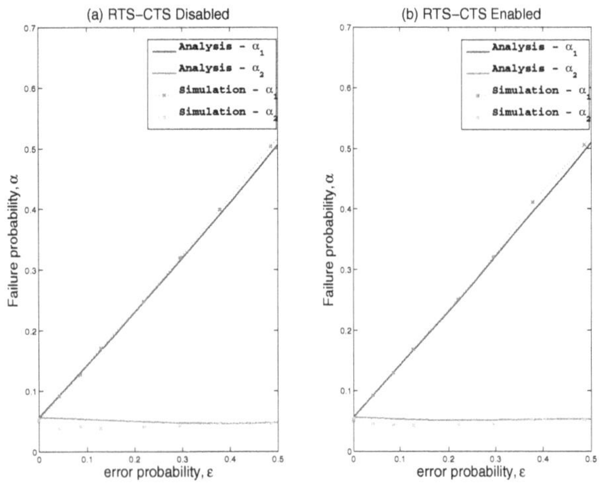

Figure 4: Probability that an attempt by the AP fails for each class when there are 2 classes and the AP buffer is very large. Class 1 has error probability ε and Class 2 has error probability 0. TCP Reno is used with W_{max} = 45 packets (a) RTS-CTS Disabled (b) RTS-CTS Enabled

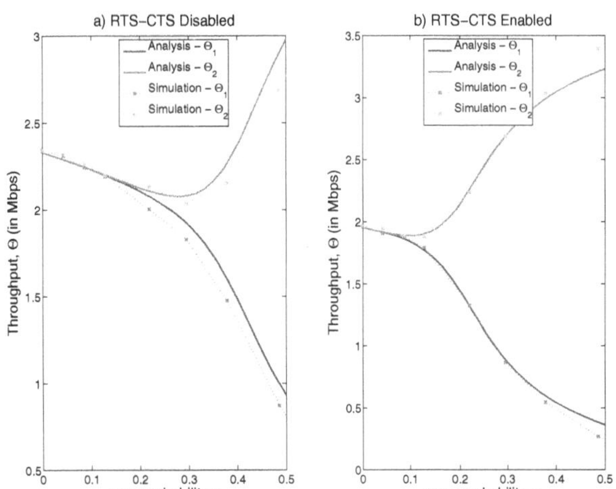

Figure 5: TCP Throughput obtained by each class when there are 2 classes and AP buffer is very large. Class 1 has error probability ε and Class 2 has error probability 0. TCP Reno is used with W_{max} = 45 packets (a) RTS-CTS Disabled (b) RTS-CTS Enabled

TCP congestion window which takes into account not only the packet losses due to channel errors but also losses due to buffer overflow to determine the value of $\boldsymbol{\eta}$. The evolution of the congestion windows of different connections are not any more independent of each other. To model the dependence, we extend the TCP Reno model. In addition to A12 and A13, we make the following approximation:

A14 As in [4], we assume *immediate feedback*, i.e., the time between the data packet being served at the AP and the corresponding TCP ACK being received at the server is ignored. This is justified since the average number of active STAs is small and, thus, the STAs can transmit the ACKs within a short period of time. A consequence of this assumption is that there is a tail drop loss only when the congestion window increases at the server. When there is no window growth, every packet served by the AP is immediately replaced in the buffer by a new packet due to immediate acknowledgment. When the buffer is full and the window grows by one, two packets arrive at the tail of the buffer but the packet served by the AP leaves space only for a single packet, resulting in tail drop loss. This leads us to the approximation that *only those connections which have had no losses due to channel errors in the round can have window growth and therefore, can lose a packet due to buffer overflow.*

The process $\{\boldsymbol{W}_k, k \geq 0\}$ evolves as follows: Let $S_k \subset \{1, 2, \ldots, M\}$ be the set of connections that lose packets due to channel errors in round k.

$$\Pr\{S_k = S | \boldsymbol{W}_k = (w_1, w_2, \ldots, w_M)\} = \prod_{i \in S} (1 - (1 - p_i)^{w_i}) \times \prod_{i \in S^c} ((1 - p_i)^{w_i}) \quad (24)$$

The congestion windows of the connections in S_k are halved and thus, these connections do not have tail drop losses in round k. If $l \in S_k$, then $W_{k+1}^{(l)} = W_k^{(l)}/2$. All the other connections grow their window by one and the number of packets lost due to buffer overflow is given by $\max\{\sum_{i \in S_k} W_k^{(i)}/2 + \sum_{j \in S_k^c} (W_k + 1) - B, 0\}$. Since all the connections in S_k^c need not have tail drop loss, we pick the connections that have tail drop loss randomly from S_k^c. It can be observed that the process $\{W_k, k \geq 0\}$ is a DTMC with finite state space, $\{1, 2, \ldots, \min\{B, W_{max}\}\}^M$. Thus, we can obtain the stationary expected value of $\{\boldsymbol{W}_k, k \geq 0\}$ to determine $\boldsymbol{\eta}$ from Equation 23.

6. MODEL VALIDATION AND DISCUSSION

All the simulation results are obtained using QualNet 4.5.1 with 802.11b. The parameters used in simulation and analysis are summarized in Table 1. We set the TCP Maximum Segment Size (MSS) to 1460 bytes and Maximum Receive Window, W_{max} to 65,535 bytes (45 packets) when there is no window scaling. Auto-rate fallback is disabled. Simulation results are obtained for two values of RTS Threshold: (a) When the RTS threshold is set to 1600 bytes, RTS-CTS mechanism is disabled as all the frames have sizes less than 1600 bytes (see Table 1), and (b) When the RTS threshold is set to 600 bytes, the RTS-CTS mechanism is enabled as the size of the data frames exceeds this threshold.

Consider the case where there are two STAs ($C = 2$); STA in Class 1 receives frames at an error rate of ε and STA in

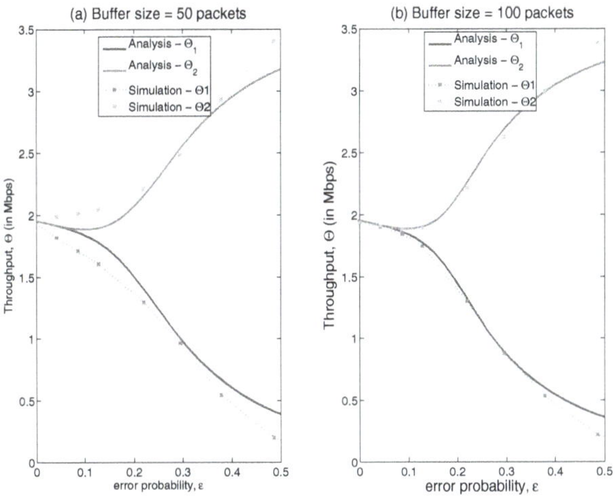

Figure 6: TCP Throughput obtained by each class when there are 2 classes. Class 1 has error probability ε and Class 2 has error probability 0. TCP Reno is used with $W_{max} = 45$ packets and RTS-CTS is enabled (a)$B = 50$ packets (b) $B = 100$ packets

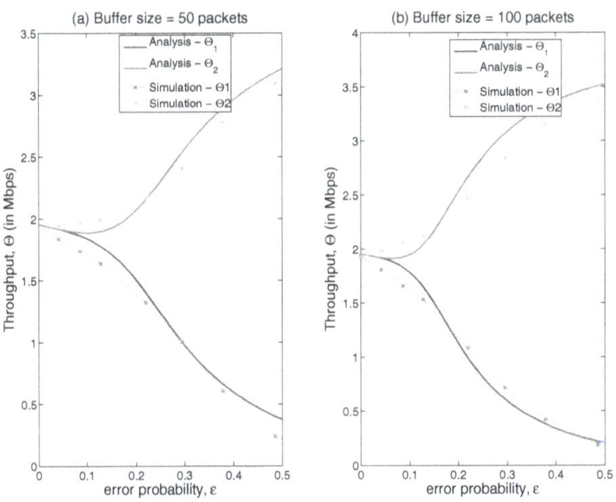

Figure 7: TCP Throughput obtained by each class when there are 2 classes. Class 1 has error probability ε and Class 2 has error probability 0. TCP Reno is used. RTS-CTS and window scaling are enabled (a)$B = 50$ packets (b) $B = 100$ packets

Class 2 correctly receives, all frames that do not collide, i.e., $\varepsilon_1 = \varepsilon$ and $\varepsilon_2 = 0$. We first consider the case when there is no window scaling and the AP buffer is large enough to avoid buffer overflows.

Fig. 4 shows plots of the probability that an attempt by the AP to transmit packets of Class 1 (α_1) and Class 2 (α_2) fails. It is seen that α_1 increases linearly with ε while α_2 remains almost constant as failure for packets of Class 2 is only due to collisions. The TCP throughput obtained by the STAs in each class is shown in Fig. 5 for the two cases.

1. It is seen that, while the aggregate network throughput decreases due to wasted time in transmitting erroneous frames, the throughputs of the two classes is equal at low error probabilities. This is because at low values of ε, the drop probability is low and both the classes will have their TCP window very close to W_{max}. Thus, the MAC level retransmission mechanism protects the TCP connections from channel losses to a large extent, and connections with sufficiently low frame error rate get the same throughput.

2. As ε increases, the STA that has a poor channel obtains a much smaller throughput as compared to the STA with a good channel. Unlike in the saturated case where unfairness occurs due to the backoff mechanism in IEEE 802.11 DCF which fails to distinguish between a collision and transmission failure due to link errors [20], unfairness in the case of TCP file downloads is due to: a) Inability of TCP to distinguish between packet losses due to congestion and packet losses due to channel errors. This leads to STAs with worse channels having a smaller TCP window and thus, fewer packets in the AP buffer. b) FIFO queuing at the AP buffer, which results in each STA getting a throughput proportional to the fraction of its packets in the AP buffer.

3. We can observe that the divergence in the throughput curves occurs at a lower value of ε when RTS-CTS is enabled as compared to the RTS-CTS disabled case. This can be justified by noting that the packet transmission attempt limit, K_l when RTS-CTS is enabled is less than the attempt limit, K_s when RTS-CTS is disabled. Thus, for the same value of ε, more number of packets of Class 1 are dropped when RTS-CTS is enabled, leading to lower throughput for Class 1.

We now consider the effect of finite buffer size at the AP for two cases: (i) window scaling option disabled, (ii) window scaling option enabled. When there is no window scaling, we set W_{max} to 45 packets and when window scaling is enabled, the scale factor is 14, which means that the window can grow up to 45×2^{14}. Fig. 6 shows the throughput plots when there is no window scaling ($W_{max} = 45$ packets) for buffer sizes of 50 and 100 packets. Note that, in this case, having a buffer size of 100 packets is same as having an infinite buffer. Fig. 7 shows the throughput plots when window scaling is enabled and $B = 50$, 100 packets. The difference in the values obtained from analysis and simulations is due to Approximation A14, where we neglect the tail drop losses of STAs that have lost packets due to channel errors.

7. CONCLUDING REMARKS

In this paper, we have addressed the problem of modeling TCP controlled file downloads by STAs with different channel conditions in an IEEE 802.11 WLAN. By combining a model for contention by DCF with a model for the TCP congestion window, we could calculate the packet failure probabilities and the throughputs obtained by the STAs. The simple TCP model that was valid for an infinite AP buffer was then extended to the case where the buffer size is finite. It was shown that STAs with poorer channel conditions get lower throughput due to two reasons: a) Inability of TCP to recognize losses due to channel errors b) FIFO queuing at the AP buffer. The latter problem can be solved by having a separate queue for each STA at the AP and serving each queue in a fair manner. Such an approach would be possible with IEEE 802.11n APs and our analysis has been found to be useful in this situation as well [24]. The results in this paper can also provide insights to the trade offs involved in designing rate adaptation algorithms. For eg., if two STAs are associated at the same PHY rate, and one of them experiences channel losses at that rate, the rate adaptation algorithm must compare the reduction in the throughput due to channel losses at that rate (small for packet error rates up to 10-20%) with the reduction in the throughput if data is transmitted to that STA at a lower rate.

8. REFERENCES

[1] Minstrel rate control algorithm. `http://madwifi-project.org/changeset/4030/madwifi/trunk/ath_rate/minstrel`.

[2] Onoe rate adaptation algorithm. `http://madwifi-project.org/svn/madwifi/branches/madwifi-0.9.4/ath_rate/onoe/`.

[3] Part 11: Wireless LAN medium access control (MAC) and physical layer (PHY) specification, 2007.

[4] O. Bhardwaj, G. V. V. Sharma, M. Panda, and A. Kumar. Modeling finite buffer effects on tcp traffic over an ieee 802.11 infrastructure wlan. *Comput. Netw.*, 53:2855–2869, November 2009.

[5] G. Bianchi. Performance analysis of the ieee 802.11 distributed coordination function. *IEEE Journal on Selected Area in Comm.*, 15(3):535–547, March 2000.

[6] J. C. Bicket. Bit-rate selection in wireless networks. Technical report, Master's thesis, MIT, 2005.

[7] L. Bright and P. G. Taylor. Calculating the equilibrium distribution in level dependent quasi-birth-and-death processes. *Stochastic Models*, 11(3):497–525, 1995.

[8] R. Bruno, M. Conti, and E. Gregori. Performance modelling and measurements of tcp transfer throughput in 802.11-based wlan. In *MSWiM*, pages 4–11, 2006.

[9] P. Chatzimisios, A. Boucouvalas, and V. Vitsas. Influence of channel BER on IEEE 802.11 DCF. *IEE Electronic Letters*, 39(23), Nov 2003.

[10] S. Ha, I. Rhee, and L. Xu. Cubic: a new tcp-friendly high-speed tcp variant. *SIGOPS Oper. Syst. Rev.*, 42:64–74, July 2008.

[11] B. Hajek. An exploration of random processes for engineers, January 2009.

[12] B. H. Jung, S. J. Kim, H. Jin, H. Y. Hwang, J. W. Chong, and D. K. Sung. Performance improvement of error-prone multi-rate wlans through adjustment of access/frame parameters. In *ICC*, pages 1–6, 2009.

[13] V. G. Kulkarni. *Modeling and analysis of stochastic systems*. Chapman & Hall, Ltd., London, UK, UK, 1995.

[14] A. Kumar. Comparative performance analysis of versions of tcp in a local network with a lossy link. *IEEE/ACM Trans. Netw.*, 6(4):485–498, 1998.

[15] A. Kumar, E. Altman, D. Miorandi, and M. Goyal. New insights from a fixed-point analysis of single cell ieee 802.11 wlans. *IEEE/ACM Trans. Netw.*, 15(3):588–601, 2007.

[16] G. Kuriakose, S. Harsha, A. Kumar, and V. Sharma. Analytical models for capacity estimation of ieee 802.11 wlans using dcf for internet applications. *Wireless Networks*, 15(2):259–277, 2009.

[17] M. Lacage, M. H. Manshaei, and T. Turletti. Ieee 802.11 rate adaptation: a practical approach. In *International Workshop on Modeling Analysis and Simulation of Wireless and Mobile Systems*, pages 126–134, 2004.

[18] T. V. Lakshman and U. Madhow. The performance of tcp/ip for networks with high bandwidth-delay products and random loss. *IEEE/ACM Trans. Netw.*, 5:336–350, June 1997.

[19] G. Latouche and V. Ramaswami. *Introduction to Matrix Analytic Methods in Stochastic Modeling*. ASA-SIAM, Philadelphia, 1999.

[20] E. Lopez-Aguilera, J. Casademont, and E. G. Villegas. A study on the influence of transmission errors on wlan ieee 802.11 mac performance. *Wireless Communications and Mobile Computing*, March 2010.

[21] S. Mare, D. Kotz, and A. Kumar. Experimental validation of analytical performance models for IEEE 802.11 networks. In *Proceedings of the Workshop on WIreless Systems: Advanced Research and Development (WISARD 2010)*, pages 1–8. IEEE Computer Society Press, January 2010.

[22] M. Neuts. *Matrix Geometric Solutions in Stochastic Models. An Algorithmic Approach.* John Hopkins University Press, Baltimore, 1981.

[23] J. Padhye, V. Firoiu, D. F. Towsley, and J. F. Kurose. Modeling tcp reno performance: a simple model and its empirical validation. *IEEE/ACM Trans. Netw.*, 8(2):133–145, 2000.

[24] K. Subhashini. Towards centralized control of channel allocation and scheduling in IEEE 802.11 WLANs. Master's thesis, Indian Institute of Science, 2011.

[25] Y. C. Tay and K. C. Chua. A capacity analysis for the ieee 802.11 mac protocol. *Wireless Networks*, 7:159–171, 2001.

Mitigating Mobility Signaling Congestion in LTE by Overlapping Tracking Area Lists

Sara Modarres Razavi
Department of Science and Technology
Linköping Institute of Technology
SE-601 74, Norrköping, Sweden
sarmo@itn.liu.se

Di Yuan
Department of Science and Technology
Linköping Institute of Technology
SE-601 74, Norrköping, Sweden
diyua@itn.liu.se

ABSTRACT

Avoiding signaling congestion in location management of cellular networks is becoming increasingly important as the population of user equipments (UEs) rapidly grows. Congestion can occur due to massive mobility of UEs behaving in a similar manner, such as the train movement scenario. In this paper, we explore the use of overlapping tracking area lists (TALs) in Long Term Evolution (LTE) networks for congestion mitigation that is not possible with the conventional tracking area concept. Each cell can use multiple and overlapping TALs, to be allocated to UEs requesting their TALs from the cell. We show that finding the optimal proportional use of TALs can be formulated as a linear program. Solving the linear program minimizes the maximum tracking area updates occurring between the cells. We present numerical results to illustrate the performance of the approach. The experiments demonstrate the effectiveness of overlapping TALs for mitigating mobility signaling congestion.

Categories and Subject Descriptors

C.2.1 [**Computer-communication Networks**]: Network Architecture and Design—*Wireless communication*; G.1.6 [**Numerical Analysis**]: Optimization—*Linear programming*; I.6.5 [**Simulation and Modeling**]: Model Development

General Terms

Design, Management, Performance

Keywords

location management, mass mobility congestion, tracking area list

1. INTRODUCTION

By the dramatic increase in the population of user equipments (UEs) in cellular networks, the study of avoiding signaling congestion scenarios becomes important. In Long Term Evolution (LTE) networks, the focus of location management is on LTE-idle mode UEs. The network knows the location of LTE-idle UEs within the granularity of a few cells forming a tracking area (TA). The UEs are frequently in the idle mode and have no active radio connection with the network. The excessive signaling required for tracing the LTE-idle UEs in some scenarios may significantly decrease the quality of service in some parts of the network, and could create signaling resource congestion.

The conventional TA scheme is almost the same concept as the Location Area (LA) in the circuit-switched (CS) domain and the Routing Area (RA) in the packet-switched (PS) domain in GSM, GPRS and UMTS networks [1]. A UE updates its location whenever it changes a TA. In order to place a call to an idle UE, the Mobility Management Entity (MME) pages all cells of the TA in which the UE is currently registered. Hence, the tracking area updates (TAU) are uplink signaling messages from UEs to the MME, and pagings are downlink signaling messages from the MME to UEs. One of the well-known performance consideration is the signaling overhead of TAU versus that for paging.

In densely-populated cities, there is a drastic difference in the mobility patterns between day and night. There are typically many UEs moving into a central area in the morning by jam-packed public transportation, causing signaling congestion around the center of the city. In this case, the UEs at the border of TAs may not have a sufficient quality of service during peak hours [3]. This is an undesirable situation from both the network and UEs standpoint [11, 13]. Reconfiguring TAs in order to adapt to the UEs movement behavior during pick hours typically requires temporally tearing down the cells, which is a costly process from the service perspective [14]. The current paper aims to mitigate the signaling congestion resulted from mass mobility of idle UEs without performing a TA revision.

In comparison to earlier generations of cellular networks, LTE networks allow for a more flexible configuration of TA design by the list concept introduced in 3GPP Release 8 [2]. In this scheme, instead of assigning one TA to each UE, one UE can have a list of TAs. The UE receives a TA list from a cell, and keeps the list, until it moves to a cell that is not included in its list. In LTE standards, a cell is also able to give different lists to different UEs. The UE location is known in the MME to at least the accuracy of the list allocated to that UE. In our study, we mitigate the excessive TAU signaling at the TA boundaries by assigning UEs with different and overlapping tracking area lists (TAL).

There are still many open problems unexplored in the

MSWiM'11, October 31–November 4, 2011, Miami, Florida, USA.

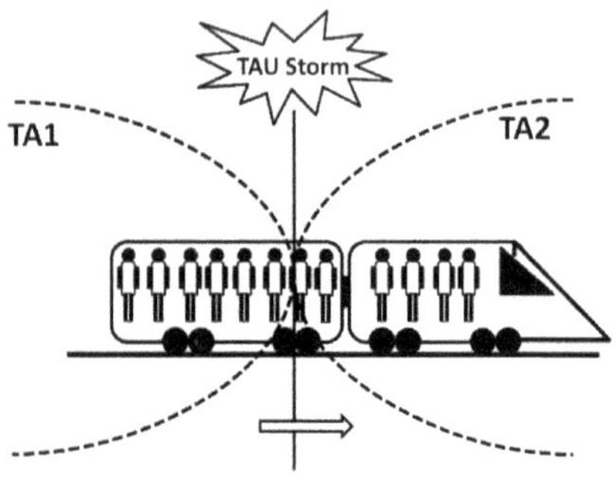

Figure 1: Example of TAU storm at the border of two TAs.

study of TAL. One main problem is how to estimate the overall signaling overhead of a fully deployed TAL scheme in a large scale network. The UEs are allowed to move in the cells contained in their TAL without any TAU. Thus, a cell contains UEs with different TALs depending on the cells from which the TALs are assigned. This diversity makes the estimation of signaling overhead difficult. The only way to calculate the exact overall signaling overhead of the network based on TAL design is to have the accurate data traces of UEs' movements. Unfortunately, the real data of UE traces is not available or too expensive to obtain. In the previous studies [15, 16], to overcome this problem, the authors generated a UE-trace scenario based on cell load and handover statistics of the network. This paper explores a new scheme in which the signaling overhead calculation is independent of UE traces. The overlapping TAL scheme in this study allows the cells to have a group of lists and to assign UEs different TALs. This feature could not be used in the framework of the TAL scheme suggested in [15] and [16].

The contributions of the paper can be summarized as follows. First, we develop the new overlapping TAL scheme. Second, we theoretically present how the scheme works with small instructive examples. Third, we present a linear programming (LP) model which minimizes the number of TAUs while keeping the overall paging overhead lower than some specified limit. The numerical results demonstrate that the proposed model is able to balance the load for representative scenarios of massive mobility signaling congestion.

2. RELATED WORKS

All the problems related to LA and RA planning and optimization can be generalized to the study of TA. Hence, there is an extensive amount of literature on the conventional TA concept [5, 17, 23, 24]. In this section we only review the limited amount of work that has been so far published on the concept of TAL, and the problem of massive mobility congestion.

Since 2007, a great attention has been given to the development of the list concept in the 3GPP standardization procedure and the telecommunication industries [13, 19, 20, 21, 25]. The TA list has been suggested to solve a couple of specific problems existing in the conventional TA scheme. One is that the TA list can prevent the frequent updates when a UE keeps hopping between two or more adjacent cells in different TAs (the so called ping-pong effect) [19]. Second, TA list can solve the problem of high uplink traffic due to simultaneous updates of a large number of UEs crossing a TA boundary (Figure 1) [13]. In this study, we propose a practical method to design an overlapping TAL scheme in order to overcome the second problem.

Introducing the concept of list gives more flexibility to the operators in their TA management. Two of the so far discussed candidate schemes are as follows: The *Overlapping TA scheme* in which one cell holds a list of overlapping TAs and a UE is assigned only to one TA in the cell's list. The other candidate is the *Multiple TAs scheme* in which a cell belongs to only one TA, but a UE can be assigned with more than one TA using a list. The objective of these two schemes is to reduce the TAU overhead, while assuming that the problem of paging is a less critical issue. The Multiple TAs scheme is considered as the more preferred scheme among the two in [20].

We have previously proposed the *TAL scheme* in which both UEs and cells are assigned a list of non-overlapping TAs [15, 16]. The assumption was that a cell gives only one common TAL to all UEs getting updated in that cell. Inevitably, this imposes restriction on the performance of TAL. It has been shown that the *TAL scheme* is potentially able to reduce the overall signaling overhead of the network compared to the conventional TA scheme [17]. However the challenge in applying this scheme is that UEs in one cell have different lists. It is already mentioned in the previous section that the perspective difference makes the estimation of signaling overhead difficult. In the current study, we are interested in a framework where there is no need of having the UE traces for calculating the exact value of signaling overhead.

The congestion mitigation problem in this paper has been previously considered in [11]. The authors of [11] use randomized time distribution of cell reselections to avoid the mass mobility signaling congestion. However, their mechanism requires additional configuration on the network side in order to provide the algorithm parameters, while our approach uses the existing feature in LTE systems.

3. TAU AND PAGING CALCULATION

Tracking area is defined as an area in which a UE may move without updating the MME. The initial design task is the formation of TAs inside a network, and the natural objective in TA planning is to reach an optimal balance between TAU and paging.

In real-life networks, splitting a site into different TAs is not a common practice. Therefore, although TA assignment is applicable to the cell level, we keep the discussion at the site level. Let us denote the set of sites in a network by $\mathcal{N} = \{1, \ldots, N\}$ and the TA of site i by t_i. A TA design can be alternatively represented by an $N \times N$ symmetric and binary matrix $\mathbf{S}$; in which element s_{ij} represents whether or not two sites are in the same TA, i.e.,

$$s_{ij} = \begin{cases} 1 & \text{if } t_i = t_j, \\ 0 & \text{otherwise.} \end{cases} \quad (1)$$

The $\mathbf{S}$ matrix has the transitive property: If sites i and j are in the same TA, and sites j and k are in the same TA, then i and k are also in the same TA.

Let u_i be the total number of UEs in site i scaled by the time proportion that each UE spends in site i. For the same time period, h_{ij} is the number of UEs moving from site i to site j. The values of u_i and h_{ij} can be assessed by site load and handover statistics of active UEs [14]. The amount of overhead of one paging and one update are denoted by c^p and c^u, respectively. The exact relationship between c^u and c^p depends on their radio resource consumption [7]. Moreover, parameter α is the call intensity factor (i.e., probability that a UE has to be paged).

The calculated C_{TAU} in Equation (2) below accounts for the total TAU overhead of the network. The TAU overhead of site i is proportional to the number of UEs moving from site i to other sites such as site j, when i and j are not in the same TA.

$$C_{TAU} = \sum_{i \in \mathcal{N}} \sum_{j \in \mathcal{N}: j \neq i} c^u h_{ij}(1 - s_{ij}) \quad (2)$$

The computed C_{PAG} in Equation (3) below is the overall paging overhead of the network. The paging overhead of site i is proportional to the number of incoming calls of the UEs in i and the number of other sites in the same TA.

$$C_{PAG} = \sum_{i \in \mathcal{N}} \sum_{j \in \mathcal{N}: j \neq i} \alpha c^p u_i s_{ij} \quad (3)$$

4. AN INSTRUCTIVE EXAMPLE

Let us consider an example of three sites, A, B, and C, which are along a train path in Figure 2. The optimum conventional TA configuration of this part of the network is one of the four possible combinations presented in Table 1. The parenthesis in the second column of the table specify different groupings of these three sites. According to Equations (2) and (3), the overall TAU and paging overheads are calculated for each of the four configurations. The configuration $\{(A,C),(B)\}$ is practically not a feasible combination, as it has been assumed that there is no direct handover from A to C in the example.

If we relax the transitivity condition of the $\mathbf{S}$ matrix, then the $\mathbf{S}^*$ matrix in (4) could become a possible configuration for the example in Figure 2.

$$\mathbf{S}^* = \begin{pmatrix} 1 & 1 & 0 \\ 1 & 1 & 1 \\ 0 & 1 & 1 \end{pmatrix} \quad (4)$$

$$C^*_{TAU} + C^*_{PAG} = c^u(h_{AC} + h_{CA}) + \alpha c^p(u_A + 2u_B + u_C) \quad (5)$$

The overall signaling overhead of $\mathbf{S}^*$ is given in (5). In the example, $h_{AC} = h_{CA} = 0$, and $C^*_{TAU} = 0$, hence the $\mathbf{S}^*$ configuration has the lowest TAU overhead compared to the configurations in Table 1. Assume that $u_A \gg u_B$ and $u_C \gg u_B$ (for example considering that there are train stations in A and C with large number of UEs standing in the platform), then the paging overhead C^*_{PAG} is almost half of the paging overhead of the fourth configuration in the table.

This example shows that the transitive property of the $\mathbf{S}$ matrix limits the performance of the conventional TA scheme. Here, we are interested to show that the TA list concept has the potential to reduce the overhead by relaxing this property, achieving an effect similar to that of S^*. However, the challenge in using the list concept is that the UEs in one site might have different lists depending on the sites from which the lists are assigned.

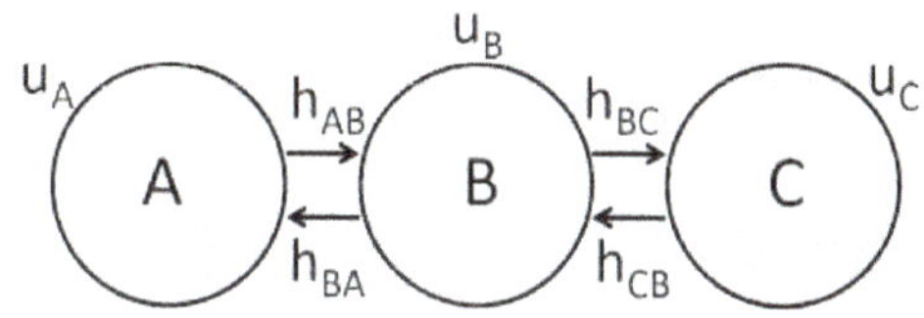

Figure 2: Three sites along a train path.

Table 1: Possible TA configurations of the ABC network with the corresponding C_{TAU} and C_{PAG}.

No.	Configuration	C_{TAU}	C_{PAG}
1	(A),(B),(C)	$c^u(h_{AB} + h_{BA} + h_{BC} + h_{CB})$	0
2	(A,B),(C)	$c^u(h_{BC} + h_{CB})$	$\alpha c^p(u_A + u_B)$
3	(A),(B,C)	$c^u(h_{AB} + h_{BA})$	$\alpha c^p(u_B + u_C)$
4	(A,B,C)	0	$2\alpha c^p(u_A + u_B + u_C)$

Table 2: The list of TALs for each site in ABC network.

A	B	C
$\{A\}$	$\{B\}$	$\{C\}$
$\{A,B\}$	$\{A,B\}$	$\{B,C\}$
$\{A,B,C\}$	$\{B,C\}$	$\{A,B,C\}$
	$\{A,B,C\}$	

Let us assume that each of the sites A, B and C are a separate TA (configuration 1) and they can assign UEs with different TALs. The TALs assigned by each site should contain that site. Table 2 presents the set of all possible TALs for each site. Site A and C are not giving TAL $\{A,C\}$ to their UEs as there is no direct handover between these two sites; this list only increases the paging overhead without any improvement in TAU. From the table, it is observable that the UEs in the ABC network have the possibility to hold one of the following six TALs: ($\{A\}$, $\{B\}$, $\{C\}$, $\{A,B\}$, $\{B,C\}$, $\{A,B,C\}$).

Let ta denote any TA of the ABC network; we define X^{ta}_{tal} as the percentage of the UEs inside each ta having tal as their list by that ta. The following three constraints are valid. Equation (6) assures that all the UEs receiving their lists from A use one of the three candidate TALs of A. The same logic for TAs B and C is represented by Equations (7) and (8), respectively.

$$X^A_A + X^A_{AB} + X^A_{ABC} = 1 \quad (6)$$

$$X^B_B + X^B_{AB} + X^B_{BC} + X^B_{ABC} = 1 \quad (7)$$

$$X^C_C + X^C_{BC} + X^C_{ABC} = 1 \quad (8)$$

According to the definition of TAL, the UEs holding the list $\{A,B,C\}$ can move in the ABC network without any TAU, and they could have also received this TAL from any of the sites. To make the calculation of signaling overhead tractable, we impose the restriction that the percentage of assigning a common TAL from different TAs are equal. This means that we require $X^A_{AB} = X^B_{AB}$, $X^B_{BC} = X^C_{BC}$ and $X^A_{ABC} = X^B_{ABC} = X^C_{ABC}$. With the mentioned assumption, there is no need to continue with the superscripts of elements (X_{tal}).

5. OVERLAPPING TRACKING AREA LIST

In the overlapping TAL scheme, the sites inside each TA hold a list of overlapping TALs. The cells update the UEs with different TALs in their lists. We consider the conventional TA design as the backbone structure, and create the lists by different combinations of TAs. The restriction explained in the previous section is applied here. It means that in the overlapping TAL scheme, the percentages of assigning a common list from different sites to the UEs in the network are equal.

The biggest difficulty in calculating the signaling overhead resulted from the TAL scheme was the challenge of having UEs with different TALs in one cell. However now with the definition of X_{tal} and the overlapping TAL scheme, the percentage of UEs holding each TAL is known. Therefore, unlike previous studies where the signaling overhead was calculated by following each UE movement pattern and call arrivals, here we can directly use the TAU and paging overhead formulas given in (2) and (3).

To overcome the signaling congestion in the train scenario, the cells (sites) along the railway path can give different TALs to the UEs inside the train. Because UEs in the train are holding different TALs, not all of them will perform TAU at the same time.

6. SOLUTION APPROACH

We aim to formulate a solution approach which finds the optimal proportional use of TALs by the sites along a railway path. The objective is to minimize the maximum TAU signaling overhead ($MTAU$), while the total paging overhead growth ($TPAG$) is bounded by a limit value in comparison to the overall paging overhead of the TA design.

For simplicity, we first consider the small example of the ABC network. In this case, we assume that A, B and C are the sites along a railway path at a peak hour. Now that it is possible to calculate the TAU and paging overheads based on Equations (2) and (3), we can minimize the maximum TAU signaling by applying the LP model in (9) and obtaining the percentages of UEs holding each TAL.

The first three constraints assure that the use of candidate TALs for each TA sum up to 100%. The next two constraints limit the TAU overhead of each site to be less or equal to $MTAU$. The last two constraints calculate the total paging overhead addition of the three sites, and keeps it lower than the *Limit*.

$$\min MTAU \tag{9}$$

subject to:

$$\begin{aligned}
X_A + X_{AB} + X_{ABC} &= 1 \\
X_B + X_{AB} + X_{BC} + X_{ABC} &= 1 \\
X_C + X_{BC} + X_{ABC} &= 1 \\
c^u h_{AB} X_A + c^u h_{BA}(X_B + X_{BC}) &\leq MTAU \\
c^u h_{BC}(X_B + X_{AB}) + c^u h_{CB} X_C &\leq MTAU \\
c^p \alpha u_A (X_{AB} + 2X_{ABC}) + & \\
c^p \alpha u_B (X_{AB} + X_{BC} + 2X_{ABC}) + & \\
c^p \alpha u_C (X_{BC} + 2X_{ABC}) &= TPAG \\
TPAG &\leq Limit \\
X_A, X_B, X_C, X_{AB}, X_{BC}, X_{ABC} &\geq 0
\end{aligned}$$

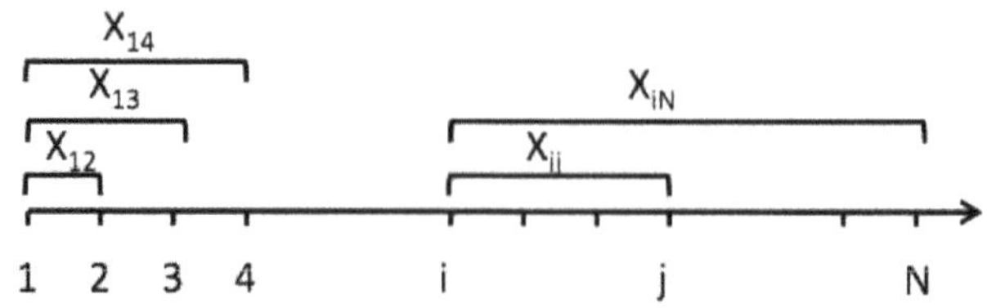

Figure 3: Definition of X_{ij} in Model (10).

Without loss of generality the terms site and TA are used interchangeably here. In general, the sites considered in a train scenario have the unique property that they only have handover with their two adjacent sites. This fact forces the TALs to contain consecutive sites. We define X_{ij} as the percentage of the list containing i, j and all the sites between them. For an illustration of an example of X_{ij} see Figure 3.

We generalize the idea in Model (9) to N sites in Model (10). Apart from the non-negativity constraints, the number of constraints of Model (10) is of $O(N)$, and the number of variables is of $O(N^2)$. This shows that the model is polynomial in size and it can be solved in polynomial time. The low computational complexity of the model is a definite advantage of our solution approach.

The first set of N constraints ensures that each site assigns all its UEs with feasible TALs. The second set of $N-1$ constraints computes the TAU signaling overhead of each site. The first sum considers UEs moving from i to j without having j in their TAL, and the second sum is for UEs moving from j to i without having i in their TAL.

The remaining constraints calculate the total paging overhead addition of all sites $\{1, .., N\}$ and bound $TPAG$ to a certain value. The paging overhead of each called UE depends on the length of its TAL, which in this case is $j - k$.

$$\min MTAU \tag{10}$$

subject to:

$$\begin{aligned}
\sum_{k=1}^{i}\sum_{j=i}^{N} X_{kj} = 1, \forall i = 1, .., N \\
c^u h_{ij} \sum_{k=1}^{i} X_{ki} + c^u h_{ji} \sum_{k=j}^{N} X_{jk} \leq MTAU \\
i = 1, .., N-1, j = i+1 \\
c^p \alpha \sum_{i=1}^{N} u_i (\sum_{k=1}^{i}\sum_{j=i}^{N} (j-k) X_{kj}) = TPAG \\
TPAG \leq Limit \\
X_{ij} \geq 0, i = 1, .., N, j = 1, .., N, i \leq j
\end{aligned}$$

7. PERFORMANCE EVALUATION

To evaluate the performance of the overlapping TAL scheme proposed in this paper, we consider two synthesized scenarios, which are constructed to resemble the reality. In both scenarios, it is assumed that 5% of the UEs are paged in every site. The overhead of a single update c^u is set ten times as much as c^p [10]. The load and handover statistics of the UEs are given for one hour period of time. The LP

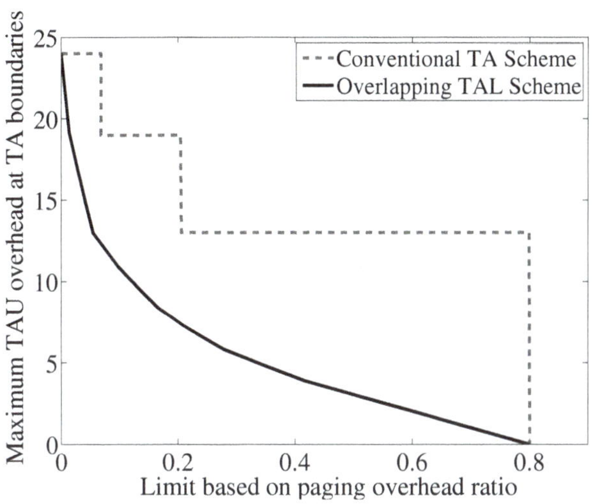

Figure 4: Comparing the performance of the two schemes for the baseline scenario.

model defined in (10) has been implemented in the CPLEX optimizer [8].

Both scenarios represent UE movement by train. We illustrate that the overlapping TAL scheme can mitigate the congestion problem in both scenarios of the network. As the LP model is polynomial in size, applying it to large-scale scenarios of this kind is scalable. The basic element of this study is TA rather than site. To compare the performance of the overlapping TAL scheme with the conventional TA scheme, we consider the TALs to be the combinations of overlapping TAs. However, it is obvious that the TALs at site level can improve the possible potentials of TALs.

Baseline Scenario

Both scenarios involve a railway path with five TAs $\{A, B, C, D, E\}$ of 150 km in length (see Figures 5 and 7). Each TA covers a large area and contains five sites. The boundaries of each TA have the lowest handovers between UEs in that region. This configuration is the optimum conventional TA design by model [22] based on the overall signaling overhead of the specified part of the network. The overall signaling overhead is 841.7 ($C_{TAU} = 122$, $C_{PAG} = 719.7$). The data excludes the mass mobility of UEs in train.

We apply Model (10) to the baseline scenario and run it for all *Limit* values from 0 to maximum. It only takes several seconds for the solver to calculate the optimum proportions of TALs for each *Limit* value. Next, we design the conventional TA configurations by minimizing the maximum $MTAU$, the same objective in Model (10), and all the *Limit* values. Note that while the *Limit* value is 0, all the five TAs are separate in the conventional TA configuration, and the equivalent solution in the overlapping TAL scheme adapts all the UEs in each TA to contain only that TA in their lists. On the other extreme, when the *Limit* value is maximum, all the five TAs are merged into one in the conventional TA configuration. The overlapping TAL, in this case, gives a list containing all the five TAs to every UE. Hence, for the two extremes of the *Limit* the optimal conventional TA scheme and the overlapping TAL coincide.

The comparison of the conventional TA scheme and the overlapping TAL for the baseline is illustrated in Figure 4. As it is explained above, the graphs coincide at the two ends. It is observable that the conventional TA scheme is changed in four discrete steps. By increasing the limit on the paging overhead, the TAs tend to merge. The behavior of the overlapping TAL scheme is smooth as the proportions of each TAL can be continuously changed. By allowing 10% addition of the paging overhead, we can reduce the maximum TAU overhead at the TA boundaries ($MTAU$) by more than 50% with the overlapping TAL scheme, while the conventional TA scheme can only reduce it by 20%. Note that even in this uncongested scenario where mass mobility is not considered, the overlapping TAL scheme can significantly perform better than the conventional TA scheme by overcoming the transitive restriction explained in Section 4.

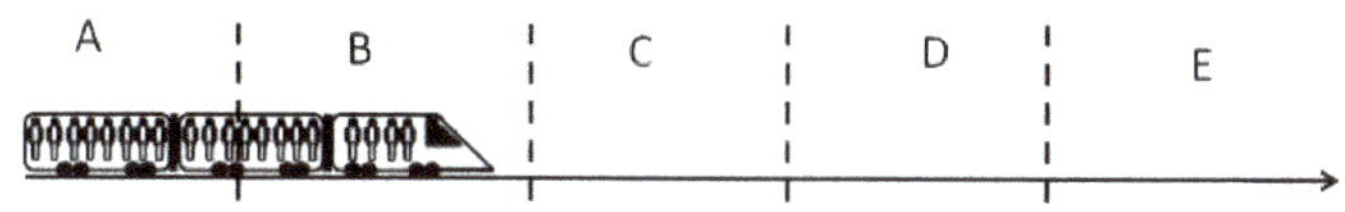

Figure 5: Congestion Scenario 1.

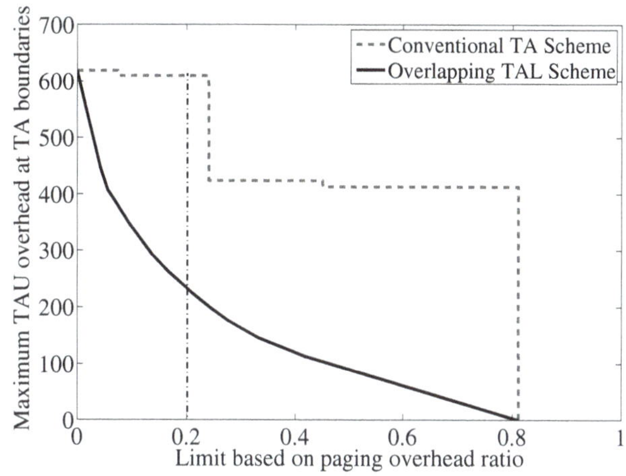

Figure 6: Comparing the performance of the two schemes for Scenario 1.

Congestion Scenario 1

In this scenario (Figure 5), a train moves from A to E having 600 passengers onboard. One example can be TGV Eurostar train which can move at a speed up to 300 km/h and may carry up to 784 passengers [13]. After a main station in C, the number of onboard passengers is reduced to 400. We assume that all passengers inside the train hold cell phones. It takes about half an hour for the train to pass the whole path (ignoring the time spent at the main station).

We aggregate the load and handover data resulted from the train and add it to the baseline scenario. All the UEs in the train passing each TA boundary will be added to the corresponding handover. The load is increased according to the proportion of time the train is spent in each TA. For example, the train spends about 6 min in A, therefore the load of A is increased by $600 * \frac{1}{10} = 60$, due to the fact that the handover and load data are given for one hour period of time.

For various *Limit* values, Figure 6 illustrates the behavior of the two schemes in minimizing the $MTAU$. Here the *Limit* refers to the additional paging overhead due to the

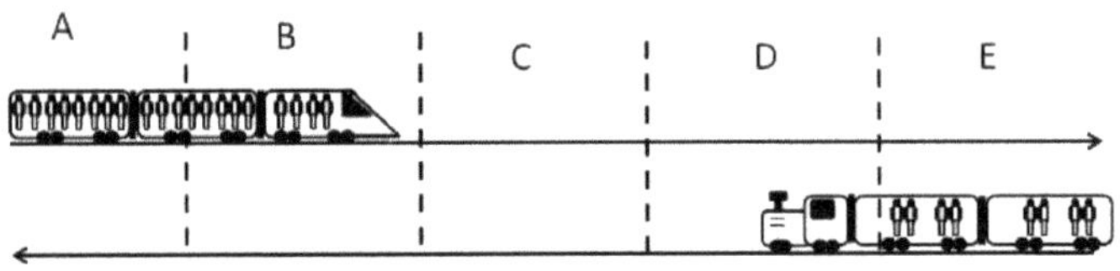

Figure 7: Congestion Scenario 2.

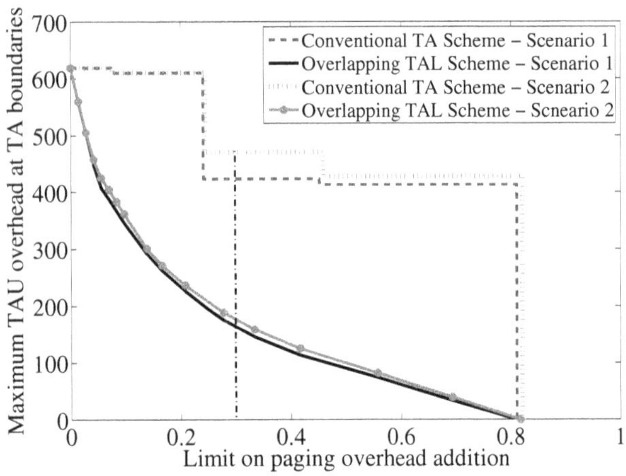

Figure 8: Comparing the performance of the two schemes for Scenarios 1 and 2.

conventional TA configurations and the proportions of the lists in the overlapping TALs compared to the original ones. By comparing the values on the y-axis in Figures 4 and 6, we can conclude that Scenario 1 is highly congested. When the *Limit* is equal to 20% of the overall paging overhead, equivalent to 143.94 in Scenario 1 (shown by the vertical line in Figure 6), the conventional TA scheme can reduce the maximum TAU signaling at the TAU boundaries by 0.01%. In the same situation, the overlapping TAL scheme reduces the $MTAU$ by 60.9%, equivalent to 242.01. This case study demonstrates the benefit of the overlapping TAL scheme in mitigating the congestion problem of the train scenario.

Congestion Scenario 2

Now, we add another train in the opposite direction to Scenario 1 (Figure 7). This train, which can be a regional train with speed reaching 160 km/h, has 450 passengers in the beginning and it stops in every TA on its way. It takes one hour for the train to pass the specified railway path. The same data aggregation is applied for this scenario.

Figure 8 illustrates the performance of the two schemes for Scenario 2. The results of Scenario 1 in Figure 6 is replicated to Figure 8 to ease comparison. Couple of observations follow from Figure 8. First by adding the second train to the scenario, the gap between the conventional TA schemes of Scenario 1 and 2 in some parts is more than ten times bigger than the corresponding gap for the overlapping TAL scheme. The proportions of TAL in reducing the maximum TAU overhead are mostly based on the largest amount of simultaneous movements of the UEs. Second, the overlapping TAL scheme performs much better than the conventional TA scheme while mass mobility in both directions is present.

Let us consider that the *Limit* is 0.3 of the overall paging overhead, equivalent to 215.91 (shown by the vertical

Table 3: Results of Scenario 1 for a certain *Limit*.

	Limit	TPAG	MTAU
Conventional TA	215.91	173.55	424.00
Overlapping TAL	215.91	215.91	163.28

Table 4: Results of Scenario 2 for a certain *Limit*.

	Limit	TPAG	MTAU
Conventional TA	215.91	175.49	471.00
Overlapping TAL	215.91	215.91	176.59

line in Figure 8). The optimal conventional TA configuration based on Model (10) is $\{(A, B, C), (D), (E)\}$, which can be represented by $X_{ABC} = 1$, $X_D = 1$ and $X_E = 1$ for both scenarios. However, the optimal TAL proportions for Scenario 1 are: $X_A = 0.27$, $X_E = 0.18$, $X_{AB} = 0.26$, $X_{CD} = 0.26$, $X_{DE} = 0.60$, $X_{ABC} = 0.39$, $X_{BCD} = 0.05$, and $X_{ABCD} = 0.08$, while the optimal TAL proportions for Scenario 2 are: $X_A = 0.29$, $X_E = 0.41$, $X_{AB} = 0.24$, $X_{CD} = 0.11$, $X_{DE} = 0.37$, $X_{ABC} = 0.42$, $X_{BCD} = 0.25$, $X_{CDE} = 0.17$ and $X_{ABCD} = 0.05$.

The overhead results of these configurations are given in Tables 3 and 4 for the two scenarios, respectively. The numerical results show that the overlapping TAL scheme is able to mitigate the first congestion scenario by 61.49% and the second congestion scenario by 62.51% more than the conventional TA scheme for the *Limit* under consideration. The fundamental flexibility of the overlapping TAL scheme in assigning different proportions of each TAL significantly improves the performance compared to the conventional TA scheme.

8. CONCLUSIONS

We have presented a low-complexity solution procedure, based on linear programming, in order to find the optimal proportional use of the lists of TAs for the proposed overlapping TAL scheme. The congestion mitigation effect is achieved by minimizing the maximum TAU signaling overhead in the network. The study shows that by assigning different TALs to the UEs involved in massive mobility, one can avoid the excessive TAU signaling messages at the TA boundaries. The transitive property of the conventional TA configuration is relaxed in the overlapping TAL scheme, which considerably improves the load balancing capability.

There are several topics deserving further research. One is to apply the overlapping TAL scheme in a large scale network to see its potentials compared to the conventional TA scheme and the previously suggested TAL algorithms. Another interesting subject worth studying is to extend the scheme by considering more data in hand, for example assuming that the destination of each UE in a train is known.

9. ACKNOWLEDGMENTS

The work has been supported by CENIIT, Linköping University, Sweden, and the Swedish Research Council.

10. REFERENCES

[1] 3GPP. TS 23.002 – v9.1.0, Network Architecture, 3G Release 9, 2009. http://www.3gpp.org/ftp/specs/archive/23_series/23.002/.

[2] 3GPP. TS 23.401 – v8.6.0, General Packet Radio Services (GPRS) enhancements for Evolved Universal Terrestrial Radio Access Network (E-UTRAN) access, 3G Release 8, 2009. http://www.3gpp.org/ftp/specs/archive/23_series/23.401/.

[3] 3GPP. TR 23.880 – v0.1.0, Stage2 for registration in densely-populated area (RED), 3G Release 8, 2007. http://www.3gpp.org/ftp/specs/archive/23_series/23.880/.

[4] 3G Americas, The Benefits of SON in LTE, White Paper, 2009.

[5] I.F. Akyildiz, J. McNair, J.S.M. Ho, H. Uzunalioglu, and W. Wang, Mobility management in next generation wireless systems, *Proc. of IEEE*, vol. 87, pp. 1347-1384, 1999.

[6] S.K. Das, S.K. Sen, R. Jayaram, and P. Agrawal, A distributed load balancing algorithm for the hot cell problem in cellular mobile networks, *Proc. of the 6th IEEE International Symposium on High Performance Distributed Computing*, pp. 254-263, 1997.

[7] P. Garcia, V. Casares, and J. Mataix, Reducing location update and paging costs in a PCS network, *IEEE Transactions on Wireless Communications*, vol. 1, pp. 200-209, 2002.

[8] ILOG CPLEX 12.2, User's Manual, 2010.

[9] S.S. Kim, A.E. Smith, and S.J. Hong, Dynamic load balancing using an ant colony approach in micro-cellular mobile communication systems, *Advances in Metaheuristics for Hard Optimization, Natrual Computing Series*, vol. 10, pp. 137-152, 2008.

[10] K. Kyamakya and K. Jobmann, Location management in cellular networks: classification of the most important paradigms, realistic simulation framework, and relative performance analysis, *IEEE Transactions on Vehicular Technology*, vol. 54, pp. 687-708, 2005.

[11] T. Mach and R. Tafazolli, Mass mobility signaling congestion avoidance mechanism using randomized time distribution of cell reselections, in *Proc. of International Conference on Telecommunications (ICT '09)*, pp. 238-242, 2009.

[12] S. Mitra and S. DasBit, On location tracking and load balancing in cellular mobile environment: a probabilistic approach, in *Proc. of the 5th International Conference on Electrical and Computer Engineering (ICECE '08)*, pp. 121-126, 2008.

[13] Mitsubishi Electric, Tracking areas sizes and tracking area list optimization, Technical Report in 3GPP TSG RAN WG3 Meeting, R3-071931, 2007.

[14] S. Modarres Razavi and D. Yuan, Performance improvement of LTE tracking area design: a re-optimization approach, in *Proc. of the 6th ACM International Workshop on Mobility Management and Wireless Access (MobiWac '08)*, pp. 77-84, 2008.

[15] S. Modarres Razavi, D. Yuan, F. Gunnarsson, and J. Moe, Exploiting tracking area list for improving signaling overhead in LTE, in *Proc. of IEEE Vehicular Technology Conference (VTC '10)*, 2010.

[16] S. Modarres Razavi, D. Yuan, F. Gunnarsson, and J. Moe, Dynamic tracking area list configuration and performance evaluation in LTE, in *Proc. of IEEE Global Communications Conference Workshop (GLOBECOM Workshop '10)*, 2010.

[17] S. Modarres Razavi, *Tracking area planning in cellular networks*, Thesis, Linköping University, 2011. http://liu.diva-portal.org/smash/get/diva2:402919.

[18] IST-2000-28088 MOMENTUM project. http://momentum.zib.de.

[19] Motorola, Long Term Evolution (LTE): A technical overview, Technical White Paper, 2007.

[20] NEC, Tracking area concepts and influence on paging load, Technical Report in 3GPP TSG RAN2 Meeting, R2-070655, 2007.

[21] QUALCOMM Europe and T-Mobile, New SON user case: tracking area optimization, Technical Report in 3GPP TSG RAN WG3 Meeting, R3-071594, 2007.

[22] D.W. Tcha, T.J. Choi, and Y.S. Myung, Location-area partition in a cellular radio network, *Journal of the Operational Research Society*, vol. 48, pp. 1076-1081, 1997.

[23] T. Winter, Mobility management and network design for UMTS, in *Proc. of IEEE International Symposium on Personal, Indoor and Mobile Radio Communications (PIMRC '04)*, 2004.

[24] V.W.S. Wong and V.C.M. Leung, Location management for next generation personal communication networks, *IEEE Network*, vol. 14, pp. 18-24, 2000.

[25] ZTE, Self optimization for tracking areas, Technical Report in 3GPP TSG RAN WG3 Meeting, R3-071415, 2007.

An Open Source Product-Oriented LTE Network Simulator based on ns-3

Nicola Baldo, Marco Miozzo, Manuel Requena-Esteso, Jaume Nin-Guerrero
Centre Tecnològic de Telecomunicacions de Catalunya (CTTC)
Av. Carl Friedrich Gauss 7, 08860 Castelldefels (Barcelona), Spain
{nbaldo, mmiozzo, mrequena, jnin}@cttc.es

ABSTRACT

In this paper we present a new simulation module for ns-3 aimed at the simulation of LTE networks. This module has been designed with a product-oriented perspective in order to allow LTE equipment manufacturers to test RRM/SON algorithms in a simulation environment before they are deployed in the field. First, we describe the design of our simulation module, highlighting its novel aspects. Subsequently, we discuss the testing methodology that we adopted to validate its output. Finally, we present some experimental result to assess its performance in terms of execution time and memory usage.

Categories and Subject Descriptors

I.6.5 [**Simulation and Modeling**]: Model Development—*Modeling methodologies*; I.6.7 [**Simulation and Modeling**]: Simulation Support Systems—*Environments*

General Terms

Algorithm, Design, Performance, Verification

Keywords

3GPP, LTE, E-UTRA, Scheduling, Simulator, ns-3

1. INTRODUCTION

3GPP Long Term Evolution (LTE) is the most promising standard for the upcoming fourth generation (4G) of mobile wireless communication systems. Very briefly, the innovative aspects included in LTE are a high speed radio interface based on OFDMA/SC-FDMA called Evolved Universal Terrestrial Radio Access (E-UTRA), an all-IP core network called Evolved Packet Core (EPC), and, most importantly, Self Organized Network (SON) capabilities, i.e., the ability of the base stations – called eNodeBs (eNBs) using the LTE nomenclature – to form a self configuring, self-optimizing and self-healing network. SONs are seen as a crucial enhancement with respect to prior mobile technologies, since they allow higher density network deployments that yield a significant increase in system capacity, and at the same time they permit a reduction in the operational expenditure of mobile network operators. In addition, the development of SON capabilities is an opportunity for LTE vendors to add value to their product and to be more competitive in the market. In this process, LTE simulators are expected to play a crucial role [1].

MSWiM'11, October 31–November 4, 2011, Miami, Florida, USA.

In this paper we describe a new software module for ns-3 that we are developing. The model partially leverages on the previous work of [2], in that the Channel model and part of the Physical layer model are reused. However, the rest of the E-UTRA model has been totally redesigned to be more representative of the LTE standard, as well as closer to real-world implementations. In particular, the LTE MAC Scheduler API published by the FemtoForum [3] has been adopted in the simulator. This interface is used by femtocell manufacturers for the implementation of scheduling and Radio Resource Management (RRM) algorithms. By introducing support for this interface in the simulator, we make it possible for LTE equipment vendors and operators to test in a simulative environment exactly the same algorithms that would be deployed in a real system.

The ultimate objective of our LTE simulator module is to become a widely accepted reference evaluation platform for LTE-based SON systems. In order to achieve this, we pay a particular attention to two important aspects. The first one is a thorough test and validation procedure that aims at guaranteeing that the output produced by the simulator represents the performance of LTE systems with a satisfactory level of accuracy. The second one is the open source nature of the simulator: by being GPL-licensed and by having a development process which is open to the community, we expect not only to foster early adoption and contributions by industrial and academic partners, but also to enhance the overall quality of the simulator by leveraging on a wide base of users and developers. The interested reader can download the source code from [4].

2. DESCRIPTION OF THE LTE SIMULATION MODEL

2.1 Design Criteria

The LTE module that we present in this paper has the objective of supporting the evaluation of the following aspects of LTE systems: 1) Radio Resource Management, 2) QoS-aware Packet Scheduling, 3) Inter-cell Interference Coordination and 4) Dynamic Spectrum Access. In order to model LTE systems to a level of detail that is sufficient to allow a correct evaluation of the above mentioned aspects, the following requirements have been considered:

- At the radio level, the granularity of the model should be at least that of the Resource Block (RB). In fact, this is the fundamental unit being used for resource allocation. Without this minimum level of granularity, it is not possible to model accurately packet scheduling and inter-cell-interference.[1]
- The simulator should scale up to tens of eNBs and hundreds of User Equipments (UEs). This rules out the use of a link level simulator, i.e., a simulator whose radio interface is modeled with a granularity up to the symbol level. This is because, to have a symbol level model, it would be necessary to implement all the PHY layer signal processing, whose huge computational complexity would severely limit simulation scalability. In fact, link-level simulators are normally limited to a single eNB and one or a few UEs.
- It should be possible within the simulation to configure different cells so that they use different carrier frequencies and system bandwidths. The bandwidth used by different cells should be allowed to overlap, in order to support dynamic spectrum licensing solutions such as those described in [5,6]. The calculation of interference should handle appropriately these configurations.
- The LTE simulation module should contain its own implementation of the FemtoForum LTE MAC Scheduler Interface Specification [3]. Neither binary nor data structure compatibility with vendor-specific implementations of the same interface should be supported; hence, a compatibility layer should be interposed whenever a vendor-specific MAC scheduler is to be used with the simulator. This requirement is necessary to allow the simulator to be independent from vendor-specific implementations of this interface specification.[2]

2.2 Implemented MAC Functionality

In this section we describe the MAC features that are implemented in the simulator. All these features are implemented after models from the literature. The reason for this choice is twofold: first, well-known models are easily understood by developers and are a convenient starting code base for the development of more advanced solutions; second, they can be used as reference algorithm when doing performance evaluation. With this latter respect, we stress that the use of a publicly available implementation of models (in particular scheduler implementations) as the reference for a performance evaluation study is beneficial to the authoritativeness of the study itself.

2.2.1 Resource Allocation Model

We now briefly describe how resource allocation is handled in LTE, clarifying how it is implemented in the simulator. The scheduler is in charge of generating specific structures called Data Control Indication (DCI) which are then transmitted by the PHY of the eNB to the connected UEs, in order to inform them of the resource allocation on a per subframe basis. In doing this, the scheduler has to fill some specific fields of the DCI structure with all the information, such as: the Modulation and Coding Scheme (MCS) to be used, the MAC Transport Block (TB) size, and the allocation bitmap which identifies which RBs will contain the data transmitted by the eNB to each user.[3] The latter bitmap can be coded in different formats; in this implementation, we considered the *Allocation Type 0* defined in [8], according to which the RBs are grouped in Resource Block Groups (RBG) of different size determined as a function of the Transmission Bandwidth Configuration[4] in use.

2.2.2 Adaptive Modulation and Coding

The Adaptive Modulation and Coding (AMC) model provided by the simulator is a modified version of the model described in [2], which in turn is inspired from [10]. Our version is described in the following. Let i denote the generic user, and let γ_i be its SINR. We get the spectral efficiency η_i of user i using the following equations:

$$\Gamma = \frac{-\ln(5 * \text{BER})}{1.5} \quad (1)$$

$$\eta_i = \log_2\left(1 + \frac{\gamma_i}{\Gamma}\right) \quad (2)$$

where BER $= 0.00005$. The procedure described in [11] is used to perform AMC. The spectral efficiency is quantized based on the table reported in [11] (rounding to the lowest value) and is mapped to the corresponding MCS scheme.

2.2.3 Round Robin (RR) Scheduler

The Round Robin (RR) scheduler is probably the simplest scheduler found in the literature. It works by dividing the available resources among the active flows, i.e., those logical channels which have a non-empty RLC queues. If the number of RBGs is greater than the number of active flows, all the flows can be allocated in the same subframe. Otherwise, the excesses flows that cannot be allocated in a subframe will be considered in the subsequent one in a circular fashion. The MCS to be adopted for each user is evaluated according to the received wideband CQIs.

2.2.4 Proportional Fair (PF) Scheduler

The Proportional Fair (PF) scheduler [7] works by scheduling a user when its instantaneous channel quality is high relative to its own average channel condition over time. Let i, j denote generic users; let t be the subframe index, and k be the resource block index; let $M_{i,k}(t)$ be MCS usable by user i on resource block k according to what the AMC model reported (see Section 2.2.2); finally, let $S(M, B)$ be the TB size in bits as defined in [8] for the case where a number B of resource blocks is used. The achievable rate $R_i(k,t)$ in

[1]The reason is that, since packet scheduling is done on a per-RB basis, an eNB might transmit on a subset only of all the available RBs, hence interfering with other eNBs only on those RBs where it is transmitting.

[2]We note that [3] is a logical specification only, and its implementation (e.g., translation to some specific programming language) is left to the vendors.

[3]For the mapping of resources to physical RBs, we adopt a *localized mapping* approach (see [7], Section 9.2.2.1); hence in a given subframe each RB is always allocated to the same user in both slots.

[4]We note that the Transmission Bandwidth Configuration, defined in [9], indicates the total number of RBs which can be used by an eNB. The allowed values are 6, 15, 25, 50, 75 and 100.

bit/s for user i on resource block k at subframe t is defined as

$$R_i(k,t) = \frac{S\left(M_{i,k}(t), 1\right)}{\tau} \tag{3}$$

where τ is the TTI duration. At the start of each subframe t, each RB is assigned to a certain user. In detail, the index $\widehat{i}_k(t)$ to which RB k is assigned at time t is determined as

$$\widehat{i}_k(t) = \underset{j=1,\ldots,N}{\operatorname{argmax}} \left(\frac{R_j(k,t)}{T_j(t)} \right) \tag{4}$$

where $T_j(t)$ is the past throughput performance perceived by the user j. According to the above scheduling algorithm, a user can be allocated to different RBGs, which can be either adjacent or not, depending on the current condition of the channel and the past throughput performance $T_j(t)$. The latter is determined at the end of the subframe t using the following exponential moving average approach:

$$T_j(t) = (1 - \frac{1}{\alpha})T_j(t-1) + \frac{1}{\alpha}\widehat{T}_j(t) \tag{5}$$

where α is the time constant (in number of subframes) of the exponential moving average, and $\widehat{T}_j(t)$ is the actual throughput achievable by the user i in the subframe t. $\widehat{T}_j(t)$ is measured according to the following procedure. First we determine the MCS $\widehat{M}_j(t)$ actually used by user j as the minimum one among the allocated RBGs. Then we determine the total number $\widehat{B}_j(t)$ of RBs allocated to user j:

$$\widehat{B}_j(t) = \left|\{k : \widehat{i}_k(t) = j\}\right| \tag{6}$$

where $|\cdot|$ indicates the cardinality of the set; finally,

$$\widehat{T}_j(t) = \frac{S\left(\widehat{M}_j(t), \widehat{B}_j(t)\right)}{\tau} \tag{7}$$

3. TESTING AND VALIDATION

To test and validate the ns-3 LTE module, several provided test suites are integrated with the ns-3 test framework. The most important ones are described in the following subsections.

3.1 Adaptive Modulation and Coding Test

The test suite `lte-link-adaptation` provides system test cases recreating scenarios with a single eNB and a single UE. Different test cases are created corresponding to different SINR values perceived by the UE. The aim of the test is to check that the chosen MCS corresponds to some known reference values. These reference values are obtained by re-implementing the model described in Section 2.2.2 in Octave. The resulting test vector is represented in figure 1. The test passes if both the following conditions are verified:

- the UE-perceived SINR calculated by the simulator corresponds to the value intended for the given test case within a tolerance of 10^{-7}. The tolerance is meant to account for the approximation errors typical of floating point arithmetic.
- the MCS index chosen by the scheduler matches exactly with the MCS index in the test vector, determined using the above described procedure.

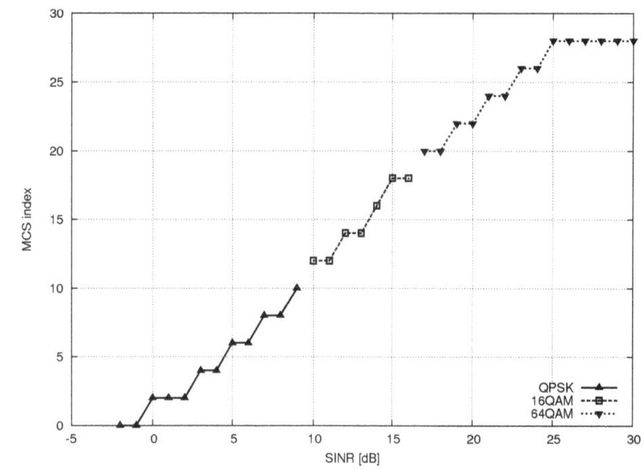

Figure 1: Test vector for the Link Adaptation test

3.2 Round Robin scheduler performance

The test suite `lte-rr-ff-mac-scheduler` creates different test cases with a single eNB and several UEs, all having the same Radio Bearer specification. In each test case, all UEs experience the same inferference and noise impairements from the eNB, i.e. same SINR; different test cases are implemented by using different distances among UEs and the eNB (therefore having different SINR values) and different numbers of UEs. The test consists on checking that the obtained throughput performance is equal among users and matches a reference throughput value obtained according to the SINR perceived within a given tolerance.

The test vector is obtained according to the values of TB size reported in table 7.1.7.2.1-1 of [8], considering an equal distribution of the physical resource blocks among the users using Resource Allocation Type 0 as defined in Section 7.1.6.1 of [8]. Let τ be the TTI duration, N be the number of UEs, B the transmission bandwidth configuration in number of RBs, G the RBG size, M the modulation and coding scheme in use at the given SINR, X the number of RBs assigned to a given user and $S(M,X)$ the corresponding TB size in bits as defined in [8]. We first calculate the number of RBGs allocated to each user as $L = \lfloor B/NG \rfloor$. The reference throughput T in bit/s achieved by each UE is then calculated as $T = S(M, LG)/\tau$. The test passes if the measured throughput matches with T within a relative tolerance of 0.1. This tolerance is needed to account for the transient behavior at the beginning of the simulation (e.g., CQI feedback is only available after a few subframes) as well as for the accuracy of the estimator of the average throughput performance over the chosen simulation time (0.4s). This choice of the simulation time is justified by the need to follow the ns-3 guidelines of keeping the total execution time of the test suite low, in spite of the high number of test cases. In any case, we note that a lower value of the tolerance can be used when longer simulations are run.

In Figure 2, the curves labeled "RR" represent the test values calculated for the RR scheduler test, as a function of the number of UEs and of the MCS index being used in each test case.

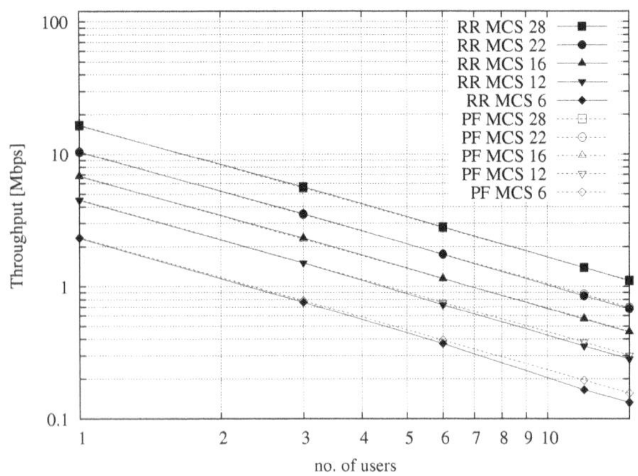

Figure 2: Test vectors for the RR and PF Scheduler in the downlink in a scenario where all UEs use the same MCS.

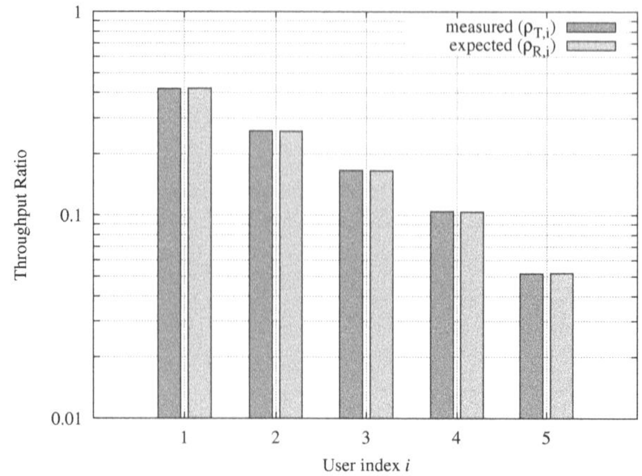

Figure 3: Throughput ratio evaluation for the PF scheduler in a scenario where the UEs have MCS index $28, 24, 16, 12, 6$

3.3 Proportional Fair scheduler performance

The test suite `lte-pf-ff-mac-scheduler` creates different test cases with a single eNB, using the Proportional Fair (PF) scheduler, and several UEs, all having the same Radio Bearer specification. The test cases are grouped in two categories that validate respectively the adaptation to the channel conditions and the fairness among UEs.

In the first category of test cases, the UEs are all placed at the same distance from the eNB, and hence all have the same SINR. Different test cases are implemented by using a different distance value and a different number of UEs. The test consists on checking that the obtained throughput performance matches with the known reference throughput up to a given tolerance. The expected behavior of the PF scheduler when all UEs have the same SINR is that each UE should get an equal fraction of the throughput obtainable by a single UE when using all the RBs. We calculate the reference throughput value by dividing the throughput achievable by a single UE at the given SINR by the total number of UEs. Using the same notation introduced in Section 3.2, the reference throughput T in bit/s is expressed as $T = S(M, B)/\tau N$.

The second category of tests aims at verifying the fairness of the PF scheduler in a more realistic simulation scenario where the UEs have a different SINR (constant for the whole simulation). In these conditions, the PF scheduler will give to each user a share of the system bandwidth that is proportional to the capacity achievable by a single user alone considered its SINR. In detail, let M_i be the modulation and coding scheme being used by each UE (which is a deterministic function of the SINR of the UE, and is hence known in this scenario). Based on the MCS, we determine the achievable rate R_i for each user i using the procedure described in Section 2.2.4. We then define the achievable rate ratio $\rho_{R,i}$ of each user i as

$$\rho_{R,i} = \frac{R_i}{\sum_{j=1}^{N} R_j} \qquad (8)$$

Let now T_i be the throughput actually obtained by the UE i, which is obtained as part of the simulation output. We define the obtained throughput ratio $\rho_{T,i}$ of UE i as

$$\rho_{T,i} = \frac{T_i}{\sum_{j=1}^{N} T_j} \qquad (9)$$

The test consists of checking that the condition $\rho_{R,i} = \rho_{T,i}$ is verified. If so, it means that the throughput obtained by each UE over the whole simulation matches with the steady-state throughput expected by the PF scheduler according to theory. In particular, the test condition can be derived from [12] as follows. From Section 3 of [12], we know that, $\forall i$, $T_i/R_i = c$, where c is a constant. By substituting the above into the definition of $\rho_{T,i}$ given previously, we get

$$\frac{T_i}{\sum_{j=1}^{N} T_j} = \frac{cR_i}{\sum_{j=1}^{N} cR_j} = \frac{R_i}{\sum_{j=1}^{N} R_j} \qquad (10)$$

which is exactly the test condition.

Figure 3 presents the results obtained from the simulator in a test case with 5 UEs $i = 1, \ldots, 5$ that are using respectively the MCS index $28, 24, 16, 12, 6$. From the figure, we note that, as expected, the obtained throughput is proportional to the achievable rate.

4. EXECUTION TIME AND MEMORY CONSUMPTION

In order to provide an evaluation of the execution time and memory consumption, a reference simulation program (`examples/profiling-reference`) has been developed. This program simulates a scenario composed by a set of eNBs, and a set of UEs attached to each eNB. All eNBs have the same number of attached UEs. Communications are performed both in downlink and in uplink using a saturation model (i.e., each RLC instance always has a PDU to transmit). The eNBs are distributed in a line with a spacing of 140m; each UE is placed at the same position of the eNB it is attached to. The total simulation time is set to 60s. Using this simulation program, we ran a simulation campaign varying the number of eNBs as well as the number of UEs per eNB. For each simulation, we measured the execution time using the `time` shell command in linux, and the memory consumption by looking at the information in

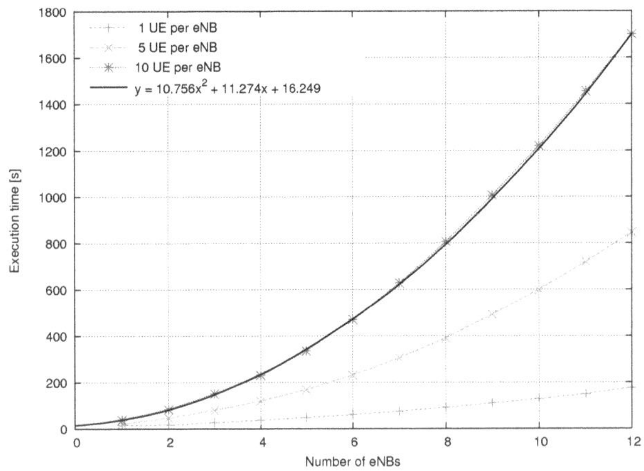

Figure 4: Execution time of the reference program.

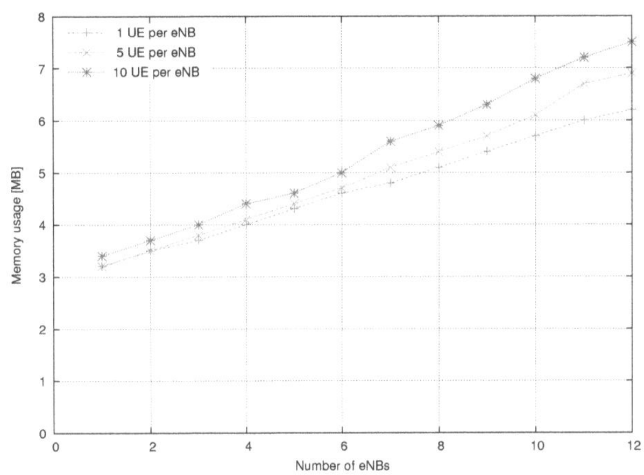

Figure 5: Memory usage of the reference program.

`/proc/{pid}/statm`. The reference hardware platform is an Intel Core2 Duo E8400 3.00GHz with 512 MB of RAM memory running a Fedora Core 10 distribution with kernel 2.6.27.5. The ns-3 build used in this experiment was configured with the options `-d optimized --enable-static`.

The results are reported in Figures 4 and 5. We note that the memory usage, as expected, primarily depends on the number of eNBs, and is in general quite low. The execution time depends significantly on both the number of eNBs and the number of UEs per eNB. For the case of 10 UEs per eNB, we also show that the experimental data can be fitted quite accurately by a quadratic function. We suggest that this behavior is due to the fact that the interference calculations, whose computational complexity is quadratic with respect to the number of eNBs, are dominant in the overall computational load.

5. CONCLUSIONS

In this paper we presented an open source product oriented simulation model for ns-3 that is aimed at the performance evaluation of LTE SONs, discussing its design, validation and run-time performance.

As a final note, we would like to mention that development on this simulator is still ongoing, with the aim of including some functionalities which are still not supported at the moment. Among these features, we would like to mention the development of pathloss models, link error models, HARQ and MIMO support, as well as the introduction of those EPC features, such as the X2 interface, that are crucial for SONs.

6. ACKNOWLEDGMENTS

We would like to acknowledge Ubiquisys, the developer of UMTS and LTE small cells, for playing a major part in instigating and funding this project.

7. REFERENCES

[1] "Ubiquisys and CTTC Develop World's First Open Source Product-Oriented LTE Network Simulator," http://www.ubiquisys.com/femtocell-media-press-releases-id-203.htm, Feb. 2011.

[2] G. Piro, N. Baldo, and M. Miozzo, "An LTE module for the ns-3 network simulator," in *Proc. of the Workshop on NS-3 (in conjunction with SimuTOOLS 2011)*, Mar. 2011.

[3] FemtoForum, "LTE MAC Scheduler Interface Specification v1.11," http://www.femtoforum.org/femto/technical.php, Oct. 2010.

[4] "The LTE-EPC Network Simulator (LENA) project," http://iptechwiki.cttc.es/LTE-EPC_Network_Simulator_(LENA).

[5] Ofcom, "Consultation on assessment of future mobile competition and proposals for the award of 800 MHz and 2.6 GHz spectrum and related issues," Mar. 2011.

[6] RealWireless, "Low-power shared access to spectrum for mobile broadband," Ofcom Project MC/073, Tech. Rep., 2011.

[7] I. T. S. Sesia and M. Baker, *LTE - The UMTS Long Term Evolution*. Ed. Wiley, 2009.

[8] 3GPP, *Tech. Specif. Group Radio Access Network; Physical layer procedures (Release 8), 3GPP TS 36.213.*

[9] ——, *Evolved Universal Terrestrial Radio Access (E-UTRA); User Equipment (UE) radio transmission and reception (Release 8), 3GPP TS 36.101.*

[10] H. Seo and B. G. Lee, "A proportional-fair power allocation scheme for fair and efficient multiuser OFDM systems," in *Proc. of IEEE GLOBECOM*, Dallas, TX, USA, Dec. 2004.

[11] 3GPP, *Tech. Specif. Group Radio Access Network; Conveying MCS and TB size via PDCCH, 3GPP TSG-RAN WG1 R1-081483.*

[12] J. Holtzman, "CDMA forward link waterfilling power control," in *Proc. of IEEE VTC Spring*, May 2000.

Throughput Modeling for Multi-rate IEEE 802.11 Vehicle-to-Infrastructure Networks with Asymmetric Traffic

Ke Xu, Benjamin T. Garrison, Kuang-Ching Wang
The Holcombe Department of Electrical and Computer Engineering
Clemson University
Clemson, SC, US
{kxu, bgarris, kwang}@clemson.edu

ABSTRACT

This paper proposes an analytical throughput model for IEEE 802.11-based vehicle-to-infrastructure networks with asymmetric traffic. Previous models assume all nodes use ad-hoc mode with uniform traffic without mobility. This model is developed using the fixed-point analysis, which assumes asymmetric traffic loads for AP and clients, and both types of traffic are nonsaturated. The mobility effect is also considered, which is modeled as a throughput scaling factor taking account of velocity and communication range. Numerical evaluation compared with Bianchi's model, verifies the proposed model and quantifies the difference caused by asymmetric traffic, and the throughput with mobile nodes is evaluated with the increase in velocity.

Categories and Subject Descriptors

C.2.6 [**Computer-Communication Networks**]: Internetworking – *Standards.* I.6.5 [**Simulation and Modeling**]: Modeling Development – *Modeling methodologies.* C.2.1 [**Computer-Communication Networks**]: Network Architecture and Design – *Wireless communication.*

General Terms

Measurement, Performance, Verification

Keywords

IEEE 802.11, performance analysis, throughput model, fixed-point analysis, vehicular networks, mobility

1. INTRODUCTION

Vehicular networks are a novel class of wireless networks that have emerged as the automotive and communication industries announce plans to bring ubiquitous broadband Internet connectivity to moving vehicles [1]. Envisioned applications include road safety, driver assistance, infotainment, and vehicle telematics utilizing a range of wireless communication methods based on Wi-Fi, dedicated short range radios (IEEE 802.11p), or 4G radios such as Mobile WiMAX and long term evolution (LTE). Depending on the applications, such networks can be realized as an ad hoc network, an infrastructure network, or a hybrid combination of the two. Vehicle ad hoc networks (VANET), also referred to as vehicle-to-vehicle networks, have been studied by many, e.g., [2-5], utilizing ad hoc multi-hop communication among cars with diverse mobility patterns. Vehicle-to-infrastructure (V2I) networks, also called vehicle-to-roadside (V2R) networks, employ statically deployed access points (APs) or base stations (BSs) to connect moving cars. Despite the higher costs to deploy and maintain the AP/BS infrastructure, industries and transportation authorities are paying high attention to V2I networks due to their higher reliability and constant availability where such infrastructure exists. In real V2I situation, the traffic load on the AP side may become extremely heavy compared to light traffic load on the client side e.g. downloads or video entertainments, resulting in asymmetric traffic in the entire network, which is different from an ad hoc network.

Radios based on the IEEE 802.11 standards (a/b/g/n/p) are most commonly considered in vehicular networks, in part due to the minimum scheduling constraints of their CSMA/CA medium access control (MAC) protocol. Recent field measurement studies have verified the feasibility and performance characteristics of these protocols in practical vehicular network conditions [6-9].

Numerous studies have explored performance models for IEEE 802.11's MAC protocol's Distributed Coordination Function (DCF) mode [10-15]. In a widely recognized work [10], Bianchi proposed a saturation throughput model for an IEEE 802.11 network based on Markov analysis, assuming N stations employing omni-directional antennas (OAs) and a common transmission probability in a single contention domain. In [13], Malone proposed a modeling framework which extended Bianchi's Markov-chain model to nonsaturated operation. The computations for this model involve a complex nonlinear system, which is time-consuming and difficult to solve. In [14], Kumar proposed a saturated renewal model for the IEEE 802.11's backoff process, which significantly simplifies the saturated performance analysis of DCF networks as compared to analysis using Bianchi's model. In [15], Zhao proposed an approximate nonsaturated model, based on Kumar's model, by scaling the transmission probability of the saturated setting with an appropriate factor. In [16], the authors modeled the mobility impact for mobile ad hoc networks with directional antennas (DAs) based on a *hit-or-miss* model, showing that such impacts can be captured with a scaling factor for the stationary throughput as well.

The objective of this paper is to derive a throughput model for IEEE 802.11-based V2I networks that can address asymmetric and nonsaturated traffic in the presence of vehicle mobility and multiple link rates. The major contributions of this paper are in two areas: (1) an analytical model based on fixed-point analysis in [14] and [15] in nonsaturated conditions is derived, which assumes AP and clients have asymmetric and nonsaturated traffic loads; this model is much easier to solve than the model in [13]; (2) the throughput accounts for both multiple link rates (other than single link rate used in conventional models) and mobility effects (as a scaling factor for throughput with static nodes). These two properties make the proposed model more suitable for measuring performance for practical V2I networks.

To derive the throughput equation for mobile nodes, three steps are involved. First, find the collision probabilities by solving fixed-point equations. Second, calculate the multi-rate throughput using the probabilities found in the first step. Third, scale the throughput based on the mobility effects. This paper is organized as follows. Section 2 describes models of the network, briefly summarizes the fixed-point analysis for uniform traffic, and gives fixed-point analysis for V2I networks in nonsaturated asymmetric traffic condition. Section 3 presents the throughput model for static nodes in V2I networks. Section 4 derives the mobility throughput scaling factor. Section 5 evaluates the theoretical results from the proposed model. The paper concludes in Section 6.

2. MODELING IEEE 802.11 DCF

The network model, introduction to fixed-point equations for uniform traffic and fixed-point analysis for V2I networks with asymmetric traffic are discussed as follows.

2.1 Network Model

An IEEE 802.11 based V2I network is envisioned to be composed of a set of APs distributed across a certain area to communicate with vehicle clients within their radio coverage. The placement of APs is similar to a typical cellular system; the AP placement problem is beyond the scope of this paper, though for modeling simplicity it is assumed that adjacent APs adopt disjoint frequency channels such that vehicles serviced by an AP do not incur interference to adjacent APs and their service vehicles. By this assumption, the throughput model will be derived for a single AP and N-1 clients connected to it. In this paper, the model assumes that all the clients are in the AP's communication range and can hear any other nodes without considering the effect of hidden terminal.

The AP covers a circular area of radius R_c, i.e., its communication range. As in [16], clients are assumed to be spatially distributed with the following probability density function (pdf):

$$\omega(r,\varphi)=\begin{cases}\dfrac{1}{2\pi R_c r} & r\in(0,R_c],\ \varphi\in[0,2\pi]\\ 0 & elsewhere\end{cases} \tag{1}$$

where r is the distance between AP and client and φ is the azimuth angle. The marginal pdfs $\omega_r(r) = 1 / R_c$ and $\omega_\varphi(\varphi) = 1 / 2\pi$ are both uniform.

All nodes adopt the IEEE 802.11 DCF MAC protocol with the RTS/CTS virtual carrier sensing mechanism. A node with arrived packets can only transmit if it senses the channel idle for a period of time equal to a distributed interframe space (DIFS). According to DCF, the time immediately following an idle DIFS is slotted, and a node is allowed to transmit only at the beginning of each slot time, which is set equal to the time to detect the transmission of a packet from any other node. The slot time depends on the physical layer, e.g. 802.11 standard sets slot time as $50\mu s$ for frequency hopping spread spectrum (FHSS) [10]. Omni-directional antennas are employed by all nodes. The IEEE 802.11 Network Allocation Vector (NAV) is used for nodes overhearing RTS/CTS to record the time they must refrain from transmission. Finally, it is assumed all nodes adopt the same transmit power P_T.

2.2 Fixed-Point Analysis for Uniform Traffic

In [15], an approximate model is developed to characterize the backoff procedure in nonsaturated homogeneous DCF networks, where all stations have the same traffic arrival rate and packet size. This model scales the transmission probability of the saturated setting with a factor to approximate the transmission probability in the nonsaturated setting. According to [15], let β ($0\le \beta \le 1$) be the transmission probability per slot for each node, and β_c be the transmission probability per slot for each node on the condition that the buffer is not empty. Letting p_a be the probability of a nonempty buffer, we have

$$\beta = p_a\beta_c \tag{2}$$

According to the decoupling approximation in [14], the collision probability γ experienced by the transmitting node can be written in terms of β:

$$\begin{aligned}\gamma &\triangleq \Gamma(\beta)\\ &=1-(1-\beta)^{N-1}\end{aligned} \tag{3}$$

Hence, β_c depends on γ, and can be expressed as $\beta_c = \beta_c(\gamma)$. Denote $R(\gamma)$ and $X(\gamma)$ as the mean number of attempts and the mean backoff time in slots, respectively. β_c can be expressed in terms of $R(\gamma)$ and $X(\gamma)$ [14]:

$$\beta_c=\beta_c(\gamma)=\frac{R(\gamma)}{X(\gamma)} \tag{4}$$

where

$$R(\gamma)=1+\gamma+\gamma^2+\cdots+\gamma^K \tag{5}$$

$$X(\gamma)=b_0+\gamma b_1+\gamma^2 b_2+\cdots+\gamma^K b_K \tag{6}$$

In Eq. 6, b_i can be expressed as

$$b_i=\begin{cases}\mathrm{CW}_0/2 & i=0\\ 2^i\cdot b_0 & 1\le i\le m-1\\ 2^m\cdot b_0 & m\le i\le K\end{cases} \tag{7}$$

where CW_0 is the minimum backoff window size (in slots), m is the maximum backoff stage and K is the retransmission limit.

Combining Eq. 2, 3 and 4, the collision probability γ is the solution of the fixed-point equation $\gamma = \Gamma(\beta)$. According to Brouwer's fixed-point theorem, $\Gamma(\beta)$ is a continuous function of γ in (0, 1) and maps it into (0, 1), and thus must have at least one fixed point.

In [15], the author proposed two models assuming different buffer sizes (a small buffer and an infinite buffer) for calculating scaling factor p_a. One fixed point exists for small buffer model but multiple fixed points exist for infinite buffer model. Although it is possible to calculate p_a for an arbitrary buffer size, for simplicity, this paper assumes all nodes in the network have small buffers.

2.3 Fixed-Point Analysis for V2I Networks with Asymmetric Traffic

In a V2I network as described in Section 2.1, there are two types of nodes, AP and client. An AP's traffic rate is much higher than each client's since traffic from all clients goes through the AP. This paper assumes all the clients have the same traffic rate. Let λ_{AP} and λ_{clt} be the traffic arrival rate of the AP and a client, respectively.

The collision probability seen by a given AP or client cannot use a homogeneous collision probability γ anymore. Let γ_{AP} *and* γ_{clt} denote the collision probability seen by the AP and a client, respectively, which can be expressed as

$$\gamma_{AP} = 1-\left(1-p_{clt}\beta_{clt}\right)^{N-1} \tag{8}$$

$$\gamma_{clt} = 1-\left(1-p_{AP}\beta_{AP}\right)\left(1-p_{clt}\beta_{clt}\right)^{N-2} \tag{9}$$

where p_{AP} and p_{clt} are probabilities of a nonempty buffer for the AP and a client; β_{AP} and β_{clt} are the transmission probability for the AP and a client, which can be calculated in terms of γ_{AP} *and* γ_{clt}:

$$\beta_{AP} = \beta_c\left(\gamma_{AP}\right) \tag{10}$$

$$\beta_{clt} = \beta_c\left(\gamma_{clt}\right) \tag{11}$$

where β_c follows Eq. 4.

The basic steps using fixed-point analysis in this subsection are as follows:

Step 1: find a fixed-point equation γ_{clt} = F(γ_{clt}, γ_{AP}, p_{clt}, p_{AP}) for collision probability γ_{clt}, where F(γ_{clt}, γ_{AP}, p_{clt}, p_{AP}) is a continuous function of γ_{clt}, γ_{AP}, p_{clt}, and p_{AP}.

Step 2: express γ_{AP}, p_{clt}, and p_{AP} in F(γ_{clt}, γ_{AP}, p_{clt}, p_{AP}) as functions of γ_{clt} .

Step 3: apply fixed-point theorem to find the solution of γ_{AP}, p_{clt}, and p_{AP} with respect to arbitrary values for γ_{clt} in (0, 1).

Step 4: plug the values of (γ_{AP}, p_{clt}, p_{AP}) obtained in Step 3 into the right-hand side of the fixed-point equation in Step 1; apply fixed-point theorem to find the solution of γ_{clt}.

From Eq. 8-11, we can rewrite γ_{clt} as

$$\gamma_{clt} = 1-\left(1-p_{AP}\beta_c\left(1-\left(1-p_{clt}\beta_c\left(\gamma_{clt}\right)\right)^{N-1}\right)\right) \cdot\left(1-p_{clt}\beta_c\left(\gamma_{clt}\right)\right)^{N-2} \tag{12}$$

Eq. 12 is the fixed-point equation for solving γ_{clt}. The next task is to find expressions of p_{AP} and p_{clt} . Applying the result of a small buffer model in [15], p_{clt} can be computed as

$$\begin{aligned} p_{clt} &= 1-\exp\left(-\lambda_{clt}Y_{clt}\right) \\ &= 1-\exp\left(-\lambda_{clt}X\left(\gamma_{clt}\right)T\right) \end{aligned} \tag{13}$$

where Y_{clt} is the mean time of a packet for client on the condition that the buffer is not empty, and $Y = X(\gamma)T$ can be found in [15]; T is the mean time that elapses for one decrement of the backoff counter, which is given by

$$T = \left(1-p_b\right)\sigma + p_s p_b\left(T_s+\sigma\right) + p_b\left(1-p_s\right)\left(T_c+\sigma\right) \tag{14}$$

where σ is the duration of one time slot; T_s and T_c are one average successful transmission duration and one average collision duration, respectively, which can be measured by

$$\begin{aligned} T_s = {} & T_{\mathrm{RTS}} + T_{\mathrm{CTS}} + T_{\mathrm{DATA}} + T_{\mathrm{ACK}} + T_{\mathrm{H}} + \mathrm{DIFS} \\ & + 3\cdot\mathrm{SIFS} + 4\cdot\delta \end{aligned} \tag{15}$$

$$T_c = T_{\mathrm{RTS}} + T_{\mathrm{CTS}} + \mathrm{DIFS} + \mathrm{SIFS} + 2\cdot\delta \tag{16}$$

where T_{RTS}, T_{CTS}, T_{DATA}, T_{ACK}, T_{H} are the times required to transmit RTS, CTS, DATA, ACK and the packet's PHY and MAC headers, respectively; DIFS and SIFS are the Distributed and Short Inter-Frame Spacing, and δ is the propagation delay.

In Eq. 14, p_b is the probability that there is at least one transmission in the considered slot time, which can be expressed as

$$\begin{aligned} p_b &= 1-\left(1-p_{AP}\beta_{AP}\right)\left(1-p_{clt}\beta_{clt}\right)^{N-1} \\ &= 1-\left(1-\gamma_{clt}\right)\left(1-p_{clt}\beta_c\left(\gamma_{clt}\right)\right) \end{aligned} \tag{17}$$

In Eq. 17, the transformation plugs in Eq. 9 to get rid of p_{AP} and β_{AP}.

In Eq. 14, p_s is the probability that a transmission is successful, which is the probability that either the AP or one client transmits successfully given that the channel is busy, therefore

$$\begin{aligned} p_s &= \frac{1}{p_b}\cdot\Big[p_{AP}\beta_{AP}\left(1-p_{clt}\beta_{clt}\right)^{N-1} + (N-1)\,p_{clt}\beta_{clt} \\ &\quad \cdot\left(1-p_{AP}\beta_{AP}\right)\left(1-p_{clt}\beta_{clt}\right)^{N-2}\Big] \\ &= \frac{1}{p_b}\cdot\left[\left(1-\frac{1-\gamma_{clt}}{\left(1-p_{clt}\beta_c\left(\gamma_{clt}\right)\right)^{N-2}}\right)\cdot\left(1-p_{clt}\beta_c\left(\gamma_{clt}\right)\right)^{N-1}\right. \\ &\quad \left. + (N-1)\,p_{clt}\beta_c\left(\gamma_{clt}\right)\left(1-\gamma_{clt}\right)\right] \end{aligned} \tag{18}$$

In Eq. 18, the transformation plugs in Eq. 9 to get rid of p_{AP} and β_{AP}.

In Eq. 12, p_{AP} is caculated by

$$\begin{aligned} p_{AP} &= 1-\exp\left(-\lambda_{AP}Y_{AP}\right) \\ &= 1-\exp\left(-\lambda_{AP}X\left(\gamma_{AP}\right)T\right) \\ &= 1-\exp\left(\frac{\lambda_{AP}X\left(1-\left(1-p_{clt}\beta_c\left(\gamma_{clt}\right)\right)^{N-1}\right)\ln\left(1-p_{clt}\right)}{\lambda_{clt}X\left(\gamma_{clt}\right)}\right) \end{aligned} \tag{19}$$

In Eq. 19, the transformation uses Eq. 7 and 8.

From Eq. 13-19, all the variables at the right side in Eq. 12 can be written in terms of γ_{clt} . The calculation of γ_{clt} follows the four steps as discussed in this subsection. After getting γ_{clt} , all the parameters in Eq. 13-19 can be computed. Notice that even if the fixed-point theorem is used multiple times (in Step 3 and 4) to obtain the solution in nonsaturated asymmetric traffic conditions, the calculation processes are much simpler than solving the nonlinear equations derived from the Markov Chain in [15].

3. THROUGHPUT ANALYSIS

In this section, the conventional single-rate throughput model is briefly introduced, and then the throughput model considering multiple link rates is derived using the probabilities calculated in Section 2.3.

3.1 Single-rate DCF Throughput Model

In [10], the author derived the computation of saturation throughput in a single-cell and single link rate IEEE 802.11 DCF network, assuming N (finite) terminals and ideal channel conditions. In [13, 14, 15] the authors followed the method in [10] to calculate normalized system throughput, which is defined as the fractional time that the channel is used to successfully transmit payload bits in a slot time. According to [10], the normalized system throughput is expressed as

$$S = \frac{p_b p_s E[P]}{T} \tag{20}$$

where $E[P]$ is the average payload size. The next subsection extends this model for multi-rate DCF networks.

3.2 Multi-Rate Throughput Model

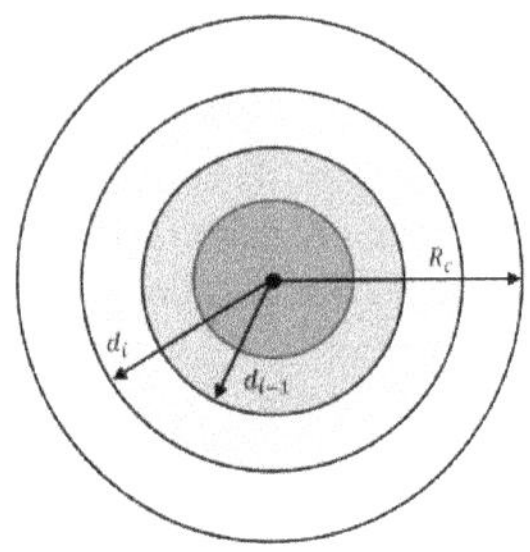

Figure 1. Link rate and distance

To account for multiple link rates in the throughput model, it is necessary to first revisit the IEEE 802.11 transmission specification. First of all, regardless of the achievable link rate between the AP and a client, RTS, CTS, and ACK packets are always transmitted using the basic rate as defined in the standard. Hence, in Eq. 15 and 16, T_{RTS}, T_{CTS}, T_{ACK}, and T_{H} remain the same. Only T_{DATA} is a function of the variable data rate based on distance between the AP and its communicating client. Let r_i be the data rate, then

$$\begin{aligned} T_s(r_i) = T_{\mathrm{RTS}} + T_{\mathrm{CTS}} + T_{\mathrm{DATA}}(r_i) + T_{\mathrm{ACK}} + T_{\mathrm{H}} \\ + \mathrm{DIFS} + 3 \cdot \mathrm{SIFS} + 4 \cdot \delta \end{aligned} \tag{21}$$

Consequently, Eq. 20 can be rewritten as

$$S = \frac{p_b p_s E[P]}{(1-p_b)\sigma + p_s p_b \left(E\left[T_s(r_i)\right] + \sigma\right) + p_b(1-p_s)(T_c + \sigma)} \tag{22}$$

where

$$\begin{aligned} E\left[T_s(r_i)\right] = T_{\mathrm{RTS}} + T_{\mathrm{CTS}} + E\left[T_{\mathrm{DATA}}(r_i)\right] + T_{\mathrm{ACK}} + T_{\mathrm{H}} + \mathrm{DIFS} \\ +3 \cdot \mathrm{SIFS} + 4 \cdot \delta \end{aligned} \tag{23}$$

$$E\left[T_{\mathrm{DATA}}(r_i)\right] = \sum_i p_{r_i} \frac{\text{DATA size}}{r_i} \tag{24}$$

Since r_i depends on the AP-client distance as illustrated in Fig. 1, p_{r_i} can be found as the fractional area where rate r_i is adopted. Assume the radio has k different data rates r_i, $i = 1 \sim k$. Let d_i be the farthest distance from the AP where r_i is feasible, $r_i < r_{i-1}$, then $d_k = R_c$,

$$p_{r_i} = \begin{cases} \dfrac{\pi(d_i^2 - d_{i-1}^2)}{\pi R_c^2} & i > 1 \\ \dfrac{\pi d_i^2}{\pi R_c^2} & i = 1 \end{cases} \tag{25}$$

and d_i is given as

$$\frac{\mathrm{P}_T G_T G_R}{d_i^{\alpha}(\mathrm{P}_I + \mathrm{P}_N)} = \mathrm{SINR}_{\mathrm{th}}(r_i) \tag{26}$$

where P_N is the noise power, P_I is the average interference power received at the node, $\mathrm{SINR}_{\mathrm{th}}(r_i)$ is the minimum required signal to interference plus noise ratio for rate r_i according to the radio specification, G_T and G_R are the transmitting and receiving antenna gains, and G_T and G_R are found according to [17]. For the V2I network, all nodes can hear the omni-directional RTS/CTS from any other node in the same cell (corrupted RTS/CTS due to channel effects are ignored here), therefore, no interfering transmissions will occur during data transmission. Hence it is reasonable to assume the interference power P_I is zero.

To derive the throughput S for a V2I network, first find the data rate probability with Eq. 25 and 26. Next, find the collision probability for one client γ_{clt} using fixed-point equations given in Section 2.3. Finally, calculate system throughput S with Eq. 17, 18, 22, and 23. Note that the throughput obtained here only considers static clients; mobility effects are discussed in Section 4.

4. MOBILITY EFFECTS ANALYSIS

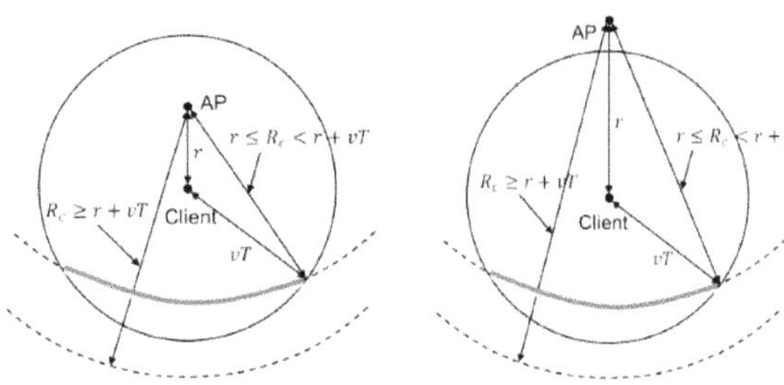

(a) when $0 < r < v \cdot \Delta T$ (b) when $r \ge v \cdot \Delta T$

Figure 2. The *hit* probability

Vehicular networks have constrained positioning and trajectories of nodes. Such topologies, however, are difficult to express exactly in an analytical model. This paper adopts a simplified throughput scaling factor which approximates the mobility impact with that of free moving vehicles in a single-cell V2I network.

In Section 3.2, the throughput equation for static clients is given by Eq. 22. In this section, mobility effects are derived as a scaling factor, using the *hit-or-miss* concept, for the throughput obtained above.

In [16], the mobility effect is modeled based on the *hit-or-miss* concept. The *hit* probability is calculated based on the probability that a receiving node moves outside the transmitting node's beam coverage, and is used to evaluate the level of throughput degradation when using a directional antenna. However, the model in [16] is limited to use directional antennas only and does not consider the cases where a mobile node moves out of the node's communication range. In this section a model, which takes into account omni-directional antenna usage and movement from inside to outside the communication range, is derived.

In our V2I network, the AP is static while clients are mobile. It is assumed that each client has a constant velocity and direction in each period ΔT. Let v be the velocity and r be the distance between AP and the desired client at the end of the previous period. At the end of the current period, the client can be anywhere on a circle centered on its prior location with radius $v \cdot \Delta T$, which is referred to as the movement circle. The mobility factor is derived based on the probability that the client remains in the AP's communication range at the end of a period, such that the modified system throughput is

$$S = M(v) \cdot S_{sta} \tag{27}$$

where S_{sta} is the system throughput with static nodes; $M(v)$ is the mobility factor associated with the node's velocity, which serves to scale S_{sta} for the mobile case.

The mobility factor depends on whether the client's initial distance from AP, r, is less or greater than the movement distance $v \cdot \Delta T$ as shown in Fig. 2a and 2b. To derive $M(v)$ we will discuss two different cases in this section.

(1) $0 < r < v \cdot \Delta T$ (Fig. 2a)

The probability that the client is still within the AP's coverage at the end of the period is denoted the *hit* probability. If $r \le R_c < r + v \cdot \Delta T$, the bold arc in Fig. 2a is the possible client's location which is within AP's coverage. Based on geometric calculation the *hit* probability is

$$P_1 = 1 - \frac{1}{\pi} \cdot \arccos\left(\frac{R_c^2 + r^2 - (v \cdot \Delta T)^2}{2 R_c r}\right) \tag{28}$$

If $R_c \ge r + v \cdot \Delta T$, the client must remain in the AP's coverage at the end of the period, thus

$$P_2 = 1 \tag{29}$$

(2) $r \ge v \cdot \Delta T$ (Fig. 2b)

If $r \le R_c < r + v \cdot \Delta T$, use the same analysis as case (1); the client's possible future locations lie anywhere on the bold arc in Fig. 2b, thus the *hit* probability is

$$P_3 = 1 - \frac{1}{\pi} \arcsin\left(\frac{R_c \sin\left(\arccos\left(\frac{R_c^2 + r^2 - (v \cdot \Delta T)^2}{2 R_c r}\right)\right)}{v \cdot \Delta T}\right) \tag{30}$$

If $R_c \ge r + v \cdot \Delta T$, the client must remain in the AP's coverage at the end of the period, so the *hit* probability is

$$P_4 = 1 \tag{31}$$

Based on these probabilities for all conditions, as well as Eq. 1, the mobility factor is found as

$$M(v) = \begin{cases} \int_0^{R_c} P_1 \cdot \omega_r(r) dr & 0 < R_c < v \cdot \Delta T \\ \int_0^{v \cdot \Delta T} \chi \cdot \omega_r(r) dr + \int_{v \cdot \Delta T}^{R_c} \eta \cdot \omega_r(r) r dr & R_c \ge r + v \cdot \Delta T \end{cases} \tag{32}$$

where

$$\chi = \begin{cases} P_1 & r \le R_c < r + v \cdot \Delta T \\ P_2 & R_c \ge r + v \cdot \Delta T \end{cases} \tag{33}$$

$$\eta = \begin{cases} P_3 & r \le R_c < r + v \cdot \Delta T \\ P_4 & R_c \ge r + v \cdot \Delta T \end{cases} \tag{34}$$

Using Eq. 32, 33, and 34, the mobility factor can be calculated, and used in Eq. 27 to obtain the system throughput with mobile clients.

5. RESULTS

Table 1. System parameters

Packet Payload	1024 bytes	MAC header	224 bits
Slot Time	20 usec	PHY header	136 bits
SIFS	10 usec	ACK	248 bits
DIFS	50 usec	RTS	296 bits
Propagation Delay	1 usec	CTS	248 bits

Table 2. Receiver sensitivity (802.11g)

54 Mbps	66 dBm	18 Mbps	83 dBm
48 Mbps	71 dBm	12 Mbps	85 dBm
36 Mbps	76 dBm	9 Mbps	86 dBm
24 Mbps	80 dBm	6 Mbps	87 dBm

This section presents numerical results obtained from the proposed models. The system parameters are chosen according to the IEEE 802.11g standard as summarized in Table 1. Receiver sensitivity for multiple link rates is based on the D-Link DWL-2100AP specifications as shown in Table 2.

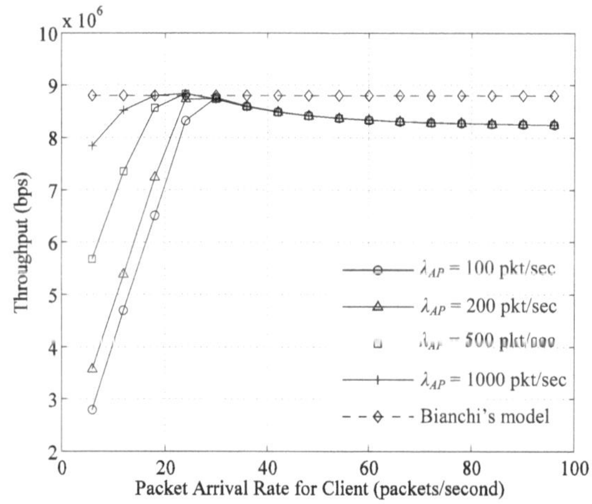

Figure 3. System throughput for different traffic loads

Fig. 3 shows the system throughput for different traffic loads when there are 40 nodes. As shown in the figure, the packet arrival rate at the AP's side is set as 100, 200, 500, and 1000 packets per second, respectively. For all four cases, the throughput reaches maximum when the client's traffic load is between 20 and 30 packets per second. After this maximum the throughput tends to be steady and flat as the client's traffic load increases, which could be explained as saturated throughput. No matter what the AP's traffic rate is (in our scenario, there are 39 clients and 1 AP, so it's reasonable to assume $\lambda_{AP} >> \lambda_{clt}$), the network has nearly the same saturated throughput. As Fig. 3 shows, the throughput obtained from the model in this paper is also compared to that from Bianchi's saturation model, and differ by 6% at saturation.

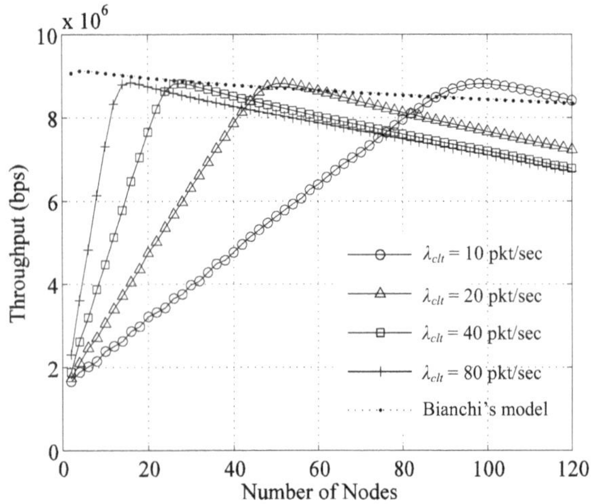

Figure 4. System throughput with stationary nodes

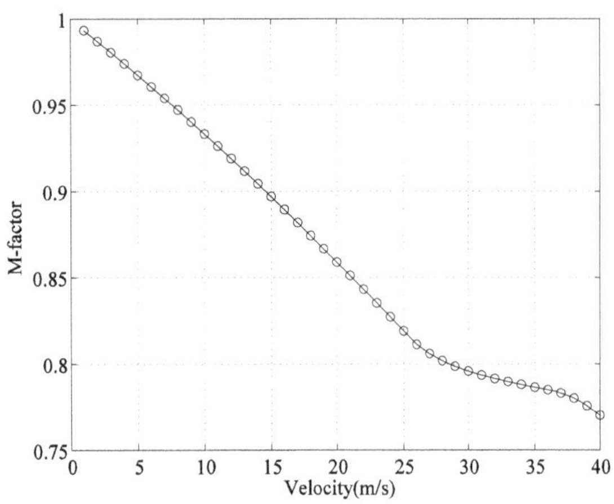

Figure 5. Mobility effects

Fig. 4 compares the system throughput with stationary nodes for λ_{clt} = 10, 20, 40, and 80 pkt/sec respectively, and λ_{AP} = 200 pkt/sec is fixed. For each curve, Fig. 4 shows the throughput increasing to the maximum and then steadily and slowly dropping as the number of nodes increases. The throughput for λ_{clt} = 80 pkt/sec climbs fastest to the maximum at which there are 16 nodes, comparing to the slowest one with λ_{clt} = 10 pkt/sec when there are 98 nodes. With more and more nodes in the network, the collision probability increases as well, which causes decrease in throughput after reaching maximum. The system throughput drops 11.34% and 10.40% when the number of nodes increases from 60 to 80, for λ_{clt} = 20 and 40 respectively. In Fig. 4, another finding is that the maximum of throughput for all four cases are nearly the same, which is around 8.8Mbps, no matter what the traffic load is for clients. Comparing to the throughput obtained from Bianchi's saturation model in Fig. 4, the maximum throughputs achieved for these four evaluated cases are close to that of Bianchi's model. With an increased number of clients in the cellular network, the throughput drops more than that from Bianchi's model, e.g. the throughput degrades 14.31% and 4.66% for λ_{clt} = 80 pkt/sec and Bianchi's model, respectively, when the number of nodes increases from 20 to 80. Such a difference represents the effects caused by asymmetric traffic in a single V2I cellular network, which is quantified by the nonlinear system of Eq. 13-19.

Fig. 5 evaluates the mobility factor derived in Section 4. When the velocity is zero, the M factor is 1 because no degradation is introduced by mobility effect. Fig. 5 shows that the M factor steadily decreases as the velocity increases. The M factor drops to 0.8590 and 0.7701 when the velocity is 20 m/s and 40 m/s, respectively.

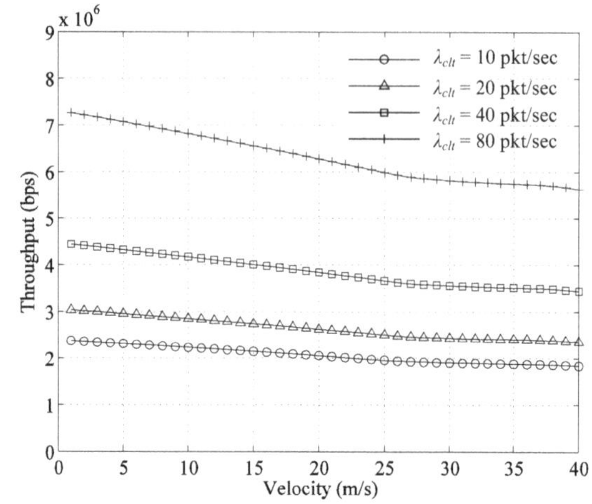

Figure 6. Throughput vs velocity (10 nodes)

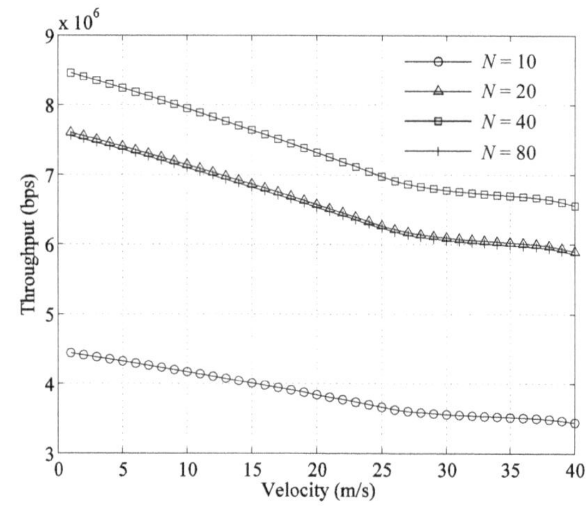

Figure 7. Throughput vs velocity (AP's traffic rate is 200 pkt/sec, client's traffic rate is 40 pkt/sec)

Fig. 6 shows the throughput versus velocity for 10 nodes (one AP and nine clients). λ_{AP} is set at 200 pkt/sec and four curves are drawn for λ_{clt} = 10, 20, 40, and 80 pkt/sec. From Fig. 4, when the number of nodes is 10, the throughput increases linearly for all four cases. Fig. 6 shows that λ_{clt} = 80 pkt/sec gives the best throughput performance, as it is 67.31%, 58.19%, and 38.84% greater than when λ_{clt} = 10, 20, 40 pkt/sec respectively, when the velocity reaches 20 m/s.

Fig. 7 shows the throughput versus velocity of different numbers of nodes when the AP's and client's traffic rates are set as 200 and 40 pkt/sec, respectively. In Fig. 7, the throughput performance is best when using 40 nodes and worst when using 10 nodes, which follows the results in Fig. 4. The throughput drops 47.52%, 10.13%, and 10.70% for 10 nodes, 20 nodes, and 80 nodes, respectively, compared to 40 nodes. Each case in Fig. 6

and Fig. 7 has the same ratio for throughput degradation following the result obtained in Fig. 5.

6. CONCLUSIONS

An analytical throughput model for IEEE 802.11-based V2I networks considering nonsaturated asymmetric traffic loads and mobility effects is proposed. Numerical results show that the maximum throughput achieved from proposed model is close to that from Bianchi's model, and degrades more than Bianchi's model when the number of nodes increases, which quantifies the impacts of asymmetric traffic loads as compared Bianchi's model. The mobility effects *M-factor* is derived as a degradation scaling factor for throughput with static clients, and the throughput for mobile clients is evaluated with the increase in vehicle's velocity.

7. ACKNOWLEDGEMEMTS

The work was supported in part by a gift from the Cisco University Research Program and a National Science Foundation grant CNS-1049947.

8. REFERENCES

[1] Networked Vehicle Association, http://www.networkedvehicle.org/. [Accessed in May. 2011].

[2] Qian, Y., Lu, K., and Moayeri, N. 2008. A secure VANET MAC protocol for DSRC applications. In *Proceedings of the Global Communications Conference* (New Orleans, LA, USA, 2008). GLOBECOM'08. IEEE, 1-5.

[3] Chung, S., Yoo, J. and Kim, C. 2009. A cognitive MAC for VANET based on the WAVE systems. In *Proceedings of the 11th International Conference on Advanced Communication Technology* (Gangwon-Do, South Korea, 2009). ICACT'09. IEEE, 41-46.

[4] Xie, X., Wang, F., Li, K., Zhang, P., and Wang, H. 2009. Improvement of multi-channel MAC protocol for dense VANET with directional antennas. In *Proceedings of the IEEE Wireless Communications & Networking Conference* (Budapest, Hungary, 2009). WCNC'09. IEEE, 1-6.

[5] Menouar, H., Filali F., and Lenardi, M. 2006. A survey and qualitative analysis of mac protocols for vehicular ad hoc networks. *IEEE Trans. Wireless Commun.* 13, 5 (Oct. 2006), 30-35.

[6] Sibecas, S., Corral, C., Emami, S. and Stratis, G. 2002. On the suitability of 802.11a/RA for high-mobility DSRC. In *Proceedings of the 55th Vehicular Technology Conference Spring 2002* (Birmingham, AL, USA, 2002). VTC Spring'02. IEEE, 229-234.

[7] Eriksson, J., Balakrishnan, H., and Madden, S. 2008. Cabernet: Vehicular content delivery using WiFi. In *Proceedings of the 14th Annual International Conference on Mobile Computing and Networking* (San Francisco, CA, USA, 2008). MobiCom'08. ACM, 199-210.

[8] Giannoulis, A., Fiore, M. and Knightly, E. W. 2008. Supporting vehicular mobility in urban multi-hop wireless networks. In *Proceedings of the 6th International Conference on Mobile Systems, Applications, and Services* (Breckenridge, CO, USA, 2008). MobiSys'08. ACM, 1-13.

[9] Hadaller, D., Keshav, S., Brecht, T., and Agarwal, S. 2007. Vehicular opportunistic communication under the microscope. In *Proceedings of the 5th International Conference on Mobile Systems, Applications, and Services* (San Juan, Puerto Rico, 2007). MobiSys'07. ACM, 206-219.

[10] Bianchi, G. 2000. Performance analysis of the IEEE 802.11 Distributed Coordination Function. *IEEE J. Sel. Areas Commun.* 8, 3 (Mar 2000), 535-547.

[11] Cali, F., Conti, M., and Gregori, E. 2000. Dynamic tuning of the IEEE 802.11 protocol to achieve a theoretical throughput limit. *IEEE/ACM Trans. Netw.*, 8, 6 (Dec. 2000), 785-799.

[12] Ho, T., and Chen, K. 1996. Performance analysis of IEEE 802.11 CSMA/CA medium access control protocol. In *Proceedings of the 7th IEEE International Symposium on Personal, Indoor and Mobile Radio Communications* (Taipei, Taiwan, 1996). PIMRC'96. IEEE, 407-411.

[13] Malone, D., Duffy, K., and Leith, D. 2007. Modeling the 802.11 distributed coordination function in nonsaturated heterogeneous conditions. *IEEE/ACM Trans. Netw.* 15, 1 (Feb. 2007), 159-172.

[14] Kumar, A., Altman, E., Miorandi, D., and Goyal, M. 2007. New insights from a fixed point analysis of single cell IEEE 802.11 WLANs. *IEEE/ACM Trans. Netw.* 15, 3 (Mar. 2007), 588-601.

[15] Zhao, Q., Tsang, D., and Sakurai, T. 2009. A simple and approximate model for nonsaturated IEEE 802.11 DCF. *IEEE Trans. Mobile Comput.* 8, 11 (Nov. 2009), 2539-2553.

[16] Hsu, J., and Rubin, I. 2005. The D-ALOHA protocol for MANETs using beamforming directional antennas. In *Proceedings of the IEEE 62nd Semiannual Vehicular Technology Conference* (Dallas, TX, USA, 2005). VTC Fall'05, IEEE, 670-674.

[17] Babich, F., and Comisso, M. 2009. Throughput and delay analysis of 802.11-based wireless networks using smart and directional antennas. *IEEE Tran. Commun.* 57, 5 (May 2009), 1413-1423.

On the Reliability of Safety Message Broadcast in Urban Vehicular Ad hoc Networks

Saeed Bastani[†,‡] Bjorn Landfeldt[†] Lavy Libman[†,‡]
[†]School of Information Technologies, University of Sydney, NSW 2006, Australia
[‡]Networks research group, NICTA, 13 Garden St., NSW 1430, Australia
sbastani@it.usyd.edu.au, bjorn.landfeldt@sydney.edu.au, lavy.libman@sydney.edu.au

ABSTRACT
Road safety is one of the most important emerging applications envisioned for Vehicular Ad hoc Networks (VANETs). Generally, such applications involve the broadcast of safety messages, consisting of beacons transmitting vehicles' state (e.g. position and velocity) with a regular period, as well as emergency messages warning about unexpected critical events. From the perspective of safety, the application performance depends foremost on two metrics: for the event-driven warning messages, the probability of message reception; and for periodic messages, the variability of the inter-reception time (IRT), which ultimately determines the freshness of the information received by the driver. In this paper, we develop an analytical model to compute the above metrics in an urban traffic scenario. Focusing on a road segment linked to a signalized junction as a basic building block of urban traffic systems, we apply a novel road traffic density model to investigate the dynamics of the reliability metrics and characterize the region(s) on the road segment according to the achieved safety level. Our numerical study shows that in broadcast mode, the hidden terminal effect is the driving factor determining the reliability of transmissions. Furthermore, the impact of hidden terminals has the greatest effect in road sections where vehicles have high velocity, leading to the poorest performance in regions where reliable reception is needed the most in order to minimize the risk of accidents.

Categories and Subject Descriptors
C.2.1 [**Network Architecture and Design**]: Wireless Communication

General Terms
Performance, Theory

Keywords
Vehicular Traffic Density; Vehicular Ad hoc Networks; Safety Applications;

1. INTRODUCTION
Vehicular Ad hoc Networks (VANETs) have recently attracted significant research and development efforts from different players including academia, standards bodies, and industry, leading to the emergence of DSRC/WAVE [1][2] as a reference standard, numerous dissemination/forwarding schemes [3], and several active projects [4]. Inter-vehicle communications open the door for a plethora of applications and services to provide safety and comfort through wireless vehicle-to-vehicle communications. Comfort applications are expected to improve the passengers' travel experience and optimize traffic efficiency, while the safety applications aim at minimizing accident levels. Safety applications can be categorized further according to their use of either periodic or event-driven messages. The first category has an informative nature, as messages are disseminated among vehicles regularly to inform drivers about local parameters such as speed and position. Event-driven messages, on the other hand, are broadcast by vehicles whenever they experience or detect a hazard or otherwise notable event. Single hop broadcast is the fundamental mechanism of periodic safety message dissemination in VANETs. In case of event-driven warning messages, even though broadcast over multiple hops may be used, the ultimate message dissemination performance depends foremost on the performance of each single hop broadcast involved.

The reliability of message broadcasts in a safety application is key to its credibility and ultimate acceptance by the drivers as the end users. Depending on the type and purpose of a safety message, a subset of parameters describes the reliability of the safety application. For event-driven messages, the reliability of the safety application is determined by the successful packet reception probability and geographical coverage of message dissemination. On the other hand, in case of periodic messages, inter-reception time (IRT) of messages is a good candidate metric for describing the application reliability. The IRT metric incorporates the variability of message reception time and packet reception probability into a single parameter. Intuitively, from a recipient vehicle perspective, a high probability of message reception from neighbor vehicles leads to high overall awareness by the recipient about its neighborhood. Furthermore, a high frequency message reception enhances the information freshness a recipient maintains at any time instance, and, in turn, promotes timely reaction to undesired events as they occur. Correspondingly, from a sender point of view, the higher the chance that the neighbor vehicles receive its message successfully and timely, the better the achieved safety level will be.

The key factor impacting the reliability metrics mentioned above is traffic density. In static wireless networks, due to the deterministic distribution of nodes throughout the network area, it is straightforward to characterize the traffic behavior and thus reliability metrics. On the other hand, in vehicular ad hoc networks (as in mobile ad hoc networks in general), characterizing reliability involves taking into account dynamic topology changes due to vehicles' mobility, which in turn is affected by microscopic and macroscopic traffic parameters [5][6][7]. These parameters include, but are not limited to, traffic regulations on junctions and road segments, driver behavior, traffic flow, road capacity, *etc.* In vehicular networks, the analysis of the reliability of a safety

[‡]NICTA is funded by the Australian Government as represented by the Department of Broadband, Communications and the Digital Economy and the Australian Research Council through the ICT Centre of Excellence program.

MSWiM'11, October 31–November 4, 2011, Miami, Florida, USA.

application is even more complicated than in other mobile ad hoc networks, due to the impact of unexpected drivers' behavior and variable traffic flow on vehicles' mobility [7]. Moreover, it is also expected that the reliability varies significantly between highway and urban VANET scenarios. This is partly due to the fact that traffic density is homogeneous under free and stable traffic flow regimes, which dominate highways, whereas a mixture of different traffic densities can be observed simultaneously in an urban scenario as simple as a road segment linked to a signalized junction. Moreover, an urban traffic network must be seen as a 2-dimensional network, compared to a 1-dimensional highway network. This makes the dynamics of traffic density more complicated in urban scenarios, resulting in more complex reliability behavior, especially near junctions.

In line with the above, we are motivated to apply a traffic density model of a simple urban traffic scenario comprising a signalized junction and road segments linked to that junction to analytically describe the spatial-temporal behavior of safety messages' reliability throughout an urban road segment. In the proposed analytical model, the probability of successful message transmission associated with both periodic and warning messages contending for a shared channel are calculated. Additionally, we determine the distribution function corresponding to the IRT of beacon messages. In this work we are not interested in the per-message channel access delay, as this is in the order of a few milliseconds and thus not a key factor impacting the requirements of a safety application [8]. Transmission failures, on the other hand, are of high importance, since in the case of periodic messages a single transmission failure causes the reception time to exceed a beacon interval as large as 100ms [8].

The remainder of this paper is organized as follows. Section 2 provides an overview of related work. In section 3, we develop a general Markovian analytical framework to characterize the reliability metrics of safety message broadcast, and apply it with an urban traffic density model in section 4. A numerical evaluation of the model is presented in section 5, and finally, section 6 concludes the paper.

2. RELATED WORK

Safety message broadcast has been studied by means of simulation in a group of works conducted by ElBatt *et al.* [8] and Torrent-Moreno *et al.* [9][10]. A few common observations that apply to these studies are: (i) Either periodic or warning messages are addressed, but not both; (ii) In spite of performance evaluation under various traffic densities, they do not address the impacts of non-uniform and heterogeneous traffic densities attributed to urban traffic systems; (iii) Per-packet delay is in the order of a few milliseconds, which is satisfactory for most safety applications. ElBatt *et al.* [8] suggested a reliability metric termed *packet inter-reception time*, defined as the elapsed time between two consecutive successfully received packets by a vehicle. The authors conducted extensive simulations to measure successful packet delivery and packet inter-reception times under high and low traffic densities in a 2-direction highway scenario with 4 lanes in each direction. In this paper, we adopt the reliability metrics proposed in [8], develop a general analytical framework for their calculation and apply it in the context of an urban scenario. In [9], Torrent-Moreno *et al.* addressed the failure of transmission coordination associated with the IEEE802.11 Distributed Coordination function (DCF) and argued that the level of coordination failure is intensified by the hidden terminal effect. We take into account the fact that the hidden terminal effect is in turn influenced by traffic density and distribution, and study the impact of hidden terminals on safety message reliability under a road network's heterogeneous traffic distribution. Torrent-Moreno *et al.* [10] studied the effects of broadcast storm on the channel. To alleviate the effects, they proposed an adaptive mechanism using transmission power adjustment and tuning of beacon transmission intervals. Yousefi *et al.* [11] studied the delivery ratio and delay of beacon messages with varying packet transmission intervals and packet sizes. Their simulation methodology considered 1-hop broadcast in a stationary large highway scenario and fixed transmission range. According to their simulation results, packet delay was in the order of a few milliseconds, which generally does not constitute a bottleneck for safety applications. They also showed that the packet reception rate decreases significantly when increasing the distance of the receiver from the transmitter, a phenomenon previously observed in [8].

Beyond the simulation studies mentioned above that are specific to vehicular environments, many analytical models have been proposed in the literature to address the performance and reliability of IEEE 802.11 DCF. For the most part, the proposed approaches are variations of the Markov-based performance evaluation method presented by Bianchi [12] and Cali *et al.* [13]; for instance, the implications of an error prone channel was modeled in [14], while retransmission retries and seizing phenomenon were taken into account in [15] and [16], respectively. This framework was extended to IEEE 802.11e QoS differentiation by Engelstad *et al.* [17], who also investigated the channel and application layer performance metrics with respect to non-saturation traffic. In [18], Lyakhov *et al.* studied the performance of IEEE 802.11 networks operating in broadcast mode. They assumed Poisson packet arrival and applied Markov chains to analytically express the mean notification time of broadcast packets. Ma *et al.* [19] studied saturation throughput, delay, and packet delivery rate while taking into consideration the impact of backoff counter freezing in a scenario termed Continuous Freeze Process (CFP). Apart from the above investigative works, efforts have also been made to enhance IEEE 802.11 MAC with respect to the specific requirements of VANETs applications. Recently, Bononi *et al.* [29] proposed an improvement to IEEE 802.11 MAC to support reliable and fast multi-hop broadcast using a dynamic virtual backbone infra-structure.

Most relevant to our work in this paper are studies performed on safety message broadcast within the DSRC framework. Ma *et al.* studied different aspects of safety message broadcast in a series of works. They investigated saturation throughput and packet delivery ratio in [20] and saturation delay in [21]. In a recent work [22], *Ma et al.* analyzed the broadcast performance of safety messages. They considered the impacts of an error-prone realistic channel, hidden terminals, and mobility issues in deriving performance metrics. In their analytical models, they assumed a 1-D highway scenario with vehicles placed on the road according to a Poisson point process with a predetermined density. In [23], Vinel *et al.* modeled IEEE 802.11p VANET as a D/M/1 queuing system and roughly estimated the mean beacon transmission delay and the beacon reception probability. Vinel *et al.* also studied the successful beacon reception probability in [24] and [25] under saturated and unsaturated beacon rates. The common features of the above works are: (i) Beacon messages are treated as random arrivals, and fail to incorporate the periodic nature of beacon messages and consequently fail to characterize the distribution of the message IRT; (ii) A homogeneous and uniform traffic density is assumed, which is not applicable to urban scenarios.

Our work differs from the above mentioned works in several ways: (i) Instead of solely adopting a simulation approach, we develop an analytical model to generalize the analysis of safety messages' reliability; (ii) We take into consideration the mutual impact of both types of safety messages, i.e. event-driven and periodic; (iii) We study the reliability of safety massage dissemination in an urban traffic scenario, involving road segments connected to a signalized junction, and using a corresponding heterogeneous density model developed in our earlier work [26].

3. RELIABILITY OF SAFETY MESSAGE BROADCAST

3.1 Assumptions

We make the following assumptions when addressing IEEE 802.11p one hop broadcast communications: (i) the ready-to-send and clear-to-send (RTS/CTS) exchange is disabled, and (ii) ACKs are not transmitted after successful reception. Consequently, no retransmissions are performed and the backoff window size is not adjusted based on the existing load on the channel. Moreover, according to the DSRC standard, it is mandatory that vehicles transmit periodic beacons and event-driven emergency messages on the same channel (channel 178) [1]. Thus, a message can potentially collide with a message of its own type or another type if they are transmitted simultaneously by more than one vehicle.

We assume that the backoff window of event-driven messages W_e is smaller than that of the periodic beacon messages W_b. This is in agreement with IEEE 802.11p standard and implies that event-driven messages are of higher priority than periodic beacons. Additionally, we assume that nodes experience a non-saturated arrival of event-driven messages as a Poisson arrival process with rate λ. Beacon messages, on the other hand, arrive on a periodic basis with inter-arrival time of α time slots. Arrival of a new beacon message cancels out old beacons, since a new beacon is assumed to always contain the most updated vehicle state. As a result, the impact of queuing delay on beacon messages is eliminated.

3.2 Analytical Model

To study the reliability metrics mentioned above, we need to calculate the probability of a vehicle's transmission attempt in a generic time slot. To that end, we propose a Markov-chain model for the backoff process corresponding to a combination of periodic beacons and event-driven messages.

We thus extend the Markov chain model proposed in [22] to account for the periodic nature of beacon transmissions. The chain, shown in Figure 1, consists of a combination of

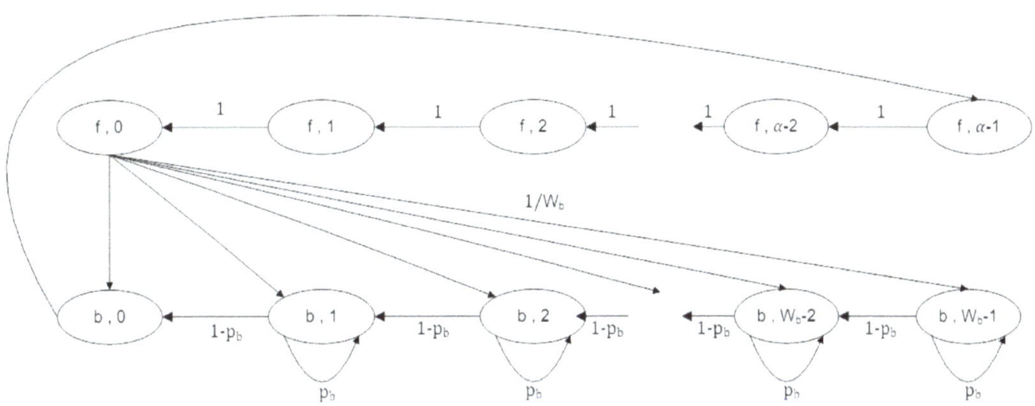

Figure 1. Markov chain model of backoff process for periodic beaconing safety messages

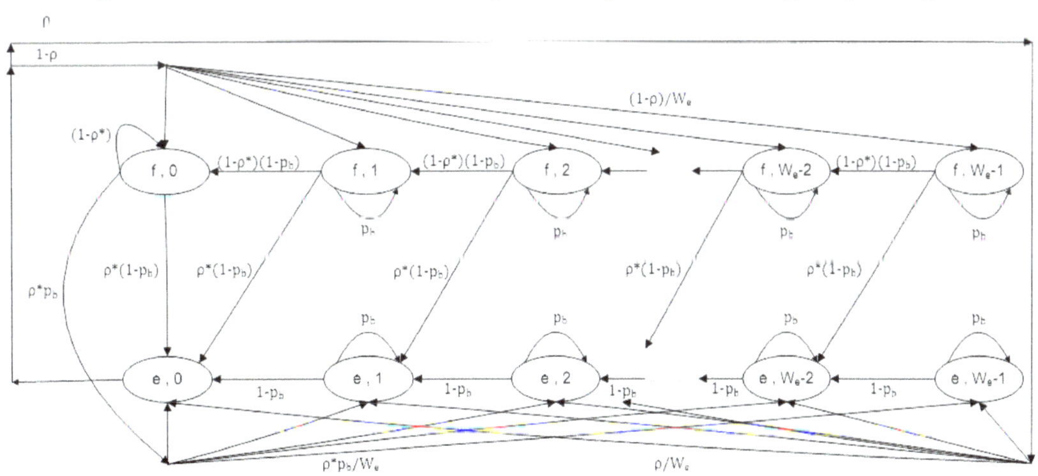

Figure 2. Markov chain model of backoff process for event-driven safety messages

deterministic post backoff (upper stage) and stochastic backoff processes (lower stage). Whenever a vehicle completes its current channel contention and transmission attempt, it enters the post backoff stage of length equal to beaconing period $\alpha = \frac{I_b}{\sigma}$, where I_b and σ are the beacon period and time slot duration, respectively. With probability $\frac{1}{W_b}$, a backoff state $(b\,, k)$ is selected and channel contention starts. The probability of a transmission attempt is equivalent to the probability that the backoff process enters state $(b\,, 0)$.

To this end we solve the Markov chain shown in Figure 1 to calculate the probability that a vehicle transmits a beacon in a generic time slot, denoted by τ^b. It is easy to verify that the chain steady-state transition and normalization conditions result in:

$$s_{b,k} = \frac{W_b - k}{W_b} s_{b\,,0} \qquad k \in (0\,, W_b - 1) \tag{1}$$

$$s_{f\,,k} = s_{b\,,0} \qquad k \in (0\,, W_b - 1) \tag{2}$$

$$\sum_{k=0}^{W_r-1} s_{b\,,k} + \sum_{k=0}^{W_r-1} s_{f\,,k} = 1 \tag{3}$$

and solving for $s_{b\,,0}$ (equivalent to the probability of transmission attempt τ^b), we obtain:

$$\tau^b = s_{b\,,0} = \frac{2}{W_b + 1 + 2\alpha} \tag{4}$$

The Markov chain corresponding to the backoff process of event-driven messages is shown in Figure 2. This is a simplified model of Engelstad *et al.* [17], customized for broadcast transmission mode. Here, $(f\,, k)$ are post backoff states during which the queue is empty and the node has to wait for a new message to arrive. $(e\,, k)$ represent backoff states where there exists a message for transmission. In this case, with probability ρ, the backoff process is immediately invoked by entering one of the backoff states $(e\,, k)$ chosen randomly. With probability $1 - \rho$ the node enters a post backoff stage. While being in a state $(f\,, k)$, if a new message arrives with probability ρ^* (different from ρ) and the channel is sensed idle, the contention process is immediately triggered by directly entering the state $(e\,, k-1)$ in the backoff stage. If the channel is sensed busy, the countdown process is blocked, otherwise a transition to state $(f\,, k-1)$ takes place.

Applying steady-state conditions recursively through the chain, it is easy to show that:

$$s_{f\,,k} = \frac{1-\rho}{W_e} \cdot \frac{1-(1-\rho^*)^{W_e - k}}{1-p_b} \cdot \frac{s_{e\,,0}}{\rho^*} \qquad k \in (1\,, W_e - 1) \tag{5}$$

$$s_{f\,,0} = \frac{1-\rho}{W_e} \cdot \frac{1-(1-\rho^*)^{W_e}}{(\rho^*)^2} \cdot s_{e\,,0} \tag{6}$$

$$s_{e\,,k} = \frac{(W_e - k)}{W_e \cdot (1-p_b)} \cdot s_{e\,,0} + \frac{(W_e - k) \cdot \rho^* \cdot p_b}{W_e \cdot (1-p_b)} \cdot s_{f\,,k} - s_{f\,,k} \tag{7}$$

and the normalization condition implies

$$\sum_{k=0}^{W_e - 1} (s_{e\,,k} + s_{f\,,k}) = 1. \tag{8}$$

Using (5)-(8), $b_{e\,,0}$ and thus the probability of event-driven message transmission τ^e in a generic time slot we obtain:

$$\frac{1}{\tau^e} = 1 + \frac{W_e - 1}{4(1-p_b)} + \frac{p_b}{(1-p_b)^2} \cdot \frac{1-\rho}{{W_e}^2} \cdot \frac{\rho^{*2} W_e (W_e - 1) + (1-\rho^*)^{W_e} (2\rho^* W_e - 2\rho^* + 2) + 2(\rho^* - 1)}{\rho^{*2}} + \frac{1-\rho}{W_e} \cdot \frac{1-(1-\rho^*)^{W_e}}{\rho^{*2}} \tag{9}$$

3.2.1 *Probability of busy channel (p_b)*

The probability of sensing a channel busy event is equivalent to the probability that at least one vehicle is transmitting a message, either a beacon or an event-driven message. This probability can be expressed as:

$$p_b = 1 - \left((1 - \tau^b) \cdot (1 - \tau^e) \right)^{N_r} \tag{10}$$

where τ^b and τ^e are the probabilities of transmission attempts corresponding to beacon and event-driven messages, described by (4) and (9), respectively. N_r is the number of vehicles in the transmission range (R) of the vehicle under investigation. For uniform traffic distribution (highway scenario) with density β, $N_r = 2\beta R$. For non-uniform traffic distribution (urban scenario), we later give an expression for calculating N_r using the density functions proposed in section 4.

3.2.2 *Probability of successful transmission (p_s)*

Without loss of generality, we address the probability of successful transmission separately for beacon and event driven messages while the mutual impacts are taken into consideration. Denote by p_s^b and p_s^e, the probability of successful transmission of beacon and event-driven messages, respectively. To obtain these probabilities, we account for simultaneous transmissions in the transmission range of a vehicle and transmission(s) from hidden nodes within the hidden area of the sender vehicle. Intuitively, p_s^b and p_s^e are equivalent to the probabilities that exactly one node attempts transmission and no hidden node transmits a message which overlaps in time with the transmission performed by the sender vehicle. Consequently:

$$p_s^b = \tau^b \cdot (1 - \tau^b - \tau^e)^{N_r - 1 + \widetilde{N}_h \cdot \frac{T_h^b}{p_b T_s^b + (1-p_b)\sigma}} \tag{11}$$

$$p_s^e = \tau^e . (1 - \tau^b - \tau^e)^{N_r - 1 + \widetilde{N}_h \cdot \frac{T_h^e}{p_b T_s^e + (1-p_b)\sigma}} \tag{12}$$

In (11) and (12), $\widetilde{N}_h$ is the average per vehicle hidden terminal effect on vehicles within the transmission range of a sender vehicle. For uniform traffic distribution with density β, $\widetilde{N}_h = \beta R$. For non-uniform traffic distribution, the hidden terminal nodes fall within the ranges $(x_v - 2R\,, x_v - R)$ and $(x_v + R\,, x_v + 2R)$ on the left and right side of a candidate sender vehicle positioned at x_v; we describe the calculation of $\widetilde{N}_h$ in greater detail in section 4.

T_h in (11) and (12) is the period during which a transmission from a vehicle may overlap with the transmission from a hidden node; hence, $T_h^r = 2T_s^b$ and $T_h^e = 2T_s^e$, where T_s^b and T_s^e are packet transmission time corresponding to beacon and event-driven messages, respectively. The subscript s in T_s^b and T_s^e is introduced to distinguish between duration of a successful message transmission and duration of a message collision. The ratios $\frac{T_h^b}{p_b T_s^b + (1-p_b)\sigma}$ and $\frac{T_h^e}{p_b T_s^e + (1-p_b)\sigma}$ in (11) and (12) are introduced due to the fact that if a node in the hidden area of the sender vehicle starts transmission, the channel will be sensed busy by the remaining vehicles in the hidden area who thus remain silent.

3.2.3 *Calculating ρ^* and ρ*

In the proposed Markov model for event-driven messages, ρ^* is the conditional probability for a new event-driven message to arrive in the queue within a generic slot time, given that at the beginning of the slot the queue was empty. Note that a generic slot can have different lengths due to blocking of the backoff process due to the channel being busy. If the channel is idle (with probability $1 - p_b$), the slot length is σ (nominal slot duration). If a successful beacon or event-driven message transmission occurs on the channel with probability P_s^b and P_s^e, the corresponding generic slot time will be of length T_s^b and T_s^e, respectively. Otherwise, with probability $p_b - P_s^b - P_s^e$, the slot duration is T_c (collision duration). Therefore, for a Poisson arrival process with rate λ, ρ^* can be expressed as:

$$\rho^* = 1 - \left((1-p_b)\cdot e^{-\lambda.\sigma} + P_s^b \cdot e^{-\lambda \cdot T_s^b} + P_s^e \cdot e^{-\lambda \cdot T_s^e} + (p_b - P_s^b - P_s^e)\cdot e^{-\lambda T_c}\right) \quad (13)$$

Note that P_s^b and P_s^e are different from p_s^b and p_s^e described by (11) and (12). More specifically, P_s^b (P_s^e) is the probability of the event that a successful beacon (event-driven) message transmission occurs, taking into account all vehicles within the transmission range of a vehicle, but neglecting the impact of hidden vehicles, since the transmitting vehicle does not have any knowledge about its hidden peers. Accordingly, we explicitly eliminate the effect of hidden terminals as follows:

$$P_s^b = \frac{N_r p_s^b}{(1-\tau^b-\tau^e)^{\tilde{N}_h \cdot \frac{T_h^b}{p_b T_s^b + (1-p_b)\sigma}}} \quad (14)$$

$$P_s^e = \frac{N_r p_s^e}{(1-\tau^b-\tau^e)^{\tilde{N}_h \cdot \frac{T_h^e}{p_b T_s^e + (1-p_b)\sigma}}} \quad (15)$$

To calculate ρ, we need to determine the channel service time, which is the time it takes a head-of-line message to access the channel and complete its transmission (whether successfully or not). To that end, we follow the approach proposed in [17] to derive the Z-transform of the Markov chain in Figure 2:

$$\mathrm{D(z)} = \frac{z^{\frac{(P^e+H^e)}{\sigma \cdot R_d}}}{W_e} \cdot \sum_{k=0}^{k=W_e-1} H_{\mathrm{state}}^k(z) = \frac{z^{\frac{(P^e+H^e)}{\sigma \cdot R_d}}}{W_e} \cdot \frac{1-(H_{\mathrm{state}}(z))^{W_e}}{1-H_{\mathrm{state}}(z)} \quad (16)$$

where $H_{state}(z)$ is the Z-transform of each state, expressed as:

$$H_{state}(z) = (1-p_b)\cdot z + P_s^b \cdot z^{\frac{T_s^b}{\sigma}} + P_s^e \cdot z^{\frac{T_s^e}{\sigma}} + (p_b - P_s^b - P_s^e)\cdot z^{\frac{T_c}{\sigma}} \quad (17)$$

The average service time of an event-driven message can be obtained by calculating the derivative of (16) in $z = 1$ (denoted by $D'(1)$ and measured in number of slots). Correspondingly, ρ is obtained as:

$$\rho = 1 - \exp(-\lambda \cdot D'(1) \cdot \sigma) \quad (18)$$

Expressions (9), (10), (11), (12), (13), and (18) are viewed as a system of equations to be solved numerically in order to obtain the values of τ^e , p_b , p_s^b , p_s^e, ρ^* and ρ.

3.2.4 Distribution of IRT

Define $p_I(\gamma)$ to be the complementary cumulative probability of the inter-reception time of beacon messages, namely, the probability that the inter-reception time I of messages from a specific sender is greater than γ. This is equivalent to the probability of at least $\left\lfloor \frac{\gamma}{I_b} \right\rfloor$ consecutive messages (where I_b is the beacon interval) to end in failure, i.e.:

$$p_I(\gamma) = \sum_{I=\gamma}^{\infty} (1-p_s^b)^{\left\lfloor \frac{I}{I_b} \right\rfloor} = \frac{(1-p_s^b)^{\frac{\gamma}{I_b}}}{1-(1-p_s^b)^{\frac{1}{I_b}}} \quad (19)$$

It is also informative to calculate the probability of an event that at least one beacon is received by a vehicle from a sender within a duration γ. Denote by $p_n(\gamma)$ the probability of such an event. We obtain:

$$p_n(\gamma) = \sum_{n=1}^{\left\lfloor \frac{\gamma}{T_b} \right\rfloor} (1-p_s^b)^{\left\lfloor \frac{\gamma}{T_b} \right\rfloor - n} . (p_s^b)^n = \frac{p_s^b\left((1-p_s^b)^{\left\lfloor \frac{\gamma}{T_b} \right\rfloor} - (p_s^b)^{\left\lfloor \frac{\gamma}{T_b} \right\rfloor}\right)}{1-2p_s^b} \quad (20)$$

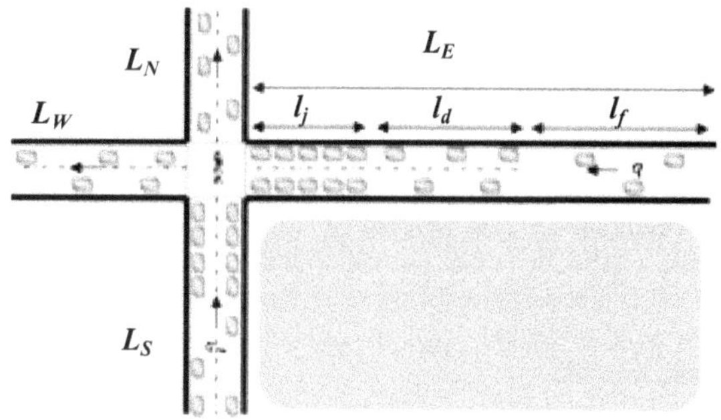

Figure 3. Urban traffic scenario

4. URBAN TRAFFIC MODEL

Our work in this paper is focused on urban traffic systems. In the following, we describe the essential features of a traffic density model for a signalized junction and the road segments linked to it, which is a basic building block of urban traffic (see Figure 3); we refer the interested reader to [26] for further details.

Relevant to our work in this paper is the traffic mobility on the main road segment denoted by L_E in Figure 3. To characterize the dynamics of traffic mobility on the road segment under consideration, we conducted traffic simulations in Paramics [27]. Our observations of traffic density behavior shows that during a red traffic light phase, three regions with different traffic densities coexist along the road segment: 1) a jam traffic density (l_j) caused by vehicles building up a queue, 2) a growing traffic density (l_d) caused by vehicles decelerating as they approach a queue ahead, and 3) an almost constant light traffic density (l_f) caused by vehicles driving in free/stable flow traffic state. During the green phase, a fourth region associated with a different traffic density emerges as vehicles in front of the previously formed queue gradually discharge the queue. By analogy, we identified that the shape and dynamics of the three density regions formed during a red phase and partly during a green phase are analogous to those of a logistic curve [28]. Correspondingly, we proposed two logistic functions to describe traffic density on the road segment under consideration and during a traffic light cycle comprising a red phase followed by a green phase. Assuming the junction is located at position $x = 0$ and t_{red} is the duration of the red phase, the traffic density functions are expressed as:

$$K(x,t) = \begin{cases} A + \frac{K_j - A}{1+exp(B(x-M))} \\ where: \\ A = \frac{l}{HD}, \quad K_j = \frac{l}{l_{hj}}, \\ M = q \cdot t \cdot l_{hj} + {}^{d_b}/_2 \\ B = \frac{2R_d}{V^2}\left(ln\left(\frac{K_j-K_1}{K_1-A}\right) - ln\left(\frac{K_j-K_2}{K_2-A}\right)\right) \end{cases} \quad (21)$$

$$K(x,t) = \begin{cases} max\left(A, A + \frac{K_j - A}{1+exp(B(x-M))} - D_{rate}\right) & s.t \ x < \frac{l_{hj}\cdot(t-t_{red})}{t_r} \\ A + \frac{K_j - A}{1+exp(B(x-M))} & o.w \end{cases} \quad (22)$$

where $K(x,t)$ is the traffic density at position x on the road segment at time instant t, during a red phase (21) or green phase (22). A is the lower bound traffic density associated with stable/free flow traffic state; K_j is the upper bound traffic density associated with jam traffic density (if $A = 0$ then K_j is called the carrying capacity); B is the growth rate of traffic density from lower bound (A) to upper bound (K_j); M is the reflection point of the logistic curve; l_{hj} is the average jam headway distance which is a known parameter; q is traffic arrival in *vehicles/hour*. V is the speed limit of the road segment; R_d is average deceleration of a vehicle; $d_b = \frac{V^2}{2R_d}$ is the distance it takes for a vehicle with average deceleration and velocity V to fully stop; K_1 and K_2 denote the traffic densities corresponding to the start and end position of the braking distance; l is the number of lanes in a multi-lane road segment; D_{rate} in (22) is the queue discharge rate and is described as:

$$D_{rate} = \begin{cases} \dfrac{l \cdot a \cdot (t' - \frac{x}{l_{hj}} t_r) \cdot t_r}{l_{hj}\left(l_{hj} + a \cdot (t' - \frac{x}{l_{hj}} t_r) \cdot t_r\right)} & s.t\ x < \dfrac{l_{hj} \cdot t'}{t_r} \\ 0 & o.w \end{cases}$$

where a is the average acceleration of a vehicle and t' is the relative time instant with respect to the start of the green phase.

4.1 Number of Nodes in Transmission Range (N_r) and Average Number of Hidden Nodes ($\widetilde{N}_h$)

The number of vehicles within the transmission range of a vehicle at position x_v at time instant t is determined by calculating the integral over the density function $K(x,t)$ described by (21) and (22), corresponding to red and green phases of the traffic light cycle:

$$N_r = \int_{\max\ (L_{min}, x_v - R)}^{\min\ (L_{max}, x_v + R)} K(x,t) \cdot dx \qquad (23)$$

where L_{min} and L_{max} are the coordinates of the start and end positions of the road segment under consideration, and x_v is the position of the vehicle.

To determine $\widetilde{N}_h$, we consider the fact that due to the non-uniform traffic distribution, the average per-vehicle number of hidden nodes affecting packet reception is no longer simply half the total number of hidden nodes in the hidden terminal region, as it is in the uniform case. To that end, focusing on the transmission range of the sender vehicle, denote by x_m^r the median position such that half of the total number of vehicles to the right of the sender within its transmission range is located to each side of x_m^r; similarly, define x_m^l to be the median point of vehicles to the left of the sender. In other words, if x_v is the position of the sender vehicle, then, x_m is the point on the road segment that minimizes the following objective function:

$$x_m^{(r,l)} = \text{argmin}_x \left(\frac{N_x}{N} - \frac{1}{2}\right) \quad s.t\ |x - x_v| \le R \qquad (24)$$

where $N_x = \int_{x_v}^{\min\ (L_{max}, x_v + x)} K(x,t) \cdot dx$ and $N = \int_{x_v}^{\min\ (L_{max}, x_v + R)} K(x,t) \cdot dx$ are used for calculating x_m^r, while $N_x = \int_{\max\ (L_{min}, x_v - x)}^{x_v} K(x,t) \cdot dx$ and $N = \int_{\max\ (L_{min}, x_v - R)}^{x_v} K(x,t) \cdot dx$ are used for the calculation of x_m^l.

To calculate $x_m^{(r,l)}$ described by (24), we propose a simple algorithm as follows.

Algorithm 1: median position to the right (left) of a sender

Input: $K(x,t), x_v, L_{min}, L_{max}, R$

Initialize: assign a small positive value (≤ 1) to Δx

$$x_m^r \leftarrow x_v \ (x_m^l \leftarrow x_v)$$

$$N_{right} \leftarrow \int_{x_v}^{\min\ (L_{max}, x_v + R)} K(x,t) \cdot dx$$

$$\left(N_{left} \leftarrow \int_{\max\ (L_{min}, x_v - R)}^{x_v} K(x,t) \cdot dx\right)$$

1: loop

2: $x_m^r \leftarrow x_m^r + \Delta x \ (x_m^l \leftarrow x_m^l - \Delta x)$

3: $N_{x_m^r} \leftarrow \int_{x_v}^{\min\ (L_{max}, x_v + x_m^r)} K(x,t) \cdot dx,$

$\left(N_{x_m^l} \leftarrow \int_{\max\ (L_{min}, x_v - x_m^l)}^{x_v} K(x,t) \cdot dx\right)$

4: until $\left|\frac{N_{x_m^r}}{N_{right}} - \frac{1}{2}\right| \le \varepsilon \ \left(\left|\frac{N_{x_m^l}}{N_{left}} - \frac{1}{2}\right| \le \varepsilon\right)$

5: return $x_m^r \ (x_m^l)$

ε in Algorithm 1 is a very small number and the expressions in parentheses correspond to x_m^l.

Correspondingly, the average per-vehicle number of hidden nodes in the right and left direction of the sender is determined as:

$$\widetilde{N}_h^r = \int_{\min\ (L_{max}, x_v + R)}^{\min\ (L_{max}, x_v + R + x_m^r)} K(x,t) dx \qquad (25)$$

$$\widetilde{N}_h^l = \int_{\max(L_{min}, x_v - R - x_m^l)}^{\max(L_{min}, x_v - R)} K(x,t) dx \qquad (26)$$

where $\widetilde{N}_h^r$ and $\widetilde{N}_h^l$ are the average number of hidden terminals in right and left directions of the sender, respectively.

Using $\widetilde{N}_h^r$ and $\widetilde{N}_h^l$, the average per vehicle terminal nodes can be expressed as:

$$\widetilde{N}_h = \beta \widetilde{N}_h^r + (1 - \beta) \widetilde{N}_h^l \qquad (27)$$

where $0 \le \beta \le 1$ is a weighting factor and can be determined based on the direction relative to the sender the reception probability of safety message broadcast is considered. In a forward collision warning application (FCW), message reception is not important for vehicles driving ahead of a sender vehicle, thus $\beta = 0$. On the other hand, in a lane change assistance application, reception in both direction are equally important and thus $\beta = 0.5$.

5. NUMERICAL RESULTS

We numerically study the reliability model derived in section 3 within two directions. First, the numerical results of the model are derived for an 8-lane highway scenario and are validated using the results of Elbatt *et al.* [8] simulation work. Second, the results of the model are derived for a 3-lane urban scenario illustrated in Figure 3. Traffic and network parameters corresponding to these scenarios are specified in Table 1. Traffic flow associated with the urban scenario is set to a near-saturation level and determined according to the capacity of the junction; as explained in [26], this traffic flow setting corresponds to an ideal signalized junction, and facilitates predictions for under-saturated and over-saturated traffic conditions near a signalized junction.

5.1 Model Validation

As the scenario simulated in [8] only considers periodic beacons, we set the probability of transmission of event-driven messages to zero to align our model with this scenario. In the scenario, the number of vehicles within the transmission range (i.e. N_r) of a candidate sender in high and low density cases are 358 and 38, respectively. The average per-vehicle number of hidden nodes ($\widetilde{N}_h$) calculated using (27) are 179 and 18 for high and low densities. As the traffic is uniformly distributed, $\widetilde{N}_h$ is simply half the number of vehicles in the entire hidden terminal area of a node. The numerical results corresponding to [8] are shown in Figure 4. According to Figure 4(a), the probability of successful reception in the dense traffic case decreases with increasing distance from the sender. This can be justified by the fact that for nodes farther from the sender, the number of hidden terminals increases, leading to a higher number of collisions. The mean reception probability achieved by the model and simulations are 0.65 and 0.72, respectively, and the mean difference between the model and simulation results is 8% with standard deviation 4%. The results corresponding to the probability of successful reception in the low density case are depicted in Figure 4(b). Due to the light traffic density, the impacts of simultaneous transmissions and hidden terminal nodes are negligible. The mean reception probability achieved by the model and simulations are 0.96 and 0.98, respectively, and the mean difference is 2% with standard deviation 0.9%.

Table 1. Simulation Parameters

Traffic Parameters	Elbatt *et al.* [8] Scenario	High density: 1920 *vehicles/mile* Low density: 208 *vehicles/mile*
	Urban Scenario	Road length =1 *km*
		Duration of red phase =50 *s*
		Traffic flow =2740 *vehicles/hour*
		Speed limit =20 *m/s*
		Jam traffic headway distance = 6 *m*
DSRC Parameters	Transmission range R =150 *m*	
	Packet length =100 *bytes*	
	Signal bandwidth = 10 *MHz*	
	Channel Data Rate = 6 *Mbit/s*	
	Slot time (σ) = 13 μs	
	Propagation delay = 1 μs	
	Preamble length = 40 μs	
	Contention window size W_b = 32 Contention window size W_e = 16	
	Arrival rate λ =1 message/s (event-driven msg.)	
	Beacon period (T_b) = 100 *ms*	

We applied the mean probabilities of successful reception measured by Elbatt *et al.* and calculated by the model to measure the distribution of IRT in high and low density scenarios. Figure 4(c) represents the complementary cumulative probability as a function of the IRT. The results show that in the low density case, a message is almost always received in less than 200ms. On the other hand, in the high density case, this increases to 400ms for some messages. Furthermore, in the low density case, the probability that the inter-reception time for a message to be above 100ms is quite small. This means that the vast majority of messages arrive in time. The mean difference between results achieved by the model and simulations in high and low density scenarios are 1% and 0.02%, respectively. We thus conclude that the proposed model fits very well with the simulation results of Elbatt *et al* [8].

5.2 Urban Scenario Performance

Our results for the urban scenario are shown in Figures 5 and 6. Figure 5(a) shows traffic density in *vehicles/meter* along the road segment during a red phase, and Figure 5(b) depicts the average per-vehicle number of hidden nodes potentially affecting a vehicle on the road segment. Figure 6 depicts the complementary cumulative probability as a function of IRT. Observe that, by increasing the queue length at the junction, the average per vehicle terminal node increases at positions behind the queue. A maximum number of hidden nodes is observed at positions 301-334 with magnitude 37 at time 50 *sec* (end of the red phase). In addition to the magnitude increase with time, the area with high number of hidden nodes also widens and expands to distances farther from the junction. As the queue length grows larger than R (transmission range), the average per vehicle terminal nodes also increases in positions close to the junction. This is shown by the rising curve near the junction from time instant 40 *sec* to 50 *sec*.

It follows from Figure 5(c) that the probability of successful transmission of a vehicle is significantly dependent on the average per vehicle number of hidden nodes. Comparing figures 5(b) and 5(c) reveals that in areas with large number of hidden nodes, the probability of successful transmission is low. In positions 301-334 and at time instant 50 *sec*, for instance, the average successful transmission probability is 0.86, which is the lowest among all positions at the same time instant. In addition, we observe that the density of vehicles within the transmission range of a sender has a very small impact on the probability of successful transmission. At time instance 50 *sec*, the highest number of vehicles within transmission range of a sender is 104, which is observed at position 150m. The number of hidden nodes seen at this position is a small number 3. Correspondingly, the probability of successful transmission is 0.95 at this position, which stresses the fact that the hidden terminal effect is the predominant driving factor determining the achievable successful transmission rate.

We continue the numerical study with the distribution of IRT shown in Figure 6. For three IRT values 100ms, 300ms, and 1s, we calculated the probability of inter-reception time using (19) and depicted the results in Figure 6(a), 6(b), and 6(c) respectively. Again, the worst-case IRT probabilities occur at positions 301-334m with average magnitudes 0.13, 0.002, and 1×10^{-9} corresponding to 100ms, 300ms, and 1s, respectively.

Our results above were given for the red phase of a traffic light. During the green phase, in the first few seconds of the phase, the probability of unsuccessful transmission and probability of high inter-reception time are exacerbated due to a slow initial discharge

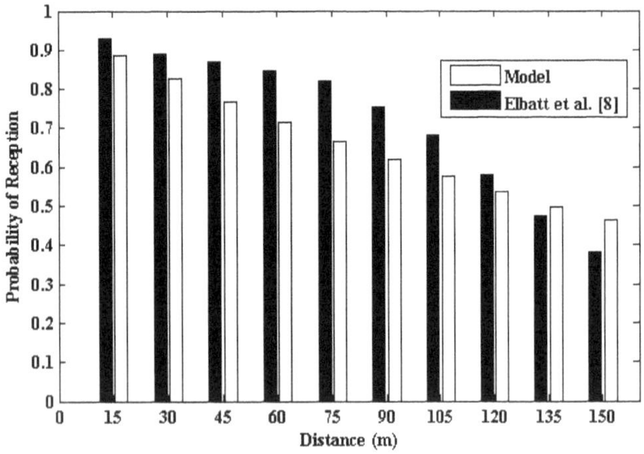

(a)

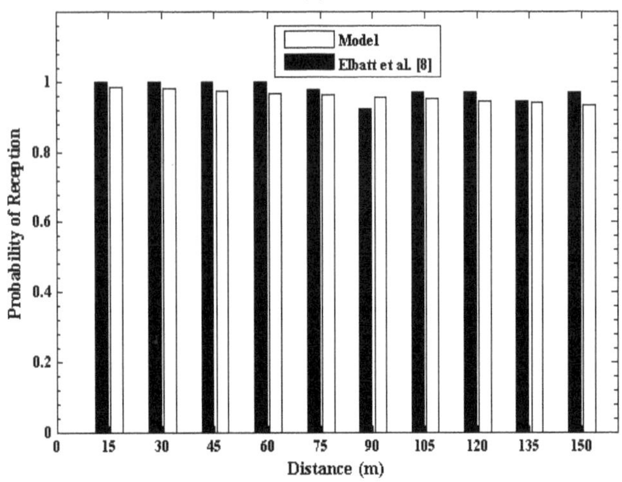

(b)

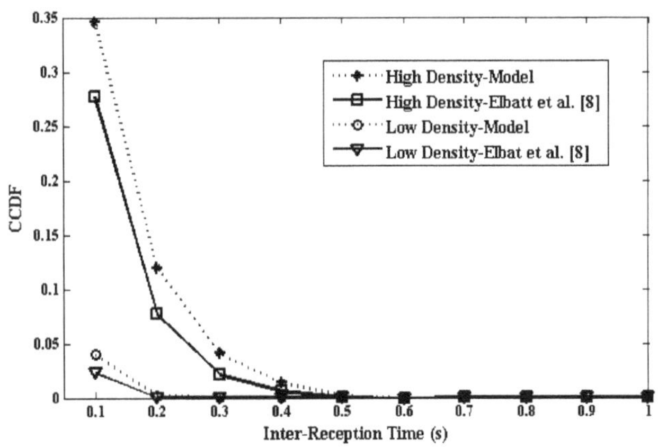

(c)

Figure 4. Comparison of the model and Elbatt *et al.* highway scenario, (a) probability of successful reception in high traffic density, (b)) probability of successful reception in low traffic density, (c) distribution of IRT

rate of the queue, thus more positions will experience a high average per-vehicle hidden terminal level. Afterwards, with increasing acceleration, the queue discharges faster and the reliability metrics improve.

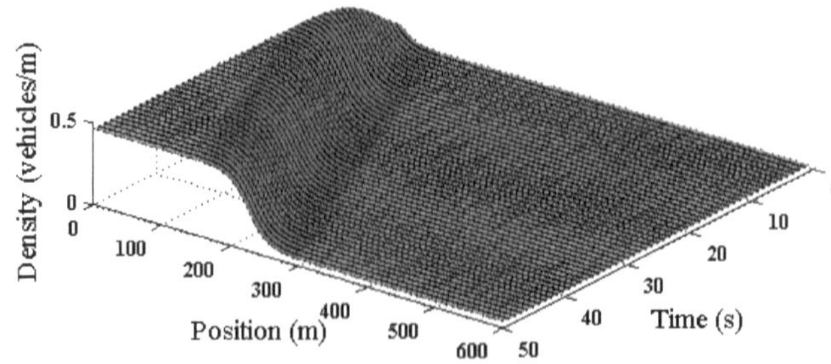

(a)

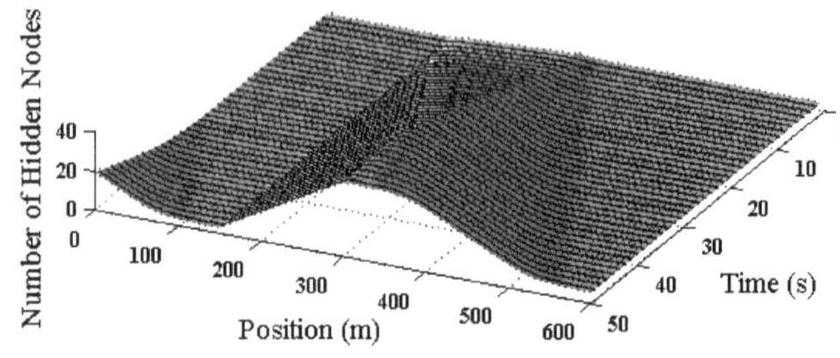

(b)

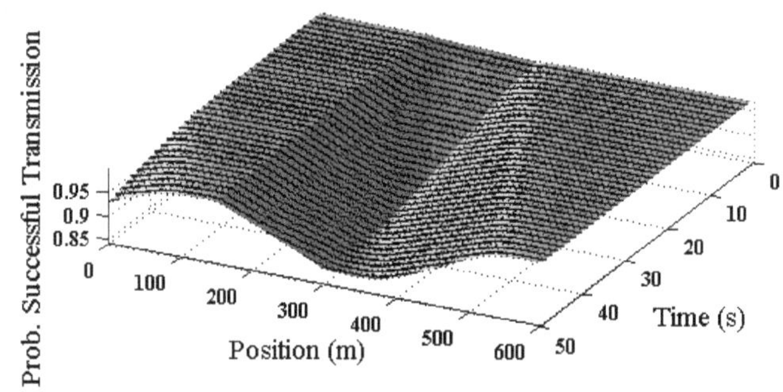

(c)

Figure 5. (a) traffic density, (b) average per vehicle hidden terminal nodes, (c) probability of successful transmission.

5.3 Discussion

The impact of the hidden terminal effect on VANETs has previously been studied in [8][11]. The studies were carried out by simulating free-flow, uniformly distributed vehicular traffic and capturing the packet delivery characteristics. In the validation of our model in section 5, we verified and corroborated these results. In our study of the reliability performance in the urban scenario, based on a realistic vehicular mobility model around a signalized junction, we find strong evidence that these results can be generalized to cover urban settings as well. One would intuitively suspect that the setting of a signalized junction, characterized by high variations of traffic density, will lead to poor packet delivery performance due to increased overlapping of transmitters. However, our results show that this impact is less significant than the hidden node problem, which is dominant in the urban non-uniform scenario as well. Moreover, we observe that the worst case results take place far from the traffic light queue, where the traffic is either in free-flow or decelerating from high speed. It is notable that, from a traffic safety point of view, these positions are arguably the most important for timely alerting of dangerous traffic conditions, and their lower reliability may correspondingly lead to an increased risk of serious incidents.

Our experiments were carried out at near-saturation traffic conditions, using DSRC communication with moderate (100 byte) payload length. Increasing the traffic load further, or increasing the packet size, will result in lower performance but generally will not change the negative impact dominance of the hidden nodes. It is possible to mitigate the impact of vehicular density by selecting appropriate radio data rates, but the hidden node problem will still remain a serious issue using the current 802.11p specifications.

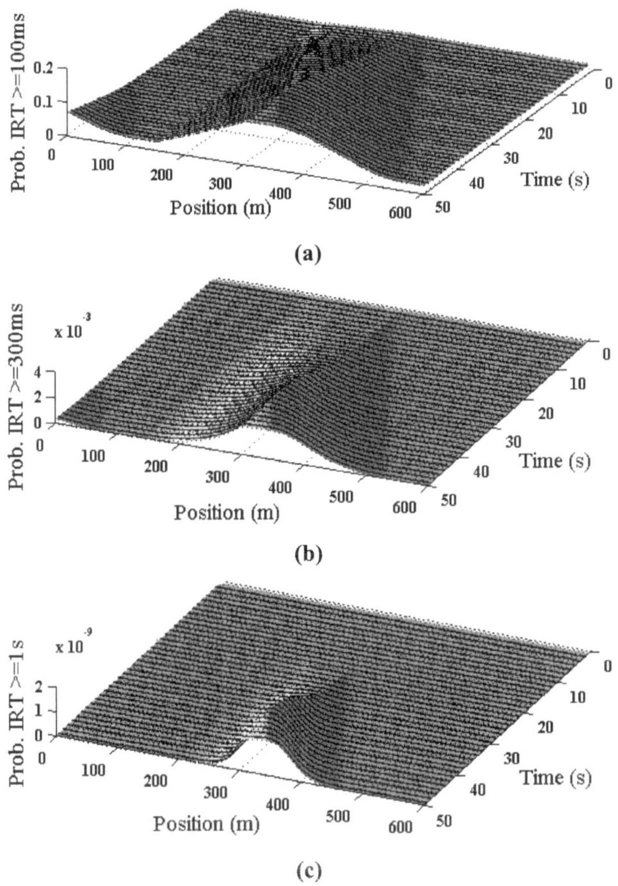

Figure 6. IRT distribution, (a) 100ms, (b) 300ms, (c) 1s

6. CONCLUSION

Vehicular networks hold great promise in delivering novel road traffic applications that can make road transportation significantly safer than today, limiting risks and catastrophic effects of incidents. It is paramount to these applications that the underlying wireless networks provide stringent packet delivery and delay characteristics. Previous work in this area has focused on investigating the performance of wireless networks in free-flow scenarios, where vehicles are uniformly distributed. This does not hold true in urban settings, where traffic is regulated by signalized junctions. In this paper, we address the urban case by studying the performance of safety messages using a realistic vehicular mobility model which captures the heterogeneous node densities at and around signalized junctions. In line with previous work, we construct Markov models for capturing the delivery probability for event-driven warning messages and the packet inter-reception time for periodic beacons. Combining the Markov models with the urban vehicular density model, for the first time we are able to accurately capture the resulting performance characteristics in a non-uniform density setting. Through a numerical evaluation, we demonstrate that the major impact of the hidden node problem on the reliability performance of safety messages extends to the urban case as well. Importantly, we find that the impact is most significant in the same road sections where the vehicle velocities are highest. These findings call for further work on mechanisms to mitigate the effect of hidden nodes in order to ensure the viability of DSRC-based safety applications.

7. ACKNOWLEDGMENTS

This work was partially sponsored by the Australian Research Council under grant DP0987782.

8. REFERENCES

[1] Standard Specification for Telecommunications and Information Exchange Between Roadside and Vehicle Systems — 5 GHz Band Dedicated Short Range Communications (DSRC) Medium Access Control and Physical Layer Specifications, ASTM E2213 -03, 2010.

[2] IEEE Draft Standard for Information Technology — Telecommunications and information exchange between systems — Local and metropolitan area networks — Specific requirements — Part 11: Wireless LAN Medium Access Control (MAC) and Physical Layer (PHY) specifications, Amendment 6: Wireless Access in Vehicular Environments. IEEE Std 802.11p, July 2010.

[3] K. C. Lee, U. Lee, and M. Gerla. Survey of routing protocols in vehicular ad hoc networks. *Advances in Vehicular Ad-Hoc Networks: Developments and Challenges*, IGI Global, October 2009.

[4] S. Zeadally, R. Hunt, Y.S. Chen, A. Irwin, and A. Hassan. Vehicular ad hoc networks (VANETs): status, results, and challenges. *Telecom Systems*, Springer Netherlands, 2010.

[5] R. Rajaraman. Topology control and routing in ad hoc networks: A survey. *ACM SIGACT News*, 33(2):60-73, 2002.

[6] M. Fiore, J. Harri, F. Filali, and C. Bonnet. Understanding vehicular mobility in network simulation. *IEEE International Conference on Mobile Adhoc and Sensor Systems (MASS)*, Pisa, Italy, October 2007.

[7] J. Harri, F. Filali, and C. Bonnet. Mobility models for vehicular ad hoc networks: A survey and taxonomy. *IEEE Communications Surveys Tutorials*, 11(4):19–41, 2009.

[8] T. ElBatt, S. K. Goel, G. Holland, H. Krishnan, and J. Parikh. Cooperative collision warning using dedicated short range wireless communications. *3rd ACM international workshop on vehicular ad hoc networks (VANET)*, Los Angeles, CA, USA, September 2006.

[9] M. Torrent-Moreno, S. Corroy, F. Schmidt-Eisenlohr, and H. Hartenstein. IEEE 802.11-based one-hop broadcast communications: Understanding transmission success and failure under different radio propagation environments. *9th ACM/IEEE International Symposium on Modeling, Analysis and Simulation of Wireless and Mobile Systems (MSWiM)*, Torremolinos, Spain, October 2006.

[10] M. Torrent-Moreno and J. Mittag. Adjusting transmission power and packet generation rate of periodic status information messages in VANETs. *3rd ACM international workshop on vehicular ad hoc networks (VANET)*, Los Angeles, CA, USA, September 2006.

[11] S. Yousefi, M. Fathy, and A. Benslimane. Performance of beacon safety message dissemination in Vehicular Ad hoc

NETworks (VANETs). *Journal of Zhejiang University SCIENCE A*, 2007.

[12] G. Bianchi. Performance analysis of the IEEE 802.11 distributed coordination function. *IEEE Journal on Selected Areas in Communications,* 18(3):535-547, 2000.

[13] F. Cali, M. Conti, and E. Gregory. IEEE 802.11 wireless LAN: Capacity analysis and protocol enhancement. *IEEE INFOCOM*, San Francisco, CA, USA, March-April 1998.

[14] Q. Ni, T. Li, T. Turletti, and Y. Xiao. Saturation throughput analysis of error-prone 802.11 wireless networks. *Wireless Communications and Mobile Computing*, 5(8):945-956, 2005.

[15] P. Chatzimisios, V. Vitsas, and A.C. Boucouvalas. Throughput and delay analysis of IEEE 802.11 protocol. *5th IEEE international workshop on networked appliances* (IWNA), Liverpool, UK, October 2002.

[16] V.M. Vishnevsky and A.I. Lyakhov. IEEE 802.11 wireless LAN: Saturation throughput analysis with seizing effect consideration. *Cluster Computing*, 5(2):133-144, 2002.

[17] P.E. Engelstad and O.N. Osterbo. The delay distribution of IEEE 802.11e EDCA and 802.11 DCF. *International Performance, Computing, and Communications Conference (IPCCC)*, Phoenix, AZ, USA, April 2006.

[18] A. Lyakhov, V. Vishnevsky, and P. Poupyrev. Analytical study of broadcasting in 802.11 ad hoc networks. *2nd International Conference on e-Business and Telecommunication Networks* (ICETE), Reading, UK, October 2005.

[19] X. Ma and X. Chen. Performance analysis of IEEE 802.11 broadcast scheme in ad hoc wireless LANs. *IEEE Transactions on Vehicular Technology,* 57(6):3757-3768, 2008.

[20] X. Ma and X. Chen. Saturation performance of IEEE 802.11 broadcast networks. *IEEE Communications Letters,*11(8):686-688, 2007.

[21] X. Chen, H. Refai, and X. Ma. Saturation performance of IEEE 802.11 broadcast scheme in ad hoc wireless LANs. *IEEE Vehicular Technology Conference* (VTC), Baltimore, MD, USA, September-October 2007.

[22] X. Ma, X. Chen, and H. Refai. Delay and broadcast reception rates of highway safety applications in vehicular ad hoc networks. *IEEE INFOCOM, Mobile Networks for Vehicular Environments Workshop*, Anchorage, AK, USA, May 2007.

[23] A. Vinel, V. Vishnevsky, and Y. Koucheryavy. A simple analytical model for the periodic broadcasting in vehicular ad-hoc networks. *4th IEEE Broadband Wireless Access Workshop*, New Orleans, LA, USA, December 2008.

[24] A. Vinel, D. Staehle, and A. Turlikov. Study of beaconing for car-to-car communication in vehicular ad-hoc networks. *IEEE VehiMob Workshop*, Dresden, Germany, June 2009.

[25] A. Vinel, Y. Koucheryavy, S. Andreev, and D. Staehle. Estimation of a successful beacon reception probability in vehicular ad-hoc networks. *International Wireless Communications and Mobile Computing Conference (IWCMC)*, Leipzig, Germany, June 2009.

[26] S. Bastani, B. Landfeldt, and L. Libman. A Traffic Density Model for Radio Overlapping in Urban Vehicular Ad hoc Networks. *IEEE Conference on Local Computer Networks (LCN)*, Bonn, Germany, October 2011.

[27] Quadstone Paramics, 2008. Available at: http://www.paramicsonline.com.

[28] S. E. Kingsland. Modeling nature: episodes in the history of population ecology. *University of Chicago Press*, ISBN 0-226-43728-0.

[29] L. Bononi, M. Di Felice, and S. Pizzi. Dba-MAC: Dynamic Backbone-Assisted Medium Access Control Protocol for Efficient Broadcast in VANETs. *Journal of Interconnection Networks*, 10(4):321-344, 2009.

VLOCI2: Improving 2D Location Coordinates using Distance Measurements in GPS-equipped VANETs

Farhan Ahammed[1,2], Javid Taheri[1], Albert Zomaya[1]
[1]School of Information Technologies
The University of Sydney
NSW 2006, Australia
{Farhan.Ahammed, Javid.Taheri, Albert.Zomaya}@sydney.edu.au

Max Ott[2]
[2]NICTA Australia
Australian Technolgy Park
Level 5, 13 Garden Street
Eveleigh NSW 2015, Australia
Max.Ott@nicta.com.au

ABSTRACT

The problem of further improving the accuracy of GPS-provided coordinates of vehicles within the same VANET (Vehicular Ad-Hoc Network) is addressed. VANETs can be used to increase the accuracy of each vehicle's computed location by allowing vehicles to share information wth each other. This paper looks at improving the coordinates provided by the GPS devices, given that erroneous measurements are present in the system. The algorithm VLOCI2 is an extension of the VLOCI algorithm by providing location improvement on multi-laned roads. VLOCI2 is shown to perform efficiently when erroneous distance measurements are present in the environment/computations.

Categories and Subject Descriptors

D.1.3 [**Programming Techniques**]: Concurrent Programming—*Distributed Programming*

General Terms

Algorithms, Design, Performance, Reliability

Keywords

vehicular ad hoc networks, localization, GPS, distance measurements, location improvement/refinement

1. INTRODUCTION

VANETs are mobile ad hoc networks where the nodes are vehicles. The vehicles are equipped with wireless communication devices allowing them to share real-time information. With this information vehicles and drivers will have up-to-date information regarding the state of traffic, allowing them to avoid congested and other abnormally affected areas.

MSWiM'11, October 31–November 4, 2011, Miami, Florida, USA.

Localization is the name given to the class of algorithms which allow each node in the network to discover its location. Additionally, *localization improvement* algorithms can be implemented allowing the nodes to further refine their computed locations after aggregating other measured data.

Many vehicles are nowadays equipped with GPS devices. GPS devices are accurate to within 10 metres [10] resulting in reported incorrect locations. Obtaining more accurate coordinates allows the vehicles to construct more precise models of their local traffic conditions.

With increased accuracy and better models, accidents can be prevented. VANETs can be used to provide information about vehicles not within line-of-sight. Increasing the accuracy of each vehicle's computed location will improve safety as vehicles can make more informed decisions on how to react in dangerous situations.

One of the most important issues facing the vehicles is that measured data will always be corrupted by noise. Any localization and localization improvement algorithm should not be expected to produce exact results and *robustness* is a desired quality in such algorithms.

It is proposed that VANETs can be used to improve on the position estimates provided by GPS devices. Every vehicle uses the VANET to share their position estimate with neighbouring vehicles. It is also assumed the vehicles can measure the distance between them and other vehicles. This can be achieved using either specialised sensors/equipments [7,8] or with the wireless communication devices [15].

An overview of previous work found in literature is presented in Section 2. Section 3 introduces the notation and defines the problem addressed in this paper. The VLOCI2 algorithm, used to solve the location improvement problem, is described in Section 4. Section 5 describes simulations used to test VLOCI2. Discussions of the results and concluding remarks are presented in Sections 6 and 7 respectively.

2. RELATED WORK

The idea of 'location improvement' is not extensively addressed in literature, nor has many work considered the situation where erroneous data is present in the system. Most work assume some nodes within the network are GPS equipped, or that some *anchor nodes* know their exact location so the aim for the remaining nodes is to discover their location. Those works that incorporate the use of GPS assume the accuracy of the GPS devices is sufficient.

Priyantha et al. proposed a technique called *anchor-free localization* (AFL) of providing localization to wireless sensor networks (WSN) [17]. Their algorithm is decentralised, where the only information each node collects is the relative distance to its neighbouring nodes.

Barani and Fathy [2] looked at the problem where not all vehicles are equipped with GPS devices or cannot receive signals from the GPS satellites. In their approach, the vehicles attempt to find three neighbours within one or two-hop distance. Their measure of error is based on node denisty.

Shekofteh et al. use tabu search with simulated annealing to provide localization [20]. The nodes take distance measurements to the anchor nodes, which are modelled to contain Gaussian noise. A 'noise factor' is introduced to control the magnitude of the noise. With 10% of the nodes designated anchor nodes, their simulation results show the algorithm does provide localization to the non-anchor nodes.

Moore et al. addresses the issue of localization where the distance measurements are corrupted by noise [13]. They address the problem of discovering the coordinates of the nodes of a graph given the edge length, and with no initial coordinate information. Their simulations produced location errors that fluctuated between values of 0.5 and 5.

Other solutions [4, 6, 9, 11, 19] are applied in situations where a subset of nodes, usually termed anchor nodes or base stations, have knowledge of their own positions. The remaining nodes then communicate with the anchor nodes to determine their locations as they have no other method to estimate their locations.

Ouyang et al. considered the problem of using terrestrial TOA measurements to improve initial GPS measurements [14]. Upon receiving the initial GPS measurements, a ranging technique (e.g. TOA) is used to measure the distance to nearby reference stations. The idea of obtaining a single GPS estimate and single distance measurements is common with this work. However Ouyang et al. assumes the anchor nodes know their exact positions which are not provided by GPS.

McGuire and Plataniotis [12] developed the Cramér-Rao, Weinstein-Weiss and Ziv-Zakai lower bounds for localization in wireless networks. In their models the nodes perform localization taking measurements from fixed base stations (anchor nodes) with known locations. The locations of the anchor nodes, regardless of how knowledge of their location is obtained, is deemed to be accurate, and is not modelled to contain error. The distance measurements are the only quantities modelled to contain noise.

3. PROBLEM STATEMENT

Let G be a network of nodes. The set of all nodes in G is denoted $V(G)$. The number of nodes in the network is $|V|$. Given any particular node $n_i \in V$, any other node that n_i can send and receive messages from are deemed its neighbour.

Each node n_i is located at $p_i = (x_i, y_i)$ (true location) and, due to inaccurate measurements, calculates that it is located at $\widehat{p_i} = (\widehat{x}_i, \widehat{y}_i)$ (its computed/estimated location). The words 'location' and 'position' are also used interchangeably. The location error for n_i is the distance between its true and computed location $\delta_i = \|p_i - \widehat{p}_i\|$. It is assumed every node is equipped with a GPS device such that their initial position estimate is a value $(\widehat{x}_i^0, \widehat{y}_i^0)$ where $x_i - 10 < \widehat{x}_i^0 < x_i + 10$ and $y_i - 10 < \widehat{y}_i^0 < y_i + 10$.

The true distance between nodes n_i and n_j is denoted $d_{i,j}$, while the distance measured by n_i between itself and node n_j is given as $\widehat{d_{i,j}} = \varepsilon \cdot d_{i,j}$, for some $\varepsilon \in \mathbb{R}$. It is also assumed that $\widehat{d_{i,j}} \neq \widehat{d_{j,i}}$.

A metric used to gauge the performance of the localisation algorithm of a network is the *network location error*. For each node n_i, its location error is already defined as δ_i. The network location error is the average of every node's location error. The *location error* of the network (E_V) is thus defined as $E_V = \frac{1}{n}\sum_{n_i \in V} \delta_i = \frac{1}{n}\sum_{n_i \in V} \|p_i - \widehat{p}_i\|$

The goal of *localization* is for every node $n_i \in V(G)$, located at p_i, to compute its position $\widehat{p}_i$ such that the network location error is minimised.

The problem of *Location Improvement* (Definition 1) is referred to in this paper. The aim is to find a method of adjusting every node's current estimated position $\widehat{p}_i$ such that the location error is reduced.

DEFINITION 1 (LOCATION IMPROVEMENT). *Let V be a network of nodes. Assume every node n_i has computed its estimated position $\widehat{p}_i$. The problem of* Location Improvement *is to find a function ϕ which modifies some or all of the nodes' estimated positions (producing $\widehat{p}_i' = (\hat{x}_i', \hat{y}_i')$) such that the location error of the modified network is minimised.*

4. METHODS AND ALGORITHM

4.1 Network Topology

The network is modelled as a straight multi-laned highway of length L metres with a set speed limit. The vehicles travel from one end to the other. GPS coordinates are set to a random coordinate within the range $(x_i \pm 10, y_i \pm 10)$, where (x_i, y_i) is the true location for vehicle n_i.

4.2 Accuracy of distance measurements

The accuracy of the distance measurement devices refers to the difference between $\widehat{d_{i,j}}$ and $d_{i,j}$. Two techniques used to measure distance are time-of-arrival and received signal strength [5]. The accuracy of these techniques can be modelled such that as the distance to be measured increases, the accuracy of the measurements taken also decreases [3, 16]. The model used for distance measuring reflects this behaviour. The statement 'the distance measurement error is α' means that $\widehat{d_{i,j}} = E \cdot d_{i,j}$ for all $n_i, n_j \in V$, where $E \sim N(1, \alpha^2)$; $N(1, \alpha^2)$ refers to the Gaussian distribution with mean 1 and standard deviation α.

Using this model, nodes further away have a larger probability of producing more erroneous data. The further away a neighbouring node is (i.e. the larger the value of $d_{i,j}$), the larger the error range of the measured distance. Measuring the distance to nodes within close range is more accurate than measuring the distance to those further away.

4.3 Computing position estimates

Each node broadcasts information containing their computed position to their neighbours and additionally are able to measure the distance to their neighbours. Using this information each node calculates a new location estimate.

Define the mapping $\phi : \mathbb{N} \times \mathbb{R}^{3N} \to \mathbb{R}^2$. Node n_i uses ϕ to produce new coordinates $(\hat{x}_i', \hat{y}_i')$ from the measured

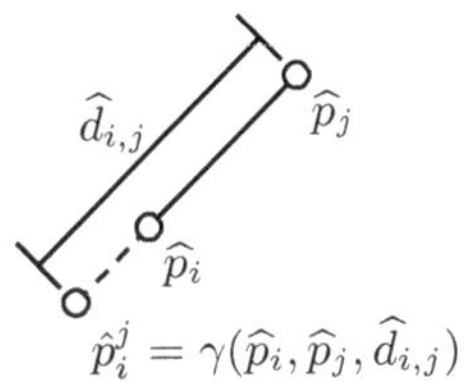

Figure 1: Calculating a new (x, y) position assuming a neighbour's position is correct

distances and provided estimated locations:

$$(i, \widehat{d}_{i,1}, \ldots, \widehat{d}_{i,N}, \hat{x}_1, \ldots, \hat{x}_N, \hat{y}_1, \ldots, \hat{y}_N).$$

ϕ is defined by two other functions: $\gamma : \mathbb{R}^5 \rightarrow \mathbb{R}^2$ and a weight function $w : \mathbb{R} \rightarrow \mathbb{R}$.

The function γ is used to calculate where node n_i is located assuming that n_j's estimated position is correct. Figure 1 shows an example of how a position $\hat{p}_i^j$ is calculated based on the values of $\widehat{p}_i$, $\widehat{p}_j$ and $\widehat{d}_{i,j}$. The values of $\widehat{p}_i$, $\widehat{p}_j$ are used to define the angle between two nodes while the new coordinate $\hat{p}_i^j$ is positioned such that the distance from the neighbouring node $\widehat{p}_j$ is the measured distance $\widehat{d}_{i,j}$.

Computing $\hat{p}_i^j$ from $\{\widehat{p}_i, \widehat{p}_j, \widehat{d}_{i,j}\}$ is straightforward. Thus given a set of nodes $\{n_1, \ldots, n_N\}$, node n_i can construct the set $\{\hat{p}_i^j \mid j = 1, \ldots, N\}$.

4.4 The weight function

To compute the weighted average, a weight function w needs to be defined. As the measured distances become more inaccurate the further away neighbouring vehicles are, thus measurements obtained from nearby vehicles are more reliable than measurements from vehicles further away.

The weight function used for the VLOCI2 algorithm is the same as that used in the VLOCI algorithm [1]. An inverse exponential function of the form $w(x) = Ae^{-x^2/B}$ was used to define the weights given to vehicles based on distance.

4.5 'Smart' lane placement

Further accuracy may be achieved if the vehicles were placed at the centre of the lanes they are located in. A further step is added to the VLOCI2 algorithm. For vehicle n_i, if $\hat{p}'_i$ is located in the same lane as $\widehat{p}_i$, then the estimated position is modified so the x-coordinate is the midpoint between the x-coordinates of $\hat{p}'_i$ and $\widehat{p}_i$ and the y-coordinate is the centre of the corresponding lane (Figure 2a).

This is to help prevent the vehicle from 'drifting' away from the lane it is located in. All the vehicles are working with erroneous data, and have no accurate information to use as a reference. When some vehicles begin to drift away, the others tend to follow to maintain the me distances between each other

In the alternate case, where $\hat{p}'_i$ and $\widehat{p}_i$ are located in different lanes, the estimated position is modified to be the midpoint of the line segment with endpoints $\{\hat{p}'_i, \widehat{p}_i\}$ (Figure 2b).

4.6 VLOCI2 overview

An iteration performed by node n_i is the process of using the location improvement procedure explained to update its estimated position. The number of iterations (I) hence determines the number of times every vehicle updates its estimated position.

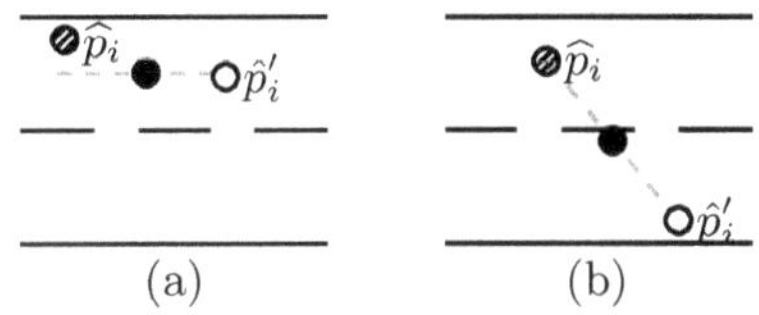

Figure 2: Smart lane placement. Either the new estimated position is on (a) a different lane or (b) on the same lane as the current estimated position.

The VLOCI2 algorithm is (a) distributed, as every vehicle runs the algorithm independently and collaborates with their neighbours, and (b) scalable, since the number of vehicles that can participate is not restricted. A brief summary of the localization algorithm is provided below:

1. Transmit a message containing $\widehat{p}_i$.
2. Wait to receive similar messages from neighbour nodes.
3. For each node a message was received from, measure the distance to it.
4. Repeat I times:
   (a) Node n_i knows the values of $\{\widehat{p}_j, \widehat{d}_{i,j}\}$ for each of its neighbours. Apply function ϕ, using weight function w, followed by smart lane placement, to compute a new estimated position.
   (b) Broadcast the new estimated position.
   (c) Wait to receive new messages from neighbour nodes.

5. SIMULATION RESULTS

Simulations were performed to assess the effectiveness of VLOCI2 for different values of α, the error of the distance measurements, given different network densities.

5.1 Simulation parameters

The network itself was modeled using the Paramics Modeler v6.7.1 [18]. The communication range of the vehicles were set to 100 metres. When the distance measurement error is set to $\alpha = 10\%$, distance measurements to vehicles further than 100m away will be within the range 100 ± 10m. The computed positions will be less accurate than the provided GPS coordinate.

A total of 8000 vehicles traveled across the highway. Once a vehicle reached the end, it did not re-enter the highway The vehicle density is defined such that at 100%, 8000 vehicles are released per hour. The simulations were run until all vehicles have traveled across the highway.

Table 1 summarises the simulation parameters used and their values.

5.2 Results

Figures 3 and 4 shows how VLOCI2 performs with differing levels of accuracy in distance measurements with the vehicle density set to 10% and 25% respectively.

How VLOCI2 performs with accurate distance measurements (ie. $\alpha = 0$) with different vehicle densities and further iterations is shown in Figure 5.

How the VLOCI2 progresses as the vehicle density and distance measurement accuracy varies is shown in Figure 6.

Figure 7 shows the percentage of vehicles that placed themselves in the correct lane and those in incorrect lanes.

Parameter	Value
Highway length (L)	3 km
No. lanes	4
Lane width	3.65 m
Speed limit	120 km/h
Dist. measurement accuracy (α)	0%, 5%, 10%, 15%, 20%
No. Dist. Meas. taken (D)	5
Weight function f coefficients	$A = 100, B = 300$
Vehicle density (% per min)	10%, 15%, 20%, 25%
No. iterations (I)	10
No. tests per scenario	2000

Table 1: The simulation parameters used.

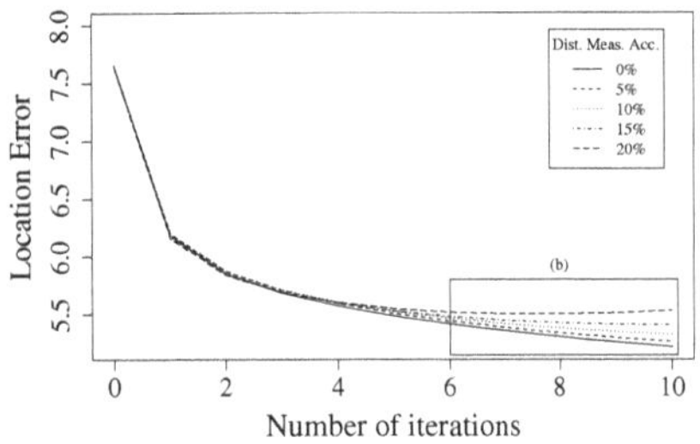

(a) Iterations 1–10

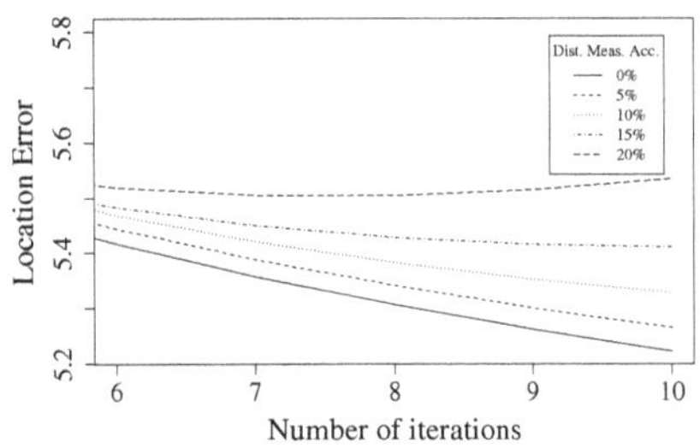

(b) Iterations 6–10

Figure 3: Effects of distance measurement error (Vehicle density 10%)

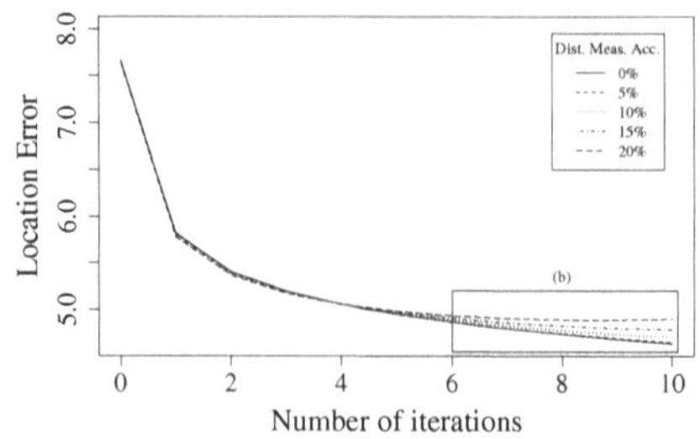

(a) Iterations 1–10

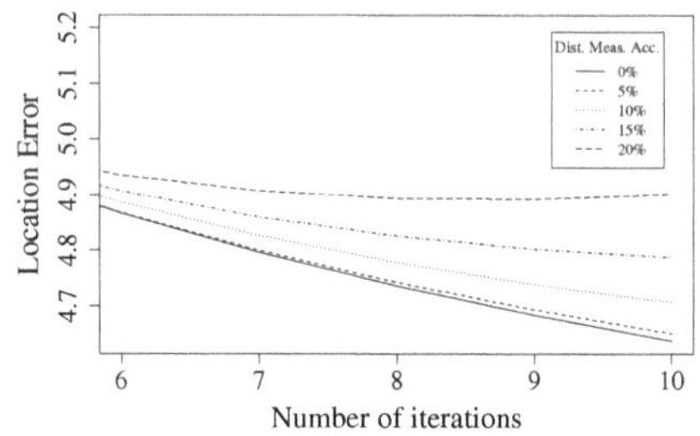

(b) Iterations 6–10

Figure 4: Effects of distance measurement error (Vehicle density 25%)

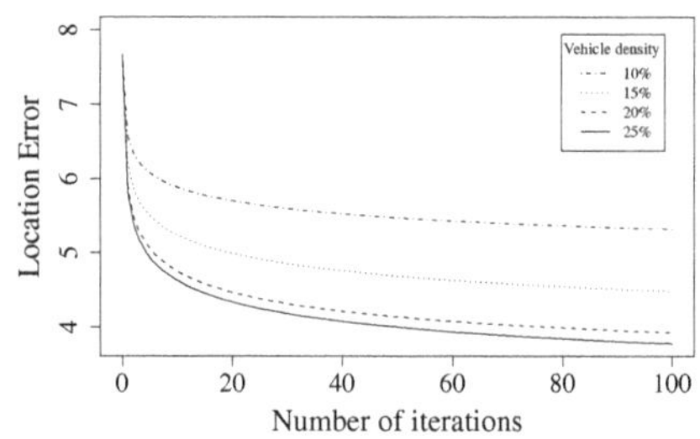

Figure 5: The reduction in location error when all distance measurements are accurate ($\alpha = 0$)

6. DISCUSSION AND ANALYSIS

The results show that VLOCI2 does cause the location error to reduce. The worst cases occurred when the vehicle density was at 10% and $\alpha = 20\%$. For most cases, the location error decreased by more than 35%.

Figure 6 shows how, regardless of distance measuring error, an increase in the vehicle density results in further reduction in location error. While it is not true that the location error strictly decreases after each iteration it nevertheless does decrease after 10 iterations in all tested scenarios.

Figure 6 also shows that as the distance measurement accuracy (α) increases, the amount of reduction in location error decreases. In all cases the location error improved by at least 20%.

In each scenario tested, every vehicle completed 10 iterations in the VLOCI2 algorithm. In Figures 3 and 4 the location error either strictly decreased or reached the minimum value by the eighth iteration.

When the code was run on a computer running Linux (Fedora 10, Kernel v2.6) with two 3GHz Intel processors, 10 iterations took 0.04ms to complete. Assuming a vehicle is equipped with a 1GHz processor, it can be estimated that it will take 0.12ms to complete the 10 iterations. If a vehicle travels at 120km/h then in 0.12ms it travels 0.004m. Similarly, at 50km/h, a distance of 0.00167m is traveled during the same amount of time. During the time it takes to complete the 10 iterations, the vehicles travel a distance less than one metre.

Figure 7 shows that when the vehicle density was greater than 10%, at least 25% of all vehicles placed themselves in the correct lane. As the vehicle density increased, so did the number of vehicles placing themselves in the correct lane. The number of vehicles placing themselves two or three lanes away from the correct lane decreases.

Since the width of two adjacent lanes is 7.3 metres and VLOCI2 was able to reduce the location error to under 7 metres, but not under 3.65 metres (the width of a single lane), the majority of vehicles were placed in either the correct or adjacent lanes.

7. CONCLUSIONS

The VLOCI2 algorithm, is shown to improve the location error of VANETs where the vehicles, traveling on a four-lane highway, receive an initial inaccurate GPS coordinate and use imprecise distance measurements.

With accurate distance measurements and largest vehicle

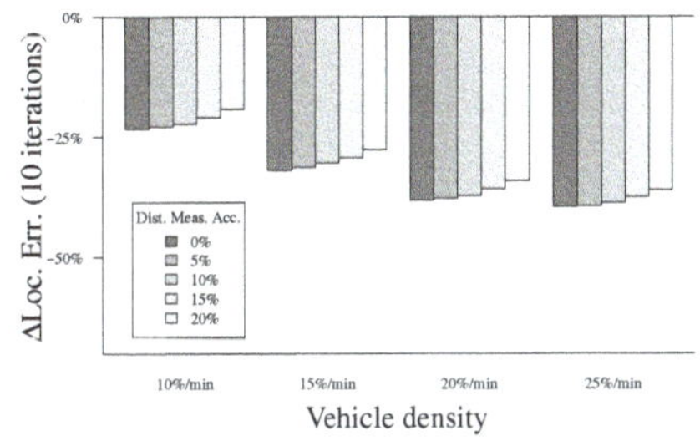

Figure 6: Effects of distance measurement error (200 nodes)

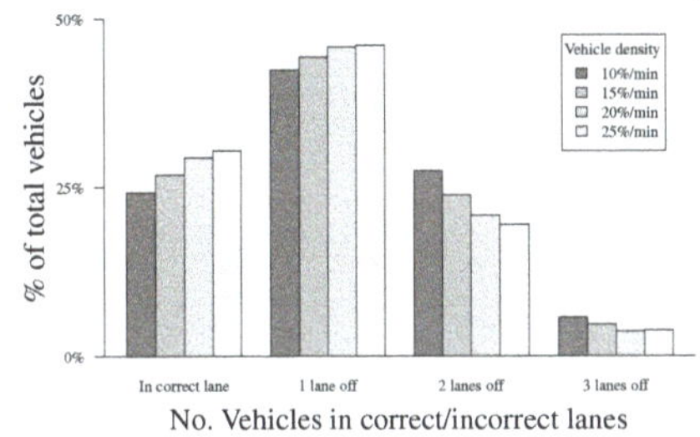

Figure 7: The % of vehicles that placed themselves in the correct lanes.

density, the location error can be reduced to values less than 4.7 metres, while in the worst case scenario tested, the location error can still be reduced to values than 5.6 metres—an improvement of almost 50%.

Additionally, at least 75% of all vehicles place themselves in either the correct lane, or an adjacent lane, while the number of vehicles placing themselves in lanes further away does decrease as the distance measurement accuracy improves or the network density increases.

8. ACKNOWLEDGEMENTS

The trace files used was provided by Saeed Bastani (School of IT, The University of Sydney, Australia).

9. REFERENCES

[1] F. Ahammed, J. Taheri, A. Zomaya, and M. Ott. VLOCI: Using distance measurements to improve the accuracy of location coordinates in gps-equipped vanets. *Mobile and Ubiquitous Systems, 7th International ICST Conference on (MobiQuituous 2010)*, 2010.

[2] H. Barani and M. Fathy. An algorithm for localization in vehicular ad-hoc networks. *Journal of Computer Science*, 6(2):168–172, 2010.

[3] G. Bellusci, G. J. M. Janssen, J. Yan, and C. C. J. M. Tiberius. Modeling distance and bandwidth dependency of TOA-based UWB ranging error for positioning. *Research Letters in Communications*, 2009(1):1–4, Jan 2009.

[4] A. Benslimane. Localization in vehicular ad hoc networks. *Systems Communications, 2005. Proceedings*, pages 19–25, Aug 2005.

[5] A. Boukerche. *Algorithms and Protocols for Wireless, Mobile Ad Hoc Networks*. Wiley-IEEE Press, 2008.

[6] X. Cheng, A. Thaeler, G. Xue, and D. Chen. TPS: a time-based positioning scheme for outdoor wireless sensor networks. In *INFOCOM 2004. Twenty-third AnnualJoint Conference of the IEEE Computer and Communications Societies*, volume 4, pages 2685–2696, 7-11 2004.

[7] W. J. Fleming. Overview of automotive sensors. *Sensors Journal, IEEE*, 1(4):296–308, Dec 2001.

[8] F. Fölster and H. Rohling. Data association and tracking for automotive radar networks. *Intelligent Transportation Systems, IEEE Transactions on*, 6(4):370–377, Dec 2005.

[9] A. A. Kannan, G. Mao, and B. Vucetic. Simulated annealing based localization in wireless sensor network. In *Local Computer Networks, 2005. The IEEE Conference on*, pages 2 pp. –514, Nov. 2005.

[10] A. LaMarca, Y. Chawathe, S. Consolvo, J. Hightower, I. Smith, J. Scott, T. Sohn, J. Howard, J. Hughes, F. Potter, J. Tabert, P. Powledge, G. Borriello, and B. Schilit. Place lab: Device positioning using radio beacons in the wild. In *Pervasive Computing*, volume 3468/2005 of *Lecture Notes in Computer Science*, pages 116–133. Springer, May 2005.

[11] L. Lazos, R. Poovendran, and S. Čapkun. ROPE: robust position estimation in wireless sensor networks. *Information Processing in Sensor Networks, 2005. IPSN 2005. Fourth International Symposium on*, pages 324–331, April 2005.

[12] M. L. McGuire and K. N. Plataniotis. Accuracy bounds for wireless localiztion methods. In *Localization Algorithms and Strategies for Wireless Sensor Networks*. IGI Global, 2009.

[13] D. Moore, J. Leonard, D. Rus, and S. Teller. Robust distributed network localization with noisy range measurements. In *SenSys '04: Proceedings of the 2nd international conference on Embedded networked sensor systems*, pages 50–61, 2004.

[14] R. W. Ouyang, A. K.-S. Wong, and K. T. Woo. Gps localization accuracy improvement by fusing terrestrial toa measurements. In *Communications (ICC), 2010 IEEE International Conference on*, pages 1–5, May. 2010.

[15] K. Pahlavan and A. H. Levesque. *Wireless Information Networks*. Wiley, second edition, 2005.

[16] S. Powell and J. Shim, editors. *Wireless Technology: Applications, Management, and Security*. Springer-Verlag, Sep 2009.

[17] N. B. Priyantha, H. Balakrishnan, E. Demaine, and S. Teller. Anchor-free distributed localization in sensor networks. *1st International Conference on Embedded Networked Sensor Systems (SenSys-03)*, 2003.

[18] I. Quadstone. The PARAMICS transportation modeling suite. URL=www.paramicsonline.com, Last Accessed Jan 19, 2011.

[19] C. Savarese, J. M. Rabaey, and J. Beutel. Locationing in distributed ad-hoc wireless sensor networks. *Acoustics, Speech, and Signal Processing, 2001. Proceedings. (ICASSP '01). 2001 IEEE International Conference on*, 4:2037–2040, 2001.

[20] S. K. Shekofteh, M. B. Khalkhali, M. H. Yaghmaee, and H. Deldari. Localization in wireless sensor networks using tabu search and simulated annealing. In *Computer and Automation Engineering (ICCAE), 2010 The 2nd International Conference on*, volume 2, pages 752–757, Feb. 2010.

A Markov Routing Algorithm for Mobile DTNs based on Spatio-Temporal Modeling of Human Movement Data

Arezu Moghadam
Columbia University
New York, NY
arezu@cs.columbia.edu

Tony Jebara
Columbia University
New York, NY
jebara@cs.columbia.edu

Henning Schulzrinne
Columbia University
New York, NY
hgs@cs.columbia.edu

ABSTRACT

Store-carry-forward communication, which is set as the heart of all routing protocols for mobile disruption-tolerant networks (DTNs), exploits nodes' mobility to bring messages closer to their destinations by exchanging messages across mobile nodes when they meet in close proximity. Understanding the subtle characteristics of human mobility leads to better service and application provisioning for mobile DTNs. We use GPS traces collected from multiple mobile users to empirically study different aspects of human mobility. Various Markov models (first, second and third-order) are estimated from users' mobility data. Based on empirical evidence, second-order Markov models are deemed sufficient to estimate mobile users' future locations accurately. These Markov models permit the design of a new routing algorithm for mobile DTNs capable of more efficiently routing data objects to their destination locations. The relay selection in this routing algorithm is based on mobile users' absorption times to the destination location. Simulations show that the proposed routing algorithm consumes less energy than legacy epidemic routing algorithms without excessive transmission delays.

Categories and Subject Descriptors

C.2 [COMMUNICATION NETWORKS]: DTNs

General Terms

Experimentation

Keywords

Algorithms, Mobile DTNs, Routing, Design, Performance

1. INTRODUCTION

An accurate understanding of human mobility is important for urban planning, traffic forecasting, and monitoring of biological and mobile virus spreading patterns [19]. Human mobility is fundamental to wireless networking research as it affects other network characteristics such as traffic and connectivity. Mobility impacts the types of applications and routing algorithms that could be provided for mobile networks in general. Knowledge of the mobility and its impact on the performance of the applications and routing algorithms depends on the types of the mobile applications and the environments in which those applications must operate. In cellular networking, for instance, the knowledge of speed and trajectories of mobile users facilitates the design of the cellular hand-off algorithms for QoS purposes. Modeling human mobility allows wireless cellular carriers to provide better service and application provisioning in 3G/4G networks [26, 29]. In mobile disruption-tolerant networks (DTNs), store-carry-forward communication which is integral to many routing protocols exploits nodes' mobility to bring messages closer to their destinations by exchanging messages with other mobile nodes when they are in adequate proximity. Therefore, long-term knowledge of future locations of mobile users is a defining factor in the routing of the messages to their final destinations in mobile DTNs. In MANETs, on the other hand, short-term knowledge of mobility is required to predict immediate connectivity for end-to-end data transmissions.

Synthetic and trace-based mobility models are two major classes of mobility models that are used to validate new routing algorithms and applications for cellular, ad-hoc and mobile DTNs. Unfortunately, many current synthetic mobility models do not accurately reflect real-world mobile user behavior [6]. In addition, trace-based mobility models are based on traces that either represent a specific class of mobile users or are extracted from cell tower traces [16] and do not provide sufficient spatial accuracy to capture fine mobility details from the users' movement. Since these cellular networking algorithms are only interested in movement of a node leaving or entering a cell, they do not require precise location data within the range of any particular cell. Conversely, for infrastructure-less ad-hoc networks and mobile DTNs, where no cell tower or access point is implicated, fine-grained mobility modeling plays a more crucial role as data exchange highly depends on device-to-device wireless proximity among mobile nodes. Therefore, the proposed algorithms for mobility predictions for cellular handoffs are not adequate for evaluating mobile ad-hoc and mobile DTN algorithms which require more fine-grained human mobility. In MANTEs or mobile DTNs the choice of mobility traces, synthetic models and analytic techniques are essential components of any evaluation platform for different algorithms or applications [4, 9].

Given the aforementioned importance of precision and

MSWiM'11, October 31–November 4, 2011, Miami, Florida, USA.

fine-granularity of real-world traces for mobility models, GPS traces [3] were collected by tracking a diverse group of mobile users in an urban area. Using these traces, we employ a Markov modeling approach to investigate different statistical aspects of human mobility and to extract mobile users' mobility patterns. Using this analysis and modeling, we develop a new routing algorithm for mobile DTNs. The routing decision in our Markov routing algorithm is based on the absorption times of mobile users to different destination locations. Our routing algorithm selects the mobile users with less absorption times to the destination as data carriers. These absorption times are calculated based on the properties of the Markov chains that we extract from our GPS traces. In addition, due to the probabilistic nature of the Markov models, we use our extracted Markov chains to generatively sample movement data to provide a meaningful simulation environment to evaluate our routing algorithm. Our evaluations show that our proposed Markov routing algorithm has a more efficient use of energy as it produces less delay per consumed unit of energy comparing to the legacy algorithms. The main objectives of this research are to explore how fine-grained mobility models can improve the performance of routing algorithms in mobile DTNs and facilitate service provisioning by wireless carriers in general. We leverage mobility data and the ability to predict mobile users' future locations to develop intelligent data distribution methods in mobile DTNs. For example, software updates, advertisements, finance news, and other commercial data could be distributed among mobile users more efficiently with such modeling and prediction tools.

This paper is organized as follows. Section 2 briefly reviews previous work and discusses the inadequacies of mobility models currently being used to design and evaluate routing protocols. Section 3 describes the characteristics of our database of movement traces [3] that distinguishes it from previously introduced mobility traces. Section 4 explains the details of the mobility models that can be extracted from the location traces. Section 5 focuses on second-order Markov models in particular which seem particularly well-suited to the location data sets being considered. Section 6 details the design, implement and evaluation of a Markov-based routing algorithm for mobile DTNs. Finally, Section 7 concludes with a brief discussions.

2. RELATED WORK

There are two classes of mobility models that are used in simulations of mobile networks. These models are either produced synthetically or they are based on real-world traces. Synthetic mobility models are usually based on random creation and movement of mobile users. Trace-based models are used either to set the parameters of the synthetically created random models or as actual traces to represent the population's behavior. In this section we describe the insufficiencies of each of these models.

2.1 Synthetic mobility models

Many routing algorithms and applications that are designed for mobile ad-hoc networks or DTNs are validated through simulations. All simulation-based studies use some class of mobility models in their experiments. Research verifies that the performance results of an ad-hoc network protocol drastically change as a result of changing the mobility model simulated [8].

All synthetic mobility models are either a variation of the random way point (RWP) mobility model or they have been implemented by modifying some parameters of the RWP model [8]. In this model each node moves from its current location to a new location by randomly choosing an arbitrary direction and speed from a given range. Some of these synthetic models try to add some realistic assumptions to the RWP model. For example, SWIM [21] is based on RWP by adding the restriction that mobile users travel most often to the places close to their home. Community-based mobility model [23, 18] is based on the fact that mobile users are grouped and assigned to different locations based on their social relationships with others. This model is oblivious to the fact that social interactions are not the only defining factor of human movements. Working day movement model [13] and city section mobility model [11] emulate the RWP model based on the street map of a city. In the Gauss-Markov mobility model [20], a node's next location is generated by its past locations and velocity. In the probabilistic version of the RWP model [10], the last step made by the random walk influences the next one based on some probability distribution. The main problem with these synthetic mobility models is their simplified assumptions to make the model more tractable and easy to implement. This leads to significant deviation of the simulation results from the real-world implementation of the protocols and routing algorithms for ad-hoc networks and mobile DTNs. For instance, The most common problem with simulation studies using random way point model is a poor choice of velocity distribution [32], e.g., uniform distribution U (0,Vmax). Such velocity distributions (which seem to be common with NS-2 simulations) lead to a situation where at the stationary state each node stops moving.

2.2 From real-world traces to mobility models

Unlike the synthetic models, trace-driven simulations are realistic to evaluate the performance of different mobile DTN algorithms and applications. Although, trace-driven simulations do not require a mobility model, model-based simulation allows researchers to explore different scenarios in their simulations by providing a larger parameter space. Different studies have been conducted to extract mobility models from real-world movement traces. Minkyong et al. [22] have derived the distribution of speed and pause times of mobile users using the WiFi access point traces collected at Dartmouth College [17]. Based on their trace-driven calculations, speed and pause time each follows a log-normal distribution. Rhee et al. [24] have studied human walks based on GPS traces involving 44 volunteers. They show that many statistical features of human walks, such as distribution of consecutive displacements and pause time distributions, follow truncated power-laws. They show these traits are similar to those of Levy walks [27]. In a another study, Gonzalez et al. [16] used the mobility traces of 100,000 individuals over a six-month period that were logged in cell towers to analyze human mobility distributions. By investigating these traces, they also suggest that human motion follows a truncated Levy flight which is a random walk for which step size follows a power-law distribution.

Although the statistical results of these mobility models are significant, they are not adequate. One problem with these models is their precision as, they have been extracted either from WiFi [22] or cell tower traces whose granularity is as large as the AP or cell tower's range [16, 28]. Another problem with some of these models is that they are inferred from traces that characterize either a specific class [22] or a small number of mobile users [24]. Therefore, generalizing these models to a more general subset of the human population yet has to be verified. Brockmann et al. [7] have generalized Levy statistics to humans, documenting that the distribution of distances between consecutive sightings of nearly half-a-million bank notes is fat tailed power-law. But because each consecutive sighting of bank notes reflects the composite motion of two or more individuals, it is not clear whether the observed distribution reflects the motion of individual users. Giradin et al. [15] have used a method to uncover the presence and movements of tourists from the georeferenced photos they generate and post on flickr.

The existence of patterns in human movement could facilitate service provisioning in mobile networks for a wide range of applications from prevention of human and electronic viruses to resource management in mobile communications. Therefore the recent research efforts in human mobility are focused on the possibility of forecasting the mobility of individuals [31, 12]. By measuring the entropy of each individual's trajectory Song et al. [28] found a 93% potential predictability in user mobility across the whole user base. Gonzalez et al. [16] also conclude that human movement follows certain patterns as users tend to return to the locations they visited before. Our goal in this paper is to characterize the inherent patterns in mobile users' movement and to use this knowledge to design a more intelligent routing algorithm for mobile DTNs.

SOLAR [14] routing algorithm for mobile DTNs bases its routing decision on delivery probability of each mobile node to the destination. The delivery probability for each mobile user is calculated by extracting "mobility profile" of that user which is the frequency of previous visits to a set of predetermined hot-spots or hubs. SOLAR tries to route data to these hubs by selecting relays which have higher probability of visiting those hubs. In our algorithm, however, pattern has a broader definition and is not limited to visiting a set of predetermined hubs with a predefined, constant size. Furthermore, SOLAR's performance has been evaluated using RWP mobility model, while we design and evaluate our algorithm using movement traces which are extracted from real-world mobile users.

3. CHARACTERISTICS OF OUR MOBILITY TRACES

We are using the movement traces collected by Sense Networks [3] in our mobility model studies. Sense Networks has created a research platform that utilizes GPS to gather traces of a diverse class of mobile users. In order to encourage users to contribute their movement data, Sense Networks has implemented a mobile social networking application called Citysense [1]. By installing and using this application, mobile users agree to contribute their own geolocation data to the Sense Networks' database. To address privacy concerns, Sense Networks has developed a new approach to data ownership by mobile users. Users are the primary owners of any historical data that they create within the system and they can choose to delete their data anytime they wish. Citysense's GPS traces are collected from pings that are sent to mobile devices every half an hour with the precision of 20 feet (GPS pings precision) which

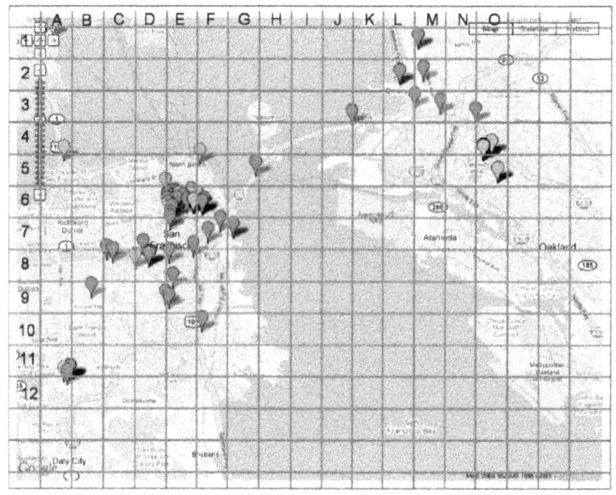

Figure 1:
Map grid of San Fransisco with number of GPS pings at each grid.

which guarantees the temporal and spacial granularity of these traces compared to the traces collected from APs or cell towers. Device-to-device contact traces, similar to the reality-mining database [2], can also be generated from the GPS traces by calculating the distance of the mobile users from each other, time of the contact, and considering the range of wireless interfaces.

3.1 Extracting Different Features of Human Movement

The first step in analyzing the movement traces is representing mobile users' GPS traces in a meaningful format. Our representation of the movement data is based on focusing on certain geographical areas, here northern California. As shown in Figure 1, we have divided this area into smaller activity areas or grids of size 300ft × 300ft which is as twice as the range of the WiFi wireless interface to guarantee the possibility of wireless communication between mobile users in the same grid. Each mobile user has a certain mobility characteristic. For example, some mobile users such as taxi drivers are random roamers and tend to visit many different grids in short amount of time. Some other mobile users have a more limited activity area by spending most of their times in very few locations such as home, office, and neighborhood grocery shops. In contrast with the random trajectories predicted by the Levy flight and random walk [6], human trajectories show a high degree of temporal and spatial regularity, meaning, humans tend to revisit the locations they visited in the past [16]. Many researchers have tried to analyze the existence or nonexistence of such patterns in human movement. We define the existence of regularity or pattern as the existence of few locations that are visited more frequently by mobile users. These patterns in human movement can be identified by calculating the prominent locations where mobile users spend most of their times (e.g. 85% of their time).

As we explained earlier we divide the users' movement map to different regions or grids of activity. Users spend some percentage of their times in each region based on their own mobility characteristics. This percentage can be also interpreted as the probability of the user being present at that specific location. We first identify the predominant ac-

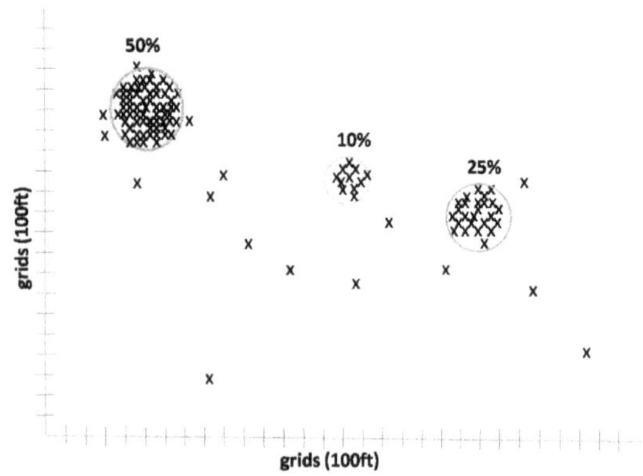

Figure 2:
A conceptual illustration of fitting Gaussian distributions on a user's movement data. In this example user has 3 predominant locations where she spends 50%, 10%, and 25% of her time.

tivity locations of each user by fitting Gaussian distributions on individual users' daily movement data as shown in Figure 2. Based on our traces, for the majority of the mobile users, 15% of the GPS pings on average are randomly scattered across the users' movement map and therefore they do not contribute in the pattern extraction. This Gaussian fitting verifies that the majority of human subjects in our sample population, on average, spend more than 85% of their time in few locations. The number of these predominant locations are few, less than 6, during workdays and maximum 15 locations during weekends for the majority of the users. We also identified some random roamers with more uniformly scattered GPS pings across a large number of regions. The number of these users is less than 10% of our sample population and we exclude them from our pattern extraction analysis.

4. MODELING USERS' MOVEMENTS BY N-GRAMS

The design goal for our movement model is to determine the possibility of predicting each user's most probable future locations after learning their past behavior. In our model, the level of granularity of the time window to learn usersŠ behavior is days of a week. In other words, our model produces users' movement model for each day of a week. The size of this time window highly depends on the type of application. For example, in traffic management applications a meaningful time window to learn users' behavior should be narrowed down to hours of a day. Even for more personalized mobile applications a user might have different movement patterns across different hours of a day for example; a user might spend the entire time window of 9am to 5pm in her office while having a totally mobile lifestyle after work hours. We make this assumption about granularity in time without any loss of generality in our model. The statistical methods that we implement to produce our model could be supplied with different traces of data with different levels of granularity to produce the corresponding output model correctly.

In on our model human movement is represented as a sequence of consecutively visited grids. Each element of this sequence is the grid in which the user was located at the time of the GPS ping. These GPS pings are uniform in time. Using this representation, the problem of predicting a user's future move converts to the problem of predicting the next item in the sequence. N-grams that are used in various areas of natural language processing (NLP) and genetic sequence analyzing are a type of probabilistic approach for predicting the next item in sequences of words or DNA. We have adopted n-grams in the sequence analysis of mobile users' movement.

An n-gram is a sequence of n items. The items in sequence can be phonemes, syllables, letters, words or grids according to the application. An n-gram of size 1 is referred to as a *unigram*; size 2 is a *bigram*; size 3 is a *trigram*; and size 4 or more is simply called an *n-gram*. The models built from n-grams are (n)-order Markov models.

Based on this definition, in unigram modeling of human movement, future locations of the users are estimated independently from the knowledge of their current or past locations. Therefore, the probability of finding a user in a specific location is simply calculated as the percentage of the time that user spent in that location in the past. The unigram model is too naive as it does not capture any trajectory in the human mobility. Trajectory in the mobility means a user's next location depends at least on the user's current location. This observation leads us to use bigram model which is in fact a subset of second-order Markov models or Markov chains.

Although using history seems intuitively to be helpful in learning human movement, it is not immediately clear how far in the past we should examine the locations in order to produce more accurate future predictions. Therefore, we also study trigram models of users' movement which as well as the current location of the user considers the previous location.

4.1 Bigram Statistics

Human movement traces can be converted to a large sequence of consecutively visited grids. In our model each element in this sequence corresponds to a grid number where user was located at the time of the GPS ping. Therefore, transition from one element in the sequence to the next occurs every 30 minute which is the time elapsed between two pings.

We consider a larger subset of each user's movement trace, here 80%, as our training data, and the remaining 20% to test the model. We implement a program that learns the user's daily bigram table. The program scans the train sequence element by element, like a sliding window, from the beginning and stores two grid sequences as it moves along until the last grid in the train dataset is reached. On the output, it creates the bigram table which stores the frequency of each double grids observed in the trace. Figure 3 shows a section of a bigram table as an example. Training the bigram model is the same as creating the bigram tables and testing the model is equivalent to calculating the likelihood of the observed sequence in the test data.

4.1.1 Calculating Likelihood of the Movement Trace

Calculating the likelihood of the test sequence is equivalent to calculating the probability of this sequence as:

Tuples of Grids	5039907665	5038663466	5038414624	5038414623	5060063904	5053345115	...
5039907665	1370	10	230	10	0	30	...
5038663466	30	130	110	0	0	0	...
5038414624	220	110	3420	120	0	60	...
5038414623	10	0	50	0	14	0	...
5060063904	0	0	12	110	0	0	...
5053345115	0	50	0	13	176	343	...
. .	. .	. .	. .	. .	. .	. .	... ...

Figure 3:
A snippet of a bigram table that is calculated from a user's movement data during weekend. Bigram tables are calculated from the 80% portion of the movement data that is considered as train data.

Triple of Grids		5039907665	5038414624	5050607875	5053345115	5038414623	...
5039907665	5039907665	1180	110	30	30	10	...
5039907665	5038414624	0	230	0	0	0	...
5038414624	5038414624	220	2820	10	60	110	...
5038414624	5039907665	110	100	0	0	0	...
5039907665	5050607875	0	20	10	0	0	...
5050607875	5038414624	0	30	133	0	44	...
. .	. .	. .	. .	. .	. .	. .	... ...

Figure 4:
A snippet of a trigram table that is calculated from a user's movement data during weekend. Trigram tables are calculated from the 80% portion of this movement data that is considered as train data.

$$P(G_1, G_2, G_3, ..., G_i, G_{i+1}, ..., G_N) = P(G_1) \times P(G_2 \mid G_1) \times ... \times P(G_N \mid G_{N-1}) \quad (1)$$

The elements of this multiplication are calculated by referring to the corresponding rows of the bigram table.

$P(G_i \mid G_{i+1})$ is the element of the bigram matrix located at row i and column $i+1$, which is the frequency of observing tuple (G_i, G_{i+1}) in the training data, divided by the total number of items in the train data. $P(G_i)$ is the unigram probability of finding the user in grid i. Hence, unigram statistics can also be calculated from the bigram matrix. Consequently, unigram likelihood of observing sequence $(G_1, G_2, G_3, ..., G_i, G_{i+1}, ..., G_N)$ is calculated as:

$$\mathcal{L}(G_1, G_2, G_3, ..., G_i, G_{i+1}, ..., G_N) = P(G_1) \times P(G_2) \times ... \times P(G_i) \times ... \times P(G_N) \quad (2)$$

Therefore, the log-likelihood of the sequence can be written as: $\log \mathcal{L}(G_1, G_2, G_3, ..., G_i, G_{i+1}, ..., G_N) = \sum_{i=1}^{n} \log P(G_i)$

4.2 Trigram Statistics

In our model we refer to trigram as a triple of grids. Trigram modeling is used in statistical natural language processing (NLP) to calculate the probability of observing a specific word after two other preceding words. We use this method in our model as the probability of a user visiting a specific grid after visiting two preceding grids and we write this probability as $P(G_i \mid G_{i-1}, G_{i-2})$.

Trigram tables are created in a similar mechanism as bigram tables from the train dataset. But this time the algorithm for scanning the train dataset counts for triplets of grids and stores their frequency in the trigram table. Here, each tuple of grids indexes a row in the trigram table and the following grid after the tuple indexes the corresponding column in that row. A partial view of one of the trigram tables of a mobile user is represented in figure 4.

After extracting the trigram tables, likelihood of the test movement trace is calculated in a similar fashion as the bigram's multiplication. Each $P(G_i \mid G_{i-1}, G_{i-2})$ item of this multiplication is calculated by referring to the trigram table in a similar method to the bigram statistics.

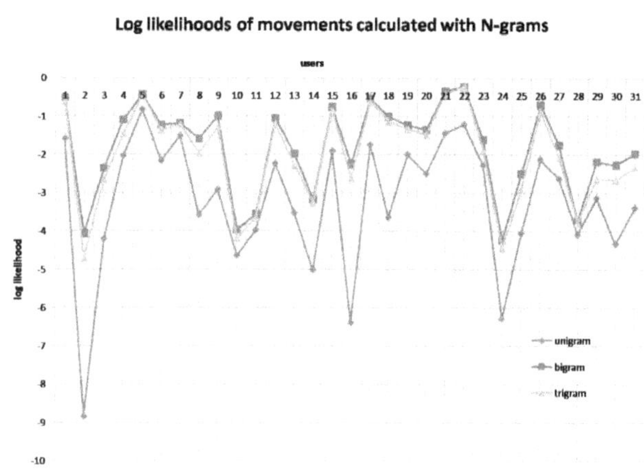

Figure 5:
Log likelihoods calculated for each user's test data based on their workday n-gram tables.

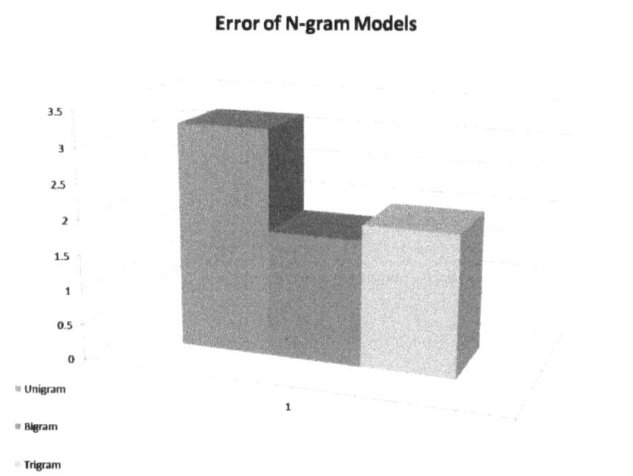

Figure 6:
Comparison between different n-gram models based on their average errors in calculating log likelihoods of users' test data.

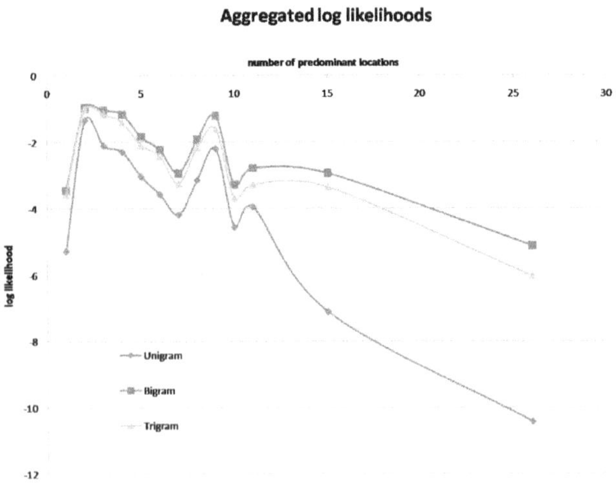

Figure 7:
Log likelihoods calculated for aggregated users' data from Figure 5. In this figure users are classified together based on their predominant locations.

4.3 Evaluation of the N-gram Models

In this section, we examine the accuracy of the n-gram models to explain patterns in mobile users' movement. We intuitively expect bigram models to be more accurate than unigrams as they take into account the current location of the user to estimate the future location. But whether trigrams are more accurate than bigrams is not immediately obvious. Figure 5 plots the log-likelihoods of test movement traces of a subset of our population consisting of 31 users after learning their unigram, bigram and trigram models. We first train the n-gram models based on each user's train data then we calculate the log-likelihood of the test data for each of the n-gram models as explained in the previous sections. Since the log-likelihoods are negative we define the amount of the error produced by each model to be the negative of its log-likelihood. As we observe in Figure 5, the unigram model which does not consider any history to predict users' future locations produces more error which is expected. Moreover, as we can see bigram and trigram models both produce similar error values and they are also on average 40% more accurate than the unigram model. The average produced error by each n-gram model is shown in Figure 6. Figure 7 displays the aggregated statistics of Figure 5 pivoted on the number or prominent visited locations by mobile users. Since bigram and trigram produce similar error values and also because trigram adds one more degree of complexity to the model without any significant gain over bigram, we use bigram model in the rest of our analysis to model mobile users' movement. The bigram model is conceptually the same as the second-order Markov model or Markov chains as it considers the current state to predict the future states. Therefore, we model human mobility by Markov chains and we investigate different characteristics of this model in the next section.

5. SPECIFYING A MARKOV CHAIN FOR HUMAN MOVEMENT

Our previous discussion in defining and evaluating the n-gram tables led us to study the patterns in human movement by Markov chains. A Markov chain is a discrete random process with the Markov property. A discrete random process means a system which is in a certain state at each "step", with the state changing randomly between steps. The steps are often thought of as time, i.e., time between consecutive GPS pings in our system. Users might change the state by moving from one grid to another one between two GPS pings by walking, driving or any other means.

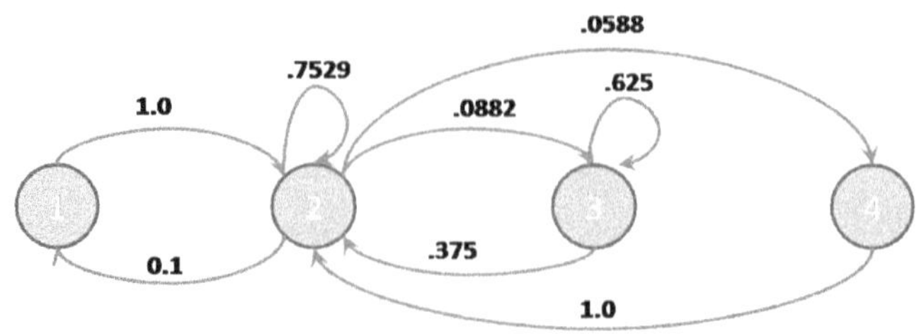

Figure 8:
An example of a user's Markov chain which is extracted from a user's workday movement trace. This Markov chain is ergodic as higher powers of its transition matrix converges to a stationary distribution.

We describe a Markov chain for each individual user's movement as follows: we define a set of locations or states, $S = \{s_1, s_2, ..., s_r\}$. The user starts in one of these states or locations and transitions successively from one state to another with some probability. Each transition corresponds to each GPS ping and is called a step. If the chain is currently in state s_i, then it transitions to state s_j at the next step with a probability denoted by p_{ij}. The probability p_{ij} indicates the frequency of user's transitions from state s_i to the state s_j. Based on the definition of Markov chains, this probability does not depend on the states the chain was before the current state i. The Markov chain is defined with the resulting transition matrix P, with p_{ij} as its elements. Transition matrix P describes each user's daily movement profile. This movement profile can be used to classify users in different movement categories. Figure 8 shows an example Markov Chain calculated from one user's movement database on Monday. In the following sections we first enumerate the important properties of the Markov chains which are extracted from users' movement traces. Then, we design a routing algorithm that uses these properties to select the most suitable forwarders to route data to a destination location.

5.1 Characteristics of the Markov Chain

Our goal is to investigate the possibility of existing patterns in human movement. One necessary condition for the existence of any pattern is the re-occurrence of an incident. Here, pattern can be interpreted as whether humans revisit the locations they visited before. In order to deduce any such properties from mobile users' movements we first enumerate the characteristics of the Markov chains we derive from mobile users' movement traces.

5.1.1 Ergodicity of the Markov Chain

A stationary process (or strictly stationary process or strongly stationary process) is a stochastic process whose joint probability distribution does not change when shifted in time or space. As a result, statistical properties such as cumulative distribution function (CDF), the mean and variance, if they exist, also do not change over time or position. Claiming that human movement is strictly stationary means with any shift in time the process (CDF, means and variance) should stay consistent. In order to prove the human movement is statistically stationary we should have data that extends across a broad spectrum in time which is practically impossible at the time of this research. The data that we have based our analysis upon is rather *seasonal* (a snapshot of human movement in time). Therefore, here, we emphasize the necessary conditions for this process to be stationary and we explore if our seasonal movement data sustains those conditions.

If a Markov chain is stationary or time-homogeneous, so that the process is described by a single, time-independent matrix p_{ij}, then the vector π is called a stationary distribution (or invariant measure) if its entries π_i sum to 1 and it satisfies:

$$\pi_j = \sum_{i \in S} \pi_i p_{ij} \tag{3}$$

Furthermore, the chain converges to the stationary distribution regardless of where it begins. Such π is called the equilibrium distribution of the chain. Such π is called the equilibrium distribution of the chain. This stationary distribution can be calculated as:

$\pi_j = \lim_{n\to\infty} p_{ij}^{(n)}$

The class of the Markov chains that have a unique stationary distribution is called *ergodic*. Since ergodicity of the Markov chain is a necessary and sufficient condition for the chain to have stationary distribution and for existence of a pattern in human movement, it is important that this property to be investigated using the real-world mobility traces. Here, we investigate if the Markov chains that are extracted from mobile users' traces are ergodic. In other words, we check to see if rows of the transition matrices converge to a stationary distribution, π.

5.1.2 Ergodicity of Human Movement

In order to investigate the ergodicity of human movement, we calculated the daily transition matrix for each individual user and evaluated its convergence over time. In our analysis we detected two types of Markov chains from users' traces. The first type is extracted for users who have large movement traces in the database that extends over a long period of time. This type of Markov Chain turns out to be ergodic and an example is represented in Figure 8. The second type of Markov chain, however, does not converge to an equilibrium state. After further investigations we realized these types of chains contain some absorbing state. An absorbing state is a state from which there is a zero probability of transition to another state. After further investigation, we realized the main reason for existence of these absorbing states is the abrupt termination of the GPS pings at a certain location. The traces show these Markov chains belong to the users who have a short movement history in the database or the users have decided to erase certain parts of their traces from the database.

6. A MARKOV BASED ROUTING ALGORITHM FOR MOBILE DTNS

Our routing algorithm for mobile DTNs is based on our second-order Markov analysis of human mobility to route a data object to a specific location. Most of the routing algorithms for mobile DTNs are designed to route data from a source to a specific destination where this destination is usually another mobile user [5]. There are scenarios, however, in which data needs to be routed to a group of mobile users located in a specific location rather than a certain individual. In these scenarios mobile users need to be informed of an event based on the region they are located for example, when an emergency notification needs to be transmitted to mobile users in a certain geographical location or in a social networking scenario, when friends who are hanging out in a certain part of the city need to be notified by other friends about a particular social event in that neighborhood. In such scenarios, the intended recipients are identified based on their locations rather than their individual identities.

6.1 Routing Objective

In mobile DTN scenarios mobile users rely on other mobile users to transmit their data to the intended locations or recipients. Carrying other users' data is costly for data carriers in terms of communication resources. The fact that these data carriers should spend their communication resources for the benefits of others is a burden on practical deployment of mobile DTN routing protocols. The most valuable communication resource is energy. Therefore, a reasonable evaluation metric for the routing algorithms is how much energy they spend to achieve their routing goals.

Epidemic routing is proven to provide the highest delivery rate and the lowest delay to transmit data to its destination [30] in mobile DTN scenarios. Epidemic routing, however, has the least efficient usage of the communication resources, especially energy, as it floods all mobile users with all the data. On the other side of the spectrum, however, we could consider an algorithm in which sender does not transmit the data to any other mobile user. Therefore, data could reach the destination only if sender arrives at that destination. This algorithm has the least energy consumption but naturally the longest expected transmission delay. Considering these two extreme cases we can define scenarios in between where some percentage of wireless contacts is excluded from any data exchange. The evaluation objective of our Markov routing algorithm is to investigate where our algorithm is located in this performance spectrum. In other words, we investigate the trade-off between energy and delay by our algorithm or the amount of delay per unit of consumed energy.

6.2 Algorithm Design

The Markov chains we extract from users' daily movement traces explain the daily movement profile of each user and the daily rates by which they transition from one location to another. Using these rates, we can calculate the expected number of steps or transitions until each Markov chain arrives at a specific state. Since the time between each two transitions is fixed we can map the number of transitions to the expected amount of time that is needed until a user reaches a certain destination after starting at some other location. This expected time which is called *absorption time* is a determining factor in our routing algorithm. By comparing the absorption times of the mobile users to a certain destination, the source of a data object is capable of selecting the best possible carrier for that data. Because of the importance of the absorption times in our routing algorithm, next we briefly explain how to calculate these values for a general Markov chain.

6.2.1 Calculating the Absorption Time

If we represent the Markov chain with n number of states as $(X_n)_{n\geq 0}$ and transition matrix P then the *absorption time* at a subset A of these states is the random variable $H^A : \Omega \rightarrow \{0, 1, 2, ...\}$ given by:

$H^A(\omega) = inf\{n \geq 0 : X_n(\omega) \in A\}$

The probability of the Markov chain $(X_n)_{n\geq 0}$ ever ends at A after starting from state i is: $h_i^A = P_i(H^A < \infty)$.

Therefore, the mean time taken for $(X_n)_{n\geq 0}$ to reach A is given by

$K_i^A = E_i(H^A) = \sum_{n<\infty} nP(H^A = n) + \infty P(H^A = \infty)$.

We can write this equivalently as:

$$H_i^A = P_i(\text{hit } A),\ K_i^A = E_i(\text{time to hit } A). \qquad (4)$$

These quantities can be calculated explicitly by means of linear equations associated with the transition matrix P. These systems of linear equations solves for the vector of mean absorption times k^A and the vector of mean absorption probabilities h^A. The system of linear equations to solve for the absorption times is a set of recursive equations that allows us to solve for the mean times that the chain ends up at certain states.

$$\begin{cases} k_i^A = 0, & \text{for } i \in A \\ k_i^A = 1 + \sum_{j \notin A} p_{ij} K_j^A, & \text{for } i \notin A \end{cases} \qquad (5)$$

6.2.2 Algorithm Mechanics

As we explained previously, we present users' daily movement models as bigram tables or transition matrices. Our algorithm's decision making process to select data carriers is based on absorption times of the mobile users to the destination. The algorithm calculates each user's mean absorption times from each location to another by solving the system of linear equations as discussed in Section 6.2.1. In a similar format as bigram tables, these absorption times also could be presented as matrices of size $n \times n$, where n is the number of locations or states of the Markov chain. Each element T_{ij} of these matrices at row i and column j shows the mean absorption time from location i to location j for a specific daily activity of a certain user. All users have the knowledge of their own absorption time tables.

When some users meet at some location i, there are two subsets of users. One subset of users have some data that needs to be transmitted to some destination location j. We call this subset source and we represent it with s. The second subset are users who could be potential forwarders of the data because they could reach location j with some probability. We call this second subset of users as relays and we symbolize the subset with r. Users in the subset of relays notify the users in subset of source of their mean absorption times to the destination location j or T_{ij}^r. Source users compare these values with their own absorption times to destination j or T_{ij}^s. After these comparisons source decides to transmit the data to the relay if: $T_{ij}^r \leq w_j T_{ij}^s$, where w_j is a positive value that weighs the importance of the destination location j. The default value for w_j is one and the larger this value, the more critical the destination location, and the less selectively the algorithm acts to choose the potential relay. By critical destination location we mean data needs to get there as fast as possible and therefore more transmissions should occur. In other words, the larger w_j implies a wider filter to choose the relay because choosing more relays means increasing the chance of data delivery to the critical destination.

6.3 Evaluation of the Algorithm by Monte Carlo Simulations

We have evaluated the performance of our Markov based routing algorithm by Monte Carlo simulations. In order to have a realistic evaluation of this algorithm, we have implemented a mobility generator module that generates mobility of the mobile users based on Sense Network's traces. The input to this mobility module is the transition matrices of the Markov chains that are learned from Sense Network's traces as explained in Section 5. This mobility generator module uses Monte Carlo methods to generate mobility of the users in the simulation scenario by sampling from these supplied transition matrices.

6.3.1 Simulation Setup

Our simulation scenarios consist of 50 mobile users who move between 100 locations. For each user we choose 10 mostly visited locations based on their GPS traces. As we explained before, we ignore the locations in which users have very few pings as they do not have any significance in our final results. We regenerate each user's mobility based on the transition matrices that we have extracted from Sense Network's traces. These transitions matrices define the inputs to the mobility generator engine.

Based on the Monte Carlo guidelines [25], we generate a random movement sample as the input, feed this input to a deterministic routing algorithm simulator engine, and then evaluate the output. We select the initial location of each user and generate the following transitions of the user from one location to another by sampling from the user's corresponding transition matrix. Therefore, each iteration in the simulation can be interpreted as generating the current state of the Markov chain or as a user's current location. After each transition, the current states of all the users are supplied to the routing algorithm simulator as an input. Based on this input if some users meet at some locations, the simulator engine emulates the behavior of the routing algorithm as explained in Section 6.2.2 to determine whether any data exchange should happen in those locations. The simulation scenario continues until the data object reaches its final destination. After the simulation ends the amount of the transmission delay and the number of total transmissions are recorded. The transmission delay is defined as the number of transitions in the Markov chain until it ends at some absorbing state. At each simulation round one of the users is uniformly selected as the source of the data object which needs to be transmitted to a randomly selected destination. We assume the amount of energy to transmit a single data object per wireless contact is constant across all mobile devices. We repeat these simulations 100 times and we aggregate the results to decrease the impact of the randomness in the final averaged results.

6.3.2 Evaluation of Results

Since epidemic routing produces the least amount of transmission delay and highest amount of Energy consumption, our main comparison base is epidemic. We define a perfor-

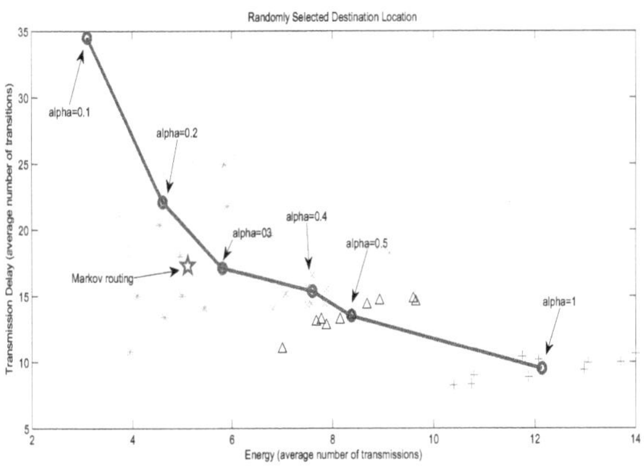

Figure 9:
Transmission delay vs. consumed energy for a randomly selected destination.

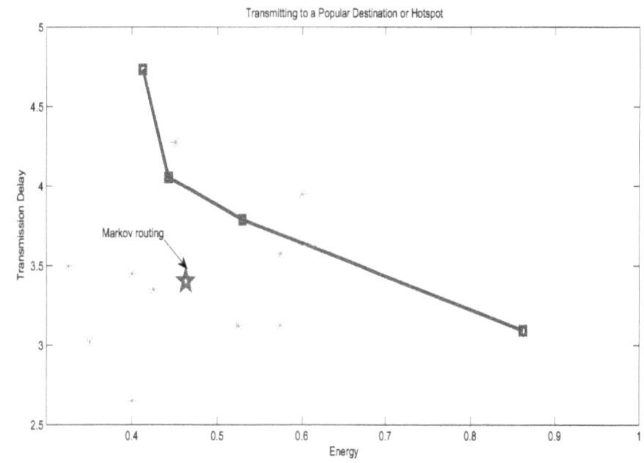

Figure 10:
Transmission delay vs. consumed energy for a popular destination or hotspot.

mance spectrum that consists of epidemic and other routing algorithms which are similar to epidemic but deliberately exclude a percentage of wireless contacts from data transmission. Therefore, we define a family of epidemic algorithms and we call them α-epidemic, where *alpha* is the percentage of the wireless contacts participating in the data exchange. When α is 1 we have a fully epidemic algorithm. The less the amount of α the more energy-aware is the algorithm but it produces larger amounts of transmission delay. Our simulation goal is to investigate where our Markov based algorithm falls in this performance spectrum compared to these algorithms with different values of *alpha*. In other words, if the class of α-epidemic algorithms has a total performance curve in the form of delay $= f(\text{energy})$, the Markov based algorithm is more energy efficient if its performance curve lies beneath the class of α-epidemic's, meaning delay $< f(\text{energy})$. In other words, an algorithm has a better performance if it produces less delay per unit of consumed energy.

The simulation results are represented in Figure 9. We change α from 1, meaning fully epidemic, to zero, meaning the sender does not transmit data over any wireless contact, and the data object reaches the destination when the sender itself arrives at the destination. The abscissa measures the consumed energy in terms of the average number of transmissions until the data object reaches its destination while the y-axis measures the average delay of these transmissions in terms of elapsed time units. The solid curve is calculated from the average values of delay and energy produced by each method of the α-epidemic family. In other words, this curve represents the delay $= f(\text{energy})$ relationship for the α-epidemic family. The point on the figure that is highlighted with an arrow is the average value of the delay and consumed energy that resulted from the Markov based routing algorithm. As we can observe in this figure, selecting the relays based on their absorption times to random destinations does not gain much over the performance much because both resulted delay and energy values are very close to the α-epidemic where $\alpha = 0.3$ or when the α-epidemic algorithm which randomly selects 30% of the encountered mobile users as potential relays. After further investigations in the movement traces, we realized that all users have large mean absorption times to the majority of the locations. The reason is the majority of the locations are not visited often by many mobile users. For example some user's home location in the suburbs of San Francisco is barely visited by other users. Having large absorption times to many locations means our algorithm cannot really take advantage of the absorption times in its routing decisions.

In order to observe the impact of the absorption times in the routing decisions, we need to select our destinations from the popular locations. Popular locations or *hot-spots* are locations or neighborhoods that are visited more often by a larger number of users for example, downtown of San Francisco. Since these locations are popular naturally more users have less mean absorption times to these locations. Figure 10 illustrates the results of these series of simulations for the class of α-epidemic v.s. Markov based algorithms. As we observe in this figure, all algorithms show improved results in terms of consuming less energy and achieving less delay. The reason for this change of trend is because the destination is a hot-spot there is a higher chance that it is visited more frequently by most of the users. But more importantly is the significant improvement of the Markov based routing compared to the family of α-epidemic. The solid line in the figure again shows the performance of the family of α-epidemic. The dots on the figure are the simulation results of our Markov based routing algorithm with the average point emphasized with an arrow. As we observe in this figure, for almost all simulation rounds of the Markov based algorithm we have delay $< f(\text{energy})$ compared to α-epidemics. Moreover, the average point of the Markov based algorithm falls considerably below the α-epidemic's curve. In our simulations, the delay of the Markov based routing is as low as the delay of an α-epidemic which transmits to 70% of the contacts. But for the same amount of delay Markov based algorithm consumes 45% less energy. Also, the energy consumption of the Markov routing is similar to the energy consumption of an α-epidemic algorithm which drops 75% of the wireless contacts. Therefore, by sacrificing less than 10% in the speed of the transmission our Markov based algorithm gains 45% more energy when transmitting to popular destinations or hot-spots.

7. CONCLUSION

We studied different aspects of human mobility and investigated how the knowledge of this mobility could affect the performance of the routing algorithms in mobile DTNs. We used Sense Networks' GPS traces that are collected from a large population of mobile users to conduct our analysis. We used n-grams to extract meaningful patterns for human movement. We realized that bigram models that are second-order Markov models are suitable candidates to explain users' movements and to predict their future locations. We used these extracted Markov chains to introduce and evaluate a new routing algorithm based on users' absorption times to different locations. Our simulations are based on traces that are produced by regenerative sampling from the Markov chains we extract from Sense Network's real-world traces. The techniques and methods that we have developed for this evaluation can be reused by feeding different mobility traces for future research. Our simulation results which is based on real-world traces shows our Markov based routing algorithm consumes considerably less energy to carry data objects to the popular destinations than the family of α-epidemic routing. Although consuming moderate amounts of energy, our routing algorithm achieves superior performance over α-epidemic family in terms of transmission delay. Our algorithm on average produces a transmission delay which is as low as an α-epidemic routing's that transmits data to 70% of the wireless contacts. However, our algorithm's energy consumption is as less as the α-epidemic which excludes 75% of the wireless contacts from data transmission.

8. REFERENCES

[1] Citysense. http://www.citysense.com/.

[2] MIT media lab reality mining. http://reality.media.mit.edu.

[3] Sense networks. http://www.sensenetworks.com/.

[4] M. Abdulla and R. Simon. The impact of the mobility model on delay tolerant networking performance analysis. *Simulation Symposium, 2007. ANSS '07. 40th Annual*, pages 177 – 184, March 2007.

[5] Arezu Moghadam and Henning Schulzrinne. Interest-aware content distribution protocol for mobile disruption-tolerant networks. In *10th IEEE International Symposium on a World of Wireless, Mobile and Multimedia Networks, Kos, Greece*, June 2009.

[6] J.-Y. L. Boudec and M. Vojnovic. The random trip model: stability, stationary regime, and perfect simulation. *IEEE/ACM Trans. Netw.*, 14(6):1153–1166, 2006.

[7] D. Brockmann and F. Theis. Money circulation, trackable items, and the emergence of universal human mobility patterns. *Pervasive Computing, IEEE*, 7(4):28–35, Oct.-Dec. 2008.

[8] T. Camp, J. Boleng, and V. Davies. A survey of mobility models for ad hoc network research. *Wireless Communications and Mobile Computing (WCMC): Special Issue on Mobile Adhoc Networking: Research, Trends and Applications*, 2:483–502, 2002.

[9] A. Chaintreau, P. Hui, C. Diot, R. Gass, and J. Scott. Impact of human mobility on opportunistic forwarding algorithms. *IEEE Transactions on Mobile Computing*, 6(6):606 – 620, 2007.

[10] C.-C. Chiang. *Wireless network multicasting*. PhD thesis, 1998.

[11] V. Davies. Evaluating mobility models within an ad hoc network. Master's thesis, 2000.

[12] N. Eagle and A. Pentland. Eigenbehaviors: Identifying structure in routine. In *Proc. Roy. Soc. A*, 2006.

[13] F. Ekman, A. Keränen, J. Karvo, and J. Ott. Working day movement model. In *MobilityModels '08: Proceeding of the 1st ACM SIGMOBILE workshop on Mobility models*, 2008.

[14] J. Ghosh, S. J. Philip, and C. Qiao. Sociological orbit aware location approximation and routing (solar) in MANET. *Ad Hoc Netw.*, 5:189–209, March 2007.

[15] F. Girardin, F. Calabrese, F. D. Fiore, C. Ratti, and J. Blat. Digital footprinting: Uncovering tourists with user-generated content. *Pervasive Computing, IEEE*, 7(4):36–43, Oct.-Dec. 2008.

[16] M. C. Gonzalez, C. A. Hidalgo, and A.-L. Barabasi. Understanding individual human mobility patterns. *Nature*, 453(7196):779–782, June 2008.

[17] T. Henderson, D. Kotz, and I. Abyzov. The changing usage of a mature campus-wide wireless network. In *MOBICOM*, pages 187–201, 2004.

[18] L. Hu and L. Dittmann. Heterogeneous community-based mobility model for human opportunistic network. *Wireless and Mobile Computing, Networking and Communication, IEEE International Conference on*, 0:465–470, 2009.

[19] J. Kleinberg. The wireless epidemic. *Nature, 2007.*, 449:287 Ú 288, September 2007.

[20] B. Liang and Z. Haas. Predictive distance-based mobility management for pcs networks. *INFOCOM '99. Eighteenth Annual Joint Conference of the IEEE Computer and Communications Societies. Proceedings. IEEE*, 3:1377–1384 vol.3, Mar 1999.

[21] A. Mei and J. Stefa. SWIM: A simple model to generate small mobile worlds. *CoRR*, abs/0809.2730, 2008.

[22] K. Mikyong, D. Kotz, and S. Kim. Extracting a mobility model from real user traces. In *Proceedings of INFOCOM 2006. 25th IEEE International Conference on Computer Communications.*, April 2006.

[23] M. Musolesi and C. Mascolo. A community based mobility model for ad hoc network research. In *REALMAN '06: Proceedings of the 2nd international workshop on multi-hop ad hoc networks: from theory to reality*, New York, NY, USA, 2006. ACM.

[24] I. Rhee, M. Shin, S. Hong, K. Lee, and S. Chong. On the levy-walk nature of human mobility. In *IEEE INFOCOM: 27th IEEE International Conference on Computer Communications. Proceedings.* ACM, 2008.

[25] R. Y. Rubinstein and D. P. Kroese. *Simulation and the Monte Carlo Method (Wiley Series in Probability and Statistics).*

[26] C. Schindelhauer. Mobility in wireless networks. In *32nd Annual Conference on Current Trends in Theory and Practice of Informatics, Czech*, 2006.

[27] M. F. Shlesinger, J. Klafter, and Y. M. Wong. Random walks with infinite spatial and temporal moments. *J. Statist. Phys.*, 27(3):499–512, 1982.

[28] C. Song, Z. Qu, N. Blumm, and A.-L. Barabási. Limits of predictability in human mobility. *Science*, 327(5968):1018–1021, 2010.

[29] M. Sricharan and V. Vaidehi. A pragmatic analysis of user mobility patterns in macrocellular wireless networks. *Pervasive and Mobile Computing*, 4(5):616–632, October 2008.

[30] A. Vahdat and D. Becker. Epidemic routing for partially connected ad hoc networks. In *Technical Report CS-200006, Duke University.*, April 2000.

[31] P. Wang, M. C. González, C. A. Hidalgo, and A.-L. Barabasi. Understanding the spreading patterns of mobile phone viruses. *Science*, 324(5930):1071–1076, May 2009.

[32] J. Yoon, M. Liu, and B. Noble. Random waypoint considered harmful. In *IEEE INFOCOM 2003. Twenty-second Annual Joint Conference of the IEEE Computer and Communications Societies (IEEE Cat. No.03CH37428)*, volume 2, pages 1312–1321. IEEE, 2003.

Modelling Inter-contact Times in Social Pervasive Networks

Andrea Passarella, Marco Conti, Chiara Boldrini
IIT-CNR
Pisa, Italy
{a.passarella,m.conti,c.boldrini}@iit.cnr.it

Robin I.M. Dunbar
Institute of Cognitive & Evolutionary Anthropology
University of Oxford, Oxford, UK
robin.dunbar@anthro.ox.ac.uk

ABSTRACT

Thanks to the diffusion of mobile user devices (e.g. smartphones) with rich computing and networking capabilities, we are witnessing an increasing integration between the cyber world of devices and the physical world of users. In this perspective, a possible evolution of pervasive networking (hereafter referred to as *social pervasive networks*, SPNs) consists in closely mapping human social structures in the network of the devices. Links between devices would correspond to social relationships between users, and communication events between devices would correspond to communications between users. It can be shown that fundamental convergence properties of SPN forwarding protocols are determined by the distributions of inter-contact times between the individual nodes (i.e. the time elapsed between two successive communication events between the nodes). Individual pairs inter-contact times are hard to completely charaterise, while the distribution of the *aggregate* inter-contact times is often a much more convenient figure. However, the aggregate distribution is not always representative of the individual pairs distributions. Therefore using it to characterise the properties of SPN forwarding protocols might not be correct. In this paper we provide an analytical model based on fundamental models of human social networks from the anthropology literature, which shows the exact dependence between the two in heterogeneous SPNs. Moreover, we use the model to i) study cases in which analysing the aggregate distribution is not enough, and ii) find sufficient conditions that guarantee that studying the aggregate distribution is enough to characterise the properties of SPN forwarding protocols.

Categories and Subject Descriptors

C.4 [**Performance of Systems**]; C.2.1 [**Network Architecture and Design**]

General Terms

Performance, Human Factors

Keywords

pervasive networks, social networks, modelling

MSWiM'11, October 31–November 4, 2011, Miami, Florida, USA.

1. INTRODUCTION

Last generation smartphones, tablets and similar pervasive devices feature extremely rich networking, computing, and sensing capabilities. It is nowadays argued in the research community that the penetration of this class of devices in the mass market is - for the first time - providing concrete grounds and a real opportunity for a massive deployment of pervasive networking applications [21]. Moreover, the fact that pervasive devices are almost constantly carried by users pushes towards the convergence of the "cyber" world, formed by the users' networked pervasive devices, and the "physical" world, formed by the users interacting with each other. In particular, an emerging design paradigm for pervasive networks consists in using off-line models and online information about the users' social behaviour to design, for example, routing [3, 19], data dissemination [4, 2], and mobile social networking [24] solutions. According to this paradigm, the human social plane (i.e., the structure and properties of social relationships between users) is translated into the cyber world to optimise the behaviour of pervasive networking systems.

The networking environment we consider in this paper, referred to as *social pervasive networks* (SPN), is a possible evolution of the pervasive networking paradigm enabled by this tight integration of the cyber and physical worlds. Assuming that the diffusion of pervasive technologies will enable, in principle, communication between any two users anytime and anywhere, the resulting network might in fact be formed by edges that correspond to communication channels activated because of a social relationship between two users, and only when those users communicate due to their social relationship. In other words, the network and the communication events between the devices might closely map the corresponding human social network and the interaction patterns of the users. Multi-hop communications will still occur in this type of networks, for example to enable diffusion of information among groups of people that not necessarily have mutual social relationships. Besides being a natural design approach, another advantage of such a design paradigm would be that activated communication channels will naturally inherit the trust level existing between their users, which is typically hard to assess in pervasive networks. Note that there is significant evidence suggesting that human social networks are almost invariant with respect to the specific technology that mediates social interactions [28].

Therefore, results in the anthropology domain that describe the properties of human social networks are already a solid starting point to investigate the properties of SPNs.

Within this scenario, the specific focus of this paper is to study some fundamental properties of inter-contact times between users. In SPNs, *contacts* are communications between two users due to a social interaction, and *inter-contact times* are time intervals between two consecutive contacts. Inter-contact times play a fundamental role for SPNs, as they have shown to do for a related networking environment, opportunistic networks [27]. In opportunistic networks, face-to-face contacts between users are exploited to forward messages. The foundational results presented in [9] highlight the impact of the distribution of inter-contact times on the convergence of opportunistic network routing protocols. Unlike in SPN, contacts in opportunistic networks require physical co-location of users. However, results in [9] hold for any network where messages can be exchanged only upon contacts between nodes, and therefore they apply also to SPNs. Specifically, [9] shows that when all inter-contact times between individual users follow a power law distribution with shape less than 2, then a large family of forwarding protocols (termed "naïve" protocols) diverge, i.e., yield infinite delay. In naïve protocols, nodes do not exploit any information describing the status of the network when taking forwarding decisions, and are only aware of the id of the destination. These protocols are attractive because they are very lightweight and simple to implement and analyse, and have been widely used in the literature [33, 16, 15, 31]. Note that the algorithm used in [15] to derive fundamental results on the capacity of opportunistic networks also falls in this category of forwarding protocols. Naïve forwarding protocols are attractive also for SPNs due to their simplicity and light cost. For example, they could be used for information dissemination and gossiping applications, where information should be spread to a large subset of users who will not necessarily share direct social relationships, and who will thus require multi-hop communication.

Although results in [9] apply to the distributions of *individual pairs* inter-contact times, it has been common in the literature [20, 6, 7, 29, 8, 5] to characterise opportunistic networks through the *aggregate* distribution of inter-contact times, i.e., the distribution of *all* inter-contact times between any two pairs considered altogether. Actually, using the aggregate distribution instead of all the distributions of individual pairs would be very convenient also in SPNs for a number of reasons:

- From a scalability standpoint, it is much less costly to compute, distribute, and store the parameters of a unique distribution than the parameters of all individual pairs distributions.
- From a statistical accuracy standpoint, much fewer samples are required to estimate with sufficient accuracy a unique aggregate distribution than all individual pairs distributions.
- From a privacy standpoint, it is much less sensitive to collect and distribute information about the aggregate distribution than about each individual pair, as from the former it is much harder to track individual users' behaviours.

Unfortunately, the aggregate distribution is in general not representative of the individual pairs distributions. Theoretically, the only case when it is representative is a completely homogeneous network, where all pairs inter-contact times are identically distributed, and thus the aggregate distribution is exactly the same as the distributions of individual pairs. However, for the reasons highlighted above, it is sensible to ask whether there are other cases in *heterogeneous* SPNs where studying the aggregate distribution is sufficient to characterise the convergence properties of forwarding protocols. To this end, it is necessary to have a clear understanding of the dependence between the individual pairs distributions and the aggregate distribution. Recently, [25] has analytically characterised this dependency for the case of opportunistic networks. In this paper we focus - instead - on the totally different scenario of SPN, where contacts do not require users mobility and physical co-location, but are driven by the structure of human social networks.

This paper provides the following contributions. We provide an analytical model showing the dependence between the inter-contact times distributions of individual pairs and the aggregate inter-contact time distribution in heterogeneous SPNs. Moreover, we highlight several cases of heterogeneous networks where considering the aggregate distribution is *not* sufficient to draw correct conclusions on the convergence properties of naïve forwarding protocols in SPNs. Specifically, we show cases where the aggregate distribution presents a power law, while all individual pairs distributions present a light tail. We highlight that, under certain conditions, this is also the case of one of the key datasets used in the anthropology literature to derive structural properties of human social networks [30]. Finally, we derive sufficient conditions for concluding that studying the aggregate distribution *is* sufficient to characterise the convergence properties of SPN naïve forwarding protocols.

The rest of the paper is organised as follows. In Section 2 we review the state-of-the-art relevant for this paper. Section 3 describes the models of human social networks available in the anthropology literature as the basis of our work. Section 4 presents the model showing the dependence between the inter-contact time distributions of individual pairs and the aggregate inter-contact time distribution. In Section 5 we use the model to analyse relevant cases of heterogeneous social pervasive networks. Finally, Section 6 concludes the paper.

2. RELATED WORK

This paper is mainly related to two bodies of work. The first one consists of the anthropology literature about models of human social networks. This body of work is described in detail in Section 3. The second body of work consists of the literature about the study of inter-contact times in opportunistic networks.

Results in [9] have demonstrated the fundamental impact of inter-conctat times on the convergence properties of opportunistic network routing protocols. As mentioned already, authors show that when the inter-contact times of individual pairs present a power law with shape less than 2, a large family of routing protocols yield infinite expected delay. [9] also analyses real traces of face-to-face inter-contact times, both originally presented in the paper and collected by others [23, 17, 32, 13]. Assuming that the network is homogeneous, authors focus on the distribution of aggregate inter-contact times, finding a good fit with a Pareto distribution with shape less than 2. These results challenge the actual applicability of popular routing protocols.

This view has been softned, to a certain extent, in [20], where authors have analysed the same traces of [9] (and, in addition, a proprietary GPS trace), noticing that the aggregate inter-contact times distribution actually presents an exponential cut-off in the tail. For what concerns the dependence between aggregate inter-contact times and the inter-contact times of individual pairs, [20] provides an initial result deriving analytically the dependence between them when the contact rates between individual pairs are known. In addition, [20] does not spend too much effort on the issue of heterogeneity, after noticing that, for a subset of the pairs in their traces, invidual inter-contact times are power law.

Results in [9, 20] had a very important impact on the subsequent literature, although not much attention has been put on the critical issue of heterogeneity. The fact that aggregate inter-contact times in popular traces present a power law has typically resulted in assuming that all distributions of individual pairs are power law. One of the most important examples is the area of mobility models. Most of the recent proposals (e.g., [6, 5, 22, 29]) aim at generating inter-contact times of individual pairs and/or aggregate inter-contact times following a power law. Similarly, other papers try to highlight which characteristics of reference mobility models generate a power law in inter-contact times [7, 8].

Authors of [10] analyse mathematically the dependence between inter-contact times of individual nodes and aggregate inter-contact times in a more general setting with respect to the model in [20]. They re-analyse the same traces used in [9, 20] showing that the distributions of inter-contact times of individual pairs are definitely heterogeneous. They propose a model to describe how heterogeneity impacts on the distribution of aggregate inter-contact times. However, as highlighted in [25], they miss to consider an important aspect, thus deriving an imprecise model. [25] presents the most precise model, as far as we know, to describe the dependence between the inter-contact time distributions of individual pairs and the aggregate inter-contact time distribution in opportunistic networking environments.

To the best of our knowledge, this is the first paper in the literature that analyses this dependence in social pervasive networks, considering models of interactions between users derived in the anthropology literature. With respect to [25], this results in a totally different model for describing the heterogeneity of inter-contact times of individual pairs. Moreover, in this paper, in addition to studying cases in which the aggregate inter-contact times distributions cannot be used to analyse the convergence of forwarding protocols, we also provide sufficient conditions under which focusing only on the aggregate distribution is enough.

3. HUMAN SOCIAL NETWORKS

Before presenting our analysis, it is worth describing our reference model for the structure of human social networks, which is based on the concept of ego network. An ego network is the network seen from the standpoint of a single individual (ego). It includes only other people (alters) the ego has social relationships with (represented by an edge in the ego network).

Ego networks have been extensively studied in the anthropology literature [11, 12, 18, 30, 35], resulting in a detailed model of their structure (Figure 1). [35] has shown that ego networks can be represented as a series of concentric layers centred around the ego. Starting from the inner-most layer, layers are characterised by a decreasing level of *intimacy* with the ego. On the other hand, the *size* of the layers (the number of alters within the layer) increases with a factor approximately equal to 3. Extensive studies have identified four layers, i.e. the support clique, the sympathy group, the band and the active network, with size approximately equal to 5, 15, 45 and 150 [18, 12, 11]. The size of the active network (150) is usually referred to as the *Dunbar's number*, and represents the maximum number of alters an ego can - on average - maintain social relationships with [18]. This is a limit related to cognitive capabilities of the human brain [12]. Many more alters can be outside the active network, corresponding to people known to the ego, but with whom the ego does not establish any significant social relation. These alters are usually not represented in the model. Note that this hierarchical structure depends very little on the communication means supporting social relationships [28].

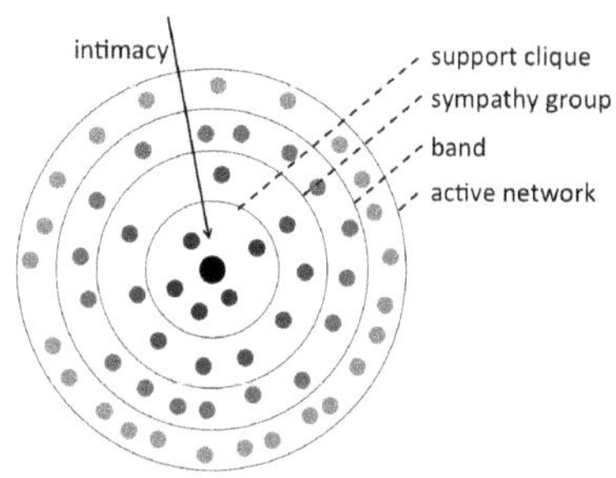

Figure 1: Ego-network's hierarchical structure.

Authors of [18] have also shown that the *emotional closeness* of the ego with a given alter is the key parameter determining the position of the alter in the layers. Moreover [18, 30] show that there is a strong correlation between the emotional closeness and the frequency of communication between the ego and the alter. Therefore, it follows that the structure of the ego network depicted in Figure 1 naturally determines the contact rates between the ego and alters in its social network. Specifically, contacts are more frequent with alters in the inner-most layer (usually referred to as *strong ties*), while the frequency progressively declines for external layers, resulting in *weaker* ties. This property is one of the starting points of the analysis presented in Section 4.

Finally, it is worth pointing out that, for our purposes, focusing on ego networks is sufficient. In general a social network contains more information than the set of ego networks of its members, as the latter does not capture correlations. However, it is straightforward to note that inter-contact times between any pair of users can be fully described by looking at ego networks only, because they only depend on the relationship between these two users, which is captured by the ego-network model.

4. INTER-CONTACT TIMES MODEL

In this section we study, through an analytical model, the dependence between the distributions of the individual pairs and aggregate inter-contact times, in a network where contacts can be described with the ego-network model presented in Section 3.

An important requirement of our model is to represent heterogeneous networks in which the distributions of inter-contact times between individual pairs are not iid. We take heterogeneity into account in the definition of the model for contact *rates* (the reciprocal of the average inter-contact times). We assume that the contact rates are random variables (r.v.) following a known distribution (hereafter Λ_p denotes the contact rate of the generic pair p). In addition, we assume that individual pairs inter-contact times are distributed according to a known *type* of distribution (e.g., Pareto, exponential, ...). For each pair p, the parameters of the inter-contact times distribution are a function of Λ_p, i.e., the parameters are set such that the average inter-contact time is equal to $1/\Lambda_p$. This allows us to model heterogeneous environments in which not all individual inter-contact times are identically distributed, and to control the type of heterogeneity through the r.v. describing the contact rates.

Therefore, three distributions play a key role in our analysis, i.e. i) the distributions of individual pairs inter-contact times (whose CCDF is hereafter denoted as $F_\lambda(x)$), ii) the distribution of individual pairs contact rates (whose density is hereafter denoted as $f(\lambda)$), and iii) the distribution of the aggregate inter-contact times (whose CCDF is hereafter denoted as $\mathcal{F}(x)$).

4.1 Modelling human networks contact patterns

As a first step in the model, we describe how we account for the human social network structures described in Section 3. This is taken into consideration in the definition of the distribution of the contact rates. Figure 2 provides a schematic representation of a generic distribution. As, in any given ego network, contacts with alters in inner shells occur more frequently than contacts with alters in outer shells, contact rates with peers in the inner-most shell should be drawn from the tail of the distribution, while contact rates with peers in the outer-most shell should be drawn from the head. Based on this observation, we divide the possible range of rates in L sectors, where L is the number of layers of an ego network, and layer 1 denotes the inner-most layer. The challenge is to meaningfully identify in the contact rate distribution the boundaries of the sectors corresponding to each layer or, in other words, to define the sectors of the contact rate distribution from where to draw contact rate samples for alters in any given social layer. The average number of relationships in each layer $n_l, l = 1, \ldots, L$, and the total number of relationships N can be derived from the results presented in Section 3 [18, 12, 11]. We can thus compute the fraction of relationships in each layer as n_l/N (note that $n_L = N$). If we denote with $\lambda_0, \ldots, \lambda_L$ the values of λ that identify the sectors of the contact rates distribution corresponding to the layers, the values of $\lambda_i, i = 1, \ldots, L$ can be computed as the $(1 - \frac{n_i}{N})$-th percentiles of the rates distribution (note that λ_L and λ_0 are the minimum and maximum possible values of λ, respectively). Therefore, contact rates with a peer in layer $l = 1, \ldots, L$ are drawn from the sector identified by λ_l, λ_{l-1}. It thus follows that the density of contact rates for relationships in layer l is as follows

$$f_l(\lambda) = \begin{cases} 0 & \lambda < \lambda_l \vee \lambda > \lambda_{l-1} \\ C_l f(\lambda) & \lambda_l \leq \lambda \leq \lambda_{l-1} \end{cases} \tag{1}$$

where C_l is a constant such that $\int_0^\infty f_l(\lambda) d\lambda = 1$, i.e. $C_l = [G(\lambda_{l-1}) - G(\lambda_l)]^{-1}$, $G(\lambda)$ being the CDF of Λ.

Note that we only consider the distribution of contact rates for alters with a contact rate greater than 0. In principle, the distribution of contact rates presents a significant mass probability in 0, corresponding to the fact that an ego "knows" alters also outside the active network layer, but relationships are so weak that the contact rate is zero.

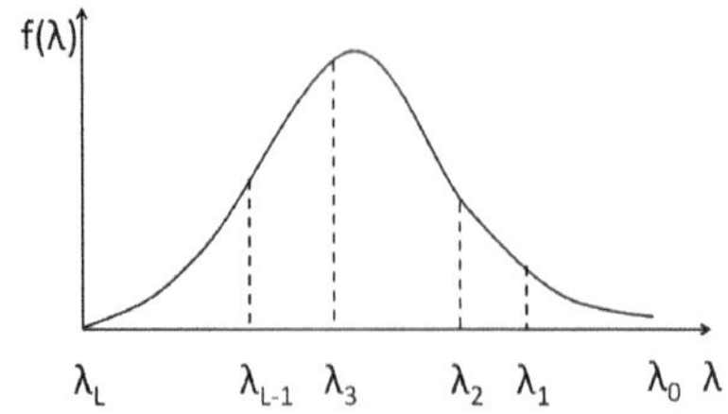

Figure 2: A representative contact rates distribution in human social networks

4.2 General inter-contact times model

The starting point of our model is a result originally presented in [20] (and recalled in Lemma 1), which describes the dependence between the distributions of the individual pairs and aggregate inter-contact times, *when the contact rates are known a priori*. Let us assume that P pairs are present in the network, that $n_p(T)$ contact events between pair p occur during an observation time T. Let us denote with $N(T)$ the total number of contact events over T, with θ_p the contact rate of pair p, with θ the total contact rate ($\theta = \sum_p \theta_p$), and with $F_p(x)$ the CCDF of inter-contact times of pair p. Then, the following lemma holds.

LEMMA 1. *In a network where P pairs of nodes exist for which inter-contact times can be observed, the CCDF of the aggregate inter-contact times is:*

$$\mathcal{F}(x) = \lim_{T \to \infty} \sum_{p=1}^{P} \frac{n_p(T)}{N(T)} F_p(x) = \sum_{p=1}^{P} \frac{\theta_p}{\theta} F_p(x) \tag{2}$$

Lemma 1 is rather intuitive. The distribution of aggregate inter-contact times is a mixture of the individual pairs distributions. Each individual pair "weights" in the mixture proportionally to the number of inter-contact times that can be observed in any given interval (or, in other words, proportionally to the contact rate).

In this paper we significantly extend this result, by i) assuming that contact rates are random variables, thus unknown a priori, and ii) exploiting an anthropology model for describing contacts between humans. Specifically, we can derive the following Theorem[1].

THEOREM 1. *In a social pervasive network where contact rates are determined by the hierarchical structure of ego networks, the CCDF of the aggregate inter-contact times is:*

$$\mathcal{F}(x) = \sum_{l=1}^{L} \frac{p_l C_l}{\sum_{l=1}^{L} p_l E[\Lambda_l]} \int_{\lambda_l}^{\lambda_{l-1}} \lambda f(\lambda) F_\lambda(x) d\lambda \tag{3}$$

where p_l is the probability that a social relationship of any given user is in layer l of its ego network, and Λ_l denotes the contact rates between an ego and its alters in layer l (i.e., its density is as in Equation 1).

While in [26] we provide the complete proof of Equation 3, here we briefly discuss its physical meaning. First of all, Equation 3 can be seen as the weighted sum of components

[1] Unless otherwise stated, proofs are provided in the [26].

related to the individual layers of ego networks. Specifically, by defining $\mathcal{F}_l(x)$ as follows:

$$\mathcal{F}_l(x) = \frac{C_l}{E[\Lambda_l]} \int_{\lambda_l}^{\lambda_{l-1}} \lambda f(\lambda) F_\lambda(x) d\lambda \tag{4}$$

we can write $\mathcal{F}(x)$ as

$$\mathcal{F}(x) = \sum_{l=1}^{L} \frac{p_l E[\Lambda_l]}{\sum_{l=1}^{L} p_l E[\Lambda_l]} \mathcal{F}_l(x) \tag{5}$$

In [26] we show that $\mathcal{F}_l(x)$ is actually the CCDF of the aggregate inter-contact times over layer l only. Equation 5 highlights an intuitive result. Each such component ($\mathcal{F}_l(x)$) "weights" in the aggregate proportionally to the fraction of pairs falling in the layer (p_l), and to the average contact rates of the layer (i.e., to the average number of inter-contact events that is generated by a pair in that layer).

Besides a more formal derivation shown in [26], the form of the individual layer CCDF in Equation 4 has a more intuitive derivation, starting from the result in Lemma 1. Specifically, it can be obtained by considering a modified network in which we assume that all rates $\lambda \in [\lambda_l, \lambda_{l-1}]$ are possibly available (for pairs in layer l), each with a probability $f_l(\lambda)d\lambda$. $\mathcal{F}_l(x)$ is thus the aggregate over all the resulting individual pairs inter-contact times distributions. As the number of such distributions becomes infinite and is indexed by Λ_l (a continuous random variable), the summation in Equation 2 becomes an integral over λ. Moreover, the weight of each distribution (θ_p in Equation 2) becomes $\lambda \cdot p(\lambda) = \lambda f_l(\lambda) d\lambda$, while the total rate ($\theta$ in Equation 2) becomes $\int_0^\infty \lambda f_l(\lambda) d\lambda = E[\Lambda_l]$. The expression in Equation 4 follows immediately.

5. STUDY OF REPRESENTATIVE SOCIAL PERVASIVE NETWORKS

The following analysis starts from the result in Theorem 1, and is divided in two parts. In the first part (Section 5.1) we show relevant cases where the aggregate inter-contact times distribution is not representative of the distributions of the individual pairs, and thus studying all individual pairs distributions is necessary to characterise the convergence properties of SPN forwarding protocols. Specifically, we identify by analysis cases where the individual distributions present a light tail, while the aggregate turns out to follow a power law. Interestingly, under the conditions discussed in Section 5.1, we show that the human social network used to collect the data at the basis of [30] falls in this category. In the second part (Section 5.2), we derive analytically sufficient conditions for concluding that the aggregate inter-contact times distribution (instead of all the individual pairs distributions) can be used to characterise the convergence properties of SPN forwarding protocols. Specifically we prove that it is sufficient that the distribution of even a single pair follows a power law for the aggregate distribution to also be power law. We thus conclude that when the aggregate is *not* power law, then all individual pairs distributions must present a light tail, and therefore the network does not satisfies the conditions for forwarding protocols divergence found in [9].

It is important to note that, to carry on our analysis, it is sufficient to study the aggregate inter-contact times distribution over individual layers only, provided by Equation 4. It is, in fact, sufficient that one such aggregate presents a heavy tail for the whole aggregate to be heavy tailed. Thus, Equation 4 is the key starting point for the following analysis.

5.1 Networks where analysis of individual pairs distributions is needed

The first network we consider is one where individual pairs distributions are exponential, and the contact rates follow a gamma distribution. Being all individual pairs distributions exponential, this network does not satisfy the conditions for divergence of forwarding protocols found in [9]. Note that assuming exponential individual pairs distributions is relevant. At least in the case of face-to-face contacts, real traces analysis has shown that the distributions of a high fraction of pairs are exponential [14, 10].

Considering contact rates following a gamma distribution is motivated by the analysis of the dataset used to derive the properties of ego networks described in [30], at the basis of the model described in Section 3. The dataset collects information about 251 ego networks. Each relationship in each ego-network provides a sample of contact rate, for a total of over 20000 samples. We fit the resulting empirical distribution to reference distributions (i.e., gamma, exponential and Pareto) using the Maximum Likelihood (ML) method [34], and compare the fitted distributions against the data using the Akaike Information Criterion (AIC, [1]). Figure 3 shows a visual comparison of the samples obtained from [30] and the ML fittings of the considered contact rates distributions (ML estimators of the parameters are provided in Table 1). As for the gamma and exponential distributions we consider the standard definitions with shape α and rate b, and with rate b, respectively. As for the Pareto distribution we consider the CCDF form $F(\lambda) = \left(\frac{b}{b+\lambda}\right)^\alpha$, $\alpha > 0$, $\lambda > 0$.

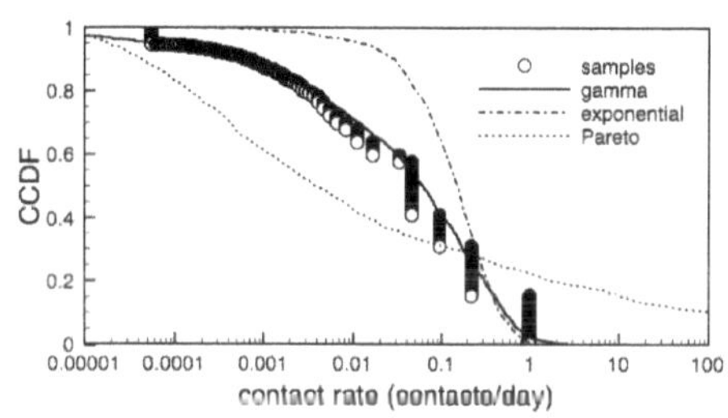

Figure 3: Fitting distributions

The intuition from Figure 3 is that the gamma distribution is the best fit for our dataset. This is confirmed by the AIC test, whose values are shown in Table 1. Remember that in AIC tests the best alternative is the one with the lowest AIC value [1].

Distribution	Best fit parameters	AIC value
Gamma	$\alpha = 0.34$, $b = 1.03$	-50280.62
Exponential	$b = 4.86$	-23505.08
Pareto	$\alpha = 0.16$, $b = 5.5\text{x}10^{-5}$	-31289.34

Table 1: AIC values for the tested distributions.

Lemma 2 and Theorem 2 characterise the distribution of the aggregate inter-contact times in a network where individual pairs follow an exponential distribution, and contact rates follow a gamma distribution.

LEMMA 2. *When contact rates follow a gamma distribution and individual inter-contact times an exponential dis-*

tribution, the CCDFs of inter-contact times aggregated over individual layers ($\mathcal{F}_l(x)$) all decay, for large x, faster than a power law with exponential cutoff, but the CCDF corresponding to the outer-most layer, which decays as a power law. Specifically, if the contact rates follow a gamma distribution with shape α and rate b, the following relations hold true, for large x:

$$\begin{cases} \mathcal{F}_l(x) \leq \frac{Re^{-\lambda_l(b+x)}}{x} & l = 1, \ldots, L-1 \\ \mathcal{F}_L(x) \simeq \frac{K}{x^{\alpha+1}} \end{cases} \tag{6}$$

where R and K do not depend on x.

THEOREM 2. *In a social pervasive network where individual pairs inter-contact times are exponentially distributed and contact rates follow a gamma distribution, the distribution of the aggregate inter-contact times features a heavy tail. Specifically, the following relation holds true:*

$$f(\lambda) = \frac{\lambda^{\alpha-1} b^{\alpha} e^{-b\lambda}}{\Gamma(\alpha)}, F_\lambda(x) = e^{-\lambda x}$$
$$\Rightarrow \mathcal{F}(x) \simeq \frac{K}{x^{\alpha+1}} \text{ for large } x$$

where K does not depend on x.

PROOF. This follows immediately from Lemma 2, by recalling the relationships between $\mathcal{F}_l(x)$ and $\mathcal{F}(x)$ in Equation 5, and noting that $\mathcal{F}_L(x)$ dominates over all the other components for large x. □

To validate the analytical results, we compare the result of Theorem 2 with simulations. Specifically, we simulate an ego-network with 150 alters divided in layers according to the model presented in Section 3. Ego and each alter meet with exponential inter-contact times, with rates drawn from a gamma distribution. Sectors of the distribution corresponding to the layers are defined as described in Section 4.1. Each simulation run reproduces an observation of the network for a time interval T, defined according to the following algorithm. For each alter a, we first generate 100 inter-contact times, and then compute the total observation time after 100 inter-contact times, T_a, as the sum of the pair inter-contact times. T is defined as the maximum of $T_a, a = 1, ..., 150$. To guarantee that all alters are observed for the same amount of time, we generate additional inter-contact times for each alter until T_a reaches T. Simulations have been replicated 20 times with independent seeds, and confidence intervals (with 99% confidence level) have been computed.

Figure 4 shows a very good agreement between the analytical and the simulation models. Recall that the analysis predicts that the tail of the aggregate inter-contact times distribution decays as $\frac{1}{x^{\alpha+1}}$ where α is the shape parameter of the contact rates distribution. Figure 4 shows that - as also found in the analysis - the lower the shape of the contact rates distribution, the heavier the tail of the aggregate inter-contact times. This results from the fact that lower shape parameters result in a higher mass of probability of contact rates around 0, i.e., in an increasing probability of very long inter-contact times.

Theorem 2 and Lemma 2 provide two interesting insights. First, the presence of aggregate inter-contact times with a heavy tail distribution does not necessarily mean that routing protocols risk divergence in SPNs, as such a heavy tail can emerge starting from exponentially distributed individual pairs. Therefore, when the contact rates follow a gamma

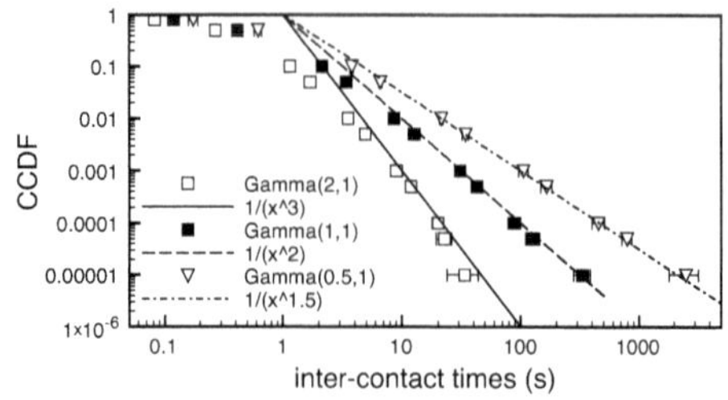

Figure 4: Aggregate inter-contact times with gamma contact rates

distribution, looking at the distribution of aggregate inter-contact times is not sufficient to check whether routing protocols may diverge or not. Instead, the distributions of individual pairs inter-contact times must be analysed. Second, the power law of $\mathcal{F}(x)$ appears because of the power law of the inter-contact times aggregated over the outer-most layers, $\mathcal{F}_L(x)$. Due to the shape of the gamma distribution, in the outer-most layers contact rates can be arbitrarily close to 0, thus resulting in arbitrarily large inter-contact times. Intuitively, this suggests a more general behaviour: Whenever the distribution of the contact rates is such that rates arbitrarily close to 0 can be drawn, the distribution of the aggregate inter-contact times presents a heavy tail.

The case of contact rates following a gamma distribution is not the only one in which these two properties hold true. Actually, the case where rates follow a Pareto distribution also present similar properties (still assuming that individual inter-contact times follow an exponential distribution). Lemma 3 and Theorem 3 analyse this case.

LEMMA 3. *When contact rates follow a Pareto distribution whose CCDF is in the form $F(\lambda) = \left(\frac{b}{b+\lambda}\right)^{\alpha}$, $\lambda > 0$ and individual inter-contact times are exponential, the CCDFs of inter-contact times aggregated over individual layers ($\mathcal{F}_l(x)$) all decay, for large x, at least as fast as a power law with exponential cutoff, but the CCDF corresponding to the outer-most layer, which decays as a power law. Specifically, the following relations hold true for large x:*

$$\begin{cases} \mathcal{F}_l(x) \leq \frac{Re^{-\lambda_l x}}{x} + \frac{Qe^{-\lambda_{l-1} x}}{x} & l = 1, \ldots, L-1 \\ \mathcal{F}_L(x) \simeq \frac{K}{x^2} \end{cases} \tag{7}$$

where R, Q and K do not depend on x.

THEOREM 3. *When contact rates follow a Pareto distribution whose CCDF is in the form $F(\lambda) = \left(\frac{b}{b+\lambda}\right)^{\alpha}$, $\lambda > 0$ and individual inter-contact times are exponential, the CCDF of the aggregate inter-contact times decays, for large x, as a power law with shape equal to 2. Specifically, the following relation holds true*

$$F(\lambda) = \left(\frac{b}{b+\lambda}\right)^{\alpha}, F_\lambda(x) = e^{-\lambda x}$$
$$\Rightarrow \mathcal{F}(x) \simeq \frac{K}{x^2} \text{ for large } x$$

where K does not depend on x.

PROOF. This comes immediately from Lemma 3 by noticing that the slowest decaying component of $\mathcal{F}(x)$ is the one corresponding to the outer-most layer, and using the corresponding expression from Equation 7. □

Also in this case, we validate the results by comparing analysis and simulation. Figure 5 shows that also in this case the analytical results capture very well the shape of the tail of the aggregate inter-contact times distribution obtained in simulation. Note that, as predicted by the model, the aggregate distribution decays as a power law with shape equal to 2, irrespective of the parameters of the contact rates distribution.

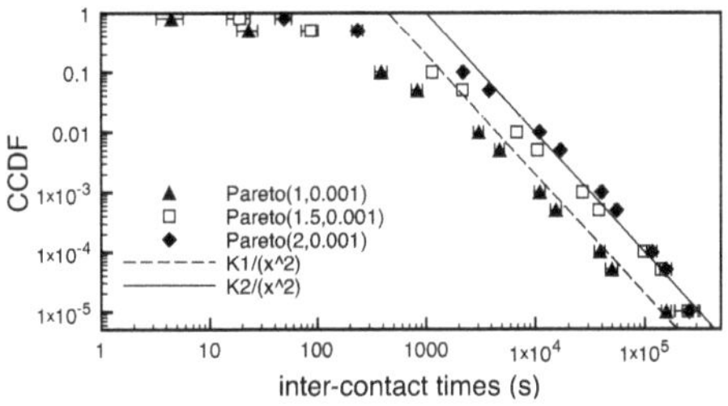

Figure 5: Aggregate inter-contact times with Pareto contact rates.

As anticipated, also in this case considering the aggregate inter-contact times distribution is not sufficient to characterise the convergence of forwarding protocols. Note that, also in this scenario, contact rates can be arbitrarily close to 0. This confirms the intuition that when contact rates present this property, the aggregate distribution of inter-contact times presents a heavy tail.

5.2 The effect of individual power-law inter-contact times distributions

In this section we study analytically the effect of even a single heavy tail individual inter-contact times distribution on the distribution of the aggregate inter-contact times. Specifically, we assume that the inter-contact times distribution of a given pair p presents a heavy tail, i.e. we assume that $F_p(x)$ is as follows:

$$F_p(x) \simeq x^{-\eta} \text{ for large x}$$

For the rest of the individual pairs distributions we consider the same assumptions used in Section 4, and, in addition, we assume that they are not power law. In other words, the distribution of pair p is the only one in the network presenting a heavy tail. Finally, we assume that the number of pairs is finite. Then, the following lemma holds true.

LEMMA 4. *In a network with a finite number of pairs, where there exists one pair whose individual inter-contact times distribution follows, for large x, a power law with shape η, the distribution of the aggregate inter-contact times, for large x, follows a power law at least as heavy as $x^{-\eta}$, i.e.*

$$\exists\, p \text{ s.t. } F_p(x) \sim x^{-\eta} \text{ for large } x \Rightarrow$$

$$\mathcal{F}(x) \geq Cx^{-\eta} \text{ for large } x \text{ and for some constant } C > 0$$

Figure 6 provides a concrete example of the result in Lemma 4. Specifically, we first consider an ego-network such that the distribution of the aggregate inter-contact times does not present a heavy tail. As shown in [26], this can be obtained, for example, by using exponentially distributed inter-contact times, and sampling the contact rates from a slightly different Pareto distribution with respect to the one considered in Section 5.1. Specifically, it is necessary to consider a CCDF in the form $F(\lambda) = \left(\frac{b}{\lambda}\right)^{\alpha}, \alpha > 0,\ \lambda > b$. In this case, the tail of the resulting aggregate distribution presents a power law with an exponential cut-off. In the Figure, the percentiles obtained by simulation are marked with white squares, and the corresponding analytical curve (derived in [26]) is plotted with a solid line. Then, we added to the same ego network one more alter, whose inter-contact times with the ego follow a Pareto distribution with shape $\eta = 1.1$ (while the scale parameter, defining the minimum inter-contact time, is set to 1 day, as this was also the minimum inter-contact time found in the traces used in Section 5.1). In the figure, percentiles obtained by simulation are marked with dark diamonds, while the corresponding analytical curve predicted by Lemma 4 is plotted with a dashed curve. According to Lemma 4, i) the existence of even a single pair whose inter-contact times are power law implies that the tail of the aggregate distribution is also heavy, and ii) the tail of the resulting aggregate distribution can be lower bounded by a power law with shape equal to $\eta = 1.1$. Figure 6 confirms both results obtained in Lemma 4.

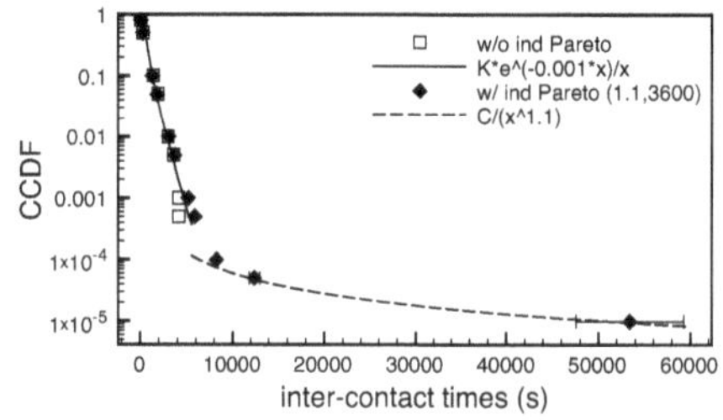

Figure 6: Aggregate inter-contact times with and without a single Pareto ICT pair.

Lemma 4 allows us to immediately identify the cases where considering the aggregate inter-contact times distribution is sufficient to characterise the convergence properties of forwarding protocols in SPNs. Specifically, the following theorem holds true.

THEOREM 4. *In a network with a finite number of pairs, if the distribution of the aggregate inter-contact times does not present a heavy tail, then no individual inter-contact times distribution can present a heavy tail.*

PROOF. This comes straightforwardly from Lemma 4. If even a single individual inter-contact times would present a heavy tail, then the aggregate distribution would also present a heavy tail. □

In practical terms, Theorem 4 tells that when the aggregate inter-contact times distribution does not present a heavy tail, it is not necessary to study all the distributions of individual inter-contact times to check the conditions on the convergence of forwarding protocols found in [9], because no individual inter-contact times distribution can follow a power law. This result is dual to those found in Section 5.1, which tell that, instead, when the aggregate distribution presents a heavy tail, a detailed analysis of the individual pairs distributions is necessary.

6. CONCLUSION

In this paper we studied fundamental properties of inter-contact times in social pervasive networks, basing our analysis on reference models of human social networks available in the anthropology literature. Social pervasive networks are a possible evolution of current pervasive networks, where the communication network maps directly the human social network of users, and communication between devices occurs when users communicate as an effect of their social tie. In social pervasive networks, forwarding actions occur upon such contact events between users. In this scenario, it is fundamental to characterise the properties of inter-contact times between individual users (i.e., the time between two consecutive contact events), as this has been shown to play a key role in determining convergence properties of forwarding algorithms when forwarding occurs upon contacts only. A complete characterisation of all individual inter-contact times distribution might be impractical to achieve. From this standpoint, using the aggregate distribution of inter-contact times would be much more convenient. Unfortunately, in general the aggregate distribution is not representative of the distributions of individual pairs. A clear understanding of the dependence between the two is thus needed.

In this paper we provided a mathematical model based on models of human social networks available in the anthropology literature to formally characterise the dependence between the individual pairs distributions and the distribution of aggregate inter-contact times. The model highlights the importance of the network heterogeneity (captured through the distribution of contact rate) in determining the shape of the aggregate distribution. We have used the model to study relevant networks in which, unfortunately, focusing on the aggregate distribution is not enough, and all individual pairs distributions must be analysed to characterise the convergence properties of forwarding algorithms. Interestingly, the model allowed us to also find sufficient conditions to be sure that considering the aggregate distribution is enough.

Beyond the specific applications presented, the contribution of the paper consists in providing a clear understanding of the dependence between the different distributions of inter-contact times in SPNs, and a practical tool to understand which statistics must be used to correctly study the convergence properties of SPN forwarding protocols.

Acknowledgements

This work was funded by the European Commission under the SCAMPI (258414) FIRE project and EINS FIRE NoE. The data were collected as part of the TESS project, funded by EPSRC. R.I.M. Dunbar's research is also funded by the British Academy Centenary Research project.

7. REFERENCES

[1] H. Akaike. A new look at the statistical model identification. *IEEE Tran. on Aut. Control*, 19(6):716 – 723, 1974.

[2] S. Allen, M. Chorley, G. Colombo, and R. Whitaker. Self adaptation of cooperation in multi-agent content sharing systems. In *IEEE SASO*, 2010.

[3] C. Boldrini, M. Conti, and A. Passarella. Exploiting users' social relations to forward data in opportunistic networks: The HiBOp solution. *Pervasive and Mobile Computing*, 4(5):633 – 657, 2008.

[4] C. Boldrini, M. Conti, and A. Passarella. Design and performance evaluation of ContentPlace, a social-aware data dissemination system for opportunistic networks. *Comput. Netw.*, 54(4):589–604, March 2010.

[5] C. Boldrini and A. Passarella. HCMM: Modelling spatial and temporal properties of human mobility driven by users' social relationships. *Computer Communications*, 33(9):1056–1074, 2010.

[6] V. Borrel, F. Legendre, M. D. de Amorim, and S. Fdida. Simps: using sociology for personal mobility. *IEEE/ACM Trans. Netw.*, 17(3):831–842, 2009.

[7] H. Cai and D. Y. Eun. Crossing over the bounded domain: from exponential to power-law inter-meeting time in manet. In *ACM MOBICOM*, pages 159–170, 2007.

[8] H. Cai and D. Y. Eun. Toward stochastic anatomy of inter-meeting time distribution under general mobility models. In *ACM MobiHoc*, pages 273–282, 2008.

[9] A. Chaintreau, P. Hui, J. Crowcroft, C. Diot, R. Gass, and J. Scott. Impact of human mobility on opportunistic forwarding algorithms. *IEEE Trans. Mob. Comput.*, 6(6):606–620, 2007.

[10] V. Conan, J. Leguay, and T. Friedman. Characterizing pairwise inter-contact patterns in delay tolerant networks. In *Autonomics*, 2007.

[11] R. Dunbar and M. Spoors. Social networks, support cliques, and kinship. *Human Nature*, 6(3):273–290, 1995.

[12] R. I. M. Dunbar. The social brain hypothesis. *Evolutionary Anthropology: Issues, News, and Reviews*, 6(5):178–190, 1998.

[13] N. Eagle and A. (Sandy) Pentland. Reality mining: sensing complex social systems. *Personal Ubiquitous Comput.*, 10:255–268, March 2006.

[14] W. Gao, Q. Li, B. Zhao, and G. Cao. Multicasting in delay tolerant networks: a social network perspective. In *ACM MobiHoc*, 2009.

[15] M. Grossglauser and D. N. C. Tse. Mobility increases the capacity of ad hoc wireless networks. *IEEE/ACM Trans. Netw.*, 10:477–486, August 2002.

[16] Z. J. Haas and T. Small. A new networking model for biological applications of ad hoc sensor networks. *IEEE/ACM Trans. Netw.*, 14:27–40, February 2006.

[17] T. Henderson, D. Kotz, and I. Abyzov. The changing usage of a mature campus-wide wireless network. In *ACM MOBICOM*, 2004.

[18] R. Hill and R. Dunbar. Social network size in humans. *Human Nature*, 14(1):53–72, 2003.

[19] P. Hui, J. Crowcroft, and E. Yoneki. Bubble rap: Social-based forwarding in delay tolerant networks. *IEEE Trans. Mob. Comput.*, 2010.

[20] T. Karagiannis, J.-Y. L. Boudec, and M. Vojnovic. Power law and exponential decay of intercontact times between mobile devices. *IEEE Trans. Mob. Comput.*, 9(10):1377–1390, 2010.

[21] N. Lane, E. Miluzzo, H. Lu, D. Peebles, T. Choudhury, and A. Campbell. A survey of mobile phone sensing. *IEEE Comm. Mag.*, 48(9):140 –150, 2010.

[22] K. Lee, S. Hong, S. J. Kim, I. Rhee, and S. Chong. Slaw: A new mobility model for human walks. In *INFOCOM*, pages 855–863, 2009.

[23] M. McNett and G. M. Voelker. Access and mobility of wireless pda users. *ACM MC2R*, 9, 2005.

[24] E. Miluzzo, N. D. Lane, K. Fodor, R. Peterson, H. Lu, M. Musolesi, S. B. Eisenman, X. Zheng, and A. T. Campbell. Sensing meets mobile social networks: the design, implementation and evaluation of the cenceme application. In *ACM SenSys*, 2008.

[25] A. Passarella and M. Conti. Characterising aggregate inter-contact times in heterogeneous opportunistic networks. In *IFIP Networking*, 2011.

[26] A. Passarella, M. Conti, C. Boldrini, and R. Dunbar. Modelling inter-contact times in human Social Pervasive Networks. Technical report, IIT-CNR-08, 2011, http://cnd.iit.cnr.it/andrea/docs/mswim11-tr.pdf.

[27] L. Pelusi, A. Passarella, and M. Conti. Opportunistic networking: data forwarding in disconnected mobile ad hoc networks. *IEEE Comm. Mag.*, 44(11):134 –141, 2006.

[28] T. Pollet, S. Roberts, and R. Dunbar. Use of social network sites and instant messaging does not lead to increased social network size, or to emotionally closer relationships with offline network members. *Cyberpsychology, Behavior, and Social Networking*, 2010.

[29] I. Rhee, M. Shin, S. Hong, K. Lee, and S. Chong. On the levy-walk nature of human mobility. In *INFOCOM*, pages 924–932, 2008.

[30] S. Roberts and R. Dunbar. Communication in social networks: Effects of kinship, network size, and emotional closeness. *Personal Relationships*, 2010.

[31] T. Spyropoulos, K. Psounis, and C. S. Raghavendra. Efficient routing in intermittently connected mobile networks: the multiple-copy case. *IEEE/ACM Trans. Netw.*, 16:77–90, February 2008.

[32] J. Su, A. Chin, A. Popivanova, A. Goel, and E. de Lara. User mobility for opportunistic ad-hoc networking. In *IEEE MSCA*, 2004.

[33] A. Vahdat and D. Becker. Epidemic routing for partially connected ad hoc networks. Technical report, Duke University, 2000.

[34] L. Wasserman. *All of statistics: a concise course in statistical inference*. Springer Verlag, 2004.

[35] W. X. Zhou, D. Sornette, R. A. Hill, and R. I. M. Dunbar. Discrete hierarchical organization of social group sizes. *Proceedings of the Royal Society B: Biological Sciences*, 272(1561):439–444, 2005.

Modeling and Analysis of Composite Network Embeddings*

Ben Baumer
City University of New York
bbaumer@gc.cuny.edu

Prithwish Basu
Raytheon BBN Technologies
pbasu@bbn.com

Amotz Bar-Noy
City University of New York
amotz@sci.brooklyn.cuny.edu

ABSTRACT

In a *composite network*, a piece of information traveling through links in a social network may have to travel over multiple links in an associated communication network. In this paper, we propose a model of composite networks that consists of two networks and an embedding between them, and several composite metrics that characterize information flow under a particular type of embedding. We present analytic results for the scaling behavior of "constrained composite stretch" of a path, "constrained composite diameter" of a graph, and "constrained composite broadcast time" of a tree, under random uniform embeddings onto various communication network structures. We validate our analytical results on composite stretch using two data sets consisting of a friendship social network geographically spread across Western Europe and a historical deployment of a military chain of command. We also present a randomized model of field deployment consistent with real-world data, and use simulations over this model to explore the distribution of constrained composite broadcast time. Finally, we show that our analytical bounds for composite broadcast time agree well with the simulation results.

Categories and Subject Descriptors

G.2.2 [**Discrete Mathematics**]: Graph Theory—*Network problems; Trees; Path and circuit problems*; C.2.1 [**Computer-Communication Networks**]: Network Architecture and Design—*Network topology; Wireless communication*

*Research was sponsored by the Army Research Laboratory and was accomplished under Cooperative Agreement Number W911NF-09-2-0053. The views and conclusions contained in this document are those of the authors and should not be interpreted as representing the official policies, either expressed or implied, of the Army Research Laboratory or the U.S. Government. The U.S. Government is authorized to reproduce and distribute reprints for Government purposes notwithstanding any copyright notation here on.

MSWiM'11, October 31–November 4, 2011, Miami, Florida, USA.

General Terms

Theory, Networks

1. INTRODUCTION

The past few years have seen explosive growth of online social networks along with proliferation of the communication networks such as the Internet, wireline telephone networks, cellular telephone networks, and even multi-hop wireless ad hoc and sensor networks, in specialized application scenarios such as disaster relief and battlefield operations. Human beings usually share information with folks they know, and then that information gets forwarded along various links in the social network, either verbatim (e.g., directives from a leader flowing through the *chain of command*) or after modifications (e.g., propagation of rumors, gossip, or news from the field to the newspaper reader). This results in a social network *constraining* or *guiding* the spread of information according to its topology. Such constraints can result in a piece of information having to traverse much longer paths in the underlying communication network between the producer and the ultimate consumer of the information. This *stretch* is justified because either the intermediaries could play a critical role in interpreting/massaging the information or they serve as important links in the acquaintance chain, without whom the producer and the ultimate consumer would not know each other.

The total time taken for a piece of information to spread through the entire social network in the aforementioned constrained fashion is often different from the time taken to just multicast the information on the underlying communication network. Additionally, a piece of information may traverse a communication link several times during the process. While this is not an issue for lightweight content such as text, it can be a major issue for multimedia content, especially in an ad hoc network/DTN setting where accessing multimedia content directly from a server over a flaky network is not reliable.

In this paper, we present a systematic analytical study of the constrained information flow problem – in particular, we model a pair of networks (social and communication) as a *composite graph* – structures that result from embedding the former network into the latter using *mapping functions* that can map a social network node to a communications network node when the former *uses* the latter as his/her communication portal.

Interactions between different genres of networks are becoming increasingly observable in both natural and man-made systems. Examples include peer-to-peer overlay networks [13], power grid networks and communication networks [5] or the interaction between the programmer social network and interconnected web of software modules developed by them [12].

We analytically study how the "constrained composite path stretch" and "constrained composite broadcast time" metrics scale with the sizes of both networks under consideration. One application scenario where this can be useful is a random deployment of workers or soldiers in a disaster relief site and their leader attempting to disseminate instructions and directives along his/her chain of command (essentially a social network of sorts). These messages trace a logical path in the social network which translates to a longer physical path on the underlying communication network.

1.1 Related Work

Related work in this area falls in three major categories: graph embedding, overlay networks, and network science approaches to study composite networks. Graph embedding has received attention in the parallel computing domain where the problem is to map a *task graph* onto a multiprocessor interconnection network (also known as *host graph*) [3, 14, 8], and in the ubiquitous computing domain where the problem is to map heterogeneous task graphs on non-regular networks such as mobile ad hoc networks [2], while attempting to determine the optimal mapping (or task to processor assignment) function such that metrics such as delay-to-task-completion, edge dilation (or stretch), node/edge congestion, etc. are minimized. Instead of the aforementioned "optimization" approaches, in this paper, we follow the "scaling law analysis" approach where both the graphs and the mapping function are given, and we characterize how a different set of appropriate "constrained" metrics such as composite path stretch, composite diameter, and broadcast time scale as a function of composite graph attributes.

There has been a significant amount of work on overlay networks in the past decade [13]. Seminal works such as CAN [15] and CHORD [16] attempt to design good distributed hash tables for P2P applications – the main endeavor there was to design good logical distributed data structures for storing (key,value) pairs *overlaid* on top of the Internet, so that efficient insertion and retrieval of hashed content is feasible from any part of the network. While this is a good example of a composite network, its similarities with our approach are slim. While overlay networks attempt to design good overlay graphs for the purpose of optimization of insertion/lookup overhead, in our problem space, the social network graph is given, and we are interested in a different set of information flow metrics. Moreover, unlike the Internet which is a complete graph, our underlying network is a multi-hop network, in general.

Recently, network science researchers have begun studying various flavors of composite network structures. Kurant and Thiran propose the Layered Complex Network model [11] for studying load in transportation networks – they considered two-layer graphs where the physical graph models the transportation network and the logical graph models traffic flow between various cities; they use computational methods to determine levels of load on various transportation sectors in Europe. In comparison, our approach is analytical and we also study different metrics. A recent analytical line of research considers interdependent networks such as power grid and communication networks [5] – they use percolation theory to determine the fraction of nodes whose removal is likely to generate cascading failures in such networks. Leicht and D'Souza show that percolation thresholds of composite networks is lower than the individual networks, when considered separately [12]. While these approaches are all analytical, they study a different graph metric, i.e., degree of failure tolerance.

There is a large body of work pertaining to the embedding of one metric space into another – in particular, normed spaces such as d-dimensional Euclidean space $\mathbb{R}^d$) – with "low-distortion". This has been summarized well in [9]. This entails establishing the necessary and sufficient conditions on the properties of the two spaces for finding such embedding functions that yield a particular distortion, and in many cases finding the best embedding function [1]. A related idea of finding embeddings is popular in geographic routing – virtual coordinates are assigned to nodes in a hyperbolic space, and such an embedding guarantees that a greedy algorithm on the virtual coordinate space yields a route between every source and destination, if one exists [10].

The focus of this paper is not *to find* the best embedding function that yields a low distortion – rather, it is *to analyze* the distortion (or stretch) of an information flow that results from a *random* embedding of the nodes of the first graph onto the second graph, in distribution or in expectation.

1.2 Our Contribution

A summary of our contributions is as follows: 1) novel models and metrics for constrained information flow in composite networks; 2) mathematical analysis of scaling laws for constrained composite path stretch when a social network path is randomly mapped onto a general graph under both one-to-one and many-to-one mappings; 3) scaling laws for constrained composite broadcast time of a tree social network (chain of command) randomly mapped onto different communication networks; 4) validation of a subset of these results using the FOAF (friend of a friend) data set embedded on a geometric graph as well as a historical deployment of a chain-of-command.

2. COMPOSITE GRAPH MODELS

For two graphs G_1 and G_2, we define the *composite graph* $\mathcal{G}$ to be the 3-tuple (G_1, G_2, R), where $R \subseteq V(G_1) \times V(G_2)$ is an *embedding relation* between the vertex sets $V(G_1)$ and $V(G_2)$ of the two graphs, respectively.

In this paper, we focus on analyzing *random embedding relations*, where vertices in G_1 are mapped to vertices in G_2 via some random process π. In particular, we study two cases:

1. Each vertex in G_1 is mapped to a vertex in G_2 that has been sampled uniformly at random *with replacement*. This is the many-to-one scenario, where many "social network" nodes can get mapped to the same communication network node.

2. Each vertex in G_1 is mapped to a vertex in G_2 that has been sampled uniformly at random *without replacement*. This is the one-to-one scenario, where a communication network node can host at most one social network node.

In general, every element of R may have multiple attributes associated with it but in this preliminary study we only consider a binary relation. Note that this relation may be time-varying as an information object may be replicated or may even move from one communication node to another over time. This is a topic of future research.

We first analyze *constrained composite path stretch*, a metric that is useful for measuring how many physical communication hops are spanned by a logical information flow under a given embedding of the logical flow on a physical network.

Throughout this paper, let $\mathcal{G} = (G_1, G_2, R)$ be a composite graph, with $V_i = V(G_i)$ the vertex set of graph i. Unless otherwise noted, $P_k = P_{uv} = \{u = v_0, v_1, ..., v_k = v\}$ is a path of length k in G_1, and $d_{G_2} : V_2 \times V_2 \rightarrow \mathbb{R}$ is a shortest path distance metric in G_2.

DEFINITION 2.1 (CSTRETCH). *The constrained composite path stretch of P_{uv} in $\mathcal{G}$ is defined as:*

$$cstretch_{G_2}(P_{uv}) = \sum_{i=0}^{k-1} \max_{\substack{x,y \in V_2: \\ (v_i,x)\in R \wedge (v_{i+1},y)\in R}} \{d_{G_2}(x,y)\}$$

CStretch characterizes the scenario with a stringent requirement that the information needs to traverse the nodes in the path P_{uv} *in order* and in the process need to traverse the appropriately mapped nodes in G_2. This is not a farfetched scenario – in military systems, the chain-of-command (modeled by graph G_1) often mandates a piece of information to flow through the logical chain even though the ultimate recipient of the information may be in close proximity to the origin and the intermediate nodes are farther away from them. The reason behind this is that information often needs to get refined or obfuscated at each level of the logical chain before it is passed on further. Similarly, even in non-military applications (such as online social networks) information such as news or gossip is often routed along logical paths of friends who may be physically located all over the globe at large "Internet distances" from each other.

In the composite graph setting, the notion of diameter[1] can be extended to that of the *constrained composite diameter* which can be defined in terms of constrained composite path stretch.

DEFINITION 2.2 (CCD). *The contrained composite diameter of $\mathcal{G}$ is defined as*

$$ccd(\mathcal{G}) = \max_{u,v \in V_1} cstretch_{G_2}(P_{uv}).$$

The *CStretch* metric captures the extra distance in G_2 that a message has to travel in order to move through a path in G_1. We need a different metric to capture the combined stretch for a message traveling through a chain-of-command tree in a composite graph. In this context, it is more natural to consider the *constrained composite broadcast time* metric.

DEFINITION 2.3 (CBTIME). *Let T be a tree in G_1, with root u. Then the constrained composite broadcast time of T in the composite graph $\mathcal{G}$ is defined as*

$$cbtime_{G_2}(T) = \max_{v \in T} cstretch_{G_2}(P_{uv})$$

[1]Diameter is the maximum length of the shortest path between any pair of nodes in a graph. It is an important measure for communication networks because it gives us a sense of the amount of time required (in the worst case) to traverse a network.

The constrained composite broadcast time represents the stretch necessary to send a message through a chain-of-command tree that is deployed in a network topology. This may be of interest, for example, in a disaster relief situation when information needs to travel from a central director to end caregivers while relief workers are deployed in the field. In other words, it measures the time at which the last worker received the message that was broadcast through the chain of command.

3. COMPOSITE STRETCH ANALYSIS

In this section, we characterize the distribution of the constrained composite path stretch of P_k over uniform random embeddings into G_2. We first prove some general results that apply to any graph G_2, and then illustrate scaling laws for a few well-known graph families.

3.1 Theoretical Results

For any graph $G = (V, E)$, let D_G be the geodesic graph distance matrix between all pairs of vertices $v_i, v_j \in V$. That is, each entry d_{ij} in D_G represents the shortest path distance from v_i to v_j in G. Then we note that the sum of the geodesic distances$\Delta_G = \sum_{v_i,v_j \in V} d_{ij}$, is a constant depending only on the structure of G.

LEMMA 3.1. *Let G be a graph with $|V| = n$, and let X be a random variable denoting the geodesic distance between two vertices of G chosen uniformly at random. Then:*

$$\mathbb{E}[X] = \begin{cases} \frac{\Delta_G}{n(n-1)}, & \text{when sampling without replacement} \\ \frac{\Delta_G}{n^2}, & \text{when sampling with replacement} \end{cases}$$

PROOF. The case where sampling is done with replacement is clear: since there are n^2 pairs of vertices from which to choose, the expression given is the average distance. If sampling is done without replacement, thenΔ_G double-counts the distance for each of the $\binom{n}{2}$ unique pairs of vertices. Note that the n diagonal entries in D_G contribute nothing toΔ_G. □

COROLLARY 3.1. *There is no asymptotic difference in $\mathbb{E}[X]$ between sampling vertices with or without replacement.*

PROOF. From the preceding Lemma, it follows that the ratio of $\mathbb{E}[X]$ when sampling without replacement to $\mathbb{E}[X]$ when sampling with replacement is $1+\frac{1}{n} \rightarrow 1$ as $n \rightarrow \infty$. □

Next, we show that the expected stretch of a link is independent of the choices of vertices already mapped, regardless of whether sampling is done with or without replacement.

LEMMA 3.2. *Let $v_1, v_2, ..., v_i$ be a sequence of vertices chosen uniformly at random from V (with or without replacement), and let X_i be the random variable giving the distance between v_i and v_{i-1}. Then $\mathbb{E}[\mathbb{E}[X_{i+1}|v_1, v_2, ..., v_i]] = \mathbb{E}[X_2]$.*

PROOF. While the statement may be obvious for the case of sampling with replacement, we exercise more care for the case where sampling is done without replacement, and prove the statement combinatorially. For the RHS, select one vertex uniformly at random and color it red (call it v_1). Then select another from the remaining and color it blue (v_2). The RHS counts the expected distance between these two vertices. We now argue that the LHS counts the same. To see this, first color one vertex blue (call it v_{i+1}), and another vertex red (v_i). Now color $i-1$ other vertices green ($v_{i-1}, ..., v_1$). The LHS counts the expected distance between the blue vertex and the red vertex. □

This leads us to a general theorem about the expected composite stretch of a path.

THEOREM 3.1. *For a path P_k embedded uniformly at random into any graph G_2 (with the sampling performed with or without replacement),*

$$\mathbb{E}[cstretch^{\pi}_{G_2}(P_k)] = k \cdot \mathbb{E}[X] ,$$

where X is the random variable giving the distance between two randomly chosen vertices in G_2.

We emphasize the expectation is being taken over the uniform random embedding R_π. But as we saw in Lemma 3.1, for a specific G_2, if the sampling method of R_π is known, then the expected distance $\mathbb{E}[X]$ is a constant.

Composite Diameter: In addition to the average-case, we also want to describe the worst-case *cstretch* for a random embedding. It is easy to see that if R_π samples vertices with replacement, then each successive link in any path can simply bounce back and forth between the furthest two vertices in G_2. Thus, $ccd(\mathcal{G}) = diam(G_1) \cdot diam(G_2)$. However, when R_π samples vertices without replacement, the problem is an instance of MAX-TSP, which is MAX SNP-hard [7]. However, a greedy approximation heuristic works well in practice.

3.2 Examples

Theorem 3.1 shows that the expected stretch of a path is equal to the length of the path times a constant depending only on the structure of G_2 and the distribution of the random embedding. In what follows, we present examples of some well-known graph families, and illustrate how their structure affects the distribution of *cstretch*.

d-dimensional Discrete Lattice: Let $D^d_n = \{0, 1, ..., n-1\}^d$ be the d-dimensional discrete lattice on n^d points, and consider a composite graph with $G_2 = D^d_n$. On this graph topology, geodesic distance is equivalent to the ℓ_1-norm (Manhattan distance) between two points in D^d_n. Thus, $d_{G_2}(v, w) = \sum_{i=1}^{d} |v_i - w_i|$, and summing all n^{2d} of these pairs gives

$$\Delta_{G_2} = \sum_{v,w \in V} \sum_{i=1}^{d} |v_i - w_i| = \frac{dn^{2d+1}}{3}\left(1 - \frac{1}{n^2}\right)$$

It follows from Lemma 3.1 and Theorem 3.1 that under a random uniform embedding with replacement into the d-dimensional discrete lattice,

$$\mathbb{E}[cstretch^{\pi}_{G_2}(P_k)] = \frac{kdn}{3}\left(1 - \frac{1}{n^2}\right)$$

Note that in this case it is also straightforward to fully explicate the distribution of X. For any $1 \le i \le d$, let $X_i = |v_i - w_i|$. Then the probability mass function for X_i is

$$p_{X_i}(\delta) = \begin{cases} \frac{1}{n} & \text{if } \delta = 0 \\ \frac{2(n-\delta)}{n} & \text{otherwise} \end{cases},$$

since each coordinate can take on any of n values, and there are $n-\delta$ ways to achieve each value of δ between 0 and $n-1$.

Since the X_i's are independent and identically distributed, we can extract (among other things), the second moment of X:

$$\text{Var}[X] = d \cdot \frac{(n^2-1)(n^2+2)}{18n^2}.$$

We can infer from this that the expected stretch is not likely to deviate from its mean.

For the discrete lattice, we have that $diam(G_2) = d(n-1)$, so as mentioned above, the *ccd* for P_k is $k(n-1)$. For the non-trivial "without replacement" scenario, we implemented a Greedy approximation heuristic, and verified that *ccd* for both without and with replacement scenarios are $O(n^2)$.

Cycle: Let C_n be the cycle of length n, and consider uniform discrete mappings from P_k onto C_n. Clearly, the maximum distance between two vertices in C_n is $\lfloor \frac{n}{2} \rfloor$. But, for each possible distance x between 0 and $\frac{n}{2}$, there are exactly n such pairs for $x = 0, \frac{n}{2}$, and exactly $2n$ such pairs otherwise. It is thus straightforward to show that

$$\Delta_{C_n} = \begin{cases} \frac{n(n^2-1)}{4} & \text{if } n \text{ is odd} \\ \frac{n^3}{4} & \text{if } n \text{ is even} \end{cases}.$$

Application of Lemma 3.1 and Theorem 3.1 then reveal that for random uniform embeddings onto C_n,

$$\mathbb{E}[cstretch^{\pi}_{C_n}(P_k)] = k \cdot \left(\frac{n}{4} + o(1)\right).$$

Greedy is optimal on C_n, since if n is odd, it finds $n-1$ pairs at distance $\lfloor \frac{n}{2} \rfloor = diam(G_2)$ from each other, which is optimal by definition. On the other hand, if n is even, it picks *all* $\frac{n}{2}$ pairs at distance $\frac{n}{2} = diam(G_2)$ from each other, and another $\left(\frac{n}{2} - 1\right)$ pairs at the next greatest distance $\left(\frac{n}{2} - 1\right)$.

Balloon graph: Next, we consider a graph family with some interesting properties. Let $B_{n,m}$ be a balloon graph consisting of a string (line graph) of length m, connected to a balloon (clique) of size $n-m$, for any $0 \le m < n$. For clarity, we specify that vertices $\{v_0, ..., v_m\}$ make up the string, while vertices $\{v_m, ..., v_{n-1}\}$ make up the balloon. Note that for any two indices $0 \le i < j \le n-1$ in this graph, we have that

$$d_{B_{n,m}}(v_i, v_j) = \begin{cases} j - i & \text{if } i < j < m \\ m + 1 - i & \text{if } i < m \le j \\ 1 & \text{if } m \le i < j \end{cases}$$

In particular, note that $diam(B_{n,m}) = m+1$. In computing the distance matrix, we distinguish three cases based on the indices of the two vertices chosen:

1. If $i \le j \le m$, then both vertices lie in the string, which is D^1_{m+1}. This contributes$\Delta_{D^1_{m+1}}$ towards$\Delta_{B_{n,m}}$.

2. If $m \le i \le j$, then both vertices lie in the balloon, and it is clear that on the complete graph $K_n, \Delta_{K_n} = n^2 - n$, since every pair of vertices are connected by an edge, but there are n ways to choose the same vertex twice.

3. If $i < m < j$, then one vertex lies in the string, and the other lies in the balloon. Consider any vertex w_j in the balloon. Its distance from the set of vertices in the string is simply $m+1, m, m-1, ..., 2$. Thus, the

contribution to $\Delta_{B_{n,m}}$ is

$$2(n-m-1)\sum_{i=2}^{m+1} i = m(m+3)(n-m-1)\,.$$

Adding these three quantities yields

$$\Delta_{B_{n,m}} = -\frac{2}{3}m^3 + (n-2)m^2 + \left(n-\frac{4}{3}\right)m + n^2 - n$$

The reader may verify that setting $m = 0$ corresponds to the special case where the balloon graph is itself a clique, while setting $m = n-1$ yields the special case where $B_{n,n-1} = D_n^1$.

By Theorem 3.1 and Lemma 3.1, the expected stretch for a path of length k onto $B_{n,m}$ is thus:

$$\mathbb{E}[cstretch^{\pi}_{B_{n,m}}(P_k)] = k \cdot \left(1 + O\left(\frac{m^2}{n}\right)\right)$$

Random Geometric Graph: Lastly, we consider the composite stretch when P_k is mapped onto a random geometric graph $G_2 = RGG(n, r(n))$, where $r(n)$ is the radius of communication. That is, G_2 consists of n vertices placed uniformly at random in $[0,1]^2$, wherein any two vertices are connected with an edge if and only if the Euclidean distance between them is at most $r(n)$. Gupta and Kumar [6] showed that a radius of connectivity of $r(n) = \sqrt{\frac{\ln n + c(n)}{\pi n}}$ ensures asymptotic connectivity in the RGG with high probability if and only if $c(n) \to +\infty$. In all of our discussions on RGG in this paper, we assume that the radius of connectivity is at least this large, i.e., $r(n) = \Omega(\sqrt{\ln n/n})$.

As before, Theorem 3.1 still applies, so it remains only to characterize the distribution of the random variable X giving the geodesic distance between two vertices in $RGG(n, r(n))$ selected uniformly at random. Note that in contrast to the previous examples we have considered, we now have two sources of randomness: 1) the randomized construction of the RGG; and 2) the random uniform embedding. If the Euclidean distance between two vertices in an RGG is δ, then recent results confirm that with high probability, the geodesic distance X differs from its minimum of δ/r by at most a constant [4].

THEOREM 3.2. *With high probability, the expected geodesic distance in $RGG(n, r(n))$ satisfies*

$$\frac{\Delta(2)}{r(n)} \leq \mathbb{E}[X] \leq \kappa(n) \cdot \frac{\Delta(2)}{r(n)}\,,$$

where $\Delta(2) \approx 0.5214054331$ is a known constant, and $\kappa(n) \geq 1$ is $O(1)$

PROOF. Let v, w be two vertices in $RGG(n, r(n))$ selected uniformly at random, and set $\delta = ||v - w||_2$. Clearly, $X \geq \delta/r$. Conversely, if $\delta = \Omega(\log^{3.5} n/r^2)$, then by a result from [4], $X = O(\delta/r)$.

Taking expectation yields the result, since $\mathbb{E}[\delta] = \Delta(2)$ is a known result [17]. □

Synthetic analysis suggests that $\kappa(n) < 1.3$ for $n > 1000$. Therefore, as before, we can easily bound (from above) the expected composite stretch.

COROLLARY 3.2. *For $r(n)$ sufficiently large (i.e., greater than the critical connectivity threshold), the composite stretch of a path P_k on a random geometric graph $RGG(n, r(n))$ satisfies with high probability:*

$$\mathbb{E}[cstretch^{\pi}_{RGG}(P_k)] = k \cdot \kappa(n) \cdot \frac{\Delta(2)}{r(n)} = O\left(k \cdot \sqrt{\frac{n}{\ln n}}\right)$$

G_1	G_2	$\mathbb{E}[cstretch]$	$\max[cstretch]$
P_k	D_n^d	$\frac{kdn}{3}(1-n^{-2})$	$kd(n-1)$
	C_n	$k \cdot (\frac{n}{4} + o(1))$	$k \cdot \lfloor \frac{n}{2} \rfloor$
	$B_{n,m}$	$k \cdot \left(1 + O\left(\frac{m^2}{n}\right)\right)$	$k(m+1)$
	$RGG(n, r(n))$	$O\left(k\sqrt{\frac{n}{\ln n}}\right)$	

Table 1: Summary of Path Stretch Metrics for Uniform Random Embeddings of P_k

3.3 Average vs. Worst-Case Analysis

Thus far, we have characterized both the average case (expected *cstretch*) and the worst case (*ccd*) for a random uniform embedding of a path onto several graph families. For both the lattice and the cycle, these quantities were of the same order of magnitude. At this point a natural question arises: Are there graphs for which the ratio of the maximum *cstretch* to the average *cstretch* of P_k is *not* $O(1)$? Indeed, the balloon graph is one such graph. As the diameter of $B_{n,m}$ is $m+1$, the maximum stretch is $diam(G_1) \cdot (m+1)$. If we let $\phi(B_{n,m})$ be ratio of the maximum *cstretch* to the mean *cstretch*, we can see that:

$$\phi(B_{n,m}) = \frac{diam(G_1)(m+1)}{diam(G_1)\left(1 + O\left(\frac{m^2}{n}\right)\right)} = O\left(\frac{n}{m}\right)$$

In particular then, for $m = \sqrt{n}$, the ratio of the maximum stretch to the mean stretch for the balloon graph $B_{n,m}$ is $O(\sqrt{n})$. Explicit calculations reveal that for $m = \sqrt{n}$, in fact $\mathbb{E}[X] \to 2$ as $n \to \infty$.

More interesting is the fact that this gap appears to be mainly an artifact of the difference between sampling with and without replacement. The results of our Greedy algorithm for CCD without replacement suggest that with $m = \sqrt{n}$, the CCD and expected *cstretch* are of the same order to magnitude.

Table 1 summarizes our theoretical results.

4. COMPOSITE BROADCAST TIME

In this section, we analytically characterize the expected composite broadcast time for tree topologies. Social networks for information dissemination commonly have tree structures (more on this in section 5), hence this analysis can be useful for specific communication network deployment scenarios. Let T_k be a k-node tree of height h and maximum (out)degree δ, for some $1 \leq \delta < k$. We assume that T_k exists in some G_1, and examine the constrained composite broadcast time for sending a message from the root to each of the other nodes.

Star Topology: We begin with the special case where T_k is a k-star. First, we introduce some notation. Let

$$p_k = \frac{1}{\binom{n-1}{k}}\left(\underbrace{0, ..., 0}_{k}, 1, \binom{k}{k-1}, ..., \binom{n-2}{k-1}\right) \in \mathbb{R}^n$$

be a column vector, and note that $||p_k||_1 = 1$. The ith entry in p_k represents the probability that the ith largest among n values is returned, when this value is the maximum among a subset of size k chosen uniformly at random. Furthermore, let $f : \mathbb{R}^{m \times n} \to \mathbb{R}^{m \times n}$ be the function that sorts the rows of a matrix in ascending order from left to right. That is,

$$D = \begin{bmatrix} d_1 \\ d_2 \\ \vdots \\ d_m \end{bmatrix} \Rightarrow f(D) = \begin{bmatrix} sort(d_1) \\ sort(d_2) \\ \vdots \\ sort(d_m) \end{bmatrix},$$

where d_i is the ith row of D. Finally, $v_m = \frac{1}{m}(1, ..., 1) \in \mathbb{R}^m$.

THEOREM 4.1. *For any graph G_2, the broadcast time of a star of size k satisfies*

$$\mathbb{E}[cbtime_{G_2}(S_k)] = v_n^T \cdot f(D_{G_2}) \cdot p_k \,.$$

PROOF. Let d_i be the ith row of D_{G_2}, and suppose that the root of the star S_k is mapped to node i in G_2. The broadcast time of S_k is the maximum *cstretch* from among its k children. But since the jth entry of p_k is the probability that the jth largest value in d_i will be returned, the inner product $\langle sort(d_i), p_k \rangle$ gives the expected value of the maximum of the k *cstretch*es. Multiplication on the left by v_n^T simply averages these n values over all n rows. □

Note that this is consistent with Theorem 3.1 for the special case where $k = 2$. Theorem 4.1 allows us to compute the broadcast time of a k-star for a variety of graph families, and we later use these as building blocks for bounds on general trees. Moreover, Theorem 4.1 improves on the trivial upper bound of $diam(G_2)$ for the broadcast time of a star. A better bound can be derived by considering the average eccentricity of G_2. The eccentricity ε of a vertex in a graph is defined as the maximum geodesic distance between that vertex and any other.

COROLLARY 4.1. *For any graph G_2, the broadcast time of a star of size k satisfies*

$$\mathbb{E}[cbtime_{G_2}(S_k)] \leq \frac{1}{n} \sum_{v \in V_2} \varepsilon(v)$$

PROOF. Substituting p_{n-1} in place of p_{k-1} returns the average eccentricity of the vertices in G_2. □

Corollary 4.1 provides a better bound than the diameter, but is not nearly as good as when using Theorem 4.1 directly. To illustrate how Theorem 4.1 can be used for a specific G_2, we provide an upper bound on the broadcast time of a star, when G_2 is the line lattice above.

COROLLARY 4.2. *For $G_2 = D_n^1$, the line lattice, the broadcast time of a star of size k satisfies*

$$\mathbb{E}[cbtime_{G_2}(S_k)] \leq \frac{k}{k+1} \cdot n \,.$$

PROOF. The maximum product on the right certainly occurs at $d_1 = (0, 1, 2, ..., n-1)$, which is already sorted. Thus,

$$\begin{aligned} \langle d_1, p_k \rangle &= \frac{1}{\binom{n-1}{k}} \sum_{j=k+1}^{n} (j-1) \binom{j-2}{k-1} \\ &= \frac{k}{\binom{n-1}{k}} \sum_{j=k+1}^{n} \binom{j-1}{k} \\ &= \frac{k}{\binom{n-1}{k}} \binom{n}{k+1} = \frac{k}{k+1} \cdot n \end{aligned}$$

□

Tree topology: For any tree T_k with maximum degree δ, let δ_i be the maximum outdegree among nodes at height $1 \leq i \leq h$ in T_k.

OBSERVATION 4.1 (CBTIME: LOWER BOUND).

$$\mathbb{E}[cbtime_{G_2}(T_k)] \geq \mathbb{E}[cstretch_{G_2}(P_h)]$$

PROOF. The lower bound represents the expected *cstretch* of a path of length h, which is the longest in T_k. No other single path from the root to a leaf could have expectation longer than this, so the expectation for the tree must be at least this large. □

OBSERVATION 4.2 (CBTIME: UPPER BOUND).

$$\mathbb{E}[cbtime_{G_2}(T_k)] \leq \sum_{\ell=1}^{h} \mathbb{E}[cbtime_{G_2}(S_{\delta_\ell})]$$

PROOF. The upper bound represents the sum (over all h levels of T_k) of the expected composite broadcast time for the largest star graph S_{δ_ℓ} at each level ℓ. This the maximum expected broadcast time, since no path from the root to a leaf could take longer than this. □

Combining Theorem 3.1 with Observations 4.1 and 4.2 yields the following bounds on the expected broadcast time of a general tree.

COROLLARY 4.3. *For any tree T_k of height h and maximum outdegree δ,*

$$h \cdot \mathbb{E}[X] \leq \mathbb{E}[cbtime_{G_2}(T_k)] \leq h \cdot \mathbb{E}[cbtime_{G_2}(S_\delta)]\,,$$

where X is the r.v. giving the expected geodesic distance between two vertices in G_2.

PROOF. The lower bound is an application of Theorem 3.1 to Observation 4.1, while the upper bound follows from Observation 4.2 and the fact that $\delta = \max_{1 \leq \ell \leq h} \delta_\ell$. □

5. SIMULATION BASED EVALUATION

In this section, we use a variety of simulation-based approaches to study how the composite stretch and composite broadcast time metrics behave under various choices of real (or at least realistic) social and communication networks. We first study the case where the social network (or G_1) is a hierarchical network like a chain of command that exists in military missions or in disaster relief operations. Our second case study is that of a social network that is richer than a tree as in friendship relationships.

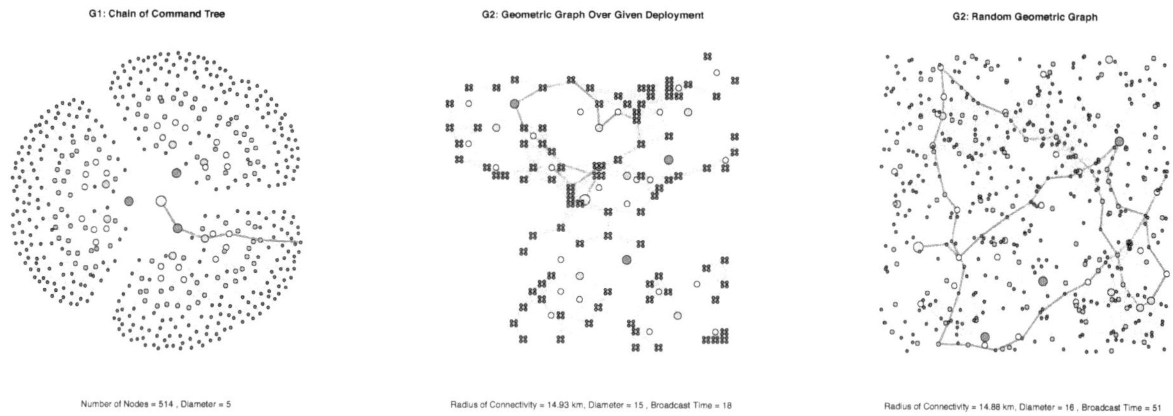

(a) G_1 (b) G_2 (c) G_2 under S2

Figure 1: Examples of *chain of command* networks. (a) a chain of command social network; (b) historical deployment of the chain of command shown in (a); (c) one random deployment of the chain of command shown in (a) under scenario S2.

5.1 Chain-of-Command

Figure 1(a) denotes the chain of command hierarchy from within a representative brigade reporting structure in the US Army. Commands move from the highest ranked (root, shown in pink) node to the lower ranked nodes in the tree until they reach the lowest ranked (leaf, shown in blue) nodes. Though not depicted in Figure 1(a), each node in G_1 occupies a physical location in space. In Figure 1(b), we show the geometric graph G_2 constructed from these physical locations by adding a communication edge between two nodes if they lie within a prescribed radio transmission range (i.e. - the critical radius of connectivity $r(n)$ outlined above) of each other. The coordinates of the nodes in G_1 are specified according to a historical deployment scenario over a $124 \times 148 km^2$ area. G_2 is not an RGG but only a geometric graph with radius of communication $14.93km$.

The composite graph $\mathcal{G}^* = (G_1, G_2, R)$ that combines Figures 1(a) and 1(b), along with the identity mapping R, is a realistic composite network structure for a military or disaster relief deployment. In this instance, the broadcast time is 18, though eccentricity in G_2 of the root node of G_1 is 9. Thus, the constraints imposed on the information flow by the chain of command require a message to travel through twice as many hops as was mandated by the actual deployment. A path that produces the broadcast time is highlighted in red in Figures 1(a) and 1(b).

In order to put this broadcast time of 18 in context, we simulated three different randomized scenarios, each of which could produce $\mathcal{G}^*$ as a singular outcome:

S1 Instead of R being the identity mapping, R is a random permutation.

S2 Instead of G_2 being a geometric graph over the actual coordinates of deployment, each node in G_1 was assigned a random 2D coordinate drawn from a square region of the same area as in S1. Therefore, G_2 is a *random* geometric graph (RGG). Fig. 1(c) shows an example of such a deployment.

S3 Instead of G_2 being a geometric graph over the actual coordinates of deployment, the coordinates were generated according to a random model that is a function of the chain of command tree G_1, with R as the identiy mapping. Details of the model are given below.

In the actual deployment, note that the lowest ranked nodes are collocated in bunches of four, hence Fig. 1(b) appears sparser than Fig. 1(c). Moreover, the broadcast time jumps from 18 to 51 for this particular RGG. This is because in the real deployment, there is strong correlation between the location of a node and its rank in the command hierarchy (even though the maximum *cstretch* is as high as 18), which does not exist in random deployments. Inspired by this observation, we constructed a correlated random deployment model for S3.

Details of Deployment Model for S3: Our model places each child in an equi-spaced, but randomly-oriented, ring around its parent, wiht a random jigger applied in both the horizontal and vertical directions. This process is recursively applied down the chain of command tree, which we assume has height h. Let v_i be the node in G_1 at distance h_i from the root node, and having n_i children. Then the location of v_i's children are determined as follows:

1. Find the n_i roots on unity $\omega_1, ..., \omega_{n_i}$ and associate one with each of the n_i children.
2. Draw a uniform random variable $u \in [0, 1]$.
3. Set the target distance from parent to child to be $\rho_i - a(h - h_i)^2$, where a is a parameter determined from analysis of the actual deployment data. [In our case $a = 1.85$.].
4. For each $j \in 1, ..., n_i$, set the target location $c_j = \rho_i \cdot e^{2\pi i u} \cdot \omega_j$, and draw two random coordinates x_j and y_j from normal distributions with mean $\Re(c_j)$ and $\Im(c_j)$, respectively, and standard deviation $b\bar{\rho}_{h_i}$. Here b is a parameter determined from the data (we assume a model with a constant coefficient of variation, $b = 0.293$), and $\bar{\rho}_{h_i}$ is the mean distance between parent and child at height h_i.
5. Return $(x_{v_i}, y_{v_i}) + (x_j, y_j)$.

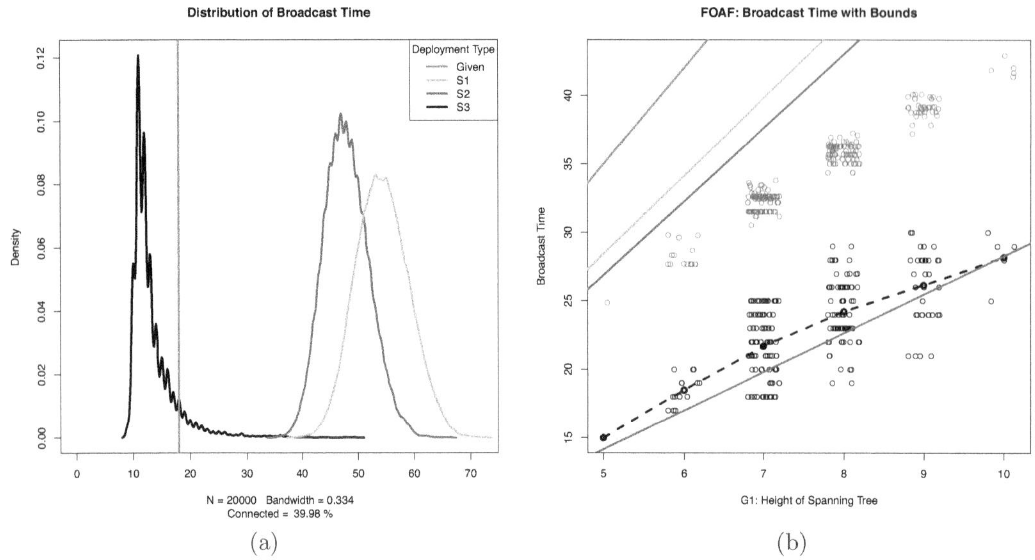

Figure 2: Composite Broadcast time: (a) comparison of *CBtime* for various deployments of the chain of command shown in Figure 1(a); (b) *CBtime* in the composite graph specified in Figure 3. Various upper and lower bounds derived in Section 4 are illustrated.

Fig. 2(a) plots the simulated distribution of broadcast time for each of the aforementioned scenarios. That the red vertical line (corresponding to $\mathcal{G}^*$) falls within the distribution of S3 (more precisely, in the 92nd percentile), suggests that our correlated deployment model is a useful one. Moreover, the other two scenarios, both of which were agnostic to the structure of the chain of command tree, performed comparatively poorly. Empirically, the probability of obtaining a broadcast time as low as that of $\mathcal{G}^*$ via either S1 or S2 appears to be negligible. This illustrates the potential efficiency dangers inherent to composite networks.

5.2 Friend-Of-A-Friend (FOAF)

Now we consider a small social network data set that was extracted from the Semantic Web Billion Triples Challenge (BTC) program. The data contains unique identifiers and indicates the existence of friendship relations between pairs of users – as per the *friend of a friend (FOAF) ontology*. It also contains the geographic coordinates of users.

One can imagine the IDs in the data set communicating with their *friends* over some underlying communication network. In this scenario, the communication would typically happen over the wired Internet that connects various users on the map; however, we used this node distribution data to study the composite stretch of the FOAF social network on a geographically distributed multi-hop network assuming a geometric graph model as described in Sec. 5.1 – in particular, we place a node at each location in the data set and construct a graph using a transmission radius that is large enough to connect most nodes in the network (akin to the notion of critical radius in case of a random geometric graph). In Figure 3, we show a 237 node social network (G_1), alongside the geometric graph imposed over the same set of vertices using the connectivity radius $r(n) = 5.1$ degrees of latitude/longitude.

Since G_1 is a rich social network, all nodes can act as sources of information which can flow along random spanning trees of G_1. Hence we pick all nodes in G_1 one by one; and assign them as roots of random spanning trees. We then measure *cbtime* for each root node and plot the results in Fig. 2(b) (black jitter-spaced hollow circles). The dashed black line shows the average *cbtime* as a function of the height of the spanning tree. That this is approximately linear accords with Corollary 4.3.

On the same figure (Fig. 2(b)) we plot the upper and lower bound formulas for $\mathbb{E}[cbtime]$ that we derived in Sec. 4. The blue lines indicate the bounds of Corollary 4.3, while the green line shows the weaker bound of Corollary 4.1 and the red line shows the trivial diamater bound. The jittered blue circles indicate the strongest bound for each spanning tree, derived from Observation 4.2. We observe that the lower bound is reasonably tight whereas the upper bounds based on Theorem 4.1 and average eccentricity are looser, but much better than the trivial diameter upper bound. The upper bound obtained from Observation 4.2 is tighter, since it takes the local structure of the spanning tree into account, rather than simply the globally maximum outdegree (46 in this case). In summary, this validates our analytical results.

6. CONCLUSION AND DISCUSSION

In this paper, we presented an analytical modeling framework (both models and metrics) for studying the composite stretch (or elongation) suffered by information on a multi-hop communication network when it is *constrained* by social network relationships. We derive scaling laws (in expected value sense) for composite stretch for random embeddings of common social network structures such as trees on a variety of multi-hop communication network models, ranging from simple linear networks to random geometric graphs.

We also analyze the constrained broadcast time metric, which measures the time taken to broadcast (or gossip) information along the edges of a social network to *all* nodes in that network while being constrained by the underlying

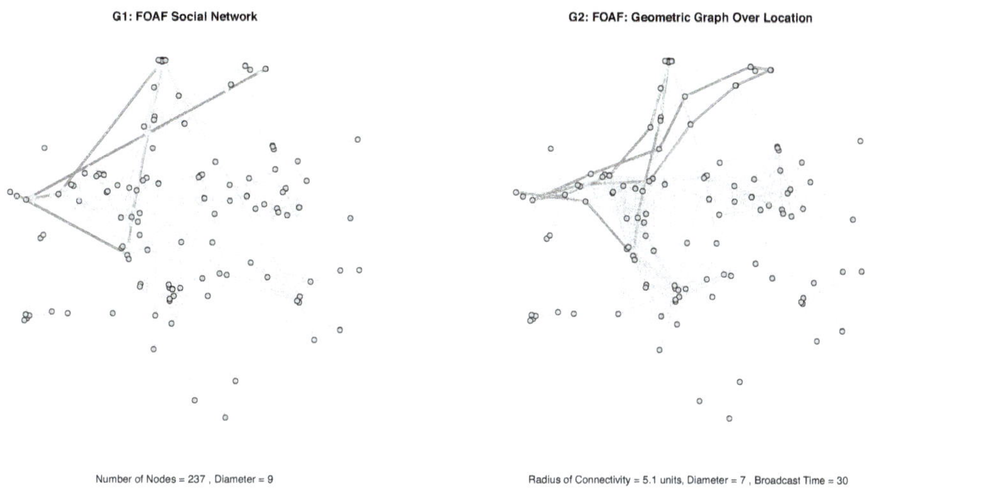

Figure 3: Social Network and Random Geometric Graph (constructed relative to the specified locations)

communication network structure. We derived analytical bounds for composite broadcast time and validated them by simulations using a friendship social network data set. We also show, using simulations based on a historical military deployment data set how the *stretch* of an information flow in a chain-of-command network is both non-optimal, but far superior to a random deployment scenario.

This is the first step toward an ambitious research program directed toward studying composite networks. Other composite metrics of interest include "composite load," which measures the number of times a piece of information needs to traverse a particular edge in G_2. Additionally, other interesting variants of the constrained stretch metric exist – suppose one is only allowed to direct communication links that exist between friends (since these are supposed to be trusted) – this will result in a routing which is likely to have a higher stretch since shorter communication paths between friends (through non-friends) are disallowed. How can one characterize this highly constrained stretch? Such insights should help us *design* better communication networks (or facilitate intelligent deployment) that are suited to particular demands of the overlaid social networks of users.

7. REFERENCES

[1] Y. Bartal. On approximating arbitrary metrices by tree metrics. In *Proceedings STOC '98*, STOC '98, pages 161–168, New York, NY, USA, 1998. ACM.

[2] P. Basu, W. Ke, and T. D. C. Little. Dynamic task-based anycasting in mobile ad hoc networks. *Mob. Netw. Appl.*, 8:593–612, October 2003.

[3] S. Bokhari. On the Mapping Problem. *IEEE Transactions on Computers*, 30(3), 1981.

[4] M. Bradonjic, R. Elsässer, T. Friedrich, T. Sauerwald, and A. Stauffer. Efficient broadcast on random geometric graphs. In *Proceedings of SODA '10*, pages 1412–1421. Citeseer, 2010.

[5] S. V. Buldyrev, R. Parshani, G. Paul, H. E. Stanley, and S. Havlin. Catastrophic cascade of failures in interdependent networks. *Nature*, 464:1025–1028, April 2010.

[6] P. Gupta and P.R. Kumar. Critical power for asymptotic connectivity. In *Proceedings of the 37th IEEE Conference on Decision and Control*, volume 1, pages 1106–1110. IEEE, 1998.

[7] G. Gutin and A.P. Punnen. *The traveling salesman problem and its variations*, volume 12. Kluwer Academic Pub, 2002.

[8] C. C. Hui and S. T. Chanson. Allocating Task Interaction Graphs to Processors in Heterogeneous Networks. *IEEE Transactions On Parallel And Distributed Systems*, 8(9), September 1997.

[9] P. Indyk and J. Matousek. Low-Distortion Embeddings of Finite Metric Spaces. In *in Handbook of Discrete and Computational Geometry*, pages 177–196. CRC Press, 2004.

[10] R. Kleinberg. Geographic Routing Using Hyperbolic Space. In *Proceedings of IEEE INFOCOM*, INFOCOM '07. IEEE, 2007.

[11] M. Kurant and P. Thiran. Layered Complex Networks. *Physical Review Letters*, 96(138701), 2006.

[12] E. A. Leicht and R. M. D'Souza. Percolation on interacting networks. *arXiv*, 2009. `http://arxiv.org/abs/0907.0894`.

[13] E. K. Lua, J. Crowcroft, M. Pias, R. Sharma, and S. Lim. A survey and comparison of peer-to-peer overlay network schemes. *IEEE Communications Surveys and Tutorials*, 7(2):72–93, 2005.

[14] R. Monien and H. Sudborough. Embedding one Interconnection Network in Another. *Computing Suppl.*, 7:257–282, 1990.

[15] S. Ratnasamy, P. Francis, M. Handley, R. Karp, and S. Shenker. A scalable content addressable network. In *Proceedings of SIGCOMM '01*, pages 161–172, 2001.

[16] I. Stoica, R. Morris, D. Karger, M. F. Kaashoek, and H. Balakrishnan. Chord: A scalable peer-to-peer lookup service for internet applications. In *Proceedings of SIGCOMM '01*, pages 149–160, 2001.

[17] E. Weisstein. Hypercube Line Picking, 2010. `http://mathworld.wolfram.com/HypercubeLinePicking.html`.

SMOOTH: A Simple Way to Model Human Mobility*

Aarti Munjal
Dept. of Math. and Computer Sciences
Colorado School of Mines
Golden, Colorado 80401
amunjal@mines.edu

Tracy Camp
Dept. of Math. and Computer Sciences
Colorado School of Mines
Golden, Colorado 80401
tcamp@mines.edu

William C. Navidi
Dept. of Math. and Computer Sciences
Colorado School of Mines
Golden, Colorado 80401
wnavidi@mines.edu

ABSTRACT

In addition to being realistic, a mobility model should be easy to understand and use. Unfortunately, most of the simple mobility models proposed thus far are not realistic and most of the realistic mobility models proposed thus far are not simple to use. The main contribution of this work is to present SMOOTH, a new mobility model that is realistic (e.g., SMOOTH is based on several known features of human movement) and is simple to use (e.g., SMOOTH does not have any complex input parameters). We first present SMOOTH. We then validate that SMOOTH imitates human movement patterns present in real mobility traces collected from a range of diverse scenarios. In addition, we compare SMOOTH with the other mobility models developed based on these mobility traces. Thus, with SMOOTH, we provide researchers with a tool that allows them to leverage the statistical features present in real human movement in a simple and easy to understand manner.

Categories and Subject Descriptors

C.2.1 [**Network Architecture and Design**]: Store and forward networks

General Terms

Performance,Theory

Keywords

mobility model; mobile ad hoc networks; simulation credibility; performance evaluation; realistic mobility.

1. INTRODUCTION

Several mobility models have been developed in the past [3]. Broadly, these mobility models can be categorized into *synthetic* and *trace-based* mobility models. The synthetic traces generated by a synthetic mobility model are not necessarily *realistic*. A *trace-based* mobility model, on the other hand, may not be *simple* to understand and use. To be widely adopted by the research community, a mobility model should be easy to comprehend and use. For example, many researchers know the RWP (Random Waypoint) mobility model is unrealistic [26], but it continues to be used due to its simplicity. In addition, a simple structure makes it easy to analyze the model mathematically [19]. The contribution of our work is in developing a *simple* mobility model that is *realistic*; our model is realistic since it generates synthetic traces that match well with real human movement. Other *simple* mobility models proposed thus far do not mimic real human movement [3, 21] and other *trace-based* mobility models proposed thus far are not *simple* to use [14]. Thus, the main contribution of this work is to provide a *simple* mobility model, SMOOTH, that is *realistic*.

To validate that our mobility model, SMOOTH, imitates real human movement, we compare the synthetic traces generated by SMOOTH with the real mobility traces collected from a wide range of scenarios [21, 22]. The statistical features present in these mobility traces have been previously analyzed in [8, 13, 14, 16]. (See Section 3 for details on these statistical features.) We simply tune SMOOTH to mimic the real human walk patterns. While real human walks are difficult to predict accurately, several statistical features present in these walks make their movement patterns predictable to some extent [24]. Thus, to develop a mobility model that can generate synthetic traces similar to real human movement, it is necessary to identify these features. None of the mobility models proposed thus far models all of the known statistical features in a simple manner.

The rest of the paper is organized as follows. In Section 2, we briefly describe the recent work that has been done in this area. Section 3 describes SMOOTH, our new mobility model, in detail. In Section 4, we validate the realism of SMOOTH by comparing the statistical features present in the synthetic traces generated by SMOOTH with the real movement traces. We also show that SMOOTH fits the statistical features present in the real movement traces better than three other trace-based mobility models. In Section 5, we compare the performance of three DTN routing protocols when using four different trace-based mobility models. Lastly, we present our conclusions in Section 6.

*This work supported in part by NSF Grant CNS-0905513. Research group's URL is *http://toilers.mines.edu*. This site gives details on obtaining SMOOTH.

MSWiM'11, October 31–November 4, 2011, Miami, Florida, USA.

2. RELATED WORK

While synthetic mobility models are simple to use, they may not generate synthetic traces that mimic real human movement. Thus, the research community has recently focused on developing trace-based models (i.e., models that are constructed from real mobility traces). In trace-based models, real human movement traces are collected from various scenarios (e.g., campus, conferences, offices). The collected traces are then analyzed to identify the key characteristics of real human walks. Using this information, a model is constructed and is called a *trace-based* mobility model. The model is expected to generate synthetic traces that match well with the real human walk traces analyzed.

Different trace-based models developed are found in [7, 10, 12, 14, 18, 20, 21, 27]. We refer readers to [1] for a survey of trace-based mobility models. In [16], the authors developed a simple mobility model, SWIM, that generates synthetic traces representing the social behavior of humans as they move. The authors validated the synthetic traces generated by SWIM using real movement traces collected from three different experiments. The synthetic traces generated by SWIM seem to match the real traces well. The movement traces used for validation, however, are only from a campus and a conference scenario; thus, these traces may not capture the movement patterns from other scenarios, e.g., cities.

In [21], the authors developed a mobility model, TLW (Truncated Levy Walk), that is based on the real GPS traces collected from five different outdoor sites. The GPS traces used in TLW are from diverse scenarios (two campuses, one state fair, one amusement park, and one metro city); however, the movement of mobile nodes is completely random and, thus, TLW does not capture the regularity of trips commonly found in human walks. Another mobility model from the same research group, SLAW [14], was developed based on the same set of traces used in [21]. The synthetic traces generated by the SLAW mobility model seem to capture the social behavior of humans as they move. These synthetic traces do, however, heavily depend on an input parameter called *hurst*. Lee et al. in [14] and our previous work in [17] show that the *hurst* value has a significant effect on the performance metrics of a routing protocol. Unfortunately, it is difficult to estimate the *hurst* value for a given data set [4]. In addition, due to the inherent complexity in techniques used for estimating *hurst* from a given data set and the possible biased trends that might be present in the measurements, one might find it difficult to use SLAW in the way one intends.

3. SMOOTH: OUR SIMPLE AND REALISTIC MOBILITY MODEL

In the evaluation of real human walks, researchers have found several statistical features that exist [2, 6, 11, 12, 21, 23, 24]. Specifically,

- **feature1:** The flights (i.e., a straight-line distance covered between two consecutive locations) distribution of mobile nodes follows a truncated power-law (TPL).[1]
- **feature2:** The ICTs (inter-contact times) (i.e., the amount of time between two successive contacts of the same pair of nodes) distribution of mobile nodes follows TPL.
- **feature3:** The pause-times (i.e., the amount of time a node pauses at a location) distribution of mobile nodes follows TPL.
- **feature4:** The distribution of mobile nodes is non-uniform in the network.
- **feature5:** Mobile nodes do not move randomly in the network; instead, their movement patterns can be predicted to some extent due to the regularity present in their movement. (See the next two features for details.)
- **feature6:** A mobile node explores a new location with probability (say *prob_explore*) inversely proportional to the total number of distinct locations it has visited so far.
- **feature7:** A mobile nodes visits a previously visited location with probability given by (1-*prob_explore*).

We capture all seven of these features in SMOOTH. First, we describe SMOOTH in detail. We then show how SMOOTH meets **feature1** - **feature7** of real human walks.

3.1 Initial Placement of Clusters and Mobile Nodes

When evaluating real GPS traces (e.g., traces in [13]), it is evident that mobile nodes do not move randomly over the simulation area. Specifically, a mobile node visits a few locations more frequently than other locations. Such movement patterns of mobile nodes represent the social behavior of humans and create communities for mobile nodes to visit. In SMOOTH, we represent communities by clusters, where a popular cluster is visited more often than a non-popular cluster. For example, students spend more time in an academic building (including classrooms) than they spend in a food court. Similarly, within a building, there are places that are visited more often than others (e.g., classrooms are visited more often than a conference room). Thus, one of the input parameters to SMOOTH is the total number of clusters, C, available in the network. (For a complete list of inputs to SMOOTH, see Section 3.2.) To simulate the level of popularity of a cluster in the network, we first assign a real number, *p_cluster*, to every cluster, such that collectively, for C clusters in the network:

$$\sum_{i=1}^{C} p\_cluster(i) \approx 1 \qquad (1)$$

where *p_cluster(i)* is the probability of cluster i. Thus, a popular cluster is represented by a cluster with a large value of *p_cluster*.

We do not define cluster boundaries in SMOOTH; instead, every cluster in SMOOTH is represented by a single *(x,y)* coordinate called *c_landmark*, where *c_landmark(i)* is the landmark that represents the *(x,y)* coordinates for cluster i. In other words, *c_landmark(i)* can be viewed as the hub[2] of cluster i. Every cluster landmark is placed uniformly over the simulation area such that no two landmarks are within each other's transmission range (i.e., R). To define cluster boundaries in a network, grids or other techniques that define the area of a cluster can be used; however, we later show that our simple scheme without boundaries accurately

[1]The truncated power-law distribution follows power-law upto a certain time after which it is truncated by an exponential cut-off.

[2]The hub may not be the *center* of the cluster.

models the community structures that exist in real traces of human walks.

For initial placement of a mobile node in the network, it selects a *c_landmark(i)* with a probability *p_cluster(i)*. The node is then placed within 0.5R (transmission range)[3] of the chosen landmark. Since a cluster with a high value of *p_cluster* is selected more often than others, the initial placement of mobile nodes in the network is concentrated in a few regions and, thus, leads to a non-uniform placement of mobile nodes in the network. The placement of a mobile node within 0.5R of a cluster's landmark ensures that all the mobile nodes that belong to the same cluster do not have the same start position. Next, we discuss the movement patterns of mobile nodes generated by SMOOTH and show how the patterns created satisfy all seven statistical features (**feature1** - **feature7**) found in real human walks.

3.2 Movement Patterns in SMOOTH

We provide the step-by-step pseudocode of SMOOTH in Algorithm 1. Further details on the algorithm follow. In SMOOTH, we first place a landmark for every cluster (Step 1). A mobile node chooses a cluster (i.e., a community) with probability *p_cluster* (Step 3), i.e., a bigger cluster is chosen more often than a smaller cluster since bigger clusters are more popular. Thus, the initial distribution of mobile nodes is non-uniform in the network (**feature4**).

While moving, the locations visited by a mobile node are not completely independent; instead, a mobile node visits a few locations (e.g., home, office, gym etc.) more often than other locations. A recent study in [23] done on more than a million mobile phone users suggests that, while moving, the following are true:

1. A mobile node n chooses to explore a *new* location with probability proportional to the number of distinct locations visited so far. Specifically,

$$prob\_explore(n) = aD_n^{-b} \quad (2)$$

   where
   - D_n is the total number of distinct locations visited by mobile node n so far, and
   - $a \approx 0.6$ *and* $b = 0.21 \pm 0.02$ are the constants derived from the empirical data analyzed in [23]. See [23] for details on these parameters and the data used.

   As D_n increases in equation 2, the probability to visit a new location, i.e., *(prob_explore(n))* decreases and, thus, the probability to visit one of the previously visited locations, i.e., *(1-prob_explore(n))* increases. In other words, as time continues, it becomes easier to predict the movement pattern of a mobile node.

2. The selection of one of the previously visited D_n locations is proportional to the total number of times the mobile node n has visited this location before. Specifically,

$$p(l) \propto n(l) \quad (3)$$

where

- $p(l)$ denotes the probability of selecting location l and
- $n(l)$ is the total number of times a mobile node has visited location l so far.

[3]In this work, we assume that every mobile node has the same transmission range, i.e., R.

After the initial placement of mobile nodes is done, we determine each mobile node's next location. For this purpose, we generate a uniform random number between 0 and 1 (Step 7). Based on the values for U and *prob_explore(n)*, a mobile node n then either chooses to visit a new location (Step 11; **feature6**) or returns to one of the locations it has visited before (Step 9; **feature7**). All the flights made by a mobile node due to Steps 9 and 11 are power-law distributed (**feature1**). (A random variable Z is power-law distributed if $P[Z > z] \sim z^{-\alpha}$ for $\alpha > 0$.) For a previously visited location, the flight length continues to be power-law distributed. For a new location to visit, we first generate the length of the flight (i.e., *flight*) for a mobile node using a power-law flight distribution (see the list of input parameters for details). Using the mobile node's current location (x_1, y_1), we then calculate its next location (x_2, y_2) as follows:

1. Generate angle uniformly in (0, 360).
2. $x_2 = x_1 + flight$*cos(angle).
3. $y_2 = y_1 + flight$*sin(angle).

Pause-times are also power-law distributed (Step 13), i.e., mobile nodes pause for a long time at a few locations visited and pause for a short time at the majority of locations visited (**feature3**). (See the list of input parameters to SMOOTH for details.) Since the initial placement of the mobile nodes is not completely independent (i.e., mobile nodes are concentrated in bigger clusters with higher probability), and since the selection of future locations to visit depends on the locations visited so far, a few mobile nodes meet each other more often than others (**feature2**). Due to this regularity, the amount of randomness in the movement patterns of mobile nodes is negligible and, thus, satisfies **feature5**.

Algorithm 1 SMOOTH: pseudocode

```
1.  Place landmarks for every cluster (see Section 3.1 for details).
2.  for each mobile node do
3.     Select a cluster i with probability p_cluster(i) (see Section
       3.1 for details).
4.  end for
5.  while (t ≤ sim_time) do
6.     for each mobile node n do
7.        Generate U ~ (0,1)
8.        if (U > prob_explore(n)) then
9.           Visit a previously visited location, such that equation
             3 is satisfied.
10.       else
11.          Select a new location to visit, such that the length of
             the flight is power-law distributed.
12.       end if
13.       At each location, pause for a pause-time distributed by
          power-law.
14.    end for
15. end while
```

The input parameters to SMOOTH follow.

1. *sim_time* is the total simulation time.
2. *nodes* is the total number of mobile nodes in the network.
3. *area* is the size of the simulation area.
4. *range (R)* is the transmission range of a mobile node.[4]
5. *clusters (C)* is the total number of clusters (to be plotted) in the network.

6. (α, f_{min}, f_{max}) are the parameters for the flights distribution (i.e., the parameters describe the asymptotic behavior of the flights distribution).

7. (β, p_{min}, p_{max}) are the parameters for the pause-time distribution (i.e., the parameters describe the asymptotic behavior of the pause distribution).

As discussed previously, every mobile node chooses one of the clusters' landmarks as its initial location using the probability distribution defined by *p_cluster(i)* (see Section 3.1). *prob_explore* for the mobile nodes is then calculated (given by equation 2). Using this probability and a random number, the mobile node either chooses to move towards a new location or stay at the same location (since this is the only location it has visited so far). If it chooses to visit a new location, we calculate its flight length (i.e., power-law distributed with input parameters α, f_{min}, f_{max}), where $\alpha > 0$. The node then starts moving towards the new location with a speed that is proportional to its flight length. The speed for a mobile node in SMOOTH is based on work done by [21]. Specifically, Rhee et. al calculate two constants (based on real trace data) that can be used to model human speed. Thus, the time taken to cover a flight (as proposed in [21]) is given by:

$$t = k * f_l^{1-\rho} \tag{4}$$

where

- t is the time taken to cover a flight of length f_l.
- k, ρ are the constants that depend on the flight length, i.e., f_l. Specifically, for $f_l < 500$, k=18.72 and ρ=0.79; for $f_l \geq 500$, k=1.37 and ρ=0.36.

Thus, the speed for a mobile node is calculated as f_l/t. After the mobile node reaches its destination, it stays there for a pause-time distributed by power-law (with the input parameters β, p_{min}, p_{max}), where $\beta > 0$. As the mobile node moves, we track its flights' lengths and ICTs (inter-contact times) with other mobile nodes.

3.3 Power-law Flights and ICTs

To illustrate that SMOOTH preserves the statistical features of real human movement, we simulated the two scenarios listed in Table 1. Due to space constraints, we only include results for Scenario I; we note, however, that results for Scenario II are similar. Figure 1 shows the cluster landmarks and the initial placement of mobile nodes w.r.t. the cluster landmarks (in red) for Scenario I. As shown in the figure, the initial placement of mobile nodes is concentrated near the landmarks. Figure 2 shows the locations visited by the mobile nodes during the simulation of Scenario I. Since the mobile nodes have a higher probability to move to locations visited earlier in simulation, the visited locations are concentrated near the landmarks. We also extracted the distributions of flights and ICTs among the mobile nodes during simulation; Figure 3 shows the CCDFs (complementary cumulative distribution functions) of flights and ICTs generated by SMOOTH for Scenario I. The figure illustrates that the flights and ICTs are distributed by truncated power-law (satisfies **feature1** and **feature2**).

[4]Many mobility models do not require the transmission range. SMOOTH needs the transmission range to associate a mobile node with the landmark of a community.

Parameter	Scenario1	Scenario2
sim_time (hours)	3	3
nodes	20	10
area (m^2)	1000x1000	800x800
range (m)	100	100
clusters	10	5
(α, f_{min}, f_{max})	(2.2, 5, 500)	(1.75, 1, 500)
(β, p_{min}, p_{max})	(2.1, 10s, 6h)	(1.8, 1s, 3h)

Table 1: Example scenarios simulated on SMOOTH.

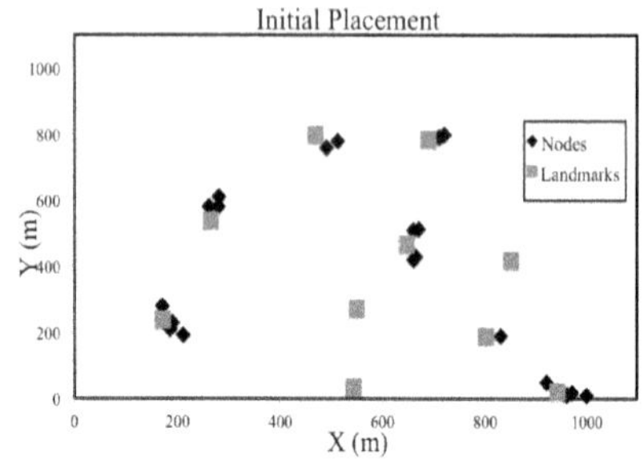

Figure 1: Initial placement of mobile nodes for Scenario I listed in Table 1.

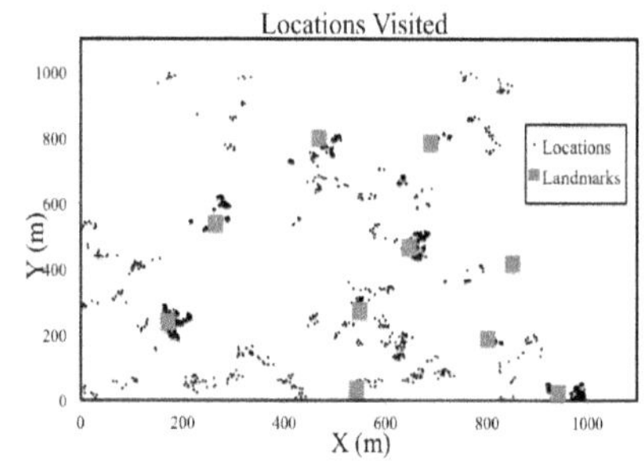

Figure 2: Locations visited by mobile nodes for Scenario I listed in Table 1.

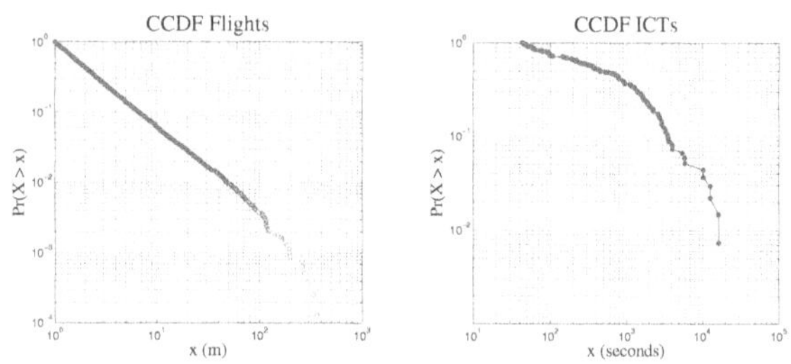

Figure 3: CCDFs of flights and ICTs extracted from SMOOTH for Scenario I listed in Table 1.

4. VALIDATION OF SMOOTH

SMOOTH was created via several known features of real human movement. In this section, we validate our mobility model by using data collected from real human movement. For this purpose, we use the real GPS traces collected from five different sites that represent diverse scenarios (two campuses, one metro city, one state fair, and one amusement park) [13]. To validate that SMOOTH generates synthetic traces that can mimic real human movement in diverse scenarios, we compare the synthetic traces generated by SMOOTH with the traces collected via GPS. Table 2 lists these five scenarios and their respective characteristics. (The values in each column have been rounded for convenience.) A brief description of two columns follows:

- **Locations** denote the total number of locations visited by the GPS users of this site.
- **Hours** denote the total duration (in hours) of the trace data collected from this site.

Since the location data for users was collected using GPS devices, the data is bound to have some noise. For data cleanup, we used techniques similar to [21]. Specifically, we used three methods, namely *rectangular, angle, and pause*, to aggregate the flights. For example, the GPS traces might contain several short flights made in the same direction; thus, the angle-based method joins all the flights made by a user into a single flight, if no pause occurs between consecutive flights and the angle is $< \theta$. See [21] for further details on these methods.

Scenario	Locations	Area	Nodes	Hours
Kaist	10600	10000x18000	34	1400
NCSU	3300	2500x2500	20	310
NYC	1100	20000x20000	10	300
State Fair	700	1100x1000	18	50
Disney	4000	1000x800	18	380

Table 2: Characteristics of the GPS traces used for validating SMOOTH.

To validate that SMOOTH imitates real human movement, we compare the statistical features extracted from the synthetic traces generated by SMOOTH with the real walk traces. Specifically, in this section, we compare the CCDFs of flights extracted from SMOOTH with the CCDFs extracted from real traces for every site listed in Table 2. We also plot CCDFs of flights extracted from two other mobility models that are based on the same set of traces. These models are described in Section 2 and summarized in the following:

TLW*[21]*

TLW refers to the Truncated-Levy walk mobility model. While the flights and ICTs distribution extracted from synthetic traces generated by this mobility model are power-law distributed, movements patterns of mobile nodes are completely random. Thus, TLW does not represent the regularity present in real human walks.

SLAW [14]

SLAW refers to the Self-Similar Least-Action Walk mobility model. An analysis of SLAW concludes that SLAW meets **feature1** - **feature5**; however, it does not model **feature6** - **feature7**[5], as the selection of a future location to visit by a mobile node is *independent* of the locations visited in the past. From the analysis of the real GPS traces used in SLAW, the authors concluded that the distribution of waypoints can be modeled as a *self-similar*[6] process. The distribution of waypoints in SLAW is controlled by a parameter called the *hurst* parameter. The synthetic mobility traces generated by SLAW are expected to be similar to the real GPS traces used in its development; however, using the SLAW mobility model can be extremely complex. For example, the waypoint placement technique used in SLAW requires a complete understanding of the *hurst* parameter. As discussed in Section 2, the *hurst* parameter is highly difficult to estimate from a given data sample [4]. In addition, any trends, noise or other forms of corruption in data (e.g., biased data) might lead to misleading results.

[5]Features 6-7 were published in September 2010 [23]. SLAW was published in 2009.

4.1 CCDFs of Flights

For validation of SMOOTH, we generated traces for all the scenarios listed in Table 2; however, due to space constraints, we only plot the CCDFs of flights extracted for three of these scenarios (i.e., the Disney, the New York city, and the Kaist scenarios). The input simulation parameters to all three mobility models correspond to the values listed in Table 2.

Since the values used for parameters a and b in equation 2 were based on data collected for 3 million users, we analyzed the real traces listed in Table 2 to estimate appropriate values of a and b for our study. Specifically, for every GPS trace, we first calculated *prob_explore* and then estimated values of a and b for every trace using the methods listed in [23]. Table 3 lists *prob_explore* and (a, b) parameter values estimated for every trace listed in Table 2.

Figure 4 illustrates the CCDFs of flights obtained from real traces and the three mobility models: SLAW, TLW, and SMOOTH. Tables 3 and 4 list SMOOTH's input parameter values to extract the CCDFs of flights shown in Figure 4.[7] For TLW and SLAW, the input parameter values that provide the best match for the flight distributions in Figure 4 are used. As shown, SMOOTH provides a better match to the real traces compared to both SLAW and TLW. To further illustrate that SMOOTH accurately models the locations visited by the mobile nodes during simulation, we plot the real map that represents the locations visited by the GPS users in the Kaist scenario (listed in Table 2). Figure 5 represents the real map (from trace data) and the synthetic maps of locations visited by the mobile nodes during simulation of SMOOTH and SLAW. We simulated SMOOTH and SLAW for all five real scenarios listed in Table 2; however, due to space constraints, we only include the real and synthetic maps for one scenario (i.e., Kaist).

Scenario	*prob_explore* (avg, conf. interval)	a	b
Kaist	0.78, [0.75-0.81]	0.80	0.01
NCSU	0.84, [0.8-0.88]	0.9	0.02
NYC	0.89, [0.83-0.95]	0.9	0.004
State Fair	0.89, [0.88-0.9]	0.9	0.01
Disney	0.83, [0.81-0.83]	0.9	0.01

Table 3: Analysis of the GPS traces for parameters.

[6]A process is called *self−similar* if the aggregated processes (i.e., the processes obtained by averaging the original process over non-overlapping blocks) are highly correlated. Specifically, the autocorrelation function of a self-similar process is non-summable (i.e., ∞). For further details, see [15].

[7]We encourage the reader to view Figures 4 and 6-7 in color.

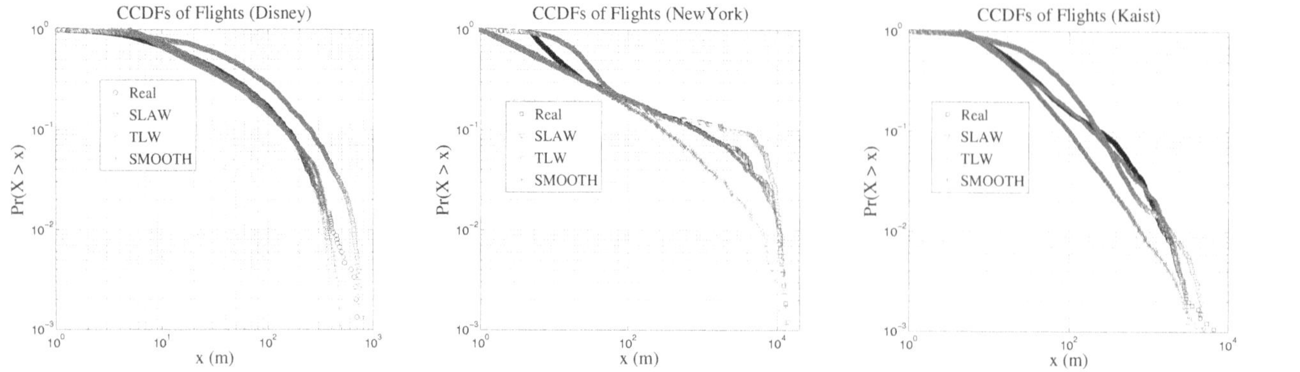

Figure 4: CCDFs for flights extracted from the real trace and the synthetic traces generated by SLAW, TLW, and SMOOTH for the Disney, NewYork and Kaist scenarios listed in Table 2.

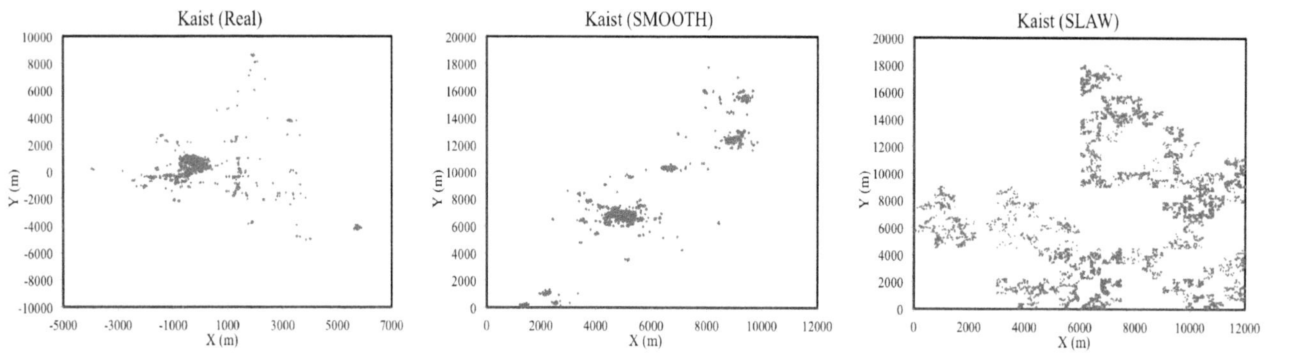

Figure 5: Synthetic maps of locations visited by mobile nodes in the real traces for Kaist (see Table 2) and SMOOTH. Synthetic map of potential locations to visit for SLAW.

Parameter	*Kaist*	*NYC*	*Disney*
α	1.97	1.45	1.4
β	1.9	1.5	1.7
f_{min} *(m)*	5	1	5
f_{max} *(m)*	3600	14000	800
p_{min} *(sec)*	10	10	10
p_{max} *(sec)*	16200	40200	9000
clusters	98	117	52
range (m)	100	100	100

Table 4: Input parameter values of SMOOTH for the results shown in Figure 4. One can use SMOOTH with these parameters and those listed in Tables 2-3 to realistically simulate one of these three scenarios.

4.2 Contact Information

Since the GPS traces used in the development of SLAW and TLW do not capture contact information of the participants, we could not validate ICTs information extracted from SMOOTH using the set of real traces listed in Table 2. Thus, for validating the contact information extracted from synthetic traces generated by SMOOTH, we used another set of real traces collected from a Bluetooth network of several imotes during two Infocom conference scenarios [8, 22].

In addition to ICTs, these two real traces are analyzed for the following two metrics:

- *Contact Duration (CD)* is the duration of contacts between a pair of mobile nodes and, thus, determines the amount of data that can be transferred during the contact.

- *Contact Number (CN)* is the number of times a pair of mobile nodes connected and, thus, determines the number of times mobile nodes can act as relays for each other.

In SLAW, a mobile node randomly replaces a cluster after every trip; thus, locations visited by a mobile node are independent of locations visited in the past. Also, we note that the movement patterns in TLW are random and, thus, do not represent the social behavior (e.g., contacts) among nodes. SWIM [16], on the other hand, seems to capture the social behavior among mobile nodes reasonably well. The selection of a location by a mobile node is based on its distance from the mobile node's current location and its popularity (i.e., the number of times a location is visited). Thus, in this section, we analyze the ICTs, CDs, and CNs generated by SMOOTH for the two Infocom scenarios and compare these with the respective distributions generated by SLAW, TLW, and SWIM. A brief description of the SWIM mobility model follows:

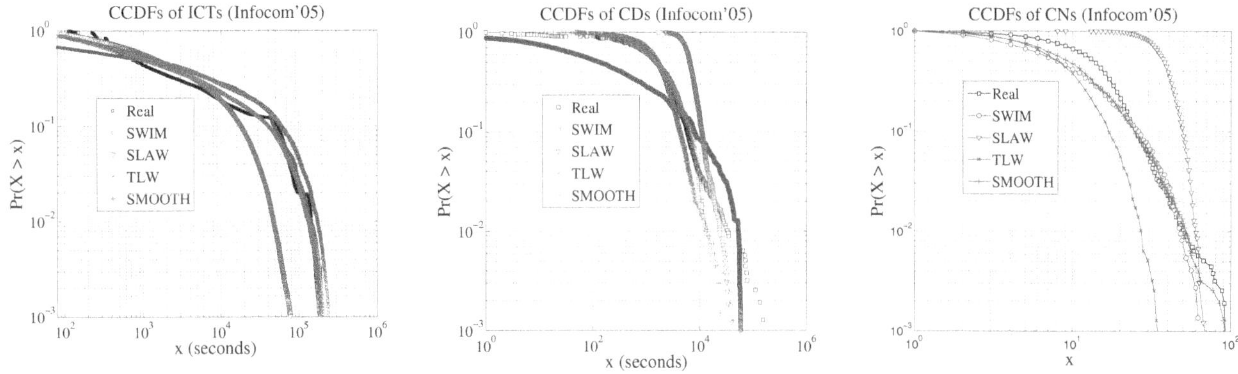

Figure 6: CCDFs of ICTs, CDs, and CNs extracted from the Infocom'05 real trace with the corresponding CCDFs extracted from the synthetic traces generated by SWIM, SLAW, TLW, and SMOOTH.

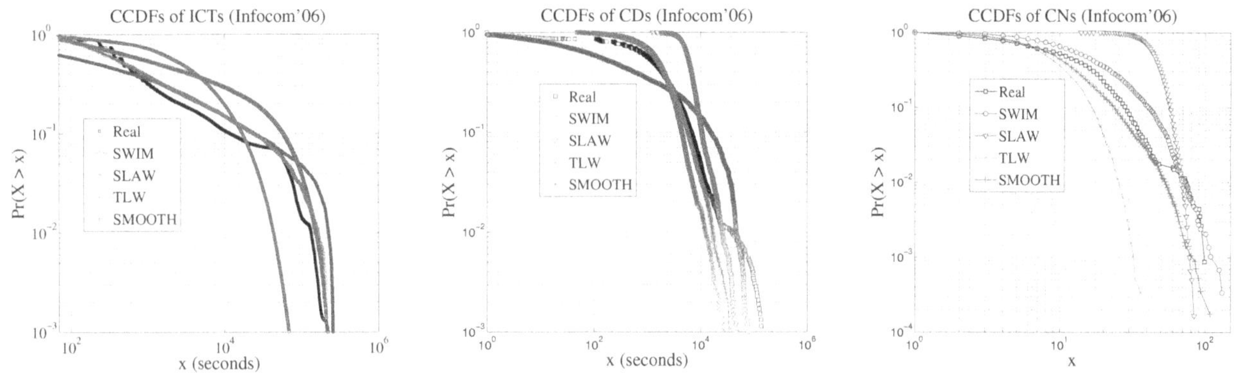

Figure 7: CCDFs of ICTs, CDs, and CNs extracted from the Infocom'06 real trace with the corresponding CCDFs extracted from the synthetic traces generated by SWIM, SLAW, TLW, and SMOOTH.

SWIM [16]

In SWIM, every possible destination is assigned a calculated weight that increases with its popularity and decreases with its distance from the current GPS location of the mobile node. The network area is divided into cells of the same size. Mobile nodes are assigned a home location (i.e., a cell) at the start of the simulation and the next location (i.e., a cell) is visited based on input parameter values (i.e., either based on the cell's popularity or the distance from the mobile node's current location).

Table 5 lists the statistical details of the real Infocom traces used in this work. In [9], the Infocom'05 and Infocom'06 real traces were analyzed using K-means clustering and the following observations were made:

- In Infocom'05 scenario, there were 41 nodes; four groups with 10, 6, 4, and 4 members each were identified.
- In Infocom'06 scenario, there were 78 nodes; four groups with 10, 4, 5, and 15 members each were identified.

Using real trace data, we simulated SMOOTH in a similar manner as [16] to simulate SWIM. Specifically, we simulated Infocom'05 and Infocom'06 scenarios with SMOOTH as follows:

- To simulate the Infocom'05 scenario, we divided 24 nodes into 4 clusters. We first placed 4 cluster landmarks and then associated 10, 6, 4, and 4 nodes with the cluster landmarks (as discussed in Section 3.1), such that the group memberships were retained. The remaining 17 nodes were randomly distributed over the simulation area.
- We simulated the Infocom'06 scenario with 4 clusters of 10, 4, 5, and 15 mobile nodes. The remaining 44 nodes were randomly distributed over the simulation area.

In TLW, since the distribution of mobile nodes is independent of each other, we started each simulation with mobile nodes distributed randomly over the simulation area. For SWIM, we used the same set of parameter values given in [16]. Since SLAW requires the *hurst* value for simulating a scenario, we estimated the *hurst* value by analyzing synthetic traces generated by SMOOTH. Specifically, the *hurst* values obtained for the Infocom'05 and Infocom'06 scenarios were 0.71 and 0.79, respectively. These *hurst* values were then used as an input to the SLAW mobility model. Table 5 lists the parameter values used for simulating the Infocom scenarios on SMOOTH and TLW.

Figures 6 and 7 represent the CCDFs of ICTs, CDs, and CNs extracted from the real Infocom traces and the synthetic traces generated by SWIM, SLAW, TLW, and SMOOTH. Using the results shown in Figures 6 and 7, we make the following observations:

- Due to the random replacement of a cluster after every trip and no truncation of flights made by mobile nodes

Scenario	*Infocom'05*	*Infocom'06*
Duration (days)	3	3
Nodes	41	78
Area (m^2)	500x500	500x500
Range (m)	25	25
(α, f_{min}, f_{max})	(2.1, 5, 100)	(2.1, 5, 100)
(β, p_{min}, p_{max})	(2.1, 10s, 12h)	(2.1, 10s, 12h)
Clusters	4	4

Table 5: Input parameter values of SMOOTH for Infocom'05 and Infocom'06 scenario.

in SLAW, the ICTs in SLAW are lower than the real traces and the CDs and CNs are relatively higher.

- The truncation of flights in TLW (i.e., f_{min}, f_{max}) avoids long random flights; thus, the CCDFs of CDs in TLW are closer to the real traces than the CCDFs of CDs in SLAW and SWIM. On the other hand, the distribution of CNs in TLW is often much lower than the real traces due to random movements.
- Figure 8 shows the pause-time distribution used in [16] to tune SWIM to real Infocom traces. We note that the duration of each Infocom trace is 3 days (=72 hours) and the upper bound to pause (p_{max}) is 12 hours. We also note that the number of pause-times at the upper bound (i.e., equal to 12 hours) generated by SWIM is extremely high. In other words, with this choice of input parameters to SWIM, long pause-times are generated, and those long pause-times do not represent a realistic scenario. In summary, the CDs generated by SWIM are higher than in real traces (if outliers in the real traces are ignored).
- SMOOTH provides a better match to all three CCDFs when compared to the other three mobility models. Thus, SMOOTH simulates a scenario more realistically than the other mobility models used in this study.

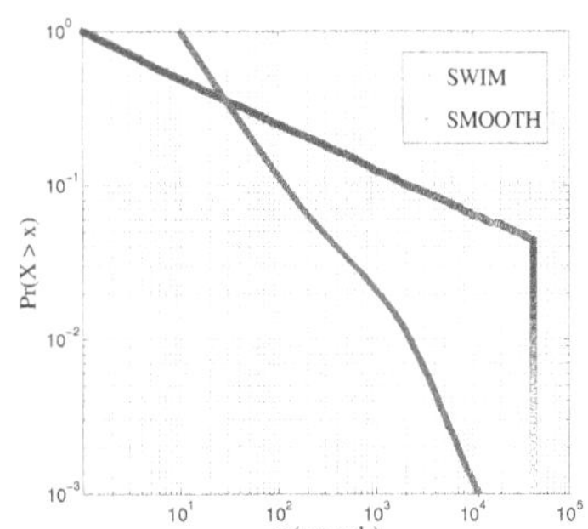

Figure 8: Pause-time distributions used to tune SWIM and SMOOTH for the Infocom'05 and Infocom'06 scenarios.

5. PERFORMANCE EVALUATION OF DTN ROUTING PROTOCOLS

The movement patterns of mobile nodes significantly affect the performance of a routing protocol. Thus, in this section, we investigate how a routing protocol performs on SWIM, SLAW, TLW, and SMOOTH. For this purpose, we simulated three DTN routing protocols (i.e., Epidemic Forwarding [25], LET [5], and ECT [14]). Broadly, these routing protocols can be categorized into: memoryless and memory-full protocols. In a memoryless routing protocol (e.g., Epidemic), a mobile node forwards a packet to its neighbor if the neighbor does not already have a copy of it. In a memory-full routing protocol (e.g., LET and ECT), on the other hand, a mobile node uses its past contact information (with other mobile nodes) in making its routing decisions. Specifically, a mobile node keeps timestamps that denote its past meeting times with the other mobile nodes. LET [5] refers to a metric called *Last Encounter Time*, in which a mobile node chooses the neighbor node that has most recently met the destination to relay a packet for that destination. ECT was introduced in [14] and refers to a metric called *Expected Contact Time*. ECT is computed as the difference between LET and ICT for a pair of mobile nodes and denotes the expected time remaining before their next encounter. Thus, a neighbor node with a lower value of ECT is chosen as a relay for the packet, as it is more likely to meet the destination sooner than other mobile nodes. Using the Infocom traces and the four mobility models discussed in this paper (SWIM, SLAW, TLW, and SMOOTH), we compare these three routing protocols for the following two metrics:

- *Success Rate* denotes the ratio of packets delivered to the total generated packets.
- *Cost* denotes the number of duplicates generated per packet.

5.1 Simulation Set Up

We simulate all three routing protocols for both Infocom'05 and Infocom'06 scenarios. Due to space constraints, we only include results for the Infocom'05 scenario in this paper; we note, however, that the results for the Infocom'06 scenario are similar. For the Infocom'05 trace, we simulate all three routing protocols on a 3 hour trace. Every 4 seconds, a random source and destination pair is chosen and a packet is generated. The nodes maintained full historical information, since we only evaluate the effect of the nodes' mobility on the performance of a routing protocol. Specifically, the entries of past contacts were not deleted even when they became stale. To avoid any transient effects of simulation, we do not generate any packet during the last hour of simulation. The effect of success rate and cost is evaluated against the TTL (Time to Live) value for a packet. Epidemic Forwarding represents the best case for packet success rate and the worst case for cost. Since a packet is forwarded to every neighbor who does not already have a copy of it, the probability to deliver a packet increases and, thus, Epidemic Forwarding acts as a benchmark for the success rate and cost.

5.2 Simulation Results

Figures 9 and 10 show the results of our simulations. We note that due to lower ICTs, SLAW overestimates the performance (i.e., higher success rate) for all three routing protocols when compared to the real traces. TLW, on the other hand, has a lower success rate when compared to the real traces due to random movements. The success rate (and hence the cost) for SWIM in LET and ECT is lower than the real traces, as SWIM does not accurately model the contact information between a pair of mobile nodes. Finally, we note that SMOOTH matches the success rate of the real traces quite well. (Confidence intervals for Figures 9 and 10 are really small and, thus, not included in figures.)

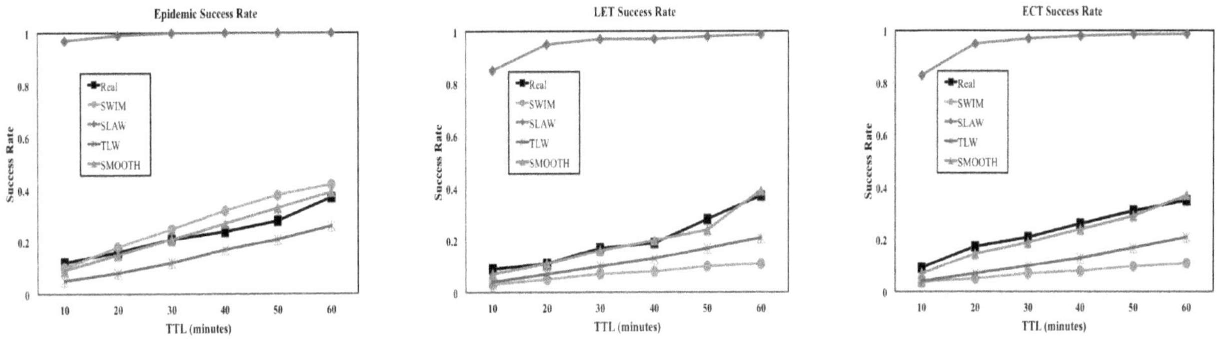

Figure 9: Average success rate for the Epidemic, LET, and ECT routing protocols on the real Infocom'05 trace and the synthetic traces generated by SWIM, SLAW, TLW, and SMOOTH.

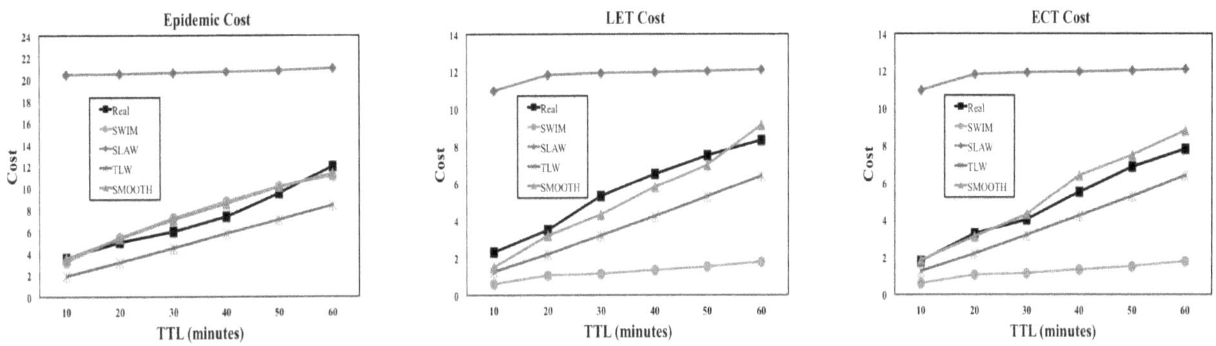

Figure 10: Average cost for the Epidemic, LET, and ECT routing protocols on the real Infocom'05 trace and the synthetic traces generated by SWIM, SLAW, TLW, and SMOOTH.

To further illustrate that SMOOTH helps mobile nodes make better decisions regarding packet forwarding, we compared SMOOTH with other mobility models for routing delays. Figure 11 shows the delay measurements obtained by simulating the ECT routing protocol for the Infocom'05 scenario. Routing delay is calculated for the delivery of the first 50 packets. Routing delay for each TTL value is an average of 20 simulation runs, with 95% confidence intervals calculated. As shown in Figure 11, SLAW overestimates the performance of the ECT protocol (similar to overestimating success rate and cost); in addition, SWIM and TLW have longer delays when compared to the real trace. SMOOTH, on the other hand, has routing delay values close to the real trace. In summary, results from Figures 9, 10, and 11 validate that SMOOTH provides an accurate prediction of the contact information for every pair of mobile nodes.

6. CONCLUSIONS

In this work, we present SMOOTH, a simple mobility model based on several known features of real human movement. To be widely adopted by the research community, a mobility model should be both *simple* and *realistic*. We validate SMOOTH using several real traces collected from sites that represent diverse scenarios. Specifically, real traces used in validating SMOOTH include: two campuses, two outdoor sites (an amusement park and a state fair), a metro city, and two conference scenarios. We also compare our model with three other mobility models that are based on the same set of traces used in this study. Based on the results obtained in this work, we derive the following conclusions:

1. Compared to the other three mobility models, we note that SMOOTH provides a better match to the statistical features present in real human movement, as SMOOTH accurately models **feature1 - feature7**.
2. SMOOTH can be easily tuned to simulate an example scenario. Specifically, input values of SMOOTH listed in Table 5 can be used to simulate a small conference scenario. In addition, parameter values listed in Tables 2-4 can be used to simulate a scenario similar to Kaist, NewYork, and Disney.
3. We validate SMOOTH with real GPS traces of human walks for a wide range of scenarios. Specifically, SMOOTH is validated for scenarios with a number of *nodes* from (10 to 78), simulation *area* size from (500x500 to 20000x20000), number of *clusters* from (6 to 117), and from sparse (e.g., the NewYork scenario) to dense (e.g., Infocom'06 scenario) networks.
4. SMOOTH provides a better estimate for ICTs, CDs, and CNs between a pair of mobile nodes compared to SWIM, SLAW, and TLW and, thus, clearly differentiates among protocols that make routing decisions from contact history of mobile nodes.

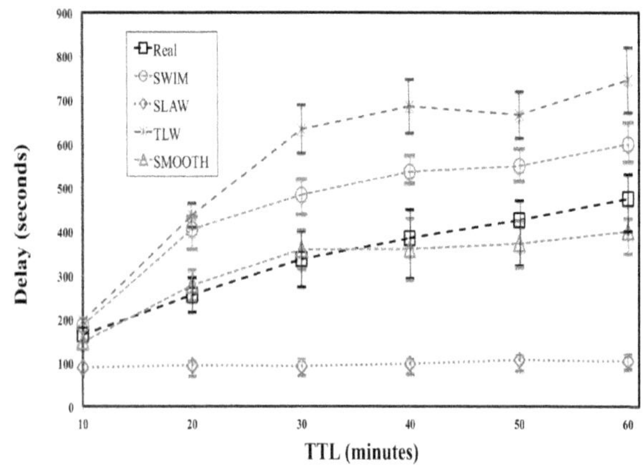

Figure 11: Average routing delays for ECT on the real Infocom'05 trace and the synthetic traces generated by SLAW, SWIM, SMOOTH, and TLW.

7. REFERENCES

[1] N. Aschenbruck, A. Munjal, and T. Camp. Trace-based mobility modeling for multi-hop wireless networks. *Elsevier Computer Communications*, pages 704–714, 2011.

[2] D. Brockmann, L. Hufnagel, and T. Geisel. The scaling laws of human travel. *Nature*, pages 462–465, 2006.

[3] T. Camp, J. Boleng, and V. Davies. A survey of mobility models for ad hoc network research. *Wireless Communication and Mobile Computing*, pages 483–502, 2002.

[4] R. G. Clegg. A practical guide to measuring the *hurst* parameter. In *Proceedings of UK Performance Engineering Workshop, University of Newcastle*, 2006.

[5] H. D.-Ferriere, M. Grossglauser, and M. Vetterli. Age matters: Efficient route discovery in mobile ad hoc networks using encounter ages. In *Proceedings of MobiHoc*, pages 257–266, 2003.

[6] M. C. Gonzalez, C. A. Hidalgo, and A.-L. Barabási. Understanding individual human mobility patterns. *Nature*, pages 779–782, 2008.

[7] W. Hsu, T. Spyropoulos, K. Psounis, and A. Helmy. Modeling time-variant user mobility in wireless mobile networks. In *Proceedings of Infocom*, pages 758–766, 2007.

[8] P. Hui, A. Chaintreau, J. Scott, R. Gass, J. Crowcroft, and C. Diot. Pocket switched networks and human mobility in conference environments. In *Proceedings of SIGCOMM Workshop on Delay-Tolerant Networking*, pages 244–251, 2005.

[9] P. Hui, E. Yoneki, S. Chan, and J. Crowcroft. Distributed community detection in delay tolerant networks. In *Proceedings of MobiArch*, pages 817–824, 2007.

[10] A. Jardosh, E. M. Belding-Royer, K. C. Almeroth, and S. Suri. Towards realistic mobility models for mobile ad hoc networks. In *Proceedings of MobiCom*, pages 217–229, 2003.

[11] T. Karagiannis, J.-Y. L. Boudec, and M. Vojnovic. Power law and exponential decay of inter contact times between mobile devices. In *Proceedings of MobiCom*, pages 1377–1390, 2007.

[12] M. Kim, D. Kotz, and S. Kim. Extracting a mobility model from real user traces. In *Proceedings of Infocom*, pages 1–13, 2006.

[13] K. Lee, S. Hong, S. J. Kim, I. Rhee, and S. Chong. Demystifying Levy walk patterns in human walks. *Technical Report, NCSU*, 2008.

[14] K. Lee, S. Hong, S. J. Kim, I. Rhee, and S. Chong. SLAW: A mobility model for human walks. In *Proceedings of Infocom*, pages 855–863, 2009.

[15] W. Leland, M. Taqqu, W. Willinger, and D. Wilson. On the self-similar nature of Ethernet traffic. *IEEE/ACM Transactions on Networking*, pages 1–15, 1994.

[16] A. Mei and J. Stefa. SWIM: A simple model to generate small mobile worlds. In *Proceedings of Infocom*, pages 2106–2113, 2009.

[17] A. Munjal, T. Camp, and W. Navidi. Constructing rigorous MANET simulation scenarios with realistic mobility. In *Proceedings of European Wireless*, pages 817–824, 2010.

[18] A. Munjal, T. Camp, and W. C. Navidi. Mobicom 2010 poster: SMOOTH: A simple way to model human walks". *Mobile Computing and Communications Review 14*, pages 34–36, 2010.

[19] W. Navidi, T. Camp, and N. Bauer. Improving the accuracy of random waypoint simulations through steady-state initialization. In *Proceedings of Modeling and Simulation*, pages 319–326, 2004.

[20] A. D. Nguyen, P. Sénac, V. Ramiro, and M. Diaz. STEPS - An approach for human mobility modeling. In *Proceedings of Networking*, pages 254–265, 2011.

[21] I. Rhee, M. Shin, S. Hong, K. Lee, and S. Chong. On the Levy walk nature of human mobility. In *Proceedings of Infocom*, pages 924–932, 2008.

[22] J. Scott, R. Gass, J. Crowcroft, P. Hui, C. Diot, and A. Chaintreau. CRAWDAD trace cambridge/haggle. *http://crawdad.cs.dartmouth.edu/cambridge/haggle*, 2006.

[23] C. Song, T. Koren, P. Wang, and A.-L. Barabási. Modelling the scaling properties of human mobility. *Nature Physics*, pages 818–823, 2010.

[24] C. Song, Z. Qu, N. Blumm, and A.-L. Barabási. Limits of predictability in human mobility. *Science*, pages 1018–1021, 2010.

[25] A. Vahdat and D. Becker. Epidemic routing for partially connected ad hoc networks. *Technical Report, Duke University*, 2000.

[26] J. Yoon, M. Liu, and B. Noble. Random waypoint considered harmful. In *Proceedings of Infocom*, pages 1312–1321, 2003.

[27] J. Yoon, B. Noble, M. Liu, and M. Kim. Building realistic mobility models from coarse-grained traces. In *Proceedings of MobiSys*, pages 177–190, 2006.

Packets Travelling in Non-homogeneous Networks

Omer H. Abdelrahman
Electrical & Electronic Engineering Dept.
Imperial College, London SW7 2BT, UK
o.abd06@imperial.ac.uk

Erol Gelenbe
Electrical & Electronic Engineering Dept.
Imperial College, London SW7 2BT, UK
e.gelenbe@imperial.ac.uk

ABSTRACT

This paper considers a probability model for travel of a packet from a source node to a destination node in a large non-homogeneous multiple hop network with unreliable routing tables. Use of a random model is justified by the lack of precise information that can be used in each step of the packet's travel, and randomness can also be useful in exploring alternate paths when a long sequence of hops has not resulted in the packet's arrival to the destination. The packet's travel may also be impeded if certain routers on its path prove to be unreliable, or the packet may be dropped from a buffer or destroyed due to packet loss. The packet also has a limited time-out that allows the source to retransmit a dropped or lost packet. Because the network itself may be extremely large, we consider packet travel in an infinite random non-homogeneous medium, with events that may interrupt, destroy or stop the packet from moving towards its destination. We derive a numerical-analytical solution allowing us to compute the average travel time of the packet from source to destination, as well as to estimate its energy consumption. Two interesting applications are then presented. In the first one a wireless network where areas which are remote from the source and destination nodes may have poor wireless coverage so that the packet losses become more frequent as the packet "unknowingly" (due to poor routing tables for instance) meanders away from the source and destination node. The second application is related to defending a destination node against attacks that take the form of packets that carry a virus or a worm that can be detected via deep packet inspection at intermediate nodes, and as the packet approaches the destination node it is more frequently inspected and dropped if it is a threat.

Categories and Subject Descriptors

C.2 [**Computer Communication Networks**]: Network Architecture and Design—*Distributed networks*
; C.4 [**Performance of Systems**]: Modeling techniques;
G.3 [**Mathematics of Computing**]: Probability and Statistics—*Markov processes*

General Terms

Performance, Reliability, Theory

Keywords

Brownian motion, diffusion process, multi-hop networks

1. INTRODUCTION

Search for recognisable objects in large random or imprecisely known environments has been studied in robotics, biology, physics, transportation and communication networks [12, 22, 21, 18, 19]. This paper focuses on the travel of a packet from a given source to a destination which is at distance D from the packet, but whose whereabouts are unknown, or imprecisely known. Thus the packet may not have precise information about which direction it should pursue as it moves from one hop to the next [14, 8, 10], and we suppose that the destination node is recognised only when the packet gets close to it, typically one hop away. Furthermore, when the network is infinitely large the packet may become lost in remote areas. The packet can also be destroyed by its own finite life-time, by a failure or error in the communication system or by buffer overflow in an intermediate node.

Random walk models for networks were analysed in [20], and in [15] the mean and variance of the hitting time were obtained for a torus-lattice network graph when the next node visited is selected at random among all neighbours, showing that in such networks the probability distribution of packet delivery time is approximately geometrically distributed. For multiple independent unbiased random walks on a connected network [23] it was shown that the mean first passage time converges to the shortest path between the source and the destination as the number of travellers approaches infinity. A random walk over a wireless network with uniform node distribution and circular symmetry was evaluated in [1]. In [4], the probability of the unbiased traveller visiting a particular node in a given step was derived for a 2-D grid topology, but unbiased routing was shown to achieve poor performance [2] because the random walker may "orbit" around the target node for a long time before attaining it.

In [9] it was shown that the time it takes a packet to travel from a source node to a destination node in an infinitely large and unreliable network is finite on average, provided that a time-out mechanism is inserted to destroy the ongoing packet after a predetermined time, and replace it with a new packet that starts at the same source and proceeds at random and independently of its predecessor. Since the network is infinite, the time-out also protects the packet from spending an unreasonably long time in remote areas from which it may never return. By using a randomly different travel path, the new packet takes a distinct path from its previous incarnation, increasing its chances of reaching the destination.

The use of Brownian motion approximations simplifies the analysis for very large networks, and the time dependent solution for the passage time using Brownian motion was considered in [3] where, differently from our work, losses and retransmissions occur at specific distances from the destination. In [11] N packets are simultaneously, but independently, sent out in the quest for the same destination node. Both the total travel time and the energy expended were obtained using a Brownian motion model of the travel process.

While most previous work has dwelt on a spatially homogeneous network, here we will consider a packet travelling in a *non-homogeneous* multi-hop network. One motivation for this work is the case where some parts of the network may be particularly faulty or degraded while the rest of the network is operating properly. Thus the network's operational quality may be quite good close to the source node, but it may become less reliable when the packet moves far away from it. This paper is motivated by two interesting applications. The first case relates to a wireless network where remote areas, away from where the source and destination nodes are located, perhaps have poor wireless coverage so that the packet losses become more frequent as the packet "unknowingly" (due to poor routing tables for instance) meanders away from the source and destination node. The second one is related to defending a destination node against attacks that take the form of packets that may carry a virus or a worm that can be detected via deep packet inspection at some intermediate nodes. Thus as a packet approaches the destination node it may be inspected by intermediate protecting nodes and dropped if it is viewed as a threat; however the source of the packet will use a time-out to attempt sending the attacking packet forward again and it is interesting to see whether the attacker will eventually be successful. Another example of non-homogeneous wireless network occurs when the packet progresses faster as it approaches the destination, for instance when directional information such as a radio signature becomes stronger as the packet approaches the destination node. However we will not examine this case in detail in this paper.

In the sequel we will model a packet's motion towards a destination node in an infinite random non-homogeneous network, with packet drops that will stop the packet's progress resulting in a subsequent time-out retransmission of the packet from the source. We obtain an exact expression for the average time and energy that it takes the packet to eventually find the destination node, based on a non-homogeneous Brownian motion model enhanced with some useful point processes representing the relaunch of an aborted or interrupted search. We develop an analytical solution technique based on a finite but unbounded number of internally homogeneous segments, yielding the average travel time and the energy expended. Then the results are applied to the two cases of interest that we have outlined.

2. AN ANALYTICAL MODEL FOR NON-HOMOGENEOUS PACKET TRAVEL

Following the approach in [11] let Y_t be the packet's distance from the destination node at time $t \geq 0$. Clearly the packet starts at distance $Y_0 = D$ and the travel process ends at some time T defined by:

$$T = \inf\{t : Y_t = 0\}$$

The random variable $s(t)$ represents the state of the packet at time $t \geq 0$ where $s(t) \in \{\mathbf{S}, \mathbf{W}, \mathbf{P}, \ ...\}$, and $s(t)$ is in:

- **S**: if the travel is proceeding and the packet's distance from the destination is $Y_t > 0$. The probability density function of the position Y_t is represented by $f(z,t)dz = P[z < Y_t \leq z + dz,\ s(t) = \mathbf{S}]$.
- **W**: its life-span has ended, and so has its travel. This can happen because the packet was destroyed or became lost, and the source was informed via the time-out. After an additional exponentially distributed delay of parameter μ, meant to avoid mistakes in assuming that the packet is lost, a new packet is placed at the source and a new travel immediately begins. We write $W(t) = P[s(t) = \mathbf{W}]$.
- **L**: if the packet is destroyed or lost, and the travel is interrupted until a new packet can be sent out. The time spent in this state is exponentially distributed with parameter r which is the same parameter as that of the "time-out" or life-span, since the source realises that the packet is lost or destroyed via the life-span or time-out effect. At the end of this exponentially distributed time, the packet is handled just as if it has been lost, and we denote $L(t) = P[s(t) = \mathbf{L}]$.
- **P**: the packet has reached its destination and the travel process ends. However, as an artefact to construct an indefinitely repeating recurrent process, after one time unit the travel process restarts at the source and a new packet is sent out. We will use the notation $P(t) = P[s(t) = \mathbf{P}]$.

Notice that the above process repeats itself indefinitely, and $E[T]$ is the average time that it takes from any successive start of the travel until the first instance when state **P** is reached again. Let $P(t)$ be the probability that the model we have just described is in state **P** at time $t \geq 0$, and let $P = \lim_{t\to\infty} P(t)$. Then:

$$P = \frac{1}{1 + E[T]}, \quad E[T] = P^{-1} - 1$$

During the packet's travel in state **S** while $\{Y_t = z > 0\}$ the following events can occur in the time interval $[t, t + \Delta t[$:

- With probability $\lambda(z)\Delta t + o(\Delta t)$ the packet is destroyed or lost, and enters state **L**. From that state it enters state **W** after an exponentially distributed delay of parameter r.
- With probability $r\Delta t + o(\Delta t)$ the packet's life-span runs out and it enters state **W**. Note that $1/r$ is the average life-span. As indicated earlier, when it

enters state $\mathbf{W}$, after an additional delay of average value $1/\mu$, the packet is replaced with a new one at the source.

The average rate per unit time at which the packet approaches the destination node when it is at distance z is denoted by $b(z)$, and the variance of the distance travelled in the interval $[t, t+\Delta t[$ is denoted by $c(z)\Delta t$ so that:

$$b(z) = \lim_{\Delta t \to 0} \frac{E[Y_{t+\Delta t} - Y_t | Y_t = z]}{\Delta t},$$
$$c(z) = \lim_{\Delta t \to 0} \frac{E[(Y_{t+\Delta t} - Y_t)^2 - (E[Y_{t+\Delta t} - Y_t])^2 | Y_t = z]}{\Delta t}$$

When $b(z) < 0$, on average the packet gets closer over time to the destination node, but $b(z) \geq 0$ is also possible. Note that P is a fictitious state that we use to create a recurrent random process which indefinitely repeats itself. The packet's distance to destination in a non-homogeneous medium is represented by the probability density function $f(z,t)$ representing the position of the packet at time $t \geq 0$, and assume that it satisfies the position dependent diffusion equation [5, 16]. Such models have been used previously to represent traffic in communications systems and transportation systems [17, 6, 7, 13].

The model for the non-homogeneous medium is simplified to a finite but unbounded number of "segments" that have different parameters for the Brownian motion describing the packet's movement as a function of its distance to the destination node, while within each segment the parameters are the same. The first segment is in the immediate proximity of the destination node, starting at distance $z = 0$. Each segment may have a different size, and we assume that there are a total of $m < \infty$ segments. By choosing as many segments as we wish, and letting each segment be as small as we wish (all segments need not be of the same length), we can approximate as closely as needed any physical situation that arises where the packet's motion characteristics vary over the distance of the packet to the destination node. If the point $0 \leq Z_k < \infty$ is the boundary between the k-th and $(k+1)$-th segments with $Z_0 = 0$, with the last segment going from Z_{m-1} to $+\infty$, with m and Z_{m-1} being finite but unbounded, we can take as many segments as needed, and they are all finite except the last segment. Thus for $1 \leq k \leq m$, the k-th segment represents the range of distances $Z_{k-1} \leq z < Z_k$; let $S_k = Z_k - Z_{k-1}$ denote its size and write $\{f(z,t), b(z), c(z), \lambda(z)\} = \{f_k(z,t), b_k, c_k, \lambda_k\}$. We use n to denote the segment number in which the source node is located, i.e. $Z_{n-1} < D < Z_n$. The differential equation for the *stationary* solution of the location dependent diffusion equation for segment $k \neq n$ is then:

$$0 = \frac{c_k}{2}\frac{d^2 f_k(z)}{dz^2} - b_k \frac{df_k(z)}{dz} - (\lambda_k + r) f_k(z) \qquad (1)$$

while the equation for the segment where the source is located is:

$$-[P + \mu W]\delta(z - D) = \frac{c_n}{2}\frac{d^2 f_n(z)}{dz^2} - b_n \frac{df_n(z)}{dz} - (\lambda_n + r) f_n(z) \qquad (2)$$

We also have:

$$rL = \sum_{k=1}^{m} \lambda_k \int_{Z_{k-1}}^{Z_k} f_k(z)dz \qquad (3)$$

$$\mu W = r[L + \sum_{k=1}^{m} \int_{Z_{k-1}}^{Z_k} f_k(z)dz] \qquad (4)$$

$$P = \lim_{z \to 0^+} [\frac{c_1}{2}\frac{df_1(z)}{dz} - b_1 f_1(z)] \qquad (5)$$

and the normalisation condition:

$$1 = P + W + L + \sum_{k=1}^{m} \int_{Z_{k-1}}^{Z_k} f_k(z)dz \qquad (6)$$

Result 1 Let u_k, v_k be, respectively, the positive and negative real roots of the characteristic polynomial of the stationary equation for the k-th segment:

$$u_k, v_k = \frac{b_k \pm \sqrt{b_k^2 + 2c_k(\lambda_k + r)}}{c_k}$$

Then the total average travel time, which is obtained by solving for P so that $E[T] = P^{-1} - 1$, is given by:

$$E[T] = (\frac{1}{r} + \frac{1}{\mu}) \times$$
$$[\sqrt{\frac{b_n^2 + 2c_n(\lambda_n + r)}{b_1^2 + 2c_1(\lambda_1 + r)}} \frac{\overline{A}_n \overline{G}_n e^{u_n S_n} - \overline{B}_n \overline{F}_n e^{v_n S_n}}{\overline{G}_n e^{u_n(Z_n - D)} + \overline{F}_n e^{v_n(Z_n - D)}} - 1] \qquad (7)$$

where the remaining parameters are computed as follows. Define:

$$\alpha_k^- = \frac{c_k u_k - c_{k-1} v_{k-1}}{c_k(u_k - v_k)}, \quad \beta_k^- = \frac{c_k u_k - c_{k-1} u_{k-1}}{c_k(u_k - v_k)}$$
$$\alpha_k^+ = \frac{c_k u_k - c_{k+1} v_{k+1}}{c_k(u_k - v_k)}, \quad \beta_k^+ = \frac{c_k u_k - c_{k+1} u_{k+1}}{c_k(u_k - v_k)} \qquad (8)$$

Then set $\overline{A}_1 = 1$ and $\overline{B}_1 = -1$ and for $2 \leq k \leq n$ compute:

$$\begin{bmatrix} \overline{A}_k \\ \overline{B}_k \end{bmatrix} = \begin{bmatrix} \alpha_k^- & \beta_k^- \\ 1 - \alpha_k^- & 1 - \beta_k^- \end{bmatrix} \begin{bmatrix} e^{u_{k-1} S_{k-1}} & 0 \\ 0 & e^{v_{k-1} S_{k-1}} \end{bmatrix} \begin{bmatrix} \overline{A}_{k-1} \\ \overline{B}_{k-1} \end{bmatrix} \qquad (9)$$

Then set $\overline{F}_m = 0$ and $\overline{G}_m = e^{v_m Z_m}$, and start another computation at $k = m - 1$ for $n \leq k \leq m - 1$ with:

$$\begin{bmatrix} \overline{F}_k \\ \overline{G}_k \end{bmatrix} = \begin{bmatrix} \alpha_k^+ & \beta_k^+ \\ 1 - \alpha_k^+ & 1 - \beta_k^+ \end{bmatrix} \begin{bmatrix} e^{-u_{k+1} S_{k+1}} & 0 \\ 0 & e^{-v_{k+1} S_{k+1}} \end{bmatrix} \begin{bmatrix} \overline{F}_{k+1} \\ \overline{G}_{k+1} \end{bmatrix} \qquad (10)$$

Proof The general solution has the form:

$$f_k(z) = \begin{cases} A_k e^{u_k z} + B_k e^{v_k z}, & Z_{k-1} \leq z \leq \min(D, Z_k) \\ F_k e^{u_k z} + G_k e^{v_k z}, & \max(D, Z_{k-1}) \leq z \leq Z_k \end{cases}$$

Thus there are $2m + 2$ constants to be determined from (a) the boundary conditions at 0 and $+\infty$, (b) the continuity condition of the probability density function at D and at the boundaries between segments, and (c) conditions obtained by integrating the defining differential equation around D

and the boundaries between segments. First consider the case $Z_{k-1} \le z \le \min(D, Z_k)$; to ensure continuity of the probability density function at $z = Z_{k-1}$ we have:

$$f_k(Z_{k-1}) = f_{k-1}(Z_{k-1}) \quad (11)$$

or equivalently

$$A_k e^{u_k Z_{k-1}} + B_k e^{v_k Z_{k-1}} = A_{k-1} e^{u_{k-1} Z_{k-1}} + B_{k-1} e^{v_{k-1} Z_{k-1}}$$

Furthermore, integrating the differential equation (1) from $z = Z_{k-1} - \epsilon$ to $z = Z_{k-1} + \epsilon$ and taking the limit as ϵ tends to 0 yields:

$$\frac{c_k}{2}\frac{df_k(Z_{k-1})}{dz} - \frac{c_{k-1}}{2}\frac{df_{k-1}(Z_{k-1})}{dz} = [b_k - b_{k-1}] f_k(Z_{k-1}) \quad (12)$$

or

$$A_k u_k e^{u_k Z_{k-1}} + B_k v_k e^{v_k Z_{k-1}} = \frac{2b_k - c_{k-1} v_{k-1}}{c_k} \times A_{k-1} e^{u_{k-1} Z_{k-1}} + \frac{2b_k - c_{k-1} u_{k-1}}{c_k} B_{k-1} e^{v_{k-1} Z_{k-1}}$$

Solving (11) and (12), we can write A_k and B_k in terms of A_{k-1} and B_{k-1} as:

$$\begin{aligned} A_k e^{u_k Z_{k-1}} &= \alpha_k^- A_{k-1} e^{u_{k-1} Z_{k-1}} + \beta_k^- B_{k-1} e^{v_{k-1} Z_{k-1}} \\ B_k e^{v_k Z_{k-1}} &= [1 - \alpha_k^-] A_{k-1} e^{u_{k-1} Z_{k-1}} \\ &\quad + [1 - \beta_k^-] B_{k-1} e^{v_{k-1} Z_{k-1}} \end{aligned}$$

where α_k^- and β_k^- are defined in (8). From the boundary condition $\lim_{z \to 0^+} f_1(z) = 0$ we have $B_1 = -A_1$; therefore the stationary solution of the differential equation for $Z_{k-1} \le z \le \min(D, Z_k)$ can be expressed as follows:

$$f_k(z) = A_1[\overline{A}_k e^{u_k(z - Z_{k-1})} + \overline{B}_k e^{v_k(z - Z_{k-1})}] \quad (13)$$

where the constants $\overline{A}_k$ and $\overline{B}_k$ are computed recursively using the matrix multiplication in (9). Next consider a segment k where $z \ge D$, and write the constants F_k and G_k in terms of F_{k+1} and G_{k+1} by solving boundary conditions similar to (11) and (12) at $z = Z_k$:

$$\begin{aligned} F_k e^{u_k Z_k} &= \alpha_k^+ F_{k+1} e^{u_{k+1} Z_k} + \beta_k^+ G_{k+1} e^{v_{k+1} Z_k} \\ G_k e^{v_k Z_k} &= [1 - \alpha_k^+] F_{k+1} e^{u_{k+1} Z_k} + [1 - \beta_k^+] G_{k+1} e^{v_{k+1} Z_k} \end{aligned}$$

Since $f(z)$ is a probability density function we must have $\lim_{z \to \infty} f_m(z) = 0$ which implies that $F_m = 0$, thus the solution for $\max(D, Z_{k-1}) \le z \le Z_k$ is given by:

$$f_k(z) = G_m[\overline{F}_k e^{-u_k(Z_k - z)} + \overline{G}_k e^{-v_k(Z_k - z)}] \quad (14)$$

where $\overline{F}_k$ and $\overline{G}_k$ are computed using (10). Note that $\overline{G}_m = e^{v_m Z_m}$ yields the desired solution for the last segment, that is $f_m(z) = G_m \overline{G}_m e^{-v_m(Z_m - z)} = G_m e^{v_m z}$. In order to determine A_1 and G_m, consider the n-th segment and apply the continuity condition of $f_n(z)$ at $z = D$ so that:

$$\begin{aligned} &G_m[\overline{F}_n e^{-u_n(Z_n - D)} + \overline{G}_n e^{-v_n(Z_n - D)}] \\ &\quad = A_1[\overline{A}_n e^{u_n(D - Z_{n-1})} + \overline{B}_n e^{v_n(D - Z_{n-1})}] \quad (15) \end{aligned}$$

Also, integrating the differential equation (2) from $z = D - \epsilon$ to $z = D + \epsilon$ and taking the limit as ϵ tends to 0 yields:

$$\begin{aligned} \frac{2[P + \mu W]}{-c_n} &= G_m[\overline{F}_n u_n e^{-u_n(Z_n - D)} + \overline{G}_n v_n e^{-v_n(Z_n - D)}] \\ &- A_1[\overline{A}_n u_n e^{u_n(D - Z_{n-1})} + \overline{B}_n v_n e^{v_n(D - Z_{n-1})}] \quad (16) \end{aligned}$$

From (5), the probability P is given by:

$$P = \frac{c_1}{2}(u_1 - v_1)A_1 = \sqrt{b_1^2 + 2c_1(\lambda_1 + r)}\ A_1 \quad (17)$$

Substituting (4) into (6) yields:

$$P + \mu W \left(\frac{1}{r} + \frac{1}{\mu}\right) = 1 \quad (18)$$

Now solving the system of linear equations (15)–(18) we can determine A_1 and G_m:

$$\begin{aligned} A_1 &= \eta\left[\overline{G}_n e^{u_n(Z_n - D)} + \overline{F}_n e^{v_n(Z_n - D)}\right] \\ G_m &= \eta\left[\overline{A}_n e^{u_n S_n} e^{v_n(Z_n - D)} + \overline{B}_n e^{v_n S_n} e^{u_n(Z_n - D)}\right] \quad (19) \end{aligned}$$

where

$$\eta = \frac{r\mu/(r+\mu)}{\sqrt{b_n^2 + 2c_n(\lambda_n + r)}}\left\{\overline{A}_n \overline{G}_n e^{u_n S_n} - \overline{B}_n \overline{F}_n e^{v_n S_n} - \sigma[\overline{G}_n e^{u_n(Z_n - D)} + \overline{F}_n e^{v_n(Z_n - D)}]\right\}^{-1},$$

$$\sigma = [1 - \frac{r\mu}{r+\mu}]\sqrt{\frac{b_1^2 + 2c_1(\lambda_1 + r)}{b_n^2 + 2c_n(\lambda_n + r)}}$$

Substituting A_1 in (17) yields P from which the average travel time follows directly. □

Remark 1 With n being the index of the discretisation segment that includes the source node at D, it is interesting to see that $E[T]$ only depends on a set of parameters that are computed for values of $k = 1$, $k = n$, and on two sets of algebraic iterations between $k = 1$ and $k = n$ and $k = m$ down to $k = n$.

Remark 2 When the source node is in the pen-ultimate segment we have $m = n$, and:

$$E[T] = \frac{r+\mu}{r\mu}[\sqrt{\frac{b_n^2 + 2c_n(\lambda_n + r)}{b_1^2 + 2c_1(\lambda_1 + r)}}\overline{A}_n e^{u_n(D - Z_{n-1})} - 1] \quad (20)$$

For a homogeneous medium $m = n = 1$ and:

$$E[T] = (\frac{1}{r} + \frac{1}{\mu})[e^{u_1 D} - 1]$$

as we would expect from [11].

Remark 3 As in [11], if energy is consumed only when a packet is actually being forwarded through the network, while during wait times for retransmissions the packet (which remains stored at the source until final successful delivery) consumes only negligible energy, then the average energy consumption $E[J]$ until the packet reaches its destination is:

$$E[J] = (1 + E[T])\sum_{k=1}^{m}\int_{Z_{k-1}}^{Z_k} f_k(z)dz \quad (21)$$

2.1 Greater loss in remote areas

An example of practical interest occurs when a packet that moves far away from its initial point and from the destination node, has a greater chance of being lost or destroyed. This could represent a multi-hop wireless network deployed in a very large area; as the packet moves to remote areas

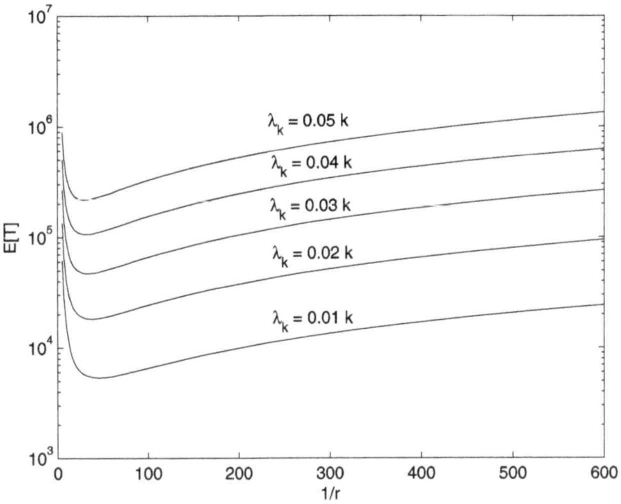

Figure 1: $E[T]$ (logarithmic scale) versus the average time out $1/r$ when loss rates increases linearly and $m = 100$.

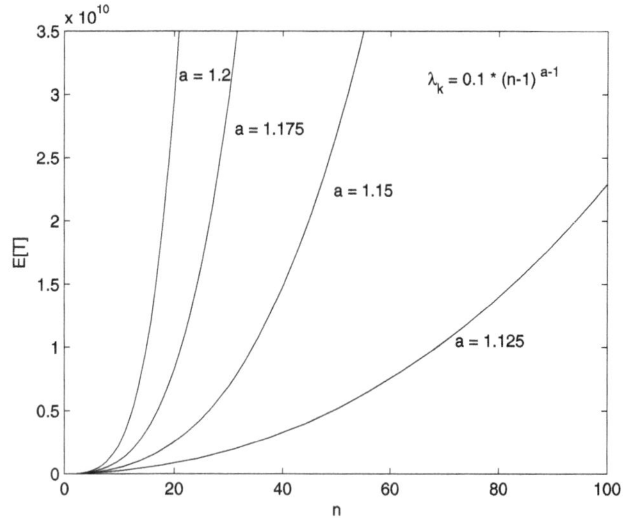

Figure 2: $E[T]$ versus n when $\lambda_k = 0.1(n-1)^{a-1}$ for different values of a; $b_k = -1$, $c_k = 0$, $S_k = D/(n-1)$, $\mu = 0.1$, $r = 0.02$, and $D = 100$.

far from the region where the source and destination are located, the nodes that the packet might visit are less likely to handle it and more likely to just discard it. This can also represent a network where there are fewer nodes in remote areas and inter-node communications in such areas are less reliable. As an example consider 100 segments with $S = 1$ and a loss rate that increases with distance: $\lambda_k = k\ell$, $\ell > 0$, $1 \leq k \leq 99$, $\lambda_{100} = 100\ell$. If average speed of the packet's motion and its second moment remain constant with $b_k = 0$ and $c_k = 1$ for $1 \leq k \leq 100$, the results with $D = 10$ in Figure 1 show that a relatively short time-out is needed to optimise the average travel time, but that the resulting optimum is nevertheless very large.

3. RETARDING AN ATTACKING PACKET

Another interesting case arises when the packet that we are modelling contains some form of attack on the destination node, such as a virus or a worm. Also, we suppose that the network protects this particular node by introducing a capability at intermediate nodes to detect the contents of the packet and to drop it. However the sender will then, after a time-out, send the attacking packet again. The question is then whether it is possible to block the attack indefinitely or whether to the contrary the attacking packet will eventually reach the destination node that is being defended.

We examine this problem in the context of a wired network that uses shortest path routing. Thus if the distance D refers to the number of hops from source to destination, and if the routers are operating properly, we will have $b = -1$ and $c = 0$ throughout the network. More generally if there is no uncertainty in routing $c_k = 0$ and $b_k < 0$, and it can be shown that the total average travel time does not depend on the network's parameters for $z > D$:

$$E[T] = \frac{r+\mu}{r\mu}\,[e^{\frac{\lambda_n+r}{|b_n|}D}\, e^{\sum_{k=1}^{n-1}\left(\frac{\lambda_k+r}{|b_k|}-\frac{\lambda_n+r}{|b_n|}\right)S_k} - 1] \quad (22)$$

Furthermore if the routers are perfect and always provide shortest distance routing we have $b_k = -1$ and:

$$E[T] = \frac{r+\mu}{r\mu}\,[e^{(\lambda_n+r)D}\, e^{\sum_{k=1}^{n-1}(\lambda_k-\lambda_n)S_k} - 1] \quad (23)$$

Now let us introduce a non-homogeneous packet drop effect by choosing an integer n to create an acceleration in the packet drop effect and let $S_k = D/(n-1)$ so that:

$$E[T] = \frac{r+\mu}{r\mu}\,[e^{(r+\frac{\sum_{k=1}^{n-1}\lambda_k}{n-1})D} - 1] \quad (24)$$

which yields the following result.

Result 2 If $\lim_{n\to\infty} \frac{\sum_{k=1}^{n-1}\lambda_k}{n-1} = +\infty$ then the packet will never reach the destination node. Otherwise it will reach it in a time which is finite on average, and with probability one.

Figure 2 illustrates Result 2 by showing that even with a small excess, represented by $a > 1$, above the $O(n)$ rate of increase for the loss rate λ_k the attacking packet's progress will be indefinitely impeded by the drops, despite the subsequent time-outs.

3.1 A neighbourhood with traps

As a final and related example suppose that routers in the neighbourhood of the destination node within a distance S contains "traps" that can identify the attacking packet and drop it. Accidental drops of the packet (due to transmission errors or buffer overflows) may also occur at a lower rate. Thus we take $m = n = 2$, so that $E[T]$ is obtained from (20) with $\lambda_2 = \lambda$ and $\lambda_1 = \lambda + \gamma$, $\gamma > 0$:

$$E[T] = \frac{r+\mu}{r\mu}[\sqrt{\frac{b_2^2 + 2c_2(\lambda+r)}{b_1^2 + 2c_1(\lambda+\gamma+r)}}\bar{A}_2 e^{u_2(D-S)} - 1]$$

Figure 3 shows the manner in which $E[T]$ sharply increases with γ, for S ranging between 10 and 15, $D = 100$, $b_2 = b_1 = 0.25$, $c_1 = c_2 = 1$, $\lambda = 0$. Also $\mu = 1/10$ and r is set to the value that minimises $E[T]$ when $\gamma = 0$ and $S = 10$. Figure 4, with $S = 10$ and the same set of parameters, shows that even small increases (more negative) in

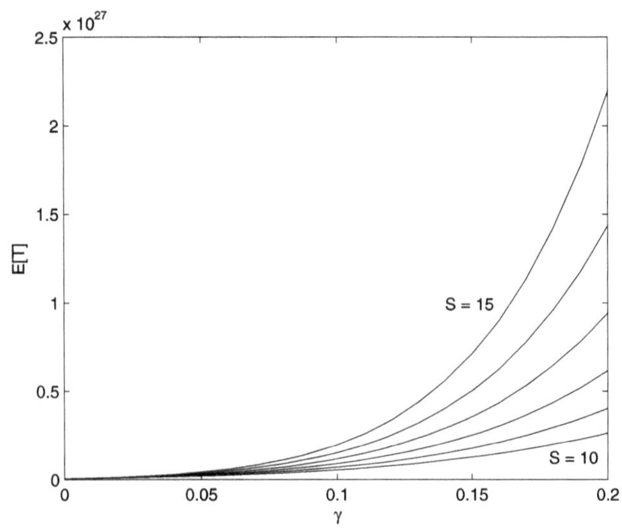

Figure 3: Average travel time $E[T]$ versus $\gamma = \lambda_1 - \lambda_2$ for $S = 10$ to 15 with a step size of 1.

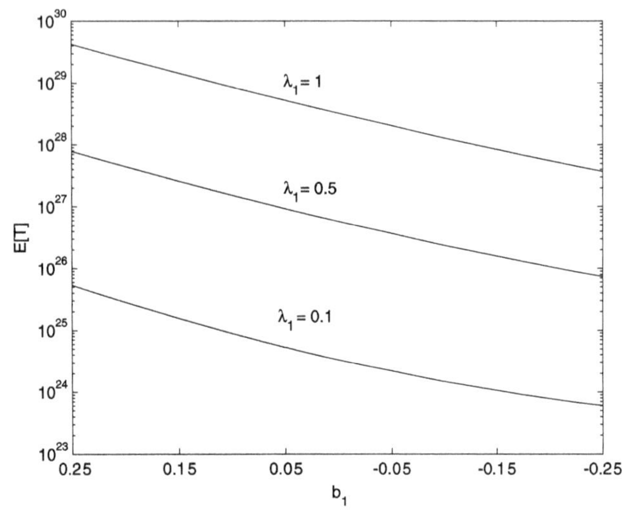

Figure 4: $E[T]$ (logarithmic scale) versus b_1 for different values of λ_1.

average speed at which the packet approaches its objective can reduce average travel time by an order of magnitude, yet $E[T]$ is still very large.

Figures 5 and 6 question how S and λ_1 may be selected together in order to maximise the protection offered to the destination node. If we keep the same set of parameters as previously but take λ_1 to be inversely proportional to S in Figure 5 so that the average number of protecting nodes $\lambda_1 \approx 1/S$, and the ratio of time rate to spatial rate will remain constant for any fixed value of b_1 which is the speed of motion. Figure 5 shows that there is indeed an optimum size of protection space $S = S^*$ that maximises the delay before the attacking packet can reach the destination node, and that it varies with the speed b_1 of the packet inside the protected neighbourhood. As the speed increases, the optimum size of the neighbourhood gets smaller. This may be counterintuitive but it follows from the fact that we have taken $\lambda_1 \approx 1/S$: a smaller size implies a higher "rate of pro-

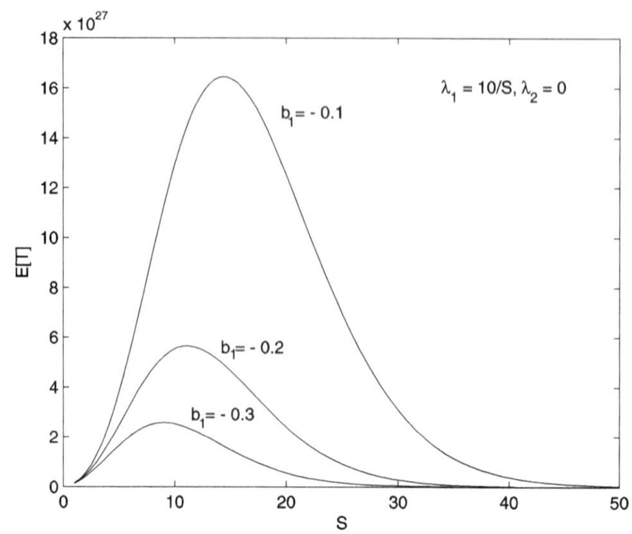

Figure 5: Average travel time $E[T]$ versus S when $\lambda_1 = 10/S$ for different values of b_1. The optimum protection area needed becomes smaller so that λ_1 increases when the packet's speed increases.

tection" and hence more frequently occurring destructions of the packet which compensate for the higher speed of the packet. However the corresponding maximum values of $E[T]$ do become smaller as the packet's speed increases. In Figure 6 we set $b_1 = b_2 = 0.25$ and Λ is varied in $\lambda_1 = \Lambda/S^2$. The results are similar to the previous ones.

For the examples of Figures 5 and 6 the average energy expenditure is closely proportional to $E[T]$ because

$$\sum_{k=1}^{m} \int_{Z_{k-1}}^{Z_k} f_k(z)dz \cong 1$$

so that we omit showing the numerical results for the energy.

4. CONCLUSIONS

We have presented a Brownian motion model to represent a packet's travel to a destination node in a very large non-homogeneous network. A mixed analytical-numerical method has been developed to compute the average packet travel time and the energy it consumes. We observe that the degree of non-homogeneity of the network will significantly affect the average travel time and energy consumed. The role of time-outs to optimise these quantities has been exhibited, and two examples have been detailed. In the first example, packet losses (for instance due to insufficient wireless network coverage) increase as the packet reaches areas which are remote from the source and destination nodes. In the second example we model an attacking packet which may be detected and destroyed as it approaches the destination node, but in turn the attacking packet may progress more rapidly as it approaches the destination node, for instance because a directional routing being used may become more accurate. Comparing the increasing speed of approach of the packet with the possible steeper defenses of the destination node, we observe that there may be conditions whereby despite the use of time-outs the attacking packet may never make it to the destination node, while in other circumstances the attack will be successful.

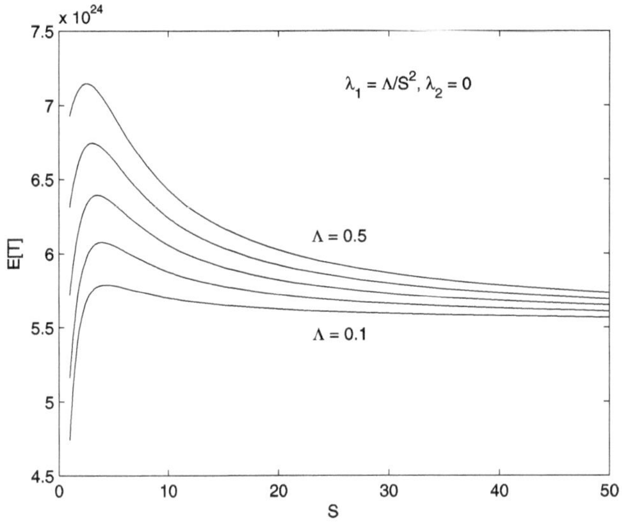

Figure 6: Average travel time $E[T]$ versus S when $\lambda_1 = \Lambda/S^2$ for $\Lambda = 0.1$ to 0.5. The protection area needed to maximise the travel time decreases as Λ increases.

5. REFERENCES

[1] R. Beraldi. Biased random walks in uniform wireless networks. *IEEE Transactions on Mobile Computing*, 8(4):500–513, April 2009.

[2] R. Beraldi, Q. Leonardo, and B. Roberto. Low hitting time random walks in wireless networks. *Wireless Communications and Mobile Computing*, 9:719–732, May 2009.

[3] T. Czachórski, K. Grochla, and F. Pekergin. Diffusion approximation model for the distribution of packet travel time at sensor networks. In *Wireless Systems and Mobility in Next Generation Internet*, volume 5122 of *LNCS*, pages 10–25. Springer Berlin / Heidelberg, 2008.

[4] D. Dhanapala, A. Jayasumana, and Q. Han. Performance of random routing on grid-based sensor networks. In *Proceedings of 6th IEEE Consumer Communications and Networking Conference (CCNC'09)*, Las Vegas, NV, USA, January 2009.

[5] A. Einstein. *Investigations on the Theory of Brownian Motion*. Dutton & Dover, New York, 1926.

[6] E. Gelenbe. On approximate computer system models. *Journal of the ACM*, 22:261–269, April 1975.

[7] E. Gelenbe. Probabilistic models of computer systems. part II: Diffusion approximations, waiting times and batch arrivals. *Acta Informatica*, 12:285–303, 1979. 10.1007/BF00268317.

[8] E. Gelenbe. Travel delay in a large wireless ad hoc network. In *Proceedings of 2nd Workshop on Spatial Stochastic Models for Wireless Networks (SpaSWiN'06)*, Boston, MA, USA, April 2006.

[9] E. Gelenbe. A diffusion model for packet travel time in a random multihop medium. *ACM Transactions on Sensor Networks*, 3(2):1–19, June 2007.

[10] E. Gelenbe. Steps towards self-aware networks. *Communications of the ACM*, 52(7):66–75, July 2009.

[11] E. Gelenbe. Search in unknown random environments. *Physical Review E*, 82(6):061112, December 2010.

[12] E. Gelenbe and Y. Cao. Autonomous search for mines. *European Journal of Operational Research*, 108(2):319–333, July 1998.

[13] E. Gelenbe, X. Mang, and R. Onvural. Diffusion based call admission control in ATM. *Performance Evaluation*, 27 & 28:411–436, 1996.

[14] Y.-B. Ko and N. H. Vaidya. Location-aided routing (LAR) in mobile ad hoc networks. *Wireless Networks*, 6:307–321, July 2000.

[15] I. Mabrouki, G. Froc, and X. Lagrange. On the data delivery delay taken by random walks in wireless sensor networks. In *Proceedings of 5th International Conference on Quantitative Evaluation of Systems (QEST'08)*, pages 17–26, St Malo, France, September 2008.

[16] J. Medhi. *Stochastic Models in Queueing Theory*. Academic Press, New York, 1991.

[17] G. F. Newell. *Applications of Queueing Theory*. Monographs on Applied Probability and Statistics, Chapman and Hall Ltd., London, 1971.

[18] G. Oshanin, O. Vasilyev, P. L. Krapivsky, and J. Klafter. Suvival of an evasive prey. *Proceedings of the National Academy of Sciences*, 106(33):13696–13701, 2009.

[19] F. Rojo, J. Revelli, C. E. Budd, H. S. Wio, G. Oshanin, and K. Lindenberg. Intermittent search strategies revisited: effect of the jump length and biased motion. *Journal of Physics A: Mathematical and General*, 43(34):345001, 2010.

[20] S. Shakkottai. Asymptotics of search strategies over a sensor network. *IEEE Transactions on Automatic Control*, 50(5):594–606, May 2005.

[21] B. Tilch, F. Schweitzer, and W. Ebeling. Active brownian particles with internal energy depots modelling animal mobility. *BioSystems*, 49:17–92, 1999.

[22] B. Tilch, F. Schweitzer, and W. Ebeling. Directed motion of brownian particles with internal energy depot. *Physica A: Statistical Mechanics and its Applications*, 273:294–314, 1999.

[23] S.-P. Wang and W.-J. Pei. First passage time of multiple brownian particles on networks with applications. *Physica A: Statistical Mechanics and its Applications*, 387(18):4699–4708, 2008.

Analytical Models of Short-Message Reliability in Mobile Wireless Networks*

Debessay Fesehaye K.
Dept. of Computer Science,
Univ. of Illinois at U-C
Urbana, IL, USA
dkassa2@illinois.edu

Klara Nahrstedt
Dept. of Computer Science,
Univ. of Illinois at U-C
Urbana, IL, USA
klara@illinois.edu

Guijun Wang
Boeing Research & Technology
WA,USA
guijun.wang@boeing.com

ABSTRACT

Applications like Twitter which use chat-like short messaging systems (SMS) have been widely used in public, political, military, emergency, humanitarian and other fields. Such applications usually involve servers (controllers) which control and forward messages from a sending client to a receiving client. The use of mobile wireless networks for such messaging systems has been increasing at a fast pace. To cope with this increase, there need to be efficient communication protocols and algorithms. To design such protocols and algorithms requires extensive analysis and understanding of the behavior of the communicating nodes under a given mobility scenario. One of the good metrics to understand the performance of such protocols is the reliability of message delivery.

In this paper we present analytical models of the average reliability of short (chat-like) message delivery in mobile wireless networks as a multivariate function of the transmission range, movement area dimensions, number of servers (base stations) and message deadline (lifetime) under moderate realistic assumptions which can be easily relaxed and extended.

Simulation results show that our analytical models give very good estimation of the average reliability of message delivery.

Categories and Subject Descriptors

C.4 [**Analytical models of reliability**]: Miscellaneous; D.2.8 [**Systems analysis**]: Metrics—*reliability measures*

General Terms

Systems

Keywords

Reliability models, short messaging, wireless networks

*This work is supported by Boeing Co 404269.

MSWiM'11, October 31–November 4, 2011, Miami, Florida, USA.

1. INTRODUCTION

With the growth of wireless and mobile network technologies many interesting applications have emerged. Some of the most common and popular applications such as Twitter [4] and SMS [12] are used to transfer short chat-like messages. Such applications have proven to be very useful in various public, political, military and emergency fields where transfer of big files is either not possible, costly, or unnecessary.

Some of the nodes in such chat-like short messaging systems are servers (controllers) which are responsible for the control and transfer of messages from the sender clients to the receiver clients. Even though there are many messaging and presence protocols for short messaging systems, in this paper we focus on protocols like the Extensible Messaging and Presence Protocol (XMPP) [7]. In the XMPP protocol, all senders and receivers register at their respective controller (server) with their names or IDs. A server broadcasts presence information of the subscribed clients to all clients and other servers. A client creates its roster by getting subscription approval from other clients. The client can then send messages to its respective peer. A moving client can subscribe to the nearest server (smart selection) which can route its messages to the intended receiver or from which it receives its messages.

With the growth of such short-messaging applications [8, 12], understanding their performance in a given network scenario becomes crucial to design efficient protocols and to efficiently dimension the network under which they operate. An important performance metric for such applications is the average reliability of message delivery from one client node to another client node via some controller nodes (servers) in given network.

In mobile wireless networks, the value of the average message reliability becomes low either because the nodes are moving outside the coverage area or due to congestion and interference which results in increased delay and packet losses. In chat-like short messaging systems the decrease in average message reliability is mainly due to nodes moving outside the coverage area or due to clients moving far away from their servers. This is because the messages generated by such systems are too short to congest the underlying dedicated physical or overlay network. Besides, the message processing and forwarding delays at the servers are negligible in such short messaging systems. So in this paper we focus on the reliability of short messaging systems where the reliability loss is mainly due to nodes moving outside the coverage area.

Performance metrics such as reliability can be studied using simulation or analytical models. Analytical models have the advantage that they are faster than simulation. Analytical models can also give detailed insight to the applications and protocols studied via the closed-form expressions which show how reliability evolves as a function of each parameter.

In this paper we present analytical models for the reliability of short message exchange between nodes via servers in mobile wireless networks as a multivariate function of the transmission range, movement area dimensions, number of servers (base stations or access points) and message deadline under moderate assumptions which can easily be relaxed and extended. In our analysis, messages are buffered at all servers until their deadline expires or until the intended mobile receiver(s) gets them from any of the servers before they expire. Any other message replication scheme can be used with our modeling approach. The fact that some (multimedia) applications tolerate some packet losses (reliability loss) makes our reliability models specially useful. This is because our analytical models give closed form expressions of what parameters give the average desired reliability values.

The rest of the paper is organized in such a way that we first present some related works, how they differ from our scheme and why they are not sufficient in Section 2. Following this we describe our analytical models of reliability in Section 3. We then present some numerical results and a summary in Sections 4 and 5.

2. RELATED WORK

There have been some works in the literature on the reliability of message delivery. Models of reliability in distributed publish-subscribe systems have been presented in [5, 6]. In this study the authors assumed that publishers (senders) are not mobile. They also assumed a specific network topology with fixed number of servers, transmission range and deadline parameters. Hence they didn't give closed form expressions of reliability as a function of the number of servers, transmission range, area dimensions and deadline parameters which are the main factors which affect the reliability of message delivery. This is specially useful to plan on how many servers with a specific transmission range are needed to obtain a specific reliability value in a given area and mobility scenario. A design and performance analysis of soft hand-off scheme for CDMA cellular systems was presented in [11]. The authors aimed to decrease both the number of dropped hand-off calls and the number of blocked calls without degrading the quality of communication service and the soft hand-off process. The authors also have given continuous-time Markov chain models to analyze their design. Their paper focused more on hand-off and a channel resource shortage condition.

3. THE ANALYTICAL MODELS

In this section we present simple and advanced analytical models of reliability. We first discuss the notations and assumptions we use in our modeling. We next present some closed form expressions for the number m of servers needed to give full wireless coverage in the movement area and for the expected maximum tolerable delay (message life time) τ. The value of τ is the expected maximum length of time before a moving client gets its message. We finally present the simple and advanced analytical models for the average reliability of short message delivery.

3.1 Notations

We use the following notations to derive the closed form expression for the average reliability. We use the terms server, base station, and access point interchangeably. This is because the servers (message) controllers are more like access points or base stations in that the mobile senders cannot send and the mobile receivers cannot receive messages unless they are in range of the servers which are connected either using reliable wireless, wired or overlay network.

Table 1: Parameters and Definitions

Parameters	Description
A	area within which nodes move
n	number of clients which move in the area
m	number of servers needed to get full coverage
m_a	number of available servers
t	transmission range of a server
R	desired reliability
d	deadline of message delivery
c	speed of light
n_c	number of clients connected
l	length in meters of the rectangular area
w	width in meters of the rectangular area
v	speed of a node (client)

3.2 Assumptions

Unless and otherwise specified, in this paper we make the following assumptions which can be relaxed.

- The n clients are uniformly distributed (located) in the area A.
- The m servers (BS) should be uniformly placed in the area A to get full coverage.
- Each server is placed at the center of each square cell as shown in Figure 1.
- The servers are placed in such a way that the overall coverage of the movement area is maximized.
- The movement of the nodes is uniformly distributed in the area.
- The area within which the nodes move and the servers are placed is considered to be rectangular. For other shapes of the area, the rectangle circumscribing the shape is considered to obtain the worst case reliability value.
- All messaged from the mobile senders are buffered at all static servers until they are delivered to their respective mobile receivers or until they expire.

We next present the algorithms and derivations for the reliability of message delivery for short messaging systems in mobile wireless networks.

3.3 Number of Servers for Full Coverage

In this section we present some expressions for the total number of servers needed to give full wireless communication coverage for all mobile nodes in the movement area. Let us consider the mobility area as given in Figure 1 with n clients which want to communicate with each other. We then use the following simple scheme to derive the relationships between the short message transfer reliability R and the variables n_s, t, l, w and d shown in Table 3.4.

- We first surround the general movement area for the mobile network with a rectangle as shown in Figure 1.

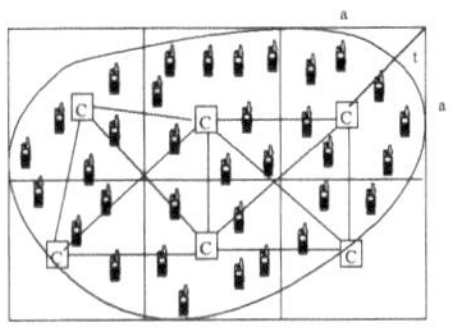

Figure 1: The mobility area with n mobile nodes and m servers (controllers C)

- Placing all m servers in the area we partition the movement area into smaller (equal) square areas where each square is covered by one server. For cases where the rectangular area cannot be divided into an integral number of square cells, a fraction of some of the squares in the last column and/or row of the rectangular area can be considered.
- Taking one square, as shown in Table 2 with the dimension a, we derive a as a function of the transmission range t as shown in Equation 1.

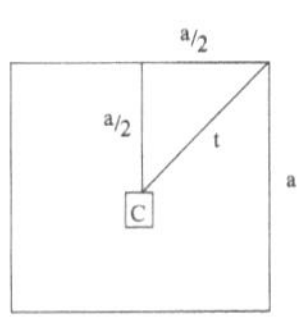

$$\begin{aligned}(\frac{a}{2})^2 + (\frac{a}{2})^2 &= t^2\\ 2(\frac{a}{2})^2 &= t^2\\ \Rightarrow a &= \hat{a}(t)\\ &= \sqrt{2}t.\end{aligned} \qquad (1)$$

Table 2: Square area covered by one server

There are m such squares. Hence we have

$$A = ma^2 = 2mt^2. \qquad (2)$$

Which implies that the total number of servers m with transmission range t needed to get full coverage assuming that is

$$m = \hat{m}(A, t) = \frac{A}{2t^2}. \qquad (3)$$

- The nodes in the given area with the furthest distance are the once at two opposite corners of the rectangle. If the rectangular area has a length of l and a width of w, the furthest distance between two nodes is $\sqrt{w^2 + l^2}$. Denoting the speed of light with c, the maximum total queuing, transmission and processing delay with D, we get a reliability R of 1.0 (full reliability) of message delivery with m servers if the deadline

$$d > \frac{\sqrt{w^2 + l^2}}{c} + D. \qquad (4)$$

We call this a deadline constraint. For short chat-like message delivery which is the main focus of this paper the value of D can be negligible.

- With n mobile clients sending and receiving messages and with m servers having a range of t, a full reliability of short message delivery can be achieved. To have a full coverage, the number of squares (servers) l_m in the horizontal (length) dimension of the rectangular area where nodes can move is given by

$$l_m = \hat{l}_m(l, t) = \lceil \frac{l}{a} \rceil = \lceil \frac{l}{\sqrt{2}t} \rceil = \lceil \frac{\sqrt{2}l}{2t} \rceil \qquad (5)$$

and the number w_m of squares along the vertical (width) dimension is given by

$$w_m = \hat{w}_m(w, t) = \lceil \frac{w}{a} \rceil = \lceil \frac{w}{\sqrt{2}t} \rceil = \lceil \frac{\sqrt{2}w}{2t} \rceil. \qquad (6)$$

Now the total number of servers in this square layout needed for the full coverage is

$$m = \hat{m}(l, w, t) = l_m \times w_m. \qquad (7)$$

It should be noted that Equation 3 assumes that $\frac{l}{a}$ and $\frac{w}{a}$ are integers.

If we have k less servers than m with a range of t, then the messages sent by the clients at the k cells at a given time t will not be transmitted, as the clients are out-of range of any server. However the messages sent by the other clients which are in range of a server are received before they expire (before the deadline d) if there are sufficient number of servers and if the destination clients which are out-of range of a server move sufficiently fast. This is because all messages transmitted by the nodes are buffered at all servers before they expire or before they are consumed by their respective receiver.

3.4 Expected Maximum Message Delay

Under this buffering assumption discussed above, the message life time or deadline τ depends on the number n_s of servers, transmission range t, node speed v and coverage area dimensions l and w and can be expressed by a function $\hat{\tau}$ as follows. Let's assume that the decision of how many servers to place in a given region of a coverage area is made based solely on how much of the area can be best covered without giving special preference to some regions of the coverage area. If some preferences need to be given to some regions, then the whole coverage area can be partitioned into sub areas each of which has no preference over its regions. To achieve this a simple relationship can be drawn on how many servers to initially place in each sub area based on some weights or user specified policies. Our scheme can then be applied to each sub area and extended to the coverage area as a whole.

To derive the deadline or message lifetime τ, we will first find the maximum possible distance between a client and its nearest server and then use some work in the literature [1] to get the expected transmission time (deadline τ) from the uncovered cell to the covered cell. We call this variable a radius r of circle centered at the client and passing through one of the nearest servers as also used in [1].

As can be seen from Figure 2(a), with the assumption that there is no preference in any specific region of the coverage area, if the number of missing servers k is less than or equal to half of the total number m of the servers required to give a full coverage of the area, then the maximum distance or radius r between a client and the nearest server (BS) is $a = \sqrt{2}t$. If $k > m/2$, the value of r depends on whether or not l_m and w_m are both even, one of them even or both odd numbers. So we have two cases A and B as described below.

Case A: l_m and w_m are odd or one is even

This case is shown in Figures 2(b) and 2(c). In this case if only the outer most round of the cells is uncovered, $r = 2t$, which is the distance from one corner of the uncovered cell to the nearest corner of the nearest covered cell in the next inner round of the coverage area. So as long as only

$$k \leq \lfloor \frac{m}{2} \rfloor + \frac{2(l_m + w_m) - 2(2)}{2}$$

servers are missing $r = 2t$. Here $2(l_m + w_m) - 2(2)$ is the total number of cells in the outer most round of the rectangular coverage area. In general, if the i outer most rounds of cells are uncovered, $r = i \times 2t = 2it$. For ρ such rounds, $1 \leq i \leq \rho$ where

$$\rho = \lfloor min\left(\frac{l_m}{2}, \frac{w_m}{2}\right) \rfloor.$$

Hence for this case denoting

$$S_{i-1} = \sum_{j}^{i-1} \left(\frac{2(l_m + w_m) - 2(2)(2j - 1)}{2}\right)$$

$$S_i = \sum_{j}^{i} \left(\frac{2(l_m + w_m) - 2(2)(2j - 1)}{2}\right),$$

for $1 \leq i \leq \rho$ we have

$$r = \begin{cases} \sqrt{2}t & \text{if } 0 < k \leq \lfloor \frac{m}{2} \rfloor \\ 2t & \text{if } \lfloor \frac{m}{2} \rfloor < k \leq \lfloor \frac{m}{2} \rfloor + S_1 \\ 2it & \text{if } \lfloor \frac{m}{2} \rfloor + S_{i-1} < k < \lfloor \frac{m}{2} \rfloor + S_i. \end{cases}$$

Case B: Both l_m and w_m are even numbers

This case is demonstrated by Figure 2(a). The only difference between this case and Case A above occurs when the two covered cells at the top left corner and bottom right corner of each round of the coverage area as shown in Figure 2(a) are removed. When the servers in these two covered cells are missing in the top most uncovered round i of the coverage area, then the shortest distance between the furthest client at the top left corner of round i to the nearest covered cell (top left corner of the coverage area) is the hypotenuse of a right angled triangle whose dimensions are ia and $(i+1)a$ as shown in Table 3. Using the Pythagoras theorem and after some algebra we get the result in Equation 8.

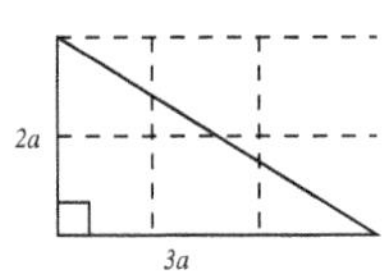

$$\begin{aligned} r &= a\sqrt{2i^2 + 2i + 1} \\ &= \sqrt{2(2i^2 + 2i + 1)}t \\ &= r_i. \end{aligned} \quad (8)$$

Table 3: The corner cases for an even $\times$ even number of squares in the area with $i = 2$

Hence we have

$$r = \begin{cases} \sqrt{2}t & \text{if } 0 < k \leq \lfloor \frac{m}{2} \rfloor \\ 2t & \text{if } \lfloor \frac{m}{2} \rfloor < k \leq \lfloor \frac{m}{2} \rfloor + S_1 - 2 \\ r_1 & \text{if } \lfloor \frac{m}{2} \rfloor + S_1 - 2 < k \leq \lfloor \frac{m}{2} \rfloor + S_1 \\ 2it & \text{if } \lfloor \frac{m}{2} \rfloor + S_{i-1} < k \leq \lfloor \frac{m}{2} \rfloor + S_i - 2 \\ r_i & \text{if } \lfloor \frac{m}{2} \rfloor + S_i - 2 < k < \lfloor \frac{m}{2} \rfloor + S_i. \end{cases}$$

In both of the above cases, with $k = m - m_a$, $r = \hat{r}(l, m, t, m_a)$ for some function $\hat{r}$.

The length L_p of the trajectory (path) to cover a distance of r is a function denoted as $P(r)$. Then the average time τ in seconds it takes a receiving client moving at a speed of v m/s to get its message from the nearest server is given as

$$\tau = \hat{\tau}(l, w, t, v, m_a) = \frac{P(r)}{v}. \quad (9)$$

For a random way point (RWP) mobility model, a movement of node (client) from uncovered cell to a covered cell (a cell with a server) can be assumed to be a two dimensional random walk (2DRW) [9]. For a 2DRW, a node takes $n_s = (r/\ell)^2$ steps of length ℓ to cover a distance r from uncovered cell to a covered cell [9]. Hence, using Wald's Equation [10], the expected value of the total length L_p of the path, covered by a client to get its message from the nearest server, is given by

$$L_p = \hat{L}_p(l, m, t, m_a) = \ell\frac{r^2}{\ell^2} = \frac{r^2}{\ell}. \quad (10)$$

In our experiments we consider the step size $\ell = a = \sqrt{2}t$ which is the dimension of each square cell.

The expected maximum delay τ of message delivery also depends on the the total pause time τ_p^t of the moving nodes. In the case of RWP mobility scenario, for a pause time of τ_p at every step of the path of length L_p and expected mobility step (transition) length of ℓ, the value of τ_p^t is given by

$$\tau_p^t = \frac{L_p}{\ell}\tau_p. \quad (11)$$

Once τ_p^t and the furthest length of the path L_p from the nearest server are obtained using Equations 11 and 10, the average time τ it takes a receiving client moving at a speed of v m/s to get its message from the nearest server is given as

$$\tau = \hat{\tau}(l, w, t, v, m_a) + \tau_p^t = \frac{L_p}{v} + \frac{L_p}{\ell}\tau_p. \quad (12)$$

The delay Equation 12 excludes the time it takes for the messages sent to propagate to all servers and other delay components due to congestion or interference. This is because as described using Equation 4 the propagation delay is negligible. Besides for short messaging systems the delay due to congestion is not significant. Otherwise, Equation 12 can be modified to take these delays into account.

3.5 Simple Reliability Model

The discussion in the above section implies that as long as $\tau \leq d$, all messages with a lifetime or deadline of d sent from clients in the covered cells are successfully received. Here as described in the notations of section 3.1 above, d is the required message deadline which is an important quality of service (QoS) parameter. Hence, on average to satisfy the deadline d of all messages transmitted from clients at cells with server coverage using buffering of messages and mobility of receivers, we need at least $m_a = \hat{m_a}(l, w, t, v, d)$. Here

$\hat{m}_a$ is an inverse of the $\hat{\tau}$ function. In the rest of this paper, we are more interested in this scenario where nodes can take advantage of buffering and mobility to receive messages sent to them.

A reliability value can also be obtained for scenarios where nodes in some cells can never get a message sent to them regardless of buffering and how fast they move. In this case, for a RWP mobility scenario, if there are κ cells whose clients can never receive messages sent to them because their $\tau > d$, then the average reliability decreases further by

$$R_d = \frac{(\kappa/m)(n/m)(m-k)}{n} = \frac{(m-k)\kappa}{m^2} = \frac{m_a\kappa}{m^2} \quad (13)$$

where κ/m is the probability that a message sent is destined in one of the κ cells with no server coverage and with $\tau > d$. More details of this case are left for future work.

Now given $m_a \geq \hat{m}_a(l, w, t, v, d)$, let's denote the distribution of the number of mobile nodes at cell (BS or server) i at time instant ω with $N_i(\omega)$ and the distribution of the rate at which client j of cell i sends packets at time ω with $R^i_j(\omega)$. The reliability R of short message delivery at time ω is given by

$$R = \hat{R}(l, w, t, m_a, d, v) = \frac{\sum_{i=1}^{m_a}\sum_{j=1}^{N_i(\omega)} R^i_j(\omega)}{\sum_{i=1}^{m}\sum_{j=1}^{N_i(\omega)} R^i_j(\omega)}. \quad (14)$$

Assuming that the clients are uniformly distributed in the area, the average number of clients per server is $f = \frac{n}{m}$. So if at any given time one server is omitted or is dis-functional, f clients are disconnected or lost. This in turn implies that out of all the messages sent, f of them are lost on average due to lack of server nearby the sender. This is assuming that all clients send messages at the same (uniform) rate and move at a uniform (the same) speed. These uniformity assumptions can be relaxed by partitioning the node set and area into nodes with similar patterns, and then aggregating the results. Therefore, if k servers are omitted, the reliability R of message delivery is given by

$$\begin{aligned} R &= \frac{n_c}{n} = \frac{n-kf}{n} \\ R = \hat{R}(l, w, t, d, v, m_a) &= \frac{n-(m-m_a)\frac{n}{m}}{n} = \frac{m_a}{m}. \\ \rightarrow m_a &= mR \\ &= \frac{A}{2t^2}R = \frac{lw}{2t^2}R. \end{aligned} \quad (15)$$

We call the above model a *simple model* as it is a simplified under-estimation of the actual reliability. The *simple model* under-estimates reliability as it does not take into account areas outside the square cells which are covered by each server.

By taking into account the areas outside each square covered by the server, we get a more accurate reliability model as described below. We call this model of reliability an *advanced model* of reliability.

3.6 Advanced Reliability Model

In this paper we approximate the server coverage area by a square cell as shown in Table 2. The same modeling approach can be used for scenarios where the server coverage area is approximated by a hexagon. For $m_a \leq \lceil m/2 \rceil$, there is no difference in reliability values between the hexagon and square cell layouts.

In the rest of the paper we consider a RWP mobility model and hence assume uniform distributions. The extensions to any other mobility models and distributions is straightforward. Hence Equation 14 can be similarly extended in the advanced reliability model for any other distribution.

In this advanced model, to derive the reliability taking into account the extra space covered by the servers outside their square areas, we first consider when half of the servers are missing, we then derive two cases where the number of missing servers is less and more than half of the total number m of servers required to give full coverage.

3.6.1 *With half the number of servers needed to cover the area ($m_a = \lceil m/2 \rceil$)*

In this scenario, the number of remaining servers (covered squares) $m_a = \lceil m/2 \rceil$. Denote the number of missing servers with k. Also denote the total extra fraction of area which is covered by the servers outside their squares by E. The reliability R_1 for this scenario is then given by

$$\begin{aligned} R_1 &= \hat{R}_1(l, w, t, d, v, m_a) = \frac{n - k\frac{n}{m} + E\frac{n}{m}}{n} \\ &= \frac{m-k+E}{m} = \frac{m_a+E}{m}. \end{aligned} \quad (16)$$

As described in the above *simple model*, the average number of messages per unit time sent in each square under the above mentioned assumptions is $\frac{n}{m}$. This assumes that the nodes under the RWP mobility scenario are uniformly distributed. As extensively studied in the literature [2, 3], the distribution of the nodes moving following the RWP mobility scenario is not always uniform. As reported in this studies, the node distribution in the RWP mobility can be approximated by a uniform distribution under a high node speed v and high pause time τ^t_p. So in our simulation experiments, we will use a high pause time in order to approximate uniform distribution. Otherwise the average number of messages sent in a given square cell may be a certain fraction $\hat{f}$ of n.

When half of the servers needed for full coverage are missing as can be seen in Figures 2(a) to 2(c),

$$k = \lfloor \frac{m}{2} \rfloor.$$

The fraction E is obtained by adding the edge E_e, side E_s and middle E_m extra area fractions as

$$E = E_e + E_s + E_m. \quad (17)$$

To find the values of E_e, E_s and E_m, we first derive expressions for the numbers n_e, n_s and n_m of edge, side and middle remaining servers which give extra coverage to the squares with missing servers. For a rectangular area where nodes can move, there are three distinct cases which give different expressions for these values. These cases are given as follows:

1. When the values of both l_m and w_m are even as shown in Figure 2(a). In this case the number of servers (BS) or covered square cells n_p on the periphery of the movement area is given by

   $$n_p = 2(l_m/2) + 2(w_m/2 - 1) = l_m + w_m - 2.$$

   Also $n_e = 2$, $n_s = n_p - n_e$ and $n_m = \lceil m/2 \rceil - n_p$.

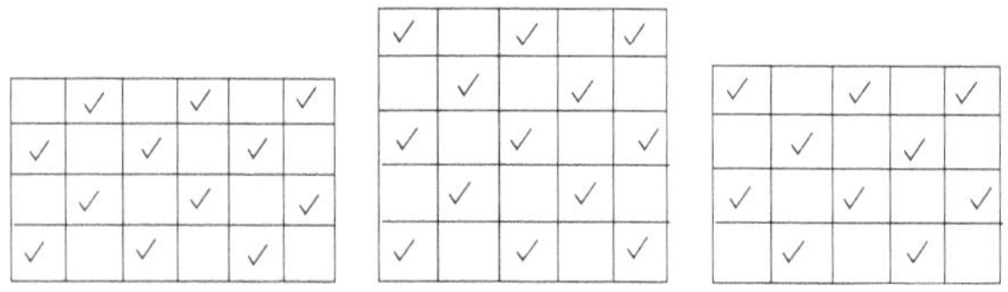

(a) Even by Even (b) Odd by Odd (c) Odd by Even

Figure 2: The marked (covered) even by even, odd by odd and even by odd number of squares

2. When the values of both l_m and w_m are odd as shown in Figure 2(b). In this case the number of available servers (BS) or covered square cells n_p on the periphery of the movement area is given by

$$n_p = 2\lceil l_m/2\rceil + 2(\lceil w_m/2\rceil - 2) = l_m + w_m - 2.$$

Also $n_e = 4$, $n_s = n_p - n_e$ and $n_m = \lceil m/2\rceil - n_p$.

3. When one of the values of l_m and w_m is odd and the other is even as shown in Figure 2(c). In this case the number of servers (BS) or covered square cells n_p on the periphery of the movement area is given by

$$n_p = 2\lceil l_m/2\rceil - 1 + 2(\lceil w_m/2\rceil - 1) = l_m + w_m - 2.$$

Also $n_e = 4$, $n_s = n_p - n_e$ and $n_m = \lceil m/2\rceil - n_p$.

We next derive the fractions E_e, E_s and E_m of areas covered by each BS outside its square. As shown in Figure 3 each BS in the middle of the movement area covers 4 sectors from the other squares whose BS is missing. Each BS at each edge square covers 2 sectors from the squares of the missing BS and each BS on the sides of the movement area covers 3 sectors. The area of each square in the rectangular movement area is given by $A_{sqr} = a^2 = 2t^2$ as shown in Equation 2. The area of each sector as shown in Figure 3 is given by $A_{sct} = \pi t^2/4 - t^2/2$. Hence, the fraction of nodes moving and exchanging messages in each sector of each square is

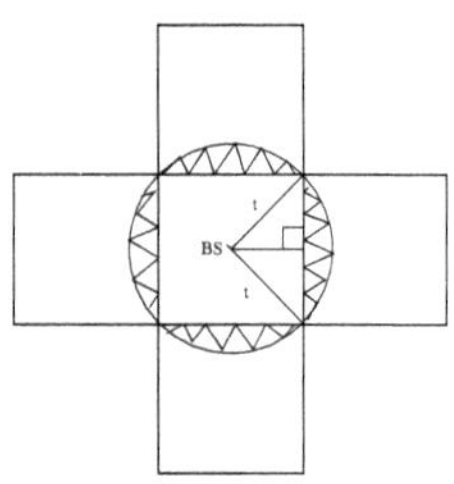

Figure 3: The sectors covered by a middle server

$$\phi = \frac{A_{sct}}{A_{sqr}} = \frac{\pi t^2/4 - t^2/2}{2t^2} = \frac{\pi/2 - 1}{4}. \qquad (18)$$

Hence as discussed above, we have the following equations.

$$E_e = n_e \times 2 \times \phi \qquad (19)$$

$$E_s = n_s \times 3 \times \phi \qquad (20)$$

$$E_m = n_m \times 4 \times \phi. \qquad (21)$$

We next present the reliability for the case where less than half of the number of servers needed to cover the rectangular area.

3.6.2 *With less than half of the number of servers needed to cover the area ($m_a < \lceil m/2\rceil$)*

In this case the number m_a of remaining BS is less than half the total number $\lceil m/2\rceil$ of BS required to give full coverage. Hence, we have the following three cases.

Edge Case: $0 < \lceil m/2\rceil - m_a \le n_e$
In this case there will be $\lceil m/2\rceil - m_a$ more BS missing from the movement area. These first missing servers are chosen to be the once at the edges. One justification for this is that a BS at a square edge covers only 2 sectors of the other uncovered squares around it. And this is the smallest number of sectors of other square areas a BS covers when compared with the side and middle covered cells. The reliability of this case decreases from when there are about half BS proportionally. Hence the reliability R_2 of this case is given as the reliability R_1 of the case $m_a = \lceil m/2\rceil$ with half of the BS missing minus the reliability loss $R_L^e(l, w, t, d, v, m_a)$ due to the additional $\lceil m/2\rceil - m_a$ missing edge BS. This reliability loss is due to the uncovered edge squares and the sectors of the neighboring cells which were supposed to be covered by the BS at these squares. Hence we have

$$R_L^e(l, w, t, d, v, m_a) = \frac{(\lceil m/2\rceil - m_a)(1 + 2\phi)n/m}{n} = \frac{(\lceil m/2\rceil - m_a)(1 + 2\phi)}{m} \qquad (22)$$

and the reliability R_2 of this case with more than half of the BS missing is given as

$$R_2 = \hat{R}_2(l, w, t, d, v, m_a) = R_1 - R_L^e(l, w, t, d, v, m_a) \qquad (23)$$

where R_1 is given by Equation 16.

Side Case: $n_e < \lceil m/2\rceil - m_a \le n_p$
In this case more than the edge BS are missing. Hence the reliability R_2 of the above case is further reduced by the reliability loss R_L^s due to the additional missing side BS to get the new reliability R_3 of this case. Hence we have

$$R_L^s(l, w, t, d, v, m_a) = \frac{(\lceil m/2\rceil - m_a - n_e)(1 + 3\phi)n/m}{n} = \frac{(\lceil m/2\rceil - m_a - n_e)(1 + 3\phi)}{m} \qquad (24)$$

and the reliability R_3 of this case with less than half of the BS missing with more than edge BS losses is given as

$$\begin{aligned} R_3 &= \hat{R}_3(l, w, t, d, v, m_a) \\ &= \hat{R}_2(l, w, t, d, v, \lceil m/2\rceil - n_e) - R_L^s(l, w, t, d, v, m_a). \end{aligned}$$

Middle Case: $n_p < \lceil m/2\rceil - m_a \le \lceil m/2\rceil$
In this case we have even more BS missing from the movement area. In addition to more than half of the missing BSs and the missing periphery BSs, we have $\lceil m/2\rceil - m_a - n_p$ middle BSs missing resulting in an additional $R_L^m(l, w, t, d, v, m_a)$ reliability loss. Hence we have

$$R_L^m(l, w, t, d, v, m_a) = \frac{(\lceil m/2\rceil - m_a - n_p)(1 + 4\phi)n/m}{n} = \frac{(\lceil m/2\rceil - m_a - 2)(1 + 4\phi)}{m} \qquad (25)$$

and the reliability R_4 of this case with less than half of the BS missing with more than periphery BS losses is given as

$$\begin{aligned} R_4 &= \hat{R}_4(l, w, t, d, v, m_a) \\ &= \hat{R}_3(l, w, t, d, v, \lceil m/2\rceil - n_p) - R_L^m(l, w, t, d, v, m_a). \end{aligned}$$

3.6.3 *With more than half of the number of servers needed to cover the area* ($m_a > \lceil m/2 \rceil$)

With more than half of the total number m of the BSs needed to cover rectangular movement area, we have an additional gain of reliability. With each additional ($m_a > \lceil m/2 \rceil$) BS covering each square area we need to exclude the sectors of each newly covered square which were previously covered by the neighboring BS in the case when $m_a = \lceil m/2 \rceil$. This is to avoid double-counting of the newly covered square areas.

We next present expressions for the total numbers n_e^r, n_s^r and n_m^r of remaining (uncovered) edge, side and middle squares to be covered by the additional BSs. There are 4 corner squares in a rectangle. If n_e of them are covered by half of the BS needed to provide full coverage, then $n_e^r = 4 - n_e$. Similarly $n_p^r = 2l_m + 2w_m - 4 - n_p = 2(l_m + w_m - 2) - n_p$ and $n_m^r = \lfloor m/2 \rfloor - n_p^r$ where $n_s^r = n_p^r - n_e^r$.

Edge Case: $0 < m_a - \lceil m/2 \rceil \leq n_e^r$

In this case, we have an additional $m_a - \lceil m/2 \rceil$ covered squares and 2 covered sectors with each square which have to be excluded to avoid double-counting. The placement of the additional BSs starts at the square corners as they are the areas with less sectors covered. For instance if we take the middle cell missing, four sectors from its square area may be covered by the neighboring servers (BS). However only two sectors from the edge cell area may be covered by the two neighboring servers. Hence the reliability gain $R_G^e(l, w, t, d, v, m_a)$ is given as

$$\begin{aligned} R_G^e(l, w, t, d, v, m_a) &= \frac{(m_a - \lceil m/2 \rceil)(1 - 2\phi)n/m}{n} \\ &= \frac{(m_a - \lceil m/2 \rceil)(1 - 2\phi)}{m} \end{aligned} \quad (26)$$

and the reliability R_5 of this case with only less than half of the BS missing is given as

$$R_5 = \hat{R}_5(l, w, t, d, v, m_a) = R_1 + R_G^e(l, w, t, d, v, m_a). \quad (27)$$

Side Case: $n_e^r < m_a - \lceil m/2 \rceil \leq n_p^r$

In this case, we have an additional $m_a - \lceil m/2 \rceil - n_e^r$ covered side squares and 3 covered sectors with each square. Hence the reliability gain $R_G^s(l, w, t, d, v, m_a)$ is given as

$$\begin{aligned} R_G^s(l, w, t, d, v, m_a) &= \frac{(m_a - \lceil m/2 \rceil - n_e^r)(1 - 3\phi)n/m}{n} \\ &= \frac{(m_a - \lceil m/2 \rceil - n_e^r)(1 - 3\phi)}{m} \end{aligned} \quad (28)$$

and the reliability R_6 of this case with only less than half of the BS missing is given as

$$\begin{aligned} R_6 &= \hat{R}_6(l, w, t, d, v, m_a) \\ &= \hat{R}_5(l, w, t, d, v, \lceil m/2 \rceil + n_e^r) + R_G^s(l, w, t, d, v, m_a). \end{aligned}$$

Middle Case: $n_p^r < \lceil m_a - m/2 \rceil \leq n_p^r + n_m^r$

In this case, we have an additional $m_a - \lceil m/2 \rceil - n_p^r$ covered middle squares and 4 covered sectors with each square. Hence the reliability gain $R_G^s(l, w, t, d, v, m_a)$ is given as

$$\begin{aligned} R_G^m(l, w, t, d, v, m_a) &= \frac{(m_a - \lceil m/2 \rceil - n_p^r)(1 - 4\phi)n/m}{n} \\ &= \frac{(m_a - \lceil m/2 \rceil - n_p^r)(1 - 4\phi)}{m} \end{aligned} \quad (29)$$

and the reliability R_6 of this case with only less than half of the BS missing is given as

$$\begin{aligned} R_7 &= \hat{R}_7(l, w, t, d, v, m_a) \\ &= \hat{R}_6(l, w, t, d, v, \lceil m/2 \rceil + n_p^r) + R_G^m(l, w, t, d, v, m_a). \end{aligned}$$

4. SIMULATION VALIDATION

To validate our analytical models of short message transfer reliability we have modified the NS2 simulation package to simulate senders and receivers of messages along with servers which control the message transfers. In the experiment presented in this paper we have used a star topology similar to the one shown in sample Figure 4 with one server at the center and all other servers connecting to it. In this figure, S represents sender node, R represents receiver node and C represents session controller (server) nodes.

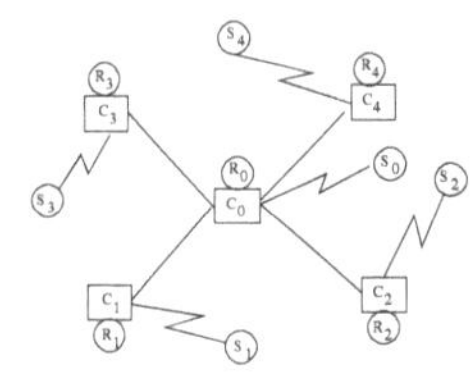

Figure 4: Sample network topology

In this simulation setup we consider a $1061mx1061m$ rectangular area with $l = m = 1061m$, where the nodes can move. We use 25 to 50 senders and 25 to 50 receivers. The message lifetime (deadline) d ranges from 60s to 120s and beacon interval is 0.1s. We use a random way point mobility model with the average speed v of nodes ranging from 1m/s to 50m/s. We use a higher pause time τ_p^t of up to to 90s for the case where receivers are mobile to emulate a uniformly distribution of the nodes in the mobility area. The transmission range t for our experiments is 150m. The results in this section are for a fixed packet (chat message) size of 24+IPHDRLEN Bytes, where $20 \leq$ IPHDRLEN ≤ 60. The chat message is generated at a uniform rate of $1packet/5s$. We next present the numerical results for cases when the number m_a of servers is 5 and 9.

4.1 Experiments with 5 servers

For the first case with 5 servers, Figures 5(a) and 5(b) show that the *Advanced Reliability Model* gives accurate estimation of the average reliability when 50 senders are mobile and 50 receivers are static. Figure 5(b) shows that the average reliability of the advanced analytical model (Adv Model) coincides with the simulation average (Sim Avg).

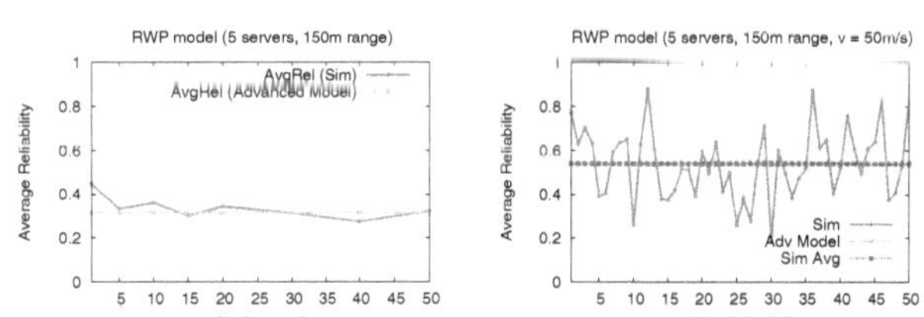

(a) With variable speed (b) With a speed of 50m/s

Figure 5: Our Model vs simulation results

4.2 Experiments with 9 servers

Figure 6(a) shows mobility results when 50 senders are mobile and 50 receivers are static. Figure 6(b) shows results where 25 senders and 25 receivers are mobile. Both experiments consider 9 servers and show that our analytical model gives accurate estimation of the average reliability of message delivery.

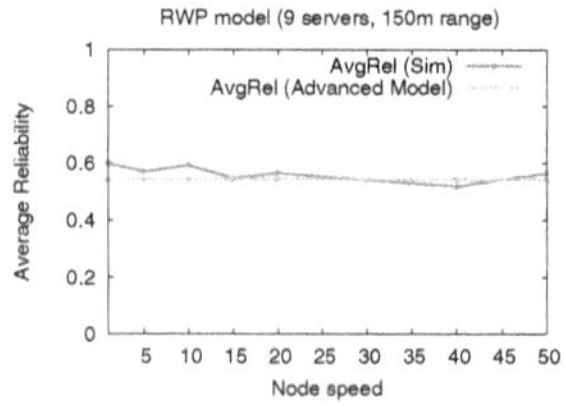

(a) With variable sender speed

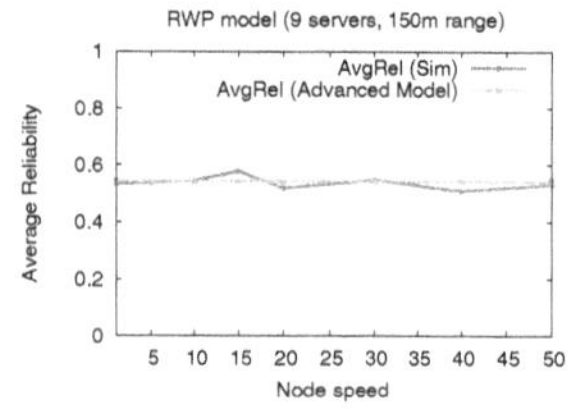

(b) With variable sender and receiver speeds

Figure 6: Our Model vs simulation results

Figure 7 shows the case of 9 servers with 50 senders and 50 receivers mobile at a speed of 30m/s. From the plot it can be seen that the average reliability using the analytical model (Adv Model) almost coincides with the average of the simulation (Sim Avg).

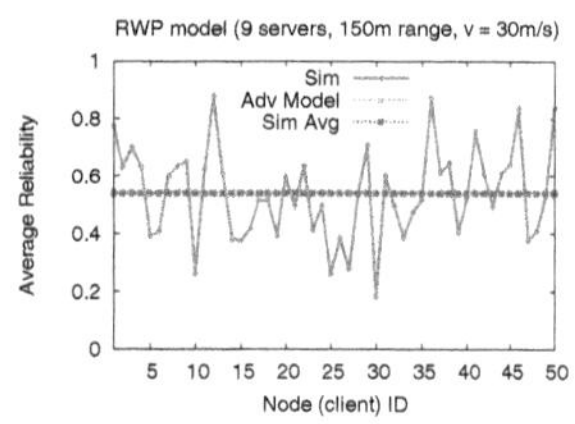

Figure 7: Our Models vs simulation results with a sender and receiver speeds of 30m/s

In all our experiments the number of servers m_a was such that maximum message deadline d didn't exceed the average maximum lifetime τ_p^t as calculated with Equation 12. This allowed moving receivers to eventually receive the messages sent to them before the deadline expires. This explains why the average reliabilities presented in Figures 6(a) and 6(b) for cases where receivers are mobile and static are almost similar. As discussed in Section 3.4, for cases where the message deadline exceeds, Equation 13 can be used to account for the loss (reduction) in reliability.

5. SUMMARY AND FUTURE WORK

In this paper we present closed form expressions for the reliability of message delivery from one mobile node to another mobile node as multivariate function of the dimensions of the rectangular movement area, client speed, message lifetime, the number of servers and their transmission range. Simulation results show that our analytical model gives accurate estimation of the average message reliability.

Making the simple extension of our work to cases where message life time expires and when the cell shapes are hexagon is left for future work. We also plan to validate our analytical model of the reliability with real mobile network experiments using Android phones.

6. REFERENCES

[1] Bettstetter, C., Hartenstein, H., and Pérez-Costa, X. Stochastic properties of the random waypoint mobility model. *Wirel. Netw. 10* (September 2004), 555–567.

[2] Bettstetter, C., Resta, G., and Santi, P. The node distribution of the random waypoint mobility model for wireless ad hoc networks. *IEEE Transactions on Mobile Computing 2* (July 2003), 257–269.

[3] Blough, D. M., Resta, G., and Santi, P. A statistical analysis of the long-run node spatial distribution in mobile ad hoc networks. *Wirel. Netw. 10* (September 2004), 543–554.

[4] Cheng, A., Evans, M., and Singh, H. Inside Twitter: An In-Depth Look Inside the Twitter World. *Sysomos Inc., A Marketwire Company* (June 2009).

[5] Pongthawornkamol, T. *Reliability and timeliness analysis of content-based publish/subscribe systems.* PhD thesis, University of Illinois at Urbana-Champaign, Urbana, Illinois, 2010.

[6] Pongthawornkamol, T., Nahrstedt, K., and Wang, G. Probabilistic QoS modeling for reliability/timeliness prediction in distributed content-based publish/subscribe systems over best-effort networks. In *ICAC '10* (New York, NY, USA, 2010), ACM, pp. 185–194.

[7] Saint-Andre, P. Extensible Messaging and Presence Protocol (XMPP): Instant Messaging and Presence (draft-ietf-xmpp-3921bis-12). *XMPP Standards Foundation*, 12 (2010).

[8] Tseng, Y.-C., Lin, T.-Y., Liu, Y.-K., and Lin, B.-R. Event-driven messaging services over integrated cellular and wireless sensor networks: prototyping experiences of a visitor system. *IEEE Journal on Selected Areas in Communications 23*, 6 (2005), 1133–1145.

[9] Weisstein, E. W. Random Walk–2-Dimensional. *MathWorld–A Wolfram Web Resource.*

[10] Weisstein, E. W. Wald's Equation. *MathWorld–A Wolfram Web Resource.*

[11] Xiaomin, M., Yun, L., and Kishor, T. S. Design and Performance Analysis of a New Soft Handoff Scheme for CDMA Cellular Systems. *IEEE TRAN. ON VEHICULAR TECH. 55*, 5 (2006), 1603–1612.

[12] Zerfos, P., Meng, X., Wong, S. H., Samanta, V., and Lu, S. A study of the short message service of a nationwide cellular network. In *IMC '06* (New York, NY, USA, 2006), ACM, pp. 263–268.

Retransmission-based Available Bandwidth Estimation in IEEE 802.11-based Multihop Wireless Networks

Nguyen Van Nam
Université Claude Bernard Lyon 1 - IFI
IRD, UMI 209 UMMISCO, MSI Team
LIP (UMR ENS Lyon - INRIA - CNRS - UCBL)
Lyon, France
nguyen.van.nam@ens-lyon.fr

Isabelle Guerin-Lassous
Université Claude Bernard Lyon 1
LIP (UMR ENS Lyon - INRIA - CNRS - UCBL)
Lyon, France
isabelle.guerin-lassous@ens-lyon.fr

Victor Moraru
Institut de la Francophonie pour l'Informatique (IFI)
IRD, UMI 209 UMMISCO, MSI Team
Hanoi, Vietnam
victor.moraru@auf.org

Cheikh Sarr
Université de Thies
UFR Sciences et Technologies
Thies, Senegal
csarr1979@hotmail.com

ABSTRACT

Estimating the available bandwidth in IEEE 802.11-based multihop wireless networks is a very difficult task due to the medium sharing among contending nodes and collisions between hidden stations. Several methods have been proposed so far for these networks to compute the available bandwidth on wireless links. If some recent solutions such as ABE and IAB now take into account collisions and their impact on the mean backoff, none considers the packet retransmissions due to collisions although these retransmissions have an impact on the available bandwidth. In this article, we propose a new available bandwidth estimation for multihop wireless networks called RABE (Retransmission-based Available Bandwidth). This method integrates the average number of retransmission attempts in the available bandwidth estimation. RABE is evaluated by simulation and the obtained results show that RABE can achieve a mean error ratio of 17% in comparison with the real measurement. Furthermore RABE is at least two times more accurate than ABE and ten times more accurate than IAB.

Categories and Subject Descriptors

C.2.1 [**Computer-Communication Networks**]: Network Architecture and Design—*Wireless Communication*; C.2.5 [**Computer-Communication Networks**]: Local and Wide-Area Networks; C.4 [**Performance of Systems**]: Measurement Techniques

General Terms

Measurement, Performance

Keywords

Available bandwidth estimation, Multihop wireless networks, IEEE 802.11, Quality of service

1. INTRODUCTION

Multihop wireless networks can be deployed in environments where networking architectures with fixed and wired infrastructure are costly and/or difficult to deploy. Mesh wireless networks are such multihop wireless networks that can be considered as an alternative to access networks [2]. In order to offer applications with good performance, similar to traditional access networks, QoS mechanisms often require an estimation of available resources, like for instance the available bandwidth. The available bandwidth of a wireless link can be defined as the maximal throughput that can be transmitted from the sender to the receiver of the link without disrupting ongoing close flows [7].

Several solutions have been proposed for estimating available bandwidth in multihop wireless networks. Many solutions assume that the used wireless technology is IEEE 802.11 with the distributed access mode (DCF - Distributed Coordination Function) [6], because 802.11 wireless cards are very popular and easy to use. We make the same assumption in our work: all the nodes are equipped with a single 802.11 wireless interface card communicating with the DCF mode.

Active estimation methods, like the ones described in [1, 4], use probing packets to derive available bandwidth on paths. Such an approach can take time because it requires to send probing flows before sending data flows and can also impact the rates and the traffic profiles of on-going flows. Passive methods, like the ones proposed in [3, 7, 8], estimate the available bandwidth without sending extra packets but only by observing what happens locally, *i.e.* on each node and on its neighbourhood. This approach requires an evaluation of different parameters like, for instance, the available bandwidth per node, the collision probability on each node and the synchronization of idle periods between senders and receivers of wireless links. But none of the proposed passive solution do not take into account the fact that, in case of collisions, packets are retransmitted with 802.11 DCF and that packets are dropped when the retransmission limit is reached. These packets retransmissions and drops clearly impact the available bandwidth.

In this paper, we present a novel passive available bandwidth estimation method that integrates the impact of the retransmission attempts on the available bandwidth. Our solution, called RABE for *Retransmission-based Available Bandwidth Estimation*, takes into account, in its estimation, the bandwidth wasted by extra waiting times and medium occupancy due to retransmissions. This estimation requires to compute the collision probability and the mean number of retransmission attempts. Note that we assume, throughout this article, a perfect physical layer and that packet loss is only due to MAC collisions.

The paper is organized as follows: Section 2 discusses the importance of retransmissions on the available bandwidth estimation. In Section 3, we describe the solution we propose, called RABE. This estimation method is based on the average number of retransmission attempts. In Section 4, we evaluate the performance of RABE by simulations carried out with ns2. We also compare RABE with other solutions under different topologies and traffic. Finally, Section 5 concludes our work.

2. MOTIVATION: PACKET RETRANSMISSIONS

Recent works, dedicated to the available bandwidth evaluation in multihop wireless networks, can be divided into two categories:

- *Active estimation methods* In these methods [4, 1], probing packets (or flows) are injected into the network in order to estimate the available bandwidth on paths and/or links. This estimation is based on the packet delay or the inter-packets delay measured on probing packets at destination nodes. The main drawbacks of this approach are the impact of the probing packets on the existing flows and a use of bandwidth that is already a scarce resource.
- *Passive estimation methods* These methods rely on information that can be locally sensed or decoded. These methods are not fully passive insofar as information can be exchanged with the nodes' neighbours. As these exchanges consume additional bandwidth, it is important to limit them (in time and to the neighbourhood). But on the other hand, they can bring up-to-date and useful information for the available bandwidth estimation. Very often, as they take profit of broadcast packets used in other protocols, like Hello packets sent in routing protocols, these methods are considered as passive.

Since the proposed solution in this article is a passive approach, we only consider here the main recent passive estimation techniques known in the literature. In Adaptive Admission Control (AAC) [3], each node measures the busy periods duration to deduce the fraction of idle periods on a measurement interval, and then deduces the available bandwidth per node. The available bandwidth of a link is computed as the minimal available bandwidth between the two end nodes of the link. Collisions are not taken into account in this solution.

Among passive estimation methods, Available Bandwidth Estimation (ABE) was the first to integrate the impact of collisions into the available bandwidth estimation [7]. To this aim, the authors compute the collision probability of Hello packets thanks to the Hello packets sent by several routing protocols for multihop wireless networks. Then, they derive the collision probability of data packets from the collision probability of Hello packets via off-line computations and Lagrange interpolated polynomials. One weakness of this technique is that it assumes that Hello packets are periodically sent and thus periodically received, which can not be true with a 802.11 DCF access. Moreover, this solution assumes that the distribution of idle periods for each node is uniform and independent.

To overcome the assumption on the independence of the idle periods distributions, Improved Available Bandwidth (IAB) [8] takes into account the common medium occupation periods between two end nodes of each link, and thus the independent occupation periods. They are computed thanks to the sensing busy state during which one end node senses the medium busy while its neighbor senses the medium idle. This computation assumes a uniform distribution of nodes in the network.

In passive methods, like ABE [7] and IAB [8], the authors consider that a communication can happen on a wireless link whenever the sender and the receiver are both idle. This means that, at the first transmission of the packet, the transmission has succeeded or failed. However, this is not true with the retransmission mechanism used in IEEE 802.11 DCF. This mechanism is shown on Fig. 1: a packet may be retransmitted several times until the transmission succeeds or until the packet is dropped due to the retransmission limit.

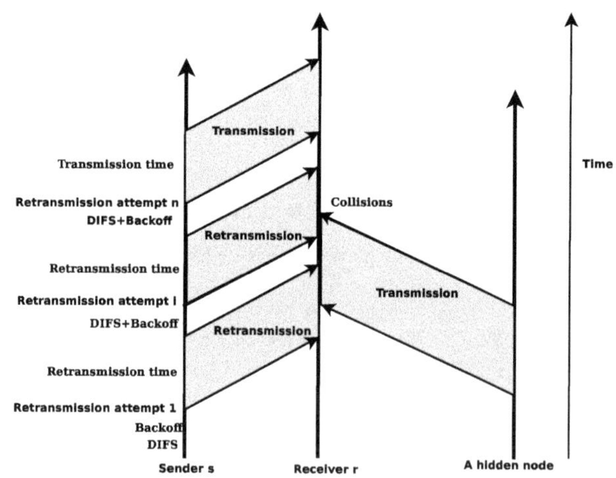

Figure 1: The communication model on a wireless link with IEEE 802.11 DCF

Let us consider the asymmetrical hidden scenario given in Fig. 2. By using NS2, we now study this simple scenario

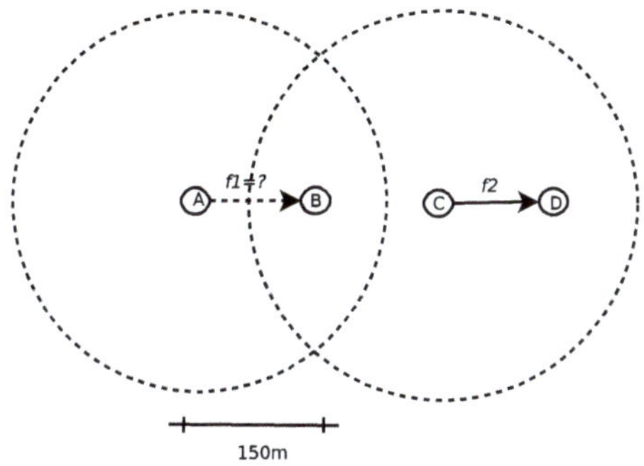

Figure 2: Asymmetrical hidden terminals scenario

that consists of four nodes A, B, C, D. The nodes have the same configuration given in Table 1.

SIFS	10 μs
DIFS	50 μs
Transmision range	200 m
Carrier sensing range	250 m
Physical rate	2 Mb/s
Packet size	1 kbytes
Retransmision limit	7
Contention window size (min, max)	31, 1023
The rate of the flow f_2	250, 500, 750, 1000 kb/s

Table 1: The nodes' configuration in the simple scenario

The distribution of the number of retransmission attempts at the sender A of the flow f_1 in function of the rate of the flow f_2 is shown in Fig. 3. The results in Fig. 3 show that, when the rate of the flow f_2 is low (250 and 500 kb/s) , the maximal number of retransmission attempts for sending the packets of the flow f_1 is 2. Next, at higher rates of the flow f_2, more retransmissions are needed to send a frame, due to a higher number of collisions. Especially, when the rate of the flow f_2 is 1000 kb/s, for all sent frames, the number of retransmission attempts is maximal meaning that all packets have been dropped.

Retransmissions imply additional medium occupancy (with the packets retransmission) and additional waiting times (with an increase of the backoff time). Moreover, when the retransmission attempts limit is reached, the packet is dropped. It means that the link is saturated.

3. RABE: RETRANSMISSION-BASED AVAILABLE BANDWIDTH ESTIMATION

In this section, we describe a new solution, called RABE (Retransmission-based Available Bandwidth Estimation), that provides an estimation of the available bandwidth per link. This estimation is based on the following parameters: the available bandwidth per node, the collision probability, the average number of retransmission attempts, the extra back-off times due to retransmissions and the packet loss ratio.

The next sub-sections describe how RABE computes these different parameters. In the following description, RABE is applied on a link (s, r).

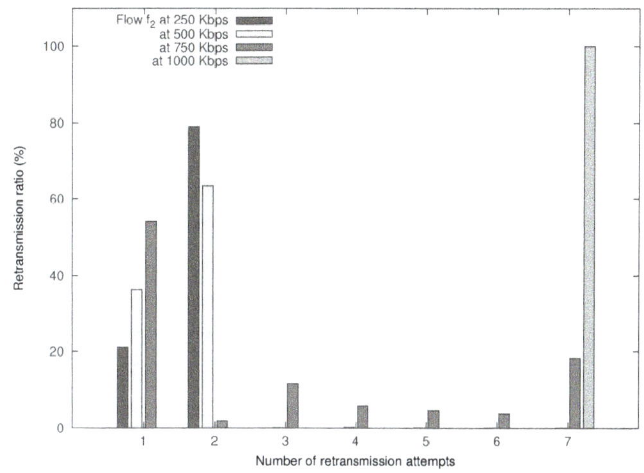

Figure 3: The distribution of the number of retransmission attempts

3.1 Estimation of the available bandwidth per node

As defined in [7], we consider that, given the fraction of idle periods of a node over a measurement interval, denoted κ^i, the available bandwidth per node for any node n can be evaluated as:

$$C_n = \kappa_n^i * C_{max} \quad (1)$$

where C_{max} is the physical rate of the node. Such a measurement takes into account the bandwidth used in the transmission range as in the carrier sensing range.

3.2 Estimation of the collision probability

3.2.1 Classification of collisions

Estimating the collision probability in multihop wireless networks is a tricky part. In multihop networks, as described in [5], following the geometric relationship of the stations, the collisions can be divided into four categories: collisions due to coordinated stations (senders in a cell draw the same backoff and access to the medium at the same time), collisions due to information asymmetry (corresponding to the asymmetrical hidden nodes scenario), collisions due to near hidden terminals and collisions due to far hidden terminals (corresponding to the symmetrical hidden nodes scenario). Several works on available bandwidth estimation, like in IAB [8] for instance, consider only the collision probability due to coordinated stations in a cell. But in [5], the authors show that the collision probability of this first category is negligible. In ABE [7], the collision probability is derived from on-line experiments on Hello packets to compute the collision probability of Hello packets and from off-line computations to derive the collision probability of data packets of a given size. One weakness of this technique is that it assumes that Hello packets are periodically sent and thus periodically received, which can not be true with a CSMA/CA access. In [5], the computations on the collision probabilities of the different categories require to gather neighbourhood information and to handle cliques operations. These tasks are complicated to integrate into an available bandwidth estimation solution that must provide accurate results in almost real-time. To simplify the computation of the collision probability, we consider unconditional probabilities. The results provided in Section 4 show that this assumption is acceptable.

In our method, we consider two types of collisions on a wireless link:

- Collisions between packets sent by two hidden emitters (EE): An example of this type of collision is given in Fig. 4. The two emitters A and C are hidden from each other. The receiver B is in the interference range of the sender C. In this case, the packets sent by A and the packets sent by C collide at B.
- Collisions between a packet sent by an emitter and a packet sent by a hidden receiver (ER): An example of this type of collisions is given in Fig. 5. The emitter A and the receiver D are hidden from each other. The receiver B is in the interference range of the other receiver D. In this case, the packets sent by A collide with the packets sent by D at B. Note that, in this case, as D is a receiver on the link (C,D), then D can only send control packets like acknowledgements or CTS for instance.

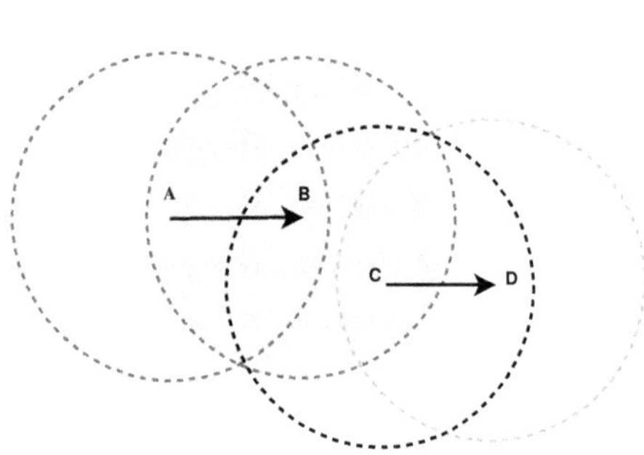

Figure 4: Collisions between packets sent by two hidden emitters (EE)

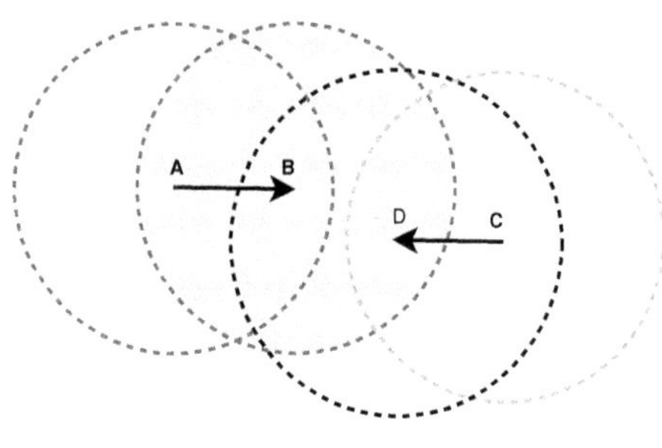

Figure 5: Collisions between packets sent by an emitter and packets sent by a hidden receiver (ER)

If the RTS/CTS mechanism is disable, then the collision type EE corresponds to collisions between data packets, while the collision type ER corresponds to collisions between data packets and acknowledgement packets. Hereafter, we consider that the RTS/CTS mechanism is disable. But the analysis can be easily modified to integrate this mechanism.

3.2.2 *Unconditional collision probability estimation*

First, we estimate the collision probability between data packets sent by hidden emitters (EE). For simplicity, we assume that, on the hidden node, packets generation follows a Poisson process with the following parameters:

- the rate of the packets generation is λ^{DATA} (expressed in packets per time unit),
- the packet duration is T^{DATA} (time units per packet),
- $\lambda^{DATA} * T^{DATA} < 1$.

At the receiver r, we have the following probabilities due to the emissions of the hidden node:

- The probability that r is busy due to packets reception is $\rho_h = \lambda^{DATA} * T^{DATA}$
- The probability that there is no packet arrival at r during the time interval T (time units) is:

$$P[N(t+T) - N(t) = 0] = \frac{e^{-\lambda^{DATA} * T}(\lambda^{DATA} * T)^0}{0!} = e^{-\lambda^{DATA} * T}$$

  where $N(t+T) - N(t)$ is the number of packets arrivals during the time interval $(t, t+T]$.

We assume that, at the node s, the emitter on the link (s, r):

- the inter-arrival times of generated packets are exponentially distributed with the rate λ_s (packets per time unit),
- the packet duration is T_s (time units per packet),
- $\lambda_s * T_s < 1$.

Then, the probability that the link (s, r) is busy due to packets transmission of node s is $\rho_s = \lambda_s * T_s$ and the probability that there is no packet arrival from node s on the link (s, r) during T (time units) is $e^{-\lambda_s * T}$.

There is a collision (of type EE) on the link (s, r) at the receiver r if:

- during the transmission of the node s on the link (s, r) with a packet duration T_s, there is another emission of a node hidden to s. The probability for this to happen is:

$$p_s^{EE} = \rho_s * (1 - e^{-\lambda^{DATA} * T_s}) \quad (2)$$

- during the emission of a node hidden to s with a packet duration T_{DATA} by reception, there is a transmission from node s. The probability for this to happen is:

$$p_h^{EE} = \rho_h * (1 - e^{-\lambda_s * T^{DATA}}) \quad (3)$$

Therefore, the unconditional collision probability of type EE occurring at the receiver r of the wireless link (s, r) is estimated as:

$$p^{EE} = 1 - (1 - p_s^{EE}) * (1 - p_h^{EE}) \quad (4)$$

Secondly, we estimate the collision probability between a packet sent by an emitter and a control packet sent by a hidden receiver (of type ER). We assume that acknowledgement packets arrivals follow a Poisson process of intensity λ^{ACK}. Since the size of acknowledgements is small, we assume that the probability that there is a packet transmission from the emitter s during the transmission of an acknowledgement from a hidden node is negligible. We only consider the probability that there is an acknowledgement emission by a hidden node during a packet transmission of s. This probability, p^{ER} is estimated as:

$$p^{ER} = p_s^{ER} = \rho_s * (1 - e^{-\lambda^{ACK} * T_s}) \quad (5)$$

Finally, the unconditional collision probability on the receiver r is:

$$p = 1 - (1 - p^{EE}) * (1 - p^{ER}) \quad (6)$$

3.3 Estimation of the average number of retransmission attempts

In this section, we estimate the average number of retransmission attempts given the collision probability computed previously. Note that the probability of failure of the first retransmission attempt is the unconditional collision probability. Then, the probability of failure for the other retransmission attempts is conditional and can be estimated in the following way. Given the unconditional collision probability p on the receiver r and the maximal number of packet retransmission attempts M, the probability of k retransmission attempts of a packet at the sender s is calculated as in [7, 8]:

$$P(X = k) = \begin{cases} p^k(1-p), & 0 < k < M \\ p^M, & k = M \\ 0, & k > M \end{cases} \quad (7)$$

The average number of retransmission attempts (including the transmission itself) can be written as:

$$\begin{aligned} \overline{n} &= 1 + \sum_{k=1}^{M-1} kP(X = k) + Mp^M \\ &= 1 + p(1-p) + 2p^2(1-p) + ... + (M-1)p^{M-1}(1-p) \\ &\quad + Mp^M \\ &= 1 + p - p^2 + 2p^2 + ... + (M-1)p^{M-1} - \\ &\quad (M-1)p^M + Mp^M \\ &= 1 + p + p^2 + \quad + p^{M-1} + p^M \end{aligned}$$

$$\overline{n} = \frac{1 - p^{M+1}}{1 - p} \quad (8)$$

The evolution of the average number of packet retransmission attempts according to the collision probability is shown in Fig. 6. In 802.11 DCF, there are two retransmission limits: Short Retry limit (7 retransmissions) that is used for packets of short size and Long Retry Limit (4 retransmissions) that is used for packets of long size [6]. From Fig. 6, we see that the retransmissions fail if the collision probability at the receiving side is more than 0.965 for Short Retry Limit and more than 0.89 for Long Retry Limit.

The retransmission mechanism impacts the available bandwidth of a link, since it increases the time needed to transmit a packet at the sender side and packets are lost due to the collision and when the retransmission attempt limit is reached. In the next two-subsections, we estimate this impact.

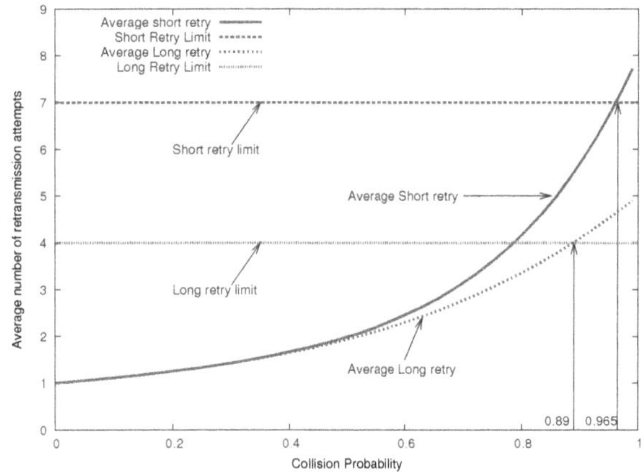

Figure 6: Evolution of the average number of packet retransmission attempts

3.4 Taking into account the extra time to send a packet

In this section, we estimate the waste of the bandwidth due to the additional time induced by retransmissions. Retransmission and backoff mechanism can be seen as a way for the sender to adapt its transmission rate to the reception rate of the receiver. This rate adaptation can be considered as a kind of synchronization.

- At the sender s, due to the retransmissions and the backoff mechanism, the total duration to send a packet is:

$$\overline{n} * (DIFS + T) + \overline{backoff} \quad (9)$$

where T includes, in addition to the time required to transmit a packet, a SIFS plus the time required to send the acknowledgement. $\overline{backoff}$ is the average backoff time estimated as follows (see [7, 8]):

$$\overline{backoff} = \frac{1 - p - 2^N p^{N+1}}{2 - 4p} CW_{min} - \frac{1}{2} \quad (10)$$

Therefore, the available bandwidth of the sender is decreased by a factor $\tau_s = \frac{DIFS + backoff + T}{\overline{n} * (DIFS + T) + \overline{backoff}}$ where $backoff$ is the average backoff time when the minimal contention window size is used.

The available bandwidth at the sender s is then estimated as:

$$AB^s_{RABE} = \tau_s * C_s \quad (11)$$

- At the receiver r, a packet is received with success at the MAC layer if the receiver is idle. The available bandwidth at the receiver r is then estimated as

$$AB^r_{RABE} = C_r \quad (12)$$

3.5 Estimation of the packet loss ratio

Packets are dropped when the retransmission attempt limit is reached. As RABE is based on the average number of retransmission attempts, we need to estimate the packet loss ratio according to this average number of retransmission attempts.

Given the retransmission attempts limit M and the average number of retransmission attempts $\overline{n}$, the maximal number of retransmissions is $(M-1)$ and the average number of retransmission failures is $(\overline{n}-1)$ (obviously, we do not take into account the successful transmission). The packet loss ratio is then estimated as $\frac{\overline{n}-1}{M-1}$. Therefore, the available bandwidth of the link must be scaled with the following factor:

$$K = \begin{cases} 1 - \frac{(\overline{n}-1)}{(M-1)} = \frac{M-\overline{n}}{M-1}, & \overline{n} \leq M \\ 0, & \overline{n} > M \end{cases} \quad (13)$$

In Eq. 13, when $\overline{n} > M$, we consider that the wireless link is saturated since all the packets sent from the sender to the receiver are dropped due to the retransmission limit that is reached.

3.6 Estimation of the available bandwidth of a wireless link

Finally, given the available bandwidth AB^{s}_{RABE} at the sender s and the available bandwidth AB^{r}_{RABE} at the receiver r, we consider that the available bandwidth on the wireless link (s, r) is:

$$AB_{RABE} = K * min\{AB^{s}_{RABE}, AB^{r}_{RABE}\} \quad (14)$$

Denoting $K_s = K * \tau_s$ and $K_r = K$, the available bandwidth of a wireless link can be rewritten as

$$AB_{RABE} = min\{K_s * C_s, K_r * C_r\} \quad (15)$$

4. SIMULATIONS

In this section, we use the simulator NS2 to evaluate the performance of RABE. We randomly generate topologies in an area of 1000m x 1000m. We present the results obtained on two topologies. The first topology, denoted 50/80/CBR topology, consists of 50 nodes and 80 one-hop point-to-point flows generated according to a CBR/UDP traffic model. The second one, denoted 100/135/Poisson topology, consists of 100 nodes and 135 one-hop point-to-point Poisson flows (the packets of each flow are generated according to a Poisson process). Note that the network may be disconnected, which is not a drawback for our study. An example of a generated network is shown in Fig. 7. The nodes are configured with the same parameters as the ones given in Table 1. Moreover, we configure the flows with the same mean transmission rate (denoted as x).

We now drop two nodes at the center of the topology: the sender s is located with the coordinates $(300m, 500m)$ and the receiver r is located with the coordinates $(450m, 500m)$. Since the communication range is $200m$, the two nodes s and r can communicate directly. Moreover, we keep the network configurations so that some flows are in the sensing range of the sender (and/or the receiver) and some others are not. This means that s and r have common and hidden nodes, which implies common and independent busy periods.

Our aim here is to evaluate the available bandwidth on the link (s, r) according to the load variation in the network. To this end, we simply vary the value of x in order to change the background traffic. For each simulation, we also measure the real collision probability and the real available bandwidth of the link (s, r).

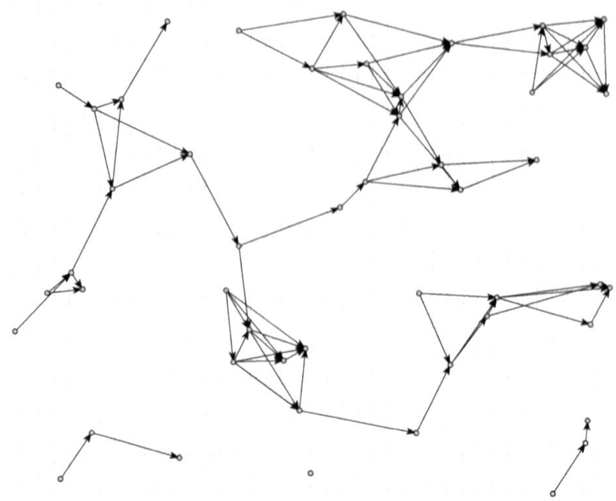

Figure 7: A 50/80/CBR topology

4.1 Estimating the available bandwidth with RABE

In this section, we describe how RABE is implemented in the simulator. Each node records, on a measurement interval, the following parameters: the number of data packets sent and received, the number of acknowledgements sent and received, the number of collisions that are perceived and the transmission times of data packets and acknowledgements. Based on these parameters, each node calculates its available bandwidth as defined in Eq. 1.

To implement RABE, we need to know the number of data packets and acknowledgements arrivals at the receiving side of the link (λ^{DATA} and λ^{ACK}). However, some packets may collide and then can not be decoded by the receiver. As the node can not decide if the collided packet is a data packet or an acknowledgement, then we apply the following approximation: the number of collisions on data packets (acknowledgements respectively) is derived as the total number of collisions multiplied by the percentage of received and decoded data packets (acknowledgements respectively). With this approximation, we consider that the ratio between data packets and acknowledgements is the same on decoded packets as on collisions. At the end, λ^{DATA} (λ^{ACK} respectively) is computed as the number of data packets (acknowledgements respectively) received and decoded plus the estimated number of collided data packets (acknowledgements respectively).

To estimate the collision probability if a flow was emitted on the link (s, r), we also need to estimate the number of generated data packets at the sending side s (λ_s). By assuming that the packet size of the flow to be emitted is known (and denoted T_s), then we apply the following estimation:

$$\lambda_s = \frac{C_s}{T_s} \quad (16)$$

where C_s is the available bandwidth per node computed by the sender s. The parameters λ_s and T_s must be sent to the receiver r.

4.2 Measuring the real available bandwidth

To measure, by simulation, the real collision probability and the real available bandwidth on the link (s,r), a flow $f_{(s,r)}$ is transmitted on the link (s,r). For each value of x, the rate of this flow is increased step by step. If one of the other flows existing in the network sees its rate decreased by more than 5%, then the rate increase of flow $f_{(s,r)}$ is stopped. The achieved rate by $f_{(s,r)}$ is considered as the available bandwidth on the link (s,r).

We also record the number of collisions perceived ($n_{collision}$) and the number of packets received ($n_{received}$) at the receiver r to compute the real collision probability as follows:

$$p_{real} = \frac{n_{collision}}{n_{received} + n_{collision}} \quad (17)$$

Note that the collision probability p estimated by Equation 6 may differ from p_{real} since p does not take into account collisions due to coordinated nodes, in the same communication area, that access to the medium at the same time.

4.3 Simulation results

For each value of x, simulations are repeated ten times on the same topology. For each simulation, the available bandwidth on the link (s,r) is computed after each measurement interval and the final available bandwidth on this link is the average of the values obtained on the measurement intervals. The plotted values are the average on the different simulations performed for a value of x. The obtained average values are also associated with a confidence interval of 95%.

Firstly, we compare the collision probability estimated by RABE and the real collision probability according to the flows mean rate. This comparison is given on Figs. 8 and 9. On the 100/135/Poisson topology, the difference between the two collision probabilities is small whatever the flows mean rate. On the 50/80/CBR topology, this difference is a bit higher even though acceptable. This can be explained by the assumption on the packets generation following a Poisson process. In this case, the collision probability estimated by RABE is lower than the real collision probability.

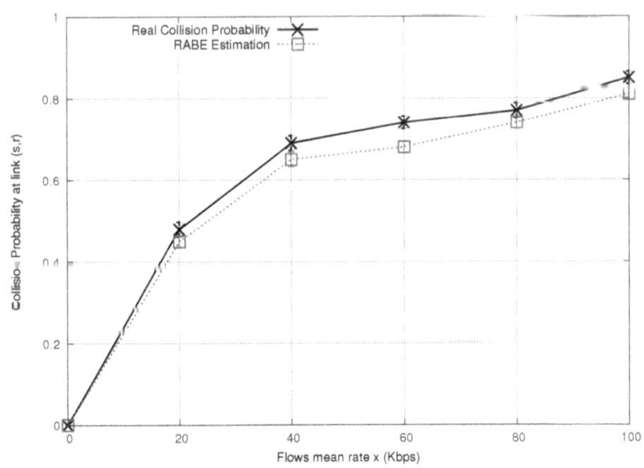

Figure 8: Comparison of the collision probability estimated by RABE with the real collision probability on a 50/80/CBR topology

Secondly, we compare RABE with the real available bandwidth, as with other available bandwidth estimations, namely

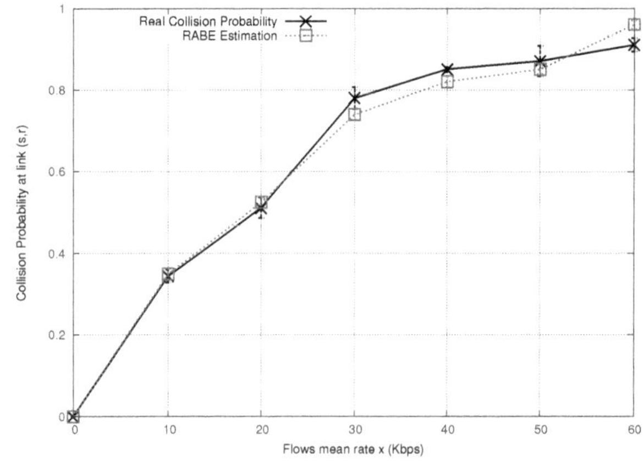

Figure 9: Comparison of the collision Probability estimated by RABE with the real collision probability on a 100/135/Poisson topology

ABE and IAB. The measurement interval is the same for the three solutions and equal to 1s. Figs. 10 and 11 show the obtained results on the two tested topologies according to the flows mean rate. We see that, in both scenarios, IAB is not accurate since the provided estimation is always very far from the real value as soon as x is higher than 20 kb/s. This can be easily explained by the fact that only collisions due to coordinated stations are taken into account and the estimation is almost insensitive to collision.

We see also that ABE under-estimates the available bandwidth. This can be explained by the assumption on the uniform and independent medium occupancy distribution and by the fact that the available bandwidth is scaled in the ABE estimation even if there is no collision.

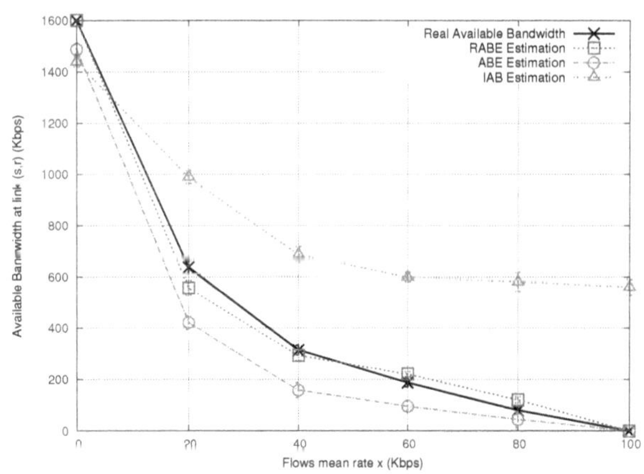

Figure 10: Available bandwidth estimation with RABE, ABE and IAB on a 50/80/CBR topology. Comparison with the real available bandwidth

From these results, we can see that RABE provides the most accurate estimation among the three tested solutions, and this whatever the scenario and the flows mean rate. The accuracy of the solutions are given in Figs. 12 and 13. In the 50/80/CBR topology, the mean estimation error of RABE is 17.49%, the one of ABE is 36.7% and the one of IAB is 205,8%. Based on the mean error ratio we can con-

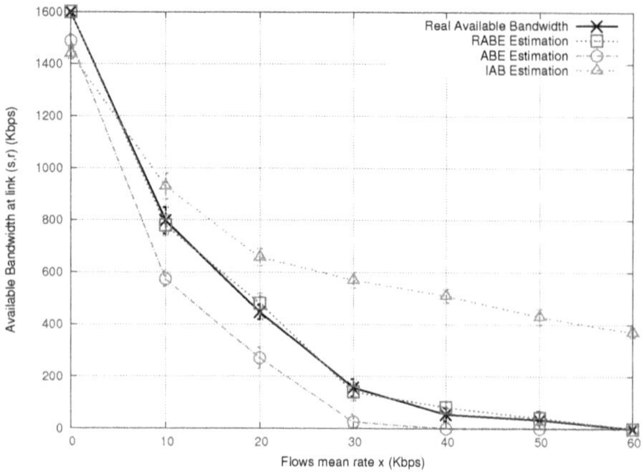

Figure 11: Available bandwidth estimation with RABE, ABE and IAB on a 100/135/Poisson topology. Comparison with the real available bandwidth

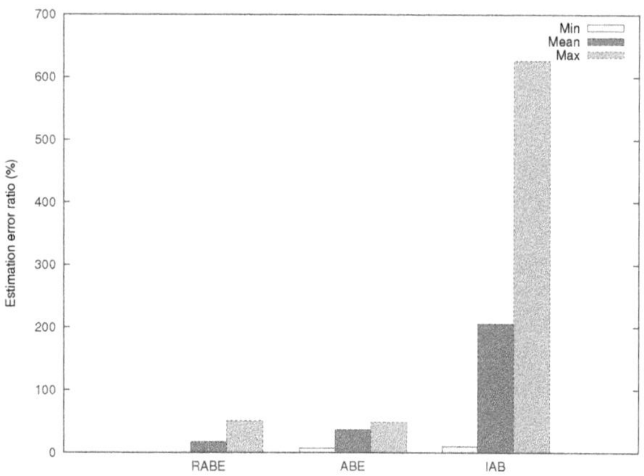

Figure 12: Estimation error (in percentage) for ABE, IAB and RABE on the 50/80/CBR topology

clude that RABE is two times more accurate than ABE and ten times more accurate than IAB. In the 100/135/Poisson topology, the mean estimation error of RABE is 15.79%, while it corresponds to 59.76% and 397% for ABE and IAB respectively. In other words, RABE is four times more accurate than ABE and twenty six times more accurate than IAB. In some simulations, the estimation error can achieve 1200% for IAB.

5. CONCLUSION

In this paper, we propose RABE, a novel available bandwidth estimation method in IEEE 802.11-based multihop wireless networks. As opposed to previous solutions, this method takes into account the retransmission mechanism of 802.11 DCF that impacts the available bandwidth. RABE is based on the available bandwidth per node, the collision probability on the link to evaluate and the mean number of retransmission attempts. From this mean number, RABE estimates the additional time needed to retransmit a packet that collides and the packet loss ratio due to the retransmission limit.

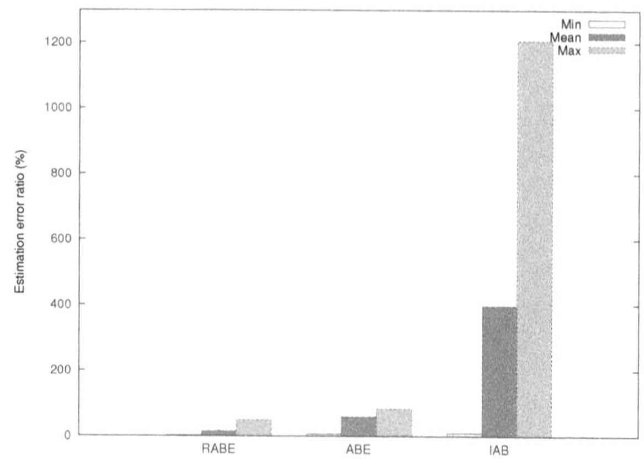

Figure 13: Estimation error (in percentage) for ABE, IAB and RABE on the 100/135/Poisson topology

Our evaluation, carried out by simulation, shows that RABE is more accurate than the other solutions, namely ABE and IAB, in different topologies and under different traffic models and loads. We think that RABE is a clear improvement compared to previous solutions.

6. REFERENCES

[1] Andreas Johnsson, Bob Melander, and Mats Björkman. Bandwidth Measurement in Wireless Network. Technical report, Mälardalen University, Sweden, Mar. 2005.

[2] J. Camp, V. Mancuso, O. Gurewitz, and E. Knightly. A Measurement Study of Multiplicative Overhead Effects in Wireless Networks. In *Proc. of Infocom*, Phoenix, AZ, USA, 2008.

[3] R. de Renesse, M. Ghassemian, V. Friderikos, and A. H. Aghvami. *Adaptive Admission Control for Ad Hoc and Sensor Networks Providing Quality of Service.* King College London, May 2005.

[4] Frank Y. Li, Mariann Haugea, Andreas Hafslund, Oivind Kure and Pal Spilling. Estimating Residual Bandwidth in 802.11-based Ad Hoc Networks: An empirical Approach. In *Proceedings of The Seventh International Symposium on Wireless Personal Multimedia Communications (WPMC 2004)*, Albano Terme, Italy, Sept. 2004.

[5] M. Garetto, T. Salonidis, and E. W. Knightly. Modeling per-flow throughput and capturing starvation in CSMA multi-hop wireless networks. In *In Proc. of IEEE Infocom*, 2006.

[6] L.S.Committee. *ANSI/IEEE Std 802.11:Wireless LAN Medium Access Control(MAC) and Physical Layer (PHY) Specifications.* IEEE Computer Society, 1999.

[7] C. Sarr, C. Chaudet, G. Chelius, and I. Guérin Lassous. Bandwidth Estimation for IEEE 802.11-Based Ad Hoc Networks. *IEEE Transactions on Mobile Computing*, 7(10):1228–1241, 2008.

[8] H. Zhao, E. Garcia-Palacios, J. Wei, and Y. Xi. Accurate available bandwidth estimation in ieee 802.11-based ad hoc networks. *Comput. Commun.*, 32:1050–1057, April 2009.

Modeling Network Traffic in Mobile Networks Implementing Offloading *

Andrey Krendzel
Centre Tecnològic de Telecomunicacions de Catalunya (CTTC)
PMT, Av. C.F. Gauss 7, 08860, Castelldefels
Barcelona, Spain
andrey.krendzel@cttc.cat

Marc Portoles-Comeras
Centre Tecnològic de Telecomunicacions de Catalunya (CTTC)
PMT, Av. C.F. Gauss 7, 08860, Castelldefels
Barcelona, Spain
marc.portoles@cttc.cat

Josep Mangues-Bafalluy
Centre Tecnològic de Telecomunicacions de Catalunya (CTTC)
PMT, Av. C.F. Gauss 7, 08860, Castelldefels
Barcelona, Spain
josep.mangues@cttc.cat

ABSTRACT

The increasing demand of mobile data traffic is starting to stress the networks of mobile network operators (MNOs). Techniques for offloading traffic (partially or totally) from the network of the MNO are currently being designed and deployed at various points of the network, which depend on the goal of each MNO. Such techniques divert traffic to offloading networks, such as Wi-Fi access networks, femtocells, etc., or directly to the Internet. This paper presents a generic analytical approach for studying the impact that offloading techniques might have on the networks of MNOs. The model is generic in the sense that it is independent of the specific offloading technology used and may be of use to provide bounds on network dimensioning. Based on previous measurements found in the literature, our model assumes that user activity periods and periods characterizing offloading are heavy-tailed. We model them as strictly alternating independent ON/OFF processes. Therefore, the non-offloaded traffic (i.e., that traffic still being served by the MNO on a regular basis) is modeled as the product of these two processes. We prove that the resulting process is long-range dependent, with heavy-tailed ON/OFF-period durations, and its characteristic parameters can be derived from those of the initial processes. We also evaluate the network resources required to serve the traffic resulting from the aggregation of many such sources. We conclude that offloading does not always mean reduction of resource consumption, and that the distribution of offloading periods turns out to be the main design parameter to deploy effective offloading strategies in the networks of MNOs. Extensive simulations, in which the Hill's estimator was used to obtain the main parameters of the resulting process, confirm the results of our analytical study.

*This work was supported in part by the Spanish Ministry of Science and Innovation under grant number TEC2008-06826 (ARTICO), by the Catalan Regional Government under grant 2009SGR-940, and by the Spanish Ministry of Industry, Tourism and Commerce under grant number TSI-020100-2010-103 (WIMSAT).

MSWiM'11, October 31–November 4, 2011, Miami, Florida, USA.

Categories and Subject Descriptors

C.2.1 [**Network Architecture and Design**]: [Wireless communication]; I.6.m [**Simulation and Modeling**]: Miscellaneous

General Terms

Performance, Theory

Keywords

mobile networks; traffic offloading; modeling; ON/OFF process; heavy-tails; long-range dependence; network dimensioning

1. INTRODUCTION

Mobile data growth is starting to stress the infrastructure of mobile network operators (MNOs). Furthermore, all forecasts predict further increases in the near future [1]. The problem of handling all such data is further exacerbated by the increasing rates offered by new access technologies (e.g., LTE, LTE-A), as the core will have to handle very high volumes of aggregated traffic. A number of techniques are being conceived to offload the core network of MNOs by diverting packets to other networks at different points of the architecture or by confining local traffic to the local network (e.g., enterprise network). Some examples of such techniques are LIPA, SIPTO, I-WLAN, IFOM, and MAPCON [2], [2], [3], [4].

Up to now, the effort has been mainly devoted to design and, in some cases, to measure the potential gains of using Wi-Fi for offloading. As for the characterization of offloading, [5] analyzes the Wi-Fi connectivity patterns of 100 iPhone users during two and a half weeks, and, based on this, estimates the potential benefits of using Wi-Fi for offloading traffic from the 3G core network. Wi-Fi connectivity patterns are shown to follow a heavy-tail distribution. In the same direction, [6] presents an experimental study of the availability of 3G and WLAN coverage from moving vehicles in three different cities. The authors also deploy a system in their testbed that exploits Wi-Fi for offloading when available and then compute the gains obtained.

However, up to our knowledge, there has been no prior work analytically studying such offloading gains. An eventual analytical framework is expected to help in taking network design decisions when deploying offloading techniques. This paper makes a contribution in this direction.

The theoretical framework proposed in this paper is based on traffic models found in the literature. In this sense, and based on previous measurements [5], we characterize the duration of offloading periods as heavy-tailed. On the other hand, [8] and references therein present well-known models of the behavior of single flow traffic, which describe traffic burstiness in terms of long-range dependence and heavy-tails. In the same way, we characterize user activity according to these models.

Therefore, the main contributions of this paper are:

- Theoretical framework for studying mobile traffic offloading, in which both user activity and offloading periods are modeled by means of strictly alternating ON/OFF processes whose ON and OFF durations are characterized by heavy-tailed distributions. Furthermore, the process describing the non-offloaded traffic (i.e., that traffic still needing to be served by the MNO on a regular basis) is modeled as the product of the two previous processes. This framework is agnostic of the specific offloading technique applied as well as the node where offloading is applied.

- Derivation of the exact parameters describing the asymptotic behavior of the non-offloaded traffic, which is shown to be long-range dependent with ON/OFF durations following a heavy-tailed distribution. We also obtain the parameters characterizing this process as a function of those of the original processes.

- Based on the characteristics of the aggregated non-offloaded traffic, this paper provides performance bounds of resource consumption in the network of the MNO when deploying offloading strategies. Furthermore, we illustrate the influence of the different design parameters on the dimensioning of the network before and after applying offloading. We conclude that offloading does not always entail a gain in terms of network resources and that the appropriate design of offloading periods is key in network dimensioning.

- Validation of the analytical results by means of simulations.

The paper is organized as follows. Fundamental concepts on network traffic modeling used in this paper are given in Section 2. In Section 3, we propose a model for the non-offloaded traffic of a single source that eventually reaches the network of the operator. In Section 4, we discuss on the applicability of our model in the context of offloading techniques. Section 5 is devoted to the main results of our study, in particular, the derivation of the parameters characterizing the asymptotic behavior of the non-offloaded traffic. In Section 6, we validate the main results by means of simulations. In Section 7, we study the implications of deploying an offloading strategy on the provisioning of resources in the network of the MNO. Finally, Section 8 concludes the paper.

2. PRELIMINARIES

Let $X(t)$ be a strictly alternating ON/OFF process where ON and OFF periods follow the cumulative distribution functions denoted by $F_{on}(t)$ and $F_{off}(t)$, respectively. Assume now that as $t \to \infty$ the complementary distribution function is

$$F_{on}^{c}(t) \sim c_1 t^{-\alpha_{on}} L_1(t) \text{ and } F_{off}^{c}(t) \sim c_2 t^{-\alpha_{off}} L_2(t), \quad (1)$$

where $c_j > 0$ is a constant and $L_j(t) > 0$, is a slowly varying function at infinity [7] ($j = 1$ for ON- and $j = 2$ for OFF-period), and $a(t) \sim b(t)$ means that $a(t)/b(t) \to 1$ as $t \to \infty$. The ON and OFF sequences are heavy-tailed whenever $1 < \alpha_{on} < 2$ and $1 < \alpha_{off} < 2$, respectively. In such a case, the process $X(t)$ presents heavy-tailed renewal cycles (i.e., one ON period and the following OFF period) with

$$1 < \alpha_{min} = min\{\alpha_{on},\alpha_{off}\} < 2.$$

The literature presents two fundamental results related to this type of processes. First, the authors of [17] show how this type of processes presents long-range dependence (LRD). Specifically, they prove how the the asymptotic behavior of the covariance function $\gamma_x(t)$ of process $X(t)$ by time t presents the following shape,

$$\gamma_x(t) \sim \sigma_x^2 t^{1-\alpha_{min}} L(t) \quad as \quad t \to \infty, \quad (2)$$

where $0 < \beta = \alpha_{min} - 1 < 1$.

Second, the authors of [7] provide a proof of the fundamental relationship between self-similarity of traffic and the aggregation of a large number of heavy-tailed ON/OFF sources. More specifically, the authors denote as $V(t)$ the variance of the process resulting from the aggregation of several heavy-tailed ON/OFF sources and they prove that,

$$V(t) \sim \sigma^2 t^{3-\alpha_{min}} L^*(t) \quad as \quad t \to \infty, \quad (3)$$

where σ is a constant and $L^*(t)$ is a slowly varying function at infinity. Then, given this result, the authors show that the fluctuations from the mean level of the aggregation of several heavy-tailed ON/OFF sources are given by a fractional Brownian motion (FBM) process (i.e., self-similar traffic) with Hurst parameter (H),

$$\frac{1}{2} < H = \frac{3-\alpha_{min}}{2} < 1. \quad (4)$$

It is also relevant to the procedure that we follow in this paper to consider the following relations [7]. For the stationary process modeling the ON/OFF behavior of a single source ($X(t)$), the probability that at time t it is in an ON-period is calculated as $P(time\ t\ is\ on) = p_{on} = \mu_{on}/(\mu_{on} + \mu_{off})$ for all t, where μ_{on}, μ_{off} are the mean durations of an ON-period and an OFF-period respectively. Then, the covariance function of the process can be expressed as

$$\gamma_x(t) = p_{on}(\pi_{11}(t) - p_{on}), \quad (5)$$

where $\pi_{11}(t) = P(time\ t\ is\ on \mid time\ 0\ is\ on)$. The renewal equation for $\pi_{11}(t)$ is:

$$\pi_{11}(t) = G_{on}^{c}(t) + \int_0^t F_{on}^{c}(t-u)dH_{12}(u), \quad (6)$$

with $G_{on}^{c}(t) = P(\text{remaining life of the first ON interval} > t \mid time\ 0\ is\ ON)$, and where $H_{12}(u)$ is the function corresponding to the inter-renewal (the pair of ON- and OFF-periods) distribution $F_{on} * F_{off}$. ($H_{12} = \sum_{k=1}^{\infty}(F_{on} * F_{off})^{*k}$).

Its density $h_{12}(u)$, when defined, is the probability density that the end of an OFF-period occurs at time u given that at time $t = 0$ the process is in ON state. To understand (6) note that if at $t = 0$ the process is in ON state, then t can be ON either if it belongs to the same ON-period as time 0 or, if there is a subsequent OFF-period, that there is an OFF/ON transition at some time u and t belongs to this subsequent ON-period. The asymptotic behavior of the function $G_{on}^c(t)$ when $t \to \infty$ can be derived as [7]

$$G_{on}^c(t) \sim l_1 t^{-\alpha_{on}+1} L_1(t), \tag{7}$$

where l_1 is a constant.

For a detailed study of heavy-tails, LRD, and self-similar traffic modeling the reader is also referred to [12], [15], [16].

3. TRAFFIC MODEL FOR OFFLOADING

In the same way as in previous literature, the activity of a single source is modeled as a strictly alternating ON/OFF process [7]. In our model, we introduce the effect of offloading over such source by means of an additional strictly alternating ON/OFF process. The resulting traffic sent to the network of the MNO in a regular way (i.e., the non-offloaded traffic) from a single source is then modeled as the product of the two above processes. Therefore, the aggregation of several such resulting processes models the traffic still needing to be handled by the operator in the conventional way, which is the one to be used for the purposes of network dimensioning. The following subsections introduce the details and notation of all above processes.

3.1 Model of user activity

As in previous literature, the activity of a single source/user is modeled as a strictly alternating ON/OFF process , where ON periods are i.i.d., OFF periods are i.i.d., and ON and OFF periods are independent. Furthermore, previous measurements have also shown that such periods follow heavy-tailed distributions (e.g., due to file size distributions, web pages) [8] (and references therein). This is also assumed in our model. Therefore, the process $Y(t)$ is a stationary binary time series $\{Y(t), t \geq 0\}$ such that [7]

$$Y(t) = \begin{cases} 1 & \text{activity period} \\ 0 & \text{idle period} \end{cases} \tag{8}$$

A representation of such a process is presented in Figure 1. During the activity period, the user is transmitting and receiving packets. On the other hand, no traffic is exchanged with any network during idle periods, e.g., user reading time of downloaded content. Notice that $Y(t)$ only describes source/user behavior and is independent from the network through which the traffic is sent, i.e., no offloading considerations have been made in this process.

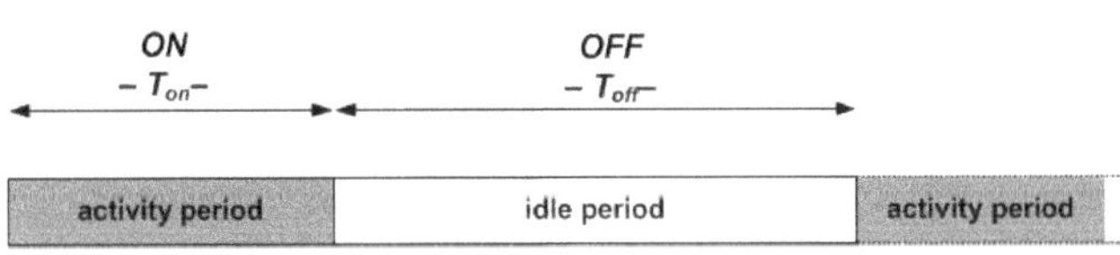

Figure 1: User activity is modeled as a strictly alternating ON/OFF process, $Y(t)$

3.2 Model of offloading periods

Previous measurements carried out in real networks have shown that smartphone connection and disconnection periods to Wi-Fi access points follow heavy-tailed distributions [5]. This observation led us to also model offloading periods for a single source as strictly alternating ON/OFF process. In the same way as before, we also assume that ON periods are i.i.d., OFF periods are i.i.d., and ON and OFF periods are independent. Therefore, the process $X(t)$ is a stationary binary time series $\{X(t), t \geq 0\}$ such that

$$X(t) = \begin{cases} 1 & \text{user flow sent through MNO network} \\ 0 & \text{user flow sent via offloading network} \end{cases} \tag{9}$$

A representation of such a process is presented in Figure 2. During ON periods, the traffic generated by the source (if any) would be sent towards the network of the MNO in a regular way (i.e., traffic is not offloaded). On the other hand, during OFF periods all traffic generated by the source would be routed through the offloading network (e.g., smartphone under the coverage of a Wi-Fi access point or a femtocell).

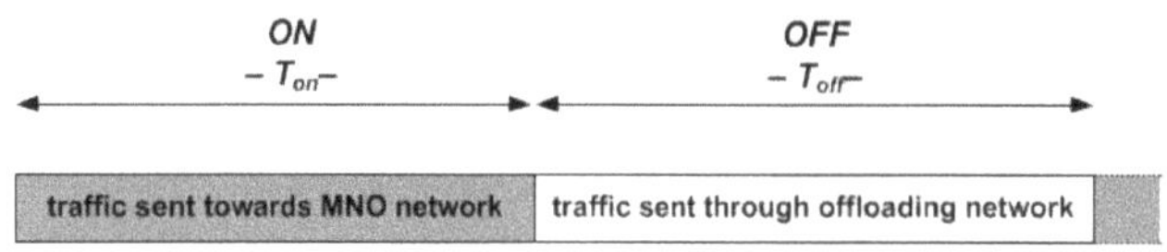

Figure 2: Offloading periods are modeled as a strictly alternating ON/OFF process, $X(t)$

3.3 Model of non-offloaded traffic from a single source

The traffic generated by a single source that is treated in a conventional way by the MNO (i.e., non-offloaded traffic) can be modeled as the product of the previously defined processes. That is,

$$Z(t) = X(t)Y(t). \tag{10}$$

Figure 3 represents such a process, which is also a strictly alternating ON/OFF process, with ON and OFF periods following heavy-tailed distributions and whose characteristic parameters can be derived from those of the original ones, as shown below. During ON periods, the traffic being generated by the source (i.e., user activity in ON state) is forwarded to the network of the MNO as usual. On the other hand, during OFF periods, either there is no activity from the source or traffic is being sent through the offloading network (e.g., through Wi-Fi, femtocells).

3.4 Problem formulation and high-level view of steps followed to solve it

The goal of this paper is to evaluate the potential benefits that offloading techniques could bring to operators. In other words, their interest is to compare what is the resource consumption in their network without applying offloading and when applying offloading on a large scale (see Section 7). Hence, operators are interested in the behavior of the process modeling the aggregation of several $Z(t)$ processes. Therefore, our problem is to obtain an analytical model describing the behavior of such an aggregated traffic and the resource consumption it entails.

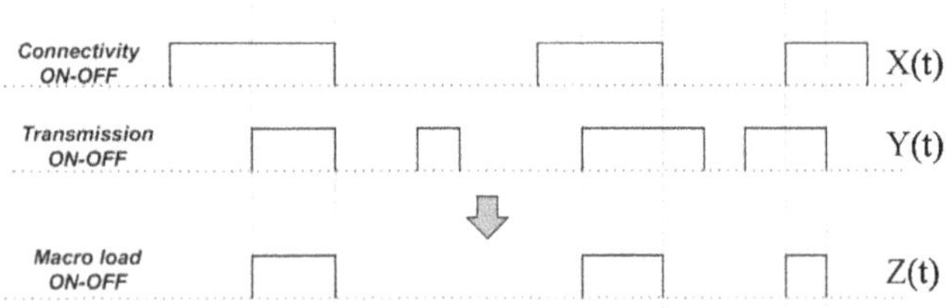

Figure 3: Non-offloaded traffic from a single source is modeled as the product of two strictly alternating ON/OFF processes, $Z(t) = X(t)Y(t)$

A high level view of the steps needed for solving the problem follows. Since, as explained above, $Z(t)$ has ON and OFF periods that follow heavy tailed durations, this process is long-range dependent. Therefore, the aggregation of several such processes is self-similar and can be characterized by means of the Hurst parameter H. This can be done by means of the same techniques presented in [7]. In turn, parameter H can be derived from the parameters characterizing the heavy tailed behavior of the ON and OFF periods of $Z(t)$. Besides, such parameters can be obtained from those of processes of the original processes $X(t)$ and $Y(t)$. Therefore, once the parameters of the original processes are known, by following the above steps, we can characterize the behavior of the aggregated non-offloaded traffic, and hence, the resources needed in the network of the MNO to serve it.

4. APPLICABILITY OF THE MODEL

There are various ways in which offloading could be deployed in a network. Currently, the most popular one is the use of Wi-Fi (instead of 3G), when available, for all the traffic exchanged by the user [2]. With the advent of femtocells, some parts of the network of the operator could also be offloaded when the user is at home or at an enterprise, as her traffic would be routed through the femtocell.

In this paper, we are interested in modeling the traffic that eventually reaches the network of the operator in a regular way (i.e., the non-offloaded traffic). Our goal is to analytically study the benefits (if any) that offloading generates in the network of the MNO in terms of resource consumption. In this sense, the traffic pattern generated towards the network of the MNO in the two above scenarios (i.e, Wi-Fi or femtocells) is the same. That is, all traffic generated by the user when outside coverage of a Wi-Fi access point (or a femtocell) is sent to the network of the MNO as usual. On the other hand, traffic is fully offloaded when under the coverage of the access point. Our model captures this traffic pattern disregarding the technology in use for offloading, and in this sense, it is agnostic from the specific technology in use.

As mentioned in Introduction, there are many other offloading techniques (e.g., I-WLAN, MAPCON, IFOM). As a group, the main qualitative difference with respect to those explained in the previous paragraph is that offloading is done on a per-flow basis, i.e., some flows are offloaded and some other flows are not when under the coverage of a Wi-Fi access point or a femtocell. In a more generic case, there may be intermediate nodes in which offloading decisions are taken (e.g., the gateway of a company towards the network of the MNO). However, in terms of traffic pattern generated towards the network of the MNO, the resulting flow would behave in the same way as in previous cases. Therefore, our model might still be used for those flows that are offloaded, no matter where the offloading point is in the network (e.g., terminal, femtocell/access point, intermediate node). Hence, the total aggregated traffic toward the network of the MNO would result from the aggregation of offloaded flows (output of our model) with regular flows that are not offloaded at all.

5. MAIN RESULTS

LEMMA 1. *Let $Z(t)$ be the process resulting from the product of processes $X(t)$ and $Y(t)$. Let these two latter processes have i.i.d. ON periods, i.i.d. OFF periods, and independent ON and OFF periods with heavy-tailed distributions. The complementary distribution function for the ON-periods of process $Z(t)$ satisfies*

$$F_{on}^{cz}(t) \sim \widetilde{c_1^z} t^{-\alpha_{on}^z} \widetilde{L_1^z}(t), \tag{11}$$

where

$$\alpha_{on}^z = \alpha_{on}^x + \alpha_{on}^y - 1, \tag{12}$$

and α_{on}^x and α_{on}^y are the parameters characterizing the heavy tails of the ON and OFF periods of processes $X(t)$ and $Y(t)$, respectively. Hence, as $\alpha > 1$,

$$\alpha_{on}^z > max\{\alpha_{on}^x, \alpha_{on}^y\}. \tag{13}$$

Additionally, the complementary distribution function for the OFF-periods of process $Z(t)$ satisfies

$$F_{off}^{cz}(t) \sim \widetilde{c_2^z} t^{-\alpha_{off}^z} \widetilde{L_2^z}(t), \tag{14}$$

where

$$\alpha_{off}^z = min\{\alpha_{off}^x, \alpha_{off}^y\}. \tag{15}$$

In (11), (14) $\widetilde{c_j^z} > 0$ is a constant and $\widetilde{L_j^z}(t) > 0$ is a slowly varying function at infinity.

PROOF. *(Lemma)* First, we prove the first part of Lemma 1 concerning the complementary distribution function of the ON-periods of process $Z(t) = X(t)Y(t)$.

The complementary distribution function of the duration of the ON periods of process $Z(t)$, $F_{on}^{cz}(t)$, can be expressed as

$$\begin{aligned} F_{on}^{cz}(t) &= 1 - F_{on}^z(t) = P(ON_Z > t) = \\ &= P(ON_X > t)P(ON_Y > t). \end{aligned} \tag{16}$$

Let us consider all possible options for the transitions from OFF to ON periods of process $Z(t)$. There are three possible situations. First, processes $X(t)$ and $Y(t)$ were in an OFF period and at a certain instant they both switched to ON period (we refer to this case as $off/off \rightarrow on/on$ transition). Second, $X(t)$ was in an ON period and $Y(t)$ in an OFF period, and at a given random instant $Y(t)$ switches to an ON period ($on/off \rightarrow on/on$). And third, $X(t)$ was in an OFF period and $Y(t)$ in an ON period, and at a given instant $X(t)$ switches to an ON period ($off/on \rightarrow on/on$).

Then, for the first type of transition, the duration of the ON-period of process $Z(t)$ is the minimum of the ON durations of $X(t)$ and $Y(t)$, i.e., the intersection of the ON periods of both processes. When the first type of transition happens, the probability that the duration of the ON period of process $Z(t)$ is greater than t can be expressed as the product of the complementary distribution functions of $X(t)$ and $Y(t)$, i.e., $F^{cx}_{on}(t)F^{cy}_{on}(t)$. As for the second type of transition, the duration of the ON-period of process $Z(t)$ is calculated as the minimum between the ON duration of process $Y(t)$ and the remaining lifetime of the ON period of process $X(t)$, and the probablity of thiss duration being greater than t can be expressed as $G^{cx}_{on}(t)F^{cy}_{on}(t)$. Similarly, for the third type of transition, we have $F^{cx}_{on}(t)G^{cy}_{on}(t)$.

By the law of total probability, the function $F^{cz}_{on}(t)$ can then be expressed as

$$\begin{aligned} F^{cz}_{on}(t) &= P[off/off \to on/on]F^{cx}_{on}(t)F^{cy}_{on}(t) + \\ &+ P[on/off \to on/on]G^{cx}_{on}(t)F^{cy}_{on}(t) + \\ &+ P[off/on \to on/on]F^{cx}_{on}(t)G^{cy}_{on}(t), \end{aligned} \quad (17)$$

where $P[off/off \to on/on]$, $P[on/off \to on/on]$, $P[off/on \to on/on]$ are the probabilities of occurrence of each of the above transitions.

Considering that the ON periods of $X(t)$ and $Y(t)$ are heavy-tailed (see eq. (1)), the complementary distribution functions of their ON-periods are

$$F^{cx}_{on}(t) \sim c^x_1 t^{-\alpha^x_{on}} L^x_1(t) \; ; \; F^{cy}_{on}(t) \sim c^y_1 t^{-\alpha^y_{on}} L^y_1(t), \quad (18)$$

respectively. Furthermore, and in accordance with (7), function $G^c_{on}(t)$ for the ON-periods of the processes $X(t)$ and $Y(t)$, respectively, is

$$G^{cx}_{on}(t) \sim l^x_1 t^{-\alpha^x_{on}+1} L^x_1(t) \; ; \; G^{cy}_{on}(t) \sim l^y_1 t^{-\alpha^y_{on}+1} L^y_1(t). \quad (19)$$

Taking into account (18) and (19), expression (17) can be expressed as:

$$\begin{aligned} F^{cz}_{on}(t) &\sim c^{'}_1 t^{-(\alpha^x_{on}+\alpha^y_{on})} L^{'}_1(t) + \\ &+ c^{''}_1 t^{-(\alpha^x_{on}+\alpha^y_{on})+1} L^{''}_1(t) + \\ &+ c^{'''}_1 t^{-(\alpha^x_{on}+\alpha^y_{on})+1} L^{'''}_1(t), \end{aligned} \quad (20)$$

where $c^{'}_1, c^{''}_1, c^{'''}_1$ are positive constants and $L^{'}_1(t), L^{''}_1(t), L^{'''}_1(t)$ are positive slowly varying functions at infinity. The asymptotic behavior of terms with $t^{-(\alpha^x_{on}+\alpha^y_{on})+1}$ decrease significantly slower than that with $t^{-(\alpha^x_{on}+\alpha^y_{on})}$ as $t \to \infty$. In other words, since $t^{-(\alpha^x_{on}+\alpha^y_{on})} = o(t^{-(\alpha^x_{on}+\alpha^y_{on})+1})$ as $t \to \infty$, the asymptotic behavior of $F^{cz}_{on}(t)$ can be expressed as

$$F^{cz}_{on}(t) \sim \widetilde{c_1} t^{-(\alpha^x_{on}+\alpha^y_{on})+1} \widetilde{L_1}(t). \quad (21)$$

Hence, $\alpha^z_{on} = \alpha^x_{on} + \alpha^y_{on} - 1$, and, as α^x_{on} and α^y_{on} are greater than 1, $\alpha^z_{on} > max\{\alpha^x_{on}, \alpha^y_{on}\}$. This concludes the proof of the first part of Lemma 1.

The second part of Lemma 1 concerning the complementary distribution function of the OFF-periods of process $Z(t)$ is proved in a similar way. But, in this case, we characterize transitions of $Z(t)$ from an ON period to an OFF period.

Thus, there are three possible transitions from ON to OFF depending on processes $X(t)$ and $Y(t)$. We refer to each of them as $on/on \to off/off$, $on/on \to off/on$, and $on/on \to on/off$, respectively.

Let us now focus on the complementary distribution function for the OFF period of $Z(t)$. The main difference with that of its ON period is that when one of the two processes ($X(t)$ and $Y(t)$) is in an OFF period, the state of the other process does not matter, as $Z(t)$ will be in an OFF period. Therefore, there may be multiple combinations that generate an OFF period of $Z(t)$. So, for the first type of transition the probability that the OFF period lasts more than t is $F^{cx}_{off}(t) + F^{cy}_{off}(t) + o(t^{-\alpha^x_{off}}) + o(t^{-\alpha^y_{off}})$, where the small o terms represent the less likely combinations that eventually generate an OFF period. As for the second type of transition, we have $F^{cx}_{off}(t) + o(t^{-\alpha^x_{off}})$. And for the third one, such probability is $F^{cy}_{off}(t) + o(t^{-\alpha^y_{off}})$.

By the law of total probability, function $F^{cz}_{off}(t)$ can be expressed as follows

$$\begin{aligned} F^{cz}_{off}(t) &= P[on/on \to off/off] \times \\ &\times (F^{cx}_{off}(t) + F^{cy}_{off}(t) + o(t^{-\alpha^x_{off}}) + o(t^{-\alpha^y_{off}}) + \\ &+ P[on/on \to off/on](F^{cx}_{off}(t) + o(t^{-\alpha^x_{off}})) \\ &+ P[on/on \to on/off](F^{cy}_{off}(t) + o(t^{-\alpha^y_{off}})). \end{aligned} \quad (22)$$

Since the OFF periods of processes $X(t)$ and $Y(t)$ have heavy-tailed durations (and in accordance with eq. (1)), their complementary distribution functions satisfy

$$F^{cx}_{off}(t) \sim c^x_2 t^{-\alpha^x_{off}} L^x_2(t) \; ; \; F^{cy}_{off}(t) \sim c^y_2 t^{-\alpha^y_{off}} L^y_2(t). \quad (23)$$

Then, taking into account (23), expression (22) can be expressed as

$$F^{cz}_{off}(t) \sim c^{'}_2 t^{-\alpha^x_{off}} L^{'}_2(t) + c^{''}_2 t^{-\alpha^y_{off}} L^{''}_2(t), \quad (24)$$

where $c^{'}_2, c^{''}_2$ are positive constants and $L^{'}_2(t), L^{''}_2(t)$ are positive slowly varying functions at infinity.

Hence $\alpha^z_{off} = min\{\alpha^x_{off}, \alpha^y_{off}\}$. This concludes the proof of the second part of Lemma 1. □

THEOREM 1. *Let $Z(t)$ be the process resulting from the product of two strictly alternating ON/OFF processes $X(t)$ and $Y(t)$ with i.i.d. ON periods, i.i.d. OFF periods, and independent ON and OFF periods with heavy-tailed distributions. Then, $Z(t)$ is long-range dependent with covariance function:*

$$\gamma_z(t) \sim \sigma^2_z t^{1-\alpha^z_{min}}, \quad (25)$$

where

$$\begin{aligned} \alpha^z_{min} &= min\{\alpha^z_{on}, \alpha^z_{off}\} = \\ &= min\{\alpha^x_{on} + \alpha^y_{on} - 1, \alpha^x_{off}, \alpha^y_{off}\}. \end{aligned} \quad (26)$$

PROOF. *(Theorem 1)* The proof goes from the expression (2) and Lemma 1 (the expressions (12),(15)). □

6. VALIDATION OF MAIN RESULTS

Taking into account the fact that both processes $X(t)$ and $Y(t)$ are heavy-tailed, we use the Pareto distribution with parameters α and K (the scale parameter) to model both the ON- and OFF-durations of the processes. In our simulations we use the Hill's estimator [10] that provides a good way to estimate tail-index α of Pareto-like tails [11]. In accordance with the methodology presented in [11], the ON- and OFF-durations were simulated with 10^{10} states and 15000 largest sequences were considered out of them in the estimation procedure. For a case study initial data were formed on a basis of statistical information provided in the experimental studies on the performance of 3G mobile data offloading through Wi-Fi networks [5] (for the process $X(t)$) and the burstiness of web traffic caused by accesses of HTTP clients

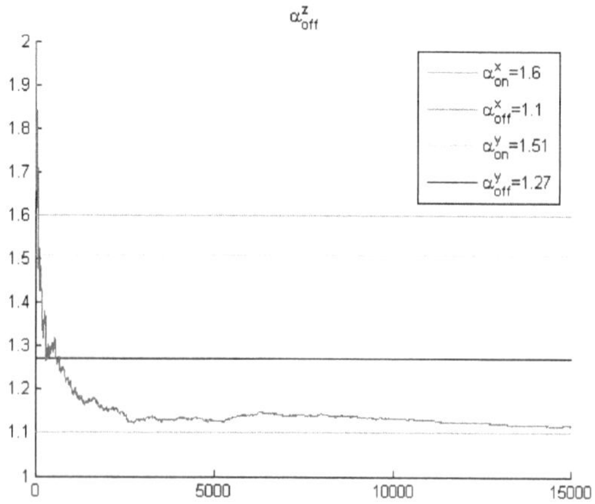

Figure 4: Estimation of the tail-index α^z_{min} by means of Hill's estimator when $\alpha^z_{min} = \alpha^z_{off}$

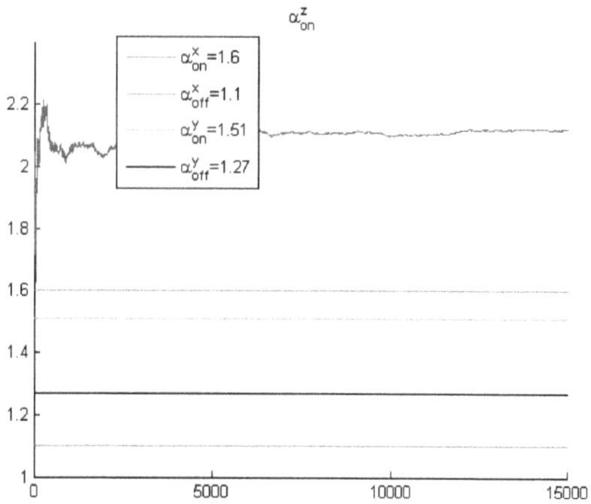

Figure 5: Estimation of the tail-index α^z_{on} by means of Hill's estimator

to the heterogeneous contents (text, images, PDF files, etc.) [9] (for the process $Y(t)$).

Thus, for the process $X(t)$ we use the following initial data: $\alpha^x_{on} = 1.6$, $\alpha^x_{off} = 1.1$, $K^x_{on} = 12.5$, $K^x_{off} = 14.5$, $p^x_{on} = 0.25$, $\mu^x_{on} = 40\ min$, $\mu^x_{off} = 120\ min$.

For the process $Y(t)$ the input data are: $\alpha^y_{on} = 1.51$, $\alpha^y_{off} = 1.27$, $K^y_{on} = 6.1$, $K^y_{off} = 5.96$, $p^y_{on} = 0.39$, $\mu^y_{on} = 18\ s$, $\mu^x_{off} = 28\ s$.

In accordance with the simulations the tail-index of the process $Z(t)$ is $\alpha^z_{min} \sim 1.1$ as illustrated in Figure 4. This agrees with Theorem 1 proved in Section 5 according to which $\alpha^z_{min} = min\{\alpha^z_{on}, \alpha^x_{off}, \alpha^y_{off}\} = min\{2.11, 1.1, 1.27\}$. In this case (26), $\alpha^z_{min} = \alpha^z_{off}$.

Figure 5 shows simulations to estimate the parameter α^z_{on}, which in accordance with Lemma 1 (12) should be equal to 2.11 ($\alpha^x_{on} + \alpha^y_{on} - 1$). The result of the simulations matches well with the analytical result.

The extensive simulations using Hill's estimators were carried out for different sets of initial data and in all cases values of the tail-index α^z_{min} were estimated correctly in accordance with the analytical study presented in Section 5. As an example, for the data set $\alpha^x_{on} = 1.17$, $\alpha^x_{off} = 1.8$, $\alpha^y_{on} = 1.23$,

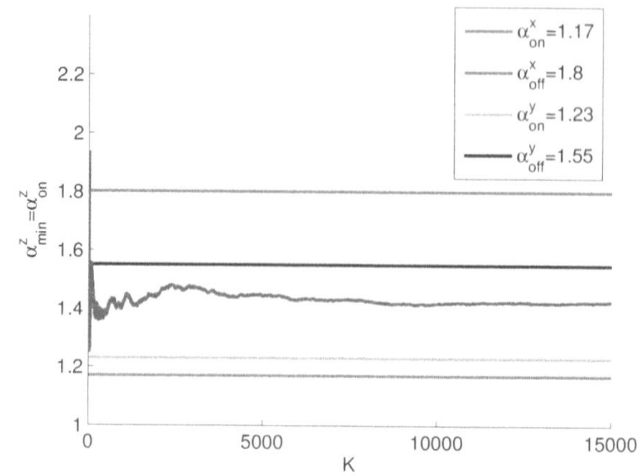

Figure 6: Estimation of the tail-index α^z_{on} by means of Hill's estimator when $\alpha^z_{min} = \alpha^z_{on}$

$\alpha^y_{off} = 1.55$, we have a case when in accordance with (26) $\alpha^z_{min} = \alpha^z_{on} = 1.4$. As seen from Figure 6 it is perfectly agree with the result of the simulations.

7. PERFORMANCE BOUNDS IN OFFLOADING SCENARIOS

This section adopts the perspective taken in [13] and [14] in order to evaluate the system-wide implications of the findings in Section 5. Specifically, we study which are the large-scale implications of implementing an offloading strategy on the provisioning of resources in a Mobile Network.

In particular, Norros [13] derives an approximation of the capacity that a system requires under fractional Brownian traffic in order to provide a target level of QoS, expressed as a bound (ϵ) on the amount of traffic lost. Let us consider an arrival traffic model such as

$$A(t) = \lambda t + \sqrt{a\lambda} B_H(t), \tag{27}$$

where λ is the average amount of traffic that arrives to the network, a is a variance coefficient and $B_H(t)$ is a normalized FBM process with Hurst parameter H.

Given $A(t)$, the capacity needed to guarantee that the probability of system buffer overflow is bounded $P(W > w) < \epsilon$ can be approximated by ([13]),

$$C \sim \lambda + (H^H (1-H)^{1-H} \sqrt{-2ln\epsilon})^{\frac{1}{H}} a^{\frac{1}{2H}} w^{\frac{-(1-H)}{H}} \lambda^{\frac{1}{2H}}. \tag{28}$$

As Theorem 1 reveals, when applying offloading to a heavy-tailed ON/OFF source, the distribution of the resulting sequence of ON/OFF durations may still keep heavy-tail behavior but with a different behavior of the tail. At large scale, when aggregating a large number of sources, the result is that the burstiness of the arrival flow has changed.

Let us denote as H_z and H_y, the Hurst parameters of the aggregate arrival process to a system when we apply an offloading strategy (H_z) and when it is not applied (H_y). Taking into account Theorem 1 and the relation $H = (3 - \alpha_{min})/2$ we make the following observations:

$$\begin{cases} H_z > H_y, & if\ \alpha^z_{min} = \alpha^z_{off} \text{ and } \alpha^x_{off} < \alpha^y_{min} \\ H_z \leq H_y, & otherwise \end{cases} . \tag{29}$$

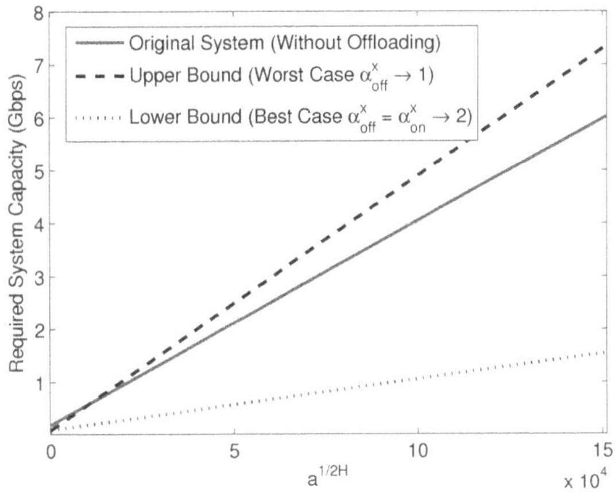

Figure 7: Bounds on Resource needs vs. variance coefficient with 0.5 of offloaded traffic

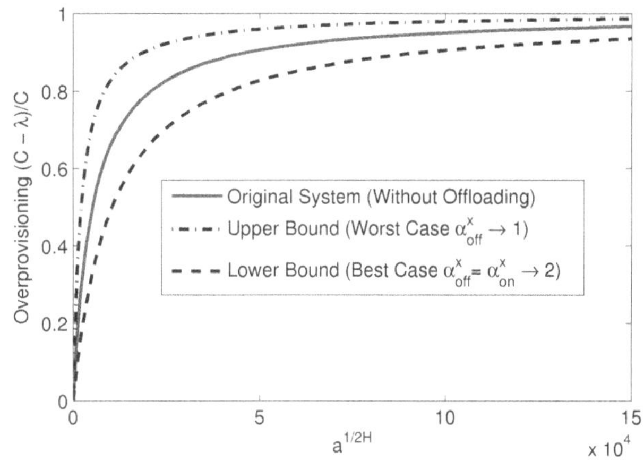

Figure 8: Normalized overprovisioning of the system taking a fluid model approximation

From a network dimensioning perspective relation (29) has a high relevance. Specifically, the expression (29) states that special care should be taken to control the distribution of the duration of offloading periods. In particular, when disconnection (offloading) durations present higher heavy-tailness than the original system (i.e., $\alpha^x_{off} < \alpha^y_{min}$), the net result is that the offered load to the Mobile Network will present a higher degree of burstiness (i.e., $H_z > H_y$).

In order to illustrate this fact let us consider the following example. Assume we have a network without the offloading service dimensioned to support an offered load of $\lambda_y = 200\ Mbps$ with an outage probability of $\epsilon = 10^{-5}$ given a buffering capacity of $w = 100\ kbytes$. Assume further that the system aggregates heavy-tailed ON/OFF sources with pareto-like ON and OFF durations with parameters $\alpha_y^{on} = 1.3$ and $\alpha_y^{off} = 1.8$.

Now assume that we introduce an offloading strategy in the previous system. Taking into account relation (29) we can define a worst case scenario when $\alpha_x^{off} \to 1$. We can also define a best case scenario when both α_x^{off} and α_x^{on} do not have a heavy-tailed distribution (i.e., $\alpha_x^{on} = 2$ and $\alpha_x^{off} = 2$). Applying (28), Figure 7 plots the amount of resources that the system needs to offer the quality of service specified ($\epsilon = 10^{-5}$). Specifically, the figure shows the capacity needed in the original system together with the worst and best case scenarios when we offload a 50% of the traffic.

The figure illustrates two important observations. First, as explained above, the parameter α^x_{off} has a capital importance as we can end up in an extreme case where we start offloading data but we need to increase the amount of resources of the network to maintain a certain level of QoS (due to an increase of data burstiness). Second, introducing a fine-grained control of the distribution of offloading periods can be highly beneficial as it can also lead to decreasing the burstiness of data traffic to the core of the network.

An alternative way to present this last observation is computing the amount of overprovisioning that the system needs to keep up with a target QoS. In this context, overprovisioning is understood as the resources beyond the average load that the operator must put in place to serve bursts of traffic. Figure 8 plots this for the same scenario as in the previous figure. In particular, it plots, for a fluid model, the excess of resources over the total capacity that the network operator needs to implement to maintain the objective probability of loss (ϵ). As the figure shows, in the worst case scenario even though we reduce the amount of load to the system, we need to increase the overprovisioning margin to keep up with the QoS. The contrary happens in the best case scenario where in addition to data traffic reduction the operator can also relax overprovisioning needs.

Finally, let us consider Figure 9, which plots the relation C_z/C_y between the resources needed after and before offloading with respect to the tail index of offloading periods (α^x_{off}). The figure illustrates the influence of the different design parameters on the dimensioning of the network after applying offloading. In particular, the figure focuses on illustrating the strong dependence of the effectiveness of offloading on the actual distribution of offloading periods. Taking as a reference the dashed line in the figure indicating the amount of traffic offloaded (i.e., $P(X(t)$ is in ON state), it can be seen that when the distribution of offloading periods exhibits large tails (i.e., α^x_{off} is low) there is a high probability that the network operator needs to increase resources to maintain QoS. Another interesting observation is that the reduction on the number of resources is best (and cannot do better than) when $\alpha^x_{off} \geq min(\alpha^y_{off}, \alpha^x_{on})$.

As a summary, this section illustrates that the main consequence of Theorem 1 is that the distribution of offloading periods turns out to be the main design parameter in order to implement effective offloading strategies in Mobile Networks.

8. CONCLUSION

This paper presents an analytical framework for modeling traffic in mobile networks implementing offloading. The framework is agnostic of both the specific offloading technique and the point of the network in which it is applied (e.g., terminal, access point). The non-offloaded traffic generated by a single source ($Z(t)$), i.e., the traffic served in a regular way by the operator, is modeled as the product of two strictly alternating ON/OFF processes with ON and OFF independent heavy-tailed durations. One of these processes models the activity of the user ($Y(t)$) and the other one models the offloading periods ($X(t)$).

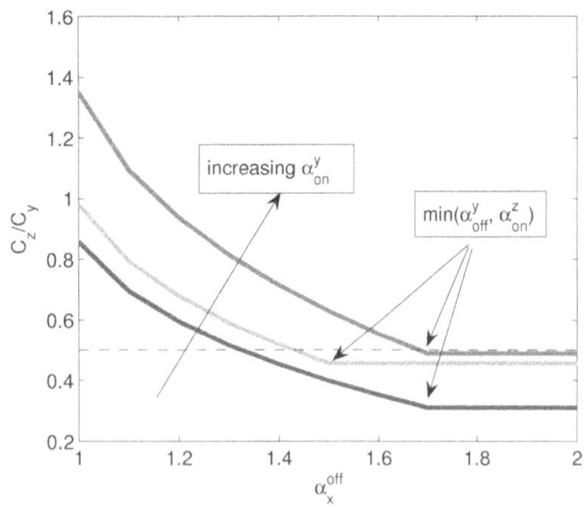

Figure 9: Illustration of the relation between the parameters of the system and the required capacity before and after implementing offloading

By studying its asymptotic behavior, we prove that the ON/OFF process $Z(t)$ has ON and OFF periods with durations that follow a heavy-tailed distribution, hence making it long-range dependent. We also derive the tail-indices describing the heaviness of the ON and OFF periods of $Z(t)$ as a function of tail-indices of the two original processes. The analytical results were validated by means of simulations.

By studying the characteristics of the aggregation of several sources of non-offloaded traffic, we provide performance bounds of the resource consumption in the network of the mobile operator, which can be used when dimensioning the network. Furthermore, we quantitatively evaluate the benefits of offloading by comparing the required resources before and after deploying offloading for providing a given quality of service.

One of the main conclusions of our research is that offloading does not necessarily entail less resource consumption in the network of the operator. Under certain conditions, and due to an increase of the burstiness of the non-offloaded traffic, the amount of network resources to offer a given level of QoS is increased. In this context, the tail-index of offloading periods (α^x_{off}) is the most important design parameter to make non-offloaded traffic less heavy-tailed, hence reducing the resources needed in the network of the operator.

9. REFERENCES

[1] Cisco Visual Networking Index: Global Mobile Data Traffic Forecast Update, 2010-2015, Whitepaper, February 2011

[2] White paper "Analysis of Traffic Offload: WiFi to Rescue", WirelessE2E LLC, September, 2010.

[3] 3GPP TR 23.861, "Multi Access PDN Connectivity", Rel'10, March, 2011.

[4] 3GPP TR 23.829 "Local IP Access and Selected IP Traffic Offload (LIPA-SIPTO)", Rel'10, March, 2011.

[5] K. Lee, Y. Yi, J. Lee, I. Rhee, S. Chong, "Mobile Data Offloading: How Much Can WiFi Deliver?", in proc. of ACM CoNEXT 2010, USA, December, 2010.

[6] A. Balasubramanian, R. Mahajan, A. Venkataramani, "Augmenting Mobile 3G Using WiFi", in proc. MobiSys'2010, San Francisco, USA, June 2010.

[7] M. Taqqu, W. Willinger, and R. Sherman, "Proof of a fundamental result in self-similar traffic modeling", Computer Communications Review 26:5-23, 1997.

[8] S. Stoev, G. Michailidis, and J. Vaughan, "On Global Modeling of Network Traffic", INFOCOM 2010, The 29th Conference on Computer Communications, San Diego, California, March 2010.

[9] E. Casilary et al, "Modeling of Individual and Aggregate Web Traffic", HSNMC 2004, LNCS 3079, pp. 84-95, 2004.

[10] B. M. Hill, "A simple general approach to inference about the tail of a distribution", Annals Statis. 3, pp. 1163-1174, 1975.

[11] J.A. Hernandez, I.W. Phillips, J. Aracil, "Discrete-Time Heavy-Tailed Chains, and Their Properties in Modeling Network Traffic", ACM Transactions on Modeling and Computer Simulation, Vol. 17, No. 4, Article 17, September, 2007.

[12] S.I. Resnick, "Heavy-tail modeling and teletraffic data". Ann. Statis. 25, 5, 1805-1869, 1995.

[13] I. Norros, "On the use of fractional Brownian motion in the theory of connectionless networks". IEEE J. Select. Areas Commun. 13, 6, 953-962, 1995.

[14] I. Norros, "A storage model with self-similar input", Queueing Systems Volume 16, Numbers 3-4, pp. 387-396, 1994.

[15] J. Beran, "Statistics for Long-Memory Processes". Chapman and Hall, London, 1994.

[16] T.Mikosch, G. Samorodnitsky, "Scaling limits for workload process", Operations Research Industrial Engineering report, Cornell University, April, 2006.

[17] D. Heath, S. Resnick, G. Samorodnitsky, "Heavy-tails and long-range dependence in ON/OFF processes and associated fluid models", Mathematics of operations research, February 1998.

Performance Repeatability of Low Power Wireless Sensor Network Protocols: A Multi Testbed Study

Taewoo Kwon
Computer Science and Engineering
395 Dreese Hall
The Ohio State University
Columbus, OH, 43210, USA
kwonta@cse.ohio-state.edu

Emre Ertin
Electrical and Computer Engineering
205 Dreese Labs
The Ohio State University
Columbus, OH, 43210, USA
ertin.1@osu.edu

Anish Arora
Computer Science and Engineering
395 Dreese Hall
The Ohio State University
Columbus, OH, 43210, USA
anish@cse.ohio-state.edu

ABSTRACT

Predicting and bounding performance is a fundamental requirement in wireless sensor network (WSN) protocol development. To support a fast development cycle principled methods are required for experimentation with wireless protocols in testbed environments that can provide repeatable performance guarantees in the target environment. In this paper we propose a method to achieve performance repeatability across test and target environments, that relies on analytical prediction of expected protocol performance as a function of RF environment parameters and forwarding protocol. For the validation of the proposed method, we present analytical, simulation, and experimental results for one-dimensional networks deployed in indoor and outdoor propagation environments, and for two-dimensional networks on four major indoor WSN testbeds to validate the performance of the proposed method on achieving repeatable protocol behavior across diverse set of RF environments.

Categories and Subject Descriptors

C.2.1 [**Computer Systems Organization**]: Computer-Communications NetworkArchitecture and Design[Wireless Communication]; C.3 [**Computer Systems Organization**]: Special-Purpose and Application-Based Systems—*Real-time and Embedded Systems*, C.4 [**Computer Systems Organization**]: Performance of Systems—*Modeling Techniques*

General Terms

Algorithms, Experimentation, Measurement, Performance

Keywords

Wireless Sensor Network,Testbed, Spatial Scaling, Protocol Performance Repeatability

MSWiM'11, October 31–November 4, 2011, Miami, Florida, USA.

1. INTRODUCTION

Predicting and bounding performance is a fundamental requirement in wireless sensor network (WSN) protocol development. Researchers deploying low-power wireless networks have observed that the network behavior in the field diverged substantially from that seen in testbeds, which made it difficult to provide performance guarantees using testbed experiments. To support a fast development cycle principled methods are required for experimentation with wireless protocols in testbed environments that can provide repeatable performance in the target environment. In this paper we propose a method to achieve performance repeatability across test and target environments, that relies on analytical modeling of protocol performance as a function of RF environment parameters and adjusting transmit power levels for matching performance measured through a user specified metric.

The link quality (*e.g.* SNR, PRR) is a function of the RF propagation environment, and makes direct impact on protocol behavior (*e.g.* link selection), which in turn determines the protocol performance. In our previous study [1], we proposed a generic feature *link usage spectrum* for summarizing protocol behavior. Link usage spectrum is the probability distribution with which the network protocol selects links of different length from among all the available links in the network at hand. and proposed to adjust power levels for matching the link usage spectrum as close as possible between the two different RF environments. The hypothesis of this method is that the link usage spectrum is a gross predictor of the performance of (a rich class of) network protocols. With this hypothesis, a network protocol will perform comparably in two network settings if the respective link usage spectrums of the protocol match closely in these settings.

Contributions of the paper. In this paper, we extend our previous work to provide a generally applicable analytical tool for adjusting transmit power levels for replicating performance across the testbed and the deployment environments by matching any user-specified performance index (*e.g.* the number of transmissions, packet forwarding distance, delay, goodput, energy efficiency). Broadly speaking, wireless network protocol performance is dependent upon the quality of the links that are chosen by each node from its available linkset, in terms of the mean and the variance of their SNR. Our tool exploits the insight that scalar network wide performance metrics (such as expected latency) aggregate link usage probabilities with the conditional link SNR statistics into a single performance index.

Armed with this insight, we derive analytical expressions for the expected value of a performance metric given the RF environment and the networking protocol. Based on the mathematical modeling of link usage spectrum, we derive closed form expressions for the moments of the link SNRs, conditioned on the link being chosen by the protocol. We then show how the conditional moments of link SNRs and the link usage spectrum can be used to evaluate the expected value of arbitrary protocol performance metrics.

We validate our tool through various simulation and experimental studies. It is desirable to validate our method in large scale networks. However it is very difficult to execute such experiments in multiple RF environments. Especially, most of the indoor testbeds can not accommodate large topologies. Therefore, we execute and present two canonical studies: micro and macro study. By combining the two, we show the proposed analytical tool will work for general topologies. Micro study uses one-dimensional chain topology with 20 TelosB nodes. The micro study is designed to adopt more granularity in choice of lengths of links than macro study. We executed the macro study in four major WSN testbeds: KanseiGenie [2], Motelab [3], Tutornet [4], and TWIST [5]. Macro study used two-dimensional grid topology with 16 TelosB (KanseiGenie) or TmoteSky (Motelab, Tutornet, TWIST) nodes.

For micro study, we first present simulation studies that show the validity of the analytical approximations employed. Then, we use outdoor deployment and indoor testbed experiments to show that the analytically calculated power scaling factors result in repeatable average protocol behavior and performance across the two environments, despite the difference in node spacing as well as radio propagation parameters. For macro study, we compare analytically derived link usage with the experimentally observed usage in four major WSN testbeds. Then, we compare analytically predicted average performance with the experimentally observed performance using two metrics (the number of transmissions and the progress to destination)with data from the the four testbeds.

Road map. In Section 2, we present the background and discuss related work. In Section 3, we provide a theoretical study to characterize the wireless protocol performance. In Section 4, we validate the performance of our tool. In Section 4.1, we present micro study and in Section 4.2, we provide macro study. In Section 5, we summarize our observations and discuss future work.

2. RELATED WORK

Spatial scaling is an important problem to achieve the performance repeatability between a testbed and a deployment. Naik et al. studied the spatial scaling problem within the *same* RF environment [6]. [6] compared the performance metrics of two WSN applications in two different scales in the KanseiGenie testbed [2] by using different transmission power. We studied the spatial scaling problem across different environments [1]. [1] defined the concept of "link usage spectrum" and tried to reproduce the spectrum of the source network as closely as possible in the target network. [1] argued that even if the available links to two networks are different but the distribution of the chosen links are comparable, the protocol behavior of the two will likely be comparable. This approach is successful in achieving comparable wireless protocol behavior, but does not target a specific performance metric for predicting and repeating mean performance.

Modeling of the expected performance is a general problem for network protocol analysis, and the solution of this problem can contribute towards achieving the performance repeatability. Chiasserini et al. [7] proposed a Markov model of a sensor network protocol performance whose nodes are duty-cycled. [7] successfully incorporated the interference model of sensor network, but it has a limitation on modeling realistic data collection protocol, since it assumes static routing, whereas in general data collection protocols use dynamic link selection schemes. Seada et al. and Zuniga et al. [8, 9] proposed $PRR \times d$ (the forwarding distance) for a metric for geographic routing and predicted average protocol performance for optimizing the protocol parameters. [8] and [9] used *numerical* simulations for estimating average protocol performance. [9] quantified the difference between their local optimum metric, PRR×d and the global optimal metric ETX. They observed that there is only narrow difference between ETX and PRR×d on the energy performance ($\simeq \frac{r}{t}$), where r is the delivery and t is the total number of transmissions, for different densities and network sizes.

Designing protocols with robust performance across different environments is a related research thrust. For example, Lin et al. [10] proposed a new link metric called *competence* capturing long term variation of link quality and designed a distributed route maintenance framework based on feedback control systems. [10] tried to achieve the performance repeatability by achieving a stable performance through choosing more stable links in long term, and equipping extra schemes (*e.g.* Retransmission, Transmission Power Control) to guarantee certain level of link reliability. In contrast, our focus is on achieving comparable performance for an *arbitrary* protocol by tuning the testbed parameters such as power attenuation and scaling.

3. WIRELESS PROTOCOL PERFORMANCE CHARACTERIZATION

In this section we present an analytical tool for determining power scaling to repeat WSN protocol performance across testbed and deployment environments. Wireless protocol performance is determined by the subset of links that are utilized by the wireless protocol and the quality of these chosen links. The quality of the each link in turn is characterized by the RF channel between the terminals (environment model) and the bit-error-performance of their wireless transceivers (radio model).

In the following we derive link usage and link quality statistics for a given network protocol by combining the RF environment and radio receiver models. Then we combine these statistics to assess the expected performance of the wireless protocol. We consider a wireless network $\mathcal{W} = (\{l_j\}, \eta, \sigma)$ with link set $\{l_j\}_{j=1}^{M}$ and the RF environment (η, σ), where η is the path loss exponent (PLE) and σ is the standard deviation of the received signal strength, employing a network protocol $\mathcal{P}$. We note $\mathcal{W}$ is a probabilistic object, referring to the ensemble of link set realizations. For each realization of the wireless network $\mathcal{W}$, network protocol $\mathcal{P}$ chooses a subset of the link set for forwarding of data. We assume $\mathcal{P}$ is a possibly randomized rule for selecting links based on a metric which combines link length $c(l_j)$ and link quality $Q(y(l_j))$ to score the link l_j, and the SNR value of the link, $y(l_j)$. As an example we consider two-dimensional grid networks, where the link lengths $c_j \equiv c(l_j)$ and the progress to destination $d_j \equiv d(l_j)$ are constrained to the finite set $\{(a\tau, b\tau) : 1 \le a, b \le N\}$, where τ is the minimum node spacing. We index elements of the above finite set from 1 to M ($=N^2$). The link usage spectrum $L(\mathcal{W}, \mathcal{P}, i)$ is the discrete probability distribution over the length of links and the progress of links to the destination induced by the network protocol $\mathcal{P}$.

$$L(\mathcal{W}, \mathcal{P}, i) = \text{Prob}[c(l) = c_i, d(l) = d_i] \qquad (1)$$

, where i is the index of $(a\tau, b\tau)$, $c_i = \sqrt{(a\tau)^2 + (b\tau)^2}$, $d_i = c_i * cos(|tan^{-1}(\frac{b}{a}) - \theta|)$, θ is the angle between source and destination. The link usage spectrum is a universal summary statistics of the network behavior. The fundamental importance of link usage spectrum stems from the fact that many network wide metrics can be calculated as the average over link realizations weighted by the link usage spectrum. The link usage spectrum can be calculated *in situ* empirically as averages over node link usages or can be derived directly from RF environment and physical layer model.

3.1 Calculating the expected performance metrics

In this subsection, we derive an equation to calculate the expected performance metric (Result 1). Let $f(y_i, d_i)$ be an arbitrary performance metric over the link quality, SNR (y_i), and the progress to destination (d_i). We assume the performance of a wireless link is measured by $f(y_i, d_i)$. Then the expected performance metric $E[f|W, \mathcal{P}]$ is computed as the average over the linkset l_i, weighted by the link usage spectrum $L(W, \mathcal{P}, i)$ (Result 2):

$$E[f|W, \mathcal{P}] = \sum_{i=1}^{M} E(f(y_i, d_i|l_i \uparrow))L(W, \mathcal{P}, index(l_i)) \qquad (2)$$

Conditional moments of the SNR of the chosen links (Result 3) can be used to approximate $E(f(y_i, d_i|l_i \uparrow))$.

RESULT 1. *The expected performance metric, $E(f(y_i, d_i|l_i \uparrow))$, of the link l_i can be calculated as below:*

$$E(f(y_i, d_i|l_i \uparrow)) \simeq f(E(y_i|l_i \uparrow), d_i)$$

$$+\frac{f''(E(y_i|l_i \uparrow), d_i)}{2}\left(E(y_i^2|l_i \uparrow) - (E(y_i|l_i \uparrow))^2\right)$$

, where y_i is SNR (dB) values of link l_i, d_i is the progress to destination of l_i

PROOF. We review the straightforward argument based on the Taylor expansion of the function $f(y, d)$ with respect to its first argument.

$$E[f(y_i|l_i \uparrow)] = E[f(\mu_{y_i} + (y_i - \mu_{y_i})|l_i \uparrow)]$$

Using the Taylor expansion,

$$f(x) = \sum_{k=0}^{n} \frac{f^{(n)}(a)}{n!}(x-a)^n$$

, we have

$$\simeq E[f(\mu_{y_i}|l_i \uparrow) + f'(\mu_{y_i}|l_i \uparrow)(y_i - \mu_{y_i}|l_i \uparrow)$$

$$+\frac{1}{2}f''(\mu_{y_i}|l_i \uparrow)(y_i - \mu_{y_i}|l_i \uparrow)^2]$$

$$\simeq f(E(y_i|l_i \uparrow)) + \frac{f''(E(y_i|l_i \uparrow))}{2}VAR(y_i|l_i \uparrow)$$

Finally,

$$= f(E(y_i|l_i \uparrow)) +$$

$$\frac{f''(E(y_i|l_i \uparrow))}{2}\left(E(y_i^2|l_i \uparrow) - (E(y_i|l_i \uparrow))^2\right)$$

□

Using Result 1, we can calculate the most common performance metrics: the expected total number of transmissions and the expected progress to destination. As an example, by choosing $f(y, d) = \frac{1}{PRR(y) \times d}$, we can calculate the expected number of transmissions per unit distance. By multiplying the progress to the destination with the calculated expected delay per unit distance, we can calculate the expected end-to-end number of transmissions of the chosen links.

3.2 Calculating the moments of the conditional SNR

In this subsection, we derive an equation (Result 3) to calculate the moments of the SNR of a link conditioned on being chosen over all other links. We revisit the link usage spectrum derivation (Result 2), which was presented in [1], for further extension in Result 3.

RESULT 2. *For a protocol $\mathcal{P}$ which uses $PRR \times d$ as the metric for choosing forwarding links, the probability of choosing link l of length over all other links is expressed as follows:*

$$L(\mathcal{W}, \mathcal{P}, index(l))$$

$$= P\left[PRR(y(l)) \cdot d(l) = \max_{i=1...M}\{PRR(y_i) \cdot d_i\}\right]$$

$$\simeq \sum_{k=0}^{M-1} \int_{a_k}^{a_{k+1}} \frac{1}{\sigma\sqrt{2\pi}} \left[\prod_{i=1}^{k}\left(\frac{1}{2} + \frac{1}{2}erf\left(\frac{g(\beta_i) - \mu_{y_i}}{\sqrt{2}\sigma}\right)\right)\right]$$

$$\left[\prod_{i=k+1}^{M}\left(\frac{1}{2} + \frac{1}{2}erf\left(\frac{y(l) - \mu_{y_i}}{\sqrt{2}\sigma}\right)\right)\right] e^{-\frac{(y(l) - \mu_{y(l)})^2}{2\sigma^2}} dy(l),$$

where $\beta_i = d_i/d(l)$, $a_i = PRR^{-1}(\min\{\beta_i, \frac{1}{\beta_i}\})$, and the links are enumerated such that $a_1 \le a_2 \le ... \le a_{M-1}$, with $a_0 = -\infty, a_M = \infty$ and y_i is the SNR experienced by link l_i, and

$$g(\beta_i) = \begin{cases} a_i, & \text{if } \beta_i \ge 1 \\ \infty, & \text{if } \beta_i < 1 \end{cases}$$

The proof is given in [1].

RESULT 3. *When an application uses $PRR \times d$ as the metric for choosing forwarding links, the expected value of the m-th moment of the SNR of a link l conditioned on the fact that it is chosen by the forwarding metric is given by:*

$$E(y(l)^m|l \uparrow) \simeq$$

$$\frac{1}{w}\sum_{k=0}^{M-1} \int_{a_k}^{a_{k+1}} \frac{y(l)^m}{\sigma\sqrt{2\pi}} \left[\prod_{i=1}^{k}\left(\frac{1}{2} + \frac{1}{2}erf\left(\frac{g(\beta_i) - \mu_{y_i}}{\sqrt{2}\sigma}\right)\right)\right]$$

$$\left[\prod_{i=k+1}^{M}\left(\frac{1}{2}+\frac{1}{2}erf\left(\frac{y(l)-\mu_{y_i}}{\sqrt{2}\sigma}\right)\right)\right]e^{-\frac{(y(l)-\mu_{y(l)})^2}{2\sigma^2}}dy(l)$$

, where $w = L(W,\mathcal{P},index(l))$, $\beta_i = d_i/d(l)$, $a_i = PRR^{-1}(\min\{\beta_i,\frac{1}{\beta_i}\})$, *and the links are enumerated such that* $a_1 \leq a_2 \leq ... \leq a_{M-1}$, *with* $a_0 = -\infty, a_M = \infty$ *and* y_i *is the SNR experienced by link* l_i, *and*

$$g(\beta_i) = \begin{cases} a_i, & if\ \beta_i \geq 1 \\ \infty, & if\ \beta_i < 1 \end{cases}$$

PROOF.

$$E(y(l)^m|l\uparrow) = \int_{-\infty}^{\infty} y(l)^m\ d_{y(l)}(y(l)|l\uparrow)dy(l)$$

From Result 2:
$d_{y(l)}(y(l)|l\uparrow)$

$$= \frac{1}{w}\sum_{k=0}^{M-1}\int_{a_k}^{a_{k+1}}\frac{1}{\sigma\sqrt{2\pi}}\left[\prod_{i=1}^{k}\left(\frac{1}{2}+\frac{1}{2}erf\left(\frac{g(\beta_i)-\mu_{y_i}}{\sqrt{2}\sigma}\right)\right)\right]$$

$$\left[\prod_{i=k+1}^{M}\left(\frac{1}{2}+\frac{1}{2}erf\left(\frac{y(l)-\mu_{y_i}}{\sqrt{2}\sigma}\right)\right)\right]e^{-\frac{(y(l)-\mu_{y(l)})^2}{2\sigma^2}}dy(l),$$

Therefore,
$E(y(l)^m|l\uparrow)$

$$\simeq \frac{1}{w}\sum_{k=0}^{M-1}\int_{a_k}^{a_{k+1}}\frac{y(l)^m}{\sigma\sqrt{2\pi}}\left[\prod_{i=1}^{k}\left(\frac{1}{2}+\frac{1}{2}erf\left(\frac{g(\beta_i)-\mu_{y_i}}{\sqrt{2}\sigma}\right)\right)\right]$$

$$\left[\prod_{i=k+1}^{M}\left(\frac{1}{2}+\frac{1}{2}erf\left(\frac{y(l)-\mu_{y_i}}{\sqrt{2}\sigma}\right)\right)\right]e^{-\frac{(y(l)-\mu_{y(l)})^2}{2\sigma^2}}dy(l),$$

□

3.3 Tuning power for matching performance

In this subsection, we present a procedure to calculate the optimal power attenuation to achieve the consistent performance repetition across two different environments. Equipped with the analytical expressions for expected network performance metrics we can define optimal power scaling rule for consistent performance across testbed and deployment environments. Consider a wireless network $\mathcal{W}$ with inter-node distances $\{d_j\}_{j=1}^m$ and its scaled version $\tilde{\mathcal{W}}$, with inter-node distances $\{\tilde{d_j} = \alpha d_j\}_{j=1}^m$ in RF environments characterized by log-normal scale model parameters (n,σ) and $(\tilde{n},\tilde{\sigma})$ respectively. We assume that transmit power in the network $\tilde{\mathcal{W}}$ is variable through: $\tilde{P_t} = \tilde{P_0}+\beta$ where β is the power attenuation or amplification in the scaled network. Since the scaled vector in general reduces the node distances for convenient testing, β in general is a negative value indicating power attenuation. As a result the scaled network realizations $\tilde{\mathcal{W}}(\beta)$ depends on β[1]. The optimal power attenuation is defined as:

$$E[f|W,\mathcal{P}] = E[f|\tilde{\mathcal{W}}(\beta),\mathcal{P}]$$

[1]We could also introduce spatial variations in β across the nodes to influence the width of the resulting spectrum, for a better match.

4. VALIDATION

In this section, we present analytical, simulation, and experimental studies of repeating protocol performance across testbed and deployed environments. .

Physical & link layer

In our experiments we use 802.15.4 transceivers embedded in popular sensor network platforms such as TelosB and MicaZ. CC2420 radio is compatible with the 2.4GHz 802.15.4 standard. 802.15.4 standard wireless physical layer employs block direct-sequence spread spectrum code with 2MChip/s chip rate and250 kbps data rate to achieve processing and coding gain. The transmitter modulates the carrier using offset quadrature phase shift keying (O-QPSK) with half-sine shaping which is equivalent to minimum shift keying (MSK) modulation which has the following Bit Error Rate:

$$\text{BER} = Q(\sqrt{2y/\text{PG}/\text{CG}}) = \frac{1}{2}\text{erfc}(\sqrt{y/\text{PG}/\text{CG}}) \quad (3)$$

, where y gives the SNR. The processing gain (PG) for 802.15.4 is given by $10\log(2/0.25)$=9 dB. The coding gain (CG) depends on the increased Hamming distance between the codes and is a function of the SNR itself. For low a packet error rate region coding gain can be approximated as 2 dB [11].

Thus, the Packet Reception Rate equation is

$$PRR = \left(1-\frac{1}{2}\text{erfc}(\sqrt{x})\right)^{8*packet\_size}$$

We note that the bit-error-rate approximation given in Equation 3 assumes coherent demodulation using carrier phase information. Practical transceiver designs use non-zero IF and noncoherent demodulation. The non-ideal receiver structures can be approximated with SNR reduction or equivalently increase in the noise floor (P_0) causing only a horizontal shift in the PRR curve.

We present to studies. The micro study uses a dense set of nodes in one dimension to provide granularity in the choice of link lengths. The macro study uses a coarse two dimensional topology to provide granularity in the choice of link direction towards source.

4.1 Micro study: chain one-dimensional topology

4.1.1 Experimental setup

We set up a chain topology with total 20 TelosB sensor nodes, as done in [1]. Here, we define D2 as the chain topology with the node separation of 6 ft, and D1 as the same topology with node separation of 3 ft. Each node is elevated about 4 inches from the ground. TelosB mote is equipped with CC2420 radio and USB serial for communication. The node 0 is set to be the source, and the node 19 is set to be the destination. Every two seconds, the source node produces and sends a packet. For each case, we gathers about 1,000 packets generated by the source. We logged all the paths that each packet has gone through, and only counts the body parts (i.e. excludes the first and the last hops of the paths) to calculate the average link length, because usually the first and the last hops are composed of the shorter links than the remainder links of the path. The metric used for these experiments are as follows: the average link length and the number of transmissions.

Analytical prediction of expected performance requires knowledge of the RF environmental parameters: PLE (η) of indoor and outdoor and standard deviation of RSSI (dB) (σ). We measured the RSSIs of links in a corridor in the second floor of Dreese Lab and on the top of a parking garage building. For the indoor experiment, we measured RSSIs at 20 different distances ($1 \sim 20$ unit, where 1 unit = 3 ft) within the maximum communication range with the highest transmission power (0 dBm). For the outdoor experiment, 10 measurements were taken from the distance of $1 \sim 10$ unit distances where 30 ft seems to be the maximum communication range with the same transmission power, 0 dBm. Table 1 presents the summary of measured RF propagation environment. We also used reported radio sensitivity of -94 dBm to adjust for the noise power P_0.

Metrics	Indoor	Outdoor
η	1.7555	2.2776
σ	5.0 dB	4.2 dB

Table 1: Log Normal Model Variables for Indoor and Outdoor RF Environments

4.1.2 Messaging layer and results

We test a messaging layer protocol, CTP [12]. CTP is a tree-based collection protocol. Nodes generate routes to roots using a routing gradient. In micro study, CTP uses ETX as the routing metric. ETX implicitly favors long links over short links because each node selects the path with the minimum number of expected transmissions.

Experimental design

TelosB mote uses 2.4 GHz frequency and provides 8 different transmission power levels: 31 (0 dBm), 27 (-1 dBm), 23 (-3 dBm), 19 (-5 dBm), 15 (-7 dBm), 11 (-10 dBm), 7 (-15 dBm), 3 (-25 dBm) [13]. We also can attach various attenuators according to attenuation levels (1 dB, 3 dB, and etc.) to modify transmitted and receive power. We used a 3 dB attenuator to construct a reasonable communication environment among 20 sensor nodes in a compact indoor area. We tested 0 dBm for D2, -1 dBm and -6 dBm for D1 at outdoor environment. For scaling node distances by half (D2 $\rightarrow$ D1), while keeping the PLE constant (outdoor) the required transmit power attenuation is give by $-10 \times 2.2776 \times log_{10}2 = -6.8514$ dB [6]. For indoor tests, we experimented -18 dBm at 3 ft spacing (D1) to determine the feasibility of mapping the performance across networks with different node spacings and PLEs.

Results

Figure 1 shows the analytical and Monte Carlo simulation results of the link usage spectra for indoor and outdoor environments with specified transmission powers for outdoor (D2 and D1) and indoor (D1) case. We observe outdoor D2 at 0 dBm, outdoor D1 at -6 dBm and indoor D1 at -18 dBm produce similar link usage spectra.

Figure 2 shows the analytical average SNR values of each links chosen over all other links for each test environments. In figure 2, we compare the conditional mean SNR of link of length d given it was chosen by the forwarding protocol over all other links and the unconditional mean SNR of a link length d. We observe that $E[SNR|d_f = d] > E[SNR|d]$.

Figure 3 shows the analytical average PRR values of each links chosen over all other links for each test environments.

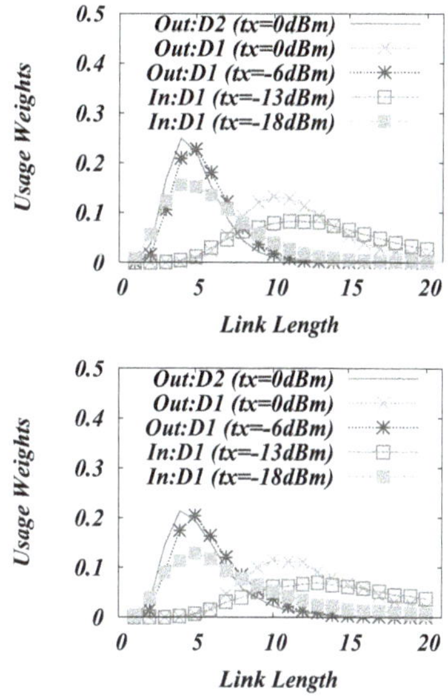

Figure 1: Link Usage Spectrum. (1) Analytical with Result 2, (2) Monte-Carlo Simulation

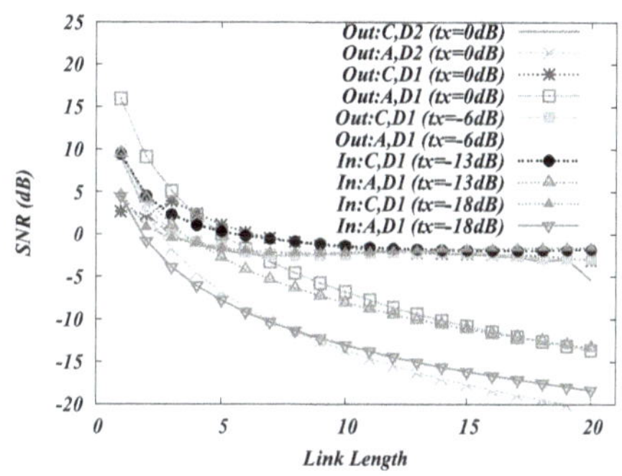

Figure 2: SNR of the Chosen/All Links. Analytical (Result 3). C:Chosen, A:All

The conditional mean PRR given the event $d_f = d$, is calculated by taking the expected conditional SNRs of links that are chosen. Again we see the average PRR of the chosen links is much higher than the average PRR of all links in figure 3. ($E[PRR|d_f = d] > E[PRR|d]$ as discussed in Section 2).

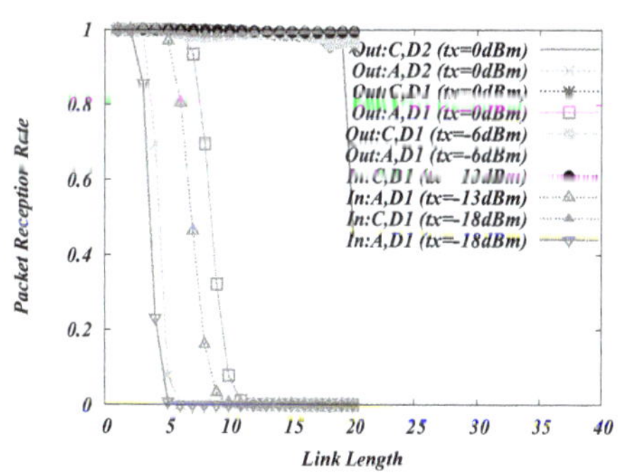

Figure 3: PRR of the Chosen/All Links. Analytical (Result 1)

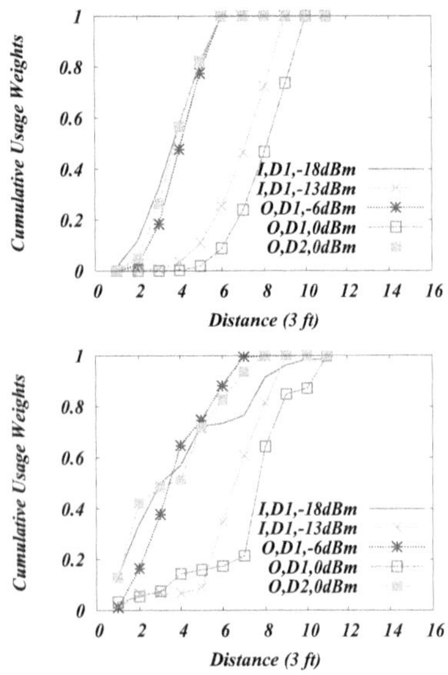

Figure 4: Cumulative Link Usage Spectrum. (1) Analytical (Result 2), (2) Experiment

Figure 4 compares the cumulative link usage spectra analytically calculated (1) and experimentally collected (2). Experimental results validate the correctness of the analytical study.

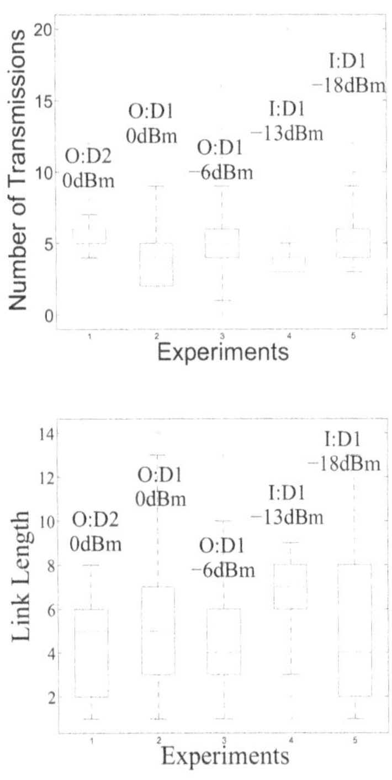

Figure 5: (1) Average Number of Transmissions (Experimental), (2) Average Link Length (Experimental)

Next, we validate analytical predictions in deployment experiments using two major performance metrics, link length and end-to-end delay. Figure 5(1) shows the average link length taken from experiments. We can see the strong similarity between outdoor D2 (0 dBm), D1 (-6 dBm), and indoor D1 (-18 dBm) as predicted with the analytical study (details shown in table 2). Figure 5(2) shows the average end-to-end delay (the number of transmissions) taken from experiments. We also observe the strong similarity between three test cases.

Number of Transmissions	O:D2 0dBm	O:D1 0dBm	O:D1 -6dBm	I:D1 -13dBm	I:D1 -18dBm
Analytical	4.6350	2.4141	4.1452	2.5046	4.3041
Experimental	5.4039	3.2049	5.2481	3.712	5.1616

Table 2: Comparisons of Performance Metrics: End-to-End delay Indoor and Outdoor

Link Length	O:D2 0dBm	O:D1 0dBm	O:D1 -6dBm	I:D1 -13dBm	I:D1 -18dBm
Analytical	4.4085	8.0155	4.9081	7.9520	5.1609
Experimental	4.1735	7.7749	4.1842	6.9329	4.709

Table 3: Comparisons of Performance Metrics: Link Length Indoor and Outdoor (Experimental : calculated without the first & last hop)

Table 2 shows the average number of transmissions taken from experiments with the specified transmission powers. Table 3 shows the average link lengths taken from experiments and compares them with the analytical results. We assume the topology is infinite to calculate the expected link length analytically. However, because the topology in experiments is limited, the link lengths of marginal hops are usually shorter than those of the intermediate hops. Table 2 and 3 show strong similarity in the performance of three matching networks (outdoor D2 with 0 dBm, outdoor D1 with -6 dBm, and indoor D1 with -18 dBm) in experimental results. This is consistent with the analytically calculated optimal attenuation factors for matching the end-to-end delay performance from the outdoor D2 to the outdoor D1 $\beta_1 = -6.9$ dB, from the outdoor D1 to the indoor D1 $\beta_2 = -12.3$ dB and from the outdoor D2 to the indoor D1 $\beta_3 = -19.2$ dB. Overall, experimental results show little more transmissions and little shorter link length than analytically predicted. Because, the last hop to destination should be shorter than other hops (edge effect), and the first hop usually tends to be shorter than other hops, we can consider little (about 1) more transmission reasonable.

4.2 Macro study: grid two-dimensional topology

4.2.1 Experimental setup

We set up grid topologies in four major WSN testbeds (KanseiGenie, Motelab, Tutornet, TWIST) with total 16 TelosB or TmoteSky sensor nodes. TelosB and TmoteSky motes are equipped with CC2420 radio and USB serial for communication. The node (1,1) is set to be the source, and the node (4,4) is set to be the destination. Every two seconds, the source node produces and sends a packet. For each experiment with a specific transmission power, we gathers about 1,000 packets generated by the source. We logged the path that each packet has gone through. The metric used for these experiments are as follows: the progress to destination and the number of transmission.

To perform analytical study to be compared with the experimental results, we measure PLE (η) and standard deviation of RSSI (σ) in four testbeds and present in table 4.

Metrics	KanseiGenie	Motelab	Tutornet	TWIST
η	1.809	3.995	4.800	2.697
σ	5.0 dB	8.5 dB	8 dB	9 dB

Table 4: Log Normal Model Variables for RF Environments of Testbeds

4.2.2 *Messaging layer and results*

We test a messaging layer protocol, CTP. We customized CTP, which uses ETX as the default routing metric, to use $PRR \times d$ as the routing metric.

Experimental design

TelosB and TmoteSky motes use 2.4 GHz frequency and provide 8 different transmission power levels. We carefully choose 4×4 (16) nodes, which form grid, in four testbeds. KanseiGenie has natural grid topology and can have the maximum 16×16 grid topology. Motelab and Tutornet do not have natural grid topology, therefore we try to choose 16 nodes to form grid-like topology. TWIST has natural grid shape topolgoy, but walls divide nodes and inter-node distance are not the same for row and column. We test multiple transmission powers (0 dBm, -5 dBm, -7 dBm, -10 dBm, -15 dBm) to find sets of transmission powers which make CTP perform comparably with given topologies in four testbeds.

Results

To validate Result 2 (calculates the link usage spectrum) in the two-dimensional topology, we present cumulative usage weights of links with the progress to destination (how much progress can be achieved with links). Figure 6(1) shows the analytical results and figure 6(2) shows the experimental results of KanseiGenie. Figure 7 shows the results of Motelab. Figure 8 shows the results of Tutornet. Figure 9 shows the results of TWIST. Overall, analytical results match experimental results in all four testbeds. However, there are gaps between analytical and experimental results for all testbeds. There can be two reasons. First, for Result 2, we assume infinite topology by which we collect the averaged distribution of link selection. But, with testbed experiments, we test 4 × 4 grid topology with limited possible path length. Good long links, bad medium links, and good short links combination contributes to the specific experimental results of KanseiGenie and Motelab. Second, we consider interference from external 802.15.4 traffic, or from 802.11 traffic as the source of variance of link selection, especially for lower power cases.

To validate how Result 1 (calculates the expected performance) performs in the two-dimensional topology, we present the analytical results of the end-to-end delay (the total number of transmissions) predicted by Result 1 and compared them with the experimental results in figure 10. Overall, analytical results match experimental results in all four testbeds. The analytical and the experimental studies on the progress to destination metric are not presented due to space limitation.

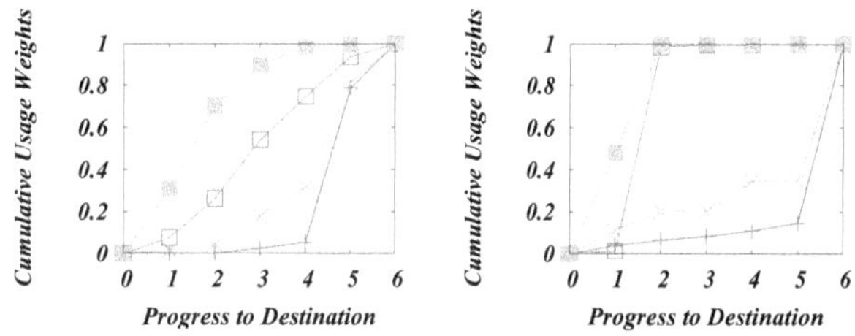

Figure 6: Cumulative Usage Weights at KanseiGenie. (1) Analytical, (2) Experimental. Transmission Power = (+:0, ×:-5, □:-10, ■:-15) dBm

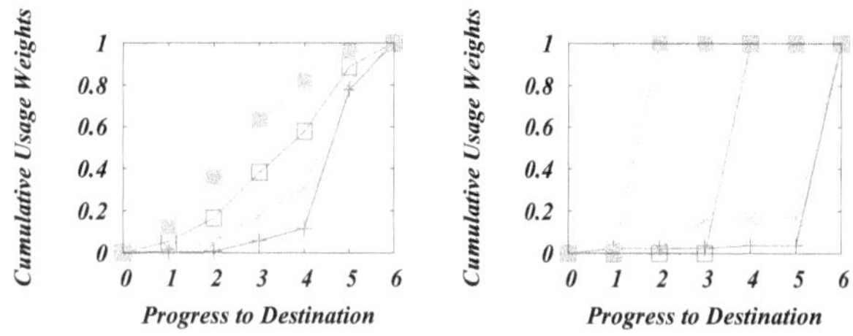

Figure 7: Cumulative Usage Weights at Motelab. (1) Analytical, (2) Experimental. Transmission Power = (+:0, ×:-5, □:-10, ■:-15) dBm

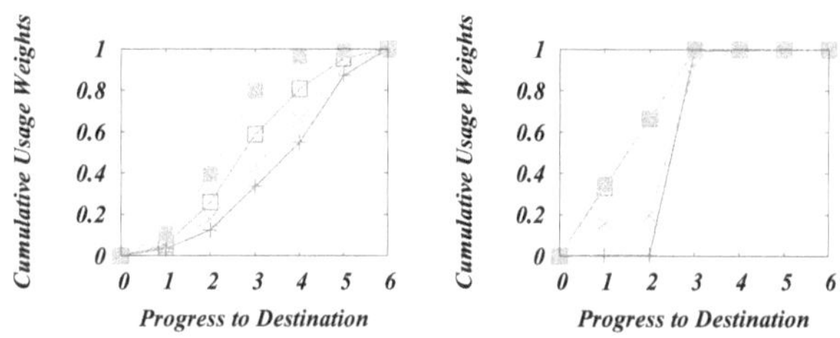

Figure 8: Cumulative Usage Weights at Tutornet. (1) Analytical, (2) Experimental. Transmission Power = (+:-5, ×:-7, □:-10, ■:-15) dBm

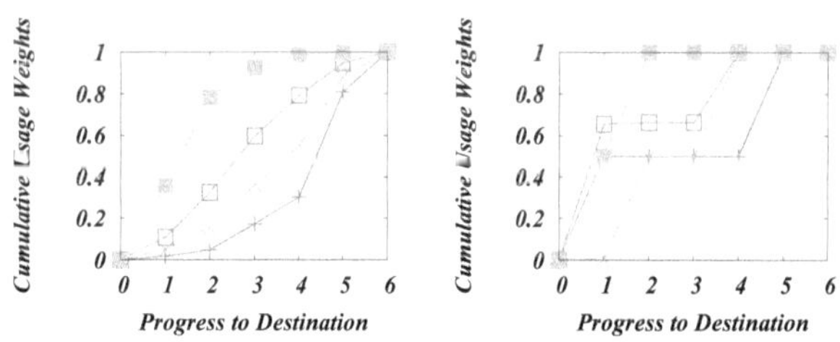

Figure 9: Cumulative Usage Weights at TWIST. (1) Analytical, (2) Experimental. Transmission Power = (+:-5, ×:-10, □:-15, ■:-25) dBm

According to figure 10, we can find a set of transmission powers which achieve comparable performance (the number of transmissions) in four testbeds. Table 5 shows examples of performance matching across four testbeds. Overall, experimental results show a little more transmissions than analytically predicted. Because, the last hop to destination should be shorter than other hops (edge effect), we can consider this reasonable.

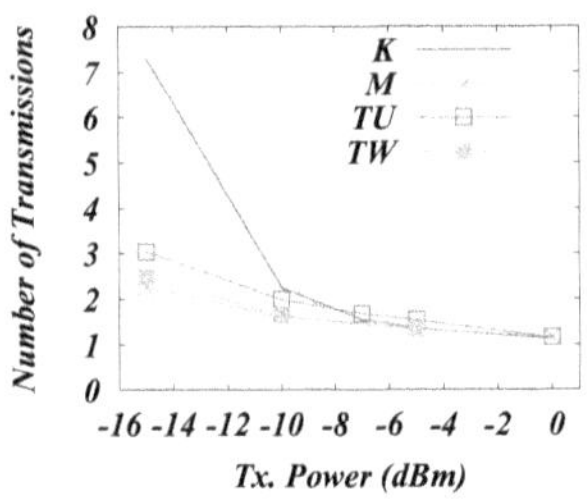

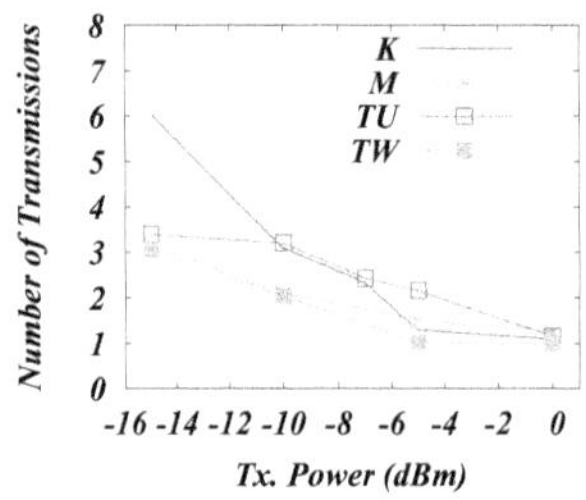

Figure 10: Number of Transmissions. (1) Analytical, (2) Experimental. K: KanseiGenie, M: Motelab, TU: Tutornet, TW:TWIST

	KanseiGenie	Motelab	Tutornet	TWIST
Tx. Power	-7dBm	-10dBm	-7dBm	-10dBm
Analytical	1.55	1.61	1.68	1.64
Experimental	2.33	2.09	2.43	2.04
Transmissions	-10dBm	-15dBm	-10dBm	-15dBm
Analytical	2.25	2.26	2.00	2.49
Experimental	3.10	3.17	3.20	3.06

Table 5: Total Ttransmissions matching Examples

5. CONCLUSIONS AND FUTURE WORK

In this paper, we prosented a technique for achieving repeatable performance across different networks deployed in diverse set of RF environments. Our technique accommodates network pairs whose signal propagation characteristics and inter-node spacings may be different. The method is executed in three steps: First, the link usage distribution is calculated analytically/ Second, conditional moments of the link quality are calculated for each link chosen by the forwarding protocol. Third, link usage and link quality statistics are aggregated to compute expected protocol performance using a user-specifed metric for each network and to select the optimal transmission power adjustment to have same expected protocol performance across the networks. We provide a multi testbed case study to validate the proposed method.

It is important to acknowledge the limitations of our tool, as presented. The tool targets consistency to a single chosen performance metric. Our experiments however indicate that this limitation is not particularly restrictive in practice, since attempting to achieve optimal consistency with respect to one metric seems to result in achieving good consistency with respect to other performance metrics as well. Moreover, the tool requires the forwarding protocol to be specified in order to calculate the link usage probabilities. Alternatively, link usage and conditional link SNR moments can be computed empirically from *in situ* observations of packet source-destination pair.

In the future, we will study how to redress the limitations of the link usage spectrum, focusing in particular on providing a continuous version of the concept which allows network nodes to be placed at random points in a geometric space. We will explore predictable performance in the target networks, using knobs other than the transmission power control and taking into account metrics other than those related to the forwarding link selection alone. We will also study the variability of protocol behavior in one network and account for the preservation of that variability in the target network, over and above the average behavior that we have focused on in this work. We will study how to extend our technique to adopt more user-specified performance indexes (e.g. goodput, energy efficiency, delay, etc.), and reflect more complex RF environment factors (e.g. macro and micro fading effects).

6. REFERENCES

[1] T. Kwon, E. Ertin, and A. Arora, "Reproducing consistent wireless protocol performance acorss environments," *Adhocnets*, 2010.

[2] KanseiGenie Testbed. http://kansei.cse.ohio-state.edu/KanseiGenie/.

[3] Motelab Testbed. http://motelab.eecs.harvard.edu/.

[4] Tutornet Testbed. http://enl.usc.edu/projects/tutornet/.

[5] TWIST Testbed. http://www.twist.tu-berlin.de.

[6] V. Naik, F. Ertin, H. Zhang, and A. Arora, "Wireless testbed bonsai," *2nd International Workshop on Wireless Network Measurement*, 2006.

[7] C.-F. Chiasserini and M. Garetto, "Modeling the performance of wireless sensor networks," in *In IEEE Infocom*, 2004.

[8] K. Seada, M. Zuniga, A. Helmy, and B. Krishnamachri, "Energy-efficient forwarding strategies for geographic routing in lossy wireless sensor networks," *Sensys*, 2004.

[9] M. Zuniga, K. Seada, B. Krishnamachri, and A. Helmy, "Efficient geographic routing over lossy in wireless sensor networks," *ACM Transactions on Sensor Networks*, vol. 4, pp. 111–143, 2008.

[10] S. Lin, G. Zhou, K. Whitehouse, Y. Wu, J. A. Stankovic, and T. He, "Towards stable network performance in wireless sensor networks," in *Proceedings of the 2009 30th IEEE Real-Time Systems Symposium*, RTSS '09, (Washington, DC, USA), pp. 227–237, IEEE Computer Society, 2009.

[11] S. Lanzisera and K. Pister, "Theoretical and practical limits to sensitivity in ieee 802.15.4 receivers," *IEEE International Conference on Electronics, Circuits and Systems*, 2007.

[12] TEP 123. http://www.tinyos.net/tinyos-2.x/doc/html/tep123.html.

[13] E. Miluzzo, X. Zheng, K. Fodor, and A. Campbell, "Radio characterization of 802.15.4 and its impact on the design of mobile sensor networks," *EWSN*, 2008.

Multidimensional Modeling and Analysis of Wireless Users Online Activity and Mobility: A Neural-networks Map Approach

Saeed Moghaddam, Ahmed Helmy

Computer and Information Science and Engineering (CISE) Department, University of Florida, Gainesville, FL

{saeed, helmy}@cise.ufl.edu

ABSTRACT

User online behavior and interests will play a central role in future mobile networks. We introduce a systematic method for large-scale multi-dimensional modeling and analysis of online activity and mobility for thousands of mobile users across 79 buildings over a variety of web domains. We propose a modeling approach based on kind of neural-networks, called self-organizing maps (SOM), for discovering, organizing and visualizing different mobile users' trends from billions of WLAN records. We find surprisingly that users' trends based on domains and locations can be accurately modeled using a self-organizing map with clearly distinct characteristics. We also find many non-trivial correlations between different types of web domains and locations.

Categories and Subject Descriptors

C. Computer Systems Organization; C.2 Computer-Communication Networks; C.2.1 Network Architecture and Design; Subjects: Wireless communication; Nouns: Internet. I. Computing Methodologies; I.6 Simulation and Modeling; I.6.5 Model Development; Subjects: Modeling methodologies.

General Terms

Measurement, Experimentation, Design, Performance

Keywords

Self-organizing map, data-driven, trend, wireless.

1. INTRODUCTION

Wireless mobile networks are evolving and integrating with every aspect of our lives. Laptops, handhelds and smart phones are becoming ubiquitous providing almost continuous Internet access. This creates a tight coupling between users and mobile networks where various characteristics of user online behavior and mobility affects network performance significantly. Hence, there is a compelling need for analysis and modeling of mobile users online activity and mobility. Such behavioral modeling will aid in the understanding of users trends and the load distribution on the network, and thus inform the design of important classes of applications, including modeling and scenario generation for network simulations, network capacity planning, web caching and behavior-aware networking protocols [1] to name a few. However, such behavioral analysis on extensive traces of Internet access is difficult, as it is large-scale, computationally costly and time-consuming. Such traces usually contain billions of records and therefore, the analysis may even be impossible when the dataset exceeds certain limits of size and complexity. Moreover, accommodating different aspects of users behavioral patterns (e.g. mobility, website visitation, time, application) is not a trivial task. As a result, it is imperative to establish systematic and scalable methods for analysis and modeling of massive multi-aspect/ multi-dimensional datasets of mobile users' online activity and mobility.

Much of the previous mobility or web usage modeling focused on individual behavior and one aspect. While individual behavior is important, investigating group behaviors and trends from different aspects is more challenging and involved. On the other hand, the conventional approach to group data analysis, i.e., direct clustering of data items based on some similarity measure, is not applicable for massive multi-aspect analysis. The common clustering algorithms are either computationally intensive, e.g., hierarchical clustering [2], or very unstable, e.g., k-means [2] for huge amount of data. Moreover, although the resulting clusters provide some insight on similarities between data items (and among features, if co-clustering [3] is used), they do not reveal intuitively how correlated they are. Therefore, they are not effectively useful for discovering correlations when there exist many data items, features and aspects.

In this paper, we focus on group behavioral modeling, and study behavioral correlations based on groups of mobile users and their trends according to the website and location visitation patterns. Our approach to solve the problem and address the above limitations is to apply self-organizing maps (SOM) [4]. A self organizing map is an artificial neural network which is trained using unsupervised learning to generate a discretized low-dimensional representation (a map) of the input space of the data samples. Unlike other artificial neural networks, self-organizing maps use a neighborhood function to preserve the topological properties of the input space. The topology-preserving mapping keeps more similar data groups closer together in the final map, which makes SOM useful for providing low-dimensional views of high-dimensional data. These views can reveal the semantic behind major user trends and the correlation between different features.

In this study, we apply SOM in a novel way on a dataset provided by processing of extensive *netflow*, DHCP and WLAN session logs for more than 22 thousand mobile users in a Wireless LAN spanning over 79 buildings and including over 700 APs, that we have collected. This dataset, including billions of records, represents by far the largest set of traces analyzed in any study of mobile networks to date. Using this technique, we extract minor and major trends in mobile users' website/location visitation

MSWiM'11, October 31-November 4, 2011, Miami, Florida, USA.

patterns and discover important correlations between different web domains and locations. We show how to apply this technique methodically to the collected large-scale multi-dimensional dataset with minimal computational complexity to facilitate its meaningful analysis. Our method is systematic and can be generally applied to discover important features of Internet behavior from other similar traces. It can also be applied for any other aspects of Internet usage, e.g., time, application in addition to the web domains and locations.

We report three major findings in this paper. First, mobile users behavior based on web domain and location visitations can be accurately modeled by a small set of neurons which can be further clustered into smaller number of major trends with clearly distinct characteristics. For example, a major trend represents Mac users who frequently visit 'mac' and 'apple' websites and have strong interest in 'washingtonpost' and 'cnet' too. Second, web domains/locations in similar categories tend to be modeled by an adjacent set of neurons. For example, most of the advertisement/marketing domains or fraternities are modeled together by neighboring neurons. Third, many nontrivial correlations exist between different kinds of web domains and locations. For example, we found that Music Practice Center is highly correlated to the Health Science Telephone Vault while they are located in two different campuses of USC. Considering these findings, we need a new model for mobile user behavioral patterns to accommodate future mobile applications, as we discuss in our applications section.

Our work has the following key contributions:

1. We propose an effective approach for multi-dimensional modeling and analysis of one of the largest set of mobile network usage traces (including billions of records) and show how self-organizing maps can be applied to model minor and major trends in wireless users online activity and mobility.
2. We conduct domain-specific and location-based analysis of mobile users' behaviors using the feature maps extracted from the SOM and show how this method can be effectively applied to discover correlations among different domains and locations.
3. We suggest feature clustering technique on top of the SOM as a quantitative way for finding domains/locations correlations.
4. We propose an extension on top of the SOM for multi-aspect (concurrent) modeling of users behaviors based on their online activity and mobility.

The rest of the paper is organized as follows. In Section 2, we review the related work. In Section 3, we briefly address challenges of data collection/processing and then explain our modeling approach. Section 4 presents our data analysis technique and Section 5 provides our case study using campus traces. Section 6 discusses applications and Section 7 concludes.

2. RELATED WORK

The rapid growth of wireless communication technologies has led to a widespread interest in analyzing the traces to understand user behavior. The scope of analysis includes WLAN usage and its evolution across time [5, 6], traffic flow statistics [7], user mobility [8, 9], user association patterns [10] and encounter patterns [11]. Some previous works [8, 11] attempt to understand user behaviors empirically from data traces. The two main trace libraries for the networking communities can be found in the archives at [12] and [13]. None of the available traces provides large-scale *netflow* information coupled with DHCP and WLAN sessions to be able to map IP addresses to MAC addresses and detect locations. We analyze wireless data around three orders of magnitude above any existing study, providing richer semantics, finer granularity and potentially more accurate models. In addition, our work includes novel data analysis techniques to address the challenges provided by this large-scale multi-dimensional data.

There are several noticeable examples of utilizing the data sets for context specific study. Mobility modeling is a fundamentally important issue, and several works focus on using the observed user behavior characteristics to design realistic mobility models [14-16]. They have shown that most widely used existing mobility models (mostly random mobility models, e.g., random waypoint, random walk; see [17] for a survey) fail to generate realistic mobility characteristics observed from the traces. Realistic mobility modeling is essential for protocol performance. It has been shown that user mobility preference matrix representation leads to meaningful user clustering [18]. Several other works focus on classifying users based on their mobility periodicity [19], time-location information [20], or a combination of mobility statistics. The work on the *TVC* model [15] provides a realistic mobility model for protocol and service performance analysis. Our work is complementary to *TVC* and can extend *TVC* dramatically to incorporate dimensions of load, interest and website visitation preferences. In [7] it is shown that the performance of resource scheduling and TCP vary widely between data-driven and non-data-driven model analysis. Using multi-dimensional modeling, our method can develop new mobility-aware Internet-usage models, and utilize the realistic profiles to enhance the performance of networking protocols. Our new application of self-organizing map technique may be extended to incorporate online activity, location and mobility, and provides user profiles that may be used in a myriad of networking applications.

One network application for multi-dimensional modeling is profile-based services. *Profile-cast* [1] provides a one-to-many communication technique to send profile-aware messages to those who match a *behavioral profile*. Behavioral profiles in [1] use location visitation preference and are not aware of online activity. Other previous works also rely on movement patterns. Our multi-dimensional modeling of mobile users, however, provides an enriched set of user attributes that relate to social behavior (e.g., interest, community as identified by web access, application) that has been largely ignored before.

While our previous work on co-clustering model provides a high level approach for analysis of Internet usage (at cluster level) [21], our new approach in this study based on the SOM provides a tool for low level (feature level) analysis of online activities and mobility. Using the new approach, not only we can find clusters of similar features (i.e., domains or locations), but also we can discover many correlations between the features themselves. The new approach provides us with a full set of feature maps that visualize the characteristics of each of the features and can be applied for individual or group-based analysis of them. Moreover, in addition to the correlated features, we can easily find the anti-correlated ones and discover the anomalies in a set of features too (for example anti-correlation between visiting 'mac' and 'mcafee' websites in addition to correlation between 'windows' and 'mcafee' which gives us a sense of their security). These tasks can be performed either mathematically or by visual inspection of maps, which is not provided by the co-clustering model.

Another difference between the new SOM approach and the co-clustering model is the fact that it attempts to find *'trends'* inside the mobile society instead of focusing on individual behaviors. This way the behavior of each user, which is usually a mixture of several trends, can be modeled more accurately as individual behaviors can affect multiple trends in the learning process of the SOM. Plus, it provides trends topology that can be effectively applied for finding major trends out of the minors. Furthermore, the learning process is incremental which means that by getting new sample behaviors we do not need to rerun the algorithm as for the co-clustering model, an important feature for very large datasets, especially when it comes to the design of distributed versions of the algorithm to get run on the mobile devices. We have extended this approach for 3D analysis of users, their activities and mobility, for the first time.

In our preliminary work [22], we briefly showed that the SOM technique has the potential to reveal some correlations in the Internet usage of wireless users. Here, we show that this technique coupled with our proposed feature clustering technique for uni-aspect modeling and our extension on top of the SOM for multi-aspect modeling can be effectively applied based on individual or multiple aspects (i.e., web domains, locations or both) to reveal correlations, anti-correlations and anomalies. While existing models do not accommodate any of these features, our approach provides a solid foundation to model multiple important aspects of users behavioral patterns for realistic design of future mobile network.

3. MODELING APROACH

Realistic modeling of large wireless networks requires three main phases to collect, process and model multi-dimensional large datasets with fine granularity. In the first phase, extensive datasets are collected using the network infrastructure that may be augmented using online directories (e.g., buildings directory, maps) and the web services (e.g., whois lookup service). Data processing is the second phase to cross-correlate obtained information from different resources (e.g., IP and MAC addresses), in which multiple datasets are manipulated, integrated and aggregated. The final phase is data modeling which includes uni-aspect and multi-aspect modeling of users' behaviors based on their web domain and location visitation patterns.

3.1 Data Collection

We collect different types of extensive traces via network switches (in USC campus) including netflows, DHCP and wireless session logs. An IP flow is defined as a unidirectional sequence of packets with some common properties (e.g., source IP address) that pass through a network device (e.g., router) which can be used for flow collection. Network flows are highly granular; flow records include the start and finish times (or duration), source and destination IP addresses, port numbers, protocol numbers, and flow sizes (in packets and bytes) (see Table 1). The source and destination IP addresses can be used to identify user device MAC addresses using DHCP log and the websites accessed respectively. The DHCP log contains the dynamic IP assignments to MAC addresses and includes date and time of each event. This information is needed to get a consistent mapping of dynamically assigned IP addresses to the device MAC addresses. The wireless session log collected by each wireless access point (AP) includes the 'start' and 'end' events for device associations (when they visited or left that specific AP) which can be used to derive the location of users at any time.

3.2 Data Processing

The variety and scale of different collected traces introduces one of the main challenges with respect to data processing. The size of the underlying data is very large and therefore, with a naïve approach the required time for this task would be in the order of months. For example, the netflow dataset gathered from USC campus includes around 2 billion flow records for each month in 2008 which equals to 2.5 terabytes of data per year. To circumvent the problem, we first compress the data via substituting similar patterns with binary codes and creating mapping headers to be used in the analysis step; then get the data exported into a database management system (MySQL) and design customized stored procedures for data integration (mapping source IPs to MAC addresses (user IDs) and destination IPs to domain names). In the last step, we aggregate the integrated data based on user ID, domain name, location and month and calculate the total online time for each resulting record.

3.3 Data Modeling

The data modeling phase includes two major parts. In the first part, we employ the self-organizing map to learn minor trends of users within the wireless society. The user trends may be learned based on website or location visitation preferences separately (using a common approach) or together (based on our proposed multi-aspect extension of the method). In any case, in the second part, we apply clustering technique on map nodes to discover major trends inside the community.

3.3.1 Trend Modeling

The SOM technique [4] provides a powerful yet intuitively understandable tool for unsupervised learning and data visualization. The SOM is defined as a set of nodes which develop a mapping of high-dimensional input vectors (which may represent website or location visitation preferences) onto a discrete output space (the "map") such that each region on the map represents an area of the input space. This mapping preserves the topology of the input space in a way that local similarity of input patterns is reflected by proximity on the map. Therefore, it can be effectively applied in capturing the properties of the input space of users' behaviors and organizing their trends in an ordered fashion. In a self-organizing map, a weight vector of the same dimension as the input data vectors and a position in the 2-D map space are associated with each node (or neuron in neural networks). The usual arrangement of nodes is in the form of a hexagonal or rectangular grid. SOM training, i.e., the iterative adjustment of the weight vectors to acquire a desired mapping, is performed by successive presentation of all input data where each presentation leads to the adjustment of weights to the presented data. The training is based on two principles:

a) Competitive learning: the weight vector most similar to a

Table 1 - Netflow Sample

Start Timestamp	Finish Timestamp	Source IP	Source Port	Dest IP	Dest Port	Protocol Num	ToS	Packet Count	Flow Size
0618.00:00:07.184	0618.00:00:07.184	128.125.253.143	53	207.151.245.121	64209	17	0	1	469
0618.00:00:07.184	0618.00:00:07.472	207.151.241.60	52759	74.125.19.17	80	6	0	4	1789
0618.00:00:07.188	0618.00:00:07.188	193.19.82.9	31676	207.151.238.90	43798	17	0	1	103

data vector is modified so that it is even more similar to it (the corresponding node is called Best Matching Unit or BMU). This way the map learns the position of the input data.

b) Cooperative learning: not only the most similar weight vector, but also its neighbors are moved towards the data vector. This way the map self-organizes.

The neighborhood function h regulates the weight changes based on the map distance between BMU and the neuron being adapted. In the case of a Gaussian shaped neighborhood function, the expression of h is given by:

$$h(i,j) = \exp\left(-\frac{dist_{map}(i,j)^2}{2r(n)}\right)$$

where $dist_{map}(i,j)$ measures the distance on the map between two neurons and $r(n)$ is a global parameter that controls the "width" of the neighborhood function. According to this expression, the amount of the changes is maximum for the BMU and decreases for nodes that are far from it. The value of $r(n)$ decreases with the number of iteration; a relatively large radius during the initial iterations allows the map to quickly organize the neurons, while a smaller value toward the end determines localized changes in a way that different parts of the map become sensitive to different input features. The learning rate of the map decreases monotonically with the number of iterations to ensure convergence.

In this way, *each neuron can learn a minor trend that represents a set of similar input data vectors*. This is one of the major advantages of SOMs with respect to clustering techniques. While a clustering technique attempts to partition the input space (e.g. users' behaviors) by assigning each sample (e.g., a user) to a cluster, the SOM technique attempts to learn trends inside the input space form the samples. Note that each input data vector (e.g., a user) affects a set of neighboring neurons (trends) and therefore the input space is not distinctly partitioned by the neurons (unlike cluster assignment in conventional clustering techniques). This is much more in consent with the natural human behaviors with no clear-cut distinctions.

Map Creation - The side lengths of the map grid are determined based on the ratio of two biggest eigenvalues of the training data. For initializing the SOM, first, linear initialization along two greatest eigenvectors is attempted, but if the eigenvectors cannot be calculated, random initialization is used instead. After the initialization, the SOM is trained by normalized input data. The normalization of the input features is very important in determining what the map will be like. If the ranges of value for some features are much bigger than the others, those features will probably dominate the map organization completely and the resulting map will not be useful. The computational complexity of SOM algorithm scales linearly with the number of data samples and quadratically with the number of map units.

3.3.2 Trend Clustering

One way to visualize the resulting map after the training phase is to create U-matrix (unified distance matrix). The U-matrix shows the distance between the weight of each node and the assigned weights of its neighbors after the learning process. Figure 1(a) shows an instance of U-matrix with interpolated shading of colors. Small U-values (Blue areas in the figure) indicate homogenous neighborhoods and large ones (Red areas) depict heterogeneous neighborhoods. As large U-values mean large distances between the neighboring nodes, they can be interpreted as borders between clusters of neurons, i.e., trends. In order to find these borders (clusters), k-means clustering algorithm can be applied. Because k-means result depends on the initial choice of cluster centroids, the algorithm is run multiple times for a given k and then the best result is selected based on the sum of the squared errors. Because the captured minor trends are already very well organized on the map, each resulting cluster maps into a contiguous area of neurons, representing a major trend (Figure 1(b)). Clustering of trends instead of original data reduces the required computational time for any kind of clustering technique as the size of input is decreased. This is very important when dealing with massive amount of data. In addition, as the weight vectors are local averages of the data, the clustering result is less sensitive to random variations in the input data.

3.3.3 Multi-Aspect Modeling

The SOM technique in its original form is suitable for uni-aspect modeling of trends, i.e., based on either web domain or location aspect. However, in multi-aspect modeling the goal is concurrent modeling of trends based on all aspects together not separately. While uni-aspect modeling is good for intra-aspect analysis, multi-aspect modeling provides an opportunity for inter-aspect analysis. In multi-aspect modeling, instead of one general usage vector per user, a set of localized vectors exists (i.e., a set of usage vector for different locations or vice versa). Therefore, the regular SOM learning method is not applicable. Our proposed approach to accommodate this situation is to consider a usage matrix for each user (representing website usage at different locations) and a weight matrix for each map unit and then get the map trained. This way each map unit can capture a multi-aspect trend. This is an extensible approach and can be applied for more than two aspects.

4. DATA ANALYSIS

We conduct uni-aspect and multi-aspect analysis considering web domain and location visitation aspects. For each type of analysis, we propose two qualitative and quantitative approaches. The qualitative approach relies on visual inspection of extracted feature maps and is useful for discovering correlations, anti-correlations and anomalies among the features. The quantitative approach is based on our proposed feature clustering technique that applies a mathematical correlation function.

4.1 Feature Map Analysis

The feature maps are extracted from the SOM and show what kind of values the weight vectors of the map units have for each feature. In other words, a feature map shows the projection of the SOM for the corresponding feature (which can be a web domain, a location, or a web domain at a specific location in multi-aspect case). The value of each unit for the feature is presented with a color. Figure 2 shows a group of resulting feature maps in our study. By visual inspection of the feature maps, we can find many different interesting facts about the trends and features as follows:

a) Comparison of feature maps with clustered SOM discovers the semantic behind each cluster of trends representing a major trend. For a cluster area, features whose maps looks red in the same area disclose the main captured trends by the cluster.

b) Similar feature maps reveals correlations between the corresponding features. The correlation can be partial or complete. If the maps seem highly similar, there exist rather complete correlation, but if they are partially similar, the correlation among features will also be partial. In our case, correlation between a set of features means that they have the same visitation pattern.

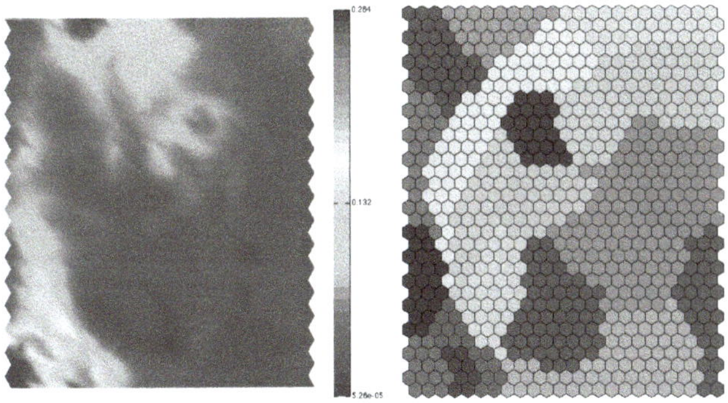

(a) U-matrix (b) Clustered SOM

Figure 1. U- matrix and clustered SOM for WLAN Internet usage in USC campus (for domains).

c) Anomalies in a set of feature maps uncover anomalies regarding the corresponding features. In our case, for example, if for a category of web domains (e.g. marketing domains) all but one feature maps looks similar; the different one brings out an anomaly.

d) Feature maps which look inverted (i.e. red areas in one are blue in the other) disclose anti-correlations. Again, anti-correlations can also be rather complete or partial.

4.2 Feature Clustering

Taking the projection of all weight vectors (or weight matrix in multi-aspect case) on each feature, we proposed to construct a description vector for the corresponding feature referred to as feature vector. Applying hierarchical clustering on the feature vectors, we can cluster features based on their correlation using the following *correlation* distance function:

$$D(i,j) = 1 - \frac{(v_i - \overline{v_i})(v_j - \overline{v_j})'}{\sqrt{(v_i - \overline{v_i})(v_i - \overline{v_i})'}\sqrt{(v_j - \overline{v_j})(v_j - \overline{v_j})'}}$$

where v_i and v_j are feature vectors. This procedure can also be interpreted as a quantitative way for comparing the feature maps.

5. CASE STUDY

In our case study, we conduct a campus-wide case study on the data we collected from the University of Southern California (USC) in 2008 based on the approach and techniques explained in the previous section.

5.1 Data Processing Details

The *netflow* and DHCP traces from the USC campus (over 700 access points covering 79 buildings) were processed to identify mobile user IDs (MAC addresses), and destinations, or 'peers' (usually web servers) using IP address prefixes. Over a billion records (for Mar. 2008) were considered. Then, the IP prefixes (first 24 bits) were filtered using a threshold of 100,000 flows [the reason for using 24 bits filter is that popular websites usually use an IP range instead of a single IP address]. For the filtered IP prefixes, their domains were resolved. Among the resolvable domains, the top 100 active ones were identified and all the users interacting with those domains were considered for the analysis.

5.2 Modeling Results

For domain specific and location-based modeling, two separate matrices were created associating the user IDs with web domains and user IDs with locations using the corresponding total online time (per minute). For our analysis, we had 22,816 users, 100 domains and 79 buildings. The data for each matrix is scaled using row-normalization of log the online time values. The two input matrices trained two SOMs of 32 by 24 nodes separately. Figure 1 (a) shows the U-matrix for domains and Figure 1 (b) represents the corresponding SOM clustered into 20 clusters. For multi-aspect modeling, we chose the highest active 40 domains and 20 locations and created a 3D matrix associating the user IDs, web domains and locations using again the corresponding online time and trained a 32 by 24 matrix-based SOM (see [23] for location and multi-aspect maps).

5.3 Analysis Results

5.3.1 Domain Specific Analysis

The feature maps were created for all the domains. Figure 2 and Figure 3 show several examples of resulting maps for different types of web domains. Inspection of the feature maps reveals many interesting facts. The following are some examples based on the presented feature maps here.

a) Figure 2 shows feature maps for advertisement and marketing domains. All these maps (except the right one) show a red area almost at the same neighborhood (right-bottom corner). This shows the major trend captured by the cluster depicted by orange at the same area in Figure 1(b) is toward this kind of web domains.

b) High similarity between feature maps in Figure 2 shows that the corresponding domains for advertisement and marketing are highly correlated. We can also observe high correlations between the following groups of domains from Figure 3: i) security related domains, i.e., 'mcafee' and 'hackerwatch'; ii) 'itunes' and 'netflix' (online media); iii) 'mac', 'apple', 'washingtonpost' and 'cnet' (showing a strong trend of Mac users toward 'cnet' and 'washingtonpost'); iv) Windows related domains, i.e., 'microsoft', 'windowsmedia' and 'microsoftoffice2007'. In the figure, we can see that 'itunes' is in one hand partly correlated to 'netflix' and on the other hand is partly correlated to 'mac' and 'apple'. This may show the facts that i) Mac users dominantly use iTunes for online media and ii) Netflix costumers shop in iTunes store too.

c) Different patterns of maps for 'doubleclick' among advertisement and marketing domains show an example of anomalies within a category of web domains. These anomalies might disclose different advertisement and marketing approaches taken by 'doubleclick'.

d) Figure 3 reveal anti-correlation between Mac and Windows related domains. As can be noticed, the bight (red) area for 'apple' and 'mac' is almost dark (blue) for 'windowsmedia', 'microsoftoffice2007' and 'microsoft'. We can also find anti-correlation between security related domains (i.e., 'mcafee' and 'hackerwatch') and 'mac' and 'apple', but partial correlation between them and Windows related domains.

We also applied our proposed feature clustering technique on top of the SOM for web domains and created 20 clusters. Table 2 shows some of the resulting clusters. As can be seen in the table, the two discussed categories of Apple and Microsoft correlated domains are clustered into two distinct clusters (Clusters A and B).

5.3.2 Location-Based Analysis

Similar to the domain-specific analysis, we can simply find the semantic behind each major trend for location visitations (see [23]). Inspection of the feature maps for locations reveals many interesting correlations too. Figure 4 shows high correlations between social and professional fraternities. As can been seen, fraternities in the first row are highly correlated. We can also

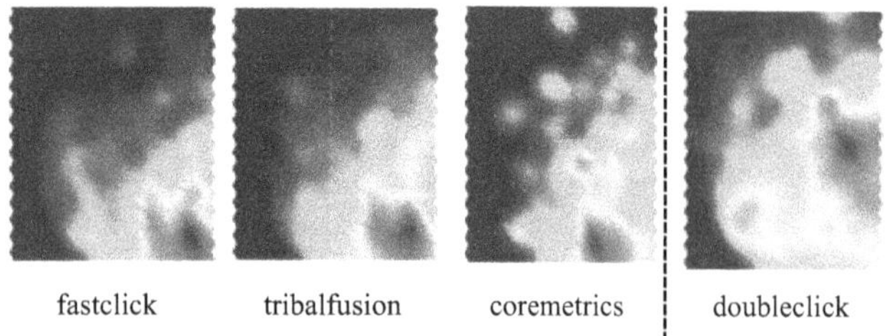

Figure 2. Feature maps for advertisement / marketing domains (important notice: maps need to be viewed in color)

observe high correlation between ATO and ARC. The feature map for PGD (playground) shows that both groups are partially correlated with the playground duplex too.

Similar to domain-specific analysis, discovered correlations among locations are not just between buildings of the same types. Figure 5 shows four pairs of highly correlated buildings that are not in the same category. As can be seen, the Music Practice Center (PIC) is highly correlated to Health Science Telephone Vault (HSV). The interesting point about these two buildings is the fact that they are located in two different campuses of USC and so relatively far from each other. However this is not the case for the Woman's Association (YMC) which is next to the Hall Building (HSH) and probably use the hall for their gatherings very frequently. We can also see the residents of housing complex TRH frequently go to the Healthcare Consultation Center (HCT). Also, fraternity PKT and sorority KAT are highly correlated which may reveal the fact that many of their members are in a relationship.

We also employed feature clustering on the locations SOM and created 20 clusters. Figure 6 shows clustered heatmap of pair-wise correlation matrix for all the buildings. Darker blocks along the main diagonal in the figure show the fact that buildings within each cluster are highly correlated together but not much to the rest. To analyze the clusters, we studied all the buildings and based on their actual context categorized them into 10 categories including: housing, auditorium, (outdoor) activity, sorority,

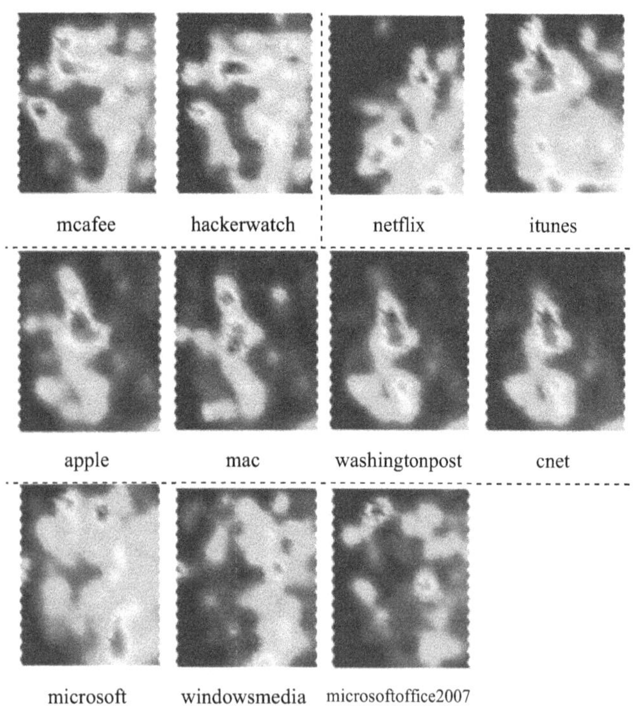

Figure 3. Feature maps for various types of domains

fraternity, school, health, music, cinema and service. These categories are available to the left of each abbreviation in the figure. As can be seen, many of the buildings in the same category are clustered together. For example many of fraternities and all sororities are placed in cluster 1 (cluster IDs are available at the right side of heatmap). We can also observe that 4 building in ''activity'' category and 7 ones in ''health'' category are clustered into clusters 5 and 8 respectively ("activity" category includes buildings with different activity context including sports, religion, social and shopping). We can also see that all of the discussed correlated buildings in Figure 4 and Figure 5 are also clustered into the same clusters.

5.3.3 Multi-Aspect Analysis

Figure 7 shows some examples of highly correlated domains at specific locations. Maps at the first row reveal a non-trivial correlation between visitation of 'yahoo' at ANH housing complex and 'live' at KAT sorority. We can also observe partial correlation between these patterns and visitation of 'mozilla' and 'google' at ATO fraternity. The second row shows two other examples of multi-aspect correlations: i) visitation of 'youtube' at LUC (cinema) and 'live' at ASC (Communication & Journalism school); ii) visitation of 'usc' at ACO (sorority) and 'yahoo' at PKF (fraternity).

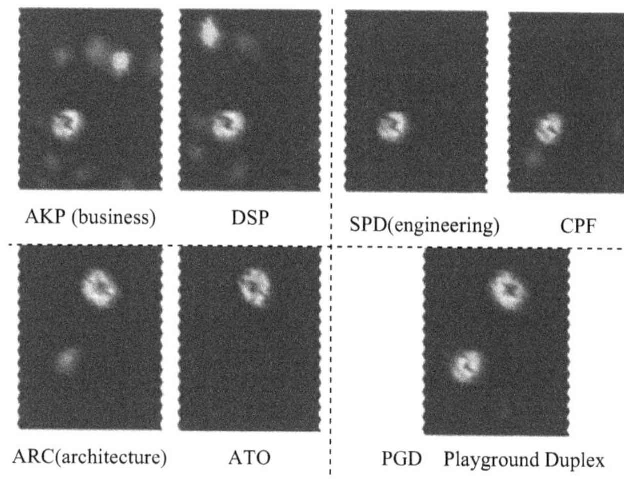

Figure 4. Feature maps for social & professional fraternities

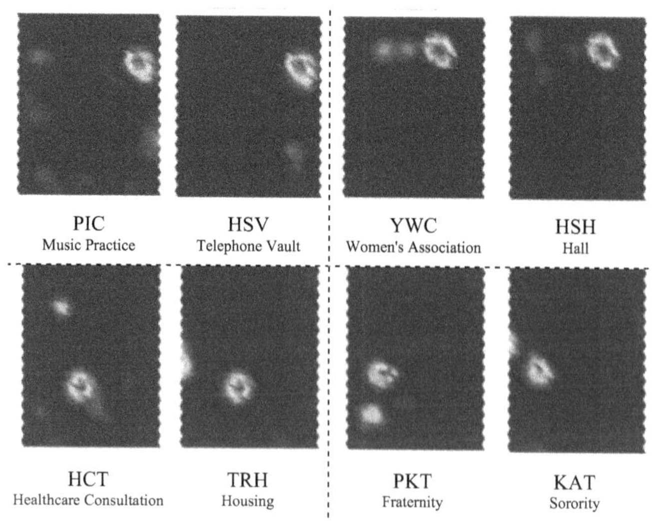

Figure 5. Feature maps for various types of locations

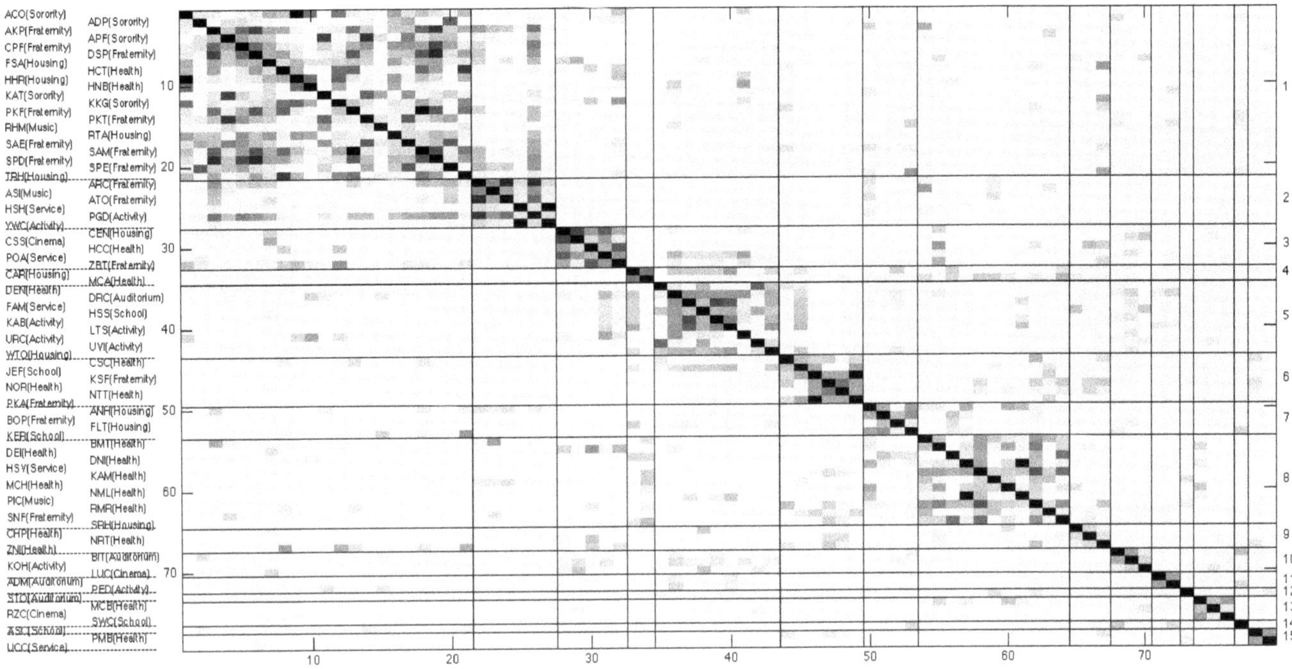

Figure 6. Clustered heatmap for all buildings as features. Rows and columns represent building IDs and lines indicate cluster borders. Numbers at the right show cluster IDs and descriptions at the left include building abbreviation and category for each row. Darker colors mean more correlation.

One point in multi-aspect analysis is the fact that inspecting many feature maps for all the combination of aspects (in our case 800 combinations of 40 domains and 20 locations) is rather difficult. This was actually one of our motivations for designing the feature clustering technique. By employing this technique, we can easily cluster all the maps and then use the visual inspection of feature maps for detailed analysis. Figure 8 shows the top-left quarter of acquired clustered heatmap for multi-aspect analysis (80 clusters in total) (see [23] for the complete map).

6. APPLICATIONS

The systematic realistic mining method proposed in this paper can be applied with any set of wireless data to reveal significant facts that may be used in several important applications in mobile networking research. Here, we briefly address three such major applications:

1- Simulation of online activity and mobility for mobile users: To provide realistic input to the simulations, realistic models of users' behaviors are required, along with scenarios of events and dynamics of mobility and Internet access. The parameters of online activity and mobility along with trend characteristics and correlations in the simulation can be easily derived from our model in this paper. None of the existing models captures such characteristics across website access. Developing such a simulation tool for mobile Internet is part of our future work. Similarly, we plan to conduct an extensive study on the spatio-temporal parameters for mobile traffic modeling in future.

Table 2 – Feature clustering result on web domains

Cluster	Domain	Cluster	Domain
A	apple mac cnet washingtonpost itunes earthlink	B	microsoft windowsmedia microsoftoffice2007 mcafee hackerwatch quiettouch
C	google mozilla nih	D	live hotmail net hamachi
E	veoh secureserver	F	comcast fastwebnet
G	torrentbox rr	H	smartbro aster fastres opendns

2- Interest-based protocols and services: A new class of protocols and services center around user-interest and similarity, including profile-cast, participatory sensing, trust establishment, location-based services, crowd sourcing, alert notification and targeted announcements and ads. Website and location visitation patterns can remarkably enhance the accuracy of interest inference and provide much needed granularity for these protocols and services. The interest models developed based on our analysis can help both the informed design of such efficient protocols and the realistic evaluation thereof.

3- Network planning and web caching: Load distribution on the network is imperative for network capacity planning and management issues, and is definitely related to web and location visitation patterns and their characteristics. Also, the caching of web objects for mobile users can only be efficient if informed by the history of visitation patterns.

7. CONCLUTION

This study is motivated by the need for developing realistic models and efficient protocols for the future mobile Internet. We provided a systematic method to model and analyze the largest wireless trace to date, with billions of records of online activity and mobility from a campus network, including thousands of users. Novel modeling and analysis were conducted utilizing self-organizing maps and our proposed extensions to the technique for feature clustering and multi-aspect trend modeling based on web domains and locations. We have shown that mobile online activity and mobility can be modeled with an organized map of trends that can be effectively used to find correlations. The details of our study enable the parameterization of new and realistic models for mobile Internet with applications in several areas of networking, including mobile web caching, simulation and evaluation of protocols, interest-aware services and network

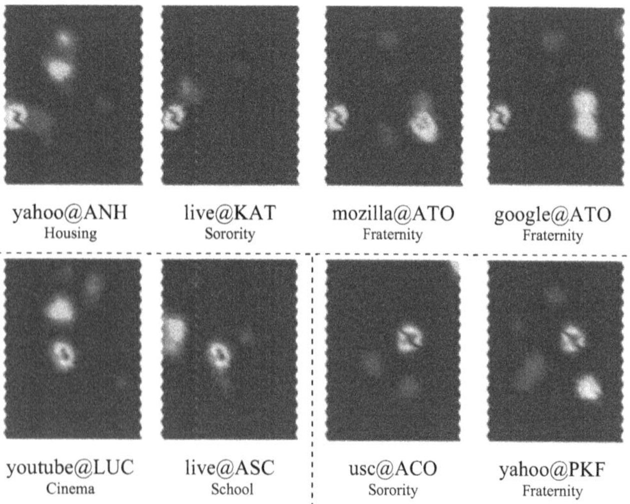

Figure 7. Multi-aspect feature maps for domains-locations

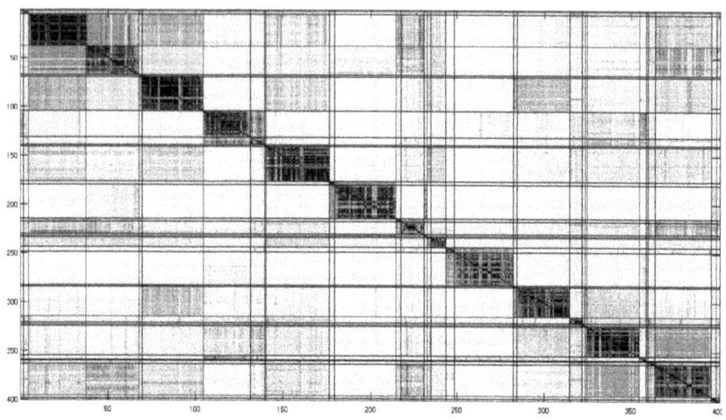

Figure 8. Clustered heatmap for multi-aspect feature analysis (zoomed on top-left quarter of the map)

planning. We hope for our modeling method to provide an example for large-scale data-driven modeling of mobile networks in the future. With more measurements from mobile and sensor networks becoming available, we expect our method to facilitate analysis of many other large multi-aspect/multidimensional datasets in future studies.

8. ACKNOWLEDGMENT

This work was partially funded by NSF award number 0832043.

9. REFERENCES

[1] Hsu, W., Dutta, D. and Helmy, A. Profile-cast: Behavior-aware mobile networking. *ACM SIGMOBILE Mobile Computing and Communications Review*, 12, 1 (Jan 2008), 52-54.

[2] Jain, A. K. and Dubes, R. C. *Algorithms for clustering data.* Prentice-Hall, Englewood Cliffs, NJ, 1988.

[3] Dhillon, I. S. and Guan, Y. Information theoretic clustering of sparse co-occurrence data. In *Proceedings of the Third IEEE International Conference on Data Mining* (2003). IEEE Computer Society.

[4] Kohonen, T. Self-organized formation of topologically correct feature maps. *Biological cybernetics*, 43, 1 (Jan 1982), 59-69.

[5] Kotz, D. and Essien, K. Analysis of a campus-wide wireless network. *Wirel. Netw.*, 11, 1-2 (Jan 2005), 115-133.

[6] Henderson, T., Kotz, D. and Abyzov, I. The changing usage of a mature campus-wide wireless network. *Computer Networks*, 52, 14 (Oct 2008), 2690-2712.

[7] Meng, X., Wong, S. H. Y., Yuan, Y. and Lu, S. Characterizing flows in large wireless data networks. In *Proceedings of the ACM MobiCom 2004* (Philadelphia, PA, USA, 2004). ACM.

[8] Hsu, W. and Helmy, A. On modeling user associations in wireless LAN traces on university campuses. In *Proceedings of the IEEE Int'l Workshop on Wireless Network Measurements(WiNMee)* (Apr, 2006).

[9] Balazinska, M. and Castro, P. Characterizing mobility and network usage in a corporate wireless local-area network. In *Proceedings of the ACM MobiSys 2003* (San Francisco, CA, 2003). ACM.

[10] Papadopouli, M., Shen, H. and Spanakis, M. Characterizing the duration and association patterns of wireless access in a campus. In *Proceedings of the 11th European Wireless Conference* (Nicosia, Cyprus, Apr, 2005).

[11] Hsu, W. and Helmy, A. On nodal encounter patterns in wireless LAN traces. In *Proceedings of the IEEE Int'l Workshop on Wireless Network Measurements(WiNMee)* (Apr, 2006).

[12] MobiLib: Community-wide library of mobility and wireless networks measurements (Investigating user behavior in wireless environments). Available: http://nile.cise.ufl.edu/MobiLib/.

[13] Kotz, D. and Henderson, T. Crawdad: A community resource for archiving wireless data at dartmouth. *IEEE Pervasive Computing*(Dec 2005), 12-14.

[14] Lelescu, D., Kozat, U. C., Jain, R. and Balakrishnan, M. Model T++: an empirical joint space-time registration model. In *Proceedings of the 7th ACM MOBIHOC* (Florence, Italy, May, 2006). ACM.

[15] Hsu, W.-J., Spyropoulos, T., Psounis, K. and Helmy, A. TVC: Modeling spatial and temporal dependencies of user mobility in wireless mobile networks. *IEEE/ACM Trans. Netw.*, 17, 5 (Oct 2009), 1564-1577.

[16] Kim, M., Kotz, D. and Kim, S. Extracting a mobility model from real user traces. In *Proceedings of the IEEE INFOCOM 2006* (Barcelona, Spain Apr, 2006).

[17] Bai, F. and Helmy, A. A Survey of Mobility Modeling and Analysis in Wireless Adhoc Networks, Wireless Ad Hoc and Sensor Networks, Springer, 2006.

[18] Hsu, W., Dutta, D. and Helmy, A. Mining behavioral groups in large wireless LANs. In *Proceedings of the ACM MobiCom 2007* (Montral, Qubec, Canada, 2007). ACM.

[19] Kim, M. and Kotz, D. Periodic properties of user mobility and access-point popularity. *Personal Ubiquitous Comput.*, 11, 6 (Aug 2007), 465-479.

[20] Eagle, N. and Pentland, A. Reality mining: sensing complex social systems. *Personal and Ubiquitous Computing*, 10, 4 (May 2006), 268.

[21] Moghaddam, S., Helmy, A., Ranka, S. and Somaiya, M. Data-driven co-clustering model of Internet usage in large mobile societies. In *Proceedings of the ACM MSWiM 2010* (Bodrum, Turkey, Oct, 2010). ACM.

[22] Moghaddam, S. and Helmy, A. Internet usage modeling of large wireless networks using self-organizing maps. In *Proceedings of the IEEE MASS SCENES Workshop 2010* (Nov, 2010).

[23] Moghaddam, S. and Helmy, A. Spatio-temporal modeling of wireless users Internet access patterns using self-organizing maps. CoRR abs/1008.4904: (2010).

TWEET: An Envelope Detection Based Broadband Ultrasonic Ranging System

Prasant Misra[1], Diethelm Ostry[2], Navinda Kottege[3], Sanjay Jha[1]
[1]School of Computer Science and Engineering, The University of New South Wales, Sydney, Australia.
CSIRO ICT Centre, {[2]Sydney, [3]Brisbane}, Australia.
{pkmisra, sanjay}@cse.unsw.edu.au {Diet.Ostry, Navinda.Kottege}@csiro.au

ABSTRACT

Fine-grained location information at long range can benefit many applications of embedded sensor networks and robotics. In this paper, we focus on range estimation - an important prerequisite for fine-grained localization - in the ultrasonic domain for both indoor and outdoor environments, and make three contributions. First, we evaluate the characteristics of broadband signals, and provide useful statistics in their design and engineering to achieve a good trade-off between range and accuracy. Second, to overcome the inaccuracy due to correlation sidelobes, we propose a signal detection technique that estimates the envelope of the correlated pulse using a simple least-square approximation approach, and undertake a simulation study to verify its ranging efficiency on linear chirps. Third, leveraging on the insights obtained from our initial study, we present the design and implementation of `TWEET`: a mote-based ultrasonic broadband ranging system based on linear chirps using the CSIRO Audio nodes, which comprises of a Fleck-3z mote along with audio codecs and a Blackfin DSP. Our evaluation results indicate that the system is precise enough to support source localization applications: a reliable operational range of 20 m (outdoor) and an average accuracy of < 2 cm with a 95% confidence interval of 2 cm.

Categories and Subject Descriptors

C.3 [**Special-Purpose and Application-Based Systems**]: Signal processing systems

General Terms

Algorithms, Design, Experimentation

Keywords

Linear Chirp, Envelope Detection, Ultrasonic Ranging

1 Introduction

A general acoustic source localization problem involves locating the sound source using acoustic sensors. It has many military and scientific applications, which include, tracking the movement of vehicles and personnel, battlefield surveillance, population estimation of various species, environmental and habitat monitoring, etc [1]. The reliability of the target location information is directly dependent on the accuracy of the position knowledge of the sensing node, where a minor estimation bias can result in localization errors that scale with range to the target. Sensing entities, with the knowledge of their precise location can provision themselves as a positioning infrastructure to provide navigational support to robots and autonomous vehicles, enable automatic system calibration and collaborative sensing, provide valuable assistance in identification of micro-climate zones in precision agriculture and facilitate search/rescue operations.

Despite significant progress in the related areas of location sensing systems, many of these outdoor systems operate in the audible range [2–7], and therefore, may not suitable for a general ranging and localization strategy, especially for covert operations. Working in the ultrasonic domain provides the advantage of inconspicuous location sensing, but its applications have been restricted to only indoor environments that support dense deployment and require short coverage range [8–10].

There are various challenges in using ultrasound as a physical layer technology. It is reported to be more sensitive to atmospheric absorption than audible domain signals that may lead to a reduced coverage range. There are less available commercial off-the-shelf (COTS) components for ultrasonics, which increases the implementation time-line of potential applications. Besides, there are other very important factors limiting sound performance, such as reflections (from the environment, ground etc.), variation in air density caused by thermal effects leading to variation in sound speed, and also, propagation effects caused by non-uniformities in the atmosphere (due to wind/turbulence).

With respect to detection techniques and algorithms, the measurement of the time-of-arrival (TOA) of sound signals have been reported to provide high accuracy ranging [2–7]. Generally, for broadband systems, TOA is estimated by identifying the first peak of the matched filter implemented by correlating the band-limited filtered version of the ranging signal with its own reference copy and counting the elapsed time samples. Depending on the propagation channel conditions, this mechanism introduces an uncertainty in estimating the correct correlation peak due to the presence of numerous sidelobes that may attain similar heights, and hence, can contribute to ranging inaccuracy.

In this paper, we address many of these problems and demonstrate that ultrasound is also a good candidate for long distance ranging for both indoor as well as outdoor environments. Following are the main contributions:

- We use broadband ultrasonic chirps as a ranging signal, and study the impact of signal length and bandwidth on range and resolution respectively along with their related trade-offs.
- We propose a high accuracy ranging mechanism using a signal detection mechanism using the simple idea of estimating the envelope of the correlated (compressed) pulse that makes the role of sidelobes irrelevant, and counters the effect of noise through the least-square curve fitting approach.
- We present the design and implementation of `TWEET`: a mote-based ultrasonic broadband ranging system using the CSIRO Audio nodes, which comprises of a Fleck-3z mote along with audio codecs and a Blackfin DSP. It attains an operational range of 20 m (outdoor) with an average accuracy of < 2 cm with a 95% confidence interval of 2 cm.

We also illustrate practical challenges in the design and execution of our system, the lessons and experiences of which will be helpful to other engineers working on similar projects.

The rest of the paper is organized as follows: A review of related work is discussed in Section 2. Section 3 presents a general broadband signal analysis followed by a comparative study between linear/non-linear chirps and PN signal. Section 4 discuses the system model of the proposed ranging system, and verifies the efficacy of the detection algorithm in simulation. Section 5 describes the design and implementation of the `TWEET` system, followed by evaluation results in Section 6. A comparison among existing systems is outlined in Section 7, and finally, a discussion of future work and conclusion is provided in Section 8.

2 Related Work

TOA-based acoustic systems fall into two categories depending on the available system bandwidth: *narrowband* and *broadband*. Broadband systems utilize a signal with relative bandwidth B_r ($= \frac{bandwidth}{centerfrequency}$) in excess of 20%. We refer our readers to our previous work in [11] where we have provided a detailed comparison of existing narrowband systems. The system proposed by Kushwaha et al. in [2], Hazas et al. in [12], AENSBox [3] and BeepBeep [4] are existing broadband systems. They share a common cross-correlation based signal detection technique, however, they differ in their signal design, synchronization schemes and methods to improve the received signal-to-noise ratio (SNR).

Kushwaha's system was based on the Mica2 platform with an attached custom 50 MHz DSP and an external speaker. The ranging signal was a Gaussian windowed linear chirp of 50 Hz-5 kHz. It employed a message time stamping technique. The SNR of the received signal was enhanced by adding a series of consecutive position-modulated chirps at the same phase and averaging these measurements.

The AENSBox system comprised of a custom designed acoustic sensor array that utilized beamforming to improve the received SNR, and time synchronization services to prevent clock skew and drifting. The ranging signal was a 2048-chip code modulated using binary phase shift keying (BPSK) on a 12 kHz carrier spread over 6-18 kHz. It differed from most of its predecessors in the use of separate synchronization service that maintained metrics to convert from one system clock to another on demand, rather than a synchronous radio and audio pulse. This approach is beneficial in scenarios where the audio range is greater than the radio range.

The BeepBeep system used a 50 ms linear chirp of 2-6 kHz. It used a two-way sensing scheme (different from the round-trip time measurements) to avoid clock synchronization and was implemented on COTS mobile devices.

As these system operated in the audible range, they were not suitable for a general ranging strategy. Therefore, Hazas et al. proposed a broadband ultrasonic localization system that was implemented on custom designed Dolphin devices. The 25 ms ranging signal was generated using a 50 kHz carrier wave modulated by Gold codes (of length 511 bits) using BPSK. The sensitivity of the receiver was improved by using a transducer with a greater surface area (10 mm radius) rather than the general 5 mm transducer applied on the transmitter. The reported ranging results showed millimeter level accuracy that is comparable to the uncertainty in hand-measured distances, but it was only targeted for very short range (< 3 m) indoor applications. We compare the results of our `TWEET` system and related characteristics to some of the related work in Section 7.

3 Choice of Signal Waveform

Based on prior work in the field of acoustical localization in air, three classes of signal waveforms were identified: single tone sinusoidal, linear chirp and pseudo-noise (PN). Chirps are frequency modulated signals, wherein a sinusoidal wave of constant amplitude sweeps the desired bandwidth (B) within a certain time period (T) in a linear or non-linear (for example following quadratic or logarithmic laws [1]) manner. PN signals are (phase) modulated sinusoidal waveforms mixed with pseudo-random sequences.

The time period (T) and bandwidth (B) of the signal have a key role in delivering the desired coverage range and resolution. From observations and reasons reported in our previous work [11], we do not consider single tone sinusoidal signals. Broadband signals (chirps/PN), which utilize correlation based processing share a common theory on the BT relationship. In the following subsection, we explain it for the linear chirp signal and then present our comparison studies for the remaining waveforms.

3.1 Linear Chirp Analysis

A linear chirp is represented by the bandpass signal:

$$s(t) = \begin{cases} \cos(2\pi(f_0 t \pm \mu\frac{t^2}{2})) & \frac{-T}{2} < t < \frac{T}{2} \\ 0 & \text{elsewhere} \end{cases} \quad (1)$$

where, f_0 is the center frequency in Hz, and μ =B(Hz)/T(s) is the chirp rate that sweeps linearly from (f_0−B/2) to (f_0+B/2). The $\pm$ term defines its sweep direction. When the signal in (1) is passed through its matched filter, the following output is generated [13]:

$$g(t) = \frac{T}{2}\cos(2\pi f_0 t)\left[\frac{\sin(\pi\mu t(T-|t|))}{\pi\mu T t}\right] \quad (2)$$

where $\frac{-T}{2} < t < \frac{T}{2}$.
This is the approximate autocorrelation of the linear chirp $s(t)$ and it provides two important results:

- The *peak value* (that signifies the energy of the signal) occurs at $t = 0$ and is proportional to T.
- The *correlation envelope*, expressed as $\left[\frac{\sin(\pi\mu t(T-|t|))}{\pi\mu T t}\right]$ is approximately $\left[\frac{\sin(\pi\mu t T)}{\pi\mu T t}\right]$ for $t \ll T$, with its first zeros at $t = \pm 1/(\mu\text{T}) = \pm(1/\text{B})$; and is inversely proportional B.

This gives the important relationship that *an increase in T increases the size of the post-correlation signal, and an increase in B gives better resolution by narrowing the envelope of the correlation peak.*
In the following subsections, we compare the features of various types of broadband chirps based on B and T.

3.2 Chirp Waveform Comparison

For studying the change of bandwidth, four types of chirp (each of category: linear, quadratic and logarithmic) with constant time-period of 1 second and varying frequency range (and thus bandwidth) were designed: Chirp-1 [20-25 kHz], Chirp-2 [20-30 kHz], Chirp-3 [20-35 kHz] and Chirp-4 [20-40 kHz]. Figure 1(a) shows the ratio between the height of the first sidelobe [P1] to the correlation peak [P0] denoted as [P1/P0] for linear, quadratic and logarithmic chirps for the different types (chirp-[1/2/3/4]). These peaks are related to the envelope of the correlation output, which is an important factor. A lower ratio of [P1/P0] signifies a narrower correlation envelope and is best supported by the highest bandwidth signal (i.e. Chirp-4). The linear and logarithmic chirps have similar correlation envelopes; however, the envelope cover of the quadratic chirp is even narrower. This suggests that although B does not define the acoustic pressure level of the chirp, a higher bandwidth signal is preferable due to its narrower correlation envelope that can improve the resolution of the range measurement.
Similarly, for the studying the change in time-period, four types of chirp were designed with constant bandwidth of 20 kHz and varying time-periods: Chirp-1 [1s], Chirp-2 [0.5s], Chirp-3 [0.1s] and Chirp-4 [0.05s]. Figure 1(b) shows that [P1/P0] is constant for all the different chirps, which suggests that the correlation envelope remains constant with the change in T, and hence is independent of the time-period, irrespective of the chirp type (linear, quadratic, logarithmic)

3.3 Discussion

As T controls the peak value of the correlated signal, one may consider choosing a longer signal duration that has higher energy to travel longer distances. However, a longer ranging signal increase the system reaction time, wherein the pulse repetition frequency has to be kept low; which means that the entire system is required to wait for one signal and all its echoes to decay before transmitting the next pulse. Second, it increases the overall system cost, in terms of processing time, energy consumption and storage, thereby making its implementation difficult on resource-deficient sensor

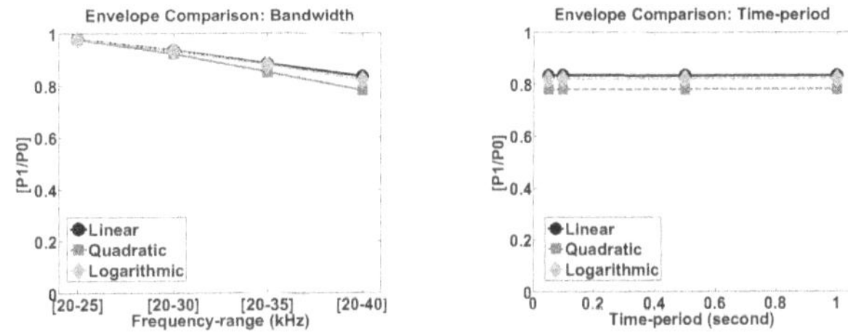

(a) (b)

Figure 1: Linear vs. Quadratic vs. Logarithmic (Chirps). Effect of change in (a): Bandwidth (B) and (b): Time-period (T) on the characteristic of the Correlation Envelope.

motes. B determines the width of the correlation envelope, and therefore, determines the range resolution. Although, working in the ultrasonic domain provides the flexibility to use band of signal frequencies above 20 kHz, higher frequencies components are more vulnerable to atmospheric absorption. This limits the use of an ultra-wide bandwidth ultrasonic signal. The appropriate choice of B and T depends on the application requirements, but from the study presented in the previous subsection, it appears reasonable to choose a broadband signal of the *highest bandwidth* (for best detection accuracy) and *shortest time-period* (for long range incurring the least cost).
With regards to the choice between linear/non-linear chirps and PN signal, we generated a 20-40 kHz/1 s pulse for each category, and compared them on the basis of their individual envelope cover (height of the first [P1] and second sidelobe [P2] to the correlation peak [P0]) and spectral complexity. For a PN signal of certain B and T, we observed that the correlation (peak and sidelobes) properties vary across different pseudo-random numbers, and therefore, we calculated the running average across 1000 randomly chosen PN codes.
Table 1 summarizes the overall statistics. A lower value of [P1/P0] and [P2/P0] signifies a narrower signal envelope, and is best supported by chirp waveforms than PN signals. With regards to the spectral complexity (observed by a spectrogram), the linear chirp is the simplest of all. In this paper, we do not make use of the spectral characteristic of this waveform; however, we intend to take advantage of this feature to improve detection accuracy in our future work. Hence, based on these requirements, we decide to use a *linear chirp* as our ranging signal.

Table 1: Chirps vs. PN Signal Characteristics

Signal Type	[P1/P0]	[P2/P0]	Complexity
Chirp (Linear)	0.8332	0.4358	Low
Chirp (Logarithmic)	0.8201	0.3961	High
Chirp (Quadratic)	0.7802	0.3030	High
Pseudo-noise	0.8106	0.6211	High

In the following section, we discuss the system model and present a simulation study of the ranging performance of linear chirps with our proposed detection algorithm.

4 System Model

We represent an indoor environment using a reverberation geometrical acoustic model [1]. For the mathematical formulation, we adopt the following notation: $s(t)$ and $d(t)$ represent the signal emitted by the transmitter(Tx) and received at the receiver(Rx) respectively; the respective impulse responses of the transmitter, environment (channel) and receiver are represented by $h_{tx}(t)$, $h(t)$ and $h_{rx}(t)$; and the white Gaussian noise in the channel is denoted by $v(t)$.

$s(t)$ is a broadband signal in the form of a linear chirp and is transmitted at $t = 0$ by Tx. The signal $d(t)$ received at Rx is the convolution:

$$d(t) = s(t) * h_{tx}(t) * h(t) * h_{rx}(t) + v(t) * h_{rx}(t)$$

Assuming $h_{tx}(t)$ and $h_{rx}(t)$ are of unit magnitude (i.e. neither the transmitter nor the receiver change the signal characteristics):

$$d(t) = s(t) * h(t) + v(t) \tag{3}$$

$h(t)$ is modeled as the sum of $M+1$ impulses corresponding to the direct path with propagation delay τ_0 and M other possible paths between Tx and Rx as:

$$h(t) = \sum_{i=0}^{M} A_i \delta(t - \tau_i)$$

where A_i is the amplitude of the i-th ray and $\delta(t - \tau_i)$ represents the delay in propagation from Tx to Rx. Ray $i = 0$ is defined here as the direct sound ray from the source to the receiver, and rays $i > 0$ are defined as reflected rays. $\tau_i = d_i/c$, where d_i is the distance traveled by ray i and c is the speed of sound under room conditions.

4.1 Signal Detection and Post-processing

The received signal $d(t)$ is processed using a matched filter implemented by correlating it with a reference signal $s(t)$ (i.e. a locally stored copy of the original emitted signal), and result in:

$$\begin{aligned} y(t) &= [d(t) \star s(t)] \\ y(t) &= [s(t) * s(-t)] * h(t) + v(t) * s(-t) \end{aligned} \tag{4}$$

$y(t)$ has its earliest component $[s \star s](t - \tau_0)$ (where: $\star$ implies correlation), whose peak can be used to determine τ_0 (direct path signal) with considerable precision, provided the other multipath components of $d(t)$ are sufficiently weak and/or separated in time from $t = \tau_0$. The noise term $v(t) * s(-t)$ may shift the peak at τ_0 from its actual time-line, which may result in an inaccurate estimate of the range information.

The signal detection scheme discussed in existing work provide resistance to multipath and low-noise signals [2–4, 12] by using a peak detection approach, which under the condition of a direct line-of-sight (DLOS) between the transmitter and receiver, identifies the first tallest correlation peak that exceeds a preset threshold. However, from our study and observations, we noticed two potential problems. First, the correlation peak is surrounded by numerous sidelobes (i.e. adjoining peaks). Second, the correlation plot obtained from processing the band-limited signal is highly oscillating within its envelope cover. Both these conditions provide an inaccurate estimate of the correct detection peak in the vicinity of similar peaks of approximately equivalent heights under noisy conditions. Therefore, we propose a simple envelope detection mechanism that makes the role of sidelobes irrelevant and counters the effect of noise through the least-square curve fitting approach. In addition, it also provides the benefit of finer resolution that can be fractions of a sampling period.

The envelope detection approach estimates the maximum value of the envelope of the compressed (correlated) pulse that should give the best estimate of its position. A simple least-squares approximation technique has been used to find the envelope of the rectified signal, rather than the standard approach of calculating the magnitude of the analytical signal. The algorithm identifies the position of the local peak (t_2, y_2) that is greater than its left and right neighbor peaks at (t_1, y_1) and (t_3, y_3), finds the parabola that passes through these points exactly, and finally, calculates the time coordinate of the maximum of this parabola ($t = t_{peak}$). Therefore, fitting a parabola to these three points requires solving the following system of three linear equations for the three unknown $[a, b, c]$:

$$\begin{aligned} y_1 &= a t_1^2 + b t_1 + c \\ y_2 &= a t_2^2 + b t_2 + c \\ y_3 &= a t_3^2 + b t_3 + c \end{aligned}$$

The corresponding representation in matrix form is:

$$\mathbf{A}\hat{\mathbf{x}} = \mathbf{B}$$

where $\hat{\mathbf{x}} = [a\ b\ c]^T$ is the matrix of unknown parameters, and:

$$\mathbf{A} = \begin{bmatrix} t_1^2 & t_1 & 1 \\ t_2^2 & t_2 & 1 \\ t_3^2 & t_3 & 1 \end{bmatrix}$$

$$\mathbf{B} = [y_1\ y_2\ y_3]^T$$

Thus, $\hat{\mathbf{x}} = \mathbf{A}^{-1}\mathbf{B}$, where $\mathbf{A}^{-1}$ is the inverse matrix of $\mathbf{A}$. The maximum of the envelope occurs at $t_{peak} = -b/(2a)$.

In case of low-noise signals, there are more peaks surrounding the highest peak as shown in Figure 2(a). The parabolic curve fitting using least-square approximation technique does the best to pass as near as possible to all the adjacent peaks, and thus, provides resistance to noise on the data points. Thus, we obtain the following three equations for the three unknowns $[a, b, c]$:

$$\sum_{i=1}^{n} y_i t_i^2 = a \sum_{i=1}^{n} t_i^4 + b \sum_{i=1}^{n} t_i^3 + c \sum_{i=1}^{n} t_i^2 \tag{5}$$

$$\sum_{i=1}^{n} y_i t_i = a \sum_{i=1}^{n} t_i^3 + b \sum_{i=1}^{n} t_i^2 + c \sum_{i=1}^{n} t_i \tag{6}$$

$$\sum_{i=1}^{n} y_i = a \sum_{i=1}^{n} t_i^2 + b \sum_{i=1}^{n} t_i + cn \tag{7}$$

This linear system can be solved (as explained before) for $[a, b, c]$ to provide an estimate for the position of the peak: $t_{peak} = -b/(2a)$. t_{peak} is the estimate of the pulse position, and thus, provides a range estimate.

4.2 Simulation Results

We simulated a custom environment in MATLAB and evaluated the performance of the proposed ranging algorithm. The simulator was designed to construct a virtual

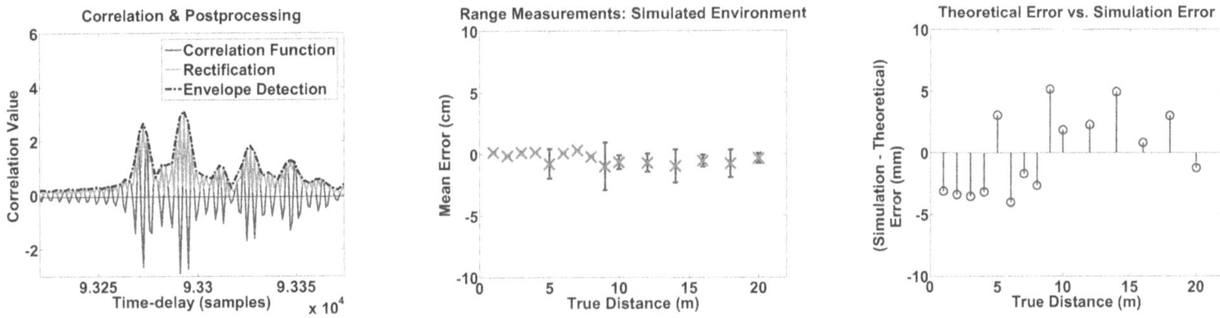

(a) (b) (c)

Figure 2: (a): Signal Detection and Post-processing. Simulation Results for Range Estimation. (b): Mean error with vertical bars representing 95% confidence intervals. (c): Theoretical error vs. Simulation error.

2D rectangular room with (top,left) and (bottom,right) coordinates as: $(-5, \zeta/2)$ and $(\zeta+5, -\zeta/2)$ respectively, where ζ is the distance between Tx and Rx and is varied from 1-20 m for every set of measurements. Tx and Rx were placed at positions $(0, 0)$ and $(\zeta, 0)$. It was configured to generate fixed number of reflection points at random positions in the enclosed geometry and was programmed as per the described system model.

Measurements were taken at different positions inside the room for distances between 1-20 m. Every simulation was run for a length of 1000 iterations. The simulator was configured for 5 reflection points with attenuation factor of 0.9, and the transmitted signal of [20-40] kHz/50 ms (sampled at 96 kHz) was added with white Gaussian noise (SNR=10).

Figure 2(a) shows the output from the correlation function, the result of rectification and envelope detection of the correlated pulse. Figure 2(b) shows the distance estimation accuracy obtained from the simulation measurements. We observe that the magnitude of the mean errors is consistently less than 1cm for distances upto 20 m. Figure 2(c) plots the difference between errors derived from simulation and their respective theoretical prediction [13] for distance measurements of Figure 2(b). A positive error denotes that the error incurred from simulated measurements are more than their predicted theoretical value, and vice-versa for negative errors. It shows that there is an average difference of 3mm between the simulated and predicted errors for all the measurement points, with a maximum deviation of 5 mm. Therefore, we conclude that our detection methodology is precise enough for fine-grained ranging.

5 TWEET: System Implementation

We, initially, developed a proof-of-concept PC based ranging system consisting of various COTS devices and custom designed units to experimentally verify the feasibility of our proposed scheme before incorporating them into an embedded design. Based on its understanding, we improved on its various stages to finally assemble all the different components onto a single mote-based ranging system named as the TWEET. In this section, we describe its hardware platform and software architecture. We refer our readers to [1] where we have reported our experiences in working with the PC-based system and its ranging performance.

5.1 System Design: Hardware & Software

TWEET has been implemented on CSIRO Audio nodes using its wireless sensor network platform: the Fleck-3z. Its main components include the Atmega1281 microcontroller, 1 MB external flash memory and a low-bandwidth Atmel RF212 radio transceiver operating in the 900 MHz band. It supports a flexible range of digital I/Os and a daughter board interface, which allows the use of expansion boards to enhance its base functionality. The architecture relies heavily on the SPI bus, where the microcontroller acts as the SPI master and communicates with the remaining system components over the SPI interface. Figure 3 shows the system architecture of the implemented TWEET system.

The audio signal processing daughter board (designed by CSIRO) was used for ultrasonic ranging. It includes four TI TLV320AIC3254 audio codecs, each providing two audio I/O channels along with internal functionalities such as programmable gain amplifiers and software configurable filtering; micro SD flash memory card slot, and a connector to hold the CM-BF537E digital signal processor module manufactured by Bluetechnix. The CM-BF537E module combines a (Analog Devices) Blackfin DSP running at up to 600 MHz, a 32 MB RAM and an Ethernet interface. The DSP communicates with the Fleck microcontroller through a serial interface. The current consumption of this daughter board is in excess of 200 mA in the active state, and so (in its present implementation) Fleck-3z mote controls power to this board ensuring that the relatively high power consumption is only incurred during audio transmission and acquisition. There are two simple daughter boards that provide connector access to the audio I/O ports and an Ethernet socket.

The transmitting front-end of the beacon mote consisted of a power amplifier driving a tweeter (speaker) transducer (VIFA 3/4" tweeter module MICRO). It was chosen due to its small size ([2 × 2 × 1] cm) and good high-frequency response compared to existing broadband transducers or piezoelectric ceramic/piezo film transducers (reported in existing literature). The amplifier is a light-weight portable unit with a maximum output power of 0.5 W. It is powered by batteries and has a tuneable gain controller.

The receiving front-end of the listener mote consisted of the Knowles microphone (SPM0404UD5 [1]) fixed to the pre-amplifier PCB designed by CSIRO. The surface mount wideband ultrasonic sensor was chosen due to its omni di-

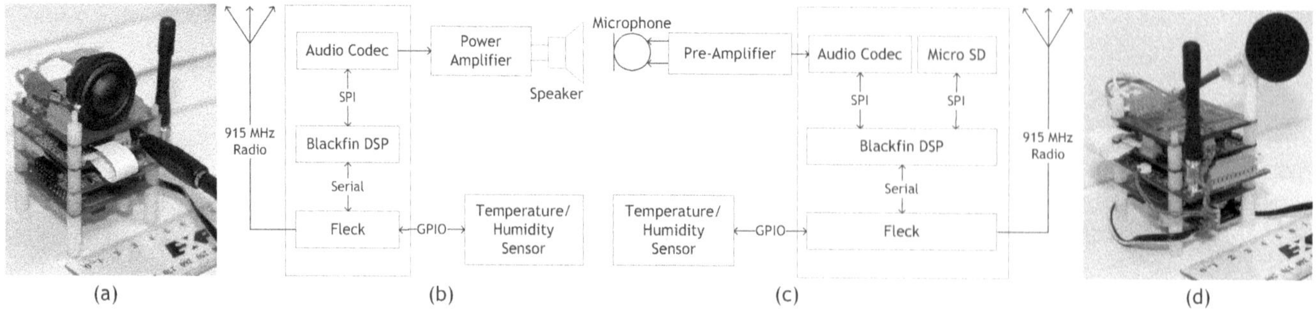

Figure 3: The TWEET Ranging System: Architecture of the (a)-(b) Beacon and the (c)-(d) Listener.

rectionality, high sensitivity and SNR within an extremely small form factor of $[4.72 \times 3.76 \times 1.40]$ mm. The small PCB has been designed to operate in close proximity with the microphone in order to minimize the susceptibility of the low amplitude microphone output signals to corruption by electrical noise. Figure 4 shows the frequency response characteristics for both the transmitting and receiving front-ends, which have an approximately 20 dB acoustic pressure level above the noise floor for frequencies between [20-40] kHz.

As temperature compensation is required for range measurements, a small form-factor PCB ($[2.5 \times 2.5]$ cm) was designed to mount the Sensirion SHT15 temperature and humidity sensor (along with a filter cap and necessary discrete components such as capacitors and pull-up resistors), and was controlled by the Fleck-3z microcontroller via a GPIO digital interface. It consumes < 5 mA of current, thus allowing it to be powered directly from the digital I/O ports of the mote.

Fleck-3z runs the TinyOS-2.x OS. The software performs the tasks of maintaining and executing a schedule of system operations, maintaining a persistent log of system actions and status, sampling from attached on-board/external sensors, controlling the operation and power switch of the audio signal processing daughter board. The software for the Blackfin DSP is responsible for configuring (sampling rate, gain, etc) and enabling the audio codec ICs, managing the incoming and outgoing digital audio stream, transferring data/information to the micro-SD card, command exchange from the Fleck-3z via serial interface, such as start/stop audio playback/recording, interrogate operational status, etc.

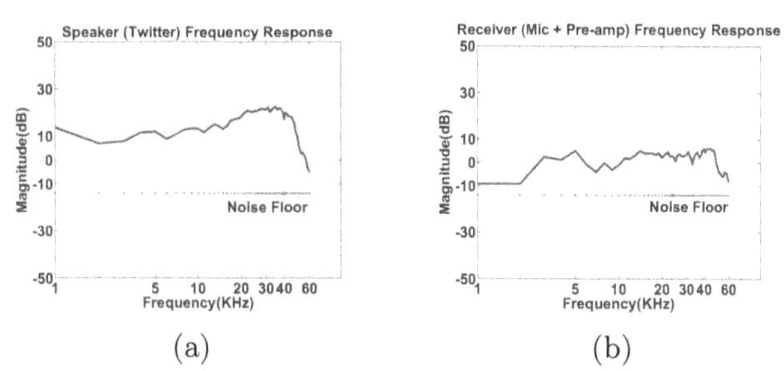

(a) (b)

Figure 4: The TWEET System: Frequency response of the (a): Speaker and (b): Receiver front-end.

5.2 Ranging Methodology

For the TWEET system, a [20-25] kHz/50 ms ranging signal was chosen and the audio codes were configured to sample at 64 kHz. Although the audio codecs could support a maximum sampling rate of 192 kHz that can generate a signal upto 96 kHz, our system tests revealed that there was a significant drop in the audio output of approximately 30-40dB beyond 25 kHz frequency range. This limited our choice of the ultrasonic bandwidth to only 5 kHz.

TWEET uses the time-difference-of-arrival (TDOA) of RF and ultrasound signals to measure the beacon-to-listener distances. The beacon periodically transmits a RF message containing the measured ambient temperature and humidity. At the start of each RF message, the beacon transmits a broadband ultrasonic linear chirp pulse. The fast propagating RF signal leads its synchronous ultrasound pulse and reaches the listener almost instantaneously, which then measures the TDOA between them. The TOA of the ultrasound pulse is measured by cross-correlating the received chirp pulse with a copy of the reference signal stored in the receiver, and then, the range estimate is computed by the envelope detection technique (Section 4.1). Since the speed of sound has a relatively large sensitivity to temperature variations than relative humidity and atmospheric pressure, the final distance estimate is computed by the corresponding speed of sound obtained by averaging the ambient temperature measured at the beacon (sent in the RF message) and the listener (measured at the TOA of the ultrasound pulse). After carefully estimating the various system induced time-delays, a final calibration was performed to eliminate a random lag of 6.42 cm and was subtracted from the final result.

6 Experiments and Results

To evaluate the performance of the TWEET system under different multipath conditions, we conducted ranging experiments in the following environments:

Case-A - Indoor, Low multipath: A quiet lecture theatre ($[25 \times 15 \times 10]$ m) with a spacious podium at one end of the large room.

Case-B - Indoor, High mutipath: A quiet meeting room ($[7 \times 6 \times 6]$ m) with a big wooden table in the center and other office furnitures.

Case-C - Outdoor, Very low multipath: A less frequently used urban walkway, and the weather being sunny with occasional mild breeze.

In all our experiments, the beacon mote was fixed while

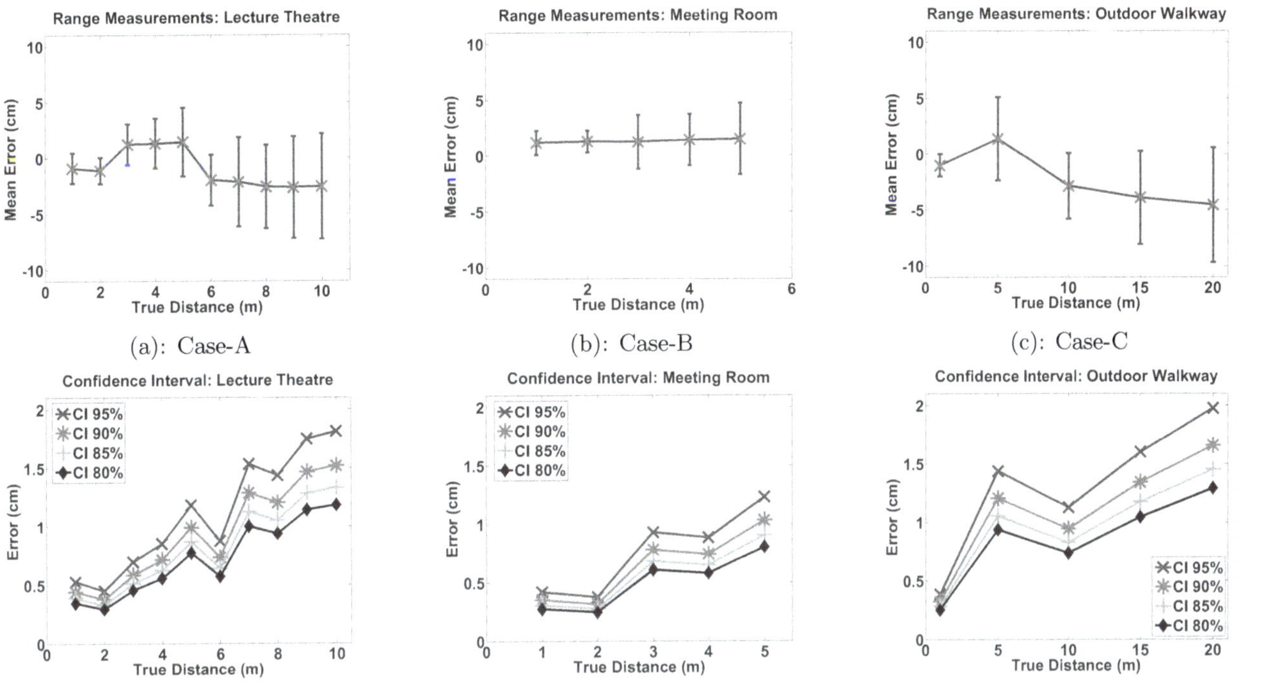

(a): Case-A (b): Case-B (c): Case-C

(d): Case-A (e): Case-B (f): Case-C

Figure 5: TWEET: Accuracy in terms of mean error, deviation (shown in blue vertical lines), and confidence intervals for the different experimentation cases.

we performed a controlled and careful movement of the listener mote along the direct LOS using a measuring tape and markers for establishing the correct ground distance. The beacon was calibrated to chirp at 70 dB. The speed of sound used in distance calculation was according to the model: $c_{air} = 331.3 + 0.6\theta$ (θ: air temperature in ^{o}C). For every setting (i.e., different distances under different test cases), the experiments were repeated 30 times. The metrics used to evaluate the system were accuracy (difference between the ranging results and the true distance) and confidence interval for the measured errors.

The accuracy and confidence measurements for the case-A setting are shown in Figure 5(a) and (d), where we can observe that our system yields accurate and stable ranging results in a (less severe) multipath environment. The mean ranging error is within ±[1-2] cm with a 95% confidence interval of < 2 cm. High percentage of experiments recorded less than 2 cm accuracy, however, the performance deteriorates for distance measurements at [9-10] m when the listener mote approaches close to the walls. Even then, the error deviation from the mean is < 5 cm. Figure 5(b) shows the accuracy for case-B, where the mean ranging error is 1.5 cm and the maximum deviation is 2.5 cm at its maximum measured distance of 5 m. Due to the multipath dominated environment, the reported error levels (for each measurement) is quite fluctuating, even for shorter distances. Nevertheless, the system was able to record a confidence interval of < 1.5 cm (Figure 5(e)).

The ranging statistics for case-C has been shown in Figure 5(c) and (f). The system shows similar performance as reported for indoor case-A/B for distances < 10 m with a maximum mean error and deviation of ≈ 2.5 cm and an estimated confidence interval of 2 cm. We also observe that

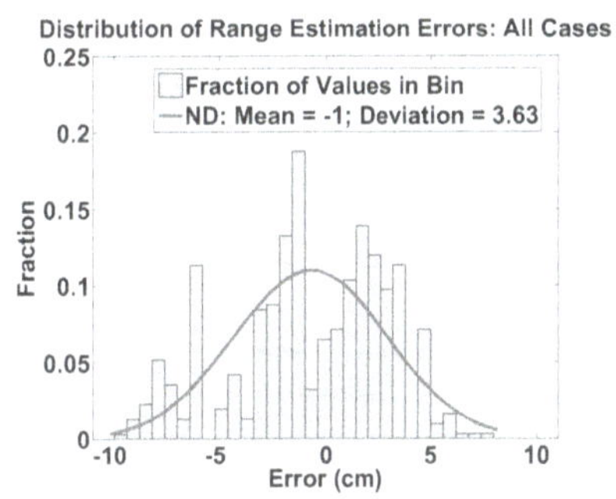

Figure 6: TWEET: Overall System Performance.

the the ranging error increases and shows a larger dynamic range for distance measurements at 15 m and 20 m, which is primarily due to the a lower SNR of the received signal caused by attenuation and non-uniformities in the atmosphere caused by wind. The measurements at 5 m shows a sudden drop in accuracy and an increased deviation, which is due to a strong breeze at that instant, but the system quickly recovers and attains a stable mean error.

For all the measurements, we observe that the absolute distance estimation error increases with the increase in separation between the transmitter and the receiver. In practice, distance information is not known a priori, therefore we have shown the distribution of all ranging errors in the different test environments (Figure 6) so as to provide an overall system snapshot. From the figure, we can see that the overall performance of TWEET is accurate with a mean error of 1 cm and deviation of 3.63 cm with the best performance obtained in indoor spacious facilities.

Table 2: Resource Usage

System	Power (W)	Approx. Processing Cost
TWEET	2.5	2*FFT + 1*IFFT
AENSBox	3.6	5*FFT + 4*IFFT

7 Discussion

Our work has similarities to the linear chirp based system designed by Kushwaha [2]; but instead of using an additional Gaussian window to compensate for high correlation sidelobes, we use a simpler and effective envelope detection method. Our ranging precision (after temperature compensation) is generally 3× better, where for ranges between [10-20] m, our standard deviation is 5 cm compared to [15-25] cm reported by Kushwaha et al. Further, Kushwaha's system achieved a maximum detection range of 30 m at a SPL of 105 dB at 10 cm (i.e. measured at the near-field of the speaker). Such near-field measurements may not provide the correct SPL representation, as they are dominated by the physical dimension of the speaker membrane and the volume of displaced air. Our experiences with two different speakers of different sizes used with our system (wherein we supplied the same power) showed different SPL values when measured in the near-field (10 cm), but were the same at 1 m. In contrast, `TWEET` attained at maximum range of 30 m, but its operational range was 20 m at SPL of 70 dB (SPL measured at 1 m).

Our system reports, approximately, the same level of accuracy as the BeepBeep [4] system for ranges under 10 m. However, the authors do not provide any ranging analysis for distances over [10-12] m, which is where our system is more useful. There is no mention of the SPL for this system, therefore, comparing the ranges would be unfair.

The AENSBox system [3], calibrated to 105 dB SPL at 10 cm, attains a maximum range of 60 m with a mean error of 1.73 cm and deviation of 1.76 cm. It is 3× better than the `TWEET` system in maximum operational range, but the accuracy is comparable to ours with a mean of 1 cm and deviation of 3.63 cm. This additional benefit in range comes at the cost of using a longer (1/3 s) ranging signal and an expensive matched-filter. However, under the condition of duty-cycling (Figure 7(a)), our system scores over AENSBox in consuming 80% lesser power. Table 2 compares the various resource usage between these two systems, where again, `TWEET` proves economical in saving power and processing cost.

A comparison of the signal envelope (an important factor for precision) among `TWEET`, AENSBox and BeepBeep has been shown in Figure 7(b), where we consider the signal length of the various systems to be 1 s (for easy analysis). We observe that `TWEET` has a narrower signal envelope than the other systems, which implies that if the same SPL is generated from all, then our system would be able to attain comparable detection range, but with better accuracy.

8 Future Work and Conclusion

The current limitation of the presented `TWEET` system is that its ranging accuracy degrades under dense multipath in cluttered environments, such as long and narrow office corridors, tunnels, small office rooms etc; and therefore, we

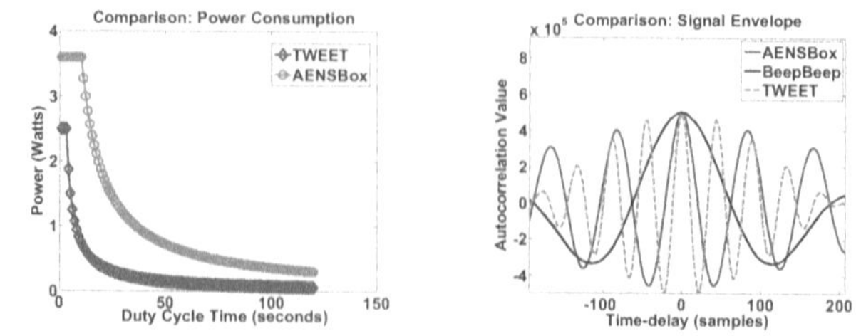

(a) (b)

Figure 7: System Comparison.

are currently investigating into robust algorithms that can improve system reliability under these extreme conditions. Our final objective is to develop an efficient localization system for indoor environments on mote-class devices.

To summarize, we presented a study of the impact of pulse length and bandwidth of broadband signals on range and resolution respectively. Leveraging on this understanding, we evaluated the ranging performance of linear chirps, and proposed an easy and low-cost signal detection technique based on estimating the envelope of the correlated pulse using a simple least-square approximation approach to counter the effect of noise and multipath. Finally, we presented the design and implementation of `TWEET`: a mote-based ultrasonic broadband ranging system using linear chirps for indoor/outdoor source localization.

9 References

[1] Prasant Misra, Diet Ostry, Sanjay Jha, and Navinda Kottege. An envelope-detection based broadband ultrasonic ranging system for improved indoor/outdoor positioning. Technical Report UNSW-CSE-TR-1106, 2011.

[2] M. Kushwaha, K. Molnar, J. Sallai, P. Volgyesi, M. Maroti, and A. Ledeczi. Sensor node localization using mobile acoustic beacons. In *MASS*, pages 9pp.–491. IEEE, 2005.

[3] Lewis Girod, Martin Lukac, Vlad Trifa, and Deborah Estrin. The design and implementation of a self-calibrating distributed acoustic sensing platform. In *SenSys*, pages 71–84. ACM, 2006.

[4] Peng Chunyi, Shen Guobin, Zhang Yongguang, Li Yanlin, and Tan Kun. Beepbeep: a high accuracy acoustic ranging system using cots mobile devices. In *SenSys*, pages 1–14. ACM, 2007.

[5] Jingbin Zhang, Ting Yan, John A. Stankovi, and Sang H. Son. Thunder: towards practical, zero cost acoustic localization for outdoor wireless sensor networks. *SIGMOBILE Mob. Comput. Commun. Rev.*, 11(1):15–28, 2007.

[6] Kwon YoungMin, K. Mechitov, S. Sundresh, Kim Wooyoung, and G. Agha. Resilient localization for sensor networks in outdoor environments. In *ICDCS*, pages 643–652. IEEE, 2005.

[7] Borriello Gaetano, Liu Alan, Offer Tony, Palistrant Christopher, and Sharp Richard. Walrus: wireless acoustic location with room-level resolution using ultrasound. In *MobiSys*, pages 191–203. ACM, 2005.

[8] Andy Harter, Andy Hopper, Pete Steggles, Andy Ward, and Paul Webster. The anatomy of a context-aware application. In *MOBICOM*, pages 59–68. ACM, 1999.

[9] Nissanka Bodhi Priyantha. *The Cricket Indoor Location System*. PhD thesis, Massachusetts Institute Of Technology, 2005.

[10] Andreas Savvides, Chih-Chieh Han, and Mani B. Strivastava. Dynamic fine-grained localization in ad-hoc networks of sensors. In *MobiCom*, pages 66–179. ACM, 2001.

[11] P. Misra, D. Ostry, and S. Jha. Improving the coverage range of ultrasound-based localization systems. In *WCNC*. IEEE, 2011.

[12] M. Hazas and A. Hopper. Broadband ultrasonic location systems for improved indoor positioning. *IEEE Transactions on Mobile Computing*, 5(5):536–547, 2006.

[13] E. Charles Cook and Marvin Bernfeld. *Radar Signals: An introduction to theory and application*. Academic Press, 1967.

Quantitative Uncertainty-Based Incremental Localization and Anchor Selection in Wireless Sensor Networks

†Zhiheng Xie, ‡Mingyi Hong, †Hengchang Liu, †Jingyuan Li
‡Kangyuan Zhu and †John A. Stankovic
†Computer Science Department, University of Virginia
‡System Engineering Department, University of Virginia
{zx3n, mh4tk, hl4d, jl3sz, kz3y, stankovic}@virginia.edu

ABSTRACT

Previous localization solutions in wireless sensor networks mainly focus on using various techniques to estimate node positions. In this paper, we argue that quantifying the uncertainty of these estimates is equally important in practice. By using the quantitative uncertainty of measurements and estimates, we can derive more accurate estimates by better fusing the measurements, provide confidence information for confidence-based applications, and know how to select the best anchor nodes so as to minimize the total mean square errors of the whole network. This paper quantifies the estimation uncertainty as an error covariance matrix, and presents an efficient incremental centralized algorithm—INOVA and a decentralized algorithm—OSE-COV for calculating the error covariance matrix. Furthermore, we present how to use the error covariance matrix to infer the confidence region of each node's estimate, and provide an optimal strategy for the anchor selection problem. Extensive simulation results show that INOVA significantly improves the computation efficiency when the network changes dynamically; the confidence region inference is accurate when the measurement number to node number ratio is more than 2; and the optimal anchor selection strategy reduces the total mean square error by four times as much as the variation-based algorithm in best case.

Categories and Subject Descriptors

I.6.4 [**Computing Methodologies**]: SIMULATION AND MODELING—*Model Validation and Analysis*

General Terms

Algorithms

Keywords

wireless sensor network, localization, uncertainty

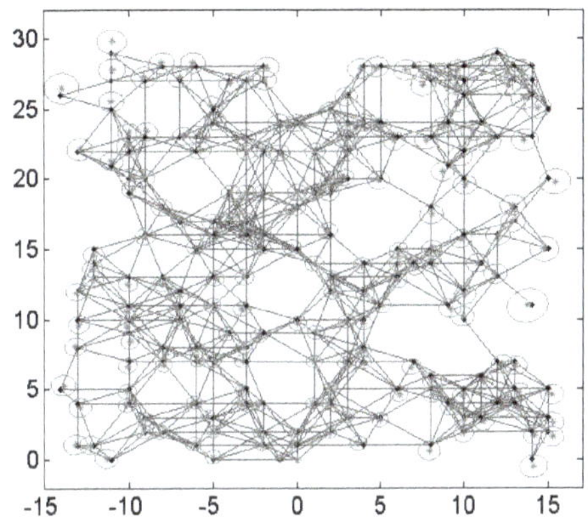

Figure 1: Confidence-based Localization (at 90% confidence level)

1. INTRODUCTION

Localizing nodes in a wireless sensor network is an important problem that has produced many solutions. Typically, each solution makes various assumptions and provides different levels of accuracy. However, most of these solutions suffer from two common drawbacks: they do not quantify the uncertainty of measurements, and they do not provide quantitative uncertainty for their location estimates. The former problem causes the estimation to be non-optimal, while the latter results in the impossibility for inferring the confidence information at the node level. A preferred localization scheme is shown in Figure 1, where not only the location estimates (the red "*"), which fuse measurements according to their quantitative uncertainty, are provided, but also the confidence region (the red ellipses) for each node is rendered.

In this paper, we emphasize the importance of quantitative uncertainty for three reasons. First, by taking the quantitative uncertainty of the measurements into account, we can achieve better localization results. For example, in Figure 2(a), Nodes 1, 2 and 3 are the anchor nodes, Node R is to be localized and its real location is at the blue point. The arrow lines are the measurements, and the dashed circles represent the uncertainty of these measurements. If we do not take the uncertainty into account and just average these three measurements (trilateration and triangulation fall into this category), the estimate of Node R would be $E1$; but if we weight the accurate measurements more and the inaccurate ones less (by using the relative measurement graph

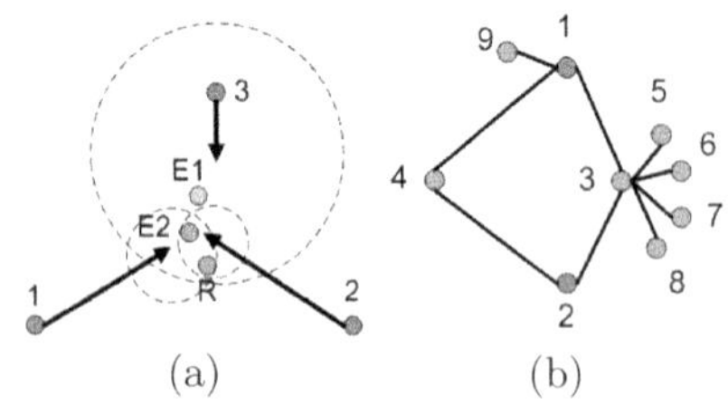

Figure 2: (a) Uncertainty Based Localization; (b) Anchor Node Selection

model as this paper does), the estimate would be $E2$, which is more accurate than $E1$.

Second, the quantitative uncertainty can provide confidence regions for confidence-based applications, which would take different actions based on different confidence-levels. Take a firefighter retreat selection application for example [16]. If a WSN is dynamically deployed in a building on fire to help select a safe retreat route (by sensing the surrounding environment), the system would only choose a safe route if it is within a 95% confidence-level region.

A third usage of quantitative uncertainty is to guide the selection of anchor nodes. Anchor nodes are those whose global locations are precisely known, e.g. either by GPS or manual input. Many WSN localization solutions (including our method) rely on a few anchor nodes to deduce other non-anchor node positions. However, in a large scale WSN, which nodes should be set as anchors so that the entire network's localization accuracy can be significantly improved? If the quantitative correlated uncertainty (e.g. the covariance matrix) for all nodes' location estimation is provided, the nodes with both large estimation variation and high correlations with other nodes should be selected. Take Figure 2(b) for example. Given the quantitative uncertainty, we would choose Node 3 as an anchor because it has both large variation (although not largest) and high correlation to other nodes, otherwise we may intuitively choose Node 4 since it is farthest away from the anchor nodes 1 and 2.

This paper models the quantitative uncertainty as the error covariance matrix for both measurements and estimates, and uses the Relative Measurement Graph (RMG) [2, 3] to calculate the location estimates and the covariance matrix. This model assumes that a set of relative position measurements (both the relative distances and angles) and their corresponding error covariance matrices are known. A RMG can be built, where the vertex set is all the sensor nodes, the directed edge set is all the node pairs which have relative measurements, and the edge function is to map each edge to its corresponding measurement and the error covariance matrix. Then, the location estimates and the corresponding covariance matrix for all these estimates can be derived by the Best Linear Unbiased Estimator (BLUE) [13].

Although the BLUE method mathematically solves the localization with quantitative uncertainty, its high computational complexity ($O(mn^2d^3)$, m is the measurement number, n is the node number and d is the variable dimension) and one-time localization scheme prevents it from being used in a large network, where measurements are incrementally generated and the number of nodes dynamically changes. Additionally, it does not provide any distributed algorithm to calculate the error covariance matrix, which is required by a peer-to-peer architecture.

This paper not only solves the above calculation problems, but also shows how to make use of a covariance matrix to infer the confidence region and help anchor selection. In summary, the contributions of this paper are: first, a more efficient incremental centralized algorithm–INOVA is developed, which reduces the computational complexity from $O(mn^2d^3)$ to $O(n^3d^3)$, and keeps a fixed computational time as measurements are incrementally generated. Second, by extending the location estimation distributed algorithm—OSE [2], a decentralized algorithm for calculating the error covariance matrix, named OSE-COV, is provided. The advantage of OSE-COV is that it reuses the scheme of the OSE algorithm, so that the location estimates and the corresponding error covariance matrix can be simultaneously obtained by using the same protocol. Third, we show that the location estimates derived by BLUE follow the multivariate normal distribution, and thus the confidence region can be derived based on the estimate and the corresponding covariance matrix. Finally, an optimal anchor selection algorithm is derived. It decides which node should be selected as an anchor so that the total mean square error of the whole network is minimized. Note that all above work is generic, and can be used for other problems, such as time synchronization [12](1D) or motion consensus [2] (3D) problems.

2. OPERATIONAL SCENARIO

Since the details of our solution are quite mathematical, we begin by presenting an operational scenario. Consider an expedition team that moves through a large forest or other wild area dropping sensor nodes. Once deployed, each of these nodes communicates with neighbors and using a technique such as TDOA [23, 21] and AOA [19, 21] to create a set of measurements and the corresponding error covariance matrices (each node can infer the error covariance matrix based on its device's accuracy property which is pre-studied, e.g. the error variation could be constant or proportional to the measured distance, and x-axis's variation could be independent with y-axis's) for the relative position between them. All these measurements and the corresponding error covariance matrices are wirelessly transmitted to the base station. Then using the RMG model and the INOVA algorithm, the location estimates and the corresponding error covariance matrix for all these nodes are obtained. If the above centralized architecture is not available in some place, the distributed algorithms OSE and OSE-COV can be used to estimate the location and the covariance matrix, respectively.

After deployment, nodes may continuously generate more relative measurements, and some may move, leave or join the network at run time, then INOVA can quickly respond to these dynamic behaviors, and update the estimates in real time. If some confidence-based application needs to know nodes' real location region at some specific confidence level, a confidence region picture like Figure 1 (the red part) can be rendered based on the inference theory in Section 6. Furthermore, based on each node's estimation error covariance matrix, we can judge whether the current estimation accuracy is good or not. If it is not good enough, we can set one or more nodes to be anchor nodes by using the optimal anchor selection strategy in Section 7, so that the total mean square error of the network is guaranteed to be minimized after each selection.

After the basic RMG model described in Section 3, we

provide our centralized algorithm INOVA and the distributed algorithm OSE-COV in Sections 4 and 5, respectively. The confidence region inference and the anchor selection problem are presented in Sections 6 and 7. The evaluations of all above conclusions are shown in Section 8, followed by the related work and the conclusion in Sections 9 and 10.

3. RELATIVE MEASUREMENT GRAPH

Estimating each node's vector-valued variable based on a set of noisy linear relative measurements, such as problems of time synchronization [12], 2D or 3D WSN localization [2], or motion consensus [2], can be solved by the relative measurement graph model. Assume the vector-valued variable for each node is $x_i \in \mathbb{R}^d, i = 1, 2, \cdots, n$, and a set of independent relative measurements about these variables are obtained, denoted as $\zeta_{uv} = x_u - x_v + \epsilon_{uv} \in \mathbb{R}^d, u, v \in \{1, 2, \cdots, n\}$, where $\epsilon_{uv} \in \mathbb{R}^d$ is the random error vector with zero mean and a covariance matrix $P_{uv} = E[\epsilon_{uv}\epsilon_{uv}^T] \in \mathbb{R}^{d\times d}$, representing the measurement uncertainty. By stacking all the node variables into one vector $\mathbf{x} = [x_1^T, x_2^T, \cdots, x_n^T]^T \in \mathbb{R}^{nd}$, all the measurements into one vector $\mathbf{z} = [\zeta_1^T, \zeta_2^T, \cdots, \zeta_m^T]^T \in \mathbb{R}^{md}$ and all the measurement errors into a vector $\epsilon = [\epsilon_1^T, \epsilon_2^T, \cdots, \epsilon_m^T]^T \in \mathbb{R}^{md}$, the measurement equations can be expressed as

$$\mathbf{z} = \mathcal{A}^T \mathbf{x} + \epsilon \tag{1}$$

where $\mathcal{A} = \mathbf{A} \otimes \mathbf{I}_d$, $\mathbf{A}$ is the incidence matrix of the relative measurement graph G, $\mathbf{I}_d$ is $d \times d$ identity matrix, and $\otimes$ denotes the Kronecker product. The relative measurement graph $G = (V, E, F)$ is a directed graph, constructed by including all the nodes as its vertex set V, all the node pairs which have measurements as its directed edge set E (the direction is from u to v if the node pair has the measurement of ζ_{uv}), and a function $F : E \to (Z, P)$ which maps an edge $e_{uv} \in E$ to the corresponding measurement $\zeta_{uv} \in Z$ (Z is the set of all measurements) and the error covariance matrix $P_{uv} \in P$ (P is the set of all error covariance matrices). The incidence matrix $\mathbf{A}$ is a $n \times m$ matrix, where each row corresponds to each node, each column corresponds to each edge. The element a_{ue} of A is 1 if Edge e leaves from Node u, -1 if e is directed toward u, and 0 if Node u is not involved in Edge e. For example, the left figure in Figure 3 can be expressed as

$$\underbrace{\begin{bmatrix} \zeta_1 \\ \zeta_2 \\ \zeta_3 \\ \zeta_4 \\ \zeta_5 \end{bmatrix}}_{\mathbf{z}} = \underbrace{\begin{bmatrix} I & -I & 0 & 0 \\ I & -I & 0 & 0 \\ 0 & I & 0 & -I \\ 0 & I & -I & 0 \\ 0 & 0 & -I & I \end{bmatrix}}_{\mathcal{A}^T} \underbrace{\begin{bmatrix} x_1 \\ x_2 \\ x_3 \\ x_4 \end{bmatrix}}_{\mathbf{x}} + \underbrace{\begin{bmatrix} \epsilon_1 \\ \epsilon_2 \\ \epsilon_3 \\ \epsilon_4 \\ \epsilon_5 \end{bmatrix}}_{\epsilon}$$

By partitioning $\mathbf{x}$ into an anchor node variable vector $\mathbf{x}_r$, which includes all anchor nodes' constant positions, and a non-anchor node variable vector $\mathbf{x}_b$, which includes all non-anchor nodes' position variables, and partitioning $\mathcal{A}$ into an anchor node incidence matrix and a non-anchor node incidence matrix, Equation 1 can be rewritten as

$$\overline{\mathbf{z}} = \mathcal{A}_b^T \mathbf{x}_b + \epsilon \tag{2}$$

where $\overline{\mathbf{z}} = \mathbf{z} - \mathcal{A}_r^T \mathbf{x}_r$. For the example in Figure 3, since Nodes 1 and 4 are anchors, the equation can be written as

$$\underbrace{\begin{bmatrix} \zeta_1 - x_1 \\ \zeta_2 - x_1 \\ \zeta_3 + x_4 \\ \zeta_4 \\ \zeta_5 - x_4 \end{bmatrix}}_{\overline{\mathbf{z}}} = \underbrace{\begin{bmatrix} -I & 0 \\ -I & 0 \\ I & 0 \\ I & -I \\ 0 & -I \end{bmatrix}}_{\mathcal{A}_b^T} \underbrace{\begin{bmatrix} x_2 \\ x_3 \end{bmatrix}}_{\mathbf{x}_b} + \underbrace{\begin{bmatrix} \epsilon_1 \\ \epsilon_2 \\ \epsilon_3 \\ \epsilon_4 \\ \epsilon_5 \end{bmatrix}}_{\epsilon}$$

Obtaining an estimate of $\mathbf{x}_b$ becomes a classical estimation problem, which can be solved by BLUE. If G is a weakly connected graph (where there always exist undirected paths between any node pair), the best estimate for $\mathbf{x}_b$ in the linear combination space of all the measurements is uniquely determined by $\hat{\mathbf{x}}_b$, the solution of the following linear system

$$\mathcal{L}\hat{\mathbf{x}}_b = \mathbf{b} \tag{3}$$

where $\mathcal{L} = \mathcal{A}_b \mathcal{P}^{-1} \mathcal{A}_b^T$ (called *G's Kirchhoff Matrix*), $\mathbf{b} = \mathcal{A}_b \mathcal{P}^{-1} \overline{\mathbf{z}}$, and $\mathcal{P}$ is a block diagonal matrix consisting of all the covariance matrices P_i of $\epsilon_i, i = 1, 2, \cdots, m$. The covariance matrix of the estimation errors is given by

$$\mathbf{\Sigma} = \mathcal{L}^{-1} \tag{4}$$

Although the above centralized solution is elegant, it has several drawbacks. First, the time complexity is high. Assuming there are n nodes and m measurements, the anchor number is constant, and the dimension of the variable vector is d, the time complexity for calculating the covariance matrix $\mathbf{\Sigma}$ is $T(\mathbf{\Sigma}) = O(n^2md^3)$, since $T(\mathcal{L}) = T(\mathcal{P}^{-1}) + T(\mathcal{A}_b\mathcal{P}^{-1}) + T((\mathcal{A}_b\mathcal{P}^{-1})\mathcal{A}_b^T) = md^3 + nmd^3 + n^2md^3 = O(mn^2d^3)$, and $T(\mathcal{L}^{-1}) = n^3d^3$ ($n <= m$). The time complexity for calculating nodes' location estimates is $T(\hat{\mathbf{x}}_b) = O(n^2md^3)$, since $T(\hat{\mathbf{x}}_b) = T(\mathcal{L}^{-1}) + T(\mathbf{b}) + T(\mathcal{L}^{-1}\mathbf{b}) = (md^3 + nmd^3 + n^2md^3 + n^3d^3) + nmd^3 + n^2d^3 = O(n^2md^3)$. So the time complexity for calculating both $\hat{\mathbf{x}}$ and $\mathbf{\Sigma}$ simultaneously is $T(\hat{\mathbf{x}}_b, \mathbf{\Sigma}) = T(\hat{\mathbf{x}}_b) = md^3 + 2nmd^3 + 2n^2d^3 + n^3d^3 = O(mn^2d^3)$. The computational complexity depending on the measurement number m makes this solution not scale well, since m would be a very large number as the node number increases (the possible different edge number of a graph is of the order of $O(n^2)$, and the same edge may be measured multiple times in practice).

Second, this method follows a one-time calculation scheme, which means it calculates $\hat{\mathbf{x}}_b$ and $\mathbf{\Sigma}$ for a fixed RMG, if the RMG is changed, all the calculations must be re-executed. However, in practice, it is very likely that new measurements are continuously generated, and sensor nodes may leave or join the network dynamically. Therefore, if the network size is large, the high one-time calculation time complexity would prevent it from providing the latest estimation in real time.

4. THE INCREMENTAL NODE-VOLTAGE ANALYSIS METHOD

To overcome the drawbacks of high computational complexity and one-time calculation scheme of the above method, we propose a new calculation scheme, called the incremental node-voltage analysis (INOVA) method, which derives the same results as BLUE, but reduces the time complexity from $O(mn^2d^3)$ to $O(n^3d^3)$, and can incrementally calculate $\hat{\mathbf{x}}_b$ and $\mathbf{\Sigma}$ as new measurements are generated, or nodes join or leave the network dynamically. Although this

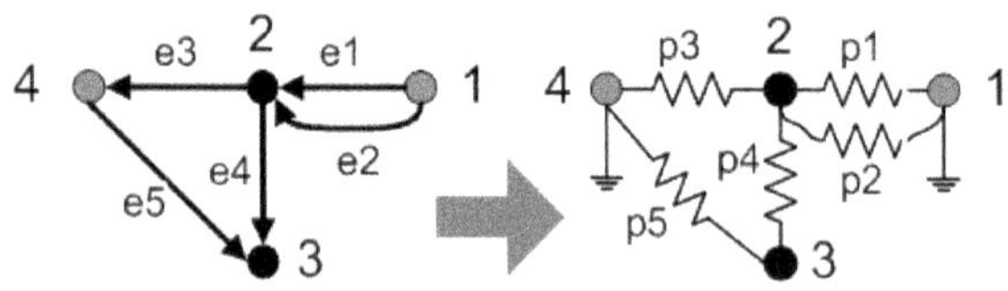

Figure 3: The Analogy Between RMG and The Generalized Electrical Network. The red nodes are anchor nodes.

method can be derived by analyzing the structure of matrices in Equation 3, we choose to deduce it based on the electrical network analogy for the following two reasons: 1) it can provide more insights of the decomposition operations; and 2) our subsequent distributed algorithm in Section 5 will also use this analogy.

It has been shown in [3] that a RMG $G = (V, E, F)$ is analogous to a generalized electrical network $\mathcal{G} = (V, \mathcal{E}, \mathcal{F})$, where V is the node set the same as in G, $\mathcal{E}$ is the edge set the same as E, but with no directions, and $\mathcal{F} : \mathcal{E} \rightarrow \mathcal{R}$ is an edge function that assigns each edge e_i a matrix valued resistance R_i which is numerically equal to the measurement error covariance matrix P_i in RMG. The generalized current J is defined as a $d \times d$ matrix associated with an edge with certain direction, and satisfies the Kirchhoff's Current Law, i.e. for each node in a generalized electrical network, the net generalized current flowing out or into that node is $\mathbf{0}$. The generalized voltage potential difference U_{uv} between two nodes u and v is defined as a $d \times d$ matrix $U_{uv} = R_{uv} \times J_{uv}$. Similarly, it satisfies Kirchhoff's Voltage Law, i.e. for any loop in a generalized electrical network, the sum of the generalized voltage potential difference in the clockwise or the counterclockwise direction is $\mathbf{0}$. Like the traditional electrical network, each node is associated with a voltage potential if there are currents or voltages imposed to the generalized electrical network. For the anchor nodes in $\mathcal{G}$, they are considered being connected to the ground, and thus their voltage potentials are always $\mathbf{0}$.

One important conclusion from [3] is that each node's estimation covariance matrix Σ_{ii} (the ith diagonal block element of network's covariance matrix $\mathbf{\Sigma}$) is numerically equal to the effective resistance R_i^{eff} between that node and the ground in $\mathcal{G}$. Hence, to calculate Σ_{ii}, we can impose an identity generalized current $\mathbf{I}$ to that node, then Node i's voltage potential U_{ii} is numerically equal to R_i^{eff}, i.e. $U_{ii} = R_i^{eff} \times \mathbf{I} = R_i^{eff} = \Sigma_{ii}$. We call the generalized electrical network imposed by a current $\mathbf{I}$ on Node i as *Node i's Current Graph, denoted as $\mathcal{G}^i$*. Therefore, to calculate Σ_{ii}, we can use node-voltage analysis method from electrical theory, i.e. set n_b voltage potential variables $\mathbf{U}_i = [U_{i1}^T, U_{i2}^T, \cdots, U_{in_b}^T]^T$ for every non-anchor node in $\mathcal{G}^i$, and then build the balance equations based on Kirchhoff's Current Law for each node.

$$\underset{dn_b\times dn_b}{\mathbf{L}}\;\underset{dn_b\times d}{\mathbf{U}_i} = \underset{dn_b\times d}{\mathbf{H}_i} \tag{5}$$

where $\mathbf{L} = [L_{uv}] \in \mathbb{R}^{dn_b \times dn_b}, u, v \in \{1, 2, \cdots, n_b\}$,

$$\underset{d\times d}{L_{uv}} = \begin{cases} \sum\limits_{\{u,q\}\in\mathcal{E}} P_{uq}^{-1} \text{ (or } P_{qu}^{-1}) & \text{if } u = v \\ -\sum P_{uv}^{-1} \text{(or } P_{vu}^{-1}) & \text{if } u \neq v \wedge \{u, v\} \in \mathcal{E} \\ \mathbf{0} & \text{if } \{u, v\} \notin \mathcal{E} \end{cases} \tag{6}$$

$$\underset{dn_b\times d}{\mathbf{H}_i} = [\underset{d\times d}{\mathbf{0}_1}, \underset{d\times d}{\mathbf{0}_2}, \cdots, \underset{d\times d}{\mathbf{I}_i}, \cdots, \underset{d\times d}{\mathbf{0}_{n_b}}]^T$$

The subscript of $\mathbf{H}_i$'s element denotes the index of that element. Actually, it is not hard to show that $\mathbf{L}$ in Equation 5 is equal to $\mathcal{L}$ in Equation 3. In addition, if we consider all nodes' $\Sigma_{ii}, i = 1, 2, \cdots, n_b$ together, we can get the following equations:

$$\underset{dn_b\times dn_b}{\mathcal{L}}\;\underset{dn_b\times dn_b}{\mathbf{U}} = \underset{dn_b\times dn_b}{\mathbf{H}} \tag{7}$$

where $\mathbf{U} = [\mathbf{U}_1, \mathbf{U}_2, \cdots, \mathbf{U}_{n_b}]$ and $\mathbf{H} = [\mathbf{H}_1, \mathbf{H}_2, \cdots, \mathbf{H}_{n_b}] = \mathbf{I}$ (size $dn_b \times dn_b$). Obviously, the solution is $\mathbf{U} = \mathcal{L}^{-1} = \mathbf{\Sigma}$. Another observation is that *the covariance matrix Σ_{ij} of Nodes i and j is numerically equal to the voltage potential U_{ij} of Node i in Node j's Current Graph $\mathcal{G}^j$, or the voltage potential U_{ji} of Node j in Node i's Current Graph $\mathcal{G}^i$ (since Σ is a symmetric matrix).*

Moreover, we can also directly build $\mathbf{b} = [b_i]$ in Equation 3 as follows.

$$b_i = \sum_{\substack{(u,v)\in E \\ u=i \vee v=i}} P_{uv}^{-1}[(-1)^{s_{uv}}\zeta_{uv} + c_u + c_v] \tag{8}$$

where $s_{uv} = 0$ if $u = i$, and $s_{uv} = 1$ if $v = i$; $c_u = x_u$ if Node u is an anchor node, otherwise $c_u = \mathbf{0}$ (size $d \times 1$); similarly, $c_v = x_v$ if Node v is an anchor node, otherwise $c_v = \mathbf{0}$.

We can directly build $\mathcal{L}$ and $\mathbf{b}$, and thus $\mathbf{\Sigma}$ and $\hat{\mathbf{x}}_b$ can be calculated. If the network is fixed, everything is fine. But what if the network changes? Do we need to redo all these calculations? There are two types of dynamic changes for a WSN: new measurements being generated and nodes quitting or joining the network. For the former problem, when a new measurement ζ_{uv} is generated, only two Nodes u and v are involved. So we only need to modify the balance equations for these two nodes if they are not anchor nodes, i.e. modifying four elements in $\mathcal{L}$ and two elements in $\mathbf{b}$ as follows (if any of them is an anchor, nothing needs to be done for that node).

OPERATION 1. *When a new measurement ζ_{uv} is generated, do $L_{uv}- = P_{uv}^{-1}, L_{vu}- = P_{uv}^{-1}, L_{uu}+ = P_{uv}^{-1}, L_{vv}+ = P_{uv}^{-1}, b_u+ = P_{uv}^{-1}\zeta_{uv} + c_v, b_v- = P_{uv}^{-1}\zeta_{uv} - c_u$*

For the event of a node leaving or joining the network, it only impacts that node and its neighbors. Especially for node joining, the new node should have measurements with the existing nodes in order to keep the graph weakly connected, otherwise, our system would not include this node. The corresponding operations are:

OPERATION 2. *If a node joins the network (with some measurements to some already existing nodes), a new column and a new row with all $\mathbf{0}$'s are appended to $\mathcal{L}$, and one element $\mathbf{0}$ is appended to $\mathbf{b}$. Then Operation 1 is taken for this node's new measurements.*

OPERATION 3. *If Node u leaves the network, for each of Node u's neighbors q, if the measurement is ζ_{qu}, then do $L_{qq}- = P_{uq}^{-1}$, $b_q+ = \zeta_{qu} - c_u$; if the measurement is ζ_{uq}, then do $L_{qq}- = P_{qu}^{-1}$ and $b_q- = \zeta_{uq} + c_u$. After that, delete Row u and Column u in $\mathcal{L}$, and the uth element of* $\mathbf{b}$.

In summary, the initial inputs of INOVA are the measurement set Z, the corresponding error covariance matrix set P and the anchor nodes' locations $\mathbf{x}_r$. The outputs are nodes' location estimates $\hat{\mathbf{x}}_b$ and the covariance matrix $\mathbf{\Sigma}$. After the initial localization, if there are some dynamic changes at

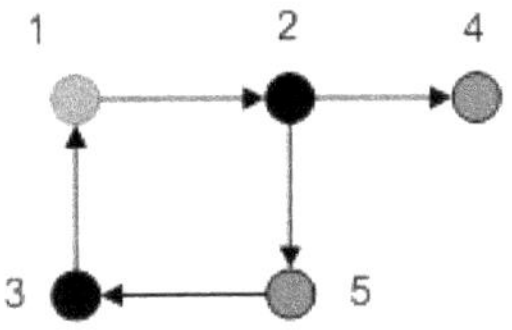

Figure 4: Node 1's Two-hop RMG

run time, the inputs are the set of new measurements, the set of nodes which leave, or the set of newly joining nodes with the corresponding measurements, and the outputs are the updated $\hat{\mathbf{x}}_b$ and $\mathbf{\Sigma}$. The whole algorithm is described as follows:

1. Calculate $P_i^{-1}, i = 1, 2, \cdots, m$. $(T(P_i^{-1}) = md^3)$.
2. Scan all measurements once, build up $\mathcal{L}$ and $\mathbf{b}$ by using Equations 6 and 8, respectively. $(T(\mathcal{L}) = 4md$, $T(\mathbf{b}) = 2md^2)$
3. Calculate $\mathbf{\Sigma} = \mathcal{L}^{-1}$. $(T(\mathbf{\Sigma}) = n_b^3 d^3)$.
4. Calculate $\hat{\mathbf{x}}_b = \mathcal{L}^{-1}\mathbf{b}$. $(T(\hat{\mathbf{x}}_b) = n_b^2 d^3)$.
5. If a new measurement is generated, do Operation 1 $(T(opt1) = O(d^3))$; if a new node joins the network with constant measurements, do Operation 2 $(T(opt2) = O(d^3))$; if a node having constant number of neighbors leaves the network, do Operation 3 $(T(opt3) = O(d^3))$. Then repeat Steps 3 and 4 $(T(\mathbf{\Sigma}) + T(\hat{\mathbf{x}}_b) = n_b^3 d^3)$.

The time complexity for INOVA to localize a fixed network is $T(INOVA) = O(n^3d^3)$, and the time complexity for one dynamic change is also $O(n^3d^3)$ (recall that n is the node number, m is the measurement number and d is the variable dimension). However, for the BLUE method, either for a fixed network or responding to a dynamic change, it all needs $O(mn^2d^3)$. The improvement of INOVA is significant for two reasons: first, in a large dense network, m is much larger than n, because the possible edge number is n^2, and multiple measurements may be taken for the same node pair. Second, if we look into the details of these two processes, we find the key difference is how to build $\mathcal{L}$. In BLUE, it always needs $O(mn^2d^3)$ time; while in INOVA, initially, it needs $O(n^3d^3)$ time, but after that, when the network dynamically changes, the time is constant. Therefore, INOVA has much better performance when the topology changes (see Section 8.1).

5. A DISTRIBUTED ALGORITHM

In some scenarios (e.g. in wild areas), the centralized architecture may not be available, and thus a distributed scheme is required. Prabir Barooah et al. proposed a distributed algorithm OSE to estimate each node's location [2]. In OSE, each node assumes its two-hop neighbor's estimations are correct, iteratively estimate its own location based on its two-hop RMG, and exchanges the estimates among neighbors. After a number of iterations, each node's estimate converges to the estimate calculated by BLUE. The essence of OSE is to use a method, called Asynchronous Filtered Weighted Additive Schwarz (AFWAS) [2], to solve linear Equation 3 for the whole network in a distributed way.

However, OSE can only estimate node positions but not the covariance matrix. Therefore, we propose a similar distributed scheme called OSE-COV to estimate the covariance matrix $\mathbf{\Sigma}$ of the whole network. The differences of OSE-COV are: first, OSE-COV uses INOVA for its local calculation instead of BLUE; second, OSE-COV is to solve n_b linear systems of Equation 5 (while OSE only solves one), and each node initially does not know how many linear systems there are.

In OSE-COV, the tasks of each node u are to discover the existence of all the other nodes, and estimate its voltage potential Σ_{uk} in Node k's current graph $\mathcal{G}^k$, $\forall k = 1, 2, \cdots, n$. Each node u maintains its two-hop RMG $G_u(2)$ (it could be also extended to more than 2 hops, but the communication cost would be large). In each iteration, Node u ($u = 1, 2, \cdots, n$) updates its known node list $S_u^{(i)}$ by merging its two-hop neighbors' $S_w^{(i)}$ ($w \in V_u(2)$), solves $\hat{\Sigma}_{uk}^{(i)}, \forall k \in S_u^{(i)}$ locally based on $G_u(2)$ by using node-voltage analysis and assuming their two-hop neighbors' $\hat{\Sigma}_{wk}^{(i-2)}$ are correct, and broadcasts $S_u^{(i)}$, $\hat{\Sigma}_u^{(i)}$ and its neighbor's estimates to its one-hop neighbors. Let $G_u(1) = (V_u(1), E_u(1), F_u(1))$ be Node u's one-hop RMG, where $V_u(1)$ consists of u and all its one-hop neighbors, $E_u(1)$ consists of all the edges (from the set E) between these nodes, and $F_u(1)$ is the corresponding mapping from edges to measurements; $G_u(2) = (V_u(2), E_u(2), F_u(2))$ is Node u's two-hop RMG; $A_u(1) = V_u(1) - \{u\}$ is the set of Node u's one-hop neighbors; $A_u(2) = V_u(2) - V_u(1)$ is the set of Node u's two-hop neighbors; $S_u^{(i)} = \{k |$Node u knows the existence of Node k at ith iteration$\}$ is Node u's known node list; $\hat{\Sigma}_u^{(i)}$ ($d \times d|S_u^{(i)}|$) denotes Nodes u's estimates for $\Sigma_{uk}, \forall k \in S_u^{(i)}$; $X[i]$ denotes the ith element of set X; and we define a node as a voltage anchor node in $\mathcal{G}^k$ if its voltage potential is known in Node k's current graph. The whole algorithm is described as follows.

1. Initially, by broadcasting twice, each node $u \in V$ gets its two-hop RMG $G_u(2)$. Node u assumes its neighbors $A_u(1) \cup A_u(2)$ know the existence of $V_u(2)$, thus sets their known node lists as $S_v^{(0)} = V_u(2)$ and $S_w^{(-1)} = V_u(2)$, where $v \in A_u(1)$ and $w \in A_u(2)$, and picks up arbitrary estimates $\hat{\Sigma}_v^{(0)}, v \in A_u(1)$ and $\hat{\Sigma}_w^{(-1)}, w \in A_u(2)$ for them.
2. At the ith iteration, if Node u is an anchor, it updates its known nodes list $S_u^{(i)} = S_{A_u(1)}^{(i-1)} \cup S_{A_u(2)}^{(i-2)}$, where $S_{A_u(1)}^{(i-1)} = \bigcup_{v \in A_u(1)} S_v^{(i-1)}$ and $S_{A_u(2)}^{(i-2)} = \bigcup_{w \in A_u(2)} S_w^{(i-2)}$, and updates its estimate $\hat{\Sigma}_u^{(i)} = \mathbf{0}$ ($d \times d|S_u^{(i)}|$). If Node u is not an anchor, it updates its known nodes list $S_u^{(i)} = S_{A_u(2)}^{(i-2)}$. Then Node u considers all its two-hop neighbors as voltage anchors (i.e. assumes their estimates $\hat{\Sigma}_w^{(i-2)}, \forall w \in A_u(2)$ are correct), and combines all their estimates $\Sigma_w, w \in A_u(2)$ into one matrix $Cov_u^{(i)} = [(Cov_u^{(i)})_{jk}]$ as follows:

$$(Cov_u^{(i)})_{jk} = \begin{cases} \Sigma_{(A_u(2)[j])(S_u^{(i)}[k])} & \text{if } S_u^{(i)}[k] \in S_{A_u(2)[j]}^{(i-2)} \\ \text{arbitrary value} & \text{otherwise} \end{cases} \quad (9)$$

Then, Node u builds up $|S_u^{(i)}|$ groups of current balance equations as Equation 5 for $\mathcal{G}^k, \forall k \in S_u^{(i)}$, which can be written together as

$$\mathcal{L}_u(2)\mathbf{U}_u^{(i)} = \mathbf{C}_u^{(i)} - \mathcal{A}_{ub}(2)\mathcal{P}_u(2)^{-1}\mathcal{A}_{ur}(2)^T Cov_u^{(i)} \quad (10)$$

where $\mathcal{L}_u(2)$ ($d|V_u(1)| \times d|V_u(1)|$), $\mathcal{A}_{ub}(2)$ ($d|V_u(1)| \times d|E_u(2)|$), $\mathcal{A}_{ur}(2)$ ($d|A_u(2)| \times d|E_u(2)|$), and $\mathcal{P}_u(2)$ ($d|E_u(2)| \times d|E_u(2)|$) are the Kirchhoff Matrix, non-anchor node incidence matrix, anchor node incidence matrix and the measurement error diagonal matrix of $G_u(2)$, respectively. $\mathbf{U}_u^{(i)}$ ($d|V_u(1)| \times d|S_u^{(i)}|$) is the variable matrix, where the block element of ith row and jth column represents the estimate of Node $V_u(1)[i]$'s voltage potential in Node $S_u^{(i)}[j]$'s Current Graph, i.e. $\hat{\Sigma}_{(V_u(1)[i])(S_u^{(i)}[j])}$. $\mathbf{C}_u^{(i)}$ ($d|V_u(1)| \times d|S_u^{(i)}|$) is defined as:

$$(\mathbf{C}_u^{(i)})_{jk} = \begin{cases} \mathbf{I} & \text{if } V_u(1)[j] = S_u^{(i)}[k] \\ \mathbf{0} & \text{otherwise} \end{cases} \quad (11)$$

After solving $\mathbf{U}_u^{(i)}$, Node u only keeps the row corresponding to itself, denoted as $\mathcal{Y}_u$ ($d \times d|S_u^{(i)}|$), and then updates its estimates as $\hat{\Sigma}_u^{(i)} = \lambda \mathcal{Y}_u + (1-\lambda)\hat{\Sigma}_u^{(i-1)}$ (if the size of $\hat{\Sigma}_u^{(i-1)}$ is less than $\mathcal{Y}_u$, append $\hat{\Sigma}_u^{(i-1)}$ with the missing elements in $\mathcal{Y}_u$), where $0 < \lambda \leq 1$ is a pre-defined parameter (we use $\lambda = 0.9$ in this paper). Then, the new estimates $\hat{\Sigma}_u^{(i)}$ plus the known node list $S_u^{(i)}$, and the estimates $\hat{\Sigma}_v^{(i-1)}$ plus their known node lists $S_v^{(i-1)}, v \in A_u(1)$, previously received from its one-hop neighbors are broadcasted to all its one-hop neighbors.

3. When each node u receives broadcasts from its one-hop neighbors, it updates corresponding $S_v^{(i)}$, $S_w^{(i-1)}$, $\hat{\Sigma}_v^{(i)}$ and $\hat{\Sigma}_w^{(i-1)}$, where $v \in A_u(1)$ and $w \in A_u(2)$. Node u waits for a timeout or a predefined number of messages, and then it begins the $(i+1)$th iteration.

The above distributed algorithm can be stopped after a timeout or a predefined number of iterations. Take Figure 4 for example. Figure 4 shows the two-hop RMG $G_1(2)$ of Node 1. Nodes 4 and 5 are Node 1's two-hop neighbors, and considered as the voltage anchors in $G_1(2)$. Assume at the ith iteration, Node 1 receives the known node lists of Nodes 2, 3, 4 and 5 as $S_2^{(i-1)}$, $S_3^{(i-1)}$, $S_4^{(i-2)}$ and $S_5^{(i-2)}$, respectively, and their corresponding covariance matrix estimates as $\hat{\Sigma}_2^{(i-1)}$, $\hat{\Sigma}_3^{(i-1)}$, $\hat{\Sigma}_4^{(i-2)}$ and $\hat{\Sigma}_5^{(i-2)}$, respectively. Assume $S_4^{(i-2)} = (1,2,3,4,5)$, $S_5^{(i-2)} = (1,2,3,4,5,6)$, $\hat{\mathbf{\Sigma}}_4^{(i-2)} = [\hat{\Sigma}_{41}^{(i-2)}, \hat{\Sigma}_{42}^{(i-2)}, \hat{\Sigma}_{43}^{(i-2)}, \hat{\Sigma}_{44}^{(i-2)}, \hat{\Sigma}_{45}^{(i-2)}]$ and $\hat{\mathbf{\Sigma}}_5^{(i-2)} = [\hat{\Sigma}_{51}^{(i-2)}, \hat{\Sigma}_{52}^{(i-2)}, \hat{\Sigma}_{53}^{(i-2)}, \hat{\Sigma}_{54}^{(i-2)}, \hat{\Sigma}_{55}^{(i-2)}, \hat{\Sigma}_{56}^{(i-2)}]$. This means by the $i-2$th iteration, Node 4 knows the existences of Nodes 1, 2, 3, 4 and 5, and estimates its voltage potentials in Current Graphs $\mathcal{G}^1$, $\mathcal{G}^2$, $\mathcal{G}^3$, $\mathcal{G}^4$ and $\mathcal{G}^5$ as $\hat{\Sigma}_{41}^{(i-2)}$, $\hat{\Sigma}_{42}^{(i-2)}$, $\hat{\Sigma}_{43}^{(i-2)}$, $\hat{\Sigma}_{44}^{(i-2)}$ and $\hat{\Sigma}_{45}^{(i-2)}$, respectively. It is similar for Node 5, although Node 5 knows one more node (Node 6) than Node 4. Then Node 1 updates its known node list as $S_1^{(i)} = (1,2,3,4,5,6)$, builds $\mathbf{C}_1^{(i)}$ and $Cov_1^{(i)}$ as:

$$\mathbf{C}_1^{(i)} = \begin{bmatrix} \mathbf{I} & \mathbf{0} & \mathbf{0} & \mathbf{0} & \mathbf{0} & \mathbf{0} \\ \mathbf{0} & \mathbf{I} & \mathbf{0} & \mathbf{0} & \mathbf{0} & \mathbf{0} \\ \mathbf{0} & \mathbf{0} & \mathbf{I} & \mathbf{0} & \mathbf{0} & \mathbf{0} \end{bmatrix}$$

$$Cov_1^{(i)} = \begin{bmatrix} \hat{\Sigma}_{41}^{(i-2)} & \hat{\Sigma}_{42}^{(i-2)} & \hat{\Sigma}_{43}^{(i-2)} & \hat{\Sigma}_{44}^{(i-2)} & \hat{\Sigma}_{45}^{(i-2)} & \mathbf{0} \\ \hat{\Sigma}_{51}^{(i-2)} & \hat{\Sigma}_{52}^{(i-2)} & \hat{\Sigma}_{53}^{(i-2)} & \hat{\Sigma}_{54}^{(i-2)} & \hat{\Sigma}_{55}^{(i-2)} & \hat{\Sigma}_{56}^{(i-2)} \end{bmatrix}$$

After that, Node 1 solves $\mathbf{U}_1^{(i)}$ for Nodes 1, 2 and 3 based on Equation 10, abandons all other nodes' solutions except for itself (the 1st row of $\mathbf{U}_1^{(i)}$, denoted as $\mathcal{Y}_1$), and updates $\hat{\Sigma}_1^{(i)} = \lambda \mathcal{Y}_1 + (1-\lambda)\hat{\Sigma}_1^{(i-1)}$. Then Node 1 broadcasts $S_1^{(i)}$, $S_2^{(i-1)}$, $S_3^{(i-1)}$, $\hat{\Sigma}_1^{(i)}$, $\hat{\Sigma}_2^{(i-1)}$ and $\hat{\Sigma}_3^{(i-1)}$ to all its one-hop neighbors.

Since OSE-COV does not require synchronization between nodes, it works under unreliable links. The convergence of OSE-COV can be proved by using the AWAS framework [9]. Another benefit of OSE-COV is that it has the same communication scheme as OSE. Hence at each iteration, all the exchange information of OSE and OSE-COV can be broadcasted together, and the location estimates and the covariance matrix estimates can be calculated simultaneously. To calculate the communication costs of OSE-COV, we assume 4 bytes are needed to represent a real number, 4 bytes for a node's address, and one node has v one-hop neighbors and w two-hop neighbors. Then for each iteration, Node u needs to broadcast $4(v+1)$ bytes for its own address and its neighbors' addresses, maximum $4(v+1)n$ bytes for the known node list of itself and its one-hop neighbors, maximum $4(v+1)nd^2$ bytes for itself and its one-hop neighbors covariance matrix estimates, $3v$ bytes for time stamps of these estimates. Therefore, the number of packets for one broadcast is

$$N_{tx}^{COV}(u) = \frac{(4nd^2 + 4n + 7)v + 4nd^2 + 4n + 4}{max\_payload} \quad (12)$$

If we combine OSE and OSE-COV together, we only need to add $4d(v+1)$ bytes for the location estimates of Node u and its one-hop neighbors.

6. CONFIDENCE REGION INFERENCE

As described in Section 1, some confidence-based applications require not only the position estimates, but also the corresponding confidence regions. Since a node's position estimate from BLUE is a linear combination of all measurements, based on the large number theorem from probability theory, it is likely that it follows a multivariate normal distribution. If this conclusion is correct, then we can deduce its confidence region at some confidence level by using the properties of the multivariate normal distribution.

One version of central limit theorem, named Lindeberg's condition [1], says that for a sequence of independent random variables $X_k, k = 1, 2, \cdots, n$, with $E(X_k) = \mu_k$ and $Var(X_k) = \sigma_k^2$, denoting $s_n^2 = \sum_{k=1}^{n} \sigma_k^2$, if $\max_{k=1,2,\cdots,n} \frac{\sigma_k^2}{s_n^2} \rightarrow 0$, as $n \rightarrow \infty$, i.e. none of σ_k^2 dominates s_n^2, then $Z_n = \sum_{k=1}^{n}(X_k - \mu_k)/s_n$ converges to a standard normal distribution $\mathcal{N}(0,1)$ when $n \rightarrow \infty$.

In BLUE, each node's position estimate is $\hat{x}_u = \sum_{k=1}^{m} W_k^u \zeta_k$, where W_k^u is the weight matrix of the measurement ζ_k Node u (or the generalized current of edge e_k in $\mathcal{G}^u$). Its corresponding covariance matrix is $\Sigma_{uu} = \sum_{k=1}^{m} W_k^u P_k (W_k^u)^T$. Let $y_k^u = W_k^u \zeta_k$, and then $\Sigma_{uu} = \sum_{k=1}^{m} COV(y_k^u)$. Define

$$Z^u = \Sigma_{uu}^{-\frac{1}{2}} \sum_{k=1}^{m} (y_k^u - E(y_k^u)) \quad (13)$$

If P_k is a diagonal matrix and $\forall k = 1, 2, \cdots, m, COV(y_k^u)$ does not dominate Σ_{uu}, then Z^u tends to be a standard multivariate normal random vector, i.e. $Z^u \backsim \mathcal{N}_d(\mathbf{0}, \mathbf{I})$, since each element $z_i^u, i = 1, 2, \cdots, d$ of Z^u tends to be a standard normal random variable under Lindeberg's condition.

In BLUE, the latter condition ($COV(y_k^u)$ does not dominate) is satisfied, because in practice usually the uncertainty of measurements generated by one device is similar. For the first condition, based on our simulation, even if P_k is not a diagonal matrix, Z^u still tends to be $\mathcal{N}_d(\mathbf{0},\mathbf{I})$. Consequently, we can derive that

$$\begin{aligned} Z^u &= \Sigma_{uu}^{-\frac{1}{2}}(\sum_{k=1}^{m}(y_k^u) - \sum_{k=1}^{m}E(y_k^u)) = \Sigma_{uu}^{-\frac{1}{2}}(\hat{x}_u - x_u) \\ \implies & \hat{x}_u \backsim \mathcal{N}_d(x_u, \Sigma_{uu}) \end{aligned} \quad (14)$$

This means each node estimate from BLUE follows the multivariate normal distribution which is centered at the real position with the covariance matrix Σ_{uu}. Then we can further infer (see [1]) that

$$(\hat{x}_u - x_u)^T\Sigma_{uu}^{-1}(\hat{x}_u - x_u) \backsim \chi_d^2(\text{with } d \text{ d.f.}) \quad (15)$$

Therefore, the solid ellipsoid $\{x_u : (\hat{x}_u - x_u)^T\Sigma_{uu}^{-1}(\hat{x}_u - x_u) \le \chi_d^2(\alpha)\}$ is the $100(1-\alpha)\%$ confidence region of x_u, where $\chi_d^2(\alpha)$ denotes the upper (100α)th percentile of the χ_d^2 distribution. For instance, after estimating node positions based on BLUE, Node u's estimate is $\hat{x}_u = [1,2]^T$ and the covariance matrix is $\Sigma_{uu} = \begin{bmatrix} 1.6 & 0.25 \\ 0.25 & 1.2 \end{bmatrix}$. Then its 90% confidence region for x_u is the solid ellipsoid $\{x_u : \left(\begin{bmatrix} 1 \\ 2 \end{bmatrix} - x_u\right)^T \begin{bmatrix} 1.6 & 0.25 \\ 0.25 & 1.2 \end{bmatrix}^{-1} \left(\begin{bmatrix} 1 \\ 2 \end{bmatrix} - x_u\right) \le 9.21\}$, since $\chi_2^2(0.01) = 9.21$.

7. OPTIMAL ANCHOR SELECTION

Anchor nodes are defined as the nodes which have accurate position information under a global coordinate frame. Any relative measurement based localization scheme must rely on one or more anchor nodes. Not only the anchor number, but also which nodes are selected to be anchors would impact the localization accuracy of the whole network. Taking Figure 2(b) for example, if Nodes 1 and 2 are already anchor nodes, obviously setting Node 4 as an anchor node is better than setting Node 9, since the node further away from the anchor nodes usually has large uncertainty, and thus more uncertainty is eliminated by setting Node 4 as an anchor. Additionally, selecting Node 3 is better than selecting Node 4, since it highly correlates to a large number of other nodes. Although the uncertainty of Node 3 is not as much as Node 4, the total uncertainty eliminated by setting Node 3 as an anchor is more (since the uncertainty of Nodes 5, 6, 7 and 8 is also reduced). Although this example provides some intuitive sense on how to select a good anchor, in large networks, anchor selection becomes much more complicated.

Without giving any quantitative uncertainty, the most straight forward strategy is to choose the node which has the max-min distance from all the anchor nodes, i.e. the node which has the maximum distance of the set of nodes' minimum distances to all the anchors is selected to be an anchor. This strategy assumes that the measurement uncertainty monotonically increases as the measured distance increases (e.g. the ToA method in [14]). For example, in Figure 2(b), assume the measured distances of node pairs are $Z = \{\zeta_{13} = 4.1, \zeta_{23} = 4, \zeta_{14} = 5.1, \zeta_{24} = 5, \zeta_{19} = 1, \zeta_{35} = 1, \zeta_{36} = 1, \zeta_{37} = 1, \zeta_{38} = 1\}$. Then the set of nodes' minimum distances to all anchors is $D_{min} = \{D_1 = 0, D_2 = 0, D_3 = 4, D_4 = 5, D_5 = D_6 = D_7 = D_8 = D_9 = 1\}$. Hence Node 4 should be selected.

When the quantitative uncertainty—the covariance matrix of all estimates—is given, the metric for selecting the best anchor node becomes selecting the node as an anchor so that the resultant trace of the covariance matrix is the smallest, that is

$$K = \underset{i=1,2,\cdots,n}{\arg\min}\, trace(\mathbf{\Sigma}_i^{New}) \quad (16)$$

Where $\mathbf{\Sigma}_i^{New}$ denotes the new covariance matrix after selecting Node i as an anchor node. The trace of the covariance matrix represents the sum of mean square errors of all the node estimates. Hence the above metric satisfies the minimum mean square error(MMSE) requirement. To reduce $trace(\mathbf{\Sigma})$ as much as possible, one naive method is to select the node which has the maximum variation, i.e. selecting Node $k = \underset{i=1,2,\cdots,n}{\arg\max}\, trace(\Sigma_{ii})$. By using this method, the $trace(\mathbf{\Sigma})$ is reduced by at least as much as $trace(\Sigma_{kk})$. Although this method is very simple, it does not take the correlation between nodes into account. Actually, some node which is highly correlated to many other nodes should have the priority to be the anchor node, as Node 3 in Figure 2(b). Following we give the optimal anchor selection strategy which satisfies Equation 16 and its proof. To select multiple anchors, we can just repeat this strategy.

THEOREM 1. *If only one node can be set as an anchor,* $trace(\mathbf{\Sigma}^{New})$ *is minimized only when Node* $K = \underset{i=1,2,\cdots,n}{\arg\max}\, trace\left(\sum_{i=1}^{n_b}(\Sigma_{ik}\Sigma_{kk}^{-1}\Sigma_{ki})\right)$ *is selected.*

Proof. To set Node k as an anchor is equivalent to get a new measurement between Node k and the global origin, and the measurement error covariance matrix $P_k = \mathbf{0}$. Assume $\mathbf{\Sigma}(t)$ and $\mathbf{\Sigma}(t+1)$ are the covariance matrices before and after selecting Node k as an anchor node, respectively. Then, $\mathbf{\Sigma}(t+1) = \mathcal{L}^{-1}(t+1) = (\mathcal{A}_b(t+1)\mathcal{P}^{-1}(t+1)\mathcal{A}_b^T(t+1))^{-1}$, where $\mathcal{A}_b(t+1) = [\ \mathcal{A}_b(t) \quad H_k\ ]$, $H_k = [\mathbf{0}_1, \mathbf{0}_2, \cdots, \mathbf{I}_k, \cdots, \mathbf{0}_{n_b}]^T$ and $\mathcal{P}(t+1) = \begin{bmatrix} \mathcal{P}(t) & \mathbf{0} \\ \mathbf{0} & P_k \end{bmatrix}$. Note P_k should be equal to $\mathbf{0}$, however, we first assume $P_k \to \mathbf{0}$, but $P_k \ne \mathbf{0}$, and P_k^{-1} exist. Then

$$\begin{aligned} &\mathcal{L}(t+1) = [\ \mathcal{A}_b(t) \quad H_k\ ] \begin{bmatrix} \mathcal{P}(t) & \mathbf{0} \\ \mathbf{0} & P_k \end{bmatrix}^{-1} \begin{bmatrix} \mathcal{A}_b^T(t) \\ H_k^T \end{bmatrix} \\ &= \mathcal{A}_b(t)\mathcal{P}^{-1}(t)\mathcal{A}_b^T(t) + H_kP_k^{-1}H_k^T = \mathbf{\Sigma}^{-1}(t) + H_kP_k^{-1}H_k^T \end{aligned}$$

By using the formula

$$(\mathbf{A} + \mathbf{B}\mathbf{D}^{-1}\mathbf{C})^{-1} = \mathbf{A}^{-1} - \mathbf{A}^{-1}\mathbf{B}(\mathbf{D} + \mathbf{C}\mathbf{A}^{-1}\mathbf{B})^{-1}\mathbf{C}\mathbf{A}^{-1}$$

we obtain

$$\begin{aligned} \mathbf{\Sigma}(t+1) &= \mathcal{L}^{-1}(t+1) \\ &= \mathbf{\Sigma}(t) - \mathbf{\Sigma}(t)H_k(P_k + H_k^T\mathbf{\Sigma}(t)H_k)^{-1}H_k^T\mathbf{\Sigma}(t) \\ &\overset{P_k=\mathbf{0}}{=} \mathbf{\Sigma}(t) - \mathbf{\Sigma}(t)H_k(H_k^T\mathbf{\Sigma}(t)H_k)^{-1}H_k^T\mathbf{\Sigma}(t) \\ &= \mathbf{\Sigma}(t) - \mathbf{\Delta}_k \end{aligned}$$

Therefore, $trace(\mathbf{\Sigma}(t+1)) = trace(\mathbf{\Sigma}(t)) - trace(\mathbf{\Delta}_k)$. To minimize $trace(\mathbf{\Sigma}(t+1))$ is equivalent to maximize $trace(\mathbf{\Delta}_k) = trace\left(\sum_{i=1}^{n_b}(\Sigma_{ik}\Sigma_{kk}^{-1}\Sigma_{ki})\right)$. **Q.E.D.**

To calculate $trace(\mathbf{\Sigma}(t+1))$ for all k's, the time complexity is $O(d^3n^2)$, and to select the maximum $trace(\mathbf{\Sigma}(t+1))$,

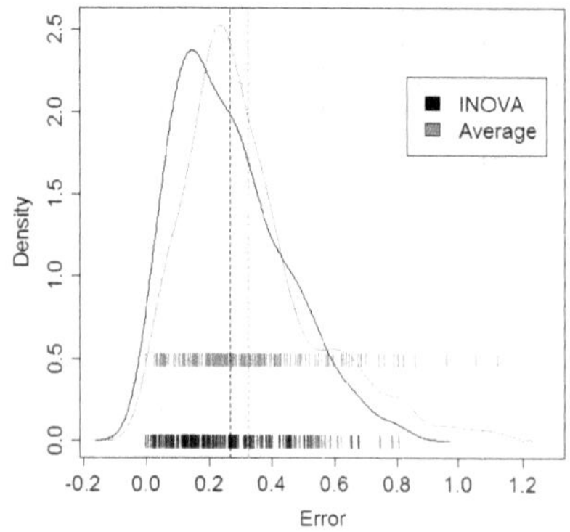

Figure 5: Error Density of INOVA and Average-Based Methods

it takes $O(n)$ time. Therefore, the total time complexity is $O(d^3n^2)$. One benefit of this method is that it is compatible with the distributed algorithm in Section 5. Since each node maintains one row of the covariance matrix, i.e. $\Sigma_{ki}, i = 1, 2, \cdots, n$, and $\Sigma_{ki} = \Sigma_{ik}$. Hence nodes can calculate $trace(\mathbf{\Delta}_k)$ locally, and then the whole network selects the node with maximum $trace(\mathbf{\Delta}_k)$ as the anchor node.

8. EVALUATION

In this section, we evaluate the localization accuracy and time complexity of INOVA, the convergence of the distributed algorithm–OSE-COV, the correctness of the confidence region inference, and the performance of the optimal anchor selection strategy. All the evaluations are based on Matlab simulation on a PC with dual Intel(R) Core(TM)2 CPU 6600 @2.40GHz, 2G memory and the Windows XP SP3 operating system.

As the default configuration in Figure 1, 200 nodes are randomly deployed in a square area with 30 by 30 grids (each grid is a 1×1 square), and 1000 relative measurements (the blue lines) are randomly generated among them. The relative measurement ζ_{ij} between Nodes i and j is determined by an angel measurement θ_{ij} and a distance measurement r_{ij}, where θ_{ij} and r_{ij} are random variables, and $\theta_{ij} = \theta^*_{ij} + \epsilon^{\theta}_{ij}$ and $r_{ij} = r^*_{ij} + \epsilon^r_{ij}$. θ^*_{ij} and r^*_{ij} are the true angle and distance between Nodes i and j. $\epsilon^{\theta}_{ij} \backsim \mathcal{N}(0, \sigma^2_{\theta})$ and $\epsilon^r_{ij} \backsim \mathcal{N}(0, \sigma^2_r)$, where $\sigma_{\theta} = 10°$ and $\sigma_r = 0.15r^*_{ij}$. Therefore, the measurement between a node pair i and j is $\zeta_{ij} = [r_{ij}\cos\theta_{ij}, r_{ij}\sin\theta_{ij}]^T$, and the error covariance matrix is

$$P_{ij} = \begin{bmatrix} y^2\sigma^2_{\theta_{ij}} + \sigma^2_r\cos^2\theta^*_{ij} & -xy\sigma^2_{\theta_{ij}} + \sigma^2_r\sin(2\theta^*_{ij})/2 \\ -xy\sigma^2_{\theta_{ij}} + \sigma^2_r\sin(2\theta^*_{ij})/2 & x^2\sigma^2_{\theta_{ij}} + \sigma^2_r\sin^2\theta^*_{ij} \end{bmatrix}$$

where $x = r^*_{ij}\cos\theta^*_{ij}$ and $y = r^*_{ij}\sin\theta^*_{ij}$. The parameters of node number, measurement number and the area size can be changed during the simulation process.

8.1 Localization Accuracy and Efficiency

We first compare the localization accuracy of INOVA with the traditional average-based methods, and then evaluate the execution time of INOVA and BLUE for calculating both position estimates and the covariance matrix. As described in Section 1, INOVA and BLUE should have better localization accuracy than the traditional average-based localization schemes, such as triangulation or multilateration. This is because INOVA better fuses the measurements by quantitatively differentiating the measurement uncertainty. Figure 5 shows the error density of INOVA versus average-based methods (AVG). The error is defined as the Euclidean distance of the estimated location and the real location. From the plot, we can see that the peak of INOVA error is at the left side of AVG, and the average error of AVG is 0.325, while that of INOVA is 0.267, which is 18% better.

Figure 6 shows the execution time for INOVA and BLUE as new measurements are incrementally generated. The result shows that initially, to localize a network with 500 nodes and 2000 measurements, BLUE takes 13.5 seconds, while INOVA only needs 8.8 seconds. After that, 200 measurements are incrementally generated in each iteration. The execution time for BLUE increases very quickly, while INOVA almost keeps a constant time. When there are 4000 measurements in the network, the computational time of INOVA is only 0.5 second, which is 70 times as fast as BLUE. Similarly, in Figure 7, which fixes the measurement number at 2000, but incrementally adds 50 nodes each time, the execution time of BLUE shows a large rate of rise, while INOVA only increases slightly. When there are 1100 nodes in the network, the computational time of INOVA is 5 seconds, which is 12 times as fast as BLUE.

This experiment shows that in a large network where measurements are continuously generated or nodes dynamically join or leave the network, INOVA can provide real-time estimates, while BLUE fails. Based on this result, we can even apply INOVA to a mobile network. When a node moves, this node is considered to leave the network, and a new node joins the new position with new measurements. Since INOVA has very low execution time, it can respond to these mobile behaviors very quickly.

8.2 Convergence Speed of OSE-COV

In this section, we evaluate the convergence speed for the OSE-COV distributed algorithm. The network in Figure 1 is used, which has 200 nodes and 1000 measurements, and Node 1 at the position $[0, 0]^T$ is set to be the anchor node. Figure 8 shows the convergence degree during 100 iterations. The y-axis is the difference ratio, which is defined as $\sum_{i=1}^{dn_b}\sum_{j=1}^{dn_b}(\sigma_{ij} - \hat{\sigma}_{ij})^2 / \sum_{i=1}^{dn_b}\sum_{j=1}^{dn_b}\sigma^2_{ij}$, where σ_{ij} denote the element of the ith row and jth column of $\mathbf{\Sigma}$ obtained by BLUE or INOVA, while $\hat{\sigma}_{ij}$ denotes the corresponding element estimated by OSE-COV. The result shows that OSE-COV converges as the iteration number increases. Actually, to further improve the convergence speed, we can initialize each node's estimate with a better value, i.e. nodes start to execute OSE-COV only when its neighbors have the estimates.

8.3 Confidence Region Inference Accuracy

Based on the analysis in Section 6, each node's estimate $\hat{x}_u$ is a multivariate normal random variable, and $(\hat{x}_u - x_u)^T\Sigma^{-1}_{uu}(\hat{x}_u - x_u) \backsim \chi^2_d$(with d d.f.). To test the correctness, we first examine the normality of the estimates. Figure 9 shows the chi-square plot for all node's estimates in Figure 1. Chi-square plot shows the relationship between the statistic distance $(\hat{x}_u - x_u)^T\Sigma^{-1}_{uu}(\hat{x}_u - x_u)$ and the quantiles of chi-square distribution. The closer the result is to the line $y = x$, the more likely the sample x_i's are from a multivariate normal distribution. The plot in Figure 9 is very closed to $y = x$, and thus our inference is correct.

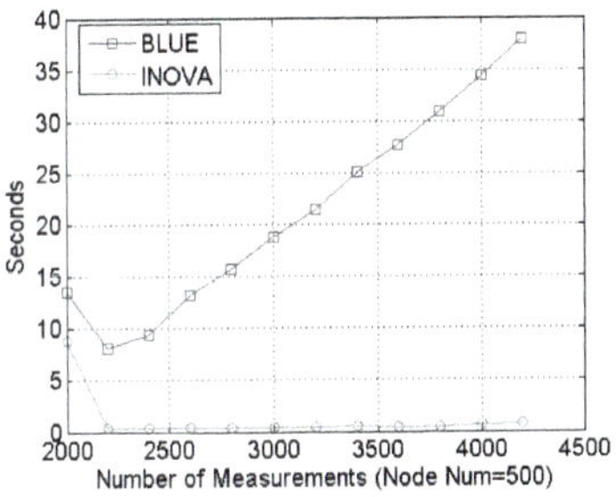

Figure 6: Execution Time for Incremental Measurements

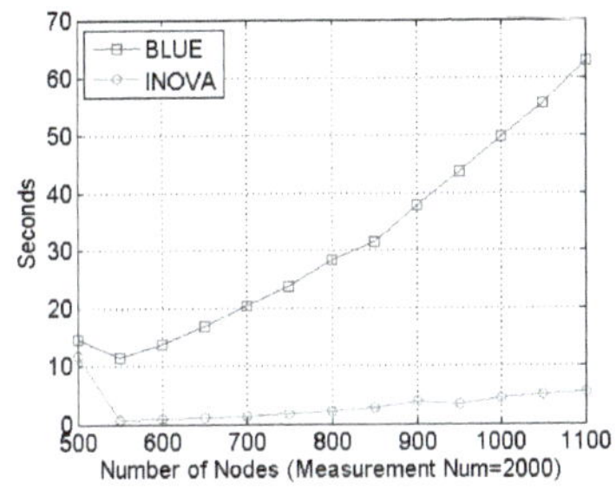

Figure 7: Execution Time for Incremental New Nodes

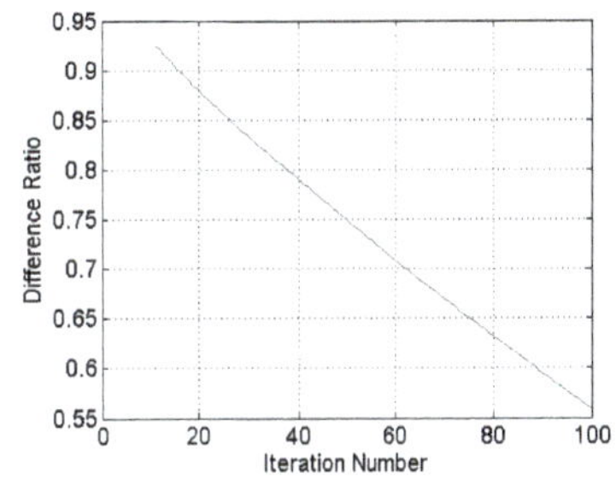

Figure 8: Convergence Speed for COV-OSE Distributed Algorithm

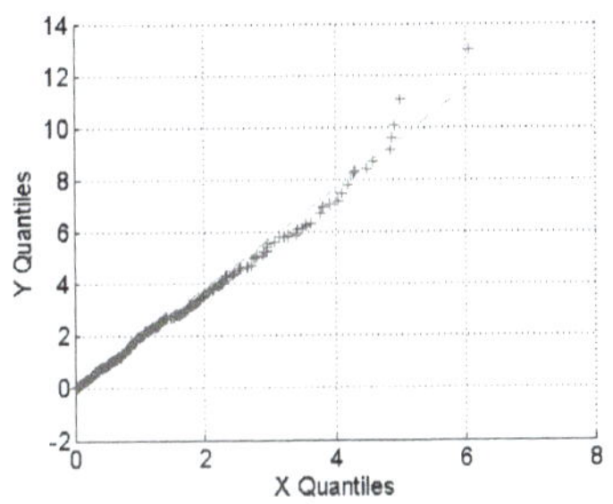

Figure 9: Chi-square Plot for All Node's Estimates

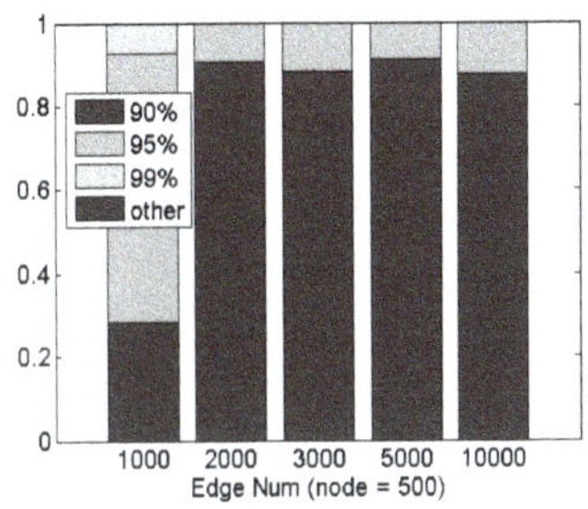

Figure 10: Confidence Region Accuracy

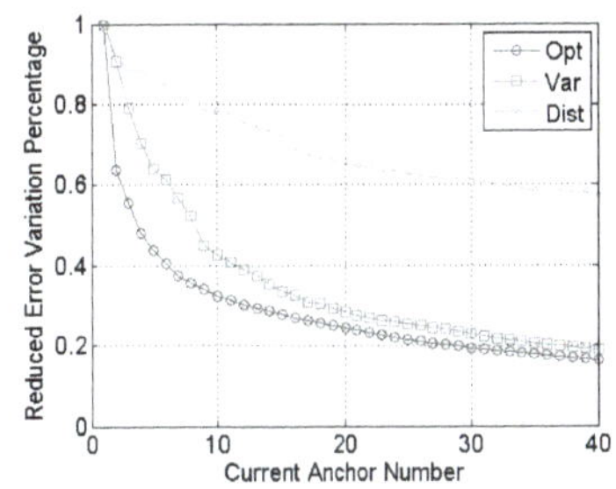

Figure 11: Anchor Selection Strategies Comparison

To show the correctness of our second inference (about the confidence region), we use three confidence levels–90%, 95% and 99%, and the corresponding chi-square quantiles are $\chi_2^2(0.1) = 4.61, \chi_2^2(0.05) = 5.99, \chi_2^2(0.01) = 9.21$. Based on the theorem in Section 6, 90% of node's real positions should fall in the ellipse $(\hat{x}_u - x_u)^T \Sigma_{uu}^{-1} (\hat{x}_u - x_u) \leq 4.61$, 95% of nodes' real positions should be within the ellipse $(\hat{x}_u - x_u)^T \Sigma_{uu}^{-1} (\hat{x}_u - x_u) \leq 5.99$, and 99% of them should be in $(\hat{x}_u - x_u)^T \Sigma_{uu}^{-1} (\hat{x}_u - x_u) \leq 9.21$. Figure 10 shows the statistical results for 5 different networks, which all have 500 nodes, but with 1000, 2000, 3000, 5000 and 10000 measurements, respectively. Except the first network (the ratio of measurement number to node number is low), all others have the consistent results with our inference. We also studied other networks of different sizes. It seems when the ratio of measurement number to node number is bigger than 2, the results are quite consistent and stable. Therefore, in practice, if the network is dense, we can use the covariance matrix to infer the confidence region accurately. Note that all the measurements are not from a multivariate normal distribution (as described at the beginning of Section 8). Therefore, there is no requirement on the population where the measurements are from.

8.4 Anchor Selection Strategies Comparison

In this section, we evaluate the performance of the three anchor selection strategies described in Section 7—the distance-based strategy (Dist), the variation-based strategy (Var) and the optimal strategy (Opt). The experiment shows how much percentage of total error variation is reduced as the anchor nodes are set one after another by using one specific strategy. In Figure 11, the x-axis is the current anchor number, and the y-axis is the percentage of the current total error variation to that at the beginning. The results show that Dist performs worst, while Opt is the best. We also observe that the first few selections reduce the total error variation a lot, where the reduced error percentage of Opt is four times as much as Var. But as more and more anchors are selected, the reduction rate decreases. This is because the nodes with largest uncertainty are selected at the beginning, leaving the nodes with little uncertainty. Another observation is that as more and more nodes are set to be anchors, method Var trends to have the similar performance as Opt. This is because when multiple anchors are selected, the anchors Var selected are very likely to be similar as Opt, although the selection order may differ.

9. RELATED WORK

In the wireless sensor networks field, localization approaches can be divided into two classes: range-based and range-free solutions. Range-based approaches use special devices, such as ultrasound [23, 25] or radio[6, 15], and various techniques, such as Time of Arrival (ToA) [15, 17], Time Difference of Arrival (TDoA) [6, 5, 23, 25], Angle of Arrival (AOA) [4, 19], or Phase Difference of Arrival (PDoA) [18] to estimate the distances or bearing among nodes, and then apply a triangulation, lateration or multilateration algorithm for final location estimations.

On the other hand, range-free localization solutions do not directly measure the distances or angles among nodes, instead, they make use of proximity information or events to infer node positions. When the density of sensor nodes is high, a number of solutions [20, 22] make the assumption of the shortest path (hop counted) between two nodes approximating the Euclidean distance of them, and thus can localize nodes with a few anchors. Other solutions make use of events to localize nodes, since these events are associated with the coordinate information [26, 27], distance information [24], or node sequence information [30]. Other works [11, 31] use relative RSSI difference to iteratively shrink node possible positions and finally localize the nodes.

However, no matter which technique is used, these methods suffer from several drawbacks: first, they do not differentiate the uncertainty of measurements, or at least do not well quantify the uncertainty [30, 29, 10, 28]. Therefore, their deduction is not theoretically optimal. Second, they do not provide the confidence of the estimates, so their estimates cannot be used for confidence-based applications. Third, there is no way for them to guide the anchor selection.

Prabir Barooah et al. present a quantitative uncertainty-based localization scheme, the Relative Measurement Graph model, which takes the quantitative uncertainty of measurements into account, and provides the covariance matrix for the estimates [2, 3]. However, their work is not adaptive to the incremental measurements generation or dynamic node number changes. Hence, their approach cannot provide real time estimation for these dynamic behaviors. Additionally, they only focus on the location estimation, but not the covariance matrix. So they do not provide any distributed algorithm for covariance matrix calculation, and do not make use of covariance matrix for confidence region inference and anchor selection.

In robotics field, there is a quantitative uncertainty-based localization solution called SLAM [7, 8], which can optimally (or approximate optimally) localize both a robot and landmarks simultaneously, make use of incremental measurements, and provide confidence of the estimates. SLAM uses the state space model, and periodically makes predictions and updates on the robot's location estimate as well as the landmark location estimates as the robot moves on. Different from SLAM, our work does not assume a state space model and also does not require a mobile agent. Additionally, we provide a distributed way for estimating both node locations and the corresponding covariance matrix. Therefore, our work is more suitable for general WSN scheme.

10. CONCLUSION

This paper emphasizes the importance of the quantitative uncertainty in WSN localization. The quantitative uncertainty is modeled as the error covariance matrix of the estimates, and can be used to infer the confidence region for location estimates and assist anchor selection so as to reduce the total mean square estimation error. To calculate the covariance matrix, we present a centralized algorithm, called INOVA, which has low time complexity and is adaptive to the dynamic changes of a network. In addition, a distributed algorithm—OSE-COV is presented for peer-to-peer architectures.

11. REFERENCES

[1] R. B. Ash and C. A. Doléans-Dade. *Probability and Measure Theory (2nd ed.)*. San Diego: Harcourt/Academic Press, 2000.

[2] P. Barooah and J. Hespanha. Estimation on graphs from relative measurements. *Control Systems Magazine, IEEE*, 27(4):57 -74, 2007.

[3] P. Barooah and J. Hespanha. Estimation from relative measurements: Electrical analogy and large graphs. *Signal Processing, IEEE Transactions on*, 56(6):2181 -2193, 2008.

[4] H.-l. Chang, J.-b. Tian, T.-T. Lai, H.-H. Chu, and P. Huang. Spinning beacons for precise indoor localization. In *SenSys*, 2008.

[5] X. Cheng, H. Shu, Q. Liang, and D. H.-C. Du. Silent positioning in underwater acoustic sensor networks. *IEEE Transactions on Vehicular Technology*, 2008.

[6] X. Cheng, A. Thaeler, G. Xue, and D. Chen. Tps: A time-based positioning scheme for outdoor wireless sensor networks. In *INFOCOM*, 2004.

[7] H. Durrant-Whyte and T. Bailey. Simultaneous localisation and mapping (slam): Part i the essential algorithms. *IEEE Robotics And Automation Magazine*, 2006.

[8] H. Durrant-Whyte and T. Bailey. Simultaneous localisation and mapping (slam): Part ii - state of the art. *IEEE Robotics And Automation Magazine*, 2006.

[9] A. Frommer, H. Schwandt, Daniel, and D. B. Szyld. Asynchronous weighted additive schwarz methods. *Electronic Transactions on Numerical Analysis*, 5:48–61, 1997.

[10] D. K. Goldenberg, P. Bihler, M. Cao, J. Fang, B. D. O. Anderson, A. S. Morse, and Y. R. Yang. Localization in sparse networks using sweeps. In *MobiCom*, 2006.

[11] T. He, C. Huang, B. M. Blum, J. A. Stankovic, and T. Abdelzaher. Range-free localization schemes for large scale sensor networks. In *MobiCom*, 2003.

[12] R. K. Jeremy, R. M. Karp, J. Elson, C. H. Papadimitriou, and S. Shenker. Global synchronization in sensornets. In *LATIN*, pages 609–624, 2004.

[13] S. Kay. *Fundamentals of Statistical Signal Processing: Estimation Theory.* Prentice-Hall, 1993.

[14] Y. Kwon, K. Mechitov, S. Sundresh, W. Kim, and G. Agha. Resilient localization for sensor networks in outdoor environments. In *ICDCS*, 2005.

[15] S. Lanzisera, D. T. Lin, and K. S. J. Pister. Rf time of flight ranging for wireless sensor network localization. In *WISES*, 2006.

[16] H. Liu, J. Li, Z. Xie, S. Lin, K. Whitehouse, J. Stankovic, and D. Siu. Automatic and robust breadcrumb system deployment for indoor firefighter applications. In *MobiSys*, 2010.

[17] J. Liu, Y. Zhang, and F. Zhao. Robust distributed node localization with error management. In *MobiHoc*, 2006.

[18] M. Maróti, P. Völgyesi, S. Dóra, B. Kusý, A. Nádas, A. Lédeczi, G. Balogh, and K. Molnár. Radio interferometric geolocation. In *SenSys*, 2005.

[19] D. Niculescu and B. Nath. Ad hoc positioning system (aps) using aoa. In *INFOCOM*, 2003.

[20] D. Niculescu and B. Nath. Dv based positioning in ad hoc networks. *Journal of Telecommunication System*, 2003.

[21] G. Oberholzer, P. Sommer, and R. Wattenhofer. Spiderbat: Augmenting wireless sensor networks with distance and angle information. In *IPSN*, 2011.

[22] L. I. On, R. Nagpal, H. Shrobe, and J. Bachrach. Organizing a global coordinate system from local information on an ad hoc sensor network. In *IPSN*, 2003.

[23] N. B. Priyantha, A. Chakraborty, and H. Balakrishnan. The cricket location-support system. In *MobiCom*, 2000.

[24] K. Römer. The lighthouse location system for smart dust. In *MobiSys*, 2003.

[25] A. Savvides, C.-C. Han, and M. B. Srivastava. Dynamic fine-grained localization in ad-hoc networks of sensors. In *MobiCom*, 2001.

[26] R. Stoleru, T. He, J. A. Stankovic, and D. Luebke. A high-accuracy, low-cost localization system for wireless sensor networks. In *SenSys*, 2005.

[27] R. Stoleru, P. Vicaire, T. He, and J. A. Stankovic. Stardust: A flexible architecture for passive localization in wireless sensor networks. In *SenSys*, 2006.

[28] W. Xi, Y. He, Y. Liu, J. Zhao, L. Mo, Z. Yang, J. Wang, and X. Li. Locating sensors in the wild: pursuit of ranging quality. In *SenSys '10*, pages 295–308, New York, NY, USA, 2010. ACM.

[29] Z. Yang and Y. Liu. Quality of trilateration: Confidence based iterative localization. In *ICDCS*, 2008.

[30] Z. Zhong and T. He. Msp: multi-sequence positioning of wireless sensor nodes. In *SenSys*, 2007.

[31] Z. Zhong and T. He. Achieving range-free localization beyond connectivity. In *SenSys*, 2009.

Performance of mac80211 Rate Control Mechanisms

Wei Yin [1 2 3], Peizhao Hu [2], Jadwiga Indulska [1 2], and Konstanty Bialkowski [1]
[1] The University of Queensland, School of ITEE, Australia
[2] National ICT Australia (NICTA)
[3] National University of Defence Technology, China
{Firstname.Lastname}@nicta.com.au

ABSTRACT

The 802.11 wireless networking standard has several transmission rates which can be adaptively selected by the MAC layer to cater for various channel conditions. Such dynamic adaptations can improve per-hop performance in wireless networks by enhancing the data throughput, reducing the channel usage time and decreasing the number of failed transmissions. This in turn has an impact on the quality of service (QoS) provided for communicating applications. The new mac80211 framework, which is now part of the Linux operating system, defines functionalities and interfaces that can be used by wireless network drivers to access physical-layer information which has impact on high-level protocols/algorithms. In this paper we present a comprehensive evaluation of the per-hop performance of two rate control mechanisms used by the mac80211 framework in Linux: Minstrel and PID. The evaluation results show that Minstrel outperforms PID. We discuss the PID's performance problems and propose an enhancement to PID to address them. The evaluation of the enhancement confirms substantial improvement of PID's performance in a range of conducted and over-the-air communication scenarios.

Categories and Subject Descriptors

D.2.8 [**Software Engineering**]: Metrics—*performance measures*

General Terms

Performance

1. INTRODUCTION

Wireless channels are subject to various types of interference which introduces variability to the wireless channel quality — adaptive selection of transmission rates according to the dynamic channel quality can improve per-hop performance by utilising the variety of transmission rates available in the IEEE 802.11 standards. A number of algorithms have been proposed to adaptively select transmission rates according to various metrics. Based on the type of metrics used, rate control algorithms can be classified into three groups: ACK (Acknowledgement), SNR (Signal to Noise Ratio) and BER (Bit Error Rate) based mechanisms. ACK based mechanisms adapt rates according to the results of successful transmissions (e.g., packet loss ratio and average transmission time), which are estimated based on received acknowledgements. For example, RRAA [1], AMRR [2], SampleRate [3], Onoe [4], Minstrel [5] and PID [6] are ACK based. SNR based mechanisms measure the SNR to select transmission rates. These mechanisms include RBAR [7], CHARM [8] and FARA [9]. BER based mechanisms use a more fine-grained metric, which uses the average number of bit errors in each frame. An example of such a mechanism is SoftRate [10]. It is difficult to accurately measure the SNR metric and there are limitations in calculating the BER using conventional network cards therefore the ACK based mechanisms are commonly used in the current wireless networks.

In earlier Linux kernels, four ACK based mechanisms, including Onoe, AMRR, SampleRate and Minstrel, were implemented on the Madwifi driver. In [11], we reported on evaluation of these rate control mechanisms and showed that Minstrel has the best performance and it is also not worse than the SNR or BER based mechanisms. The recently defined mac80211 framework introduced functionalities and interfaces that can be used by drivers to access physical-layer information, which was previously provided by a proprietary implementation of HAL (Hardware Abstraction Layer) in Madwifi driver. As a result the functionality of rate control is now part of the framework rather than being dependent on individual drivers. The mac80211 framework is used by a number of new wireless network card drivers, including ath5k — a replacement of the MadWifi driver.

In this paper we present a comprehensive evaluation of the per-hop performance of two rate control mechanisms used by the mac80211 framework in Linux: Minstrel and PID, and propose an enhancement to PID that improves its performance significantly. In addition, as packet/frame loss ratio is used as a metric in many rate adaptation algorithms we analyse the pros and cons of using this metric. Although we have evaluated Minstrel in Madwifi previously, this paper presents the first comprehensive evaluation of Minstrel on the mac80211 framework.

The evaluation is carried out in a range of conducted and over-the-air experiment scenarios. We first measure the performance of these two rate control mechanisms in static channel conditions and follow it with a performance investigation for various dynamic channel conditions caused by

MSWiM'11, October 31–November 4, 2011, Miami, Florida, USA.

interference and mobility (repeating pattern, fading channel, or progressive increase/decrease of channel link quality). For the interference scenario, we analyse throughput while varying three parameters: the duration, the interval and the strength of the interference. For the fading channel, a well-known Rayleigh model is used to generate the channel conditions, whereas for the progressive increase/decrease of link quality we model the channel quality changes of two communication stations moving towards or away from each other. The responsiveness of rate control mechanisms to channel changes is also studied. Furthermore, there are over-the-air experiments that confirm our observations in the conducted testbed.

The remainder of the paper is organized as follows. In section 2 we briefly describe the two rate control mechanisms. The performance comparison of Minstrel and PID carried out in a conducted testbed for repeatable experiments is described in section 3, followed by the validation of the comparison results by over-the-air experiments in section 4. A detailed discussion of PID shortcomings, a description of a proposed PID enhancement and performance evaluation of the enhanced PID (PIDE) are presented in section 5. The lesson learned about using packet/frame loss ratio as a metric for rate control mechanisms is presented in section 6. The paper concludes in section 7.

2. RATE CONTROL ON MAC80211

In this section we briefly describe the rate control mechanisms used by the mac80211 framework.

2.1 Minstrel

The core of the Minstrel [5] rate control algorithm is a retry chain, which proposes four candidate rates (i.e., $r0$, $r1$, $r2$ and $r3$) to attempt in case re-transmission is necessary. Each of these candidate rates has a corresponding counter (i.e., $c0$, $c1$, $c2$, and $c3$), which specifies the maximum number of retry a candidate rate is allowed before giving up. For example, the algorithm will first attempt to use rate $r0$ for the transmission; if the transmission fails $c0$ times, then rate $r1$ is attempted for the re-transmission; in the case when rate $r1$ also fails $c1$ times, then rate $r2$ and $r3$ are attempted in the same manner until it reaches the base rate. Minstrel uses this retry chain to define the retry preferences (shown in Table 1).

Table 1: Retry preferences

Attempt	Lookaround rate		Normal
	$RR < BTR$	$RR > BTR$	rate
$r0$	BTR	RR	BTR
$r1$	RR	BTR	NBTR
$r2$	BPR	BPR	BPR
$r3$	BR	BR	BR

As an attempt to determine the optimal rate for a given channel condition, Minstrel dedicates 10% of its data traffic to probe the performance statistics of other rates by randomly selecting a rate (as *Lookaround rate*) that are not currently in use. For this 10% data traffic, the retry preferences are the best throughput rate (BTR), the random rate (RR), the best probability rate (BPR) and the base rate (BR) if the randomly selected rate (RR) is lower than the current best throughput rate (BTR); otherwise they are the random rate, the best throughput rate, the best probability rate, and the base rate. For other 90% of data traffic (considered as *Normal rate*), the retry preferences are the best throughput rate, next best throughput rate (NBTR), the best probability rate and the base rate.

To obtain the statistics for the retry chain, for every rate adaptation period of 100 ms Minstrel calculates the measured throughput and the probability of success in transmissions for each rate. The calculation of the probability of success is based on an Exponential Weighted Moving Average (EWMA), which controls the balance of influence of both the old and new packet delivery statistics, as shown in Eqn (1). The throughput of each rate is then computed based on the weighted probability of success P_{new} and the maximum number of packets that can be sent.

$$P_{new} = (1 - \alpha) * P_{this\_interval} + \alpha * P_{previous} \quad (1)$$

- P_{new} is the weighted probability of success for this interval, which will be used by the rate selection process.
- α is a smoothing factor (or the *scaling value*) in the EWMA mechanism.
- $P_{this\_interval}$ is the probability of success of this interval before the rate selection, and it is calculated as the ratio of the number of packets sent successfully to the number of packets sent.
- $P_{previous}$ is the weighted probability of success for the last interval used to select last transmission rate.

In Eqn (1), the smoothing factor α can be manually configured at Minstrel's instantiation time, and defaults to 75%. The α value controls the influence of current and historical measurements on the next rate selection. A value of 0% means using only the latest measurements and a value of 99% means using the old results, with a tiny influence from the new measurements.

2.2 PID

The PID [6] rate control algorithm conducts rate control based on a proportional integral derivative (PID) controller. In essence, the controller is a control loop feedback mechanism that tries to minimise the difference (i.e., *error* in control system's term) of the current and target frame loss ratio (i.e., $FLR_{current}$ and FLR_{target} respectively) as a result of switching to a new transmission rate. By default, the FLR_{target} is set to 14% for all rates.

To determine the appropriate transmission rate, the controller computes an adjustment value, adj, as follow.

$$adj = \gamma * (1 + sharpening) * (e_{current} - e_{last}) + \alpha * e_{current} + \beta * e_{avg} \quad (2)$$

where $e_{current}$ is the current error, and its value is calculated as $FLR_{target} - FLR_{current}$. e_{avg} is the average of recent errors, while e_{last} is the last error. In addition, there are four tuneable parameters. *sharpening* is a smoothing factor (non-zero when fast response is needed), whereas α, β and γ are the corresponding *proportional*, *integral* and *derivative* coefficients.

Using Eqn. 2 PID computes the adjustment value, adj, at the end of each rate adaptation period and decides on whether to switch to a new transmission rate. When adj is positive, the new rate R_{new} is set to the highest rate, in the range of $R_{current} \leq R_{new} \leq (R_{current} + adj)$, and its error (i.e., the difference between the target and respective frame loss ratio) is no more than the error of the current rate, $R_{current}$. When adj is negative, the new rate R_{new} is set to the lowest rate, in the range of $(R_{current} + adj) \leq R_{new} \leq R_{current}$, and its error is no more than the error of the current rate. No rate adaptation is required, if adj is equal to zero.

3. CONDUCTED EXPERIMENTS

To fairly compare the performance of Minstrel and PID, a fully controlled testbed is used and the results are also validated using over-the-air experiments (discuss in the next section). In this section, we describe the controllable evaluation platform and the results of performance evaluation carried out in this platform for five selected scenarios.

3.1 Controllable Platform

Fig. 1 shows the experiment setup. As shown we connect a traffic source to a traffic sink using co-axial cables and vary the link quality/channel conditions using a variable attenuator (Vaunix LabBrick LDA-602). The wireless signals are transmitted along the co-axial cables [12], rather than being sent over the air by antennas. Therefore, the testbed can produce repeatable evaluation experiments, i.e., the rate control mechanisms are compared for the same parameters (e.g., path loss) and offered load. During the experiments, the attenuation value varies according to the specifications defined in each evaluation scenario.

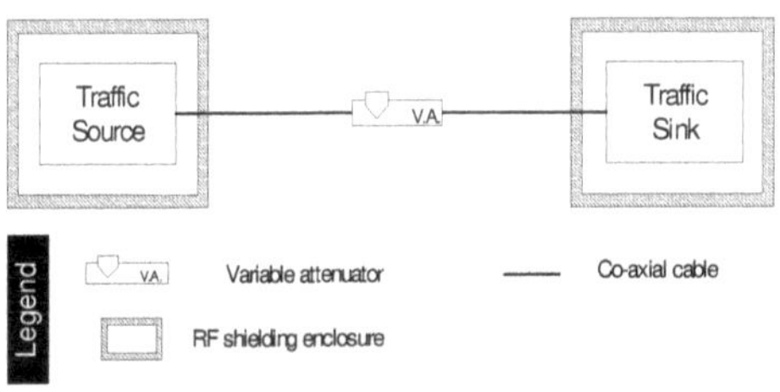

Figure 1: Experiment setup.

Each traffic source and traffic sink runs on a single board computer. Each computer is equipped with a Wistron CM9 Atheros wireless card. The computers run the Linux operating system (with kernel version 2.6.35 and the corresponding mac80211 framework). The debugfs system for Minstrel and PID are modified to allow access to necessary information. In addition to the programmable attenuator, a 25 dB fixed attenuator is attached to each traffic node, and each node's transmission power is set to 16 dBm. This prevents damage of the radios and ensures that within the operation range of the variable attenuator (0–63 dB), a full range of link qualities is possible (i.e., minimal attenuation corresponds to full throughput and maximum attenuation corresponds to a disconnected link).

To ensure consistent results, experiment nodes are enclosed within RF shielding boxes (JRE 4400), which ensures 85 dB of isolation to external sources. Further to this, all experiments are performed using IEEE 802.11a. This has two benefits, the first being that the 5 GHz frequency band is currently much less used than 2.4 GHz, and secondly it means that all transmission rates use the same family of modulation and coding methods. After configuration, each setup is verified manually using various tools.

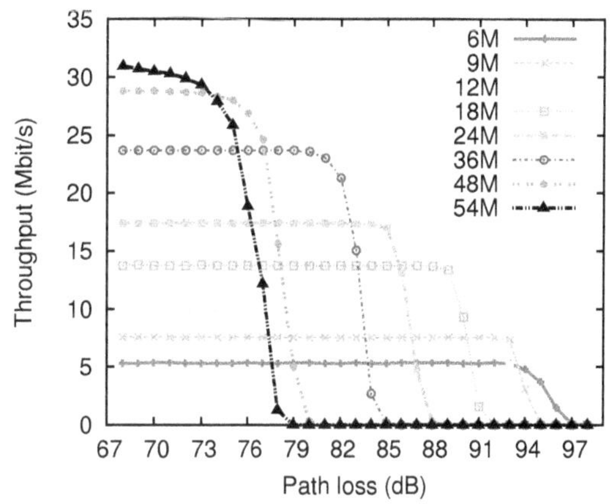

Figure 2: Fixed rates performance.

All transmitted data packets are captured at the receiver using a packet sniffing tool (tcpdump). Each measurement contains information regarding the achievable throughput and the number of successful packets at each rate. A parser library (Banjax) is used to analyse the measurement files. In the experiments, we use UDP traffic; the UDP packet size is set to the iperf default size. Experiments using TCP packets are not discussed in this paper, as congestion control in TCP makes it difficult to determine whether the poor performance is due to the congestion control mechanism or rate control at the MAC layer.

3.2 Evaluation Scenarios and Results

In this section we discuss the results of our extensive evaluation of the Minstrel and PID performance. The evaluation was carried out in five representative communication scenarios, such as static channel, dynamic channel, fading channel, progressive increase/decrease of channel quality and a sudden channel quality change.

3.2.1 Static channel conditions

Under static channel conditions, the rate control mechanisms should converge at an optimal rate and minimise hopping between different rates. To evaluate this scenario, we set a constant effective path loss between the two traffic nodes for a duration and observe the throughput and rate selection in relation to the given effective path loss. During the experiment the transmission power of the traffic source is set to 16 dBm, while iperf offered load is set to saturated.

To compare the rate control mechanisms to the optimum transmission rate, all fixed transmission rates are also evaluated in the same channel conditions as shown in Fig. 2. This experiment is performed to investigate the highest throughput rate (optimum rate) for each path loss, e.g., the optimum rate for 70 dB is 54 Mbit/s, because it achieves the highest throughput.

Fig. 3 shows the throughput achieved by the two rate control mechanisms as a function of the path loss. Higher path loss means lower received signal strength. In general, Minstrel shows better throughput compared with PID, except for very good or bad channel conditions. The number of successful packet arrivals at the receiver for each rate is plotted in Fig. 4. It shows that Minstrel almost selects the highest throughput rate and achieves throughput similar to the optimum rate for each scenario, except 79 dB. For

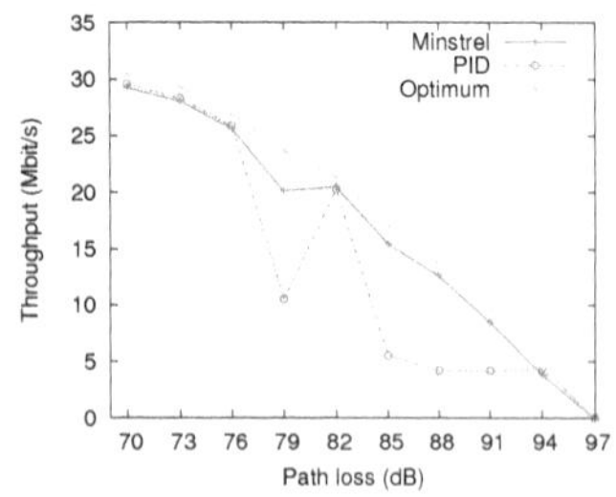

Figure 3: Auto rate performance.

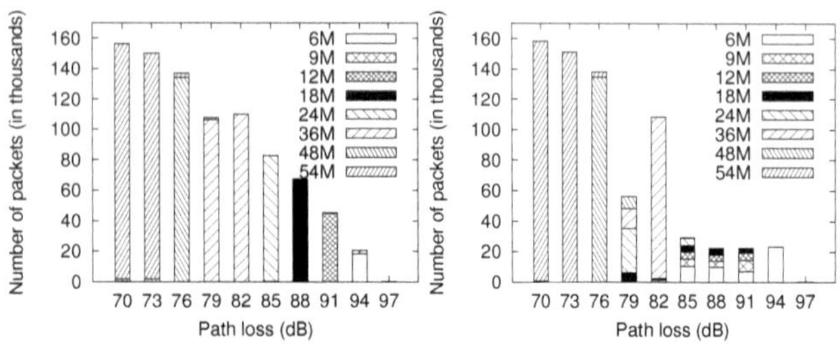

(a) Minstrel (b) PID

Figure 4: # of packets succeeded at each rate.

this path loss, a small number of packets are transmitted at 48 Mbit/s, which may fail and increase the retransmission count.

As shown in Fig.3, PID performs well in extreme channel conditions in which either the highest rate of 54 Mbit/s or the base rate achieves the best throughput, which matches the optimum rate. However, in other channel conditions (e.g., 79, 85, 88 and 91 dB) PID performs poorly. This is because PID fails to select a rate that can achieve the highest throughput as shown in Fig. 4(b). To further investigate this problem, we show in Fig. 5 the rate selection history of PID for the static channel conditions scenario with path loss value at 85 dB. For clarity we show only the first 30 seconds of a 60 seconds experiment of the repeating pattern. From this repeating pattern, we can see that PID tends to select a higher rate than the channel is capable to support; this results in higher frame loss ratio (FLR) and PID then falls back to the base rate and gradually increases the rate again as shown. By analysing Figs. 2 and 5, we discover that this instability (repeating pattern in rate selection) is the result of the FLR of a selected rate falls below the target FLR defined in PID; thus PID tends to increase rate. The instability effect occurs whenever a rate reaches the path loss regions (e.g., between 82-84dB for 36 Mbit/s), as shown in Fig. 2, where the rate's throughput experiences sharp fall. In these path loss regions, it is likely that the FLR is above the defined target FLR in PID, which causes PID to drop rate. This is a problem in rate control mechanisms if they do not stop switching to a higher rate, even if the current channel conditions cannot support it.

3.2.2 Impact of Interference

In this section, we study the impact of interference on rate control mechanisms. In wireless networks, one of the significant effects of interference is the drop of channel quality (i.e., decrease in the signal to noise ratio). In our controlled platform, we emulate the presence of interference by increasing in the effective path loss, which in effect also decreases the signal to noise ratio.

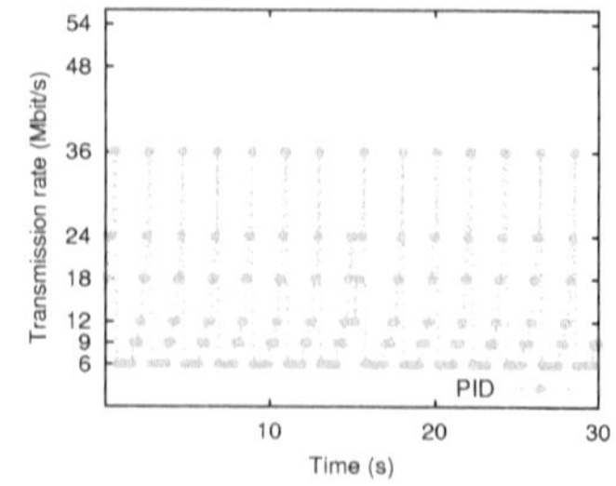

Figure 5: Repeating pattern in rate selection in PID.

To this end, we define an evaluation scenario in which link quality changes periodically in a repeating pattern (i.e., to emulate the presence of interference), and we measure how well a rate control algorithm performs when the duration, interval, and strength of interference vary.

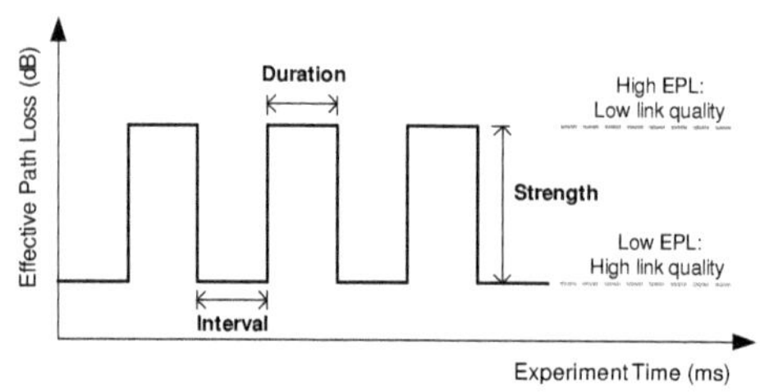

Figure 6: Emulation model for channel changes.

As shown in Fig. 6, the changes in link quality are emulated by adjusting the effective path loss (EPL); that is, low EPL means high/good link quality in which data transmission at 54 Mbit/s is possible, while high EPL means low/bad link quality allowing data transmission at rates lower than 54 Mbit/s. By varying the interval, duration and strength of the effective path loss, we can emulate various changes of link quality.

Fig. 7 shows the results when varying different interference parameters. In the figures, we also include the optimum rate at each interference configurations that is calculated using Fig. 2. Details of each experiment are discussed below.

Duration

In this experiment, we vary the duration exponentially from 1 to 1000 ms. The path loss during interference is set to 85 dB (for which the highest throughput rate is 24 Mbit/s, as shown in Fig. 2). Between the two interference there is a clear channel in which a 54 Mbit/s traffic is supported. The interval between interference is set to 100 ms.

As shown in Fig. 7(a), the throughput drops when the interference duration increases; that is, when interference occurs only lower rates are supported, therefore the longer the duration the lower the throughput can be achieved. One observation is that Minstrel always outperforms PID and achieves almost the maximum throughput when duration is less than 100 ms. We argue the poor performance of PID is due to the instability of rate selection.

Interval

This is similar to the previous experiment, but we vary the interval between interference from 1 to 1000 ms and fix the

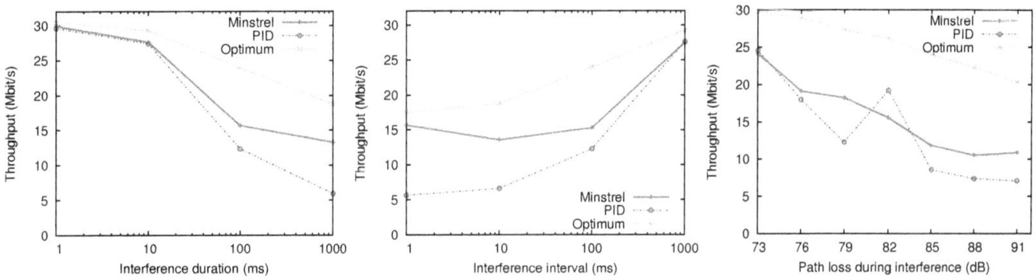

(a) Duration (b) Interval (c) Strength

Figure 7: Auto rates performance for various interference parameters.

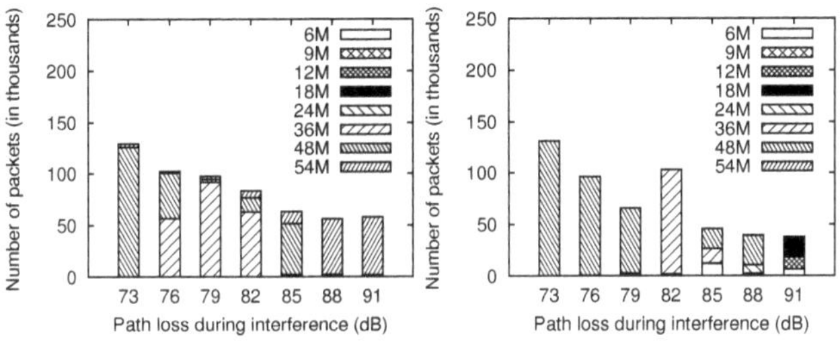

(a) Minstrel (b) PID

Figure 8: # of packets succeeded at each rate.

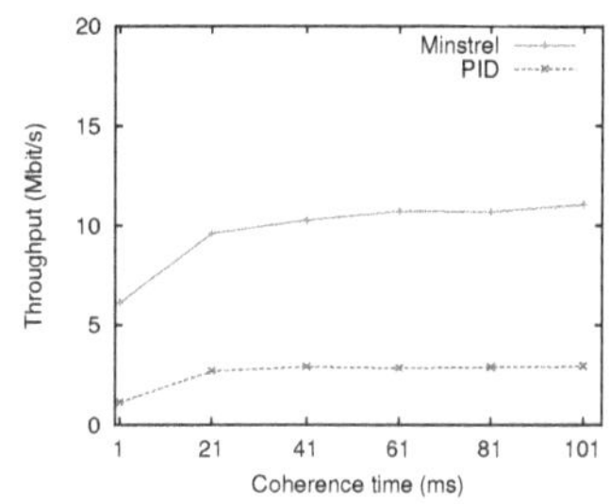

Figure 9: Performance in Rayleigh model.

interference duration to 100 ms. The path loss during interference is set to 85 dB, while there is clear channel between interference. Fig. 7(b) shows the results for this experiment. The throughput increases when we increase the interval between interference. In this experiment, Minstrel again constantly outperforms PID. Similar to the previous scenario, instability of rate selection in PID could be the cause.

Strength

In this experiment we vary the path loss from 73 to 91 dB to emulate different strength of interference. A clear channel (capable of supporting 54 Mbit/s) is used in between the interference. The channel condition varies between two path losses (very good channel and lower quality channel) and each lasts for 15 ms, for every 15 ms (i.e., the typical minimum coherent time in residential urban area with vehicles passing by [13]).

From Fig. 7(c) we can see that for Minstrel the throughput decreases as the emulated interference becomes stronger. In general PID follows similar behaviour, except for one occasion when path loss is at 82 dB (correlate to Fig. 3). Fig. 8(b) shows that PID selects 36 Mbit/s for all its transmissions at path loss of 82 dB. The reason being Minstrel and PID adapt rates based on the performance of each rate over the previous rate adaptation periods. Within each rate adaptation period, half of the time when there is interference, using optimal rate will gain full throughput, while using any higher rate will only achieve half of the throughput. In this case, rate selection is depending on whether full throughput of the optimal rate is better than half of the throughput of higher rates. From Fig. 2 we see that at 82 dB, 36 Mbit/s with full throughput is clearly better.

3.2.3 Fading Channel

Rayleigh fading [14] is one of best-known models which capture the effect of signal propagation in wireless environments. It represents the statistical time-varying nature of the received signal envelope of a flat fading signal when there are objects in the environment that scatter the signal before it reaches the receiver. The probability density function of the received signal power is described:

$$p(\gamma) = \frac{\gamma}{\sigma}e^{-\frac{\gamma^2}{2\sigma^2}}(0 \leq \gamma \leq \infty) \qquad (3)$$

where $\frac{\gamma^2}{2}$ is the instantaneous power and σ^2 is the average power of the received signal. The Rayleigh distribution is controlled by σ, the root mean square value of the received signal. We can generate various sets of Rayleigh distribution random numbers with different σ by using the inverse transform function with the uniform distribution.

By using this Rayleigh fading model, we investigate how a rate control algorithm behaves under channel fading in a fully controlled environment. Using Eqn. 3, we generate 60000 samples of attenuation value which are to be played back on our programmable attenuator. The average path loss of the model is set to 85 dB. To study how the two mechanisms cope under different mobility pattern, we vary the coherence time, which refers to the time period during which the channel remains constant. In [13], Camp and Knightly discuss that typical coherence time in residential urban area caused by passing vehicles is roughly between 15 and 100 ms. As our attenuators can get up to 1 ms granularity we perform experiments with coherence time varying from 1 to 101 ms to include cases of faster fading patterns. Fig. 9 shows the throughput achieved by both mechanisms in this scenario. Minstrel performs much better than PID.

3.2.4 Channel quality progressive increase/decrease

Node mobility is common in wireless networks. As a wireless node moves toward/away from another node, there are changes in link quality. To compare how the rate control

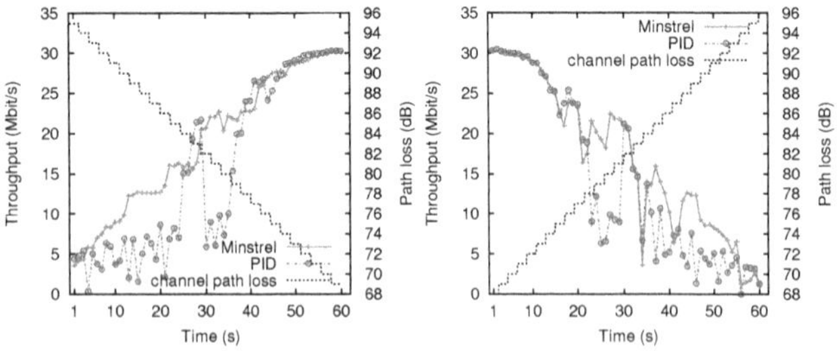

(a) Quality linearly increase (b) Quality linearly decrease

Figure 10: Throughput when channel quality changes linearly.

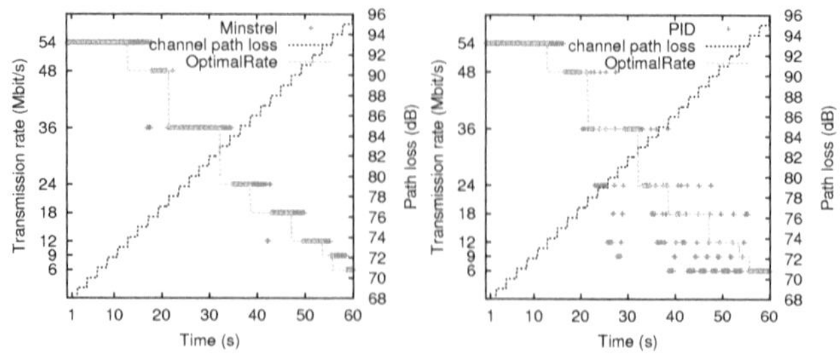

(a) Minstrel rate selection (b) PID rate selection

Figure 11: History of rate selection when channel quality linearly decrease.

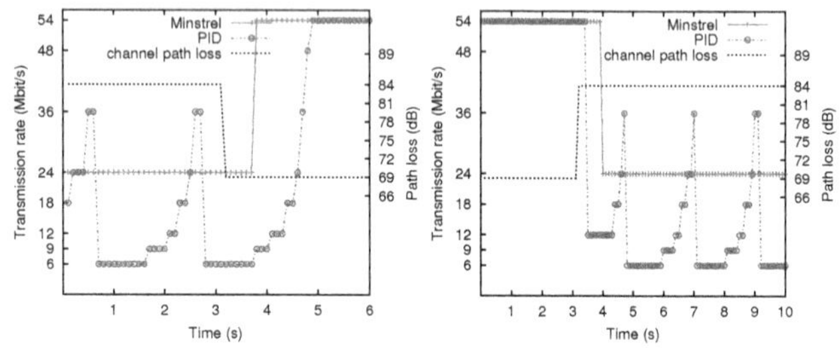

(a) Decrease attenuation (b) Increase attenuation

Figure 12: Rate selection when channel quality sudden changes.

mechanisms perform under these scenarios, we designed a scenario in which channel quality progressively increase or decrease.

In our conducted platform, we emulate these two scenarios by gradually decreasing/increasing the attenuation (corresponding to the increase/decrease of channel quality) between two nodes. In the experiment we vary the attenuation between 68 and 95 dB over a period of 60 s. During the experiment, we record, in every second, the achieved throughput and the history of rate selection.

Fig. 10 shows the results of these two experiments. There are several important observations: (i) in both cases Minstrel outperforms PID whenever rate adaptation is possible and necessary; the instability in rate selection of PID being the reason for such poor performance; (ii) Minstrel performs relatively well in the case when channel quality is gradually increased, but it experiences problems when channel quality falls. There are seven signs of sharp throughput drops, as shown in Fig. 10(b). After a detailed analysis, we discover that the path loss values of these throughput drops correlate exactly with the path loss regions where the throughput of a fixed rate drops rapidly, as shown in Fig. 2. We call these regions the *rate transition regions*. As an example the drop between the time 32 to 34 s, the path loss value increases from 82 to 84 dB. At these path loss, there should be a transition in rate selection, from 36 to 24 Mbit/s. However, for Minstrel a great portion of the outgoing packets are still sent at 36 Mbit/s, rather than using the optimal rate, as shown in the history of rate selection when channel quality decreases (in Fig. 11(a)); some of them will fail and retry using lower rates, hence results in poor throughput. One reason for why Minstrel still selects 36 Mbit/s after the channel quality changed is due to the way Minstrel calculates its metric. As discussed in Section 2, the smoothing factor α (in Eqn. 1) defaults to 75% to increase the weight on history measurements in its calculation. Before the channel changes, the rate of 36 Mbit/s achieves highest throughput. It will take time for the 36 Mbit/s measurement to have no impact on the rate selection. A comprehensive study on the impact of α value on the Minstrel performance (on the Madwifi driver) is presented in [15]. Another reason is that Minstrel wrongly assumes that a lost frame takes the same time as a successfully delivered frame to transmit, thus it overestimates the link capacity. Due to the combination of these reasons, Minstrel uses 36 Mbit/s even within the rate transition regions. The other throughput drops can be explained in the same manner.

3.2.5 *Responsiveness due to sudden channel quality changes*

Responsiveness is one of key aspects of a rate control mechanism; that is, how fast a rate control mechanism can select the optimal rate for the given channel conditions. To create such a scenario, we start the experiment with one path loss value; after a period of time, we either increase or decrease the path loss value to emulate sudden channel quality changes. From the point when attenuation is changed, we measure the time it takes for rate control mechanisms to converge on an appropriate rate for the new channel conditions.

Fig. 12 shows traces for these experiments. In the case when the channel quality suddenly increases (i.e., decrease in path loss) as shown in Fig. 12(a), we notice that Minstrel will take approximately 500 ms to stabilise at 54 Mbit/s for the new channel conditions. For PID, we again observe the repeating pattern as before. Obviously PID does not converge at a rate. While it may look that PID converges at 54 Mbit/s at around 5 s, 54 Mbit/s is in fact already the highest rate supported by the IEEE 802.11a standard; otherwise, the repeating pattern will continue. This is a fundamental issue of the PID rate selection mechanism — it fails to verify whether the higher rate should be used. As expected, we see a similar repeating pattern for PID in the case when channel quality decreases, as shown in Fig. 12(b). On the other hand, Minstrel takes around 600 ms to converge at 24 Mbit/s after the channel quality changes.

4. OVER-THE-AIR EXPERIMENTS

In the last section, we discussed the evaluation of both rate control mechanisms in a conducted platform in which all RF signals are sent over coaxial cables rather than over-

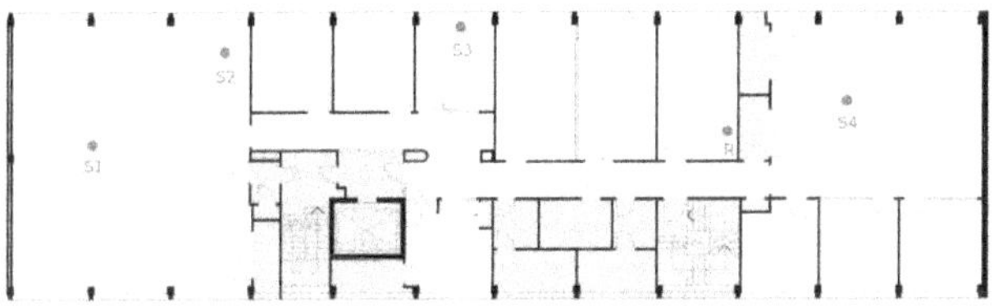

Figure 13: Experimental floor plan.

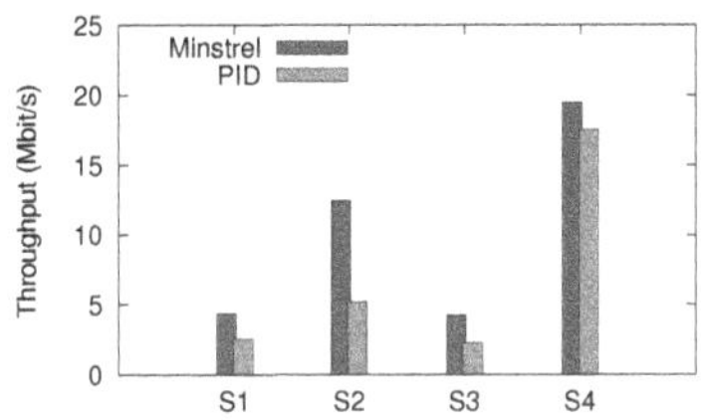

Figure 14: Performance of semi-controlled experiments.

the-air. The advantage is that we have a full control of the experiments — the experiments are repeatable; hence we can analyse the cause of a particular observation in the measurements. To validate these conducted evaluation results, we put together an over-the-air setup (i.e., RF signals are sent via antenna) and present our findings from two experiment configurations.

Fig. 13 shows the random placement of four senders (i.e., S1, S2, S3, and S4), which send saturated traffic (one at a time) to the receiver (i.e., R) in our office environment. With this five nodes setup, we perform two experiments: *Semi-controlled* configurations in which we select a wireless channel and experiment time that have minimum level of external interference; and *In-the-wild* configurations in which we perform our experiments using a wireless channel that is shared with wireless access points around the office and at random office hours. The latter experiments will potentially test a rate control mechanism's ability to adapt to uncontrolled (but realistic) channel quality changes.

4.1 Semi-controlled Configurations

The aim of this over-the-air experiment is to recreate the static channel scenario we have in the conducted experiment setup. We want to validate our findings in a more realistic environment; that is, RF signals are sent via the antenna and from senders at different random locations (hence different propagation delays and different path loss to the receiver). To ease our analysis, we first run our experiments at midnight over the weekends and use a wireless channel that is not shared with other wireless access points in the vicinity. Each experiment runs for one hour, and the measurements are averaged.

Fig. 14 presents the throughput results for both mechanisms measured from different locations (refer to Fig. 13) and shows that Minstrel clearly outperforms PID at all locations. Since these experiments are performed at off-peak time and using channel that is not shared, the channel conditions are relatively stable. We investigate how both rate control mechanisms determine what rate to use under the given channel conditions. Fig. 15 shows the number of packets sent and succeeded at each rate for these experiments. We can see Minstrel is very stable in terms of picking on the

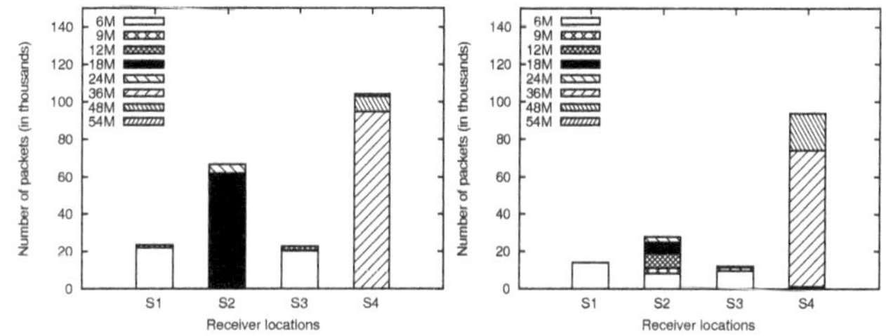

(a) Minstrel (b) PID

Figure 15: # of packets succeeded at each rate.

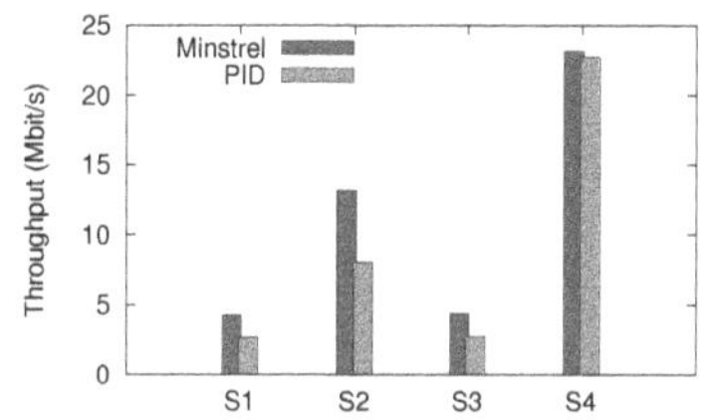

Figure 16: Performance of *in-the-wild* experiments.

optimum rate for the given channel conditions. In contrast, these experiments confirm the instability of rate selection in PID.

4.2 In-the-wild Configurations

Having the experiment results in the semi-controlled configurations in which we compared the two mechanisms, our next goal is to investigate how the mechanisms perform in a completely uncontrolled settings; i.e., the *In-the-wild* experiments. In these experiments, we perform random tests during random office hour; therefore there will be people moving around in the office (or even using the microwave at a location between these nodes). In addition, we use a wireless channel that is shared with four university's wireless access points around the office; in terms of traffic pattern, they are completely out of our control. We run each experiment for one hour and average the measurements.

The results of these experiments are presented in Fig. 16. Although difference in the performance is not very significant at some locations, overall we still see that Minstrel outperforms PID. Due to the complexity involved in modelling the exact channel changes, we cannot provide the most definitive reason for the performance results. However, we strongly believe that the instability in rate selection will be one of the significant drawbacks, which causes poor performance for PID.

5. WAYS TO IMPROVE PID

Based on the previous experiments we can conclude that Minstrel outperforms PID. In this section, we discuss the potential reasons for the PID's poor performance, present a proposal for improving PID, and show that this PID enhancement significantly improves its throughput.

5.1 Issues in PID

As we have emphasised a number of times, the most significant problem with PID is its instability in rate selection, as first shown in Fig. 5 and demonstrated in subsequent experiments.

As discussed in Section 2, the idea of PID is based on control system's principle of a feedback loop. That is, defining a target frame loss ratio (FLR) — e.g., the target FLR in PID is fixed at 14% — let the rate control mechanism to adapt its rate to meet that target by minimising the difference between the current FLR and the target FLR. It is very straight forward; such that PID drops the sending rate when FLR is greater than 14%, while increases the sending rate when FLR is less than the target FLR.

During the adaptation process PID will increase the rate whenever the current FLR is below the target threshold, regardless the current channel conditions which may not be sufficient to support the higher rate. The consequence of this is the FLR of the higher rate is higher than the target threshold and this causes PID to drop the rate again. Hence, this results in oscillation in rate selection as we can see in Fig. 5.

One solution to this problem is to implement a verification mechanism that probes the achieved throughput of the proposed rate compared to the current throughput. The goal is to maximise throughput therefore if the proposed rate achieves higher throughput than the current sending rate the algorithm should select the proposed rate. Otherwise, the rate adaptation requests should be ignored. The same rule applies for requests either to increase rate or to decrease rate.

5.2 The Making of PIDE

As an enhancement of PID (PIDE) we implemented the proposed verification mechanism. The mechanism uses information passed by the PID to detect requests on rate adaptation at the end of each rate adaptation period. When a new rate is proposed (at the end of the last adaptation period) and the proposed rate has not been verified, PIDE then sends n frames (e.g., current implementation uses three frames) using the proposed rate in the current rate adaptation period. Other frames within the adaptation period will continue to use the current rate. Each frame is associated with a status that recordes a number of statistics regarding the transmission (e.g., retry count). PIDE collects these statistics and maintains a table of performance information for each rate. At the end of the current adaptation period, PIDE compares the achieved throughput to decide whether to use the proposed rate in the next rate adaptation period. The achieved throughput tp is calculated as

$$tp = (1 - FLR) * (1s\ /T) \quad (4)$$

where FLR is the frame loss ratio, $1s$ is one second, and T is calculated as $DIFS + T_{DATA} + SIFS + T_{ACK}$ [16]; and T_{DATA} and T_{ACK} are the respective transmission times of the DATA and ACK frames.

In addition to the verification mechanism, we also modify the way frame loss ratio is calculated in PID. When calculating the FLR, PID counts multiple failures of frame transmission attempts as one failed frame; this potentially underestimates the frame loss ratio.

5.3 Evaluation of PIDE

To show the effectiveness of our proposed enhancement, we revisit the same set of conducted and over-the-air experiments, as discussed in Sections 3 and 4. Due to the space limitation, we only discuss some of them in the paper.

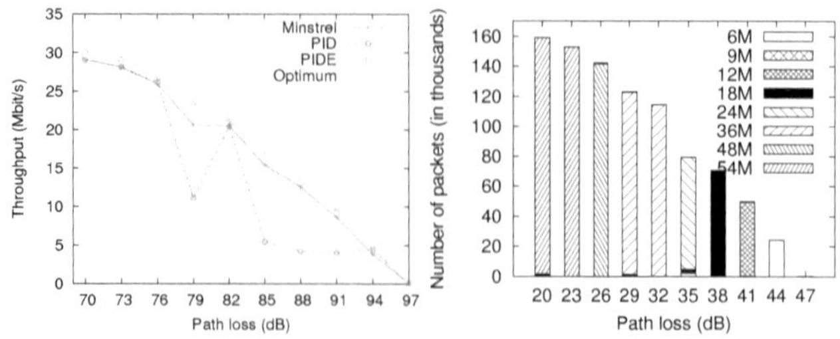

(a) Performance under different path loss (b) # of packets succeeded at each rate for PIDE

Figure 17: Performance in static channel.

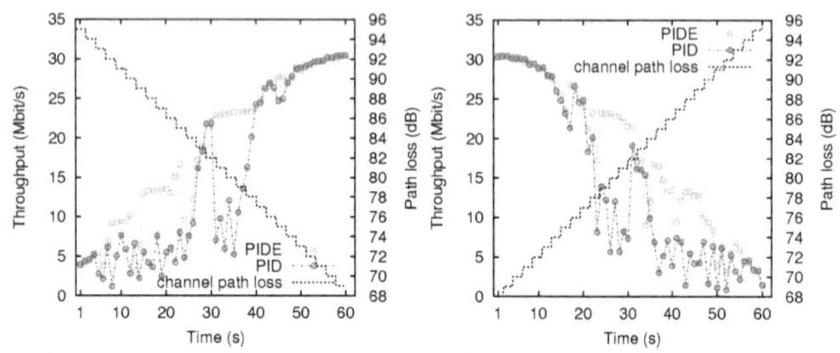

(a) Quality linearly increase (b) Quality linearly decrease

Figure 18: Performance when channel quality linearly changes

5.3.1 Static channel conditions

Fig. 17(a) shows that PIDE achieves much better performance than PID when the channel conditions are relatively static; on some occasions, PIDE even outperforms Minstrel and achieves throughput that is very close to the optimum. When comparing the history of rate selection as in Figs. 4(b) and 17(b), it is very obvious that PIDE solves the instability problem in PID. Rather than hoping around between different rates, as in Fig. 4(b), PIDE is able to quickly converge at rates that can achieve high throughput.

5.3.2 Channel quality progressive increase/decrease

Fig. 18 shows the achieved throughput for PID and PIDE as the link quality linearly increases (i.e., decrease in attenuation)/decreases (i.e., increase in attenuation). Although PIDE does not solve the significant throughput drops in both cases, it is still very obvious that PIDE by far performs better than PID in these scenarios. The reason being PIDE selects the optimal rate quicker and stays at the optimal rate most of the time, as shown in the history of rate selection when channel quality decreases (in Fig. 19(b)); in contrast PID has fewer time slots in which it uses the optimal rate, as shown in Fig. 19(a). Another observation is that in terms of the optimal rate usage PIDE and Minstrel are very similar, as shown in Figs. 11(a) and 19(b). Fundamentally, this is because the verification mechanism in PIDE solves the problem of rate hoping in PID.

5.3.3 Responsiveness to sudden channel quality changes

In the experiments to test the responsiveness of PIDE due to sudden channel quality changes, PIDE shows significant improvement over PID, as it converges at an optimal rate in the matter of around 200 ms in both cases when changes of channel conditions happen. Another observation is that in

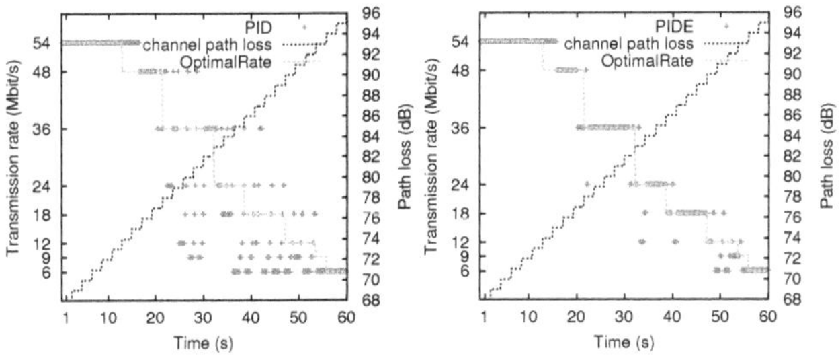

(a) Rate selection for PID. (b) Rate selection for PIDE.

Figure 19: History of rate selection channel quality decreases.

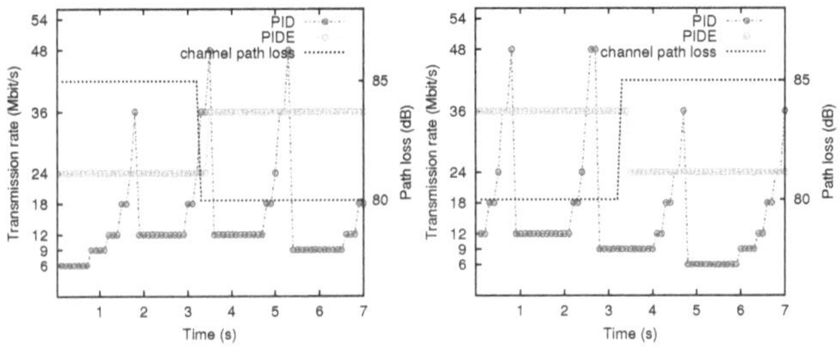

(a) Rate selection of PIDE. (b) Rate selection of PID.

Figure 20: Rate selection when channel quality suddenly changes.

these experiment scenarios PIDE shows better responsiveness than Minstrel, which is shown when comparing Figs. 12 and 20.

5.3.4 *Over-the-air experiments*

To validate results of our conducted experiments, we also revisit the two over-the-air experiment configurations. Fig. 21 shows the performance results in both configurations. As shown in the figures, the results confirm our findings from the conducted experiments, showing that PIDE performs much better than PID and outperforms Minstrel. The absolute throughput difference for the same pair of nodes can be explained by the frequency-selective fading, as described in [9] and [17]. Figs. 15(a) (refers to Minstrel) and 22 (refers to PID and PIDE) show the number of packets sent and succeeded at each rate (for the semi-controlled experiments), which again confirm that PIDE with the verification mechanism is much better in selecting the optimal rate and staying at this rate. Similar pattern in rate selection is also observed for the *in the-wild* experiments.

6. DISCUSSION

Based on the experience we gained from this exercise of evaluating and improving the PID rate control mechanism, we can summarise three lessons learned in terms of using frame loss ratio (FLR), or packet loss ratio (PLR), as a metric in designing rate control mechanisms.

First, usually when the FLR of a rate is below a threshold, the mechanism should increase the rate. But one condition is that the proposed rate should be verified to check whether the new rate provides better throughput (or is still below the target FLR, depending on the goal of the network) before switching rate; at a minimum whether the current channel

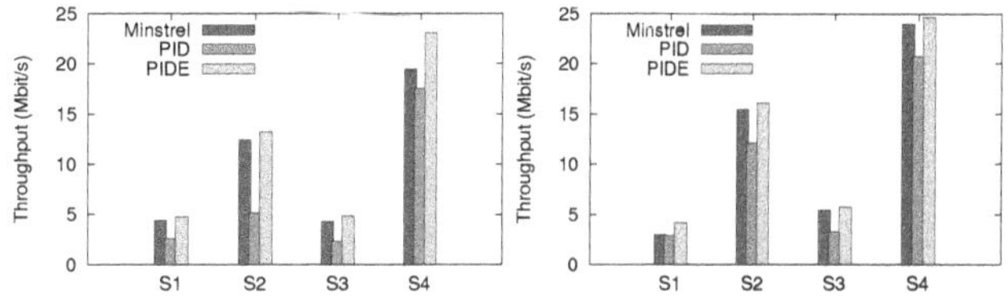

(a) Semi-controlled (b) In-the-wild

Figure 21: Over-the-air performance.

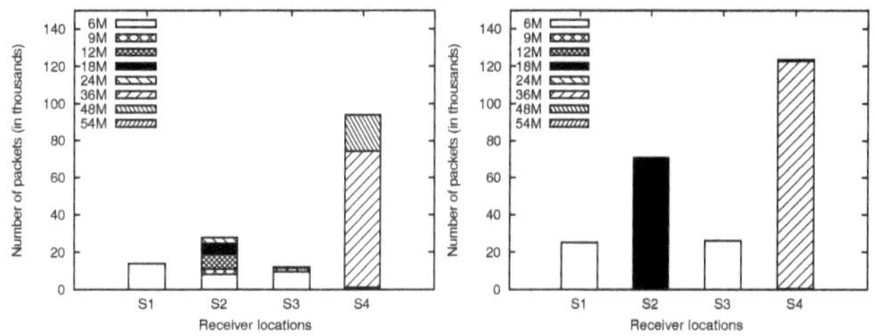

(a) PID (b) PIDE

Figure 22: # of packets succeeded at each rate.

conditions can support such a rate increase. This will prevent oscillation in rate selections, as we have seen in PID.

Second, rate control mechanisms usually decrease the rate when the FLR of a rate is above the target threshold (e.g., in PID this target FLR is set to 14%). This is based on the assumption that a rate with some degree of FLR achieves lower throughput than its adjacent lower rate. But this might not be the case in some scenarios. Fig. 23 shows two examples of the relation between throughput and FLR of two adjacent fixed rates at different path loss values; that is, 12 and 18 Mbit/s in Fig. 23(a), and 24 and 36 Mbit/s in Fig. 23(b). These figures show two evidences of how PID using 14% as the target FLR for all rates failed under the above assumption. At the path loss of 92 dB in Fig. 23(a) and 84 dB in Fig. 23(b), we can see that the FLRs at these path loss values are both above 20% for the rate of 18 and 36 Mbit/s, respectively. Therefore, PID will drop the rate under the above assumption. However, if we look closely at the throughput that each respective fixed rate is achieving, we will notice that even with FLR above 20% they still outperform their adjacent lower rates. Hence, we conclude that the above assumption does not hold true in some cases.

Third, assuming a unique FLR threshold for all rates is dangerous. If the goal is to achieve maximum throughput, from Fig. 2 we should see that the absolute maximum throughput for a given path loss is on the envelop that connects the maximum throughput points of all the fixed rates. Hence the optimal maximum FLR of each fixed rate are the points when a higher rate switches to a lower rate to maintain highest throughput. When we compared the FLR of these throughput points in Fig. 2, we found that they are all different values. Therefore, PID using a fixed target FLR of 14% for all rates fails to achieve maximum throughput. In literature, authors of RRAA [1] also notice that there is no single optimal FLR threshold for all the rates. In addition, in [11] we have shown that other similar mechanisms (e.g., AMRR [2], Onoe [4]) perform poorly when compared to Minstrel, which uses the achieved throughput as the metric for rate adaptation.

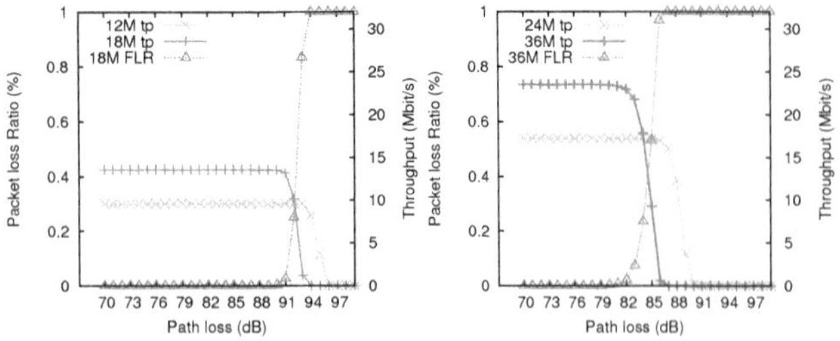

(a) Evidence 1 (b) Evidence 2

Figure 23: Relation between throughput and FLR.

In summary, we are questioning whether FLR (or PLR) is a good metric to be used in rate control mechanisms. It can be seen from our analysis that directly measuring the achieved throughput and using it as the metric for rate adaptation leads to higher throughput.

7. CONCLUSION

In this paper we presented the performance evaluation of two MAC layer rate control mechanisms implemented on the mac80211 framework: Minstrel and PID. These rate control mechanisms are used in millions of computers running the Linux operating system. We showed that PID has poor performance compared to Minstrel; we then proposed and evaluated a PID enhancement (PIDE) that improves the performance. In addition, we discussed a lesson learned in this research, i.e., frame loss ratio is not the best metric for rate adaptation.

We compared the rate control mechanisms in a conducted testbed that can provide the same experiment environment (i.e., the same effective path loss) for each experiment. We evaluated the mechanisms in five communication scenarios: (i) static channel conditions, (ii) interference with repeating pattern, (iii) Rayleigh model fading channel, (iv) progressive increase/decrease of channel quality, and (v) sudden change of channel quality. The results from the conducted testbed were also validated by over-the-air experiments.

Acknowledgement

NICTA is funded by the Australian Government as represented by the Department of Broadband, Communications and the Digital Economy and the Australian Research Council through the ICT Centre of Excellence program; and the Queensland Government.

8. REFERENCES

[1] S. H. Y. Wong, H. Yang, S. Lu, and V. Bharghavan, "Robust rate adaptation for 802.11 wireless networks," in *Proceedings of MobiCom 2006*. Los Angeles, California, USA: ACM, 2006, pp. 146–157.

[2] M. Lacage, M. H. Manshaei, and T. Turletti, "Ieee 802.11 rate adaptation: a practical approach," in *Proceedings of MSWiM 2004*. Venice, Italy: ACM, 2004, pp. 126–134.

[3] J. C. Bicket, "Bit-rate selection in wireless networks," MIT Master Thesis, 2005.

[4] "Onoe specification," Madwifi Project.

[5] "Minstrel specification," Madwifi Project.

[6] "Pid specification," LinuxWireless Project.

[7] G. Holland, N. Vaidya, and P. Bahl, "A rate-adaptive mac protocol for multi-hop wireless networks," in *Proc. of ACM MobiCom*, 2001.

[8] G. Judd, X. Wang, and P. Steenkiste, "Low-overhead channel-aware rate adaptation," in *Proceedings of MobiCom '07*. Montreal, Quebec, Canada: ACM, 2007, pp. 354–357.

[9] H. Rahul, F. Edalat, D. Katabi, and C. G. Sodini, "Frequency-aware rate adaptation and mac protocols," in *Proceedings of MobiCom 2009*. Beijing, China: ACM, 2009, pp. 193–204.

[10] M. Vutukuru, H. Balakrishnan, and K. Jamieson, "Cross-layer wireless bit rate adaptation," in *Proceedings of ACM SIGCOMM 2009*. Spain, Barcelona: ACM, 2009, pp. 3–14.

[11] W. Yin, K. Bialkowski, J. Indulska, and P. Hu, "Evaluation of madwifi mac layer rate control mechanisms," in *Preceedings of IWQoS 2010*, Beijing, China, June 2010.

[12] K. Bialkowski and M. Portmann, "Design of test-bed for wireless mesh networks," in *IEEE Antennas and Propagation International Symposium*, Toronto, Canada, July 2010.

[13] J. Camp and E. Knightly, "Modulation rate adaptation in urban and vehicular environments: cross-layer implementation and experimental evaluation," *IEEE/ACM Trans. Netw.*, vol. 18, pp. 1949–1962, December 2010.

[14] V. K. Garg, *Wireless Communications and Networking*. ELSEVIER, June 2007.

[15] W. Yin, P. Hu, J. Indulska, and K. Bialkowski, "A method to improve adaptability of the minstrel mac rate control algorithm," in *Proceedings of UIC 2010*, Xi'an, China, October 2010.

[16] "Ieee standard 802.11-2007," IEEE Computer Society, p. 625, 2007.

[17] D. Halperin, W. Hu, A. Sheth, and D. Wetherall, "Predictable 802.11 packet delivery from wireless channel measurements," in *Proceedings of SIGCOMM 2010*, New Delhi, India, 2010, pp. 159–170.

Seamless Rate Adaptation for Wireless Networking

Hao Cui*
University of Science and Technology of China
Hefei, 230027, P.R. China
hcuipro@mail.ustc.edu.cn

Chong Luo
Microsoft Research Asia
Beijing, 100080, P.R. China
chong.luo@microsoft.com

Kun Tan
Microsoft Research Asia
Beijing, 100080, P.R. China
kuntan@microsoft.com

Feng Wu
Microsoft Research Asia
Beijing, 100080, P.R. China
fengwu@microsoft.com

Chang Wen Chen
University at Buffalo, The State University of New York
Buffalo, NY 14260-2000, USA
chencw@buffalo.edu

ABSTRACT

This paper aims at designing a *Seamless Rate Adaptation* for wireless networking which achieves smooth rate adjustment in a broad dynamic range of channel conditions. Conventional rate adaptation can only achieve a stair-case rate adjustment. Even when combining with hybrid ARQ, it suffers from an irreconcilable conflict between throughput and dynamic range. We tackle this problem from a new perspective by relying on modulation, instead of channel coding, for rate adaptation. We propose rate compatible modulation (RCM), in which modulation signals are incrementally generated from information bits through weighted mapping. Rate adaptation is achieved through varying the number of modulated signals. As more signals are transmitted, information bits gradually accumulate energy. The weights in bit-to-symbol mapping are delicately designed to ensure fine-grained energy accumulation so that smoothness and efficiency can both be achieved. We design and implement a rate adaptation system, called SRA and evaluate its performance through a software radio testbed. Results show that, under highly dynamic channel conditions, SRA achieves over 80% throughput gain over 802.11a adaptive modulation and coding, and achieves 28.8% and 43.8% gain over HARQ systems implemented with Turbo code and Raptor code. We believe that the concept of rate compatible modulation opens up a fresh research avenue toward the wireless rate adaptation problem.

Categories and Subject Descriptors

C.2.1 [**Computer-Communications Networks**]: Network Architecture and Design- *Wireless Communications*

*This work was carried out when H.Cui was a research intern at Microsoft Research Asia.

MSWiM'11, October 31–November 4, 2011, Miami, Florida, USA.

General Terms

Design, Performance, Experimentation

Keywords

rate adaptation, modulation

1. INTRODUCTION

In wireless networking, rate adaptation is critical to the system performance by exploiting the dynamic bandwidth. Though it is not specified in current 802.11 standards, all 802.11 systems need to dynamically change the rate for transmission. In the standards, a set of rates is mandated in physical layer for selection. They are implemented by combining different modulation and channel coding schemes. According to where the rate selection is performed, rate adaptation can be categorized as sender rate adaptation and receiver rate adaptation.

Most existing rate adaptation schemes adopt sender rate adaptation, which is also known as adaptive modulation and coding (AMC). Specifically, the receiver observes channel condition and delivers a feedback to the sender for rate selection. These schemes mainly focus on which metric should be used for channel estimation. Usually, the metrics for evaluating channel condition are signal to noise ratio(SNR) [7, 8], frame loss rate [9, 23], interference-free bit error rate [22] and symbol-level dispersions [15]. However, there always exists a dilemma in sender rate adaptation no matter what metric is used. That is, in time varying wireless networks, instant channel estimation cannot guarantee accuracy, while accurate channel estimation needs a long term observation. Either inaccurate or delayed estimation may lead to degradation from achievable throughput. In addition, the rate set provided by physical layer is discrete. Even a perfect channel state is known at the sender, it can only achieve stair-case rate adjustment. The gap between adjacent rates cannot be fully utilized.

Hybrid ARQ (HARQ) is a complementary to sender rate adaptation in a form of channel coding, which fills this gap in some sense. It defines an acknowledgement mechanism to issue retransmission when error occurs. The retransmission is either a repetition (type I HARQ) or incremental redundancy (type II HARQ). The rate can be smoothly changed by increasing the retransmission times. Since the throughput of type II HARQ is much higher than type I HARQ, we only discuss type II HARQ in this paper and use HARQ as type II HARQ unless otherwise specified. In HARQ, the incremental redundancy is usually encoded by either a rate compatible code, e.g. punctured Turbo code [13] or fountain code, e.g. Raptor code [19]. However, both of them only have a limited rate dynamic range. To cover a typical range, from 0.5 bit/s/Hz to 6 bit/s/Hz, one way is to use AMC. This will fall back to the dilemma. The other way is to fix the modulation and use a low rate mother code for puncturing. The problem in punctured Turbo code is that a low rate mother code encounters throughput loss at high end [14]. In Raptor code, which is originally designed for binary erasure channel (BEC), a 0.9 outer LDPC code is needed before inner LT code for Gaussian channel [21]. In addition, the overhead of LT code in Gaussian channel is larger than that in BEC. Thus, the outer coding rate and larger overhead make it saturate at a lower transmission rate than punctured Turbo code. We will show this through simulation.

To tackle these issues, we need a receiver rate adaptation approach, which can be seamless. By seamless, we mean that the rate adjustment is rateless, smooth and spread on the typical range. The sender continuously generates and transmits coded symbols for a data block. The receiver collects symbols until the data block can be correctly decoded and delivers back an ACK for the transmission of next block. In addition, since we cannot guarantee an accurate channel estimation, the system performance should be insensitive to the channel estimation error.

In order to achieve the desired seamless adaptation, we propose a rate compatible modulation (RCM) which is motivated by the analysis of conventional modulation. RCM incrementally generates symbols with fine-grained bit energy allocation. Collecting symbols can progressively accumulate the transmission power until meeting the requirement, which is rateless. Unlike conventional modulation, the generated symbols are directly placed on each dimension of the constellation to modulate signals, so that the symbol locates in the same domain with channel noise. This makes it possible for smooth degradation. The constellation in RCM is fixed and very dense (23×23 QAM). By fixing the constellation, there is no need to make decision on modulation selection and avoid the dilemma as mentioned above. Dense constellation contains large amount information and it will not be easily saturated. Therefore, it covers a broad dynamic range.

Based on RCM, we design and implement a rate adaptation system, called SRA (seamless rate adaptation), on orthogonal frequency division multiplexing (OFDM) physical layer, and evaluate its performance on a software radio platform. Results show that, while SRA excels others in all four test scenarios, it shows greatest advantage in non-line-of-sight scenario, achieving over 80% gain over conventional 802.11 rate adaptation which is pure AMC, 28.8% and 43.8% improvement on Turbo code and Raptor code based HARQ with AMC.

The rest of the paper is organized as follows. Section 2 provides the recent works on receiver rate adaptation. Section 3 presents the motivation and design of RCM, then introduces the modulation and demodulation process and its complexity. Section 4 evaluates RCM through Matlab simulations. Section 5 describes the implementation and evaluation of SRA system on software radio testbed. Finally, we conclude this paper in Section 6.

2. RELATED WORK

One representative work of receiver adaptation is adaptive demodulation (ADM) [5]. It employs rateless coding and fixed 16QAM constellation. In particular, source bits are encoded by a Raptor code, then every four rateless bits form a symbol and is mapped to 16QAM constellation with Gray code. Upon receiving a symbol, the receiver collects reliable bits by adjusting the decision regions for demodulation. These rateless bits are then accumulated for decoding. This scheme has two problems. First, it obviously has a saturation rate at 4 bits/Hz. Second, the authors neglect the fact that rateless code can also be used in channels other than BEC. In a noisy channel, unreliable bits also contain some amount of information. Simply discarding them would dramatically decrease channel throughput.

A recent work automatic rate adaptation (ARA) [6] shares the same goal with our work to achieve feedback-free rate adaptation. In particular, ARA adds a minimum distance transformer (MDT) after existing coding and modulation to perform linear combinations for modulated symbols. Although ARA requires no change to existing modulation and coding, its performance is adversely affected by the limitations of existing modules. In addition, ARA performs linear combination through a dense random matrix. This dramatically increases complexity in the decoding. The authors have suggested Lattice decoding which has a complexity of $O(N^3)$ [10].

3. RATE COMPATIBLE MODULATION

Rate compatible indicates that the constellation is fixed and the transmitted symbols of high rate modulation are used by all low rate modulation. Let N be the bit block length for modulation, and K be the number of modulation signals for successful transmission. Then the modulation rate is $\frac{N}{K}$. First, K_0 signals are transmitted to keep a high rate modulation. If low rate modulation is needed for transmission, we just increase the number of signals step by K_c. Therefore, the achievable modulation rate are:

$$R = \frac{N}{K} = \frac{N}{K_0 + i \cdot K_c} \quad i = 0, 1, \ldots$$

3.1 Motivation

In conventional modulation, the minimum Euclidean distance between adjacent constellation points is called free distance. Under the same noise amplitude, the bit error rate (BER) exponentially decreases with increasing free distance [18]. When the transmission power is fixed, high rate modulation conveys more data bits per transmission, but its free distance is small. Although low rate modulation conveys less data bits per transmission, its free distance is large. So switching between different modulations is to choose an appropriate free distance for resisting the immediate channel noise. In fact, the free distance is equivalent to another modulation parameter, i.e. average bit energy. Denote E_s as the average symbol energy of the constellation and K as the number of bits contained in each symbol, then the average bit energy is $E_b \triangleq E_s/K$. The free distance is proportional to E_b [18]. It is possible to change the free distance by tuning the number of transmitted symbols. However, this is

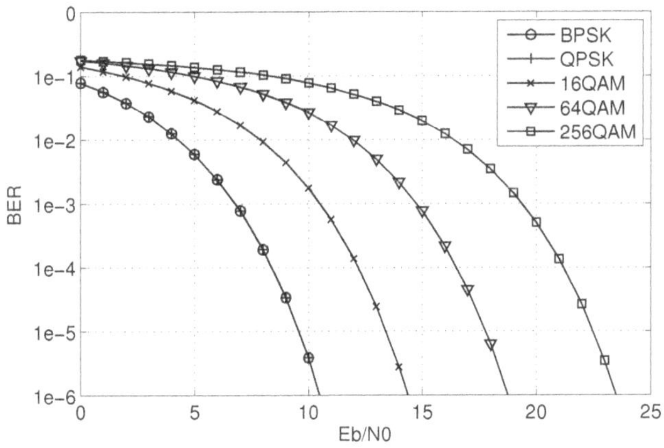

Figure 1: BER of different modulation schemes

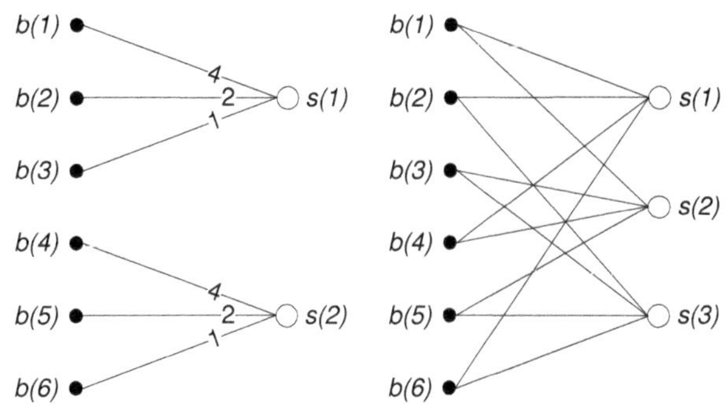

(a) Conventional (b) Heuristic

Figure 2: Bit-to-symbol mapping

not efficient for conventional modulations. To do this, the bit energy should be accumulated from a low point which means a high rate modulation. The problem in conventional modulation is that the bit energy is not equally allocated in the symbols of a higher rate modulation. Unequal bit energy allocation leads to unequal BER. Since the BER of a modulation is calculated by averaging over all bits, it will be dominated by the bits with higher BER. Fig. 1 shows the BER vs. bit energy to noise power ratio (E_b/N_0) for different modulations. We see that even under the same average bit energy, the BER of high rate modulation is higher than that of low rate modulation. That's why rate adaptation based on conventional modulation can only be achieved by switching modulation schemes.

Therefore, to realize the rate compatibility in modulation, we need an even bit energy allocation, which can increase the free distance by accumulating the received symbols and achieve the same effect of switching between different modulations. Fig. 2 presents our heuristic idea. Assume that there are 6 bits for modulation. Fig. 2(a) shows the mapping of conventional modulation 8 PAM (or one dimension of 64 QAM). It takes each 3 bits as one group and then generates a symbol by weighted sum. $\{1, 2, 4\}$ is taken as weight for identical bit to symbol mapping. However, this mapping cannot equally increase the bit energy with increasing retransmissions, since the bit energy ratio is always 1:4:16. A heuristic method is to replace the mapping by what is shown in Fig. 2(b), each bit is sampled by multiple symbol and the sampling weight vector for each bit is with equal norm. Only such a mapping can get evenly increasing bit energy

3.2 Design

Now the problem becomes how to map bits to symbols. Since the mapping is a weighted bipartite graph, it can be described by its incidence matrix from graph theory. Denote $\mathbf{G}$ with dimension $M \times N$ as the incidence matrix which is named mapping matrix in this paper. Let the row represents the bits and the column represents the symbols. If there exists one edge between a bit and a symbol, the corresponding element in $\mathbf{G}$ is the weight on the edge, or the corresponding element will be zero. In the following, we first describe the constraints on $\mathbf{G}$. Then we will show how to construct such a mapping matrix.

3.2.1 Constraints on Mapping Matrix

Constraint 1: $\mathbf{G}$ *should be regular in rows.*

We desire a fixed constellation, which implies a fixed symbol alphabet. As the RCM symbols are generated by weighted sum operation, the symbol alphabet is determined by the weights. Therefore, we require $\mathbf{G}$ to be regular in rows, i.e. each row contains the same number of non-zero entries drawn from a fixed weight set.

Let L be the number of non-zero entries in each row, and $\mathcal{W}_s = \{w_1, w_2, ...w_L\}$ be the fixed weight set, then each symbol $u(\cdot)$ can be computed by:

$$u(m) = \sum_{l=1}^{L} w_l b(m_l)$$

where $b(\cdot)$ is the bit block and m_l is the corresponding bit index of weight w_l in the mth row.

Although the weight set is fixed, the weights should be randomly placed in each row. Such a random property can create many symbols, which enables rateless.

Constraint 2: $\mathbf{G}$ *should be as regular as possible in columns.*

All weights in a column are associated with its corresponding bit. The allocated energy to bit $b(n)$ is actually the norm of each column vector:

$$E_{b(n)} = C \cdot \sum_{r=1}^{R_n} w_{n_r}^2$$

where R_n is the number of non-zero weights in the nth column, n_r indicates which weight is used and C is a normalization factor. Equal energy increase can be achieved if the columns are regular in norm.

It is trivial to get regular columns when $M = N$. We require the matrix to be regular when $M = N/2$ or even $M = N/4$. A positive-negative symmetric weight set can easily satisfy the regularity when $M = N/2$.

Constraint 3: *The weight set $\mathcal{W}_s$ should create diverse symbol values.*

From information theoretic perspective, each received symbol carries a certain amount of information upper bounded by the channel capacity and the entropy of the transmitted symbol. In order to avoid rate saturation at high channel SNR, the entropy of modulated symbols should be high enough.

Using diverse weights can increase the diversity of symbols, and consequently increase symbol entropy. We set a threshold for symbol entropy at 4 bit/dimension, which is equivalent to 256QAM rate.

3.2.2 Construction of Mapping Matrix

We find that the weight set $\{\pm1, \pm2, \pm4, \pm4\}$ satisfies all the constraints outlined above. Now, we present three steps to construct the generation matrix $\mathbf{G}$ such that it has good properties at varying values of M. First, we construct three elementary matrices A_1, A_2 and A_4. Each elementary matrix has dimension $N/8 \times N/4$. The structure of A_1 is shown as follows. Matrices A_2 and A_4 have the same structure, but the non-zero values are replaced with $+2/-2$ or $+4/-4$.

$$A_1 = \begin{bmatrix} +1 & -1 & & & & \\ & & +1 & -1 & & \\ & & & \ddots & & \\ & & & & +1 & -1 \end{bmatrix}$$

Second, we form an $N/2 \times N$ matrix $\mathbf{G}_0$ by stacking random permutations of the three elementary matrices as follows:

$$\mathbf{G}_0 = \begin{bmatrix} \pi(A_4) & \pi(A_4) & \pi(A_2) & \pi(A_1) \\ \pi(A_2) & \pi(A_1) & \pi(A_4) & \pi(A_4) \\ \pi(A_4) & \pi(A_4) & \pi(A_1) & \pi(A_2) \\ \pi(A_1) & \pi(A_2) & \pi(A_4) & \pi(A_4) \end{bmatrix}$$

where $\pi(\cdot)$ denotes randomly permutated columns of a matrix. By using different permutation choices, we may construct virtually unlimited number of matrix $\mathbf{G}_0$.

The third and final step in constructing $\mathbf{G}$ is to stack all the randomly generated $\mathbf{G}_0'$s. In practice, we only stack two matrices to form an $N \times N$ matrix, and repeatedly use it when the channel condition is poor. This matrix can be communicated between the sender and the receiver before transmission. When a symbol is repeatedly transmitted, multiple versions can be combined through maximal ratio combining.

3.3 Modulation

A data packet from upper layer is first divided into bit blocks by length N. Let $\mathbf{b} = \{b(n)\}_{n=1}^N$ be one bit block. Based on the above mapping matrix $\mathbf{G}$, we can generate symbols by

$$\mathbf{u} = \mathbf{G} \cdot \mathbf{b}$$

where $\mathbf{u} = \{u(m)\}_{m=1}^M$ is the generated symbol block for transmission. Then the symbol is directly used to modulate the signal amplitude. Since the weight set $\{\pm1, \pm2, \pm4, \pm4\}$ is used, the signal amplitude ranges from $-11d$ to $11d$ and is Bernoulli distributed, where d is a scale factor and is also the minimum distance between constellation points. To fully utilize the constellation plane, i.e. I (in phase) and Q (quadrature phase), each two consecutive symbols compose one modulation signal by:

$$u(2k) + \sqrt{-1} \cdot u(2k+1), \ (k = 0, 1, \ldots, M/2 - 1).$$

Therefore, we are actually using a 23×23 QAM constellation. When the transmission power is normalized to 1 based on its distribution, $d \approx 0.1644$ which is similar to 256 QAM.

3.4 Demodulation

We use the standard AWGN channel model to design our demodulation algorithm. Denote $\mathbf{u}'$ as the received symbols. We have

$$\mathbf{u}' = \mathbf{G} \cdot \mathbf{b} + \mathbf{e}$$

where $\mathbf{e} = \{e(m)\}_{m=1}^M$ is a Gaussian noise vector with $e(m) \sim \mathcal{N}(0, \sigma^2)$.

The demodulation process is to find the optimal solution to the following problem:

$$\hat{\mathbf{b}} = \arg \max_{\mathbf{b} \in GF(2^N)} P(\mathbf{b}|\mathbf{u}') \qquad (1)$$

The transmitted symbol is generated by a non-identical mapping, we cannot perform symbol-wise demodulation that is to find the closest constellation point to the received symbol. From [4], the problem (1) can be solved by belief propagation. However, binary digits constraint $\mathbf{b} \in GF(2^N)$ is introduced in this problem. The computation of convolution by FFT [4] is expensive when performing on binary digits. We modify the FFT to convolution operations in a FPGA manner, where only non-zero elements are taken into computation. The details of our demodulation algorithm is shown in Appendix A.

During the belief propagation process, we assume that the channel SNR is correctly estimated. Later, in Section 4, we will show through simulation that our algorithm is not sensitive to SNR estimation error. In Section 5, the evaluation over software radio testbed will demonstrate that our demodulation algorithm performs well when the real channel is not AWGN.

3.5 Complexity

We shall analyze the complexity of RCM in two aspects: modulation complexity and demodulation complexity. Let R and L be the number of nonzero elements in each column and row of the generation matrix respectively. Constructing the generation matrix is a one-time task and the complexity is $O(R \cdot N)$. As each RCM symbol is the weighted sum of L bits, it can be generated through looking up a pre-defined table of size 2^L. Therefore, the modulation complexity for generating M symbols is $O(M)$. Our demodulation procedure is based on belief propagation algorithm. The only difference is that we use convolution and deconvolution to deduce the confidence of each bit from the received RCM symbol. In each iteration, the computation cost of belief deduction for each bit is bounded by $O(W \cdot R)$ which is a constant, where $W = 1 + \sum_{i=1}^L |w_i|$. Therefore, the demodulation complexity is $O(W \cdot R \cdot N)$, where W and R are constant for the RCM. Therefore, RCM has a linear modulation and demodulation complexity.

4. SIMULATION

The Matlab simulation serves two purposes. One is to evaluate the rate adaptation capability of RCM. The other is to choose appropriate parameters for the practical system design.

4.1 Reference Schemes

The rate adaptation capability is evaluated against ideal AMC and HARQ with AMC based on both Turbo codes and Raptor codes. For all schemes, we assume immediate feedback from the receiver. The sender stops the transmission when an ACK is received or the maximum symbol number is reached.

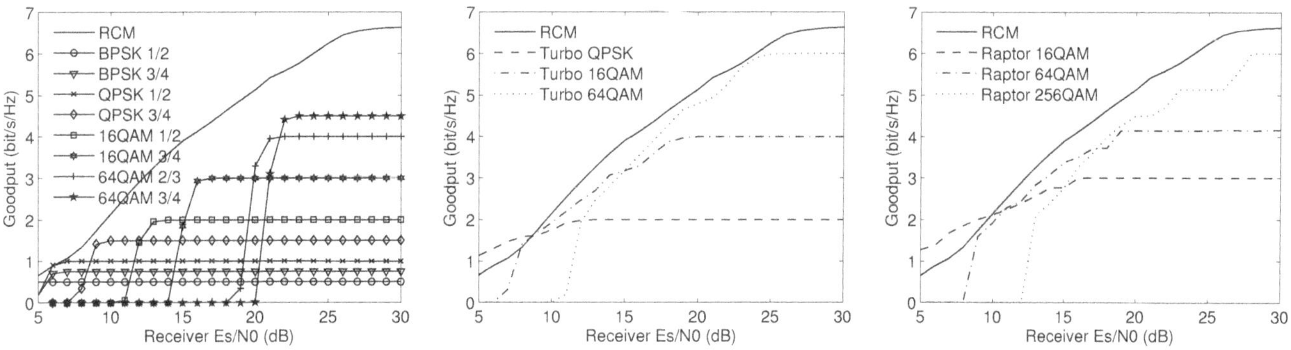

(a) Comparison with Ideal AMC (b) Comparison with HARQ-CTC (c) Comparison with HARQ-Raptor

Figure 3: Comparing RCM's rate adaptation capability with existing schemes

4.1.1 Ideal AMC

Conventional rate adaptation is exemplified by the AMC based on 802.11a standard [1]. The modulations are BPSK, QPSK, 16QAM and 64QAM and the channel code is convolutional code. In total, there are eight different combinations. By ideal, we mean that the sender knows exactly what the channel condition is and can make the optimal rate selection. Therefore, ideal AMC is the upper bound of all conventional rate adaptation schemes.

4.1.2 HARQ with AMC based on Turbo code

The first implementation of HARQ is based on convolutional Turbo code (HARQ-CTC). We use 1/3 Turbo code as mother code. As in WCDMA and LTE [3], the code length is set to 1200 bits, and the component encoder is based on recursive convolutional code with polynomial (13, 15) in octal. In order to achieve smooth rate adaptation, the puncturing period is set to 8. We use the same puncture pattern as presented by Rowitch and Milstein [14]. Rates corresponding to $\frac{8}{8+l}$ $(l = 0, 2, ...16)$ are used.

HARQ-CTC cooperates with three modulation schemes: QPSK, 16QAM, and 64QAM. Different modulation and coding schemes create 27 combinations and 21 identical rates. The decoder we used at the receiver is soft input Viterbi decoder with 8 iterations.

4.1.3 HARQ with AMC based on Raptor code

The second implementation of HARQ is based on Raptor code (HARQ-Raptor). We use an LDPC code with rate 0.9 as the outer code. Rateless bits are generated using the optimal degree distribution at BEC channel [17]. According to WiMax PHY [2], the code length is set to 1800 bits before LDPC coding and therefore 2000 bits before LT coding. Using a longer block may improve the performance but the complexity and transmission delay will also be increased.

HARQ-Raptor combines with three modulations schemes, 16QAM, 64QAM and 256QAM. As explained earlier, the highest achievable rate of Raptor code is well below 1. Therefore, we do not use QPSK modulation as the rate saturates too early. The transmission step size is set to 50 modulation signals, which translates to 200 rateless bits in 16QAM, 300 rateless bits in 64QAM and 400 rateless bits in 256QAM. The maximum number of bits generated for each data block is 16000, creating a minimum code rate of 1/8. At the receiver, belief propagation decoding is deployed.

4.2 Achievable Rates

In this simulation, RCM uses block size $N = 400$ and the retransmission step is 10 modulation signals. The rate adaptation granularity is actually the same as HARQ-Raptor, as the block length is shorter. We run all schemes on every integer SNR from 5dB to 30dB. For each SNR, a total of 10^7 bits are transmitted. We use goodput as evaluation metric [11], which is the rate of correctly received bits. We separately compare RCM with the other three schemes and show the performance of RCM in each comparison.

Fig. 3(a) shows the achievable rates of each combination of modulation and coding in 802.11a. The envelope of these combinations is the performance of ideal AMC. RCM is well above ideal AMC. Due to the coarse-granularity of AMC rate adaptation, RCM shows great advantage around the SNRs where switching occurs. For example, at 20dB, AMC still suggests to use (16QAM, 3/4) scheme whose rate is only 3 bit/s/Hz. In contrast, RCM can achieve 4.65 bit/s/Hz which is over 53% gain.

Fig. 3(b) shows the performance of HARQ-CTC with AMC. From the curves of each modulation, we see that the dynamic range is limited when only HARQ-CTC is adopted, i.e. the single curve of each modulation. To cover the typical rate range, it needs to cooperate with AMC, i.e. the envelope of three modulations. RCM has a significantly larger dynamic range than HARQ-CTC. In addition, the achievable rate of RCM is above the envelope of using 16QAM and 64QAM. When the channel SNR is below 7dB, HARQ-CTC with QPSK has a slightly better performance. However, it contributes a very small portion to the overall performance.

Fig. 3(c) shows the rate of HARQ-Raptor with AMC. Surprisingly, the dynamic range of HARQ-Raptor is even smaller than that of HARQ-CTC. We find that, although the Raptor code rate can be arbitrarily small, the additional parity bits transmitted at low SNR can not improve the decoding performance. From the curves, we see that the highest achievable rate of Raptor code is around 0.75. Therefore, the rate saturates at 4.2 bit/s/Hz when combined with 64QAM and at 6 bit/s/Hz when combined with 256QAM.

This simulation confirms that HARQ need to be combined with AMC to cover a large dynamic range. In contrast, the proposed RCM scheme has been using the same set of parameters from 5dB to 30dB, and the overall goodput is higher than the "Oracle" combined HARQ and AMC scheme which knows exactly when to switch the modulations.

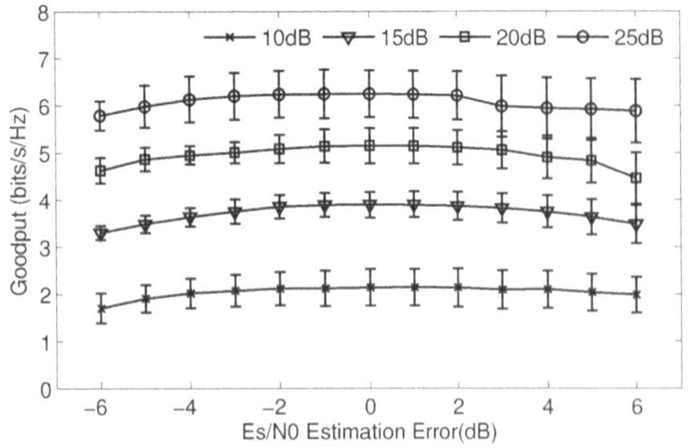

Figure 4: Sensitivity to SNR estimation errors

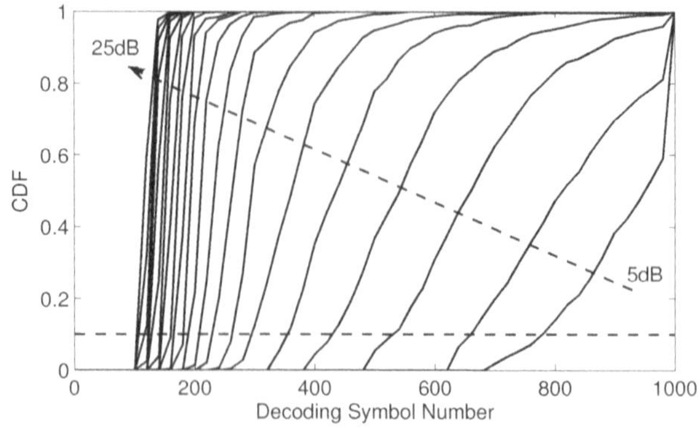

Figure 5: RCM demodulation threshold

SNR (dB)	5	6	7	8	9	10	11
Threshold	760	640	520	420	340	280	260
SNR (dB)	12	13	14	15	16	17	18
Threshold	220	200	180	180	160	160	140
SNR (dB)	19	20	21	22	23	24	25
Threshold	140	120	120	120	100	100	100

Table 1: Demodulation threshold at different channel SNRs

4.3 Sensitivity to SNR Estimation Error

As proposed earlier, precise channel SNR is hard to obtain in practical system and an ideal rate adaptation should be insensitive to channel estimation error. Therefore, we evaluate whether RCM demodulation is sensitive to inaccurate SNR input. We implemented the SNR estimation algorithm proposed by Rahul et al. [12], and found that 98% of SNR estimation error is within -6dB to 6dB.

We run four sets of simulations at four typical SNR values: 10dB, 15dB, 20dB and 25dB. In each set of simulation, we perform RCM demodulation at imprecise SNR estimation. The estimation error varies from -6dB to 6dB step by 1dB. Fig. 4 shows the mean and standard deviation of achieved goodput. Data presented are averaged over 1000 data blocks of length $N = 400$. It can be found that RCM demodulation is not sensitive to imprecise SNR input. The performance loss is not significant even when the input SNR is 6dB lower than the actual value.

4.4 Demodulation Threshold

In a practical system, a receiver should start RCM demodulation after a minimum number M_0 of modulation signals have been received. A small M_0 will increase the system complexity, but a large one may degrade the achievable rate. We create a lookup table of demodulation threshold with respect to channel SNR.

The table is created by the following process. We run a large number of simulations on each integer SNR value from 5dB to 30dB, and plot the cumulative distribution function (CDF) of successful decoding at different M's. Fig. 5 shows the CDF from 5 to 25dB. The curves for 26dB and above are very close to each other and are not shown here. Then we take number of symbols corresponding to the 10% percentile as the demodulation threshold as shown in Table 1. These settings are adopted in our system implementation.

5. IMPLEMENTATION AND EVALUATION

We design and implement a rate adaptation system, called SRA, based on the proposed rate compatible modulation and OFDM physical layer.

5.1 Implementation

The protocol of SRA can be depicted by Fig. 6. At sender end, information bits are block-wise mapped to RCM symbols. Then, the RCM symbols are pair-wisely transmitted as in-phase and quadrature-phase components of QAM modulation. For each block of information bits, the sender progressively transmits modulation signals with a certain step until an acknowledgement is received or the maximum number is achieved. At receiver end, the RCM symbols are accumulated for demodulation after the symbol number exceeds a pre-defined threshold. This threshold is shown in Table 1. If the demodulated bits pass the CRC check, an acknowledgement is delivered to the sender. Otherwise, more RCM symbols need to be collected for another demodulation.

Based on the above protocol, we implement SRA on OFDM PHY defined in IEEE 802.11a, which is illustrated in Fig. 7. Specifically, the channel is divided into 64 subcarriers and 48 of them are used to transmit modulation signals. To reduce the overhead of PLCP header, we use 150 OFDM symbols in each PLCP frame for data transmission. Therefore, the total number of modulation signals in each transmission is 7200. Since RCM is a block based modulation, the data from MAC layer will be divided into bit blocks before transmission. In current SRA system, 400 bits block size is adopted. The progressive transmission step is 10 modulation signals and the maximum number of modulation signals is set to 800 which is equal to the lowest rate 0.5 bits/s/Hz in 802.11a for PLCP header transmission. Therefore, we transmit 720 bit blocks in parallel pipeline style and need to maintain a 36KB buffer at both sender and receiver end.

5.2 Evaluation Methodology

We perform trace-driven evaluation on the software radio platform SORA [20]. SORA is a programmable software radio platform built on commodity general purpose PCs. We perform trace-driven experiments instead of real-time experiments mainly for two reasons. First, as wireless channel is changing over time, it is not possible to fairly compare several schemes under exactly the same channel condition. Second, SORA does not allow for parallel RCM demodulation at present, so our implementation is not real-time.

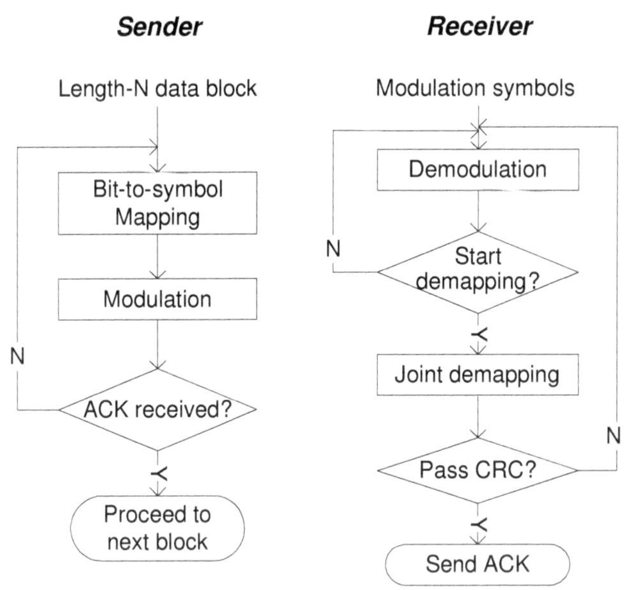

Figure 6: Data transmission flow charts at the sender and the receiver in SRA system

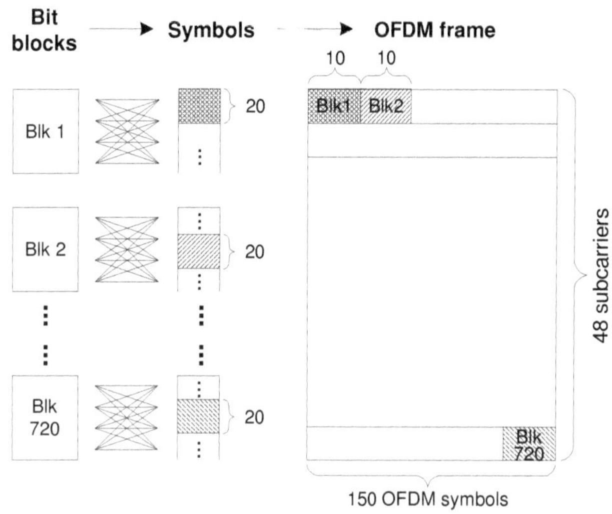

Figure 7: SRA implementation on OFDM

bits/s/Hz	LOS	LOS_M	NLOS	NLOS_M
SRA	5.50	4.71	3.78	2.89
System1	4.04	3.03	2.27	1.56
System2	5.06	4.02	2.93	2.43
System3	4.71	3.77	2.63	2.29

Table 2: Average Rate of each scenario

Real-time demodulation implementation will be one of our future works.

The channel condition is traced by transmitting uncoded BPSK symbols. This is valid because Sen et al. [16] have demonstrated that the dispersions on transmitted symbols are not influenced by the modulation scheme and is only the function of channel condition. We create four distinct channel conditions by combining static/mobile and Line-of-Sight (LOS)/Non-Line-of-Sight (NLOS). In all scenarios, the sender location is fixed. The receiver keeps motionless in static settings and moves at walking speed in mobile settings. For each scenario, one minute data is traced including the fading gain and the noise on 4 pilots and 48 carriers.

In each scenario, our SRA system is compared against with the three reference systems. System1 is an AMC scheme based on 802.11a physical layer. The eight modulation and coding combinations are the same as in the Simulation section. System2 is HARQ-CTC combined with AMC. According to the simulation results shown in Fig. 3, HARQ-CTC should switch to QPSK modulation when the SNR is below 9dB, and should switch to 64QAM when the SNR is above 15dB. In the SNR region between 9dB and 15dB, 16QAM modulation is used. System3 is HARQ-Raptor combined with AMC. The switching of modulation in all these three systems is determined by the channel estimation based on previous transmission.

In all systems, we use the SNR estimation algorithm proposed by Rahul et al. [12]. Similar to the setting in simulation, we assume that both channel SNR feedback and ACK are immediately available.

5.3 Results

In all four scenarios, SRA significantly outperforms the other schemes. Fig. 8 shows the SNR statistics of the four scenarios. Fig. 9 presents the CDF (cumulative distribution function) of goodput for all schemes. Table 2 summarizes the average goodput of SRA and the three reference systems.

In LOS static scenario, the channel is very stable and the SNR is in the range of 17dB to 25dB. All schemes perform fairly well. The poor performance of System1 in this scenario is mainly due to two reasons. First, 802.11a only provides a discrete rate set. The achievable rates are stair-shaped. By looking into the trace results, we find that System1 selects rate 3.0 (16QAM, 3/4) and 4.5 (64QAM, 3/4) in most time slots, so the average goodput is around 4 bit/s/Hz. Second, convolutional code cannot achieve the optimal coding performance so the achievable rates will be lower than using LDPC code or Turbo code under the same channel condition. In the other three scenarios, the performance gain of SRA over System1 becomes more significant. This is due to the fact that the channel varies dramatically and the imprecise estimation (prediction) misleads the rate selection algorithm. In NLOS mobile scenario, the gain of SRA over System1 is over 80%.

Comparing Fig. 8(a)(b)(c), we find that the dynamic range is gradually extended. The gain of SRA over System2 and System3 also gradually increases. This is because the advantage of using a fixed modulation in SRA becomes prominent when the other two systems have to switch between different settings. In NLOS scenario, SRA achieves 28.8% gain over System2 and 43.8% gain over System3.

In the last NLOS mobile scenario, the channel SNR varies between 2dB and 17dB. As shown in simulations, when channel SNR is below 8dB, SRA has a little performance loss with respect to HARQ schemes using low rate modulations. Despite of this, SRA manages to achieve around 18% gain over System2 and around 26% gain over System3 in average throughput.

The CDF of goodput shown in Fig. 9 presents the details

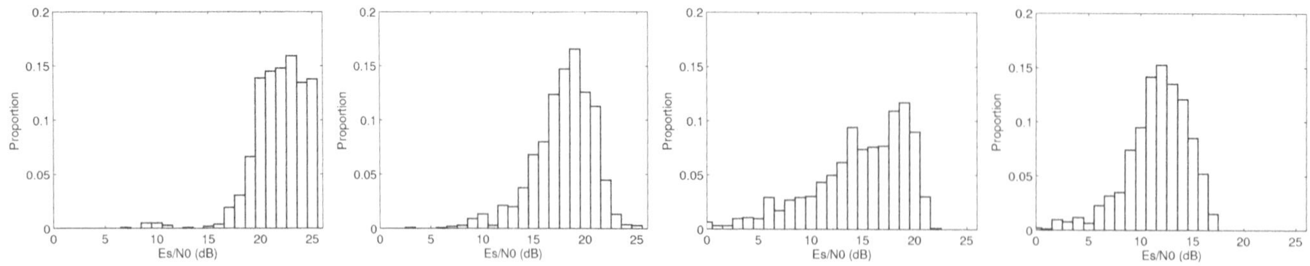

(a) LOS (b) LOS Mobile (c) NLOS (d) NLOS Mobile

Figure 8: PDF of Es/N0

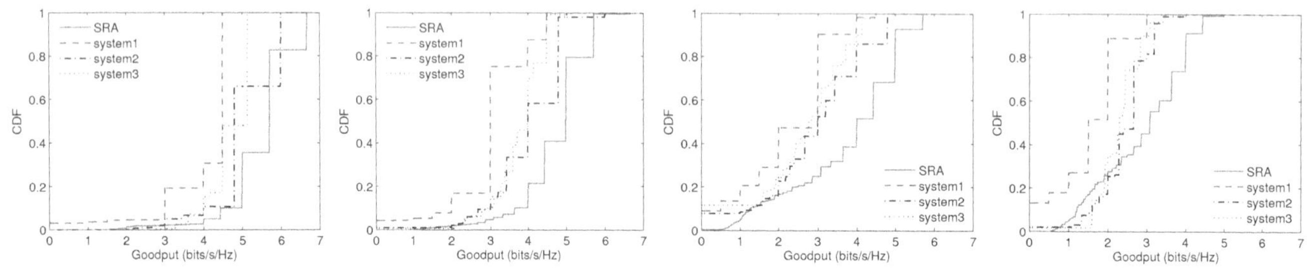

(a) LOS (b) LOS Mobile (c) NLOS (d) NLOS Mobile

Figure 9: CDF of Goodput

of rate selection. The results show that SRA successfully achieves our design goal of smooth and efficient rate adaptation over a broad dynamic range.

6. CONCLUSION

We have described a novel rate compatible modulation scheme that successfully achieves smooth and efficient rate adaptation over a broad dynamic range of channel conditions. By imposing the rate compatible restriction on modulation, instead of channel coding, we successfully extend the dynamic range of rate adaptation without compromising the spectral efficiency. The seamless adaptation is achieved through an incremental symbol generation process with fine-grained bit energy allocation.

We carry out extensive simulations and obtain the achievable rates of proposed RCM for the typical operation range of channel SNR. Results show that RCM covers a much larger dynamic range than HARQ systems based on rate compatible channel code. We have also designed and implemented a rate adaptation system SRA over OFDM. The evaluation on SORA testbed confirms the advantage of SRA when channel condition varies dramatically. In NLOS scenario, SRA outperforms the combined HARQ with AMC scheme by 35% on average. These promising results validate the practical use of the proposed RCM in real wireless environments.

7. REFERENCES

[1] IEEE Standard for Information technology - Telecommunications and information exchange between systems - Local and metropolitan area networks - Specific requirements. Part 11: Wireless LAN Medium Access Control (MAC) and Physical Layer (PHY) Specifications. *IEEE Std 802.11-2007*, 2007.

[2] IEEE Standard for Local and metropolitan area networks - Part 16: Air Interface for Broadband Wireless Access Systems. *IEEE Std 802.16-2009*, May 2009.

[3] 3rd Generation Partnership Project; Technical Specification Group Radio Access Network; Evolved Universal Terrestrial Radio Access (E-UTRA);Multiplexing and channel coding. *3GPP TS 36.212*, Sep 2010.

[4] D. Baron, S. Sarvotham, and R. G. Baraniuk. Bayesian compressive sensing via belief propagation. *Signal Processing, IEEE Transactions on*, 58(1):269 –280, Jan. 2010.

[5] J. D. Brown, S. Pasupathy, and K. N. Plataniotis. Adaptive demodulation using rateless erasure codes. *Communications, IEEE Transactions on*, 54:1574–1585, 2006.

[6] A. Gudipati and S. Katti. Automatic rate adaptation. In *ACM Hotnets'10*, 2010.

[7] G. Holland, N. Vaidya, and P. Bahl. A rate-adaptive mac protocol for multi-hop wireless networks. In *Proc. of ACM MobiCom*, pages 236–251, 2001.

[8] G. Judd, X. Wang, and P. Steenkiste. Efficient channel-aware rate adaptation in dynamic environments. In *Proc. of ACM MobiSys*, pages 118–131, 2008.

[9] R. T. Morris, J. C. Bicket, and J. C. Bicket. Bit-rate selection in wireless networks. Technical report, Masterąŕs thesis, MIT, 2005.

[10] W. H. Mow. Universal lattice decoding: principle and recent advances. *Wireless Communications and Mobile Computing*, 3:553 – 569, Aug. 2003.

[11] D. Qiao, S. Member, S. Choi, and K. G. Shin. Goodput analysis and link adaption for ieee 802.11a wireless lans. *IEEE Trans. on Mobile Computing*, 1:278–292, 2002.

[12] H. Rahul, F. Edalat, D. Katabi, and C. Sodini. Frequency-aware rate adaptation and mac protocols. In *Proc. of ACM MobiCom*, pages 193–204, 2009.

[13] D. Rowitch and L. Milstein. On the performance of hybrid fec/arq systems using rate compatible punctured turbo (rcpt) codes. *Communications, IEEE Transactions on*, 48(6):948 –959, June 2000.

[14] D. N. Rowitch and L. B. Milstein. On the performance of hybrid fec/arq systems using rate compatible punctured turbo (rcpt) codes. *Communications, IEEE Transactions on*, 48(6):948–959, Jun 2000.

[15] S. Sen, N. Santhapuri, R. R. Choudhury, and S. Nelakuditi. Accurate: constellation based rate estimation in wireless networks. In *Proceedings of the 7th USENIX conference on Networked systems design and implementation*, NSDI'10, pages 12–12, Berkeley, CA, USA, 2010. USENIX Association.

[16] S. Sen, N. Santhapuri, R. R. Choudhury, and S. Nelakuditi. Accurate: constellation based rate estimation in wireless networks. In *Proceedings of the 7th USENIX conference on Networked systems design and implementation*, NSDI'10, pages 12–12, Berkeley, CA, USA, 2010. USENIX Association.

[17] A. Shokrollahi. Raptor codes. *Networking, IEEE/ACM Transactions on*, 14:2551–2567, 2006.

[18] B. Sklar. *Digital communications: fundamentals and applications*. Prentice-Hall, Inc., Upper Saddle River, NJ, USA, 1988.

[19] E. Soljanin, N. Varnica, and P. Whiting. Punctured vs rateless codes for hybrid arq. In *Information Theory Workshop, 2006. ITW '06 Punta del Este. IEEE*, pages 155 –159, 2006.

[20] K. Tan, J. Zhang, J. Fang, H. Liu, Y. Ye, S. Wang, Y. Zhang, H. Wu, W. Wang, and G. M. Voelker. Sora: high performance software radio using general purpose multi-core processors. In *NSDI'09*, pages 75–90. USENIX Association, 2009.

[21] A. Venkiah, C. Poulliat, and D. Declercq. Jointly decoded raptor codes: analysis and design for the biawgn channel. *EURASIP J. Wirel. Commun. Netw.*, 2009:16:1–16:11, January 2009.

[22] M. Vutukuru, H. Balakrishnan, and K. Jamieson. Cross-layer wireless bit rate adaptation. In *Proc. of ACM SIGCOMM*, pages 3–14, 2009.

[23] S. H. Y. Wong, H. Yang, S. Lu, and V. Bharghavan. Robust rate adaptation for 802.11 wireless networks. In *Proc. of ACM MobiCom*, pages 146–157, 2006.

APPENDIX

A. RCM DEMODULATION ALGORITHM

Algorithm 1 Demodulation

Initialization:
  $p_j = p(b_j = 1) = 0.5$; $q_{ji}^{(0)}(0) = 1 - p_j$; $q_{ji}^{(0)}(1) = p_j$
Iterations: {T is the maximum number of iterations}
while $t < T$ **do**
  Horizontal Processing:
  for $i = 1$ **to** M **do**
    $r_{ij}^{(t)}(0) = p(b_j = 0|u_i') = p(U_j = u_i')$
    $r_{ij}^{(t)}(1) = p(b_j = 1|u_i') = p(U_j = u_i' - w_l)$
    {where $u_i' = \sum_{l=1}^{L} w_l \cdot b_{i_l} + e_i$ is the received symbol}
    {$U_j = u_i' - w_l \cdot b_j$ and $j = i_l$}
    Distribution $P(U_j) = \{\bigotimes_{i_k, 0<k\leq L, k\neq l} P(w_k \cdot b_{i_k})\} \otimes P(e_i)$
    {where $p(w_k \cdot b_{i_k} = 0) = q_{ji}^{(t-1)}(0)$, $p(w_k \cdot b_{i_k} = w_k) = q_{ji}^{(t-1)}(1)$}
    {$p(e_i) \sim N(0, \sigma^2)$, $\otimes$ is convolutional product}
  end for
  Vertical Processing:
  for $j = 1$ **to** N **do**
    $q_{ji}^{(t)}(0) = C_{ji} \cdot (1 - p_j) \cdot \prod_{k \in T_j \setminus i} r_{kj}^{(t)}(0)$
    $q_{ji}^{(t)}(1) = C_{ji} \cdot p_j \cdot \prod_{k \in T_j \setminus i} r_{kj}^{(t)}(1)$
    {C_{ji} is the normalization factor which ensures $q_{ji}^{(t)}(0) + q_{ji}^{(t)}(1) = 1$}
    {T_j is the index set of the check nodes connected with variable node j}
  end for
end while
Output: {Hard Decision}
$b_j = 0$, if $q_j^{(T)}(0) \geq q_j^{(T)}(1)$; Otherwise, $b_j = 1$
  {$q_j^{(T)}(0) = C_j \cdot (1 - p_j) \cdot \prod_{k \in T_j} r_{kj}^{(T)}(0)$}
  {$q_j^{(T)}(1) = C_j \cdot p_j \cdot \prod_{k \in T_j} r_{kj}^{(T)}(1)$}

Author Index